Decade Map Volume
To Accompany the Neotectonic Maps,
Part of the Continent-Scale Maps
of North America

Neotectonics of North America

Edited by

D. Burton Slemmons
P.O. Box 81050
Las Vegas, Nevada 89180-0695

E. R. Engdahl
National Earthquake Information Center
U.S. Geological Survey
MS 967, Box 25046, Denver Federal Center
Denver, Colorado 80225

Mark D. Zoback
Department of Geophysics
Stanford University
Stanford, California 94305

David D. Blackwell
Department of Geological Sciences
Southern Methodist University
Dallas, Texas 75275

1991

Acknowledgment

Publication of this volume, one of the synthesis volumes of *The Decade of North American Geology Project* series, has been made possible by members and friends of the Geological Society of America, corporations, and government agencies through contributions to the Decade of North American Geology fund of the Geological Society of America Foundation.

Following is a list of individuals, corporations, and government agencies giving and/or pledging more than $50,000 in support of the DNAG Project:

Amoco Production Company
ARCO Exploration Company
Chevron Corporation
Cities Service Oil and Gas Company
Diamond Shamrock Exploration Corporation
Exxon Production Research Company
Getty Oil Company
Gulf Oil Exploration and Production Company
Paul V. Hoovler
Kennecott Minerals Company
Kerr McGee Corporation
Marathon Oil Company
Maxus Energy Corporation
McMoRan Oil and Gas Company
Mobil Oil Corporation
Occidental Petroleum Corporation

Pennzoil Exploration and Production Company
Phillips Petroleum Company
Shell Oil Company
Caswell Silver
Standard Oil Production Company
Oryx Energy Company (formerly Sun Exploration and Production Company)
Superior Oil Company
Tenneco Oil Company
Texaco, Inc.
Union Oil Company of California
Union Pacific Corporation and its operating companies:
    Union Pacific Resources Company
    Union Pacific Railroad Company
    Upland Industries Corporation
U.S. Department of Energy

© 1991 by The Geological Society of America, Inc.
All rights reserved.

All materials subject to this copyright and included in this volume may be photocopied for the noncommercial purpose of scientific or educational advancement.

Copyright is not claimed on any material prepared by government employees within the scope of their employment.

Published by The Geological Society of America, Inc.
3300 Penrose Place, P.O. Box 9140, Boulder, Colorado 80301

Printed in U.S.A.

This book does not list Cataloging-in-Publication data because it is an accompanying volume to the Neotectonic Maps, part of the Continent-Scale Maps of North America, and the Library of Congress does not issue CIP data for map sets. Therefore, libraries should shelve this volume under "neotectonics" and cross reference it under "seismology, heat flow, or crustal stress of North America."

Cover Photo: Pleasant Valley, central Nevada, showing 4- to 6-m high Pearce fault scarp from the M = 7.5 to 7.8 earthquake of 1915.

10 9 8 7 6 5 4 3 2 1

Contents

Preface vii

Foreword ix

1. Introduction 1
   D. B. Slemmons

SEISMICITY

2. Seismicity Map of North America Project 21
   E. R. Engdahl and W. A. Rinehart

3. Seismicity of the Aleutian Arc 29
   J. J. Taber, S. Billington, and E. R. Engdahl

4. Seismicity of continental Alaska 47
   R. A. Page, N. N. Biswas, J. C. Lahr, and H. Pulpan

5. An overview of western Canadian seismicity 69
   G. C. Rogers and R. B. Horner

6. Seismicity of Washington and Oregon 77
   R. S. Ludwin, C. S. Weaver, and R. S. Crosson

7. Northern California seismicity 99
   R. A. Uhrhammer

8. The seismotectonic fabric of central California 107
   D. P. Hill, J. P. Eaton, W. L. Ellsworth, R. S. Cockerham,
   F. W. Lester, and E. J. Corbett

9. Seismotectonics of southern California 133
   L. K. Hutton, L. M. Jones, E. Hauksson, and D. D. Given

10. **The seismicity of Nevada and some adjacent parts of the Great Basin** 153
    A. M. Rogers, S. C. Harmsen, E. J. Corbett, K. Priestley, and
    D. dePolo

11. **Seismicity of the Intermountain Seismic Belt** 185
    R. B. Smith and W. J. Arabasz

12. **Seismicity of the Rio Grande rift in New Mexico** 229
    A. R. Sanford, L. H. Jaksha, D. J. Cash

13. **Seismotectonics of the central United States** 245
    B. J. Mitchell, O. W. Nuttli, R. B. Herrmann, and W. Stauder

14. **The seismicity and seismotectonics of eastern Canada** 261
    J. Adams and P. Basham

15. **Earthquake activity in the northeastern United States** 277
    J. E. Ebel and A. L. Kafka

16. **Seismicity of the southeastern United States; 1698 to 1986** 291
    G. A. Bollinger, A. C. Johnston, P. Talwani, L. T. Long,
    K. M. Shedlock, M. S. Sibol, M. C. Chapman

17. **Seismotectonics of Middle America** 309
    J. W. Dewey and G. Suárez

18. **Tectonic implications of upper-crustal seismicity in Central America** 323
    R. A. White

STRESS

19. **Tectonic stress field of North America and relative plate motions** 339
    M. D. Zoback and M. L. Zoback

20. **Crustal stresses in Canada** 367
    J. Adams and S. Bell

21. **Stress indicators in Alaska** 387
    C. H. Estabrook and K. H. Jacob

22. **State of stress and active deformation in Mexico and western Central America** 401
    M. Suter

THERMAL ASPECTS

23. **Heat-flow patterns of the North American continent; A discussion of the Geothermal Map of North America** 423
    D. D. Blackwell, J. L. Steele, and L. S. Carter

24. **Terrestrial heat flow in Canada** 437
    A. M. Jessop

25. **Heat flux in the Canadian Cordillera** 445
    T. Lewis

26. *An overview of heat flow in southwestern United States and northern Chihuahua, Mexico* 457
   M. Reiter, M. W. Barroll, and J. Minier

27. *Subsurface temperatures in the northern Great Plains* 467
   W. D. Gosnold, Jr.

NEOTECTONICS

28. *Late Quaternary glacial isostatic recovery of North America, Greenland, and Iceland; A neotectonics perspective* 473
   J. T. Andrews

Index 487

Preface

The Geology of North America series has been prepared to mark the Centennial of The Geological Society of America. It represents the cooperative efforts of more than 1,000 individuals from academia, state and federal agencies of many countries, and industry to prepare syntheses that are as current and authoritative as possible about the geology of the North American continent and adjacent oceanic regions.

This series is part of the Decade of North American Geology (DNAG) Project, which also includes seven wall maps at a scale of 1:5,000,000 that summarize the geology, tectonics, magnetic and gravity anomaly patterns, regional stress fields, thermal aspects, and seismicity of North America and its surroundings. Together, the synthesis volumes and maps are the first coordinated effort to integrate all available knowledge about the geology and geophysics of a crustal plate on a regional scale.

The products of the DNAG Project present the state of knowledge of the geology and geophysics of North America through the 1980s, and they point the way toward work to be done in the decades ahead.

A. R. Palmer
General Editor for the volumes
published by The Geological
Society of America

J. O. Wheeler
General Editor for the volumes
published by the Geological
Survey of Canada

Foreword

This volume provides a nucleus of ideas, syntheses, and discussions to accompany the continent-scale maps of the Decade of North American Geology (DNAG) Project that relate to neotectonics. Initial plans in April 1986 included preparation of texts related to four maps that embraced the topics of seismicity, stress, thermal aspects, and neotectonics of North America. Three of the four maps have been completed. Texts related to these maps, plus two chapters dealing with regional neotectonics are included in this volume.

The three completed maps each consist of four sheets covering North America at a scale of 1:5,000,000.

Seismicity Map of North America, 1988, compiled by E. R. Engdahl, and prepared for publication by W. A. Rinehart, Geological Society of America CSM-004. Construction of the seismicity map was a team effort that began in 1984 at the Seismological Society of America meeting in Anchorage, Alaska, and culminated in publication of the final map with data contributed by more than 100 researchers. The 17 chapters that discuss the seismicity of North America by region were edited by E. R. Engdahl. Compilation of the map data base involved integration of earlier felt reports, dating back to 1534, with modern improved earthquake epicenters derived from data recorded by an increasing number of sophisticated seismographic programs. Thus the colorfully displayed map and the text provide the most thoroughly researched source material available for studying the historical seismicity of North America during the period 1534 to 1985.

Stress Map of North America, 1991, compiled by M. D. Zoback and M. L. Zoback, Geological Society of America CSM-005. The stress map and text were edited by M. D. Zoback and include the most comprehensive and complete data base that has been prepared for the current state of crustal stress in North America. The data demonstrate well-defined and regional stress orientations and relative stress magnitudes. They also reflect a compressional midplate region and extensional regions in the western Cordillera, southern Great Plains, and Texas-Louisiana Gulf Coast. The data are based on four general categories of tectonic stress indicators, including earthquake focal mechanisms, stress-induced wellbore breakouts, in-situ measurements at depth, and young geologic data such as volcanic alignments and fault-slip events.

Thermal Aspects Map of North America, 1991, compiled by D. D. Blackwell, J. L. Steele, and L. S. Carter, Geological Society of America, CSM-006. The geothermal map and accompanying text include a summary of the heat flow sites and values for North America and surrounding oceanic and island areas where data are available. The heat flow values are contoured in areas where data are sufficient. In addition to the locations of major hot spring systems, explored geothermal systems and Quaternary volcanoes are shown. The data base for the map was compiled with the cooperation of 11 heat flow investigators.

A *Neotectonic Map of North America* and accompanying texts were the last part of the neotectonics project to be started. Due to the extensive work needed to complete the map for many key regions and organizational problems, only a neotectonic overview, in the Introduction to this volume, and a review of late Quaternary isostatic recovery of North America, Greenland, and Iceland, with a neotectonic perspective, are included here. The map is indefinitely delayed.

In addition to the above maps and supporting chapters in this volume, several other geological and geophysical maps produced by the Geological Society of America provide a complement to the understanding of the North American neotectonic setting. They include *The Geologic Map of North America,* still in compilation by J. C. Reed, Jr., J. O. Wheeler, B. E. Tucholke, and A. R. Palmer; the *Gravity Anomaly Map of North America,* 1987, compiled by the Gravity Anomaly Map Committee; and the *Magnetic Anomaly Map of North America,* 1987, compiled by the Committee for the Magnetic Anomaly Map.

The digital data bases for the neotectonic maps and the gravity and magnetic anomaly maps are available on a CD-ROM, *Geophysics of North America,* produced by the National Geophysical Data Center, Boulder, Colorado.

This volume and the three related maps are the Centennial contribution by the Geophysics and the Structural Geology and Tectonics Divisions of the Geological Society of America to the DNAG Program.

The progress and completion of this volume and associated maps were strongly influenced by the unfailing help of A. R. (Pete) Palmer, the Centennial Science Program Coordinator of the Society. Words are inadequate to express the gratitude of the editors and contributors for his continued and careful support and vigorous encouragement at all stages of development of this project.

D. B. Slemmons
Volume editor
August 1991

Chapter 1

Introduction

D. Burton Slemmons
P.O. Box 81050, Las Vegas, Nevada 89180-0695

INTRODUCTION

The three topics covered in this volume—seismicity, stress, and heat flow—are essential for an understanding of the neotectonics of North America. These topics are discussed in three groups of chapters that support associated map compilations (seismicity—Engdahl and Rinehart, 1988; stress—Zoback and Zoback, 1991; thermal aspects—Blackwell and others, 1991) prepared on a standardized continent-wide base to provide information on the neotectonic setting of North America. Each of the three topics in this volume was organized separately and internally edited. The seismicity chapters, edited by E. R. Engdahl, show general regional earthquake patterns and indicate the locations, styles, and relative frequencies of earthquakes from neotectonic faulting and fault-related seismogenic folding. The chapters on stress, edited by Mark D. Zoback, highlight regional differences in stress type and orientation, and are important for determining likely earthquake focal mechanisms and styles of neotectonic deformation. The chapters on geothermal aspects, edited by David D. Blackwell, provide an indication of neotectonic activity developed over much longer time frames, particularly for volcanic and geothermal phenomena.

The original intent was to include a fourth section and an accompanying neotectonics (strict sense) map, the first compilation of North American faults, folds, and other neotectonic features. Problems of compiling and analyzing the overwhelmingly large accumulation of new paleoseismic data, particularly in the western United States, prevented completion of this task in time for inclusion in this volume, with the exception of this chapter, which emphasizes midplate neotectonics (Slemmons), and a chapter on glacial rebound (Andrews). Accordingly, in this chapter I highlight some major contributions to the study of active tectonics, paleoseismicity, and seismotectonics in the midplate region of North America.

Neotectonic studies in North America have increased at an exponential rate in recent years due to advances in geological, paleoseismological, and geophysical techniques, and to increasing population growth in seismically active, potentially hazardous areas. Awareness of earthquake hazards is generally high in the western United States, Alaska, and Mexico, where the frequent occurrence of large earthquakes periodically reminds residents, builders, and planners of the potential dangers posed by neotectonic processes. Equally significant are the less frequent, although severely damaging, earthquakes in midplate settings in the central and eastern parts of North America. These events are especially important because the seismic energy for larger earthquakes is attenuated less in the central and eastern parts of North America compared with the geologically younger and more heterogeneous rocks of western North America; therefore, similar-sized events are felt much more widely in midplate settings (Bollinger and others, 1991). Until recently, this region, which includes about two-thirds of the population of United States, received relatively little attention from neotectonic geologists. Intraplate tectonic settings were generally considered to be inactive, yet an examination of the historical record of "stable" continental regions in North America shows that there have been many moderate- to large-magnitude earthquakes of $M > 6$ to 8, and a small number of great earthquakes of $M > 8$. A repeat of one of the four New Madrid earthquakes of A.D. 1811–1812 of $M > 8$, or the Charleston, South Carolina, earthquake of A.D. 1886 of $M \geq 7.5$, for example, would cause severe and widespread damage, injury, and loss of life. Accordingly, examples of North American intraplate or midplate neotectonic activity (Fig. 1 and Table 1) are given special attention in the latter part of this chapter.

Terminology and Time Frames of Neotectonic Processes

Bates and Jackson (1987, p. 675) defined tectonics as "a branch of geology dealing with the broad architecture of the outer part of the Earth, that is, the regional assembling of structural or deformational features, a study of their mutual relations, origin, and historical evolution." Neotectonics can be broadly described as tectonic events and processes that have occurred in post-Miocene time (approximately the past 5–6 m.y.), although many studies and applications use more restrictive time frames. Mörner (1990), editor of the International Quaternary Association (INQUA) Neotectonics Commission *Bulletin,* noted that the start

---

Slemmons, D. B., 1991, Introduction, *in* Slemmons, D. B., Engdahl, E. R., Zoback, M. D., and Blackwell, D. D., eds., Neotectonics of North America: Boulder, Colorado, Geological Society of America Decade Map Volume 1.

of the time span for neotectonics in the European-African region may be defined on the basis of various tectonic changes at the base of the Miocene (22 Ma), at the Messinian (5–6 Ma), or at the new and current tectonic regime that appears to have been initiated in the Mediterranean region (2.5–3.0 Ma). Mörner (1990) concluded that 2.5–3.0 Ma "may well be chosen as a natural base for the neotectonics of the Mediterranean region." The Quaternary Period (approximately 1.65 Ma at the end of the Olduvai subchron or 2.48 Ma at the Gauss-Matuyama Chronozone boundary, as discussed in Morrison, 1991) is used as the interval for many neotectonic studies in the western United States (e.g., Jennings and others, 1975).

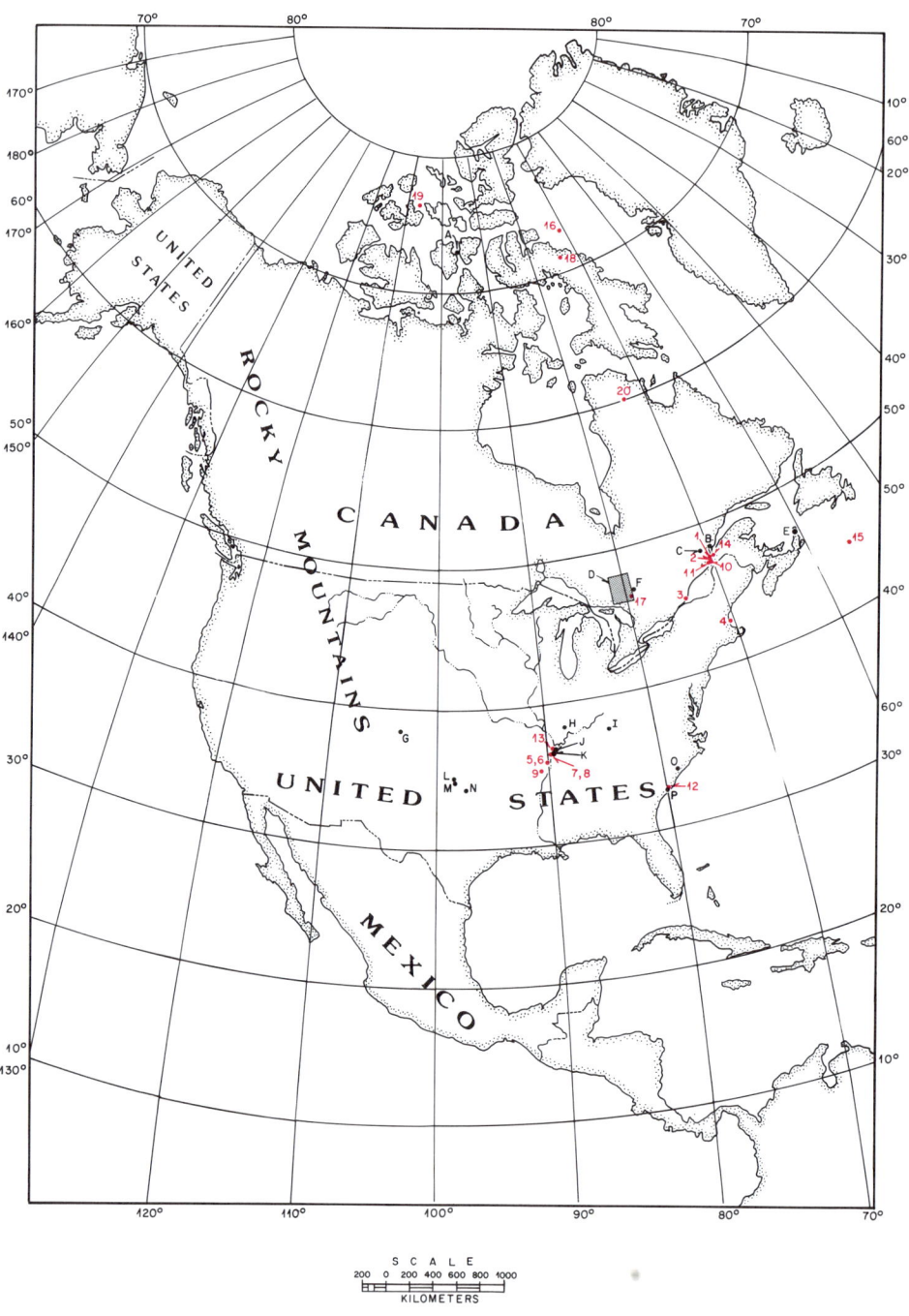

Figure 1. Map showing 20 historical earthquake epicenters of $M_s > 6$ (denoted by number and listed in Table 1) and 15 localities with paleoseismic surface or near-surface faulting (with lengths greater than 5 km and/or displacement more than 0.5 m) or recurrent liquefaction events from paleoseismic earthquakes (denoted by letter and listed in Table 2).

In addition to the use of neotectonics as a term with specific time connotations, it is also widely used in conjunction with "active tectonics," "seismotectonics," and "paleoseismicity." These terms are defined differently by different authors and are not always synonymous with neotectonics, because some neotectonic processes and structures are not necessarily seismogenic, and many neotectonic structures may no longer be active.

"Active tectonics" was defined by Wallace (1986, p. 5) as "tectonic movements that are expected to occur within a future time span of concern to society." He further commented that the present is a moving instant that progresses ever forward in time and is viewed differently by the layperson and the geologist. By defining active tectonics in terms of the future, it is possible to avoid the problem of selecting a single time period that represents the hazard and risk of future potential earthquakes and surface deformation. He concluded that events of the past 10 to 20 ka to as much as 100 to 200 ka are most useful in predicting future activity. The shorter time frame (the Holocene, or 10^4 yr) may be used by the task group of the Inter-Union Commission on the Lithosphere for preparation of a world map of active faults planned for completion in 1994.

The Holocene may be suitable for evaluating well-studied faults with high rates of activity, such as faults of the San Andreas fault system. However, longer periods of time may be more ap-

TABLE 1. HISTORICAL EARTHQUAKES OF M >6 IN THE MIDPLATE PROVINCE OF NORTH AMERICA, 1834–1990*

No.	Year	Lat. (°N)	Long. (°W)	Mag	Comments, References	No.	Year	Lat. (°N)	Long. (°W)	Mag	Comments, References
1†	1534	47.7	70.1	6.67	La Malbaie (Charlevoix), Quebec; compressive stress inferred.	12	1886	32.90	80.0	7.56	Charleston, South Carolina; Bollinger (1983) prefers a dip-slip mechanism and Talwani (1982) prefers a N26E right-slip with a steep dip to the WNW. P and R for the earthquake area.
2†	1663	47.60	70.10	6.67	La Malbaie (Charlevoix), Quebec; compressive stress inferred. P and R inferred by Doig (1990) for the Charlevoix events with an indirectly determined magnitude of M ≥7.5	13	1895	37.0	89.4	6.79	Charleston, Missouri; m_b = 6.2, Nuttli and Brill, 1981; compressive stress inferred. Parallel to the main right-slip fault of the New Madrid area. P and R for the zone.
3	1732	45.50	73.60	6.12	Mid-St. Lawrence River Valley, Canada; compressive stress inferred.						
4	1755	72.7	70.3	6.33	Cape Ann, Massachusetts; compressive stress inferred.	14†	1925	47.76	69.84	6.86	La Malbaie (Charlevoix), Quebec; compressive stress inferred. P (Doig, 1990)
5	1811	36.0	90.0	8.20	New Madrid; Ms = 8.5; Nuttli, 1983; right-slip for southern section of fault. P and R shown for parts of the zone.	15	1929	44.39	56.006	7.38	Grand Banks, Canada; tectonic or submarine landslide origin is uncertain.
6	1811	36.0	90.0	....	New Madrid; Ms = 8.3, Street and Nuttli, 1984; may be an aftershock of event 5. P and R for parts of the zone.	16	1933	73.3	70.7	7.70	NE Baffin Island; may be located in oceanic crust; normal-fault mechanism is preferred by Hasegawa (1991).
7	1812	36.3	89.6	8.09	New Madrid; Ms = 8.4, Nuttli, 1983; thrust central arm. P and R for parts of the zone.	17	1935	46.874	79.051	6.35	Timiskaming, Ontario; compression stress and thrust fault inferred.
8	1812	36.5	89.6	8.30	New Madrid; Ms = 8.8, Nuttli, 1983; rightslip on northern arm. P and R for parts of the zone.	18	1963	71.25	72.94	6.11	NE Baffin Island, Canada; compressive stress inferred.
9	1843	35.5	90.5	6.37	NE Arkansas; mb = 6.0, Nuttli and Brill, 1981.	19	1972	76.80	106.49	6.28	Svedrup Basin, Arctic; type faulting is strikeslip with a slight normal component.
10†	1860	47.5	70.10	6.08	La Malbaie (Charlevoix), Quebec; compressive stress inferred.	20	1989	60.12	73.60	6.3	Ungava, Quebec; Adams and others, 1991; reverse fault in compressive stress regime. Probably no postglacial activity.
11†	1870	47.40	70.50	6.55	La Malbaie (Charlevoix), Quebec; compressive stress inferred. P (Doig, 1990).						

*Modified from the Electric Power Research Institute, 1989, with data prepared by Johnston, Kanter, and Metzer; magnitudes based on seismic moment magnitude, with moment magnitudes quoted in two decimal places to indicate a derived, not a measured quanity. Epicenters indicated in Figure 1. Symbols include P for paleoseismic activity and R for recurrent activity.
†The Malbaise-Charlevoix area appears to have recurrent historic and paleoseismic activity. Doig (1990) correlated the thickness of paleoseismic silting events in Lake Tadoussac, Charlevoix, Quebec, with the magnitude of historic earthquakes.

propriate for many faults of the Basin and Range province or parts of the central and eastern United States, where active faults have much longer recurrence intervals, or have short bursts of activity after long periods of inactivity.

"Paleoseismicity" refers to the study of the age, frequency, and size of prehistoric, or "fossil," earthquakes and earthquake effects. Paleoseismic activity was originally identified by the presence of seismites, or geologic evidence of deformation of sediments by earthquake shaking, by Seilacher (1969) and was redefined by Vittori and others (1991) to include other geologic structures that are genetically related to earthquake events. This definition of paleoseismicity is independent of the timing or age of activity, and therefore may not necessarily be neotectonic. Most recent paleoseismicity studies are aimed at identifying recent surface faulting, deformation, and earthquake-induced liquefaction events by use of geomorphic and exploratory-trenching techniques. Paleoseismicity studies extend the history of past earthquakes beyond the very short historical record into the much longer geologic record.

"Seismotectonics" is the study of the history and relations between earthquakes and causative tectonic processes. The term "present seismotectonic regime" is generally used for the period of time that began with the origin of the current tectonic forces and processes. Accordingly, it may have similarities in scope to neotectonics and active tectonics, but unless data on prehistoric (paleoseismic) activity are included, it may be based only on historical earthquake activity.

PREVIOUS AND CURRENT NEOTECTONIC STUDIES IN NORTH AMERICA

North American Neotectonic Maps

Prior to the Decade of North American Geology (DNAG) series of publications, there were no comprehensive compilations of neotectonic settings and activity for North America on a standardized map base. One of the goals of the DNAG project was to gather data from available sources on the seismicity, stress, heat flow, and neotectonic structures for North America and publish this data on a standardized, 1:5,000,000 scale base. Other related DNAG map compilations on the same base include the *Geologic Map of North America* (Reed and others, 1992), the *Gravity Anomaly Map of North America* (Gravity Anomaly Map Committee, 1987), and the *Magnetic Anomaly Map of North America* (Committee for the Magnetic Anomaly Map of North America, 1987).

Supporting texts for three neotectonic compilation maps, seismicity, stress, and heat flow, are included in this volume. The seismicity map (Engdahl and Rinehart, 1988) depicts historical (preinstrumental and instrumental) events. The duration of the historical record varies regionally, but is limited to a few hundred years, a small fraction of the Holocene or Quaternary. Because the recurrence interval for activity on individual active faults varies between about 100 years to more than 100 ka, the historical seismicity record of large events represents only a partial record of neotectonic activity. The stress map (Zoback and Zoback, 1991) includes many historical measurements, including stress-induced wellbore breakouts, earthquake focal mechanisms, and hydraulic fracturing alignments, as well as geologic data (e.g., volcanic alignments) that represent much longer time frames. The heat flow (geothermal) map (Blackwell and others, 1991) shows general patterns that are controlled by tectonic activity that may be measured in tens to hundreds of millions of years, although features of late Cenozoic age are emphasized. Accordingly, this map is partly, but not strictly, related to the Neogene definition of neotectonics.

Regional Map Compilations

There are currently no comprehensive neotectonic maps of North America. The most comprehensive continental-scale tectonic map America is the *Tectonic Map of North America* (1:5,000,000 scale) compiled by King (1969a), along with a general text (King, 1969b). Although this compilation map does not emphasize the more recent Cenozoic tectonic processes and structures, it provides an overall structural framework for North America compiled on a map base similar to the DNAG maps described in this volume.

An extensive literature exists for specific neotectonic structures or regions, particularly for the tectonically active parts of western United States. Regional map compilations of neotectonically active regions, states, and structures prepared since the early 1970s at scales of 1:500,000 to 1:1,500,000. These include the following maps, along with references for supporting texts.

Alaska. An active fault map for Alaska at a scale of 1:2,500,000 (Plafker and others, unpublished), will be published in 1992 in the DNAG volume *Geology of Alaska*. The map will show active faults with dates and rates of activity and regions of tectonic uplift and subsidence.

Arizona. The Arizona Geological Survey has published several neotectonic maps at scales of 1:500,000 to 1,000,000. Maps and supporting texts include those by Menges (1983), Menges and Pearthree (1983, 1989), Morrison and Menges (1981), Pearthree and others (1983), and Scarborough and others (1983, 1986).

Basin and Range and Rio Grande Rift Provinces. An active fault map at a scale of 1:1,000,000 was prepared for this region by Nakata and others (1982). The map depicts historical, Holocene, and Quaternary faults of this large neotectonic province.

California. The active fault map for California by Jennings and others (1975) was published at a scale of 1:750,000 by the California Division of Mines and Geology. It includes age classifications for historic, Quaternary, and pre-Quaternary faults. Jennings is currently updating the fault-activity map of California at a scale of 1:750,000 with publication planned for 1992. Wallace (1990) organized and edited a comprehensive discussion of the San Andreas fault system, including a map of the main active faults of both the San Andreas fault system and the adjoining Mojave block at a scale of 1:1,500,000.

Colorado. Witkind (1976) compiled a map of active faults of Colorado at a scale of 1:500,000. The active fault map by Kirkham and Rogers (1981), compiled at a scale of 1:1,000,000 for the Colorado Geological Survey in a report on earthquake potential in Colorado, depicts Neogene and Quaternary faults in Colorado, with descriptions of historic, Holocene, late Pleistocene (post-Wisconsin, or post-125 ka), Pleistocene (post-2 Ma), and late Cenozoic (post–middle Miocene or 20 Ma) faulting.

Idaho, Montana, and Wyoming. Witkind (1975a, 1975b, and 1975c) compiled maps of these three states, as well as Colorado, at a scale of 1:500,000 for the U.S. Geological Survey. The times of most recent offset (historic, Holocene, late Quaternary, Pleistocene, and post–middle Miocene) are shown on the maps.

Utah. Anderson and Miller (1979) compiled a map of Quaternary faults in Utah at a scale of 1:1,000,000. Hecker (unpublished) is updating this map to include faults, folds, and volcanic rocks with five age intervals of most recent tectonic displacement. It is planned for publication by the Utah Geological Survey as an open-file map in 1991 and as a colored map in 1992.

Other Studies

In addition to numerous individual publications, there are several important compilations of recent papers relating to neotectonic issues and problems. Although none provides a comprehensive overview of the neotectonics of North America, they do provide an up-to-date nucleus for many key neotectonic topics. These include the following.

Active Tectonics (Wallace, 1986) provides 16 excellent summaries of relevant topics in neotectonic studies, including papers relating to active tectonics along the western continental margin, epeirogenic and intraplate movements, evaluation of active faulting hazards, active faults related to folding, alluvial responses to active tectonics, coastal tectonics, tectonic geomorphology, seismological and paleoseismological techniques, geodetic measurement, near-field tectonic geodesy, methods for dating, and volcanic-hazard assessment.

The greatly increased use of paleoseismic methods for assessments of earthquake hazards and understanding of global dynamics resulted in a U.S. Geological Survey (USGS) workshop and redbook *Directions in Paleoseismology* (Crone and Omdahl, 1987). The 44 papers present state-of-the-art discussions of Quaternary dating techniques, recognition of paleoseismic events in the geologic record, determination of Quaternary slip rates, and other related topics.

The USGS workshop and redbook *Fault Segmentation and Controls of Rupture Initiation and Terminations* (Schwartz and Sibson, 1989) presented 27 investigations of the physical properties that control the nucleation and termination of ruptures along fault zones, including evidence that some fault segments have behaved characteristically through several seismic cycles. The book considers fault segmentation, geometric features of fault zones related to nucleation and termination of earthquake ruptures, segmentation of Basin and Range province normal and strike-slip faults, implications of the characteristics of end-points of historical surface ruptures on segmentation, effects of restraining bends, surface faulting at stepovers, fault-slip estimation from geologic, geodetic, and seismologic observations, and the persistence of fault segments over geologic time.

The San Andreas Fault System (Wallace, 1990) covers a spectrum of 10 topics related to this broad fault system, including geomorphic expression, plate-tectonic development, Quaternary deformation, seismicity, earthquake history, present-day crustal movements, cyclic deformation, lithospheric structures, stress, and heat flow.

Neotectonics in Earthquake Evaluation (Krinitzsky and Slemmons, 1990) includes seven papers that discuss methods for estimating earthquake size, use of probabilistic methods, seismic-hazard assessments in the central and eastern United States, implications of the Meers fault on seismic potential in the central United States, neotectonics in the southeastern United States, and fault segmentation.

PLATE TECTONIC RELATIONS

Plate tectonics provides a widely accepted model for neotectonic settings and crustal behavior. This model suggests that the Earth's upper crust is composed of several large plates and numerous smaller "microplates" that move gradually with respect to each other. Nearly all tectonic activity occurs at or near plate boundaries, and plates are thought to be internally stable or undeformed. Although nearly all North American neotectonic activity (including most active faulting and folding, areas of high seismicity and volcanic activity, regions with perturbations in stress fields, and variable heat flow) occurs near plate boundaries, there are several recognized localities of Neogene to historic tectonic activity within the stable continental plate, referred to as the "Midplate Province" in Zoback and Zoback (this volume). The Midplate province is characterized by low rates of seismicity, with generally low-magnitude earthquakes ($M < 6$) and infrequent higher-magnitude earthquakes ($M = 6$ to 8.3), uniform and nearly consistent patterns of compressional stresses, and normal heat flow. Earthquakes are widely scattered in diffuse to well-aligned zones; regional and local areas are separated by regions with little or no seismicity.

INTERPLATE BOUNDARY REGIONS

The western one-fourth of the North American continent is characterized by strong tectonic interaction between the west-southwest–moving North American plate and the adjoining Pacific, Juan De Fuca, Cocos, and Caribbean plates, which have different directions and velocities of absolute movement. The plate and microplate boundaries vary from sharply defined and narrow zones to broad, diffuse zones that extend hundreds to thousands of kilometers from the main boundary. An example of the latter type of boundary is the Cordilleran extension province (Zoback and Zoback, this volume), which includes the Basin and Range subprovince, the regions surrounding the Colorado Plateau, the Rio Grande rift, and much of the central Rocky Mountains and Denver basin regions. Most of the methods used to evaluate active faults and determine earthquake potentials or hazards for active faults were developed in interplate regions (Wallace, 1986; Slemmons and dePolo, 1986; dePolo and Slemmons, 1990).

Interplate boundaries fall into three types, subduction, extensional, and transform, depending on the directions and geometries of plate interactions. Provinces and subprovinces of these types are defined and discussed for the main North American stress fields by Zoback and Zoback (1980, 1989, and this volume). Unlike midplate regions, stress orientations in intraplate settings are typically more diverse, and the direction of maximum horizontal compressive stress does not show the remarkable correlation with stress-release patterns that is typical in the eastern and central United States. Parts of some provinces are influenced by complex interactions between the plates. Within the classic Basin and Range subprovince in the Cordilleran extension province, the Walker Lane belt of western Nevada, for example, has a component of superimposed shear that appears to be related to the San Andreas transform system. Zoback and Zoback note that oblique subduction in the Aleutian subduction and Aleutian back-arc provinces is partitioned into separate zones of dip-slip and strike-slip deformation.

It is apparent from the seismicity map (Engdahl and Rinehart, 1988, and this volume), that interplate boundaries and adjoining tectonically active parts of the North American plate interior have higher than average earthquake activity, commonly in belts and zones that may relate to the distribution of more recently active faults. Epicenters shown in green on the seismicity map denote the deeper-focus earthquakes, generally at subduction zones.

The heat flow of interplate boundary zones is also more variable than in the stable plate interior (Figs. 1, 2, and 3 of Blackwell and others, this volume); heat flow in the western United States is somewhat higher than in the Midplate province, and has a greater variability and larger extremes. Regions of high heat flow include recently active volcanic, geothermal, and some tectonically active areas. Regions of very low heat flow are above subducting oceanic plates.

MIDPLATE, OR STABLE CONTINENTAL INTERIOR PROVINCES

The Midplate province of North America includes the eastern three-fourths of the continent. It is characterized by generally low levels of historical seismicity (Table 1 and Engdahl and Rinehart, 1988), although there are some zones of higher seismicity, such as New Madrid, Missouri, and the Charlevoix–La Malbaise–Saugenay region of the St. Lawrence Valley, as indicated on the seismicity map of Engdahl and Rinehart (1988) and described in Engdahl and others (this volume). Compilations of regional seismicity and tectonic relations for this region include the eastern United States (Barosh, 1990; Electric Power Research Institute, 1989), Canada (Adams and Basham, 1989; Hasegawa, 1988, 1991), the central United States (Johnston and Nava, 1990), and the southeastern United States (Johnston and others, 1984; Talwani, 1990).

The Midplate province is within the eastern part of the North American stable continental plate and has a nearly uniform compressional stress field, maximum horizontal compression being oriented in an east-northeast direction. This stress orientation correlates well with both the direction of absolute plate velocity and ridge push from the mid-Atlantic Ridge (Fig. 2, Zoback and Zoback, this volume). Within this province, heat flow is generally normal to below normal (50–60 Mwm^{-2}), but these are scattered areas of higher than average heat flow in regions of high crustal radioactivity, and lower than average values in areas of low heat-flow production or where there is regional redistribution of heat by large-scale ground-water flow (Figs. 1 and 2, Blackwell and others, this volume).

Until the last decade or so, the Midplate province was believed to have little or no potential for large earthquakes, active faulting, or neotectonic deformation. The great New Madrid earthquakes of 1811–1812, and the Charleston, South Carolina, earthquake of 1886 were considered by many workers to be special cases within an otherwise tectonically inactive province. During the past two decades these epicentral areas received special and intensive study, in part due to the need for comprehensive earthquake-hazard studies for vital engineering structures, particularly for nuclear power facilities. Investigation of the Charleston earthquake led Wentworth and Mergner-Keefer (1981) and Wentworth (1983) to speculate that there is a likelihood of other undiscovered seismic sources in the eastern United States. Although the issue is not resolved at this time, subsequent studies have not identified other major neotectonic structures along the eastern United States seaboard, and the liquefaction studies at the Charleston source area show repeated paleoseismic events with a short recurrence interval. This suggests that the Charleston source area may be a unique source for large events on the eastern seaboard.

The possibility of more widespread intraplate neotectonic activity was suggested by the recent recognition of the seismic potential of the spectacular Meers fault scarp along the Wichita frontal fault system of Oklahoma. The scarp was formed mainly by two late Holocene faulting events; the scarp is up to 5 m in height in a fault zone of at least 40 km length (Ramelli and Slemmons, 1990). A greater rupture length of 70 km is possible if the 30-km-long extension to the northwest reported by Cetin (1991) was a synchronous tectonic event. Ramelli and Slemmons (1990) noted that the activity on the Meers fault had two implications: (1) it presents a potential source of strong ground motion in a region that is generally thought to be tectonically stable and devoid of potential for large-magnitude earthquakes, and (2) the active section of the Meers fault may be part of a much longer zone that links the Amarillo-Wichita-Arbuckle uplift through the

Texas panhandle to other active structures, such as the Criner fault (Swan, 1989; Kelson and others, 1990a, 1990b), the Washita Valley fault (VanArsdale and others, 1989), and/or the Holocene and historically active Monroe uplift in northern Louisiana and the Wiggins uplift in Mississippi (Schumm, 1986).

The great historical 1811–1812 earthquakes of the New Madrid seismic zone (M = 8.1–8.3) and the large 1886 Charleston earthquake (M = 7.6; magnitudes determined by the Electric Power Research Institute [1989]), as well as the paleoseismic Meers scarp (M = 6.75–7.5, to 7.5–8; magnitudes estimated by Ramelli and Slemmons [1990]), and other recently studied examples of midplate tectonic activity, suggest that there is more widespread and significant intraplate deformation than is generally assumed for stable continental plates. Evidence for midplate activity now includes 20 historical earthquakes with surface-wave magnitude (M_s) > 6, and 15 faults, or epicentral areas, with significant Quaternary surface faulting, deformation, or liquefaction. The earthquake epicenters and neotectonic features are shown in Figure 1 and listed in Tables 1 and 2. Figure 1 and Tables 1 and 2 do not include smaller-scale faults of postglacial age that have maximum displacements of a small fraction of a meter, are less than a kilometer in length, and may be related to postglacial rebound (e.g., Oliver and others, 1970; Adams, 1989). The distributed occurrence of neotectonic features in North America suggests that neotectonic activity may include many older reactivated tectonic structures. Other unresolved issues include the following.

• Are there other seismogenic zones and faults similar to the 1811–1812 New Madrid seismic zone and the Meers fault that have not been recognized?

• Are any of the Neogene or older faults interconnected to form long, reactivated, neotectonic fault zones?

• Is the temporal clustering of activity that appears to be present at New Madrid or the Meers fault typical or commonplace in midplate settings?

• Is the earthquake activity of the New Madrid region unique to the Reelfoot rift, or is there a potential for large earthquakes on other structures of the Midplate province?

• Has the paucity of Quaternary and late Tertiary deposits limited recognition of many neotectonic features? Has the repeated and nearly continentwide Pleistocene glaciation removed much of the neotectonic evidence?

Neotectonic processes within stable continental plate and midplate regions are not well understood. Given the brief (and probably nonrepresentative) historical earthquake record, and the complex interplay between stresses and older faults and zones of weakness, it is difficult to resolve midplate neotectonic processes and the effects of specific geologic structures.

Major Active or Reactivated Fault Zones

Many of the historical and paleoseismic events appear to have occurred along reactivated parts of larger-scale faults, arches, and other major tectonic features of the Midplate province. In contrast, some smaller-scale earthquakes, such as the 1989 Ungava surface-faulting event in Quebec, occurred on smaller-scale structures that do not appear to have had earlier Quaternary offsets. The large range in size of features—from the great New Madrid earthquakes and the dramatic Meers fault scarp to small-scale postglacial "pop-ups" (Oliver and others, 1970; Adams, 1989)—raises the question of whether there is a continuum between major neotectonic structures such as the Reelfoot rift–New Madrid seismic zone and smaller-scale surface faults like that of Ungava, Quebec, or the even smaller-scale, postglacial faults.

MAJOR ACTIVE INTRAPLATE FAULTS

Reelfoot Rift–New Madrid Seismic Zone (earthquakes 5, 6, 7, and 8, Fig. 1, and paleoseismic event A, Figs. 1 and 2, and Tables 1 and 2)

The great earthquakes of 1811–1812 and the evidence that these events reactivated a major tectonic structure in North America provide the strongest indication for a major neotectonic structure in the Midplate province. Johnston and Nava (1990) emphasized the importance of New Madrid zone, including its possible extensions, with possibly unique factors that make the Reelfoot complex susceptible to a high rate of seismicity and generation of major earthquakes.

The New Madrid earthquakes of 1811–1812 not only included the three largest-magnitude historical earthquakes within the stable continental plate of North America, but may have included the three largest-magnitude events in the conterminous United States. The three main shocks were felt throughout most of the eastern half of United States. These three earthquakes (December 16, 1811, January 23, 1812, and February 7, 1912) are estimated to have had surface-wave magnitudes (M_s) between 8.4 and 8.8 (Nuttli, 1983; Street and Nuttli, 1984) and moment magnitudes (M_w) of 8.2, 8.1, and 8.3, respectively (Electric Power Research Institute, 1989).

The earthquakes were located along a major North American structural zone marked by prominent gravity and magnetic anomalies and by subsidence and warping of the Mississippi embayment in the Mesozoic and early Cenozoic. The zone appears to be a reactivated older structure, and has prominent complex and branching fault patterns to the north (Fig. 3), some of which also may be neotectonic.

Although the evidence is not conclusive for surface faulting from the 1811–1812 events, Russ (1979, 1982), Russ and others (1978), and Zoback (1979) demonstrated clearly a tectonic origin for the Reelfoot scarp and uplift of the Tiptonville dome (Fig. 4). Recent tectonism is also supported by the near-surface faulting and folding reported by Sexton and Jones (1986), and by faulting with associated folding, which appears to extend into the Holocene flood-plain deposits along the Mississippi River. Late Holocene uplift of the Tiptonville dome, deformation along the Reelfoot scarp, and associated subsidence of Reelfoot Lake are reported to have destroyed all of the bottomland hardwood trees

TABLE 2. CHARACTERISTICS OF PALEOSEISMIC STRUCTURES OR PALEOLIQUEFACTION EVENTS
IN THE MID-PLATE PROVINCE OF NORTH AMERICA

Event	Approximate Date of Last Event	Estimated Paleoseismic Magnitude	Latitude (°N)	Longitude (°W)	Recurrent Activity	Recurrence Interval (RI)	Comments	References
A	Holocene	Unknown	72	96.5	Probable	<<10,000 yr?	Regional uplift and scarps on Holocene marine terraces; Boothia Arch/Bell Arch, Northwest Territories	Dyke and others, 1991
B	Holocene	6 to 7.5	48.23	69.68	Yes	75 to 120 yr	Paleoseismic silting events in Lake Tadoussac, Quebec	Doig, 1991
C	Late Holocene	>5.9	48.17	70.84	Yes	3,300 yr (3 events) 10,000 yr	Paleoliquefaction; Ferland, Saguenay, Quebec	Tuttle and others, 1990
D	8,000 B.P.	>6	47 ± 2	80 ± 2	Yes	One per 600 yr until 8,000 B.P.	Six distributed paleoscarps; may include Event F; (Glacial Lake Barlow-Ojibway, Ontario)	Adams, 1989, and personal communication, 1991
E	Post-Sangamonian	Undetermined, ca. 7?	45	36	Yes	10,000 yr?	Paleoseismic offset of marine platform; Aspy fault, Nova Scotia	Grant, 1990
F	Holocene	≥6.3	47	79	Yes	1,100 yr (events at 400 B.P. and 1,500 B.P.)	Paleoseismic silting events in Lac Tee	Doig, 1991
G	Quaternary	6.5 to 7+	38.4	103.3	Yes	>100,00 yr?	Scarp on Quaternary surfaces; Cheraw fault, Las Animas Arch, Colorado	Kirkham and Rogers, 1981; Witkind, 1976
H	1,500 to 7,500 B.P.	≥6.2 to 6.7	38.4	88	Probable	>14,000 yr?	Paleoliquefaction events; Wabash Valley, Illinois/Idaho	Obermeier and others, 1991
I	Holocene?	6 to 7	37.9	84	Yes?	>100,000 yr?	Offset and folding of Quaternary terraces; Kentucky River fault, Kentucky	VanArsdale, 1986
J	~1,500 B.P.	>6.2	36.5	89.2	Yes	~468 yr	Paleoliquefaction	Saucier, 1991
K	~600 B.P.	Variable, 1811 event was ca. 8.4	36.5	89.5	Yes	600 to 700 yr events may be of variable magnitude with temporal clustering	Paleo-scarp and liquefaction; Reelfoot scarp, Missouri	McKeown, 1982; Russ and others, 1978; Russ, 1982; Sexton and Jones, 1986
L	1,100 B.P.	>6.75	35.0	98.7	Yes	>10,000 yr?	Paleo-offset drainages and Quaternary stratigraphic units; Meers fault, Oklahoma	Cetin, 1991
M	1,100 B.P.	6.75 to 8	34.8	98.5	Yes	Holocene RI = 1,800 yr and late Quaternary RI >100,000 yr with temporal clustering	Paleo-scarps and terrace offsets, Meers fault, Oklahoma	Luza and others, 1987; Madole, 1988; Ramelli, 1988; Ramelli and Slemmons, 1990

TABLE 2. CHARACTERISTICS OF PALEOSEISMIC STRUCTURES OR PALEOLIQUEFACTION EVENTS IN THE MID-PLATE PROVINCE OF NORTH AMERICA (continued)

Event	Approximate Date of Last Event	Estimated Paleoseismic Magnitude	Latitude (°N)	Longitude (°W)	Recurrent Activity	Recurrence Interval	Comments	References
N	Quaternary	Unknown	34.2	97.3	Unknown	Unknown	Quaternary strat. offsets; Criner fault, Oklahoma	Swan and Kelson, 1990; Kelson and others, 1990a, b
O	Holocene	>6.2	33.3	79	Unknown	Unknown	Paleoliquefaction event; Myrtle Beach, South Carolina	Amick and others, 1990; Obermeier and others, 1990
P	Holocene	>7.6+	32.9	80.0	Yes	Four events to 7,200 B.P. (<1,800 yr)	Historic and paleo-liquefaction; Charleston, South Carolina	Amick and others, 1990; Obermeier and others, 1990; Weems and others, 1986

within the impounded area (VanArsdale and others, 1991). An exploratory trench of the Reelfoot scarp (Russ and others, 1978) showed two pre-1811–1812 liquefaction events at the Reelfoot scarp during the past 2000 years, suggesting a recurrence interval of about 600 years, although the pre-1811–1812 events may have been of much lower magnitudes than the historical events.

The epicentral area for the 1811–1812 New Madrid earthquakes (Fig. 4) appears to be defined by low-magnitude earthquakes during the period July 1974 to March 1987 (Johnston and Nava, 1990) and by a zone of major liquefaction deposits (Obermeier, 1984; Obermeier and others, 1990). The earthquake sequence of 1811–1812 includes three main events (Fig. 4). (1) The two large December 16, 1811, earthquakes of about M_s = 8.2 (earthquakes 5 and 6, Table 1) are inferred to be mainly right-slip events in a narrow zone extending northeastward from Marked Tree to the southern part of the Lake County uplift; there is an adjacent zone of extensive liquefaction (Obermeier, 1984; Obermeier and others, 1990). (2) Activity continued on January 23, 1812, with an earthquake of about M_s = 8.1 (earthquake 7, Table 1). This event is believed to be associated with reverse-slip and thrust faults, and associated uplifts between New Madrid and Tiptonville dome–Lake County uplift–Ridgley Ridge; major uplift was above two southwestward-dipping reverse faults that are marked by the north-northwest–oriented zone of diffuse epicenters between New Madrid and Dyersburg. The deformation and uplift at Tiptonville dome and the Lake County uplift, the formation of the Reelfoot scarp, the subsidence at Reelfoot Lake, and the reverse-slip focal solutions (Herrmann and Canas, 1978) are features of this zone. (3) The highest-magnitude earthquake of M_s = 8.3 occurred on February 7, 1812 (earthquake 8, Table 1). It affected at least part of the Lake County uplift and may be associated with the right-slip prong of epicenters that extends northeastward from New Madrid toward Cairo, Missouri (Fig. 2).

Wesnousky and Leffler (1991) prepared many exploratory trenches near the Bootheel lineament between New Madrid and Marked Tree (Fig. 2), a possible neotectonic fault marked by an alignment of extensive 1811–1812 liquefaction deposits (Schweig and Jibson, 1989). The trenches and natural exposures reveal deposits of about 5 to 10 ka that show the major 1811–1812 liquefaction, but no evidence for earlier Holocene events. This suggests that the two paleoliquefaction events at the Reelfoot scarp were from epicenters at or north of the Tiptonville dome, and there was no pre-1811–1812 liquefaction between New Madrid and Marked Tree.

A possible northeastward extension of the zone of historical seismicity (earthquake 13, Fig. 2 and Table 1) and paleoliquefaction (events at J, Fig. 2 and Table 2) was proposed by Saucier (1991) on the basis of two liquefaction events at an archaeological site near Towosahgy, an event estimated at about A.D. 500 (<100 yr prior to A.D. 539), and a probable second event after A.D. 539, but before A.D. 991. The site is about 30 km northeast of the exploratory trench at the Reelfoot scarp and near the epicenter of the 1895 Charleston, Missouri, earthquake (earthquake 13, Fig. 1 and Table 1); estimated magnitudes are (body wave) m_b = 6.2 (Saucier, 1991) to M = 6.8 (Electric Power Research Institute, 1989). Saucier (1991) commented that the two events at Towosahgy could correlate with the liquefaction events observed in the trench at the Reelfoot scarp, and that the activity could be temporally clustered, with an average recurrence interval as short as 468 years, shorter than the 600–700 year interval estimated by McKeown (1982). This suggests that the most recent prehistoric events were from epicenters north of the Reelfoot scarp. These events may have been of lower magnitude than the historical earthquakes; they apparently did not cause liquefaction of Holocene sediments in the southern part of the seismic zone. This may explain the lack of pre-1811–1812 Holocene liquefaction deposits in the exploratory trenches described by Wesnousky and Leffler (1991), and supports the field observations for Holocene liquefaction between the Reelfoot scarp (Russ and others, 1978), Towosahgy (Saucier, 1991), and the Wabash Valley (Obermeier and others, 1991), and those noted by Kerr (1991).

Recent mapping of sand blows and vented surface sand (Fig.

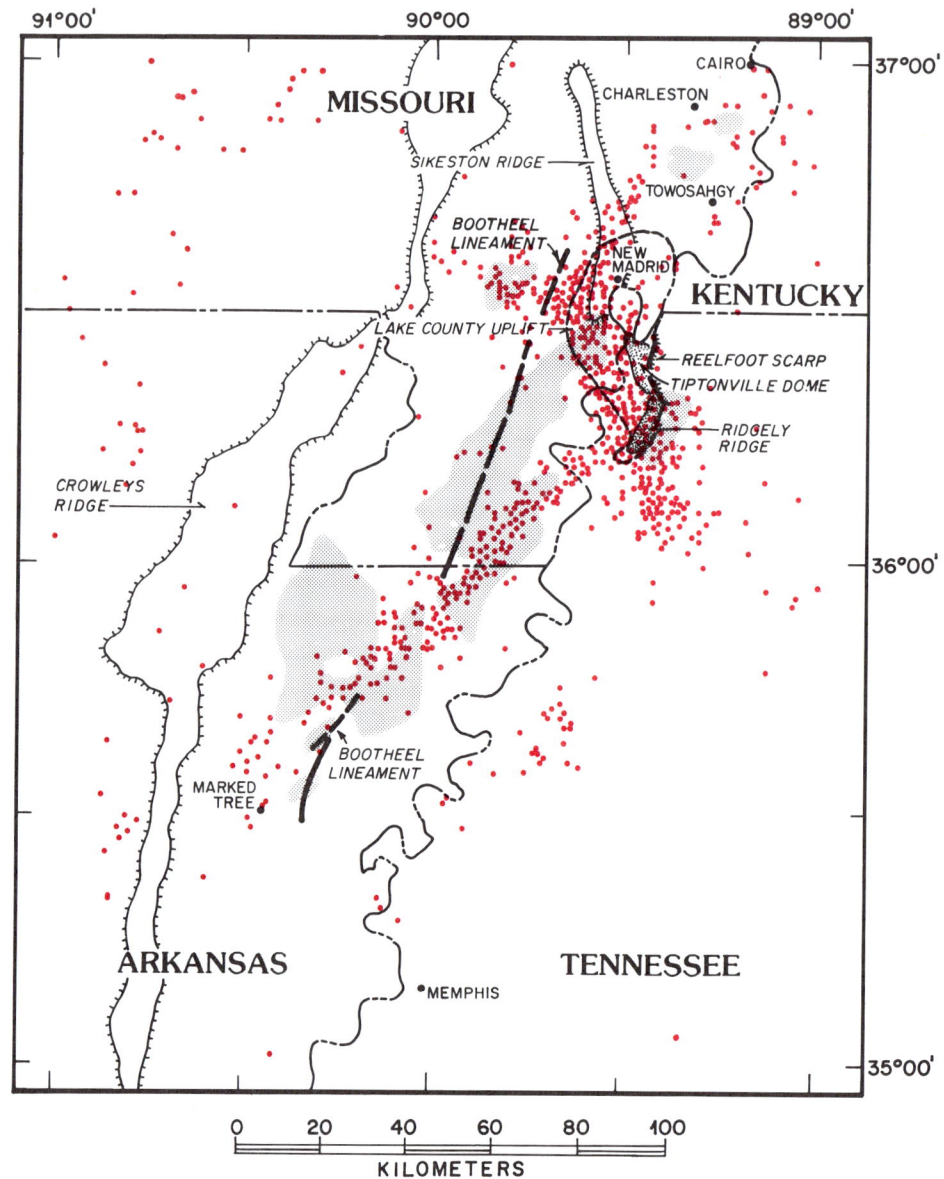

Figure 2. Seismicity of the New Madrid seismic zone from 1974 to 1987 showing earthquakes of M > 1. Adapted from Johnston and Nava (1990) and Reiter (1990). The earthquakes are mainly concentrated in two zones: a north-northwest-trending zone of diffuse reverse faulting and a narrow, nearly southwest-trending zone of right-lateral faulting. Shaded areas are regions of alluvium covered by > 25% liquefaction deposits (Obermeier, 1984). The Bootheel lineament, shown with the light dashed line, is from Schweig and Jibson (1989).

5) indicates that there were extensive effects of the 1811–1812 earthquakes north of New Madrid. This, along with the pattern of recent seismicity (Fig. 2), suggests that the 1811–1812 epicentral zone may have extended northward from New Madrid toward other fault systems (Fig. 3).

The evidence for earthquake recurrence in the New Madrid seismic zone is very strong, but the location of epicenters for prehistoric events is not well known. Although the recurrence interval for events at the Reelfoot scarps could be as few as 600 years (McKeown, 1982; Talwani and Collinsworth, 1988), the prehistoric earthquakes appear to be of much lower magnitude than the 1811–1812 events (Kerr, 1991). This rate of activity is much higher than the average Cenozoic or Quaternary rates, which suggests that there is currently a temporal clustering of activity.

Figure 3 shows part of the complex structural interaction with zones to the north, including northeastward to the Wabash Valley fault zone, northwestward to the St. Genevieve–Cottage

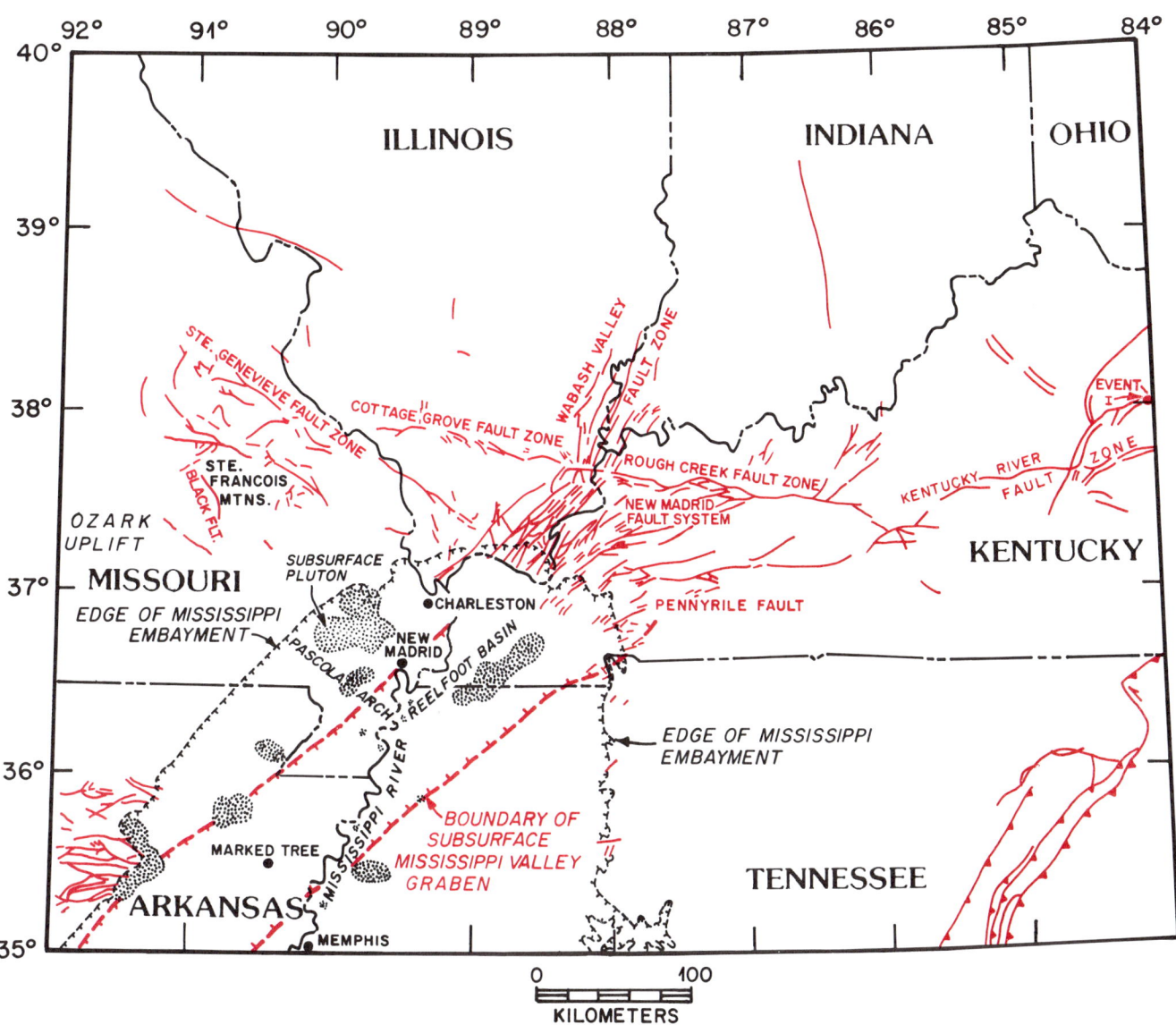

Figure 3. Regional tectonic features of the New Madrid seismic zone–Reelfoot rift zone and surrounding area showing the northern Mississippi embayment (light red hachures), generalized faults and fault systems (fine red lines), plutons (black stipple), and rift boundaries (heavy black lines). From Braile and others (1982), Heyl and McKeown (1978), and Hildebrand and others (1984).

Grove fault zones, and eastward to the Rough Creek fault zone, and possibly farther eastward to the Kentucky River fault zone. Possible extension of the Reelfoot rift–New Madrid seismic zone westward toward the Ouachita fold belt along the linear section of the Mississippi River is difficult to assess due to recent fluvial deposition and erosion.

Geologic and seismologic studies of the New Madrid rift lead to the conclusion that the zone is along a major older tectonic structure that was reactivated and is capable of large to great earthquakes and large-scale surface deformation. The aftershock patterns define major active faults and conjugate faults and associated uplifts. Several issues have not been resolved. (1) Does activity continue, perhaps at a lower rate, into adjacent, interconnected structural zones? (2) Are there other similar zones in other parts of the Midplate province? (3) Is the zone isolated, and does it have a unique character? According to Johnston and Nava (1990, p. 52–53), the neotectonic setting has four factors that combined make the zone more susceptible to large earthquakes:

First... it is a major, throughgoing crustal structure... Second, the rift is oriented ideally with respect to the regional stress regime... for the ratio of shear-to-normal stress to be maximized on preexisting fault systems... Third, the major Mesozoic-Cenozoic reactivation of the Reelfoot rift is tectonically relatively young, and its disruption has not had time to heal... Fourth, and most speculative, is the observation that the Reelfoot rift complex is saturated with water from the largest of the North American drainage systems.

Wabash Valley, Indiana (Event H, Fig. 1 and Table 2)

The north-northeast–striking Wabash Valley fault zone (Figs. 3 and 4) is well established by geologic studies, and appears to connect south-southwestward with the Reelfoot rift–New Madrid seismic zone (Braile and others, 1984). Neotectonic activity along the zone is indicated by historical seismicity and paleoliquefaction deposits. Obermeier and others (1991) used data from several tens of sand blows in a 60-km-long zone along the Wabash River in southern Indiana to demonstrate that they were caused by a substantial earthquake. They believe that all or nearly all of the sand dikes and vented sand from paleo-sand blows probably developed during a single event. Radiocarbon dating of contemporaneous archeological sites yields an age between 1500 and 7500 B.P., according to Obermeier and others (1991). The magnitude of the causative earthquake probably exceeded $m_b =$

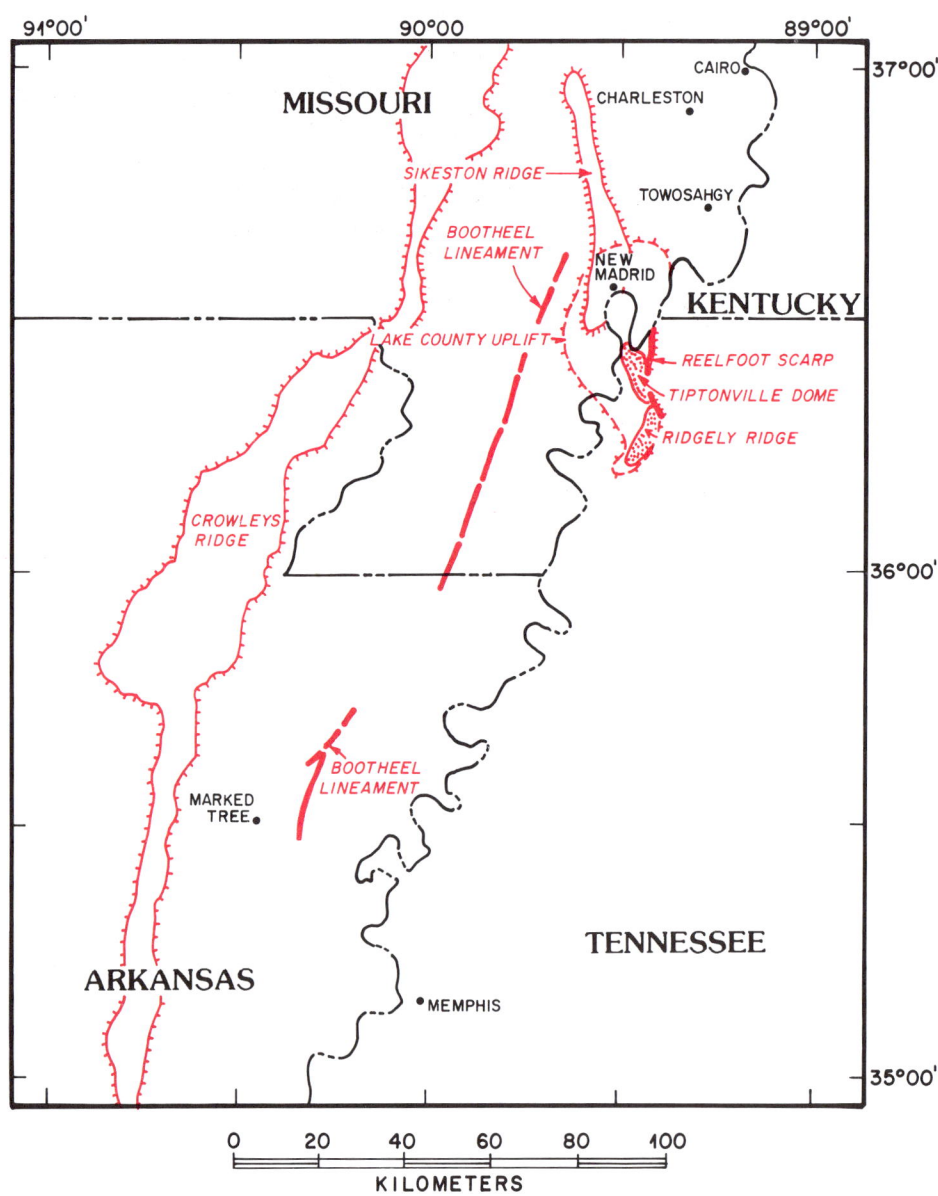

Figure 4. Map of Reelfoot scarp, Reelfoot Lake, and deformed areas. Generalized structural and geomorphic features include the disturbed Holocene Mississippi River Valley sediments at Reelfoot Lake, Tiptonville dome, Lake County uplift, and Ridgley ridge. Adapted from Russ (1982).

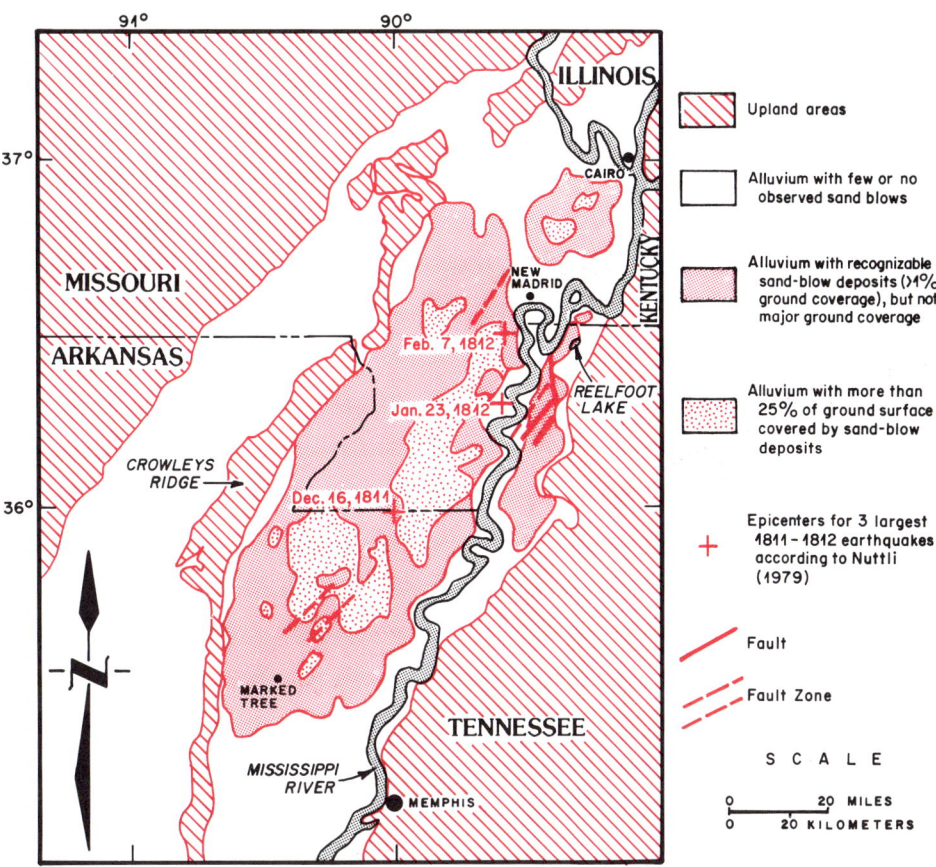

Figure 5. Map of areas covered by vented sand (Obermeier, 1984; Obermeier and others, 1990) and estimated locations of the strongest 1811–1812 earthquakes epicenters (from Obermeier and others, 1990; Nuttli, 1979).

6.3, and could have approached that of the 1886 Charleston, South Carolina, earthquake.

Kentucky River Fault System (Event I, Fig. 1 and Table 2)

The Kentucky River fault system is east of, and on strike with, the Rough Creek fault system, which appears to connect with the northern end of the Reelfoot rift system (Fig. 3). The Kentucky River fault system forms the north-bounding fault system of the Rome trough, a major Paleozoic structure of eastern Kentucky. It appears to be a system of very weak neotectonic activity. Exploratory trenches and boreholes at several sites along the fault system revealed reverse fault offsets of up to 1.1 m vertical displacement in Pleistocene terrace sediments, as well as associated folding (VanArsdale, 1986). No recurrent activity is noted at any of the sites; however, the timing of the offset varies and there may be more than one Quaternary event in different parts of the system (VanArsdale, 1986). The earthquake epicenters for this fault system are sparse and diffuse (Fig. 16 of Barosh, 1990; Fig. 6 of Johnston and Nava, 1990).

Wichita Frontal Fault System and Meers-Duncan-Criner Fault Zone (events L, M, and N of Figs. 1 and 7 and Table 2)

The recognition of the seismic potential for the Wichita frontal fault system in a region that is historically aseismic was not only a great surprise to geologists and seismologists, but raised the question of whether other major seismic sites have been overlooked in the Midplate province. The unexpected nature of this discovery, and its recentness (1983), augmented by the historical record of three great earthquakes in the New Madrid region, is leading geologists and seismologists to a fresh look at the potential of many other regions within the Midplate province.

The fault zone extends more than 600 km from the Amarillo-Wichita uplift eastward to the Duncan and Criner faults at the southern end of the Ouachita fold belt, and includes the Wichita frontal fault zone at the southwestern edge of the Anadarko basin (the deepest intracontinental basin in United States) and the Meers fault (Fig. 6). Evidence of neotectonic activity and/or seismicity is only sporadic along the 600 km long

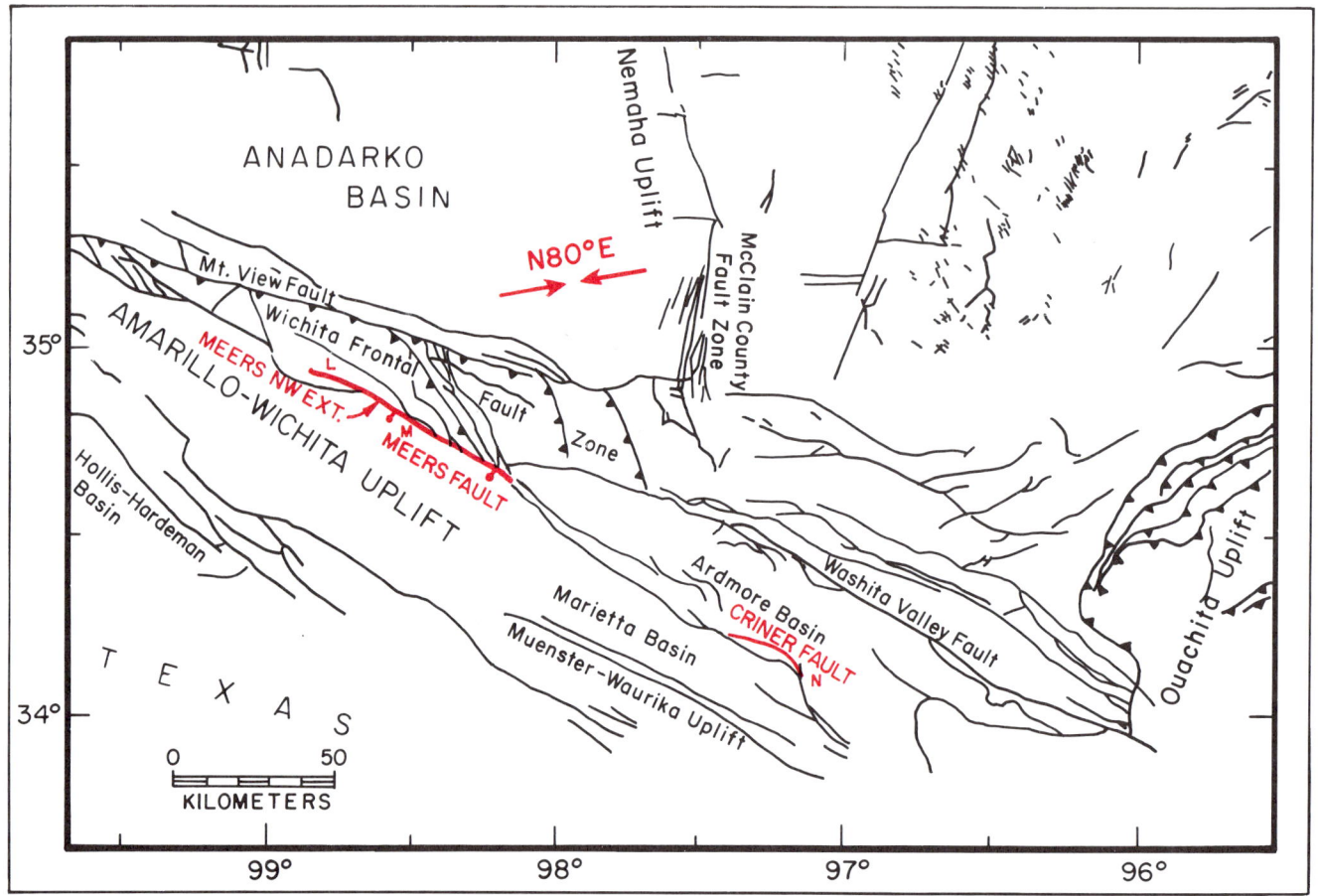

Figure 6. Map of the main structural features of southwestern Oklahoma (from Ramelli and Slemmons, 1990). Red overprint for known Holocene and Quaternary surface faults, including the Meers fault (Event M of Table 2, from Ramelli, 1988), northwest extension of the Meers fault (Event L of Table 2, from Cetin, 1991), and Criner fault (Event N, from Swan and Kelson, 1990). The N80°E direction is the orientation of the maximum horizontal stress axis (Dart, 1987).

zone, although the ductile character of some of the stratigraphic units of this zone may obscure geomorphic evidence for activity (Ramelli, 1988; Luza and others, 1987). Three parts of the zone show recognized Holocene or Quaternary activity: the Meers fault, the northwest extension of the Meers fault, and the Criner fault.

Gilbert (1983) first recognized the seismic potential of the Meers fault along part of the Wichita frontal fault zone by observing on aerial photographs a prominent paleoseismic scarp trending N63°W– in southwestern Oklahoma (event M, Fig. 1 and Table 2). The Meers fault scarp is the best-expressed late Quaternary fault scarp in the Midplate province. Its location is more than 600 km east of active faults of the Rio Grande rift. Gilbert observed that, although the fault is in a currently aseismic area, the youthful-appearance of the 5-m-high, 27-km-long paleoseismic fault scarp suggests that the fault could be the source of a future large earthquake. Studies by Ramelli (1988), Ramelli and Slemmons (1990), and Swan (1989) documented at least two late Holocene events with ruptures of several meters each and a late Holocene rupture length of 40 km. Fluvial terrace studies by Madole (1988) and exploratory trenching by Luza and others (1987) confirmed the recentness of the late Holocene faulting. The scarp is prominent and well defined in the relatively resistant Post Oak Conglomerate (Fig. 7), but is subdued and appears to be characterized by monoclinal deformation in the Hennessey Shale.

Field observations show that the left-lateral strike-slip component on the Meers fault is three to five times greater than the vertical offset, although the exploratory trench study of Kelson and others (1990a, 1990b) suggests a lower ratio of about 1.3 to 1.4, including two events with cumulative displacement of about 2 m each at about 1,100 and 2,800 B.P. This dominant left-slip component is consistent with the N80°E maximum horizontal stress orientation of Dart (1987).

The recent activity of the Meers fault presents several significant problems for analyses of seismic hazards of this and other similar midplate regions, and led Ramelli and Slemmons (1990) to pose the following questions.

• What is the frequency and how great is the potential for a damaging earthquake along this fault?

• To what extent have potential sources of damaging earthquakes been overlooked or underestimated in stable continental interior regions?

• Is this a one-of-a-kind feature, or does it suggest that conventional methods for evaluating seismic potential in stable continental interior regions may be inadequate and misleading?

Swan and Kelson (1990) identified Pleistocene offset along the Criner fault (event N, Figs. 1 and 7 and Table 2) and along the Meers-Duncan-Criner fault zone, about 120 km southeast of the Meers scarp (Swan and Kelson, 1990; Kelson and others, 1990a). No active faults were identified from aerial reconnaissance in the gap between the Criner fault and the Meers scarp. This observation suggests that the active fault segments are intermittently distributed along the fault system, or that plastic failure conceals some or most of the active segments. The nearby Washita Valley fault that is subparallel to the Criner fault has features that suggest Quaternary activity, but the activity is pre-Wisconsin in age (VanArsdale and others, 1989).

Cetin (1991) observed a 30-km-long northwestward extension of the Meers fault (event L, Figs. 1 and 6 and Table 2). He reported an offset in a buried soil with ^{14}C dates of 1090 ± 80 B.P. and 760 ± 70 B.P. that suggests a late Holocene faulting event that may be synchronous with the activity at the main Meers scarp. In addition, variations in lateral offsets of drainage lines suggest that there may have been many other Quaternary events.

St. Lawrence Valley–Charlevoix–La Malbaise–Saguenay zone (earthquakes 1, 2, 10, 11, 14, Fig. 1 and Table 1; events B and C, Table 2)

This zone is one of the most seismically active parts of the Midplate province (the other is the New Madrid region), and has major historic and paleoliquefaction activity. The Charlevoix–La Malbaise part of the St. Lawrence Valley zone is the most active region in eastern Canada (Hasegawa, 1988, 1991). Hasegawa noted that the St. Lawrence Valley structure is a reactivated part of a Paleozoic rift, and the Charlevoix part of the zone appears to have been affected by a 350 Ma meteorite impact. The zone was reactivated about 200 Ma with the opening of the present Atlantic Ocean, and also may be affected by the retreat of the ice sheet about 10 ka. The seismologic character is not discussed in this chapter (see discussions and references in Adams and Basham, 1989; Hasegawa, 1988, 1991), but two nearby localities, Lake Tadoussac and Ferland, Quebec, are included, mainly to document the paleoseismic activity.

Lake Tadoussac, in the Charlevoix, Quebec, region (event C, Fig. 1 and Table 2). Lake Tadoussac is a post-Wisconsin age lake and is about 40 km north of the edge of the main zone of post-1924 seismicity in the Charlevoix seismic zone

Figure 7. Low-sun-angle aerial photograph of the Meers fault in Post Oak Conglomerate. Scale 1" = 500 ft.

of the St. Lawrence River. Doig (1990) correlated six large historical earthquakes with thickness of associated silting events observed in cores from approximately the uppermost 1 m of sediments in Lake Tadoussac. He correlated the thickness of silt layers with historic earthquakes and determined similar paleoseismic events. Assuming a correlation between sediment thickness and earthquake magnitude, the thickness of the silts indicates estimates of M = 6 to 7 for prehistorical events, and an estimated average recurrence interval of 120 years from 300 B.C. to A.D. 800, and 75 years for the period A.D. 1500 to present (event 35, Fig. 1 and Table 1). Although the magnitude and frequency of paleoseismic earthquakes may be modified by future studies, these observations indicate that the zone had large and frequent prehistoric earthquakes.

Ferland, Sauguenay region, Quebec (event C, Table 2). Tuttle and others (1990) reported liquefaction and ground failures about 25–30 km north of the M = 5.9 Saguenay earthquake of 1988, and observed evidence for two earlier Holocene paleoseismic events after the deposition of the sediments (about 10 ka). They concluded that there is a 3,300 year recurrence interval for moderate to large earthquakes in this region.

Charleston, South Carolina (earthquake 12, Fig. 1; events O and P, Fig. 1 and Table 2)

The 1886 Charleston, South Carolina, earthquake had estimated m_b = 6.6–7.1 (Bollinger, 1977; Nuttli, 1983a) and m_s = 7.56 (Electric Power Research Institute, 1989). The present very low rate of seismicity makes it difficult to establish conclusively the earthquake source, although several faults and low earthquake activity are associated with an underlying Cenozoic graben (Talwani, 1990) (Fig. 8). Although the seismotectonic structure is not clear, the excellent record of recurrent paleoliquefaction events is firmly established.

The Charleston event was the largest historical earthquake along the Atlantic seaboard; it was accompanied by extensive liquefaction, effects extending to at least 35–40 km from the epicenter (Amick and others, 1990; Obermeier and others, 1990). Many studies indicate that there have been many paleoseismic events in this region (Gohn and others, 1984; Talwani and Cox, 1985; Obermeier and others, 1985, 1989, 1990; Weems and others, 1986; Talwani and Collinsworth, 1988). Weems and others (1986) indicated that there were four events dating back to 7,200 B.P., and they estimated an average recurrence rate of 1,800 years or less. Talwani (1990) estimated a maximum recurrence interval of about 1,000 to 2,000 years. Studies by Amick and Gelinas (1991) and Obermeier and others (1990) indicate that for the Charleston area, the recurrence interval for moderate to large earthquakes over the past 2,000 years was about 600 years. One or more pre-1886 events with epicenters north of Charleston, South Carolina (event O, Fig. 1 and Table 2), were reported by Obermeier and others (1990). The magnitudes of the pre-1886 events are not well established, but were at least M =

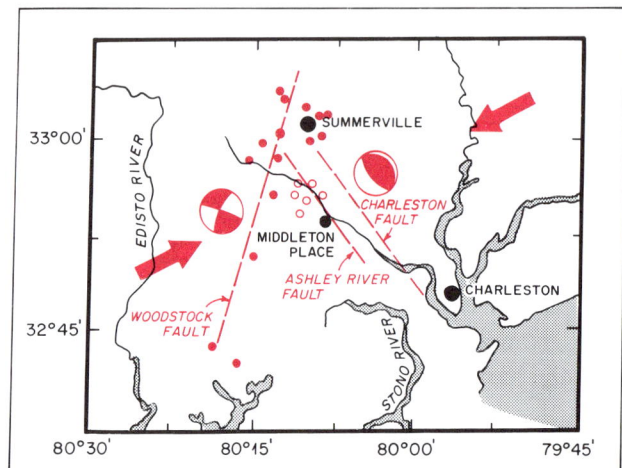

Figure 8. Faults of the Charleston, South Carolina, area (modified from Talwani, 1990).

5.5 (Obermeier and others, 1990). Some of the sand-blow craters appear to have dimensions and characteristics that, to Obermeier and others (1990), indicated that some of the paleoliquefaction events may have had stronger ground motion than for the same sites in 1886.

Aspy fault, northern Nova Scotia, Canada (event D, Fig. 1 and Table 2)

The Aspy fault is represented by a northerly-trending, east-facing scarp, 500 m high, that separates the Cape Breton Highlands Plateau of lower Paleozoic metamorphic rocks and granite from the Aspy Lowland of Carboniferous sedimentary rocks Grant (1990). The scarp has a 15 m offset of an emerged Sangamon (125 ka) intertidal rock platform. The large 15 m offset of the old uplifted marine terrace indicates that there were multiple surface-fault offsets along the scarp, and an average fault slip rate of about 0.12 mm/yr for post-Sangamon time along this trace.

NEOTECTONIC ACTIVITY AT MAJOR INTRAPLATE FOLDS AND ARCHES

Boothia arch or uplift-Bell arch, Northwest Territories, Canada (Event A, Fig. 1 and Table 2)

Dyke and others (1991) observed many northerly-trending, east-facing scarps or lineaments on Dixon Island and at Flexure Bay, Prince of Wales Island, about 115 km north of Dixon Island. The lineaments are sharp and linear and were interpreted by Dyke and others (1991) as possible active faults along the easern side of the seismically active (Hasegawa, 1988, 1991) Boothia arch near and on Prince of Wales Island. The main deformation is in a zone about 600 km long (trending north-south) and 100 km wide; there is post-9,300 B.P. deformation of Holocene marine terraces and about 60 to 100 m of uplift. Locally the scarps are up to 1 m high, with lengths of up to 1 km. The observed short length of the scarps, at least in part, may be due to concealment beneath the waters of Peel Sound and Franklin Straight. The scarps offset the glaciated surface, and the east faces of the scarps are devoid of glacial polish. They have steeply inclined striae that suggest a postglacial fault origin. The scarps are back-facing relative to the main western boundary of Boothia arch.

The Cheraw fault along the Las Animas arch, Colorado (event G, Fig. 1 and Table 2)

The Cheraw fault is in the eastern part of Colorado, about 100 km east of the Rocky Mountains (Witkind, 1976; Kirkham and Rogers, 1981) and in the Midplate province (Zoback and Zoback, this volume). The fault is aligned along a prominent gravity gradient that extends more than 100 km along the Las Animas arch. The curvilinear trace of the fault strikes N45°E, and offsets Rocky Flats Alluvium by 12 m; the northwest side is down. The 12 m height of the scarp indicates recurrent faulting events, although there are no detailed field studies of the scarp. The length of the scarp is greater than 50 km. The most recent offset is pre-Holocene, but younger than late Pleistocene. It is marked by prominent sag ponds that may reflect ground-water movement along fault zones. The fault type is unknown, although the gently curvilinear trace suggests that the fault plane is steep or vertical.

Minor Faults

Ungava, Quebec, from the earthquake of December 25, 1989 (earthquake 20, Figs. 1 and 9 and Tables 1 and 2)

Minor faults with surface faulting are present within the stable continental interiors of many continents. Within Australia, for example, five midplate earthquakes of M = 6 to 7 developed surface scarps with associated thrust faulting of up to 37 km length (McCue and others, 1987). Within the Midplate province of North America, the only example of this type is the clearly expressed historic surface faulting at Unvaga (Fig. 9). The Unvaga earthquake of M_s = 6.3 had left reverse-slip surface faulting of more than 8.5 km length, with maximum displacement of 1.8 to 2 m (Adams and others, 1990, 1991; Drysdale and others, 1990). The shallow-focus fault rupture is in well-exposed glaciated Precambrian gneiss, and is not located at a major structure. Adams (1991, oral commun.) indicated that, although there may be some evidence for paleoscarps, there is no definitive evidence for Holocene paleoseismicity, and the earthquake does not appear to have reactivated a major earlier zone of weakness. Other examples probably include event D of Table 2.

POSTGLACIAL ISOSTATIC REBOUND

A neotectonics perspective of postglacial isostatic rebound in North America, Greenland, and Iceland is presented by Andrews (this volume). The effects are mainly in areas that were previously covered by glacial ice or large Pleistocene pluvial lakes. Deglaciation of the main ice sheets started about 14 to 15 ka, and in many regions most of the glaciers had largely disappeared by about 10 ka. Maximum vertical elevation changes are about 300 m, with an exponential decrease in uplift rate after deglaciation. Initial rates exceeded 10 m/100 years, but present rates are 1 to 2 m/100 years. In addition to uplift and associated tilts, Andrews (this volume) cites nine references that report evidence of postglacial faulting on faults that vary from small (Oliver and others, 1970; Adams, 1989) to a vertical separation of about 30 m (Dyke and others, unpublished). In Fennoscadia, many scarps of great length (up to 150 km) and significant height (10 m) developed during the final stages of deglaciation, but similar very large offsets have not been described in North America.

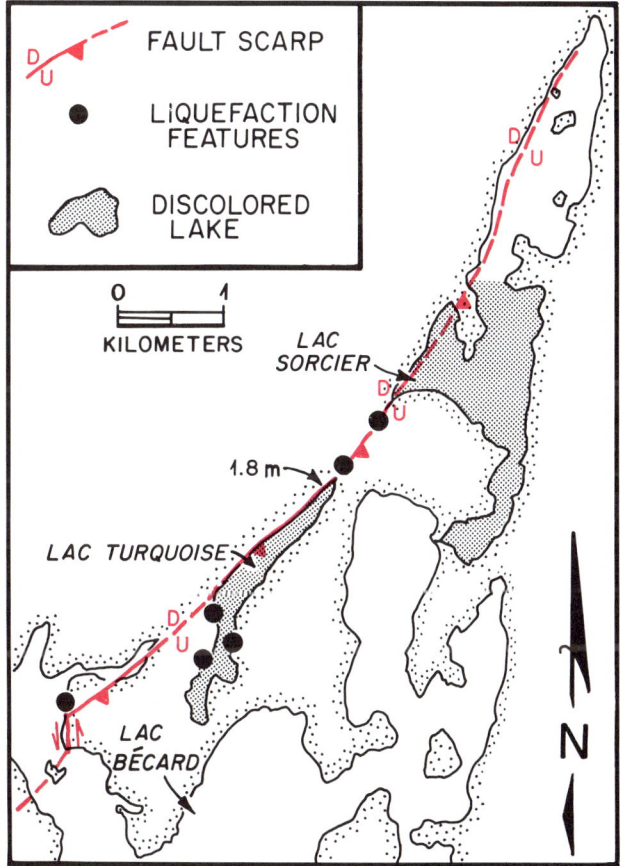

Figure 9. Map of the 1989 Ungava, Quebec, surface-faulting event (lat 60°12′N, long 73°60′W) (after Adams and others, 1990).

REFERENCES CITED

Adams, J., 1989, Postglacial faulting in eastern Canada: Nature, origin and seismic hazard implications: Tectonophysics, v. 163, p. 323-331.

Adams, J., and Basham, P., 1989, The seismicity and seismotectonics of Canada east of the Cordillera: Geoscience Canada, v. 16, no. 1, p. 3-16.

Adams, J., Wetmiller, R. J., Hasegawa, H. S., Lamontagane, M., and Drysdale, J., 1990, Surface faulting and lakeshore deformation caused by the 1989 Ungava earthquake: Seismological Research Letters, v. 61, no. 3-4, p. 143.

Adams, J., Wetmiller, J. D., and Hasegawa, H. S., 1991, The first surface rupture from an earthquake in eastern North America, in Current research, Part C: Geological Survey of Canada Paper 91-1C, p. 9-15 (in press).

Amick, D., and Gelinas, R., 1991, Paleoliquefaction investigations along the Atlantic seaboard: Results of recent studies and plans for future investigations: Eos (Transactions, American Geophysical Union), v. 72, no. 17, p. 195.

Amick, D., Maurath, G., and Gelinas, R., 1990, Characteristics of seismically induced liquefaction sites and features located in the vicinity of the 1889 Charleston, South Carolina earthquake: Seismological Research Letters, v. 61, no. 2, p. 117-130.

Anderson, L. E., and Miller, D. G., 1979, Quaternary fault map of Utah: Long Beach, California, Fugro Inc., 35 p., scale: 1:1,000,000.

Barosh, P. J., 1990, Neotectonic movement and earthquake assessment in the eastern United States, in Krinitzky, E. L., and Slemmons, D. B., eds., Neotectonics in earthquake evaluation: Geological Society of America Reviews in Engineering Geology, v. 8, p. 77-109.

Bates, R. L., and Jackson, J. A., 1987, Glossary of geology: Alexandria, Virginia, American Geological Institute, 786 p.

Blackwell, D. D., Steele, J. L., and Carter, L. C., 1991, Geothermal map of North America: Geological Society of America map CSM-007, 4 sheets, scale 1:5,000,000 (in press).

Bollinger, G. A., 1983, Speculations on the nature of seismicity at Charleston, South Carolina, in Gohn, G. S., ed., Studies related to the Charleston, South Carolina earthquake of 1886—Tectonics and seismicity: U.S. Geological Survey Professional Paper 1313, p. T-1-T-11.

Bollinger, G. A., Sibol, M. S., and Chapman, M. C., 1991, The areal distribution of earthquake intensity levels—eastern United States versus western United States: Geological Society of America, Abstracts with Programs, v. 23, no. 1, p. 9.

——, 1991, Distribution of earthquake intensity levels, eastern United States versus western United States: Geological Society of America Abstracts with Programs, v. 23, p. 9.

Braile, L. W., Hinze, W. J., Keller, G. R. and Lidiak, E. G., 1982, The northeastern extension of the New Madrid seismic zone, in McKeown, F. A., and Pakiser, L. C., eds., Investigations of the New Madrid, Missouri, earthquake region: U.S. Geological Survey Professional Paper 1236, p. 175-184.

Braile, L. W., Hinze, W. J., Sexton, J. L., Keller, G. R., and Lidiak, E. G., 1984, Tectonic development of the New Madrid seismic zone, in Gori, P. L., and Hays, W. W., eds., Proceedings of the symposium on the New Madrid seismic zone: U.S. Geological Survey Open-File Report 84-770, p. 204-233.

Cetin, H., 1991, The northwest extension of the Meers fault, Oklahoma: Geological Society of America Abstracts with Programs, v. 23, no. 4, p. 10.

Committee for the Magnetic Map of North America, 1987, Magnetic anomaly map of North America: Geological Society of America map CSM-003, 4 sheets, scale 1:5,000,000.

Crone, A. J., and Omdahl, E. M., eds., 1987, Directions in paleoseismology: U.S. Geological Survey Open-File Report 87-673, 456 p.

Dart, R. L., 1987, South-central United States well-bore breakout-data catalog: U.S. Geological Survey Open-File Report 87-405, 95 p.

dePolo, C. M., and Slemmons, D. B., 1990, Estimation of earthquake size for earthquake hazards: Geological Society of America Reviews in Engineering Geology, v. 8, p. 1-28.

Doig, R., 1990, 2300 yr history of seismicity from silting events in Lake Tadoussac, Charlevoix, Quebec: Geology, v. 18, p. 820-823.

——, 1991, Effects of strong seismic shaking in lake sediments and earthquake recurrence interval, Temiscaming, Quebec: Canadian Journal of Earth Sciences, v. 23, p. 930-937.

Drysdale, J. A., Lamontagne, M., Wetmiller, R. J., Adams, J., Lapointe, S. P., and Cajka, M. G., 1990, Investigations of the Ungava earthquake (poster): Seismological Research Letters, v. 61, no. 3-4, p. 149.

Dyke, A. S., Morris, T. S., and Green, D.E.C., 1991, Postglacial tectonic and sea level history of the central Canadian Arctic: Geological Survey of Canada Bulletin 397, 56 p.

Electric Power Research Institute, 1989, Methods for assessing maximum earthquakes in the central and eastern United States: Palo Alto, California, Geomatrix Consultants, Inc., and Center for Earthquake Research and Information, Review Draft, Chapters 2 and 3, Appendices A, B, and C, EPRI Project RP-2556-12.

Engdahl, E. R., and Rinehart, W. A., compilers, 1988, Seismicity map of North America: Geological Society of America map CSM-004, 4 sheets, scale 1:5,000,000.

Gilbert, M. C., 1983, The Meers fault of southwestern Oklahoma; evidence for possible strong Quaternary seismicity in the midcontinent [abs.]: Eos (Transactions, American Geophysical Union), v. 64, no. 18, p. 313.

Gohn, G. S., Weems, R. E., Obermeier, S. F., and Galinas, R. L., 1984, Field studies of earthquake-induced liquefaction-flowage features in the Charleston, South Carolina area—Preliminary report: U.S. Geological Survey Open-File Report 84-670, 26 p.

Grant, D. R., 1990, Late Quaternary movement of Aspy fault, Nova Scotia: Canadian Journal Earth Sciences, v. 27, p. 984-987.

Gravity Anomaly Map Committee, 1987, Gravity anomaly map of North America: Geological Society of America map CSM-002, 4 sheets, scale 1:5,000,000.

Hasegawa, H. S., 1988, Seismogenesis in eastern Canada: Seismological Research Letters, v. 59, no. 4, p. 219-225.

——, 1991, Four seismogenic environments in eastern Canada: Tectonophysics, v. 186, p. 3-17.

Herrmann, R. B., and Canas, J., 1978, Focal mechanism studies in the New Madrid seismic zone: Seismological Society of America Bulletin, v. 68, no. 4, p. 1095-1102.

Heyl, A. V., and McKeown, F. A., 1978, Preliminary seismotectonic map of central Mississippi Valley and environs: U.S. Geological Survey Miscellaneous Field Studies Map MF-1011, scale 1:500,000.

Hildebrand, T. G., Kane, M. F., and Hendricks, J. D., 1984, Magnetic basement in the upper Mississippi embayment region—A preliminary report, in McKeown, F. A., and Pakiser, L. C., eds., Investigations of the New Madrid, Missouri, earthquake region: U.S. Geological Survey Professional Paper 1236, p. 39-53.

Jennings, C. W., Strand, R. G., Rogers, T. H., Stinson, M. C., Burnett, J. L., Kahle, J. E., Streitz, R., and Stinson, R. A., 1975, Fault map of California, with locations of volcanoes, thermal springs, and thermal wells: California Division of Mines and Geology, California Map Series, 1 sheet, scale 1:750,000.

Johnston, A. C., and Nava, S. J., 1990, Seismic-hazard assessment in central United States, in Krinitzsky, E. L., and Slemmons, D. B., eds., Neotectonics in earthquake evaluation: Geological Society of America Reviews in Engineering Geology, v. 8, p. 47-58.

Johnston, A. C., Reinbold, D. J., and Brewer, S. I., 1984, Seismotectonics of the southern Appalachians: Seismological Society of America Bulletin, v. 75, p. 291-312.

Kelson, K. I., Swan, F. M., and Coppersmith, K. J., 1990a, Implications of late Quaternary faulting along the Meers and Criner faults to assessments of seismic hazards in the central and eastern U.S.: Geological Society of America Abstracts with Programs, v. 22, no. 1, p. 10.

Kelson, K. I., Swan, F. H., and Wesling, J. R., 1990b, Late Quaternary displacements on the Meers fault, southern Oklahoma: Geological Society of America Abstracts with Programs, v. 22, no. 1, p. 10-11.

Kerr, R. A., 1991, A fruitless search for great midwest quakes: Science News, v. 139, no. 24, p. 1497.

King, P. B., compiler, 1969a, Tectonic map of North America: U.S. Geological Survey, scale 1:5,000,000.

—— , 1969b, The tectonics of North America: U.S. Geological Survey Professional Paper 628, 95 p.

Kirkham, R. M., and Rogers, W. P., 1981, Earthquake potential in Colorado: A preliminary evaluation: Colorado Geological Survey Bulletin, v. 43, 1 sheet, scale: 1:500,000 and 171 p.

Krinitzsky, E. L., and Slemmons, D. B., eds., 1990, Neotectonics in earthquake evaluation: Geological Society of America Reviews in Engineering Geology, v. 8, 160 p.

Leffler, L. M., and Wesnousky, S. G., 1991, Paleoseismic study of the southern New Madrid seismic zone by using liquefaction features as seismic indicators: Eos (Transactions, American Geophysical Union), v. 72, no. 17, p. 196.

Luza, K. V., Madole, R. F., and Crone, A. J., 1987, Investigation of the Meers fault, southwestern Oklahoma: Oklahoma Geological Survey Special Publication 87-1, 75 p.

Madole, R. F., 1988, Stratigraphic evidence of Holocene faulting in the midcontinent; the Meers fault, southwestern Oklahoma: Geological Society of America Bulletin, v. 100, p. 392–401.

McCue, K., Barlow, B. C., Denham, D., Jones, T., Gibson, G., and Michael-Leiba, M., 1987, Another chip off the old Australian block: Eos (Transactions, American Geophysical Union), v. 68, p. 609–612.

McKeown, F. A., 1982, Overview and discussion, in McKeown, F. A., and Pakiser, L. C., eds., Investigations of the New Madrid earthquake region: U.S. Geological Survey Professional Paper 1236, p. 1–14.

Menges, C. M., 1983, The neotectonic framework of Arizona: Implications for the regional character of basin-range tectonism: Arizona Bureau of Geology and Mineral Technology, Open-File Report 83-19, 109 p.

Menges, C. M., and Pearthree, P. A., 1983, Map of neotectonic (latest Pliocene–Quaternary) deformation in Arizona: Arizona Bureau of Geology and Mineral Technology, Open-File Report 83-22, 2 sheets, scale 1:500,000 and 48 p.

—— , 1989, Late Cenozoic tectonism in Arizona and its impact on regional landscape evolution, in Jenney, J. P., and Reynolds, S. J., eds., Geologic evolution of Arizona: Arizona Geological Society Digest, v. 17, p. 649–680.

Mörner, N., 1990, Neotectonics and structural geomorphology: General introduction: International Quaternary Association Neotectonics Commission Bulletin, no. 13, p. 87.

Morrison, R. B., 1991, Introduction, in Morrison, R. B., ed., Quaternary nonglacial geology; Conterminous U.S.: Boulder, Colorado, Geological Society of America, The Geology of North America, v. K-2, p. 1–12.

Morrison, R. B. and Menges, C. M., 1981, Neotectonic maps of Arizona: Arizona Geological Society Digest, v. 13, p. 179–183.

Nakata, J. K., Wentworth, M., and Machette, M. N., 1982, Quaternary fault map of the Basin and Range and Rio Grande rift provinces, western United States: U.S. Geological Survey Open-File Report 82-579, 2 sheets, scale 1:1,000,000.

Nuttli, O., 1979, Seismicity of the central United States, in Hatheway, A. W., and McClure, C. R., Jr., eds., Geology in the siting of nuclear power plants: Geological Society of America Reviews in Engineering Geology, v. 4, p. 67–93.

—— , 1983, Average seismic-source parameter relations for mid-plate earthquakes: Seismological Society of America Bulletin, v. 73, p. 519–535.

Nuttli, O. W., and Brill, K. G., Jr., 1981, Catalog of central United States earthquakes since 1800 of $m_b \geq 3.0$, Part 2 and Appendix B-2, in Barstow, N. L., Brill, K. G., Jr., Nuttli, O. W., and Pomeroy, P. W., eds., An approach to seismic zonation for siting nuclear electric power generating facilities in the eastern United States: Washington, D.C., U.S. Nuclear Regulatory Commission NUREG/CR-1577, p. 97–143, B2-1–B2-31.

Obermeier, S. F., 1984, Liquefaction potential in the central Mississippi Valley, in Gori, P. L., and Hays, W. W., eds., Proceedings: Symposium on the New Madrid earthquakes: U.S. Geological Survey Open-File Report 84-770, p. 391–446.

Obermeier, S. F., Gohn, G. S., Weems, R. E., Galinas, R. L., and Rubin, M., 1985, Geologic evidence for recurrent moderate to large earthquakes near Charleston, South Carolina: Science, v. 227, p. 408–411.

Obermeier, S. F., Weems, R. E., and Jacobson, R. B., 1987, Earthquake-induced liquefaction features in the coastal South Carolina region: U.S. Geological Survey Open-File Report 87-504, 20 p.

Obermeier, S. F., Weems, R. E., Jacobsen, R. B., and Gohn, G. S., 1989, Liquefaction evidence for repeated Holocene earthquakes in the coastal region of South Carolina: New York Academy of Sciences Annals, v. 558, p. 183–195.

Obermeier, S. F., and 10 others, 1991, Evidence of strong earthquake shaking in the Wabash Valley from prehistoric liquefaction features: Science, v. 251, p. 1061–1063.

Oliver, J., Johnson, T., and Dorman, J., 1970, Postglacial faulting and seismicity in New York and Quebec: Canadian Journal of Earth Sciences, v. 7, p. 579–590.

Pearthree, P. A., Menges, C. M., and Mayer, L., 1983, Distribution, recurrence, and possible tectonic implications of late Quaternary faulting in Arizona: Arizona Bureau of Geology and Mineral Technology Open-File Report 83-20, 51 p

Ramelli, A. R., 1988, Late Quaternary tectonic activity of the Meers fault, southwestern Oklahoma [M.S. thesis]: Reno, University of Nevada, 123 p.

Ramelli, A. R., and Slemmons, D. B., 1990, Implications of the Meers fault on seismic potential in the central United States, in Krinitzski, E. L., and Slemmons, D. B., eds., Neotectonics in earthquake evaluation: Geological Society of America Reviews in Engineering Geology, v. 8, p. 59–75.

Reed, J. C., Wheeler, J. O., Tucholke, B. E., and Palmer, A. R., 1992, Geologic map of North America: Geological Society of America map CSM-001, 4 sheets, scale: 1:5,000,000 (in press).

Reiter, L., 1990, Earthquake hazard analysis: Issues and insights: New York, Columbia University Press, 254 p.

Russ, D. P., 1979, Late Holocene faulting and earthquake recurrence in the Reelfoot Lake area, northwestern Tennessee: Geological Society of America Bulletin, Part I, v. 90, p. 1013–1018.

—— , 1982, Style and significance of surface deformation in the vicinity of New Madrid, Missouri, in McKeown, F. A., and Pakiser, L. C., eds., Investigations of the New Madrid, Missouri, earthquake region: U.S. Geological Survey Professional Paper 1236-H, p. 95–114.

Russ, D. P., Stearns, R. G., and Herd, D. G., 1978, Map of exploratory trench across Reelfoot scarp, northwestern Tennessee: U.S. Geological Survey Miscellaneous Field Studies Map MF—985.

Saucier, R. T., 1991, Geoarchaeological evidence of strong prehistoric earthquakes in the New Madrid (Missouri) seismic zone: Geology, v. 19, p. 296–298.

Scarborough, R. B., Menges, C. M., and Pearthree, P. A., 1983, Map of Basin and Range (post 15 M.Y.A.) exposed faults, grabens, and basalt-dominated volcanism in Arizona: Arizona Bureau of Geology and Mineral Technology Open-File Report 83-21, 2 sheets, scale: 1,500,000 and 25 p.

Scarborough, R. B., Menges, C. M., and Pearthree, P. A., 1986, Map of late Pliocene–Quaternary (post-4 m.y.) faults, folds, and volcanic outcrops in Arizona: Tucson, Arizona Bureau of Geology and Technology map 22, scale 1:000,000.

Schumm, S. A., 1986, Alluvial river response to active tectonics, in Wallace, R. E., ed., Active tectonics, studies in geophysics: Washington, D.C., National Academy of Science, 266 p.

Schwartz, D. P., and Sibson, R. H., eds., 1989, Fault segmentation and controls of rupture initiation and termination: U.S. Geological Survey Open-File Report 89-315, 447 p.

Schweig, E. S., III, and Jibson, R. W., 1989, Surface effects of the 1811–1812 New Madrid earthquake sequence and seismotectonics of the New Madrid seismic zone, western Tennessee, northeastern Arkansas, and southeastern Missouri, in Vineyard, J. E., and Wedge, W. K., eds., Gateway to the future . . . frontiers in geoscience: Rolla, Missouri Department of Natural Resources, Division of Geology and Land Survey, Geological Society of

America 1989 field trip guidebook, p. 48–68.
Seilacher, A., 1969, Fault-graded beds interpreted as seismites: Sedimentology, v. 13, p. 155–159.
Sexton, J. L., and Jones, P. B., 1986, Evidence for recurrent faulting in the New Madrid seismic zone from Mini-Sosi high-resolution reflection data: Geophysics, v. 51, no. 9, p. 1760–1788.
Slemmons, D. B., and dePolo, C. M., 1986, Evaluation of active faulting and associated hazards, in Wallace, R. E., ed., Active tectonics: Washington, D.C., National Academy Press, p. 45–62.
Street, R., and Nuttli, O. W., 1984, The central Mississippi Valley earthquakes of 1811–1812, in Hays, W. W., and Gori, P. L., eds., Proceedings of the symposium on The New Madrid Seismic Zone: U.S. Geological Survey Open-File Report 84-770, p. 33–63.
Swan, F. H., 1989, Preliminary results of paleoseismic investigations along the Meers Valley fault, southwestern Oklahoma [abs.]: Geological Association of Canada Program with Abstracts, v. 14, p. 128–129.
Swan, F. H., and Kelson, K. I., 1990, Segmentation of the Meers-Duncan-Criner fault zone, Oklahoma: Implications to active tectonics in the mid-continent: Geological Society of America Abstracts with Programs, v. 22, no. 7, p. A18.
Talwani, P., 1982, Internally consistent pattern of seismicity near Charleston, South Carolina: Geology, v. 10, p. 654–658.
—— , 1990, Neotectonics in the southeastern United States with emphasis on the Charleston, South Carolina area, in Krinitzsky, E. L., and Slemmons, D. B., eds., Neotectonics in earthquake evaluation: Geological Society of America Reviews in Engineering Geology, v. 8, p. 111–129.
Talwani, P., and Collinsworth, K., 1988, Recurrence intervals for intraplate earthquakes in eastern North America from paleoseismological data: Seismological Research Letters, v. 59, no. 5, p. 207–211.
Talwani, P., and Cox, J., 1985, Paleoseismic evidence for recurrence of earthquakes near Charleston, South Carolina: Science, v. 229, p. 379–381.
Tuttle, M., Law, K. T., Seeber, L., and Jacob, K., 1990, Liquefaction and ground failure induced by the 1988 Saguenay, Quebec, earthquake: Canadian Geotechnical Journal, v. 27, p. 580–589.
VanArsdale, R. B., 1986, Quaternary displacement on faults within the Kentucky River fault system of east-central Kentucky: Geological Society of America Bulletin, v. 97, p. 1382–1392.
VanArsdale, R., Ward, C., and Cox, R., 1989, Post-Pennsylvanian reactivation along the Washita Valley fault, southern Oklahoma: U.S. Nuclear Regulatory Commission, NUREG/CR-5375, 48 p.
VanArsdale, R. B., Stahle, D. N., and Cleaveland, M. K., 1991, Tectonic deformation revealed in bald cypress tree-rings of Reelfoot Lake, Tennessee: Eos (Transactions, American Geophysical Union), v. 72, no. 17, p. 196.
Vittori, E., Sylos Labini, S., and Serva, L., 1991, Paleoseismology: Review of the state-of-the-art: Tectonophysics, v. 193, p. 9–32.
Wallace, R. E., 1986, Overview and recommendations, in Wallace, R. E., ed., Active tectonics: Washington, D.C., National Academy Press, p. 3–19.
—— , 1990, The San Andreas fault system, California: U.S. Geological Survey Professional Paper 1515, 283 p.
Weems, R. E., Obermeier, S. F., Pavich, M. J., Gohn, G. S., Rubin, M., Phipps, R. L., and Jacobson, R. B., 1986, Evidence for three moderate to large prehistoric Holocene earthquakes near Charleston, South Carolina: Proceedings of the Third U.S. National Conference on Earthquake Engineering, Volume 1: Earthquake Engineering Research Institute, p. 3–13.
Wentworth, C. M., 1983, Regenerate faults of small Cenozoic offset; probable earthquake sources in the southeastern United States: U.S. Geological Survey Professional Paper 1313-S, 20 p.
Wentworth, C. M., and Mergner-Keefer, M., 1981, Reverse faulting along the eastern seaboard and the potential for large earthquakes, in Beavers, J. E., ed., Earthquakes and earthquake engineering; the eastern United States: Ann Arbor, Michigan, Ann Arbor Science, p. 109–128.
Wesnousky, S. G., and Leffler, L., 1991, Search for paleoearthquakes in the New Madrid seismic zone: Seismological Research Letters, v. 62, no. 1, p. 38.
Witkind, I. J., 1975a, Preliminary map showing known and suspected active faults in Idaho: U.S. Geological Survey Open-File Report 75-278, 1 sheet, scale 1:500,000 and 71 p.
—— , 1975b, Preliminary map showing known and suspected active faults in Wyoming: U.S. Geological Survey Open-File Report 75-279, 1 sheet, scale 1:500,000 and 36 p.
—— , 1975c, Preliminary map showing known and suspected active faults in Montana: U.S. Geological Survey Open-File Report 75-285, 1 sheet, scale 1:500,000 and 36 p.
—— , 1976, Preliminary map showing known and suspected active faults in Colorado: U.S. Geological Survey Open-File Report 76-154, 42 sheets, scale 1:500,000.
Zoback, M. D., 1979, Recurrent faulting in the vicinity of Reelfoot Lake, northwestern Tennessee: Geological Society of America Bulletin, v. 90, p. 1019–1024.
Zoback, M. D., and Zoback, M. L., 1980, State of stress of the conterminous United States: Journal of Geophysical Research, v. 85, p. 6113–6156.
—— , compilers, 1991, Stress map of North America: Geological Society of America map CSM-005, 4 sheets, scale 1:5,000,000 (in press).
Zoback, M. L., and Zoback, M. D., 1989, Tectonic stress field of the conterminous United States, in Pakiser, L. C., and Mooney, W. D., eds., Geophysical framework of the continental United States: Geological Society of America Memoir 172, p. 523–539.

ACKNOWLEDGMENTS

I am especially grateful to Doug Clark for many discussions, and extensive editorial assistance. I also wish to thank my wife, Ruth Slemmons, for her encouragement and editorial assistance. Peg O'Malley's scientific expertise was invaluable for all of the illustrations. John Adams contributed information on Canadian neotectonics, and also served as the technical reviewer. I also thank Arch Johnston for many stimulating discussions of the New Madrid seismic zone.

MANUSCRIPT ACCEPTED BY THE SOCIETY JULY 26, 1991

The Geology of North America
Decade Map Volume 1
1991

Chapter 2

Seismicity Map of North America Project

E. R. Engdahl
National Earthquake Information Center, U.S. Geological Survey, MS 967, Box 25046, Denver Federal Center, Denver, Colorado 80225

W. A. Rinehart
National Geophysical Data Center, National Oceanic and Atmospheric Administration, 325 Broadway, Boulder, Colorado 80303

INTRODUCTION

A highlight of the Decade of North American Geology (DNAG) program is the publication of new continent-scale maps of North America. The map set includes a Geological Map, Gravity Anomaly Map, Magnetic Anomaly Map, Seismicity Map, Stress Map, Neotectonic Map, and Thermal Aspects Map. The maps are wall size (four 42-in × 55-in sheets per map), in color, and made on a common base map at a scale of 1:5,000,000.

A GSA-sponsored workshop was held at the 1984 Seismological Society of America meeting in Anchorage, Alaska, to discuss construction of the Seismicity Map. Consensus was reached that the goal of the Seismicity Map would be to provide a useful resource on North American seismicity for seismologists, earth scientists at the postgraduate level, and industry, and to accurately portray North American seismotectonic features using the entire earthquake history from the early 1500s (pre-instrumental) through the modern period. It was agreed that construction of the map database would require the rationalization of hundreds of thousands of earthquake hypocenters from global, national, regional, and local catalogs, and from other source materials, and that because modern data are more useful than historic data for resolving seismotectonic features, a scheme of data selection and representation would have to be devised that could reveal details of the seismotectonic fabric of North America yet preserve a perspective of historical earthquake occurrence. The problem was further complicated by the highly variable monitoring capability of seismic networks and by the different types and rates of occurrence of earthquake activity throughout North America. The scheme developed at the workshop utilizes regional magnitude-completeness thresholds that vary with time for data selection and a system of symbols and colors to portray the seismicity data on the map. It was decided that further details of seismotectonic features shown on the map would be addressed on a regional basis in contributions to the accompanying volume that would include small-scale maps and cross sections.

The goal of this chapter is to present further details of the rationale used to construct the Seismicity Map (Engdahl and Rinehart, 1988) and to assess some of the more interesting aspects of earthquake monitoring in North America.

EARTHQUAKE DATA BASES

Catalog information on the occurrence of earthquakes usually includes parameters such as the origin time, location, focal depth, and magnitude or intensity of individual events. In the case of historic pre-instrumental earthquakes, these parameters must be determined entirely from intensity data and, hence, are less accurate and more nonuniform spatially than modern earthquake data. Presently, institutions around the world and in North America have varying degrees of responsibility for the collection and processing of earthquake data and for the catalog reporting of earthquake parameters on global, national, regional, and local scales.

On the global scale, since 1928 U.S. government agencies have been reporting the parameters of earthquakes worldwide within a few months of their occurrence through the Preliminary Determination of Epicenters (PDE) and earlier programs. The PDE program, now part of the U.S. Geological Survey's National Earthquake Information Center (NEIC), routinely attempts to report parameters for all earthquakes greater than magnitude 4.5 worldwide and 3.5 within the conterminous United States. Two years after publication of the PDE reports, the International Seismological Centre (ISC) (prior to 1964 the International Seismological Summary [ISS]) revises the PDE parameters and determines new events using a significantly larger database. Global earthquake parameters have also been available from other sources, such as the Bureau Central International de Séismologie (BCIS), special catalogs of larger earthquakes that often provide useful new information on the magnitudes of earlier events, and technical articles in the professional journals.

For events occurring prior to 1964, the global catalog of earthquake parameters is less complete, hypocenters are subject to greater uncertainty, and magnitude information is available for

---

Engdahl, E. R., and Rinehart, W. A., 1991, Seismicity Map of North America Project, *in* Slemmons, D. B., Engdahl, E. R., Zoback, M. D., and Blackwell, D. D., eds., Neotectonics of North America: Boulder, Colorado, Geological Society of America, Decade Map Volume 1.

only the larger events. In the early 1960s, with the advent of routine computer processing of earthquake data, the installation of the Worldwide Standardized Seismograph Network (WWSSN), and the installation of standard seismographs in Canada, the reporting of earthquake parameters was sharply advanced in North America. The database for those areas in North America not monitored by regional networks relies heavily on the global earthquake parameters reported since about 1964.

On the national level, the U.S. Geological Survey (USGS) and the National Oceanic and Atmospheric Administration (NOAA) have compiled data on earthquake occurrence in the publications *Earthquake History of the United States* and the annual *United States Earthquakes*. In addition, special compilations have been used to construct state seismicity maps and for hazard analysis. The Geological Survey of Canada has had a similar effort.

On regional and local scales, the databases are somewhat more complicated. The University of California at Berkeley, with regional network coverage in California, has provided a stable, long-term database on the seismicity of northern and central California since 1910, and the California Institute of Technology has cataloged the seismicity of southern California since 1932. It was not until the 1960s in Canada and the 1970s in the United States that regional networks in other parts of North America started to routinely compute the parameters of smaller earthquakes. Many of the regional institutions started upgrading their historical catalogs in the 1960s and 1970s, and some groups such as the California Division of Mines and Geology, the Electric Power Research Institute, and private consulting firms combined data from several regional networks into single catalogs. To our knowledge, the database assembled for the North American Seismicity Map is the first attempt to construct a single catalog from all available earthquake catalogs in North America.

DATABASE CONSTRUCTION

One of the most formidable problems in assembling the North American database was the elimination of duplicate entries resulting from the same earthquake being listed in more than one catalog. The choice of preferred source parameters is not often obvious and may require special knowledge about the effects of source structure on the set of network observations used to locate the earthquake. For example, we often find that the parameters of larger earthquakes based on globally distributed stations are significantly different from parameters for the same earthquakes determined by regional data alone. It is also common to find overlap in regional network coverage so that parameters for the same earthquake are independently estimated by different networks. These earthquake locations can be significantly different because of the effects of earth structure or because one of the networks may be poorly positioned to locate the earthquake. To further complicate the problem, the differences are not systematic but can vary between regions and in time.

Resolution of the problem required development of a catalog hierarchy and subjective estimates of probable errors in origin time, location, and magnitude for network source parameters on a region-by-region basis. An index map and table on the seismicity map define the hierarchy of data contributors for each region. In the most general sense, preferred earthquake parameters for individual earthquakes were ranked in the order of local, regional, national, and global estimates, but hierarchies in regions of overlap between networks often required discussion with regional experts. The probable errors in source parameters were used to develop "association windows" between catalogs for the identification of duplicate entries.

The difficulty of determining whether the parameters for an earthquake in one catalog correspond to different parameters for the same earthquake in another catalog depends on the source of the differences in these parameters. In some cases, the reason for a discrepancy is obvious. An origin time error commonly found in data prior to 1964 is simply the difference between local time and Greenwich Mean Time. There are also many one-hour errors due to changes between time zones, daylight saving time, and war time (during World War II). One-minute errors in origin time are also common. Simple errors in origin time such as these are easy to identify, but other differences in reported parameters that make association between catalogs difficult require more careful individual attention. For example, locations for historical earthquakes are often quoted to only the nearest degree in some catalogs but to 1/4, 1/2, or 1/10 degree in others, resulting in large differences between locations. Even in modern catalogs, differences in regional location estimates for the same earthquake are sometimes found to be as large as 100 km because of the effects of source structure or poor network coverage. Finally, very large differences in magnitude, well beyond those that might be expected in the estimation of magnitude by different methods, are often found between catalogs for earthquakes that appear to be the same. We had to resolve these difficulties by experimentation with origin time, location, and magnitude association windows that varied from one pair of catalogs to another and with region and time. For small association windows, the association of earthquakes between catalogs for some previously determined hierarchy could be accomplished automatically. However, larger association windows were invariably needed to process historical data or to search for gross errors in modern data. In the latter case, the DNAG database required the tedious examination of long lists of possible associations of earthquakes between catalogs on the basis of origin time, location, and magnitude. In this process, there are always some duplicate entries that cannot be resolved without the assistance of data contributors. This assistance was provided with varying degrees of thoroughness.

A number of magnitude estimates are often reported for the same earthquake. These magnitudes may be determined by different methods and/or from different wave types that are not always easily related to one another. Magnitude scales frequently used include those based on estimates of the seismic moment (M_w), on the maximum amplitudes of body waves (m_b) and surface waves (M_s), on the maximum amplitudes of local record-

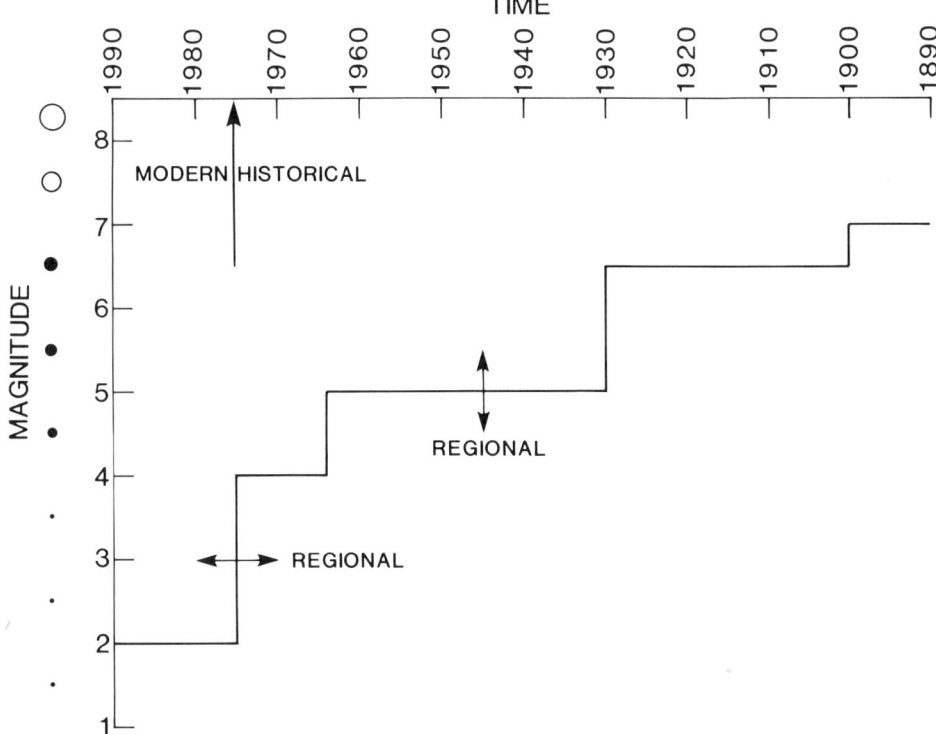

Figure 1. Generalized magnitude completeness thresholds with time. Boundaries vary regionally. On the map, shades of color are used to distinguish modern from historical data in each region; symbol definition and scaling shown help to preserve seismotectonic perspective.

ings (M_L), and on signal duration (M_D). For earlier, noninstrumental earthquakes, only maximum intensities are known, and these must be converted to a magnitude. In this case, we have converted maximum reported intensities to a magnitude (MI) using MI = 2/3 I_o + 1 (Richter, 1958) for the western United States and Canada and MI = 0.5 * (I_o + 3.5) (Nuttli, personal communication) east of the Rockies. To facilitate the assessment of reported magnitudes, we carry M_s, m_b, and up to five reported magnitudes for each event in the database. For data selection and plotting, we use the largest reported magnitude of each event unless some other hierarchy has been suggested by a regional contributor. Any errors introduced by this approach are not serious, as we plot only in magnitude interval classes of one unit, and although the larger events are reviewed carefully, we are not attempting to produce definitive magnitude estimates for each event.

DATA SELECTION

It was recognized at the outset that to plot the entire historical record without accounting for changes in monitoring capability with time or for the greater accuracy of modern data would severely limit the usefulness of the seismicity map. As an obvious example of a change in monitoring capability, the expansion of regional networks and significant advances in processing of seismic data in the 1970s significantly lowered the magnitude threshold for seismicity data over large regions of the United States. From a historical perspective, we would like to know the locations, even though approximate, of larger-magnitude earthquakes relative to source zones that are well defined by low-magnitude modern seismicity data.

A solution to the problem was suggested by a method developed in a seismic hazard study of Canada (Basham and others, 1982, 1985). The method requires the identification of spatially distinct earthquake source zones and the determination of historical time periods over which earthquakes at different magnitude levels are assumed to be completely reported within each of these zones. A list of earthquakes could then be selected for each zone that would be complete at the magnitude completeness threshold for each time period. We extend this concept to the case of regional network coverage as shown in Figure 1, which displays generalized magnitude completeness thresholds with time from which we will make our selection of earthquakes. This approach differs from Basham and others (1982, 1985) in that generally we are concerned with completeness of seismicity within regions defined by regional network coverage and/or by special seismicity catalogs rather than by source zones that are seismotectonically distinct.

The benefits of such a scheme are numerous. It provides a natural selection of seismicity data that emphasizes larger earthquakes in the historical period and smaller earthquakes in the modern period. By proper choice of symbol definition and scaling, as shown in Figure 1, we can view the relative levels of activity by scanning only the large symbols on the map, yet also see the fine detail provided by modern data by narrowing the field of vision. We further enhance the representation by using a brighter shade of each color for modern data. This enables users to easily draw their own conclusions about the relationship of larger historical earthquakes to modern source zones.

REGIONAL NETWORK COVERAGE

Figure 2 displays approximate regional/national network boundaries or source regions in North America over which magnitude completeness thresholds with time have been assigned. Especially active source zones, such as Mammoth (MAM) and Yellowstone (YSTON), have been isolated because of the considerable overlap in network coverage in those zones. This often appears to artificially partition the adjacent regional network coverage (as, for example, the coverage of YSTON by the MBMG network). However, in nearly every case, more than one catalog has contributed to the data base for a particular source region. For regions in the conterminous United States not covered by regional networks, we use a magnitude threshold of 3.5 since 1975, by which time many regional networks had been installed and the U.S. Geological Survey had initiated a project to recover source parameters of all U.S. earthquakes to that level.

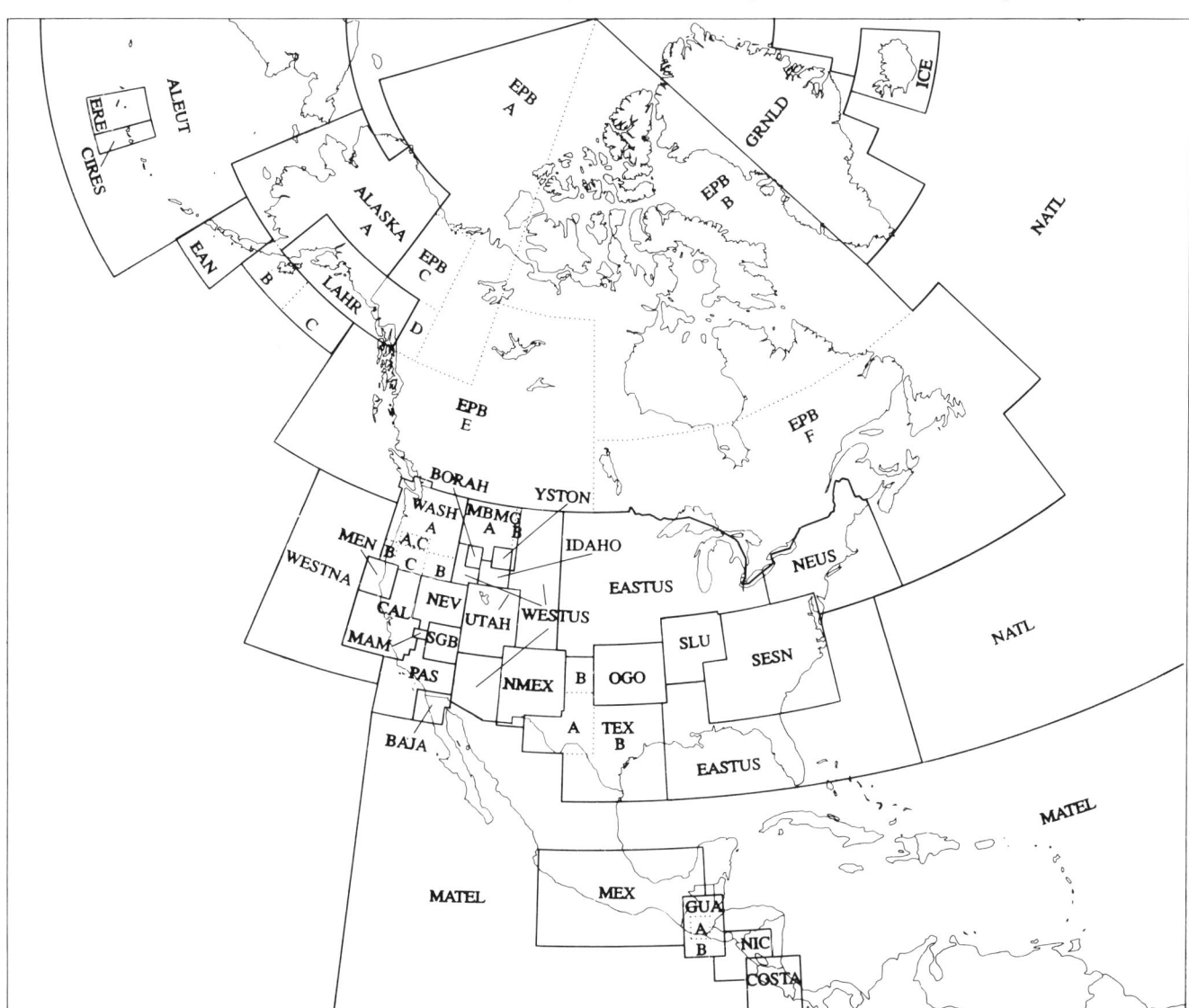

Figure 2. Definition of approximate regional/national network boundaries or source regions in North America over which magnitude completeness thresholds have been estimated. Except in regions of major overlap, abbreviations are identical to those used for catalogs in the Earthquake Information System of the U.S. Geological Survey's National Earthquake Information Center. Some of these catalogs now use new abbreviations: e.g., GSC (Geological Survey of Canada) rather than the old name EPB (Earth Physics Branch).

Magnitude threshold estimates with time are shown by region in Figure 3. The monitoring capability in some regions has been rigorously analyzed (e.g., SESN), but in most regions, completeness thresholds with time are based entirely on the subjective judgment of regional experts. In some instances, the magnitude threshold estimates with time vary spatially within the region (e.g., WASH and EPB). In these cases, the reader will have to refer to the original publications for a complete description. The period prior to the 1920s is known to be incomplete for all regional compilations. However, because of the importance of some of the moderate-size earlier earthquakes, we have, unless otherwise specified, set the magnitude threshold for the earlier period low enough to include all known earthquakes of magnitude 5¾ (maximum intensity 8) or greater in the conterminous United States. To provide some measure of homogeneity on the map, we plot only earthquakes of magnitude 2.5 or above in regions such as southern Nevada and California, where the magnitude-completeness threshold might be as low as magnitude 2 in the modern period. Throughout the map, separation of modern and historical data usually occurs in the 1970s, when magnitude completeness thresholds were significantly lowered by the installation of regional networks.

FOCAL DEPTH DISTRIBUTION

Except for earthquakes in subduction zones, no attempt has been made to portray focal depth distribution on the seismicity map. Although some improvement has been realized through expanded coverage by regional networks, focal depth remains what it has been historically—one of the most poorly determined focal parameters for shallow earthquakes in North America.

In North American subduction zones, the map attempts to distinguish shallow depth earthquakes from intraplate events at depth (i.e., mantle earthquakes in subducted slabs). Table 1 summarizes by region the criteria for plotting intraplate subduction-zone earthquakes at depth. In the Alaskan-Aleutian arc, the effect of the high-velocity descending slab produces a bias between teleseismic and regionally determined focal depth estimates so that different depth ranges must be used.

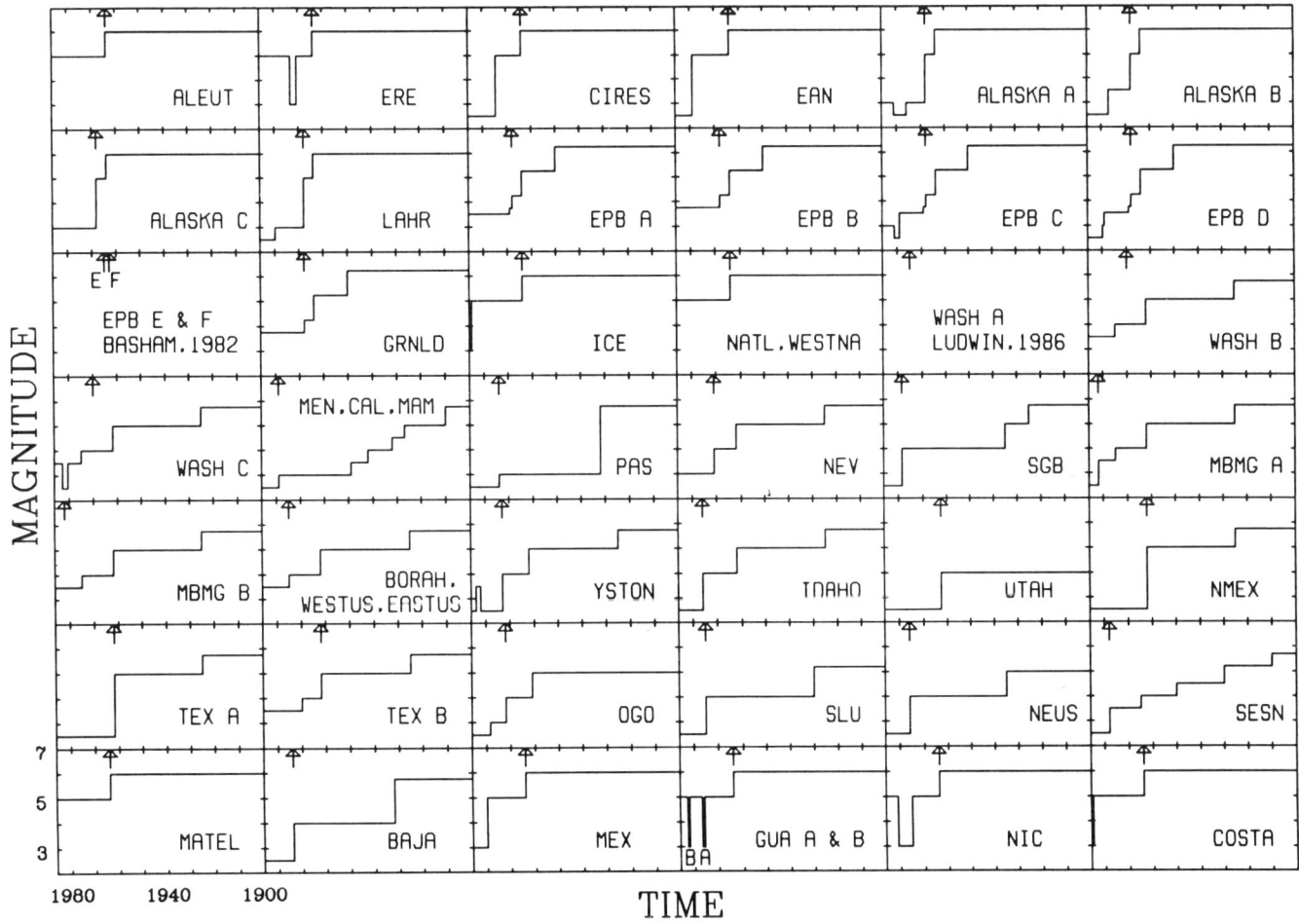

Figure 3. Regional magnitude completeness thresholds with time for North American networks or source zones. The period prior to the 1920s is incomplete but includes all known earthquakes of maximum intensity 8 or greater. The separation of modern and historical data (indicated by the arrows) usually occurs in the 1970s. Where the thresholds are spatially dependent over a region, only a reference is given.

Over some of the regions in Table 1, focal depth is not uniformly resolved because of poor network coverage. Hence, deeper than normal earthquakes in northern and western Alaska and offshore Middle America were, for lack of better information, assumed to be shallow depth.

SEISMICITY MAP

A seismicity map for North America was produced by the scheme previously described. Since detailed regional interpretations are published in separate chapters, we will attempt to describe only general characteristics of seismicity patterns revealed by this map.

Regional patterns of seismicity in North America can be interpreted within the context of plate tectonic models. With modern high-resolution seismicity data, it is possible to relate these patterns to local and global tectonic processes and to associate earthquakes with geologically mapped faults or other geologic structures. For example, northwest-trending lineations of epicenters associated with right-lateral faults in California are superimposed on a background of more widely scattered epicenters and clusters of epicenters. Moreover, the rate of seismicity differs between zones that are approximately delineated by major structural provinces.

It is also quite evident that without the historical data the perspective of earthquake occurrence in time is lost or distorted. For example, sections of the San Andreas ruptured by major earthquakes in 1857 and 1906 are now relatively quiet, even at the lower magnitudes. On the other hand, some source zones, such as the region off Cape Mendocino and the Mammoth, California, area, have clearly been highly seismic over the entire historical record. Larger historical earthquakes fold into the seismic patterns quite well, and inferences on their relationship to modern earthquake patterns (or lack thereof) are made possible.

Overall, the map achieves the desired objective. It displays fine details of the seismicity revealed by modern high-resolution data yet preserves information about historical earthquake occurrence. The result is an accurate portrayal of the seismotectonic fabric of North America over the period for which we have a recorded history.

DISCUSSION

Construction of the DNAG Seismicity Map of North America has revealed important information about the state of earthquake monitoring in North America. Prior to 1950, the coverage of North American seismicity was highly nonuniform. Except for the conterminous United States, instrumental earthquake coverage was probably no better than magnitude 6 over large regions of the continent. From 1950 to 1963, Basham and others (1982) considered that all of Canada was covered down to magnitude 5.5, which may also be true for most of the rest of the continent. The modern period of global catalog data, beginning about 1964, provided a more uniform data set for earthquakes magnitude 5 and above in North America.

TABLE 1. CRITERIA FOR PLOTTING INTRAPLATE SUBDUCTION ZONE EARTHQUAKES AT DEPTH

Region	Depth
ALEUT ERE CIRES EAN ALASKA LAHR	h>60 km (teleseismic) h>50 km (regional)
WASH	h>35 km
MATEL MEX GUA NIC COSTA	h>60 km

With the expansion of modern telemetered networks in the 1970s, and with greater timing accuracy provided by the later introduction of digital recording, significant progress was realized in the resolution of earthquake hypocenters. Detailed knowledge of particular regions now comes from lower magnitude earthquake data collected by regional and local seismic networks. Most of these networks are installed to better understand the regional seismotectonics and details of structure. In addition, the seismicity may be used as data for studies of earthquake prediction or of the seismic effects of nuclear explosions. Many networks are permanent installations, but others are short lived, having served the purpose for which they were installed (e.g., ERE for monitoring seismic effects of nuclear explosions on Amchitka Island in the central Aleutians).

The Alaska-Aleutian arc is one of the most seismically active subduction zones in North America. Attempts to monitor Alaska with regional networks have required enormous efforts to install instrumentation and process the data. The Aleutian Islands also present special logistic difficulties; although a number of networks have been installed there, only two remain operational at present.

Since the early 1960s, the Canadian national network has provided a uniform monitoring capability over its territories that has significantly lowered the completeness thresholds over large areas of that country. The Geodetic Institute of Denmark has improved the national catalog for Greenland with the help of Canadian data, and Iceland provided a systematic compilation of earthquakes recorded by its regional network.

For the most part, the western United States, including the Rocky Mountain region, is reasonably well monitored by existing networks. Areas monitored to only about the magnitude 3.5 level include Oregon, large parts of Idaho, and Arizona. These are regions of low to moderate seismicity in which funding agencies have not been inclined to support the installation of permanent networks (although some short-term studies have been made). The offshore regions, especially west of about 126° west longitude, near Cape Mendocino, obviously pose a special problem because they are outside of landbased network coverage and are complete only to relatively high teleseismic magnitude thresholds.

There is often partial overlap in network coverage over large areas of the western United States. Exchange of data is closely coordinated between the United States and Canada in the Pacific Northwest. Networks in California and Nevada apparently share phase data and in some cases the actual data stream, but individual catalogs often extend beyond their network boundaries, and there is considerable uncertainty in the choice of preferred hypocenters. To some extent, there is a duplication of effort, but this is probably desirable to insure continuity in the overall catalog. The California networks have been especially effective in support of research on earthquake sources.

Large parts of the central United States are among the most poorly monitored in that country. The few permanent stations are widely spaced and provide a monitoring capability to only about the magnitude 3.5 level.

Earthquake monitoring in the eastern United States is well organized. Central collection points of data from numerous sources, including Canada, exist for both the northeastern and southeastern regions. Network operators remain in close communication, the best use is made of available resources, and duplication of effort is avoided.

Regional network coverage in Middle America has been uneven. Although many parts of Middle America have operating networks, the data are not generally available or are available for only short time periods. The eastern Caribbean needs a special effort to bring together the numerous sources of network data in that region.

The users of regional network data in North America could certainly benefit from more consistent estimates of magnitudes and more careful analysis of magnitude completeness thresholds. The types of magnitudes commonly used vary, and relationships between them are not well known. Some standardization seems essential. In most regions, network magnitude thresholds with time are only subjective estimates. A more rigorous examination of the catalog data with time over well-defined source regions seems desirable.

Finally, there needs to be a coordinated effort to correct and update the DNAG database for all of North America. This will require better definition of network coverages with time, determination of more accurate regional magnitude completeness thresholds, and some attempt to rationalize the differences between reported magnitudes.

REFERENCES CITED

Basham, P. W., Wiechert, D. H., Anglin, F. M., and Berry, M. J., 1982, New probabilistic strong seismic ground motion maps of Canada; A compilation of earthquake source zones, methods, and results: Earth Physics Branch Open-File Report 82-33, 205 p.

——, 1985, New probabilistic strong seismic ground motion maps of Canada: Bulletin of the Seismological Society of America, v. 75, p. 563–595.

Engdahl, E. R. and Rinehart, W. A., 1988, Seismicity Map of North America: Geological Society of America, Centennial Special Map CSM-4, scale 1:5,000,000.

Richter, C. F., 1958, Elementary Seismology: San Francisco, California, W. H. Freeman and Co., 768 p.

MANUSCRIPT ACCEPTED BY THE SOCIETY OCTOBER 12, 1988

ACKNOWLEDGMENTS

We thank G. Reagor for invaluable assistance in the preparation of catalog data for entry in the U.S. Geological Survey computer-based Earthquake Information System. W. L. Ellsworth, P. W. Basham, J. W. Dewey, and J. N. Taggart are thanked for helpful reviews and suggestions. We also gratefully acknowledge the wisdom of the Anchorage workshop attendees and the numerous contributors of data to this map.

Printed in U.S.A.

Chapter 3

Seismicity of the Aleutian Arc

J. J. Taber*
Lamont-Doherty Geological Observatory of Columbia University, Palisades, New York 10964
S. Billington
U.S. Bureau of Mines, Building 20, Denver Federal Center, Denver, Colorado 80225
E. R. Engdahl
National Earthquake Information Center, U.S. Geological Survey, MS 967, Box 25046, Denver Federal Center, Denver, Colorado 80225

INTRODUCTION

The Alaska-Aleutian Arc extends 3,600 km from the Gulf of Alaska to Kamchatka. It is a classic convergent margin, exhibiting all the basic elements of the subduction of an oceanic plate. This chapter will emphasize the seismic aspects of the Pacific-North American plate interaction along the Alaska Peninsula and the Aleutian Arc. The seismicity of the other parts of Alaska is discussed by Page and others (this volume). Some elements of this convergent margin remain remarkably the same over most of the arc while other features change greatly. Many of the variations in the subduction zone may be related to the transition in the overriding plate from continental to oceanic composition and thickness, or to the shift in plate convergence direction from near normal to parallel to the arc. Based on these differences, the arc can be divided into three regions (Fig. 1): (1) the far western Aleutians where plate motion becomes parallel to the arc, (2) the central Aleutians where there is oceanic-oceanic convergence that varies from oblique to nearly perpendicular to the arc, and (3) the eastern Aleutians where the convergence is near perpendicular but the overlying plate is of continental thickness. The western Aleutians will be discussed separately, while most aspects of the central and eastern Aleutians will be considered together.

Two types of data will be used in the discussion of seismicity. The overall pattern of seismicity is best examined using a uniform data set such as the Preliminary Determination of Epicenters (PDE) catalog for events greater than or equal to magnitude 5 (Fig. 1). Focal mechanisms for these larger events give information about broad-scale stresses within the arc. Detailed knowledge of particular sections of the arc comes from data collected by local seismic networks. Four networks have operated or are currently running in the Aleutians (Fig. 2), in addition, there is a Soviet station on Beringa Island, and a station did exist

on Mednyy Island in the far western Aleutians. There were short-term networks in the Amchitka region of the central Aleutians and on Unalaska in the eastern Aleutians, and there still are networks in the Adak region of the central Aleutians and in the Shumagin Islands region of the eastern Aleutians. These networks, whose data show as high-density regions on the DNAG seismicity map, provide improved hypocentral parameters and structural information. Stress variations are calculated from composite focal mechanisms and frequency spectra for microearthquakes. Temporal variations not visible in the data set of larger earthquakes provide information about the episodic nature of subduction. Network results will be discussed throughout the text.

The installation of a microearthquake monitoring network on Amchitka and nearby islands in the central Aleutians to monitor seismic effects from the nuclear explosions MILROW and CANNIKIN provided the first opportunity to study the details of seismicity and structure in the Aleutian subduction zone (Engdahl, 1971). The completed network of eight stations operated continuously from October 1970 through April 1973. Earthquakes of magnitude 3 and greater that occurred during that time period are plotted on the DNAG seismicity map.

The Adak network, located on Adak and seven nearby islands in the central Aleutians, consists of 14 stations. Data from the Adak seismograph network have been used for a number of studies, the results of which generally fall into two classes: studies on seismotectonics and structure (with emphases on both shallow and intermediate-depth structure), and studies on earthquake prediction. Epicenters of earthquakes located using data from the Adak network are shown on the DNAG map for earthquakes of magnitude 3 or greater that occurred from August 1974 through December 1985.

The location of the 17-station Shumagin network was chosen based on the likelihood of the region being a seismic gap and on the existence of the Shumagin Islands. The islands extend out toward the trench, allowing better than usual depth resolution there. Earthquakes of magnitude 2.5 or more from 1979 to 1985

---

*Present address: Institute of Geophysics, Victoria University, P.O. Box 600, Wellington, New Zealand

Taber, J. J., Billington, S., and Engdahl, E. R., 1991, Seismicity of the Aleutian Arc, *in* Slemmons, D. B., Engdahl, E. R., Zoback, M. D., and Blackwell, D. D., eds., Neotectonics of North America: Boulder, Colorado, Geological Society of America, Decade Map Volume 1.

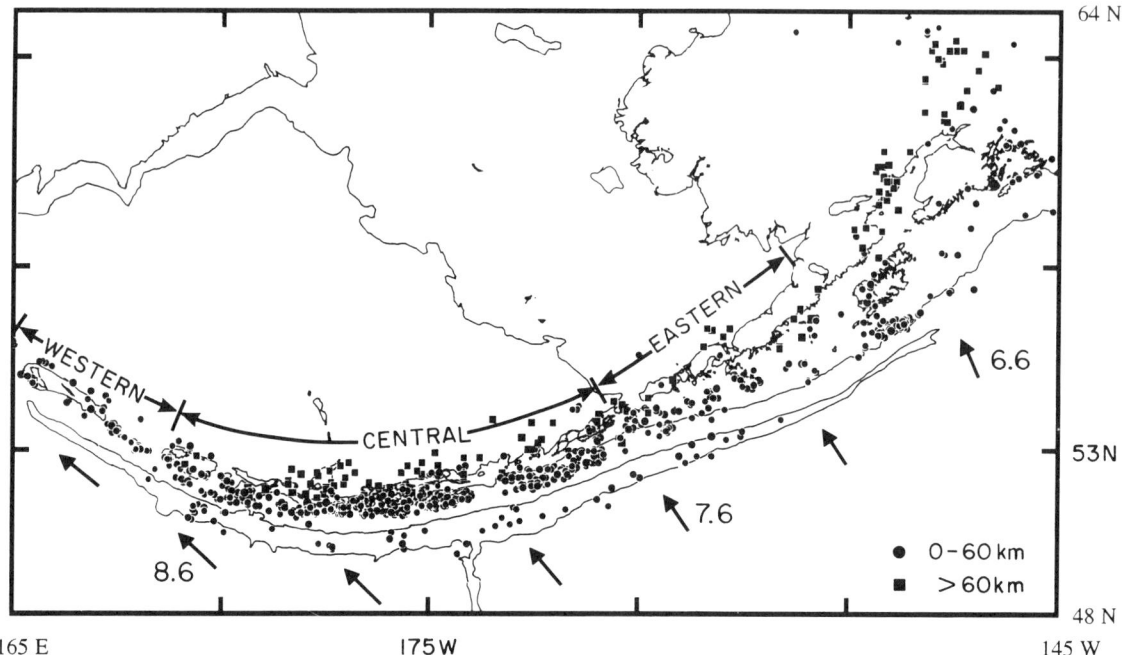

Figure 1. PDE epicenters, $m_b \geq 5$, for the period 1968 to 1986. Symbols define depth classes and are scaled by magnitude. Bathymetric contours are at 400 and 5,200 m. Pacific–North America plate convergence vectors (RM2, Minster and Jordan, 1978) are indicated by arrows, with rates in cm/yr. See Figure 5 for names of physiographic features.

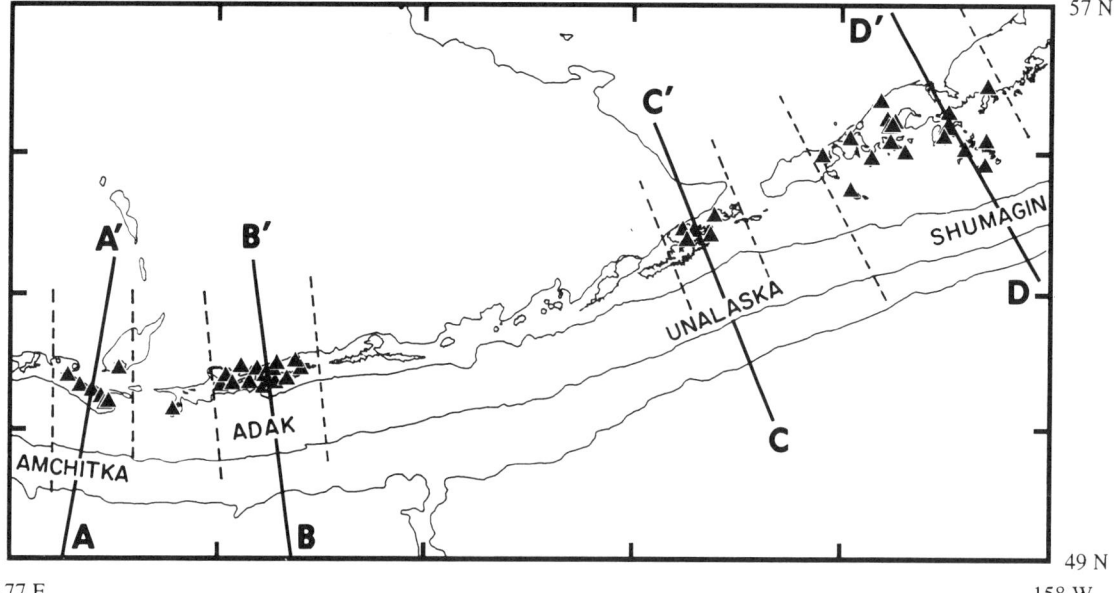

Figure 2. Location of seismic networks in the Aleutians. The Unalaska and Amchitka networks are no longer operating. Triangles are station locations. Dotted lines show widths of cross sections; solid lines show length and angle of cross sections in Figure 10.

are plotted on the DNAG map. The smallest and shortest-lived network operated on Unalaska Island from 1980 to 1982. Those data are not included on the DNAG map but will be mentioned here.

The seismicity within any section of the arc can be divided into four distinct regions based on location and focal mechanism (Fig. 3): (1) trench and outer rise, (2) main thrust zone, (3) Wadati-Benioff zone, or (4) overriding plate. Each of these regions responds to the driving force of plate convergence in a different manner. Occasional outer rise events oceanward of the trench may be responding to the state of stress on the plate interface or to the bending of the plate itself. The main thrust zone is the seismically active contact between the Pacific and North American plates. This region, which is the location of great interplate earthquakes, exhibits the largest variation along the arc. The zone of interplate seismicity extends arcward from near the trench to a depth of 40 to 60 km, below which plate convergence is accommodated aseismically. The seismogenic zone then shifts to within the subducting plate, and the dip of the plate steepens to 30 to 60°. For the purpose of this chapter, this latter region will be referred to as the Wadati-Benioff zone (WBZ). However, the WBZ is typically defined as the entire dipping zone of seismicity from the trench to the deepest earthquakes and thus includes the main thrust-zone earthquakes and events that are within the slab and directly below the main thrust zone. The location of the intermediate-depth seismicity in the WBZ relative to the Quaternary volcanoes is remarkably similar along the arc. The depth to the base of this zone varies considerably along the arc, reaching a maximum of nearly 300 km. Velocity anomalies outlining the plate are still measurable to greater depths (Engdahl and Gubbins, 1987; Boyd and Creager, 1990), but the pressure-temperature regime is such that deformation occurs aseismically. There is also seismicity above the main thrust zone in the overlying North American plate. The maximum trenchward extent of this upper plate seismicity varies from the 400- to the 4,000-m bathymetric contour along the arc, and may mark the trenchward extent of the rupture zones of great earthquakes. The arcward boundary of upper plate seismicity usually terminates near the volcanic arc.

OUTER RISE AND TRENCH

The rate of seismicity near the trench and outer rise is low compared to the other parts of the subduction zone (Fig. 1). Along most of the arc a spatial separation exists between trench events and events on the main thrust zone and in the upper plate. Since the amount of trench seismicity is limited, local studies offer much better resolution than studies based on globally recorded events. The results of an ocean-bottom seismometer (OBS) microseismicity study in the central Aleutians (Frohlich and others, 1982) are probably representative of much of the arc. Events were recorded at depths less than 20 km and extended only 10 km landward and 60 km oceanward of the trench axis. Relocated

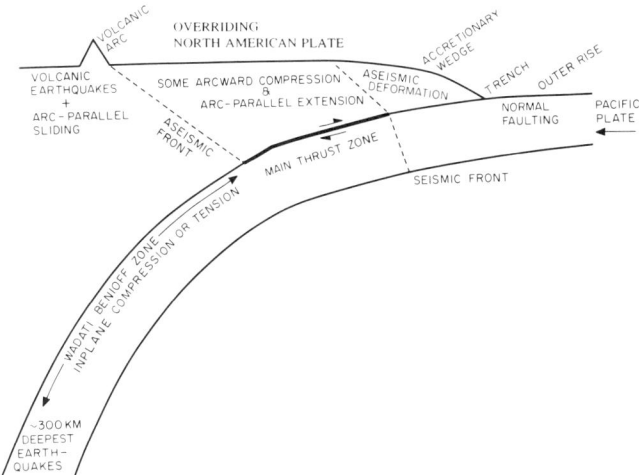

Figure 3. Diagram depicting major elements of the Aleutian subduction zone. Seismicity occurs within the Pacific plate near the trench axis and the outer rise and arcward of the main thrust zone. Plate-boundary earthquakes occur on the MTZ. Upper (North American) plate seismicity occurs at shallow depths between the seismic and aseismic fronts or beneath the volcanic arc.

larger events show the same spatial extent and a similar but narrower range of subcrustal depths beneath the trench and outer rise (Engdahl and Billington, 1986; Engdahl and others, 1989). The OBS study also confirmed the gap in seismicity between trench events and the main thrust zone in the Adak Island region. Focal mechanisms of moderate-sized events recorded globally, as well as locally recorded microseismicity, are consistent with normal faulting, and the tension axis is approximately perpendicular to the trench (Stauder, 1968b; Frohlich and others, 1980; Ekström and Engdahl, 1989). These authors infer that the earthquakes are caused by high tensile stresses generated in the Pacific plate as it bends beneath the continental margin. The limited range of the seismicity implies that the bending strength of the lithosphere is exceeded in only a narrow region.

Some of the extensional events may also be the result of movement on the main thrust zone. An event on March 30, 1965 ($M_s = 7.5$) that occurred trenchward of the 1965 Rat Island thrust earthquake and followed it by less than two months, may have ruptured through most of the oceanic lithosphere (Abe, 1972). Such large extensional stresses may be generated in the shallow part of the slab after the compressional stresses are relieved on the plate interface during a thrust earthquake. Normal faulting events also occurred in the trench after the 1957 thrust earthquake (Stauder and Udias, 1963). Extensional events may continue to follow a thrust earthquake for 10 to 15 years (Sykes, 1971), and then at some later time, compression may occur temporarily at the outer rise. Compressional events have occurred near the trench just before great thrust earthquakes in other subduction zones (Christensen and Ruff, 1983), but there are as yet no documented examples of this in the Aleutians.

MAIN THRUST ZONE

Most of the strain generated by plate convergence is released during great earthquakes along what will be referred to in this chapter as the main thrust zone (MTZ). The dip of this contact between the Pacific and North American plates varies from a maximum of 25 to 30° in the central Aleutians to a minimum of 4 to 5° in the Gulf of Alaska. Since the downdip edge only varies in depth from 40 to 60 km, the decrease in dip requires a widening of the MTZ. The change in arc-trench separation, from a minimum of 170 km in the central Aleutians to a maximum in the Gulf of Alaska of 570 km (Jacob and others, 1977), is almost entirely due to variations in the width of the MTZ. The thickness and exact geometry of the zone are not well determined in most regions, but recent studies near Adak, using relocated earthquakes with well resolved focal depths (Engdahl and Billington, 1986; Engdahl and others, 1989), show that north of the trench the plate dips gently beneath the terrace, then steepens and descends gradually into the mantle. This result is also consistent with the dip of shallow thrust planes determined from focal-mechanism solutions (Ekström and Engdahl, 1989). The thickness of the interplate thrust-earthquake zone defining the plate boundary is no more than 10 to 15 km.

The earthquake cycle and seismic gaps

Earthquake activity follows a distinct cycle within the MTZ. The cycle begins with a period of relatively low seismicity rate lasting an average of about 80 years, but with considerable variation across the arc (Jacob, 1984). This is followed by a major moment-releasing event that has been as large as M_w = 9.2. The aftershock sequence, including possible trench events, lasts months to several years, depending on the size of the mainshock, and then decreases to low background levels.

Based on the extent of their aftershock zones, great earthquakes appear to rupture adjacent sections of the arc (Stauder, 1972). Many of the boundaries between rupture zones can be defined by tectonic features (Figs. 4 and 5), but the boundaries, and thus the size of the earthquakes, are not necessarily the same for each cycle (e.g., Rundle and Kanamori, 1987; Boyd and Lerner-Lam, 1988). There may be a correlation between tectonic complexity and the length of the rupture, with simple regions having the longest ruptures (Sykes, 1971). This may explain why the historic record shows only moderate-sized events near the tip of the Alaska Peninsula in the transition region from oceanic to continental upper plate.

There are few gaps between earthquakes in the same cycle. The most recent cycle consisted of earthquakes in 1938, 1946, 1957, 1964, and 1965 rupturing most of the Alaska-Aleutian Arc, while the previous cycle of activity near the turn of the century seems to have consisted of a greater number of smaller (i.e., magnitude 7.2 to 7.8) events (Boyd and Lerner-Lam, 1988). The notable exceptions to rupture in the last cycle are the Shumagin and Komandorsky Islands regions and perhaps the region near Unalaska. Large earthquakes were last reported in the Komandorsky Islands in 1849 and 1858 (Sykes and others, 1981). The earlier of the two shocks also generated a tsunami that was recorded locally and on islands in the South Pacific. The estimated magnitude for both events was 7.5 ± .7 (Kondorskaya and Shebalin, 1977). The generation of a tsunami requires some vertical slip during the earthquake (i.e., either thrust or normal faulting); therefore, the earthquakes probably were confined to the eastern portion of the gap where thrusting is still occurring. It is not known what the repeat time might be for this section of the arc.

The state of stress in the Unalaska region is still unresolved. Aftershocks from the 1957 event in the region near Unalaska were located only in a narrow band close to the arc, suggesting that the shallower trenchward region remained locked (House and others, 1981). However, a relocation of the events (Boyd, 1985) showed that most of the 1957 aftershocks along the entire rupture zone were concentrated in a narrow band, and thus, the seismicity pattern in the Unalaska region may be the same as the rest of the aftershock zone. The tsunami source areas for both the 1957 and adjacent 1946 event are also poorly determined (Ruff, 1985; Hatori, 1981). Slip may have occurred within the Unalaska region, but the speed of rupture and the amount of slip may have been different than elsewhere along the rupture zone (Boyd and Jacob, 1986).

The Shumagin Islands region is another candidate for a seismic gap. To be a seismic gap, a region must not have ruptured for at least 30 years, and there must be evidence of previous major events (McCann and others, 1979). Historic earthquakes in 1788 and 1847, according to felt reports, probably ruptured a significant portion of the Shumagin gap (Fig. 4). The most recent cycle appears to have consisted of several M_s ≈ 7.5 shocks. The instrumental location for an M_s = 7.5 event and its aftershocks in 1917 is within the Shumagin Islands region (Boyd and others, 1988; Estabrook and Boyd, 1989). This event may have ruptured the eastern portion of the gap, while events in 1898 and 1948 ruptured the central and western portions. Given the time between the 1847 and 1917 earthquakes and the relatively small amount of slip calculated for the 1917 event (120 cm), at least some of the plate convergence must be occurring aseismically (Estabrook and Boyd, 1989). Aseismic slip has also been proposed to explain the lack of significant strain accumulation in the Shumagin Islands over the period from 1980 to 1987 based on trilateration measurements (Savage and Lisowski, 1986; Lisowski and others, 1988). The largest events (such as in 1788) may occur when rupture begins outside the Shumagin gap. Nishenko and Jacob (1990) have pointed out that it has been long enough for the probability of rerupturing the 1938 zone to also be high, so the possibility exists for an earthquake similar in size to the 1788 event.

An M_w = 8.0 earthquake in 1986 occurred within the 1957 rupture zone (M_w = 9.1), only 29 years after the previous great earthquake. This is only one-third the 86-year average recurrence time for the Aleutian Arc (Jacob, 1984), and thus, the calculated

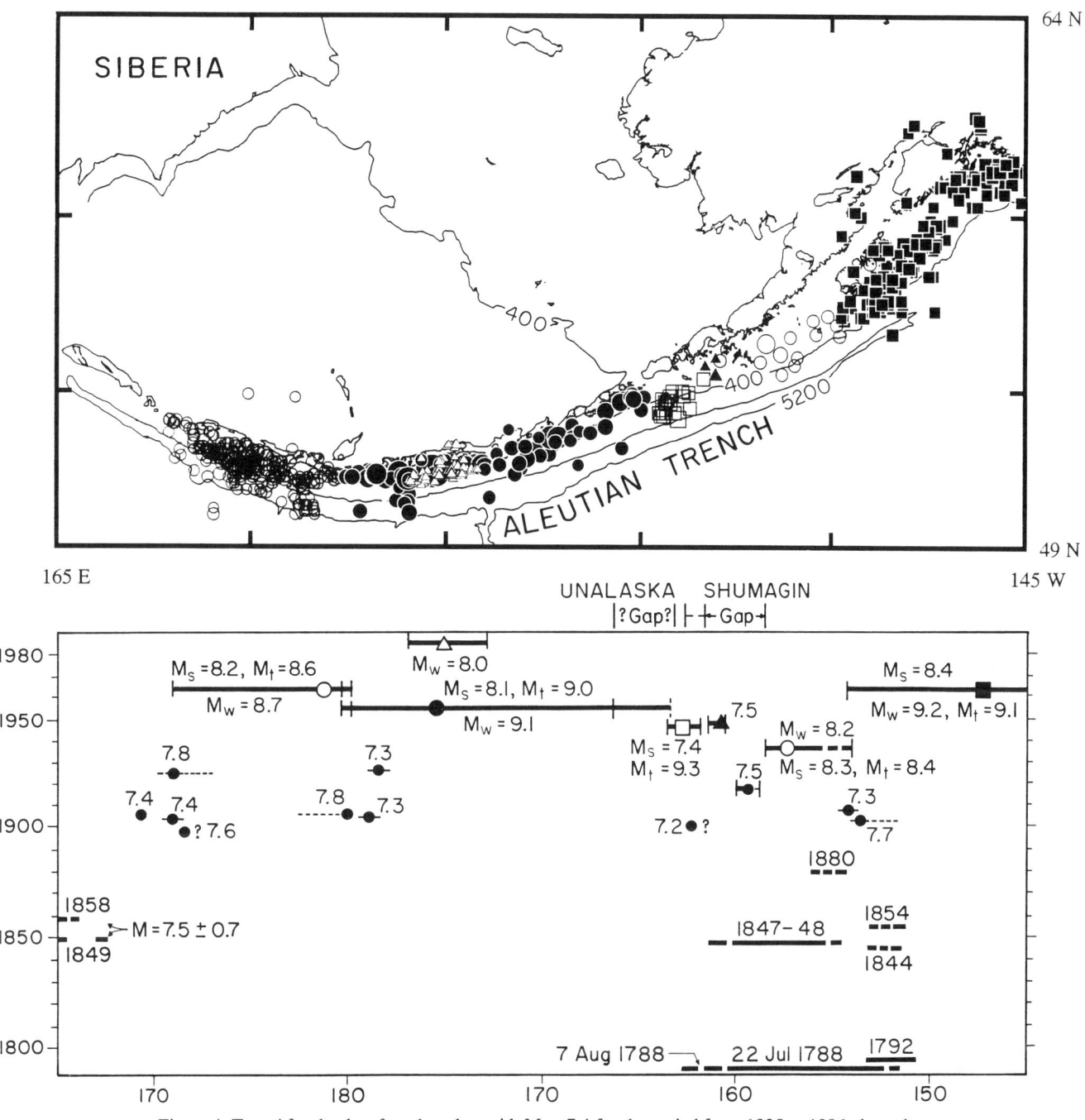

Figure 4. Top: Aftershocks of earthquakes with M ≥ 7.4 for the period from 1925 to 1986 along the Alaska-Aleutian Arc (modified from Sykes, 1971, with new locations from Boyd and others, 1988, and Boyd, personal communication, 1988). Symbol type matches symbols on rupture lines in lower figure. Bottom: Space-time diagram showing inferred rupture zones, magnitudes, and locations of mainshocks for known events of M ≥ 7.2 from 1784 to 1986 (after Davies and others, 1981, with new data from Boyd and Lerner-Lam, 1988, and Estabrook and Boyd, 1989). Dashes denote uncertainties in size of rupture zones. Large shocks in 1925 and 1965 that involve normal faulting in the trench are omitted. Absence of events before 1898 along several portions of the arc reflect a lack of historic record in those areas. M_s, M_w, and M_t refer to surface, moment, and tsunami magnitudes, respectively. Magnitude is M_s if not labeled.

probability of occurrence was very low. Such earthquakes show that it may not be possible to predict over the long term the occurrence of events that are small relative to the last great earthquake in the region. A very large event does not release strain uniformly over the rupture area, and thus, parts of a rupture zone may be able to rupture again, long before the expected recurrence interval based on a uniform release of strain. An alternative explanation is that the repeat time of individual segments varies significantly along the arc and that the arc average is not meaningful when predicting the recurrence of a particular segment.

Seismicity variations

A plot of shallow seismicity since 1967 shows several along-arc variations that can be correlated with major tectonic features and edges of great earthquake rupture zones (Figs. 1 and 5). These changes are most evident in the MTZ seismicity, with a less noticeable pattern in the trench events. West of 170°E there is a low level of seismicity both north and south of the islands (Cormier, 1975; Newberry and others, 1986). Near 170°E (the western edge of the 1965 aftershock zone) the seismicity rate increases, and the earthquakes are more evenly scattered across the arc. This boundary is coincident with the westernmost Quaternary volcanoes. These changes mark the transition from arc-parallel sliding to more normal subduction. Continuing eastward, Bowers Ridge intersects the eastern edge of the 1965 zone. At this point the trenchward extent of the subduction zone seismicity shifts toward the volcanic arc, and the arc itself is offset (Spence, 1977). Gravity signatures and dredge samples suggest that both Bowers and Shirshov Ridges are remnant island arcs (Scholl and others, 1975), although their relation to the current arc is unclear (Spence, 1977). The level of activity is high within most of the 1957 and 1965 aftershock zones, with notable gaps where the Amlia fracture zone is being subducted in the center of the 1957 aftershock zone, and in the Unalaska gap at the eastern edge of the 1957 zone. There is another shift of the volcanic axis at the Amlia fracture zone. The 1957 event ruptured across the Amlia fracture zone and finally stopped near the transition from oceanic to continental overlying plate near 164°W. Assuming that the offset in volcanic axis corresponds to an offset in the position of the subducting slab, it is surprising that the rupture front propagated across the Amlia fracture zone.

The arc-trench separation begins to widen dramatically where the overlying plate thickens to continental dimensions and the seismic front shifts back to follow the 400-m bathymetric contour instead of the 4,000-m contour. It has been suggested that the widening is due at least partially to the increased sediment supply (Jacob and others, 1977). Moving eastward, the level of activity decreases again within the Shumagin gap and the 1938 zone. The low rate of small- and moderate-sized earthquakes in the Unalaska area, Shumagin area, and just east of the Amlia fracture zone relative to adjacent regions is statistically

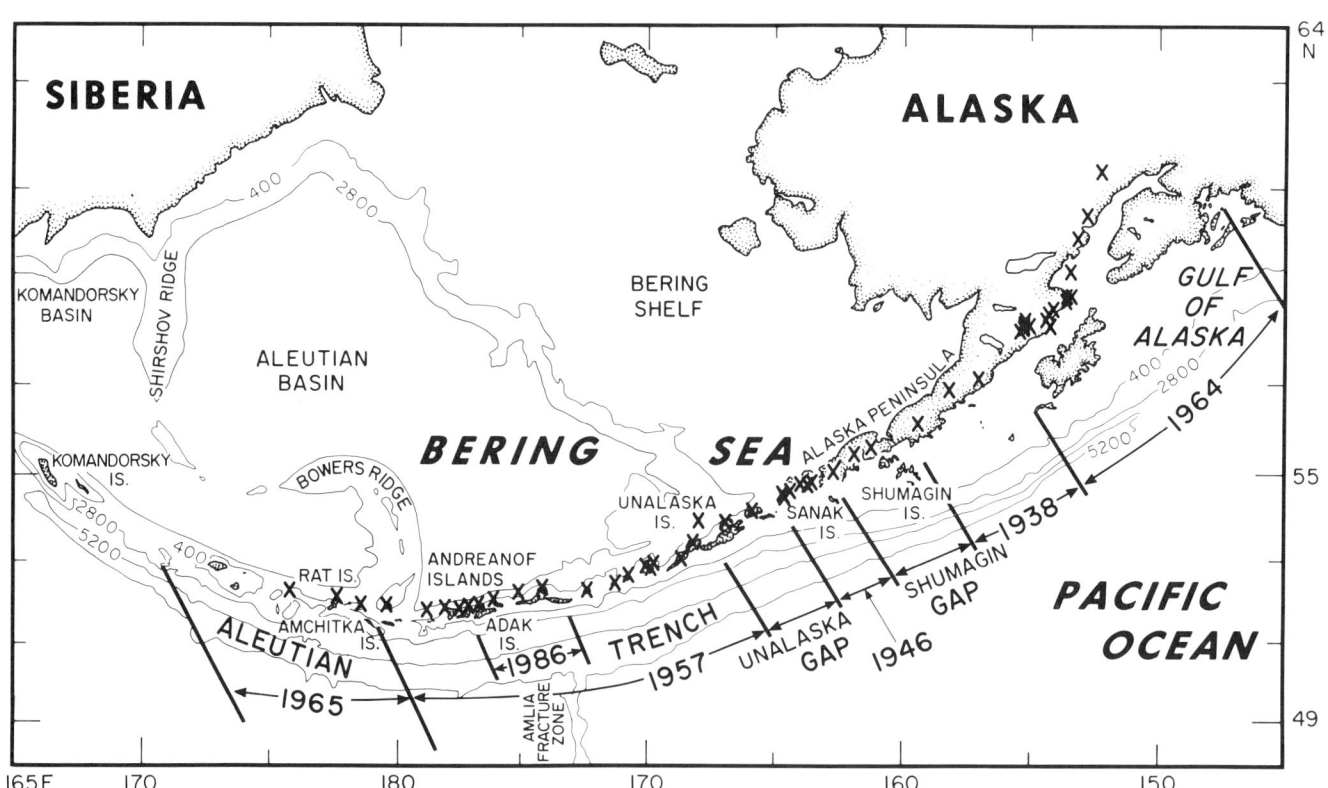

Figure 5. Location map of the Alaska-Aleutian Arc (after Davies and others, 1981). Historically active volcanoes shown as X's (from Simkin and others, 1981). Bathymetry in meters. The aftershock zones of the last major earthquake in each region are designated by year of occurrence.

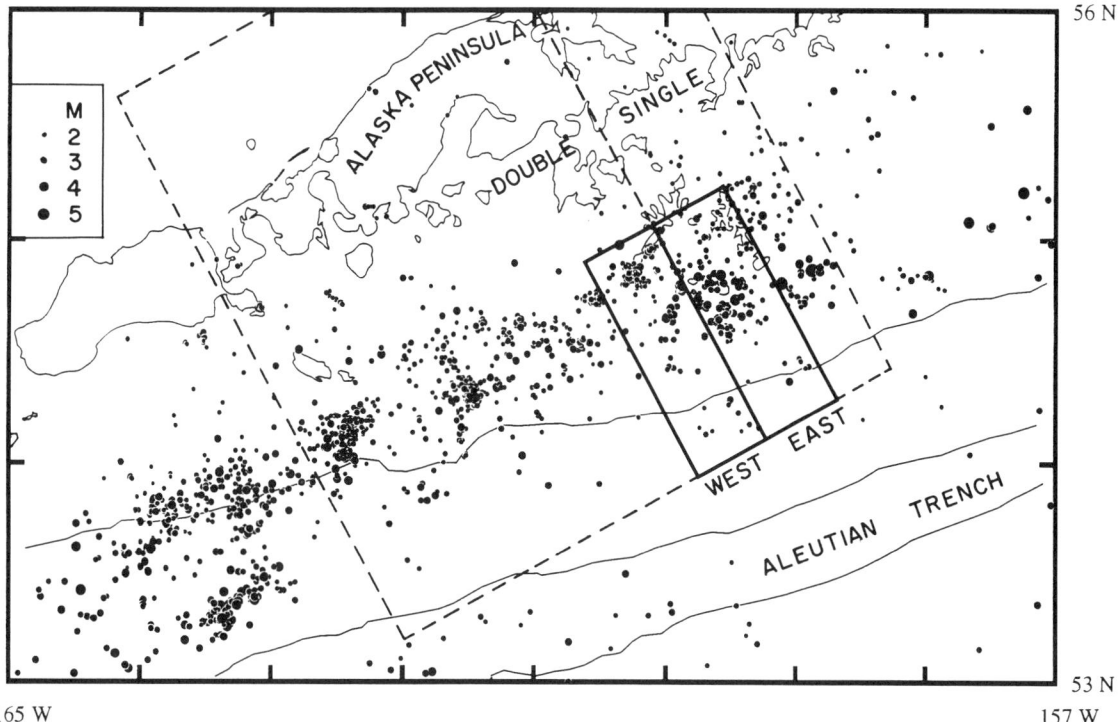

Figure 6. Map view of shallow seismicity, recorded by the Shumagin network from 1979 to 1987, M > 2.0. Events are scaled by magnitude. Dashed boxes show regions and orientation of deep cross sections in Figure 8. Small solid boxes show regions plotted in the thrust zone cross sections in Figure 7.

significant at the 99 percent level (Boyd and Jacob, 1986) using the technique of Habermann (1984). The Amlia fracture zone area has been interpreted as aseismic (House and Jacob, 1983) based on its tectonic regime and the fact that the 1957 earthquake propagated across it, whereas the low rate in the Shumagin and Unalaska areas has been used as evidence for locking of the MTZ (Davies and others, 1981; Hauksson and others, 1984) since historic great earthquakes have occurred in the region.

Focal mechanisms from microseismicity (Billington and others, 1981; Reyners and Coles, 1982; Taber and Beavan, 1986) and from larger events (House and Jacob, 1983; Ekström and Engdahl, 1989) during the interseismic part of the cycle also show thrusting along the plate interface. Thus, motion along most of the plate boundary does not cease completely between great earthquakes. Some regions, however, appear to experience no thrusting during the interseismic period. Since a local network was installed in 1971 near the 1964 great earthquake rupture zone (Fig. 5), there have been few events located near the MTZ, and no thrust-type focal mechanisms have been calculated (Page and others, this volume). Elsewhere along the Alaska-Aleutian Arc, occasional events occurring in the lower plate just beneath the MTZ exhibit lateral compression in the subducting slab (Boyd and Jacob, 1986; House and Jacob, 1983). This may be a geometric effect due to a change in curvature as the plate subducts (e.g., Frank, 1968; Laravie, 1975).

Along-arc variations are evident in the microseismicity (M<5) recorded by the local networks. Although these variations may be temporally dependent, their general features may delineate structurally controlled arc segments of the kind that define the rupture extents of large and great earthquakes (e.g., Kanamori, 1981). Significant variations in the along-arc distribution of shallow seismicity have been recorded by the Shumagin seismic network between 1979 and 1987 (Fig. 6). First is the abrupt seismicity decrease just east of the Shumagin Islands, and second is a distinct offset in the trenchward extent of the currently observed microseismicity southwest of Sanak Island (island location shown in Fig. 5). The first along-arc change coincides with the edge of aftershock activity associated with the 1938 earthquake (see Fig. 4). East of the second change, which coincides with the nucleation point of the 1946 earthquake, aftershocks of the 1946 earthquake are less numerous and spatially separated from the main locus of activity to the west. The spatial correlation between the microseismicity patterns and the patterns in the distributions of previous aftershock activity suggests that these changes reflect arc segmentations that can affect the lateral propagation of rupture during great earthquakes.

Segmentation is evident both along-arc and downdip within the subduction zone. The lower part of the MTZ is steeper than the section that extends toward the trench (Fig. 7), both near the Shumagin Islands (Boyd and others, 1988) and in the Adak

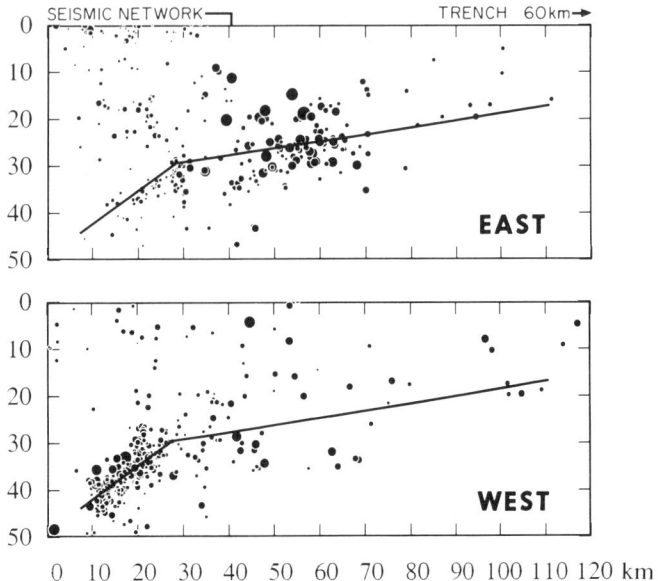

Figure 7. Cross sections of the solid boxes in Figure 6. The bent line is the suggested location of the main thrust zone. There is an abrupt change in dip near 30 km depth and much less seismicity on the steeply dipping section in the eastern block.

region (LaForge and Engdahl, 1979; Ekström and Engdahl, 1989; Engdahl and others, 1989). Within the Shumagins, seismicity is concentrated near the change in dip. The high level of activity within this narrow band of seismicity does not extend into the eastern portion of the Shumagin gap, but instead stops abruptly at the western edge of the Shumagin Islands. This change (Fig. 8) is coincident with a transition in the intraplate seismicity within the downgoing plate from a single to double WBZ (Hudnut and Taber, 1987). The lack of a double WBZ and the small number of events on the steeply dipping portion of the main thrust zone beneath the Shumagin Islands may result from a change in the degree of seismic coupling or may only be temporally significant.

Temporal variations in seismicity distribution perpendicular to the arc are also seen. The southern extent of shallow seismicity monitored by the Adak network since 1974 is relatively abrupt and corresponds to roughly the 4,000-m bathymetric contour, which is displaced northward relative to the rest of the arc in this region (Fig. 9a). This southern termination is not an artifact of the location of the Adak network stations, since it has been confirmed by ocean-bottom seismograph studies (Frohlich and others, 1982). However, prior to 1974, swarms of shallow-focus events occurred to the south under the feature called Hawley Ridge, defined by a 4,000-m closed bathymetric contour, and this region was significantly reactivated by the 1986 $M_w = 8.0$ earthquake (Fig. 9b).

The aftershock zone of the 1986 large earthquake (Engdahl and others, 1989) extended over a 250-km-long segment of the arc, primarily along the trenchward portion of the active thrust

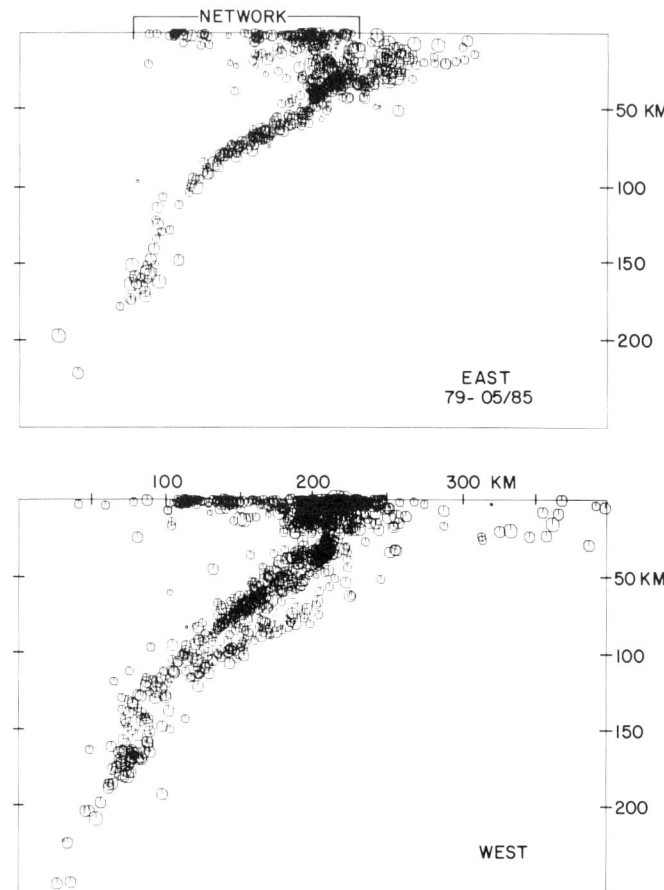

Figure 8. Cross sections of seismicity within the dashed boxes in Figure 6. Note transition from single to double WBZ. Three-dimensional ray tracing suggests that the apparent lack of seismicity in the accretionary wedge in the western area is probably due to location errors (from Hudnut and Taber, 1987).

zone (Fig. 9c), in a pattern very similar to the spatial distribution of relocated aftershocks (Boyd, 1985) of the great 1957 earthquake in this region. Before the 1986 earthquake and since at least 1964, seismicity in the main thrust zone had been characterized by localized concentrations of moderate-size earthquakes (Fig. 9b). Aftershocks of the 1986 earthquake occurred both within these concentrations and within new centers of activity. A striking feature of both the pre- and post-mainshock seismicity is the absence of lower thrust zone activity east of the 1986 mainshock. Also note that the western edge of the aftershock zone coincides with Adak canyon. This is another example of structural features correlating with the ends of rupture zones.

The largest earthquakes along the arc have nucleated near the base of the MTZ (Kelleher and others, 1973), whereas some smaller events appear to start near the shallow edge of the thrust zone and propagate downward toward higher strength regions. Stress-drop calculations within the Shumagin gap seem to show that stress increases with depth in the subduction zone (House

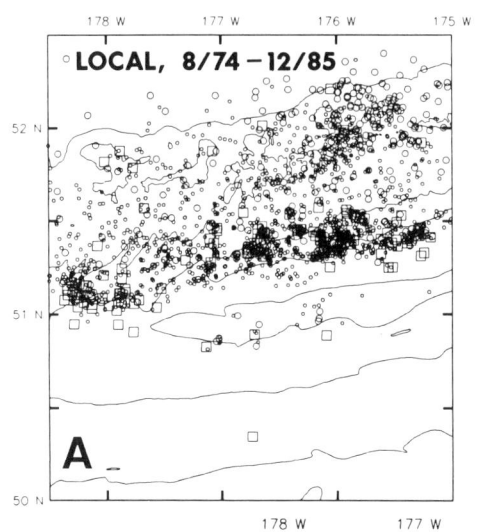

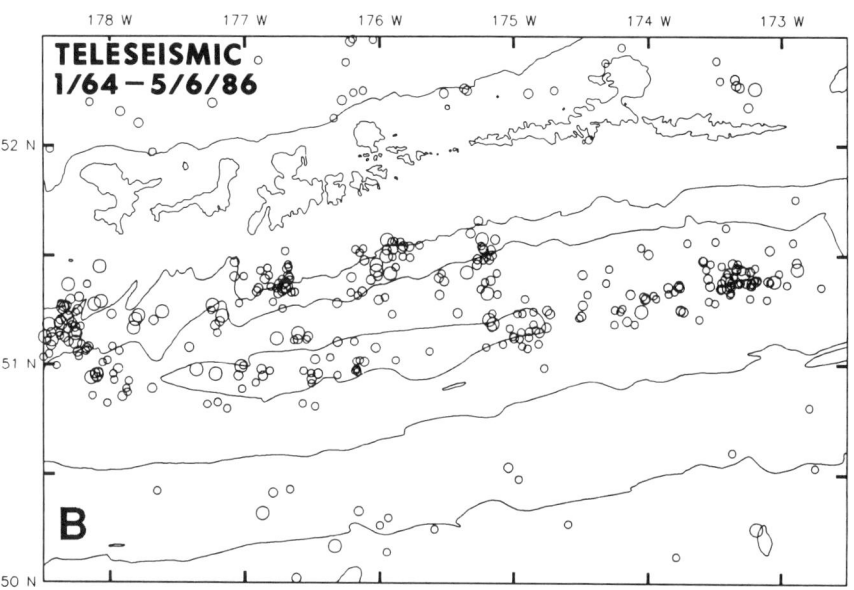

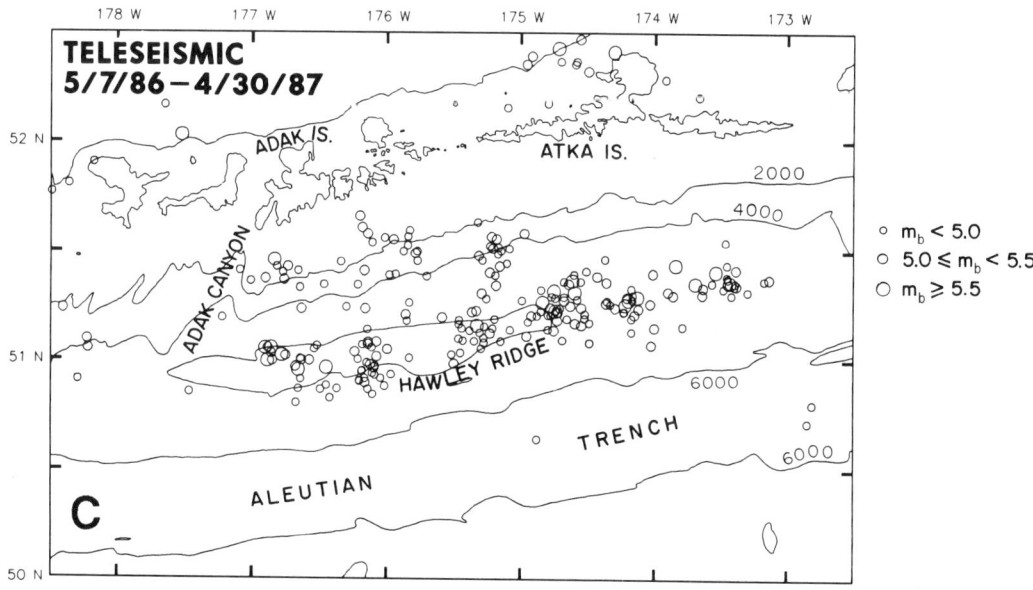

Figure 9A. Map of seismicity that occurred from August 1974 through December 1985 as recorded by the Adak network. These are events for which depth as well as epicenter could be determined and which have $M_D \geq 2.3$. Events marked with squares are those for which a teleseismic body-wave magnitude has been determined by the USGS; all other events are shown by symbols that indicate the duration magnitude determined from Adak network data. The closed contours show central Aleutian islands, bathymetry at 1,000-m intervals. (B) Hypocenters relocated using teleseismic data from 1964 to May 6, 1986. (C) Teleseismic seismicity after the 1986 earthquake. There is no relative mislocation between teleseismic and local network locations.

38 J. J. Taber and Others

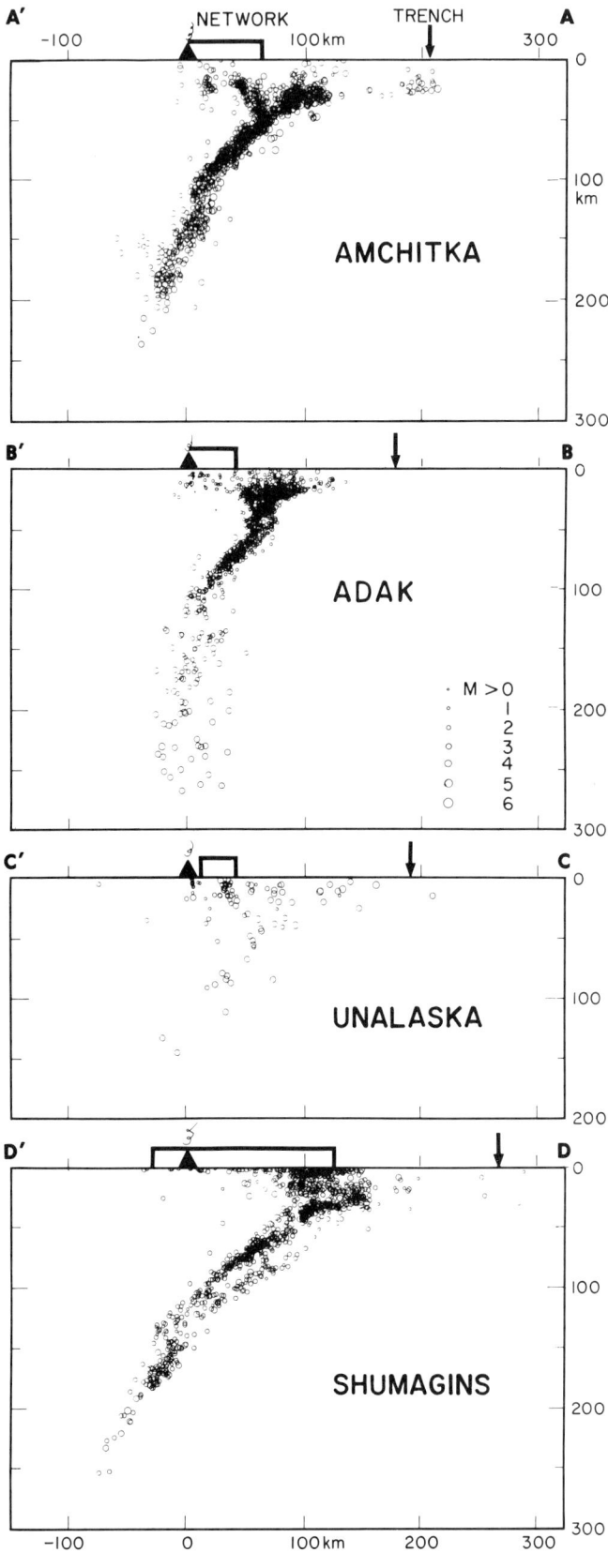

and Boatwright, 1980; Taber and Beavan, 1986). Earthquakes that initiate in the deeper high-stress regions can easily propagate upward into the shallower, lower stress part of the MTZ, whereas in the opposite case, propagation is inhibited by the increasing stress (Das and Scholz, 1983).

WADATI-BENIOFF ZONE

In contrast to the large variations in width, dip, and rate in the seismicity of the MTZ, the patterns of intermediate-depth earthquakes are very similar along the arc (Fig. 10). The variation in arc-trench separation mentioned above is almost entirely due to variations in the MTZ. Instead of the primarily episodic nature of the MTZ, earthquakes occur at a more uniform rate within the WBZ over the earthquake cycle. However, small temporal variations exist in seismicity rate and stress direction within the subducting slab. A decrease in seismicity rate between the MTZ and the WBZ is particularly noticeable in cross sections of network seismicity (Fig. 10). This area of limited seismicity marks the transition from interplate thrust events on the MTZ and intraplate compressional and tensional events in the WBZ. The Wadati-Benioff zone (WBZ) dips at an angle between 30° (Davies and House, 1979) in the Shumagins and 60° (Engdahl and Gubbins, 1987) in the Adak region. Offsets in the hypocentral locations that delineate the subducting slab occur in the same regions as offsets of the volcanic axis, suggesting that there is a distinct step in the slab itself at these points (House and Jacob, 1983).

Double seismic zone and focal mechanisms

A true double WBZ has been observed at depths of 50 to 150 km in the Shumagin Islands (Reyners and Coles, 1982; Hudnut and Taber, 1987), and a laterally segmented double zone has been proposed for the arc as a whole (House and Jacob, 1983). The only major change in the WBZ seismicity pattern

Figure 10. Cross sections of seismicity from four regions covered by local networks in the Aleutians. Location and width of sections shown in Figure 2. All hypocenters are located using just the local network, the extent of which is shown above each cross section. The cross sections are aligned relative to the volcanic axis, which is labelled as the origin on the horizontal axis. Only the highest quality solutions of the Amchitka, Adak, and Shumagin networks are used. Number of events does not reflect seismicity rate, as different time periods and selection criteria are used for each plot. Note general similarity of WBZ and position of aseismic front, even though the arc-trench distance varies considerably. The strong clustering of hypocenters at 26 km depth in Figure 10B and the systematic trenchward mislocation of the intermediate-depth earthquakes in that section are both artifacts of the flat-layered velocity model used for routine determination of Adak hypocenters (Toth and Kisslinger, 1984; Engdahl and others, 1982). In Figure 10D, note the clear double WBZ and the gap in seismicity between the thrust zone and the WBZ. The abrupt termination of seismicity at the edge of the Shumagin network is due to the quality criteria and is not a real feature.

along the arc is a possible shift of seismicity from the top to the middle of the subducting slab east of the transition from oceanic to continental upper plate near 164°W. This lateral segmentation of the subducting slab is based on the location of teleseismically recorded events relative to the volcanic arc and their focal mechanisms (House and Jacob, 1983). The teleseismic focal mechanisms generally show downdip tension in the lower, eastern plane and downdip compression in the upper, western plane. Except in the Shumagin gap, the change in stress, as exhibited by the focal mechanisms, is the primary basis for the laterally segmented model since the spatial coverage of the events is poor and the comparative cross sections depend heavily on the presumed consistency of the downgoing slab relative to the volcanic arc. The intraplate stresses are consistent with models of unbending of the subducting slab as it enters the mantle (Engdahl and Scholz, 1977; Kawakatsu, 1986), thermal changes (House and Jacob, 1982), and the sagging of the slab (Sleep, 1979).

When microseismic locations are compared to the teleseismic locations, the choice of a lateral segmentation model is not as clear. Within the Shumagin Islands, there is a transition from a double to single WBZ (Hudnut and Taber, 1987; Fig. 8). The lower plane exists only in the western portion of the Shumagin gap, while the upper plane continues as far east as can be determined by the local network data. This is the opposite of what would be predicted by the teleseismic data where the lower plane is to the east (House and Jacob, 1983). It may be that both planes exist to the east but that the b-value is much lower in the lower plane, and therefore, very few small shocks and only occasional moderate-sized events occur there. A double zone may start again to the east of the Shumagins, but the depth control there of the local data is marginal. Few deep events occurring within the Shumagin network have been teleseismically recorded since the network was installed, and none with $m_b > 5.0$, so it is difficult to compare the data sets. The depth of the easternmost teleseismic events are consistent with a single WBZ recorded by a local seismic network slightly to the east (Pulpan and Frohlich, 1985). The Unalaska data are consistent with a double zone, but there are far too few events to be conclusive. A double zone was originally proposed in the Adak region (Engdahl and Scholz, 1977), but it appears that the offset is really due to segmentation of the subducting plate (Topper, 1978). The apparent double zone there can be explained by the projection of a single, apparently slightly contorted Benioff zone onto one plane. It is possible that the systematic mislocation of intermediate-depth earthquakes by the local network may also play a role in the apparent slight contortion of the Benioff zone detected by Topper (Spencer and Engdahl, 1983).

Near Amchitka, composite focal mechanisms for events from 70 to 120 km are consistent with the teleseismic mechanisms (Engdahl and others, 1977). The mechanisms could be characterized by downdip pressure axes (or lateral extension along the strike of the arc), whereas first motions from earthquakes at greater depths suggest that the deeper portions of the slab are in tension. In the Adak Island region, Frohlich and others

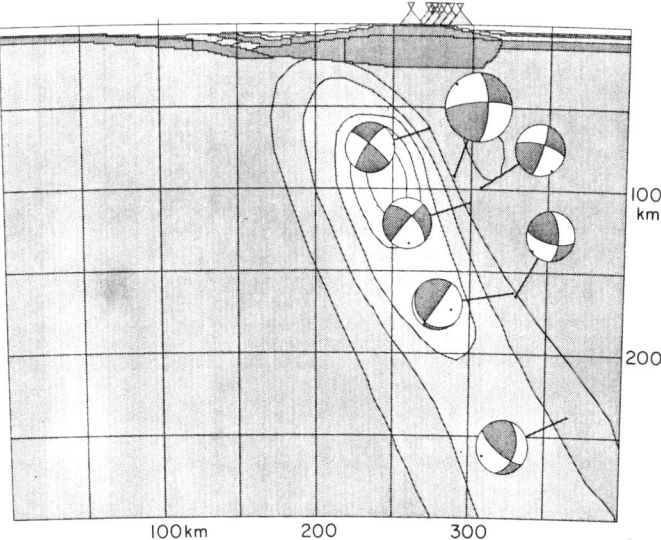

Figure 11. Double-couple component of 7 centroid moment tensor solutions near Adak (from Ekström and Engdahl, 1989) superimposed on the velocity model of Engdahl and Gubbins (1987). Mechanisms are back-hemisphere projections. Slab contours are 2 percent velocity variations. Note that events are in upper part of slab.

(1982) reported a downdip pressure axis for a composite focal mechanism for events between 70 and 100 km. More recently, centroid-moment tensor solutions for seven $M_w \geqslant 5$ intermediate-depth earthquakes (Fig. 11) have been determined by Ekström and Engdahl (1989). The null axes for six of these events are roughly arc-parallel and nearly horizontal. The shallowest of the events (61 km depth) shows primarily downdip tension, but this event may lie along an extension of the MTZ. Three events near 100 km depth have mechanisms dominated by downdip pressure axes. Two of the three deepest events (depth > 150 km) have focal mechanisms that reflect predominantly downdip tension.

Microearthquake focal mechanisms within the two planes of the WBZ in the Shumagin gap for the period from 1978 to 1986 (Fig. 12) are nearly the reverse of the teleseismic mechanisms of House and Jacob (1983). The local events exhibit downdip tension in the upper plate and in-plane compression in the lower plane. Reyners and Coles (1982) suggest that this may be related in some way to the preparation for a great earthquake, as has also been suggested by Astiz and Kanamori (1986), who examined larger intermediate-depth earthquakes in the Chilean subduction zone. Their results, however, were based on events with magnitudes greater than 7, whereas the Shumagin events have been less than magnitude 4. Along the arc as a whole, in addition to events with downdip pressure and tension axes, some larger events of $m_b > 5.2$ have horizontally directed pressure and tension axes that can be explained by a simple geometric model based on the curvature of the arc and the conservation of material (Creager and Boyd, 1990).

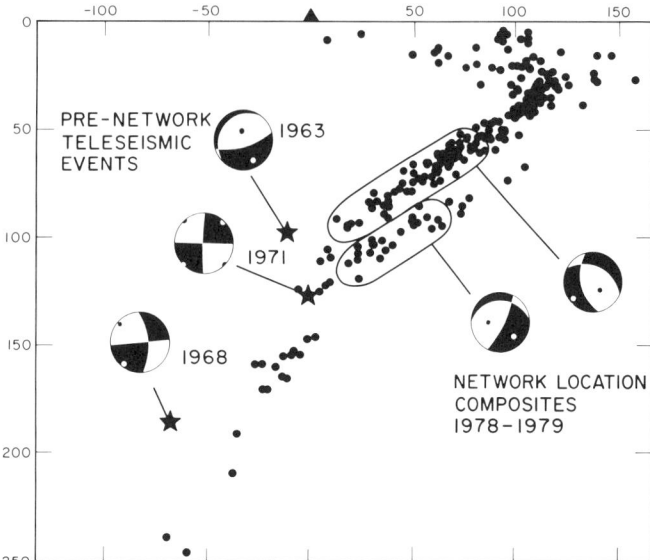

Figure 12. Composite focal mechanisms for the Shumagin network (Reyners and Coles, 1982) and mechanisms determined from teleseismic data for events that occurred within the same region (House and Jacob, 1983) but before the network was in operation. Mechanisms are front hemisphere, side projection along the plane of the cross section (line D-D′ in Fig. 2). Stars are locations of teleseismic events that may be mislocated relative to the local hypocenters. Lines around network hypocenters show events used in compositing. Note apparent reversal of P and T axes (black and white dots) between teleseimic and local mechanisms.

Temporal variations occurred within the WBZ during the 1978–1986 time period in the Shumagin region. Composite focal mechanisms from 1981 gave reversed pressure and tension axes from the Reyners and Coles 1978–1979 data (Hauksson and others, 1984). A later study (Taber and McNutt, 1984) showed that the results depended on the way in which events were composited. No real reversal in axes occurred, but the composite mechanisms were much less consistent in 1981–1986 compared to the 1978–1979 time period. 1978 through 1979 was also a time of high microseismicity rate and a reversal in surface tilt measured by repeated precision leveling. The tilt reversal was modeled as an aseismic slip event extending from 80 to 20 km in depth (Beavan and others, 1984). Slip on the plate interface might reduce the downdip tensional stresses within the subducting slab, which would explain the reduced coherency of the focal mechanism data after the proposed slip event.

Event relocations and structure

The best information available about the intermediate-depth seismicity and structure near Adak comes from recent efforts to relocate hypocenters, taking into account the presence of the subducted slab in the velocity model (Engdahl and Gubbins, 1987). Some earlier studies of intermediate-depth earthquakes reflect the systematic mislocation of intermediate-depth events by the Adak local network, in which a one-dimensional, flat-layered velocity model is routinely used (Frohlich and others, 1982; Spencer and Engdahl, 1983; McLaren and Frohlich, 1985). The dip of the WBZ as determined by the Adak network alone is steeper than that determined from teleseismic locations (Barazangi and Isacks, 1979). Shallow events have the poorest teleseismic locations, while intermediate-depth events can be well located teleseismically (Nieman and others, 1986).

Engdahl and Gubbins (1987) inverted travel-time data to teleseismic and local stations to calculate earthquake hypocenters and two-dimensional velocity structure for the central Aleutians. Their results show that intermediate-depth earthquakes there apparently occur within a narrow downdip zone near the upper surface of the descending slab, instead of being within the presumably stronger and colder inner core of the slab (Fig. 11). A slab thickness of 80 to 100 km and a downdip length of about 400 km, well below the deepest seismic activity, are indicated. The slab is characterized by seismic velocities as much as 11 percent higher than the surrounding mantle in its upper portions and 4 to 6 percent higher at depth. A sharp velocity gradient and lower velocities occur directly beneath the volcanic arc near the top of the slab.

In the Amchitka region, Engdahl (1973) relocated intermediate-depth earthquakes by ray tracing through a simplified plate model using data recorded by the local network. The relocated hypocenters define a zone about 10 km thick in the upper half of the plate, corresponding to the colder, more brittle region of the slab suggested by thermal models. However, Engdahl and others (1977) used a thermal model of the slab and teleseismic data to relocate 39 intermediate-depth events, and obtained a zone more closely resembling the results of Engdahl and Gubbins (1987) for the Adak region.

Engdahl (1977) noted that in both the Amchitka and Adak regions, intermediate-depth earthquakes in the depth range 50 to 110–120 km were uniformly distributed along the arc in a zone no more than 20 km thick, dipping steeply beneath the insular platform. In the Adak region, at least since 1974, earthquakes deeper than about 120 km occur more frequently toward the eastern edge of the Adak network coverage area than toward the western edge (Fig. 13). This may be in part a temporal effect.

A single plane of seismicity exists everywhere below 170 km and extends to 250 to 300 km. Local network data suggests a steepening of the dip below 100 km (e.g., Fig. 10d), but this seems to be a location error caused by the slab velocity anomaly (Fig. 14; Spencer and Engdahl, 1983; Hauksson, 1985; Engdahl and Gubbins, 1987). Ray paths in the slab pull the deeper events toward the trench when a standard flat-layer velocity model is used in the location process. The maximum depth of earthquakes varies considerably along the arc. Figure 13 shows the variation just within the Adak network. The general variation may be related to the age of the subducted slab (Creager and Boyd, 1990) and the decreasing convergence angle (Spence, 1977). The larger deep earthquakes seem to cluster at only a few locations along the arc (Fig. 1). It is not clear how these pockets relate to the shallow seismicity or to the overlying volcanoes. The maximum depth of earthquakes in the Aleutians is much less than many other subduction zones because of the relatively young (<65 Ma) age of the slab. Teleseismic inversions have traced the aseismic part of the slab to depths of at least 360 or 400 km in the central Aleutians (Nieman and others, 1986; Engdahl and Gubbins, 1987), and at least 300 km deeper than the deepest seismicity in the central and eastern Aleutians (Boyd and Creager, 1990).

UPPER PLATE SEISMICITY

Though much of the plate convergence seismicity is associated with the subducting oceanic slab, earthquakes do occur in the overriding plate above and arcward of the MTZ (Fig. 10). As with the MTZ, this seismicity does not extend all the way to the trench, but instead terminates at or within the accretionary prism at a point where deformation begins to take place plastically. It has been suggested that this cessation of seismicity within the upper plate may also mark the trenchward extent of rupture on the MTZ during great earthquakes (Byrne and others, 1988). This has been termed the seismic front (House and Jacob, 1983; see Fig. 3). The arcward boundary of upper plate seismicity begins near the base of the MTZ and angles up toward the surface expression of the volcanic arc. This feature has been named the aseismic front (Yoshii, 1979) and seems to occur in a similar location in most subduction zones. It may mark the base of the seismically active plate boundary. This feature is quite consistent in the cross sections in Figure 10. Between the seismic and aseismic fronts the seismicity in the upper plate is diffuse, with little indication of movement on specific faults in contrast with strike-slip regions such as California where most shallow seismicity is associated with mapped faults. The width of the zone varies with the dip and width of the MTZ. The events in this region are generally too small to be recorded by the global network, so local networks must be relied on for locations. Unfortunately, depth control is usually poor because islands are seldom located above the seismic front. Composite focal mechanisms for events within the Shumagin network show primarily horizontal compression in

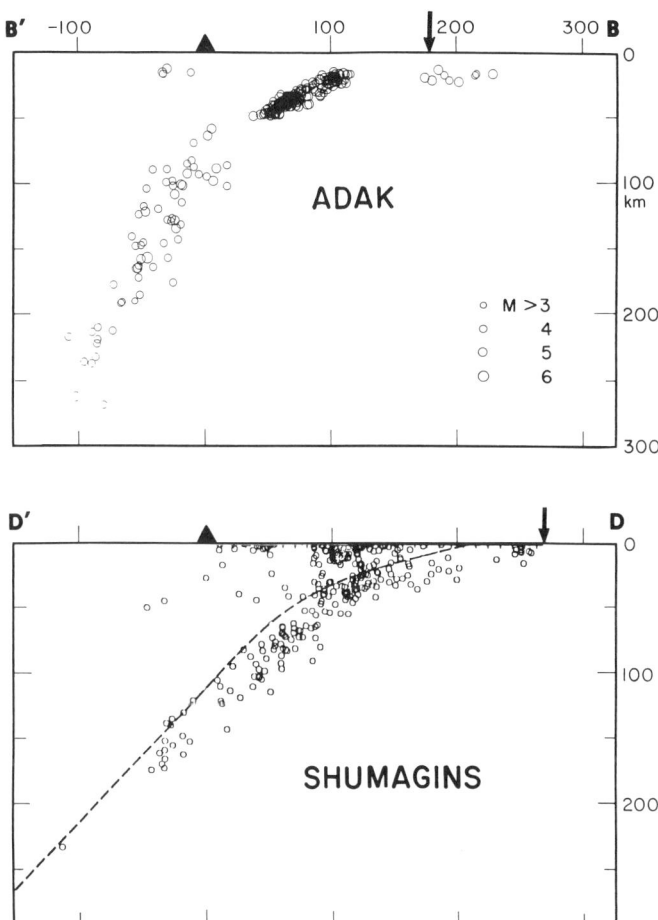

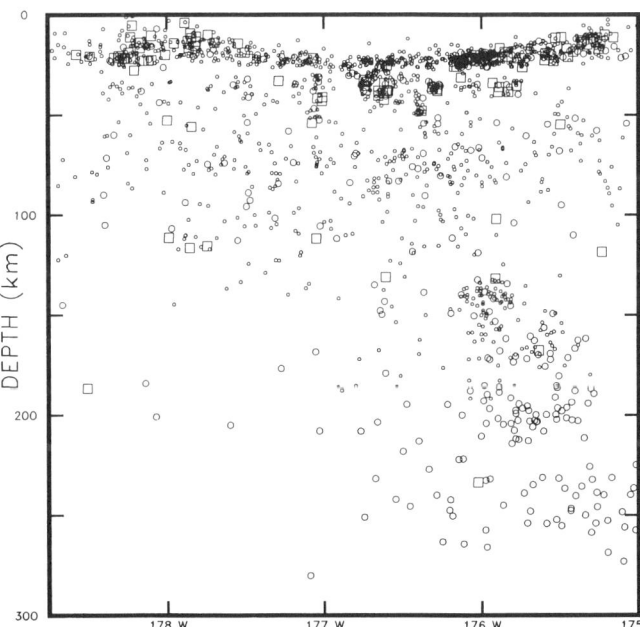

Figure 13. Plot of depth of hypocenter versus longitude, for seismicity that occurred from August 1974 through December 1985 within the Adak network. Section shows the same 1834 events as mapped in Figure 9a. Events marked with squares are those for which a teleseismic body-wave magnitude has been determined by the USGS; all other events are shown by symbols that indicate the duration magnitude determined from Adak network data. The strong clustering of hypocenters at depths near 26 km is an artifact of the flat-layered velocity model used for the routine determination of hypocenters (Toth and Kisslinger, 1984).

Figure 14. Top: Teleseismic locations for the Adak region accounting for the effect of the subducting slab (Engdahl and Gubbins, 1987). Compare to the local data in Figure 10b. Bottom: Relocation of Shumagin network data correcting for the high velocity of the slab, the top of which is shown as a dashed line (Hauksson, 1985). Compare to standard locations in Figure 10d. Note that the dip of the slab does not increase below 100 km. Triangle shows location of volcanic axis, arrow is location of trench axis.

the direction of convergence (Reyners and Coles, 1982). This is consistent with stress being transferred across the plate interface to the overlying plate.

In the central Aleutians, however, near Adak (LaForge and Engdahl, 1979) and Amchitka (Kisslinger and Engdahl, 1974), a component of along-arc (east-west) extension exists within the upper plate. This normal faulting occurs in the region where there are large transverse canyons in the fore arc. These canyons appear to be tectonically produced extensional features (Gates and Gibson, 1956). They are not connected to a land mass large enough to erode them, are consistently asymmetrical in cross section, and lie subparallel to each other at an angle to the maximum slope gradient. LaForge and Engdahl (1979) suggest that along-arc extension is occurring because of the component of plate convergence that is parallel to the arc. East of Adak, where plate convergence becomes more normal, the canyons die out. When the relative plate motion becomes nearly parallel to the boundary east of Amchitka, the canyons also disappear.

Geist and others (1988) have developed a tectonic model for the central and western Aleutians based on structural, morphological, and seismic observations. In their model, the arc massif between approximately 170°E and 170°W is broken into five blocks, which since late Miocene or early Pliocene time have moved to the west and rotated clockwise with respect to the rest of the North American plate. The rotation away from the plate to the north caused the formation of several summit basins along the arc. Transverse canyons are formed along block-block boundaries due to the left-lateral shear between blocks as well as extensional stresses caused by differential rotation of adjacent blocks and the curved geometry of the arc. As a direct consequence of these geometrical considerations, the spacing between blocks must increase and the arc extend in the along-arc direction, as observed. The fault-controlled boundaries between blocks also coincide with the ends of rupture zones of major earthquakes, as previously noted.

Large strike-slip earthquakes have also occurred at crustal depths near the volcanic arc following the 1965 Rat Islands earthquake (Stauder, 1968a) and the 1986 Andreanof Islands earthquake (Ekström and Engdahl, 1989). Mechanisms for these earthquakes suggest either right-lateral arc-parallel motion or left-lateral sliding roughly in the direction of plate convergence. The Aleutian Arc is clearly affected by a tectonically significant component of oblique convergence. Ekström and Engdahl (1989) present two observations that suggest a partitioning of strain between slip on the MTZ and internal deformation of the arc massif: (1) the deviation of slip vectors along the whole Aleutian Arc from those predicted by large-scale plate motion; and (2) the occurrence of large strike-slip earthquakes in the upper plate. A model of oblique plate interaction is proposed in which a portion of the along-arc motion occurs along a weak strike-slip shear zone in the upper plate, near the volcanic line. The rate of along-arc extension estimated from this model is 25 mm/yr, similar in magnitude to an estimate of 14 to 16 mm/yr by Geist and others (1988) based on long-term average rates of deformation. On the basis of these and other lines of evidence, Engdahl and others (1989) concluded that the sequence of strike-slip earthquakes that occurred a few weeks after the 1986 Andreanof Islands mainshock was triggered by the mainshock and manifests a partial decoupling of oblique slip in this region along a west-striking right-lateral fault with low shear strength. For a more complete discussion of shallow stresses in the Aleutians, see Estabrook and Jacob (this volume).

Small earthquakes near the volcanic arc and above the plate interface were among the shallow-focus earthquakes in the overriding North American plate that were examined by Pohlman (1982) using data from the local Adak network. This study showed that small earthquakes near the volcanic arc in the Adak region exhibit a great diversity of focal mechanisms. The study suggests that the diversity of mechanism observed for shallow crustal earthquakes, particularly in the vicinity of a dormant volcano on Adak Island, is the result of crustal heterogeneities in the region rather than the result of volcanic processes. Larger earthquakes near the volcanic arc and above the plate interface occur in the crust at depths beneath the sediment-basement interface (Engdahl and Billington, 1986).

Engdahl (1972) showed that explosion-produced seismic effects from MILROW and CANNIKIN were confined to the upper crust of Amchitka Island. They are clearly separable into those occuring near the explosion cavity, which are related to cavity deterioration and chimney development, and those somehow related to a change in stress and/or strength of the nearby crust. The extent of the explosion-related effects and other lines of evidence suggest a low level of ambient tectonic stress in the upper crust of Amchitka Island. This might be expected early in the seismic cycle, i.e., only a few years after the 1965 great earthquake (L. R. Sykes, personal communication, 1988). On the basis of this evidence and the structural stability of the insular platform surrounding Amchitka, it was also concluded that the island is apparently seismically decoupled from the active zone of interplate earthquakes 40 to 50 km below.

WESTERN ALEUTIANS

The far western Aleutians (west of 176°E) exhibit very different seismicity patterns, primarily because of the change in the convergence direction between the Pacific and North American plates (Fig. 1). Additional complexity occurs at the intersection of the arc with the Kuril-Kamchatka subduction zone. Through much of the arc, subduction occurs almost normal to the arc, but by 170°E, plate motion is nearly parallel to the plate boundary. The Aleutian trench continues, but the Quaternary volcanoes do not. Mid-Tertiary volcanic rocks do exist because there had been convergence between the Pacific and North American plates around 50 Ma, which changed to strike-slip by 30 Ma (Engebretson and others, 1985). The trench is left over from the previous episode of subduction, and the current plate motion has not obliterated it (Cormier, 1975). Based on geochemistry, the downdip end of the plate may have reached 125 to 150 km (Vlasov, 1964).

Seismicity deeper than 70 km is absent west of 177°E, probably because the plate, which is moving nearly sideways instead of downdip, is now too warm to support brittle fracture (Cormier, 1975). However, there may still be a density contrast between the slab and the surrounding material, which could produce buoyant forces causing vertical movements in the overlying crust (Newberry and others, 1986).

In map view the shallow seismicity is clustered north and south of the arc (Fig. 1). Hypocenter data from a Soviet seismic network (Fedotov and others, 1987a), which has operated a station on Beringa Island since 1962 (Fedotov and others, 1987b), shows the same clustering. The pattern of shallow seismicity changes at 171°E near the western edge of aftershocks of the 1965 M_s = 8.2 Rat Islands earthquake. Normal faulting near the trench does not extend past 172°E, while shallow thrust mechanisms continue to 167°E (Fig. 15; Newberry and others, 1986). The convergence vector is nearly parallel to the arc at this point, so the strike-slip motion is being accommodated by slip on a nearly horizontal fault plane. West of 167°E the mechanisms are primarily strike slip, but there are occasional thrust events (Cormier, 1975; Zobin and Simbireva, 1977; Newberry and others, 1986) and a group of normal faulting solutions from an earthquake swarm in 1981 south of Beringa Island (Zobin and others, 1988). The line of seismicity to the north consists of right-lateral strike-slip motion on vertical fault planes west of 167°E, but east of this point the mechanisms can be interpreted as either overthrusting from the southeast toward the northwest or near-vertical motion (Newberry and others, 1986), depending on the choice of nodal plane. Thus, some of the relative plate motion is accommodated north of the arc.

The Aleutian Arc ends at the intersection with the Kuril-Kamchatka trench. Although there is no surface expression of a plate boundary north of the intersection, seismicity and focal mechanisms suggest the existence of a small plate between the North American and Eurasian plates. In this model the joining of the Aleutian strike-slip boundary and the Kuril-Kamchatka subduction zone is a triple junction with a combination of strike-slip and thrust faulting along the plate boundary to the north (Cook and others, 1986; Newberry and others, 1986).

CONCLUSIONS

The subduction of the Pacific plate beneath North America along the Aleutian Arc creates a pattern of seismicity that can be broken into four basic parts: (1) trench and outer rise, (2) main thrust zone (MTZ), (3) Wadati-Benioff zone (WBZ), and (4) upper plate. The division is based on the location of the events

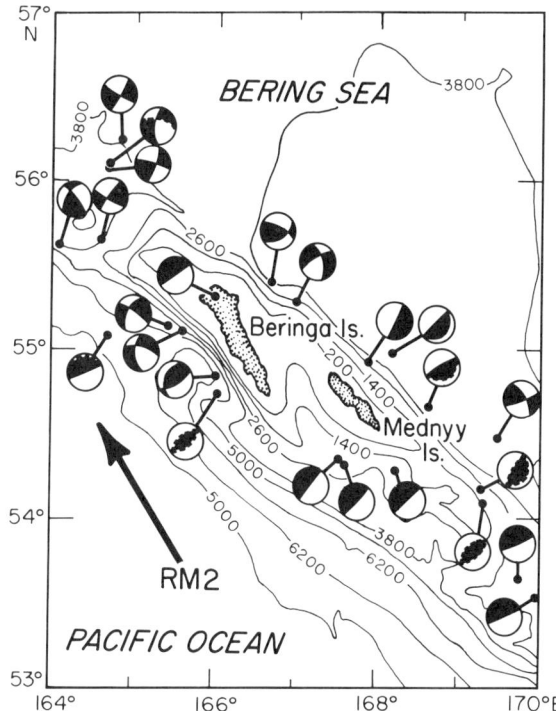

Figure 15. Focal mechanisms for the Komandorsky Islands regions. Mechanisms are shown as lower-hemisphere equal-area projections. Nodal planes dashed where poorly constrained. RM2 denotes North America–Pacific relative motion vector according to Minster and Jordan (1978). Bathymetry is in meters. (After Newberry and others, 1986.)

and direction of stresses as determined by focal mechanisms. Extensional events near the trench relate to bending of the plate and movement on the MTZ. The majority of the energy release within the subduction zone occurs along the MTZ during major plate-boundary ruptures. The ends of these major events seem to abut smoothly with only a few gaps. The gaps are areas of expected future earthquakes. WBZ seismicity occurs within the subducting Pacific plate. Focal mechanisms from events there show that the plate may be responding to mechanical and thermal stresses as it descends into the mantle. The plate can be traced at least 100 km and possibly as much as 300 km beneath the deepest seismicity, which occurs at depths of nearly 300 km. Seismicity in the upper plate is confined primarily to the region above the MTZ between the accretionary prism and the volcanic arc. Stress directions in this region can be either compression parallel to the plate convergence or, where the convergence is oblique, sliding or tension along strike.

REFERENCES CITED

Abe, K., 1972, Lithospheric normal faulting beneath the Aleutian trench: Physics of the Earth and Planetary Interiors, v. 5, p. 190–198.

Astiz, L., and Kanamori, H., 1986, Interplate coupling and temporal variations of mechanisms of intermediate-depth earthquakes in Chile: Seismological Society of America Bulletin, v. 76, p. 1614–1622.

Barazangi, M., and Isacks, B. L., 1979, A comparison of the spatial distribution of mantle earthquakes from data produced by local and teleseismic networks for the Japan and the Aleutian arcs: Seismological Society of America Bulletin, v. 69, p. 1763–1770.

Beavan, J., Bilham, R., and Hurst, K., 1984, Coherent tilt signals observed in the Shumagin seismic gap; Detection of time-dependent subduction at depth?: Journal of Geophysical Research, v. 89, p. 4478–4492.

Billington, S., Engdahl, E. R., and Price, S., 1981, Changes in the seismicity and focal mechanism of small earthquakes prior to an M_s 6.7 earthquake in the central Aleutians, *in* Simpson, D., and Richards, P. G., eds., Earthquake prediction; An international review: American Geophysical Union, p. 348–356.

Boyd, T. M., 1985, Characteristics of seismicity associated with the 1957 (M_w = 9.1) Aleutian Island earthquake [abs.]: EOS Transactions of the American Geophysical Union, v. 66, p. 298.

Boyd, T. M., and Creager, K. C., 1990, The geometry of Aleutian subduction; Three-dimensional seismic imaging: Journal of Geophysical Research (in press).

Boyd, T. M., and Jacob, K., 1986, Seismicity of the Unalaska region, Alaska: Seismological Society of America Bulletin, v. 76, p. 463–481.

Boyd, T. M., and Lerner-Lam, A., 1988, Spatial distribution of turn-of-the-century seismicity along the Alaska–Aleutian arc: Seismological Society of America Bulletin, v. 78, p. 636–650.

Boyd, T. M., Taber, J. J., Lerner-Lam, A., and Beavan, J., 1988, Seismic rupture and arc segmentation within the Shumagin Islands seismic gap, Alaska: Geophysical Research Letters, v. 15, p. 201–204.

Byrne, D. E., Davis, D. M., and Sykes, L. R., 1988, Loci and maximum size of thrust earthquakes and the mechanics of the shallow region of subduction zones: Tectonics, v. 7, p. 833–857.

Christensen, D. H., and Ruff, L. J., 1983, Outer-rise earthquakes and seismic coupling: Geophysical Research Letters, v. 10, p. 697–700.

Cook, D. B., Fujita, K., and McMullen, C. A., 1986, Present-day plate interactions in northeast Asia; North American, Eurasian, and Okhotsk plates: Journal of Geodynamics, v. 6, p. 33–51.

Cormier, V., 1975, Tectonics near the junction of the Aleutian and Kuril-Kamchatka arcs and a mechanism for middle Tertiary magmatism in the Kamchatka basin: Geological Society of America Bulletin, v. 86, p. 443–453.

Creager, K. C., and Boyd, T. M., 1990, The geometry of Aleutian subduction; three-dimensional kinematic flow model: Journal of Geophysical Research (in press).

Das, S., and Scholz, C. H., 1983, Why large earthquakes do not nucleate at shallow depths: Nature, v. 305, p. 621–623.

Davies, J., and House, L., 1979, Aleutian subduction zone seismicity, volcano-trench separation, and their relation to great thrust-type earthquakes: Journal of Geophysical Research, v. 84, p. 4583–4591.

Davies, J., Sykes, L., House, L., and Jacob, K., 1981, Shumagin seismic gap, Alaska Peninsula; History of great earthquakes, tectonic setting and evidence for high seismic potential: Journal of Geophysical Research, v. 86, p. 3821–3856.

Ekström, G., and Engdahl, E. R., 1989, Earthquake source parameters and stress distribution in the Adak Island region of the central Aleutian Islands, Alaska: Journal of Geophysical Research, v. 94, p. 15,499–15,519.

Engdahl, E. R., 1971, Explosion effects and earthquakes in the Amchitka Island region: Science, v. 173, p. 1232–1235.

—— , 1972, Seismic effects of the Milrow and Cannikin nuclear explosions: Seismological Society of America Bulletin, v. 62, p. 1411–1423.

—— , 1973, Relocation of intermediate-depth earthquakes in the central Aleutians by seismic ray tracing: Nature, Physical Sciences, v. 245, p. 23–25.

—— , 1977, Seismicity and plate subduction in the central Aleutians, *in* Talwani, M., and Pitman, W. C., III, eds., Island arcs, deep sea trenches, and back arc basins: American Geophysical Union, p. 259–271.

Engdahl, E. R., and Billington, S., 1986, Focal depth determination of central Aleutian earthquakes: Seismological Society of America Bulletin, v. 76, p. 77–93.

Engdahl, E. R., and Gubbins, D., 1987, Simultaneous travel-time inversion for earthquake location and subduction zone structure in the central Aleutian Islands: Journal of Geophysical Research, v. 92, p. 13855–13862.

Engdahl, E. R., and Scholz, C., 1977, A double Benioff zone beneath the central Aleutians; An unbending of the lithosphere: Geophysical Research Letters, v. 4, p. 473–476.

Engdahl, E. R., Sleep, N. H., and Lin, M.-T., 1977, Plate effects in North Pacific subduction zones: Tectonophysics, v. 37, p. 95–116.

Engdahl, E. R., Dewey, J. W., and Fujita, K., 1982, Earthquake locations in island arcs, *in* Engdahl, E. R., ed., Earthquake algorithms: Physics of the Earth and Planetary Interiors, v. 30, p. 145–156.

Engdahl, E. R., Billington, S., and Kisslinger, C., 1989, Teleseismically recorded seismicity before and after the May 7, 1986, Andreanof Islands, Alaska, earthquake: Journal of Geophysical Research, v. 94, p. 15,481–15,498.

Engelbretson, D. C., Cox, A., and Gordon, R. G., 1985, Relative motions between oceanic and continental plates in the Pacific basin: Geological Society of America Special Paper 206, 59 p.

Estabrook, C. H., and Boyd, T. M., 1989, The Shumagin Islands earthquake of May 31, 1917 [abst.]: EOS Transactions of the American Geophysical Union, v. 70, p. 1225.

Fedotov, S. A., Feofilaktov, V. D., Gordeev, E. I. Gavrilov, V. A., and Chebrov, V. N., 1987a, The development of seismic studies in Kamchatka: Volcanology and Seismology, no. 6, p. 11–28 [in Russian].

Fedotov, S. A., Shumilina, L. S., and Chernysheva, G. V., 1987b, Seismicity of Kamchatka and Commander Islands as derived from detailed studies: Volcanology and Seismology, no. 6, p. 29–60 [in Russian].

Frank, F. C., 1968, Curvature of island arcs: Nature, v. 220, p. 363.

Frohlich, C., Caldwell, J. G., Malahoff, A., Latham, G. V., and Lawton, J., 1980, Ocean-bottom seismograph measurements in the central Aleutians: Nature, v. 286, p. 144–145.

Frohlich, C., Billington, S., Engdahl, E. R., and Malahoff, A., 1982, Detection and location of earthquakes in the central Aleutian subduction zone using land and ocean-bottom seismograph stations: Journal of Geophysical Research, v. 87, p. 6853–6864.

Gates, G., and Gibson, W., 1956, Interpretation of the configuration of the Aleutian ridge: Seismological Society of America Bulletin, v. 67, p. 127–146.

Geist, E. L., Childs, J. R., and Scholl, D. W., 1988, The origin of summit basins of the Aleutian ridge; Implications for block rotation of an arc massif: Tectonics, v. 7, p. 327–341.

Habermann, R. E., 1984, Spatial seismicity variations and asperities in the New Hebrides seismic zone: Journal of Geophysical Research, v. 89, p. 5891–5903.

Hatori, T., 1981, Tsunami magnitude and source area of the Aleutian–Alaska tsunamis: Bulletin of the Earthquake Research Institute of Tokyo University, v. 56, p. 97–110.

Hauksson, E., 1985, Structure of the Benioff zone beneath the Shumagin Islands, Alaska; Relocation of local earthquakes using 3-D ray tracing: Journal of Geophysical Research, v. 90, p. 635–649.

Hauksson, E., Armbruster, J., and Dobbs, S., 1984, Seismicity patterns (1963–1982) as stress indicators in the Shumagin seismic gap: Seismological Society of America Bulletin, v. 74, p. 2541–2558.

House, L., and Boatwright, J., 1980, Investigations of two high stress-drop earthquakes in the Shumagin seismic gap, Alaska: Journal of Geophysical Research, v. 85, p. 7151–7165.

House, L., and Jacob, K. H., 1982, Thermal stresses in subducting lithosphere; Consequences for double seismic zones: Nature, v. 295, p. 587–589.

—— , 1983, Earthquakes, plate subduction, and stress reversals in the eastern Aleutian arc: Journal of Geophysical Research, v. 88, p. 9347–9373.

House, L., Sykes, L., Davies, J., and Jacob, K., 1981, Identification of a possible seismic gap near Unalaska Island, eastern Aleutians, Alaska, in Simpson, D. W., Richards, P. G., and Sykes, L. R., eds., Earthquake prediction; An international revview: American Geophysical Union, v. 4, p. 81–92.

Hudnut, K. W., and Taber, J. J., 1987, Transition from double to single Wadati-Benioff seismic zone in the Shumagin Islands: Geophysical Research Letters, v. 14, p. 143–146.

Jacob, K. H., 1984, Estimates of long-term probabilities for future great earthquakes in the Aleutians: Geophysical Research Letters, v. 11, p. 295–298.

Jacob, K. H., Nakamura, K., and Davies, J. N., 1977, Trench-volcano gap along the Alaskan–Aleutian arc; Facts, and speculations on the role of terrigenous sediments, in Talwani, M., and Pitman, W. C., III, eds., Island arcs, deep sea trenches, and back-arc basins: American Geophysical Union, Maurice Ewing Series 1, v. 1, p. 243–258.

Kanamori, H., 1981, The nature of seismicity patterns before large earthquakes, in Simpson, D. W., and Richards, P. G., eds., Earthquake prediction; An international review: American Geophysical Union, p. 1–19.

Kawakatsu, H., 1986, Double seismic zones; Kinematics: Journal of Geophysical Research, v. 91, p. 4811–4825.

Kelleher, J., Sykes, L., and Oliver, J., 1973, Possible criteria for predicting earthquake locations and their application to major plate boundaries of the Pacific and the Caribbean: Journal of Geophysical Research, v. 78, p. 2547–2585.

Kisslinger, C., and Engdahl, E. R., 1974, A test of the Semyenov prediction technique in the central Aleutians: Tectonophysics, v. 23, p. 237–246.

Kondorskaya, N. V., and Shebalin, N. V., eds., 1977, New catalog of strong earthquakes in the Territory of the U.S.S.R.: Moscow, U.S.S.R., Nauka, p. 347–433.

LaForge, R., and Engdahl, E. R., 1979, Tectonic implications of seismicity in the Adak Canyon region, central Aleutians: Seismological Society of America Bulletin, v. 69, p. 1515–1532.

Laravie, J. A., 1975, Geometry and lateral strain of subducted plates in island arcs: Geology, v. 3, p. 484–486.

Lisowski, M., Savage, J. C., Prescott, W. H., and Gross, W. K., 1988, Absence of strain accumulation in the Shumagin seismic gap, Alaska, 1980–1987: Journal of Geophysical Research, v. 93, p. 7909–7922.

McCann, W. R., Nishenko, S. P., Sykes, L. R., and Krause, J., 1979, Seismic gaps and plate tectonics; Seismic potential for major plate boundaries: Pure and Applied Geophysics, v. 117, p. 1082–1147.

McLaren, J. P., and Frohlich, C., 1985, Model calculations of regional network locations for earthquakes in subduction zones: Seismological Society of America Bulletin, v. 75, p. 397–413.

Minster, J. B., and Jordan, T. H., 1978, Present-day plate motions: Journal of Geophysical Research, v. 83, p. 5331–5354.

Newberry, J. T., Laclair, D. L., and Fujita, K., 1986, Seismicity and tectonics of the far western Aleutians: Journal of Geodynamics, v. 6, p. 13–32.

Nieman, T. L., Fujita, K., and Rogers, W. J., 1986, Teleseismic mislocation of earthquakes in island arcs; Theoretical results: Journal of Physics of the Earth, v. 34, p. 43–70.

Nishenko, S. P., and Jacob, K. H., 1990, Seismic potential of the Queen Charlotte-Alaska-Aleutian seismic zone: Journal of Geophysical Research, v. 95, p. 2511–2532.

Pohlman, J. C., 1982, A study of a shallow-focus earthquake cluster possibly related to volcanism [M.S. thesis]: Boulder, University of Colorado, 205 p.

Pulpan, H., and Frohlich, C., 1985, Geometry of the subducted plate near Kodiak island and the lower Cook Inlet, Alaska, determined from relocated earthquake hypocenters: Seismological Society of America Bulletin, v. 75, p. 791–810.

Reyners, M., and Coles, K., 1982, Fine structure of the dipping seismic zone and subduction mechanics in the Shumagin Islands, Alaska: Journal Geophysical Research, v. 87, p. 356–366.

Ruff, L., 1985, The 1957 great Aleutian earthquake [abs.]: EOS Transactions of the American Geophysical Union, v. 66, p. 298.

Rundle, J. B., and Kanamori, H., 1987, Application of an inhomogeneous stress (patch) model to complex subduction zone earthquakes; A discrete interaction matrix approach: Journal of Geophysical Research, v. 92, p. 2606–2616.

Savage, J. C., and Lisowski, M., 1986, Strain accumulation in the Shumagin seismic gap: Journal of Geophysical Research, v. 91, p. 7447–7454.

Savostin, L. A., Zonenshain, L., and Baranov, B., 1983, Geology and plate tectonics of the Sea of Okhotsk: American Geophysical Union Geodynamics Series, v. 11, p. 189–221.

Scholl, D. W., Buffington, E. C., and Marlow, M. S., 1975, Plate tectonics and the structural evolution of the Aleutian–Bering Sea region, in Forbes, R., ed., Contributions of the geology of the Bering Sea basin and adjacent regions: Geological Society of America Special Paper 151, p. 1–30.

Simkin, T., Siebert, L., McClelland, L., Bridge, D., Newhall, C., and Latter, J. H., 1981, Volcanoes of the world: Stroudsburg, Pennsylvania, Hutchinson Ross Publishing Co., 232 p.

Sleep, N. H., 1979, The double seismic zone in downgoing slabs and the viscosity of the mesosphere: Journal of Geophysical Research, v. 84, p. 4565–4571.

Spence, W., 1977, The Aleutian Arc; Tectonic blocks, episodic subduction, strain diffusion, and magma generation: Journal of Geophysical Research, v. 82, p. 213–230.

Spencer, C. P., and Engdahl, E. R., 1983, A joint hypocentre location and velocity inversion technique applied to the central Aleutians: Geophysical Journal of the Royal Astronomical Society, v. 72, p. 399–415.

Stauder, W., 1968a, Mechanism of the Rat Island earthquake sequence of February 4, 1965, with relation to island arcs and sea-floor spreading: Journal of Geophysical Research, v. 73, p. 3847–3858.

—— , 1968b, Tensional character of earthquake foci beneath the Aleutian trench with relation to sea floor spreading: Journal of Geophysical Research, v. 73, p. 7693–7701.

—— , 1972, Fault motion and spatially bounded character of earthquakes in Amchitka Pass and the Delarof Islands: Journal of Geophysical Research, v. 77, p. 2072–2080.

Stauder, W., and Udias, A., 1963, S-wave studies of earthquakes of the North Pacific; Part 2, Aleutian Islands: Seismological Society of America Bulletin, v. 53, p. 57–77.

Sykes, L. R., 1971, Aftershock zones of great earthquakes, seismicity gaps and earthquake prediction for Alaska and the Aleutians: Journal of Geophysical Research, v. 76, p. 8021–8041.

Sykes, L. R., Kisslinger, J. B., House, L., Davies, J. N., and Jacob, K. H., 1981, Rupture zones and repeat times of great earthquakes along the Alaska-Aleutian arc, in Simpson, D. W., and Richards, P. G., eds., Earthquake prediction: American Geophysical Union, p. 73–80.

Taber, J. J., and Beavan, J., 1986, February 14, 1983, earthquake sequence in the Shumagin seismic gap, Alaska: Seismological Society of America Bulletin, v. 76, p. 1588–1596.

Taber, J. J., and McNutt, S. R., 1984, Temporal and spatial changes in seismicity associated with subduction near the Shumagin Islands, Alaska [abs.]: EOS Transactions of the American Geophysical Union, v. 65, p. 987.

Topper, R. E., 1978, Fine structure of the Benioff zone beneath the central Aleutian Arc [M.S. thesis]: Boulder, University of Colorado, 148 p.

Toth, T., and Kisslinger, C., 1984, Revised focal depths and velocity model for local earthquakes in the Adak seismic zone: Seismological Society of America Bulletin, v. 74, p. 1349–1360.

Vlasov, G. M., ed., 1964, Geolgiya SSSR, Tom 31; Kamchatka Kuriskiy i Komandorsky Ostrova, Cast 1: Geologicheskoye opisaniye Izdatelstvo "Nedra" Moscow (Geology of the USSR, vol. 31, Kamchatka, Kuril and Komdanorsky Islands, Document N68-151812, Nat. Tech. Inf. Center, U.S. Dept. of Commerce, Springfield, Va.).

Yoshii, T., 1979, A detailed cross-section of the deep seismic zone beneath northeastern Honshu, Japan: Tectonophysics, v. 55, p. 349–360.

Zobin, V. M., and Simbireva, I. G., 1977, Focal mechanism of earthquakes in the Kamchatka–Commander region and heterogeneities of the active seismic zone: PAGEOPH, v. 115, p. 283–299.

Zobin, V. M., Ivanova, E. I., and Chirkova, V. N., 1988, Earthquake source parameters for Kamchatka and Commander Islands area: Volcanology and Seismology, v. 6, p. 279–305 [English translation, Gordon and Breach Science Publishers].

MANUSCRIPT ACCEPTED BY THE SOCIETY FEBRUARY 28, 1990

ACKNOWLEDGMENTS

Helpful comments and plotting software were provided by T. Boyd. The Russian-language references were translated by Vadim Levin. The original manuscript was much improved by reviews by N. Biswas, K. Fisher, K. Fujita, and L. Sykes. Drafting was done by K. Nagao. The research was partially supported by Department of Energy grant ER13221. Lamont-Doherty Contribution No. 4626.

Printed in U.S.A.

Chapter 4

Seismicity of continental Alaska

Robert A. Page
U.S. Geological Survey, 345 Middlefield Road, Menlo Park, California 94025
Nirendra N. Biswas
Geophysical Institute, University of Alaska, Fairbanks, Alaska 99775
John C. Lahr
U.S. Geological Survey, 345 Middlefield Road, Menlo Park, California 94025
Hans Pulpan
Geophysical Institute, University of Alaska, Fairbanks, Alaska 99775

INTRODUCTION

Alaska spans 4,800 km of the active boundary between the Pacific and North American Plates and is the site of three of the world's ten largest earthquakes of this century. The largest of the three—the magnitude M_w 9.2 earthquake that struck southern coastal Alaska in 1964—ranks second only to the M_w 9.5 Chilean earthquake of 1960. The other two earthquakes (M_w 9.1, 1957, and M_w 8.7, 1965) occurred in the Aleutian Islands.

The seismicity of Alaska is discussed in two chapters. This chapter focuses on the seismicity of continental Alaska (Fig. 1), while a companion chapter (Taber and others, this volume) treats that of the Alaska Peninsula and the Aleutian Islands.

The earliest written accounts of Alaskan earthquakes date back to at least 1786 and appear in the reports of the early explorers and traders. The instrumentally recorded seismic history commenced more than a century later, when, in 1899, a series of great earthquakes located in southeast Alaska (near Yakutat Bay) were registered on early seismographs throughout the world (Tarr and Martin, 1912). The first seismograph in Alaska was installed in 1904 at Sitka on Alaska's southeastern panhandle, while the first in mainland Alaska was established three decades later, in 1935, at the University of Alaska at College, near Fairbanks (Fig. 1). At the time of the great 1964 earthquake, College and Sitka were the only permanent seismograph stations in Alaska.

The extensive destruction caused by the 1964 earthquake and its tidal waves triggered the growth of both statewide and regional seismograph networks and stimulated national interest in earthquake prediction and hazard reduction. By 1966, the University of Alaska at Fairbanks had installed a four-station telemetered regional network in central and south-central Alaska. The next year, the U.S. Coast and Geodetic Survey, which is now part of the National Oceanic and Atmospheric Administration, established the Alaska Tsunami Warning Center and initiated, with six seismic stations, a sparse telemetered network of seismic and ocean-tide stations in continental Alaska and the Aleutian Islands for the purpose of rapidly locating large, potentially tsunamigenic earthquakes and providing tsunami warnings (Butler, 1971). These two networks markedly improved the detection and location of regional earthquakes. Over the remainder of the 1960s and during the early 1970s, stations were added to the two networks, and by the end of 1972 the networks totaled nearly 40 stations.

Regional seismograph networks flourished in the 1970s with the need to develop seismic design criteria for the trans-Alaskan oil pipeline and offshore petroleum facilities. In 1971, the U.S. Geological Survey (USGS) established a ten-station array spanning the Cook Inlet and Valdez area. By 1974, the network was expanded to 54 stations stretching from Cook Inlet to Yakutat Bay (Fogleman and others, 1988). In the middle and late 1970s, the University of Alaska initiated 4- to 17-station regional networks in northeastern Alaska (Gedney and others, 1977), in western Alaska around the Seward Peninsula (Biswas and others, 1980, 1986b), and in lower Cook Inlet and around Kodiak Island (Kienle and others, 1983; Pulpan and Frohlich, 1985). The number of operating seismograph stations peaked in the early 1980s. The university networks in northeastern and western Alaska and around Kodiak Island were terminated by 1983, and the number of stations operated by the USGS in southern coastal Alaska declined to 42 by 1986.

Important compilations of historical earthquakes in Alaska

Page, R. A., Biswas, N. N., Lahr, J. C., and Pulpan, H., 1991, Seismicity of continental Alaska, *in* Slemmons, D. B., Engdahl, E. R., Zoback, M. D., and Blackwell, D. D., eds., Neotectonics of North America: Boulder, Colorado, Geological Society of America, Decade Map Volume 1.

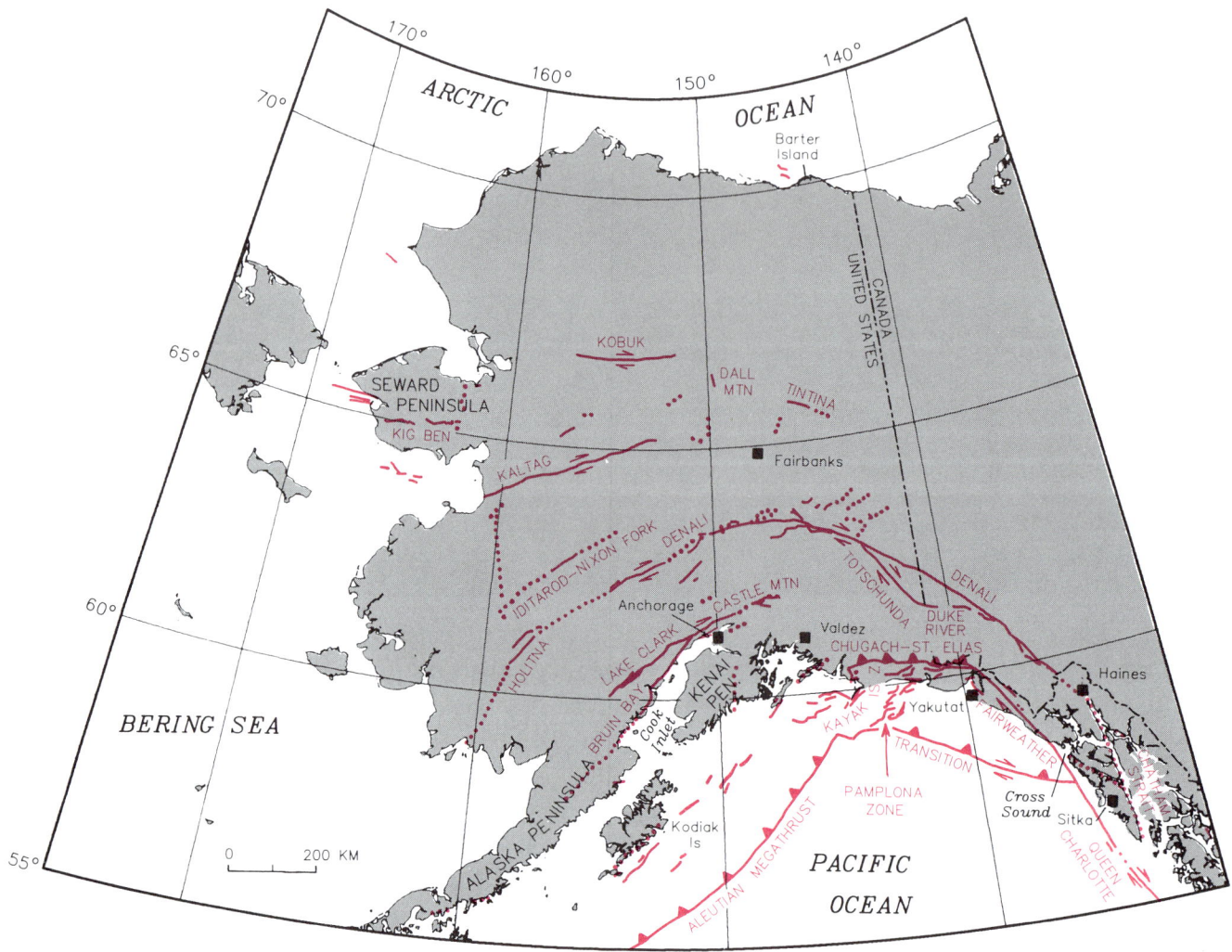

Figure 1. Map of continental Alaska showing known (solid) or suspicious (dotted) late Cenozoic faults (modified from Plafker and others, 1991). KIG = Kigluaik fault; BEN = Bendeleben fault.

include Tarr and Martin (1912) for the interval 1788 to 1909, Davis and Echols (1962) for 1788 to 1961, U.S. Coast and Geodetic Survey (1966) for 1786 to 1964, and Meyers (1976) for 1899 to 1974. More recent earthquakes as small as magnitude m_b 3.5 are routinely catalogued in the Preliminary Determination of Epicenters (PDE) reports of the USGS and its predecessor organizations and in the Bulletin of the International Seismological Centre. Smaller shocks recorded by the regional seismograph networks are catalogued in reports of the USGS and the Geophysical Institute of the University of Alaska at Fairbanks along with the larger ($m_b > 3.5$) earthquakes. Many historical earthquakes have been relocated by using modern computation techniques (e.g., Tobin and Sykes, 1966, 1968; Sykes, 1971; Boyd and Lerner-Lam, 1988) or by evaluating intensity data (e.g., McCann and others, 1980; Stover and others, 1980).

The seismicity of mainland Alaska is dominated by the intense activity that occurs along the Pacific coast (Fig. 2). Since 1899, 30 earthquakes of magnitude 7.0 or larger have originated in and adjacent to continental Alaska (Table 1). Although most of these occurred along the Pacific coast, some originated in interior Alaska and the Gulf of Alaska (Fig. 2). The largest instrumentally recorded events in the interior are of magnitude 7.3, two near Fairbanks (1904 and 1937) and one in western Alaska (1958).

Most of the shallow earthquakes in continental Alaska, with the exception of those in central western Alaska, appear to result from the interaction of the Pacific and North American lithospheric plates (Davies, 1983; Estabrook and others, 1988). The northwestward motion of the Pacific Plate relative to the North American Plate is accommodated by dextral transcurrent faulting in southeast Alaska and by underthrusting and subduction of the Pacific Plate beneath the continent along the Aleutian Trench (Fig. 3). The subduction process is manifest in the great 1964 earthquake (Fig. 2) and in the belt of subcrustal earthquakes along the west- to northwest-dipping Aleutian and the north-northeast–dipping Wrangell Wadati-Benioff zones (Fig. 3). A complexity must be added to this simple plate model in the transitional area between the transcurrent and convergent segments of the plate boundary. Here, the Yakutat terrane, which is largely transported with the Pacific Plate, impinges against the continent and resists subduction (Plafker, 1987). The horizontal compressional stress resulting from this collision is transmitted inland across the eastern half of the state and also seaward into the Pacific Plate (Lahr and others, 1988). In western Alaska, and on the Seward Peninsula in particular, the seismicity (Biswas and others, 1986b) and recent tectonism (Hudson and Plafker, 1978) reflect regional extension, the origin of which is not understood (Estabrook and others, 1988).

Crustal earthquakes pervade practically the entire state, as seen in the shallow seismicity from 1967 through mid-1988 (Fig. 4). The greatest concentration of activity is along the Pacific margin, where the major shocks occur (Fig. 2); however, a prominent belt of diffuse activity extends northward into interior Alaska. The only area of Alaska that appears to be aseismic is the North Slope, the part of Alaska lying north of the Brooks Range (Fig. 4). Numerous dense clusters of seismicity are apparent. Many of these mark aftershocks of recent large earthquakes, such as the 1979 St. Elias (No. 28, Fig. 2 and Table 1) and the 1987 and 1988 (Nos. 29 and 30) Gulf of Alaska shocks; however, other clusters are unrelated to the occurrence of a previous large shock (or at least to one that appears in the historical record). Few of the epicenters can be correlated with certainty to mapped geologic faults, even to those faults with evidence of late Cenozoic displacement. As discussed below, various authors have suggested correlations between shocks larger than magnitude 3.0 and several of the faults shown in Figures 1 and 4, including the Castle Mountain, Denali, Duke River, Totschunda, Fairweather, Queen Charlotte, Tintina, Dall Mountain, and Kobuk faults. In

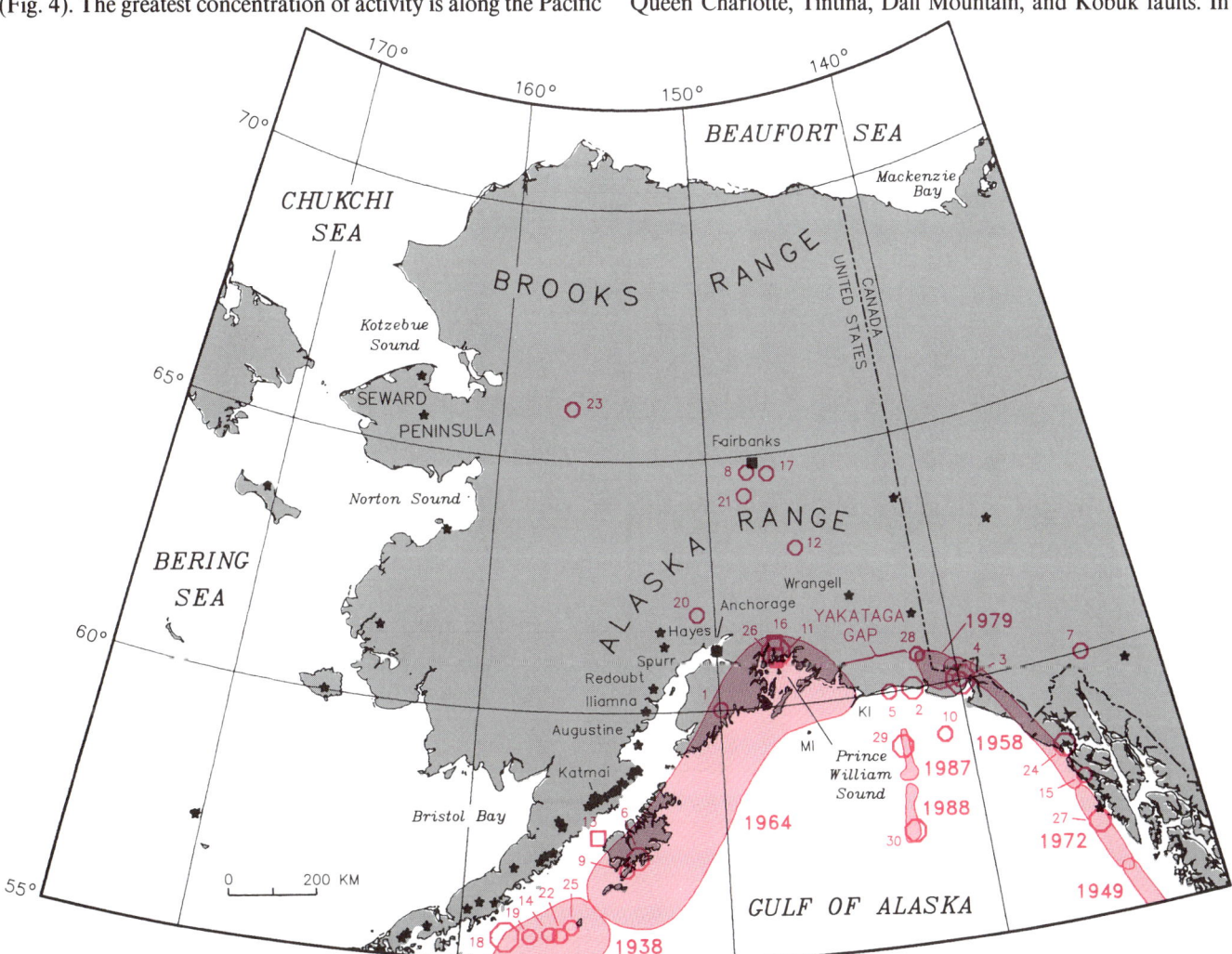

Figure 2. Epicenters of magnitude 7.0 or greater earthquakes for 1899 to 1988 (shallow depth, circles; intermediate depth, squares) and aftershock zones of large plate-boundary earthquakes for 1930 to 1988. Event numbers refer to Table 1. Stars are volcanoes active or probably active with the last 10,000 yrs. From Simkin and others (1981). KI = Kayak Island; MI = Middleton Island.

many of these cases, however, uncertainty in the epicenters makes the correlation with a specific mapped fault trace tenuous. In southern coastal Alaska, there is the added complication that many of the shallow shocks originate in the shallow part of the subducting oceanic plate and cannot be expected to correlate with mapped faults in the overriding continental plate.

The subsequent sections of this paper address the seismicity of continental Alaska in detail. For this purpose, we divide mainland Alaska into five areas: southeast, southern, central, northeast, and western Alaska (Fig. 5).

SOUTHEAST ALASKA

The tectonic framework of southeast Alaska (the panhandle of Alaska south and east of approximately 60°N, 141°W) is

TABLE 1. SOURCE PARAMETERS OF MAGNITUDE 7.0 OR GREATER EARTHQUAKES IN OR ADJACENT TO CONTINENTAL ALASKA (FIG. 2)

No.	Date			Time			Lat. °N	Long. °W	Depth km	Ref.*	Magnitude					
	Yr	Mo	Da	Hr	Mn	Sec					m_b	Ref.*	M_s	Ref.*	M_w	Ref.*
1	1899	07	14	13	32?		60§	150§		A			7.2	a		
2	1899	09	04	00	22		60	142	(<60)	B			7.9	a		
3	1899	09	10	17	04		60	140	(<60)	B			7.4	a		
4	1899	09	10	21	41		60	140	(<60)	B			8.0	a		
5	1899	09	23	12	50		60	143	(<60)	A			7.0	a		
6	1900	10	09	12	28	17.6	57.09	153.48	(33)	C			7.7	a		
7	1901	01	18	04	39?		60**	135**		A			7.1	a		
8	1904	08	27	21	56	20.3	64.66	148.08	(33)	C			7.3	a		
9	1906	12	23	17	21	11.7	56.85	153.90	(33)	C			7.3	a		
10	1908	05	15	08	31	36	59	141	(<60)	D			7.0	a		
11	1912	01	31	20	11	48	61	147½	80	D	7.0	b				
12	1912	07	07	07	57	41.0	63.07	146.14	(33)	C			7.2	a		
13	1912	11	07	07	40	24	57½	155	90	D	7.3	b				
14	1923	05	04	16	26	39.3	55.55	156.75	(0)	E			7.1	b		
15	1927	10	24	15	59	44.8	57.69	136.07	(0)	F			7.1	b		
16	1934	05	04	04	36	07	61¼	147½	80	D	7.1	b				
17	1937	07	22	17	09	28	64.6	147.1	(33)	G			7.3	b		
18	1938	11	10	20	18	41.2	55.48	158.37	(0)	E			8.3	b	8.2	c
19	1938	11	17	03	54	34.0	55.45	157.55	(0)	E			7.3	b		
20	1943	11	03	14	32	17.5	61.90	150.84	(0)	E			7.4	b		
21	1947	10	16	02	09	44	64.2	148.3	(33)	G			7.2	b		
22	1951	02	13	22	12	53.8	55.55	156.35	(0)	E			7.1	b		
23	1958	04	07	15	30	40.3	65.99	156.55	(0)	H			7.3	b		
24	1958	07	10	06	15	53.6	58.34	136.52	(0)	F			7.9	b	7.7	d
25	1964	02	06	13	07	23.1	55.72	155.95	13	G			7.0	b		
26	1964	03	28	03	36	13.9	61.05	147.48	23	G			8.4	b	9.2	c
27	1972	07	30	21	45	15.8	56.77	135.91	29	G			7.4	b	7.6	e
28	1979	02	28	21	27	07.2	60.64	141.60	13	I			7.1	f	7.5	f
29	1987	11	30	19	23	16.4	58.91	142.76	(10)	J	7		7.6	g	7.9	h
30	1988	03	06	22	35	36.4	57.23	142.78	(10)	J	7		7.6	g	7.8	h

Notes: m_b is the broad-band body-wave magnitude. M_s is the 20-second surface-wave magnitude. M_w is the moment magnitude. Depths in parentheses were assigned rather than computed.

*References: A = Kanamori and Abe (1979); B = Gutenberg (1956); C = Boyd and Lerner-Lam (1988), epicenters; T. M. Boyd (personal communication, 1989), origin times; D = Gutenberg and Richter (1954); E = Sykes (1971); F = Tobin and Sykes (1968); G = International Seismological Summary, or Bulletin of the International Seismological Centre; H = Tobin and Sykes (1966); I = Stephens and others (1980); J = Lahr and others (1988); a = Abe and Noguchi (1983); b = Abe (1981); c = Kanamori (1977); d = Abe and Kanamori (1980); e = Schell and Ruff (1989); f = Buland and Taggart (1981); g = Preliminary Determination of Epicenters (PDE), U.S. Geological Survey; h = This study. Computed from Harvard University (HRV) best double-couple moment published in PDE.

§Epicenter unreliable. Based on felt reports from Unga and Unalaska Islands (Tarr and Martin, 1912), epicenter may be near tip of Alaska Peninsula (Davies and others, 1981).

**Epicenter unreliable. No constraints or confirmation from felt reports.

dominated by right-lateral shear produced by the northwestward motion of the Pacific Plate with respect to the North American Plate (Fig. 6). South of 55.5°N, this shear is concentrated on the Queen Charlotte fault, whereas to the north it is distributed among the Queen Charlotte–Fairweather fault system, the Chatham Strait–Denali fault system, and the Transition fault zone (Lahr and Plafker, 1980). Both the geologic evidence for the displacement rate along the Fairweather fault (Plafker and others, 1978) and the historical record of large earthquakes support the view that the Queen Charlotte–Fairweather fault zone is the principal plate boundary (Tobin and Sykes, 1968; Page, 1973).

The historical record of magnitude 7 and greater earthquakes in southeast Alaska is complete since about 1918 and includes four events along the Queen Charlotte–Fairweather fault zone (1927, 1949, 1958, and 1972), all of which involved dextral slip (Stauder, 1959; Rogers, 1986; Stauder, 1960; Perez and Jacob, 1980). The 1927 shock (No. 15, Fig. 2) with a magnitude 7.1 (M_s) was located near latitude 57.7°N; although the longitude of this event is not well constrained, its focal mechanism is consistent with an origin on the Queen Charlotte–Fairweather fault system (Stauder, 1959). This event was followed by an M_w 8.1 shock in 1949 on the Queen Charlotte fault, which nucleated near 53.6°N (south of the areas shown in Figs. 2 and 6) and ruptured bilaterally, probably extending as far north as 56°N and totaling 470 km in length (Rogers, 1986). In 1958, an M_s 7.9 earthquake (No. 24) broke about 350 km of the Fairweather fault (Ben-Menahem and Toksoz, 1963; Tobin and Sykes, 1968) with as much as 6.6 m of dextral surface slip measured onshore (Tocher, 1960). Shaking from this event induced a large rockslide at the head of Lituya Bay causing a spectacular water wave that surged up and deforested the opposite shore of the fjord to an elevation of 530 m (Miller, 1960). The 1972 M_s 7.4 (No. 27) Sitka earthquake ruptured a 190-km segment of the fault system between the northern limit of the 1949 fault break and the southern limit of the 1958 break (Page, 1973), with inferred slip of 7 m in the epicentral region (Schell and Ruff, 1989). Prior to 1972, this plate-boundary segment had been identified as a seismic gap and a likely site for an earthquake (Tobin and Sykes, 1968; Kelleher, 1970; Sykes, 1971); thus, the Sitka earthquake was successfully forecast.

Although all the well-recorded historic shocks larger than magnitude 7.0 have occurred on the main plate boundary, significant seismicity occurs inland of the Queen Charlotte–Fairweather fault system (Fig. 6). North of about 59°N, a 40-km-wide band of seismicity follows the Denali fault system, including the Duke River fault; historic shocks as large as M_s 6½ (No. 10, Table 2) have originated in the vicinity of the Denali fault system in Canada (Horner, 1983). The epicenter of the M_s 7.1 shock of 1901 (No. 7, Fig. 2) lies 80 km northeast of the Denali fault system, but it is poorly constrained; within the uncertainty of the data, this event might have occurred on the main plate boundary. A broad swath of activity, first noted by Rogers (1976), lies between the Denali and the Queen Charlotte–Fairweather fault systems near 59°N and may represent the principal tectonic link connecting the two systems (Horner, 1983). Earthquakes as large as M_s 6 (No. 13, Table 2) are frequent in this area. Prominent seismicity clusters occur in the Coast Mountains, including swarm activity near 58.5°N, 133.5°W, which may be related to volcanic processes (Horner, 1983). Few earthquakes appear to be associated with the Chatham Strait fault (Rogers, 1976; Horner, 1983), although in 1987 a m_b 5.3 event (No. 44, Table 2) occurred 30 km south of Haines, in close proximity to the fault.

Seaward of the Fairweather fault, an M_s 6.7 earthquake in 1973 (No. 30, Table 2) furnishes evidence that the Pacific Plate is obliquely underthrusting the Yakutat terrane along the Transition fault zone (Fig. 6). Both relocated teleseismic and regional aftershock epicenters define a northwest-trending zone that lies along the continental slope southwest of Cross Sound at the location of the inferred Transition fault zone (Gawthrop and others, 1973; Page, 1975). Focal mechanisms of earthquakes in the 1973 sequence are consistent with underthrusting of the Pacific Plate to the northeast (Perez and Jacob, 1980).

SOUTHERN ALASKA

The convergence of the Pacific and North American Plates dominates the tectonic framework of southern Alaska. Between Yakutat Bay and the eastern end of the Aleutian Trench, or the northeast end of the Aleutian megathrust (Fig. 1), the plate boundary is complex, and the relative plate motion is distributed among at least three fault systems (Lahr and Plafker, 1980). At the northern end of the Fairweather fault, the primary component of plate motion is transferred to the Chugach–St. Elias fault, thence to the system of thrust faults disrupting the western part of the Yakutat terrane between the Pamplona and Kayak Island zones, and finally to the Aleutian megathrust (Figs. 1 and 3). Smaller components of motion are accommodated by the Transition fault at the edge of the continental margin and the Denali–Totschunda–Duke River fault system in the interior of the continent. Locally the Denali and Totschunda faults exhibit Holocene dextral slip rates as large as 1 or 2 cm/yr (Richter and Matson, 1971; Stout and others, 1973; Plafker and others, 1977). Between the primary plate boundary and the Denali fault system lie other active surface faults, including those in the Prince William Sound region that slipped in 1964 (Plafker, 1969) and the Castle Mountain fault, which exhibits Holocene oblique-slip offsets (Detterman and others, 1974). The tectonics in this part of southern Alaska are complex and transitional in character between those characterizing the transform plate margin in southeast Alaska and the convergent margin along the Aleutian arc.

West of this area of transitional tectonics, the plate boundary is relatively simple and lies along the Aleutian Trench. The classic features of a typical volcanic arc are present: a highly active, landward-dipping sheet of seismicity composing the Aleutian Wadati-Benioff zone (WBZ) and a chain of active volcanoes approximately following the 100-km isobath of the WBZ (Fig. 3).

The tectonic regime of southern Alaska gives rise to five

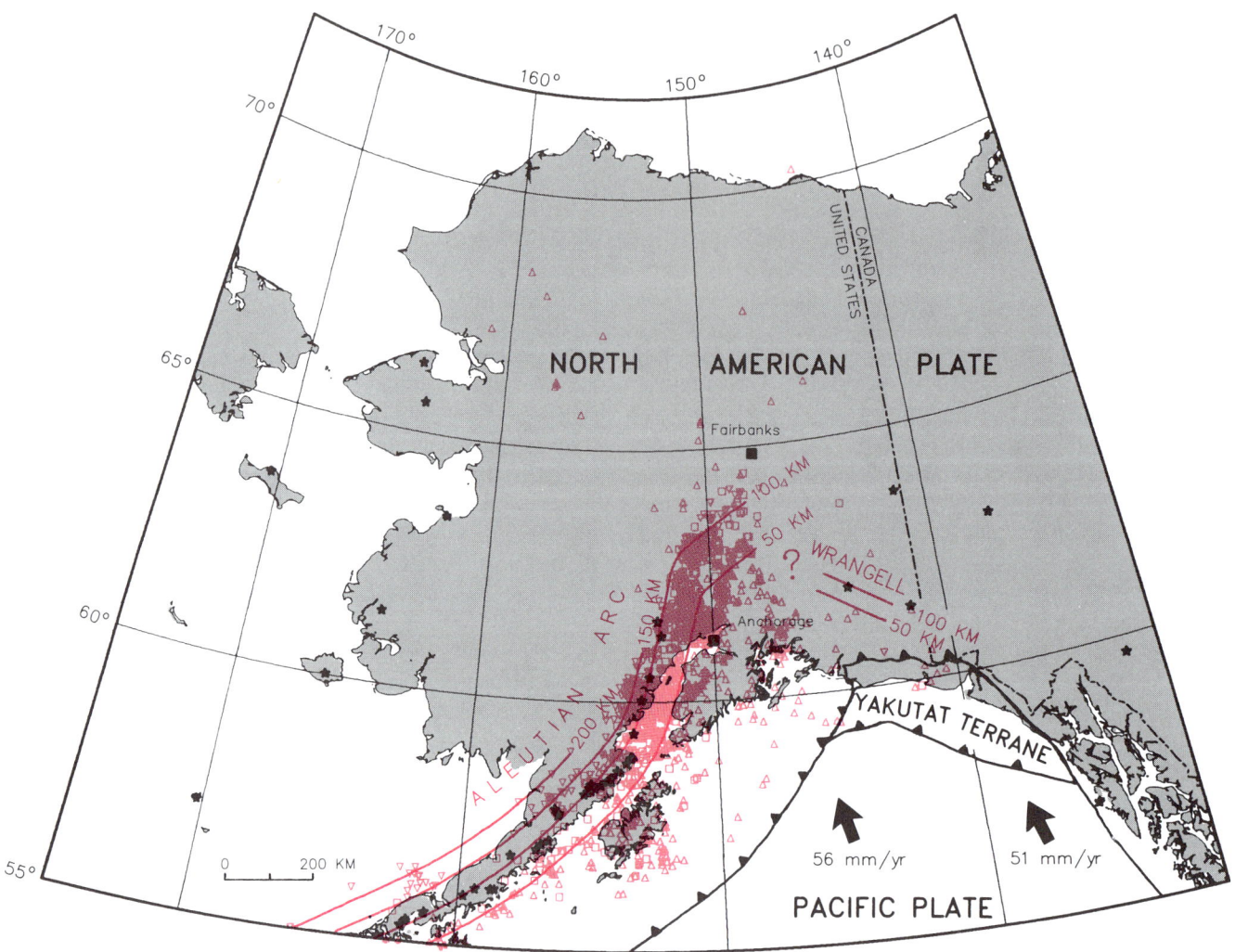

Figure 3. Epicenters of earthquakes deeper than 4 km and contours defining the tops of the Aleutian and Wrangell Wadati-Benioff zones. Earthquakes are equal to or greater than magnitude 3.0 and are from 1967 through mid-June 1988. Symbol type indicates depth range: triangle pointing up, 40 to 79 km; square, 80 to 119 km; triangle pointing down, 120 km or deeper. Source of data is the Preliminary Determination of Epicenters of the U.S. Geological Survey. Note the data are not complete or spatially homogeneous at the magnitude 3 level. Question mark indicates area of low seismicity in which Wadati-Benioff zone has not been identified and may be absent. Stars as in Figure 2. Arrows show motion of Pacific Plate relative to North American Plate (Demets and others, 1990).

distinct classes of earthquakes (Fig. 7): *main thrust zone earthquakes* along the interface between the subducting oceanic plate and the overriding continental plate, *subsea earthquakes* within the oceanic plate beneath or seaward of the trench, *Wadati-Benioff zone earthquakes* within the subducted part of the oceanic plate landward of the trench, *overriding plate earthquakes* in the overriding continental plate exclusive of those along the volcanic axis, and *volcanic axis earthquakes* in the overriding continental plate along the volcanic axis.

Main thrust-zone earthquakes

Most of the seismic energy in southern Alaska is released in major earthquakes that rupture the shallow part of the gently dipping thrust interface between the subducting and overriding plates. Beneath a certain depth, estimated to be about 40 km in the Aleutian arc (Davies and House, 1979) and about 20 km in the northern Prince William Sound region (Page and othrs, 1989), the interface slips aseismically. Since 1899 the entire length of the main seismic thrust zone west of about 140°W has ruptured in a series of major shocks.

West of Kayak Island, the main thrust zone is strongly coupled, as reflected in both the occurrence of the great 1964 earthquake (No. 26, Table 1 and Fig. 2)—the second largest shock yet to be recorded instrumentally—and in the gentle dip and large width of the plate interface. The gentle dip of the main thrust zone accounts for the large distance between the trench and

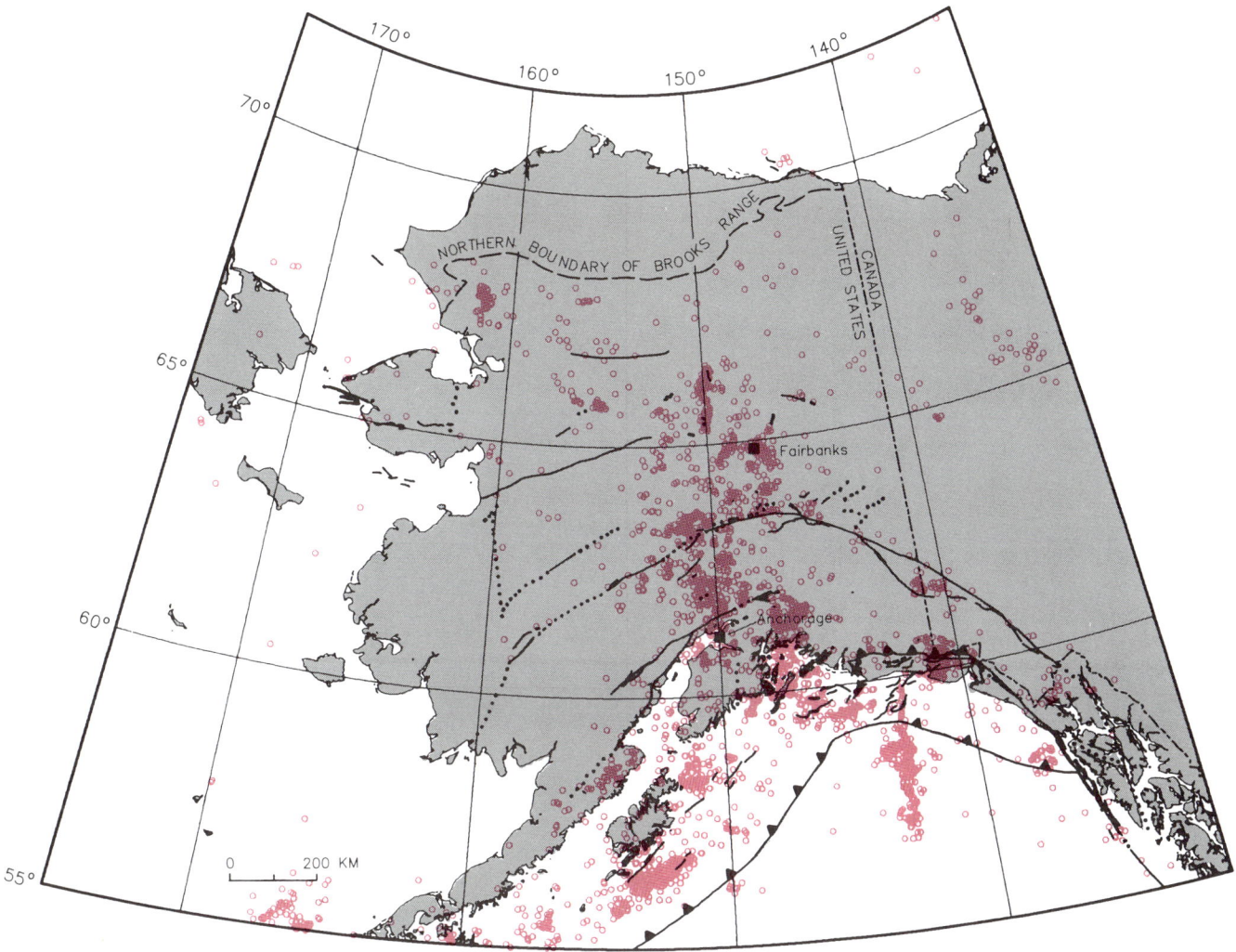

Figure 4. Epicenters of earthquakes with focal depths less than or equal to 40 km and magnitude greater than or equal 3.0 from 1967 through mid-June 1988. Earthquake data source same as for Figure 3. Note the date are not complete or spatially homogeneous at the magnitude-3 level. Faults as in Figure 1.

the 50-km WBZ isobath at the eastern end of the Aleutian arc (Fig. 3). The 1964 earthquake resulted from sinistral reverse slip on the order of 20 m, which produced extensive, large surface displacements, including as much as 10 m of uplift of Prince William Sound and the continental shelf and smaller amounts of subsidence landward on the Kenai Peninsula and Kodiak Island (Plafker, 1969; Hastie and Savage, 1970; Miyashita and Matsu'ura, 1978). The dimensions of the rupture zone were about 750 km by 200 km as deduced from the aftershock zone (Fig. 2). The origin and effects of this earthquake are discussed in the volumes of the National Academy of Sciences report, *The Great Alaska Earthquake of 1964* (Committee on the Alaska Earthquake, 1972).

Since its installation in 1971, the southern Alaska regional seismograph network has recorded a large number of shallow shocks in the Prince William Sound region (Fig. 8A). Few of these events appear to occur in the main thrust zone (Page and others, 1988); thus, in recent years the main thrust zone may have been practically aseismic. This observation suggests rapid locking of the main thrust zone after the 1964 earthquake. Estimates of the recurrence interval for 1964-type earthquakes vary widely, from a low range of about 200 to 600 yr based on average fault displacement inferred from the seismic moment and on long-term rates of relative plate motion (Sykes and Quittmeyer, 1981), to a high range of about 500 to 1,350 yr based on the uplift chronology of elevated marine terraces on Middleton Island (Plafker and Rubin, 1978).

The plate boundary east of Kayak Island ruptured in 1899 in a sequence of large earthquakes (Tarr and Martin, 1912), including events of M_s 7.9 and 8.0 (Nos. 2 and 4, Table 1 and Fig. 2). McCann and others (1980) argue that the first of these two events ruptured the plate boundary between Yakutat Bay

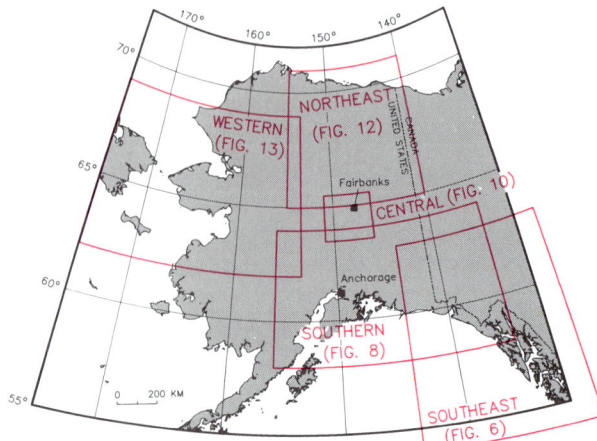

Figure 5. Index map defining geographic areas discussed in text and shown in Figures 6, 8, 10, 12, and 13.

and the 1964 zone; however, as mentioned earlier, the plate boundary in this region is distributed and does not contain just one simple thrust fault. Although this shock cannot be confidently linked to a particular fault, an onshore fault—possibly the Chugach–St. Elias fault—is implicated by the lack of a significant tsunami. The second 1899 event apparently involved deformation within the Yakutat terrane. From observations of elevated shorelines (Tarr and Martin, 1912), Thatcher and Plafker (1977) inferred as much as 10 to 20 m of reverse slip on northwest-striking, northeast-dipping thrust faults in the Yakutat Bay region.

The region between the aftershock zones of the 1958 Fairweather and the 1964 Prince William Sound earthquakes (Fig. 2) had been identified by the early 1970s as a seismic gap due to the lack of significant shocks since the turn of the century (Tobin and Sykes, 1968; Kelleher, 1970; Sykes, 1971). In 1979, the St. Elias earthquake (No. 28, Table 1 and Fig. 2) occurred at the eastern edge of this gap and involved north-northwest–directed thrusting on a subhorizontal fault about 15 km deep (Stephens and others, 1980). Whether this faulting lies within the Yakutat terrane or between the Yakutat terrane and underlying subducted Pacific Plate is uncertain. In any case, the Yakataga seismic gap, defined as the area situated between the 1964 and 1979 aftershock zones (Fig. 2), is considered to be a likely site for a major (M_s 7.8 or greater) earthquake within the next few decades (McCann and others, 1980; Lahr and others, 1980; Jacob, 1984).

Subsea earthquakes

The historical record contains little evidence of seismicity in the Gulf of Alaska seaward of the continental margin. However, in November 1987 and March 1988, two M_s 7.6 dextral strike-slip events (Nos. 29 and 30, Table 1) ruptured a 250-km-long north-south fault in the Pacific Plate south of the Yakataga seismic gap (Fig. 2). Lahr and others (1988) attribute these shocks to the combination of enhanced tensional stress in the Pacific Plate seaward of and following the great 1964 earthquake and compressional stress resulting from collision of the Yakutat terrane with North America. The orientation of the fault zone, parallel to the trend of the magnetic lineations in the northern Gulf of Alaska, suggests rupture of a zone of weakness inherited from the process of plate formation. These shocks are among the largest oceanic intraplate earthquakes ever recorded.

Several subsea shocks occurred off Kodiak Island beneath or just seaward of the Aleutian trench after the great 1964 earthquake. The one shock for which a focal mechanism is available—an m_b 5.6 event (No. 20, Table 2) four months after the main shock—displays normal faulting with a steeply dipping nodal plan subparallel to the trench (Stauder and Bollinger, 1966). Such extensional shocks are a common feature of convergent margins, especially after a large earthquake on the main thrust zone (e.g., Christensen and Ruff, 1988), and reflect tension in the oceanic lithosphere induced by plate bending or by gravitational body forces associated with sinking of the subducted portion of the plate, or by some combination of the two.

Epicenter files of historical seismicity include several shocks smaller than magnitude m_b 5 located in the Pacific Plate as much as 200 km seaward of the toe of the continental slope. The origin and mechanisms of these earthquakes are obscure.

Wadati-Benioff zone earthquakes

Oceanic plates, as they are subducted, deform in a brittle manner. The resulting seismicity within the subducted part of the oceanic plate constitutes the Wadati-Benioff seismic zone. In many studies the term "Wadati-Benioff zone" (WBZ) refers only to those shocks in the subducted plate that are landward of the main thrust zone, or equivalently, in many areas, to shocks that are deeper than about 40 km. We extend the definition of the WBZ to also include those shocks within the subducted plate that are directly beneath the main thrust zone (Fig. 7). We can broaden the definition because many of the earthquakes recorded by the southern Alaska regional seismograph network are located with sufficient precision to differentiate between shocks in the main thrust zone and those immediately below it in the underlying subducted plate (Page and others, 1989).

Wadati-Benioff zones are associated with both the northeast- to north-trending Aleutian volcanic arc and the east-southeast–trending Wrangell arc. The longer and more active Aleutian WBZ parallels the volcanic arc along the Alaska Peninsula and Cook Inlet (Fig. 3). Although the arc terminates at Hayes Volcano (Fig. 2), the WBZ continues about 350 km farther north to 64.1°N beneath the northern foothills of the Alaska Range (Gedney and Davies, 1986). The intense seismicity defining the Aleutian WBZ ends abruptly along a north-northwest–trending boundary that approximately parallels the direction of plate convergence.

The geometry of the Aleutian WBZ changes along strike as documented in studies of both teleseismically and regionally recorded earthquakes (Van Wormer and others, 1974; Lahr, 1975; Davies, 1975; Agnew, 1980; Pulpan and Frohlich, 1985). The contours defining the deeper, more steeply dipping part of the zone diverge from the Aleutian Trench in the Cook Inlet region (Fig. 3). In lower Cook Inlet near 59°N, the strike of the deep WBZ rotates 15° counterclockwise (Pulpan and Frohlich, 1985), and about 350 km farther north the strike swings about 35° clockwise (Agnew, 1980). Accordingly, the average dip of the shallow (depth < 50 km) part of the zone decreases from about 13° south of Kodiak Island to about 7° beneath Prince William Sound. Below a depth of 30 to 40 km, the dip of the WBZ increases with depth; beneath Cook Inlet, where the geometry is most precisely resolved from regionally recorded shocks, the dip steepens from about 35° at 50 km to 50° at 100 km (section AA′, Fig. 9). The maximum focal depth varies along the length of the zone from about 180 km near Iliamna Volcano to about 125 km beneath the northernmost Cook Inlet Volcano (Fogleman and others, 1988, Fig. 8) and to about 150 km beneath the Alaska Range (Gedney and Davies, 1986). Beneath continental Alaska, the Aleutian WBZ comprises a single zone of earthquakes, in contrast to the two zones of seismicity beneath the eastern Aleutian arc near the Shumagin Islands (Taber and others, this volume).

The magnitude of the largest historical shock within the Aleutian WBZ is uncertain because large uncertainties in the focal depths and epicenters of shocks earlier than the 1960s preclude ascribing shallow shocks with confidence to the WBZ rather than to the main thrust zone or overriding plate. The

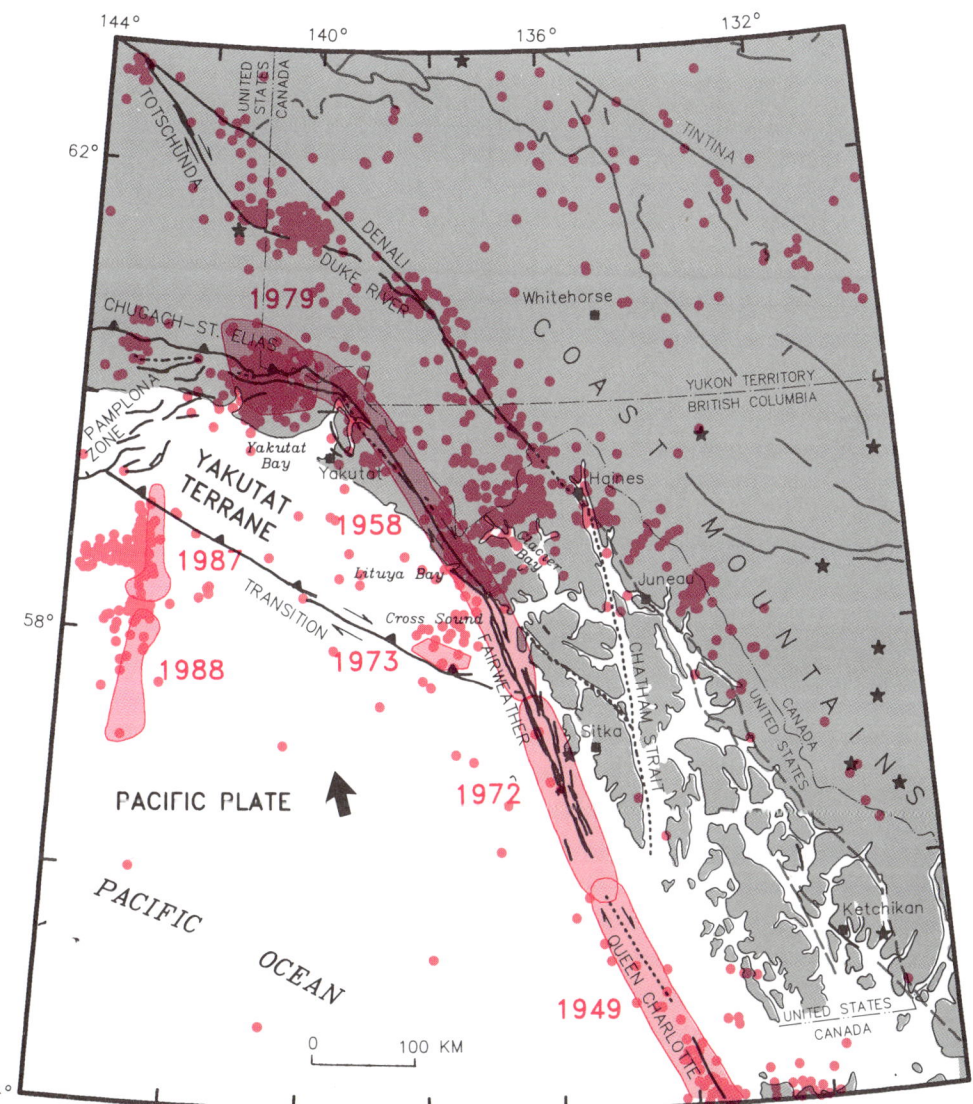

Figure 6. Epicenters of earthquakes in southeast Alaska and adjacent parts of Canada for 1982 to 1987 and principal known aftershock zones for 1930 to 1988. Epicenters determined by Geological Survey of Canada; magnitude thresholds are about M2.5 onshore and M3.0 offshore (Horner, 1990). Known (solid black) and suspicious (dotted black) late Cenozoic faults modified from Plafker and others (1991); other faults (gray), east of Denali and Chatham Strait faults, from Horner. Arrow indicates relative plate motion as in Figure 3. Stars are volcanoes as in Figure 2. Figure adapted from Horner (1990).

**TABLE 2. SOURCE PARAMETERS OF EARTHQUAKES SMALLER THAN MAGNITUDE 7.0
IN OR ADJACENT TO CONTINENTAL ALASKA** (Referred to in this article)

No.	Date Yr	Mo	Da	Time Hr	Mn	Sec	Lat. °N	Long. °W	Depth km	Ref.*	m_b	Ref.*	M_s	Ref.*
1	12	06	10	16	06	06	59	153		a			6.9	b
2	20	07	07	18	41	32	61.5	140.0	(15)	c			6	a
3	28	02	21	19	49	04	67	172		a			6.9	a
4	28	02	24	14	10	23	67	171		a			6¼	a
5	28	02	26	01	19	10	68	172		a			6½	a
6	28	05	01	18	54	41	67	172		a			6¼	a
7	29	01	21	10	30		64.24	147.98		d			6¼	a
8	29	07	04	04	28		64.24	147.88		d			6½	a
9	33	04	27	02	36	04	61¼	150¾		a			6.9	e
10	44	02	03	12	15	00	60.10	137.88	(15)	c			6½	a
11	50	05	25	08	34	32	65½	151½		f			6	f
12	50	08	26	04	39	27	65	162		f			6½	f
13	52	03	09	20	00	18	59.11	136.68	(15)	c			6	g
14	58	04	08	00	14	16.0	65.87	155.95	(0)	h			6.1	i
15	58	04	13	09	07	24.6	65.83	155.55	(0)	h			6¾	f
16	58	05	10	22	54	39.5	65.12	152.09	(0)	h			6¼–½	f
17	58	05	11	05	23	55.6	65.01	151.97	(0)	h			6¼–½	f
18	58	07	14	02	54	22.7	61.72	141.42	9	h				
19	58	08	31	23	00	17.0	63.22	144.32	17	h			6.2	i
20	64	08	02	08	36	17.3	56.18	149.90	31	j	5.6	j		
21	64	12	13	00	33	26.9	64.88	165.57	28	j	5.3	j		
22	65	04	16	23	22	18.6	64.69	160.23	(5)	j	5.8	j		
23	67	06	21	18	04	49.5	64.76	147.37	16	k	5.4	k	5.6	k
24	67	06	21	18	13	04.9	64.70	147.66	21	j	5.5	j	5.9	k
25	67	06	21	18	24	46.8	64.78	147.70	17	j	5.4	j	5.5	k
26	68	01	22	23	44	30.4	70.36	144.0	9	j	4.7	l	4.9	m
27	68	10	29	22	16	16.5	65.46	150.07	7	j	6.0	j	6.5	l
28	72	04	11	18	21	33	61.99	150.43	4	j	4.5	j		
29	72	11	28	13	35	38.5	65.65	145.6	36	j	4.2	j	5.0	l
30	73	07	01	13	33	34.4	57.86	137.42	30	j	6.2	j	6.7	l
31	79	02	23	09	42	03.7	64.96	147.84	22	j	4.3	j		
32	79	05	20	08	13	58.4	56.65	150.70	70	j	6.4	j	6.2	j
33	80	10	06	14	57	35.4	66.78	154.96	(33)	j	4.6	j	3.8	j
34	81	04	02	16	10	43.8	62.49	141.92	18	j	4.3	j		
35	81	07	12	01	27	56.4	67.70	161.44	34	j	5.3	j	5.1	j
36	81	12	30	14	00	32.6	64.59	148.01	15	j	4.9	j	4.6	j
37	83	03	30	18	06	16.7	61.55	140.09	18	j	5.4	j	4.9	j
38	83	07	12	15	10	03.7	61.03	147.18	30	n	6.2	j	6.4	j
39	83	09	07	19	22	05.0	60.98	147.32	30	n	6.2	j	6.2	j
40	84	08	14	01	02	09.3	61.82	148.98	19	o	5.6	j	5.4	j
41	85	02	14	05	04	02.2	66.24	150.02	5	j	5.0	j	5.0	j
42	85	03	09	14	08	04.6	66.22	149.98	7	j	5.8	j	6.1	j
43	86	09	15	14	48	22.7	61.50	143.90	52	p	4.6	j		
44	87	11	14	15	48	29.9	58.96	135.24	(5)	l	5.3	l	4.5	l

Notes: m_b is the short-period body-wave magnitude. M_s is the 20-second surface-wave magnitude. Depths in parentheses were assigned rather than computed.

*References: a = Gutenberg and Richter (1954); b = Abe and Noguchi (1983); c = Horner (1983); d = L. R. Sykes (unpublished data) in Pulpan (1988); e = Abe (1981); f = Annual volumes of U.S. Earthquakes; g = U.S. Coast and Geodetic Survey (1966); h = Tobin and Sykes (1966); i = Rothé (1969); j = Bulletin of the International Seismological Centre; k = Jordan and others (1968); l = Monthly Listing Preliminary Determination of Epicenters of the U.S. Geological Survey; m = Stevens and others (1976); n = Page and others (1985); o = Lahr and others (1986); p = Page and others (1989).

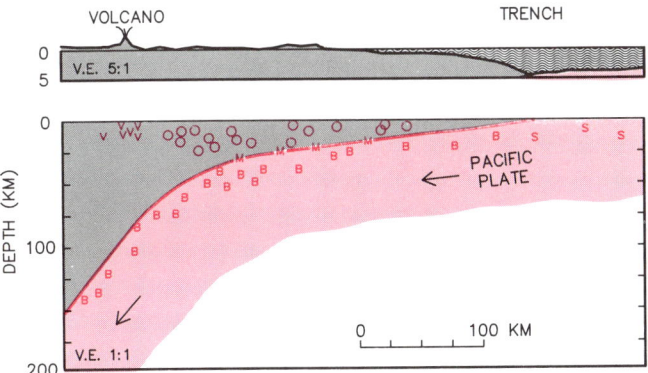

Figure 7. Schematic cross section defining terminology for earthquake classifications. Section is normal to Aleutian volcanic axis in southern Alaska. Upper profile shows topography and bathymetry with 5:1 vertical exaggeration. Lower graph shows distribution of five distinct classes of earthquakes: main thrust zone (M); subsea (S); Wadati-Benioff zone (B); overriding plate (O); and volcanic axis (V). No vertical exaggeration.

historical record suggests the possibility that shocks as large as magnitude 7 may have occurred in the WBZ, such as the 1912 m_B 7.3 shock off the southwest coast of Kodiak Island at a depth of 90 km (No. 13, Table 1). Since 1960, when more accurate hypocenters became available, the largest shocks have been in the magnitude 6 class: the two m_b 6.2 Columbia Bay shocks (No. 38 and 39, Table 2) in the shallow part of the WBZ beneath the north coast of Prince William Sound (Page and others, 1985, 1989); and an m_b 6.4 shock (No. 32, Table 2) at a depth of 70 km in the deep WBZ in the Kodiak Island region, south of the area in Figure 8.

Significant differences in the rate and character of seismicity exist between the shallow (h < 35 km) and deep parts of the Aleutian WBZ, with the higher level of activity in the deep part of the zone beneath the western Kenai Peninsula, Cook Inlet, and the volcanic arc (Fig. 8B and section AA′, Fig. 9). In both the shallow and deep parts, the seismicity since 1971 generally has been diffuse and constant but with a significant component of spatial and temporal clustering. Clustering is more pronounced in the shallow WBZ, as epitomized in aftershock sequences. Vigorous sequences, for example, followed the two 1983 Columbia Bay earthquakes in the shallow WBZ (Page and others, 1985). In contrast, few, if any, aftershocks have followed deep WBZ shocks in the Cook Inlet region. Nonetheless, spatial clustering in the deep WBZ is conspicuous (Fig. 8B). Some parts of the deep zone are very active, such as beneath Iliamna Volcano, whereas others are nearly aseismic, such as in the vicinity of the points 60.8°N, 152.5°W and 62.5°N, 150.2°W. These patterns, the causes of which are unknown, have been stable since at least the late 1960s when regional recording began.

East of the Aleutian zone lies the weakly expressed Wrangell WBZ, which dips to the north-northeast beneath the Wrangell volcanoes (Stephens and others, 1984a; Page and others, 1989). This zone extends at least 150 km along strike and to depths of at least 100 km; however, the dimensions of the zone could be greater because there are few hypocenters with which to define the width and downdip length of the zone. The thickness and downdip curvature of the Wrangell zone is similar to that of the Aleutian zone beneath Cook Inlet (compare sections AA′ and BB′, Fig. 9). The Wrangell WBZ is about two orders of magnitude less active than the Aleutian zone; the reason for this difference is not yet understood. The largest historical earthquake definitely related to this zone is an m_b 4.6 event (No. 43, Table 2) with a focal depth of 52 km.

The seismicity in the Aleutian and Wrangell zones in the depth range 20 to 45 km appears to be continuous across their common boundary; thus the two zones may represent the adjacent limbs of a buckle in the subducted plate (Page and others, 1989). Below 45 km, however, the continuity of seismicity cannot be established due to the lack of hypocenters in the Wrangell zone adjacent to the eastern end of the Aleutian zone (Fig. 3).

The stress orientation within the Aleutian WBZ, inferred from fault-plane solutions of earthquakes since 1970, is generally characterized by in-plane, downdip tension. In the subhorizontal, shallow WBZ beneath and north of Prince William Sound—the epicentral region of the great 1964 earthquake—the tensional, or least compressive, stress axis points toward the deep part of the WBZ (Page and others, 1988, 1989). Numerous solutions from the steeply dipping, deep WBZ in the Cook Inlet region suggest a dominant stress field with the tensional axis aligned in the dip direction of the subducted plate and the compressional axis aligned with the strike (Lahr, 1975; Engle, 1982; Pulpan and Frohlich, 1985). Downdip tension seems to characterize the deep WBZ north of Cook Inlet (Bhattacharya and Biswas, 1979) to its northern limit (Gedney and Davies, 1986). The stress field within the Wrangell WBZ is unknown because of the low level of seismicity (Page and others, 1989).

Overriding plate earthquakes

Shallow earthquakes are widely distributed in southern Alaska (Figs. 4 and 8A). Only since the regional seismograph network was installed, however, has it been possible to distinguish routinely and reliably the seismicity in the overriding plate from that in the main thrust zone and from that in the subducted plate. In the relatively short record acquired to date, upper-plate events occur less frequently than WBZ earthquakes (e.g., Fig. 9).

Shallow shocks located landward of the 30-km isobath on the Aleutian-Wrangell WBZ lie in the overriding continental plate (Fig. 8A). Over the past two decades, the seismicity of this region has been dominated by four concentrations: (1) a narrow belt parallel to the Duke River fault on the U.S.–Canada border, (2) a diffuse north-northeast–trending band near 150°W longitude extending between the Denali fault and Cook Inlet, (3) a band north of and paralleling the Castle Mountain fault east of 151°W, and (4) a belt along the volcanic axis west of Cook Inlet (Figs. 4 and 8A). We defer discussion of the latter to the section on volcanic axis earthquakes.

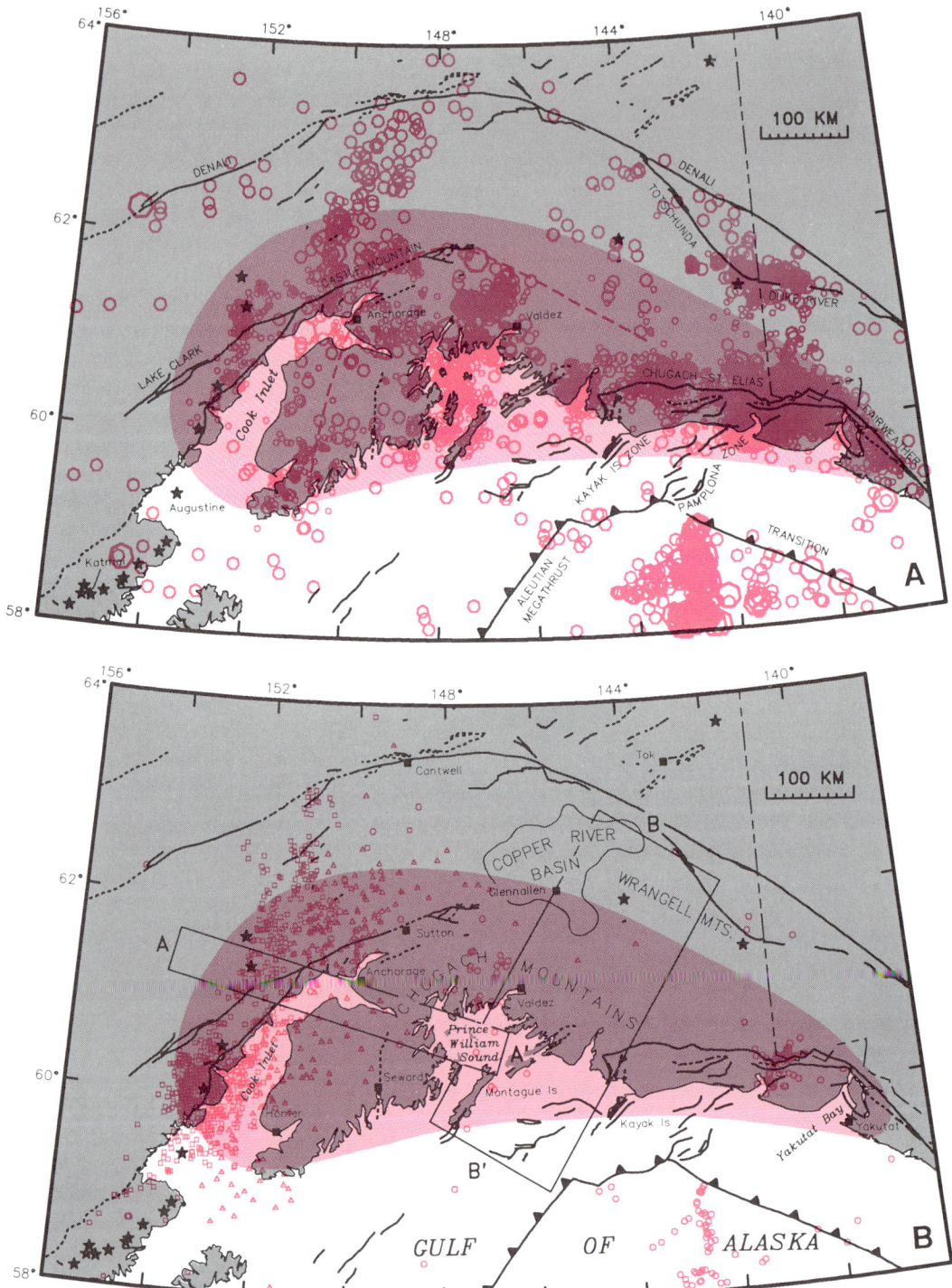

Figure 8. Epicenters of better-located earthquakes recorded by USGS southern Alaska seismograph network, shown in relation to known (solid) or suspicious (dotted) late Cenozoic faults (Plafker and others, 1991). Earthquake data are more complete and epicenters more accurate in central shaded region where station density is greatest. Stars are volcanoes as in Figure 2. Top: earthquakes shallower than 30 km for January 1979 through April 1988. Symbol size reflects equivalent m_b magnitude: small, 1.6 or smaller; medium, 1.7 to 3.7; large, 3.8 or larger. Dashed red line approximates 30-km isobath on top of Wadati-Benioff zone. Bottom: earthquakes for October 1971 through April 1988 with magnitudes greater than or equal to approximately m_b 3.8. Symbol type reflects depth: circle, 0 to 29 km; triangle, 30 to 69 km; square, 70 km or deeper. Boxes indicate search areas for hypocenters shown in Figure 9.

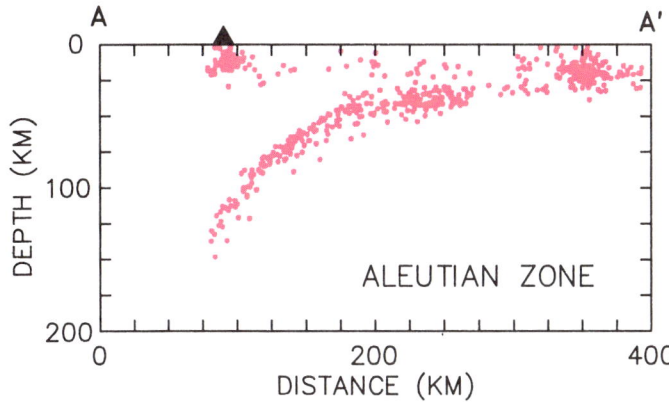

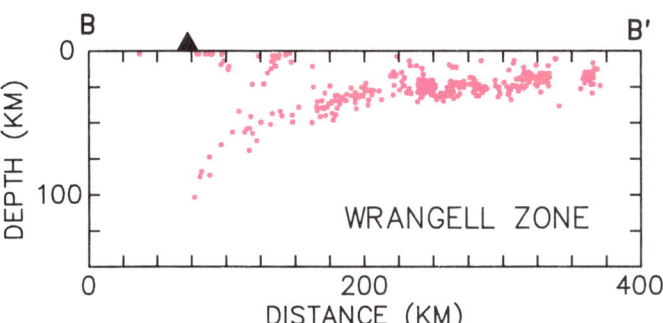

Figure 9. Sections of better-constrained hypocenters recorded by USGS southern Alaska seismic network. Section locations indicated in Figure 8B. A, view looking approximately north-northeast through Aleutian WBZ beneath upper Cook Inlet. Data from January 1979 to April 1988. B, view looking east-southeast through Wrangell WBZ. Data for October 1971 to December 1986. Figure adapted from Page and others (1989, Fig. 8).

Both historic (Horner, 1983) and recent (Fig. 4) teleseismic epicenters for shocks as large as M_s 6 (No. 2, Table 2) appear to scatter throughout the block between the Duke River and Denali faults. To a large degree, however, the apparent scatter may reflect errors in the epicenters. Recent shocks located from both regional (Horner, 1983; and Fig. 8A) and local (Power, 1988b) recordings reveal that, at least since the mid-1970s, the seismicity is mostly concentrated in the vicinity of the Duke River fault. The fault-plane solution for a m_b 5.4 shock in 1983 (No. 37, Table 2) suggests predominantly reverse slip on a southeast-striking plane (Dziewonski and others, 1983). A composite solution for a group of locally recorded shocks is also consistent with oblique reverse faulting (Power, 1988b).

The north-northeast–trending band of diffuse seismicity extending from northern Cook Inlet to the Denali fault (Fig. 8A) may mark a deformational zone that accommodates northwest-southeast compression between the block south of the Denali-Totschunda fault system and the interior of Alaska, as postulated by Lahr and Plafker (1980). The earthquakes in this broad band lie well above the Aleutian WBZ; in the northern half of the band, the shocks are shallower than 20 km, whereas the WBZ shocks are deeper than 40 km (Woodward-Clyde Consultants, 1980, Fig. 9-9). Earthquakes as large as the 1943 M_s 7.4 shock (No. 20, Fig. 2) may originate in this band; the largest shock since 1971 is a m_b 4.5 event (No. 28, Table 2). Analysis of seismic waveforms and first motions of the 1943 shock and composite fault-plane solutions for a few shallow microearthquakes in 1980 (Woodward-Clyde Consultants, 1980, 1982) suggest that reverse faulting with west- to northwest-oriented P axes may characterize this seismic belt.

The Castle Mountain fault, which passes 40 km north of Anchorage, exhibits geologic evidence of Holocene offset (Detterman and others, 1974) and possibly may have generated the large historic shock in 1933 (M_s 6.9; No. 9, Table 2). The 1984 m_b 5.6 Sutton earthquake (No. 40, Table 2) ruptured a 10-km-long buried segment of the fault and involved dextral slip (Lahr and others, 1986). Although the earthquake did not rupture the surface, well-constrained aftershock hypocenters unequivocally link this event to the fault. North of the fault lies a band of moderate seismicity (Fig. 8A) whose origin and tectonic significance are unknown.

In recent decades, only minor to moderate seismicity has occurred on the Totschunda fault and on the Denali fault west of its intersection with the Totschunda fault, although both faults bear abundant evidence of Holocene dextral displacement. The segment of the Denali fault between the Totschunda fault and 150°W longitude is thought to be capable of producing a magnitude class 8 shock (e.g., Woodward-Clyde Consultants, 1982), but no events of this size exist in the historical record. The largest event possibly associated with this segment of the Denali fault is the 1912 M_s 7.2 shock (No. 12, Table 1 and Fig. 2). More recently, a M_s 6.2 shock in 1958 (No. 19, Table 2) occurred in close proximity to the fault (Tobin and Sykes, 1966; Horner, 1983). Holocene offset on the Denali fault diminishes east of its junction with the Totschunda fault (Richter and Matson, 1971); a recent m_b 4.3 strike-slip earthquake (No. 34, Table 2) is identified with this segment of the fault (Gedney, 1985). No historic shock as large as magnitude 5 has been linked with the Totschunda fault, although an event of undetermined magnitude (No. 18, Table 2) occurred in the vicinity of the fault (Tobin and Sykes, 1966), four days after the 1958 Fairweather earthquake (Fig. 2).

Minor seismicity occurs elsewhere within the overriding plate landward of the 30-km WBZ isobath (Fig. 8A). Significant activity is observed in the vicinity of the U.S.–Canada boundary between the Duke River fault and the 1979 St. Elias aftershock zone (Fig. 2), where a fault system linking the Fairweather and Totschunda faults has been postulated to explain the rate of slip on the Denali-Totschunda fault system (e.g., Lahr and Plafker, 1980). Northeast of this activity, numerous scattered shocks lie between the 30-km isobath and the Denali and Totschunda faults. Little is known about these shocks; however, the P-wave initial motions for a group of shocks beneath the Copper River basin near 62.2°N, 145.3°W seem to reflect a roughly horizontal, north-south orientation of the compressive stress axis (Page and others, 1989).

In summary, the available fault-plane solutions for shallow

shocks in southern Alaska east of 150°W reveal strike-slip and reverse faulting. Although the solutions are few and generally not tightly constrained, they suggest a dominant horizontal compression oriented north-south. The pattern of P-axis directions appears to be consistent with the idea of the compressive stress originating from the collision of the Yakutat terrane with the North American Plate.

Seaward of the 30-km isobath, distinguishing between shocks in the overriding plate and those in the main thrust zone or in the WBZ becomes difficult because of the problem of obtaining sufficiently accurate focal depths for small shallow earthquakes recorded by a sparse regional seismograph network. In the Prince William Sound region, where the hypocenters are most reliable, only a small fraction of the shallow shocks since 1971 have been located in the overriding plate; most belong to the deeper, subhorizontal WBZ (Page and others, 1988, 1989). None of the overriding plate shocks has yet been associated with mapped faults, such as those on Montague Island that slipped in 1964.

Volcanic axis earthquakes

Small, shallow earthquakes are abundant along both the Aleutian (Matumoto and Ward, 1967; Kienle and others, 1983; Stephens and others, 1984b) and Wrangell (Page and others, 1989) volcanic axes. The spatial distribution of these shocks has been resolved most clearly in the Cook Inlet region, where they form a diffuse band, approximately 30 km wide, punctuated by prominent clusters (Fig. 8A). The volcanoes are marked by dense clusters of earthquakes shallower than 5 km, whereas elsewhere along the axis shocks fall in the depth range 5 to 20 km (Stephens and others, 1984b). For neither class of shocks are there published focal mechanisms.

Both the 1976 and 1986 eruptions of Augustine Volcano in lower Cook Inlet were preceded and accompanied by large numbers of small earthquakes (Lalla and Kienle, 1986; Reeder and Lahr, 1987; Power, 1988a); the duration of precursory seismicity in each instance was about nine months. Most of the earthquakes occurred directly beneath the volcano at depths shallower than 1 km below sea level and had magnitudes less than 2.0. (Too small to be recorded by the USGS regional seismic network, these shocks do not appear in Fig. 8A). The increase in seismicity preceding the 1986 eruption foretold its imminence (Kienle, 1986) and prompted advisories of the potential hazards that might be associated with an eruption (Kienle and others, 1986).

The largest known volcanic axis earthquakes are possibly those accompanying the June 1912 eruption of Novarupta (Katmai); however, the proximity of these widely felt shocks, which were as large as M_s 6.9 (No. 1, Table 2), to the eruptive center remains uncertain (Fenner, 1925), as do their focal depths.

CENTRAL AND NORTHEAST ALASKA

The historical record of seismicity in central and northeast Alaska (areas defined in Fig. 5), which derives mainly from teleseismic recordings of moderate and large earthquakes, includes eight shocks of M_s 6.0 or greater before 1960, three of which are about M_s 7.3 (Fig. 2). All of these shocks are located south of 66.5°N and between about 147°W and 152°W. The extent of seismicity is not accurately reflected in the scanty historical data. The many new seismographs installed in the late 60s and the 1970s record activity throughout much of central and northeast Alaska, including a concentration of activity in the area between longitudes 145°W and 153°W and south of 67°N (Fig. 4). This area includes the epicenters of the large historical shocks, but to some degree the heavy concentration of epicenters in this region reflects the detection capability of the modern seismic network. North of 67°N, regional network data show seismicity along the northeast extension of the Brooks Range, extending onto the Beaufort Shelf in the vicinity of Barter Island (Biswas and others, 1986a, Fig. 2; Estabrook and others, 1988).

The earthquakes in Alaska north of about 64°N all occur at shallow depths (less than 40 km) and are confined to the crust of the North American Plate. The characteristics of the seismicity are discussed more fully for the following areas: Fairbanks, Rampart–Dall City, and northeast Alaska.

Fairbanks area

The recent seismicity of the Fairbanks area as determined from data recorded by the regional seismograph network is broadly distributed (Fig. 10). The overall pattern of epicenters displays a general northeast-southwest grain with several noteworthy clusters and trends. Much of the seismicity falls within three zones (Fig. 10).

The Salcha seismic zone (Biswas and Tytgat, 1988) is characterized by a conspicuous northeast-trending band of epicenters about 50 km long. The epicenter of the 1937 M_s 7.3 Salcha earthquake (No. 17, Fig. 2 and Table 1) lies within the zone. This event, which was widely felt throughout central Alaska, produced extensive ground failures in the epicentral area but no documented evidence of tectonic faulting (Bramhall, 1938). The focal mechanism (Wickens and Hodgson, 1967) indicates predominantly strike-slip faulting (Fig. 10). If the nodal plane that strikes subparallel to the recent seismicity is interpreted as the fault plane, the focal mechanism represents predominantly left-lateral motion on a steeply dipping fault. Although the recent seismicity indicates a northeast-trending active fault, no fault with Quaternary displacement has yet been recognized (Cluff and others, 1974).

The Fairbanks seismic zone (Gedney and others, 1980) was the site of three M_s 5.5 to 5.9 earthquakes in 1967 (Nos. 23 to 25, Table 2), which occurred within several minutes of each other and were followed by a prominent aftershock sequence. Although widely felt, the three shocks caused only minor damage in the greater Fairbanks area (Jordan and others, 1968; Gedney and Berg, 1969). The focal mechanism solution for the first event (M_s 5.6) indicates nearly pure strike slip (Jordan and others, 1968; Gedney, 1970). Gedney and Berg (1969) and Gedney (1970)

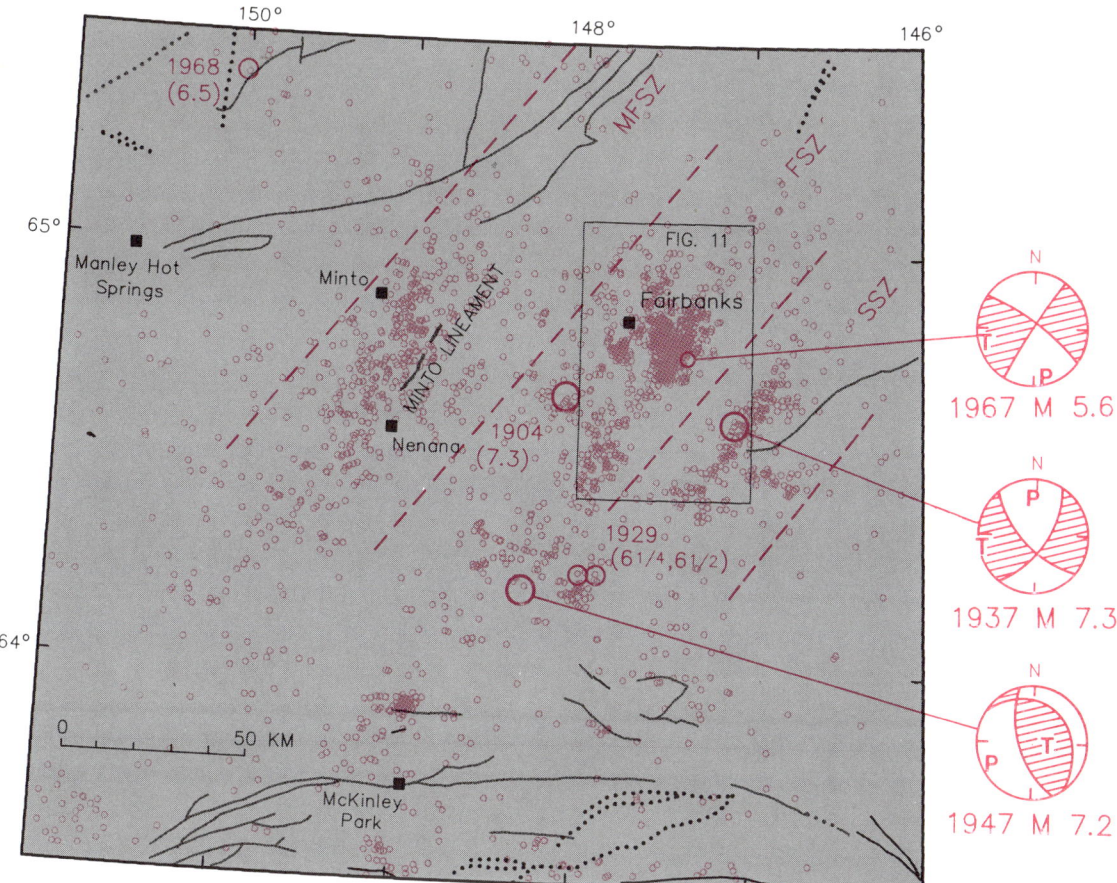

Figure 10. Epicenters of shallow earthquakes in central Alaska. Large circles are historical shocks of magnitude (M_S) 6.0 or greater, and the 1967 Fairbanks mainshock (Tables 1 and 2). Small circles represent better-located shocks with focal depths less than or equal to 40 km for 1982 to June 1985 from University of Alaska Fairbanks earthquake catalog; most shocks are larger than magnitude M_L 1.0. Seismic zones are: SSZ, Salcha; FSZ, Fairbanks; MFSZ, Minto Flats. Known (solid black) and suspicious (dotted black) late Cenozoic faults modified from Plafker and others (1991); Minto lineament and older faults (gray) from Péwé and others (1966) and from Beikman (1980), respectively. Fault-plane solutions are lower hemisphere, equal-area projections with compressional quadrants shaded. P, compressional axis; T, tensional axis. Figure adapted from Biswas and Tytgat (1988).

interpreted the northwest-striking nodal plane as the fault; this intrepretation implies dextral slip. On the other hand, Matumoto and others (1968) concluded that the northeast-striking plane was the fault based on its approximate alignment with the planar distribution of aftershock hypocenters (strike 225°, dip 75° northwest) they resolved from data recorded by an array of four temporary stations deployed near the epicentral area. This latter interpretation implies left-lateral strike-slip motion, similar to that inferred for the 1937 Salcha earthquake. No surface faulting was reported.

Since 1967, about five or six felt earthquakes have been located yearly in the Fairbanks seismic zone, along with sporadic sequences of intense microearthquake activity that last from a few days to about a month. The area of the sequences has migrated in time within the zone (Gedney and others, 1980, 1982), as illustrated in Figure 11. Area a on Figure 11 is the aftershock zone of the three 1967 earthquakes, as determined by Gedney and Berg (1969). Activity in area b increased markedly during late 1970 and early 1971 and included several hundred shocks without a single predominantly large event. In 1977, a swarm incorporating about 20 felt events occurred in area c. Two years later, an m_b 4.3 earthquake (No. 31, Table 2) triggered an aftershock sequence (area d). A minor swarm of events occurred in area e in October 1981. Separated from these areas, a sequence of about 80 aftershocks (area f) followed an m_b 4.9 earthquake in 1981 (No. 36, Table 2). In April 1983, activity revisited area d and extended into the northwest part of area a. In October 1983 and again in November 1985, the same part of area a generated two swarms. All these clusters may represent the release of strain energy on spatially concentrated crustal fractures that respond to a regional north- to northwest-directed maximum compressive stress (Gedney and others, 1980) resulting from the collision of the Pacific and North American Plates along the Pacific margin (Davies, 1983; Estabrook and others, 1988).

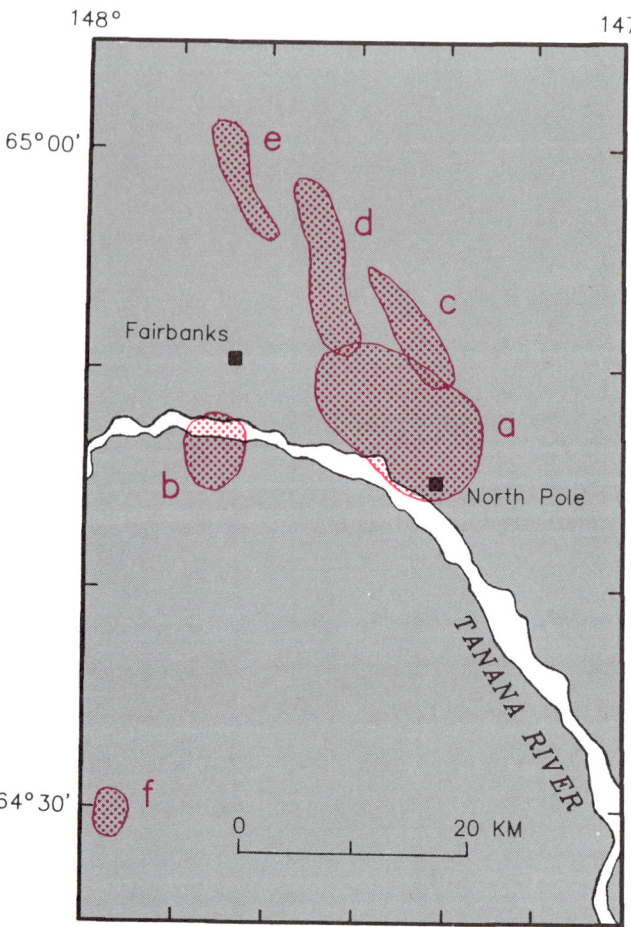

Figure 11. Areas of warm and aftershock activity in Fairbanks region, 1967 to 1985. Figure modified from Biswas and Tytgat (1988).

The Minto Flats seismic zone (Biswas and Tytgat, 1988) comprises a more diffuse distribution of earthquakes than the other two zones and includes occasional felt earthquakes of about M_L 3.0 to 4.0 (Fig. 10). Some of the apparent scatter in the epicenters may be an artifact of the weak network geometry of the seismic stations recording the shocks. Lying along the axis of the diffuse seismic band is the Minto lineament, which has been mapped as a fault by Péwé and others (1966) and characterized as active in Quaternary (Brogan and others, 1975) and Holocene (Plafker and Jacob, 1986) time, but more recently this lineament has been categorized as a geomorphic feature rather than a fault (G. Plafker, personal communication, 1989). In any case, the abundant seismicity betrays active faulting at depth in this region.

Unlike the 1937 earthquake, discussed previously, the other four historical shocks larger than M_s 6.0 in the Fairbanks area are not obviously correlated with conspicuous lineations in the recent seismicity. The epicenters of the 1904 (No. 8, Fig. 2 and Table 1) and two 1929 (Nos. 7 and 8, Table 2) shocks, all of which predate the College seismograph station, are uncertain by several tens of kilometers. In contrast, the location of the 1947 M_s 7.2 earthquake (No. 21, Fig. 2, and Table 1) is constrained within 10 to 20 km by estimates of aftershock epicenters determined from the College seismograms (St. Amand, 1948); these data suggest that the focus of the mainshock could lie slightly west of the teleseismic location plotted in Figure 10. The fault-plane solution for this event (Wickens and Hodgson, 1967) suggests thrust faulting in apparent contrast to the strike-slip faulting that otherwise seems to characterize the Fairbanks area.

Rampart–Dall City area

The Rampart–Dall City area, which lies about 150 km northwest of Fairbanks, experienced magnitude class 6 earthquakes in 1968 and 1985. The first shock (M_s6.5; No. 27, Table 2) struck near the village of Rampart and caused landslides and ground fissures at several places in and around the epicentral area (Gedney and others, 1969). As deduced from the spatial distribution of the ensuing aftershocks (Fig. 12) and the mainshock fault-plane solution, the earthquake was caused by sinistral slip on a 50-km-long, nearly north-south, vertical rupture (Huang and Biswas, 1983). After 1968, shocks smaller than M_L 5 continued in and around the Rampart seismic zone.

The second large shock (M_s 6.1; No. 42, Table 2) occurred north of the Rampart zone, near the abandoned Dall City and close to the active Dall Mountain fault (Brogan and others, 1975). This shock was preceded about three weeks by a nearby M_s 5.0 event (No. 41, Table 2). The aftershocks of the two Dall City earthquakes define an elongate epicentral pattern that subparallels the Rampart zone but apparently crosscuts the more northerly striking Dall Mountain fault (Fig. 12). Based on the fault-plane solutions, the Dall City events, like the Rampart earthquake, involved sinistral slip on a steeply dipping, north-northeast-striking fault (Estabrook and others, 1988). The exact relation of the seismicity to the Dall Mountain fault, which Brogan and others (1975) suggest is a normal fault, is unclear.

East of the Rampart–Dall City area, a band of recent epicenters (not shown in Fig. 12) including an M_s 5.0 earthquake in 1972 (No. 29, Table 2), parallels the west-northwest-striking Tintina fault (Estabrook and others, 1988, Fig. 4), along which hundreds of kilometers of Cretaceous and early Tertiary dextral offset is inferred (e.g., Gabrielse, 1985). To date, only one segment of the fault has been classified as active based on geomorphic evidence (Brogan and others, 1975). The strike-slip focal mechanism obtained for the 1972 shock (Fig. 12) suggests that the dextral sense of slip continues today.

Northeast Alaska

North of 66.5°N, the seismicity of northeast Alaska (Fig. 12) is characterized by widespread but relatively weak activity as far north as the northern boundary of the Brooks Range; a broad zone of diffuse activity extending from the northeast Brooks Range, across the coastal plain and onto the Beaufort Shelf; and a notable quiescence beneath the North Slope (Gedney and others, 1977; Biswas and Gedney, 1979; Estabrook and others, 1988). These features are most clearly resolved in the earthquakes re-

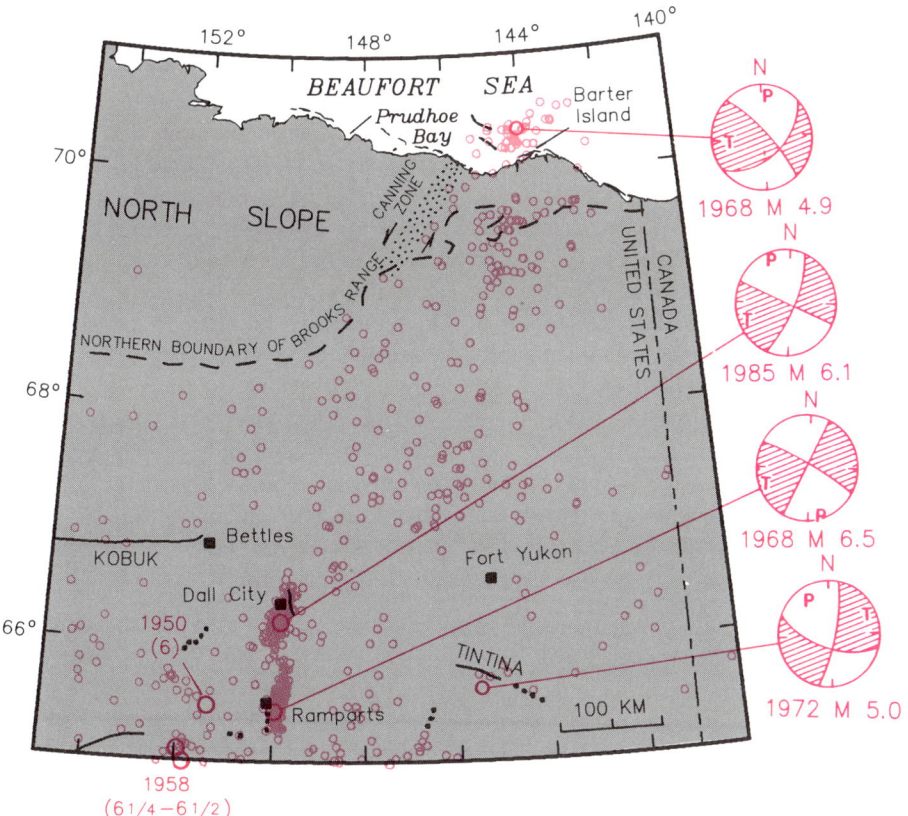

Figure 12. Epicenters of earthquakes in northeast Alaska. Large circles are historic shocks of magnitude M_S 6.0 or greater, and 1968 Barter Island and 1972 Tintina fault earthquakes (Table 2). Small circles south of 67°N are shocks for 1939 to March 1988 from Global Hypocenter Database, Version 1.0, National Earthquake Information Center of the U.S. Geological Survey; most shocks are larger than magnitude 3. Small circles north of 67°N are shocks from University of Alaska Fairbanks earthquake catalog for September 1975 to August 1978, when a regional network was operating north of 67°N; most shocks are in the magnitude interval M_L 1.0 to 4.0. Known (solid) or suspicious (dotted) late Cenozoic faults modified from Plafker and others (1991); Canning displacement zone from Grantz and others (1983). Fault-plane solution conventions as in Figure 10. Figure adapted from Biswas and Tytgat (1988).

corded by the regional seismograph networks operated in the mid-1970s (Fig. 12). Between 67°N and the northern limit of the Brooks Range, no strong concentrations of shocks are observed. There is little doubt that seismicity is widely distributed throughout this region (Estabrook and others, 1988), although the weak geometry of the seismograph network introduces scatter into the epicenters. The density and precision of the available data are insufficient to identify active faults. The swath of seismicity that follows the northeast Brooks Range and crosses the Beaufort coast appears to be bounded on the west by the Canning displacement zone, a young, active, mainly strike-slip shear zone along which the northeast Brooks Range is moving northward and upward relative to the lowland to the west (Grantz and others, 1983).

The most prominent concentration of seismicity in northeast Alaska, seen in maps of both teleseismically (Fig. 4) and regionally (Fig. 12) recorded shocks, lies offshore on the Beaufort Shelf near Barter Island (around 70.3°N, 144°W) and includes events as large as M_S 4.9 (No. 26, Table 2). West- to northwest-striking faults and monoclines deform the sea floor in the vicinity of the earthquakes (Grantz and others, 1983); however, correlations between specific shocks and geologic structures are not possible with the quality of the available earthquake data. Two focal mechanisms have been proposed for the 1968 M_S 4.9 shock: oblique normal faulting, possibly reflecting continuation of the local tectonic processes that have deformed the young shelf sediments (Fujita and others, 1983); and predominantly strike-slip faulting (Fig. 12), consistent with sinistral slip on a northeast-striking plane (Biswas and others, 1986a). The latter mechanism is consistent with the sense of shear along the Canning displacement zone.

WESTERN ALASKA

The historical record of seismicity includes four earthquakes of magnitude 6.0 or greater in western Alaska, of which the largest is the 1958 Huslia earthquake (No. 23, Fig. 2 and Table

1), and four in the Chukchi Sea (Fig. 13). The M_S 7.3 Huslia shock produced extensive failures in surficial unconsolidated deposits within an elongate northeast-striking zone that possibly betrays a buried culprit fault (Davis, 1960). Predominantly normal slip on a plane subparallel to the trend of the failure zone is inferred from the published fault-plane solutions (Wickens and Hodgson, 1967; Estabrook and others, 1988). Two magnitude 6 aftershocks followed (Nos. 14 and 15, Table 2). The second largest shock in Figure 13 is the M_S 6.9 Chukchi Sea earthquake of 1928 (No. 3, Table 2); three magnitude 6 shocks (Nos. 4 to 6, Table 2) followed this event. All three of these events are poorly located. The remaining magnitude 6 earthquake (No. 12, Table 2) occurred on the southern Seward Peninsula in 1950; little is known about this shock.

Regional seismic monitoring in the vicinity of the Seward Peninsula from 1977 to 1982 (Biswas and others, 1986b) affords a more detailed view of the seismicity in western Alaska (Fig. 13). In the magnitude range 2.0 to 5.0, activity seems to be widespread throughout this part of Alaska and the adjacent offshore regions and confined to shallow (i.e., crustal) depths. The most complete and accurate data are from the area 64 to 67°N, 161 to 168°W. The mapped seismicity forms a diffuse cloud punctuated by several clusters; no prominent linear trends are apparent. The broad distribution of activity suggests that seismic deformation is distributed over many active faults and not concentrated on one or two major fault systems. To some degree, location errors contribute to the apparent scatter in the mapped epicenters. Clusters occur in the epicentral regions of some of the larger historical events, such as the 1950 and 1964 earthquakes. Activity also concentrates along the eastern end of the late Cenozoic offshore faults west of Teller, but not along the more recently active Bendeleben fault. Diffuse seismicity is mapped both north and south of the Kigluaik fault; it is not clear whether any of this activity actually centers on the fault. The Bendeleben

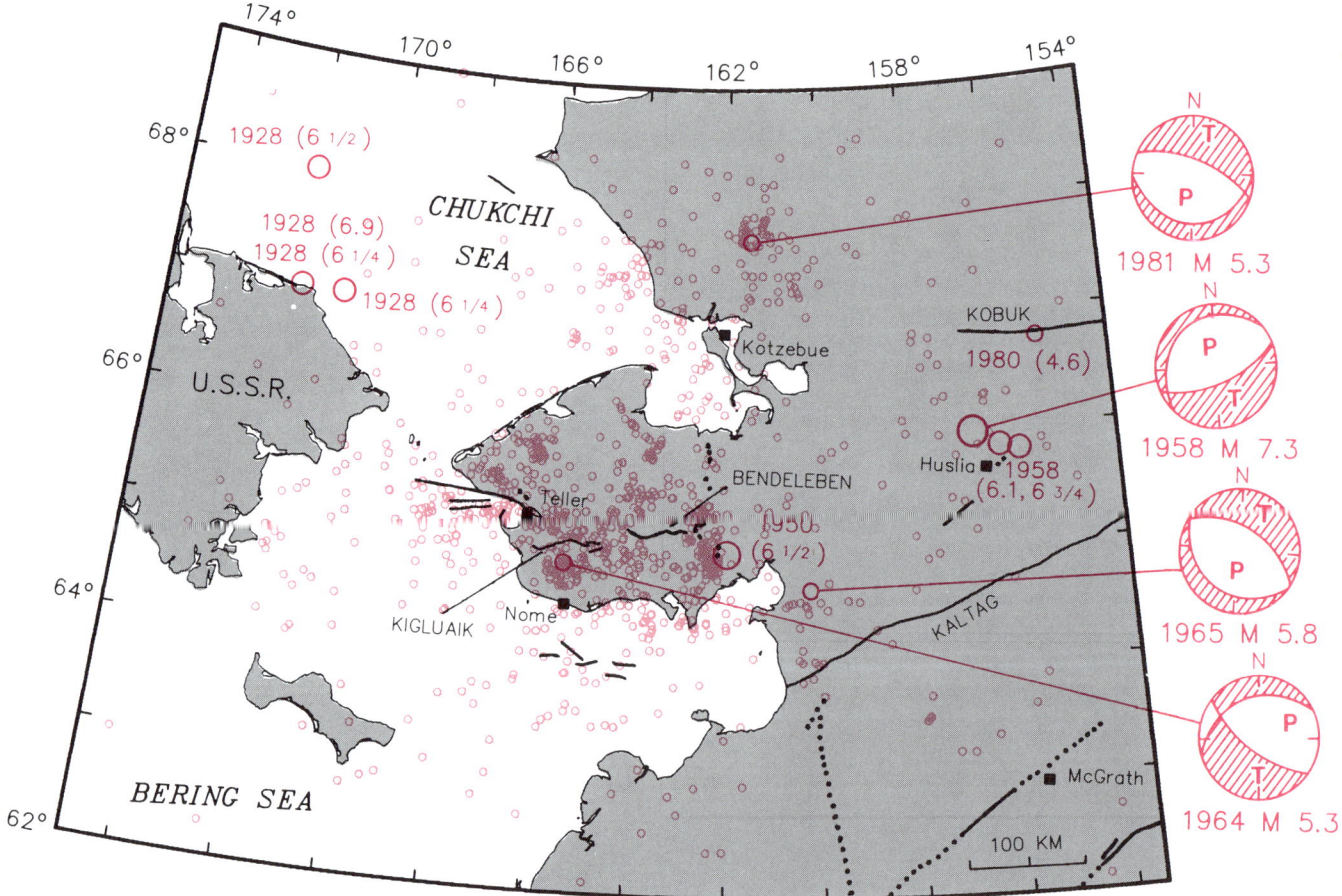

Figure 13. Epicenters of earthquakes in and around Seward Peninsula, western Alaska. Large circles represent historical shocks of magnitude M_S 6.0 or greater and smaller shocks with fault-plane solutions (Tables 1 and 2). Small circles are shocks from University of Alaska Fairbanks earthquake catalog for 1977 through June 1982, when a regional network was operating on the Seward Peninsula; most shocks are in the magnitude interval M_L 1.0 to 4.0. Known (solid) or suspicious (dashed) late Cenozoic faults modified from Plafker and others (1991). Sources of fault-plane solutions: 1958 (No. 23, Table 1), Wickens and Hodgson (1967); 1964 and 1981 (Nos. 21 and 35, Table 2), Biswas and others (1986b); and 1965 (No. 22, Table 2), Sykes and Sbar (1974). Alternative solutions for 1964 and 1981 shocks in Estabrook and others (1988). Fault-plane solution conventions as in Figure 10.

and Kigluaik faults are the principal mapped active faults on the Seward Peninsula, both of which exhibit Holocene normal displacement (Hudson and Plafker, 1978).

A moderate earthquake sequence, triggered by an m_b 4.6 mainshock (No. 33, Table 2), occurred in the Kobuk Trench in 1980 (Fig. 13). The proximity of the epicenters to the east-west trace of the Kobuk (or Kobuk–Alatna Hills) fault, which locally displaces late Quaternary deposits in a dextral sense (Brogan and others, 1975), suggests that it or a neighboring fault may be active today (Gedney and Marshall, 1981). The shocks are too small to yield reliable fault-plane solutions, but the observed pattern of first motions for the mainshock is generally compatible with dextral slip on an east-west fault (Gedney and Marshall, 1981; Estabrook and others, 1988).

Only a few fault-plane solutions are available for shocks in western Alaska. Although in some cases different solutions for the same shock have been obtained by different authors, there is consensus that normal faulting distinguishes seismicity on the Seward Peninsula, consistent with the late Cenozoic, north-south extension observed on east-striking normal faults and also with widespread basaltic volcanism (Hudson and Plafker, 1978). The extent of the extensional domain is uncertain. Biswas and others (1986a) surmise that roughly the western half of continental Alaska is characterized by a stress regime in which the tensional axis is nearly horizontal and the compressional axis nearly vertical. On the other hand, Estabrook and others (1988) suggest that the normal-slip Huslia earthquake and the extensional tectonics on the Seward Peninsula are localized features within a larger region of predominantly strike-slip faulting that encompasses all of northern Alaska, and they hypothesize that the localized normal faulting may be associated with the formation of pull-apart structures.

SUMMARY

In the 25 years since the devastating 1964 Prince William Sound earthquake, much progress has been made in defining the distribution of seismicity in continental Alaska and understanding its origins, particularly in the more populated parts of the state. Important to this progress has been the operation of regional seismograph networks that have provided detailed information not available from either teleseismic recordings of larger shocks or the short written history.

Earthquakes of continental Alaska can be identified with one of three tectonic elements: the Pacific Plate, the North American Plate, and the primary boundary between the two plates. Most of the earthquake energy in Alaska is released in major shocks that rupture large segments of the primary plate boundary and that accommodate most of the motion between the two plates. The sense of the slip varies along the plate boundary from dextral strike-slip in southeast Alaska to reverse-slip along the eastern Aleutian arc. During this century, the entire primary boundary north of 55°N, except for the Yakataga seismic gap (Fig. 2), has slipped in a series of six shocks ranging in size from M_s 7.1 to M_w 9.2. The Yakataga gap, which last broke in 1899, is the likely site for the next major shock.

Earthquakes within the Pacific Plate divide into two groups based on their location either seaward or landward of the primary plate boundary. Subsea shocks are relatively rare in the short historical record; however, an unusual pair of M_s 7.6 strike-slip shocks ruptured a 250-km-long fault in the northern Gulf of Alaska in 1987 to 1988. Earthquakes landward of the primary plate boundary lie in the subducted part of the plate and compose the northwest- to north-dipping Aleutian and north-northeast-dipping Wrangell Wadati-Benioff zones, which dip landward beneath their respective volcanic arcs. The Aleutian zone extends inland 350 km beyond the northernmost volcano in the Aleutian arc and to depths as great as 180 km. The extent of the Wrangell zone is not yet fully resolved; minimum values of its length along strike and its depth are 150 and 100 km, respectively. The apparent continuity of the two zones at shallow depths is consistent with the possibility that they may be adjacent limbs of a buckle in the subducted plate. The Aleutian WBZ is two orders of magnitude more active than the Wrangell zone and since 1960 has generated shocks in the magnitude 6 class in both the shallow, gently dipping and deeper, steeply dipping segments of the zone. In comparison, the largest shock in the Wrangell WBZ during this time is m_b 4.5. The stress orientation in the Aleutian WBZ inferred from fault-plane solutions since 1970 is dominantly in-plane, downdip tension and along-strike compression in the deep part of the zone; in the shallow part of the zone beneath Prince William Sound and the adjacent Chugach Mountains to the north, the inferred tension axis is in-plane and directed toward the deep part of the zone. WBZ seismicity, especially that deeper than 50 km, is more uniformly distributed in both space and time than that occurring in either the Pacific Plate seaward of the plate boundary or that in the North American Plate.

Practically the entire North American Plate composing most of continental Alaska is seismically active. The rate of activity is highest in southern Alaska and decreases northward away from the plate boundary. Only the North Slope appears to be quiet. The historical record documents events as large as magnitude M_s 7.3 or 7.4, the most northern of which occurred at 66°N. All the shocks are shallow, but only a small fraction of the observed seismicity can be correlated with known faults. Although earthquakes have been located in proximity to several faults with recognized late Cenozoic displacement, in many cases the uncertainties in the epicenters preclude unequivocal correlation with mapped fault traces. Fault-plane solutions of moderate-sized earthquakes in south-central, central, and northern Alaska typically reveal strike-slip faulting with northwesterly to northerly oriented compressional axes; whereas solutions from west-central Alaska, particularly the Seward Peninsula, generally exhibit normal faulting with northerly oriented tensional axes. Both the distribution of seismicity and the available focal mechanisms are qualitatively consistent with the hypothesis that the seismicity within the North American Plate, apart from west-central Alaska, results from stresses arising from the collision of the Pacific Plate, or more specifically the Yakutat terrane, with the North American Plate along the northern Gulf of Alaska. Furthermore, the seismicity and mechanisms support the idea that the plate boundary in southern Alaska is not composed of a single fault system but encompasses secondary faults, both seaward and landward of the primary boundary, on which a small fraction of the relative plate motion is distributed.

REFERENCES CITED

Abe, K., 1981, Magnitudes of large shallow earthquakes from 1904 to 1980: Physics of the Earth and Planetary Interiors, v. 27, p. 72–92.

Abe, K., and Kanamori, H., 1980, Magnitudes of great shallow earthquakes from 1953 to 1977: Tectonophysics, v. 62, p. 191–203.

Abe, K., and Noguchi, S., 1983, Revision of magnitudes of large shallow earthquakes, 1897–1912: Physics of the Earth and Planetary Interiors, v. 33, p. 1–11.

Agnew, J. D., 1980, Seismicity of the central Alaska Range, Alaska, 1904–1978 [M.S. thesis]: Fairbanks, University of Alaska, 88 p.

Beikman, H. M., compiler, 1980, Geologic map of Alaska: U.S. Geological Survey, scale 1:2,500,000.

Ben-Menahem, A., and Toksoz, M. N., 1963, Source mechanism from spectra of long-period seismic surface waves; 3, The Alaska earthquake of July 10, 1958: Bulletin of the Seismological Society of America, v. 53, p. 905–919.

Bhattacharya, B., and Biswas, N. N., 1979, Implications of North Pacific plate tectonics in central Alaska; Focal mechanisms of earthquakes: Tectonophysics, v. 53, p. 99–130.

Biswas, N. N., and Gedney, L. D., 1979, Seismotectonic studies of northern and western Alaska; Environmental assessment of the Alaskan continental shelf: National Oceanic and Atmospheric Administration Report on Hazards and Data Management, v. 10, p. 155–208.

Biswas, N. N., and Tytgat, G., 1988, Intraplate seismicity in Alaska: Seismological Research Letters, v. 59, p. 227–233.

Biswas, N. N., Gedney, L., and Agnew, J., 1980, Seismicity of western Alaska: Bulletin of the Seismological Society of America, v. 70, p. 873–883.

Biswas, N. N., Aki, K., Pulpan, H., and Tytgat, G., 1986a, Characteristics of regional stresses in Alaska and neighboring areas: Geophysical Research Letters, v. 13, p. 177–180.

Biswas, N. N., Pujol, J., Tytgat, G., and Dean, K., 1986b, Synthesis of seismicity studies for western Alaska: Tectonophysics, v. 131, p. 369–392.

Boyd, T. M., and Lerner-Lam, A. L., 1988, Spatial distribution of turn-of-the-century seismicity along the Alaska-Aleutian arc: Bulletin of the Seismological Society of America, v. 78, p. 636–650.

Bramhall, E. H., 1938, The central Alaska earthquake of July 22, 1937: Bulletin of the Seismological Society of America, v. 28, p. 71–75.

Brogan, G. E., Cluff, L. S., Korringa, M. K., and Slemmons, D. B., 1975, Active faults of Alaska: Tectonophysics, v. 29, p. 73–85.

Buland, R., and Taggart, J., 1981, A mantle wave magnitude for the St. Elias, Alaska, earthquake of 28 February 1979: Bulletin of the Seismological Society of America, v. 71, p. 1143–1159.

Butler, H. M., 1971, Palmer Seismological Observatory: Earthquake Notes, v. 42, p. 15–36.

Christensen, D. H., and Ruff, L. J., 1988, Seismic coupling and outer rise earthquakes: Journal of Geophysical Research, v. 93, p. 13421–13444.

Cluff, L. S., Slemmons, D. B., Brogan, G. E., and Korringa, M. K., 1974, Basis for pipeline design for active-fault crossings for the Trans-Alaska Pipeline System: Houston, Texas, Alyeska Pipeline Service Company, 115 p.

Committee on the Alaska Earthquake, 1972, The Great Alaska Earthquake of 1964. Volume 2, Seismology and geodesy: Washington, D.C., National Academy of Sciences,

Davies, J. N., 1975, Seismological investigations of plate tectonics in south-central Alaska [Ph.D. thesis]: Fairbanks, University of Alaska, 193 p.

—— , 1983, Seismicity of the interior of Alaska; A direct result of Pacific-North American Plate convergence? [abs.]: EOS Transactions of the American Geophysical Union, v. 64, p. 90.

Davies, J. N., and House, L., 1979, Aleutian subduction zone seismicity, volcano-trench separation, and their relation to great thrust-type earthquakes: Journal of Geophysical Research, v. 84, p. 4583–4591.

Davies, J., Sykes, L., House, L., and Jacob, K., 1981, Shumagin seismic gap, Alaska Peninsula; History of great earthquakes, tectonic setting, and evidence for high seismic potential: Journal of Geophysical Research, v. 86, p. 3821–3855.

Davis, T. N., 1960, A field report on the Alaska earthquakes of April 7, 1958: Bulletin of the Seismological Society of America, v. 50, p. 489–510.

Davis, T. N., and Echols, C., 1962, A table of Alaskan earthquakes, 1788–1961: Fairbanks, University of Alaska Geophysical Institute Geophysical Research Report 8, 44 p.

Demets, C., Gordon, R. C., Argus, D. F., and Stein, S., 1990, Current plate motions: Geophysical Journal International, v. 101, p. 425–478.

Detterman, R. L., Plafker, G., Hudson, T., Tysdal, R. G., and Pavoni, N., 1974, Surface geology and Holocene breaks along the Susitna segment of the Castle Mountain fault, Alaska: U.S. Geological Survey Miscellaneous Field Studies Map MF-618, 1 sheet, scale 1:24,000.

Dziewonski, A. M., Friedman, A., and Woodhouse, J. H., 1983, Centroid-moment tensor solutions for January-March, 1983: Physics of the Earth and Planetary Interiors, v. 33, p. 71–75.

Engle, K. Y., 1982, Earthquake focal mechanism studies of Cook Inlet area, Alaska [M.S. thesis]: Fairbanks, University of Alaska, 81 p.

Estabrook, C. H., Stone, D. B., and Davies, J. N., 1988, Seismotectonics of northern Alaska: Journal of Geophysical Research, v. 93, p. 12026–12040.

Fenner, C. N., 1925, Earth movements accompanying the Katmai eruption: Journal of Geology, v. 33, p. 116–139.

Fogleman, K. A., Stephens, C. D., and Lahr, J. C., 1988, Catalog of earthquakes in southern Alaska for 1985: U.S. Geological Survey Open-File Report 88-31, 112 p.

Fujita, K., Cook, D. B., and Coley, M. J., 1983, Tectonics of the western Beaufort and Chukchi Seas [abs.]: EOS Transactions of the American Geophysical Union, v. 64, p. 263.

Gabrielse, H., 1985, Major dextral transcurrent displacements along the Northern Rocky Mountain Trench and related lineaments in north-central British Columbia: Geological Society of America Bulletin, v. 94, p. 1–14.

Gawthrop, W. H., Page, R. A., Reichle, M., and Jones, A., 1973, The southeast Alaska earthquakes of July 1973 [abs.]: EOS Transactions of the American Geophysical Union, v. 54, p. 1136.

Gedney, L., 1970, Tectonic stresses in southern Alaska in relationship to regional seismicity and the New Global Tectonics: Bulletin of the Seismological Society of America, v. 60, p. 1789–1802.

—— , 1985, Stress trajectories across the northeast Alaska Range: Bulletin of the Seismological Society of America, v. 75, p. 1125–1134.

Gedney, L. D., and Berg, E., 1969, The Fairbanks earthquakes of June 21, 1967; Aftershock distribution, focal mechanisms, and crustal parameters: Bulletin of the Seismological Society of America, v. 59, p. 73–100.

Gedney, L. D., and Davies, J. N., 1986, Additional evidence for down-dip tension in the Pacific Plate beneath central Alaska: Bulletin of the Seismological Society of America, v. 76, p. 1207–1214.

Gedney, L., and Marshall, D., 1981, A rare earthquake sequence in the Kobuk trench, northwestern Alaska: Bulletin of the Seismological Society of America, v. 71, p. 1587–1592.

Gedney, L., Berg, E., Pulpan, H., Davies, J., and Feetham, W., 1969, A field report on the Rampart, Alaska, earthquake of October 29, 1968: Bulletin of the Seismological Society of America, v. 59, p. 1421–1423.

Gedney, L., Biswas, N., Huang, P., Estes, S., and Pearson, C., 1977, Seismicity of northeast Alaska: Geophysical Research Letters, v. 4, p. 175–177.

Gedney, L., Estes, S., and Biswas, N., 1980, Earthquake migration in the Fairbanks, Alaska, seismic zone: Bulletin of the Seismological Society of America, v. 70, p. 223–241.

Gedney, L. D., Estes, S. A., Biswas, N. N., and Marshall, D. L., 1982, A note on further activity in the Fairbanks, Alaska, seismic zone: Bulletin of the Seismological Society of America, v. 72, p. 1415–1417.

Grantz, A., Dinter, D. A., and Biswas, N. N., 1983, Map, cross-sections, and chart showing late Quaternary faults, folds, and earthquake epicenters on the Alaskan Beaufort shelf: U.S. Geological Survey Miscellaneous Investigations Map I-1182-C, 7 p., 3 sheets, scale 1:500,000.

Gutenberg, B., 1956, Great earthquakes 1896–1903: Transactions of the American Geophysical Union, v. 37, p. 608–614.

Gutenberg, B., and Richter, C. F., 1954, Seismicity of the Earth and associated phenomena, 2nd ed.: Princeton, New Jersey, Princeton University Press, 310 p.

Hastie, L. M., and Savage, J. C., 1970, A dislocation model for the 1964 Alaska earthquake: Bulletin of the Seismological Society of America, v. 60, p. 1389–1392.

Horner, R. B., 1983, Seismicity in the St. Elias region of northwestern Canada and southeastern Alaska: Bulletin of the Seismological Society of America, v. 73, p. 1117–1137.

Horner, R., 1990, Seismicity in the Glacier Bay region of southeast Alaska and adjacent areas of British Columbia, in Milner, A. M., and Wood, J. D., Jr., eds., Proceedings, Second Glacier Bay Science Conference, Glacier Bay Lodge, Alaska, September 19–22, 1988: Atlanta, Georgia, U.S. National Park Service Science Publications, p. 6–11.

Huang, P. Y., and Biswas, N. N., 1983, Rampart seismic zone of central Alaska: Bulletin of the Seismological Society of America, v. 73, p. 813–829.

Hudson, T., and Plafker, G., 1978, Kigluaik and Bendeleben faults, Seward Peninsula, in Johnson, K. M., ed., The United States Geological Survey in Alaska; Accomplishments during 1977: U.S. Geological Survey Circular 772-B, p. 47–50.

Jacob, K. H., 1984, Estimates of long-term probabilities for future great earthquakes in the Aleutians: Geophysical Research Letters, v. 11, p. 295–298.

Jordan, J. N., Dunphy, G. N., and Harding, S. T., 1968, Preliminary seismological report, in The Fairbanks, Alaska, earthquakes of June 21, 1967: U.S. Coast and Geodetic Survey, p. 1–32.

Kanamori, H., 1977, The energy release in great earthquakes: Journal of Geophysical Research, v. 82, p. 2981–2987.

Kanamori, H., and Abe, K., 1979, Reevaluation of the turn-of-the-century seismicity peak: Journal of Geophysical Research, v. 84, p. 6131–6139.

Kelleher, J. A., 1970, Space-time seismicity of the Alaska-Aleutian seismic zone: Journal of Geophysical Research, v. 75, p. 5745–5756.

Kienle, J., 1986, Augustine Volcano; Awake again?: EOS Transactions of the American Geophysical Union, v. 67, p. 172.

Kienle, J., Swanson, S. E., and Pulpan, H., 1983, Magmatism and subduction in the eastern Aleutian arc, in Shimozuru, D., and Yokoyama, I., eds., Arc volcanism; Physics and tectonics: Terra Scientific Publishing Company, p. 191–224.

Kienle, J., Davies, J. N., Miller, T. P., and Yount, M. E., 1986, 1986 eruption of Augustine Volcano; Public safety response by Alaskan volcanologists: EOS Transactions of the American Geophysical Union, v. 67, p. 580–582.

Lahr, J. C., 1975, Detailed seismic investigation of Pacific-North American Plate interaction in southern Alaska [Ph.D. thesis]: New York, Columbia University, 88 p.

Lahr, J. C., and Plafker, G., 1980, Holocene Pacific-North American Plate interaction in southern Alaska; Implications for the Yakataga seismic gap: Geology, v. 8, p. 483–486.

Lahr, J. C., Stephens, C. D., Hasegawa, H. S., and Boatwright, J., 1980, Alaskan seismic gap only partially filled by 28 February 1979 earthquake: Science, v. 207, p. 1351–1353.

Lahr, J. C., Page, R. A., Stephens, C. D., and Fogleman, K. A, 1986, Sutton, Alaska, earthquake of 1984; Evidence for activity on the Talkeetna segment of the Castle Mountain fault system: Bulletin of the Seismological Society of America, v. 76, p. 967–983.

Lahr, J. C, Page, R. A., Stephens, C. D., and Christensen, D. H., 1988, Unusual earthquakes in the Gulf of Alaska and fragmentation of the Pacific Plate: Geophysical Research Letters, v. 15, p. 1483–1486.

Lalla, D. J., and Kienle, J., 1986, Seismic and thermal precursors to the January 1976 eruption of Augustine Volcano, Alaska [abs.]: Auckland, New Zealand, International Volcanological Congress, February 1986, p. 251.

Matumoto, T., and Ward, P. L., 1967, Microearthquake study of Mount Katmai and vicinity, Alaska: Journal of Geophysical Research, v. 72, p. 2557–2568.

Matumoto, T., Gumper, F., and Sbar, M., 1968, A study of microaftershocks following Fairbanks earthquake of June 21, 1967 [abs.]: EOS Transactions of the American Geophysical Union, v. 49, p. 294.

McCann, W. R., Perez, O. J., and Sykes, L. R., 1980, Yakataga gap, Alaska; Seismic history and earthquake potential: Science, v. 207, p. 1309–1314.

Meyers, H., 1976, A historical summary of earthquake epicenters in and near Alaska: National Oceanic and Atmospheric Administration Technical Memorandum EDS NDSDG-1, 57 p.

Miller, D. J., 1960, Giant waves in Lituya Bay, Alaska: U.S. Geological Survey Professional Paper 354-C, p. 51–86.

Miyashita, K., and Matsu'ura, M., 1978, Inversion analysis of static displacement data associated with the Alaska earthquake of 1964: Journal of the Physics of the Earth, v. 26, p. 333–349.

Page, R., 1973, The Sitka, Alaska, earthquake of 1972; An expected visitor: Earthquake Information Bulletin, v. 5, no. 5, p. 4–9.

—— , 1975, Evaluation of seismicity and earthquake shaking at offshore sites: Houston, Texas, 7th Offshore Technology Conference, Proceedings, v. 3, p. 179–190.

Page, R. A., Stephens, C. D., Fogleman, K. A., and Maley, R. P., 1985, The Columbia Bay earthquakes of 1983, in Bartsch-Winkler, S., and Reed, K. M., eds., The United States Geological Survey in Alaska; Accomplishments during 1983: U.S. Geological Survey Circular 945, p. 80–82.

Page, R. A., Fogleman, K. A., Stephens, C. D., and Lahr, J. C., 1988, State of stress in the subducted Pacific Plate beneath Prince William Sound, southern Alaska [abs.]: Seismological Research Letters, v. 59, p. 16.

Page, R. A., Stephens, C. D., and Lahr, J. C., 1989, Seismicity of the Wrangell and Aleutian Wadati-Benioff zones and the North American Plate along the Trans-Alaska Crustal Transect, Chugach Mountains and Copper River basin, southern Alaska: Journal of Geophysical Research, v. 94, p. 16059–16082.

Perez, O. J., and Jacob, K. H., 1980, Tectonic model and seismic potential of the eastern Gulf of Alaska and Yakataga seismic gap: Journal of Geophysical Research, v. 85, p. 7132–7150.

Péwé, T. L., Wahrhaftig, C., and Weber, F., 1966, Geologic map of the Fairbanks Quadrangle, Alaska: U.S. Geological Survey Miscellaneous Geologic Investigations Map I-455, scale 1:250,000.

Plafker, G., 1969, Tectonics of the March 27, 1964, Alaska earthquake: U.S. Geological Survey Professional Paper 543-I, 74 p.

Plafker, G., 1987, Regional geology and petroleum potential of the northern Gulf of Alaska continental margin, in Scholl, D. W., Grantz, A., and Vedder, J. G., eds., Geology and resource potential of the continental margin of western North America and adjacent ocean basins: American Association of Petroleum Geologists Circum-Pacific Earth Science Series, v. 6, p. 229–268.

Plafker, G., and Jacob, K., 1986, Seismic sources in Alaska, in Hays, W. W., and Gori, P. L., eds., A workshop on evaluation of regional and urban earthquake hazards and risk in Alaska: U.S. Geological Survey Open-File Report 86-79, p. 76–82.

Plafker, G., and Rubin, M., 1978, Uplift history and earthquake recurrence as deduced from marine terraces on Middleton Island, Alaska, in Proceedings, Conference 6, Methodology for Identifying Seismic Gaps and Soon-to-Break Gaps: U.S. Geological Survey Open-File Report 78-943, p. 687–721.

Plafker, G., Hudson, T., and Richter, D. H., 1977, Preliminary observations on late Cenozoic displacements along the Totschunda and Denali fault systems, in Blean, K. M., ed., The United States Geological Survey in Alaska; Accomplishments during 1976: U.S. Geological Survey Circular 751-B, p. 67–69.

Plafker, G., Hudson, T., Bruns, T., and Rubin, M., 1978, Late Quaternary offsets along the Fairweather fault and crustal plate interactions in southern Alaska:

Canadian Journal of Earth Sciences, v. 15, p. 805–816.
Plafker, G., Gilpin, L., and Lahr, J. C., 1991, Neotectonic map of Alaska, *in* Plafker, G., and Berg, H. C., eds., Geology of Alaska: Boulder, Colorado, Geolgocial Society of America, The Geology of North America, v. G-1 (in press).
Power, J., 1988a, Seismicity associated with the 1986 eruption of Augustine Volcano, Alaska [M.S. thesis]: Fairbanks, University of Alaska, 142 p.
Power, M. A., 1988b, Mass movement, seismicity, and neotectonics in the northern St. Elias Mountains, Yukon [M.S. thesis]: Edmonton, University of Alberta, 125 p.
Pulpan, H., 1988, Seismic-hazard analysis of the Nenana Agricultural Development Area, central Alaska: Alaska Division of Geology and Geophysical Surveys Report of Investigations 88-1, 29 p.
Pulpan, H., and Frohlich, C., 1985, Geometry of the subducted plate near Kodiak Island and the lower Cook Inlet, Alaska, determined from relocated earthquake hypocenters: Bulletin of the Seismological Society of America, v. 75, p. 791–810.
Reeder, J. W., and Lahr, J. C., 1987, Seismological aspects of the 1976 eruptions of Augustine Volcano, Alaska: U.S. Geological Survey Bulletin 1768, 32 p.
Richter, D. H., and Matson, N. A., 1971, Quaternary faulting in the eastern Alaska Range: Geological Society of America Bulletin, v. 82, p. 1529–1539.
Rogers, G. C., 1976, A microearthquake survey in northwest British Columbia and southeast Alaska: Bulletin of the Seismological Society of America, v. 66, p. 1643–1655.
—— , 1986, Seismic gaps along the Queen Charlotte fault: Earthquake Prediction Research, v. 4, p. 1–11.
Rothé, J. P., 1969, The seismicity of the Earth, 1953–1965: Paris, UNESCO, 336 p.
St. Amand, P., 1948, The central Alaska earthquake swarm of October 1947: Transactions of the American Geophysical Union, v. 29, p. 613–623.
Schell, M. M., and Ruff, L. J., 1989, Rupture of a seismic gap in southeastern Alaska; The Sitka earthquake (M_S 7.6): Physics of the Earth and Planetary Interiors, v. 54, p. 241–257.
Simkin, T., and 5 others, 1981, Volcanoes of the world; A regional directory, gazetteer, and chronology of volcanism during the last 10,000 years: Stroudsburg, Pennsylvania, Hutchinson Ross Publishing Co., 232 p.
Stauder, W., 1959, A mechanism study; The earthquake of October 24, 1927: Geofisica Pura e Applicata, v. 44, p. 135–143.
—— , 1960, The Alaska earthquake of July 10, 1958; Seismic studies: Bulletin of the Seismological Society of America, v. 50, p. 293–322.
Stauder, W., and Bollinger, G. A., 1966, The focal mechanism of the Alaska earthquake of March 28, 1964, and of its aftershock sequence: Journal of Geophysical Research, v. 71, p. 5283–5296.
Stephens, C. D., Lahr, J. C., Fogleman, K. A., and Horner, R. B., 1980, The St. Elias, Alaska, earthquake of February 28, 1979; Regional recording of aftershocks and short-term, pre-earthquake seismicity: Bulletin of the Seismological Society of America, v. 70, p. 1607–1633.
Stephens, C. D., Fogleman, K. A., Lahr, J. C., and Page, R. A., 1984a, Wrangell Benioff zone, southern Alaska: Geology, v. 12, p. 373–376.
Stephens, C. D., Lahr, J. C., and Page, R. A., 1984b, Seismicity along southern coastal Alaska, October 1981-September 1982, *in* Bartsch-Winkler, S., and Reed, K. M., eds., The United States Geological Survey in Alaska; Accomplishments during 1982: U.S. Geological Survey Circular 939, p. 78–82.
Stevens, A. E., and 5 others, 1976, Canadian earthquakes; 1968: Ottawa, Canada, Earth Physics Branch Seismological Series 71, 39 p.
Stout, J. H, Brady, J. B., Weber, F. and Page, R. A., 1973, Evidence for Quaternary movement on the McKinley strand of the Denali fault in the Delta River area, Alaska: Geological Society of America Bulletin, v. 84, p. 939–947.
Stover, C. W., Reagor, B. G., and Wetmiller, R. J., 1980, Intensities and isoseismal map for the St. Elias earthquake of February 28, 1979: Bulletin of the Seismological Society of America, v. 70, p. 1635–1649.
Sykes, L. R., 1971, Aftershock zones of great earthquakes, seismicity gaps, and earthquake prediction for Alaska and the Aleutians: Journal of Geophysical Research, v. 76, p. 8021–8041.
Sykes, L. R., and Sbar, M. L., 1974, Focal mechanism solutions of intraplate earthquakes and stresses in the lithosphere, *in* Kristjansson, L., ed., Geodynamics of Iceland and the North Atlantic area: Boston, Massachusetts, D. Reidel Publishing Company, p. 207–224.
Sykes, L. R., and Quittmeyer, R. C., 1981, Repeat times of great earthquakes along simple plate boundaries, *in* Simpson, D. W., and Richards, P. G., eds., Earthquake prediction: American Geophysical Union Maurice Ewing Series, v. 4, p. 217–247.
Tarr, R. S., and Martin, L., 1912, Earthquakes at Yakutat Bay, Alaska, in September, 1899: U.S. Geological Survey Professional Paper 69, 135 p.
Thatcher, W., and Plafker, G., 1977, The 1899 Yakutat Bay, Alaska, earthquakes; Seismograms and crustal deformation: Geological Society of America Abstracts with Programs, v. 9, p. 515.
Tobin, D. G., and Sykes, L. R., 1966, Relationship of hypocenters of earthquakes to the geology of Alaska: Journal of Geophysical Research, v. 71, p. 1659–1667.
—— , 1968, Seismicity and tectonics of the northeast Pacific Ocean: Journal of Geophysical Research, v. 73, p. 3821–3845.
Tocher, D., 1960, The Alaska earthquake of July 10, 1958; Movement on the Fairweather fault and field investigation of southern epicentral region: Bulletin of the Seismological Society of America, v. 50, p 267–292.
U.S. Coast and Geodetic Survey, 1966, The Prince William Sound, Alaska, earthquake of 1964 and aftershocks: U.S. Coast and Geodetic Survey Publication 10-3, v. 1, p. 11–73.
Van Wormer, J. D., Davies, J., and Gedney, L., 1974, Seismicity and plate tectonics in south central Alaska: Bulletin of the Seismological Society of America, v. 64, p. 1467–1475.
Wickens, A. J., and Hodgson, J. H., 1967, Computer re-evaluation of earthquake mechanisms solutions 1922–1962: Ottawa, Canada, Publications of the Dominion Observatory, v. 33, no. 1, 560 p.
Woodward-Clyde Consultants, 1980, Interim report on seismic studies for Susitna Hydroelectric Project: Buffalo, New York, Report to Acres American Incorporated for Alaska Power Authority Susitna Hydroelectric Project, 273 p.
—— , 1982, Final report on seismic studies for Susitna Hydroelectric Project: Buffalo, New York, Report to Acres American Incorporated for Alaska Power Authority Susitna Hydroelectric Project, 155 p.

Manuscript Accepted by the Society November 30, 1990

ACKNOWLEDGMENTS

We honor the contributions of E. Berg and L. D. Gedney, both formerly of the University of Alaska at Fairbanks (UAF) and of C. Stephens and K. A. Fogleman of the U.S. Geological Survey, Menlo Park, in supervising the acquisition and reduction of much of the earthquake data used in this study. We also recognize the assistance of the personnel of the Alaska Tsunami Warning Center and College Observatory in the acquisition of data. We acknowledge the computational assistance of G. Tytgat and C. Rohwer of the UAF and helpful reviews of this paper by D. Hill, R. Horner, J. Taber, and R. White. The research support of NNB was partly derived from the State of Alaska funds appropriated to the Geophysical Institute of UAF.

Chapter 5

An overview of western Canadian seismicity

Garry C. Rogers and Robert B. Horner
Geological Survey of Canada, Pacific Geoscience Centre, P.O. Box 6000, Sidney, British Columbia V8L 4B2, Canada

INTRODUCTION

Western Canada, for the purposes of this chapter, refers to all regions west of the Canadian Shield. This includes the provinces of British Columbia and Alberta, the Yukon Territory, the southern parts of Saskatchewan and Manitoba, the western part of the Northwest Territories, and offshore regions in the Pacific Ocean and Beaufort Sea (Fig. 1). Of the approximately 2,000 earthquakes located by seismologists in Canada each year, about three-quarters occur in western Canada (Fig. 2). Of the 94 earthquakes magnitude six and greater that occurred in or very near Canada from 1900 to 1989, 83 were in western Canada. Most of these larger earthquakes occurred offshore, west of Vancouver Island, but 29 were on land or immediately adjacent to land.

Documentation of earthquakes in western Canada begins with a paper describing historical seismicity by Milne (1956) and one by Meidler (1961) describing historical seismicity north of 60°N. From 1951 to 1984, annual catalogs were published by the Department of Energy, Mines, and Resources (first as the Dominion Observatory, then the Earth Physics Branch, and now the Geological Survey of Canada) documenting earthquakes in Canada and briefly describing larger events. A list of earlier catalogs can be found in Milne and others (1978); a current list in recent catalogues is published by the Geological Survey of Canada (e.g., Drysdale and Horner, 1987). Beginning with 1985 to 1986, catalogs are published biannually (e.g., Wetmiller and others, 1989).

Major reviews of seismicity in western Canada have been written by Milne (1963), Milne and others (1978), and Rogers (1983b) for the coastal and Cordilleran regions south of 60°N and by Basham and others (1977) for the northern regions. Discussion of the St. Elias region can be found in Horner (1983), and of the Interior Platform or prairie region in Horner and Hasegawa (1978). Western Canada is also discussed in reviews of Canadian seismicity by Milne (1967) and Milne and others (1970). With each succeeding review the relation between earthquake occurrence and the tectonic setting has become clearer. The correlation of seismicity with the active plate tectonic regime along the west coast and offshore is striking (e.g., Riddihough and others, 1983), but not understood in the same detail everywhere. Causative factors of the seismicity within the Cordillera and the extension into the Beaufort Sea are less well understood. Several large earthquakes have occurred in areas of low seismicity within the Cordillera. In the Interior Platform or prairie region, many of the earthquakes detected today are induced by mining or hydrocarbon extraction activities, but natural seismicity does also occur.

The purpose of this paper is to discuss briefly the distribution of seismicity within western Canada and to introduce the reader to the published literature on the subject.

OFFSHORE

The most intense region of seismicity in Canada lies just to the west of Vancouver Island (Fig. 2). More than 100 earthquakes with a magnitude of 5 or greater have occurred here in the past 70 years (Ellis and Rogers, 1986). It is a region of active

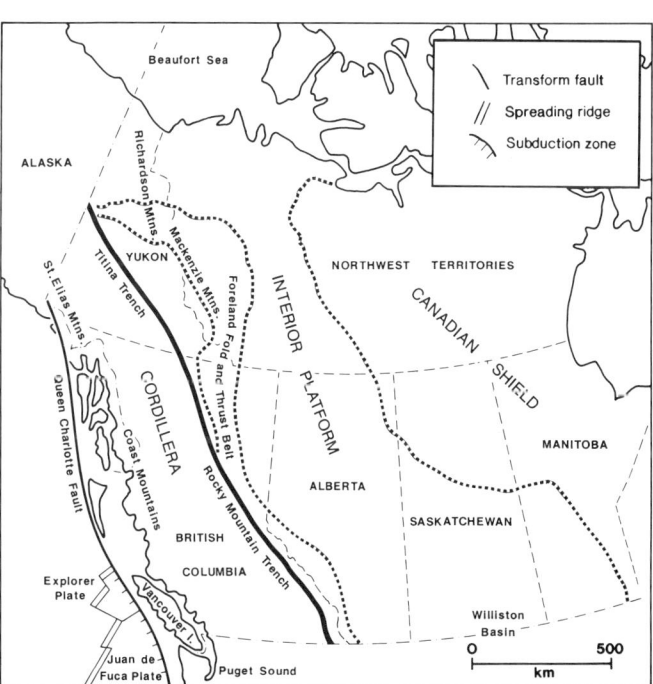

Figure 1. Major tectonic features of western Canada. Most seismicity is associated with the active plate boundaries along the west coast: the Queen Charlotte transform fault, the Cascadia subduction zone (hatched line), and the offshore fracture zones.

Rogers, G. C., and Horner, R. B., 1991, An overview of western Canadian seismicity, *in* Slemmons, D. B., Engdahl, E. R., Zoback, M. D., and Blackwell, D. D., eds., Neotectonics of North America: Boulder, Colorado, Geological Society of America, Decade Map Volume 1.

crustal spreading and thin oceanic lithosphere. Most of the seismicity occurs in fracture zones that mark the boundaries of the small Explorer and Juan de Fuca Plates (Wetmiller, 1971; Rogers, 1980b; Wahlstrom and Rogers, 1990). However, significant numbers of earthquakes also occur within the Explorer Plate, indicating internal deformation (Fig. 3). The rate of seismicity on the fracture zones correlates with the rate expected from plate motions (Hyndman and Weichert, 1983). The thin lithosphere determines the maximum depth of microearthquakes (Hyndman and Rogers, 1981) and places an upper limit of about magnitude 7 on the size of earthquake that can occur. The largest magnitude observed was 6.8 on December 17, 1980, in the Sovanco fracture zone. Focal mechanism studies indicate strike-slip faulting is most common in the larger offshore earthquakes (Chandra, 1974; Rogers, 1979, 1980c; Wahlstrom and others, 1990).

CASCADIA SUBDUCTION ZONE

The southwest margin of Canada is dominated by the Cascadia subduction zone. It is characterized by ongoing seismicity (Fig. 4) that appears to have no correlation to mapped faults. Most of the seismicity is shallow, less than 30 km in depth, but events have been detected to depths of 90 km. A cross section through southern Vancouver Island shows two suites of earthquakes (Fig. 5). The lower suite dips to the northeast and defines the position of the subducting Juan de Fuca Plate (e.g., Rogers, 1983b; Rogers and others, 1990a). The upper suite has a sharp cutoff at about 30 km depth; this probably defines an isotherm below which the crust is too hot for brittle fracture. Normal faulting predominates in the subducting plate, and down-dip tension in the plane of the plate appears to be the dominant stress

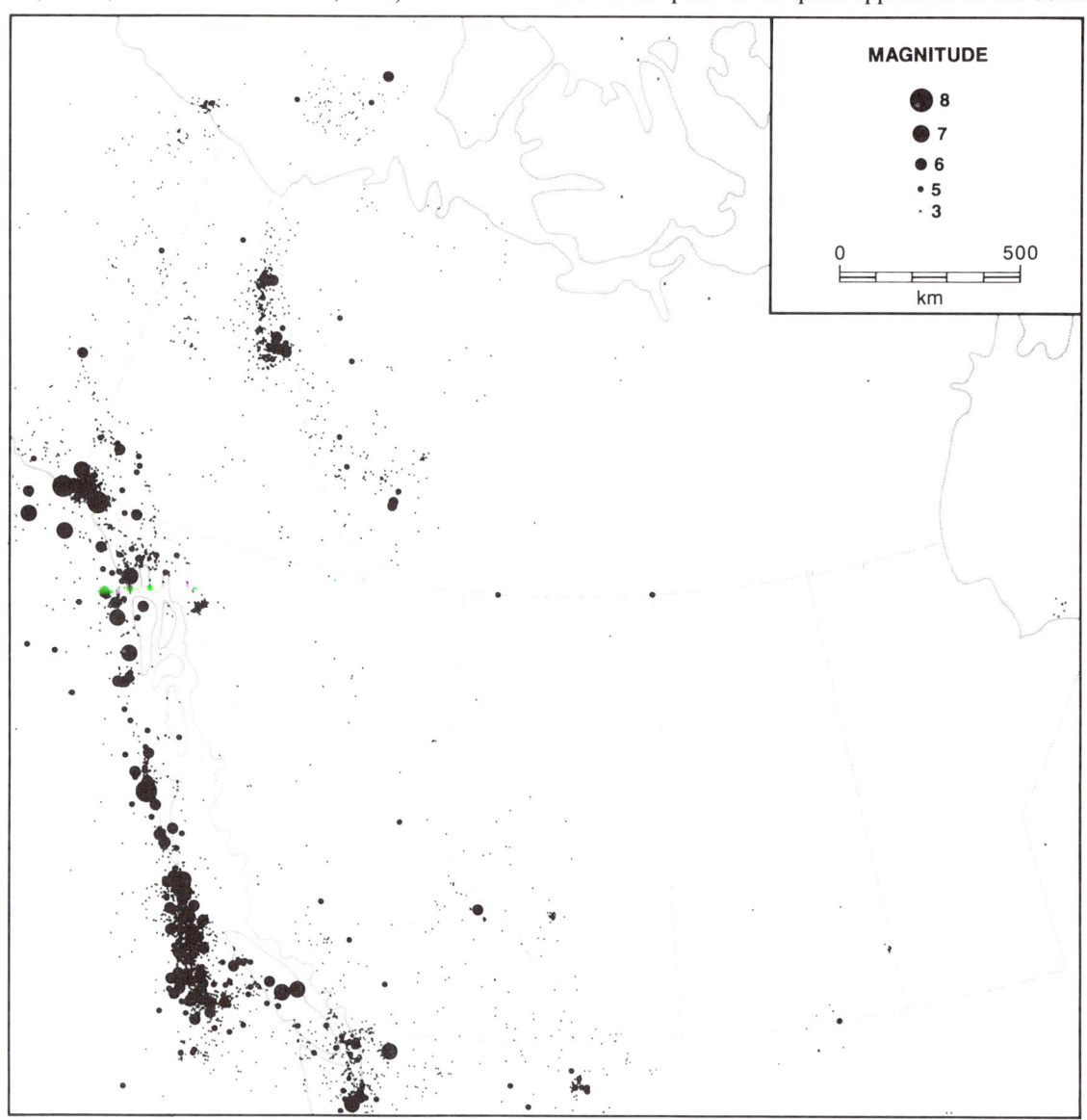

Figure 2. Seismicity of the Cordilleran region of western Canada to the end of 1986. All events of the Geological Suvey of Canada data file of magnitude 3 and greater are shown, but detection capability has varied significantly with time. Data is incomplete in the United States.

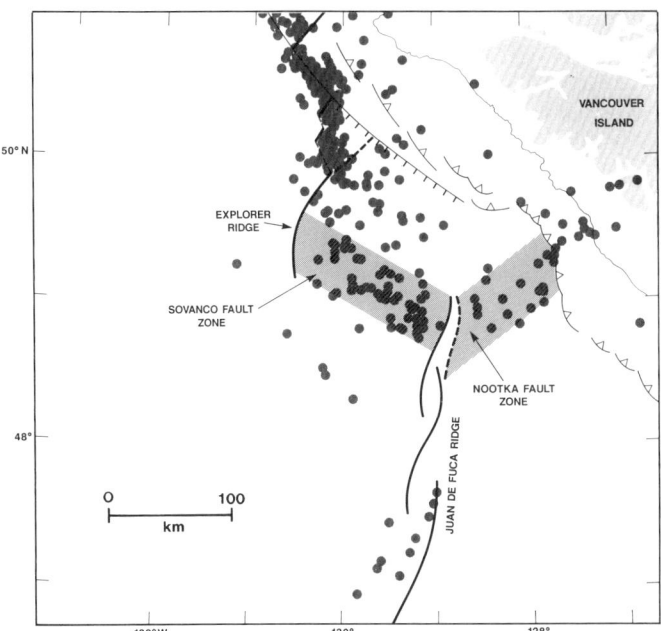

Figure 3. Relocated earthquake epicenters in the tectonically active region of young sea floor west of Vancouver Island (adapted from Wahlstrom and Rogers, 1990).

(Rogers, 1983a, b); however, considerable variability is seen in the focal mechanisms of small earthquakes immediately to the south in Puget Sound (e.g., Crosson, 1983).In the overlying plate, strike slip and thrust faulting occur consistent with right-lateral shear parallel to the margin in the region above the subduction zone (Rogers, 1979).

Thus far no earthquakes have been observed on the subduction interface (Rogers, 1988a), but the tectonic setting suggests that large subduction earthquakes may be possible (Heaton and Kanamori, 1984; Heaton and Hartzell, 1987; Rogers, 1988a; Dragert and Rogers, 1988). Cyclical deposits along the outer coast in Washington State (Atwater, 1987) and on the deep sea floor off Oregon (Adams, 1990) have been interpreted as evidence of past large subduction earthquakes occurring at intervals of several hundreds of years. The possibility of large earthquakes on the subduction interface of the Cascadia subduction zone is an area of active research and continued scientific debate (Rogers, 1988b; Heaton, 1990).

The largest earthquakes in the subducting plate have occurred under the southern Georgia Strait and Puget Sound where phase changes increase the density of the subducting lithosphere, resulting in the plate bending to descend into the asthenosphere (Rogers, 1983b). Just to the south in Puget Sound, earthquakes in this environment have ranged in size up to magnitude 7 (April 13, 1949), but the largest historical event of this type in Canada was magnitude 5.4 on May 16, 1976. The largest earthquakes in the overlying plate have been two events near magnitude 7 on December 6, 1918 (Dennison, 1919; Cassidy and others, 1988), and on June 23, 1946 (Hodgson, 1946; Hasegawa and Rogers, 1978;

Rogers and Hasegawa, 1978; Mathews, 1979; Rogers, 1980a). These earthquakes occurred in the central Vancouver Island region above the subducted Nootka fault zone (Hyndman and others, 1979), which is the boundary between the Juan de Fuca and Explorer Plates. This fault zone is identified by a broad zone of seismicity (Hyndman and others, 1979) and has been positioned beneath the continental shelf by persistent small earthquake activity within the subducted plate (Fig. 4) and the occurrence of several events with magnitude about 6 (Rogers, 1976a, 1979; Cassidy and others, 1988).

THE QUEEN CHARLOTTE TRANSFORM REGION

All major earthquakes have occurred in a narrow zone just west of the Queen Charlotte Islands, which defines the Queen Charlotte fault (Rogers, 1982). This is the site of the large historic earthquake in Canada, magnitude 8.1 on August 22, 1949, which ruptured the fault for almost 500 km (Bostwick, 1984). Earthquakes of about magnitude 7 occurred just south of the Queen Charlotte Islands on May 26, 1929, and on June 24, 1970. Mechanisms along the fault zone are right-lateral strike slip; a significant thrust component in events at the southern end of the region indicates a component of convergence there (Rogers, 1983b). The Queen Charlotte fault has experienced almost complete rupture in this century except for a small region at the southern tip of the Queen Charlotte Islands that has been identified as a seismic gap with a potential for future rupture (Rogers, 1986).

Only minor seismicity had been observed inland from the Queen Charlotte fault until a magnitude 5.2 event in northern Hecate Strait on January 12, 1990. Increased seismic monitoring in the Queen Charlotte Islands since 1982 has identified a concentration of small earthquakes on northern Graham Island and in northern Hecate Strait (Fig. 6). The seismicity does not appear to be associated with any mapped faults (Bérubé and others, 1989; Hyndman and Ellis, 1981; Rogers, 1982). Focal mechanisms of microearthquakes inland from the Queen Charlotte fault indicate north-south thrusting (Bérubé and others, 1989).

THE ST. ELIAS REGION

Convergence between the Pacific and North American Plates in the eastern Gulf of Alaska is resulting in a major orogeny with crustal shortening and uplift that involves parts of the southwest Yukon Territory and extreme northwest British Columbia. Most of the relative motion is accommodated by transform motion on the Queen Charlotte–Fairweather fault and convergence along the Chugach–St. Elias, and Pamplona fault zones to the west (Fig. 7). A number of major earthquakes have occurred on the plate margin and have been felt strongly in adjacent parts of Canada. A discussion of this seismicity is given by Page and others (this volume).

The most significant inland zone of seismicity follows the Dalton and Duke River segments of the Denali fault zone through the southwest Yukon. A microearthquake survey conducted on a 40-km segment of the Denali fault zone near Kluane Lake showed events distributed over a 15-km-wide zone with all focal depths less than 15 km (Horner, 1983). A composite P-nodal solution defined planes that are not consistent with regional geologic trends and suggested a reactivation of the fault zone due to current tectonic stress. This is consistent with the low average slip rates estimated for the Denali zone—approximately 1 mm/yr (Horner, 1983)—and the absence of obvious Holocene displacements (Clague, 1979). The rate of seismicity is about an order of magnitude lower than that along the coast. The largest historic earthquake occurred on February 3, 1944, with a magnitude of 6.5. It was located west of the Dalton fault just north of the British Columbia border (Horner, 1983).

Another significant inland seismicity zone is observed in the Glacier Bay region of southeastern Alaska. This zone straddles the British Columbia–Alaska border and may extend from the Denali fault zone at the north end of Chatham Strait, southwest across the Fairweather fault to the Transition fault zone (Fig. 7).

Horner (1983, 1989, 1990) suggests this seismicity might result from convergence across the Fairweather transform boundary. The largest historic earthquake in this zone, magnitude 6.0, occurred on March 9, 1952 (Horner, 1983).

Minor seismicity is observed between the Denali and Tintina fault zones and is confined mainly to the Yukon Territory (Fig. 7). There are very few events observed in southeast Alaska south of Glacier Bay. The Chatham Strait fault appears to be aseismic (Rogers, 1976c; Horner, 1983, 1990).

THE NORTHERN CORDILLERA

In the rest of the Cordillera north of 60°N, earthquakes are observed in a broad zone through the Mackenzie and Richardson Mountains (Figs. 1 and 2). Historical seismicity is described by Leblanc and Wetmiller (1974), and Basham and others (1977). The largest earthquake, magnitude 6.9, occurred in the Foreland Fold and Thrust belt of the Mackenzie Mountains on December 23, 1985. This earthquake was part of a sequence in the Nahanni,

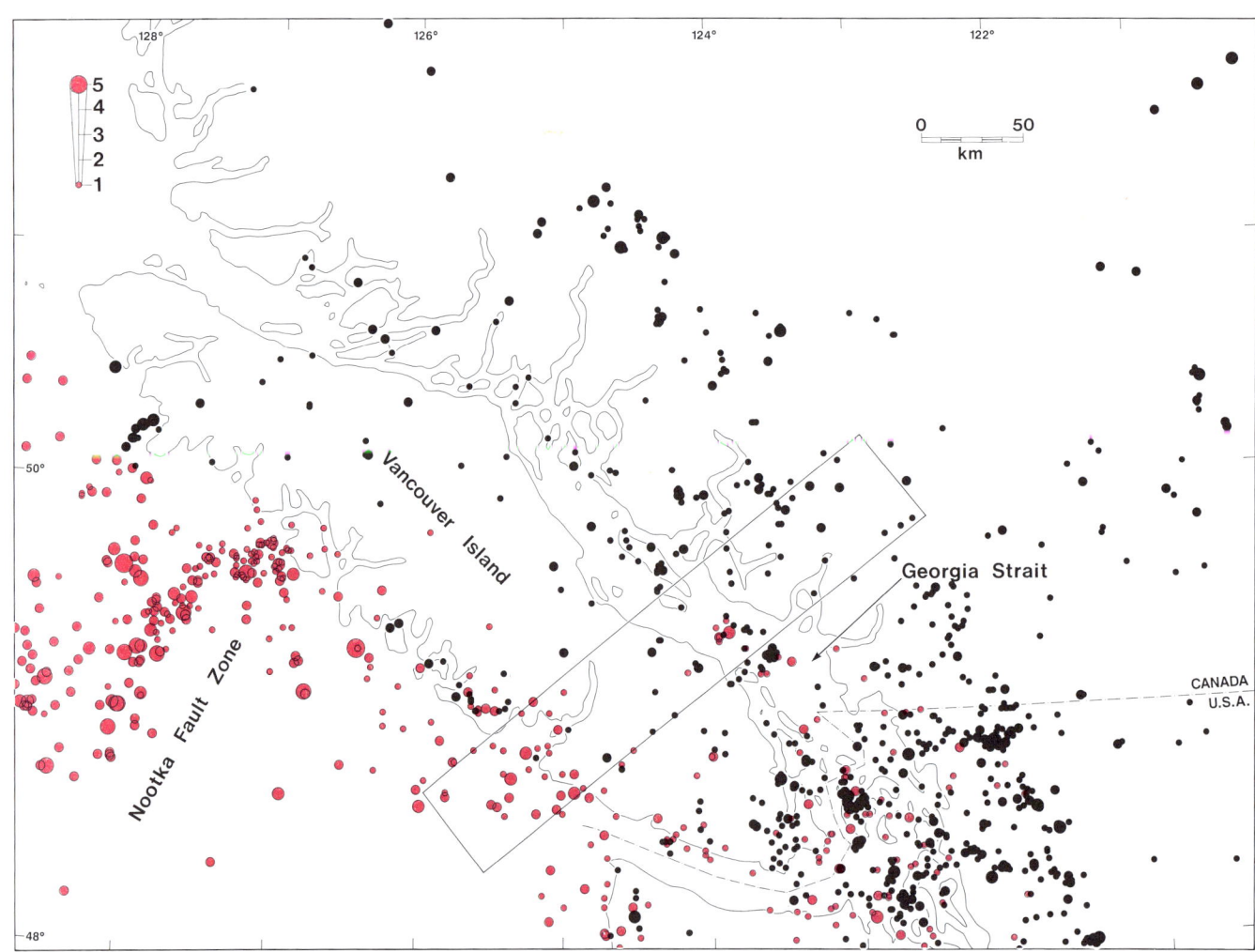

Figure 4. Earthquakes in the Cascadia subduction region 1980 to 1985. Events magnitude 1 and greater are shown. Those occurring within the oceanic plates are colored red, and those within the continental plate are colored black. The box outlines the area of the cross section in Figure 5.

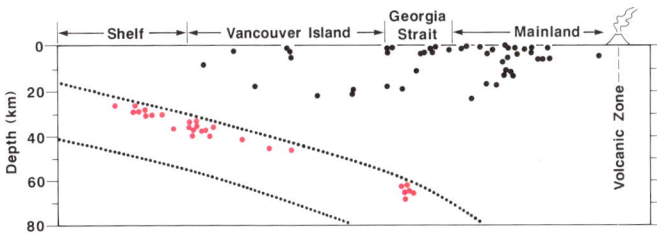

Figure 5. Cross section indicated on Figure 4 including only earthquakes well located in depth (adapted from Rogers and others, 1990a). Colors as in Figure 4.

Northwest Territories, region (62°N, 124°W) that began with a magnitude 6.6 event on October 5, 1985, and included a magnitude 6.0 earthquake on March 25, 1988. The Nahanni earthquakes are described by Weichert and others (1986a, b), Evans and others (1987), Horner and others (1987), Wetmiller and others (1988), Choy and Boatwright (1988), Lamontgne and others (1989), and Horner and others (1990). These earthquakes were the result of thrust faulting on north-striking, shallow west-dipping planes. Three other magnitude-6 earthquakes in the northern Cordillera occurred in the Richardson Mountains region; magnitude 6.2 on May 29, 1940; magnitude 6.5 on June 5, 1940; magnitude 6.6 on March 1, 1955. Elsewhere in the northern Cordillera, only minor seismicity is observed. There is some suggestion in Figure 2 of an alightment of epicenters along the Tintina Trench (Fig. 1), with increasing activity closer to the Alaska border.

The intense cluster of seismicity in the Richardson Mountains is not well understood. A regional microearthquake survey conducted over the area in 1972 (Leblanc and Wetmiller, 1974) found that the seismicity could be grouped into two zones separated by the Bonnet Plume basin; one in the Richardson Mountains north of the Peel River (approx. 66°N), the other to the south in the Mackenzie Mountains. The two 1940 magnitude 6 events occurred in the north zone, whereas the 1955 earthquake occurred in the south zone. The regional geologic strike changes from approximately east-west in the Mackenzie Mountains to north-south in the Richardson Mountains. A poorly controlled P-nodal solution for a magnitude 4.8 earthquake in the Richardson Mountains on July 26, 1972, indicates right-lateral motion on a nearly vertical fault with a north-northwest strike. Focal depths during the field survey were poorly determined but ranged from 0 to about 25 km.

THE BEAUFORT SEA

A distinct cluster of earthquakes is observed in the Beaufort Sea (Figs. 1 and 2). The largest historical earthquake was magnitude 6.5 on November 16, 1920 (Basham and others, 1977). The epicenters lie on the continental slope between the 200- and 2,500-m bathymetric contours. Hasegawa and others (1979) noted a spatial correlation with the seaward slope of an ellipti-

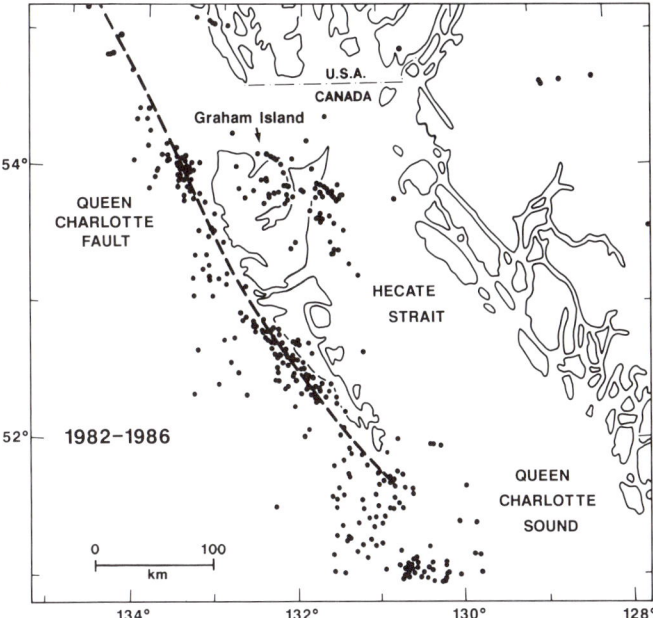

Figure 6. Earthquake epicenters in the Queen Charlotte Islands region 1982 to 1986. All events above about magnitude 1.5 and greater are shown. The detection capability improved significantly through this period.

cally NNE-oriented gravity anomaly believed to be caused by an uncompensated load of Quaternary sediments. They calculate that the maximum horizontal tensile stress under this load occurs at a depth of about 40 km, oriented normal to the continental slope. The mechanism of a magnitude 5.1 event in 1975, at a depth of about 40 km, also indicated a nearly horizontal tension axis oriented approximately normal to the continental slope (Hasegawa and others, 1979).

A more complete pattern of well-located epicenters in the Beaufort cluster since Hasegawa and others' (1979) study indicates more complexity. The cluster does not trend parallel to the gravity anomaly, nor does it correlate in any clear way with the gradient of the gravity field. Moreover, the mechanism for a 1987 event located about 250 km north of the 1975 event has its pressure axis, not its tension axis, normal to the continental slope. The depths of earthquakes in the Beaufort Sea cluster are still insufficiently resolved, mainly because the distance to the nearest stations is of the order of 200 km. Generally, the hypocenters are thought to be in the upper mantle, perhaps close to the 40-km depth determined by Hasegawa and others (1979). This conclusion is supported by the absence of Lg waves at near-coastal stations that would normally be expected from shallow events, particularly in this case where there is no intervening oceanic crust to extinguish these waves.

THE SOUTHERN CORDILLERA

South of 60°N, seismicity drops off markedly away from the coastal region to a low level throughout much of the Cordillera

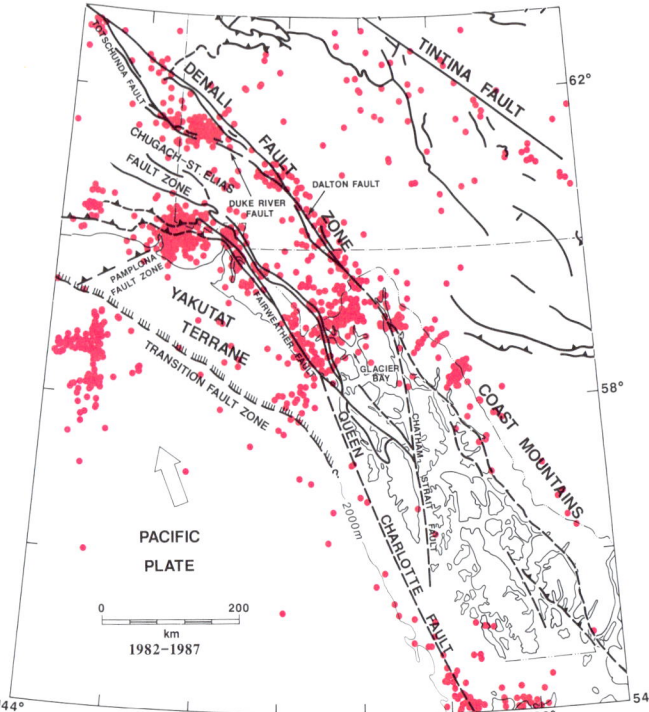

Figure 7. Earthquake epicenters from 1982 to 1987 and major faults in the tectonically active St. Elias region (from Horner, 1990). Magnitude thresholds are about 2.5 onshore and 3.0 offshore.

(Fig. 2). It is slightly higher in the Coast Mountains from southern British Columbia to the Yukon border. A number of moderate earthquakes have occurred within the Coast Mountains (e.g., Rogers, 1976b). The largest are magnitude 5.5 events on September 17, 1926, and January 31, 1942, in the southern region. Very few earthquakes are observed in northern British Columbia east of the Coast Mountains; however, there is some minor activity along the northern Rocky Mountain Trench. Minor seismicity is observed across southern British Columbia and into southwestern Alberta. There is a slightly higher level of activity on the eastern side of the Cordillera in the region of the Rocky Mountain trench and the Foreland fold and thrust belt (Fig. 1). A concentration of activity occurs in the vicinity of the Rocky Mountain trench about latitude 52° N, and this has been the most studied region (Ellis and others, 1976; Rogers and Ellis, 1979; Rogers and others, 180; Ellis and chandra, 1981; Rogers, 1981). The largest earthquake in the southern Cordillera, magnitude about 6.0, occurred here on February 4, 1918.

The only focal mechanisms in the southern Cordillera have been obtained for two earthquakes within the Foreland fold and thrust belt, a magnitude 4.8 event (Rogers and others, 1980) and a magnitude 5.4 event (Rogers and others, 1990b). Both earthquakes involved thrust faulting in the same east-west sense that formed the original structures in this belt, which is similar to the Nahanni earthquakes (Wetmiller and others, 1988) that occurred much farther to the north. Reactivation of these older structures is consistent with present-day stress directions compiled by Adams (1987).

THE INTERIOR PLATFORM

Earthquakes in the Interior Platform or prairie region south of 60°N are predominantly confined to southern Saskatchewan in a zone that extends into northeastern Montana (Fig. 1 and 2). The largest earthquake in this zone was a magnitude 5.5 event on May 15, 1909, near the Canada–United States border. It was widely felt but not well located. Horner and Hasegawa (1978) have shown a general spatial correlation between the observed seismicity and the more structurally disturbed areas of the Precambrian basement and overlying sedimentary sequence in the Williston Basin area. There is abundant evidence of basement control on structures in the Phanerozoic sequence, such as the many salt-solution features. Several of the earthquakes in southern Saskatchewan show a spatial correlation with a large, multi-stage salt-solution structure marked by a major anomaly in electrical conductivity and a steep Bouguer gravity gradient. Indirect evidence suggests the earthquakes occur in the upper crust at depths of less than 10 km. A strike-slip mechanism is proposed for a magnitude 3.7 earthquake in southern Saskatchewan on July 26, 1972 (Horner and others, 1973; Horner and Hasegawa, 1978). This is the only earthquake for which even a partial focal mechanism solution has been obtained.

In addition to the natural seismicity, induced earthquakes have been associated with potash mining in Saskatchewan (Gendzwill and others, 1982; Hasegawa and others, 1989), and oil and gas production in Alberta (Reboller and others, 1981; Wetmiller, 1986). The largest was magnitude 4.6 on March 8, 1970 (Milne, 1970). A number of small events located near Fort St. John, British Columbia (56°N, 121°W), since 1984 (Drysdale and Horner, 1987), all less than magnitude 3.5, may also be related to oil and gas production. Because of the shallow depth of these earthquakes, many have been felt with intensities as high as MM VI over small areas.

SUMMARY

Most of Canada's earthquakes occur in western Canada, but the most intense region of activity is offshore, associated with the young plate regime west of Vancouver Island. The majority of the offshore activity occurs on or near major sea-floor transform fault zones, but there is significant activity within the Explorer Plate suggesting that internal deformation is occurring.

In the subduction environment of southwest British Columbia, ongoing minor seismicity denotes the position of the subducted lithosphere beneath Vancouver Island and the adjacent mainland. Damaging earthquakes have occurred within both the overlying and subducted plates. No seismicity has been detected on the subduction interface, but the possibility of very large subduction earthquakes at intervals of several hundreds of years has been suggested. Approximately ten percent of Canada's population lives in southwest British Columbia, thus, adequate assessment of earthquake hazard here is essential.

To the north of the subduction zone the continental margin is characterized by a very narrow zone of ongoing seismicity delineating the Queen Charlotte fault. Here, strike-slip motion takes place between the Pacific and North American Plates. The Queen Charlotte fault was the site of Canada's largest known earthquake, a magnitude 8.1 event in 1949. Farther north where Pacific-America motion changes from strike-slip motion to convergence in Alaska, significant seismicity occurs in the St. Elias region of southwest Yukon and extreme northwest British Columbia, adjacent to the plate boundary.

Seismicity decreases significantly away from the coastal regions, but some seismicity occurs inland across all of the Cordillera, with concentrations of activity in the Mackenzie and Richardson Mountains in norther Canada. The prairie regions are mainly aseismic, with the exception of sporadic minor activity in southern Saskatchewan. Some induced earthquake activity associated with potash and hydrocarbon extraction is also observed in Alberta and Saskatchewan.

REFERENCES CITED

Adams, J., 1987, Canadian crustal stress data; A compilation to 1987: Geological Survey of Canada Open-File Report 1622, 130 p.
——, 1990, Paleoseismicity of the Cascadia subduction zone; Evidence from turbidites off the Oregon-Washington margin: Tectonics, v. 9, p. 569–583.
Atwater, B. F., 1987, Evidence for great Holocene earthquakes along the outer coast of Washington State: Science, v. 236, p. 942–944.
Basham, P. W., Forsyth, D. A., and Wetmiller, R. J., 1977, The seismicity of northern Canada: Canadian Journal of Earth Sciences, v. 14, p. 1646–1667.
Bérubé, J., Rogers, G. C., Ellis, R. M., and Hasselgren, E. O., 1989, A microseismicity study of the Queen Charlotte Island region: Canadian Journal of Earth Sciences, v. 26, p. 2556–2566.
Bostwick, T. K., 1984, A re-examination of the August 22, 1949, Queen Charlotte Islands earthquake [M.Sc. thesis]: Vancouver, University of British Columbia, 115 p.
Cassidy, J. F., Ellis, R. M., and Rogers, G. C., 1988, The 1918 and 1957 Vancouver Island earthquakes: Bulletin of the Seismological Society of America, v. 78, p. 617–635.
Chandra, U., 1974, Seismicity, earthquake mechanisms, and tectonics along western North America, from 42° to 61°N: Bulletin of the Seismological Society of America, v. 64, p. 1529–1549.
Choy, C. L., and Boatwright, J., 1988, Teleseismic and near field analysis of the Nahanni earthquakes in the Northwest Territories, Canada: Bulletin of the Seismological Society of America, v. 78, p. 1627–1652.
Clague, J. J., 1979, The Denali fault system in the southwest Yukon Territory; A geological hazard?, in Current Research, Part A: Geological Survey of Canada Paper 79-1A, p. 169–178.
Crosson, R. S., 1983, Review of seismicity in the Puget Sound region from 1970 through 1978; A brief summary, in Yount, J. C., ed., Earthquake hazards of the Puget Sound region, Washington State: U.S. Geological Survey Open-File Report 83-19, p. 6–18.
Dennison, F. N., 1919, The British Columbia earthquake of December 6, 1918: Bulletin of the Seismological Society of America, v. 9, p. 20–23.
Dragert, H., and Rogers, G. C., 1988, Could a megathrust earthquake strike southwestern British Columbia?: Geos, v. 17, no. 3, p. 5–8.
Drysdale, J. A., and Horner, R. B., 1987, Canadian earthquakes; 1984: Geological Survey of Canada Paper 87-19, 44 p.
Ellis, R. M., and Chandra, B., 1981, Seismicity in the Mica Reservoir (McNaughton Lake) area, 1973–1978: Canadian Journal of Earth Sciences, v. 18, p. 1708–1716.
Ellis, R. M., and Rogers, G. C., 1986, Preliminary investigation of earthquakes west of Vancouver Island: Geological Survey of Canada Open-File Report 1400, 85 p.
Ellis, R. M., Dragert, H., and Ozard, J. M., 1976, Seismic activity in the McNaughton Lake area, Canada: Engineering Geology, v. 10, p. 227–238.
Evans, S. G., Aitken, J. D., Wetmiller, R. J., and Horner, R. B., 1987, A rock avalanche triggered by the October 1985 North Nahanni earthquake, District of Mackenzie, N.W.T.: Canadian Journal of Earth Sciences, v. 24, p. 176–184.
Gendzwill, D.J., Horner, R. B., and Hasegawa, H. S., 1982, Induced earthquakes at a potash mine near Saskatoon, Canada: Canadian Journal of Earth Sciences, v. 19, p. 466–475.
Hasegawa, H. S., and Rogers, G. C., 1978, Quantification of the magnitude 7.3 British Columbia earthquake of June 23, 1946: Tectonophysics, v. 19, p. 185–188.
Hasegawa, H. S. Chou, C. W., and Basham, P. W., 1979, Seismotectonics of the Beaufort Sea: Canadian Journal of Earth Sciences, v. 16, p. 816–830.
Hasegawa, H. S., Wetmiller, R. J., and Gendzwill, D. J., 1989, Induced seismicity in mines in Canada; An overview: Pure and Applied Geophysics, v. 129, p. 423–453.
Heaton, T. H., 1990, The calm before the quake?: Nature, v. 343, p. 511–512.
Heaton, T. H., and Hartzell, S. H., 1987, Earthquake hazards on the Cascadia subduction zone: Science, v. 236, p. 162–168.
Heaton, T. H., and Kanomori, H., 1984, Seismic potential associated with subduction in the northwestern United States: Bulletin of the Seismological Society of America, v. 74, p. 933–942.
Hodgson, E. A., 1946, The British Columbia earthquake, June 23, 1946: Journal of the Royal Astronomical Society of Canada, v. 40, p. 285–319.
Horner, R. B., 1983, Seismicity in the St. Elias region of northwestern Canada and southeastern Alaska: Bulletin of the Seismological Society of America, v. 73, p. 1117–1137.
——, 1989, Low-level seismic monitoring at the Windy Craggy deposit in northwest British Columbia, in Current research, Part E: Geological Survey of Canada Paper 98-1E, p. 275–278.
——, 1990, Seismicity in the Glacier Bay region of southeast Alaska and adjacent areas of British Columbia, in Milner, A. M., and Wood, J. D., Jr., eds., Proceedings, 2nd Glacier Bay Science Symposium, Glacier Bay Lodge, Alaska, September, 1988: p. 6–11.
Horner, R. B., and Hasegawa, H. S., 1978, The seismotectonics of southern Saskatchewan: Canadian Journal of Earth Sciences, v. 15, p. 1341–1355.
Horner, R. B., Stevens, A. E., and Hasegawa, H. S., 1973, The Bengough, Saskatchewan, earthquake of July 26, 1972: Canadian Journal of Earth Sciences, v. 10, p. 1805–1821.
Horner, R. B., Lamontagne, M., and Wetmiller, R. J., 1987, Rock and roll in the N.W.T.: The 1985 Nahanni earthquakes: Geos, v. 16, no. 2, p. 1–4.
Horner, R. B., Wetmiller, R. J., Lamontagne, M., and Plouffe, M., 1990, A fault model for the Nahanni earthquakes from aftershock studies: Bulletin of the Seismological Society of America, v. 80, p. 1553–1570.
Hyndman, R. D., and Ellis, R. M., 1981, Queen Charlotte fault zone; Microearthquakes from a temporary array of land stations and ocean bottom seismographs: Canadian Journal of Earth Sciences, v. 18, p. 776–788.

Hyndman, R. D., and Rogers, G. C., 1981, Seismicity surveys with ocean bottom seismographs off western Canada: Journal of Geophysical Research, v. 86, p. 3867–3880.

Hyndman, R. D., and Weichert, D. H., 1983, Seismicity and rates of relative motion on the plate boundaries of western North America: Geophysical Journal of the Royal Astronomical Society, v. 72, p. 59–82.

Hyndman, R. D., Riddihough, R. P., and Herzer, R., 1979, The Nootka fault zone; A new plate boundary off western Canada: Geophysical Journal of the Royal Astronomical Society, v. 58, p. 667–683.

Lamontagne, M., Horner, R. B., Wetmiller, R. J., Monsees, D., and Vonk, A., 1989, Le seisme de mars 1988 de la riviere North Nahanni, T.N.-O., et ses repliques: Recherches en cours, Part E, Commission Geologique du Canada, Etude 89-1E, p. 265–268.

Leblanc, G., and Wetmiller, R. J., 1974, An evaluation of seismological data available for the Yukon Territory and the Mackenzie Valley: Canadian Journal of Earth Sciences, v. 11, p. 1435–1454.

Mathews, W. H., 1979, Landslides of central Vancouver Island and the 1946 earthquake: Bulletin of the Seismological Society of America, v. 69, p. 445–450.

Meidler, S. S., 1961, Seismic activity in the Canadian Arctic 1899–1955: Ottawa, Dominion Observatory Seismological Series 1961-3, 9 p.

Milne, W. G., 1956, Seismic activity in Canada west of the 113° meridian 1841–1955: Ottawa, Dominion Observatory Publications, v. 18, p. 113–145.

——, 1963, Seismicity of western Canada: Bulletin Bibliografico de Geofisica y Oceanografia Americanas, v. 2, p. 17–40.

——, 1967, Earthquake epicentres and strain release in Canada: Canadian Journal of Earth Sciences, v. 4, p. 1–18.

——, 1970, The Snipe Lake, Alberta, earthquake of March 8, 1970: Canadian Journal of Sciences, v. 7, p. 1564–1567.

Milne, W. G., Smith, W.E.T., and Rogers, G. C., 1970, Canadian seismicity and microearthquake research in Canada: Canadian Journal of Earth Sciences, v. 7, p. 1–11.

Milne, W. G., Rogers, G. C., Riddihough, R. P., McMechan, G. A., and Hyndman, R. D., 1978, Seismicity of western Canada: Canadian Journal of Earth Sciences, v. 15, p. 1170–1193.

Rebollar, C. J., Kanasewich, E. R., and Nyland, E., 1981, Source parameters from shallow events in the Rocky Mountain House earthquake swarm: Canadian Journal of Earth Sciences, v. 19, p. 907–918.

Riddihough, R. P., and 6 others, 1983, Geodynamics of the Juan de Fuca Plate, in Cabre, R., ed., Geodynamics of the eastern Pacific region; Caribbean and Scotia arcs: American Geophysical Union Geodynamics Series, v. 9, p. 5–21.

Rogers, G. C., 1976a, The Vancouver Island earthquake of July 5, 1972: Canadian Journal of Earth Sciences, v. 13, p. 92–101.

——, 1976b, The Terrace earthquake of November 5, 1973: Canadian Journal of Earth Sciences, v. 13, p. 495–499.

——, 1976c, A microearthquake survey in northwest British Columbia and southeast Alaska: Bulletin of the Seismological Society of America, v. 66, p. 1643–1655.

——, 1979, Earthquake fault plane solutions near Vancouver Island: Canadian Journal of Earth Sciences, v. 16, p. 523–531.

——, 1980a, A documentation of soil failure during the British Columbia earthquake of 23 June, 1946: Canadian Geotechnical Journal, v. 17, p. 122–127.

——, 1980b, Juan de Fuca Plate map; Seismicity: Canada Department of Energy, Mines and Resources, Earth Physics Branch Open-File 80-3, scale 1:2,000,000.

——, 1980c, Juan de Fuca Plate map; Fault plane solutions: Canada Department of Energy, Mines and Resources, Earth Physics Branch Open-File 80-4, scale 1:2,000,000.

——, 1981, McNaughton Lake seismicity; More evidence for an Anahim hotspot?: Canadian Journal of Earth Sciences, v. 18, p. 826–828.

——, 1982, Revised seismicity and revised fault plane solutions for the Queen Charlotte Islands region: Canada Department of Energy, Mines and Resources, Earth Physics Branch Open-File 82-23, 73 p.

——, 1983a, Some comments on the seismicity of the northern Puget Sound–southern Vancouver Island region, in Young, J. C., and Crosson, R. S., eds., Earthquake hazards of the Puget Sound region, Washington State: U.S. Geological Survey Open-File Report 83-19, p. 19–39.

——, 1983b, Seismotectonics of British Columbia [Ph.D. thesis]: Vancouver, University of British Columbia, 247 p.

——, 1986, Seismic gaps along the Queen Charlotte fault: Earthquake Prediction Research, v. 4, p. 1–11.

——, 1988a, An assessment of the megathrust earthquake potential of the Cascadia subduction zone: Canadian Journal of Earth Sciences, v. 24, p. 844–852.

——, 1988b, Seismic potential of the Cascadia subduction zone: Nature, v. 332, p. 17.

Rogers, G. C., and Ellis, R. M., 1979, The eastern British Columbia earthquake of February 4, 1918: Canadian Journal of Earth Sciences, v. 16, p. 1484–1493.

Rogers, G. C., and Hasegawa, H. S., 1978, A second look at the British Columbia earthquake of 23 June, 1946: Bulletin of the Seismological Society of America, v. 68, p. 653–676.

Rogers, G. C., Ellis, R. M., and Hasegawa, H. S., 1980, The McNaughton Lake earthquake of May 14, 1978: Bulletin of the Seismological Society of America, v. 70, p. 1771–1786.

Rogers, G. C., Spindler, C., and Hyndman, R. D., 1990a, Seismicity along the Vancouver Island Lithoprobe corridor, in Proceedings of the Lithoprobe Southern Canadian Cordillera Transect Workshop, 3–4 March, 1990: Calgary, University of Calgary, p. 166–169.

Rogers, G. C., Cassidy, J. F., and Ellis, R. M., 1990b, The Prince George, British Columbia, earthquake of 21 March 1986: Bulletin of the Seismological Society of America, v. 80, p. 1144–1161.

Wahlstrom, R., and Rogers, G. C., 1990, Relocation of earthquakes offshore Vancouver Island: Geological Survey of Canada Open-File Report 2288, 63 p.

Wahlstrom, R., Rogers, G. C., and Baldwin, R. E., 1990, P-nodal focal mechanisms for large earthquakes offshore Vancouver Island: Geological Survey of Canada Open-File Report 2268, 316 p.

Weichert, D. H., Wetmiller, R. J., and Munro, P. S., 1986a, Vertical earthquake accelerations exceeding 1G? The case of the missing peak: Bulletin of the Seismological Society of America, v. 76, p. 1473–1478.

Weichert, D. H., Wetmiller, R. J., Horner, R. B., Munro, P. S., and Mark, P. N., 1986b, Strong motion records from the 23 December 1985 M 6.9 Mahanni, N.W.T., and some associated earthquakes: Geological Survey of Canada Open-File Report 1330, 65 p.

Wetmiller, R. J., 1971, An earthquake swarm on the Queen Charlotte Island fracture zone: Bulletin of the Seismological Society of America, v. 61, p. 1489–1505.

——, 1986, Earthquakes near Rocky Mountain House, Alberta, and their relationship to gas production facilities: Canadian Journal of Earth Sciences, v. 23, p. 172–181.

Wetmiller, R. J., and 6 others, 1988, An analysis of the 1985 Nahanni earthquakes: Bulletin of the Seismological Society of America, v. 78, p. 590–616.

Wetmiller, R. J., Drysdale, J. A., Horner, R. B., and Lamontagne, M., 1989, Canadian earthquakes 1985–86: Geological Survey of Canada Paper 88-14, 25 p.

MANUSCRIPT ACCEPTED BY THE SOCIETY SEPTEMBER 26, 1990
GEOLOGICAL SURVEY OF CANADA CONTRIBUTION 42787

The Geology of North America
Decade Map Volume 1
1991

Chapter 6

Seismicity of Washington and Oregon

R. S. Ludwin
Geophysics Program, University of Washington, Seattle, Washington 98195
C. S. Weaver
U.S. Geological Survey, University of Washington, Seattle, Washington 98195
R. S. Crosson
Geophysics Program, University of Washington, Seattle, Washington 98195

INTRODUCTION

This chapter examines seismicity in Oregon and Washington, and combines historical accounts of damaging earthquakes with the detail of recent seismicity from modern instrumental locations. A revised seismic catalog has been compiled for Washington and Oregon from 1850 through 1987. This catalog provides a snapshot of the seismotectonic setting of the Pacific Northwest at a single point in geologic time. From the catalog we identify three primary elements of the seismicity distribution: (1) a dipping Benioff zone beneath western Washington and northwestern Oregon, (2) crustal earthquakes in the forearc that shallow eastward toward the volcanic arc, and (3) scattered earthquakes in the back-arc region east of the Cascade volcanic arc.

The tectonics, geology, and seismicity of the Pacific Northwest are briefly reviewed in this introduction. Later, we describe the catalog, and include a review of data sources and data-selection criteria, a discussion of the largest earthquakes known in the Pacific Northwest, and a study of temporal variations by decades for earthquakes larger than magnitude 4 since 1960. Microearthquake data available from the University of Washington catalog reveal variations of seismicity in the crust of North America and in the Juan de Fuca plate. These more detailed seismity data allow us to describe variations that exist in the subduction zone framework, and to suggest that continental, as well as subduction tectonics have a profound influence on Pacific Northwest seismicity. In a section on earthquake hazards, we comment on two significant unresolved seismological issues: the probability of an earthquake of magnitude 8 or greater on the subduction zone thrust interface, and the likelihood of another large shallow event like the 1872 North Cascades earthquake, probably the largest historical earthquake in the Pacific Northwest.

Washington and Oregon have features typical of convergent boundaries in other parts of the world. Offshore, on the continental shelf, deformed Quaternary sediments are found where the converging plates meet (Barnard, 1978). A convergence rate of 4 to 4.5 cm/yr between these two plates, with an approximately N50°E direction of convergence, has been estimated by Riddihough (1977, 1984). The regional free-air and Bouguer gravity anomalies show a pattern common in subduction zones: an elongated gravity low along the toe of the continental slope and a gravity high inland over the coastal mountains. A volcanic arc of andesitic composition, dating from the mid-Miocene, stretches from northern California to southern British Columbia and is consistent with the presence of a subducting plate about 100 km beneath the arc. Beneath western Washington, a zone of earthquakes at depths of 35 to 80 km inclines easterly from the coast (Crosson, 1983); this has been interpreted as an intraplate Benioff zone (Crosson, 1983; Taber and Smith, 1985). Most of the larger earthquakes in the Pacific Northwest since the mid-1800s have probably occurred within this deep zone (Rogers, 1983b). Figure 1 shows the regional seismicity from the NOAA catalog through 1985 for magnitudes greater than 4. Major plate boundaries (transform zones and spreading ridges) are illustrated for offshore regions, and major physiographic provinces onshore.

Onshore, the surficial geology of Oregon and Washington is complex (Fig. 1), reflecting in part the long-term convergence and accretionary plate environment. The Klamath Mountains, North Cascades, and Blue Mountain provinces are largely pre-Cenozoic terranes that include folded and metamorphosed complexes of continental basements and sediments (Misch, 1966). Volcanism in the Cenozoic was extensive and voluminous. The Western Cascades, High Cascades, and Columbia Plateau provinces experienced repeated eruptions of basaltic and/or andesitic lavas. During the Quaternary, volcanism has been more limited in extent. South of the latitude of Mt. Adams, Quaternary volcanics are exposed nearly continuously in the High Cascades in Oregon and northern California. North of Mt. Adams, Quaternary volcanism has been confined to the large stratocones of the Cascade Range. The Olympic Mountains and the Coast Range of

---

Ludwin, R. S., Weaver, C. S., and Crosson, R. S., 1991, Seismicity of Washington and Oregon, *in* Slemmons, D. B., Engdahl, E. R., Zoback, M. D., and Blackwell, D. D., eds., Neotectonics of North America: Boulder, Colorado, Geological Society of America, Decade Map Volume 1.

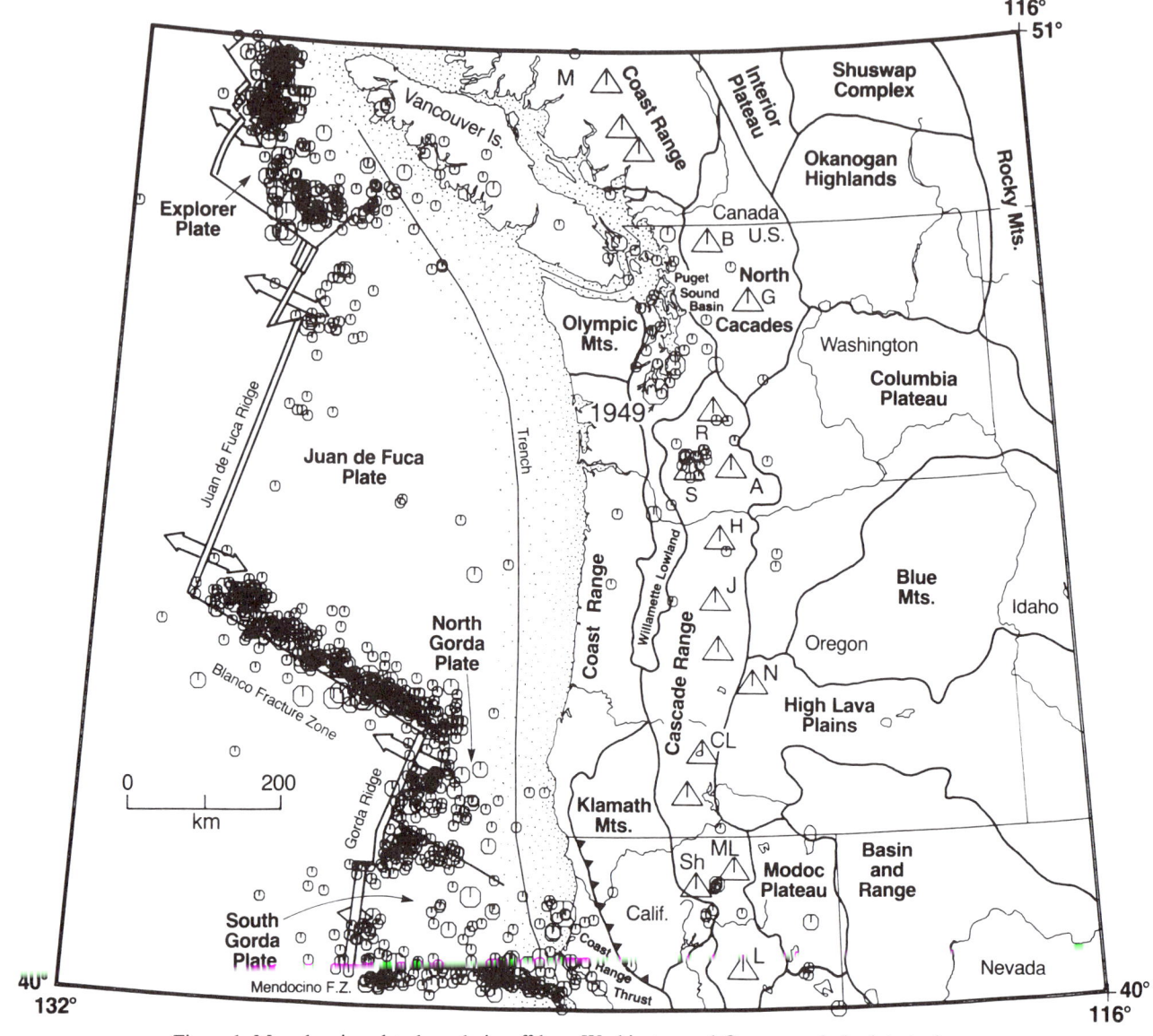

Figure 1. Map showing plate boundaries offshore Washington and Oregon, and physiotectonic provinces of Washington and Oregon. Earthquakes shown are magnitude 4 or larger events listed in the NOAA catalog through 1985. This catalog is fairly complete at this magnitude range since 1963, before that date the data is not complete. The 1949 Olympia earthquake (not included in the NOAA catalog) is also shown. Volcanos are indicated by triangles, and marked as follows: M = Meager Mountain, B = Mt. Baker, G = Glacier Peak, R = Mt. Rainier, S = Mt. St. Helens, A = Mt. Adams, J = Mt. Jefferson, H = Mt. Hood, N = Newberry Volcano, L = Lassen Peak, Sh = Mt. Shasta, ML = Medicine Lake Volcano.

Washington and Oregon consist mainly of Tertiary oceanic sediments and basalts (Cady, 1975). These sediments appear to have been accreted to the continent in the subduction process (Tabor, 1972). The Olympic Mountains have been interpreted as oceanic basement obducted onto the continental margin (Snavely, 1988). Paleomagnetic data from the Coast Range have been used to infer a significant rotation for this province compared to inland provinces (Magill and others, 1981, 1982; Beck and Engebretson, 1982). Finally, the Puget-Willamette Lowland is an extensive depression covered by thick sedimentary sequences eroded from the Olympic Mountains and Cascade Range during Miocene uplift (Tabor, 1972; Cady, 1975).

Known Pacific Northwest earthquakes with magnitudes greater than 4.0 (Fig. 1) only weakly reflect the convergence framework. Most of the events are located offshore, with major activity at the transform fault boundaries such as the Blanco fracture zone. Although there is moderate activity near the Gorda and Explorer ridges, the major ridge in the region, the Juan de Fuca, is seismically quiet. Likewise, the subduction zone and continental margin of Washington and Oregon are seismically quiet compared to the offshore transform faults. Very few earthquakes have been located in Oregon, in continuing agreement with an early observation by de Ballore (1906), who noted the hiatus, which still persists, in damaging earthquakes between Portland, Oregon, and Crescent City, California. The sparse distribution of earthquakes onshore in Washington and Oregon contrasts markedly with the distribution of seismicity reported in most active subduction zones where numerous earthquakes occur at the interface between the two plates, within the subducting plate, and within the overriding plate (Uyeda and Kanamori, 1979). For the Oregon-Washington portion of the Cascadia subduction zone, no subduction or thrust earthquakes have been identified along the plate interface at any magnitude level.

EARTHQUAKE HISTORY

Sources of data

High-quality earthquake locations (epicentral precision approximately ± 2 km) for Washington and northern Oregon are available beginning in 1970, when the installation of the modern high-gain seismic network began (Crosson, 1972). Prior to 1970, only a few earthquakes have even moderately well-constrained epicentral locations (± 10 km). However, because all known earthquakes greater than magnitude 6 in Oregon and Washington occurred prior to 1970, it is imperative to incorporate the best estimates of the epicenters of older, large events along with modern, high-quality observations of smaller earthquakes.

The earliest data available are felt reports published in newspapers, which began publication fewer than 150 years ago. Locations for the older earthquakes are estimated from maximum reported intensities (e.g., Malone and Bor, 1979). These older earthquakes, aside from the few well-located ones, are useful mainly for estimating recurrence intervals and approximate locations of larger earthquakes. By the early part of the 20th century, seismographic instruments had been installed at Victoria, British Columbia (installed 1898; Milne and others, 1978), and at Seattle (installed 1906; Poppe, 1979). Although these instruments were capable of detecting local earthquakes larger than magnitude 4.5, the historical record of the early 20th century is still based largely on felt reports. By the 1930s, enough seismic stations had been installed in North America and Europe to detect and approximately locate events larger than magnitude 6. Due to poor absolute timing, readings from these distant stations are usually not accurate enough to locate events in the Pacific Northwest with better precision than using felt reports alone. Until 1951, when the installation of three high-gain seismic stations in southern British Columbia was completed, earthquake locations and magnitudes of felt or instrumentally detected events continued to be estimated from felt reports supplemented with instrumental data from the few regional stations then operating.

Since 1960, three major improvements in instrumental coverage have allowed locations to be calculated for small- to moderate-magnitude (2.5 to 4.5) and smaller events (<2.5). First, the advent of greatly improved absolute timing and the installation of the World-Wide Seismographic Station Network (WWSSN) in the early 1960s, along with the installation of Wood-Anderson seismometers in Victoria, Spokane, and Seattle, allowed sufficient resolution to compile a complete catalog, beginning in 1963, of earthquakes magnitude 4.0 and larger (Rogers, 1983a). Second, in 1970 a multi-station telemetered seismograph network was installed that was capable of detecting and precisely locating earthquakes in Washington of magnitudes less than 4.0 (Crosson, 1974; University of Washington Geophysics Program, 1979). Throughout the 1970s, this network was expanded and modified to increase sensitivity to small earthquakes and to provide better coverage of northwestern Washington, eastern Washington, and a portion of northeastern Oregon. Third, large areas of southwestern Washington and northern Oregon were instrumented further after the 1980 eruption of Mount St. Helens. As a result of this expansion, the Mount St. Helens area, the southern Washington Cascade Range, and areas along the Washington and northern Oregon coasts were monitored for the first time with enough instruments to routinely locate microearthquake activity. Similar coverage of the central and southern Oregon Cascade Range also began in 1980, but was discontinued in 1982. During 1987, stations were installed in Oregon east of the Cascade crest as far south as Newberry Volcano. Current seismic stations, as well as stations in operation prior to 1970, are shown in Figure 2. Five velocity models are routinely used for earthquake location by the Washington Regional Seismic Network (WRSN), and the areas covered by these models are also shown in Figure 2. Summaries of dates of operation and locations of pre-1970 seismic stations are given in Tables 1 and 2. Table 3 gives the velocity models used for earthquake locations.

The catalog

For the oldest earthquakes, neither location, depth, nor magnitude can be determined accurately. Our final earthquake catalog eliminates all but the largest earthquakes that occurred before the installation of the modern seismic network beginning in 1970. For the period prior to 1970, 94 earthquakes are included in the catalog, whereas more than 1,500 earthquakes larger than magnitude 2.5 have been included for the period 1970 to 1987. For pre-1970 earthquakes, where possible, we have used the results of special studies to select the preferred location and magnitude of an event. Since the regional U.W. network began operating in 1970, we rely on it as our primary data source for Washington and northern Oregon. Washington State earthquakes before 1960 were checked against the list given by Rasmussen (1967). In addition, due to their large numbers, volcanic earthquakes at Mount St. Helens that occurred since the 1980 eruption have been eliminated from the catalog.

The final catalog is divided into six time periods. The magnitude threshold for each interval is shown in Table 4; only earthquakes equal to or larger than that magnitude were included. For

TABLE 1. PRE-NETWORK STATION HISTORY*

Sta	00	10	20	30	40	50	60	70	80	Record Location	Seis	Comments
ALB	-	-	-	-	-	8/51	X	X	X	GSC, Ottawa	SPZ	Closed 7/72-8/75
BLL	-	-	-	-	-	-	12/61-69	-	-	WWU	Benioff 3-D	Incomplete
BMO	-	-	-	-	-	-	1962	5/75	-	Teledyne in VA 62-66 USGS 66-75	SP-3D LP-3D	Array
COR	-	-	-	-	-	12/50	X	X	X	UCB pre-62 OSU 62 and later	SP-3D (pre 62) WWSSN (62)	
HBC	-	-	-	-	-	8/51	4/60	-	-	GSC, Ottawa	SPZ	replaced by PNT
KFO	-	-	-	-	-	-	62	X	84	OSU	SPZ	gain unknown
LON	-	-	-	-	-	3/58	X	X	X	UW	SP-3D (pre 62) (D)WWSSN	
NEW	-	-	-	-	-	-	6/66	X	X	USGS	SP-3D LP-3D WA-2D	
NTI	-	-	-	-	-	-	6/67	12/75	-	USGS	SP	
PNT	-	-	-	-	-	-	1/60	X	X	GSC, Ottawa	SP-3D, LP-3D	
PTD	-	-	-	-	-	-	5/64	X	11/86	OSU	SPZ	
SEA	06	X	X	31 31	- X	- X	- X	- X	- X	??? UW	Bosch-Omori 2D SP 3D (49-66) LP-2D (55-69) WA-2D since 66	
STT	-	-	-	-	-	-	64	76	-	Gerald Marshall 35765 26th South Auburn, WA 98002 (206) 838-3578 (206) 927-4411	Galitzen type NS $T_0 = 10s$	
SPO	09	X	X	X	X	X	X	X	-	Battelle in Richland, WA; 46-49 in St. Louis at S.L.U.	Weichert WA-2D Benioff 3D	
TUM	-	-	-	-	-	5/58	69	-	-	UW	SP-3D LP-3D	
Vict.	X	16	-	-	-	-	-	-	-	GSC, Ottawa	Milne LPZ in '14	
VGZ	-	16	X	39	-	-	-	-	3/82	GSC, Ottawa	Milne Milne-Shaw in '22 SPZ in '82	
VIC	-	-	-	39	X	X	X	3/78	-	GSC, Ottawa	Milne-Shaw SP Z since 6/48 SP-3D and LP-3D since 1958 WA-2D 66-3/78	replaced by PGC

*Poppe, 1979; Milne and others, 1978.
Decades: X = Operating entire decade; - = Not operating entire decade.
Record Location Abbreviations: GSC = Geological Survey of Canada; OSU = Oregon State University; SLU = St. Louis University; UCB = University of California, Berkeley; USGS = United States Geological Survey; UW = University of Washington; WWU = Western Washington University.
Seismometer Abbreviations: 2D = two components (horizontal); 3D = three components (2 horizontal, 1 vertical); SP = short period; LP = long period; SPZ = short period vertical; LPZ = long period vertical; NS = one component, horizontal, north-south; WA = Wood-Anderson.

TABLE 2. LOCATIONS OF SEISMIC STATIONS OPERATING IN THE PACIFIC NORTHWEST PRIOR TO 1969

Sta	Latitude	Longitude	Elevation (km)	Dates	Name
ALB	49°16'13.8"	124°49'18.0"	0.025	8/1951–present	Alberni, Canada
BLL	48°44'20.0"	122°29'05.0"	0.096	12/1961–1969	Bellingham, Washington
BMO	44°50'56.0"	117°18'20.0"	1.189	1962–5/1975	Blue Mountains, Oregon
COR	44°35'08.5"	123°18'11.5"	0.121	12/1950–present	Corvallis, Oregon
HBC	49°22'42.0"	123°16'36.0"	0.050	8/1951–4/60	Horseshoe Bay, Canada
KFO	42°16'00.0"	121°44'42.0"	1.439	1962–1984	Klamath Falls, Oregon
LON	46°45'0.00"	121°48'36.0"	0.853	3/1958–present	Longmire, Washington
NEW	48°15'50.0"	117°07'13.0"	0.760	6/1966–present	Newport, Washington
NTI	48°37'48.0"	116°57'47.9"	0.823	8/67–12/75	Nordman, Idaho
PNT	49°19'00.0"	119°37'00.0"	0.550	1/1960–present	Penticton, Canada
PTD	45°30'29.9"	122°42'59.0"	0.208	5/64–present	Portland, Oregon
SEA	47°39'18.0"	122°18'30.0"	0.030	1906–present	Seattle, Washington
STT	47°25'22.0"	122°18'49.2"	0.0	4/64–4/76	Seattle-Marshall
SPO	47°43'48.0"	117°20'32.0"	0.713	1909–1969	Spokane, Washington
TUM	47°00'54.0"	122°54'30.0"	0.084	5/1958–1969	Tumwater, Washington
Vict.	48°25'24.0"	123°22'00.0"	0.005	10/1898–1916	Victoria, Canada
VGZ	48°24'54.0"	123°19'24.0"	0.068	1916–1939 1982–present	Victoria, Canada
VIC	48°31'12.0"	123°24'54.0"	0.197	1939–1978	Victoria, Canada

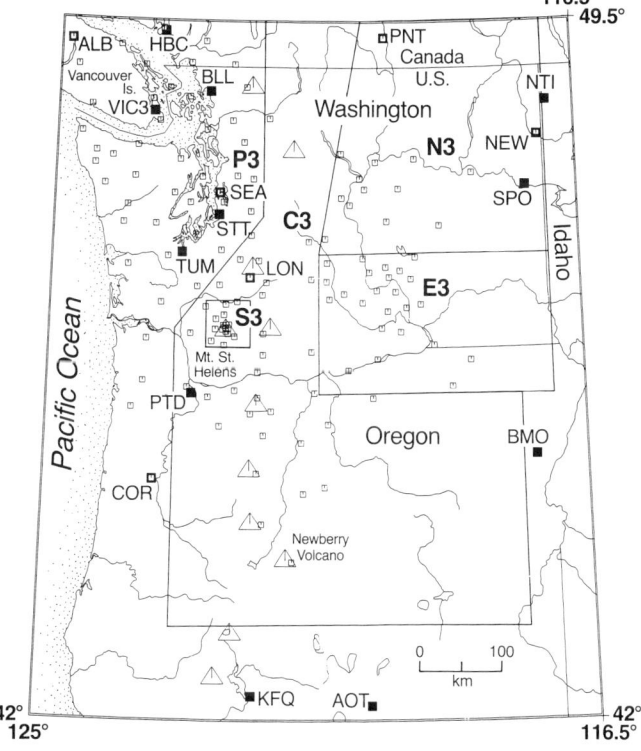

Figure 2. Map view of seismic stations in Washington, Oregon, and southern British Columbia. Stations operating at the end of 1987 are shown as small boxes. Larger filled boxes indicate stations operating prior to 1970. Several stations (e.g., LON) began operation before 1970, and continued through 1987; they are symbolized by the small box superimposed on the larger filled box. Tables 1 and 2 give operational dates of pre-1970 stations. Velocity model areas currently used by the Washington Regional Seismic Network (WRSN) are also indicated. Velocity models for the areas indicated on the figure are given in Table 3.

each period in Table 4 we have estimated the magnitude at which the catalog is complete over all of Oregon and Washington. For earthquakes earlier than 1946 we have used magnitudes estimated by Rasmussen (1967), Rogers (1983a), and Noson and others (1988).

From the catalog we have selected three subsets of earthquakes for discussion based on estimated magnitude and time period. The first subset includes all events estimated to be greater than magnitude 6 from 1870 through 1987. We believe that this selection of 8 earthquakes is complete. The second selection includes all earthquakes greater than magnitude 4 between 1960 and 1987; and although we estimate that this selection is complete for both Oregon and Washington, it is possible that some magnitude 4 events could have been missed in southeastern Oregon in the early 1960s. Finally, we have plotted all events in the catalog that have adequate instrumental hypocenters (accuracy better than about ± 5 km for hypocenters).

Earthquakes greater than magnitude 6. Locations of earthquakes believed to be larger than magnitude 6 are shown in Figure 3, and these events are largely restricted to northwestern Washington. The 1872 North Cascades earthquake is generally considered the largest earthquake known in Washington and Oregon (Milne, 1956), with an estimated magnitude of 7.4 (Malone and Bor, 1979). It was felt over an area of more than 1,010,000 km^2, including Washington, central to northern Oregon, northern Idaho, western Montana, and southern British Columbia. The earthquake was followed by an extensive aftershock sequence (Milne, 1956). Study of damage reports suggests that the maximum intensity exceeded VII and may have been as high as IX on the Modified Mercalli scale (Milne, 1956). One spectacular effect of the event was a triggered landslide that reportedly temporarily dammed the Columbia River in north-central Washington (Weston Geophysical Research Inc., 1976). Both the location and depth of the North Cascades earthquake are subjects of controversy. The location shown in Figure 3 was determined by Malone and Bor (1979) based on the intensity pattern; these authors also summarize other suggested locations of the event. Hopper and others (unpublished abstract, 1982) have suggested a shallow depth based on intensity contours and the extensive aftershock sequence. All instrumentally located earthquakes (since 1970) near the epicenter proposed by Malone and Bor (1979) are shallower than 25 km. Both the depth and location of this earthquake remain controversial because a shallow quake of magnitude 7.4 would be expected to have a rupture area large enough that primary surface faulting would be evident; but none has been noted (Shannon and Wilson, Inc., 1977).

An earthquake estimated to have a magnitude of at least 6.75 (Toppozada, 1981) occurred in 1873 near the Oregon-California border (Fig. 3). The earthquake was felt from near Portland, Oregon, to Sacramento, California, but was felt with intensity VIII only in a small area north of Crescent City (Toppozada, 1981). There is no mention of felt aftershocks accompanying the mainshock in newspaper accounts of the time. This observation suggests that either the event was sufficiently far offshore that any aftershocks were not felt, or that it was deep, possibly within the subducting Gorda plate.

The 1909 earthquake near the San Juan Islands (Fig. 3) has a magnitude estimated from felt area of 6.0 (Rogers, 1983a) and is the largest known earthquake in the northern Puget Sound basin (Rasmussen and others, 1974). This event has been relocated, based on felt reports, by Rogers (1983a), who noted that the new epicentral location is within the contour of the greatest felt intensity. Based on the lack of higher intensities in the epicen-

TABLE 3. VELOCITY MODELS FOR WASHINGTON AND OREGON

	Depth Range (km)	Velocity (km/sec)
Cascades Area (C3)		
	0.0 - 1.0	5.1
	1.0 - 10.0	6.0
	10.0 - 18.0	6.6
	18.0 - 34.0	6.8
	34.0 - 43.0	7.1
	43.0 - ∞	7.8
Southeastern Washington and Eastern Oregon (E3)		
	0.0 - 0.4	3.70
	0.4 - 8.5	5.15
	8.5 - 13.0	6.10
	13.0 - 23.0	6.40
	23.0 - 38.0	7.10
	38.0 - ∞	7.90
Northeastern Washington (N3)		
	0.0 - 0.5	5.1
	0.5 - 14.0	6.1
	14.0 - 24.0	6.4
	24.0 - 38.0	7.1
	38.0 - ∞	7.9
Western Washington (P3)		
	0.0 - 4.0	5.40
	4.0 - 9.0	6.38
	9.0 - 16.0	6.59
	16.0 - 20.0	6.73
	20.0 - 25.0	6.86
	25.0 - 41.0	6.95
	41.0 - ∞	7.80
Mt. Saint Helens Area (S3)		
	0.0 - 2.2	4.6
	2.2 - 3.4	5.1
	3.4 - 6.0	6.0
	6.0 - 10.0	6.2
	10.0 - 18.0	6.6
	18.0 - 34.0	6.8
	34.0 - 43.0	7.1
	43.0 - ∞	7.8

tral region, Rogers (1983a) has suggested that the 1909 earthquake was deep.

The 1936 Milton-Freewater earthquake is the largest known event in the eastern Washington region. Maximum intensity was reported as VII (Coffman and others, 1982), and the surface wave magnitude has been calculated as 5.75 (Gutenberg and Richter, 1954); the magnitude estimated from felt area has been calculated to be 6.4 by Noson and others (1988). Because all earthquakes in eastern Washington located since 1970 are in the crust, the 1936 event is also assumed to be crustal. Numerous aftershocks were felt. In 1979 a shallow M_c 4.3 (M_c denotes coda length magnitudes determined by the WRSN) earthquake was located in the same epicentral area as the 1936 event (Fig. 4c).

Between 1939 and 1965, four earthquakes (in 1939, 1946, 1949, and 1965) greater than magnitude 6 (magnitudes based on felt areas) occurred in the southern Puget Sound basin (Rogers, 1983a; Fig. 3). The 1949 (m_b = 7.1) and the 1965 (m_b = 6.5) earthquakes caused significant damage in the Puget Sound region (Murphy and Ulrich, 1951; Nuttli, 1952; Algermissen and others, 1965) and had instrumentally determined hypocentral depths of 54 and 60 km, respectively (Baker and Langston, 1987; Algermissen and others, 1965). Eight people were killed in the 1949 event (Ulrich, 1949), and six died in the 1965 earthquake (Algermissen and others, 1965). No aftershocks were felt or recorded after the 1949 earthquake; instrumentation available at the time would have detected events larger than magnitude 4.5. Similarly, following the 1965 earthquake, no aftershocks were felt, and an examination of seismograms recorded on stations operating within the region failed to identify any aftershocks greater than magnitude 2.5.

Earthquakes of magnitude 4 or greater, 1960 to 1987.
Fig. 4a shows earthquakes of magnitude 4 or greater from 1960 through 1987. Figures 4b through d divide the seismicity in Figure 4a into three plots, covering 1960–1969, 1970–1979, and 1980–1987. For the 1960s, when depth determination was somewhat uncertain, earthquakes are shown as shallow unless hypocenters are well confirmed as deeper than 35 km. Generally, locations in the 1960s are less reliable than during the 1970s and 80s.

TABLE 4. MAGNITUDES INCLUDED IN THE WASHINGTON-OREGON EARTHQUAKE CATALOG

Date	Minimum Magnitude	Estimated Completeness Level
Before 1917	5.75	6.5 (before 1900)
		6.0 (after 1900)
1917-1939	5.25	6.0
1940-1955	4.75	5.5
1956-1964	4.25	5.0
1965-1970	3.75	4.25
1970-present	2.5	4.0

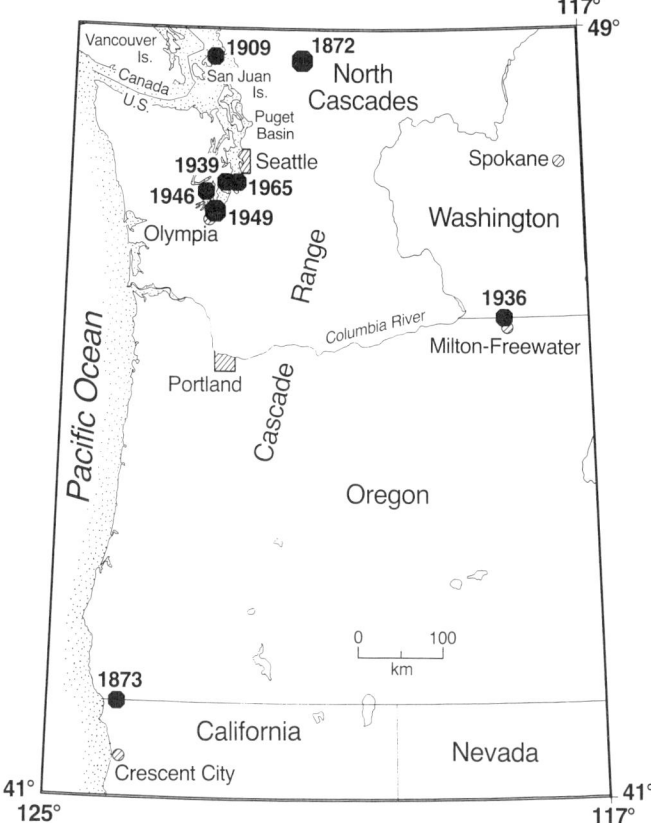

Figure 3. Map showing epicenters of largest known earthquakes in the Pacific Northwest; magnitudes estimated from felt areas to be larger than 6.

Comparing Figure 4a to Figure 3, the most striking difference is in southwestern Washington; since 1960 all four of the crustal earthquakes greater than magnitude 5 west of the Cascades have occurred here, yet there are no magnitude 6 events known in this region. The southern Oregon coast presents the contrary case, in that there are no known events greater than magnitude 4 located here since 1960 despite occurrence of the 1873 event (Fig. 3).

The decade plots (Figs. 4b through d) clearly indicate that, over several decades, the spatial distribution of earthquakes at the magnitude 4 and greater level in Washington and Oregon varies with time. A similar variation for the Portland, Oregon, area has been previously noted by Couch and others (1968). Seismic activity near Portland was concentrated in the 1960s (Fig. 4b). Only a few earthquakes of magnitude 4 or larger were located in Oregon from 1970 to 1987, all east of the Cascades (Figs. 4c and d).

During the 1960s, the four confirmed events in the Juan de Fuca plate span five degrees of latitude, ranging from the central Willamette Lowland to British Columbia. This decade includes the southernmost deep event, near the central Oregon coast (Tobin and Sykes, 1968). In the 1970s, earthquakes in the Juan de Fuca plate were more limited in latitude, spanning only two

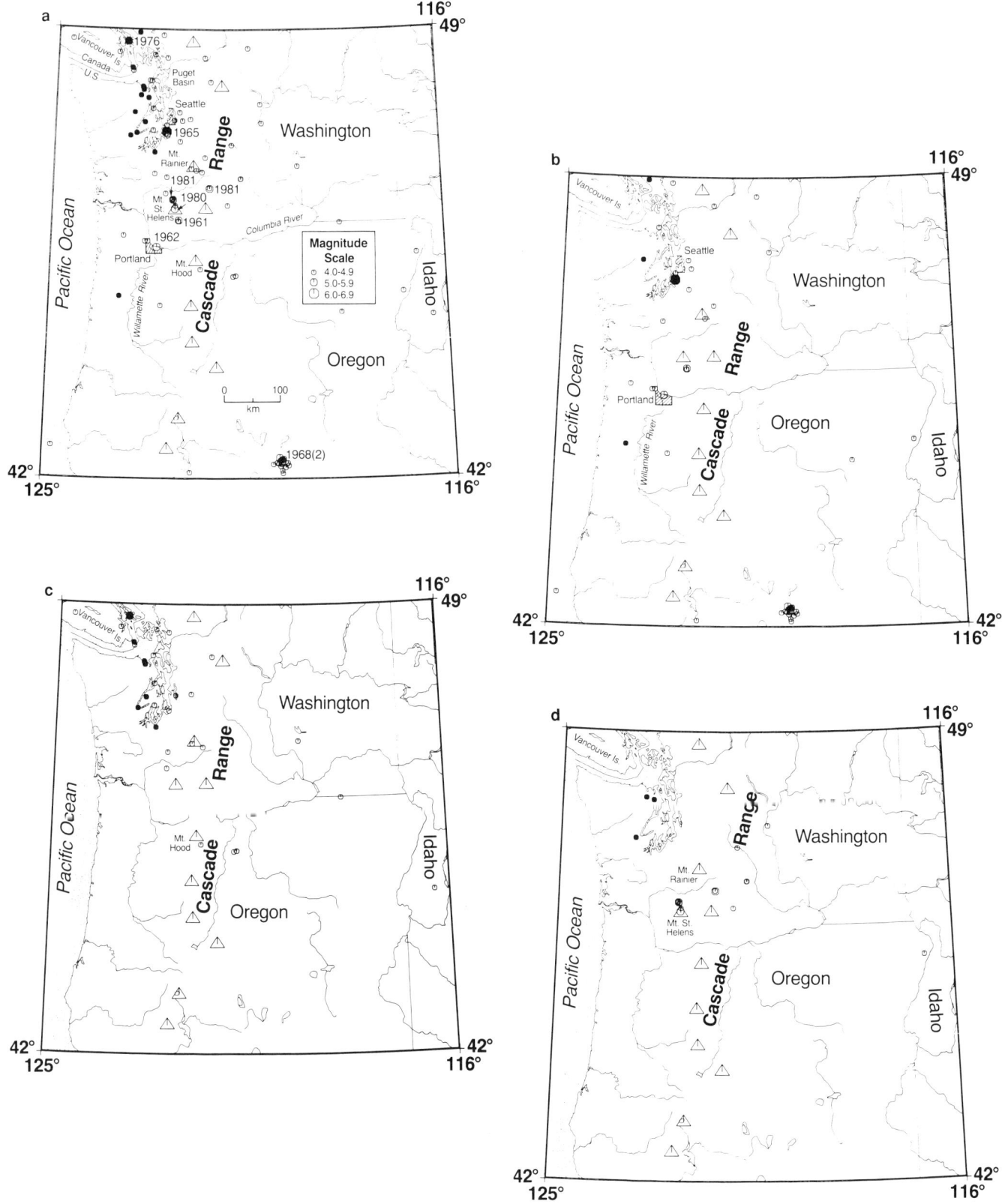

Figure 4. Earthquakes with magnitudes equal to or larger than 4. Symbols are scaled according to earthquake magnitude. Unfilled symbols show earthquakes shallower than 35 km, while filled symbols indicate earthquakes at depths of 35 km or greater. (a) 1960–1987, (b) 1960–1969, (c) 1970–1979, (d) 1980–1987.

degrees (Fig. 4c). Between 1980 and 1987, deep earthquakes were confined to the central Puget basin, within about 1° of latitude (Fig. 4d). Within the Puget basin, the rate of occurrence of subcrustal earthquakes has been relatively consistent.

Crustal earthquakes of magnitude ≥4.0 west of the Cascade arc (Fig. 4a) show similar changes in spatial distribution over the decades. Crustal seismicity in the 1960s, like the deep earthquakes during the same period, stretched from the central Willamette Lowland to southern Vancouver Island (Fig. 4b). In the 1970s, shallow activity occurred between Mount Hood in the south and British Columbia, and was confined to a narrower band longitudinally (Fig. 4c). Further concentration continued from 1980 through 1987, with all crustal seismicity west of the Cascades confined to a small area between Mount St. Helens and Mount Rainier (Fig. 4d).

The narrowing areas of seismic activity west of the Cascades, both subcrustal and crustal, most probably represent a short-term variation, and illustrate the nonstationary nature of Pacific Northwest seismicity on the scale of decades. Clearly, long-term observations are needed to produce realistic estimates of earthquake source regions in Oregon and Washington.

East of the Cascades, the most notable activity was a swarm of shallow earthquakes in the Basin and Range province in southeastern Oregon (Fig. 4b) in 1968, which included two events of magnitude 5 or greater. Other seismicity at the magnitude 4 level displays no obvious pattern, although five earthquakes were located on the eastern flank of the Cascades from 1980 through 1987.

Earthquakes magnitude 2.5 and greater, 1970 to 1987. In Figure 5 we show all instrumentally located earthquakes since 1970 that are greater than magnitude 2.5, plus earlier earthquakes that have adequate instrumental locations. For clarity, aftershocks of the 1981 Elk Lake (M_L 5.5) earthquake smaller than M_c 4.0 have been removed. Grant and others (1984) reported locations for more than 1,000 aftershocks in the 24 months following the mainshock. Two prominent differences with the patterns observed at the magnitude 4.0 and greater level can be noted. First, the area of deep earthquakes seen in Figure 5 has increased relative to the area of deep activity in Figure 4a. Hypocenters of the 1949 and 1965 earthquakes fall within the deep suite of events, and a few deep events have been located as far south as the northern Oregon Coast Range. Second, a prominent zone of earthquakes can be seen to strike north-northwest through Mount St. Helens (Fig. 5). This distribution of events includes magnitude 5+ earthquakes in 1961 and 1981 (Figs. 4b and d) and is known as the St. Helens Seismic Zone (SHZ) (Weaver and Smith, 1983). East of the Cascade Range, seismicity remains scattered, and even at smaller magnitudes there are few earthquakes located in the Mesozoic terranes of the Okanogan Highlands and Blue Mountains.

Microearthquakes and seismotectonics

Although the distribution of earthquakes changes little when events with magnitudes less than 2.5 are added, these smaller magnitude events do help to unambiguously define both crustal seismic zones and the general shape of the dipping Juan de Fuca Benioff zone. Accordingly we have plotted in Figure 6 the seismicity from Figure 5 plus the best-located earthquakes of magnitudes less than 2.5 in western Washington and northwestern Oregon from 1970 through 1987. The lineament of shallow earthquakes defining the SHZ is very prominent; a second, less distinct lineation of earthquakes is west of Mount Rainier (Fig. 6). The Puget Sound basin region is distinct in that there is not an obvious alignment of seismicity that suggests a seismic zone. Despite the fact that the SHZ is well defined by seismicity, neither it nor the lineation west of Mount Rainier correlates with mapped Quaternary surface faults. In fact, seismicity in western Washington has not been correlated with mapped surface faults, at least partly because thick Pleistocene glacial sediments in the Puget Sound region, along with thick vegetation, make active faults difficult to identify.

The structure of the Juan de Fuca plate beneath western Washington has previously been interpreted to include an upward arch of the plate (Crosson and Owens, 1987; Weaver and Baker, 1988). The arch structure represents a change in the azimuth of plate dip, with the plate dipping to the northeast beneath the northern Puget Sound basin and dipping east-southeast beneath southern Puget Sound and southwestern Washington. If this change in the geometry of the plate is not taken into account, cross-section plots of seismicity fail to show well-defined Benioff zones. Accordingly, for cross-section plots we have chosen areas on either side of the interpreted arch of the Juan de Fuca plate (Fig. 6).

With respect to the Benioff zone seismicity, the cross sections (Figs. 7a and b) show significant differences between the northern and southern areas, both in the dip of the Benioff zone and in the frequency of occurrence of subcrustal earthquakes. In the Puget basin (Fig. 7b), subcrustal earthquakes are much more frequent than beneath southwestern Washington (Fig. 7a). Referring to Fig. 6, the southern cross section (Fig. 7a) encompasses twice the area of the northern section (Fig. 7b), but shows far fewer subcrustal earthquakes. However, because the 1949 earthquake occurred in the southern section, the seismic moment release of this region for the period since at least 1870 dwarfs that of the northern section.

Finally, to define the seismotectonic provinces within the Pacific Northwest, we have plotted a representative selection of focal mechanisms (Fig. 8; Table 5). These focal mechanisms, along with distribution of seismicity (Figs. 4 through 7) and the prevailing style of tectonics inferred from geologic observations, allow us to define five distinct tectonic provinces in Oregon and Washington; the data do not allow us to subdivide large areas of Oregon and Washington that are lacking contemporary or historical seismicity (the Okanogan Highlands is one example). Our tectonic provinces are: (1) the subducting Juan de Fuca plate; (2) crustal northwestern Washington (earthquakes above 30 km); (3) crustal southwestern Washington and northwestern Oregon, including the Mount St. Helens region; (4) the Columbia Plateau

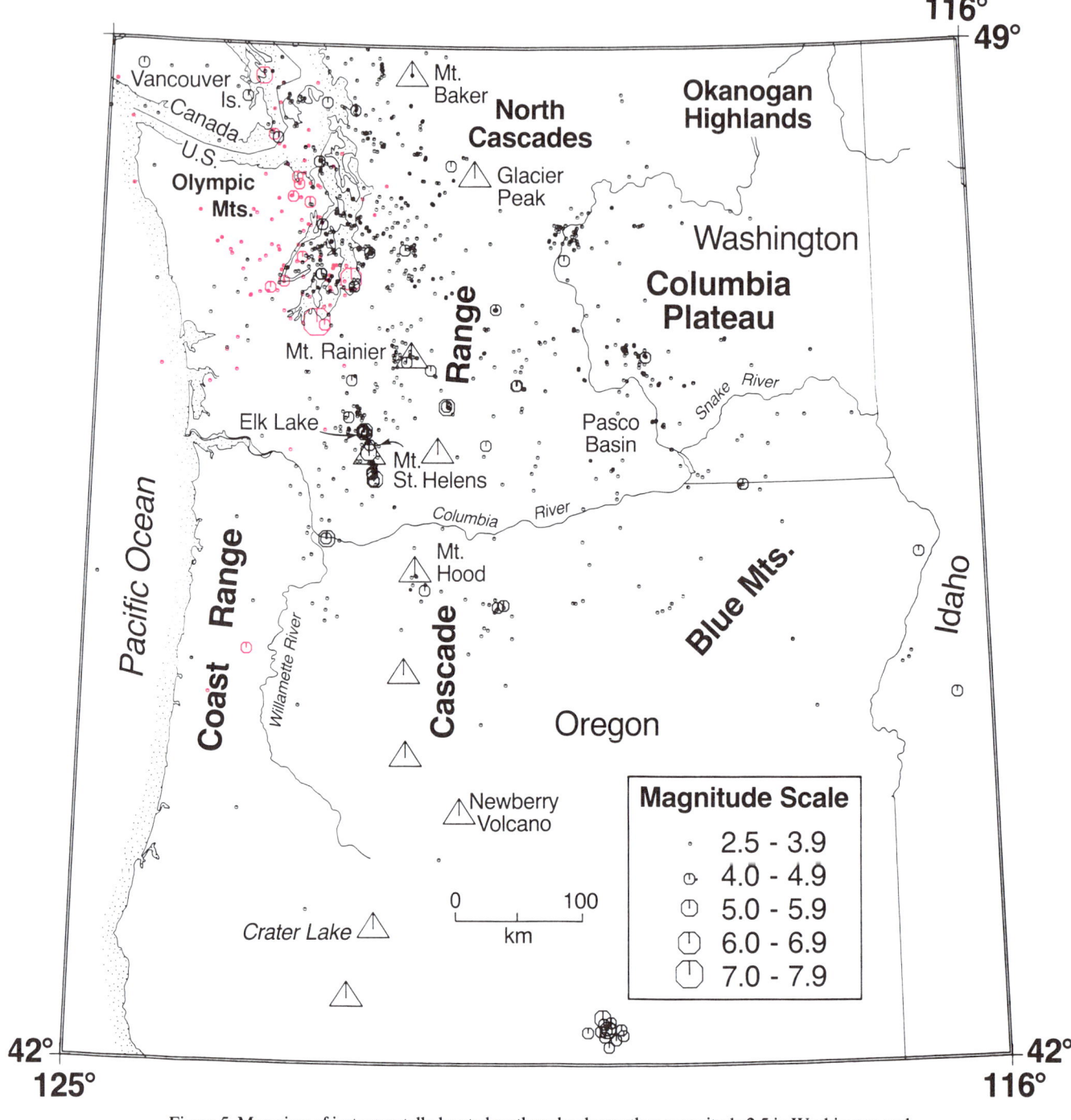

Figure 5. Map view of instrumentally located earthquakes larger than magnitude 2.5 in Washington and Oregon. Symbols are scaled proportionally to earthquake size for events larger than 4.0 (each symbol size represents one magnitude unit; 4.0–4.9, 5.0–5.9, 6.0–6.9, 7.0–7.9). From magnitude 2.5 to 3.9, all events have the same symbol size. Earthquakes shallower than 35 km have black symbols; deeper earthquakes are shown in red. All events since 1970 located by the WRSN with magnitudes larger than 2.5 are included. Some earlier earthquakes larger than magnitude 4.0 also have adequate instrumental locations. These include the 1949 and 1965 Puget Lowland earthquakes, the 1962 Portland earthquake, the Warner Valley sequence of 1968 in southern Oregon, and the Swift Reservoir earthquakes south of Mt. St. Helens in 1960 and 1961. An unusual deep earthquake in Oregon in 1963 (magnitude 4.3 [Tobin and Sykes, 1968]) is also shown. Volcanic earthquakes at Mt. St. Helens have not been plotted. Aftershocks of the 1981 Elk Lake earthquake smaller than magnitude 4 have also been omitted.

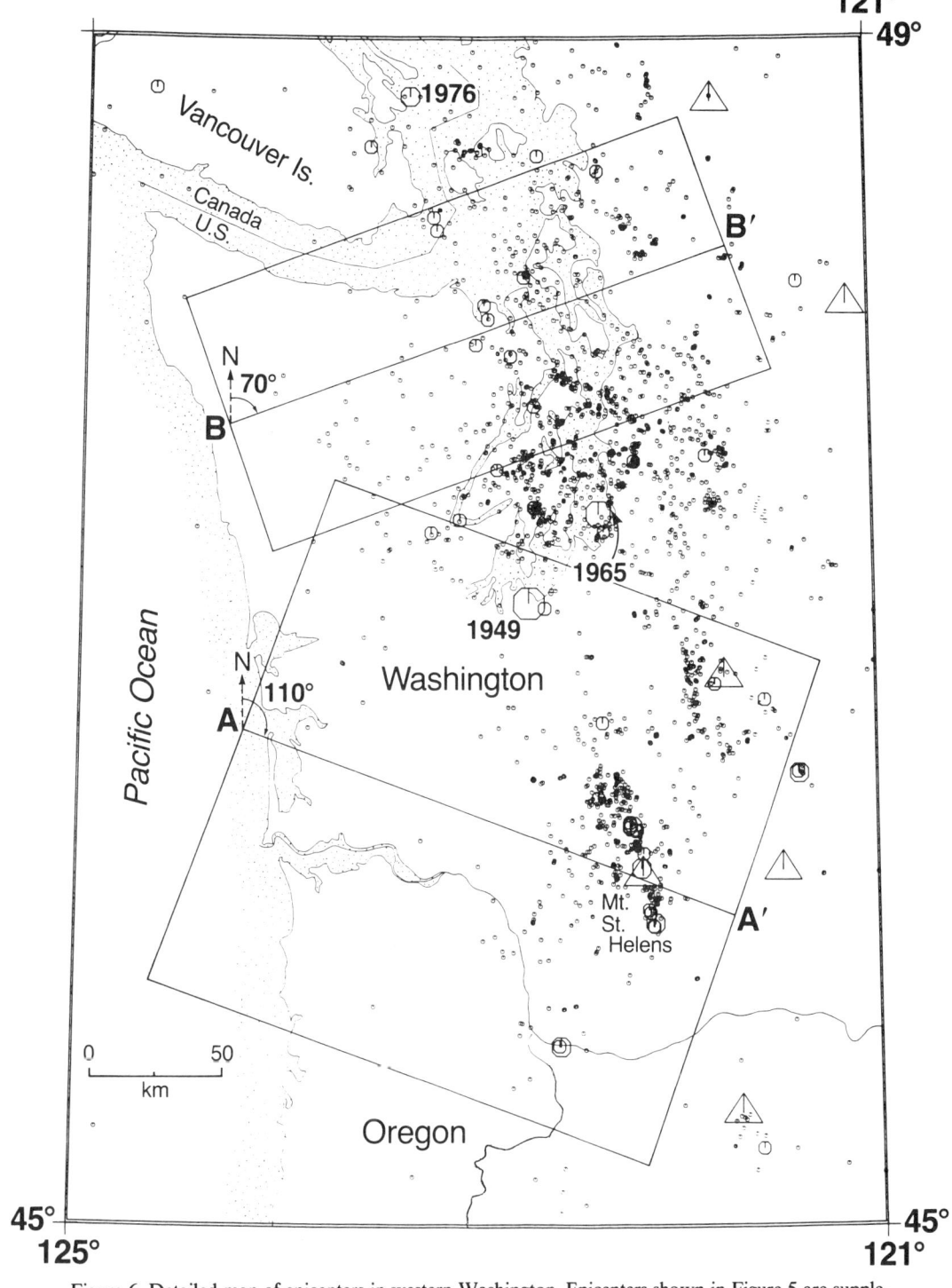

Figure 6. Detailed map of epicenters in western Washington. Epicenters shown in Figure 5 are supplemented by the addition of epicenters of best-located earthquakes regardless of magnitude since 1970. Each event since 1970 met the following criteria; at least 5 stations and 8 phases read, azimuthal gap smaller than 100°, nearest station no farther than 40 km, WRSN quality factors B or higher, and events with problem depths excluded. Earthquakes at Mt. St. Helens were omitted except for unusual earthquakes deeper than 3 km in April and May, 1980. Aftershocks of the 1981 Elk Lake earthquake (M_L 5.5) smaller than magnitude 3 were also excluded. For earthquakes larger than 4.0, symbols are scaled as in Figure 5. Earthquakes smaller than magnitude 4 have a single symbol size. Boxed areas labeled A and B correspond to cross sections 7a and b. Significant deep earthquakes in 1949, 1965, and 1976 are labeled.

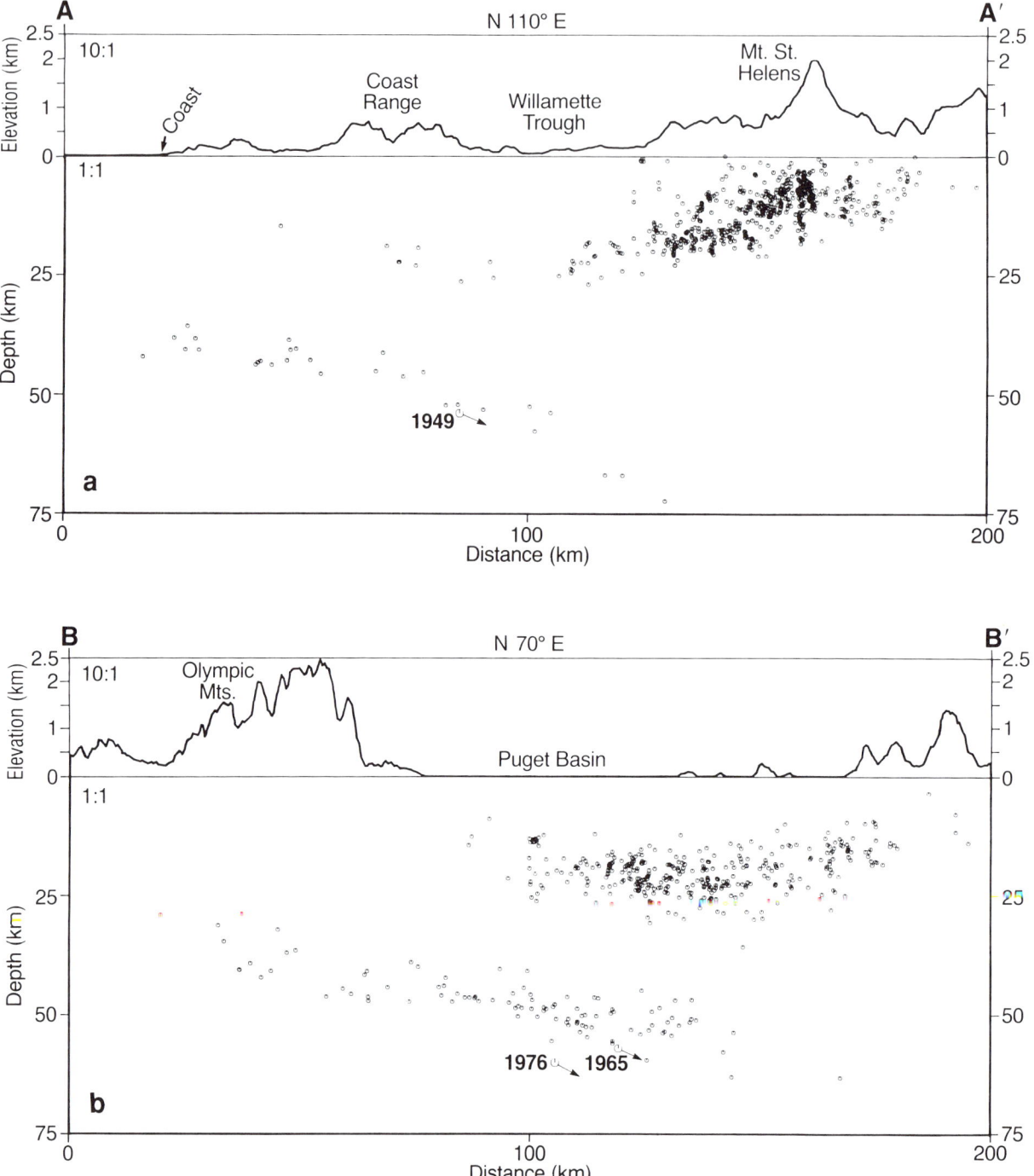

Figure 7. a. Southwestern Washington cross section A-A' (Fig. 6); 10:1 vertical exaggeration of topography; no vertical exaggeration of subsurface. Best-located (criteria in Fig. 6) earthquake hypocenters since 1970, projected onto a vertical plane striking N110°E, are shown by a single symbol size. The hypocenter of the damaging 1949 Olympia earthquake is within this distribution, and is plotted as a larger symbol with a vector representing the extensional axis from its focal mechanism (Baker and Langston, 1987) shown. The profile of Mt. St. Helens shown is prior to the May 18, 1980 explosion, which lowered the summit by 0.4 km. b. Northwestern Washington cross section B-B' (Fig. 6); 10:1 vertical exaggeration of topography; no vertical exaggeration of subsurface. Best-located (criteria in Fig. 6) earthquake hypocenters since 1970, projected onto a vertical plane striking N70°E, are shown by a single symbol size. The 1965 and 1976 earthquakes are shown as larger symbols with vectors representing extensional axes from focal mechanisms. These earthquakes were located to the south and north, respectively, of the cross-section area (see Fig. 6).

region of eastern Washington and eastern Oregon; (5) the large area of crustal extension in southeastern Oregon; and (6) the seismically quiet regions of northeastern and easternmost Washington, and central and southwestern Oregon.

Subducting Juan de Fuca plate. As noted above, Wadati-Benioff zones are clearly evident from the two cross sections (Figs. 7a and b). In the southern section (Fig. 7a), the dip of the hypocenter distribution in the subducting slab clearly steepens toward the southeast. Near the coast, the hypocenters dip at 10 to 12°, whereas west of Mount St. Helens, the dip is almost 20°. The 1949 magnitude 7.1 earthquake occurred at a depth of 54 km within the volume defined by the earthquakes (Fig. 7a). The focal mechanism determined for this event by Baker and Langston (1987) is strike-slip (Fig. 8, number 1), with the preferred fault plane striking east-west. The T-axis is oriented to the southeast with a plunge of ~20°, parallel to the dip of slab hypocenters in southwestern Washington (indicated in Fig. 7a). This suggests that the 1949 earthquake was related, at least in part, to down-dip tensional forces within the subducting plate (Baker and Langston, 1987; Weaver and Baker, 1988).

In contrast to the increase in plate dip defined by the hypocenters in Fig. 7a, the distribution of earthquakes in the subducting plate in the northern cross section (Fig. 7b) is approximately planar, dipping 10 to 12° to the northeast across the entire section. Two moderate-to-large earthquakes, which are located just north and south of the cross section, have reliable focal mechanisms determined and are included in Figure 7b. These are the 1965 m_b 6.5 south Seattle earthquake (Figs. 3, 4a, 4b, 5; Fig. 8, number 4) and the 1976 m_b 5.1 Pender Island (British Columbia) earthquake (Figs. 4a, 4c; Fig. 8, number 11). Both earthquakes had nearly identical focal mechanisms showing normal faulting on NNW-striking fault planes, and both T axes dip to the northeast at approximately 25 to 35° (Fig. 7b; Rogers, 1983b). As this dip is steeper than that defined by the hypocenters beneath Puget Sound (Fig. 7b), Rogers (1983b) suggested that these deep events occurred near the position where the Juan de Fuca plate dip is expected to steepen.

Crustal northwestern Washington. Crustal seismicity in the Puget Sound area (Fig. 7b) is generally confined to depths between 10 and 30 km. The distribution of shallow earthquakes describes a concave-upward bowl shape, with little variation in earthquake density within the selected area. In this section, the deepest crustal earthquakes occur near the northeastern limit of the subcrustal earthquakes. Crustal focal mechanisms in western Washington, from the Canadian border as far south as Mount Rainier and Goat Rocks (located south and east of Mount Rainier), typically have thrust or strike-slip mechanisms with nearly horizontal P axes oriented approximately north-south (Fig. 8, numbers 7, 9, 12, 15, 16, and 20). To the south of Mount Rainier and west of the Cascade crest, the thrust mechanisms cease, and all crustal earthquakes have strike-slip focal mechanisms (Fig. 8, numbers 2, 3, 7, 9, 13, 14, and 15).

Crustal southwestern Washington and northwestern Oregon. In contrast to the bowl-like distribution of crustal seismicity in the northern cross section, the southern cross section (Fig. 7a) describes a sort of cornucopia shape. The most westerly earthquakes are at depths of 10 to 20 km, and sparsely distributed. To the southeast, the sparse distribution of crustal earthquakes deepens to a maximum of about 25 km above the area where the dip of the subducting slab appears to steepen southeastward. After reaching the maximum depth, crustal earthquakes become more numerous, and the distribution shallows, with some hypocenters at depths of 2 km or less. The shallowest part of the distribution is near the Cascade volcanic arc.

A majority of the crustal seismicity in Figure 7a is concentrated along the St. Helens Seismic Zone (SHZ). Although the SHZ does not appear as a linear feature in this cross section because of the section orientation, it is a near-vertical strike-slip fault zone, and if viewed along strike, plots as a linear feature extending to a depth of about 12 km (Grant and others, 1984). Along the SHZ (prominent in Figs. 5 and 6), on the north side of Mount St. Helens away from the flanks of the volcano, the strike of the epicenters and focal-mechanism fault planes indicates that individual fault segments strike nearly northerly (Fig. 8, number 14). South of the volcano, the strike of epicenters and focal mechanisms suggests fault planes that strike northwest (Fig. 8, number 2). Most focal mechanisms in this area have northeast-southwest P axes (Weaver and Smith, 1983; Grant and others, 1984; Grant and Weaver, 1986). The two largest tectonic earthquakes to occur along the SHZ since 1960 are the 1961 Siouxon Peak earthquake (M_L = 5.1) and the 1981 Elk Lake earthquake (M_L = 5.5) (Fig. 8, numbers 2 and 14). Both the 1961 and 1980 mainshocks were preceded by precursory swarms that contained several magnitude 4 earthquakes and began about 8 months prior to the mainshock. After several months without activity, foreshocks occurred within several days of both of the mainshocks. Both earthquakes were followed by aftershock sequences.

Mount St. Helens is located at an apparent right-stepping offset in the SHZ, and Weaver and others (1987) have proposed that the volcano is located at a small volume of the crust that is under local extension. Directly beneath the cone of Mount St. Helens, and visible in Figure 7a, is a vertical alignment of hypocenters extending to depths of 20 km. These earthquakes, ranging in depths from 3 to 20 km, occurred in May and June of 1980, beginning just after the catastrophic eruption of May 18, 1980. Focal mechanisms for earthquakes at Mount St. Helens show preferred nodal planes that are generally either north-south or northwest-southeast; consistent with the strike of the SHZ both north and south of the volcano (Weaver and others, 1987). Detailed descriptions of seismic and volcanic activity at Mount St. Helens can be found in Malone and others (1981), Weaver and others (1981), Endo and others (1981), Swanson and others (1983), Shemeta and Weaver (1986), Qamar and others (1987), and Weaver and Malone (1987).

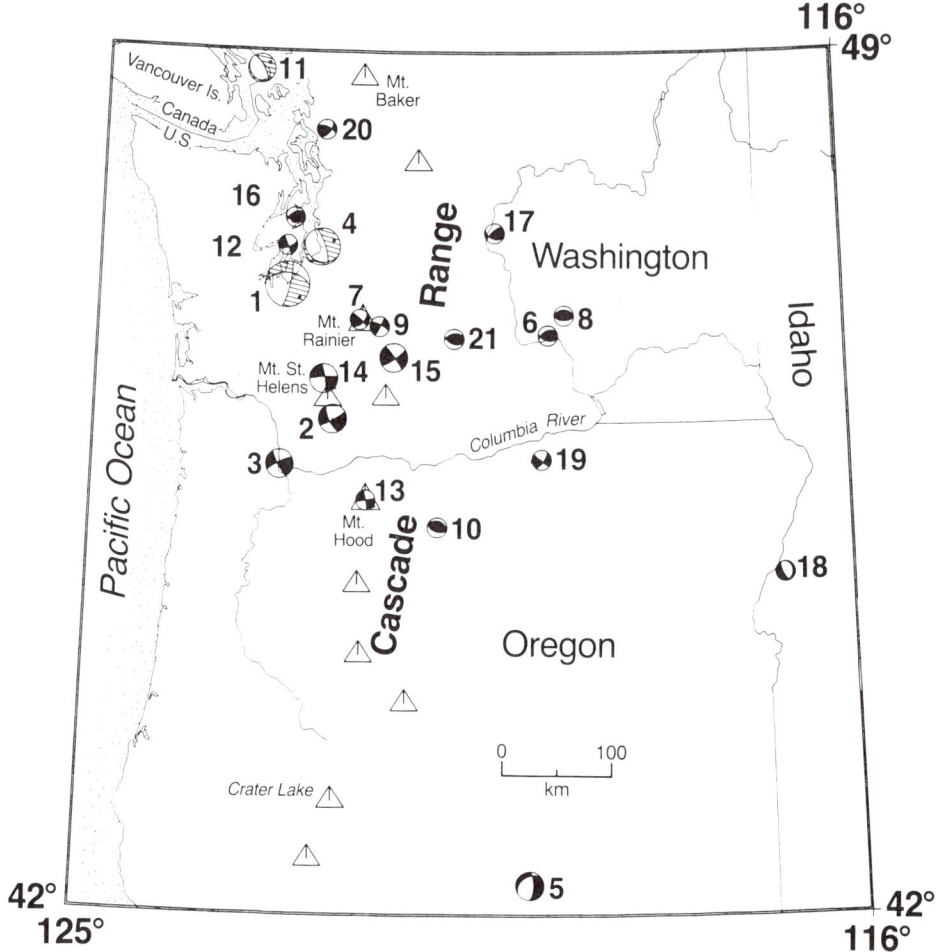

Figure 8. Map showing lower-hemisphere focal mechanisms of representative large and moderate earthquakes. Shaded quadrants represent compressional first arrivals, and numbers correspond to Table 5. Mechanisms with solid fill are for earthquakes with depths less than 35 km, while those with striped fill represent deeper earthquakes. Symbols are scaled to earthquake magnitude, each symbol size represents one magnitude range (3.0–4.9, 5.0–5.9, 6.0–6.9, 7.0–7.9).

Columbia Plateau and adjacent eastern Cascade Range in Washington. East of the Cascade crest in Washington and northern Oregon there are no earthquakes located within the Benioff zone, and all earthquakes locate within the mid to upper crust. More than 90 percent of the earthquakes occur at depths less than 8 km, and there are no events located deeper than 25 km. Although several magnitude 4+ earthquakes have been located in the Columbia Plateau east of the northern Oregon Cascade Range, the great majority of earthquakes occur around the Pasco basin (Fig. 9). The shallow depths of most earthquakes in eastern Washington mean that their hypocenters lie within the Columbia River Basalt Group, a sequence of flood basalts of Miocene age that thickens to a maximum depth of 6 km in the central Pasco basin (Figs. 5 and 9). The shallowest events frequently occur in earthquake swarms isolated in both space and time. Swarms occur in restricted volumes only a few kilometers in diameter and last a few weeks to several months. Some swarms appear spatially associated with east-west–trending anticlines, such as the steeply dipping north limb of the Saddle Mountains anticline, possibly along short segments of a high-angle reverse fault. There are few other associations of seismicity with mapped surface structures and no indication of horizontal alignments of earthquakes that would indicate major segments of near-vertical fault systems. At depths greater than 8 km, earthquakes occur as isolated events, rather than in swarms. One area with persistent earthquake activity since monitoring began in 1970 is located along the eastern boundary of the Cascade province, south of Lake Chelan (Fig. 9). Earthquakes here are scattered approximately uniformly over a 20 × 30 km region; hypocenters vary from near surface to a depth of 10 km. Despite the presence of mapped faults of Miocene age and older in the vicinity of Lake Chelan, no alignment of earthquakes has been observed coincident with these faults.

Throughout the Columbia Plateau in eastern Washington,

TABLE 5. FOCAL MECHANISMS

	Year/ Month/Day	Latitude (N)	Longitude (W)	Depth (km)	M	P AZ (°)	P PL (°)	T AZ (°)	T PL (°)	References*
1	1949/4/13	47°06.00'	122°42.00'	54.00	7.1	244	35	138	21	Baker and Langston, 1987; Murphy and Ulrich, 1951†
2	1961/9/17	46°01.36'	122°07.34'	7.20	5.1	16	21	110	10	Grant and Weaver, 1986
3	1962/11/16	45°36.48'	122°35.87'	18.00	5.1	18	0	108	14	Grant, personal communication
4	1965/4/29	47°24.00'	122°18.00'	60.00	6.5	274	60	63	26	Isacks and Molnar, 1971; Algermissen and others, 1965†
5	1968/6/3	42°15.00'	119°48.00'	20.00	5.0	233	58	111	18	Patton, 1985; Couch and Johnson, 1968†
6	1971/10/25	46°42.48'	119°32.87'	3.99	3.8	174	0	82	67	WRSN
7	1973/7/18	46°49.61'	121°48.84'	6.10	4.0	5	20	262	31	Crosson and Frank, 1975; WRSN†
8	1973/12/20	46°52.03'	119°21.17'	2.30	4.4	180	7	356	83	Malone and others, 1975 (composite); WRSN†
9	1974/4/20	46°46.43'	121°34.02'	0.20	4.9	160	10	253	15	Crosson and Lin, 1975; WRSN†
10	1976/4/13	45°09.24'	120°51.66'	15.00	4.8	198	12	17	78	Couch and others, 1976
11	1976/5/16	48°47.99'	123°21.06'	60.00	5.1	250	60	57	29	Rogers, 1983b; WRSN†
12	1978/3/11	47°25.32'	122°43.08'	24.00	4.8	27	14	293	14	Yelin and Crosson, 1982; WRSN†
13	1978/9/6	45°22.26'	121°42.53'	2.08	3.4	35	0	125	0	Weaver and others, 1982
14	1981/2/14	46°20.93'	122°14.16'	7.28	5.5	220	10	130	2	Grant and others, 1984; WRSN†
15	1981/5/28	46°31.50'	121°23.63'	3.20	5.0	11	3	280	11	Zollweg and others, 1982; WRSN†
16	1981/11/26	47°38.75'	122°37.32'	21.90	3.5	174	0	84	67	WRSN
17	1984/4/11	47°32.09'	120°11.09'	8.0	4.3	346	14	93	49	Malone and others, 1987; WRSN†
18	1984/9/19	44°46.37'	116°49.02'	10.00	3.8	45	65	251	23	Zollweg and Jacobson, 1986 (composite); WRSN†
19	1985/2/10	45°42.23'	119°38.04'	18.41	3.9	353	21	260	7	Malone and others, 1987; WRSN†
20	1985/4/30	48°21.54'	122°15.80'	18.16	3.3	348	6	253	38	Ma, 1988; WRSN†
21	1987/12/2	46°40.75'	120°40.39'	17.80	4.3	186	3	85	74	WRSN

Latitude and longitude are given in degrees and minutes.
WRSN = Washington Regional Seismic Network.
*References given are to mechanism source. Several mechanisms are composites.
†Hypocentral source, if different from mechanism source.

thrust or reverse mechanisms are common. These mechanisms suggest that movement occurs on east-west–trending faults, although no earthquake has been correlated to a specific mapped fault. P-axes of these focal mechanisms are roughly horizontal and oriented north-south, and the direction of maximum principal stress has been inferred to be north-south (Malone and others, 1975). This stress direction is consistent with the east-west–trending anticlines mapped in the area (shown in Fig. 9; Shannon and Wilson Inc., 1977). The two focal mechanisms given for the Pasco basin (Fig. 8, numbers 6 and 8) are typical. Similar thrust-type earthquakes extend into north-central Oregon, where a M_c 4.8 mainshock (Fig. 5; Fig. 8, number 10) occurred near the Deschutes River in 1976. This focal mechanism again indicates thrust faulting on east-west–striking fault planes; Couch and others (1976) inferred a north-south direction of maximum principal stress, similar to that inferred for the Columbia Plateau. The cluster of seismicity just south of Lake Chelan is the one area along the eastern Cascade front in which the orientation of the P axes consistently varies from the north-south direction inferred on the Columbia Plateau. Mechanisms in this area (Fig. 8, number 17) suggest thrust faulting on northeasterly striking fault planes, with northwest-southeast P axes, and nearly vertical T axes.

A crustal refraction profile across the Columbia Plateau through the Pasco basin allows a comparison between details of the crust and the distribution of seismicity. This refraction line (C-C' in Fig. 9) crossed the series of east-west–striking thrust faults at acute angles and passed through the center of the Pasco

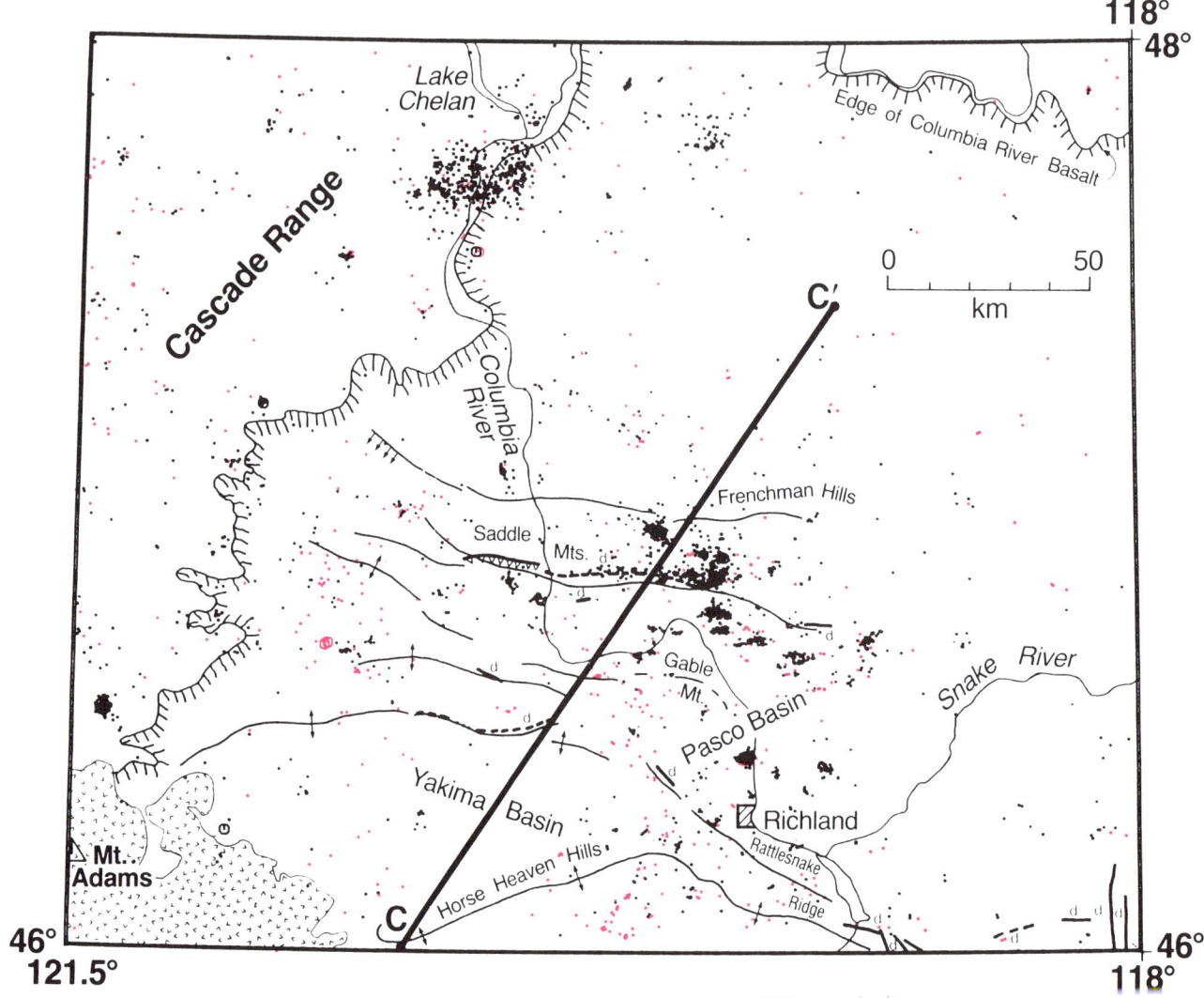

Figure 9. Map view of seismicity in eastern Washington, 1970–1987, magnitude 1.0 and larger. Earthquakes smaller than magnitude 4 are shown by a single symbol size; larger events are scaled according to magnitude (as in Fig. 5). Earthquakes shallower than 8 km are shown in black, those at depths of 8 km or more are colored red. Also shown are mapped faults of post-Miocene age (heavy lines) and axes of major anticlines (light lines; adapted from Shannon and Wilson, Inc., 1977). The western margin of the Columbia River Basalt Group is indicated by hachures; Quaternary volcanics near Mt. Adams are shown as a stippled pattern. Cross-section line C-C′ is indicated.

basin where the flows of the Columbia River Basalt Group (CRBG) are thought to be thickest.

The crustal structure (Fig. 10) interpreted by Catchings and Mooney (1988) along line C-C′ contains four features of interest with respect to the earthquake distribution:

1. A series of alternating layers of high and low velocity (5.85 and 5.3 km/s, respectively) that compose the CRBG. The high velocity represents the Miocene basalts, whereas the low velocity represents sedimentary deposits between flows (Catchings and Mooney, 1988). This integrated structure reaches its maximum depth near the center of the Pasco basin (Fig. 10).

2. A low-velocity layer beneath the CRBG that is interpreted by Catchings and Mooney as a pre-Miocene sedimentary basin.

3. The low-velocity layer is contained within a graben structure defined in the mid and upper crust (Fig. 10). The graben structure is about 70 km wide and has a vertical offset of about 4 km between the central axis and either side; the low-velocity sedimentary fill is thickest within the graben axis.

4. A lower crustal upwelling of velocity 7.5 km/s occurs beneath the entire graben structure, but reaches its highest crustal level of about 25 km directly beneath the center of the graben.

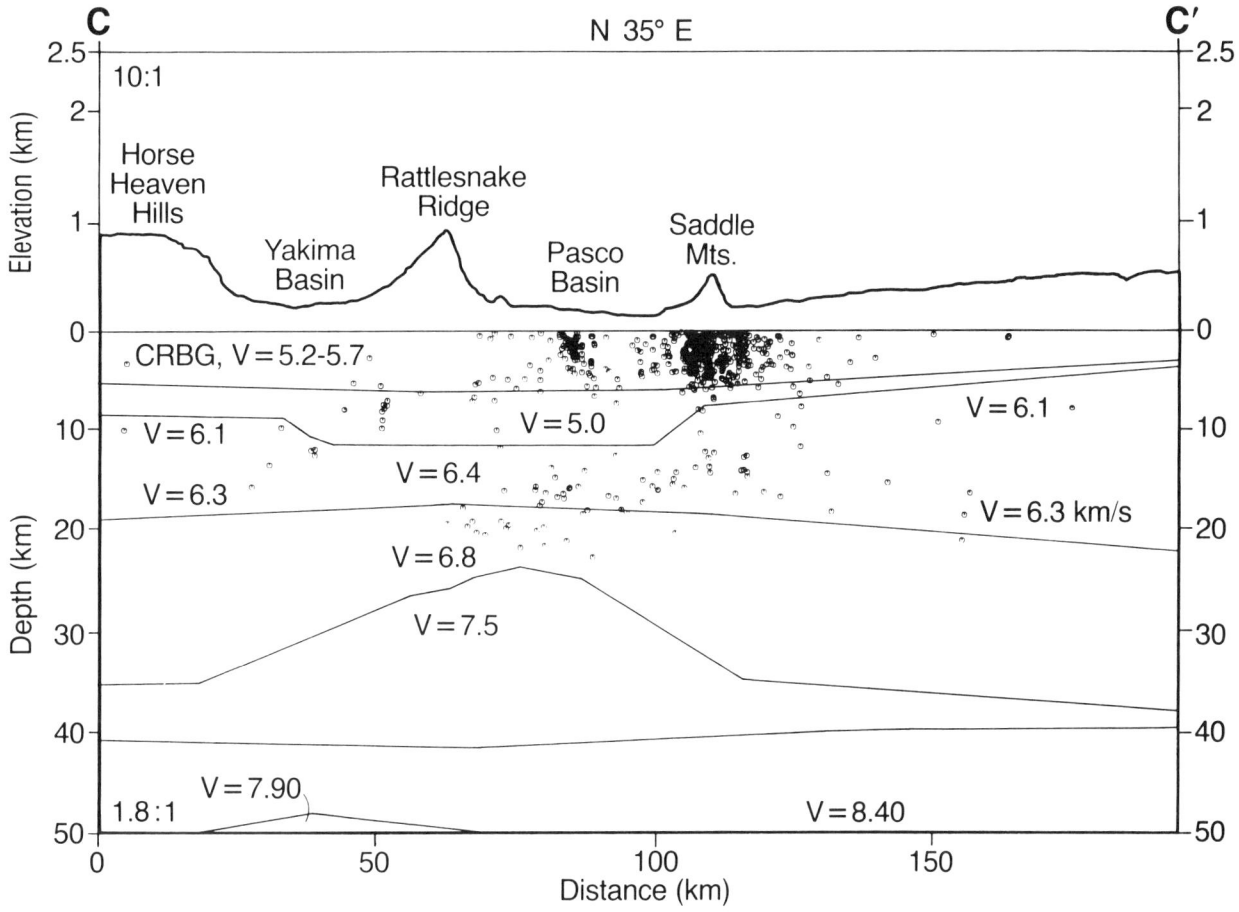

Figure 10. Eastern Washington cross section C-C' (Fig. 9); structural information from Catchings and Mooney (1988). 10:1 vertical exaggeration of topography; 1.8:1 vertical exaggeration of subsurface. The Columbia River Basalt Group is labeled CRBG. Best-located (criteria as in Fig. 7a) earthquake hypocenters since 1970, regardless of magnitude, are shown by a single symbol size. Earthquakes for 50 km on either side of line C-C' in Figure 8a are projected onto a vertical plane striking N35°E.

Taken together, these four elements are typical of the structure of continental rifts (Catchings and Mooney, 1988).

The earthquake distribution (Fig. 10) shows clearly the concentration of events within the CRBG. Most of the shallow crustal events within the CRBG are near the Saddle Mountains, which occur over the northeastern edge of the upper crustal graben. The low-velocity sediments (of velocity 5.0 km/s) within the graben are nearly devoid of earthquake hypocenters: the deeper crustal events occur in crystalline basement and sub-basement layers with velocities of 6.1 to 6.4 km/s and 6.8 km/s, respectively (Catchings and Mooney, 1988). The deepest crustal events, at about 22 km, occur above the shallowest extent of the lower crustal upwelling. Seismicity seems to be distributed unevenly about the upper crustal graben, with most of the events occurring on the northeast side of the graben. The graben, though predating the CRBG and the contemporary tectonics, may exert some influence on the shallow crust producing (or localizing) the Saddle Mountain system and contemporary seismicity, although the processes responsible for localizing the shallow thrusts and earthquakes are unknown. Likewise, there is a tendency for most of the earthquakes below the CRBG to occur beneath the northeastern half of the graben (Fig. 10).

Crustal southeastern Oregon. Inferred north-south horizontal compression in the Columbia Plateau and along the eastern Cascade Range of Washington changes dramatically to horizontal extension oriented in a westerly direction in the Basin and Range of southeastern Oregon (Zoback and Zoback, 1980). The dominant extensional earthquake sequence occurred in the Warner Valley near Adel, Oregon in 1968 (Fig. 5). This basin-and-range-type sequence (Couch and Johnson, 1968) included dual mainshocks of M_L 5.1 and 5.0, and 23 aftershocks greater than magnitude 3.5 (Couch and Johnson, 1968). The magnitude 5.1 mainshock is the largest known earthquake in Oregon east of the Cascade Range this century. The sequence was interpreted by Couch and Johnson as a typical Basin and Range earthquake swarm on a preexisting fault with Quaternary offsets in the Warner graben. The focal mechanism for the M_L 5.1 mainshock (Fig. 8, number 5) was recomputed by Patton (1985) using surface waves. It indicates a high-angle normal faulting mechanism with a T axis aligned approximately east-west. Patton (1985) noted that this mechanism matches the local geology of Warner Valley, where nearly vertical slickensides were interpreted by Couch and Johnson (1968) as suggesting recent high-angle normal faulting.

Two earthquakes of magnitude M_L 3.8 during 1984 were located in the southeastern Blue Mountains (Fig. 5; Fig. 8, number 18) near the Idaho border (Zollweg and Jacobson, 1986). A composite focal mechanism using these events as well as several smaller ones indicates normal faulting on a fault striking north-south or northwest. This normal faulting may indicate the presence of basin-and-range-style extension north of the Brothers fault zone, the geographic northern limit of the Basin and Range province.

Seismically quiet regions. No focal mechanisms are available either in easternmost Washington or in central and southwestern Oregon. Although there are very few earthquakes detected in these regions by the contemporary seismic network, the historical seismic record indicates that southwestern Oregon was the site of one of the largest earthquakes known in Oregon or Washington (Fig. 3).

SOURCES OF EARTHQUAKE HAZARD

The seismotectonic framework of the Pacific Northwest has three possible large-magnitude sources of seismic hazards: large-magnitude deep earthquakes, large shallow crustal earthquakes, and a great earthquake on the megathrust boundary. Deep, large intra-slab earthquakes beneath Puget Sound and vicinity are the most clearly identified seismic hazard. The maximum size of the deep intra-slab earthquakes is uncertain but has often been taken to be approximately 7.5 (Rasmussen and others, 1974). The largest deep earthquake known is the 1949 magnitude 7.1 event. However, similar earthquakes in other subduction zones have had magnitudes as large as M_W 8.0 (Astiz and others, 1988). The maximum size of intra-slab earthquakes probably depends at least partly on the elastic thickness of the slab. Since the elastic part of the Juan de Fuca slab is probably quite thin due to its young age and high temperature, an upper limit of 7.5 may be reasonable. Because damaging deep earthquakes are known only in the Puget Sound, it is uncertain whether these earthquakes can occur at other localities along the subduction zone.

The largest known earthquake in Washington and Oregon, the magnitude ~7.4 1872 earthquake, was probably shallow. Shallow earthquakes, with magnitudes larger than 5 and perhaps larger than 6 (Milton-Freewater earthquake of 1936), are well documented. Neither these, nor most other crustal earthquakes in Washington and Oregon, have been placed on mapped faults. Seismicity may be the only way to identify active fault zones in those parts of the Pacific Northwest that are heavily vegetated, have rugged topography and high relief, or are masked by extensive glacial deposits or alluvium. Well-located earthquakes since 1960 show considerable variation of seismic activity through time, and fault zones the length of the SHZ or longer may be revealed in the future, either through substantial earthquake sequences (as in the case of the SHZ) or by studies of microseismicity. Based on regression analysis of fault length and observed magnitude, the SHZ has been judged capable of generating an earthquake as large as magnitude 6.8 (Grant and Weaver, 1990). From the historical record, it is feasible that shallow earthquakes of magnitude 5 could occur almost anywhere in Washington or Oregon. However, based on historical and modern seismicity, the least likely areas to show such activity would be central Oregon, where significant earthquake activity is unknown, and the northeastern part of Washington in the Okanogan Highland. Despite uncertainties surrounding the location, depth, and tectonic source of the 1872 earthquake, if shallow crustal earthquakes of similar size were to be included in the hazards assessment of the major urban areas, the extent of earthquake hazard would be significantly increased. Currently, it is impossible to unequivocally rule out this possibility.

Finally, although it is the subject of active debate and investigation, the possibility of a great earthquake occurring on the Cascadia megathrust adds a new dimension to the issue of seismic hazards along the continental margin. Although a subduction zone has been recognized along the Pacific Northwest for some time (Atwater, 1970), only recently has the suggestion that great earthquakes may be possible here been seriously raised (Heaton and Kanamori, 1984; Heaton and Hartzell, 1987; Atwater, 1987). Heaton and Kanamori (1984), using a correlation between plate age and rate of subduction, argued that the Cascadia zone could sustain a great subduction earthquake of magnitude 8 or larger. Worldwide, the ability of a subduction zone to generate large thrust earthquakes is strongly correlated to convergence rate and the age of the subducted oceanic floor. The largest earthquakes globally are associated with rapid subduction of youthful oceanic crust (Ruff and Kanamori, 1980; Heaton and Kanamori, 1984). Although the Juan de Fuca plate has one of the slowest convergence rates, it is extremely young (<8 Ma at the trench).

Atwater (1987) interprets evidence of buried peats of late Holocene age from the coastal estuaries of southwestern Washington to be the consequence of coseismic deformation of the coast during large subduction earthquakes. Time coincidence of such subsidence over a broad stretch of coastline, coupled with geologic evidence for strong ground shaking, would provide convincing evidence of the existence of a subduction earthquake hazard along the Cascadia margin. However, before these earthquakes can be incorporated into the hazards assessment for the Pacific Northwest, considerable effort will be needed to narrow the uncertainty surrounding possible magnitudes and recurrence intervals for these events.

DISCUSSION

The seismicity catalog for Washington and Oregon indicates that the distribution of earthquakes in the Pacific Northwest has elements directly attributable to the subduction system: offshore seismicity at fracture zones and spreading ridges, a Wadati-Benioff zone beneath Washington and northern Oregon where downdip tensional faulting occurs within the subducting plate, and deep crustal earthquakes within the overlying North American plate. The subduction regime is also reflected in the physiographic provinces of Washington and Oregon that have been accreted to the continental margin since Tertiary time, and in the volcanism of the Cascade arc.

However, there are clearly limits to the extent of the effects of the subduction regime. East of the Cascade Range there is essentially no evidence of the contemporary subduction system that exists to the west. As an example of these limits, the extensional tectonics of the Basin and Range province in southeastern Oregon have developed since the late Miocene, and are distinct from the subduction tectonics to the west. Likewise, the tectonics of the Columbia Plateau, where evidence of post-Miocene north-south compression is abundant, would appear to be only weakly related to the northeastward convergence at the plate margin. Although the flood basalts of the Columbia River Basalt Group and the evidence of continental rifting beneath them may well be related to the subduction process, the seismicity distribution of the area east of the Cascades shows no clear connection to the subduction framework. Similarly, effects of the subduction regime are not easily linked to the 1872 North Cascades earthquake in an otherwise seismically quiet continental area. The lack of a link to subduction processes for the 1872 earthquake is emphasized by the sudden truncation of the crustal earthquake distribution just east of the Puget Sound basin (Fig. 7b).

West of the Cascade Range, there are distinctly different distributions of crustal earthquakes in northwestern Washington compared to southwestern Washington (Figs. 7a and b). Further, the persistent north-south axes of compression from strike-slip and thrust focal mechanisms calculated for crustal events in the Puget Sound basin contrast with the NNE- to NE-striking compressional axes found in the strike-slip regime of the southern Washington Cascade Range. If these observations reflect the effect of the subduction zone on the continental crust, then it is clear that there must be processes within the crust that allow this effect to be manifested so differently over such a restricted geographic area. Perhaps some of the complications seen in the crustal seismicity of northwestern Washington result from the uparched shape of the Juan de Fuca plate beneath the Puget Sound basin.

Understanding the current state of crustal strain west of the volcanic arc is important for understanding the influence of subduction processes on crustal seismicity. Unfortunately, the interpretation of strain data is ambiguous at present. Recent horizontal strain observations in Washington are limited mainly to trilateration networks in the Puget Sound basin and Olympic Mountains (Savage and others, 1981; Lisowski and others, 1987), and reoccupation of triangulation stations across the Strait of Juan de Fuca (Lisowski and others, 1989). Data for the Puget Sound region were interpreted to be consistent with the northeast compression expected from coupling across the subduction zone by Savage and others (1981). However, these early strain results were in apparent disagreement with the north-south principal compressive stress direction determined for the same region from earthquake focal mechanisms (Crosson, 1972; Yelin, 1982). As Crosson (1986) pointed out, significant time dependence of strain may complicate the interpretation. Sbar (1983) has also suggested that strain measurements need not reflect the principal axes of stress deduced from focal mechanisms since strain measurements reflect only incremental changes in stress/strain, and not the possibly larger ambient stress field. Further, as is apparent in Figure 7b, much of the central Puget Sound seismicity is at depths greater than 20 km, well below the surface strain measurements. Finally, although more recent analysis of strain in the Strait of Juan de Fuca region still suggests a northeast orientation of the direction of relative contraction (consistent with subduction zone coupling), the magnitude of strain accumulation does not appear to exceed the level of uncertainty (Lisowski and others, 1989). Thus, the picture is still unclear, and more extensive measurements over a longer time base are needed to characterize strain in this region, as well as better models to improve our understanding of stress/strain relations.

The significant temporal variations in the distribution of the earthquakes in our catalog are cause for concern in assessing the earthquake hazards of the region. It is noteworthy that two of the largest known earthquakes in the historical record, the 1872 North Cascade event and the 1873 earthquake near the Oregon-California border, occurred in areas lacking contemporary seismicity. The implication of these temporal variations for assessing earthquake hazards of the region, particularly when coupled with uncertainties surrounding the cause of the differences in the seismicity in western Washington, is that the possibility of large crustal earthquakes occurring in the major urban areas of Oregon and western Washington cannot be ruled out.

A longer baseline of seismic and geodetic data, a full three-dimensional model of the subduction zone, and a better grasp of plate rheology will help complete our understanding of the Juan de Fuca–North American plate interaction. Details of the shape and position of the subducting plate are becoming better known, and we expect that these can be combined with a more complete understanding of the crustal structure, and of the position and extent of the discrete crustal blocks that compose the western part of the North American plate, to form a more comprehensive understanding of the seismicity and tectonics of the Pacific Northwest.

CONCLUSIONS

Seismicity and tectonics of the Pacific Northwest have features consistent with continuing subduction: offshore seismicity at fracture zones and spreading ridges, a Wadati-Benioff zone beneath Washington and northern Oregon where down-dip tensional faulting occurs within the subducting plate, and active volcanism in the Cascade Range. Physiographic provinces of

Washington and Oregon, which have been accreted to the continental margin since Tertiary time, also reflect continued convergence over geologic time.

Other patterns of seismicity in Washington and Oregon are less clearly related to the subduction of the Juan de Fuca plate. The extensional tectonics of the Basin and Range province and the persistent north-south direction of compression from strike-slip and thrust crustal focal mechanisms in the Columbia Plateau, Puget basin, and the Cascade Range in central Washington suggest a crustal stress direction different from the N50°E convergence direction between the Juan de Fuca and North American plates. A link between the 1872 event and subduction tectonics is unclear, considering the location of the event within the pre-Cenozoic terrain of the North Cascades, in an area lacking contemporary seismicity. Further, large-framework features, such as the Columbia River Basalt Group of mid to late Miocene age, have not yet been convincingly related to the Cenozoic convergence tectonics.

In the Puget basin, much of the shallow microseismicity occurs in a volume above the crest of an arch in the Juan de Fuca plate (Crosson and Owens, 1987; Weaver and Baker, 1988), whereas microseismicity at depths of 35 to 60 km occurs in a slightly larger area, and large-magnitude earthquakes (1949 and 1965) at depths of 50 to 60 km have been located on the south flank of the arch. Since 1960, moderate-sized shallow crustal earthquakes have occurred in southwestern Washington on the SHZ (1961, 1981), southeast of Mount Rainier near Goat Rocks (1981), and near Portland, Oregon (1962). Farther south, much of central Oregon has been seismically quiet throughout the historical and instrumental record.

Observations of the distribution of seismicity in the Pacific Northwest indicate that the pattern varies considerably in time, making interpretations of tectonic activity and earthquake hazards difficult. The historical record, although brief (150 years), suggests that microseismic observations may not be adequate to identify seismic hazards, since two of the largest earthquakes in the region (1872 and 1873) apparently occurred in areas that are currently seismically quiet. Even at the magnitude 4 level, significant variations in spatial distribution are seen over a few decades. The catalog, which provides a snapshot of seismicity in geologic time, can provide only limited assistance in forecasting earthquake hazards in terms of human time.

REFERENCES CITED

Algermissen, S. T., Harding, S. T., Steinbrugge, L. V., and Cloud, W. K., 1965, The Puget Sound, Washington, earthquake of April 29, 1965: U.S. Department of Commerce, Coast and Geodetic Survey, 51 p.

Astiz, L., Lay, T., and Kanamori, H., 1988, Large intermediate depth earthquakes and the subduction process: Physics of Earth and Planetary Interiors, v. 53, p. 80–166.

Atwater, B. F., 1987, Evidence for great Holocene earthquakes along the outer coast of Washington State: Science, v. 236, p. 942–944.

Atwater, T., 1970, Implications of plate tectonics for the Cenozoic tectonic evolution of western North America: Geological Society of America Bulletin, v. 81, p. 3513–3536.

Baker, G. E., and Langston, C. A., 1987, Source parameters of the 1949 magnitude 7.1 south Puget Sound, Washington, earthquake as determined from long-period body waves and strong ground motion: Bulletin of the Seismological Society of America, v. 77, p. 1530–1557.

Barnard, W. D., 1978, The Washington continental slope; Quaternary tectonics and sedimentation: Marine Geology, v. 27, p. 79–114.

Beck, M. E., and Engebretson, D. C., 1982, Paleomagnetism of small basalt exposures in the west Puget Sound area, Washington, and speculations on the accretionary origin of the Olympic Mountains: Journal of Geophysical Research, v. 87, p. 3755–3760.

Cady, W. M., 1975, Tectonic setting of the Tertiary volcanic rocks of the Olympic Peninsula, Washington: U.S. Geological Survey Journal of Research, v. 3, p. 573–782.

Catchings, R. D., and Mooney, W. D., 1988, Crustal structure of the Columbia Plateau; Evidence for continental rifting: Journal of Geophysical Research, v. 93, p. 459–474.

Coffman, J. L., Von Hake, C. A., and Stover, C. W., eds., 1982, Earthquake history of the United States; revised edition: National Oceanic and Atmospheric Administration-U.S. Geological Survey Publication 41-1, 270 p.

Couch, R., and Johnson, S., 1968, The Warner Valley earthquake sequence; May and June 1968: Ore Bin, v. 30, p. 191–204.

Couch, R. W., Johnson, S., and Gallegher, J., 1968, The Portland earthquake of May 13, 1968, and earthquake energy release in the Portland area: Ore Bin, v. 30, p. 185–190.

Couch, R., Thrasher, G., and Keeling, K., 1976, The Deschutes Valley earthquake of April 12, 1976: Ore Bin, v. 38, p. 151–161.

Crosson, R. S., 1972, Small earthquakes, structure, and tectonics of the Puget Sound region: Bulletin of the Seismological Society of America, v. 62, p. 1133–1171.

—— , 1974, Compilation of earthquake hypocenters in western Washington, July 1970 to December 1972: Washington Department of Natural Resources, Geology and Earth Resources Division, 25 p.

—— , 1983, Review of seismicity in the Puget Sound region from 1970 through 1978, in Yount, J. C., and Crosson, R. S., eds., Proceedings of Workshop 14, Earthquake Hazards of the Puget Sound Region, Washington: U.S. Geological Survey Open-File Report 83-19, p. 6–18.

—— , 1986, Comment on "Geodetic strain measurements in Washington" by J. C. Savage, M. Lisowski, and W. H. Prescott: Journal of Geophysical Research, v. 91, p. 7555–7557.

Crosson, R. S., and Frank, D., 1975, The Mt. Rainier earthquake of July 18, 1973, and its tectonic significance: Bulletin of the Seismological Society of America, v. 65, p. 393–401.

Crosson, R. S., and Lin, J. W., 1975, A note on the Mt. Rainier earthquake of April 20, 1974: Bulletin of the Seismological Society of America, v. 65, p. 549–556.

Crosson, R. S., and Owens, T. J., 1987, Slab geometry of the Cascadia subduction zone beneath Washington from earthquake hypocenters and teleseismic converted waves: Geophysical Research Letters, v. 14, p. 824–827.

de Ballore, F. de Montessus, 1906, Les Tremblements de Tierre: Paris, Geographic Seismologique.

Endo, E. T., Malone, S. D., Noson, L. L., and Weaver, C. S., 1981, Locations, magnitudes, and statistics of the March 20–May 18 earthquake sequence, in Lipman, P. W., and Mullineaux, D. R., eds., The 1980 eruptions of Mount St. Helens, Washington: U.S. Geological Survey Professional Paper 1250, p. 93–107.

Grant, W. C., and Weaver, C. S., 1986, Earthquakes near Swift Reservoir, Washington, 1958–1963; Seismicity along the southern St. Helens seismic zone: Bulletin of the Seismological Society of America, v. 76, p. 1573–1587.

—— , 1990, Seismicity of the Spirit Lake area; Estimates of possible earthquake magnitudes for engineering design, *in* Schuster, R. L., and Meyer, W., eds., The formation and significance of major lakes impounded during the 1980 eruption of Mount St. Helens, Washington: U.S. Geological Survey Professional Paper (in press).

Grant, W. C., Weaver, C. S., and Zollweg, J. E., 1984, The 14 February 1981 Elk Lake, Washington, earthquake sequence: Bulletin of the Seismological Society of America, v. 74, p. 1289–1309.

Gutenberg, B., and Richter, C. F., 1954, Seismicity of the Earth, 2nd ed.: Princeton, New Jersey, Princeton University Press, 310 p.

Heaton, T. H., and Hartzell, S. H., 1987, Earthquake hazards on the Cascadia subduction zone: Science, v. 236, p. 162–168.

Heaton, T. H., and Kanamori, H., 1984, Seismic potential associated with subduction in the northwestern United States: Bulletin of the Seismological Society of America, v. 74, p. 933–941.

Isacks, B., and Molnar, P., 1971, Distribution of stresses in the decending lithosphere from a global survey of focal mechanism solutions of mantle earthquakes: Reviews in Geophysics and Space Physics, v. 9, p. 103–174.

Lisowski, M., Savage, J., Prescott, W., and Dragert, H., 1987, Strain accumulation across the Strait of Juan de Fuca and in the Olympic Mountains, Washington [abs.], *in* 1987 Symposium on Recent Crustal Movements in the Pacific Northwest: Geological Association of Canada Programme and Abstracts, p. 17.

Lisowski, M., Prescott, W., Dragert, H., and Hohldahl, S. R., 1989, Results from 1986 and 1987 GPS surveys across the Strait of Juan de Fuca, Washington and British Columbia [abs.]: Seismological Research Letters, v. 60, p. 1.

Ma, L., 1988, Regional tectonic stress in western Washington from focal mechanisms of crustal and subcrustal earthquakes [M.S. thesis]: Seattle, University of Washington, 84 p.

Magill, J. R., Cox, A. V., and Duncan, R., 1981, Tillamook volcanic series; Further evidence for tectonic rotation of the Oregon Coast Range: Journal of Geophysical Research, v. 86, p. 2953–2970.

Magill, J. R., Wells, R. E., Simpson, R. W., and Cox, A. V., 1982, Post-12-million-year rotation of southwest Washington: Journal of Geophysical Research, v. 87, p. 3761–3776.

Malone, S. D., and S. Bor, 1979, Attenuation patterns in the Pacific Northwest based on intensity data and the location of the 1872 North Cascades earthquake, Bulletin of the Seismological Society of America, v. 69, p. 531–546.

Malone, S. D., Rothe, G. H., and Smith, S. W., 1975, Details of microearthquake swarms in the Columbia Basin, Washington: Bulletin of the Seismological Society of America, v. 65, p. 855–864.

Malone, S. D., Endo, E. T., Weaver, C. S., and Ramey, J. W., 1981, Seismic monitoring for eruption prediction, *in* Lipman, P. W., and Mullineaux, D. R., eds., The 1980 eruptions of Mount St. Helens, Washington: U.S. Geological Survey Professional Paper 1250, p. 803–813.

Malone, S. D., Johnson-Browne, P., Mc Clurg, D., Qamar, A. I., Ramey, J., and Thompson, K., 1987, Annual technical report 1987, U.S. Department of Energy Contract EY-76-S-06-2225 Task agreement 39: Seattle, University of Washington Geophysics Program, 80 p.

Milne, W. G., 1956, Seismic activity in Canada west of the 113th meridian, 1841–1951: Ottawa, Ontario, Dominion Observatory Publication 18, p. 126–127.

Milne, W. G., Rogers, G. C., Riddihough, R. P., McMechan, G. A., and Hyndman, R. D., 1978, Seismicity of western Canada: Canadian Journal of Earth Sciences, v. 15, p. 1170–1193.

Misch, P., 1966, Tectonic evolution of the Northern Cascades of Washington State: Canadian Institute of Mining and Metallurgy Special, v. 8, p. 101–148.

Murphy, L. M., and Ulrich, F. P., 1951, United States earthquakes 1949: U.S. Department of Commerce Coast and Geodetic Survey, serial 745, 64 p.

Noson, L. L., Qamar, A., and Thorsen, G. W., 1988, Washington State earthquake hazards: Washington Division of Geology and Earth Resources Information Circular 85, 77 p.

Nuttli, O. W., 1952, The western Washington earthquake of April 13, 1949: Bulletin of the Seismological Society of America, v. 42, p. 21–28.

Patton, H. J., 1985, P-wave fault-plane solutions and the generation of surface waves by earthquakes in the western United States: Geophysical Research Letters, v. 12, p. 518–521.

Poppe, B. B., 1979, Historical survey of U.S. seismograph stations: U.S. Geological Survey Professional Paper 1096, 389 p.

Qamar, A., Ludwin, R., Crosson, R. S., and Malone, S. D., 1987, Earthquake hypocenters in Washington and northern Oregon, 1982–1986: Washington Department of Natural Resources, Division of Geology and Earth Resources, 78 p.

Rasmussen, N. H., 1967, Washington State earthquakes 1840 through 1965: Bulletin of the Seismological Society of America, v. 57, p. 463–476.

Rasmussen, N. H., Millard, R. C., and Smith, S. W., 1974, Earthquake hazard evaluation of the Puget Sound region, Washington State: Seattle, University of Washington, 99 p.

Riddihough, R. P., 1977, A model for recent plate interactions off Canada's west coast: Canadian Journal of Earth Sciences, v. 14, p. 384–396.

——, 1984, Recent movements of the Juan de Fuca plate system: Journal of Geophysical Research, v. 89, p. 6980–6994.

Rogers, G. C., 1983a, Seismotectonics of British Columbia [Ph.D. thesis]: Vancouver, University of British Columbia, 227 p.

——, 1983b, Some comments on the seismicity of northern Puget Sound–southern Vancouver Island region, *in* Workshop 14, Earthquake Hazards of the Puget Sound Region, Washington: U.S. Geological Survey Open-File Report 83-19, p. 19–39.

Ruff, L., and Kanamori, H., 1980, Seismicity and the subduction process: Physics of the Earth and Planetary Interiors, v. 23, p. 240–252.

Savage, J. C., Lisowski, M., and Prescott, W. H., 1981, Geodetic strain measurements in Washington: Journal of Geophysical Research, v. 86, p. 4929–4940.

Sbar, M. L., 1983, An explanation for contradictory geodetic strain and fault plane solution data in western North America: Geophysical Research Letters, v. 10, p. 177–180.

Shannon and Wilson, Inc., 1977, Geologic studies in the 1872 earthquake epicentral region, *in* Washington Public Power Supply System (WPPSS) Nuclear Projects Numbers 1 and 4 Preliminary Safety Analysis Report, Amendment 23, subappendix 2R D.

Shemeta, J. E., and Weaver, C. S., 1986, Seismicity accompanying the May 18, 1980, eruption of Mount St. Helens, Washington, *in* Keller, S.A.C., ed., Mount St. Helens, five years later: Cheney, Eastern Washington University Press, p. 44–58.

Snavely, P. D., Jr., 1988, Tertiary geologic framework, neotectonics, and petroleum potential of the Oregon–Washington continental margin, *in* Scholl, D. S., Grantz, A., and Vedder, J. G., eds., Geology and resource potential of the continental margin of western North America and adjacent ocean basins; Beaufort Sea to Baja California: Circum-Pacific Council for Energy and Mineral Resources Earth Science Series, v. 6, p. 305–335.

Swanson, D. A., and 5 others, 1983, Predicting eruptions at Mt. St. Helens, June 1980 through December 1982: Science, v. 221, p. 1369–1376.

Taber, J. J., and Smith, S. W., 1985, Seismicity and focal mechanisms associated with the subduction of the Juan de Fuca plate beneath Washington: Bulletin of the Seismological Society of America, v. 75, p. 237–249.

Tabor, R. W., 1972, Age of the Olympic metamorphism, Washington; K-Ar dating of low-grade metamorphic rocks: Geological Society of America Bulletin, v. 83, p. 1805–1816.

Tobin, D. G., and Sykes, L. R., 1968, Seismicity and tectonics of the northeast Pacific Ocean: Journal of Geophysical Research, v. 73, p. 3821–3845.

Toppozada, T. R., Real, C. R., and Parke, D. L., 1981, Preparation of isoseismal maps and summaries of related effects of pre-1900 California earthquakes: California Division of Mines and Geology Open-File Report 81-11, 182 p.

Ulrich, F. P., 1949, Reporting the northwest earthquake: Building Standards Monthly, June 1949. Reprinted 1986 *in* Thorsen, G. W., compiler, The Puget lowland earthquakes of 1949 and 1965: Washington Division of Geology and Earth Resources Information Circular 81, p. 19–23.

University of Washington Geophysics Program, 1979, Eastern Washington earthquake catalog 1969–1974; Appendix to annual technical report 1979 on earthquake monitoring of the Hanford region, eastern Washington: Seattle, University of Washington Geophysics Program, 37 p.

Uyeda, S., and Kanamori, H., 1979, Back-arc opening and the mode of subduction: Journal of Geophysical Research, v. 84, p. 1049–1061.

Weaver, C. S., and Baker, G. E., 1988, Geometry of the Juan de Fuca plate beneath Washington and northern Oregon from seismicity: Bulletin of the Seismological Society of America, v. 78, p. 264–275.

Weaver, C. S., and Malone, S. D., 1987, Overview of the tectonic setting and recent studies of eruptions of Mount St. Helens, Washington: Journal of Geophysical Research, v. 92, p. 10149–10154.

Weaver, C. S., and Smith, S. W., 1983, Regional tectonic and earthquake hazard implications of a crustal fault zone in southwestern Washington: Journal of Geophysical Research, v. 88, p. 10371–10383.

Weaver, C. S., Grant, W. C., Malone, S. D., and Endo, E. T., 1981, Post May 18 seismicity: Volcanic and tectonic implications, in Lipman, P. W., and Mullineaux, D. R., eds., The 1980 eruptions of Mount St. Helens, Washington: U.S. Geological Survey Professional Paper 1250, p. 109–121.

Weaver, C. S., Green, S. M., and Iyer, H. M., 1982, Seismicity of Mount Hood and structure as determined from teleseismic P wave delay studies: Journal of Geophysical Research, v. 87, p. 2782–2792.

Weaver, C. S., Grant, W. C., and Shemeta, J. E., 1987, Local crustal extension at Mount St. Helens, Washington: Journal of Geophysical Research, v. 92, p. 10170–10178.

Weston Geophysical Research, Inc., 1976, The 1872 earthquake; Significant data and conclusions, prepared for United Engineers and Constructors, Inc.: Weston Geophysical Research, Inc., unpaginated.

Yelin, T. S., 1982, The Seattle earthquake swarms and Puget basin focal mechanisms and their tectonic implications [M.S. thesis]: Seattle, University of Washington, 96 p.

Yelin, T. S., and Crosson, R. S., 1982, A note on the south Puget Sound basin magnitude 4.6 earthquake of 11 March 1978 and its aftershocks: Bulletin of the Seismological Society of America, v. 72, p. 1033–1038.

Zoback, M. L., and Zoback, M., 1980, State of stress in the conterminous United States: Journal of Geophysical Research, v. 85, p. 6113–6156.

Zollweg, J. E., and Jacobson, R. S., 1986, A seismic zone on the Oregon-Idaho border; The Powder River earthquakes of 1984: Bulletin of the Seismological Society of America, v. 76, p. 985–999.

Zollweg, J. E., Rathbun, A. P., and Crosson, R. S., 1982, The Goat Rocks Wilderness, Washington, earthquake of 28 May 1981, in Annual technical report 1980–1981, U.S.G.S. contract #14-08-0001-19274: Seattle, University of Washington Geophysics Program, 5 p.

Manuscript Accepted by the Society August 18, 1989

ACKNOWLEDGEMENTS

Support for this work was provided by the U.S. Geological Survey under grant 14-08-0001-G1390, and by the Geothermal, Volcano, and Earthquake Hazards Programs. Numerous scientists from the University of Washington Geophysics Program and the USGS office at the University of Washington contributed useful information and suggestions.

Chapter 7

Northern California seismicity

Robert A. Uhrhammer
Seismographic Station, 475 Earth Sciences Building, University of California, Berkeley, California 94720

INTRODUCTION

Northern California (39°N to 42°N, as shown in Figure 1) is a geologically complex region that exhibits a wide variation in its characteristic seismicity. The rate of seismicity varies in zones that are approximately delineated by the major structural provinces shown in Figure 2. The seismicity rate varies by a factor of approximately 20, from very low in the northeastern corner of California, to high in the vicinity of the Gorda Escarpment and Gorda Basin just off the coast.

This paper discusses the general characteristics of the seismicity in the region recorded by the University of California Seismographic Stations (1949–1985), the U.S. Geological Survey Calnet (1980–1986), and numerous special studies.

SOURCES

The seismicity list is from two primary sources: Bolt and Miller (1975) prior to 1973, and the *Bulletin of the Seismographic Stations* (U.C. Berkeley) for 1973 through 1985. The 1949 starting date was chosen for two reasons. First, the installation of a high-sensitivity Benioff seismograph at Mineral (MIN in Fig. 2) in 1949 allowed routine detection and location down to M_L 3.0 in the region. Second, the Wood-Anderson torsion seismographs at Mineral and Arcata (ARC) provided a uniform method for determining the local magnitude. Since 1949 the number and quality of seismographic stations has increased in northern California, due primarily to the interest of several groups in either regional monitoring or special studies. Most notably, the USGS has expanded the central California network northward during the past 20 years. This data is used as a supplement to the primary source.

RATE OF SEISMICITY

From 1949 through 1985, 1,614 earthquakes $M_L \geq 3.0$ occurred in the 141,000-km² region of northern California and vicinity (bounded by 39°N and 42°N, and 120°W and 125°W).

This seismicity list was declustered to 927 principal earthquakes ($M_L \geq 3.0$) by removing foreshocks and aftershocks. The principal earthquakes were fit to the Gutenberg-Richter relation using a maximum likelihood procedure; the result is:

$$\log N = 4.18 - 0.88 \, M_L \quad (1)$$

with an uncertainty of:

$$\sigma^2_{\log N} = 0.00131 - 0.000550 M_L + 0.000060 M_L^2. \quad (2)$$

In (1) and (2), N is the cumulative number of principal earthquakes with magnitude M_L or larger (normalized to earthquakes per year).

The average annual rate of seismicity ($r = 10^{\log N}$) with local magnitude M_L or larger is:

$M_L \geq$	r (eq/yr)	σ_r (eq/yr)
3	34.7	1.1
3.5	12.6	.31
4	4.57	.086
4.5	1.66	.070
5	.60	.085
5.5	.219	.123

For $M_L > 6$, σ_r is larger than r; thus, the rate for $M_L > 6$ cannot be reliably inferred from the 37-year seismicity sample.

Northernmost California was divided into the six dominant provinces shown in Figure 2 for the purpose of analyzing the seismicity. The provinces and their associated rates of seismicity ($M_L \geq 3$) (normalized to eq/yr/10,000 km²) are:

Province	Region	r	σ_r
Warner	1	0.5	0.13
Modoc Plateau	2	1.5	0.12
Sierra Nevada	3	2.0	0.19
Klamath Mtns.	4	0.8	0.10
Northern Coast Ranges	5	2.1	0.13
Gorda Basin	6	8.6	0.58

Uhrhammer, R. A., 1991, Northern California seismicity, *in* Slemmons, D. B., Engdahl, E. R., Zoback, M. D., and Blackwell, D. D., eds., Neotectonics of North America: Boulder, Colorado, Geological Society of America, Decade Map Volume 1.

Thus, the level of seismicity, as shown in Figure 1, varies by a factor of 20 over the region, from a low of 0.5 ± 0.13 eq/yr in the northeastern corner of the state (region 1 in Fig. 2), to a high of 8.6 ± 0.9 eq/yr in the Gorda Basin/Gorda Escarpment area off the northern coast (region 6 in Fig. 2). Both rates are normalized to earthquakes per year per 10,000 km^2 with $M_L \geq 3.0$.

PROMINENT EARTHQUAKES

Forty-eight strong earthquakes ($M_L \geq 5$; 1949–1985) listed in Table 1 and plotted in Figure 3, have occurred in northern California and vicinity. Two-thirds of these earthquakes occurred in the vicinity of the Gorda Escarpment and Gorda Basin (region 6) off the coast, and one-fourth of the earthquakes occurred in the vicinity of the Sierra Nevada (region 3). Refer to Figures 2 and 6 for place and fault names. There have been six large earthquakes ($M_L \geq 6$), and all except one (the 1966 Truckee earthquake; Ryall and others, 1968) occurred near or off the coast. The largest earthquake (M_L 6.9) occurred on November 8, 1980, in the Gorda Basin (Tera Corporation, 1981).

Four other strong earthquake sequences are worth noting. On March 20, 1950, a M_L-5.5 earthquake occurred east of Red-

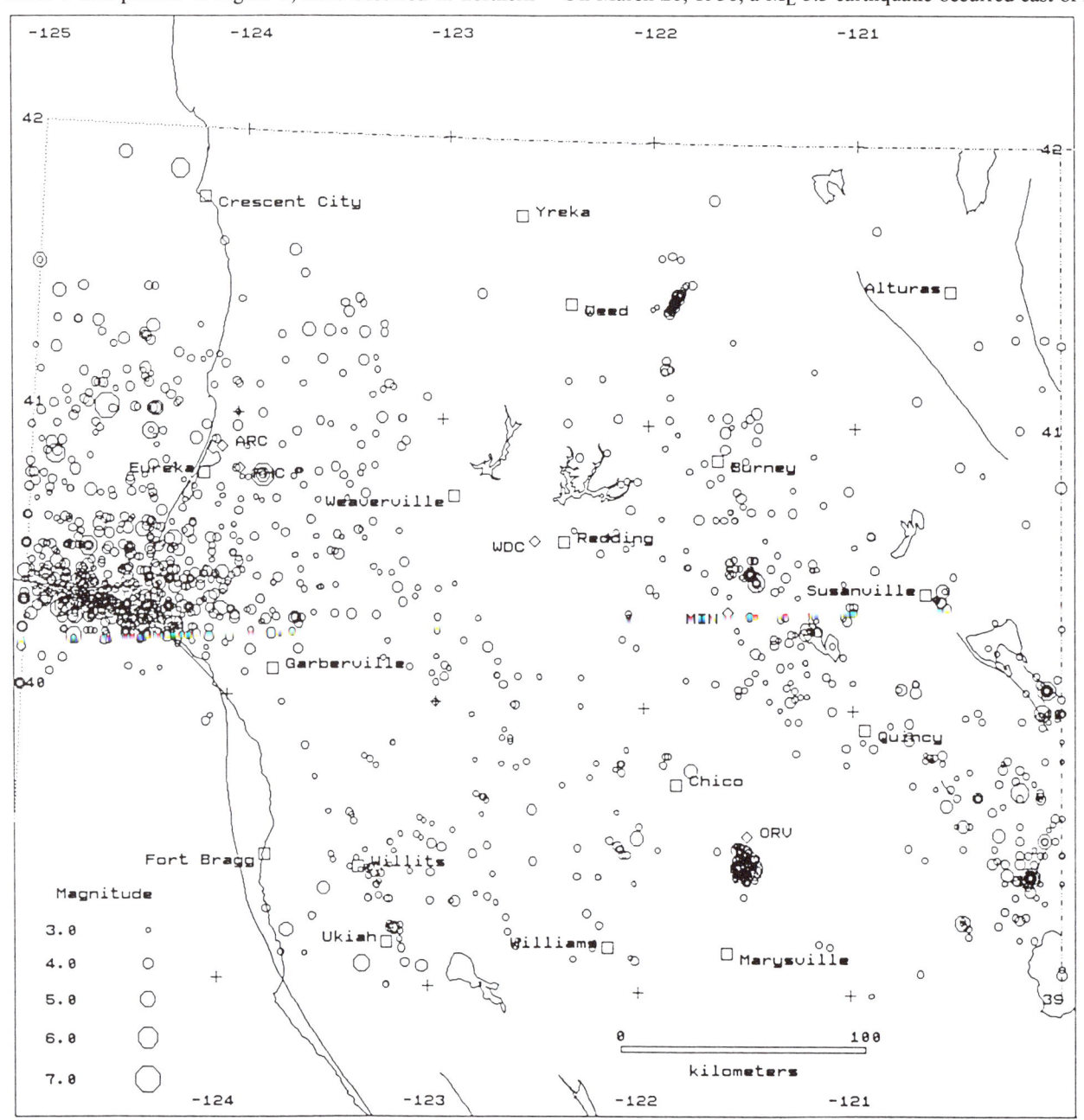

Figure 1. Northern California seismicity ($M_L \geq 3.0$; 1949–1985). 927 principal earthquakes (with foreshocks and aftershocks omitted) that occurred in the 141,000-km^2 area bounded by 39° and 42°N, and 120° and 125°W.

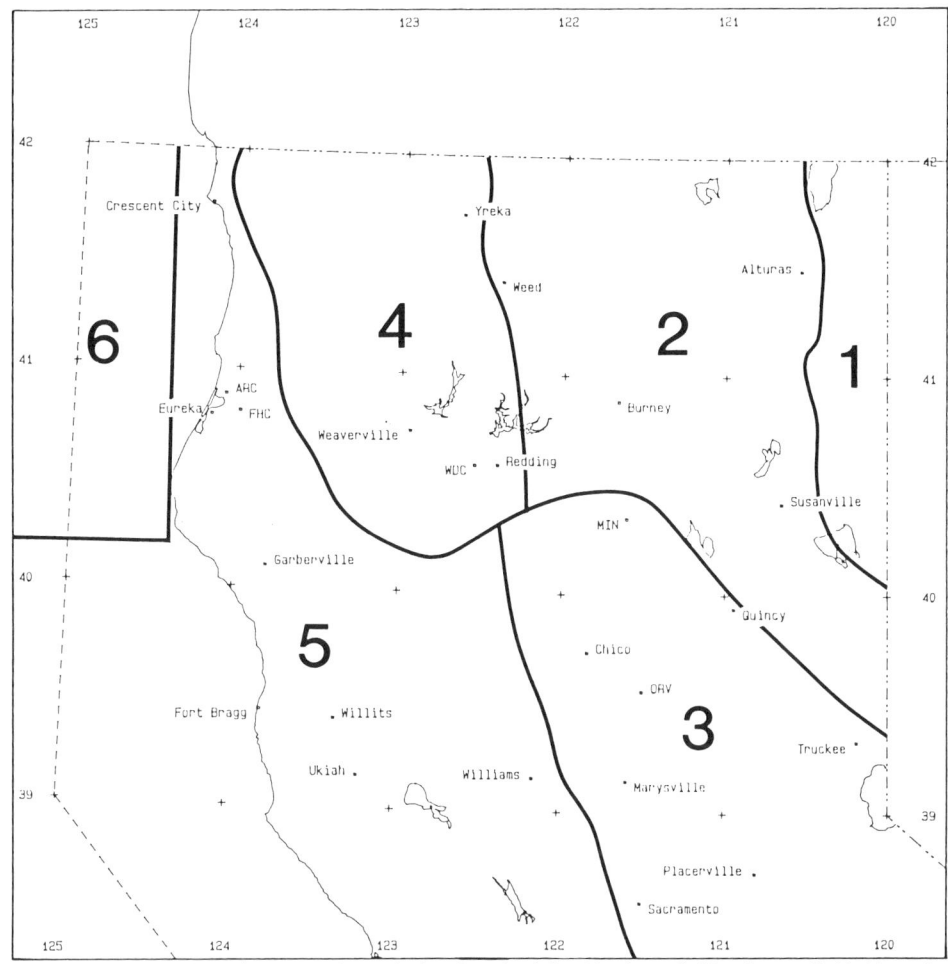

Figure 2. Structural provinces used in analyzing the seismicity. Northern California was divided into six regions: (1) Warner; (2) Modoc Plateau; (3) Sierra Nevada; (4) Klamath Mtns.; (5) Northern Coast Ranges; and (6) Gorda Basin.

ding near Lassen Peak (Klein, 1979). The area east of Lassen Peak has had numerous swarms of small earthquakes. A M_L-5.2 earthquake occurred in the northern coast ranges southwest of Ukiah (region 5) on June 6, 1962. In May and June 1968 a pair of M_L-5.1 earthquakes occurred near the California-Oregon border in the vicinity of Adel, Oregon (northeast of Yreka) (USCGS, 1969). On August 1, 1975, a M_L-5.7 earthquake occurred along the foothills of the Sierra Nevada near Oroville (ORV) (Morrison and others, 1976). This earthquake generated a lot of interest and led to a thorough reevaluation of the seismic potential along the foothills fault system in the Sierra Nevada.

GENERAL CHARACTERISTICS OF THE SEISMICITY

The seismicity of northern California has been studied by several seismologists (e.g., Byerly, 1937; Tocher, 1956; Niazi, 1964; Bolt and Miller, 1971; Simila, 1981; McPherson, 1989). In the following sections the seismicity characteristics are discussed by province. Refer to Figures 1 through 6 as appropriate. The region numbers correspond to Figure 2, and the fault names are on Figure 6.

Gorda Basin

The most seismically active region is along the Gorda Escarpment from the coast to the Gorda Ridge (near left edge of Fig. 6). The hypocentral depths approach 30 km along the coast (Simila and others, 1975) as shown in the depth cross-section (c in Fig. 5a), and a shallow, eastward-dipping Benioff zone is indicated (Smith and Knapp, 1980). The two lines of seismicity in the left part of Figure 5a at depths of 15 and 20 km are submarine hypocenters whose depth was restrained due to poor depth resolution. The largest earthquake (M_L 6.9) in northern California since 1949 occurred in the Gorda Basin on November 8, 1980, and its 100-km-long aftershock zone is indicated by the northeast-southwest trend of epicenters west of Eureka in Figure 4. Ocean-bottom seismographs indicate a high microseismicity rate of shallow earthquakes in the Gorda Basin (Asada and others, 1975).

TABLE 1. PROMINENT EARTHQUAKES IN NORTHERN CALIFORNIA 1949–1985

Date	Time (UTC)	Lat. (°N)	Long. (°W)	M_L
24 Mar 49	20:56:56	41.30	126.00	5.9
20 Mar 50	15:22:17	40.45	121.47	5.5
14 Dec 50	13:24:19	40.08	120.07	5.6
01 Apr 51	19:21:08	40.47	125.30	5.0
08 Oct 51	04:10:35	40.28	124.80	5.8
09 May 52	15:31:32	39.42	119.78	5.1
22 Sep 52	11:41:25	40.20	124.42	5.2
26 Sep 53	03:34:29	39.53	119.98	5.3
25 Nov 54	11:16:35	40.27	125.63	6.1
21 Dec 54	19:56:29	40.78	123.87	6.5
11 Oct 56	16:48:50	40.67	125.77	6.0
01 Apr 59	18:18:30	39.72	120.20	5.6
24 Jul 59	01:23:09	41.13	125.30	5.8
05 Dec 59	08:13:42	40.30	125.42	5.1
06 Jun 60	01:17:48	40.82	124.88	5.7
27 Dec 60	10:35:26	41.52	125.05	5.4
06 Apr 61	04:04:45	40.18	124.75	5.1
14 Apr 62	07:53:14	40.27	125.32	5.4
06 Jun 62	17:50:06	39.07	123.32	5.2
14 Jul 62	19:43:46	40.43	125.52	5.1
23 Aug 62	19:29:13	41.85	124.33	5.6
16 Sep 65	04:10:08	40.50	125.80	5.0
12 Sep 66	16:41:01	39.42	120.15	6.0
12 Sep 66	17:20:11	39.42	120.15	5.3
10 Dec 67	12:06:52	40.50	124.70	5.6
30 May 68	00:36:01	42.17	120.00	5.1
04 Jun 68	02:34:16	42.30	119.90	5.1
26 Jun 68	01:42:20	40.23	124.27	5.9
13 Sep 70	21:10:21	40.13	125.08	5.4
27 Feb 71	00:31:37	40.27	124.93	5.2
01 Mar 72	09:28:42	40.67	125.25	5.2
15 Jun 73	19:18:47	41.50	125.53	5.0
03 Jul 74	05:00:58	40.34	125.21	5.1
07 Jun 75	08:46:22	40.54	124.29	5.3
01 Aug 75	20:20:12	39.44	121.53	5.7
02 Aug 75	20:22:16	39.44	121.46	5.1
02 Aug 75	20:59:02	39.43	121.47	5.2
16 Nov 75	17:29:29	40.40	126.30	5.0
22 Feb 79	15:57:00	40.00	120.10	5.2
03 Mar 80	14:17:04	40.60	125.03	5.1
08 Nov 80	10:27:28	41.08	125.12	6.9
09 Nov 80	04:09:10	40.59	125.30	5.2
28 Nov 80	18:21:13	39.31	120.43	5.3
29 May 83	06:55:29	40.39	125.99	5.3
11 Jun 83	03:09:54	39.26	120.47	5.1
24 Aug 83	13:36:31	40.30	124.77	5.7
28 Feb 84	15:16:07	40.36	125.90	5.2
10 Sep 84	03:14:07	40.39	127.15	6.6

The Mendocino Escarpment and the Gorda Basin both exhibit left-lateral strike-slip focal mechanisms (see Fig. 6). The question of which plane was the fault plane in the Gorda Basin was cleared up by the occurrence of the November 8, 1980 earthquake and the orientation of its aftershock zone.

Northern Coast Ranges

Moderate seismicity is observed throughout the Northern Coast Ranges (region 5). There are two areas where the seismicity is concentrated, as shown in Figures 4 and 5b. The first is along the South Fork Mountain fault zone (the common boundary with the Klamath Mountains; see S in Fig. 5b) (Smith, 1976). The second is along the Maacama fault zone (Severy, 1978). Note that the seismicity along the Maacama Fault is very shallow; few hypocenters are deeper than 5 km. The San Andreas fault in this area has exhibited a very low rate of seismicity since the occurrence of the great San Francisco earthquake of April 18, 1906.

The prominent San Andreas and Maacama fault zones exhibit right-lateral strike-slip focal mechanisms (Bolt and others, 1968; Severy, 1978). Earthquakes occurring along the South Fork Mountain fault zone have reverse faulting mechanisms, indicating north-south compression (Bolt and others, 1968; Simila, 1981).

Klamath Mountains

This region (4) exhibits a low level of seismicity. Most of the events occur along the western side in the vicinity of the South Fork Mountain fault; very few events occur in the eastern half between Redding, Yreka, and Weaverville as indicated in Figures 1 and 4 (Bolt, 1979).

Modoc Plateau

The general trend in the Modoc Plateau (region 2) is a scattering of epicenters extending northwestward from the eastern flank of the Sierra Nevada (Bolt and Miller, 1971). Concentrations in the seismicity are observed in the vicinity of Lassen Peak (Klein, 1979).

The dominant mechanism in the central part of the Modoc Plateau (including swarms in the vicinity of Mt. Shasta and Lassen Peak) is normal faulting with east-west extension (Simila, 1981). The Honey Lake fault and the Likely fault exhibit right-lateral strike-slip mechanisms (Hannah, 1973).

Sierra Nevada

The eastern flank of the Sierra Nevada (region 3) has produced large-magnitude earthquakes in the vicinity of Verdi (1948) (California-Nevada border 30 km northeast of Truckee) and Truckee (1966) (Ryall and others, 1968). The focal mechanisms are predominantly normal with east-west extension (Ryall

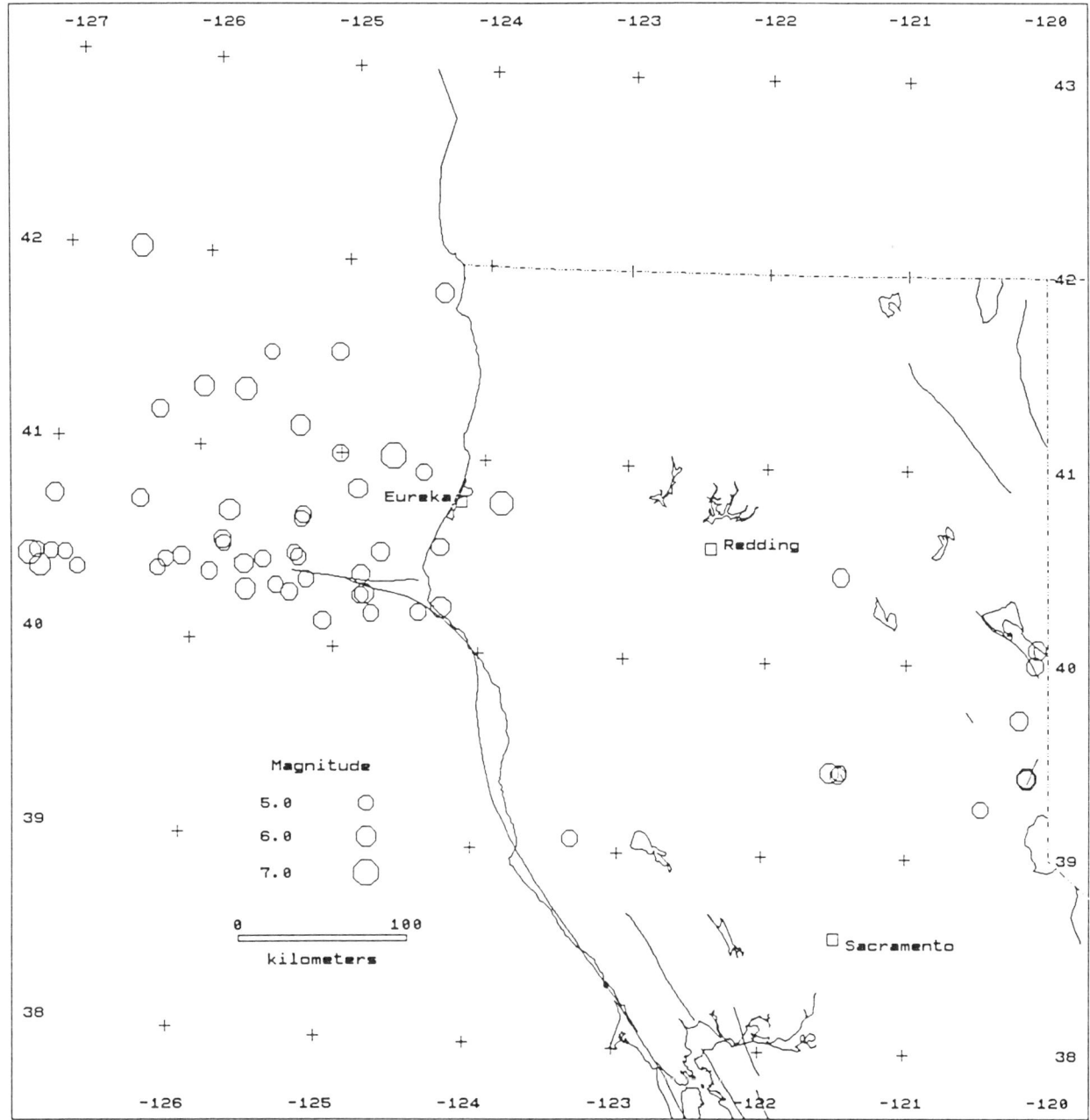

Figure 3. Prominent earthquakes ($M_L \geq 5.0$; 1949–1985). Forty-eight strong earthquakes occurred in Northern California and vicinity in the 37-year sample.

and others, 1968). The wide zone of seismicity in Figure 5c corresponds to the general trend in seismicity running northwest of Truckee in Figure 4.

The Foothills fault system along the west side of the Sierra Nevada (see Figure 4 and F in Fig. 5c) (McNally and others, 1978). The largest event since 1949 (M_L 5.7) in the Foothills occurred in August 1975 (Morrison and others, 1976) near Oroville (ORV in Fig. 2). The Foothills Fault system exhibits normal faulting with east-west extension (west side down), as indicated by the Oroville earthquake (Bufe and others, 1976). Note that the seismicity along the Foothills Fault system and vicinity extends to a depth of 40 km.

Warner

This province (region 1) exhibits the lowest level of seismicity in northern California. Only 10 earthquakes $M_L \geq 3.0$ were recorded in the 37-year sample. Note, however, that there is geologic evidence for Quaternary surface displacements along the Surprise Valley fault in the region (Jennings, 1975).

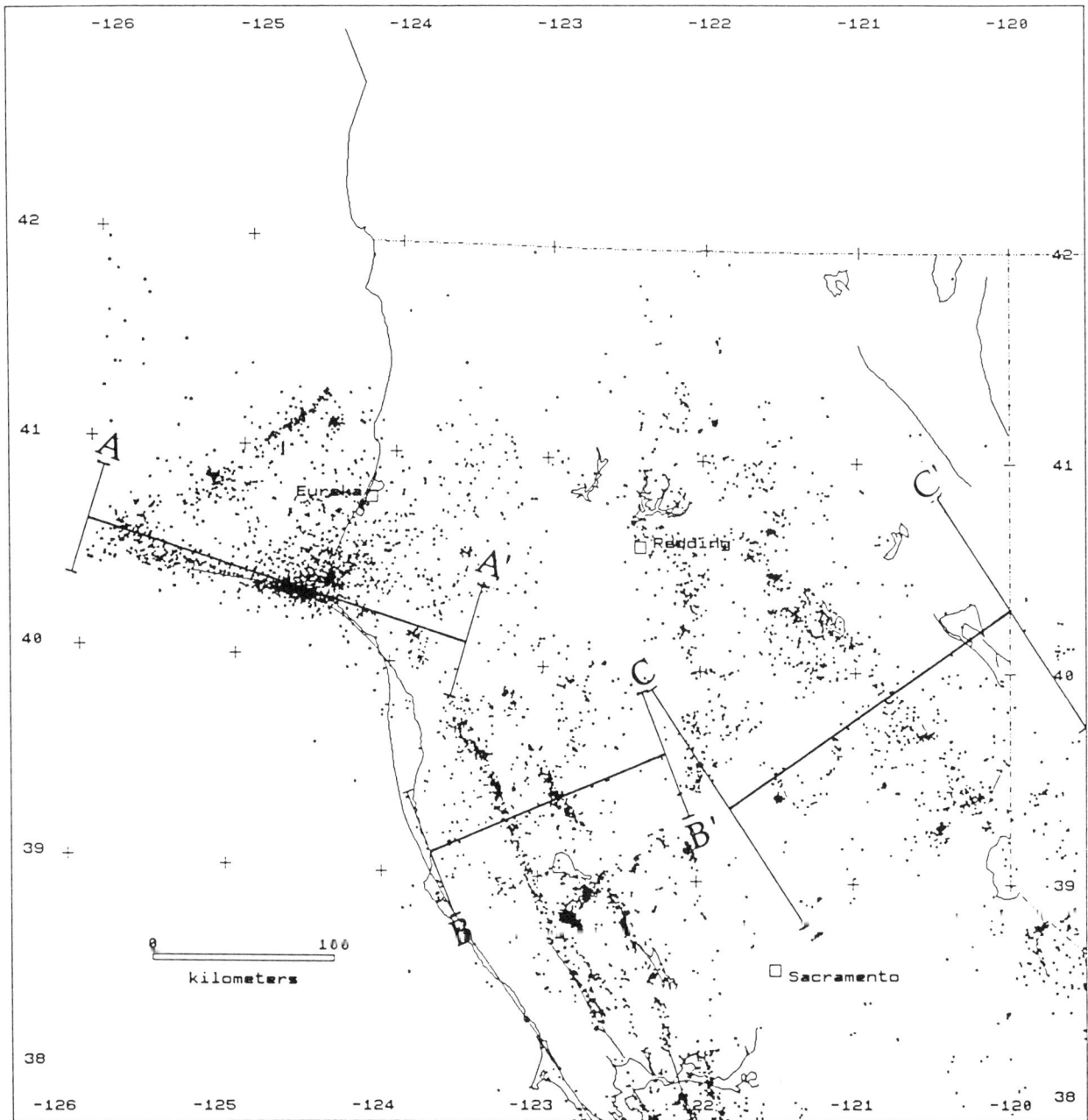

Figure 4. Seismicity from USGS Calnet earthquake catalog ($M_L > 1.3$; 1980–1986). Only well-constrained solutions are included that were recorded by six or more stations.

GORDA PLATE SUBDUCTION

Seismological evidence for the subduction of the Gorda Plate has been accumulating during the past ten years. A well-defined Wadati-Benioff zone beneath Cape Mendocino was identified by Smith and Knapp (1980). Small earthquakes were analyzed by Cockerham (1984) at depths down to 58 km beneath the eastern side of the coast ranges, from which he inferred a subducting slab 180 km long dipping 25 in a S68°E direction. Walter (1986) analyzed seven earthquakes, ranging from 52 km to 87 km in depth, which line up with the inferred slab and extend it 70 km farther east (as far as Redding). A thrust focal mechanism was determined for one of the 80-km-deep earthquakes (Walter, 1986), which is consistent with the active subduction zone tectonics.

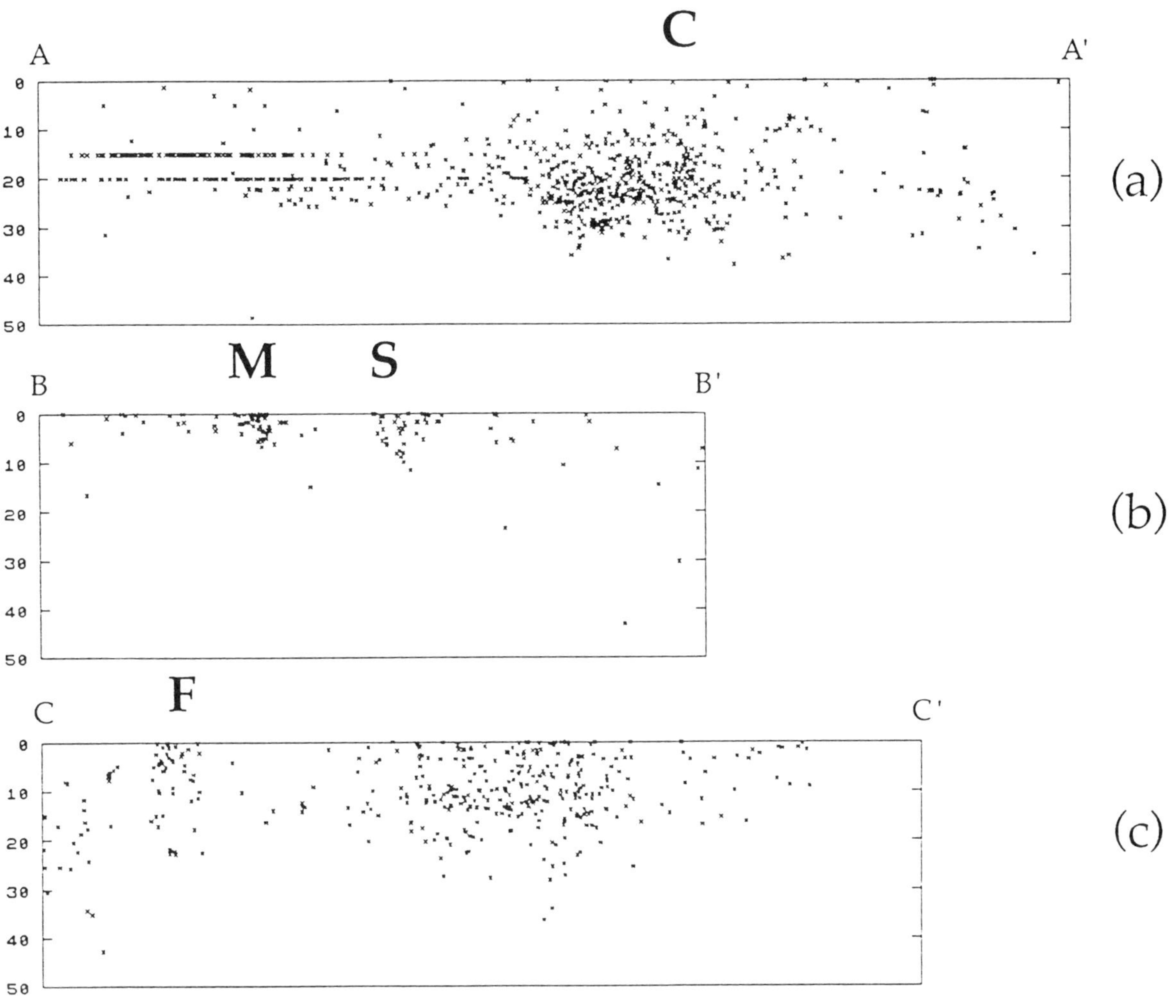

Figure 5. Depth cross-sections A-A', B-B', and C-C' (shown in Fig. 4) are shown in a, b, and c respectively. The depth range in each plot is 0 to 50 km, and there is no vertical exaggeration. The letters C, M, F, and S are the locations of the coastline, the Maacama Fault, the Foothills Fault system, and the South Fork Mountain Fault zone, respectively. A detailed description is given in the text.

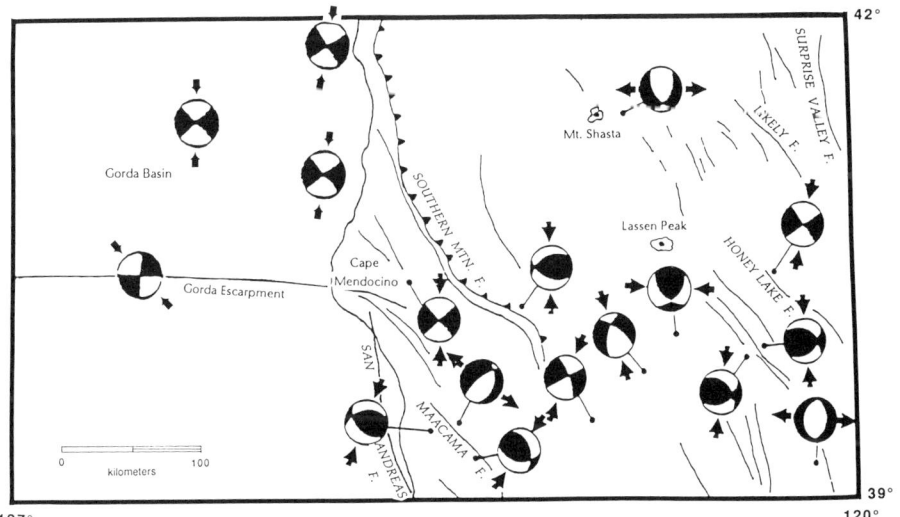

Figure 6. Focal mechanisms for selected earthquakes in northern California. The dark zones are compression, and the arrows indicate the direction of principal stress. Principal faults and geographic landmarks are named.

REFERENCES CITED

Asada, T., Bolt, B. A., Shimamura, H., Takano, K., and Moriya, T., 1975, Earthquake studies in the Gorda basin with ocean bottom seismographs: EOS Transactions of the American Geophysical Union, v. 56, p. 1084.

Bolt, B. A., 1979, Seismicity of the western United States, in Hatheway, A. W., and McClure, C. R., eds., Geology in the siting of nuclear power plants: Geological Society of America Reviews in Engineering Geology, v. 4, p. 95–107.

Bolt, B. A., and Miller, R. D., 1971, Seismicity of northern and central California, 1965–1969: Bulletin of the Seismological Society of America, v. 61, p. 1831–1847.

—— , 1975, Catalogue of earthquakes in northern California and adjoining areas; 1 January 1910–31 December 1972: Berkeley, University of California Seismographic Stations, 567 p.

Bolt, B. A., Lomnitz, C., and McEvilly, T. V., 1968, Seismological evidence on the tectonics of central and northern California and the Mendocino Escarpment: Bulletin of the Seismological Society of America, v. 58, p. 1725–1767.

Bufe, C. G., and 5 others, 1976, Oroville earthquakes; Normal faulting in the Sierra Nevada foothills: Science, v. 192, p. 72–74.

Byerly, P., 1937, Earthquakes off the coast of northern California: Bulletin of the Seismological Society of America, v. 27, p. 73–96.

Cockerham, R. S., 1984, Evidence for a 180 km-long subducted slab beneath northern California: Bulletin of the Seismological Society of America, v. 74, p. 569–576.

Hannah, J. L., 1973, Tectonic setting of the Modoc region of northeastern California: California Division of Mines and Geology Special Report 129, p. 35–39.

Jennings, C. W., 1975, Fault map of California with locations of volcanoes, thermal springs, and thermal wells: California Division of Mines and Geology California Geologic Data Map Series Map 1, scale 1:750,000.

Klein, F. W., 1979, Earthquakes in Lassen Volcanic National Park, Cailfornia: Bulletin of the Seismological Society of America, v. 69, p. 867–875.

McNally, K. C., Simila, G. W., and Von Dollen, F. J., 1978, Microearthquake activity adjacent to the Raklin pluton near Auburn, California: Bulletin of the Seismological Society of America, v. 68, p. 239–243.

McPherson, R. C., 1989, Seismicity and focal mechanisms near Cape Mendocino, northern California; 1974–1984 [M.S. thesis]: Arcata, California, Humboldt State University, 75 p.

Morrison, P. W., Jr., Stump, B. W., and Uhrhammer, R. A., 1976, The Oroville earthquake sequence of August 1975: Bulletin of the Seismological Society of America, v. 66, p. 1065–1084.

Niazi, M., 1964, Seismicity of northern California and western Nevada: Bulletin of the Seismological Society of America, v. 54, p. 845–850.

Ryall, A., Van Wormer, J. D., and Jones, A. E., 1968, Triggering of microearthquakes by Earth tides and other features of the Truckee, California, earthquake sequence of September 1968: Bulletin of the Seismological Society of America, v. 58, p. 215–248.

Severy, N. I., 1978, A microearthquake survey of the Maacama Fault zone in the northern Coast Ranges: Earthquake Notes, v. 49, p. 34.

Simila, G. E., 1981, Seismic velocity structure and associated tectonics of northern California [Ph.D. thesis]: Berkeley, University of California, 180 p.

Simila, G. W., Peppin, W. A., and McEvilly, T. V., 1975, Seismotectonics of the Cape Mendocino, California, area: Bulletin of the Seismological Society of America, v. 86, p. 1399–1406.

Smith, S. W., 1976, Seismicity north of Cape Mendocino: Earthquake Notes, v. 47, p. 20.

Smith, S. W., and Knapp, J. S., 1980, The northern termination of the San Andreas fault zone in northern California: California Division of Mines and Geology Special Report 140, p. 153–164.

Tera Corporation, 1981, Report on the November 8, 1980, Eureka earthquake and aftershocks: Berkeley, California, Tera Corporation report B-81-15.

Tocher, D., 1956, Earthquakes off the North Pacific coast of the United States: Bulletin of the Seismological Society of America, v. 46, p. 165–173.

USCGS, 1969, Abstracts of earthquake reports for the United States, April-June 1968: U.S. Coast and Geodetic Survey MSA-138.

Walter, S. R., 1986, Intermediate-focus earthquakes associated with the Gorda Plate subduction in northern California: Bulletin of the Seismological Society of America, v. 76, p. 583–588.

Manuscript Accepted by the Society February 11, 1991

ACKNOWLEDGMENT

My thanks to Jerry Eaton for proving the subset of the USGS Calnet earthquake catalog.

Chapter 8

The seismotectonic fabric of central California

David P. Hill, Jerry P. Eaton, William L. Ellsworth, Robert S. Cockerham, and Frederick W. Lester
U.S. Geological Survey, 345 Middlefield Road, Menlo Park, California 94025
Edward J. Corbett
Seismological Laboratory, University of Nevada, Reno, Nevada 89557

INTRODUCTION

The northwest-trending structural grain of central California is defined by the Coast Ranges on the west, the Sierra Nevada on the east, and the intervening Great Valley (Fig. 1) and reflects a long history of plate-boundary tectonism along this same northwest-striking trend (Dickinson, 1981). In this chapter, we look at the detailed patterns of earthquake occurrence during the six-year period from 1980 to 1985 as a clue to currently active tectonic processes in central California and the adjacent section of western Nevada.

The data for these detailed seismicity patterns derive from the 300-station network of telemetered seismograph stations operated by the U.S. Geological Survey in central and northern California in conjunction with the 73-station network in eastern California and western Nevada recorded by the University of Nevada at Reno (e.g., Vetter and Corbett, 1987). These networks, which by mid-1979 had essentially attained their current configuration (Fig. 2), provide the capability for routinely detecting and locating earthquakes of magnitude 1.5 to 2.0 and greater occurring throughout central and northern California. The much longer history of instrumentally recorded earthquakes in northern California began in 1887 with the installation of the University of California seismograph stations at Berkeley and Mount Hamilton, as documented in the UC Berkeley earthquake catalogs from 1910 onward (Bolt and Miller, 1975). This network currently includes 19 stations and, among other things, provides a stable, long-term data base for magnitudes of larger earthquakes (M> 4) occurring throughout the region (see Uhrhammer, this volume).

After briefly highlighting critical tectonic elements in central California and the historical seismicity pattern for the region, we will focus on detailed seismicity patterns recorded during the six-year interval from 1980 to 1985, relying heavily on a series of seismicity maps and cross sections. As a guide to patterns of seismogenic deformation of the brittle upper crust in central California, we also review selected, well-constrained focal mechanism data for several of the larger earthquakes in the region. We will conclude with some thoughts on the significance of these patterns in terms of current seismotectonic processes in central California.

TECTONIC SETTING

Our principal focus in this chapter is on the seismotectonic fabric of California between the Mendocino triple junction and the Transverse Ranges (see Fig. 1), and in this section we review the tectonic elements in this central section of California that are integral to the description and broad interpretation of current seismicity patterns. More detailed descriptions of the regional tectonics and structure are clearly laid out in associated DNAG volumes or, for example, in the collection of papers on the geotectonics of California edited by Ernst (1981). The chapters by Uhrhammer (this volume) and Hutton and others (this volume) address the seismotectonic fabric of California from the triple junction north and from the Transverse Ranges south, respectively.

The three lithospheric plates that control the modern tectonics of coastal California meet at the Mendocino triple junction just off Cape Mendocino (Fig. 1). North of the triple junction, oblique subduction dominates, with the eastern margin of the Gorda plate slipping beneath the North American plate in a north-northeast direction at a rate of roughly 48 mm/yr (Wilson, 1986). South of the triple junction, transform tectonics dominate, with the Pacific plate sliding past the North American plate in a north-northwest direction, subparallel with the central California coast line. Estimates of the rigid-plate motion along the section of the transform boundary through central California vary between 48 and 65 mm/yr, with the Pacific Plate moving N35° to 38° W with respect to the North American Plate (Minster and Jordan, 1987; Ness and others, 1985). Recent evidence presented by DeMets and others (1987) points strongly to the lower of these estimates.

Right-lateral strike-slip displacement along the major branches of the San Andreas fault system, which cuts diagonally

Hill, D. P., Eaton, J. P., Ellsworth, W. L., Cockerham, R. S., Lester, F. W., and Corbett, E. J., 1991, The seismotectonic fabric of central California, *in* Slemmons, D. B., Engdahl, E. R., Zoback, M. D., and Blackwell, D. D., eds., Neotectonics of North America: Boulder, Colorado, Geological Society of America, Decade Map Volume 1.

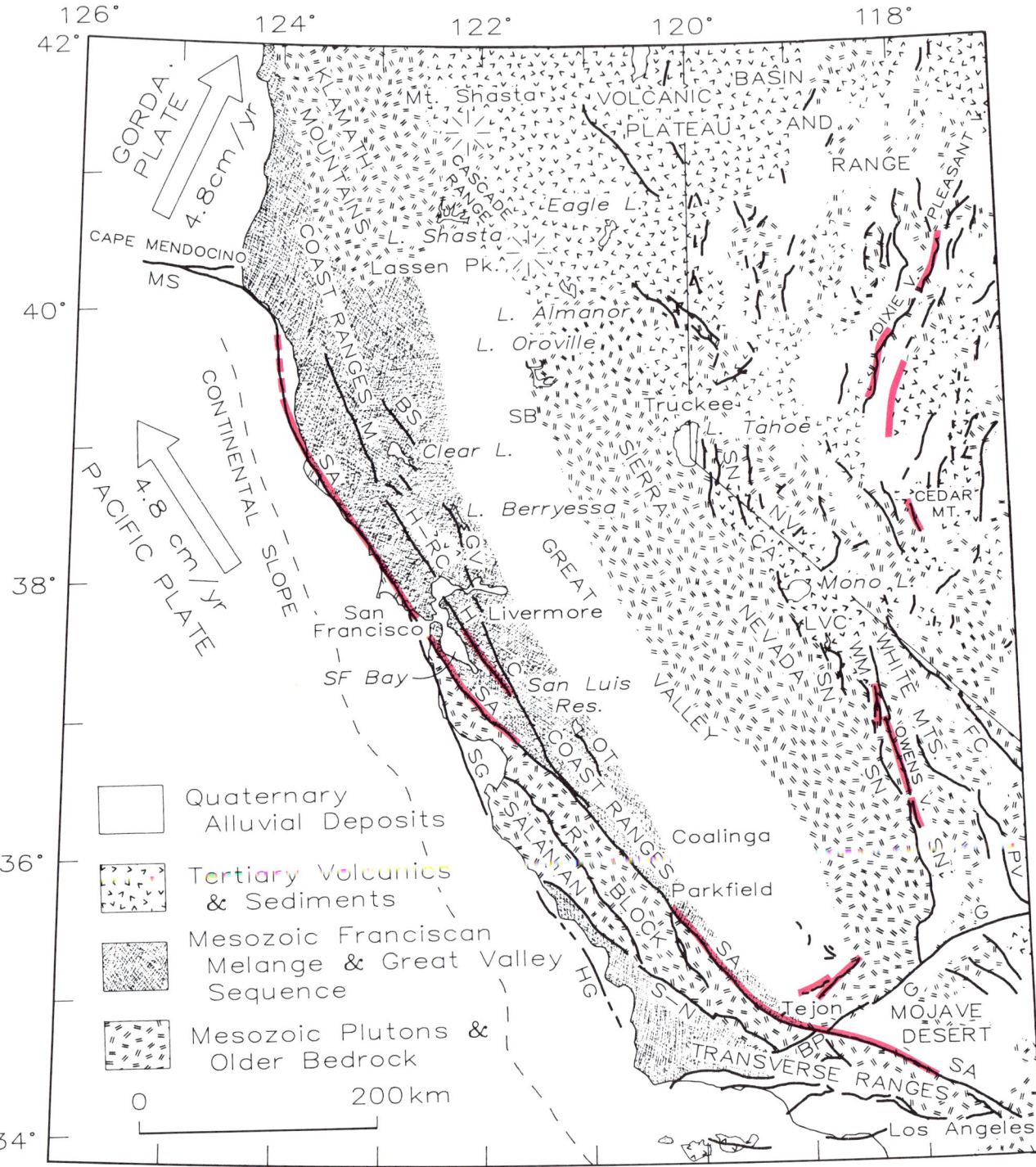

Figure 1. Map showing principal tectonic elements in central and northern California. Large arrows show motion of Pacific and Gorda plates with respect to the North American plate. Abbreviations: BS = Bartlett Springs fault; C = Calaveras fault; GV = Green Valley fault; G = Garlock fault; H-RC = Rodgers Creek fault; HG = Hosgri fault; H = Hayward fault; M = Maacama fault; OT = Ortigalita fault; PV = Panamint Valley fault; R = Rinconada fault, S-N = Sur-Nacimiento fault; SA = San Andreas fault; SB = Sutter Buttes; SG = San Gregorio fault; SN = Sierra Nevada fault zone; WW = White Wolf fault. Faults with historic surface ruptures shown in red.

through the Coast Ranges, accommodates a large part of the transform motion between the Pacific and North American plates in central California. Geological evidence for cumulative Holocene offsets across the San Andreas fault system (Wallace, 1970; Sieh and Jahns, 1984) and modern geodetic evidence from repeated trilateration measurements spanning the fault system (Savage, 1983; Prescott and others, 1981) give a slip rate of 34 ± 3 mm/yr, parallel with the average strike of the San Andreas fault in central California (N41° ±2° W). The remaining fraction of the Pacific–North American plate motion (the so-called "San Andreas discrepancy") must be accommodated by deformation of the plate margins themselves. In central California, this residual deformation is reflected in Holocene tectonism and seismicity on either side of the San Andreas fault system extending as far west as the continental margin and as far east as the Sierra Nevada and the western Basin and Range province (Atwater, 1970; Minster and Jordan, 1987).

Immediately east of the Coast Ranges, the Great Valley and the adjacent Sierra Nevada form a relatively stable crustal block that extends the length of central California, from east of the Mendocino triple junction in the north to the Transverse Ranges in the south (Fig. 1). The Mesozoic crystalline basement within this block dips gently westward from the granitic peaks of the high Sierra on the east through the metamorphic belt of the western Sierran foothills, and then beneath the westward-thickening wedge of Quaternary through Mesozoic sedimentary rocks of the Great Valley.

Major, eastward-dipping normal faults form the tectonically active eastern escarpment of the Sierra Nevada, which defines the western margin of the extensional Basin and Range province. A series of Quaternary through recent volcanic centers have erupted both silicic and basaltic lavas along the base of the escarpment, and many of the range-front normal faults show several tens of meters of Holocene displacement (Wallace, 1984a; Clark and Gilespie, 1981).

The crust beneath central California thickens eastward from between 25 and 30 km under the Coast Ranges and the San Andreas fault system to more than 50 km under the high Sierra Nevada (Eaton, 1966; Prodehl, 1970; Pakiser and Brune, 1980; Oppenheimer and Eaton, 1984). The transition from the thick Sierran crust to the 30- to 35-km-thick crust beneath the adjacent Basin and Range province on the east remains poorly resolved (Priestley and others, 1982). Both seismic-refraction and gravity data, however, suggest that the transition may be displaced as much as 50 km eastward from the topographic boundary between the Sierra Nevada block and the Basin and Range province (Oliver, 1977; Hill, 1978).

HISTORICAL SEISMICITY

Figure 3, together with the seismicity map of North America (Engdahl and Rinehart, 1988), provides a historical context for our discussion of detailed seismicity patterns in central and northern California for the 1980–1985 interval illustrated in Figure 4.

The epicenters plotted in Figure 3 are from the data base compiled by Engdahl (this volume) for the seismicity map of North America. As Engdahl (this volume) emphasizes, the catalog for earthquakes in the region is not complete for M>5 earthquakes until after about 1910, and the minimum magnitude for earthquakes earlier than 1910 plotted in Figure 3a is M=5.75. The catalog is complete, however, for M≥3.5 events during the 1952–1979 interval plotted in Figure 3b.

Epicenters of M≥5 earthquakes during the 172-yr period from 1808 through 1979 (Fig. 3a) define an irregular cluster about the triple junction off Cape Mendocino and two diffuse-trending belts of activity. The first of these belts broadly follows the San Andreas transform boundary through the Coast Ranges of central California, and the second follows the eastern escarpment of the Sierra Nevada and the western margin of the Basin and Range province. This second band of epicenters continues its general northwestward trend north of Lake Tahoe, extending past Lassen Peak toward Mount Shasta in the southern Cascade Range (refer to Fig. 1). A third, somewhat broader belt of epicenters that extends northward from the knot of epicenters near Long Valley caldera in east-central California into the Basin and Range

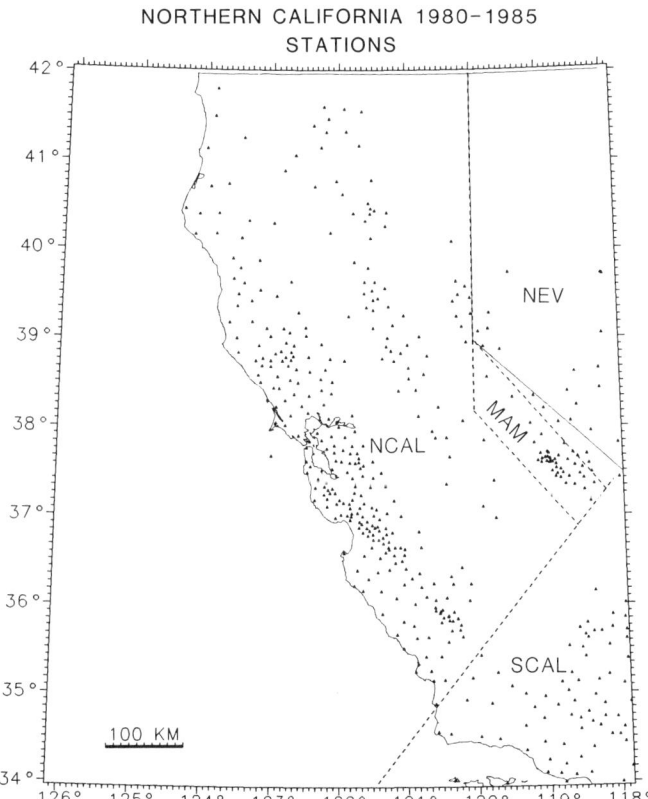

Figure 2. Telemetered seismograph stations in northern California. NCAL network telemetered to U.S. Geological Survey in Menlo Park, California; NEV network telemetered to University of Nevada, Reno; MAM stations shared between NCAL and NEV networks. Dotted lines indicate subsets of stations CA, MA, and NE used to compile frequency-magnitude statistics summarized in Figure 9 and Table 2.

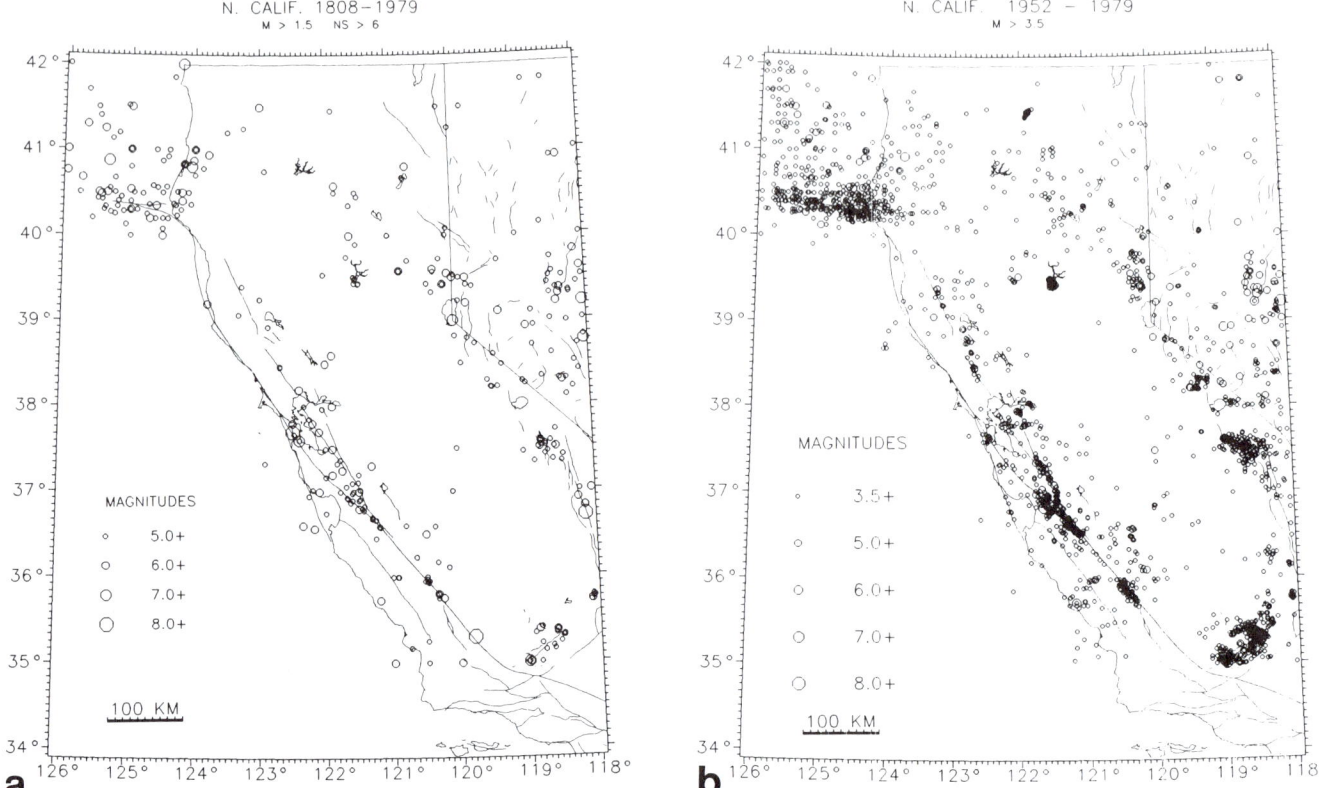

Figure 3. Historical seismicity in northern and central California. (a) M > 5 earthquakes for 1808–1979; (b) M > 3.5 earthquakes for 1952–1979. Epicenters and magnitudes from database compiled by Engdahl (this volume).

province and central Nevada is partially visible along the right margin of Figure 3. This belt is part of Wallace's (1981) eastern California–central Nevada seismic zone, and it is discussed in detail by Rogers and others (this volume). The considerably more numerous epicenters of M≥3.5 earthquakes during the 27-yr interval from 1952 to 1979 (Fig. 3b) clarify the same basic seismicity pattern more sparsely outlined by M>5 events during the 172-yr interval. These relations serve to emphasize that, in broad outline, the seismicity pattern in northern and central California remains relatively stable over periods of decades to perhaps centuries.

Some of the details that emerge from the M>3.5 seismicity in Figure 3b include: (1) linear clusters of epicenters defining short segments of the San Andreas fault system in central California, (2) a diffuse pattern of epicenters that widens northeastward from the vicinity of Clear Lake in the northern Coast Ranges across the northern one-fourth of the Great Valley (Bolt and Miller, 1971), and (3) widely scattered epicenters within the southern Sierra Nevada–Great Valley block.

A particularly notable aspect of seismicity along the San Andreas fault system in Figures 3a and b is the absence of epicenters along the section of the San Andreas fault that ruptured in 1906 with the M=8 San Francisco earthquake. A few earthquakes cluster in the vicinity of the 1906 epicenter near San Francisco and along the peninsula to the south, but the stretch of the fault that ruptured at least 150 km to the northwest with offsets as large as 6 m shows up as an aseismic zone in Figure 3. Much the same is true for the section of the San Andreas fault in southern California that ruptured in 1857 with the M=8 Fort Tejon earthquake (see Hutton and others, this volume) and the 1872 M=8 Owens Valley earthquake in eastern California. We will see in the following sections that this pattern of seismic quiescence along the rupture zones of the great historic earthquakes persists for earthquakes as small as M=1.5.

DETAILED SEISMICITY PATTERNS: 1980–1985

In broad outline, the pattern of M≥1.5 earthquakes in central California for the 6-yr period 1980 to 1985 (Fig. 4) repeats that defined by M>5 earthquakes for the 172-yr period through 1979 (Fig. 3a) and that defined by M>3.5 earthquakes for the 27-yr-period prior to 1979 (also compare the post-1978 [red] and the pre-1978 [blue] seismicity patterns for central California on the seismicity map of North America [Engdahl and Rinehart, 1988]. The resolution afforded by the combined seismographic networks (NCAL and NEV in Fig. 2), however, brings a detailed fabric into focus within the broad seismicity pattern not evident in the earlier data. We use the combined 1980–1985 epicentral data plotted in Figure 4 to emphasize some of the more obvious relations between detailed seismicity patterns and specific tectonic structures. To highlight the fabric defined by the seismicity patterns alone, we have separated the data into 1980–1982 and 1983–1985 intervals and replotted them in Figures 5a and b, respectively, without the background of faults or geographic information.

Coast Ranges and the San Andreas fault system

A network of northward branching, densely aligned epicenters dominates the seismicity pattern in the Coast Ranges. As first documented by Eaton and others (1970a, b), and illustrated in Figure 5, these dense linear clusters of epicenters coincide closely with the traces of the San Andreas fault and its major branches in central California (also see Ellsworth and others, 1982). More specifically, as suggested by Wesson and others (1973), they mark actively creeping sections of the San Andreas fault system. Wesson and others (1973) based their inference on the correlation of creep and microseismicity along the 200-km stretch of the main branch of the San Andreas fault north of Parkfield, but the relation appears to hold for the creeping stretches of the Hayward, Calaveras, Concord, and Greenville faults east of San Francisco Bay as well. The two subparallel lines of epicenters north of San Francisco Bay coincide with the Healdsburg-Rodgers Creek-Macaama fault on the west and the Green Valley-Bartlett Springs faults on the east (Cockerham and Herd, 1982), although it remains to be determined whether or not these faults show evidence of active creep. Interestingly, the northern arm of San Francisco Bay and its eastern extension (San Pablo Bay) coincide with an aseismic interruption in the linear seismicity patterns associated with the eastern branches of the San Andreas fault system.

Close inspection of the lower part of Figure 6 reveals that the densely aligned epicenters are systematically displaced 3 to 5 km west of the main branch of the San Andreas fault and a comparable distance east of the adjacent Calaveras fault. Much of this offset results from a strong P-wave velocity contrast across the faults that is not adequately modeled in routine hypocenter location calculations (Mayer-Rosa, 1973; Speith, 1981). When the hypocenter locations are carefully recalculated using an appropriate two-dimensional velocity structure, the systematic offsets are much reduced but not completely eliminated. The remaining offsets suggest that the fault planes deviate from the vertical, dipping 70° to the west beneath the trace of the San Andreas fault (Pavoni, 1973; Spieth, 1981) and dipping 80° to the east beneath the trace of the Calaveras fault (Reasenberg and Ellsworth, 1982).

The main section of the San Andreas fault north of San Francisco is nearly devoid of earthquake activity, as is the section to the south of Parkfield within the southern Coast Ranges. These sections of the San Andreas fault ruptured with the great (M=8) earthquakes of 1906 and 1857, respectively, and are currently

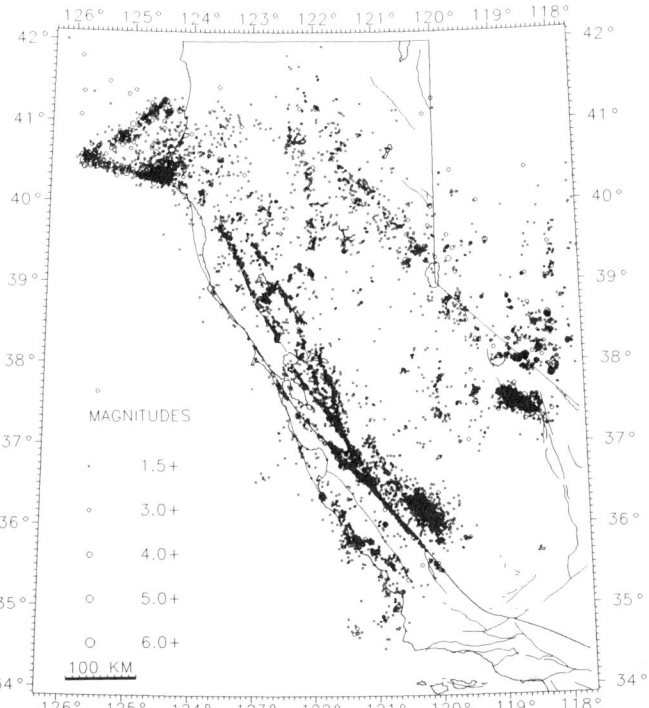

Figure 4. Northern and central California seismicity for 1980–1985. Faults are same as plotted in Figure 1.

locked, according to the results of repeated geodetic surveys across the fault (Savage, 1983; Prescott and Yu, 1986). As described by Allen (1981) and Hutton and others (this volume), a comparable quiescence applies to the 1857 break in southern California as well. This quiescence is most pronounced along the section of the fault that produced the largest offset in 1906. Fewer than a dozen small earthquakes were located along the fault north of San Francisco during the 6-yr period represented in Figures 4, 5, and 6 (the detection threshold is M = 1.5 to 2.0 along this section of the fault for most of 1980–1985). As is the case with the historic data (Fig. 3), however, a number of earthquakes occurred along the section of the fault near the probable position of the 1906 epicenter just west of San Francisco (Boore, 1977), and a comparable number of events scatter about the fault south of the epicenter where offsets were less than half of those to the north (see Thatcher and Lisowski, 1987).

Scattered clusters of epicenters in the Coast Ranges form a mottled background to the dense alignment of epicenters associated with the creeping segments of the San Andreas fault system. Aftershocks of the M=6.7 Coalinga earthquake of May 2, 1983, which was located 20 km east of the San Andreas fault at the eastern margin of the southern Coast Ranges, form the most pronounced of these clusters (see Figs. 4 and 5).

The background seismicity level within the southern Coast

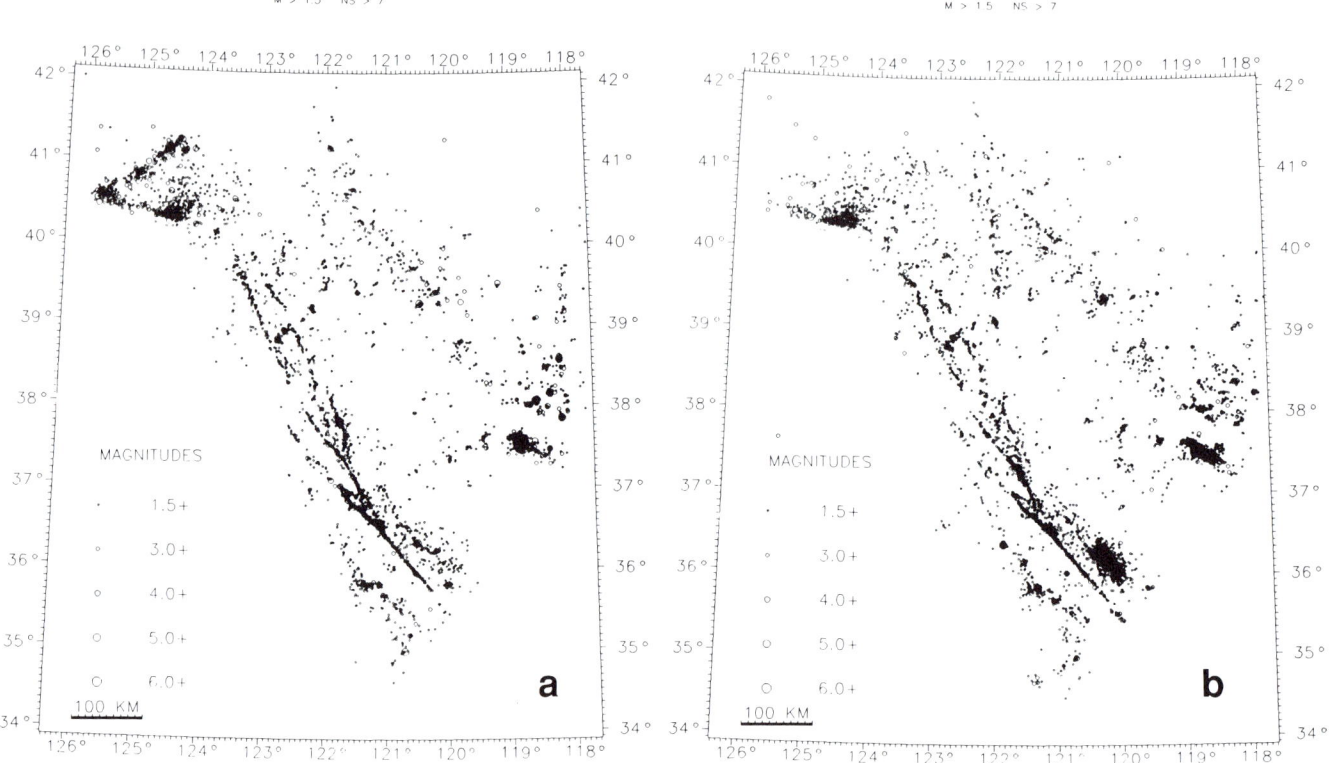

Figure 5. Northern and central California seismicity for 1980–1985 without geographic or structural reference lines. (a) 1980–1982 (pre-Coalinga); (b) 1983–1985 (post-Coalinga).

Ranges is distinctly more intense in the Franciscan block east of the San Andreas fault than in the granitic Salinian block to the west, and as emphasized in Figure 5, this is true both prior to and following the Coalinga event. Indeed, the Salinian block appears as a 50-km-wide band of relative quiescence sandwiched between higher levels of seismicity associated with the Franciscan blocks on either side. The edges of the Salinian block are marked distinctly by concentrated earthquake activity along the San Andreas fault on the east and somewhat more scattered activity along the Sur-Nacimiento fault zone on the west (compare Figs. 1 and 4). The Rinconada fault within the Salinian block appears to be nearly aseismic except, perhaps, toward the south where it approaches the Sur-Nacimiento fault zone. The small cluster of epicenters just east of the midpoint of the Rinconada fault represents a persistent spot of microearthquake activity at depths of 8 to 10 km near the San Ardo oil field (Poley, 1988). Scattered offshore seismicity in the Franciscan block west of the Sur-Naciemento fault may in part be associated with slip on the San Gregorio–Hosgri fault system that parallels the coast line (Gawthrop, 1978). A weakly defined linear trend of epicenters emerges from the diffuse seismicity in the Franciscan block north of the Coalinga aftershock zone. The northern half of this trend is subparallel with the Calvareras fault and coincides with the Ortigalita fault, the northern end of which passes beneath San Luis Reservoir (LaForge and Lee, 1982).

Background seismicity in the northern Coast Ranges (Fig. 6) forms a diffuse pattern encompassing the two subparallel lineations of epicenters along the Maacama–Healdsberg–Rodgers Creek and Green Valley–Bartlett Springs fault zones. This diffuse seismicity widens northward with the Coast Ranges toward the Mendocino triple junction and the southern end of the Klamath Mountains (refer to Fig. 1). The western half of the dense cluster of epicenters just south of Clear Lake between the subparallel lineations reflects the persistent microearthquake activity associated with the Geysers geothermal field (Oppenheimer, 1986). Notably, the seismic quiescence characteristic of the 1906 break of the San Andreas fault seems to extend into adjacent sections of the Coast Ranges for some 20 to 30 km.

Eastern California, the Sierra Nevada escarpment, and the Basin and Range province

In contrast to the densely aligned epicenter that define the San Andreas fault system, the seismicity data in Figure 4 show the northwest-trending band of earthquake activity in eastern California to be a 50-km-wide zone defined by loosely aligned clusters of epicenters along the Sierra Nevada escarpment to Lake Tahoe and beyond to the southern Cascades (largely coincident with the Sierra Nevada–Great Basin boundary zone of Van Wormer and Ryall, 1980). The even broader, north-trending

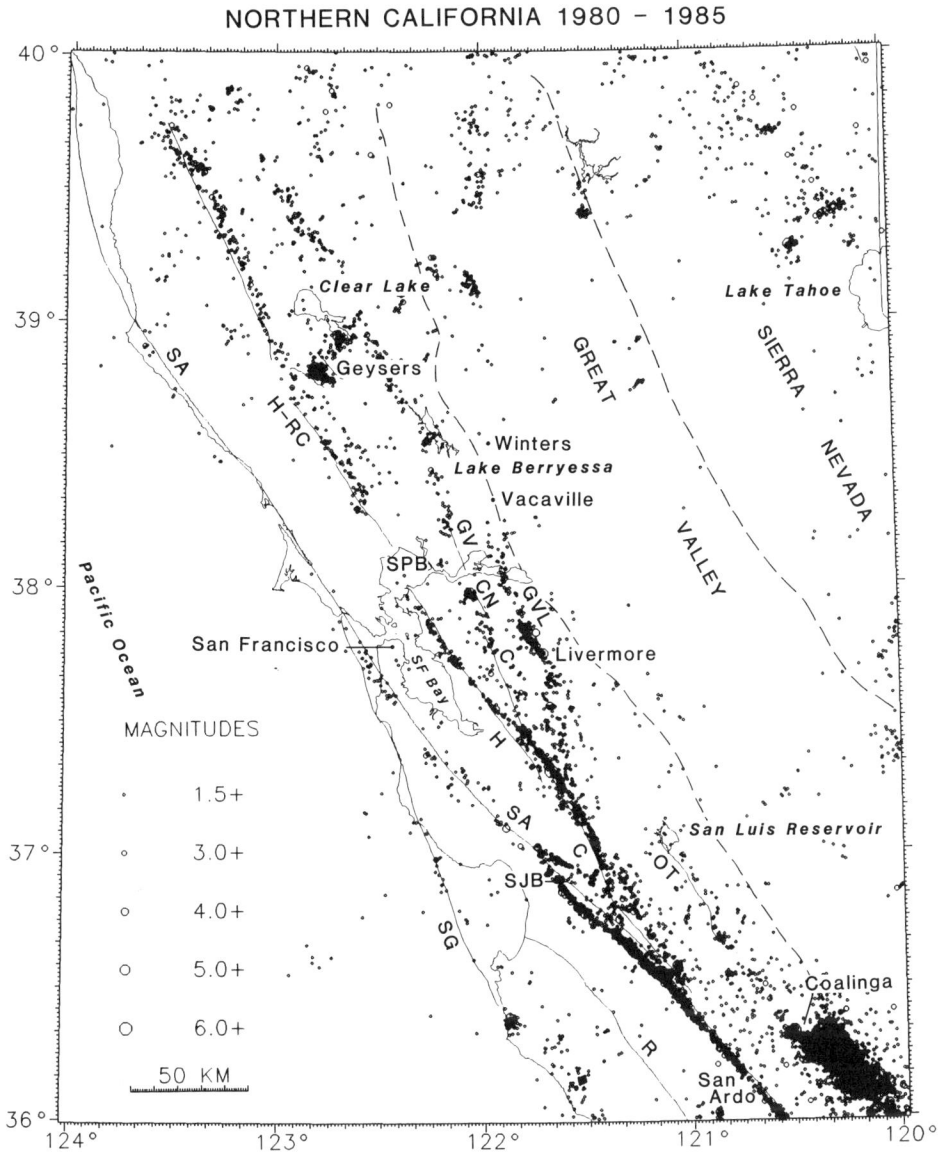

Figure 6. Enlarged map showing details in seismicity pattern along the San Andreas fault and its major branches through the central Coast Ranges for 1980–1985.

band of activity extending into central Nevada and the Basin and Range province consists of loosely grouped clusters of epicenters (see Rogers and Corbett, this volume).

The most pronounced of these seismicity clusters is located where the bands meet in the region between Long Valley caldera, the northern end of Owens Valley, and the White Mountains (Figs. 1 and 4). This site has consistently produced more M=5 to M=6 earthquakes since 1978 than any other in the conterminous United States (Hill and others, 1985; Savage and Cockerham, 1987). Furthermore, as indicated in Figure 3, it has been a persistent source of moderate earthquakes for at least the last century (Van Wormer and Ryall, 1980; Cramer and Toppozada, 1980). As is evident from the seismicity map for North America (Engdahl and Rinehart, 1988) this Long Valley–White Mountains seismicity cluster forms a conspicuous knot at the junction of the eastern California and central Nevada seismicity belts. A third belt extends eastward from the Long Valley–White Mountains cluster across southern Nevada (see Rogers and others, this volume) to join with the Intermountain seismic belt in southern Utah (see Smith and Arabasz, this volume).

Although this Long Valley–White Mountains cluster has produced numerous moderate (M=5 to M=6) earthquakes this century, it remains a seismic gap (the White Mountains seismic gap) with respect to the series of major earthquakes that have ruptured the surface along the north-trending eastern California–central Nevada seismic belt in historical time (Wallace, 1981; Hill and others, 1985). The largest of these, the M=8 Owens Valley earthquake of 1872, formed a 100-km-long rupture in the floor of

Owens Valley with right-lateral offsets of up to 11 m (Lubetkin and Clark, 1988). The northern end of this rupture reached to within roughly 20 km of the southern margin of the Long Valley–White Mountains seismicity cluster (Figs. 1 and 4). Major earthquakes to the north along this belt include the M=7.3 event near Ceder Mountain (38°26' to 38°44'N) in 1932, the sequence of four M=6.5 to M=7.1 events in the Dixie Valley–Fairview Peak region (39°10' to 39°49'N) in 1954, and the M=7.6 Pleasant Valley event (40°10' to 40°41'N) in 1915. Each of these earthquakes involved a combination of right-lateral and normal slip along generally north-striking rupture surfaces (Wallace, 1984a and b; Rogers and Corbett, this volume).

Figure 7 reveals a distinct fabric within the Long Valley–White Mountains seismicity cluster. This fabric is characterized by subparallel, north-northeast lineations of densely aligned epicenters in the Sierra Nevada block south of the caldera and by a west-northwest–trending band of epicenters within the south moat of the caldera (Fig. 7). The north-northeast lineations in the Sierran block represent aftershock activity of the four M=6 earthquakes of May 25–27, 1980, and the elongate cluster within the caldera represents recurring earthquake swarms, the largest of which occurred in January 1983, and included two M=5.2 earthquakes. The triangular-shaped cluster of epicenters to the southeast represent aftershocks of the M=5.8 Round Valley earthquake of November 1984.

None of these seismicity patterns relates in any simple way to the large north–northwest-striking normal faults that form the eastern escarpment of the Sierra Nevada. Fault plane solutions for earthquakes in these sequences show predominantly strike-slip mechanisms with left-lateral motion along north-northeast-striking planes and right-lateral motion along west-northwest planes (Cockerham and Corbett, 1987; Vetter and Ryall, 1983). None of the M=5.5 to M=6.0 earthquakes in the sequence from 1978 to 1985 have produced obvious tectonic surface ruptures.

The seismicity band extending to the northwest follows the structural trend of the Sierra Nevada escarpment as far as Lake Tahoe (Van Wormer and Ryall, 1980; Ryall and Ryall, 1981). Northwest of Lake Tahoe, however, the relation between this seismicity band and mapped structures becomes less clear. Jennings and others (1975) show a broad zone of subparallel, northwest-striking faults pervading much of northeastern California, and it appears that the belt of seismicity is concentrated along the southwestern margin of this zone, which broadly follows the topographic crest of the northern Sierra Nevada. Jennings and others (1975) also indicate, however, that displacements on the faults within this zone are largely pre-Quaternary between Lake

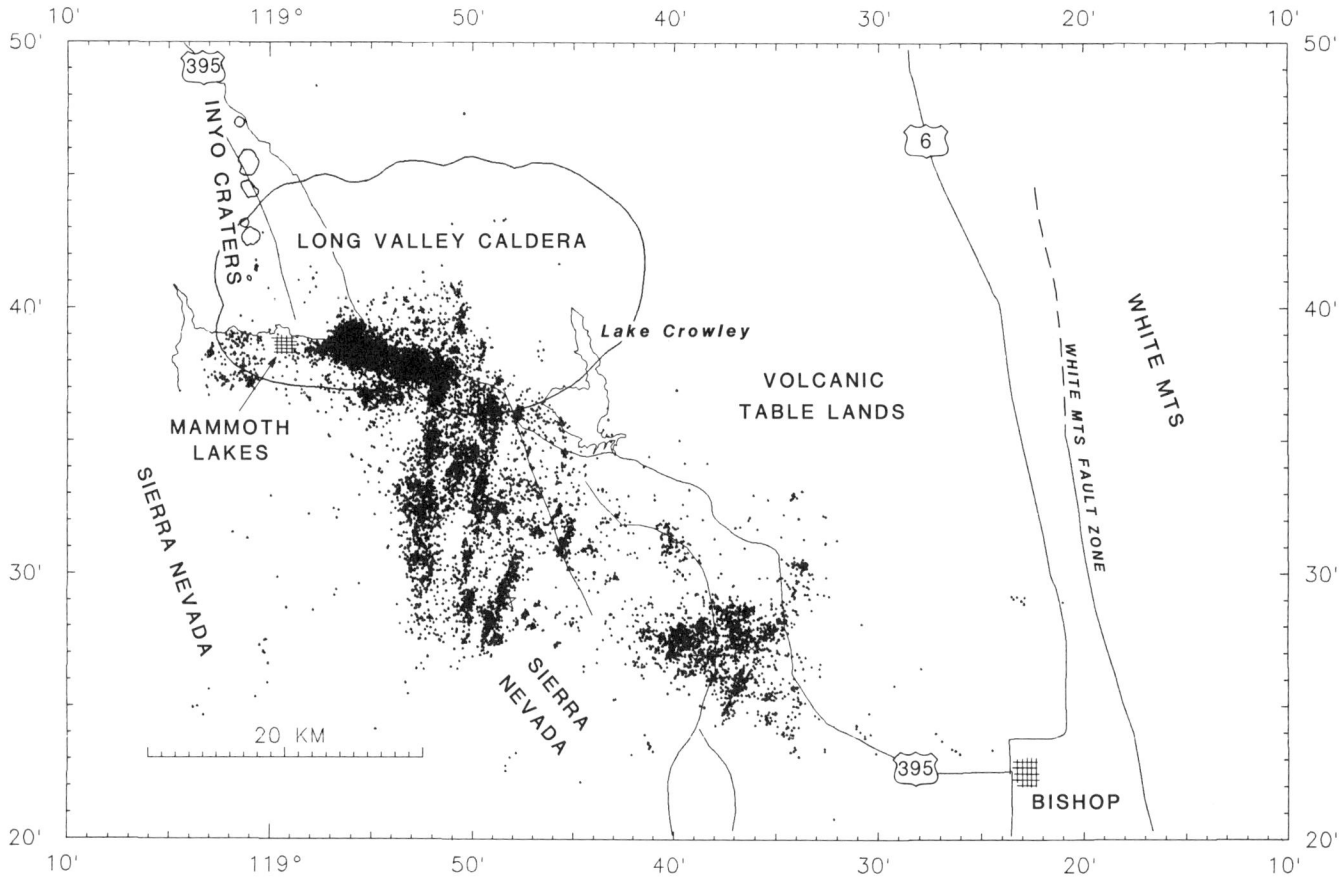

Figure 7. Enlarged map showing details in seismicity pattern within the Long Valley–White Mountains region of eastern California for 1980–1985.

Tahoe and Lassen Peak, although northwest of Lassen, Quaternary fault displacements are common, as are widespread Quaternary volcanic centers. Activity in the vicinity of Lassen itself is dominated by recurring swarms of small, shallow earthquakes associated with the geothermal system immediately south of Lassen Peak (Klein, 1979; Walter and others, 1984). Next to the Long Valley-White Mountains cluster, the most seismically active zone within this belt coincides with the cluster of epicenters 30 to 35 km northwest of Lake Tahoe. A M=5 earthquake occurred near the southwest end of this cluster on November 28, 1980 (see Figs. 4 and 5), and the relatively dense cluster of epicenters immediately to the northeast represents persistent activity (aftershocks?) near the epicenter of the 1966 Truckee earthquake (M=6.0). The latter involved left-lateral slip on a northeast-striking plane (Tsai and Aki, 1970) conjugate to the overall northwest trend of the seismicity belt.

Great Valley and Sierra Nevada foothills

Broad zones of diffuse seismicity encroach into the northern and central sections of the relatively stable Great Valley-Sierra Nevada block (Fig. 4). Earthquakes occurring within these zones are some of the deepest intraplate events within the contiguous United States, with maximum focal depths in excess of 40 km. To the north, isolated clusters of epicenters align to form a north-trending lineation along the axis of the Great Valley, beginning some 50 km east of Clear Lake and extending to Lake Shasta, beyond which it merges with the northwest-trending eastern California seismicity belt. Eaton (1986) has suggested that this lineation may be related to a structural trend that continues southward through the epicenter of the M=6.5 Vacaville/Winters earthquake of 1892, which was located 20 km east of Lake Berryessa (see Figs. 1 and 3a). A more scattered zone of seismicity extends across the central section of the Great Valley toward the Long Valley-White Mountains seismicity cluster in eastern California. The clusters of epicenters within this zone tend to be more intense beneath the Sierra Nevada foothills than beneath the Great Valley (see Wong and Savage, 1983).

Mendocino triple junction

Uhrhammer (this volume) discusses seismicity of the Mendocino triple junction in detail. Here we simply note that the interaction of the Gorda, North American, and Pacific plates results in a broad zone of persistent earthquake activity in the vicinity of Cape Mendocino and a broad area to the north and east. The M=7.0 event of November 8, 1980, which ruptured the southeastern tip of the Gorda plate seaward of the subduction zone with left-lateral displacements (see Wilson, 1986), was the largest earthquake to occur in California during the 1980–1985 period. Its aftershocks defined the pronounced northeast-striking lineation of epicenters off the coast north of Cape Mendocino evident in Figure 4.

Temporal seismicity variations

The series of maps in Figure 8 illustrates the annual seismic activity in central and northern California for the 1980–1985 period; the corresponding frequency-magnitude statistics are summarized in Figure 9. A cursory look at these annual seismicity maps affirms that broad outlines of the seismicity pattern persist from year to year, while details within the outlines fluctuate from one year to the next. The most pronounced of these changes are associated with the occurrence of moderate to large (M>5.5) earthquakes and the rapidity with which their aftershock sequences decay (see Table 1 for a list of M>5.5 earthquakes in central California during the 1980–1985 period).

Aftershock activity of the M=7 earthquake off Cape Mendocino in November 1980, for example, decayed exceptionally quickly, leaving little evidence of the occurrence of this major earthquake in the seismicity record two months later. Aftershocks of the January 1980, M=5.8 Livermore earthquake along the Greenville fault also died off within about two months of the mainshock (compare Figs. 8a and b). In contrast, the long-lived aftershock sequence of the M=6.7 Coalinga earthquake of May 2, 1983, has persisted as a dense cluster of epicenters in the southern Coast Ranges east of the San Andreas fault for more than 2½ yr (also see Fig. 5). The long-lived seismicity cluster in the Long Valley-White Mountains region has persisted throughout the 6-yr period covered in Figure 8. Ryall and Ryall (1981), however, document 1½ yr of nearly complete quiescence in the Long Valley-White Mountains region preceding the onset of this activity in October 1978.

More diffuse seismicity patterns show fluctuations from year to year as well, although some of these fluctuations reflect local changes in the configuration and resolution of the seismic networks. To provide a quantitative indication of network-dependent fluctuations in the seismicity patterns, we summarize the frequency-magnitude data for each year from 1980 through 1985 in Table 2 and Figure 9. These frequency-magnitude data are based on the complete set of earthquakes plotted in Figures 4 through 8 and, thus, they include foreshocks and aftershocks associated with larger events.

The CA data, for example, fit the Gutenberg-Richter relation ($\text{Log}N = a - bM$, where N is the cumulative number of earthquakes of magnitude greater than M) for $M \geqslant 1.5$ during the entire 6-yr period, as do the combined data for the 1982–1985 period. To the extent that the coda-duration method for estimating the magnitudes of $M<4$ events as described in Lee and Stewart (1981) provides a reliable, unbiased measure of earthquake size (see Bakun, 1984), the frequency-magnitude statistics in Figure 9 indicate that the seismicity data for 1980 through 1985 are essentially complete for most of central and northern California $M \geqslant 1.5$ earthquakes.

A significant fraction of the short-fall in the number of $M \leqslant 2.5$ events in the combined (ALL) data for 1980 and 1981 must be attributed to the Long Valley caldera seismicity cluster in

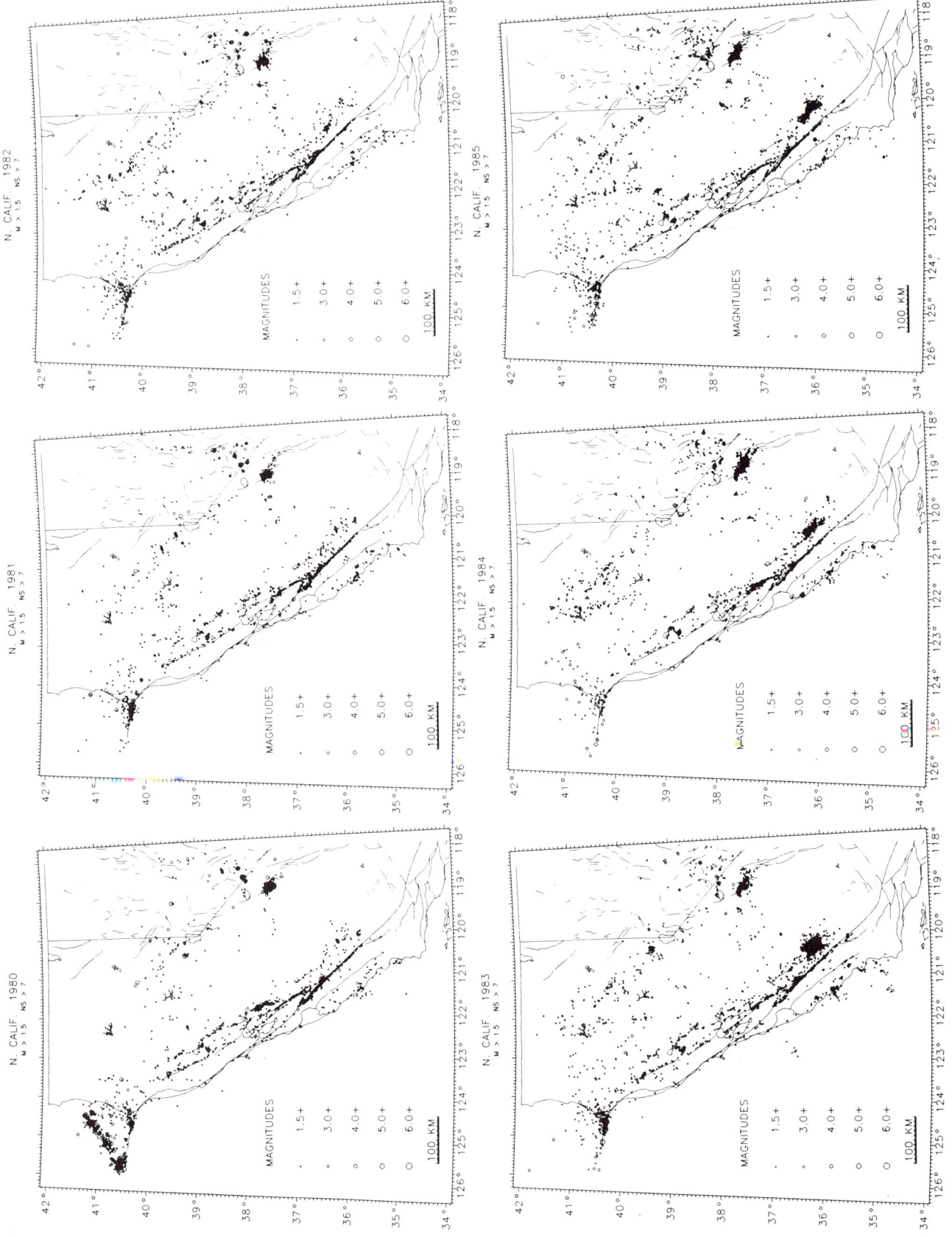

Figure 8. Variations in the 1980–1985 seismicity pattern on a year-by-year basis.

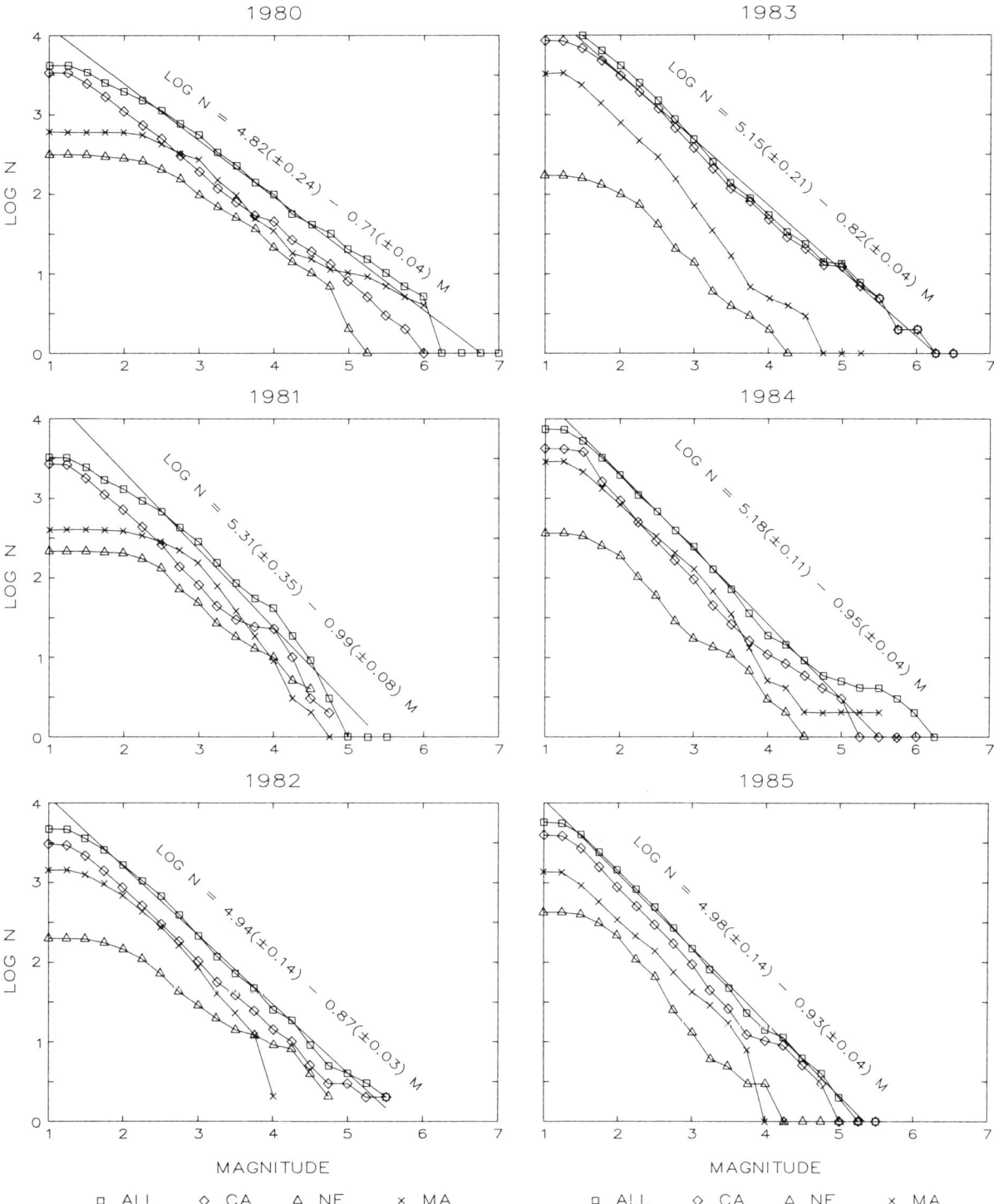

Figure 9. Annual frequency-magnitude curves for the earthquakes plotted in Figure 8. N is the cumulative number of earthquakes during a given year with magnitudes $M > M_o$, where we have taken $1 < M_o < 7$ in $\frac{1}{4}M$ steps. CA, MA, and NE refer to statistics for the regions illustrated in Figure 2. ALL refers to the combined statistics for all three regions. Linear Reg. shows the best-fitting line of the form LOG $N = a + bM$ through the combined (ALL) data over the magnitude interval between the arrows (see Table 2).

TABLE 1. LIST OF M≥5.5 EARTHQUAKES FOR 1980-1985

y	m	d	Time hr	min	Lat	Long	Depth (km)	Mag
80	01	24	1900	9.14	37°50.2'	122°46.88'	11.9	5.9
80	05	25	1633	44.13	37°35.38'	118°49.57'	6.6	6.1
80	05	25	1649	26.71	38°68.68'	118°53.97'	7.1	6.0
80	05	25	1944	48.10	37°19.86'	118°49.80'	9.9	6.1
80	05	25	2035	48.00	37°34.98'	118°50.10'	8.4	5.7
80	05	26	1857	55.76	37°31.98'	118°51.83'	6.6	5.5
80	05	27	1451	56.70	37°26.88'	118°47.10'	16.4	6.2
80	06	28	0058	42.30	37°35.10'	118°48.78'	6.7	5.9
80	11	08	1027	32.52	41°06.96'	124°39.87'	6.0	7.0
80	11	28	1821	12.06	39°16.31'	120°27.94'	22.7	5.5
81	09	30	1153	26.20	37°35.21'	118°52.24'	5.0	5.7
82	10	15	1058	39.12	32°50.50'	125°49.27'	<10	5.7
82	10	25	2226	03.67	36°19.31'	120°30.44'	11.0	5.5
83	05	02	2342	38.14	36°13.96'	120°18.57'	10.0	6.7
83	05	02	2347	28.01	36°12.65'	120°19.05'	10.2	5.6
83	07	22	0239	54.07	36°14.44'	120°24.53'	7.4	6.0
83	08	24	1336	27.90	40°21.98'	125°06.35'	17.0	5.5
83	12	20	1041	05.21	40°25.44'	125°31.35'	15.0	5.7
84	04	24	2115	18.78	37°18.56'	121°40.68'	8.4	6.2
84	09	10	0314	11.99	40°30.18'	126°49.86'	6.6	6.4
84	11	23	1808	20.30	37°27.70'	118°35.57'	12.8	5.8
84	11	23	1912	34.62	37°25.97'	118°36.71'	14.2	5.5
85	08	04	1201	55.76	36°08.69'	120°09.16'	11.8	5.7

Lat = latitude; Long = longitude; Mag = magnitude; y m d = year/month/day.

eastern California, which produced an exceptionally large number of M≥2.5 events in 1980 and 1981 prior to installation of the locally dense seismograph network in mid-1982. Similarly, based on the position of the corner in the NE frequency–magnitude curves, the completeness threshold for northeastern California and adjacent parts of western Nevada varies between M=2.0 and M=2.5.

Parameters for the "best-fitting" Gutenberg-Richter relation through the combined frequency-magnitude data (ALL) are listed in Table 2. The straight line plotted through ALL represents the outcome of linear regression applied to the magnitude interval specified in Table 2 for each year from 1980 through 1985. The resulting annual b-values, which, within the 95 percent confidence limits, agree with those estimated using the maximum likelihood method of Aki (1965), cluster around an average value of b = 0.9 for this 6-yr interval. Marginally significant deviations from this average occur in 1980 (b = 0.7 to 0.8) and in 1981 and 1985 (b = 1.0). Within 95 percent confidence limits, the associated a-values show no significant variations from an average value of a = 5.1 (Table 2). This nearly constant annual a-value for the combined data set results from mutually compensating fluctuations in annual seismicity levels between the broad central and northern California region (CA in Figs. 2 and 9) and the more concentrated activity in the Long Valley–White Mountains cluster of eastern California (MA in Figs. 2 and 9). In 1983, for example, the aftershocks of the vigorous Coalinga earthquake sequence lift the CA frequency-magnitude curve well above the MA and NE curves. In 1980 through 1982, however, the numerous events in the 1980 Mammoth Lakes earthquake sequence and the recurring earthquake swarms within Long Valley caldera elevate the MA curve so that it overlaps the CA curve. At this stage, we see no reason to attribute special meaning to a nearly constant a-value during this 6-yr period beyond a coincidence of timing and the particular combination of regions included in this seismicity study.

SEISMICITY CROSS SECTIONS AND FOCAL DEPTHS

To illustrate the distribution of earthquake focal depths in central and northern California for the 1980–1985 period, we have constructed nine cross sections combining data for the entire period. Figure 10 shows the locations of the cross sections on the 1980–1985 seismicity map; each of the cross sections plotted in Figure 11 was generated by projecting the hypocenters of all earthquakes within a given rectangle in Figure 10 onto a vertical plane parallel with the long sides of the rectangle. Each of the sections is plotted with a two-fold vertical exaggeration.

When looking at these depth sections, bear in mind that the epicentral coordinates of a hypocenter are usually better constrained than the associated focal depth. This is particularly true for earthquakes occurring near or beyond the margins of a seismic network and in areas where the station spacing is significantly greater than the earthquake focal depths. Also, bear in mind that larger errors associated with focal depths may, in some cases, artificially smear a pattern of hypocentral solutions in the vertical dimension, giving the false impression of a pipe-like structure.

The pattern of focal depths painted by the 1980–1985 earthquake data provides reasonably clear definition of the thickness of the brittle seismogenic crust beneath central California. In many areas, these data show a relatively sharp transition with depth, from densely distributed hypocenters to a seismically quiescent lower crust. The sharp transition beneath the central Coast Ranges, for example, indicates that the seismogenic crust associated with this section of the San Andreas fault system is 12 to 15 km thick, or about half the thickness of the structural crust as defined by the Mohorovičić (M) discontinuity (see Figs. 11d through g). Sibson (1982) has suggested that this transition to seismic quiescence marks the transition from brittle to ductile behavior of the crust (also see Meisner and Strehlau, 1982).

Broadly considered, the hypocentral data summarized in Figure 11 show a distinct eastward thickening of the seismogenic crust from less than 15 km beneath the Coast Ranges to between 40 and 50 km beneath the weakly seismogenic and relatively cold crust of the Great Valley–Sierra Nevada block. Farther east, the seismogenic crust thins rapidly to less than 20 km beneath the eastern Sierra Nevada–Great Basin boundary zone. All of the earthquakes associated with the Long Valley caldera–White Mountains seismicity cluster, for example, have focal depths less than 20 km; the majority are confined to the upper 15 km of the crust.

The distribution of maximum focal depths also suggests some interesting variations in the thickness of the seismogenic crust beneath the central California Coast Ranges (see the longitudinal profiles H-H′ and I-I′ in Fig. 11). The most pronounced relief occurs beneath the northern Coast Ranges between the isolated cluster of 20-km-deep earthquakes beneath the north arm of San Francisco Bay (Suisun Bay) and the Geysers geothermal area some 90 km to the north where maximum focal depths are only 5 to 6 km (Oppenheimer, 1986). On a somewhat broader scale, the focal depth data suggest an overall tendency of the seismogenic crust to thin northward beneath the northern Coast Ranges from the 12 to 15 km thickness typical of the central Coast Ranges to roughly 10 km just south of the Mendocino triple junction. In the vicinity of the triple junction, however, the seismogenic crust thickens abruptly to more than 30 km, coincident with the southern edge of the subducting Gorda plate beneath northern California (see the chapter by Uhrhammer, this volume). The landward extension of the southern edge of the Gorda plate cuts obliquely across profile A-A′ (Figs. 10, 11a), and the thickened seismogenic crust formed by the subducted section of the Gorda plate beneath the North American plate is evident in the 20- to 30-km-deep earthquakes beneath the Great Valley in this profile.

Two additional aspects of the depth distribution for earthquake foci associated with the San Andreas fault system deserve special note. The first is the clearly defined vertical alignment of hypocenters beneath the traces of the San Andreas fault and its major branches in central California evident in the transverse profiles (Figs. 11b through g). This emphasizes that the dense alignment of epicenters coincident with the creeping sections of these faults in map view represents a nearly vertically oriented, planar distribution of hypocenters over the upper 10 to 15 km of the fault surfaces. The second involves the variation of maximum focal depths beneath the central creeping section of the San Andreas fault (profile I-I′ in Fig. 12). Note, in particular, the tendency of maximum focal depths within the creeping section to increase toward the locked sections of the fault associated with the 1906 break to the north and the 1857 break to the south. Even more pronounced, however, is the rapidity with which minimum focal depths increase as one moves northward into the 1906 break and southward into the 1857 break from the creeping section of the fault (see Poley and others, 1987).

FOCAL MECHANISMS AND KINEMATICS OF THE BRITTLE CRUST

The dominant contribution to the seismogenic deformation of the brittle crust in a specified region is associated with the

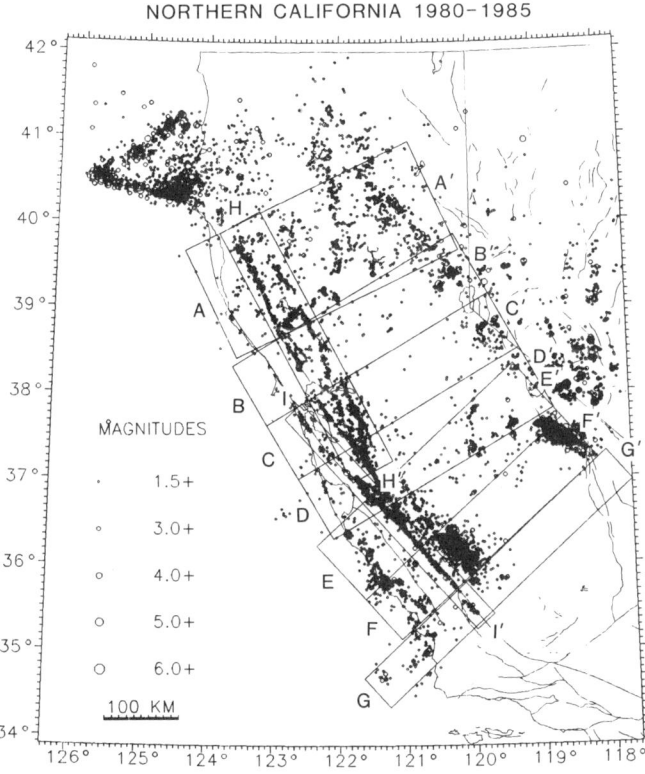

Figure 10. Key to cross section for 1980–1985 seismicity plotted in Figures 11 and 12. Hypocenters of earthquakes within a given rectangle are projected onto a vertical plane parallel with the long axis of the rectangle to generate the associated cross section.

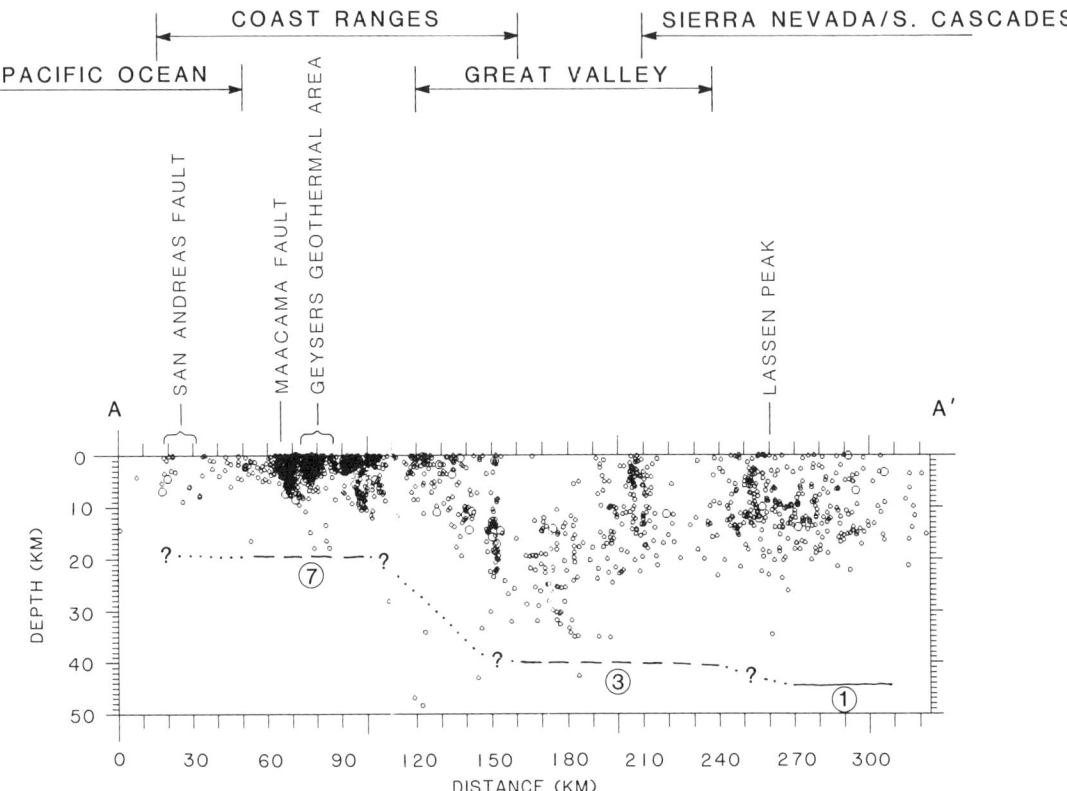

Figure 11 (above and following 3 pages). Transverse cross sections A-A' through G-G'. Crustal thickness based on published seismic-refraction and gravity interpretations: solid and dashed lines indicate reversed and unreversed seismic-refraction coverage, respectively; dotted line based on regional P-wave travel-time and potential field data. Sources: (1) Eaton (1963, 1966); (2) Holbrook and Mooney (1987); (3) Jachens and Griscom (1983); (4) Macgregor-Scott and Walter (1988); (5) Trehu and Wheeler (1987); (6) Walter and Mooney (1982); and (7) Warren (1981).

largest earthquakes within the region. Thus, over the long run, the great (M=8) plate-boundary earthquakes, such as the 1857 and 1906 events that ruptured the main branch of the San Andreas fault with offsets of 5 to 10 m, dominate the seismogenic deformation of California. The growing number of published focal mechanisms for earthquakes throughout central and northern California, however, provides an ever-clearer picture of the kinematic pattern of brittle deformation of the crust during intervals between these great plate-boundary events. In Figure 13, we show the focal mechanisms of some of the larger (M>4) of these earthquakes.

The San Andreas fault system

The overwhelming number of earthquakes occurring along the San Andreas fault or its major branches in the central California Coast Ranges have focal mechanisms that are consistent with right-lateral slip on near-vertical planes with the dextral plane parallel with the local strike of the fault trace (see, for example, Ellsworth, 1975; Ellsworth and others, 1982). This relation is emphasized in Figure 13 by the focal mechanism of the M=5.6 Parkfield earthquake of 1966, which occurred along the San Andreas fault in the southern Coast Ranges (660628 in Fig. 13), the mechanisms for the 1979 M=5.9 and 1984 M=6.2 Coyote Lake and Morgan Hill earthquakes along adjacent segments of the Calaveras fault (790806 and 840424, respectively), and the three M=4.3 to M=4.8 events along the creeping segment of the San Andreas fault (820818, 820625, and 820811). Evidently the active surfaces of the San Andreas fault system, which span the entire thickness of the brittle crust (Fig. 11), serve as pre-existing planes of weakness controlling the local slip direction during brittle failure, not only for major plate-boundary earthquakes but also for small-to-moderate earthquakes along the creeping sections of the fault.

Focal-mechanism studies of microearthquakes occurring along the subparallel Healdsburg–Rodgers Creek–Maacama and the Green Valley–Bartlett Springs faults north of San Francisco Bay also show a predominance of strike-slip mechanisms consistent with right-lateral displacements parallel with the strike of the local fault trace. They also, however, show a substantial number of mixed mechanisms. For the most part, these mixed mechanisms involve oblique-normal slip and tend to be associated with offsets (predominantly right-stepping) between the multiple en-echelon segments that characterize these less well developed

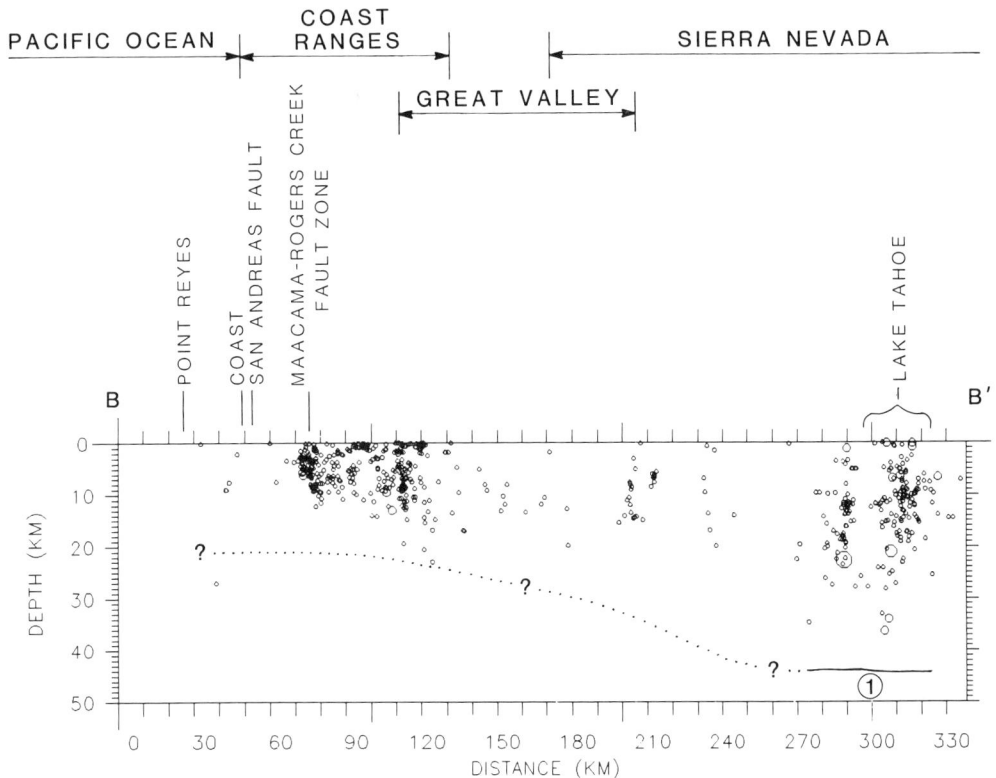

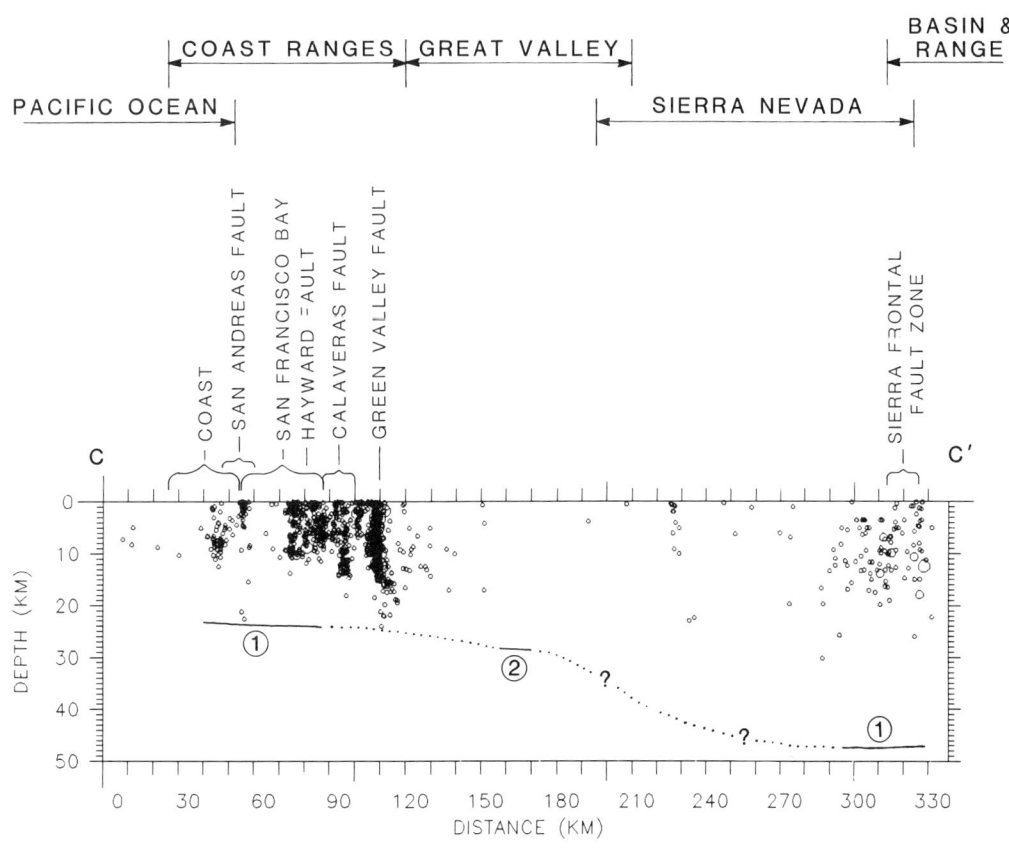

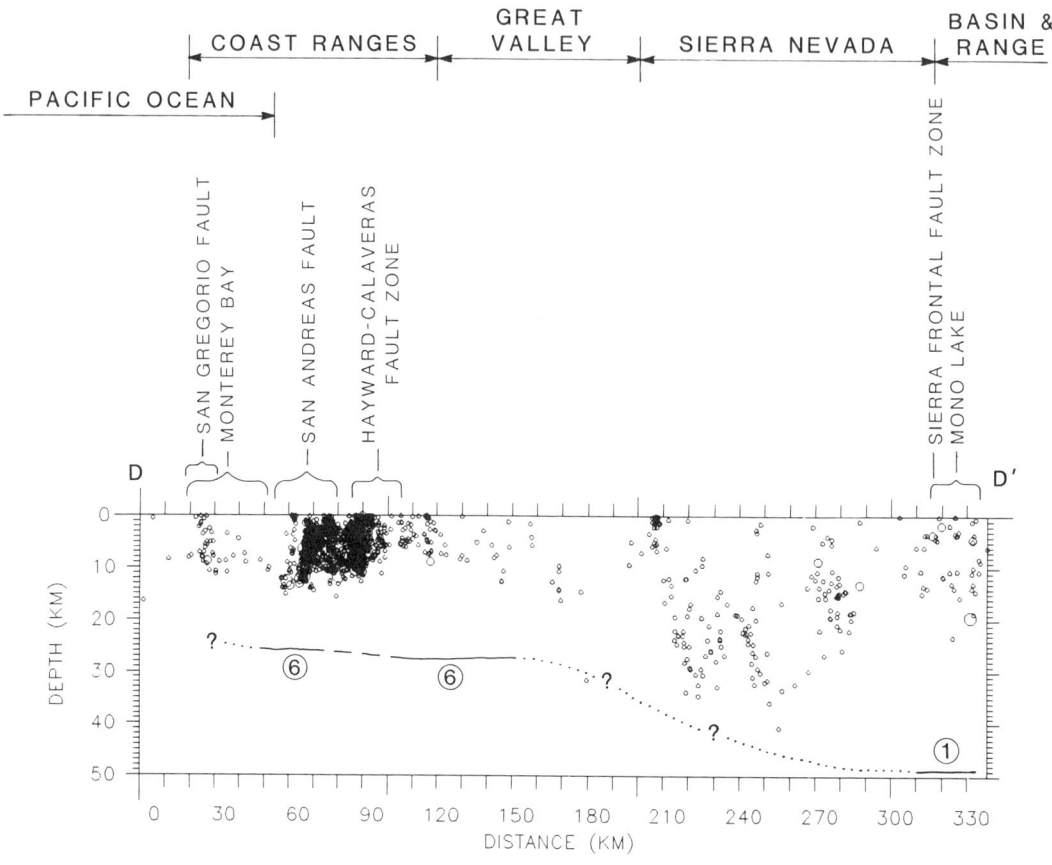
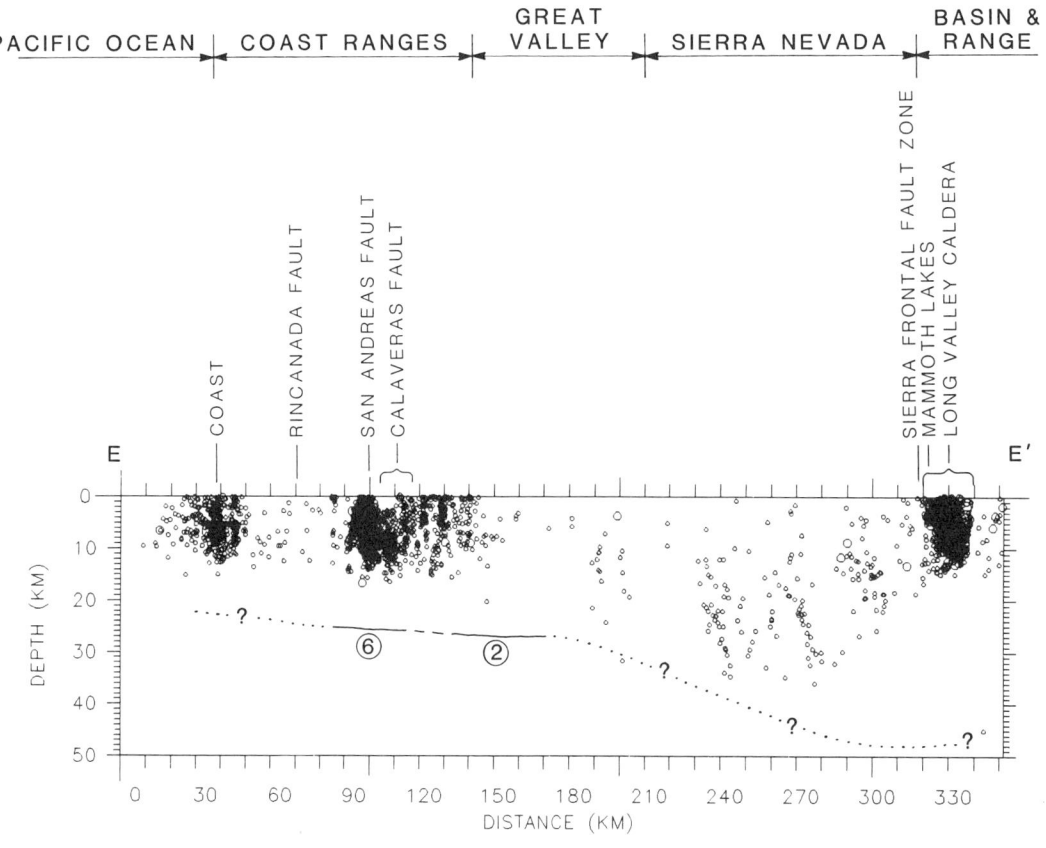

Seismotectonic fabric of central California

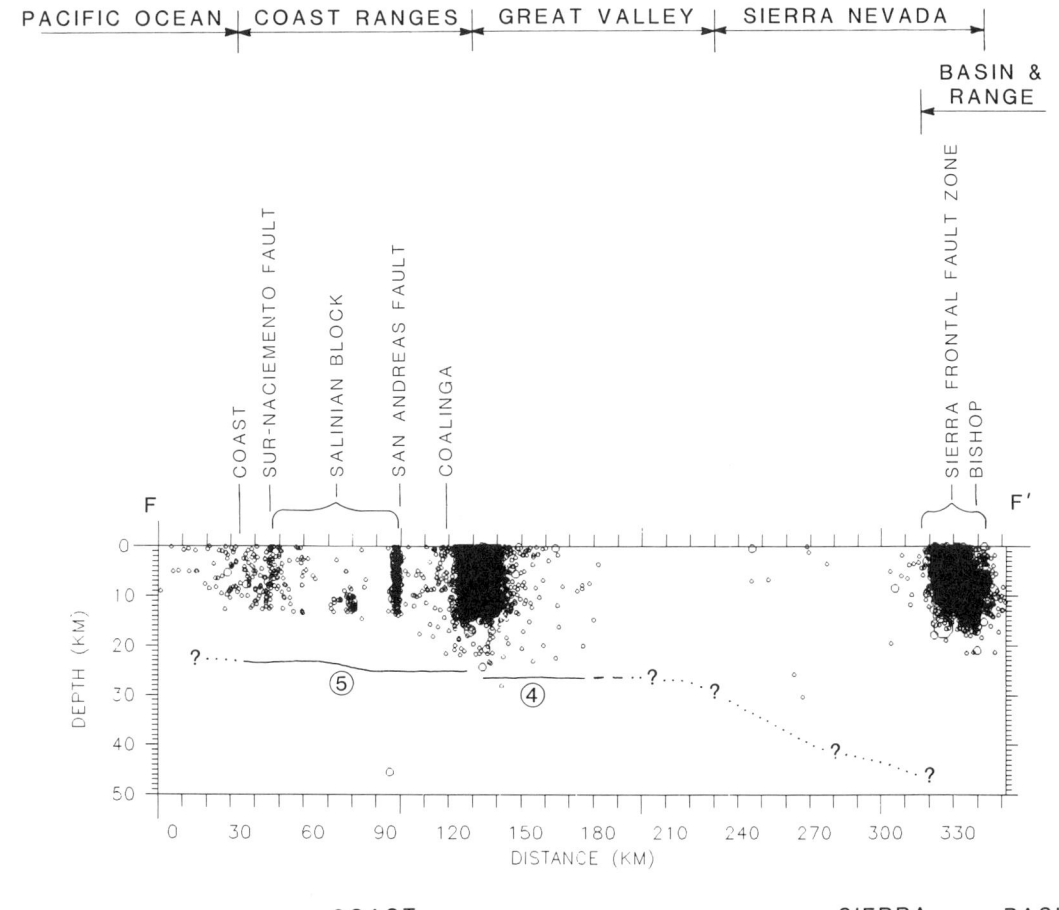

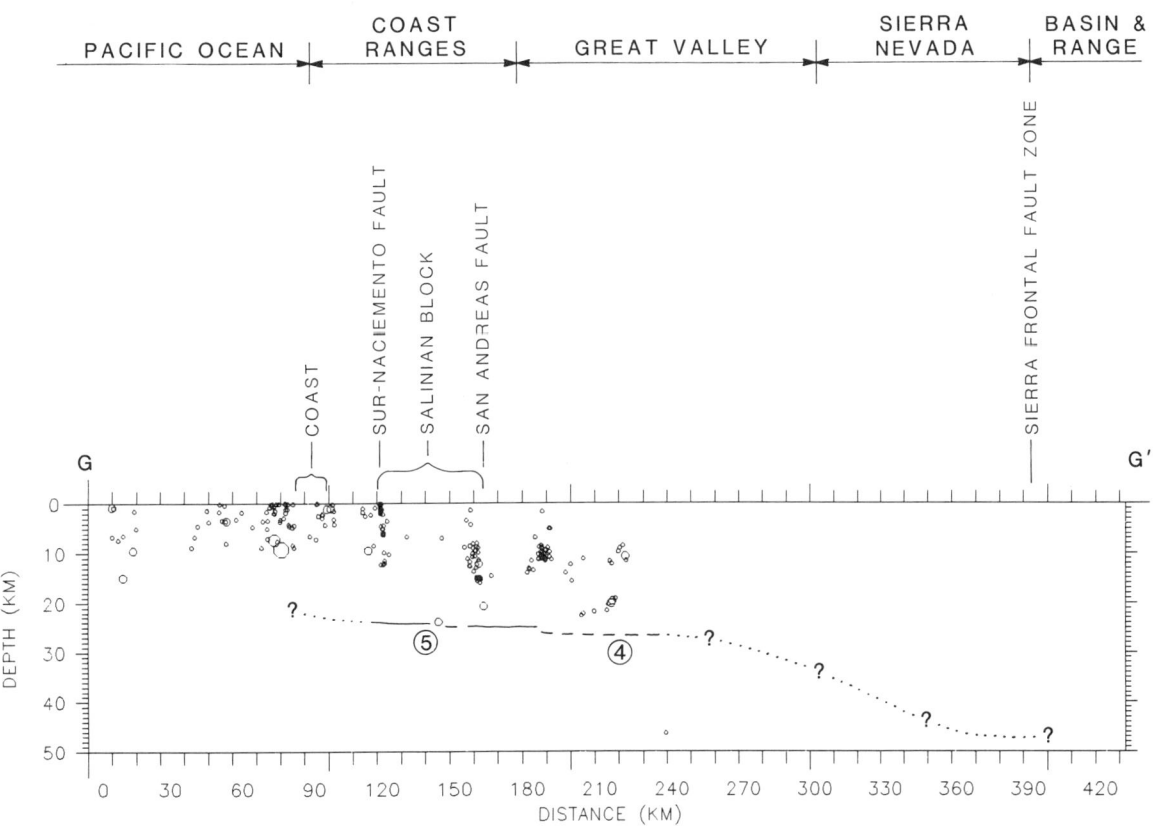

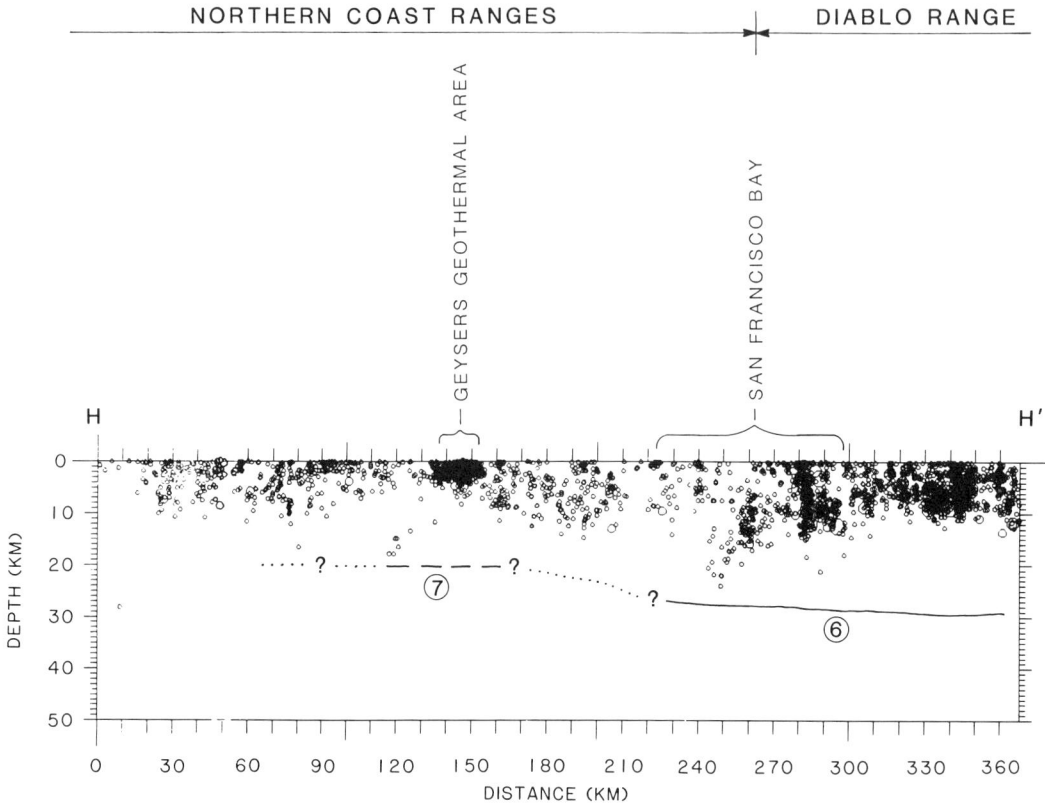
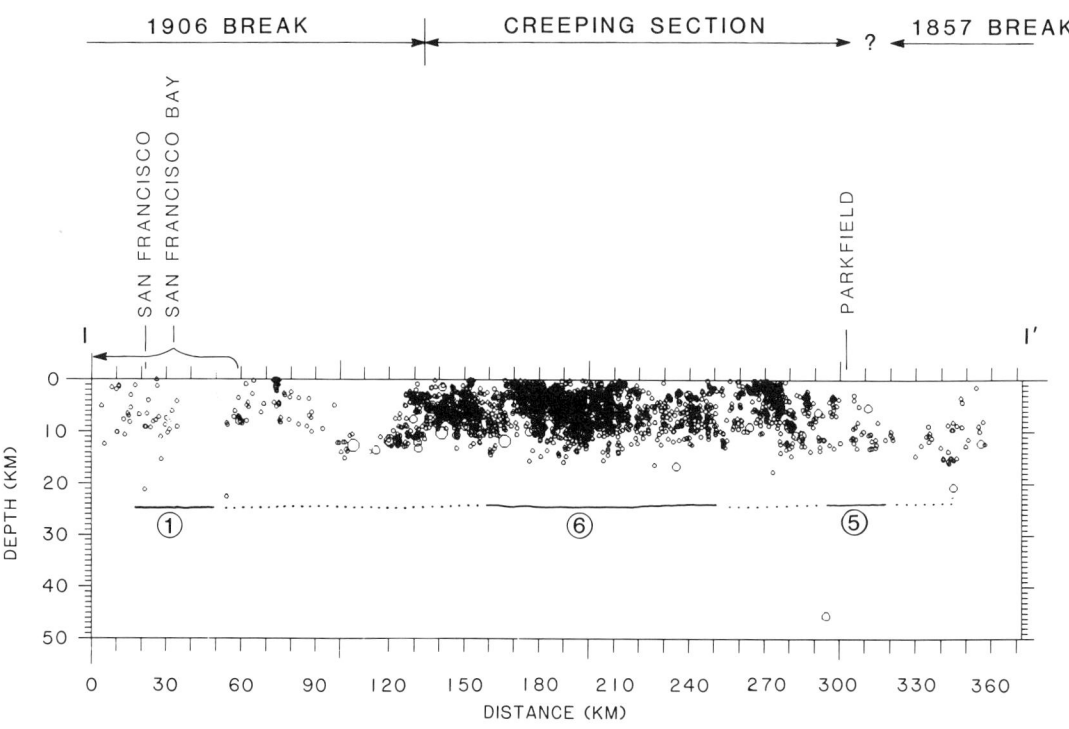

Figure 12. Longitudinal cross sections H-H' and I-I'. Symbols are the same as in Figure 11.

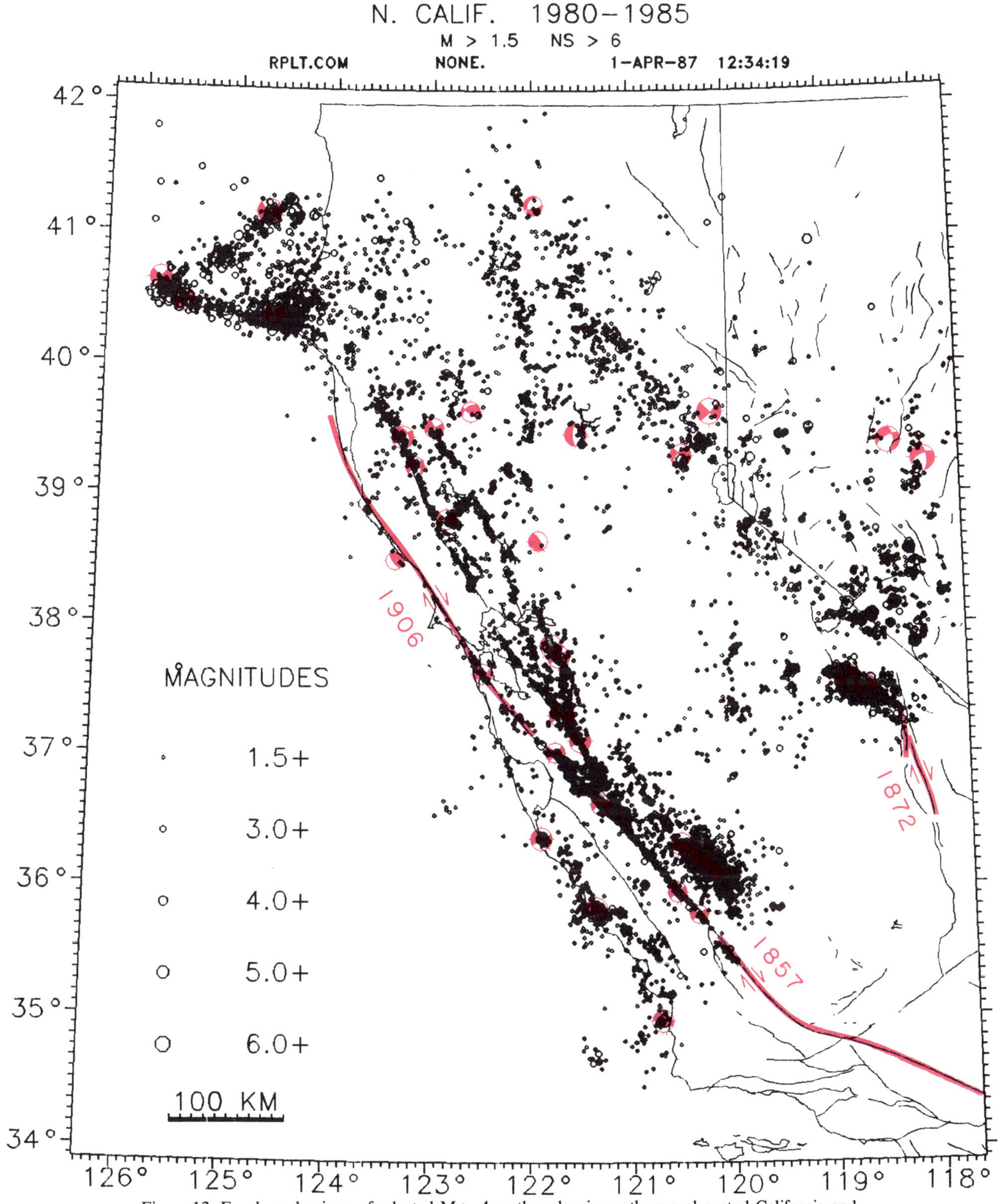

Figure 13. Focal mechanisms of selected M > 4 earthquakes in northern and central California and western Nevada (red) superimposed on the 1980–1985 seismicity pattern. Focal mechanisms shown using lower-hemisphere projections with solid red indicating compressional quadrants. Numerical code identifies events listed in Table 3 by date (yr, mo, da). Heavy lines and double arrows indicate rupture surfaces and slip directions of the three great (M = 8), historic earthquakes in California.

TABLE 2. FREQUENCY-MAGNITUDE STATISTICS FOR THE 1880-1985 DATA IN FIGURE 9

Year	Linear Regression				Maximum Likelihood	
	M-interval	R2	$a \pm 2\sigma$	$b \pm 2\sigma$	M-Interval	$b \pm c.l.$
1980	2.0–7.0	0.989	4.8 ± 0.2	0.71 ± 0.04	3.0–7.0	0.76 ± 0.06
1981	2.0–5.5	0.977	5.3 ± 0.3	0.98 ± 0.08	3.0–5.5	0.97 ± 0.11
1982	2.0–5.5	0.995	4.9 ± 0.1	0.86 ± 0.03	2.0–5.5	0.83 ± 0.04
1983	2.0–6.5	0.992	5.1 ± 0.2	0.82 ± 0.04	2.0–6.5	0.90 ± 0.03
1984	2.0–4.8	0.997	5.1 ± 0.1	0.94 ± 0.04	2.0–6.0	0.96 ± 0.04
1985	2.0–5.5	0.996	4.9 ± 0.1	0.93 ± 0.04	2.0–5.5	0.96 ± 0.05

M-Interval = Magnitude interval ; R2 = Goodness of fit; *c.l.* = confidence limits.

branches of the San Andreas system through the northern Coast Ranges (Bufe and others, 1981; Eberhart-Phillips and Oppenheimer, 1984; Oppenheimer, 1986; Dehlinger and Bolt, 1984).

Compressional deformation within the Coast Ranges

Earthquakes occurring off the main branches of the San Andreas fault within the Coast Ranges show a variety of focal mechanisms, most of which indicate a strong element of crustal shortening perpendicular to the San Andreas fault. These focal mechanisms typically involve a mix of reverse- and strike-slip displacements; the slip vector has a significant component perpendicular to the local strike of the San Andreas fault. The P-axes associated with these "off fault" earthquakes are commonly rotated clockwise with respect to the north to north-northeast P-axis orientations typical of earthquakes located along the main branches of the San Andreas fault system. Zoback and others (1987) argue that the focal mechanisms for off-fault earthquakes are consistent with the greatest compressive stress within the Coast Ranges having an orientation normal to the San Andreas fault, which they interpret in terms of a very low shear strength for the San Andreas fault system.

The M=6.7 Coalinga earthquake of May 2, 1983, is the largest of these "off fault" Coast Range earthquakes yet recorded by a modern seismograph network. It was located 20 km east of the San Andreas fault near the eastern margin of the Coast Ranges (830502 in Fig. 13), and although this earthquake produced no surface rupture, both seismic waveform data and local geodetic data show that it involved reverse slip along a northwest-striking plane subparallel with the adjacent section of the San Andreas fault (Stein and King, 1984; Eaton, 1984b). The focal mechanisms for three moderate earthquakes along the central coast (events 800529, 830829, and 840123 in Fig. 13) show a similar relation to the San Andreas fault, although the reverse component and the component of slip perpendicular to the San Andreas fault decrease systematically between events from south to north (Eaton, 1984b). Yet another example involves a moderate (M=4.3) earthquake along the eastern margin of the northern Coast Ranges in 1978 (event 780809) that also shows reverse slip on a plane subparallel with the trend of the adjacent San Andreas fault system (Eaton, 1986).

The Great Valley–Sierra Nevada block

The M=5.7 Oroville earthquake of August 1, 1975 (event 750801 in Fig. 13), the largest instrumentally recorded earthquake to occur within the relatively quiescent western foothill belt of the Sierra Nevada, involved predominantly normal displacement along a west-dipping plane extending from the mainshock hypocenter at a depth of approximately 8 km to the surface (Langston and Butler, 1976; Lahr and others, 1976). Focal mechanisms determined by Wong and Savage (1983) for several earthquakes in the 12- to 40-km-deep range beneath the central section of the Great Valley–Sierra Nevada block show a mix of strike-slip, reverse, and normal displacements. The P- and T-axes for these focal mechanisms have northerly and easterly orientations, respectively, consistent with north-south shortening and east-west extension in the mid- to lower-crust beneath this central section of the Sierra Nevada foothill belt.

Eastern Sierra Nevada and the Basin and Range

Earthquakes occurring along the eastern Sierra Nevada seismic belt and within the adjacent Basin and Range province show a mix of strike-slip and dip-slip focal mechanisms consistent with a tectonic regime influenced by both east-west extension of the Basin and Range province and dextral shear of the San Andreas transform boundary (Van Wormer and Ryall, 1980; Vetter and Ryall, 1983; Vetter and others, 1983; Zoback and Zoback, 1980). This deformation style is typified by the strike-slip mechanisms from M=5 and M=6 earthquakes in the persistent seismicity cluster northwest of Lake Tahoe (events 801128 and 660912, respectively; see Hawkins and others, 1986) and the normal-oblique mechanisms from the three largest events (M = 7.1, 6.8, and 6.6) in the 1954 Fairview Peak–Dixie Valley sequence in western Nevada (events 541216, 540824, and 540706, respectively; see Doser, 1986). While the pattern of distributed east-west extension across most of the eastern Sierra Nevada–Basin and Range region is supported by a consistent east-west to

TABLE 3. KEY TO FOCAL MECHANISMS SHOWN IN FIGURE 13

Event y m d	Lat	Long	Depth (km)	Mag	Reference
54 07 06	39.29°	118.36°	10.0	6.6	Doser (1986)
54 08 24	39.35°	118.34°	12.0	6.8	Doser (1986)
54 12 16	39.20°	118.00°	15.0	7.1	Doser (1986)
66 07 02	35.79°	120.34°	9.1	3.5	Eaton and others (1970)
66 09 12	39.44°	120.17°	10.0	6.0	Tsai and Aki (1970)
75 08 01	39.44°	121.54°	5.5	5.7	Langston and Butler (1976)
77 11 22	39.42°	123.27°	6.6	5.1	Hill and others (1989)
78 03 26	39.20°	123.14°	6.0	4.7	Hill and others (1989)
78 03 31	38.48°	123.33°	6.4	3.6	Hill and others (1989)
78 09 08	38.65°	121.89°	11.8	4.3	Eaton (1986)
78 11 12	39.49°	122.95°	10.2	4.3	Hill and others (1989)
79 04 28	37.62°	122.46°	11.6	4.2	Hill and others (1989)
79 08 06	37.10°	121.52°	7.8	5.9	Reasenberg and Ellsworth (1982)
80 01 24	37.84°	121.78°	11.9	5.9	Hill and others (1989)
80 01 27	37.75°	121.71°	12.4	5.3	Hill and others (1989)
80 05 25	37.60°	118.83°	8.0	6.1	Kramer and Toppazada (1980)
80 05 29	34.98°	120.71°	9.2	5.1	Eaton (1984a)
80 11 08	41.12°	124.66°	10.0	7.0	Hill and others (1989)
80 11 09	40.59°	125.77°	15.0	5.4	Hill and others (1989
80 11 28	39.28°	120.45°	13.9	5.4	Hawkins and others (1986)
81 09 16	40.34°	124.59°	28.9	4.8	Hill and others (1989)
82 05 29	38.80°	122.82°	4.8	4.3	Hill and others (1989)
82 06 21	41.18°	121.94°	7.3	4.3	Hill and others (1989)
82 06 25	35.96°	120.55°	9.0	4.2	Eaton and Rymer (1987)
82 08 11	36.63°	121.30°	11.6	4.8	Eaton and Rymer (1987)
82 08 18	37.02°	121.75°	12.2	4.3	Eaton and Rymer (1987)
82 09 03	39.62°	122.58°	14.6	4.2	Hill and others (1989)
82 10 25	36.32°	120.51°	10.9	5.5	Eaton and Rymer (1989)
83 05 02	36.24°	120.32°	10.0	6.7	Eaton (1984b)
83 08 29	35.84°	121.37°	6.6	5.4	Eaton (1984a)
84 01 23	36.37°	121.89°	7.7	5.2	Eaton (1984a)
84 04 24	37.32°	121.69°	8.4	6.2	Cockerham and Eaton (1984)
84 11 23	37.47°	118.60°	8.0	5.8	Barker and Wallace (1986)
85 08 04	36.14°	120.16°	11.4	5.7	Hill and others (1989)

Lat = latitude; Long = longitude; Mag = magnitude; y m d = year/month/day.

northwest-southeast orientation of focal mechanism T-axes, detailed relations between adjacent seismicity patterns, focal mechanisms, and mapped faults are often complex.

Thus, for example, the largest historic earthquake in the region, the M=8 Owens Valley event of 1872, involved nearly pure right-lateral strike-slip displacement on a plane striking slightly west of north (Beanland and Clark, 1987). Just 30 km to the northwest, however, the M=6.0 Mammoth Lakes earthquakes of May 1980 involved predominantly left-lateral slip on planes striking slightly east of north (Cramer and Toppazada, 1980; Julian and Sipkin, 1985; Lide and Ryall, 1985). Furthermore, neither of these historic earthquake sequences shows any obvious relation to adjacent range-bounding normal faults, several of which have large, purely dip-slip displacements of Holocene age.

Focal mechanisms from earthquakes of the Mammoth Lakes sequence in the Sierra Nevada block just south of Long Valley caldera are unusual in two respects. First, the systematic southwest-northeast orientation of their T-axes deviates 45° or more (in a counterclockwise sense) from the east-west to northwest-southeast T-axis orientation common to earthquakes throughout the rest of the eastern Sierra Nevada–Basin and Range region (Vetter and others, 1983). Second, two of the four M=6 Mammoth Lakes earthquakes of May 1980 showed significant nondouble couple components at long periods. The interpretation of these nondouble couple events remains controversial; Julian and Sipkin (1985) propose that they represent the abrupt opening of dikes associated with magma injection; others (Wallace, 1985; Priestley and others, 1985; Lide and Ryall, 1985) propose that they result from complex shear failure on fault planes of differing orientation.

Although little has been published on the focal mechanisms of earthquakes along the northwestern extension of the eastern Sierra Nevada seismicity belt beyond the Lake Tahoe region, Klein's (1979) work in the Lassen Peak area indicates that a mix of strike-slip and normal displacements with predominantly east-west T-axis orientations prevails here as well.

DISCUSSION AND CONCLUSIONS

The San Andreas fault system accommodates from 50 to 70 percent of the relative motion between the Pacific and North American plates by right-lateral slip on well-defined vertical fault planes that cut the entire 10- to 15-km thickness of the seismogenic crust. Slip along the central, creeping section of the San Andreas fault occurs primarily by aseismic creep accompanied by numerous small-to-moderate ($M \leq 5$) earthquakes, the vast majority of which have strike-slip focal mechanisms with the right-lateral plane parallel to the local strike of the fault. Geodetic measurements show that sections of the crust on either side of the creeping section of the fault between San Juan Bautista and Parkfield are sliding past one another as rigid blocks at a rate of about 34 mm/yr without accumulating significant shear strain (Thatcher, 1979). In contrast, most of the slip on the "locked" segments of the San Andreas fault north of San Juan Bautista and south of Parkfield accumulates with great ($M=8$) plate-boundary earthquakes that recur at intervals of 100 to 200 yr (Sieh, 1981). These locked segments of the San Andreas fault have produced only a few small earthquakes over the last several decades while geodetic measurements show that shear strain is accumulating within the brittle crust on either side of the locked segments at rates of several tenths of a microstrain per year (Prescott and Yu, 1986; Savage, 1983).

Considerable interest is currently focused on a 25-km-long segment of the San Andreas fault in the transition zone between the locked and creeping sections of the fault near Parkfield. Bakun and McEvilly (1984) documented that this segment of the fault has produced moderate ($M=5.5$ to $M=6.0$) earthquakes at a remarkably regular 21- to 22-yr interval since 1881, and they forecast that the next "Parkfield" earthquake would occur in 1988 with an estimated standard error of 5 yr. Based on this forecast, an intensive experiment is under way to obtain a detailed geophysical record of both the precursory stages to this earthquake and the rupture process associated with the earthquake itself (Bakun and Lindh, 1985).

The fraction of Pacific–North American plate motion in central California not accommodated by slip along the San Andreas fault system (the so-called "San Andreas discrepancy") involves distributed deformation of the plate margin at least as far inland as the eastern escarpment of the Sierra Nevada and the western Basin and Range province. Minster and Jordan (1987) identified two principal contributions to the San Andreas discrepancy based on their analysis of global plate motions and recent Very Long Baseline Interfereometry (VLBI) results for the western United States. The first is the result of intraplate extension across the Basin and Range province and involves the northwestward movement of the Sierra Nevada–Great Valley block with respect to the interior of the North American plate. Minster and Jordan (1987) estimate that this motion accounts for 7 to 9 mm/yr of right-lateral slip parallel to the average strike of the San Andreas fault and that it includes minor (1 to 3 mm/yr) extensional deformation perpendicular to the average strike of the San Andreas fault. The second contribution involves distributed deformation within the Coast Ranges and along the continental margin and accounts for between 5 and 13 mm/yr of right-lateral slip parallel with the San Andreas fault and 6 to 9 mm/yr of compressional deformation perpendicular to the fault. The larger of these values associated with the Coast Ranges derive from the Minster and Jordan (1987) estimate of 56 mm/yr for the relative velocity between the Pacific and North American plates. Recent evidence presented by DeMets and others (1987) for a lower (48 mm/yr) relative plate velocity, however, favors the smaller of these values for deformation associated with the San Andreas discrepancy in the Coast Ranges.

The seismicity patterns for central California in Figures 3, 4, and 5 suggest that most of the Sierra Nevada–Great Valley block contribution to the San Andreas discrepancy is concentrated along the eastern escarpment of the Sierra Nevada. The right-oblique extensional deformation characteristic of the eastern Sierra Nevada seismic belt is consistent with the sense of deformation inferred by Minster and Jordan (1987) for the Sierra Nevada–Great Valley block. The 1975 Oroville ($M=5.8$) earthquake with its normal focal mechanism in the Sierra Nevada foothills (750801 in Fig. 13) gives further evidence for a component of extensional deformation within the Sierra Nevada–Great Valley block. Because of its low seismicity rate, however, the thick seismogenic crust (40 km) of the central section of the Sierra Nevada–Great Valley block contributes little to plate-boundary deformation.

Seismicity patterns within the Coast Ranges indicate that, east of the San Andreas fault, deformation contributing to the San Andreas discrepancy is distributed throughout the Franciscan terrain along the entire length of the Coast Ranges. In contrast, deformation west of the San Andreas fault appears to be confined to the continental margin and the western margin of the Salinian block associated with the San Gregorio–Hosgri and Sur-Naciemento fault systems (see Fig. 1) south of San Francisco Bay. The Salinian block itself appears to be relatively stable.

Youthful uplift and folding within the Coast Ranges, with the fold axes commonly subparallel with the San Andreas, are clear evidence for Quaternary compressional deformation perpendicular to the San Andreas fault (Page, 1981). The strong component of convergent slip perpendicular to the San Andreas fault shown by focal mechanisms for the distributed seismicity off the San Andreas fault emphasizes that compressional deformation is currently active and involves brittle deformation of the upper 15 km of the crust. Based on their analysis of the 1983 Coalinga $M=6.7$ earthquake (event 830502 in Fig. 13), Eaton and Rymer (1989) propose that this deformation involves a wedge of eugeosynclinal Franciscan rocks being driven eastward along a low-angle, mid-crustal décollement surface between the crystalline basement and the overlying miogeosynclinal Great Valley sediments. Kinematically, this convergence across the Coast Ranges and the San Andreas fault system is related, in part, to the 3° to 5° counterclockwise (restraining) misalignment of the San Andreas fault in central California with respect to the N35°W orientation of the Pacific–North American plate boundary and, in part, to the westward movement of the Great Valley–Sierra Nevada block associated with Basin and Range spreading, both of which imply an element of decoupling of the seismogenic crust from the San Andreas transform in the lower aseismic lithosphere (Hill, 1982; Weldon and Humphreys, 1986; Minster and Jordan, 1987).

On a regional scale, the P- and T-axes orientations from focal mechanisms of central California earthquakes are broadly consistent with a stress field in which the greatest and least horizontal stresses have north-south and east-west orientations, respectively, an orientation that is also consistent with the broad interaction of the Pacific and North American plates along the San Andreas transform boundary (Zoback and Zoback, 1980; Hill, 1982). Two exceptions to this regional pattern, however, are particularly noteworthy:

1. P- and T-axes of earthquakes adjacent to the San Andreas fault system in the Coast Ranges are systematically rotated 30 to 45° in a clockwise direction from both the regional orientation and the orientation shown by most of the strike-slip mechanisms for earthquakes along the San Andreas fault (see Ellsworth and others, 1982). Zoback and others (1987) interpret this locally abrupt change in P- and T-axis orientations in terms of an exceptionally low shear strength along the San Andreas fault system.

2. P- and T-axes for focal mechanisms of the earthquakes in the dense seismicity cluster just south of Long Valley caldera in eastern California have a 45- to 90° counterclockwise orientation with respect to the regional pattern. These earthquakes are exceptional not only because their T-axes have an anomalous southwest-northeast alignment, but also because the distribution of their hypocenters and their strike-slip focal mechanisms show no obvious relation to the large range-front normal faults that dominate the area. Evidently this activity reflects westward encroachment of Basin and Range tectonics into the stable Sierra Nevada block complicated by proximity to the large, Quaternary magmatic system forming the Long Valley caldera and the Inyo-Mono Craters volcanic chain (Vetter and others, 1983).

Both historic and instrumental seismicity records indicate that the spatial distribution of earthquakes in central California changes only slowly over periods of decades to centuries (Figs. 3, 4, and 5). Indeed, while the pattern of $M \geqslant 1.5$ seismicity for the 6-yr period summarized in Figure 8 shows some fluctuation in intensity from one year to the next, its spatial outline remains nearly stationary. An important issue for earthquake prediction in this regard involves the sort of spatial-temporal variations in seismicity patterns we might see as we approach the return time for the next great earthquake along the locked section of the San Andreas fault. The frequency of moderate to large earthquakes in the San Francisco Bay area, for example, was distinctly higher in the 50 yr prior to the 1906 M=8 earthquake than in the following 50 yr (Tocher, 1959). Ellsworth and others (1981) point out that the frequency of moderate earthquakes in the San Francisco Bay area began to increase again sometime in the mid-1950s and that these temporal fluctuations are consistent with Mogi's (1981) model for the seismic cycle for great plate-boundary earthquakes. Our instrumental record of earthquake activity is too short, however, to indicate whether we can expect to see a change in the spatial seismicity pattern a short time (months to years?) prior to the next great "San Francisco" earthquake with, say, numerous small to moderate earthquakes (foreshocks) occurring along the currently aseismic, locked section of the San Andreas fault.

REFERENCES CITED

Aki, K., 1965, Maximum likelihood estimate of b in the formula logN=a-bM and its confidence limits: Bulletin of the Earthquake Research Institute, v. 43, p. 237–239.

Allen, C. R., 1981, The moder San Andreas fault, in Ernst, W. G., ed., The geotectonic development of California; Rubey Volume 1: Englewood Cliffs, New Jersey, Prentice-Hall, p. 512–534.

Atwater, T., 1970, Implications of plate tectonics for the Cenozoic tectonic evolution of western North America: Geological Society of America Bulletin, v. 81, p. 3513–3536.

Bakun, W. H., 1984, Seismic moments, local magnitudes, and coda-duration magnitudes for earthquakes in central California: Bulletin of the Seismological Society of America, v. 74, p. 439–458.

Bakun, W. H., and Lindh, A. G., 1985, The Parkfield, California, prediction experiment: Science, v. 229, p. 619–624.

Bakun, W. H., and McEvilly, T. V., 1984, Recurrence models and the Parkfield, California, earthquakes: Journal of Geophysical Research, v. 89, p. 3051–3058.

Barker, J. S., and Wallace, T. C., 1986, A note on the teleseismic body waves from the 23 November 1984 Round Valley, California, earthquake: Bulletin of the Seismological Society of America, v. 76, p. 883–888.

Beanland, S., and Clark, M. M., 1987, The Owens Valley fault zone, eastern California, and surface rupture associated with the 1872 earthquake [abs.]: Seismological Society of America, Seismological Research Letters, v. 58, p. 32.

Bolt, B. A., and Miller, R. D., 1971, Seismicity of northern and central California: Bulletin of the Seismological Society of America, v. 61, p. 1831–1848.

—— , 1975, Catalogue of earthquakes in northern California and adjoining areas 1 January 1910–31 December, 1972: Berkeley, University of California seismographic station, 576 p.

Boore, D. M., 1977, Strong-motion recordings of the California earthquake of April 18, 1906: Bulletin of the Seismological Society of America, v. 67, p. 561–577.

Bufe, C. G., Marks, S. M., Lester, F. W., Ludwin, R. S., and Stickney, M. C., 1981, Seismicity of the Geysers–Clear Lake region, in McLaughlin, R. J., and Donnelly-Nolan, J. M., Research in the Geysers–Clear Lake geothermal area, northern California: U.S. Geological Survey Professional Paper 1141, p. 129–138.

Cockerham, R. S. and Corbett, E. J., 1987, The July 1986, Chalfant Valley, California, earthquake sequence: preliminary results: Bulletin of the Seismological Society of America, v. 77, p. 280–289.

Cockerham, R. S., and Eaton, J. P., 1984, The earthquake and its aftershocks, April 24 through September 30, 1984, *in* Hoose, S. N., ed., The Morgan Hill, California, earthquake of April 24, 1984: U.S. Geological Survey Bulletin 1639, p. 15–28.

Cockerham, R. S., and Herd, D. G., 1982, Seismicity of the north end of the San Andreas fault system [abs.]: EOS Transactions of the American Geophysical Union, v. 63, p. 384.

Clark, M. M., and Gillespie, A. R., 1984, Record of late Quaternary faulting along the Hilton Creek fault in the Sierra Nevada, California: Earthquake Notes, v. 52, p. 46.

Cramer, C. H., and Toppozada, T. R., 1980, A seismological study of the May 1980 and earlier earthquake activity near Mammoth Lakes, California, *in* Sherburne, R. W., ed., Mammoth Lakes, California, earthquakes of May 1980: California Division of Mines and Geology Special Report 150, p. 91–130.

Dehlinger, P., and Bolt, B. A., 1984, Seismic parameters along the Bartlett Springs fault zone in the Coast Ranges of northern California: Bulletin of the Seismological Society of America, v. 74, p. 1785–1798.

DeMets, C., Gordon, R. G., Stein, S., and Argus, D., 1987, A revised estimate of Pacific–North America motion and implications for western North America plate boundary tectonics: Geophysical Research Letters, v. 14, p. 911–914.

Dickinson, W. R., 1981, Plate tectonics and the continental margin of California, *in* Ernst, G. W., ed., The geotectonic development of California; Rubey Volume 1: Englewood Cliffs, New Jersey, Prentice-Hall, p. 2–28.

Doser, D. I., 1986, Earthquake processes in the Rainbow Mountain–Fairview Peak–Dixie Valley, Nevada, region (1954–1959): Journal of Geophysical Research, v. 91, p. 12572–12586.

Eaton, J. P., 1963, Crustal structure from San Francisco, California, to Eureka, Nevada, from seismic-refraction measurements: Journal of Geophysical Research, v. 69, p. 5789–5806.

—— , 1966, Crustal structure in northern and central California from seismic evidence, *in* Geology of northern California: California Division of Mines and Geology Bulletin 190, p. 419–426.

—— , 1984a, Focal mechanisms of near-shore earthquakes between Santa Barbara and Monterey, California: U.S. Geological Survey Open-File Report 84-477, 10 p.

—— , 1984b, The May 2, 1983, Coalinga earthquakes and its aftershocks; A detailed study of the hypocenter distribution and of the focal mechanisms of the larger aftershocks, *in* Rymer, M. J., and Ellsworth, W. L., eds., Mechanics of the May 2, 1983, Coalinga earthquake: U.S. Geological Survey Open-File Report 85-44, p. 132–201.

—— , 1986, Tectonic environment of the 1892 Vacaville/Winters earthquake, and the potential for large earthquakes along the western edge of the Sacramento Valley: U.S. Geological Survey Open-File Report 86-370, 11 p.

Eaton, J. P., and Rymer, M. J., 1990, Regional seismotectonic model for the southern Coast Ranges, and the May 2, 1983, Coalinga earthquake, *in* Rymer, M. J. and Ellsworth, W. L., eds., The Coalinga California earthquake of May 2, 1983: U.S. Geological Survey Professional Paper 1487, p. 97–111.

Eaton, J. P., Lee, W. H. K., and Pakiser, L. C., 1970a, Use of microearthquakes in the study of mechanics of earthquake generation along the San Andreas fault in central California: Tectonophysics, v. 9, p. 559–582.

Eaton, J. P., O'Neill, M. E., and Murdock, J. N., 1970b, Aftershocks of the 1966 Parkfield–Cholame, California, earthquake; A detailed study: Bulletin of the Seismological Society of America, v. 60, p. 1151–1197.

Eberhart-Phillips, D., and Oppenheimer, D. H., 1984, Induced seismicity in the Geysers geothermal area, California: Journal of Geophysical Research, v. 89, p. 1191–1207.

Ellsworth, W. L., 1975, Bear Valley, California, earthquake sequence of February–March 1972: Bulletin of the Seismological Society of America, v. 65, p. 483–506.

Ellsworth, W. L., Lindh, A. G., Prescott, W. H., Herd, D. G., 1981, The 1906 San Francisco earthquake and the seismic cycle, *in* Simpson, D. W., and Richards, P. G., eds., Earthquake prediction; An international review: American Geophysical Union Maurice Ewing Series 4, p. 126–140.

Ellsworth, W. L., Olson, J. A., Shijo, L. N., and Marks, S. M., 1982, Seismicity and active faults in eastern San Francisco Bay region, *in* Hart, E. W., Hirschfeld, S. E., and Schulz, S. S., eds., Proceedings of the Conference on Earthquake Hazards in the Eastern San Francisco Bay Area: California Division of Mines and Geology Special Publication 62, p. 83–92.

Engdahl, E. R., and Rinehart, W. A., 1988, Seismicity Map of North America: Boulder, Colorado, Geological Society of America, Continent-Scale Map-004, scale 1:5,000,000.

Ernst, W. G., 1981, Summary of the geotectonic development of California, *in* Ernst, W. G., ed., The geotectonic development of California; Rubey Volume 1: Englewood Cliffs, New Jersey, Prentice-Hall, p. 601–613.

Gawthrop, W. J., 1978, Seismicity and tectonics of the central California coastal region, *in* Silver, E. A., and Normark, W. R., eds., San Gregorio–Hosgri fault zone, California: California Division of Mines and Geology Special Report 137, p. 45–56.

Hawkins, F. F., LaForge, R., and Hansen, R. A., 1986, Seismotectonic study of the Truckee/Lake Tahoe area, northeastern Sierra Nevada, California: U.S. Bureau of Reclamation Seismotectonic Report 85-4, 210 p.

Hill, D. P., 1978, Seismic evidence for the structure and Cenozoic tectonics of the Pacific coast state, *in* Smith, R. B., and Eaton, G. P., eds., Cenozoic tectonics and regional geophysics of the western Cordillera: Geological Society of America Memoir 152, p. 145–174.

—— , 1982, Contemporary block tectonics; California and Nevada: Journal of Geophysical Research, v. 87, p. 5433–5450.

Hill, D. P., Wallace, R. E., and Cockerham, R. S., 1985, Review of evidence on the potential for major earthquakes and volcanism in the Long Valley–Mono Craters–White Mountains regions of eastern California: Earthquake Prediction Research, v. 3, p. 571–594.

Hill, D. P., Eaton, J. P., and Jones, L., 1990, Seismicity: 1980–1986, *in* Wallace, R. E., ed., The San Andreas fault system: U.S. Geological Survey Professional Paper 1515, p. 114–151.

Holbrook, W. S., and Mooney, W. D., 1987, The crustasl structure of the axis of the Great Valley, California, from seismic refraction measurements: Tectonophysics, v. 140, p. 49–63.

Jachens, R. C., and Griscom, A., 1983, Three-dimensional geometry of the Gorda plate beneath northern California: Journal of Geophysical Research, v. 88, p. 9375–9392.

Jennings, C. W., and 7 others, 1975, Fault map of California with locations of volcanoes, thermal springs, and thermal wells: Sacramento, California Division of Mines and Geology, California Geologic Data Map Series Map 1, scale 1:750,000.

Julian, B. R., and Sipkin, S. A., 1985, Earthquake processes in the Long Valley caldera area, California: Journal of Geophysical Research, v. 90, p. 11155–11169.

Klein, F. W., 1979, Earthquakes in Lassen Volcanic National Park, California: Bulletin of the Seismological Society of America, v. 69, p. 867–875.

LaForge, R., and Lee, W.H.K., 1982, Seismicity and tectonics of the Ortigalita fault and southeast Diablo Range, California, *in* Hart, E. W., Hirschfeld, S. E., and Schulz, S. S., eds., Proceedings of a Conference on Earthquake Hazards in the Eastern San Francisco Bay Area: California Division of Mines and Geology Special Publication 62, p. 93–102.

Lahr, K. M., Lahr, J. C., Lindh, A. G., Buffe, C. G., and Lester, F. W., 1976, The August 1975 Oroville earthquakes: Bulletin of the Seismological Society of America, v. 66, p. 1085–1099.

Langston, C. A., and Butler, R., 1976, Focal mechanism of the August 1, 1985, Oroville earthquake: Bulletin of the Seismological Society of America, v. 66, p. 1121–1132.

Lee, W. H. K., and Stewart, S. W., 1981, Principles and applications of microearthquake networks, *in* Saltzman, B., ed., Advances in geophysics, Supplement 2: New York, Academic Press, 193 p.

Lide, C. S., and Ryall, A. S., 1985, Aftershock distribution related to the controversy regarding mechanisms of the May 1980 Mammoth Lakes, California, earthquakes: Journal of Geophysical Research, v. 90, p. 11151–11154.

Lubetkin, K. C., and Clark, M. M., 1988, Late Quaternary activity along the Lone

Pine fault, eastern California: Geological Society of America Bulletin, v. 100, p. 755-766.

Mayer-Rosa, D., 1973, Traveltime anomalies and distribution of earthquakes along the Calaveras fault zone, California: Bulletin of the Seismological Society of America, v. 63, p. 713-729.

Macgregor-Scott, N., and Walter, A. W., 1988, Crustal velocities near Coalinga, California, modeled from a combined earthquake/explosion refraction profile: Bulletin of the Seismological Society of America, v. 78, p. 1475-1490.

Meissner, R., and Strehlau, J., 1982, Limits of stresses in continental crusts and their relation to the depth-frequence distribution of shallow earthquakes: Tectonics, v. 1, p. 73-98.

Minster, J. B., and Jordan, T. H., 1987, Vector constraints on western U.S. deformation from space geodesy; Neotectonics and plate motion: Journal of Geophysical Research, v. 92, p. 4798-4804.

Mogi, K., 1981, Seismicity in western Japan and long-term earthquake forecasting, in Simpson, D. W., and Richards, P. G., eds., Earthquake prediction; An international review: American Geophysical Union Maurice Ewing Series 4, p. 43-51.

Ness, G. E., Lyle, M. W., and Lonseth, A. T., 1985, Revised Pacific, North American, Riviera, and Cocos relative motion poles; Implications for strike-slip motion along the trans-Mexican volcanic belt [abs.]: EOS Transactions of the American Geophysical Union, v. 66, p. 849.

Oliver, H. W., 1977, Gravity and magnetic investigations of the Sierra Nevada batholith, California: Geological Society of America Bulletin, v. 87, p. 1537-1546.

Oppenheimer, D. H., 1986, Extensional tectonics at the Geysers geothermal area, California: Journal of Geophysical Research, v. 91, p. 11463-11476.

Oppenheimer, D. H., and Eaton, J. P., 1984, Moho orientation beneath central California from regional earthquake traveltimes: Journal of Geophysical Research, v. 89, p. 10267-10282.

Page, B. M., 1981, The southern Coast Ranges, in Ernst, G. W., ed., The geotectonic development of California, Rubey Volume 1: Englewood Cliffs, New Jersey, Prentice-Hall, p. 330-417.

Pakiser, L. C., and Brune, J. N., 1980, Seismic models of the root of the Sierra Nevada: Science, v. 210, p. 1088-1094.

Pavoni, N., 1973, A structural model for the San Andreas fault zone along the northeast side of the Gabilan Range, in Kovach, R. L., and Nur, A., eds., Proceedings of a Conference on Tectonic Problems of the San Andreas Fault System: Stanford, California, Stanford University Geological Science, v. 13, p. 259-267.

Poley, C. M., 1988, The San Ando, California, earthquake of 24 November 1985: Bulletin of the Seismological Society of America, v. 78, p. 1360-1366.

Poley, C. M., Lindh, A. G., Bakun, W. H., and Schulz, S. S., 1987, Temporal changes in microseismicity and creep near Parkfield, California: Nature, v. 327, p. 134-137.

Prescott, W. H., and Yu, S-B, 1986, Geodetic measurement of horizontal deformation in the northern San Francisco Bay region, California: Journal of Geophysical Research, v. 91, p. 7475-7484.

Prescott, W. H., Lisowski, M., and Savage, J. C., 1981, Geodetic measurement of crustal deformation on the San Andreas, Hayward, and Calaveras faults near San Francisco, California: Journal of Geophysical Research, v. 86, p. 10853-10869.

Priestley, K., Ryall, A., and Fezie, G., 1982, Crust and upper mantle structure in the northwest Basin and Range province: Bulletin of the Seismological Society of America, v. 72, p. 911-923.

Priestley, K., Brune, J. N., and Anderson, J. G., 1985, Surface wave excitation and source mechanisms of the Mammoth Lakes earthquakes sequence: Journal of Geophysical Research, v. 90, p. 11177-11185.

Prodehl, C., 1970, Seismic-refraction study of crustal structure in the western United States: Geological Society of America Bulletin, v. 81, p. 2629-2646.

Reasenberg, P., and Ellsworth, W. L., 1982, Aftershocks of the Coyote Lake, California, earthquake of August 6, 1979; A detailed study: Journal of Geophysical Research, v. 87, p. 10637-10655.

Ryall, A. S., and Ryall, F., 1981, Spatial-temporal variations in seismicity preceding the May 1980 Mammoth Lakes earthquakes, California: Bulletin of the Seismological Society of America, v. 71, p. 747-760.

Savage, J. C., 1983, Strain accumulation in western United States: Annual Review of Earth and Planetary Sciences, v. 11, p. 11-43.

Savage, J. C., and Cockerham, R. S., 1987, Quasiperiodic occurrence of earthquakes in the 1978-1986 Bishop-Mammoth Lakes sequence, eastern California: Bulletin of the Seismological Society of America, v. 77, p. 1347-1358.

Sibson, R. H., 1982, Fault zone models, heat flow, and the depth distribution of earthquakes in the continental crust of the United States: Bulletin of the Seismological Society of America, v. 72, p. 151-163.

Sieh, K. E., 1981, A review of geological evidence for recurrence times of large earthquakes, in Simpson, D. W., and Richards, P. G., eds., Earthquake prediction; An international review: American Geophysical Union Maurice Ewing Series 4, p. 181-207.

Sieh, K. E., and Jahns, R. H., 1984, Holocene activity of the San Andreas fault at Wallace Creek, California: Geological Society of America Bulletin, v. 95, p. 883-896.

Spieth, M. A., 1981, Two detailed seismic studies in central California; Part 1, Earthquake clustering and crustal structure studies of the San Andreas fault near San Juan Bautista; Part 2, Seismic velocity structure along the Sierra foothills near Oroville, California [Ph.D. thesis]: Stanford, California, Stanford University, 162 p.

Stein, R. S., and King, G.C.P., 1984, Seismic potential revealed by surface folding: 1983 Coalinga, California, earthquake: Science, v. 224, p. 869-872.

Thatcher, W., 1979, Systematic inversion of geodetic data in central California: Journal of Geophysical Research, v. 84, p. 2283-2295.

Thatcher, W., and Lisowski, M., 1987, Long-term seismic potential of the San Andreas fault southeast of San Francisco, California: Journal of Geophysical Research, v. 92, p. 4771-4784.

Tocher, D., 1959, Seismic history of the San Francisco region: California Division of Mines and Geology Special Report 57, p. 39-48.

Trehu, A. M., and Wheeler, W. H., 1987, Possible evidence for subducted sedimentary materials beneath central California: Geology, v. 15, p. 254-258.

Tsai, Y., and Aki, K., 1970, Source mechanism of the Truckee, California, earthqake of September 12, 1966: Bulletin of the Seismological Society of America, v. 60, p. 1199-1280.

Van Wormer, J. D., and Ryall, A. S., 1980, Sierra Nevada-Great Basin boundary zone; Earthquake hazards related to structure, active tectonic processes, and anomolous patterns of earthquake occurrence: Bulletin of the Seismological Society of America, v. 70, p. 1557-1572.

Vetter, U. R., and Corbett, E. J., eds., 1987, Bulletin of the Seismological Laboratory for the period January 1 to December 31, 1982: Reno, University of Nevada Maclay School of Mines, 44 p.

Vetter, U. R., and Ryall, A. S., 1983, Systematic change of focal mechanism with depth in the western Great Basin: Journal of Geophysical Research, v. 88, p. 8237-8250.

Vetter, U. R., Ryall, A. S., and Sandera, C. O., 1983, Seismological investigations of volcanic and tectonic processes in the western Great Basin, Nevada and eastern California: Geothermal Resources Council Special Report 13, p. 333-343.

Wallace, R. E., 1970, Earthquake recurrence intervals on the San Andreas fault: Geological Society of America Bulletin, v. 81, p. 2875-2890.

—— , 1981, Active faults, paleoseismology, and earthquake hazards in the western United States, in Richards, P. G., and Simpson, D. W., eds., Earthquake prediction; An international review: American Geophysical Union Maurice Ewing Series 4, p. 209-216.

—— , 1984a, Pattern and timing of late Quaternary faulting in the Great Basin province and relation to some regional tectonic features: Journal of Geophysical Research, v. 89, p. 5763-5769.

—— , 1984b, Fault scarps formed during the earthquakes of October 2, 1915, in Pleasant Valley, Nevada, and some tectonic implications: U.S. geological Survey Professional Paper 1274-A, 33 p.

Wallace, T. C., 1985, A reexamination of the moment tensor solutions of the

1980 Mammoth Lakes earthquakes: Journal of Geophysical Research, v. 90, p. 11171–11176.
Walter, A. W., and Mooney, W. D., 1982, Crustal structure of the Diablo and Gabilan Ranges, central California: A reinterpretation of existing data: Bulletin of the Seismological Society of America, v. 72, p. 1567–1590.
Walter, S. R., 1986, Intermediate-focus earthquakes associated with Gorda plate subduction in northern California: Bulletin of the Seismological Society of America, v. 76, p. 583–588.
Walter, S. R., Rojas, V., and Kollmann, A., 1984, Seismicity of the Lassen Peak area, California: Geothermal Resources Council Transactions, v. 8, p. 523–527.
Warren, D. H., 1981, Seismic-refraction measurements of crustal structure near Santa Rosa and Ukiah, California, *in* McLaughlin, R. J., and Donnelly-Nolan, J. M., eds., Research in the Geysers–Clear Lake geothermal area, northern California: U.S. Geological Survey Professional Paper 1141, p. 167–182.
Weldon, R., and Humphreys, E., 1986, A kinematic model of southern California: Tectonics, v. 5, p. 33–48.
Wesson, R. L., Burford, R. O., and Ellsworth, W. L., 1973, Relationship between seismicity, fault creep, and crustal loading along the central San Andreas fault, *in* Kovach, R. L., and Nur, A., eds., Proceedings of the Conference on Tectonic Problems of the San Andreas Fault System, Stanford, California, Stanford University, p. 303–321.
Wilson, D. S., 1986, A kinematic model for the Gorda deformation zone as a diffuse southern boundary of the Juan de Fuca plate: Journal of Geophysical Research, v. 91, p. 10270.
Wong, I. G., and Savage, W. U., 1983, Deep intraplate seismicity in western Sierra Nevada, California: Bulletin of the Seismological Society of America, v. 73, p. 797–812.
Zoback, M. L., and Zoback, M. D., 1980, Interpretative stress map of the conterminous United States: Journal of Geophysical Research, v. 85, p. 6223–6154.
Zoback, M. D., and 11 others, 1987, State of stress of the San Andreas fault system: Science, v. 238, p. 1105–1111.

Manuscript Accepted by the Society August 7, 1989

NOTES ADDED IN PROOF

Two large earthquakes occurred in central California after the 1980–1985 time interval covered in this chapter that warrant comment as this volume goes to press: the M = 6.4 Chalfant Valley earthquake of 21 July 1986 and the M = 7.0 Loma Prieta earthquake of 17 October 1989. Both earthquakes occurred in recognized seismic gaps, although neither event was large enough to completely "fill" its respective gap. Both earthquakes together with their aftershock sequences produced distinct local changes in the seismicity patterns from those illustrated in the chapter for the 1980–1985 period.

The Chalfant Valley earthquake sequence occurred in a historically aseismic section of eastern California (the White Mountains seismic gap; see Hill and others, 1985, and Figure 7) just west of the White Mountains and roughly 20 km east-southeast of Long Valley caldera. The sequence began on 3 July with a well-developed foreshock sequence that culminated in an M = 5.9 event on 20 July. The M = 6.4 main shock occurred approximately 24 hours later (1442 UT on 21 July) followed by a rich aftershock sequence that included M = 5.8, M = 5.6, and M = 5.3 earthquakes plus thousands of smaller events (Cockerham and Corbett, 1987; also see Figure 3 in Rundle and Hill, 1988). The focal mechanism and displacement associated with the main shock indicate dominantly right-lateral slip but with a minor normal component on a plane striking north-northwest and dipping steeply (50° to 60°) to the southwest, consistent with displacement on the west-dipping White Mountains fault (Cockerham and Corbett, 1987; Gross and Savage, 1987). Earthquakes have continued to occur in the Chalfant aftershock zone through mid-1991 at a rate of tens of events per month.

The M = 7.0 Loma Prieta earthquake occurred beneath the Santa Cruz Mountains segment of the San Andreas fault at 5:04 p.m. (PDT) on 17 October 1989. This is the first major earthquake ($M \geq 6.5$) to occur along the main branch of the San Andreas fault since the great ($M \approx 8$) 1906 earthquake. Indeed, the main shock rupture is closely coincident with the southernmost 40 km of the 1906 break (in Figure 12B, the Loma Prieta rupture extends over all but the upper 4 km of the aseismic area above the U-shaped distribution of deep earthquake hypocenters in the distance range from 90 to 130 km in cross section I-I'). The Loma Prieta earthquake is perhaps the best recorded $M > 6.5$ earthquake in history (see McNally and Ward, 1990; U.S. Geological Survey Staff, 1990; Hanks and Krawinkler, 1991; Olson, 1991).

The Working Group on California Earthquake Probabilities (1988), among others, had identified this Santa Cruz Mountains segment of the San Andreas fault as having a somewhat higher probability (30% over the next 30 years) of producing an $M \geq 6.5 = 7.0$ earthquake than other major fault segments in the greater San Francisco Bay Area. In this respect, occurrence of the Loma Prieta earthquake was not entirely a surprise. The wealth of data from this earthquake, however, produced a number of important surprises that have forced a reevaluation of the nature of the San Andreas fault system through the Santa Cruz Mountains. Both seismic and deformation data, for example, show that the earthquake involved nearly equal parts of right-lateral strike-slip and reverse slip on a plane that is parallel with the local strike of the surface trace of the San Andreas fault but that dips roughly 70° to the southwest. This raises a question of just how the Loma Prieta earthquake relates to the vertical San Andreas fault that is believed to have ruptured in the great 1906 earthquake, and that has a geomorphic expression consistent with long-term, strike-slip displacement on a near-vertical fault. The strong component of reverse slip associated with this earthquake also raises a host of intriguing questions regarding the uplift history of the Santa Cruz Mountains and the recurrence of Loma Prieta-like earthquakes. For additional information on the tectonics of the San Andreas fault system, see Wallace (1990).

ADDITIONAL REFERENCES

Gross, W. K., and Savage, J. C., 1987, Deformation associated with the 1986 Chalfant Valley earthquake, eastern California: Bulletin of the Seismological Society of America, v. 77, p. 306–310.
Hanks, T. C., and Krawinkler, H., 1991, The 1989 Loma Prieta earthquake and its effects: Introduction to the special issue: Bulletin of the Seismological Society of America, v. 81 (in press).
McNally, K., and Ward, S. N., 1990, The Loma Prieta earthquake of October 17, 1989; Introduction to the special issue: Geophysical Research Letters, v. 17, p. 1177.
Rundle, J. B., and Hill, D. P., 1988, The geophysics of a restless caldera—Long Valley, California: Annual Review of Earth and Planetary Sciences, v. 16, p. 251–271.
U.S. Geological Survey Staff, 1990, The Loma Prieta, California, earthquake; An anticipated event: Science, v. 247, p. 286–293.
Wallace, R. E., ed., 1990, The San Andreas Fault system, California: U.S. Geological Survey Professional Paper, 1515, 283 p.
Working Group on California Earthquake Probabilities, 1988, Probabilities of large earthquakes occurring in California on the San Andreas Fault: U.S. Geological Survey Open File Report 88-398, p. 62.

Chapter 9

Seismotectonics of southern California

L. Katherine Hutton
Seismological Laboratory, California Institute of Technology, Pasadena, California 91125
Lucile M. Jones
U.S. Geological Survey, Pasadena, California 91106
Egill Hauksson
University of Southern California, Los Angeles, California 90089-0741
Douglas D. Given
U.S. Geological Survey, Pasadena, California 91106

Southern California straddles the boundary between the North American and the Pacific plates. The relative motion between these two plates has been determined from paleomagnetic lineations in the Gulf of California, from global solutions to known slip rates along plate boundaries, from geology, and from geodesy (Minster and Jordan, 1978; Minster and Jordan, 1978; DeMets and others, 1987) to be primarily horizontal at a rate of about 48 mm/yr (DeMets and others, 1987). This results in one of the highest levels of seismicity in the conterminous United States (e.g., Evernden, 1970). In southern California, the deformation is spread over a large area, encompassing numerous normal, strike-slip, and reverse faults. A majority of the plate motion appears to be accommodated by the San Andreas fault, with the rest distributed among the dozen or so other major faults (Weldon and Humphreys, 1986). This is in contrast to the plate boundary in northern California, where the plate motion is more concentrated near the San Andreas fault than it is in southern California (e.g., Hill and others, this volume). The diffuse deformational pattern leads to the high level of seismic activity and to a complicated tectonic structure.

On a broad scale, the North American–Pacific plate boundary in California is a transform fault that extends from the Gulf of California to Cape Mendocino (Fig. 1). The San Andreas fault and the transform plate boundary end at the Mendocino Triple Junction in northernmost California. North of Cape Mendocino, the spreading center and subduction zone of the Juan de Fuca plate lie between the North American and Pacific plates. Another spreading center lies south of southern California in the Gulf of California, creating parts of the Pacific and Rivera plates. The transform faults of that spreading system merge into the San Andreas fault system near the Imperial Valley and the Salton Sea (Fig. 1).

The relative motion of the plates in southern California is primarily accommodated on the right-lateral strike-slip faults of the San Andreas fault system (Fig. 2), which includes the San

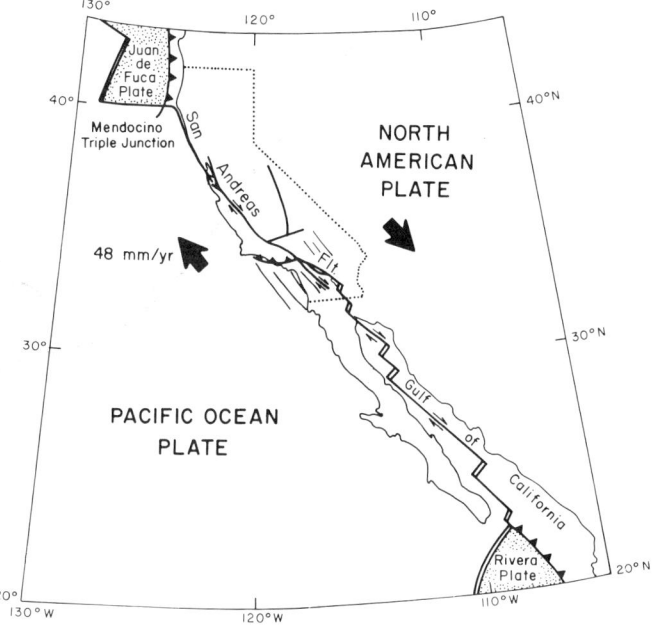

Figure 1. Schematic map of the North American–Pacific plate boundary in California.

Andreas fault itself and a number of subparallel faults, including those offshore in the Continental Borderland (Weldon and Humphreys, 1986). These faults strike northwest, except through the Transverse Ranges, where they strike west-northwest. The San Andreas fault has the highest slip rate in this system, with slip of 25 mm/yr to 35 mm/yr (e.g., Sieh and Jahns, 1984; Weldon and Sieh, 1985). South of, and truncated by, the Transverse Ranges, the San Jacinto, Elsinore, and Newport-Inglewood faults and several probable offshore faults parallel the San Andreas to the west. North of the Transverse Ranges, numerous smaller

Hutton, L. K., Jones, L. M., Hauksson, E., and Given, D. D., 1991, Seismotectonics of southern California, *in* Slemmons, D. B., Engdahl, E. R., Zoback, M. D., and Blackwell, D. D., eds., Neotectonics of North America: Boulder, Colorado, Geological Society of America, Decade Map Volume 1.

strike-slip faults strike northwest through the Mojave desert east of the San Andreas. Because the direction of plate motion trends N35°W (Minster and Jordan, 1978) while the San Andreas system faults strike at N45°W to N70°W, a significant component of convergent motion must also be accommodated. Much of this deformation occurs along the reverse faults of the Transverse Ranges (Fig. 2). These faults generally strike east-west and have slip rates on the order of millimeters per year (Ziony and Yerkes, 1985).

Studies of the earthquakes and tectonics of southern California have continued for over half a century since the establishment of the Seismological Laboratory in Pasadena in 1927 (Goodstein, 1984) and its associated seismographic stations. A continuous catalog has been maintained since 1932, when six stations were in operation. In that period, detection was complete at magnitude 3.0 and greater for the central part of the region (Hileman, 1978), except during intense aftershock sequences and swarms, when earthquakes were sometimes missed. The Southern California Seismic Network (SCSN) was expanded following the 1952 Kern County earthquake (M_S = 7.7), so that an average of 25 to 30 stations were in operation between 1952 and 1971. Further expansion occurred following the 1971 San Fernando earthquake (M_L = 6.4), when the U.S. Geological Survey (USGS) joined with the Seismological Laboratory in operating the SCSN; by the end of 1974, more than 100 stations were in operation. Data were for the most part are telemetered to the Seismological Laboratory and recorded on helicorder and develocorder. Since January 1977, the bulk of the telemetered data have been digitized and recorded using an on-line computer system. Various versions of the hardware and software, notably CEDAR

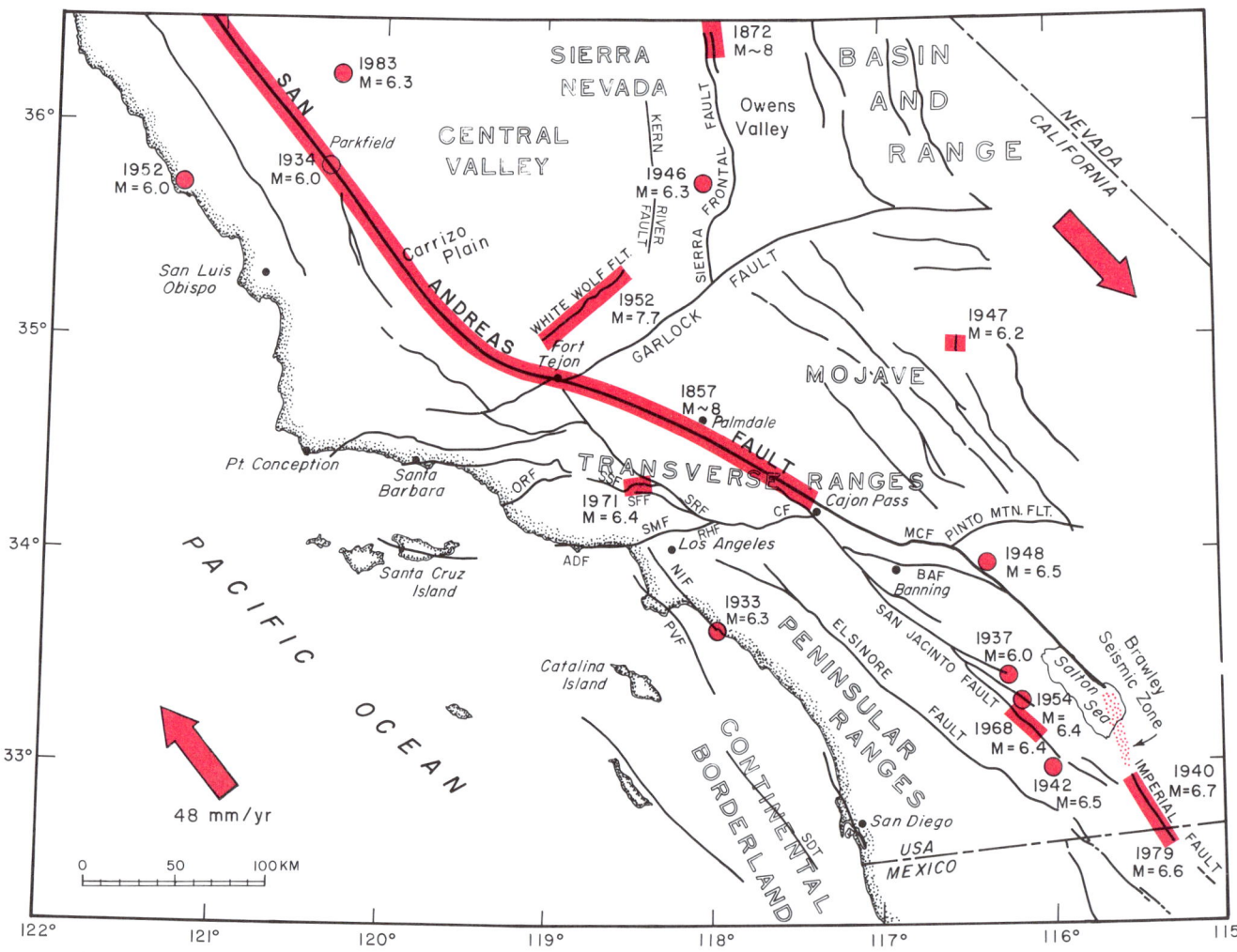

Figure 2. A map of southern California showing the major faults and physiographic regions and the M_L ≥ 6.0 earthquakes recorded by the Southern California Seismic Network from 1932 through 1986. Historic fault rupture is shown in red. Large arrows indicate the sense and magnitude of plate motion. Some fault names are abbreviated as: ADF = Anacapa Dume fault; BAF = Banning fault; CF = Cucamonga fault; MCF = Mission Creek fault; NIF = Newport-Inglewood fault; ORF = Oak Ridge fault; PVF = Palos Verdes fault; RHF = Raymond Hill fault; SDT = San Diego Trough-Bahia-Soledad fault; SFF = San Fernando fault; SMF = Santa Monica fault; SRF = Sierra Madre fault; SFF = Santa Susana fault.

Earthquake Detection and Recording; Johnson, 1979) and CUSP (Caltech-USGS Seismic Processor; Johnson, 1983; Given and others, 1987), have been in operation since then.

As of 1987, more than 250 components were recorded by the SCSN, making possible the detection of approximately 1,000 events per month. The U.S. Geological Survey and Caltech together operate 215 sites, 15 with multiple components. The SCSN also records signals from stations operated by the University of Southern California, California Department of Water Resources, and the U.S. Geological Survey in central California. At present, the earthquake catalog is largely complete above about magnitude 1.8 for the central part of the network, although many earthquakes smaller than 1.8 are recorded and analyzed (Fig. 3). As previously described in detail by Gutenberg and Richter (1954) and Allen and others (1965), the ratio of small earthquakes to large in southern California follows a magnitude frequency distribution where b is the slope of the curve. The maximum likelihood estimate of the b-value of 0.96 determined here from 6 years of $M \geqslant 1.8$ earthquakes is higher than graphical estimates of 0.88 and 0.86 (Gutenberg and Richter, 1954; Allen and others, 1965).

As has long been observed (e.g., Allen and others, 1965), the spatial pattern of microseismicity differs from the distribution of damaging earthquakes in southern California (Figs. 2, 4, and 5). Some Quaternary faults capable of major earthquakes, such as the San Jacinto fault, also have a high level of microseismicity, but other major faults like the San Andreas fault are very quiet at the microseismic level during much of the interseismic period. In spite of the recent improvements in location capability, a map of well-located earthquakes in southern California "still looks like a shotgun has been fired at it" (Allen, 1981; Figs. 4 and 5). Most $M \geqslant 6.0$ earthquakes are associated with Quaternary fault structures, but in many cases the potential for damaging earthquakes was not recognized until after the earthquake, as was true for the 1971 San Fernando ($M_L = 6.4$), the 1940 Imperial Valley ($M_L = 6.7$), and the 1952 Kern County ($M_S = 7.7$) earthquakes. Two of the $M_L \geqslant 6.0$ earthquakes, the 1946 Walker Pass ($M_L = 6.3$) and the 1983 Coalinga ($M_S = 6.5$) earthquakes, were not associated with a recognizable Quaternary fault at all. It appears that earthquakes at the $M_L = 5.0$ level and smaller can occur virtually anywhere in the seismically active regions of southern California (Fig. 4), without regard to obvious Quaternary faults. This widespread distribution is in marked contrast to that in central California, where both microseismicity and moderate earthquakes are concentrated on a few major fault structures (Hill and others, this volume).

In the 56-year catalog (1932 through 1987) of southern California earthquakes, 80 sequences have been recorded that included at least one earthquake of $M_L \geqslant 5.0$ (Table 1), giving an average rate of 1.4 $M_L \geqslant 5.0$ earthquake sequences per year. This rate has not been constant over the recording period of the SCSN. After the 1952 earthquake, the average number of earthquakes recorded decreased (Fig. 6; Hutton and others, 1979), presumably because that $M_S = 7.7$ earthquake relieved regional tectonic

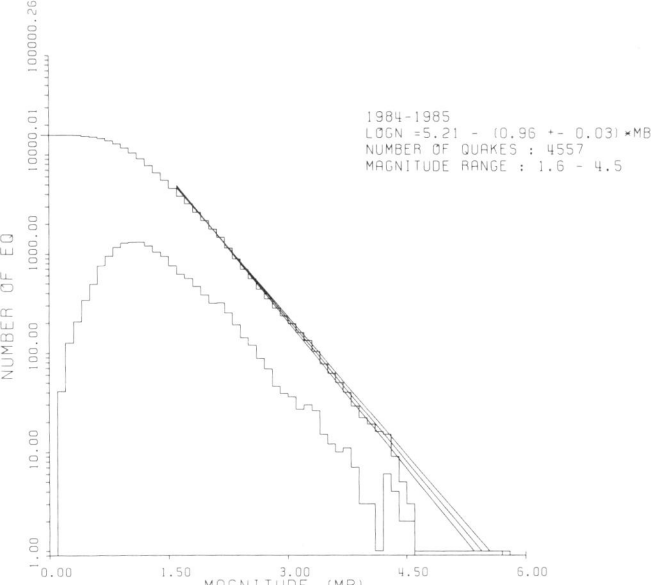

Figure 3. The number of earthquakes recorded in the central part of southern California from 1984 to 1985 versus magnitude. The lower curve is the total number of events in each 0.1 magnitude increment. The upper curve is the cumulative number of all events less than or equal to the magnitude on the x-axis. Straight lines represent b-value fits.

stress within a large area of southern California. An average of two $M_L \geqslant 5.0$ sequences per year were recorded between 1932 and 1952, but only one such sequence per year has occurred since 1953. Short-term increases in activity in 1978 and 1979 led to speculation that the long quiescence was ending (Raleigh and others, 1982). However, such increases did not continue for more than two or three years (Fig. 6).

The recent increase in station density, combined with the introduction of computerized hypocentral location of earthquakes and improved models of the velocity structure of the upper crust, has allowed more accurate resolution of the depth distribution of the earthquakes. Significant regional variations in maximum depth of earthquakes have been observed (Corbett and Hearn, 1984; Ziony and Jones, 1989; Fig. 7). The most notable variation is the group of deep earthquakes (depths of 15 to 22 km) near the Mission Creek and Banning faults. The greater density of stations and computer locations has also provided better coverage for the determination of focal mechanisms. Focal mechanisms have been determined for several hundred earthquakes over the last ten years (Fig. 8). They show a relatively consistent regional stress field with a maximum horizontal compression trending approximately north-south, in agreement with geodetic data (e.g., Savage and others, 1981) but with a wide variation in type of faulting with reverse, strike-slip, and normal faulting all in evidence.

The complicated tectonic structure of southern California leads to a diversity of seismicity patterns. In this chapter, we address the four major tectonic divisions of southern California—

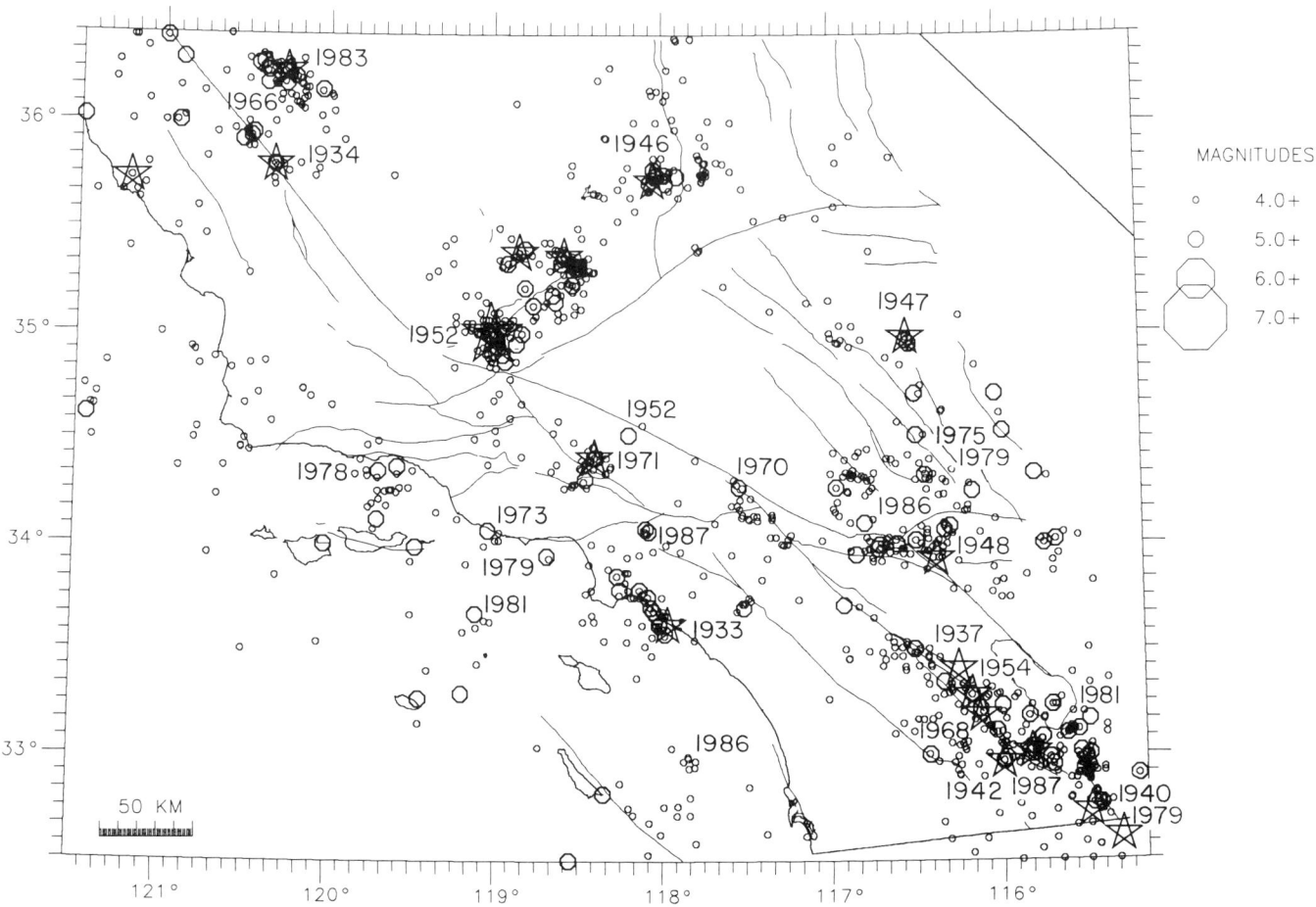

Figure 4. A map of southern California showing all $M_L \geq 4.0$ earthquakes recorded by the Southern California Seismic Network from 1932 through 1987. Earthquakes of $M_L \geq 6.0$ are shown by stars.

the San Andreas fault system, the Transverse Ranges, the Mojave desert, and the Sierra Nevada and southern Basin and Range. We discuss the seismicity as mapped in the Seismicity Map of North America (Engdahl and Rinehart, 1988) and summarize the work of numerous authors on earthquake sequences and seismotectonic structure.

SAN ANDREAS FAULT SYSTEM

At least two-thirds of the relative motion between the North American and Pacific plates in California occurs on the San Andreas fault system (Allen, 1981; Minster and Jordan, 1978; Sieh and Jahns, 1984). In southern California, this is a complex fault system with deformation spread over four major faults, including the San Andreas fault itself, and numerous minor ones (Fig. 2). The parent name of San Andreas has been associated with the easternmost member of the system on which the largest Quaternary displacements are recorded (Allen, 1981). In the following section, we discuss the San Andreas fault, the Imperial Valley fault, the San Jacinto fault, the Elsinore fault, and the faults of the coastal zone and the Continental Borderland, including the Newport-Inglewood fault.

San Andreas Fault

The San Andreas fault, which extends for 600 km in southern California, was responsible for the largest earthquake of the region, the 1857 $M \simeq 8$ Fort Tejon earthquake (Fig. 2). In northern and central California, the San Andreas is a clearly delineated feature, striking northwest approximately parallel to the direction of plate motion, N35°W (e.g., Hill and others, this volume; Fig. 1). Near Fort Tejon, along the northern edge of the Transverse Ranges, the fault changes strike to N70°W, in the so-called Big Bend. The main trace of the fault is clearly followed from there east-southeast to the northern end of the Banning region. Near Banning, the fault splits into several strands, with the Banning and the Mission Creek strands as the two most recently active features (Matti and others, 1985). Southeast of Banning,

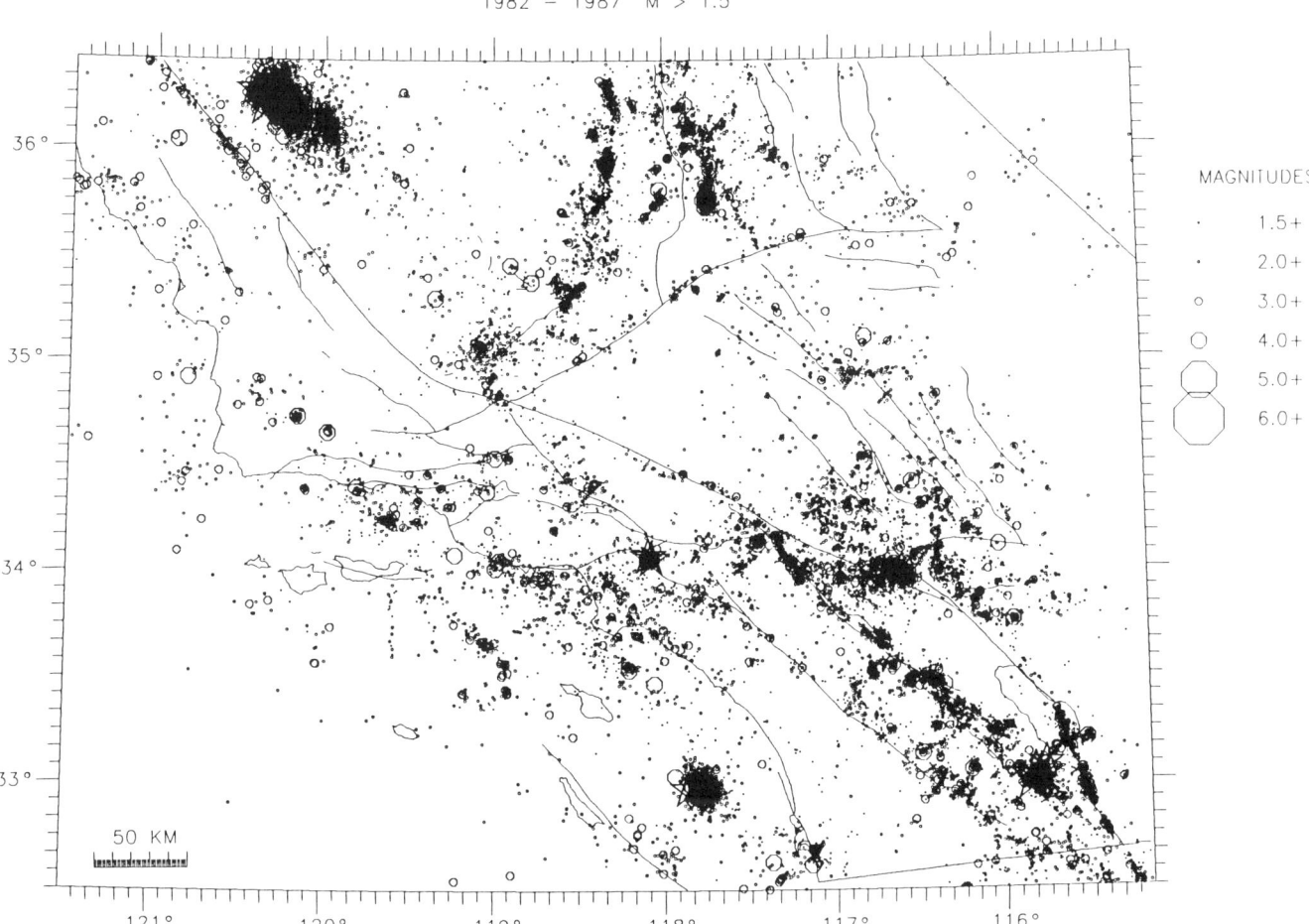

Figure 5. A map of southern California showing all $M_L \geq 1.5$ earthquakes recorded by the Southern California Seismic Network from 1982 through 1987. Earthquakes of $M_L \geq 5.0$ are shown by stars.

the multiple strands rejoin to form the Indio segment, and the strike of the fault resumes its northwest trend.

The rupture zone of the $M \simeq 8$ 1857 Fort Tejon earthquake on the San Andreas fault extended from Parkfield in central California, through the "Big Bend" near Fort Tejon, south to Cajon Pass (Sieh, 1978). The coseismic slip during the 1857 earthquake varied along strike, from 9 m near Wallace Creek and 6 m around Fort Tejon, to 3 to 4 m near Palmdale (Sieh, 1978). Jacoby and others (1988) have shown evidence from tree rings that the 1812 earthquake occurred on the San Andreas fault in the areas of Wrightwood (north of Cajon Pass) and San Bernardino (south of Cajon Pass). No large earthquake ($M \geq 7.0$) has been documented in the historic record (since 1749) for the San Andreas south of San Bernardino (the Banning and Indio sections (Allen, 1968).

The occurrence of moderate earthquakes in the Banning region led some researchers to suggest that this part of the fault is incapable of large earthquakes (Allen, 1968; Wallace, 1970); however, recent field studies have shown that the fault southeast of Banning does produce large to great earthquakes with repeat intervals of about 230 years (Sieh, 1986). This evidence suggests that the last major earthquake occurred about 1680 ± 20 years. Moreover, the hypocentral distribution of the 1986 North Palm Springs earthquakes showed that, in spite of the complicated surface geology, the rupture surface of the Banning strand at depth continues on the N60°W trend that is the local trend of the San Andreas fault farther southeast (Jones and others, 1986). This strengthens the hypothesis of Matti and others (1985) that the San Bernardino and the Coachella Valley segments of the Banning fault, which both strike N60°W, may join at depth as a continuous fault system.

The San Andreas fault has generally produced few moderate-sized earthquakes in historic times. The Banning segment of the fault is the one exception, with numerous $5.0 \leq M \leq 6.5$ events located near or on the several strands of the fault in the 1930s and 1940s (Fig. 4; Richter and others, 1958; Sykes and Seeber, 1985; Jones and others, 1986). After a hiatus in moderate earthquake activity from 1948 to 1986, an $M_L = 5.6$ event oc-

TABLE 1. MAIN SHOCKS IN SOUTHERN CALIFORNIA $M_L \geq 5.0$

Date			Time	Latitude	Longitude	Depth	Mag	Name
1933	3	11	1:54	33° 48.00	117° 58.00		6.3	Long Beach
1934	6	8	4:47	35° 48.00	120° 20.00		6.0	Parkfield
1935	9	8	17:03	32° 54.00	115° 13.00		5.0	
1935	10	24	14:48	34° 6.00	116° 48.00		5.1	
1935	12	20	7:45	33° 10.00	115° 30.00		5.0	
1937	3	25	16:49	33° 24.51	116° 15.68	10.0	6.0	
1938	5	31	8:34	33° 41.93	117° 30.63	10.0	5.5	
1938	6	6	2:42	32° 54.00	115° 13.00		5.0	
1938	9	17	14:23	35° 37.83	117° 30.80		5.0	
1938	9	27	12:23	36° 18.00	120° 54.00		5.0	
1939	6	24	13:01	36° 24.00	121° 0.00		5.5	
1939	12	28	12:15	35° 48.00	120° 20.00		5.0	
1940	5	18	5:03	34° 5.00	116° 18.00		5.4	Imperial Valley
1940	5	19	4:36	32° 44.00	115° 30.00		6.7	
1940	6	4	10:35	33° 0.00	116° 26.00		5.1	
1941	7	1	7:70	34° 22.00	119° 35.00		5.9	
1941	9	21	19:53	34° 52.00	118° 56.00		5.2	
1941	11	14	8:41	33° 47.00	118° 15.00		5.4	
1942	3	3	1:03	34° 0.00	115° 45.00		5.0	
1942	5	23	15:47	32° 59.00	115° 59.00		5.0	
1942	10	21	16:22	32° 58.00	116° 0.00		6.5	
1943	8	29	3:45	34° 16.00	116° 58.00		5.5	
1943	12	22	15:50	34° 30.00	115° 48.00		5.5	
1944	5	12	11:16	33° 59.66	116° 42.70	10.0	5.3	Kitching Peak
1945	3	20	21:55	34 15.00	116° 10.00		5.0	
1945	4	1	23:43	34° 0.00	120° 1.00		5.4	
1945	8	15	17:56	33° 13.00	116° 8.00		5.7	
1946	1	8	18:54	33° 0.00	115° 50.00		5.4	
1946	3	15	13:49	35° 43.50	118° 3.27	22.0	6.3	Walker Pass
1946	7	18	14:27	34° 32.00	115° 59.00		5.6	
1946	9	28	7:19	33° 57.00	116° 51.00		5.0	
1947	4	10	15:58	34° 59.00	116°323.00		6.2	Manix
1947	7	24	22:10	34° 1.00	116° 30.00		5.5	Morongo Valley
1947	11	18	21:59	33° 16.00	119° 27.00		5.0	
1948	2	24	8:15	32° 30.00	118° 33.00		5.3	
1948	12	4	23.43	33° 56.00	116° 23.00		6.5	Desert Hot Springs
1949	3	9	12:28	36° 1.00	121° 29.00		5.2	
1949	5	2	11:25	34° 1.00	115° 41.00		5.9	
1950	7	29	14:36	33° 7.00	115° 34.00		5.5	
1951	1	24	7:17	32° 59.00	115° 44.00		5.6	
1951	12	26	0:46	32° 49.00	118° 21.00		5.9	
1952	7	21	11:52	35° 0.00	119° 1.00		7.7	Kern County
1952	8	23	10:09	34° 31.16	118° 11.89		5.0	
1952	11	22	7:46	35° 44.00	121° 12.00		6.0	
1953	6	14	4:17	32° 57.00	115° 43.00		5.5	
1954	3	19	9:54	33° 17.00	116° 11.00		6.2	
1955	11	2	19:40	36° 0.00	120° 55.00		5.2	
1955	12	17	6:07	33° 0.00	115° 30.00		5.4	
1956	11	16	3:23	35° 57.00	120° 28.00		5.0	
1957	4	25	21:57	33° 12.98	115° 48.50		5.2	
1961	1	28	8:12	35° 46.69	118° 2.92	5.5	5.3	
1961	10	19	5:09	35° 49.89	117° 45.67		5.2	
1963	9	23	14:41	33° 42.61	116° 55.50	16.5	5.0	
1965	9	25	17:43	34° 42.75	116° 30.16	10.6	5.2	
1966	6	28	4:26	35° 54.94	120° 32.00	18.6	5.6	Parkfield

TABLE 1. MAIN SHOCKS IN SOUTHERN CALIFORNIA $M_L \geq 5.0$ (continued)

Date			Time	Latitude	Longitude	Depth	Mag	Name
1968	4	9	2:28	33° 11.39	116° 7.72	11.1	6.4	Borrego Mountain
1968	7	5	0:45	34° 7.06	119° 42.14	5.9	5.2	
1969	4	28	23:20	33° 20.60	116° 20.78	20.0	5.8	Coyote Mountain
1969	10	24	8.29	33° 17.46	119° 11.55	10.0	5.1	
1969	11	5	17:54	34° 36.54	121° 26.11	10.0	5.6	
1970	9	12	14:30	34° 16.18	117° 32.40	8.0	5.4	Lytle Creek
1971	2	9	14:00	34° 24.67	118° 24.03	8.4	6.4	San Fernando
1971	9	30	22:46	33° 2.01	115° 49.23	8.0	5.1	
1973	2	21	14:45	34° 3.89	119° 2.10	8.0	5.9	Point Mugu
1973	8	6	23:29	33° 59.16	119° 28.52	16.9	5.0	
1975	6	1	1:38	34° 30.94	116° 29.72	4.5	5.2	Galway Lake
1978	8	13	22:54	34° 20.82	119° 41.75	12.7	5.1	Santa Barbara
1279	1	1	23:14	33° 56.65	118° 40.88	11.3	5.0	Malibu
1979	3	15	21:07	34° 19.63	116° 26.68	2.5	5.2	Homestead Valley
1979	10	15	23:16	32° 36.81	115° 19.09	12.3	6.6	Imperial Valley
1980	2	25	10:47	33° 30.05	116° 30.79	13.6	5.5	
1981	4	26	12:09	33° 5.90	115° 37.90	3.7	5.7	Westmorland
1981	9	4	15:50	33° 40.26	119° 6.67	5.0	5.3	Santa Barbara Is.
1982	10	25	22:26	36° 17.49	120° 23.77	6.0	5.0	New Idria
1983	5	2	23:42	36° 14.69	120° 15.80	9.0	6.3	Coalinga
1985	8	4	12:01	36° 9.07	120° 2.91	6.0	5.8	Kewttleman Hills
1986	7	8	9:20	33° 59.91	116° 36.38	11.7	5.6	N. Palm Springs
1986	7	13	13:47	32° 58.24	117° 58.19	6.0	5.3	Oceanside
1987	10	1	14:42	34° 2.97	118° 4.75	14.4	59	Whittier Narrows
1987	11	24	13:15	33° 0.76	115° 50.31	2.8	6.5	Superstition Hills

curred on the Banning fault in 1986. Although the earthquakes on the Banning segment have been locally damaging, they have relieved little of the accumulated slip on the southern San Andreas fault (Jones and others, 1986). The only other instrumentally recorded earthquakes of moderate size to occur near the southern San Andreas fault were an $M_L = 5.0$ that occurred in 1952 near Palmdale and the 1970 $M_L = 5.4$ Lytle Creek earthquake. The Lytle Creek earthquake probably occurred on the northern San Jacinto fault (Jones, 1984).

The main trace of the San Andreas fault south of Parkfield appears to rupture with strike-slip displacement only in large earthquakes and is nearly aseismic during at least a major portion of the interseismic period of the seismic cycle. Adjacent to the main trace are secondary fault structures that often have significant compressional or extensional components (e.g., Davis, 1983; Weldon, 1986). The small to moderate-sized earthquakes recorded by the SCSN near the San Andreas occur mostly on these secondary features (Jones, 1988). Nicholson and others (1986) proposed that much of the microseismicity near the southern San Andreas fault occurs on northeast-striking faults and could be interpreted as resulting from rotations of small crustal blocks with dimensions of a few to a few tens of kilometers. Jones (1988), however, found evidence of northeast-striking fault planes only in the area east of the Indio segment of the San Andreas fault (south of Banning).

A wide range in level and type of seismic activity is evident along the San Andreas, from almost complete quiescence in the Carrizo Plain (Carlson and others, 1979) to one of the highest levels in southern California near Banning. The straight segments of the fault, near Carrizo Plain, Mojave, and Indio, have a lower level of seismicity than the two sections of the fault where large changes in strike occur, near Fort Tejon and Banning (Fig. 5; Sykes and Seeber, 1985; Jones, 1988). There is also a wide range in the depth distribution of the earthquakes: the deepest hypocenters are at Banning (22 km), and the shallowest at Indio (2 km) (Jones, 1988; Fig. 7). In general, however, the depths of the earthquakes along the San Andreas are similar to those elsewhere in southern California (Sibson, 1982): between 5 and 15 km depth.

Focal mechanism studies have found large variations in the style of seismic deformation along strike of the southern San Andreas fault (Fig. 8). From Parkfield northwestward, right-lateral, strike-slip motion parallel to the San Andreas fault is the primary mode of slip (e.g., Eaton and others, 1970). The Carrizo Plain is largely aseismic. Near Fort Tejon, reverse faulting mixed with strike-slip faulting on secondary fault structures predominates and continues south through the Mojave segment to Cajon Pass (McNally and others, 1978; Sauber and others, 1983; Pechmann, 1983; Webb and Kanamori, 1985; Jones, 1988). The most abrupt change occurs near Cajon Pass, where the reverse mecha-

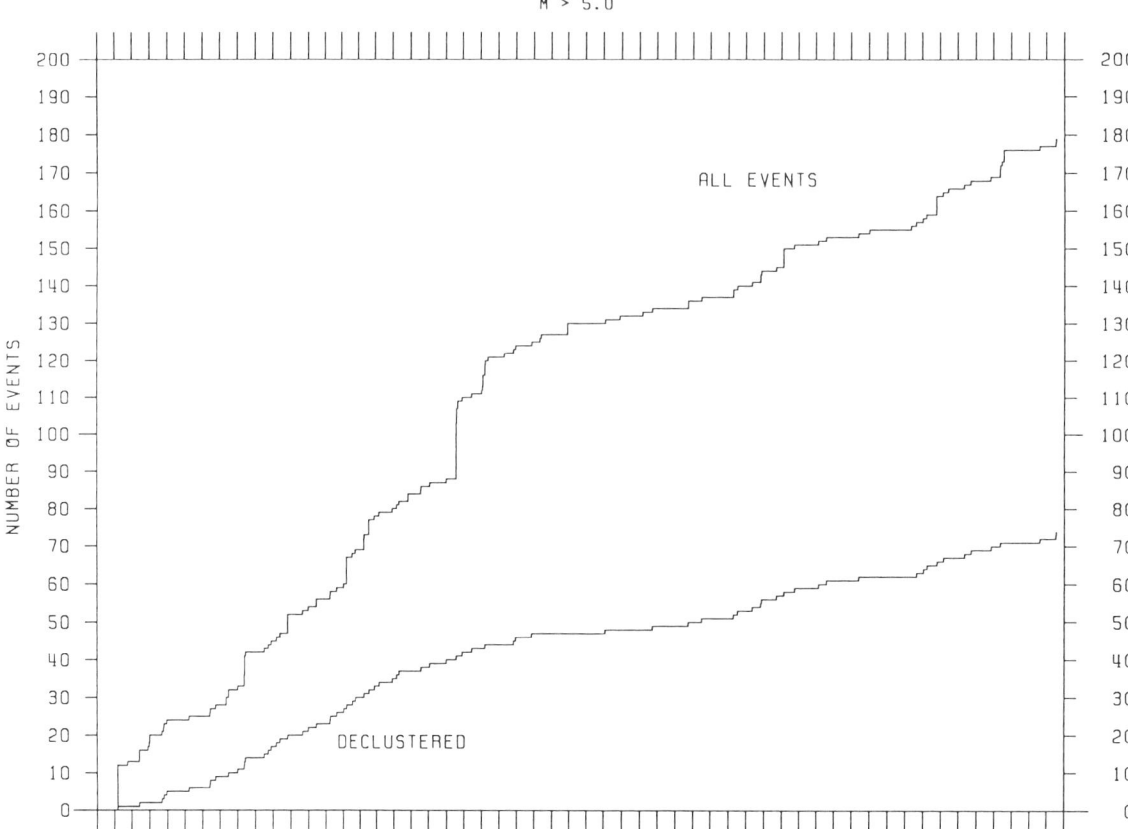

Figure 6. The cumulative number of earthquakes of $M_L \geq 5.0$ recorded in southern California versus time since 1932. The "declustered" curve is all $M_L \geq 5.0$ events except for foreshocks and aftershocks.

nisms of the Mojave segment give way within 10 km to the normal mechanisms of the San Bernardino segment (Jones, 1988). The most active and the most complex segment of the fault is the Banning segment where reverse, strike-slip, and normal faulting earthquakes occur in close proximity (Green, 1983; Webb and Kanamori, 1985; Nicholson and others, 1986; Jones, 1988). Southeast of Banning, on the Indio segment, the type of seismic slip returns to a simpler pattern of mixed normal and strike-slip faulting earthquakes (Jones, 1988).

Imperial Valley

The Imperial Valley—or, more broadly, the Salton Trough—is the zone of transition between the ocean-floor spreading regime in the Gulf of California and the right-lateral, strike-slip regime that characterizes the San Andreas fault system to the north (Larson and others, 1968; Moore and Buffington, 1968; Lomnitz and others, 1970; Elders and others, 1972; Sharp, 1982a). This major structural trough forms the northern end of the Gulf of California, which has been filled with the deltaic sediment from the Colorado River. The delta partially fills a sub-sea-level basin beneath the Coachella and Imperial valleys and forms a natural dam impounding the Salton Sea.

Seismic refraction studies (Fuis and others, 1982, 1984; Kohler and Fuis, 1986) reveal that the Salton Trough is 10 to 16 km deep and underlain by mafic ocean floor (subbasement). Minor Quaternary volcanism is a feature of the Salton Sea Geothermal Area (Robinson and others, 1976) in the northern Imperial Valley, and crustal extension and sea-floor spreading are thought to be active in oceanic crust beneath the sediment and metasediment. The Imperial fault connects the Salton Trough to the next postulated ridge segment to the southeast (Fig. 1 and 2). Lomnitz and others (1970) show that the spreading rate at each ridge segment of the Gulf of California probably decreases to the north, with some percentage of the slip carried on strike-slip faults (such as the Elsinore and San Jacinto) to the northwest.

The Imperial Valley is one of the most seismically active regions of southern California. The historical record indicates that several damaging earthquakes occurred in the Imperial Valley before 1932. Two large earthquakes have occurred on the Imperial fault since the beginning of the SCSN catalog, a $M_S = 7.1$ event in 1940 (Neumann, 1942) and an $M_L = 6.6$ earthquake in

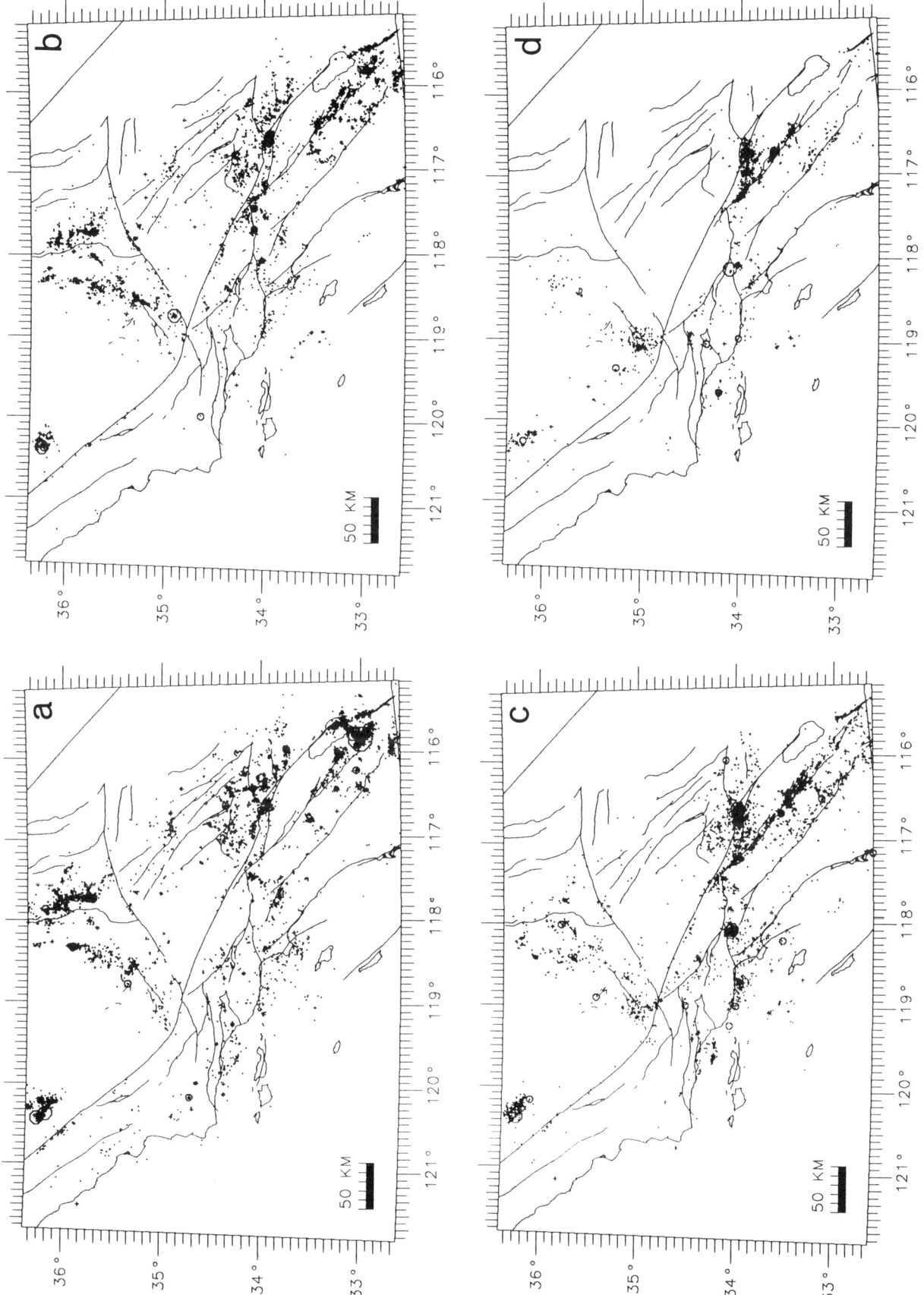

Figure 7. Four maps of southern California showing all $M_L \geq 4.0$ earthquakes recorded by the Southern California Seismic Network from 1982 to 1986 separated by depth. Earthquakes with calculated depths between 0 and 5 km (a); 5 and 10 km (b); 10 and 15 km (c); ≥ 15 km (d).

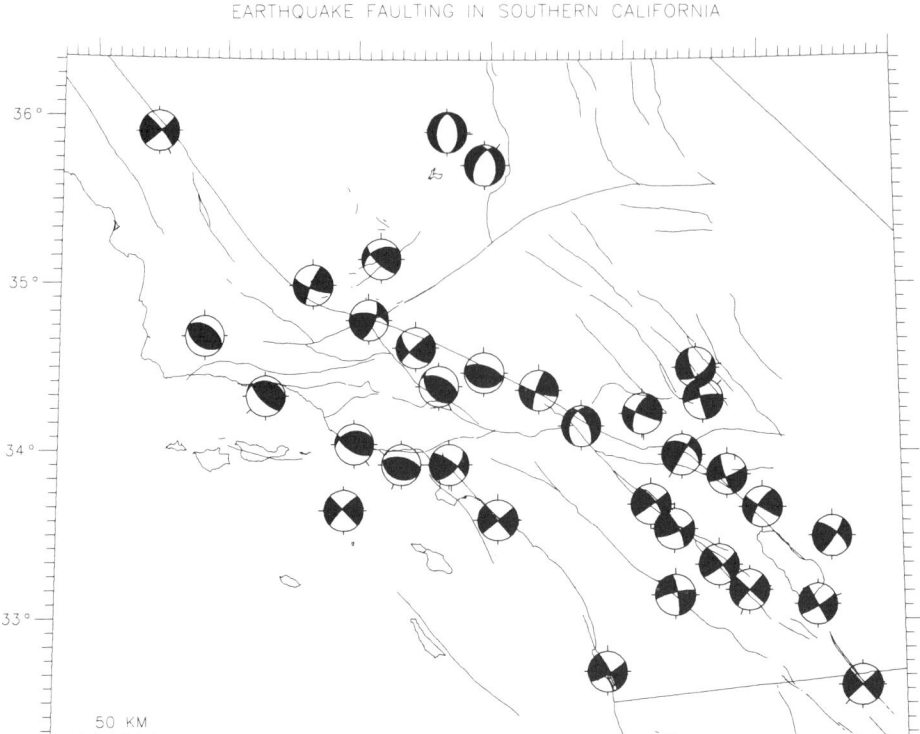

Figure 8. A map of southern California showing focal mechanisms of selected earthquakes since 1932. Focal mechanisms taken from Allen and Nordquist (1972), Corbett and Johnson (1982), Dollar and Helmberger (1985), Given (1983), Hauksson (1987), Hauksson and Saldivar (1986), Hauksson and others (1988), Heaton (1982), Hutton and others (1980), Hutton and Johnson (1981), Johnson and Hutton (1982), Jones (1988), Jones and Dollar (1986), Jones and others (1986), Magistrale and Kanamori (1988), Stein and Thatcher (1981), Stierman and Ellsworth (1976), Webb and Kanamori (1985), and unpublished data. Focal mechanisms are lower-hemisphere, equal-area projections; compressional quadrants are shaded. All $M_L \geq 5.0$ earthquakes of the last 20 years are included. The size of the mechanism scales with magnitude.

1979 (e.g., Johnson and Hutton, 1982). The epicenter of the 1940 earthquake was at the northern end of the Imperial fault, near the juncture with the Brawley Seismic Zone (Trifunac and Brune, 1970), and rupture was to the south, with the largest displacements occurring just north of the international border (Neumann, 1942; Richter, 1958; Reilinger, 1984). In contrast, the 1979 earthquake nucleated south of the border, near the juncture of the Imperial fault with the Cerro Prieto Seismic Zone, and ruptured largely to the north (Archuleta, 1982; Crook and others, 1982; Hartzell and Helmberger, 1982; Hartzell and Heaton, 1983; Langbein and others, 1982, 1983; Olsen and Apsel, 1982; Sharp, 1982b; Sharp and others, 1982; Snay and others, 1982; Slade and others, 1984; Reilinger and Larsen, 1986). In general, the 1979 earthquake reruptured the northern part of the 1940 earthquake fault rupture. Although evidence for a small amount of slip south from the hypocenter in 1979 was apparent in Mexican strong-motion data (Anderson and Silver, 1985; Silver and Masuda, 1985), this clearly did not extend as far south as the 1940 rupture did. Anderson and Bodin (1987) suggest that repeated magnitude 6 to 7 earthquakes, along with as much as 5 mm/yr of creep (Louie and others, 1985) on the Imperial fault, account for a large percentage of the motion between the Pacific and North American plates at that latitude.

At the microseismic level, the most prominent seismogenic structure in the Imperial Valley is a band of high seismicity connecting the northwestern end of the Imperial fault and the southeastern end of the San Andreas fault (Fig. 5), called the Brawley Seismic Zone (Johnson, 1979). This zone is characterized by swarms of small earthquakes ($M_L \leq 5.6$) (Johnson and Hadley, 1976; Hill, 1977; Gilpin and Lee, 1978; Johnson and Hill, 1982; Hutton and Johnson, 1981). The southern end of the Brawley Seismic Zone abuts the Imperial fault at a structural depression where a minor component of normal faulting was observed following the 1979 earthquake (Sharp and others, 1982). Focal mechanisms of small earthquakes in this area, however, exhibit nearly pure strike-slip solutions. Most of the aftershocks of the 1979 main shock occurred within the Brawley Seismic Zone rather than near the main-shock epicenter on the Imperial fault.

Faults conjugate to the Imperial and San Andreas faults play a significant role within the Brawley Seismic Zone. Evidence suggests that the largest 1979 aftershock (M_L = 5.8), which occurred 16 hours after the main shock, occurred on such a structure. In particular, a northeast-southwest trend appeared in the seismicity after that shock (Johnson and Hutton, 1982), coincident with a zone of liquifaction and other surface disturbance (Heaton and others, 1983). A similar trend appeared farther north during the 1981 Westmorland sequence (Hutton and Johnson, 1981) following the largest shock (M_L = 5.6), and an M_S = 6.2 ruptured the surface along such a trend even farther north during the 1987 Superstition Hills sequence (Magistrale and Kanamori, 1988; Johnson and Hutton, 1988).

Johnson (1979) discussed two types of Imperial Valley earthquake swarms. Those in the southern end of the Brawley Seismic Zone, near the town of Brawley, tend to occur in pairs, the members of which nucleate on the Imperial fault itself and propagate away from it on conjugate structures. The spatial and temporal arrangement suggests that they are triggered by creep events on the Imperial fault. The swarms in the northern part of the Brawley Seismic Zone also nucleate at a point within the Brawley Seismic Zone and spread from there, but do not generally happen in pairs (Johnson, 1979; Hutton and Johnson, 1981; Johnson and Hutton, 1981). Focal mechanisms of most swarm earthquakes in the Imperial Valley are strike slip (Hutton and Johnson, 1981), with a strike consistent with the trend of the nearby San Andreas and Imperial faults. Hill (1977) has postulated a fluid-control model of earthquake swarms that seems applicable to the Imperial Valley. Imperial Valley swarms typically show complex spatial-temporal patterns; the Westmorland sequence of 1981 (Hutton and Johnson, 1981; Johnson and Hutton, 1981) involved activity on at least seven distinct fault planes.

Following the 1940 earthquake, swarm activity in the Brawley Seismic Zone ceased until the mid-1970s, presumably because the regional stress dropped as a result of that M_S = 7.1 event. Heavy swarm activity began again in 1975 and occurred at intervals until the 1979 earthquake, and for a few years afterward. The activity is currently relatively low (Given and others, 1986).

Doser and Kanamori (1986) showed that most Imperial Valley seismicity occurs in the metasedimentary basement or deeper. However, some events in the 1981 Westmorland sequence may have been shallower (Hutton and Johnson, 1981). Both the 1979 and 1981 sequences also included a few events on the southeastern projection of the San Andreas fault through Niland, at the eastern boundary of the Salton Trough. These events were all small, but those with sufficient data for focal mechanisms showed normal displacement (Johnson and Hutton, 1982), which is consistent with the subsidence of the Salton Trough.

Johnson and Hutton (1981, 1986) point out that the junctures where the 1940 and 1979 earthquakes occurred are stress concentration points where the Imperial fault is most heavily loaded by nearly continuous swarm activity in the two adjacent (Brawley and Cerro Prieto) seismic zones. Similarly, activity along the Brawley Seismic Zone suggests that a major portion of the displacement observed on the Imperial fault is being transferred to the San Andreas fault to the northeast.

San Jacinto Fault

The San Jacinto fault zone is a major tectonic feature both structurally and seismically. It is a member of the San Andreas fault system and strikes northwest for more than 200 km, parallel to the plate boundary. At its northwest end it appears to splay from the San Andreas fault in the San Gabriel mountains, forming a complex surface trace of anastomosing and splaying strands to its apparent southeastern termination in the Quaternary sediments of the Salton Trough. It is predominantly a right-lateral strike-slip fault, but some segments display significant amounts of dip-slip offset (Sharp, 1967).

The San Jacinto fault zone is seismically the most active structure in southern California at all magnitude levels less than 7.0. It has produced at least nine events of $M_L \geqslant 6.0$ since 1890; four of these since the installation of the SCSN (1937, 6.0; 1942, 6.5; 1954, 6.2; 1968, 6.4) (Fig. 2). The fault zone, therefore, has an average recurrence interval of about 10 years for magnitude 6.0 and larger events, with the longest interval between consecutive events being 19 years. The last large events on the San Jacinto fault system have been the Borrego Mountain earthquake (M_L = 6.4) on April 9, 1968 (Allen and Nordquist, 1972), and the Superstition Hills earthquake (M_S = 6.6) on November 24, 1987.

The potential for a large, damaging earthquake on the San Jacinto fault is enhanced by the existence of a seismic gap. Thatcher and others (1975) estimated moment and rupture dimensions of all the $M_L \geqslant 6.0$ events on the San Jacinto fault since 1890. On the basis of their study they proposed two segments of the fault that may be gaps in seismic strain release. One of the proposed gaps, which covers the northern 40 km of the fault, may have been filled by one or both of two poorly located events in 1899 and 1907. The other gap is a 40-km-long segment in the central portion of the fault near the town of Anza. The northern 20 km of this gap also has a very low rate of background activity bounded to the northwest and southeast by persistent, deep clusters of seismicity (Sanders and Kanamori, 1984). This quiet segment is often referred to as the "Anza Gap."

The high level of microseismic activity on the San Jacinto fault stands out clearly in seismicity plots (Fig. 5). More detailed seismicity maps and cross sections reveal that the earthquakes occupy a diffuse band about the fault trace and do not sharply define individual fault planes. Background activity is generally concentrated in clusters at points of complication in the fault zone, such as fault segment junctions and terminations. Focal mechanisms (Fig. 8) indicate slip planes consistent with the strike of the fault zone, even though much of the microseismicity clearly occurs on secondary structures.

Early studies of the seismicity of the San Jacinto fault were carried out by Brune and Allen (1967), Arabasz and others

(1970), Cheatum and Combs (1973), and Hadley and Combs (1974). An apparent deepening of seismicity from southwest to northeast across the central portion of the fault was interpreted by Arabasz and others (1970) as a vertical offset in the seismogenic zone. Studies of the 1968 Borrego Mountain earthquake (Hamilton, 1970) and the 1969 Coyote Mountain earthquake (M_L 5.8; Thatcher and Hamilton, 1973) place epicenters of these two sequences to the northeast of the mapped fault trace. More recent work by Given (1983) and Sanders and Kanamori (1984) suggests that these observations are a result of a northeastward dip of the fault plane.

Most of the hypocenters along the northern and central San Jacinto fault occur between 10 and 20 km, but the bottom of the seismogenic zone shallows to the southeast. This pattern may be related to the regional heat flow, with earthquakes deep where low heat flow drives the brittle-ductile transition zone down, and shallow where heat flow is high (Sibson, 1982). To the north the cool roots of the San Jacinto Mountains depress the geotherms, allowing deep brittle failure; toward the Salton Trough, heat flow increases, allowing ductile deformation at shallower depths (Given, 1983; Doser and Kanamori, 1986).

The microseismicity suggests the possibility that the San Jacinto fault—not the San Andreas fault—is the main locus of plate motion south of the Transverse Ranges. This idea is not consistent, however, with the total geologic displacement of only 24 km on the San Jacinto fault zone (Sharp, 1967) or with estimates of the Holocene slip rate, which vary from 1.6 mm/yr (Sharp, 1981) to as much as 15 mm/yr (Brune, 1968; Sharp, 1981). Even the largest estimates only represent a fraction of the plate rate of 48 mm/yr (DeMets and others, 1987).

Elsinore fault and the Continental Borderland

Several large northwest-striking strike-slip faults, such as the Elsinore, Newport-Inglewood, and the Palos Verdes, are located west of both the San Andreas and San Jacinto faults near the coast of southern California (Fig. 2). These faults extend south of the U.S.–Mexico international border and merge with major faults in Baja California, such as the San Miguel and Agua Blanca systems. To the north, these faults are truncated at the southern margin of the Transverse Ranges, with some oblique-slip faults, such as the Whittier, part of the transition.

The region west of the San Andreas and San Jacinto faults and south of the Transverse Ranges is characterized by a moderate level of seismicity. During the last 56 years the region has experienced one M_L = 6.3 earthquake and several earthquakes in the magnitude range M = 5.0 to 6.0 (Fig. 4). Although the spatial distribution of the 1982 to 1986 seismicity is diffuse, some northwest-trending alignments of epicenters that are subparallel to major Quaternary strike-slip faults emerge from the background (Fig. 5). Most of the background seismicity consists of scattered single shocks, located adjacent to but not necessarily within the fault zones. Mechanisms are mostly strike-slip, although minor local variations of reverse or normal faulting are observed (Fig. 8). The Quaternary faults themselves are believed to rupture in infrequent moderate to large earthquakes.

On the basis of its aftershock distribution, the damaging 1933 Long Beach earthquake (M_L = 6.3) appears to have ruptured a 30-km-long segment of the Newport-Inglewood fault (Richter, 1958; Fig. 2). The focal mechanism of the main shock, derived from teleseismic data, shows pure right-lateral strike-slip motion along a N40°W striking plane (Woodward-Clyde Consultants, 1979; Fig. 8). During the last decade, several moderate-sized earthquakes (M_L = 5–6) have occurred offshore within the Continental Borderland. The 1981 Santa Barbara Island earthquake (M_L = 5.3) occurred 40 km offshore near the Santa Cruz–Catalina sea-floor escarpment. The main-shock focal mechanism showed strike-slip motion on a vertical plane, subparallel to the escarpment (Corbett, 1984). The 1986 Oceanside earthquake (M_L = 5.3) and its aftershock sequence, located 50 km offshore, showed a more complex pattern of faulting (Hauksson and Jones, 1988; Pacheco and Nabelek, 1988). That earthquake was located at the northern end of the San Diego Trough–Bahia–Soledad fault zone as mapped by Legg (1985). The focal mechanism of the main shock, as well as the mechanisms of several of the aftershocks, shows reverse faulting on a N70°W striking fault plane. Hence this sequence may have occurred on a small fault that provides for a left step in the San Diego Trough fault as it curves around the Santa Cruz–Catalina escarpment. That earthquake was unusual in having one of the largest and most protracted aftershock sequences ever recorded in southern California (Hauksson and Jones, 1988).

Detailed studies of the recent background seismicity in the region indicate that the calculation of both epicentral locations and focal depths can be influenced by heterogeneities in the local velocity structure. Hauksson (1987) showed that small earthquakes near the Newport-Inglewood fault have depths ranging from 6 to 11 km if the low velocities of the basin sediments are taken into account. Hence the depth distribution of the earthquakes in the Los Angeles basin falls between the depth distributions of the Imperial Valley and the Peninsular Ranges as reported by Doser and Kanamori (1986). The depth distribution of earthquakes in the Continental Borderland is largely unresolved but is probably relatively shallow (<15 km) (Corbett, 1984; Hauksson and Jones, 1988). The spatial distribution of seismicity can be correlated with major offshore fault systems that have been mapped using remote sensing techniques (Legg, 1980; 1985).

The presence of the small-magnitude background seismicity west of the San Andreas and San Jacinto faults indicates that tectonic stress is accumulating in the region. Hence, many of the Quaternary fault zones that crosscut the coastal zone and the Continental Borderland are potentially seismically active and capable of generating damaging earthquakes of $M_L \geq 6.0$.

TRANSVERSE RANGES

The Transverse Ranges of southern California are a series of east-trending mountain ranges and deep sedimentary basins (e.g.,

Keller, 1981; Keller and Prothero, 1987). The Santa Ynez Mountains and Point Conception form the province's westernmost extent, and the Little San Bernardino Mountains its easternmost. In between lie the Santa Monica and San Gabriel Mountains, north of Los Angeles, and the San Bernardino Mountains. The mountains cut across the predominantly northwest-trending structures of the surrounding tectonic provinces and the San Andreas fault system. The San Andreas fault itself is the only fault of the system that is not truncated by the Transverse Ranges. Anderson (1971) and Keller (1981), among others, relate the origin of the Transverse Ranges to the subduction of the Murray Transform. Most authors agree that the north-south compression arises because of the so-called Big Bend, or left step, in the right-lateral San Andreas Fault and the compressional component to the relative plate motion. Wave propagation studies in the upper crust (e.g., Hearn and Clayton, 1986a) indicate that the batholiths of the San Gabriel and San Bernardino ranges do not extend below 10 km depth; these mountains are essentially rootless.

Central and western Transverse Ranges

The central and western Transverse Ranges trend east-west from the San Andreas fault on the east to Point Conception on the west. Three major systems of seismically active, west- to west-northwest–striking reverse faults have been recognized in the Transverse Ranges (Fig. 2). The middle system is a group of Quaternary oblique reverse faults that dip steeply to moderately northward. This system of faults includes the Santa Cruz Island, the Anacapa Dume, the Santa Monica, the Raymond Hill, and the Cucamonga faults (Ziony and Yerkes, 1985). A second system of Quaternary reverse faults branches north from the Cucamonga fault and extends to the west-northwest along the Sierra Madre, San Fernando, and Santa Susana faults. This system can be traced across the Ventura basin and offshore from Santa Barbara to merge with the Hosgri fault (Bird and Rosenstock, 1984). The 1987 Whittier Narrows earthquake (M_L = 5.9) drew attention to another system of blind thrust faults buried beneath the sediments of the Los Angeles basin, south of the Raymond Hill and Cucamonga faults (Hauksson and others, 1988). The surface expression of these faults appears to be a series of anticlinal hills, but slip on faults occurs at depth. This fault system assumes major importance in studies of earthquake hazard because it passes directly under downtown Los Angeles.

The destructive 1971 San Fernando earthquake (M_L = 6.4) illustrates the convergent tectonic style of this province. It was the largest earthquake to occur in this region since the beginning of the instrumental catalog in 1932 (Fig. 2). The main shock ruptured a 20-km-long segment of the San Fernando thrust fault. Detailed aftershock studies by Whitcomb (1971) and modeling of observed ground motions by Heaton and Helmberger (1979) and Heaton (1982) reveal a main rupture surface with a 30° to 35° dip to the northeast at shallow depths, presumably the San Fernando fault, and a steeper surface with a 50° to 55° dip at depths greater than 8 km, which could be the Sierra Madre xault.

The estimated Quaternary rate of crustal convergence across the Cucamonga–San Gabriel fault system is 3 to 9 mm/yr (Ziony and Yerkes, 1985). These high slip rates, in particular along faults in the Ventura Basin, suggest that damaging earthquakes ($M_L \geq$ 6.0) could occur relatively frequently in the Transverse Ranges along this fault system (Wesnousky, 1986).

The rupture pattern of the October 1, 1987, Whittier Narrows earthquake (M_L = 5.9) was quite different from that of the San Fernando earthquake. Instead of a steeply dipping reverse fault extending to the surface, a shallowly dipping ($\simeq$ 25°), buried thrust fault produced the Whittier Narrows event. The earthquake was deep, with an aftershock zone between 11 and 16 km (Hauksson and others, 1988). No surface rupture for the Whittier Narrows earthquake has been documented (Hauksson and others, 1988). The aftershock zone was also small for this size main shock, about 4 km by 5 km. A blind thrust-fault system, the Elysian Park fault system, extending under the anticlines of the northern Los Angeles basin and never exposed at the surface, has been proposed as the causative fault (Hauksson and others, 1988). Geologic evidence suggests that the Quaternary rate of shortening across these anticlines that extend for almost 100 km across the Los Angeles basin is on the order of 5 mm/yr (Davis and Hayden, 1987). How much of this slip is released seismically is still unresolved.

Numerous moderate-sized earthquakes (M_L = 5.0–6.0) have also occurred in the central and western Transverse Ranges during the last 56 years. Most of these earthquakes show reverse faulting on approximately west- or northwest-striking planes, consistent with the dominantly north-south maximum principal stress (Fig. 8). The most destructive of these earthquakes was the 1978 Santa Barbara (M_L = 5.1) earthquake, which was followed by an extensive aftershock sequence (Corbett and Johnson, 1982; Yeats and Olson, 1984). The main shock focal mechanisms showed thrust faulting at a depth of about 13 km on a shallow-dipping plane with north-over-south movement (Corbett and Johnson, 1982). In addition to main shock–aftershock sequences, the Santa Barbara Channel area is prone to frequent earthquake swarms (Sylvester and others, 1970). Several other moderate earthquakes have occurred on the southern margin of the Transverse Ranges: the 1930 Santa Monica Bay (M_L = 5.2) earthquake (Gutenberg and others, 1932), the 1973 Point Mugu (M_L = 5.9) earthquake (Stierman and Ellsworth, 1976), and the 1979 Malibu (M_L = 5.0) earthquake (Hauksson and Saldivar, 1986).

The spatial distribution of microseismicity recorded during the period 1982 to 1986 shows a scattered pattern of epicenters, in part because most of the earthquakes are associated with dipping and geometrically complex structures (Fig. 5). In a study of the spatial distribution of seismicity from 1933 to 1977 in the central Transverse Ranges, Pechmann (1983; 1987) found similar diffuse patterns. He also determined focal mechanisms for some of the more recent smaller events and found mostly reverse faulting on east-striking planes, with a few cases of strike-slip faulting on northwest- or northeast-striking planes.

The focal depths of small earthquakes in the central and

western Transverse Ranges vary from 3 to 15 km, with a few events as deep as 20 to 25 km (Fig. 7; Ziony and Jones, 1989). Some focal mechanisms of the deeper earthquakes in the western Transverse Ranges indicate shallow south-over-north reverse movement in the northern part of the province and deeper north-over-south movement farther south (Webb and Kanamori, 1985). These focal mechanism data have been used to infer the existence of a regional decollement surface at the base of the seismogenic zone (e.g., Hadley and Kanamori, 1978; Webb and Kanamori, 1985). Much of the motion on the hypothetical detachment surface, however, is expected to be aseismic. Some authors believe there is geodetic evidence for episodic aseismic slip on such a detachment surface; for example, the "Palmdale Bulge" (Thatcher, 1979; Rundle and Thatcher, 1980) and an unusual strain episode that may have been observed in the Central Transverse Range area in 1979 (Savage and others, 1981; King and Savage, 1984; Turcotte and others, 1984; Savage and Gu, 1985; Savage and others, 1986). There is some temporal clustering of seismicity in the Transverse Ranges; for example, in 1978–1979 (Sauber and others, 1983), which could be related to episodic slip.

Such a decollement surface is also consistent with some recent models that propose aseismic subduction of ductile lithospheric material under the Transverse Ranges (Walck and Minster, 1982; Humphreys and others, 1984). Other models of aseismic subduction (e.g., Bird and Rosenstock, 1984; Hearn, 1984; Hearn and Clayton, 1986b), which do not require a decollement surface, also could explain the significant crustal shortening and associated uplift and thrust faulting that is taking place in the Transverse Ranges.

Eastern Transverse Ranges

The eastern Transverse Ranges extend from the San Andreas fault on the west to the Little San Bernardino mountains on the east (Fig. 2). Unlike the western and central Transverse Ranges, where reverse faults are common, the eastern Transverse Ranges are dominated by the San Andreas fault and its many strands that bound the region to the south (Matti and others, 1985). The most prominent reverse fault is the Banning fault (Fig. 2). This fault shows geologic evidence of reverse motion but experienced predominantly strike-slip motion in the 1986 North Palm Springs earthquake (Jones and others, 1986). Numerous moderate earthquakes have occurred in the eastern Transverse Ranges; these are described above, in conjunction with the San Andreas fault. Only one moderate earthquake, the 1979 Big Bear earthquake (M_L = 4.9) (Webb and Kanamori, 1985), is cataloged in the eastern Transverse Ranges north of the San Andreas system.

The microseismicity of the eastern Transverse Ranges occurs at a very high rate (Hileman and others, 1973; Friedman and others, 1976; Hutton and others, 1985) and has been extensively studied (e.g., Green, 1983; Webb and Kanamori, 1985; Nicholson and others, 1986; Jones, 1988). The shallow seismicity ($\leqslant$12 km depth) tends to be spatially clustered, with the clusters distributed more or less uniformly over the area. Earthquakes at depths greater than 12 km occur only south of the Mission Creek branch of the San Andreas fault, indicating that it may be a fundamental tectonic boundary. In fact, some of the deepest earthquakes in southern California occur in this area (Fig. 7). The 1986 North Palm Springs earthquake (see the San Andreas fault section above) occurred at the boundary of the area of deep seismicity (Jones and others, 1986).

From the San Andreas north, the maximum depth of brittle failure shallows toward the Mojave desert, so that the volume of earthquake activity in the eastern Transverse Ranges resembles a wedge (Fuis and Lamanuzi, 1978; Green, 1983; Corbett and Hearn, 1984; Fig. 7). Nicholson and Simpson (1985) present seismic-velocity evidence for a mid-crustal low-velocity zone in the San Bernardino Mountains, which they identify with a detachment surface corresponding to this seismogenic wedge.

The earthquakes of the eastern Transverse Ranges show predominantly thrust and strike-slip mechanisms (Fig. 8; Green, 1983; Webb and Kanamori, 1985; Nicholson and others, 1986; Jones, 1988), indicating north-south tectonic compression. This pattern of deformation is confirmed by strain measurements (Lysenga and others, 1986; Savage and others, 1986). The reverse-faulting mechanisms are most common near the Banning fault (Green, 1983; Jones, 1988); elsewhere strike-slip mechanisms predominate (Webb and Kanamori, 1985; Nicholson and others, 1986).

MOJAVE DESERT

The Mojave Desert is a large, wedge-shaped tectonic block that occupies the central portion of southern California. It is bounded on the northwest by the left-lateral Garlock fault, on the southwest by the right-lateral San Andreas fault, and for our purposes, on the east by the California border.

The late Cenozoic tectonic grain of the Mojave block is dominated by a group of at least seven right-lateral, strike-slip faults that cut through the central section of the block along a northwest-southeast trend. Because the individual faults strike more or less parallel to the plate boundary, they are usually considered part of the San Andreas fault system. These faults have a total cumulative displacement of between 26.7 and 38.4 km (Dokka, 1983). Hadley and Kanamori (1978) have suggested that these faults are the surface expression of the true plate boundary at depth. A less prominent set of east-west–striking faults exists on the north edge of the block near and parallel to the eastern termination of the Garlock fault. East-west–striking faults also occur on the south edge of the block, near the San Andreas fault. These, rather than being parallel to the bounding fault, are oriented at a high angle to it. One of these, the Pinto Mountain fault, intersects the San Andreas fault zone and appears to act as a southern boundary to the northwest-striking faults. All of these faults have been active in the Quaternary (Jennings, 1975).

Most of the Quaternary faults are concentrated in the central part of the Mojave block. This pattern is reflected in the seismicity of the area (Figs. 4 and 5), with seismic activity occurring in a diffuse, patchy swath, coincident with the band of late Cenozoic faults that separates the very quiet areas to the west and east. A majority of the earthquakes occur either near the south edge of the block, near the San Bernardino Mountains, or in a less active zone in the center of the block that includes the epicenter of the Manix earthquake of April 10, 1947 (M_L = 6.4). The background seismicity does not appear to delineate the mapped faults. In fact, focal mechanisms of the larger events generally indicate slip on more northerly striking planes than those mapped at the surface (Fig. 8).

Historically the larger earthquakes (with the exception of the Manix quake) and most of the background earthquakes have been located near the south edge of the block. This activity is continuous with the very active San Gorgonio Pass area of the San Andreas fault zone. As described in the previous section, the seismogenic bottom of this activity is as deep as 22 km in the San Andreas fault zone, shallows abruptly to about 12 km under the San Bernardino Mountains, and then continues to shallow to the north to less than 10 km under the Mojave block (Green, 1983; Corbett and Hearn, 1984). Several recent sequences have occurred in the Mojave block, including the Homestead Valley sequence of March 15, 1979 (M_L = 5.0, 5.3, 4.5, 4.8), the Goat Mountain earthquakes of November 15 and December 14, 1975 (M_L = 4.6, 4.3), and the Galway Lake earthquake of June 1, 1975 (M_L = 5.0). These earthquakes have all been extremely shallow. The Goat Mountain earthquakes and the Homestead Valley main shock (M_L = 5.3) were less than 5 km deep (Jones, 1984), and the Galway Lake main shock had a depth of 2 km (Lindh and others, 1978).

Like the background activity, some large sequences have not been clearly associated with surface faults. The largest event in the area since the beginning of the SCSN was the Manix earthquake of 1947 (M_L = 6.4), which caused 1.6 km of surface rupture on the east-west-striking Manix fault (Fig. 2; Richter, 1958). Based on the alignment of the larger aftershocks, Richter (1958) suggested that the event was on a concealed northwest-trending structure at depth; however, relatively inaccurate hypocentral locations at that time make interpretation difficult. The aftershocks of the Homestead Valley sequence delineated a cruciform pattern with one lineation between two mapped faults, both of which showed surface rupture (Hutton and others, 1980). This trend had a somewhat more northerly strike than the mapped structures, and the spreading aftershock zone eventually "crossed over" into adjacent blocks with no apparent regard for the bounding faults. The perpendicular lineation may have represented rupture on conjugate faults. This earthquake sequence is particularly perplexing because, as stated above, the hypocenters were extremely shallow.

SIERRA NEVADA AND SOUTHERN BASIN AND RANGE

North of the Garlock fault, which bounds the Mojave Desert on the north, lie the Sierra Nevada and Owens Valley. The Sierra Nevada is a high mountain range underlain by a Mesozoic granitic batholith. The eastern front of the Sierra Nevada is controlled by a normal fault that is the western edge of the Basin and Range Province (e.g., Wallace, 1984). The Owens Valley, east of the Sierra Nevada frontal fault, is a graben of the Basin and Range. At the southern edge of Owens Valley is the Coso geothermal field, a site of Quaternary volcanism (e.g., Duffield and others, 1980).

The Garlock fault is a left-lateral strike-slip fault, extending at least 265 km eastward from the San Andreas fault. The large lateral offsets across it, which vary along strike from approximately 64 km in the west to near zero in the east (Smith, 1962), may contribute to the large deflection in the San Andreas fault in southern California (e.g., Hill, 1982). Geologic evidence shows that large prehistoric earthquakes have occurred on the fault (e.g., Burke, 1979; Roquemore and others, 1982), with average recurrence intervals of about 1,000 ± 500 years (Astiz and Allen, 1983). The largest earthquake recorded on the Garlock fault since 1932 had a magnitude of M_L = 4.3. (An M_L = 5.4 earthquake occurred near the western end of the Garlock fault on June 10, 1988; the mechanism showed oblique reverse and left-lateral strike strike slip on a plane parallel to the Garlock fault, and a 70° northerly dip. The 5-km offset of this earthquake north of the Garlock fault makes it unclear at this time if it was on the Garlock or not.) Astiz and Allen (1983) showed that small to moderate earthquakes align along the western half of the Garlock fault where aseismic creep is recorded, but that seismic activity is very sparse along the eastern half. These earthquakes show left-lateral strike-slip motion. The seismic moment release on the Garlock fault in the last 50 years is at least three orders of magnitude less than the moment release implied by geologically determined slip rates, indicating that the fault is a temporal seismic gap with the potential for large earthquakes (Astiz and Allen, 1983).

One of the largest earthquakes in California (M ≃ 8) occurred in 1872 on the Owens Valley fault, just east of the Sierra Nevada frontal fault (Townley and Allen, 1939; Richter, 1958; Fig. 2). Details of this earthquake are sketchy, but geologic evidence suggests that it resulted from several meters of right-lateral oblique normal slip (Lubetkin and Clark, 1985). The largest earthquake within the Sierra Nevada was the 1946 Walker Pass earthquake (M_L = 6.3) (Chakrabarty and Richter, 1949; Dollar and Helmberger, 1985), which was notable for its large foreshock (M_L = 5.3) and rich aftershock sequence.

At the microseismic level, the southeastern Sierra Nevada and southern Owens Valley are very active. The seismicity of the Sierra Nevada is characterized by intense swarms with no large earthquakes (Jones and Dollar, 1986). These swarms form a lineation parallel to but about 15 km east of the Kern River fault (Fig. 5), called the southern Sierra lineation (Jones and Dollar, 1986). The Kern River fault appears to have been inactive for at least 3 million years (Moore and du Bray, 1978). The adjoining section of the Basin and Range, including Owens Valley and the Coso geothermal area, is also characterized by a high rate of swarm activity without an obvious spatial pattern.

The maximum depth of earthquakes in Owens Valley is similar to that elsewhere in southern California: about 15 km (Bent and Kanamori, 1986). In spite of the prevalence of extensional geologic features, the focal mechanisms of earthquakes in Owens Valley are usually strike-slip (Roquemore and others, 1982; Bent and Kanamori, 1986). The same behavior has been observed to the north of this region, in the Long Valley Caldera (Savage and Cocherham, 1984). In contrast, the earthquakes in the southern Sierra seismic lineation occur at very shallow depths (≤ 6 km), with pure normal faulting along north-south-striking planes. Jones and Dollar (1986) interpreted this lineation as a basin-and-range normal fault beginning to form within the Sierran block.

The southern Sierra seismic lineation extends southward to join with the aftershock zone of the 1952 M_S = 7.7 Kern County earthquake. This event was the largest to occur after the initiation of the SCSN. It resulted from left-lateral, oblique reverse slip on the White Wolf fault (e.g., Stein and Thatcher, 1981), which parallels the Garlock fault. Aftershocks of the 1952 earthquake are still occurring and are almost the only seismic activity in the southern San Joaquin Valley. The transition between oblique reverse faulting in the 1952 aftershock zone and normal faulting in the southernmost Sierra has not been clearly delineated.

The Death Valley region, which lies east of the Owens Valley, is remarkable for its dramatic Holocene and late Quaternary faulting and extremely low historical seismicity. The only known moderate earthquake took place in 1908 and disturbed the region with aftershocks for three weeks (Richter, 1958). Based on intensity reports, the magnitude is estimated to be about 6.5. The microseismicity level is low, with a few M $\simeq$ 3 earthquakes each year.

SUMMARY

The diverse tectonic deformation and high level of seismicity in southern California result from the broad transform plate boundary between the North America and Pacific plates. In historical time, the area has experienced two great earthquakes (M $\simeq$ 8.0): one on the San Andreas fault in 1857 and the other on the Sierra Nevada frontal fault in Owens Valley in 1872. The historic large earthquakes (M_L = 6.5 to 8.0) have occurred most frequently along the San Jacinto fault and in the Imperial Valley, but the largest event to occur since the beginning of the Southern California Seismic Network was the destructive 1952 Kern County earthquake (M_S = 7.7), which occurred along the northern margin of the Transverse Ranges. Moderate-sized and small earthquakes occur throughout southern California. In some cases they form broad zones of seismicity that coincide with major Quaternary faults. However, with the exception of the San Jacinto fault, the major faults that accommodate most of the plate boundary slip, such as the San Andreas fault, are seismically quiescent at the microseismic level during most of the interseismic period between large events. Swarms of earthquakes occur in the Imperial Valley, the Santa Barbara channel, the Sierra Nevada, and the southern Basin and Range.

Depths of earthquakes in southern California range from near the surface to 10 to 15 km deep, with earthquakes in certain areas as deep as 20 to 25 km. Southern California is primarily a transform plate boundary with right-lateral strike-slip motion, although normal- and reverse-fault plane solutions occur throughout the region. The rate of reported seismicity is approximately 1,000 earthquakes ($M_L \geq 1.0$) per month when excluding major aftershock sequences.

REFERENCES CITED

Allen, C. R., 1968, The tectonic environments of seismically active and inactive areas along the San Andreas fault system: Stanford University Publication in Geological Science, v. 11, p. 70–82.

——, 1981, The modern San Andreas fault, in Ernst, W. G., ed., The Geotectonic Development of California, Rubey v. 1: Englewood Cliffs, New Jersey, Prentice-Hall, Inc., p. 511–534.

Allen, C. R., and Nordquist, J. M., 1972, Borrego Mountain earthquake—Foreshock, mainshock, and larger aftershocks, in The Borrego Mountain Earthquake of April 9, 1968: U.S. Geological Survey Professional Paper 787, p. 16–23.

Allen, C. R., St. Amand, P., Richter, C. F., and Nordquist, J. M., 1965, Relationship between seismicity and geologic structure in the southern California region: Bulletin of the Seismological Society of America, v. 55, p. 753–797.

Anderson, D., 1971, The San Andreas fault: Scientific American, v. 225, p. 42–66.

Anderson, J. G., and Bodin, P., 1987, Earthquake recurrence models and historical seismicity in the Mexicali–Imperial Valley: Bulletin of the Seismological Society of America, v. 77, p. 562–578.

Anderson, J. G., and Silver, P. G., 1985, Accelerogram evidence for southward rupture propagation on the Imperial fault during the October 15, 1979, earthquake: Geophysical Research Letters, v. 12, p. 349–352.

Arabasz, W. J., Brune, J. N., and Engen, G. R., 1970, Locations of small earthquakes near the trifurcation of the San Jacinto fault southeast of Anza, California: Bulletin of the Seismological Society of America, v. 60, p. 617–627.

Archuleta, R. J., 1982, Analysis of near-source static and dynamic measurements from the 1979 Imperial Valley earthquake: Bulletin of the Seismological Society of America, v. 72, p. 1927–1956.

Astiz, L., and Allen, C. R., 1983, Seismicity of the Garlock fault, California: Bulletin of the Seismological Society of America, v. 73, p. 1721–1734.

Bent, A., and Kanamori, H., 1986, Indian Wells Valley earthquake swarm [abs.]: EOS Transactions of the American Geophysical Union, v. 67, p. 1084.

Bird, P., and Rosenstock, R. W., 1984, Kinematics of present crust and mantle flow in southern California: Geological Society of America Bulletin, v. 95, p. 946–957.

Brune, J. N., 1968, Seismic moment, seismicity, and rate of slip along major fault zones: Journal of Geophysical Research, v. 73, p. 777–784.

Brune, J. N., and Allen, C. R., 1967, A micro-earthquake survey of the San Andreas fault system in southern California: Bulletin of the Seismological Society of America, v. 57, p. 277–296.

Burke, D. B., 1979, Log of a trench in the Garlock fault zone, Fremont Valley, California: U.S. Geological Survey Miscellaneous Series Map MF-1028.

Carlson, R., Kanamori, H., and McNally, K., 1979, A study of microearthquake activity along the San Andreas fault from Carrizo Plains to Lake Hughes: Bulletin of the Seismological Society of America, v. 69, p. 177–186.

Chakrabarty, S. K., and Richter, C. F., 1949, The Walker Pass earthquakes and structure of the southern Sierra Nevada: Bulletin of the Seismological Society of America, v. 39, p. 93–107.

Cheatum, C., and Combs, J., 1973, Microearthquake study of the San Jacinto Valley, Riverside County, California, in Proceedings of the Conference on Tectonic Problems of the San Andreas Fault System: Stanford University Publications in the Geological Sciences, v. 13, p. 1–10.

Corbett, E. J., 1984, Seismicity and crustal structure studies of southern California; Tectonic implications from improved earthquake locations [Ph.D. thesis]: Pasadena, California Institute of Technology, 231 p.

Corbett, E. J., and Hearn, T. M., 1984, The depth of the seismic zone in the Transverse Ranges of southern California [abs.]: Earthquake Notes, v. 55, p. 23.

Corbett, E. J., and Johnson, C. E., 1982, The Santa Barbara, California, earthquake of 13 August 1978: Bulletin of the Seismological Society of America, v. 72, p. 2201–2226.

Crook, C. N., Mason, R. G., and Wood, P. R., 1982, Geodetic measurements of horizontal deformation on the Imperial Fault, in The Imperial Valley, California, Earthquake of October 15, 1979: U.S. Geological Survey Professional Paper 1254, p. 183–191.

Davis, T. L., 1983, Late Cenozoic structure and tectonic history of the western "Big Bend" of the San Andreas fault and adjacent San Emigdio Mountains [Ph.D. thesis]: Santa Barbara, University of California, 580 p.

Davis, T. L. and Hayden, K., 1987, A retrodeformable cross-section across the central Los Angeles basin and implications for seismic risk evaluation [abs.]: EOS (Transactions of the American Geophysical Union), v. 68, p. 1502.

DeMets, C., Gordon, R. G., Stein, S., and Argus, D. F., 1987, A revised estimate of Pacific–North America motion and implications for western North America plate boundary zone tectonics: Geophysical Research Letters, v. 14, p. 911–914.

Dokka, R. K., 1983, Displacement on late Cenozoic strike-slip faults of the central Mojave Desert, California: Geology, v. 11, p. 305–308.

Dollar, R. S., and Helmberger, D. V., 1985, Body wave modeling using a master event for the sparsely recorded 1946 Walker Pass, California, earthquake [abs.]: EOS Transactions of the American Geophysical Union, v. 66, p. 964.

Doser, D. I., and Kanamori, H., 1986, Depth of seismicity in the Imperial Valley region (1977–1983) and its relationship to heat flow, crustal structure, and the October 15, 1979, earthquake: Journal of Geophysical Research, v. 91, p. 675–688.

Duffield, W. A., Bacon, C. R., and Dalrymple, G. B., 1980, Late Cenozoic volcanism, geochronology, and structure of the Coso Range, Inyo County, California: Journal of Geophysical Research, v. 85, p. 2381–2404.

Eaton, J. P., O'Neill, M. E., and Murdock, T. N., 1970, Aftershocks of the 1966 Parkfield-Cholame, California, earthquake—a detailed study: Bulletin of the Seismological Society of America, v. 60, p. 1151–1197.

Elders, W. A., Rex, R. W., Meidav, T., Robinson, P. T., and Biehler, S., 1972, Crustal spreading in southern California: Science, v. 178, p. 15–24.

Engdahl, E. R. and Rinehart, W. A., 1988, Seismicity Map of North America: Geological Society of America, Centennial Special Map CSM-4, scale 1:5,000,000.

Evernden, J. F., 1970, Study of regional seismicity and associated problems: Bulletin of the Seismological Society of America, v. 60, p. 393–446.

Friedman, M. E., Whitcomb, J. H., Allen, C. R., and Hileman, J. A., 1976, Seismicity of the Southern California Region; 1 January 1972 to 31 December 1974: Pasadena, California Institute of Technology, Division of Geological and Planetary Sciences Seismological Laboratory, 52 p.

Fuis, G. S., and V. Lamanuzi, 1978, Seismicity of the eastern Transverse Ranges, southern California [abs.]: EOS Transactions of the American Geophysical Union, v. 59, p. 1051.

Fuis, G. S., Mooney, W. D., Healy, J. H., McMechan, G. A., and Lutter, W. J., 1982, Crustal structure of the Imperial Valley region, in The Imperial Valley, California, Earthquake of October 15, 1979: U.S. Geological Survey Professional Paper 1254, p. 25–49.

——, 1984, A seismic refraction survey of the Imperial Valley region, California: Journal of Geophysical Research, v. 89, p. 1165–1189.

Gilpin, B., and Lee, T. C., 1978, A microearthquake study in the Salton Sea geothermal area, California: Bulletin of the Seismological Society of America, v. 68, p. 441–450.

Given, D. D., 1983, Seismicity and structure of the trifurcation in the San Jacinto fault zone, southern California [M.S. thesis]: Los Angeles, California State University, 73 p.

Given, D. D., Norris, R., Jones, L. M., Hutton, L. K., Johnson, C. E., and Hartzell, S., 1986, The Southern California Network Bulletin, January–June, 1986: U.S. Geological Survey Open-File Report 86-598, 28 p.

Given, D. D., Hutton, L. K., and Jones, L. M., 1987, The Southern California Network Bulletin, July–December, 1986: U.S. Geological Survey Open-File Report 87-488, 39 p.

Goodstein, J., 1984, Waves in the earth; Seismology comes to southern California: History of Science, v. 14, p. 201–230.

Green, S. M., 1983, Seismotectonic study of the San Andreas, Mission Creek, and Banning fault system [M.S. thesis]: Los Angeles, University of California, 52 p.

Gutenberg, B., and Richter, C. F., 1954, Seismicity of the Earth, 2nd ed.: Princeton, New Jersey, Princeton University Press, 310 p.

Gutenberg, B., Richter, C. F., and Woods, H. O., 1932, The earthquake in Santa Monica Bay, California, on August 30, 1930: Bulletin of the Seismological Society of America, v. 22, p. 138–154.

Hadley, D. and Combs, J., 1974, Microearthquake distribution and mechanisms of faulting in the Fontana–San Bernardino area of southern California: Bulletin of the Seismological Society of America, v. 64, p. 1477–1499.

Hadley, D., and Kanamori, H., 1978, Seismic structure of the Transverse Ranges, California: Geological Society of America Bulletin, v. 88, p. 1469–1478.

Hamilton, R. M., 1970, Time-term analysis of explosion data from the vicinity of the Borrego Mountain, California, earthquake of 9 April 1968: Bulletin of the Seismological Society of America, v. 60, p. 367–381.

Hartzell, S. H., and Heaton, T. H., 1983, Inversion of strong ground motion and teleseismic waveform data for the fault rupture history of the 1979 Imperial Valley, California, earthquake: Bulletin of the Seismological Society of America, v. 73, p. 1553–1583.

Hartzell, S., and Helmberger, D. V., 1982, Strong-motion modeling of the Imperial Valley earthquake of 1979: Bulletin of the Seismological Society of America, v. 72, p. 571–596.

Hauksson, E., 1987, Seismotectonics of the Newport-Inglewood fault zone in Los Angeles basin, southern California: Bulletin of the Seismological Society of America, v. 77, p. 539–561.

Hauksson, E., and Jones, L. M., 1988, The July 1986 Oceanside (M_L = 5.3) earthquake sequence in the Continental Borderland, southern California: Bulletin of the Seismological Society of America, v. 78, p. 1885–1906.

Hauksson, E., and Saldivar, G. V., 1986, The 1930 Santa Monica and the 1979 Malibu, California, earthquakes: Bulletin of the Seismological Society of America, v. 76, p. 1542–1559.

Hauksson, E., and 17 others, 1988, The 1987 Whittier Narrows earthquake in the Los Angeles metropolitan area, California: Science, v. 239, p. 1409–1412.

Hearn, T. M., 1984, Pn travel times in southern California: Journal of Geophysical Research, v. 89, p. 1843–1855.

Hearn, T. M., and Clayton, R. W., 1986a, Lateral velocity variations in southern California: I. Results for the upper crust from Pg waves: Bulletin of the Seismological Society of America, v. 76, p. 495–509.

——, 1986b, Lateral velocity variations in southern California. II. Results for the lower crust from Pn waves: Bulletin of the Seismological Society of America, v. 76, p. 511–520.

Heaton, T. H., 1982, The 1971 San Fernando earthquake—a double event?: Bulletin of the Seismological Society of America, v. 72, p. 2037–2062.

Heaton, T. H., and Helmberger, D. V., 1979, Generalized ray models of the San Fernando earthquake: Bulletin of the Seismological Society of America, v. 69, p. 1311–1341.

Heaton, T. H., Anderson, J. G., and German, P. T., 1983, Ground failure along the New River caused by the October 1979 Imperial Valley earthquake sequence: Bulletin of the Seismological Society of America,

v. 73, p. 1161–1171.

Hileman, J. A., 1978, Part I. A contribution to the study of the seismicity of southern California [Ph.D. thesis]: Pasadena, California Institute of Technology, 230 p.

Hileman, J. A., Allen, C. R., and Nordquist, J. M., 1973, Seismicity of the southern California region, 1 January 1932 to 31 December 1972: Pasadena, California Institute of Technology, Division of Geological and Planetary Sciences Seismological Laboratory, 487 p.

Hill, D. P., 1977, A model for earthquake swarms: Journal of Geophysical Research, v. 82, p. 1347–1352.

——, 1982, Contemporary block tectonics—California and Nevada: Journal of Geophysical Research, v. 87, p. 5433–5450.

Humphreys, E., Clayton, R. W., and Hager, B. H., 1984, A tomographic image of mantle structure beneath southern California: Geophysical Research Letters, v. 11, p. 625–627.

Hutton, L. K., and Johnson, C. E., 1981, Preliminary study of the Westmorland, California, earthquake swarm [abs.]: EOS Transactions of the American Geophysical Union, v. 62, p. 957.

Hutton, L. K., Minster, J. B., and Johnson, C. E., 1979, Seismicity trends in southern California: EOS Transactions of the American Geophysical Union, v. 60, p. 883.

Hutton, L. K., Johnson, C. E., Pechman, J. C., Ebel, J. E., Given, D. W., Cole, D. M., and German, P. T., 1980, Epicentral locations for the Homestead Valley earthquake sequence, March 15, 1979: California Geology, v. 33, no. 5, p. 110–114.

Hutton, L. K., Allen, C. R., and Johnson, C. E., 1985, Seismicity of Southern California; Earthquakes of ML 3.0 and Greater, 1975 through 1983: Pasadena, California Institute of Technology, Division of Geological and Planetary Sciences, Seismological Laboratory, 142 p.

Jacoby, G. C., Sheppard, P. R., and Sieh, K. E., 1988, Irregular recurrence of large earthquakes along the San Andreas fault—evidence from trees: Science, v. 241, p. 196–199.

Jennings, C. W., 1975, Fault Map of California with Volcanoes, Thermal Springs and Thermal Wells: California Division of Mines and Geology Geologic Data Map No. 1, scale 1:750,000.

Johnson, C. E., 1979, I. CEDAR–An approach to the computer automation of short-period local seismic networks, II. Seismotectonics of the Imperial Valley of southern California [Ph.D. thesis]: Pasadena, California Institute of Technology, 332 p.

——, 1983, CUSP–Automated processing and management for large regional networks [abs.]: Earthquake Notes, v. 54, p. 13.

Johnson, C. E., and Hadley, D. M., 1976, Tectonic implications of the Brawley earthquake swarm, Imperial Valley, California, January 1975: Bulletin of the Seismological Society of America, v. 66, v. 1133–1144.

Johnson, C. E., and Hill, D. P., 1982, Seismicity of the Imperial Valley, in The Imperial Valley, California, earthquake of October 15, 1979: U.S. Geological Survey Professional Paper 1254, p. 15–24.

Johnson, C. E., and Hutton, L. K., 1981, Imperial Valley seismicity and the San Andreas Fault [abs.]: EOS Transactions of the American Geophysical Union, v. 62, p. 957.

——, 1982, Aftershocks and pre-earthquake seismicity, in The Imperial Valley, California, earthquake of October 15, 1979: U.S. Geological Survey Professional Paper 1254, p. 59–76.

——, 1986, A tectonic model for the Imperial Valley and its relation to seismic risk on the southern San Andreas Fault [abs.]: EOS Transactions of the American Geophysical Union, v. 67, p. 1200.

——, 1988, Tectonic implications of the November 24, 1987, Superstition Hills earthquakes, Imperial Valley, California [abs.]: Seismological Research Letters, v. 59, p. 48.

Jones, L. M., 1984, Foreshocks (1966–1980) in the San Andreas System, California: Bulletin of the Seismological Society of America, v. 74, p. 1361–1380.

——, 1988, Focal mechanisms and the state of stress on the San Andreas fault in southern California: Journal of Geophysical Research, v. 93, p. 8869–8891.

Jones, L. M., and Dollar, R. S., 1986, Evidence for basin-and-range extensional tectonics in the Sierra Nevada—The Durrwood Meadows swarm, Tulare County, California (1983–1984): Bulletin of the Seismological Society of America, v. 76, p. 439–461.

Jones, L. M., Hutton, L. K., Given, D. D., and Allen, C. R., 1986, The North Palm Springs, California, earthquake sequence of July 1986: Bulletin of the Seismological Society of America, v. 76, p. 1830–1837.

Keller, B., 1981, A model forming the Transverse Ranges of California by subduction of the Murray Transform: Geophysical Research Letters, v. 8, p. 305–308.

Keller, B., and Prothero, W., 1987, Western Transverse Ranges crustal structure: Journal of Geophysical Research, v. 92, p. 7890–7906.

King, N. E., and Savage, J. C., 1984, Regional deformation near Palmdale, California, 1973–1983: Journal of Geophysical Research, v. 89, p. 2471–2477.

Kohler, W. M., and Fuis, G. S., 1986, Travel-time, time-term, and basement depth maps for the Imperial Valley region, California, from explosions: Bulletin of the Seismological Society of America, v. 76, p. 1289–1303.

Langbein, J. O., Johnston, M.J.S., and McGarr, A., 1982, Geodetic observations of postseismic deformation around the north end of surface rupture, in The Imperial Valley, California, earthquake of October 15, 1979: U.S. Geological Survey Professional Paper 1254, p. 205–212.

Langbein, J., McGarr, A., Johnston, M.J.S., and Harsh, P. W., 1983, Geodetic measurements of postseismic crustal deformation following the 1979 Imperial Valley earthquake, California: Bulletin of the Seismological Society of America, v. 73, p. 1203–1224.

Larson, R. L., Menard, H. W., and Smith, S. M., 1968, Gulf of California; A result of ocean-floor spreading and transform faulting: Science, v. 161, p. 781–784.

Legg, M. R., 1980, Seismicity and tectonics of the inner Continental Borderland of southern California and northern Baja California, Mexico [M.S. thesis]: San Diego, University of California, 60 p.

——, 1985, Geologic structure and tectonics of the inner Continental Borderland offshore northern Baja California, Mexico [Ph.D. thesis]: Santa Barbara, University of California, 410 p.

Lindh, A., Fuis, G., and Mantis, C., 1978, Foreshock amplitudes and fault plane changes—A new earthquake precursor?: Science, v. 201, p. 56–59.

Lomnitz, C., Mooser, F., Allen, C. R., Brune, J. N., and Thatcher, W., 1970, Seismicity and tectonics of northern Gulf of California region, Mexico—Preliminary results: Geofisica Internacional, v. 10, p. 37–48.

Louie, J. N., Allen, C. R., Johnson, D. C., Haase, P. C., and Cohn, S. N., 1985, Fault slip in southern California: Bulletin of the Seismological Society of America, v. 75, p. 811–833.

Lubetkin, L., and Clark, M. M., 1985, Late Quaternary activity along the Lone Pine fault, eastern California, in Proceedings of Workshop XXVIII on the Borah Peak, Idaho, Earthquake: U.S. Geological Survey Open-File Report 85–290, p. 118–140.

Lysenga, G. A., Wallace, K. S., Fanslow, J. L., Raefsky, A., and Groth, P. M., 1986, Tectonic motions in California inferred from very long baseline interferometry observations, 1980–1984: Journal of Geophysical Research, v. 91, p. 9473–9487.

McNally, K. C., Kanamori, H., Pechmann, J. C., and Fuis, G., 1978, Earthquake swarm along the San Andreas fault near Palmdale, southern California: Science, v. 201, p. 814–817.

Magistrale, H., and Kanamori, H., 1988, Superstition Hills earthquakes and basement structure of the western Imperial Valley [abs.]: Seismological Research Letters, v. 59, p. 48.

Matti, J. C., Morton, D. M., and Cox, B. F., 1985, Distribution and geologic relations of fault systems in the vicinity of the Central Transverse Ranges, southern California: U.S. Geological Survey Open-File Report 85–365, 27 p.

Minster, J. B., and Jordan, T. H., 1978, Present-day plate motions: Journal of Geophysical Research, v. 83, p. 5331–5354.

Moore, D. G., and Buffington, E. C., 1968, Transform faulting and growth of the Gulf of California since the late Pliocene: Science, v. 161, p. 1238–1241.

Moore, J. G., and du Bray, E., 1978, Mapped offsets on the right-lateral Kern

Canyon fault, southern Sierra Nevada, California: Geology, v. 6, p. 205–208.

Neumann, F., 1942, United States Earthquakes, 1940: U.S. Coast and Geodetic Survey Serial 647, 74 p.

Nicholson, C., and Simpson, D. W., 1985, Changes in Vp/Vs with depth—Implications for appropriate velocity models, improved earthquake locations, and material properties of the upper crust: Bulletin of the Seismological Society of America, v. 75, p. 1105–1123.

Nicholson, C., Seeber, L., Williams, P., and Sykes, L. R., 1986, Seismicity and fault kinematics through the eastern Transverse Ranges, California—Block rotation, strike-slip faulting, and low-angle thrusts: Journal of Geophysical Research, v. 91, p. 4891–4908.

Olson, A. H., and Apsel, R. J., 1982, Finite faults and inverse theory with applications to the 1979 Imperial Valley earthquake: Bulletin of the Seismological Society of America, v. 72, p. 1969–2001.

Pacheco, J., and Nabelek, J., 1988, Source mechanisms of three moderate southern California earthquakes of July 1986: Bulletin of the Seismological Society of America, v. 78, p. 1907–1929.

Pechmann, J. C., 1983, The relationship of small earthquakes to strain accumulation along major faults in southern California [Ph.D. thesis]: Pasadena, California Institute of Technology, 175 p.

—— , 1987, Tectonic implications of small earthquakes in the central Transverse Ranges, in Recent Reverse Faulting in the Transverse Ranges, California: U.S. Geological Survey Professional Paper 1339, p. 97–112.

Raleigh, C. B., Sieh, K., Sykes, L. R., and Anderson, D. L., 1982, Forecasting southern California earthquakes: Science, v. 217, p. 1097–1104.

Reilinger, R., 1984, Coseismic and postseismic vertical movements associated with the 1940 M7.1 Imperial Valley, California, earthquake: Journal of Geophysical Research, v. 89, p. 4531–4537.

Reilinger, R., and Larsen, S., 1986, Vertical crustal deformation associated with the 1979 M = 6.6 Imperial Valley, California, earthquake—Implications for fault behavior: Journal of Geophysical Research, v. 91, p. 14044–14056.

Richter, C. F., 1958, Elementary Seismology: San Francisco, California, W. H. Freeman and Co., p. 768.

Richter, C. F., Allen, C. R., and Nordquist, J. M., 1958, The Desert Hot Springs earthquakes and their tectonic environment: Bulletin of the Seismological Society of America, v. 48, p. 315–337.

Robinson, P. T., Elders, W. A., and Muffler, L.J.P., 1976, Quaternary volcanism in the Salton Sea geothermal field, Imperial Valley, California: Geological Society of America Bulletin, v. 87, p. 347–360.

Roquemore, G. R., Smith, P. E., and Banks, E. W., 1982, Holocene earthquake activity of the eastern Garlock fault in Christmas Canyon, San Bernardino County, California [abs.]: Geological Society of America Abstracts with Programs, v. 14, p. 228.

Rundle, J. B., and Thatcher, W., 1980, Speculations on the nature of the southern California uplift: Bulletin of the Seismological Society of America, v. 70, p. 1869–1886.

Sanders, C. O., and Kanamori, H., 1984, A seismotectonic analysis of the Anza seismic gap, San Jacinto fault zone, southern California: Journal of Geophysical Research, v. 89, p. 5873–5890.

Sauber, J., McNally, K., Pechmann, J. C., and Kanamori, H., 1983, Seismicity near Palmdale, California, and its relation to strain changes: Journal of Geophysical Research, v. 88, p. 2213–2219.

Savage, J. C., and Cockerham, R. S., 1984, Earthquake swarm in the Long Valley Caldera, January, 1983—Evidence for dike inflation: Journal of Geophysical Research, v. 89, p. 8315–8324.

Savage, J. C., and Gu, G., 1985, The 1979 Palmdale, California, strain event in retrospect: Journal of Geophysical Research, v. 90, p. 10301–10309.

Savage, J. C., Prescott, W. H., Lisowski, M., and King, N. E., 1981, Strain accumulation in southern California, 1973–1980: Journal of Geophysical Research, v. 86, p. 6991–7001.

Savage, J. C., Prescott, W. H., and Gu, G., 1986, Strain accumulation in southern California, 1973–1984: Journal of Geophysical Research, v. 91, p. 7455–7473.

Sharp, R. V., 1967, San Jacinto fault zone in the Peninsular ranges of southern California: Geological Society of America Bulletin, v. 78, p. 705–730.

—— , 1981, Variable rates of late Quaternary strike-slip on the San Jacinto fault zone, southern California: Journal of Geophysical Research, v. 86, p. 1754–1762.

—— , 1982a, Tectonic setting of the Imperial Valley region, in The Imperial Valley, California, earthquake of October 15, 1979: U.S. Geological Survey Professional Paper 1254, p. 5–14.

—— , 1982b, Comparison of 1979 surface faulting with earlier displacements in the Imperial Valley, in The Imperial Valley, California, earthquake of October 15, 1979: U.S. Geological Survey Professional Paper 1254, p. 213–221.

Sharp, R. V., and 14 others, 1982, Surface faulting in the central Imperial Valley, in The Imperial Valley, California, earthquake of October 15, 1979: U.S. Geological Survey Professional Paper 1254, p. 119–143.

Sibson, R. H., 1982, Fault zone models, heat flow, and the depth distribution of earthquakes in the continental crust of the United States: Bulletin of the Seismological Society of America, v. 72, p. 151–163.

Sieh, K. E., 1978, Slip along the San Andreas fault associated with the great 1857 earthquake: Bulletin of the Seismological Society of America, v. 68, p. 1421–1448.

—— , 1986, Slip rate across the southern San Andreas and prehistoric earthquakes at Indio, California [abs.]: EOS Transactions of the American Geophysical Union, v. 67, p. 1200.

Sieh, K. E., and Jahns, R. H., 1984, Holocene activity of the San Andreas fault at Wallace Creek, California: Geological Society of America Bulletin, v. 95, p. 883–896.

Silver, P., and Masuda, T., 1985, A source extent analysis of the Imperial Valley earthquake of October 15, 1979, and the Victoria earthquake of June 9, 1980: Journal of Geophysical Research, v. 90, p. 7639–7651.

Slade, M. A., Lysenga, G. A., and Raefsky, A., 1984, Modeling of the surface static displacements and fault plane slip for the 1979 Imperial Valley earthquake: Bulletin of the Seismological Society of America, v. 74, p. 2413–2433.

Smith, G. I., 1962, Large lateral displacements on the Garlock fault, California, as measured from offset dike swarms: Bulletin American Association of Petroleum Geologists, v. 46, p. 85–104.

Snay, R. A., Cline, M. W., and Timmerman, E. L., 1982, Horizontal deformation in the Imperial Valley, California, between 1934 and 1980: Journal of Geophysical Research, v. 87, p. 3959–3968.

Stein, R. S., and Thatcher, W., 1981, Seismic and aseismic deformation associated with the 1952 Kern County, California earthquake and relationship to the Quaternary history of the White Wolf fault: Journal of Geophysical Research, v. 86, p. 4913–4928.

Stierman, D., and Ellsworth, W., 1976, Aftershocks of the February 21, 1973, Point Mugu, California, earthquake: Bulletin of the Seismological Society of America, v. 66, p 1931–1952.

Sykes, L. R., and Seeber, L., 1985, Great earthquakes and great asperities, San Andreas fault, southern California: Geology, v. 13, p. 835–838.

Sylvester, A. G., Smith, S. W., and Scholz, C. H., 1970, Earthquake swarm in the Santa Barbara Channel, California, 1968: Bulletin of the Seismological Society of America, v. 60, p. 1047–1060.

Thatcher, W., 1979, Horizontal crustal deformation from historic geodetic data in southern California: Journal of Geophysical Research, v. 84, p. 8547–8576.

Thatcher, W. and Hamilton, R. M., 1973, Aftershocks and source characteristics of the 1969 Coyote Mountain earthquake, San Jacinto fault zone, California: Bulletin of the Seismological Society of America, v. 63, p. 647–661.

Thatcher, W., Hileman, J. A., and Hanks, T. C., 1975, Seismic slip distribution along the San Jacinto fault zone, southern California, and its implications: Geological Society of America Bulletin, v. 86, p. 1140–1146.

Townley, S. D., and Allen, M. W., 1939, Descriptive catalog of earthquakes of the Pacific Coast of the United States, 1769 to 1928: Bulletin of the Seismological Society of America, v. 29, p. 1–297.

Trifunac, M. D., and Brune, J. N., 1970, Complexity of energy release during the Imperial Valley, California, earthquake of 1940: Bulletin of the Seismological Society of America, v. 60, p. 137–160.

Turcotte, D. L., Liu, J. Y., and Kulhawy, F. H., 1984, The role of an intracrustal asthenosphere on the behavior of strike-slip faults: Journal of Geophysical Research, v. 89, p. 5801–5816.

Walck, M. C., and Minster, J. B., 1982, Relative array analysis of upper mantle lateral velocity variations in southern California: Journal of Geophysical Research, v. 87, p. 1757–1772.

Wallace, R. E., 1970, Earthquake recurrence intervals on the San Andreas fault: Geological Society of America Bulletin, v. 81, p. 2875–2890.

Wallace, R. E., 1984, Patterns and timing of late Quaternary faulting in the Great Basin Province and relation to some regional tectonic features: Journal of Geophysical Research, v. 89, p. 5763–5770.

Webb, T. H., and Kanamori, H., 1985, Earthquake focal mechanisms in the eastern Transverse Ranges and San Emigdio Mountains, southern California, and evidence for a regional décollement: Bulletin of the Seismological Society of America, v. 75, p. 737–757.

Weldon, R. J., 1986, The late Cenozoic geology of Cajon Pass; Implications for tectonics and sedimentation along the San Andreas fault [Ph.D. thesis]: Pasadena, California Institute of Technology, 400 p.

Weldon, R. J., and Humphreys, E., 1986, A kinematic model of southern California: Tectonics, v. 5, p. 33–48.

Weldon, R. J., and Sieh, K. E., 1985, Holocene rate of slip and tentative recurrence interval for large earthquakes on the San Andreas fault, Cajon Pass, southern California: Geological Society of America Bulletin, v. 96, p. 793–812.

Wesnousky, S. G., 1986, Earthquakes, Quaternary faults and seismic hazards in California: Journal of Geophysical Research, v. 91, p. 12587–12632.

Whitcomb, J. H., 1971, Fault-plane solutions of the February 9, 1971, San Fernando Earthquake and some aftershocks, *in* The San Fernando, California, earthquake of February 9, 1971: U.S. Geological Survey Professional Paper 733, p. 30–32.

Woodward-Clyde Consultants, 1979, Report on the evaluation of maximum earthquake and site ground motion parameters associated with the Offshore Zone of Deformation, San Onofre Nuclear Generation Station, Los Angeles, California: Report prepared for Southern California Edison, 56 p.

Yeats, R. S., and Olson, D. J., 1984, Alternate fault model for the Santa Barbara, California, earthquake of 13 August 1978: Bulletin of the Seismological Society of America, v. 74, p. 1545–1553.

Ziony, J. I., and Jones, L M., 1989, Map showing late Quaternary faults and 1978–1984 seismicity of the Los Angeles region, California: U.S. Geological Survey Miscellaneous Series Map MF-1964, scale 1:250,000.

Ziony, J. I., and Yerkes, R. F., 1985, Evaluating earthquake and surface faulting potential, *in* Evaluating Earthquake Hazards in the Los Angeles Region— An Earth-Science Perspective: U.S. Geological Survey Professional Paper 1360, p. 43–92.

Manuscript Accepted by the Society October 12, 1988

ACKNOWLEDGMENTS

The Southern California Seismic Network has continued and prospered through the dedication of the analysts who time and locate the earthquakes; these analysts include Lee Anderson, Ann Blanchard, Steve Bryant, Sang Choi, Jane Cooper, Jeff Cronk, Shirley Fisher, Doug Ford, Wolfgang Franzen, Martin Friedman, Riley Geary, Lind Gee, Peter German, Susan Green, Susanna Gross, Gary Gutierrez, Donna Jenkins, Vic Lamanuzzi, Carl Loeffler, John Nordquist, Robert Norris, Barbara Reed, Karen Richter, Joan Rudnicki, Lisa Stach, John Strub, Violet Taylor, Marla Turner, and Kathy Watts. Understanding the pattern of microseismicity and how it relates to the older moderate earthquakes that have occurred in southern California between 1932 and 1975 would not be possible without the foresight and perseverance of such individuals as Charles Richter, Harry O. Wood, and Clarence Allen. The modern capability to process over 10,000 earthquakes per year results primarily from the extensive software written by Carl Johnson. The earthquake analysis has been supported in part by the U.S. Geological Survey Grants 14-08-0001-A0257 (Caltech) and 14-08-0001-G1158 (USC). The clarity of this paper benefited greatly from insightful reviews by David Hill, Clarence Allen, Carl Johnson, and Craig Nicholson. Contribution 4521, Division of Geological and Planetary Sciences, California Institute of Technology, Pasadena.

Chapter 10

The seismicity of Nevada and some adjacent parts of the Great Basin

A. M. Rogers and S. C. Harmsen
U.S. Geological Survey, Box 25046, Denver Federal Center, Denver, Colorado 80225
Edward J. Corbett, K. Priestley, and Diane dePolo
Seismological Laboratory, University of Nevada, Reno, Nevada 89557

INTRODUCTION

The seismicity of Nevada is distributed in several broad zones that connect with significant seismic zones in surrounding states and appear to concentrate the largest earthquakes in the Great Basin province. During the historic record, which extends over the last 140 years, a number of large, damaging earthquakes occurred in some of these zones, and those larger than magnitude 6 typically produced surface rupture. Based on geologic evidence, most of these earthquakes are believed to have occurred on steeply dipping range-front normal faults that penetrate the crust to midcrustal depths. For numerous cases, however, seismic and geodetic data suggest that strike slip and oblique slip occurred at focal depths. Microearthquake data also indicate a preference for dextral strike slip and oblique slip on northerly trending, steeply dipping faults at depths ranging from near-surface to about 15 km. In addition, some discrepancy exists between the orientation of faults inferred from seismic and geologic data. Faults show a tendency to be rotated clockwise relative to preferred nodal plane orientation. Little seismic evidence has been found for slip on low-angle detachment or listric faults in spite of abundant geologic evidence for this deformation style in the last 15 m.y.

The existence of seismic evidence for transcurrent slip on northerly trending faults is at variance with popular tectonic models for the large, young structures in the region—the basins and ranges. The seismic data are also not in accord with the abundant northwest and northeast conjugate strike-slip faults that exist in the Walker Lane belt and the margins of the southern Great Basin. The tectonic framework of the seismicity of the region is, thus, incompletely understood. Discerning between various tectonic driving mechanisms could help resolve these problems. For instance, previous discussions have raised questions regarding whether Great Basin deformation is causally related to tectonics along the continental plate margin (i.e., Slemmons, 1967; Atwater, 1970), or to processes internal to the Great Basin such as an over-thickened crust (Coney and Harms, 1984), gravitational spreading (Wernicke, 1981), or other processes related to back-arc extension (Matthews and Anderson, 1973; Coney, 1987). This summary of historical and current seismicity data of the region provides a basis for evaluating contemporary deformation in terms of generalized tectonic models of the Great Basin.

SEISMICITY OVERVIEW

The seismic patterns in Nevada (Fig. 1, 2, and 3) can be discussed in terms of several significant trends. The most prominent seismic trend is the zone of moderate-to-large earthquakes extending northward from southern California into west-central Nevada. This zone was referred to as the 118°W Meridian Zone by Slemmons and others (1965) and as the Ventura-Winnemucca zone by Ryall and others (1966). The part of the zone within Nevada has been referred to as the central Nevada seismic belt (CNSB) by Wallace (1984a). We use the latter term in this chapter. This zone includes the largest Nevada earthquakes in historic time. Since 1900, three earthquakes greater than magnitude 7.0 occurred, and seven earthquakes produced surface rupture (Tables 1 and 2; magnitudes in this chapter are M_L or M_S, unless otherwise noted).

The second zone of prominent seismicity is the Sierra Nevada–Great Basin boundary zone (SNGBZ), described by Van Wormer and Ryall (1980). The zone extends northwest near the California-Nevada border from Bishop to Reno and continues for hundreds of kilometers into northern California. Earthquakes in the zone tend to concentrate along the east flank of the Sierra Nevada. In addition to frequent moderate earthquakes, large events have also occurred in 1852, 1860, and 1872. The Mammoth earthquake series is also in this zone. Five earthquakes in the SNGBZ have produced surface rupture.

The third trend that is apparent in Figure 2 is the arcuate earthquake zone extending from the south end of the Wasatch front in Utah into and across southern Nevada to eastern California near Bishop. This earthquake trend has been referred to as the

---

Rogers, A. M., Harmsen, S. C., Corbett, E. J., Priestley, K., and dePolo, D., 1991, The seismicity of Nevada and some adjacent parts of the Great Basin, *in* Slemmons, D. B., Engdahl, E. R., Zoback, M. D., and Blackwell, D. D., eds., Neotectonics of North America: Boulder, Colorado, Geological Society of America, Decade Map Volume 1.

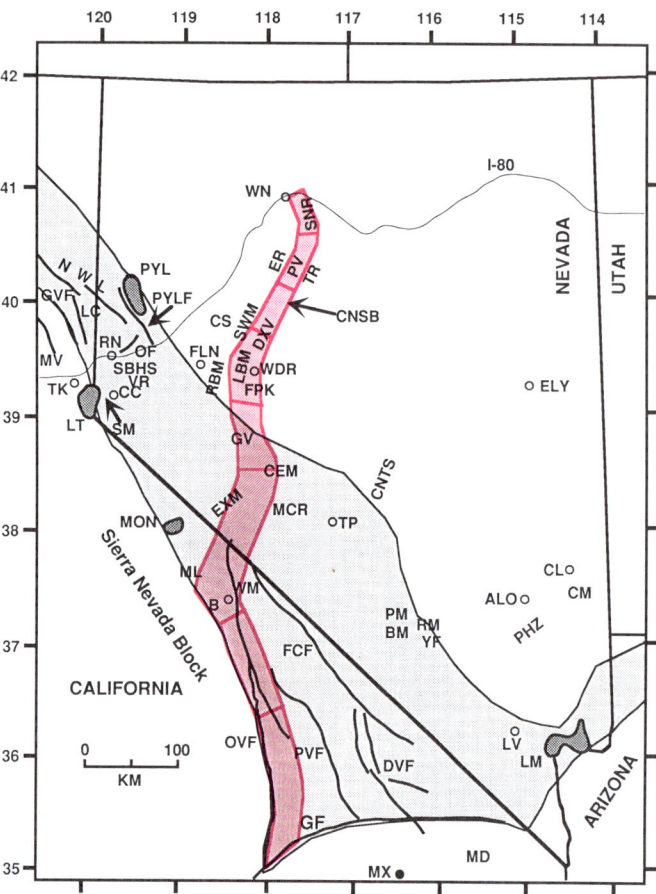

Figure 1. Map showing the location of towns, ranges, and other geographic and geologic features discussed in the text. The northwest-trending shaded zone is the Walker Lane belt as defined by Stewart (1988). The northerly trending shaded zones make up the Central Nevada Seismic Belt (CNSB) defined by Wallace (1984a). The northern most gap is the Sonoma Range gap (Thenhaus and Barnhard 1989). Breaks in the CNSB roughly denote the extent of seismic zones, fault segmentation, and gaps defined by Wallace (1984a). Other symbols are: ALO-Alamo; B-Bishop; BM-Buckboard Mesa; CC-Carson City; CEM-Cedar Mountain; CL-Caliente; CM-Clover Mountains; CNTS-Central Nevada Test Area; CS-Carson Sink; DXV-Dixie Valley; DVF-Death Valley fault; ER-East Range; EXM-Excelsior Mountain; FCF-Furnace Creek fault; FLN-Fallon; FPK-Fairview Peak; GF-Garlock fault; GV-Gabbs Valley; GV-Grizzly Valley fault; LC-Last Chance fault; LM-Lake Mead; LM-Louderback Mountains; LT-Lake Tahoe; LV-Las Vegas; MCR-Monte Cristo Range; MD-Mojave Desert; ML-Mammoth Lakes; MON-Mono Lake; MV-Mohawk Valley fault; MX-Manix earthquake; NWL-northern Walker Lane belt; OF-Olinghouse fault; OVF-Owens Valley fault; PHZ-Pahranagat Shear Zone; PM-Pahute Mesa; PV-Pleasant Valley; PVF-Panamint Valley fault; PYL-Pyramid Lake; PYLF-Pyramid Lake fault zone; RBM-Rainbow Mountain; RM-Rainier Mesa; RN-Reno; SBHS-Steamboat Hot Springs; SM-Slide Mountain; SNR-Sonoma Range; SWM-Stillwater Range; TK-Truckee; TP-Tonopah; TR-Tobin Range; VR-Virginia Range; WDR-Wonder; WM-White Mountains; WN-Winnemucca; YF-Yucca Flat.

southern Nevada Transverse Zone (NTZ) by Slemmons and others (1965) and as the east-west seismic belt by Smith (1978). The most recent maps of this area show that the NTZ may have considerable north-south expression; a north-northeasterly trending prong of activity extends from about 37.5°N and 116°W to as much as 40°N. Only small to moderate historic earthquakes have occurred in this zone, including four events of about magnitude 6 (1908 and 1916 in Death Valley, California; 1902 Pine Valley, Utah; 1966 Clover Mountains, Nevada). A fifth event, the magnitude 6.4 1947 Manix earthquake, occurred on what might be described as the southern fringe of the NTZ and produced surface rupture (Richter, 1958). Several areas of induced seismicity have affected the record in this region (see below) due to underground nuclear testing and the impoundment of Lake Mead. Excluding the Manix earthquake, the only known historic surface rupture in this zone accompanied a number of the larger underground nuclear tests. The zone might not appear to be continuous from east to west across southern Nevada if earthquakes had not been induced. A few earthquakes, however, exist in the historic record of the southern Great Basin area before underground testing or the impoundment of the lake (Meremonte and Rogers, 1987). In addition, data from regional network studies show that microearthquakes are widespread across the southern Nevada region, and many are likely unrelated to nuclear testing (Rogers and others, 1987).

All three of these trends appear to converge in the Mammoth-Bishop area, but the significance of this fact is not understood. A high percentage of Nevada earthquakes occur within these three trends, including all earthquakes over magnitude 6. Elsewhere, scattered diffuse seismicity covers the rest of Nevada at a level that is low compared to California, but high compared to other surrounding states. The apparent drop-off in seismicity at several of Nevada's borders may, however, be an artifact of a lower level of seismic monitoring in Oregon, Idaho, Utah, and Arizona.

Possible secodary seismic zones are also apparent in Figure 2. For instance, there may be one or more additional zones of seismicity that subparallel the CNSB, both on the northwest and southeast of the CNSB.

BRIEF HISTORY OF THE EARTHQUAKE RECORD IN THE GREAT BASIN

The early seismic history of the region is contained in a "Catalog of Nevada Earthquakes, 1852–1960" by Slemmons and others (1965), which they based on Indian traditions, newspaper accounts, and other historical records, as well as catalogs published by other institutions. Information on Great Basin earthquakes prior to 1916 derives mainly from catalogs developed by Holden (1898), McAdie (1907), and Townley and Allen (1939). Additional accounts of northern Nevada seismicity are given by Ryall and Douglas (1974), Ryall and Priestley (1975), and Ryall and Vetter (1982). The accuracy of pre-instrumental event locations is highly variable because the records depend on felt reports

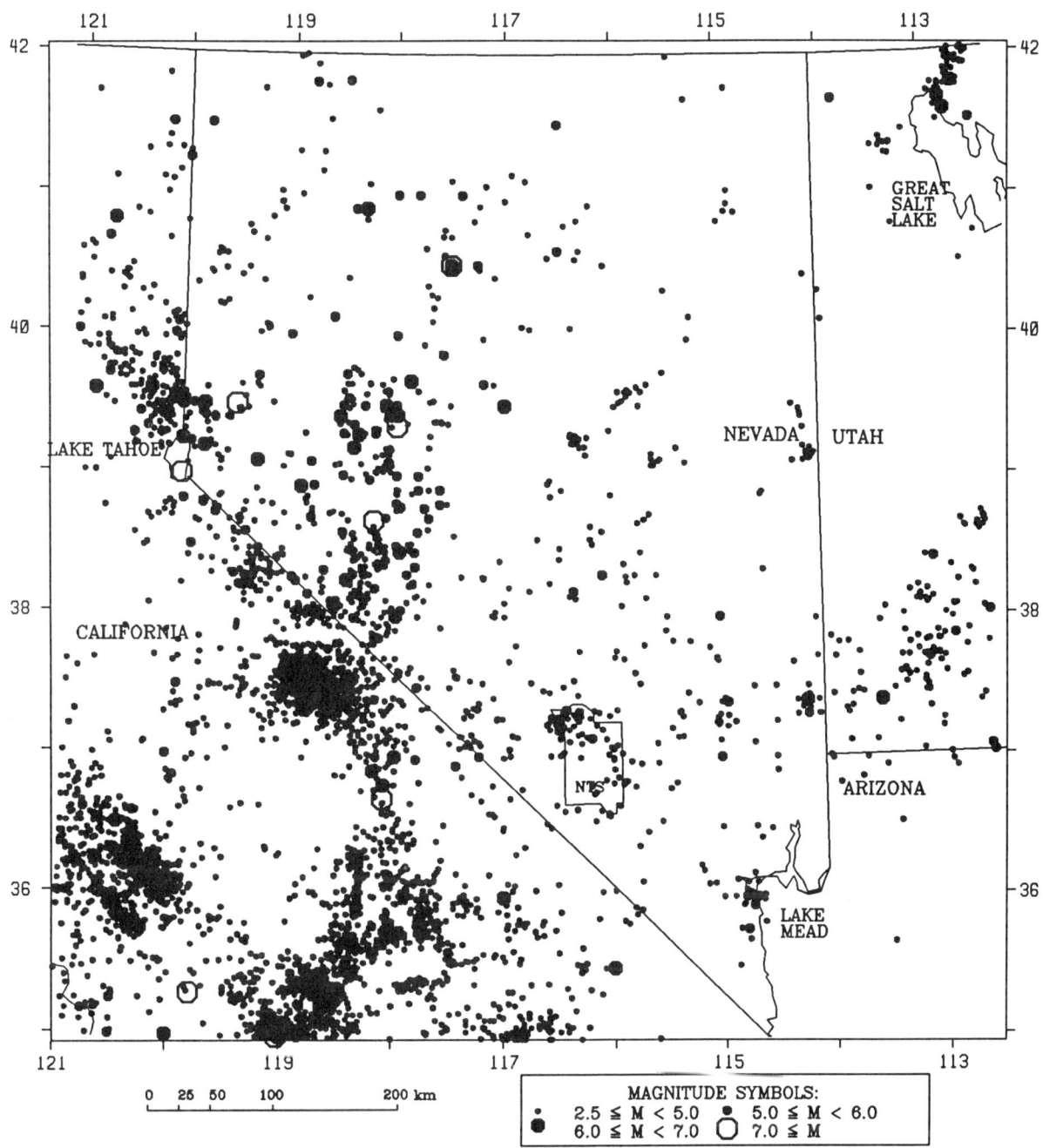

Figure 2. Map of M ≥2.5 earthquake epicenters in Nevada and surrounding areas for the time period 1850 to 1986, from Engdahl (this volume). Symbol sizes correspond to magnitudes, with smallest symbols for earthquakes having magnitudes in the range 2.5 ≤M < 4.0.

in sparsely populated Nevada, involving potential inaccuracies of several hundred km, except in cases where fault rupture was found.

The history of instrumental seismological recording in the Great Basin dates back to 1888 with the installation of a seismograph at Carson City, and 1914 with the installation of a seismograph in Salt Lake City at the University of Utah. Another station was installed in 1916 in Reno at the University of Nevada. The University of Nevada operated, semicontinuously, a two-component, horizontal Wiechert seismograph with smoked-paper recording from 1916 to 1959 (Jones, 1963, 1975). The University of California operated a three-component Sprengnether instrument on the Reno campus from 1948 until operation was taken over by the University of Nevada in 1963. Since that time, the permanent network has expanded gradually from three stations in 1962 to 74 short-period telemetered stations in 1986. With the initiation, expansion, and improvement of instrumental recording, the accuracy of earthquake locations and magnitudes

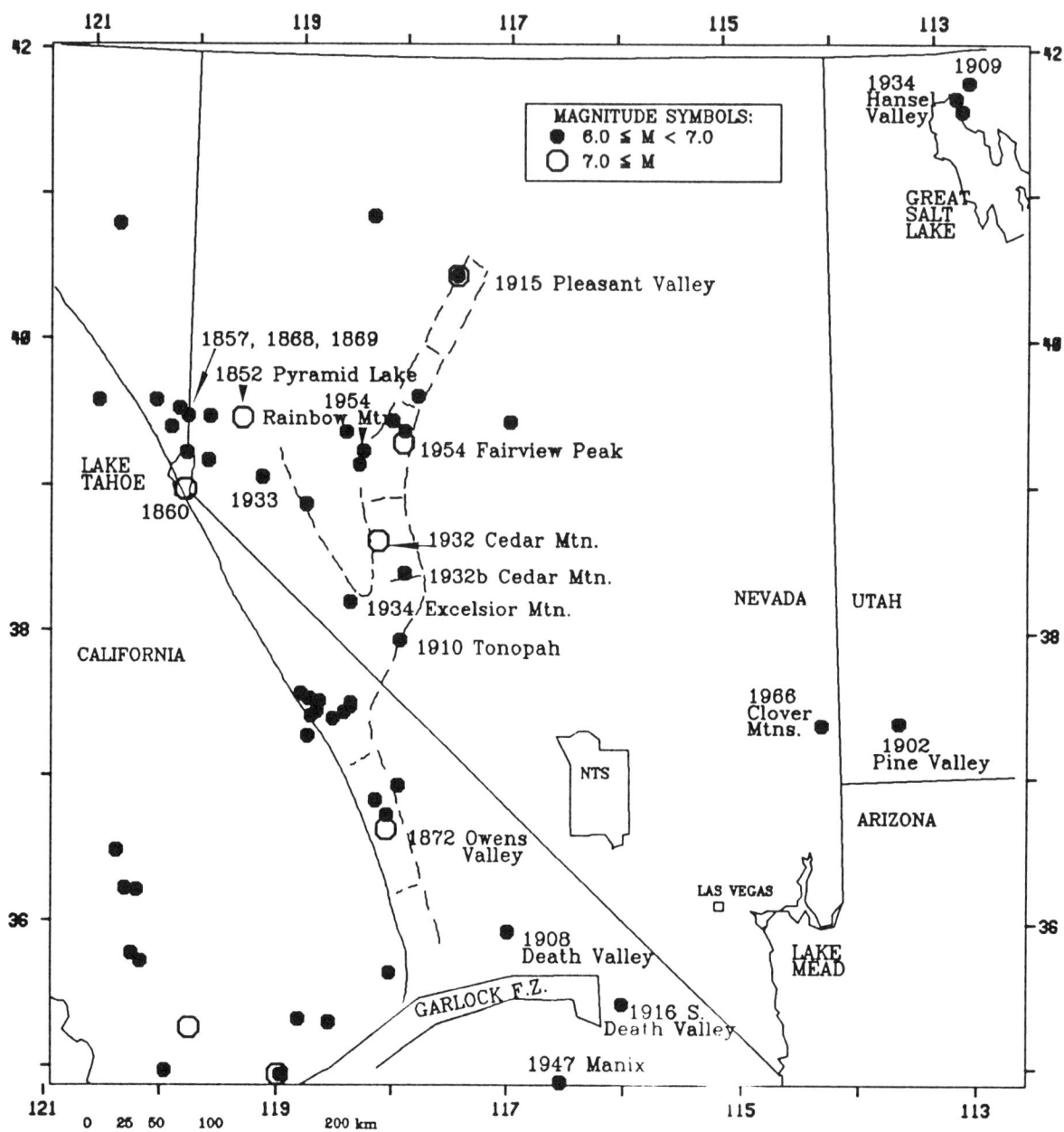

Figure 3. Map showing locations of earthquakes having magnitudes M ≥ 6.0 in Nevada and surrounding region for the period 1852 through 1986.

increased with time. Post-1916 distance and magnitude measurements were available from the Reno Wiechert station, and combined with data from Salt Lake City, Berkeley, and later Pasadena (post-1932), location accuracies improved to within a few tens of kilometers. With the installation of modern telemetered networks in the 1970s, this accuracy improved even more. The onset of digital recording in 1981 for southern Nevada and in 1984 for northern Nevada resulted in timing accuracy of a few hundredths of a second; present-day epicenters are now accurate to less than a kilometer for the best-recorded earthquakes occurring within the denser parts of the networks. The University of Nevada Seismological Laboratory has produced annual catalogs for the northern Nevada region since 1971.

Numerous networks have operated intermittently in southern Nevada. These networks were primarily deployed to study underground-nuclear-explosion aftershocks, their associated ground motions, and earthquakes induced by the filling of Lake Mead. Because these studies were conducted by several different agencies and organizations, no complete catalog of the data exists. Meremonte and Rogers (1987) collected all known earthquake hypocenter data through 1978 for southern Nevada in a comprehensive catalog, and their report serves as a reference to all known data sources for that region. Since 1978, a 54-station telemetered network has operated in southern Nevada to evaluate the seismic hazard to a proposed high-level radioactive waste repository at the Nevada Test Site; the catalogs for these data are Rogers and others (1987; hypocentral data through 1983) and Harmsen and Rogers (1987; hypocentral data for 1984 to 1986). Gawthrop and Carr (1988) have relocated many of the best-recorded earthquakes in the southwestern Great Basin using modern location techniques and station corrections. Their catalog spans 1931 to 1974.

LARGE EARTHQUAKES IN NEVADA AND VICINITY

Despite the brief historical record for the Nevada region, the area has experienced a substantial number of potentially damaging earthquakes. Since 1852, 55 earthquakes of magnitude 6 or larger have been recorded in the central and western Great Basin (Table 1 and Fig. 3). Six of these earthquakes were magnitude 7 or larger, the largest being $M \approx 8$, and as many as 12 earthquakes produced surface rupture. Thus, the western Great Basin is among the most active earthquake regions in the conterminous United States. Faulting associated with the historic events is shown in Figure 4 together with Holocene and late Quaternary faulting. Details of the largest of these earthquakes are presented in the following sections.

Pyramid Lake earthquake, about 1850

Because there were few pioneer settlements in Nevada prior to the 1850s, details of the early historic Nevada earthquakes are sketchy. The first known large earthquake produced substantial geologic effects, including open fissures with 100-ft-high water spouts, failed river banks and a changed river course, a large landslide on Slide Mountain south of Reno (Slemmons and others, 1965), and surface faulting. Slemmons and others (1965) estimate the date to have been 1852 and the location to have been about 30 km south of Pyramid Lake, while Ryall (1977) suggests that the event occurred earlier, in about 1845, in the Carson Sink–Stillwater area. Anderson and Hawkins (1984) note, on the basis of geomorphic evidence, that the Pyramid Lake fault zone may have been the source of this earthquake. This fault demonstrates evidence of recurrent Holocene dextral strike slip on a northwest-trending fault (Table 2). The dramatic earthquake effects (maximum Modified Mercalli intensity, $MMI \approx X$) and their widespread occurrence suggest that this event was at least magnitude 7.3 (Slemmons and others, 1965).

Western Nevada, 1860

A large earthquake on March 15, 1860, was felt from California to Utah and was very violent in Carson City. Its epicenter is unknown, but a Nevada location is likely, and its wide felt area (maximum MMI IX, Slemmons and others, 1965) indicates that it was a major event. Anderson and Hawkins (1984) suggested this event could also have occurred on the Pyramid Lake fault, Nevada. Based on maximum intensity and felt area (518,000 km^2), Slemmons and others (1965) estimated that this event was about magnitude 7.0.

Western Nevada, 1869

Two earthquakes occurred in this region in 1869. Sanders and Slemmons (1979) estimated, on the basis of maximum intensities (MMI IX) and isoseismal areas, that the larger of these two events was M = 6.7. Sanders and Slemmons (1979) suggested that this earthquake may have accompanied faulting on the Olinghouse fault, producing 3.65 m of sinistral slip on a northeast to east-northeast trend. If the earthquake did rupture this fault, then the length of faulting and maximum displacement indicate a maximum magnitude closer to M = 7.0 (Sanders and Slemmons, 1979).

Wonder, Nevada, earthquake, 1903

Slemmons and others (1959), on reexamination of unpublished geologic mapping by F. C. Schrader, interviews with residents, and field studies, concluded that an earthquake in the fall of 1903 ruptured as much as 19 km of the northerly trending Gold King normal fault, where the fault transects Tertiary volcanic bedrock of the Louderback Mountains. The fault trace was characterized by fissures as much as 8 m wide and 8 m deep. The fault zone was ruptured again in the 1107 GMT December 16, 1954, earthquake with displacements of "the same order and type" as the 1903 rupture (Slemmons and others, 1959). If the maximum rupture length of the 1903 earthquake is 19 km, we estimate the magnitude to be M ≈ 6.5 (Mark, 1977).

Pleasant Valley earthquake, 1915

The largest earthquake in the history of Nevada was located south of Winnemucca in Pleasant Valley. This earthquake, which had a magnitude of 7.8, occurred on October 3, 1915 (Slemmons and others, 1965). The felt area of the main shock was 1.3 million km^2, extending from Washington to Mexico and from the Pacific to Colorado, Montana, and Arizona. The maximum intensity of this event is estimated to be MMI X. Most of the damage was to adobe or stone buildings at the mining town of Kennedy and the few ranches in the epicentral region (Jones, 1915; Page, 1935). Based on newspaper reports, Ryall and Priestley (1975) and Ryall and Vetter (1982) infer that this earthquake was preceded by moderate seismicity for several decades prior to the main shock. Six or seven events of magnitude 4 to 4.5 occurred between 1872 and 1900, no events from 1901 to 1914, and two events immediately preceded the main shock in 1915. The largest of these foreshocks, M = 6.3 (maximum MMI VIII), occurred on the day of the main shock. The main shock was also followed by an intense aftershock series (Jones, 1915).

TABLE 1. EARTHQUAKES IN NEVADA AND VICINITY MAGNITUDE ≥6; 1852 THROUGH 1986

Date yr mo day	Time UT	Latitude °N	Longitude °W	Max. mag.	Region
1852		39.5	119.5	7.3*	Pyramid Lake, Nevada
1857 9 3	0305	39.5	120.0	6.2	Southwest of Reno; Nevada-California border
1860 3 15	1845	39.	120.	7.0*	West Carson City, Nevada
1868 5 30	0511	39.25	120.	6.0	West Carson City, Navada
1868 9 4	1600	37.	118.	6.3	Owens Valley, California
1869 12 27	0135	39.5	120.	6.7*	Virginia City, Nevada
1869 12 27	0940	39.5	120.	6.1	Virginia City, Navada
1872 3 26	1030	36.7	118.1	8.0*	Lone Pine, Owens Valley, California
1872 3 26	1406	36.9	118.2	6.7	Lone Pine, Owens Valley, California
1872 4 3	1215	36.9	118.2	6.6	Lone Pine, Owens Valley, California
1872 4 11	1900	37.5	118.5	6.9	Lone Pine, Owens Valley, California
1872 11 12		39.5	117.	6.0	Austin, Nevada
1875 1 24	1200	39.6	120.3	6.0	Reno, Nevada
1887 6 3	1048	39.2	119.8	6.3	Carson City, Nevada
1896 8 17	1130	36.8	118.1	6.4	Owens Valley, California
1902 11 17	1950	37.393	113.520	6.3	Pine Valley, Utah
1903		39.5	118.1	6.5*	Wonder, Nevada
1908 11 4	0837	36.0	117.0	6.5	Death Valley, California
1909 10 6	0250	41.766	112.666	6.3	Hansel Valley, Utah
1910 11 21	2323	38.	118.	6.3	Tonopah, Nevada
1914 2 18	1817	39.5	119.8	6.0	Reno, Nevada
1914 4 24	0834	39.5	119.8	6.4	Reno, Nevada
1915 10 3	0149	40.5	117.5	6.1	Pleasant Valley, Nevada
1915 10 3	0653	40.5	117.5	7.8*	Pleasant Valley, Nevada
1916 11 10	0911	35.5	116.	6.1	Southern Death Valley?, California
1918 3 12	1030	39.580	120.830	6.3	Reno, Nevada
1927 9 18	0207	37.500	118.750	6.0	Bishop, California
1932 12 21	0610a	38.68	118.21	7.2*	Cedar Mountain, Nevada
1932 12 21	0610b	38.46	117.97	6.2*	Cedar Mountain, Nevada
1933 6 25	2045	39.1	119.3	6.0	Wabuska/Yerington, Nevada
1934 1 30	2016	38.26	118.46	6.3*	Excelslor Mountain, Nevada
1934 12 3	1505	41.658	112.795	6.6	Hansel Valley, Utah
1934 12 3	1820	41.571	112.745	6.1	Hansel Valley, Utah
1941 8 30	1328	40.900	118.300	6.0	Winnemucca, Nevada
1941 9 14	1643	37.570	118.730	6.0	Mammoth Lakes, California
1941 9 14	1839	37.570	118.730	6.0	Mammoth Lakes, California
1942 5 28	0039	40.800	120.700	6.0	Northern Walker Lane, California
1946 3 15	1349	35.716	118.050	6.3	Owens Valley, California
1948 12 29	1253	39.55	120.08	6.0	Verdi, California
1954 7 6	1113	39.29	118.36	6.8*	Fallon/Rainbow Mountain/Stillwater Range, Nevada
1954 7 6	2207	39.20	118.40	6.4	Fallon/Rainbow Mountain/Stillwater Range, Nevada
1954 8 24	0551	39.35	118.34	6.8*	Fallon/Rainbow Mountain/Stillwater Range, Nevada
1954 12 16	1107	39.20	118.00	7.3*	Fairview Peak, Nevada
1954 12 16	1111	39.67	117.87	6.9*	Dixie Valley/Stillwater Range, Nevada
1959 3 23	0710	39.43	117.99	6.3	Dixie Valley/Stillwater Range, Nevada
1959 6 23	1435	38.92	118.89	6.3	Southwest Fallon, Nevada
1966 8 16	1802	37.395	114.206	6.0	Caliente/Clover Mountains, Nevada
1966 9 12	1641	39.420	120.150	6.0	North Truckee, Nevada
1980 5 25	1633	37.589	118.826	6.5	Mammoth Lake, California
1980 5 25	1649	37.620	118.900	6.0	Mammoth Lake, California
1980 5 25	1944	37.331	118.830	6.7	35 km west-southwest Bishop, California
1980 5 27	1451	37.470	118.802	6.3	South Mammoth Lake, California
1984 11 23	1808	37.456	118.602	6.2	Round Valley, west-northwest Bishop, California
1986 7 20	1429	37.540	118.439	6.2	Chalfant Valley, Bishop, California
1986 7 21	1442	37.540	118.442	6.6*	Chalfant Valley, Bishop, California

Note: An asterisk (*) in the magnitude column refers to a surface-rupturing event. Max. mag. indicates maximum reported M_L or M_S magnitude.

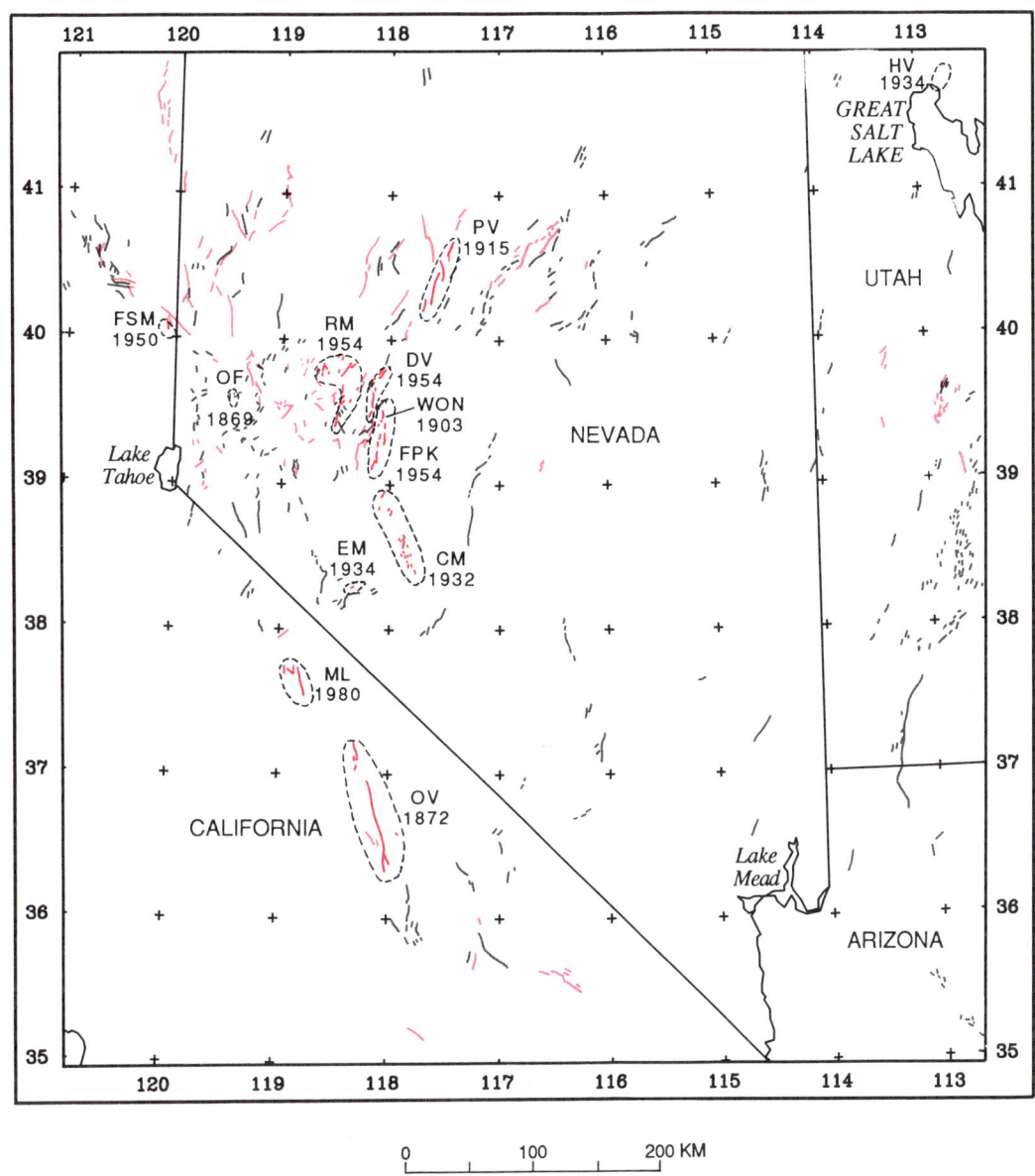

Figure 4. Map of historic (red), Holocene (shaded red), and late Quaternary (gray) faulting in Nevada and vicinity. Faulting has been adopted from Nakata and others (1982) by Thenhaus and Barnard (1989). Symbols are: CM-Cedar Mountain; DV-Dixie Valley; EM-Excelsior Mountain; FPK-Fairview Peak; FSM-Fort Sage Mountain; HV-Hansel Valley; ML-Mammoth Lakes; OV-Owens Valley; OF-Olinghouse; PV-Pleasant Valley; RM-Rainbow Mountain; WON-Wonder.

Scarps were formed along the west base of four mountain blocks in this earthquake, in some places forming new scarps and in other places breaking existing scarps (Wallace, 1984b). These blocks compose parts of the Tobin and Stillwater Ranges. The scarps formed along a zone 6 km wide and 59 km long, trending N25°E. The maximum vertical displacement on these faults is 5.8 m. Wallace (1984b) found evidence for 1 to 2 m of right slip in several locations. In a waveform modeling study, Doser (1988) found this event to be composed of two subevents that occurred on a fault striking N14°E with predominantly dip slip and some dextral slip.

Cedar Mountains earthquake, 1932

This earthquake, which occurred in the Monte Cristo Valley, west of Cedar Mountain on December 21, 1932, was M = 7.2 (Slemmons and others, 1965) and was felt over an area of 1.6 million km^2 (Coffman and von Hake, 1973). The earthquake was accompanied by about 61 km of discontinuous rifts in a belt 6 km to 15 km wide trending N21°W (Gianella and Callaghan, 1934a, 1934b). Horizontal displacements as great as 0.9 to 1.8 m and maximum vertical components of 0.5 m were measured (Gianella and Callaghan, 1934a, 1934b). The rifted zones dis-

TABLE 2. SEISMIC AND GEOLOGIC SOURCE PARAMETERS FOR EVENTS OF ABOUT MAGNITUDE 6 OR GREATER IN NEVADA AND VICINITY 1860 THROUGH 1986

Year	Region/place name	Orientation	Slip Sense	Fault Parameters Length (km)	Width (km)	Max. displ. (m)	Aftershock/ microearth-quake	Orientation of Preferred nodal plane	Nodal plane slip sense	References
1860	Pyramid Lake, Nevada (Pyramid Fault?)	N30W	Dextral, normal	~30	3.0	0.7 [V], ?[H]	--	--	--	Anderson and Hawkins (1984)
1869	S. of Pyramid Lake, Nevada (Olinghouse Fault)	N65E	Sinistral	23	--	3.65 [H]	--	--	--	Sanders and Slemmons (1970)
1872	Owens Valley, California	N20W[1]	Pred. normal, dextral[1]; pred. dextral, normal[2]	64[1]; 100[2]	1.3[1]	7 [V], 4.6 [H][1]; 1 [4.4,D] [V][2]; 4-5 [10,D] [H][2]	--	--	--	[1]Bateman (1961) [2]Beanland and Clark (1987)
1903	Wonder, Nevada	N-S	Normal	4.8-19.3	--	--	--	--	--	Slemmons and others (1959)
1915	Pleasant Valley, Nevada	N25E[1]	Pred. normal, some dextral slip on NW faults[1]	59[1]	6[1]	5.8 [V], 1-2 [H][1]	--	N14E[2]	Pred. normal; 36% dextral[2]	[1]Wallace (1984a) [2]Doser(1988)
1932	Cedar Mountain, Nevada	(E) N11E, N61E, N11W[1] (F) N21W[1]	Dextral, normal[1]	61[1]	6-15[1]	0.5 [V], 1.8 [H][1] 1.4 [V], 2.0 [H][2]; 0.3 [V], 1-2 [H][5]	(A) ~ N21W[3]	N2E to N14W[4]; (A) N27W[3]; (C) N13W[4]	100% dextral[4]; (A) Normal, 40% dextral[3]; (C) 100% dextral[4]	[1]Gianella and Callaghan (1934a,1934b) [2]Molinari (1984) [3]Gumper and Scholz (1971) [4]Doser (1988) [5]DePolo and others (1987)
1934	Excelsior Mountain, Nevada	N65E[1]	Normal, sinistral[1]	1.4[1]	--	0.13 [V][1]	(A) Diffuse[2]	N68E[4] (A) N86E[2] (A) N49E[3]	Normal[4]; (A) Sinistral[2]; (A) Pred. normal, 44% sinistral[3]	[1]Callaghan and Gianella (1935) [2]Gumper and Scholz (1971) [3]Ryall and Priestley (1975) [4]Doser (1988)
07/06/54, 1113 UT; 2207 UT	Rainbow Mountain, Nevada	N15E[3]	Normal[2]	17.7[3]	3[1]	0.3 [V][3]	NNE[5]	N34W[4]; N24W[6]; (C) N12W[6]	54% dextral, normal[4]; 54% dextral, normal[6]; (C) Pred. normal, 35-50% dextral[6]	[1]Richter (1958) [2]Byerly and others (1956) [3]Tocher (1956) [4]Fara (1964) [5]Douglas and Ryall (1972) [6]Doser (1986)
		N7E†	10% dextral, normal†							
08/24/54, 0551 UT	Rainbow Mountain, Nevada	N20E[3]	Normal[2]	22.5[3]	3[1]	0.76 [V][3]	(I) N12E[5,7]	N8W[4]; N5W[6] (C) N0W[6] (I) N34E[7]	Pred. normal, 44% dextral[4] 59% dextral, normal[6] (C) 75% dextral, normal[6] (I) Normal[7]	[1]Richter (1958) [2]Byerly and others (1956) [3]Tocher (1956) [4]Fara (1964) [5]Douglas and Ryall (1972) [6]Doser (1986) [7]Ryall and Malone (1971)

TABLE 2. SEISMIC AND GEOLOGIC SOURCE PARAMETERS FOR EVENTS OF ABOUT MAGNITUDE 6 OR GREATER IN NEVADA AND VICINITY 1860 THROUGH 1986 (continued)

Year	Region/place name	Fault Parameters					Orientation of Aftershock/microearthquake	Orientation of Preferred nodal plane	Nodal plane slip sense	References
		Orientation	Slip Sense	Length (km)	Width (km)	Max. displ. (m)				
12/16/54 1107 UT	Fairview Peak, Nevada	N10E[1]	(G) Normal, dextral[6]	49[1]	6[1]	3.7 [V], 4.3 [H][6]	N-S[2] N50E[4]	N11W[3] N10W[5]	63% dextral, normal[3] 63-73% dextral, normal[5]	[1]Richter (1958) [2]Westphal and Lange (1967) [3]Romney (1957) [4]Stauder and Ryall (1967) [5]Doser(1986) [6]Slemmons (1957b) [7]Smith and others (1972)
		N9E to N13W†	33-61% dextral, normal†				(A) NNE[2] (A) N11W and N50E[4] (A) NS-N7W[7] (A) N40E[7]	(A) N9W[7] (A) N36E[7]	(A) Normal[4] (A) Pred. normal 46% dextral[7] (A) Normal[7]	
12/16/54 1111 UT	Dixie Valley, Nevada	N17E[1]	Normal[1]	40[1]	5[1]	2.1 [V][2], 0 [H][1]	N-S[3]	N10W[4] (B) N12W to N2E[4]	Normal[4] (B) 82-96% dextral[4]	[1]Richter (1958) [2]Shawe (1965) [3]Douglas and Ryall (1972) [4]Doser (1986)
		N2E to N6E†								
1986	Chalfant Valley, California	N11W[1]	Dextral, normal[1]	15.5[1] 15.0[2]	0.5[1]	0.11 [H][1] 0.7 [V], 1.3 [H][2]	N25W[3]	N25W[3]	Dextral[3]	[1]DePolo and Ramelli (1987) [2]Gross and Savage (1987) [3]Cockerham and Corbett (1987)

(A) = Microearthquake data; (B) = 1959 Dixie Valley earthquake; (C) = aftershock data; (D) = one location near Lone Pine, Calif.; (E) = individual rifts/faults; (F) = rift zone; (G) slip varied from mostly strike slip on the south to mostly normal on the north; (I) = microearthquakes offset to the east of the rupture zone; [H] = strike-slip displacement; [V] = vertical displacement; † = modeling of geodetic data (Whitten, 1957) by Snay and others (1985) and Savage and Hastie (1969). Percentage of dextral slip is calculated as the fraction horizontal slip in the fault plane to the sum of horizontal plus downdip slip times 100%. Percentage of normal slip is the fraction of downdip slip to the sum of horizontal plus downdip slip times 100%.

played variable orientation (Table 2), but are predominantly oriented east of north, whereas the nodal plane orientations strike north to west of north. Gianella and Callaghan (1934b) found the rifting patterns and slip style to be consistent with the southward shift of the Paradise Range–Cedar Mountain block to the south relative to the Gabbs Valley–Pilot Mountains block. Based on wave-form inversion, Doser (1988) considers the main shock to be two strike-slip events of magnitude 6.7 and 6.6, respectively, occurring about 20 seconds apart. The surface breaks occurred within a larger fault zone, termed the Stewart and Monte Cristo fault zone (SMCFZ) by Molinari (1984). Molinari (1984) found scarps for the 1932 event as high as 1.4 m and as much as 2 m of dextral slip; he also found evidence for two additional Holocene events and at least three and possibly five or six surface faulting events during the latest Pleistocene and Holocene. The faulting style was principally dextral slip with tensional components and normal components in an en echelon pattern. According to Richter (1958) the faulting is consistent with dextral slip on a fault underlying Gabbs Valley, but not reaching the surface as a continuous break. Richter (1958) assumed the orientation of the buried fault was parallel to the trend of the rupture belt (i.e., N21°W).

Fallon–Stillwater–Rainbow Mountain earthquakes of July–August 1954

Three large earthquakes occurred one and a half months apart on the same fault. The first two events, a main shock (M = 6.8 and maximum MMI = IX) and its aftershock, were on July 6, 1954. The epicenter was located on the main fault zone on the east edge of Rainbow Mountain in the Stillwater Range, 24 km southeast of Fallon, and was felt over portions of California and Nevada (337,000 km^2; Slemmons and others, 1965). The third major event (M = 6.8) occurred on August 24, 1954. The maximum MMI intensity for this event was VIII (Slemmons and others, 1965) or IX (Coffman and von Hake, 1973); it was felt over 388,000 km^2, and there were instances of more severe damage than from the first shock (Slemmons and others, 1965; Steinbrugge and Moran, 1956).

These earthquakes produced discontinuous normal faulting along a NNE-trending fault bounding the eastern border of Rainbow Mountain (24 km east of Fallon) northward to the southern edge of the Carson Sink (Slemmons, 1957a). The July 6 events produced a fresh scarp 0.03 to 0.3 m high, west side up, for 17.7 km through the alluvial apron at the base of Rainbow Mountain and into the desert flats to the north (Tocher, 1956). The August 23 rupture extended the break 22.5 km northward with scarps as high as 0.8 m and increased some of the July 6 offset to as much as 0.5 m. The observed normal faulting from these events is in contrast with the focal mechanisms, which indicate predominantly dextral and dextral-oblique slip on a nodal plane striking north-northwest (Table 2). Doser (1986) concluded on the basis of wave-form modeling that the July 6 main shock was composed of two subevents. The first subevent, at a depth of 10 km, exhibited a predominance of dextral slip. The second subevent, at a depth of 7 km, was predominantly normal slip. This result could explain why strike slip was not observed at the surface. Doser (1986) finds that the August event was composed of three subevents, all best fit by dextral oblique slip on a fault trending north-northwest (Table 2). Geodetic studies also suggest that these events had substantial dextral slip (Meister and others, 1968). Snay and others (1985), in an attempt to explain the geodetic data on the basis of a dislocation model of the four historic rupture zones in this area, find their model does not adequately fit the slip associated with the Rainbow Mountain earthquakes. Their models for this Rainbow Mountain fault range from predominantly dip slip to predominantly strike slip, but in all cases the fault length and width appears too large given the seismically determined magnitude. They suggest magma movement as an alternative explanation for the displacements on this fault.

Dixie Valley–Fairview Peak earthquakes, December 16, 1954

Two large earthquakes occurred on December 16, 1954, in the eastern part of Churchhill County, Nevada. These earthquakes were the magnitude 7.3 and 6.9 Fairview Peak and Dixie Valley events, respectively (Slemmons and others, 1965). The mainshock felt area (518,000 km^2) included all of Nevada and parts of California, Oregon, Idaho, Utah, and Arizona (Cloud, 1957). In the sparsely populated epicentral zone, the mainshock maximum intensity was MM IX, where it could be estimated, but damage to structures in nearby towns was only intensity VII. The epicenter of the first shock was near a system of fault breaks on the east side of Fairview Peak and in the Louderback Mountains. The second epicenter was near fault breakage along the western edge of Dixie Valley at the base of the Stillwater Range, 48 km north of the first event (Tocher, 1957).

The earthquakes of December 16, 1954, were accompanied by offsets along many faults in four main zones of a north-trending belt 96 km long by 32 km wide, mainly along normal faults of the Basin-Range type (Slemmons, 1957b). The maximum vertical slip was 3.7 m at Bell Flat on the east side of Fairview Range. Significant strike-slip motion, amounting to 3.7 m of dextral slip, was also observed at Fairview Peak. Dip-slip displacements were prevalent in the northern part of the area, and oblique-slip or strike-slip displacements characterized the southern part.

Doser (1986) finds the seismic data for the Fairview Peak event best fit by three subevents occurring on a fault striking N10°W with predominantly dextral slip (Table 2). Snay and others (1985) find that the geodetic data in this region are consistent with slip on two separate planes for the Fairview Peak earthquake. A shallow dextral-oblique fault extends to 5 km depth and strikes N12°E, while the buried fault extends from 2 to 20 km depth, strikes N13°W, and is predominantly dip slip. Seismic and geodetic data for the Dixie Valley event are consistent with normal slip on a fault with strike of N10°W (Doser, 1986) or N6°E (Snay and others, 1985). Studies of microearthquakes in this region (Westphal and Lang, 1967; Stauder and Ryall, 1967; and Smith and others, 1972) indicated activity continuing along both north-south epicentral trends and northeast-trending zones (Table 2). Microearthquake focal mechanisms displayed both pure normal faulting along the northeast trends and oblique dextral slip along the north-trending zones.

OTHER SIGNIFICANT SEISMICITY

The Sierra Nevada–Great Basin boundary zone and the 1872 Owens Valley earthquake

This seismic zone follows the boundary region between the Sierra Nevada and the Great Basin. The earthquakes form a nearly continuous northwest-trending zone (Fig. 1). The breadth of this zone and the clustering of numerous earthquakes in several areas suggest activity on multiple faults. In the northern part of the zone near Reno (Fig. 1), clusters of earthquakes are observed in the following areas: (1) on the California-Nevada border west of Reno, where magnitude 6 events occurred twice in 1914 and once in 1948 (new work by Priestley, written communication, 1988, suggests that the largest of these events was east of Reno); a magnitude 5.6 earthquake produced surface faulting in 1950 along the Fort Sage Mountain fault in the northern Walker Lane belt (Slemmons, 1967); (2) southwest of Reno on the California-Nevada border, near the presumed location of the 1857 earthquake (Real and others, 1978); (3) north of Truckee, California, where an M_L = 6.0 shock occurred in 1966; (4) south of Reno in the Steamboat Hot Springs area; and (5) in the Virginia Range southeast of Reno. This band of activity continues to the northwest, into California, where several major seismic trends are apparent that may be associated with the Dog Valley, Mohawk Valley, Grizzly Valley, and Last Chance fault zones (Hawkins and others, 1986). Ryall (1977) suggests, on the basis of the long linear epicentral trends and their close association with the Sierra Nevadan range-front faults that the SNGBZ is a likely location for a major (M = 7+) earthquake. The largest event within the SNGBZ occurred in Owens Valley in 1872; on the basis of geologic data, it was estimated to be M = 7.5 to 7.8 (Beanland and Clark, 1987; Hanks and Kanamori, 1979), and on the basis of felt area, M = 8.0 (Oakeshott and others, 1972).

Mammoth

At the junction of the central Nevada seismic belt and the Sierra Nevada–Great Basin boundary zone (SNGBZ), earthquake swarms occur frequently in the area north of Bishop, and a major swarm near Mammoth Lakes has been in progress since October 1978. Through 1988, the Mammoth Lakes sequence has produced seven earthquakes with magnitudes greater than 6. Uplift, spasmodic tremor, and increased fumarole activity have also occurred in Long Valley caldera and likely are associated with magma injection (Ryall and Ryall, 1981, 1983; Savage and Clark, 1982; Miller and others, 1982). This zone has had the highest level of seismic activity since 1980. Savage and Cockerham (1987) suggested that the M ⩽ 5.5 events occur in a quasi-periodic fashion with an average interval of 18 months. As noted by Smith and Priestley (1988), however, the Chalfant Valley earthquake, to the north of Bishop, is an exception to this observation because aftershocks continued for more than six months after the mainshock. The causes of seismicity in this region are uncertain, but the proximity of Long Valley caldera and the intimate association of volcanic processes is a likely factor (Hill and others, 1985).

Caliente/Clover Mountains earthquake, 1966

This earthquake series deserves special mention because it is the largest earthquake (M = 5.7 to 6.1, University of California, Berkeley seismograph station) to occur in the southern Nevada Transverse Zone. The earthquake was accompanied by an aftershock swarm, which at its peak produced more than 8,000 events per day (Boucher and others, 1967). Page (1968) reports the depths of the aftershocks to be less than 9 km. Some of the aftershocks were relocated by Beck (1970) and Rogers and others (1983). In the latter study, a selected subset of the best-recorded aftershocks and the main shock were relocated using the joint-hypocenter relocation technique (Dewey and Spence, 1979). These relocated events occupied a NNE—trending zone 22 km long and 8 km wide on the southern end of the Clover Mountains. Based on the aftershock trend and several focal mechanisms for the main shock (Wallace and others, 1983; Smith and Lindh, 1978). Rogers and others (1983) infer dextral slip on a north-trending fault for this event.

Induced seismicity

Earthquakes have been induced by the activities of man at two locations in Nevada—the Nevada Test Site and Lake Mead. Tectonic stress apparently has been released as a result of underground nuclear explosions at the Nevada Test Site (NTS). This stress release is seen as surface displacements occurring at the time of the event (McKeown, 1975), as seismic energy release concurrent with the detonations, and as numerous aftershock earthquakes outside the zones of shattering. Hamilton and Healy (1969), Boucher and others (1969), Smith and others (1971), and Rogers and others (1977) presented evidence that numerous aftershocks followed the detonation of high-yield nuclear devices at NTS and the Central Nevada Test Area (CNTA); some of these earthquakes have had magnitudes ≃ 5. Based on these studies it appears that nuclear tests of at least m_b ⩾ 5 (Boucher and others, 1969) are required to trigger stress release in the form of aftershocks, although there may be exceptions, and the thresholds may vary between test areas (i.e., Pahute Mesa, Rainier Mesa, Buckboard Mesa, Yucca Flats, and Hot Creek Valley [CNTA]; Rogers and others, 1977). The tectonic energy that is released seismically at the time of the explosion and the seismic energy released in aftershocks have comparable values. The magnitude values equivalent to this energy release range between approximately the magnitude of the nuclear test and one magnitude unit less than that of the explosion (Aki and others, 1969; Aki and Tsai, 1972; Wallace and others, 1983; Lay and others, 1984; Wallace and others, 1985; Wallace and others, 1986). Bucknam (1969), McKeown and Dickey (1969), Orkild and other (1969), Maldonado (1977a, b), Snyder (1971, 1973), and Morris (1971), in studies of faults near nuclear tests, found that these events can generate vertical and horizontal displacements on existing tectonic faults that are produced with the first arrival of the explosion-generated seismic waves. As much as 100 cm of vertical displacement and 15 cm of horizontal dextral slip have been observed on faults as great as 10 km in length.

Hamilton and others (1971) noted that 95 percent of the aftershocks of four nuclear explosions having magnitudes greater than about 5.8 occurred within 14 km of ground zero, and 94 percent occurred in the upper 5 km of the crust. Rogers and others (1977) conducted similar studies for eight nuclear tests in the high-yield nuclear-test series about six years after the earlier series and found that, although the aftershocks in the latter testing series clustered more closely around ground zero, the depths of the aftershocks appeared to be significantly greater than the earlier study. Between 93 and 100 percent of the aftershocks occurred within 6 km of ground zero, and at least 95 percent of the hypocenters occurred between 4 and 10 km. These earthquakes do not clearly align with known faults in most cases, but they do sometimes exhibit north to north-northeast trends. Depth sections of the alignments suggest that nuclear-induced seismicity occurs on steeply dipping faults, and based on focal mechanisms, slip style ranges from strike slip on north-trending faults to dip slip on NNE-trending faults.

During and following the impoundment of Lake Mead, numerous earthquakes occurred; some of these events had magnitudes as great as 5 (Mead and Carder, 1941; Rogers and Lee, 1976). Rogers and Lee (1976) suggested that these earthquakes were caused by increases in pore pressure and hydrostatic head along existing faults of about 3 to 6 bars; they showed that the expected energy release resulting from this pore-pressure increase was the same order of magnitude as the total seismic energy release of earthquakes in the Lake Mead area. Earthquakes at Lake Mead occur from near-surface to about 12 km, and clusters appear to occur on steep faults, with slip style varying from strike slip on north-trending faults to dip slip on north-northeast- to northeast-trending faults. Activity rates and the maximum magnitudes of earthquakes occurring at Lake Mead have declined with time, suggesting partial relaxation of stress in the crust in the immediate vicinity of the lake. In 1988 renewed activity, including numerous felt earthquakes, occurred southwest of Hoover Dam (S. C. Harmsen, written communication, 1988).

SEISMIC GAPS

Wallace (1978) has noted that three seismic gaps may exist within the CNSB. These gaps, which are likely candidates for future large earthquakes (Figs. 1 and 3), are the Stillwater gap that lies between the 1954 and 1915 ruptures (Wallace and Whitney, 1984), the White Mountains gap between the 1872 and 1932 zones, and the southern Owens Valley gap that is south of the 1872 rupture. The White Mountains seismic gap has been identified as an area in which the next major earthquake in the western Great Basin has a high likelihood of occurring, and increased seismicity in and around the White Mountains gap may be precursory to such an event (Ryall and Ryall, 1983).

Thenhaus and Barnhard (1989) discuss the characteristics of previously recognized (Stewart, 1980) east-west zones of accommodation that bound northerly trending bands of seismicity, zones of like-age faulting, and domains of range tilt. Thenhaus and Barnhard suggest that an additional seismic gap, which they term the Sonoma Range gap, may lie between the northern terminus of the Pleasant Valley historic rupture zone and the northernmost accommodation zone (Fig. 1).

Some of these gaps, as defined by the present-day seismicity, appear shorter than the rupture-defined gaps (Figs. 1 and 2). That is, small- to moderate-magnitude earthquakes appear to extend into parts of the rupture-defined gaps. For example, there are a few scattered events in the Stillwater Gap, just north of the 1954 rupture zone. Although Gumper and Scholz (1971) first commented on the White Mountains gap, since then, earthquakes have occurred at its north end near the Excelsior Mountains, and also east of Mono Lake. The gap appears to be shortened on the south by the 1986 Chalfant Valley earthquakes (Cockerham and Corbett, 1987; Smith and Priestley, 1988). These events occurred north of the termination of the 1872 ruptures. Thus, the White Mountains seismicity gap is now only half the length of the White Mountains rupture gap. Ryall and Priestley (1975) suggested that high seismicity in the Excelsior Mountains may indicate that stress is released by a continuing series of small to moderate earthquakes and fault creep and that rupture of the northern segment of the White Mountains gap is less likely than it would be otherwise.

Inactivity on the northwest-trending Death Valley–Furnace Creek fault zone is a paradox because, although numerous Holocene scarps exist on these faults, suggesting that they are active (Hunt and Mabey, 1966; G. E. Brogan, written communication, 1979), little or no historic seismicity has occurred in this region that can be associated with these faults (Real and others, 1978; Rogers and others, 1987). Thus, this fault zone could be in the part of the seismic cycle in which the faults have experienced stress release in the recent geologic past. The alternative hypothesis—that the fault is locked and accumulating stress for the next event—appears to have less merit. Based on geologic and seismic data, Rogers and others (1983, 1987) inferred a clockwise horizontal principal-stress rotation eastward across the zone and suggested that, if the fault were locked and highly stressed, the stress orientations on both sides of the fault could be expected to be approximately equal. The inferred stress difference on either side of this structure could be indirect evidence that this fault zone is presently in a low-stress state.

Astiz and Allen (1983) suggested that the eastern section of the Garlock fault, which bounds the Great Basin, is a seismic gap. Clear evidence of Holocene displacement exists for this structure, yet very few earthquakes have occurred there in the historic record. Because the western half of the fault exhibits the youngest offsets, active creep, and low-level seismicity, while the eastern section appears locked, Astiz and Allen (1983) believe the eastern section of the Garlock has the greatest potential for a large earthquake.

FREQUENCY OF LARGE HISTORIC EVENTS

As noted by Ryall (1977), the time between large events in the historical record for Nevada ranges from 4 minutes to 43 years. Our understanding of earthquake recurrence in this region is confounded by inconsistencies between projected rates of large events from low- or intermediate-magnitude earthquake occurrence, the rate of occurrence of historic large-magnitude events, and the rate of occurrence of large-magnitude events predicted from geological studies. The wide disparity between these rates is evident in Table 3. The rates from Algermissen and others (1982) are based on their maximum-likelihood estimates (Bender, 1983) and historic seismicity in their zones 31, 32, 33; these zones represent the CNSB, but exclude the SNGBZ. This estimating procedure tends to heavily weight the intermediate-magnitude events, and reduces the influence of the large historic events, which accounts for part of the discrepancy between A and B in Table 3. There are, however, substantial differences in the rates of large events forecast on the basis of geologic data compared to those based on seismic data. The mean seismic rates (Table 3, B) exceed the mean geologic rates (Table 3, C). These differences can be qualitatively explained by a model of Great Basin earthquake occurrence discussed by Wallace (1987).

Based on studies of range-front faults in Nevada, Wallace (1987) suggests that clusters of events may occur on individual segments of a fault during an active period lasting hundreds to thousands of years. The active period is followed by quiescence for 10,000 to 30,000 years. Other segments of the same fault may or may not be active at the same time, but once active, these segments follow the same pattern. This model may extend to subzones of the Great Basin, as observed by Buckham and others (1980) in western Utah. Some subzones of that region display late Quaternary, but not Holocene faulting or the converse. Subzones can be expected to turn on for periods of hundreds to thousands of years, then become dormant while other subzones are active. This model appears consistent with the historic record of seismicity in the CNSB where several colinear fault segments became active almost simultaneously. In fact, most of west-central Nevada, where Holocene displacements have been observed on many range-front faults, may be part of a subzone in the active phase (Wallace, 1984a).

As noted by Wallace (1987), the temporal behavior of Great Basin events (here taken to be M ≥ 7.0) complicates the assessment of the seismic hazard in the region. Use of the mean

TABLE 3. RATES OF LARGE (M ≥7.0) EARTHQUAKE OCCURRENCE IN THE NEVADA SEISMIC ZONE

Source of Data (Assumption)	Return Period (years)
A. Historic seismicity (M≥7; all within the CNSB)	27
B. Projected seismic rates (CNSB)*	1,080
C. Geologic (Mean RP, Fully Cycle, SF)†	5,000-10,000
D. Geologic (Mean RP, Active Period, SF)†	1,500-3,000
E. Geologic (Quiet Period, SF)†	(No events)
F. Geologic (Mean RP, Active Period, CNSB incl. gaps))	300-1,500

Note: Abbreviations used are SF = single fault, RP = return period, and CNSB = Central Nevada Seismic Belt.
*Algermissen and others (1982). The rate for their zones 31, 32, and 33 combined is proportioned to the area of the CNSB (1.22 x 10^4 km^2).
†Wallace (1987)

rate of occurrence of large events (determined from geologic data; Table 3, C) overestimates the hazard on a fault segment if the short-term forecast extrapolates from the active period into the quiescent period and underestimates the hazard if the short-term forecast extrapolates from the quiescent period into the active period.

Estimating the rate of occurrence of large events over a region such as the CNSB is also problematic. The principal difficulty lies in the lack of enough geologic data to estimate the number of faults in the CNSB capable of producing large events, the ages of slip events on these faults, the time since the last slip event, and the duration of the active and dormant periods. Although these data are lacking for most CNSB faults, it is possible, nonetheless, to estimate a rate for large events under certain assumptions that are based on the patterns of fault behavior that Wallace has observed. If we assume, for instance, that one full cycle, including the active and dormant phases for a single fault segment, lasts 15,000 to 30,000 years and that, on average, three large events occur during the active phase (5,000 to 10,000 years), then the mean return period over a full cycle is 10,000 years (Table 3, C; Wallace, 1987). Following similar reasoning, the mean return period over the active period is 1,500 to 3,000 years (Wallace, 1987). If we assume that six segments (including gaps) exist in the CNSB that are capable of producing large events, then the mean return period for CNSB large events during the active period is 300 to 1,500 years (Table 3, F). Notably, this return-period range brackets the return period projected by the intermediate-level seismicity (Table 3, B).

Furthermore, it has also been suggested that seismic strain in some regions may be released in a "characteristic earthquake" on each major fault segment, such that a zone might have only low-level seismicity before and after a characteristic earthquake, with few or no events in the magnitude range 5 to 7 (Schwartz and Coppersmith, 1984). In this scenario the recurrence curve would be discontinuous with a rate spike above magnitude 7.

TABLE 4. SEISMIC RATES IN NEVADA

Zones*/ Area (10^4 km^2)	Area and approximate age of faulting†	Return period for M≥7.0 (years)	Annual rate of exceedance per 10^4 km^2 M≥6.4	M≥7.0	M*$_{max}$
31/ 4.71	NSZ, excluding historic rupture zone, Holocene	280	2.6×10^{-3}	7.6×10^{-4}	7.6
31, 32, 33/ 6.16	NSZ, including historic rupture zone, Holocene	210	2.6×10^{-3}	7.6×10^{-4}	7.6
31, 32, 33, 29/ 7.99	Nevada seismic zone and the eastern Sierran front, Holocene	180	2.0×10^{-3}	6.8×10^{-4}	7.6
31, 32, 33, 34/ 16.2	Western Nevada, Holocene	170	1.3×10^{-3}	3.6×10^{-4}	7.6
17, 18, 19/ 12.6	Eastern/southern Nevada, late Quaternary	640	5.8×10^{-4}	1.2×10^{-4}	7.6

*Zones, rates, and maximum magnitudes are adapted from Algermissen and others (1982).
†Based on Wallace (1984a).

Thus, the difference in seismically inferred return periods given in Table 3, A and B, could be partially explained by this model, but could also be the result of "contagion" between fault segments. That is, once rupture occurs in an active zone, other nearby fault segments rupture over short periods of time (Perkins, 1987), perhaps due to increased loading of segments adjacent to the ruptured segment.

As a means of comparing seismic rates for various areas of Nevada, rates of seismicity from Algermissen and others (1982) for several geologic subzones of the Great Basin are shown in Table 4. The logarithm of the number of events, N, versus magnitude, M, can be fit with lines of the form

$$\log N = a - bM. \quad (1)$$

These lines have slopes (b-values) for Nevada seismic zones varying from about 0.9 to 1.1, although b-values as great as 1.5 to 2.0 have been observed in the regions of induced seismicity at the Nevada Test Site (NTS) and Lake Mead (Hamilton and others, 1971; Rogers and others, 1977, Rogers and Lee, 1976). Recurrence slopes are not well determined for some regions of Nevada, particularly east and south of the CNSB, because of the low level of earthquake occurrence. The completeness of the earthquake record for Nevada is not well determined, and is likely to be a function of both time and geographic position. The record is likely most complete for the region near Reno, where the instrumental record is longest. Rogers and others (1977) estimated completeness for a large portion of Nevada, including most of the CNSB. They evaluated the time period from 1845 to 1974 and a region that included a substantial portion of southern California. Their results for this region suggest that earthquakes with magnitudes ranging between 4 and 5, 5 and 6, 6 and 7, and 7 and 8 are complete for the most recent 40, 50, 60, and 130 years, respectively.

Ryall and Priestley (1975) plotted the number of earthquakes occurring in the epicentral zones of large Nevada earthquakes versus time since the main shock and found that the logarithm of the number of events was linearly related to the time since the last large event in that region. They conclude that the aftershocks effectively die out about a century after the mainshock.

EARTHQUAKE DEPTH DISTRIBUTIONS IN THE GREAT BASIN

Accurate estimates of focal depth are an important factor in integrating seismicity with tectonic models. The most accurate depth estimates are likely to derive from dense network data with stations that are within one focal depth of the earthquake, for events within the perimeter of the network. In addition, however, accurate crustal models are required, and the location procedure should include at least five P-wave phase readings and one or more S-wave readings. Because data used to locate earthquakes are typically from stations more than one focal depth from the epicenter, earthquake focal depth is frequently poorly constrained. In general, the most accurate earthquake depths for this region have been obtained from data derived from detailed microearthquake surveys and telemetered local networks. Such studies have been conducted in Nevada by Oliver and others (1966), Westphal and Lange (1967), Stauder and Ryall (1967), Ryall and Savage (1969), Gumper and Scholz (1971), Ryall and Malone (1971), Hamilton and others (1971), Smith and others (1971), Fischer and others (1972), Papanek and Hamilton (1972), Ryall and Priestley (1975), Rogers and Lee (1976), Rogers and others (1977), Ryall and Vetter (1982), Tarr and Rogers (1986), and Rogers and others (1987). Although considerable additional study of individual active zones will be necessary before confidence can be acquired concerning earthquake depths in this region, hypocenters appear to display some consistent general characteristics.

Great Basin earthquakes are rarely deeper than 20 km. For most seismic zones in the Great Basin, more than 95 percent of the events occur in the upper 15 km (Ryall and Savage, 1969; Rogers and others, 1987). Within the upper 15 km of the crust, hypocenter concentrations display considerable variability (Fig. 5; Stauder and Ryall, 1967; Ryall and Savage, 1969; Okaya and Thompson, 1985; Richins and others, 1985). Modal depths of background microearthquakes or aftershocks may occur at any depth between about 1 and 15 km. In contrast, mainshock focal depth occurs in the range 8 to 16 km for the best-determined values (Doser and Smith, 1985; Doser, 1986, 1987, 1988; Baker and Doser, 1988). This observation has been one of the chief arguments supporting the existence of a brittle-ductile crustal boundary at this depth in the Great Basin (Anderson, 1971; Tocher, 1975; Smith and Bruhn, 1984). Because mainshock events commonly initiate near the base of the brittle upper crust between 10 and 15 km, Smith and Bruhn (1984) infer that maximum strength of the brittle crust occurs at the brittle-ductile boundary.

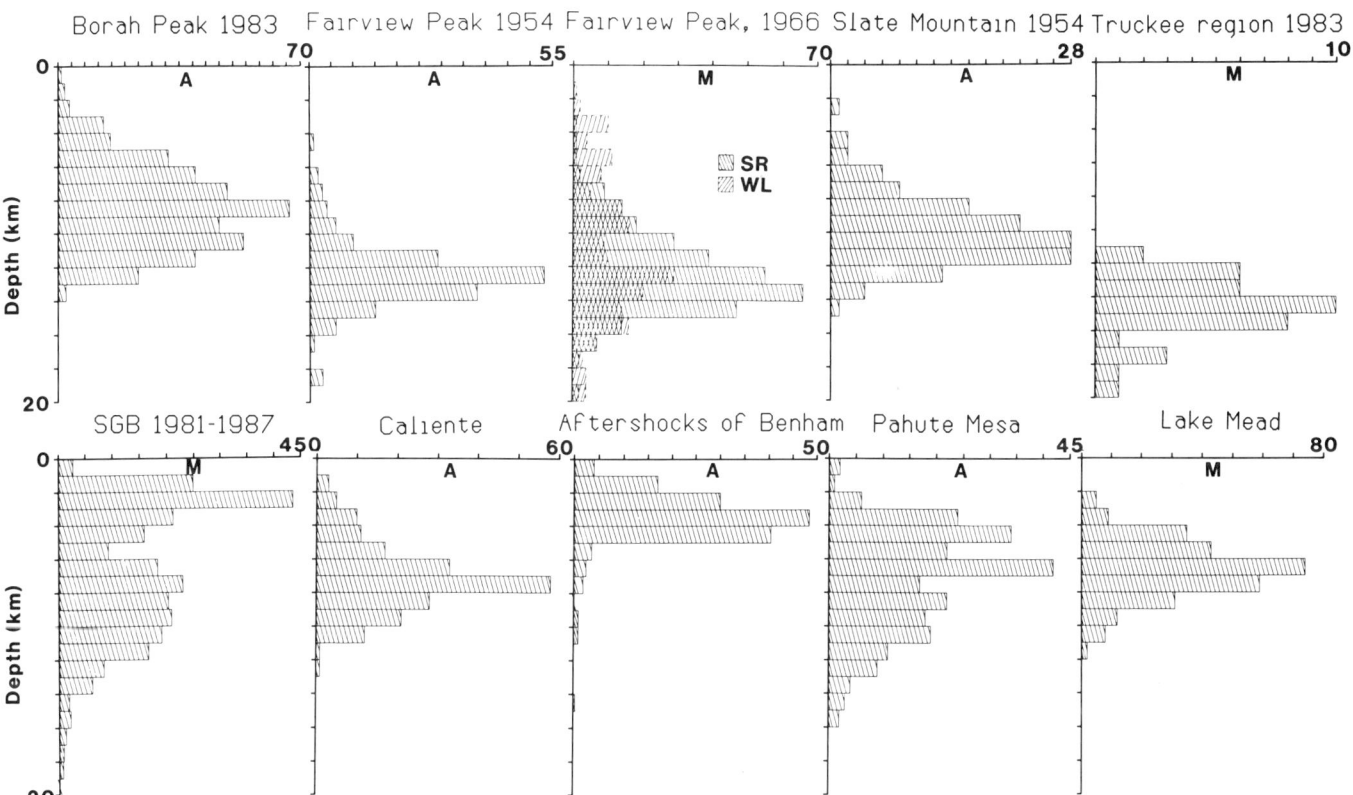

Figure 5. Depth-of-focus histograms for aftershock studies (labeled A) and microearthquake studies (labeled M) for various regions in the Great Basin. We plot numbers of earthquakes versus depth of focus for events in the range 0.0 to 20.0 km. For all but one of the data sets, less than 1 percent of the estimated depths are greater than 20 km. About 8 percent of the Fairview Peak depths computed by Westphal and Lange (1967) are greater than 20 km. References are: Borah Peak 1983—Charley Langer (written communication, 1987); Fairview Peak 1954, Slate Mountain 1954, and Caliente 1966—Ryall and Savage (1969); Fairview Peak 1966—Stauder and Ryall (SR) (1967) and Westphal and Lange (WL) (1967); Truckee region 1983—Hawkins and others (1986); southern Great Basin (SGB)—Rogers and others (1987) and Harmsen and Rogers (1987); Benham and other Nevada Test Site (NTS) nuclear test aftershocks—Hamilton and others (1971); aftershocks of Pahute Mesa (NTS) nuclear tests of 1976—Rogers and others (1977); and Lake Mead earthquakes of 1972 and 1973—Rogers and Lee (1976). Depths are relative to the mean surface level for each study area.

Hypocentral distributions for small-earthquake series fall into several categories. Some events occur in weakly tabular or cylindrical clusters over depth ranges of only several kilometers (Stauder and Ryall, 1967; Fischer and others, 1972; Hawkins and others, 1986; Arabasz and Julanders, 1986; Rogers and others, 1987). In other cases, clusters of microearthquakes appear to occur in steeply plunging cylindrical or tabular volumes of rock extending from near-surface to 10 to 15 km (Rogers and Lee, 1976; Rogers and others, 1987). Although these patterns may be questioned on the basis of known focal depth errors, the patterns have been observed in several cases using tightly clustered stations (i.e., Hamilton and others, 1971; Rogers and Lee, 1976; Rogers and others, 1987). In some cases this appearance may be an artifact of location error; nonetheless, this feature of microearthquake occurrence has been noted in studies of well-located earthquakes from dense arrays, leading Rogers and others (1987) to suggest that the cylindrical clusters may be real and represent failure in weak rock along the intersection of faults.

On the basis of seismic data such as aftershock distribution and nodal-plane dip, all large Basin and Range earthquakes appear to occur on steeply dipping faults that penetrate the upper 15 km of the crust (Romney, 1957; Doser, 1986, 1987, 1988, 1989). Some of these events yield predominantly normal-fault seismic signatures, such as the Pleasant Valley, Borah Peak, and Excelsior Mountain earthquakes, but others yield strike-slip or oblique slip signatures (Table 2). Most microearthquake and aftershock activity occurs within steeply dipping volumes that may reflect steeply dipping structures (Stauder and Ryall, 1967; Westphal and Lange, 1967; Gumper and Scholz, 1971; Rogers and Lee, 1976; Van Wormer and Ryall, 1980; Rogers and others, 1987; Doser, 1987). The main-shock hypocenter and the surface rupture appear to lie on the nodal-plane projection in many cases (Smith and others, 1985), suggesting that large Great Basin earthquakes, $M \geqslant 7.0$, occur on steeply dipping planar faults (45° to 60°) that penetrate to at least 15 km.

Despite this evidence for historic failure on steeply dipping structures, there is extensive geologic and geophysical evidence suggesting that large parts of the Great Basin have deformed in response to shear along listric and detachment faults and low-angle uncoupling zones (i.e., Anderson and others, 1983), some of which penetrate to midcrustal depths and possibly to the base of the crust. Extensive block rotations are common above the shallow parts of detachment faults. As yet, however, little evidence exists for seismic slip on low-angle faults in the Great Basin or in any other extensional regime (Jackson, 1987). Smith and Richins (1984) termed this problem the "paradox and paradigm" of Great Basin deformation. Although a small percentage of focal-mechanism nodal planes do have shallow dip (Figs. 6, 7, 8, 9), in many cases these low-angle nodal planes are likely to be auxiliary nodal planes. This conclusion is drawn on the basis of correlations between inferred principal-nodal planes and aftershock lineations and/or mapped structure in the southern Great Basin (Rogers and others, 1987).

Low-angle structures could be active aseismically. Eddington and others (1987) find that geodetically determined slip rates for extension of the Great Basin are about the same order of magnitude as the seismically inferred extensional slip rates and the slip rates estimated from geologic observations averaged over the Holocene (Minister and Jordan, 1987). Both the seismic and geologic estimates are based on the assumption of a predominance of normal slip on steeply dipping faults and nodal planes. Thus, aseismic slip on low-angle faults is not required to explain these observations.

FOCAL MECHANISMS AND TECTONICS

Figures 6 and 7 present the focal mechanism determinations that have been obtained for the largest Nevada earthquakes, as well as representative mechanisms determined for low-magnitude seismicity or microearthquakes (the data for these figures are presented in Appendix A, Tables A1 and A2). Table 2 compares fault parameters obtained from geologic and seismic data. Many of the focal mechanisms demonstrate strike slip and oblique slip (Fig. 9), even for those cases in which geologic data are commonly interpreted to indicate normal faulting. The seismic data are surprising in the context of neotectonic deformation, dominated by vertical tectonics that produced the basins and ranges. Although substantial strike-slip faulting accompanied extension, most commonly the strike of these faults is roughly northwest or northeast. Holocene and contemporary lateral faulting, by contrast, exhibit more northerly strike (Fig. 4). If west-northwest extension is active today (Zoback and others, 1981), local accomodation to boundaries of differential extension could be expected to produce strike slip on trends subparallel to this direction. Yet, the widespread geographic extent of dextral slip and dextral-oblique slip on faults of northerly trend does not confirm this expectation.

Notable differences between the geologic and seismic data are apparent in Table 2. For instance, the focal-mechanism inferred fault orientations are commonly rotated counterclockwise relative to geologic field observations. Differences in fault orientation and/or slip style have been observed across much of the Great Basin for events as far north as the Hansel Valley earthquake (northwestern Utah; Doser, 1989) and as far south as the Manix earthquake (Fig. 1; Richter, 1958). Similar differences in slip style have been observed at Pahute Mesa in response to nuclear testing. Underground nuclear explosions radiate significant seismic components of strike-slip energy that can be comparable in energy release to the explosion itself (i.e., Aki and Tsai, 1972), while producing predominantly vertical displacements on scarps at the surface.

Although differences are common between seismic and geologic observations, they seem to be in accord locally in some cases. For instance, in the southern Nevada transverse zone, Rogers and others (1987) find correlations in several locales between epicenter lineations, focal-mechanism nodal planes, and surrounding structural grain, suggesting northerly trending dextral slip and oblique and normal slip on northeasterly trending faults.

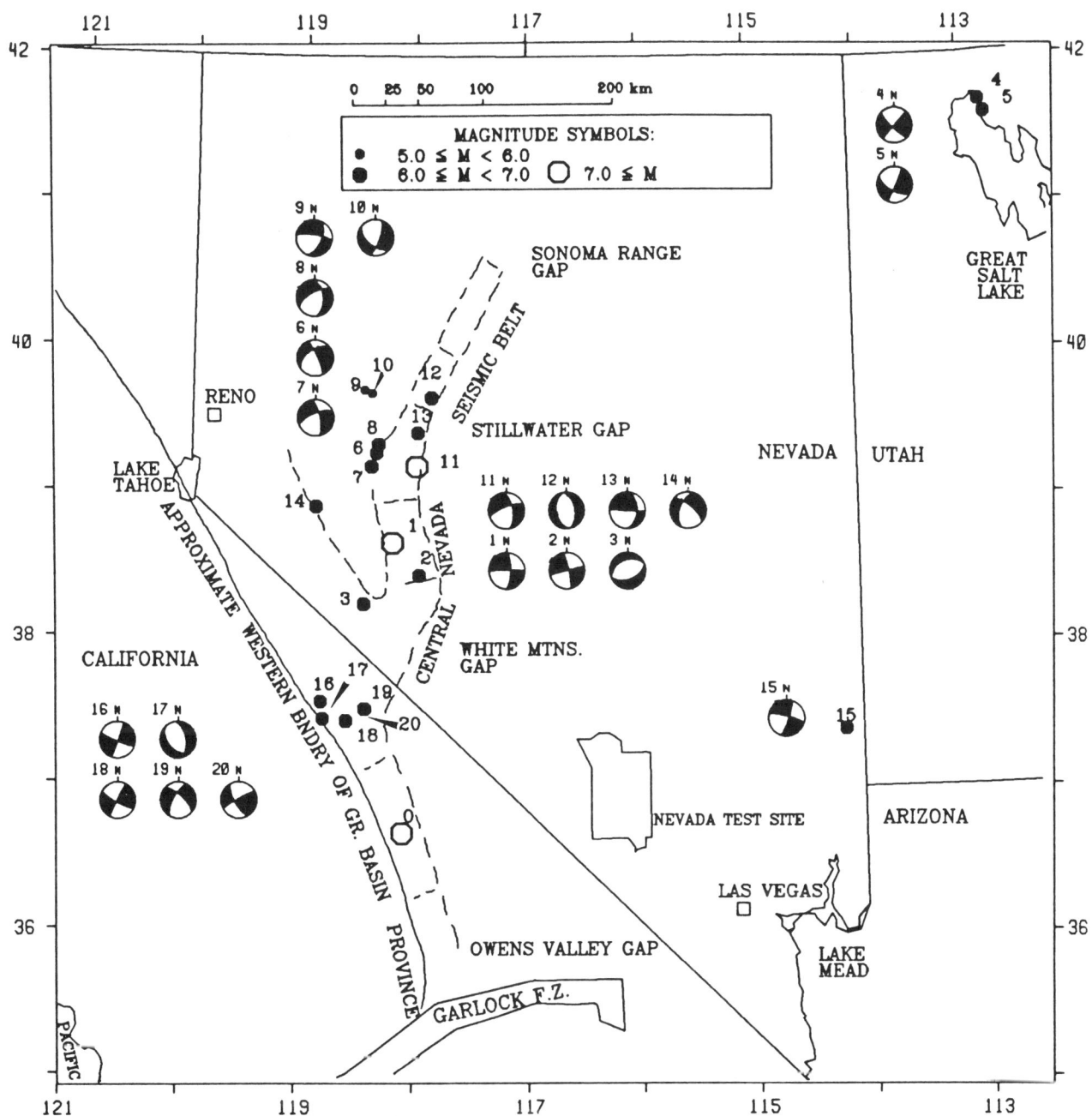

Figure 6. Map showing earthquake locations and their focal mechanisms computed by body-waveform inversion for Great Basin earthquakes having M ≥ 5.0 for the time period 1932 through 1986. In the case of Mammoth Lakes, the non-double-couple component of the solution is substantial (Sipkin, 1986). Focal mechanisms 1 to 14 were computed by Doser (1986, 1987), 15 by Wallace and others (1983), 16 and 17 by Sipkin (1986), 18 by Barker and Wallace (1986), 19 and 20 by Cockerham and Corbett (1987). Boundaries of central Nevada seismic belt from Wallace (1984a), and Garlock fault zone from Astiz and Allen (1983).

Anderson and Barnhard (1987), in a topical study of young deformation in the Sevier Valley, Utah, concluded there was a fair correspondence between late Quaternary strike-slip and normal faulting and deformation inferred from earthquake focal mechanism data (Arabasz and Julander, 1986).

Figure 10 shows the average T-axes (tension) orientations from earthquake focal mechanisms in various active areas of the Great Basin. Because both strike slip and normal slip are observed in most of these zones, the greatest horizontal principal stress and the vertical stress are inferred to be approximately equal (Zoback and Zoback, 1980). Rogers and others (1987) argue that these stress conditions are pervasive throughout the brittle crust (in the

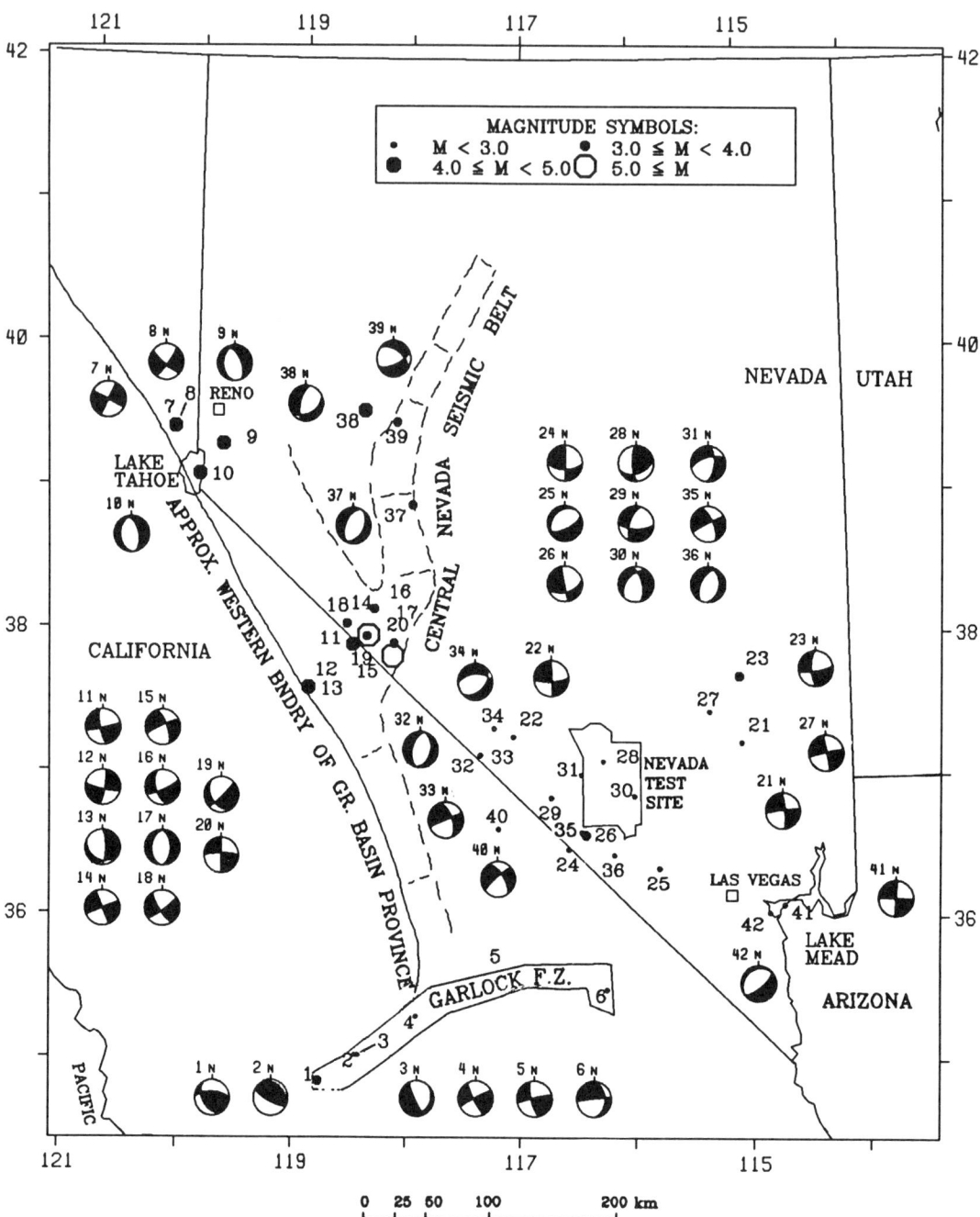

Figure 7. Map showing earthquake locations and their focal mechanisms computed from first-motion P-wave arrivals at local seismograph networks in the Great Basin and Garlock fault zone. Focal mechanisms 1 through 6 from Astiz and Allen (1983), 7 through 20 and 37 through 39 from Vetter and Ryall (1983) and Vetter (1984), 21 through 24 from Rogers and others (1987), and 25 through 36 and 40 from Harmsen and Rogers (1987).

southern Great Basin) because both styles of faulting are observed from near-surface to about 20 km. Figure 11 shows three cross sections with focal spheres in plan view, projected to the section plane at the depth of focus of each earthquake. This figure emphasizes the comingling of strike slip and normal faulting over the full depth range of earthquake occurrence. Harmsen and Rogers (1986) note that in a region where all fault orientations are available for slip, stress conditions of this type permit both slip styles with equal likelihood. In fact, based on the focal mechanisms, an apparent preference exists in the data sets for strike slip and oblique slip. It is unclear whether this preference is due to greater availability of certain fault strikes or whether it is due to a greatest horizontal principal stress generally slightly in excess of the vertical stress. Given the highly fractured nature of the Great Basin, the latter choice seems preferable.

Cenozoic strike-slip faulting in the Great Basin has been widely discussed (i.e., Shawe, 1965; Hamilton and Myers, 1966; Slemmons, 1967; Atwater, 1970; Anderson, 1971, 1973; Wright, 1976; Stewart, 1978, 1988; Hill, 1982; Wernicke and others, 1988), resulting in disparate interpretations to account for the presence of northwest-trending dextral and northeast-trending sinistral faults in an otherwise extensional environment. The origin of these faults has been attributed to (1) megashear related to movement along the continental plate boundary; (2) essentially north-south–directed compression, producing conjugate lateral faults; (3) differential extension between the northern and southern Great Basin or between smaller subzones; and (4) multiple changes in stress magnitudes and/or orientation favoring either dip-slip faulting in one mode or strike-slip faulting in the alternate mode. Combinations of some of these interpretative origins are also possible.

Some models of Great Basin extension treat strike-slip faults as secondary or even passive features that accommodate boundaries of differential extension. Yet, there is evidence that in some subregions, strike slip may be the primary deformation mode and normal faulting secondary, as in parts of the Walker Lane belt (Stewart, 1988), the Pahranagat shear zone (Shawe, 1965), and the Lake Mead region (Anderson, 1971; Ron and others, 1986). The widespread aspect of strike slip in the central Great Basin earthquake record also suggests that it may play a primary role in contemporary deformation.

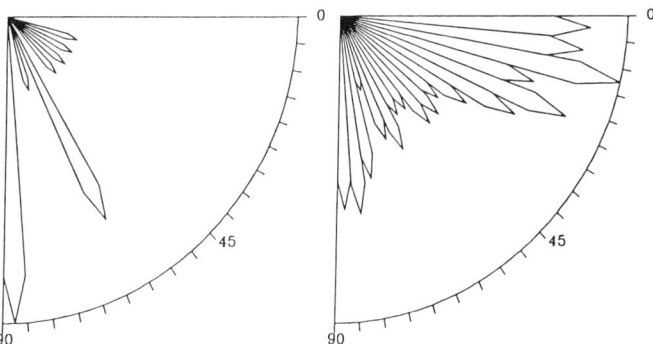

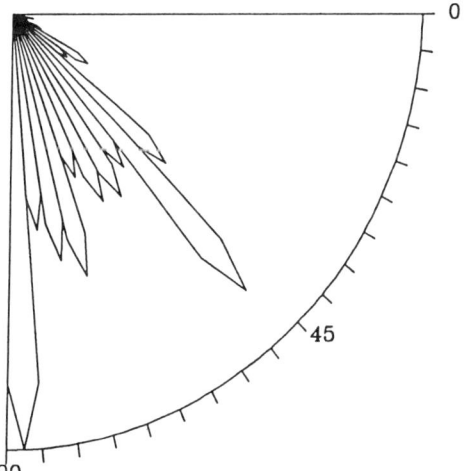

Figure 8. Rose diagram showing distribution of dip angles (in degrees) of Great Basin focal-mechanism nodal planes. The source regions for these data are shown in greater detail in Figure 9 below. We have not attempted to identify the preferred fault planes; thus, two dip angles per focal mechanism are included. The rose petal *lengths* (not areas) are proportional to the number of nodal plane dip angles in each 5° interval.

Figure 9. Rose diagrams showing distribution of rake angles (modified to lie in the range 0° to 90°) for Great Basin focal-mechanism nodal planes (same data as in Figs. 4 and 6). Figure 9(A) shows rake angle distribution for nodal planes having dip <25° and 9(B) shows rake angle distribution for nodal planes having dip ≥25°, respectively. Predominantly strike-slip mechanisms compose the majority of the data. For shallowly dipping nodal planes (of which there are only 12), normal slip predominates.

Stewart (1988) suggests that in the Walker Lane belt the coexistence of subparallel late Cenozoic normal- and strike-slip faults is best accounted for by temporally alternating extensional-mode and strike-slip mode stress systems having a common least principal stress orientation. In some subsections, northeast- and northwest-trending strike-slip zones are the result of conjugate faulting during periods of horizontal greatest principal stress orientation. Elsewhere, some northwest-trending strike-slip faults bound extended areas. In periods during which the greatest principal stress is vertical, extensional block faulting predominates on north- to northeast-trending structures; large-scale extension along northwest trends is also possible where right steps occur in northwest-trending right-lateral faults. Both Stewart (1988) and Hill (1982) account for the limited length of strike-slip faults in essentially the same manner. That is, these faults are principally the boundaries of rigid and/or extending terranes in a compressional (north-south directed) environment.

It appears that contemporary strike-slip deformation, which on the basis of seismic data seems to occur primarily along northerly trending faults, differs substantially in orientation and geographic location from neotectonic faulting. Wright (1976) notes that Cenozoic strike-slip faults are most frequently observed in the boundary zones of the Great Basin. Although the western boundary zone may be large, as noted by Stewart (1988) in his description of the Walker Lane belt, seismicity data indicate that contemporary strike slip is observed outside the Walker Lane within the CNSB, across the southern Great Basin, and as far north and east as Hansel Valley, Utah.

The discrepancies in fault orientations and slip directions inferred from seismic and geologic data, where they occur, are poorly understood. The lack of identified north-trending strike-slip faults may indicate that such faults are deep-seated and hidden or that total slip is limited to the extent that it is not readily visible at the surface. On the other hand, Slemmons (1967) has noted that many of the Pliocene and Quaternary northwest- to north-south–trending faults exhibit components of dextral slip, and likewise many of the north-south– to northeast-trending faults exhibit components of sinistral slip. Slemmons (1967) notes that this pattern of lateral faulting, in groups of faults trending northeasterly that form en echelon northerly elongations, is widespread across the Great Basin.

For some subregions of the Great Basin, evidence has been found suggesting that young wrench faulting may be obscured by one or more overlying detachments (Hardyman, 1978; Molinari, 1984). Their geologic data support a model in which the initial stage of deep-seated horizontal shear may act to produce folding and sets of Reidel shears in a hypothesized detached upper-crustal plate. In this interpretation, observed surface displacements and block faulting are a response to deep-seated lateral shear. It is also possible, of course, that differences in geologic and seismic data are due to lack of detailed geologic investigations throughout the Great Basin of the kind conducted by Molinari (1984), Angelier and others (1985), and Anderson and Barnhard (1987).

One exception to the paucity of observable north-trending dextral faults is the faulting associated with the Owens Valley 1872 earthquake. Dextral slip appears to be the primary slip sense in this earthquake (Beanland and Clark, 1987); nearby, however, spectacular young range-front faults bound the eastern side of Owens Valley, but these faults were not activated by this event. The 1872 earthquake ruptured about 100 km of a north- to north-northwest–trending fault zone on the floor of Owens Valley with 4 to 10 m of displacement. Some secondary faults, such as the Lone Pine fault, had 1 to 2 m of vertical rupture (Lubetkin and Clark, 1987a). This example demonstrates that major contemporary strike-slip and late Quaternary normal faulting events can coexist within small subregions of the Great Basin (Zoback, 1989).

Another means of coping with discordance between contemporary and neotectonic deformation styles, is to argue that the region is presently subjected to a short-lived regional-stress field (Eaton and others, 1978). In principal, either the orientation or magnitudes of the principal stresses may exhibit temporal variation. Stress changes within the Tertiary have been inferred from the geologic record in selected locales (i.e., Donath, 1962; Wright, 1976; Anderson and Ekren, 1977; Angelier and others, 1985; Frizzell and Zoback, 1987), or for the Great Basin (Zoback and others, 1981; Stewart, 1988), lending credence to this possibility. Zoback and Beanland (1986a, b) and Zoback (1989) suggest that large fluctuations in the relative magnitudes of the principal stresses are required to account for the presence of both strike-slip and normal faulting in the Owens Valley–Sierran Front region in Pleistocene to Holocene time (Lubetkin and Clark, 1987b). Intraplate stress changes may be related to stress build-up and stress release along major faults of the continental plate margin and stress transfer to intervening faults, such as the Death Valley–Furnace Creek fault system, the Garlock fault, or other faults within the Walker Lane belt. In this way, contemporary intraplate deformation would be attributed to superposition of variable plate-boundary stresses with internal stresses such as back-arc extension (Zoback and others, 1981; Coney, 1987), gravitational collapse (Wernicke and others, 1988), or other mechanisms (Spencer and Chase, 1989); the relative influence of these two stresses as a function of time would be determined by factors such as the relative motion and coupling of the Pacific and North American plates.

Although it is possible to look to the plate boundary for an explanation of complexities and inconsistencies in the Great Basin, we favor the notion that forces internal to the province are responsible for contemporary deformation (Wernicke and others, 1987; Wernicke and others, 1988; Sonder and others, 1986; Jackson, 1987). For instance, the coexistence of strike-slip and normal faulting throughout the brittle crust (at least in the southern Great Basin) requires that the greatest horizontal stress increase with depth, leading to the conclusion that crustal stresses are consistent with a basal traction acting along the base of the crust or at the brittle-ductile boundary (McGarr, 1982; Rogers and others, 1987). Gumper and Scholz (1971) reached the same conclusion regarding deformation in this region, but on the basis

---

Figure 10. Map showing directions of average tension axes for focal-mechanism data collected from local network studies (Type I) or from large ($M > 6$) earthquake waveform inversion studies (Type II). Horizontal projections of average compression axes are shown for data sets in which predominantly strike-slip mechanisms constitute at least 50 percent of the data. Lengths of compression axes equal lengths of tension axes where strike-slip events make up ≥ 80 percent of the data; compression axes lengths equal half those of the tension axes when strike-slip events make up 50 to 80 percent of the data. Circles surrounding map are lower hemisphere, equal-area projections of P-axes (darkened circles) and T-axes (open circles) for mechanisms in selected regions, numbered on the map. Average T directions are computed from maximum eigen values (Watson statistics). Dashed circles are shown at inclinations of 25°, 45°, and 65°. Regions and references are: 1, Tahoe and Truckee region (Type I, Hawkins and others, 1986); 2, Dixie Valley–Fairview Peak (Type II, Doser, 1986, 1987); 3, Mono Lakes region (Type I, Vetter, 1984); 4, Fishlake–West Fishlake (Type I, Vetter, 1984); 5, Mammoth Lakes region (Type I, Vetter, 1984); 6, Mammoth Lakes mainshocks of May 25 and 27, 1980 (Type II, Sipkin, 1986); 7, Chalfant Valley 1986 earthquakes (Type II, Cockerham and Corbett, 1987); 8, southern Great Basin region (Type I, Rogers and others, 1987; Harmsen and Rogers, 1987); 9, eastern half of Garlock fault zone (Type I, Astiz and Allen, 1983); 10, Clover Mountains mainshock of 1966 (Type II, Wallace and others, 1983); 11, eastern Basin and Range and Wasatch Front (Type I, Arabasz and Julander, 1985); 12, Hansel Valley 1934 earthquakes (Type II, Doser, 1987); 13, T azimuth of only available focal mechanism for a southern Great Basin earthquake just west of the Death Valley fault trace (event 40, Figure 5); 14, Durrwood Meadows series of 1983 to 1984 (Type I, Jones and Dollar, 1986, their Table 2). The only regional data set showing predominantly normal-slip mechanisms is Durrwood Meadows in the Sierra Nevada, and thus not within the Great Basin.

Seismicity of Nevada and some parts of the Great Basin 173

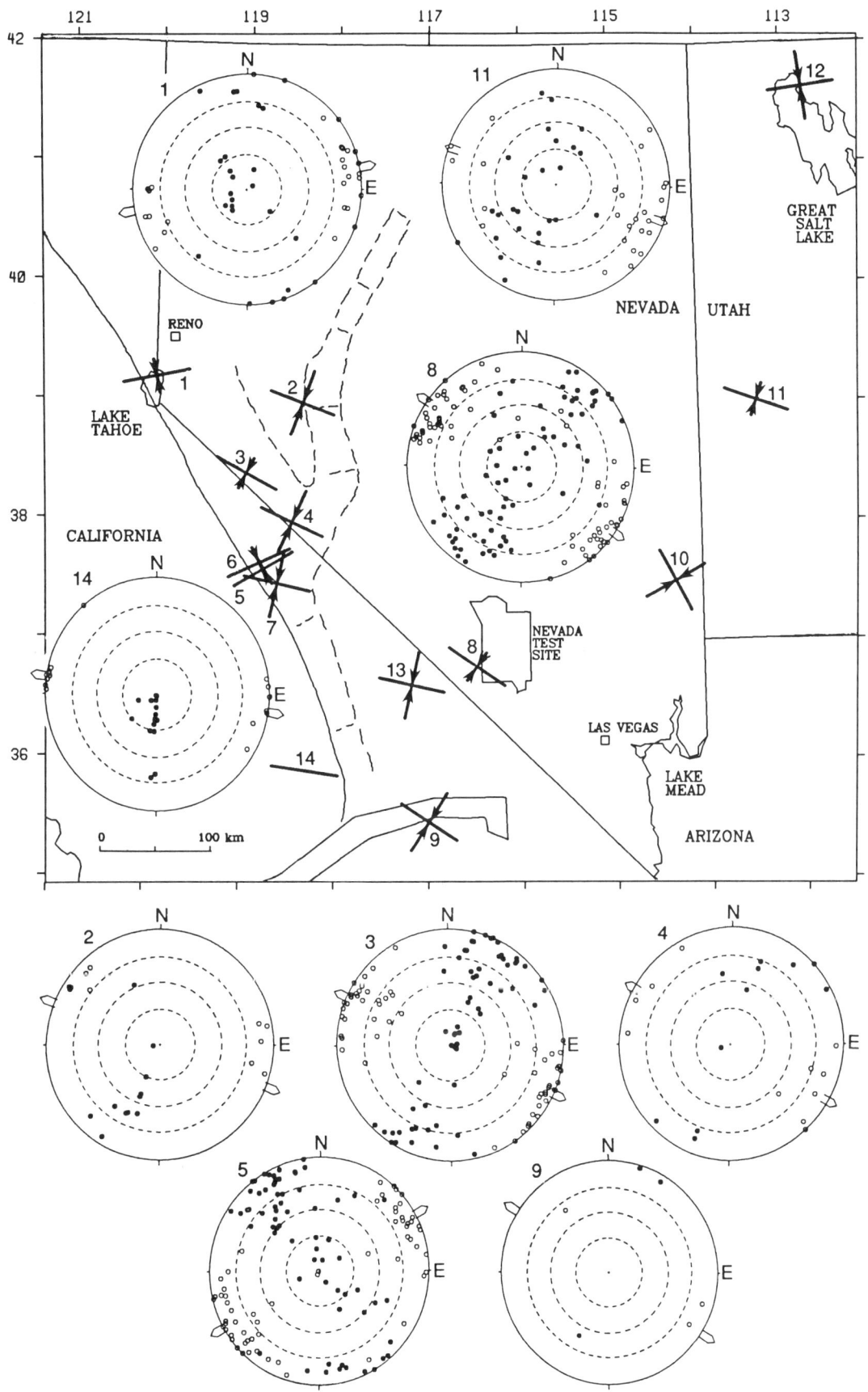

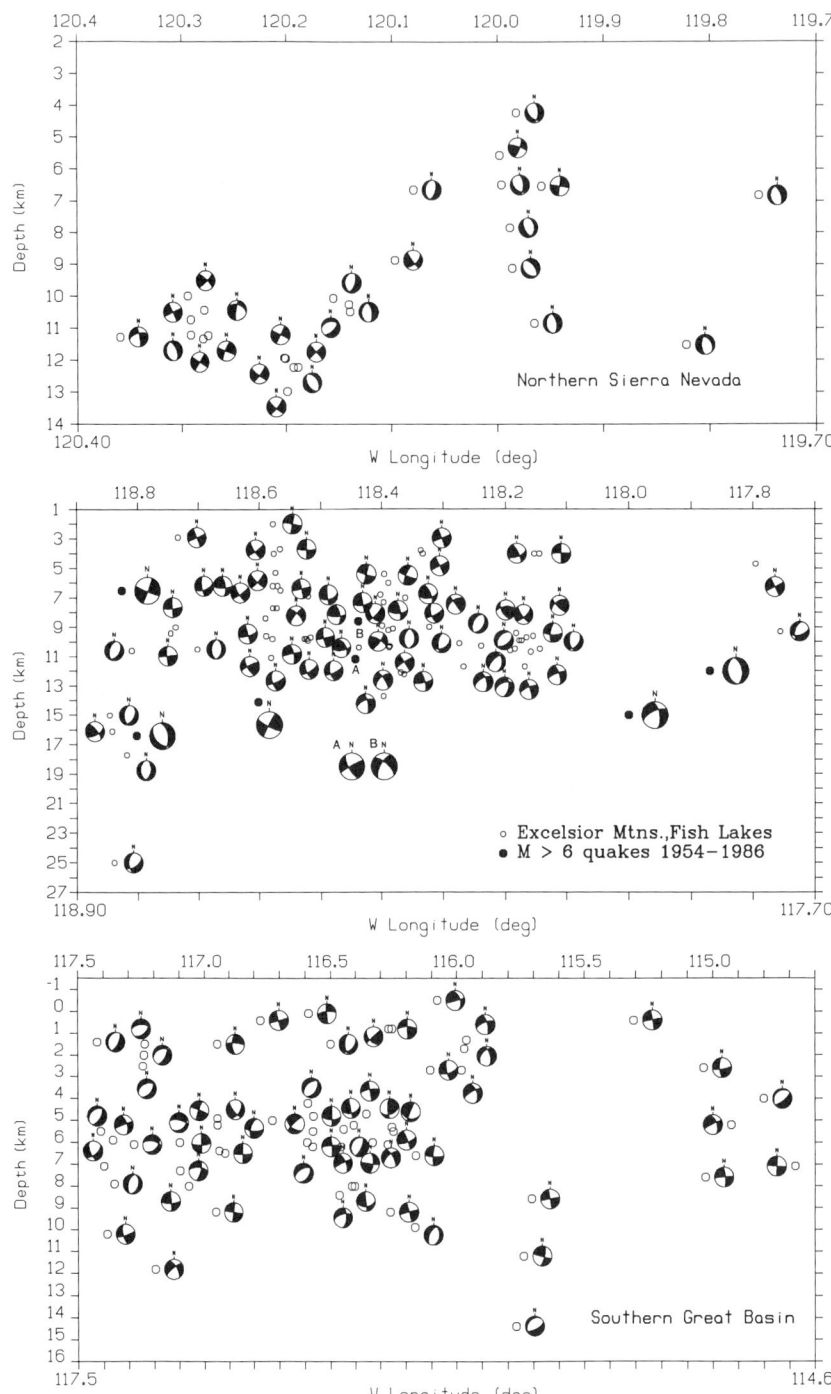

Figure 11. The east-west cross sections on which the focal mechanisms in Figures 6 and 7 have been projected. The focal hemispheres are lower hemisphere in plan view, shown at the focal depth. The larger focal spheres are events with M > 6. The solid dots labeled A and B are the locations of the events for focal mechanisms A and B. The events shown in the panels for the Northern Sierra region and the southern Great Basin are all less than magnitude 5.

of factors related to the temporal spread of volcanism, the location of the Nevada Seismic Zone, and crustal thickness between this zone and the eastern Sierra Nevada. Further, the slip direction for contemporary north-south strike slip is inconsistent with the slip direction that would be induced by remote plate-boundary influence (Atwater, 1970; Hill, 1982; Stewart, 1988), which could be taken as a contraindication for a plate-margin influence. Also, theoretical modeling of plate-boundary conditions by Sonder and others (1986) suggests that the Great Basin is too distant from the plate margin to be influenced significantly.

Some seismic (Doser, 1989; Arabasz and Julander, 1986) and geologic (Anderson and Barnhard, 1987) evidence exists for sinistral strike slip on northerly to northeasterly trending faults within the eastern Great Basin and possibly into the Colorado Plateau (Anderson and Christiansen, in preparation). On the basis of geologic and seismic data in the Sevier Valley, Utah, region, Anderson and Barnhard (1987) inferred locally active southerly transport of thin-skinned crustal blocks (upper 5 to 6 km). Mixtures of normal faulting and slip-incompatible strike-slip faulting suggest block-boundary faulting rather than a response to remote stress. This deformation style is more compatible with locally variable basal tractions than with regionally applied remote stresses.

Zoback (1989) noted that the orientation of the least-principal stress is roughly westerly along both the Colorado Plateau–Great Basin boundary and the Sierra Nevada–Basin and Range boundary, whereas within the interior Great Basin this stress is oriented northwest to N60°W. While it may be difficult to explain the origin of the geographic dependence in orientation of the least principal stress, it is equally difficult to explain the identical orientation of this stress at both province boundaries without inferring an internal deformation mechanism. A process wherein deformation interior to the Great Basin interacts with northerly trending province boundaries seems most plausible.

In a speculative vein, we suggest that the seismic data support a model of internally driven, southerly directed, lateral transport of the brittle crust in the southern Great Basin, a concept first proposed by Anderson (1984) on the basis of geologic data. Rogers and others (1987) suggested that the seismic data from the southern Great Basin were also best explained by this model. Given the data reviewed here, we suggest extension of this hypothesis to include the central Great Basin as well. Cartoons depicting this concept are shown in Figure 12. The existence of dextral motion on a series of subparallel faults across the central Great Basin and the presence of mixed mode and sinistral slip along the Colorado Plateau–Great Basin boundary suggest transport of a large subsection of the brittle crust in this region to the south. Oblique faulting may also occur as a secondary manifestation of the processes described by Hardyman (1978) and Molinari (1984) and is fundamentally related to deep-seated lateral slip. In this model, lateral deformation may extend through previously active shallow detachments in some locales, but not in all cases. In these areas, previously formed detachments may partially decouple deep-seated lateral shear from the surface. Normal

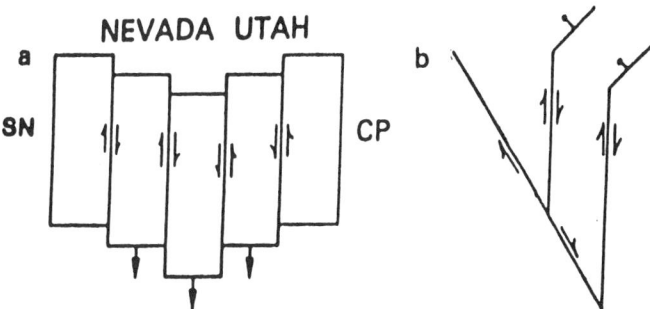

Figure 12. (a) Schematic diagram depicting dextral slip along north-trending faults in the southern (and possibly central) Great Basin, bounded by the Sierra Nevada block (SN), and sinistral slip along north-trending faults in the eastern margin of the Great Basin, bounded by the Colorado Plateau (CP). (b) Schematic diagram showing the relation between dextral slip on north-trending faults and normal slip on NNE-trending faults. The northwest-trending dextral slip fault could represent slip within the Walker Lane belt.

and oblique faulting may also occur on north-northeasterly striking faults as pull-apart or right-offsets in essentially north-trending strike-slip zones (Figs. 12b). In this model the scarcity of easily detectable northerly trending lateral faults could be attributed to limited duration of the stresses driving this process and limited total slip. On the western margin of this zone, where the transported blocks abut the Walker Lane, the dextral motion could have been taken up along the northwest trends of that zone in the past, although this is not presently the pattern. Based on the Owens Valley event, southerly directed transport would extend to the Sierra Nevada boundary. On the eastern margin of the zone, where the blocks adjoin the Colorado Plateau, one would expect predominantly sinistral motion along north- to southeast-trending faults. These suggestions are highly speculative given the quantity of data available, the simplicity of this model, and its shortcomings. For example, the model implies that west-northwest extension is presently inactive, yet satellite geodesy data suggest that the Great Basin is extending 9.7 mm/yr in the direction N56°W (Minster and Jordan, 1987). The model also implies kinematic inconsistencies in some locales that could be expected to produce compressional features such as folding and reverse faulting. As yet, this style of deformation has not been observed in the seismic record.

CONCLUSIONS

We find that seismic data from the Great Basin imply that substantial amounts of strike-slip and normal faulting are occurring at present. Furthermore, inconsistencies exist between the orientation and slip sense of faulting inferred from focal mechanisms and epicenter lineations compared with surface observations of historic fault rupture in the CNSB. Finally, although much of the inferred strike slip occurs as dextral slip on northerly trending faults, slip-incompatible dextral and sinistral slip has been noted on northerly to northeasterly trending faults near the

Colorado Plateau–Great Basin boundary. These data taken together are not wholly consistent with any previously derived models of Great Basin deformation. Although a model of southerly directed lateral transport may be most consistent with seismic slip styles and orientation, this model is also lacking in some aspects. Regardless of the adequacy of this model, the seismic data for both large and small earthquakes appear to call for some revision of hypotheses concerning contemporary deformation in the Great Basin.

Continued detailed studies of the seismicity of the Great Basin are required to help improve our understanding of deformation processes. For example, studies using high station densities centered in active zones would provide data to better define the geometry of active faults. Accurate estimates of earthquake depth are required to determine the vertical extent of strike-slip faulting; the relation between strike-slip faulting and multiple levels of normal and detachment faults that may exist within the crust is of primary importance for complete understanding of Great Basin deformation. Data from dense seismic networks would also permit more detailed study of the velocity structure of the upper crust, a factor that could act to constrain some deformation models. New geodetic data are also needed that span only the Great Basin and that are independent of deformation in other tectonic provinces. Although seismic data alone are not sufficient to resolve questions concerning Great Basin tectonics, continued collection of these and other kinds of geophysical and geologic data should help to further restrict the number of potential models.

Seismicity of Nevada and some parts of the Great Basin

TABLE A1. SELECTED GREAT BASIN FOCAL MECHANISMS 1972–1986

Origin time (UTC) Date	Time	North latitude	West longitude	Focal depth (km)	Mag	Region	Fig. 7 Index	Nodal plane Orientation St	Dp	Rk	Principal axes P Tr	Pl	T Tr	Pl	B Tr	Pl	Ref.
1978 0209	03:47	34.884	-118.771	9.3	3.0	Garlock Fault Zone	1	150.0	44.0	139.0	27.1	10.7	133.6	56.2	290.4	31.6	1
1979 0517	00:37	35.069	-118.454	6.5	2.5	Garlock Fault Zone	2	118.0	74.0	90.0	208.0	29.0	28.0	61.0	118.0	0.0	1
1981 0612	02:59	35.069	-118.425	6.5	2.5	Garlock Fault Zone	3	154.0	80.0	-110.0	41.4	51.0	260.6	32.1	157.6	19.7	1
1977 1028	23:44	35.340	-117.917	8.1	2.5	Garlock Fault Zone	4	243.0	85.0	10.0	17.2	3.5	107.9	10.6	269.3	78.8	1
1980 0306	07:45	35.573	-117.237	5.2	2.5	Garlock Fault Zone	5	257.0	85.0	15.0	30.5	6.9	122.2	14.1	275.0	74.2	1
1978 0717	14:46	35.525	-116.236	3.2	2.5	Garlock Fault Zone	6	264.0	87.0	-53.0	206.2	36.9	324.0	31.8	81.7	36.9	1
1983 0703	15:08	39.415	-120.202	11.9	4.2	Truckee, California	7	26.0	90.0	0.0	161.0	0.0	71.0	0.0	0.0	90.0	2
1983 0703	12:48	39.424	-120.199	13.0	3.7	Truckee, California	8	38.0	80.0	-12.0	353.9	15.5	84.2	1.3	178.8	74.4	2
1981 0429	11:55	39.301	-119.754	6.8	4.2	Virginia City, Nevada	9	183.0	33.0	-68.0	214.9	71.8	77.2	13.6	344.3	11.8	2
1977 0201	18:47	39.091	-119.965	10.8	4.2	Lake Tahoe	10	180.0	34.0	-80.0	236.0	77.3	82.8	11.4	351.7	5.6	2
1981 0507	01:02	37.926	-118.524	9.8	4.6	Mono Lake	11	169.0	76.0	184.0	32.7	12.6	124.3	7.1	242.9	75.5	3
1981 0809	15:52	37.618	-118.920	8.2	3.7	Mammoth Lakes	12	10.0	80.0	-10.0	325.9	14.1	235.9	0.1	145.4	75.9	3
1981 0809	16:01	37.619	-118.919	14.2	4.3	Mammoth Lakes	13	0.0	75.0	-65.0	300.0	53.3	70.6	25.8	173.1	24.1	3
1982 0430	19:42	38.173	-118.334	14.2	3.0	Excelsior Mtns.	14	158.0	90.0	180.0	23.0	0.0	113.0	0.0	0.0	90.0	3
1982 0924	07:40	37.848	-118.159	10.7	5.3	Fishlake	15	66.0	100.0	20.0	201.4	21.2	294.0	6.6	40.5	67.7	3
1982 1228	19:06	37.990	-118.383	9.1	5.0	West Fishlake	16	66.0	78.0	-34.0	19.9	32.3	118.9	13.9	228.9	54.2	3
1982 1229	02:30	37.988	-118.387	10.3	2.5	West Fishlake	17	184.0	38.0	-86.0	251.7	82.5	91.2	7.1	0.8	2.5	3
1983 0101	01:32	38.070	-118.577	4.0	3.1	Mono Lake	18	147.0	75.0	191.0	10.0	18.3	101.0	3.0	200.1	71.5	3
1983 0106	11:09	37.984	-118.397	7.3	3.8	Fishlake	19	46.0	82.0	-50.0	352.0	39.4	105.6	26.0	219.3	39.5	3
1983 0117	23:20	37.936	-118.145	4.0	3.7	Fishlake	20	90.0	100.0	0.0	225.4	7.1	134.6	7.1	0.0	80.0	3
1979 0813	18:23	37.238	-115.029	7.6†	2.7	Pahranagat (SGB)	21	355.0	80.0	-177.0	219.3	9.2	310.1	5.0	68.2	79.6	4
1979 1225	14:17	37.288	-117.062	8.0†	2.8	Sarcobatus Flat (SGB)	22	88.0	74.0	3.0	43.7	9.2	311.5	13.3	167.2	73.7	4
1982 0706	02:10	37.698	-115.037	2.6†	3.2	Hancock Summit (SGB)	23	177.5	69.0	-167.0	38.7	23.7	131.4	6.0	234.7	65.5	4
1983 0102	16:32	36.502	-116.569	5.5†	2.6	Rock Valley (SGB)	24	9.1	60.0	3.0	49.2	18.8	311.1	22.7	175.0	59.9	4
1985 0113	22:21	36.366	-115.769	14.4†	2.5	Mt. Sterling (SGB)	25	221.0	27.0	249.0	354.5	68.4	146.6	19.3	239.9	9.4	5
1985 0120	18:40	36.617	-116.444	5.4†	2.3	Lathrop Wells (SGB)	26	75.5	50.2	-4.2	39.6	29.5	294.8	24.4	172.0	50.0	5
1985 0515	10:29	37.455	-115.309	0.5†	2.3	Hancock Summit (SGB)	27	260.0	87.0	-3.8	215.1	4.8	305.1	0.6	41.8	85.2	5
1985 0530	05:46	37.113	-116.262	5.5†	1.7	Nevada Test Site (SGB)	28	177.4	74.4	50.3	296.0	19.5	47.2	45.6	190.0	38.0	5
1985 1212	11:57	36.860	-116.726	5.0†	1.5	Bare Mountain (SGB)	29	92.7	65.6	-32.7	53.5	39.8	146.1	3.2	239.9	50.0	5
1986 0116	15:24	36.873	-115.984	1.7†	2.0	Nevada Test Site (SGB)	30	220.5	44.8	315.0	204.5	58.7	100.1	8.6	5.1	29.9	5
1986 0217	20:29	37.020	-116.464	8.4†	1.5	Timber Mountain (SGB)	31	255.0	60.0	325.0	219.7	44.8	129.6	0.1	39.5	45.2	5
1986 0306	20:16	37.160	-117.359	8.0†	2.6	Ubehebe Crater (SGB)	32	15.0	40.0	270.0	105.0	85.0	285.0	5.0	15.0	0.0	5
1986 0306	20:29	37.158	-117.366	6.0†	1.8	Ubehebe Crater (SGB)	33	341.0	60.0	183.0	201.1	22.7	299.2	18.8	65.0	59.9	5

177

TABLE A1. GREAT BASIN FOCAL MECHANISMS 1972–1986 (continued)

Origin time (UTC)		North latitude	West longitude	Focal depth (km)	Mag	Region	Fig. 7 Index	Nodal plane Orientation			Principal axes						Ref.
Date	Time							St	Dp	Rk	P		T		B		
											Tr	Pl	Tr	Pl	Tr	Pl	
1986 0604	15:07	37.345	-117.236	1.5†	2.8	Scottys Junction (SGB)	34	266.0	52.8	-64.6	235.7	69.4	338.3	4.7	70.0	20.0	5
1986 0702	08:10	36.599	-116.407	5.2†	2.6	Lathrop Wells (SGB)	35	244.1	85.8	-24.7	197.3	20.3	292.6	14.1	55.1	65.0	5
1986 1206	09:20	36.462	-116.165	9.9†	2.5	Amargosa Flat (SGB)	36	206.9	55.6	-77.9	154.4	75.9	288.2	9.8	20.0	10.0	5
1977 1231	03:43	38.899	-117.984	12.4	3.1	CNSB	37	31.0	40.0	-85.0	85.1	84.0	297.5	5.1	207.2	3.2	3
1978 0215	09:25	39.552	-118.442	12.0	4.0	CNSB	38	64.0	40.0	-48.0	57.7	61.6	305.1	11.8	209.4	25.5	3
1980 0926	06:07	39.471	-118.133	11.8	3.2	CNSB	39	66.0	62.0	-132.0	284.6	52.7	184.5	7.6	88.9	36.2	3
1986 0708	03:02	36.645	-117.195	11.8†	2.4	Death Valley (SGB)	40	232.0	77.8	-27.6	187.2	28.0	282.5	9.8	30.0	60.0	5
1972 1020	23:16	36.093	-114.678	6.2	1.6	Lake Mead (SGB)	41	94.0	89.9	0.0	49.0	0.1	319.0	0.1	184.0	89.9	6
1972 1209	06:46	36.042	-114.802	3.1	2.6	Lake Mead (SGB)	42	52.0	70.0	-90.0	322.0	65.0	142.0	25.0	52.0	0.0	6

Notes: This table presents the data for selected focal mechanisms shown in Figure 7 for earthquakes recorded by various Great Basin seismograph networks. The mechanisms are based on P-wave first-motion studies. Focal depth is relative to the surface except where a dagger (†) follows the depth estimate, in which case depth is relative to sea level or 0 km. Magnitude (Mag), indicating the size of the earthquake, is ML (local) or MD (duration). The symbols are: St = strike of nodal plane, Rk = rake of slip vector; principal axes are pressure (P), tension (T), and null (B); Tr = trend of axis, Pl = plunge of axis. No inferred fault planes for these focal mechanisms are suggested in this table; however, only one of the two nodal planes is indicated above. References: 1. Astiz and Allen (1983); 2. Hawkins and others (1986); 3. Vetter and Ryall (1983); 4. Rogers and others (1987); 5. Harmsen and Rogers (1987); 6. Rogers and Lee (1976). (SGB) = earthquakes in the Southern Great Basin.

TABLE A2. SELECTED FOCAL MECHANISMS FROM THE LARGEST GREAT BASIN EARTHQUAKES

Origin time (UTC) Date Time	North latitude	West longitude	Focal depth (km)	Mag (ML)	Fig 6 Index	Nodal plane Orientation St	Dp	Rk	Principal axes P Tr	Pl	T Tr	Pl	B Tr	Pl	Ref.
1932 1221 6:10a	38.68	-118.21	14	7.2	1	1.0	72.0	-176.0	224.0	15.4	316.7	9.9	78.2	71.6	2
1932 1221 6:10b	38.46	-117.97	12	6.2	2	347.0	81.0	-179.0	211.6	7.1	302.3	5.6	70.6	80.9	2
1934 0130 20:16	38.26	-118.46	14	6.3	3	248.0	40.0	-91.0	346.1	85.0	158.7	5.0	248.8	0.6	2
1934 0312 15:05	41.658	-112.795	9	6.6	4	42.0	90.0	-12.0	356.4	8.5	87.6	8.5	222.0	78.0	3
1934 0312 18:20	41.571	-112.745	3	6.1	5	25.0	85.0	-29.0	337.5	23.8	74.9	16.3	196.1	60.6	3
1954 0706 11:13	39.29	-118.36	10	6.8	6	336.0	78.0	-140.0	203.8	36.3	100.6	17.3	349.9	48.5	1
1954 0706 22:07	39.20	-118.40	8	6.4	7	348.0	70.0	-157.0	209.2	30.1	118.6	1.0	26.9	59.9	1
1954 0824 05:51	39.35	-118.34	12	6.8	8	355.0	50.0	-145.0	200.9	49.9	300.8	8.3	37.6	38.9	1
1954 0831 22:20	39.72	-118.47		5.8	9	22.0	55.0	-160.0	235.5	37.1	334.8	12.0	79.6	50.3	1
1954 0901 05:18	39.70	-118.40		5.5	10	20.0	70.0	-42.0	336.3	43.4	77.5	11.7	179.2	44.3	1
1954 1216 11:07	39.20	-118.00	15	7.3	11	350.0	60.0	-160.0	206.5	34.3	302.2	8.3	43.9	54.5	1
1954 1216 11:11	39.67	-117.87	12	6.9	12	350.0	50.0	-90.0	260.0	85.0	80.0	5.0	170.0	0.0	1
1959 0323 07:10	39.43	-117.99		6.3	13	2.0	45.0	178.0	218.2	28.9	327.7	31.2	94.8	45.0	1
1959 0623 14:35	38.92	-118.89		6.3	14	310.0	68.0	-138.0	172.3	44.7	72.2	10.0	332.6	43.6	1
1966 0816 18:02	37.395	-114.206	7.0	6.0	15	17.0	80.0	190.0	241.1	14.1	331.1	0.1	61.6	75.9	4
1980 0525 16:33	37.589	-118.826	6.5	6.5	16	21.0	87.0	-6.0	336.0	6.4	66.3	2.1	174.5	83.3	5
1980 0527 14:50	37.470	-118.802	16.4	6.3	17	135.0	42.0	-118.0	316.6	70.6	64.5	6.2	156.6	18.3	5
1984 0112 03:18	37.456	-118.602	14.1	6.2	18	207.7	86.5	10.2	342.0	4.7	72.8	9.7	226.4	79.2	6
1986 0720 14:29	37.540	-118.439	8.6	6.2	19	205.0	75.0	-20.0	171.2	34.7	80.1	1.6	347.8	55.2	7
1986 0721 14:42	37.540	-118.442	11.2	6.6	20	155.0	60.0	-180.0	14.1	10.3	107.7	19.0	257.2	68.2	7

Notes: This table presents the data for the selected focal mechanisms shown in Figure 6 for Great Basin earthquakes having magnitude ≥5.5. These mechanisms were computed using waveform inversion techniques (except events 19 and 20, which are computed from P-wave first-motion polarities). St = strike of nodal plane, Dp = dip of nodal plane, Rk = rake of slip vector; principal axes are Pressure (p), Tension (T), and Null (B); Tr = trend of axis, Pl = plunge of axis; Mag = local magnitude. Nodal planes: no inferred fault planes for these focal mechanisms are suggested in this table; however, only one of the two nodal planes is indicated above. Ref = references: 1. Doser (1986); 2. Doser (1988); 3. Doser (1989); 4. Smith and Lindh (1978); 5. Sipkin (1986); 6. Barker and Wallace (1986); 7. Cockerham and Corbett (1987).

REFERENCES CITED

Aki, K., and Tsai, Y., 1972, Mechanism of Love-wave excitation by explosive sources: Journal of Geophysical Research, v. 77, p. 1452–1475.

Aki, K., Reasenberg, P., DeFazio, T., and Tsai, Y., 1969, Near-field and far-field seismic evidences for triggering of an earthquake by the Benham explosion: Seismological Society of America Bulletin, v. 59, no. 6, p. 2197–2207.

Algermissen, S. T., Perkins, D. M., Thenhaus, P. C., Hanson, S. L., and Bender, B. L., 1982, Probabilistic estimates of maximum acceleration and velocity in rock in the contiguous United States: U.S. Geological Survey Open-File Report 82-1033, 99 p.

Anderson, D. L., 1971, The San Andreas fault: Scientific America, v. 225, p. 52–67.

Anderson, L. W., and Hawkins, F. F., 1984, Recurrent Holocene strike-slip faulting, Pyramid Lake fault zone, western Nevada: Geology, v. 12, p. 681–684.

Anderson, R. E., 1971, Thin-skinned distension in Tertiary rocks of southeastern Nevada: Geological Society of America Bulletin, v. 82, p. 43–58.

——, 1973, Large magnitude late Tertiary strike-slip faulting north of Lake Mead, Nevada: U.S. Geological Survey Professional Paper 794, 18 p.

——, 1984, Strike-slip faults associated with extension in and adjacent to the Great Basin: Geological Society of America Abstracts with Programs, v. 16, p. 429.

Anderson, R. E., and Barnhard, T. P., 1987, Neotectonic framework of the central Sevier Valley area, Utah, and its relationship to seismicity, *in* Gori, P. L., and Hays, W. W., eds., Assessment of regional hazards and risk along the Wasatch Front, Utah: U.S. Geological Survey Open-File Report 87-585, p. F1–134.

Anderson, R. E., and Ekren, E. B., 1977, Late Cenozoic fault patterns and stress fields in the Great Basin and westward displacement of the Sierra Nevada block: Geology, v. 5, p. 388–389.

Anderson, R. E., Zoback, M. L., and Thompson, G. A., 1983, Implications of selected subsurface data on the structural forms and evolution of some basins in the northern Basin and Range Province, Nevada and Utah: Geological Society of America Bulletin, v. 94, p. 1055–1072.

Angelier, J., Colletta, B., and Anderson, R. E., 1985, Neocene paleostress changes in the Basin and Range; A case study of Hoover Dam, Nevada-Arizona: Geological Society of America Bulletin, v. 96, p. 347–361.

Arabasz, W. J., and Julander, D. R., 1986, Geometry of seismically active faults and crustal deformation within the Basin and Range–Colorado Plateau transition in Utah, *in* Mayer, L., ed., Extensional tectonics of the southwestern United States; A perspective on processes and kinematics: Geological Society of American Special Paper 208, p. 43–74.

Astiz, L., and Allen, C. R., 1983, Seismicity of the Garlock fault, California: Seismological Society of America Bulletin, v. 73, p. 1721–1734.

Atwater, T., 1970, Implications of plate tectonics for the Cenozoic tectonic evolution of western North America: Geological Society of America Bulletin, v. 81, p. 3513–3536.

Baker, M. R., and Doser, D. I., 1988, Joint inversion of regional and teleseismic waveforms: Journal of Geophysical Research, v. 93, p. 2037–2045.

Barker, J. S., and Wallace, T. C., 1986, A note on the teleseismic body waves for the 23 November 1984 Round Valley, California, earthquake: Seismological Society of America Bulletin, v. 76, p. 883–888.

Bateman, P. C., 1961, Willard D. Johnson and the strike-slip component of fault movement in the Owens Valley, California, earthquake of 1872: Seismological Society of America Bulletin, v. 51, no. 4, p. 483–493.

Beanland, S., and Clark, M. M., 1987, The Owens Valley fault zone, eastern California, and surface rupture associated with the 1872 earthquake: Seismological Resarch Letters, v. 58, p. 32.

Beck, P. J., 1970, The southern Nevada–Utah border earthquakes, August to December, 1966 [M.S. thesis]: Salt Lake City, University of Utah, 62 p.

Bender, B., 1983, Maximum likelihood estimation of b values for magnitude grouped data: Seismological Society of America Bulletin, v. 73, p. 831–851.

Boucher, G., Seeber, L., Ward, P., and Oliver, J., 1967, Microaftershock observations at the epicenter of a moderate-sized earthquake in Nevada: EOS Transactions of the American Geophysical Union, v. 48, p. 205.

Boucher, G., Ryall, N., and Jones, A. E., 1969, Earthquakes associated with underground nuclear explosions: Journal of Geophysical Research, v. 74, p. 3809–3820.

Bucknam, R. C., 1969, Geologic effects of the Benham underground nuclear explosion, Nevada Test Site: Seismological Society of America Bulletin, v. 59, p. 2209–2220.

Bucknam, R. C., Algermissen, S. T., and Anderson, R. E., 1980, Patterns of late Quaternary faulting in western Utah and an application in earthquake hazard evaluation, *in* Anderson, R. E., Ryall, A., and Smith, R. A., eds., Earthquake hazards along the Wasatch and Sierra-Nevada frontal fault zones: U.S. Geological Survey Open-File Report 80-801, p. 299–314.

Byerly, P., and 5 others, 1956, The Fallon-Stillwater earthquakes of July 6, 1954, and August 23, 1954: Seismological Society of America Bulletin, v. 46, p. 1–40.

Callaghan, E., and Gianella, V., 1935, The earthquake of January 30, 1934, at Excelsior Mountains, Nevada: Seismological Society of America Bulletin, v. 47, p. 327–334.

Cloud, W. K., 1957, Intensity distribution and strong-motion seismograph results, Nevada earthquakes of December 16, 1954: Seismological Society of America Bulletin, v. 47, p. 327–334.

Cockerham, R. S., and Corbett, E. J., 1987, The July 1986 Chalfant Valley, California, earthquake sequence; Preliminary results: Seismological Society of America Bulletin, v. 77, p. 280–289.

Coffman, J. L., and von Hake, C. A., 1973, Earthquake history of the United States: U.S. Department of Commerce, National Oceanic and Atmospheric Administration NOAA EDS Publication 41-1/COM 74 500 35, 108 p.

Coney, D. J., and Harms, T. A., 1984, Cordilleran metamorphic core complexes; Cenozoic extensional relics of Mesozoic compression: Geology, v. 12, p. 550–554.

Coney, P. J., 1987, The regional tectonic setting and possible causes of Cenozoic extension in the North American Cordillera, *in* Coward, M., and others, eds., Continental extensional tectonics: Geological Society of London Special Publication 28, p. 177–186.

dePolo, C. M., and Ramelli, A. R., 1987, Preliminary report on surface fractures along the White Mountains fault zone associated with the July 1986 Chalfant Valley earthquake sequence: Seismological Society of America Bulletin, v. 77, p. 290–296.

dePolo, C. M., Bell, J. W., and Ramelli, A. R., 1987, Geometry of strike-slip faulting related to the 1932 Cedar Mountain earthquake, central Nevada: Geological Society of America Abstracts with Programs, v. 19, p. 371.

Dewey, J. W., and Spence, W., 1979, Seismic gaps and source zones of recent large earthquakes in coastal Peru: Pure and Applied Geophysics, v. 117, p. 1148–1171.

Donath, F. A., 1962, Analyses of basin-range structure, south-central Oregon: Geological Society of America Bulletin, v. 73, p. 1–16.

Doser, D. I., 1986, Earthquake processes in the Rainbow Mountain–Fairview Peak–Dixie Valley, Nevada, region (1954–1959): Journal of Geophysical Research, v. 91, p. 12572–12586.

——, 1987, Modeling Pn1 waveforms of the Fairview Peak–Dixie Valley, Nevada, U.S.A., earthquake sequence (1954–1959): Physics of the Earth and Planetary Interiors, v. 48, p. 64–72.

——, 1988, Source mechanisms of earthquakes in the Nevada seismic zone (1915–1943) and implications for deformation in the western Great Basin: Journal of Geophysical Research, v. 93, p. 15001–15015.

——, 1989, Extensional tectonics in northern Utah, U.S.A.; The 1934 Hansel Valley sequence: Physics of the Earth and Planetary Interiors, v. 54, p. 120–143.

Doser, D. I., and Smith, R. B., 1985, Source parameters of the 28 October 1983, Borah Peak, Idaho, earthquake from body wave analysis: Seismological Society of America Bulletin, v. 75, p. 1041–1051.

Douglas, B. M., and Ryall, A., 1972, Special characteristics and stress drop for microearthquakes near Fairview Peak, Nevada: Journal of Geophysical Research, v. 77, p. 351–359.

Eaton, G. P., Wahl, R. R., Prostka, H. J., Mabey, D. R., and Kleinkopf, M. D., 1978, Regional gravity and tectonic patterns; Their relation to late Cenozoic epeirogeny and lateral spreading in the western Cordillera, in Smith, R. B., and Eaton, G. P., eds., Cenozoic tectonics and regional geophysics of the western Cordillera: Geological Society of America Memoir 152, p. 51–91.

Eddington, P. K., Smith, R. B., and Renggli, C., 1987, Kinematics of Basin and Range intraplate extension, in Coward, M., and others, eds., Continental extensional tectonics: Geological Society of London Special Publication 28, p. 371–392.

Fara, H. D., 1964, A new catalog of earthquake fault plane solutions: Seismological Society of America Bulletin, v. 54, p. 1491–1518.

Fischer, F. G., Papanek, P. J., and Hamilton, R. M., 1972, The Massachusetts Mountain earthquake of 5 August 1971 and its aftershocks, Nevada Test Site: U.S. Geological Survey Report USGS-474-149; available only from U.S. Department of Commerce National Technical Information Service, Springfield, Virginia 22161 as Report NTS-238, 20 p.

Frizzell, V. A., and Zoback, M. L., 1987, Stress orientation determined from fault slip data in Hampel Wash area, Nevada, and its relation to contemporary regional stress field: Tectonics, v. 6, p. 89–98.

Gawthrop, H. W., and Carr, W. J., 1988, Location refinement of earthquakes in the southwestern Great Basin, 1931–1974, and seismotectonic characteristics of some of the important events: U.S. Geological Survey Open-File Report 88-560, 64 p.

Gianella, V. P., and Callaghan, E., 1934a, The earthquake of December 20, 1932, at Cedar Mountain, Nevada, and its bearing on the genesis of Basin-Range structure: Journal of Geology, v. 42, p. 1–22.

—— , 1934b, The Cedar Mountains, Nevada, earthquake of December 20, 1932: Seismological Society of America Bulletin, v. 24, p. 345–384.

Gross, W. K., and Savage, J. C., 1987, Deformation associated with the 1986 Chalfant Valley earthquake, eastern California: Seismological Society of America Bulletin, v. 77, p. 306–310.

Gumper, F. J., and Scholz, C., 1971, Microseismicity of the Nevada seismic zone: Seismological Society of America Bulletin, v. 61, p. 1413–1432.

Hamilton, R. M., and Healy, J. H., 1969, Aftershocks of the Benham nuclear explosion: Seismological Society of America Bulletin, v. 59, p. 2271–2281.

Hamilton, R. M., Smith, B. E., Fischer, F. G., and Papanek, P. J., 1971, Seismicity of the Pahute Mesa area, Nevada Test Site, 8 December 1968 through 31 December 1971: U.S. Geological Survey Report USGS-474-138, 170 p; available only from U.S. Department of Commerce, National Technical Information Service, Springfield, Virginia 22161.

Hamilton, W., and Myers, W. B., 1966, Cenozoic tectonics of the western United States: Review of Geophysics, v. 4, p. 509–549.

Hanks, T. C., and Kanamori, H., 1979, A moment magnitude scale: Journal of Geophysical Research, v. 84, p. 2348–2350.

Hardyman, R. F., 1978, Volcanic stratigraphy and structural geology of Gillis Canyon Quadrangle, northern Gillis Range, Mineral County, Nevada [Ph.D. thesis]: Reno University of Nevada, 248 p.

Harmsen, S. C., and Rogers, A. M., 1986, Inferences above the local stress field from focal mechanisms; Applications to earthquakes in the southern Great Basin of Nevada: Seismological Society of America, v. 76, p. 1560–1572.

—— , 1987, Earthquake location data for the southern Great Basin of Nevada and California; 1984 through 1986: U.S. Geological Survey Open-File Report 87-596, 92 p.

Hawkins, F. F., LaForge, R., and Hansen, R. A., 1986, Seismotectonic study of the Truckee/Lake Tahoe area, northeastern Sierra Nevada, California, for Stampede, Prosser Creek, Boca, and Lake Tahoe dams: U.S. Bureau of Reclamation Seismotectonic Report 85-4, 210 p.

Hill, D. P., 1982, Contemporary block tectonics; California and Nevada: Journal of Geophysical Research, v. 87, p. 5433–5450.

Hill, D. P., Wallace, R. E., and Cockerham, R. S., 1985, A review of evidence on the potential for major earthquakes and volcanism in the Long Valley–Mono Craters–White Mountains region of eastern California: Earthquake Prediction Research, v. 3, p. 571–594.

Holden, E. S., 1898, Catalogue of earthquakes on the Pacific coast 1769 to 1897: Smithsonian Institute Miscellaneous Collections, no. 1087, 253 p.

Hunt, C. B., and Mabey, D. R., 1966, General geology of Death Valley, California; Stratigraphy and structure: U.S. Geological Survey Professional Paper 494-A, p. A1–A165.

Jackson, J. A., 1987, Active normal faulting and crustal extension, in Coward, M., and others, eds., Continental extensional tectonics: Geological Society of London Special Publication 28, p. 3–17.

Jones, A. E., 1963, Catalog of Nevada earthquakes; Part 2, Study of Wiechert records at Reno, 1916–1950: Geological Society of America Special Paper 76, 206 p.

—— , 1975, Recording of earthquakes at Reno, 1916–1951, in Van Wormer, J. D., ed., Bulletin of the Seismological Laboratory: Reno, University of Nevada Seismological Laboratory Report, 199 p.

Jones, J. C., 1915, The Pleasant Valley, Nevada, earthquake of October 2, 1915: Seismological Society of America Bulletin, v. 5, p. 190–205.

Jones, L. M., and Dollar, R. S., 1986, Evidence of Basin-and-Range extensional tectonics in the Sierra Nevada; The Durrwood Meadows swarm, Tulare County, California (1983–1984): Seismological Society of America Bulletin, v. 76, p. 439–461.

Lay, T., Wallace, T. C., and Helmberger, D. V., 1984, The effects of tectonic release on short-period P-waves from NTS explosions: Seismological Society of America Bulletin, v. 74, p. 819–842.

Lubetkin, K. C., and Clark, M. M., 1987a, Late Quaternary activity along the Lone Pine fault, eastern California: Seismological Research Letters, v. 58, p. 32.

—— , 1987b, Late Quaternary fault scarp at Lone Pine, California; Location of oblique slip during the great 1872 earthquake and earlier earthquakes, in Hill, M. L., ed., Cordilleran Section of the Geological Society of America: Boulder, Colorado, Geological Society of America Centennial Field Guide 1, p. 151–156.

Maldonado, F., 1977a, Results from fault-monitoring stations on Pahute Mesa, Nevada Test Site, from July 1973 through December 1976: U.S. Geological Survey Report USGS-474-242, 32 p.; available only from U.S. Department of Commerce National Technical Information Service, Springfield, Virginia 22161.

—— , Composite postshot fracture map of Pahute Mesa, Nevada Test Site, June 1973 through March 1976: U.S. Geological Survey Report USGS-474-243, 8 p.; available only from U.S. Department of Commerce National Technical Information Service, Springfield, Virginia 22161.

Mark, R. K., 1977, Application of linear statistical models of earthquake magnitude versus fault length in estimating maximum expectable magnitude earthquakes: Geology, v. 5, p. 464–466.

Matthews, V., III, and Anderson, C. E., 1973, Yellowstone convection plume and break-up of the western United States: Nature, v. 243, p. 158–159.

McAdie, A. G., 1907, Catalog of earthquakes on the Pacific coast 1897–1906: Smithsonian Institute Miscellaneous Collections, no. 1721, 64 p.

McGarr, A., 1982, Analysis of states of stress between provinces of constant stress: Journal of Geophysical Research, v. 87, p. 9279–9288.

McKeown, F. A., 1975, Relation of geological structure to seismicity at Pahute Mesa, Nevada Test Site: Seismological Society of America Bulletin, v. 65, p. 747–764.

McKeown, F. A., and Dickey, D. D., 1969, Fault displacements and motion related to nuclear explosions: Seismological Society of America Bulletin, v. 59, no. 6, p. 2253–2269.

Mead, T. C., and Carder, D. S., 1941, Seismic investigation in the Boulder Dam area in 1939: Seismological Society of America Bulletin, v. 31, p. 321–324.

Meister, L. J., Burford, R. O., Thompson, G. A., and Kovach, R. L., 1968, Surface strain changes and strain energy release in the Dixie Valley–Fairview Peak area, Nevada: Journal of Geophysical Research, v. 73, p. 5981–5994.

Meremonte, M. E., and Rogers, A. M., 1987, Historical catalog of southern Great Basin earthquakes 1868–1978, U.S. Geological Survey Open-File Report 87-80, 203 p.

Miller, C. D., Crandell, D. R., Mullineaux, D. R., Hoblitt, R. P., and Bailey, R. A., 1982, Preliminary assessment of potential volcanic hazards in the Long Valley-Mono Lakes area, east-central California and southwestern Nevada: U.S. Geological Survey Open-File Report 82-583, 30 p.

Minster, J. B., and Jordan, T. H., 1987, Vector constraints on western U.S. deformation from space geodesy: Neotectonics and plate motion: Journal of Geophysical Research, v. 92, p. 4798–4804.

Molinari, M. P., 1984, Late Cenozoic geology and tectonics of Stewart and Monte Cristo valleys, west-central Nevada [M.S. thesis]: Reno, University of Nevada, 124 p.

Morris, R. H., 1971, Geologic effects, in Geologic and hydrologic effects of the Handley event, Pahute Mesa, Nevada Test Site: U.S. Geological Survey Report USGS-474-95, p. 7–26; available only from U.S. Department of Commerce National Technical Information Service, Springfield, Virginia 22161.

Nakata, J. K., Wentworth, C. M., and Machette, M. N., 1982, Quaternary fault map of the Basin and Range and Rio Grande Rift Provinces, western United States: U.S. Geological Survey Open-File Report 82-579, scale 1:2,500,000.

Oakeshott, G. B., Greensfelder, R. W., and Kahle, J. E., 1972, The Owens Valley earthquake of 1872; One hundred years later: California Geology, v. 25, p. 55–62.

Okaya, D. A., and Thompson, G. A., 1985, Geometry of Cenozoic extensional faulting; Dixie Valley, Nevada: Tectonics, v. 4, no. 1, p. 107–125.

Oliver, J., Ryall, A. Brune, J. N., and Slemmons, D. B., 1966, Microearthquake activity recorded by portable seismographs of high sensitivity: Seismological Society of America Bulletin, v. 56, p. 899–924.

Orkild, P. P., Sargent, K. A., and Snyder, R. P., 1969, Geologic map of Pahute Mesa, Nevada Test Site and vicinity, Nye County, Nevada: U.S. Geological Survey Miscellaneous Geologic Investigations Map I-567, scale 1:48,000.

Pacheco, J., and Nábělek, J., 1988, Source mechanisms of three moderate California earthquakes of July 1986, Seismological Society of America Bulletin, v. 78, no. 6, p. 1907–1929.

Page, B. M., 1935, Basin-Range faulting of 1915 in Pleasant Valley, Nevada: Journal of Geology, v. 43, p. 690–707.

Page, R., 1968, Focal depths of aftershocks: Journal of Geophysical Reseerch, v. 713, p. 3897–3903.

Papanek, P. J., and Hamilton, R. M., 1972, A seismicity study along the northern Death Valley-Furnace Creek fault zone, California-Nevada boundary: U.S. Geological Survey Report USGS-474-141, 41 p.; available only from U.S. Department of Commerce National Technical Information Service, Springfield, Virginia 22161.

Perkins, D. M., 1987, Contiguous fault rupture, probabilistic hazard, and contagion observability, in Crone, A. J., and Omdahl, E. M., eds., Directions in paleoseismology; Proceedings of Conference 39: U.S. Geological Survey Open-File Report 87-673, p. 428–439.

Real, C. R., Toppozada, T. R., and Parke, D. L., 1978, Earthquake epicenter map of California showing events from 1900 through 1974 equal to or greater than magnitude 4.0 or intensity V: California Division of Mines and Geology Map Sheet 39, scale 1:1,000,000.

Richins, W. D., and 5 others, 1985, The 1983 Borah Peak, Idaho, earthquake; Relationship of aftershocks to the mainshock, surface faulting, and regional tectonics, in Stein, R. S., and Bucknam, R. C., eds., Proceedings of Workshop 28 on the Borah Peak, Idaho, earthquake: U.S. Geological Survey Open-File Report 85-290A, p. 285–310.

Richter, C. F., 1958, Elementary seismology: San Francisco, California, W. H. Freeman, 768 p.

Rogers, A. M., and Lee, W.H.K., 1976, Seismic study of earthquakes in the Lake Mead, Nevada-Arizona region: Seismological Society of America Bulletin, v. 66, no. 5, p. 1657–1681.

Rogers, A. M., Woulett, G. M., and Covington, P. A., 1977, Seismicity of the Pahute Mesa area, Nevada Test Site, 8 October 1975 to 30 June 1976: U.S. Geological Survey Report USGS-474-184; available only from U.S. Department of Commerce National Technical Information Service, Springfield, Virginia 22161.

Rogers, A. M., Harmsen, S. C., Carr, W. J., and Spence, W. J., 1983, Southern Great Basin seismological data report for 1981 and preliminary data analysis: U.S. Geological Survey Open-File Report 83-669, 240 p.

Rogers, A. M., Harmsen, S. C., and Meremonte, M. E., 1987, Evaluation of the seismicity of the southern Great Basin and its relationship to the tectonic framework of the region: U.S. Geological Survey Open-File Report 87-408, 196 p.

Romney, C. F., 1957, The Dixie Valley-Fairview Peak, Nevada, earthquakes of December 16, 1954; Seismic waves: Seismological Society of America Bulletin, v. 47, p. 301–320.

Ron, H., Aydin, A., and Nur, A., 1986, The role of strike-slip faulting in the deformation of Basin and Range provinces [abs.]: EOS Transactions of the American Geophysical Union, v. 67, p. 358.

Ryall, A. S., 1977, Earthquake hazard in the Nevada region: Seismological Society of America Bulletin, v. 67, p. 517–532.

Ryall, A. S., and Douglas, B. M., 1974, Seismicity of northwest Nevada related to the feasibility of power plant siting: Reno, Nevada, Report to Sierra Pacific Power Company, 79 p.

Ryall, A. S., and Malone, S. D., 1971, Earthquake distribution and mechanism of faulting in the Rainbow Mountain-Dixie Valley-Fairview Peak area, central Nevada: Journal of Geophysical Research, v. 76, p. 7241–7248.

Ryall, A. S., and Priestly, K., 1975, Seismicity, secular strain, and maximum magnitude in the Excelsior Mountain area, western Nevada and eastern California: Geological Society of America Bulletin, v. 86, p. 1585–1592.

Ryall, A. S., and Ryall, F., 1981, Spatial-temporal variations in seismicity preceding the May 1980 Mammoth Lakes earthquakes, California: Seismological Society of America Bulletin, v. 71, p. 747–760.

——, 1983, Spasmodic tremor and possible magma injection in Long Valley caldera, eastern California: Science, v. 219, p. 1432–1433.

Ryall, A. S., and Savage, W. U., 1969, A comparison of seismological effects for the Nevada underground test Boxcar with natural earthquakes in the Nevada region: Journal of Geophysical Research, v. 74, p. 4281–4289.

Ryall, A. S., and Vetter, U. R., 1982, Seismicity related to geothermal development in Dixie Valley, Nevada: Reno, University of Nevada Final Report on USDOE Contract DE-AC08-79-NV10054, 102 p.

Ryall, A. S., Slemmons, D. B., and Gedney, L. D., 1966, Seismicity, tectonism, and surface faulting in the western United States during historic times: Seismological Society of America Bulletin, v. 156, p. 1105–1135.

Savage, J. C., and Clark, M. M., 1982, Magnetic resurgence in the Long Valley caldera, California: Possible cause of the 1980 Mammoth Lakes earthquakes: Science, v. 217, p. 531–533.

Savage, J. C., and Cockerham, R. S., 1987, Quasi-periodic occurrence of earthquakes in the 1978–1986 Bishop-Mammoth Lakes sequence, eastern California: Seismological Society of America Bulletin, v. 77, p. 1347–1358.

Savage, J. C., and Hastie, L. M., 1969, A dislocation model for the Fairview Peak, Nevada, earthquake: Seismological Society of America Bulletin, v. 59, p. 1937–1948.

Schwartz, D. P., and Coppersmith, K. J., 1984, Fault behavior and characteristic earthquakes; Examples from the Wasatch and San Andreas fault zones: Journal of Geophysical Research, v. 89, p. 5681–5698.

Shawe, D. E., 1965, Strike-slip control of Basin-Range structure indicated by historical faults in western Nevada: Geological Society of America Bulletin, v. 76, p. 1361–1378.

Sipkin, S. A., 1986, Interpretation of non-double-couple earthquake mechanisms derived from moment tensor inversion: Journal of Geophysical Research, v. 91, no. B1, p. 531–547.

Slemmons, D. B., 1957a, Geologic setting for the Fallon-Stillwater earthquakes of 1954: Seismological Society of America Bulletin, v. 47, p. 4–9.

——, 1957b, Geological effects of the Dixie Valley-Fairview Peaks, Nevada, earthquakes of December 16, 1954: Seismological Society America Bulletin, v. 47, p. 353–375.

——, 1967, Pliocene and Quaternary crustal movements of the Basin-and-Range Province, U.S.A.: Osaka City University Journal of Geosciences, v. 10, p. 91–103.

Slemmons, D. B., Steinbrugge, K. V., Tocher, D., Oakeshott, G. B., and Gianella, V. P., 1959, Wonder, Nevada, earthquake of 1903: Seismological Society of America Bulletin, v. 49, p. 251–256.

Slemmons, D. B., Jones, A. E., and Gimlett, J. I., 1965, Catalog of Nevada earthquakes, 1852–1960: Seismological Society of America Bulletin, v. 55, p. 537–583.

Smith, B. E., Hamilton, R. M., and Jackson, W. H., 1971, Seismicity of the central Nevada Test Area, 26 September 1969–1 October 1970: U.S. Geological Survey Report USGS-474-109, 27 p.; available only from U.S. Department of Commerce National Technical Information Service, Springfield, Virginia 22161.

Smith, B. E., Coakley, J. M., and Hamilton, R. M., 1972, Distribution, focal mechanisms, and frequency of earthquakes in the Fairview Peak area, Nevada: Seismological Society of America Bulletin, v. 62, p. 1223–1240.

Smith, K. D., and Priestley, K. F., 1988, The foreshock sequence of the 1986 Chalfont, California, earthquake: Seismological Society of America Bulletin, v. 78, p. 172–187.

Smith, R. B., 1978, Seismicity, crustal structure, and intraplate tectonics of the interior of the western Cordillera, in Smith, R. B., and Eaton, G. P., eds., Cenozoic tectonics and regional geophysics of the western Cordillera: Geological Society of America Memoir 152, p. 111–114.

Smith, R. B., and Bruhn, R. L., 1984, Intraplate extensional tectonics of the eastern basin range; Inferences on structural style from seismic reflection data, regional tectonics, and thermal-mechanical models of brittle-ductile deformation: Journal of Geophysical Research, v. 89, p. 5733–5762.

Smith, R. B., and Lindh, A. G., 1978, Fault plane solutions of the western United States; A compilation, in Smith, R. B., and Eaton, G. P., eds., Cenozoic tectonics and regional geophysics of the western Cordillera: Geological Society of America Memoir 152, p. 107–109.

Smith, R. B., and Richins, W. D., 1984, Seismicity and earthquake hazards of Utah and the Wasatch Front; Paradigm and paradox, in Hays, W. W., and Gori, P. L., eds., Evaluation of regional and urban earthquake hazards and risk in Utah; Proceedings of Conference 26: U.S. Geological Survey Open-File Report 84-763, p. 73–112.

Smith, R. B., Richins, W. D., and Doser, D. I., 1985, The 1983, Borah Peak, Idaho, earthquake; Regional seismicity, kinematics of faulting, and tectonic mechanism, in Stein, R. S., and Bucknam, R. C., eds., Proceedings of Workshop 28 on the Borah Peak, Idaho, Earthquake: U.S. Geological Survey Open-File Report 85-290, v. A, p. 236–263.

Snay, R. A., Cline, M. W., and Timmermann, E. L., 1985, Dislocation models for the 1954 earthquake sequence in Nevada: U.S. Geological Survey Open-File Report 85-290, p. 531–555.

Snyder, R. P., 1971, Composite postshot fracture map of Pahute Mesa, Nevada Test Site: U.S. Geological Survey Report USGS-474-100, 17 p.; available only from U.S. Department of Commerce National Technical Information Service, Springfield, Virginia 22161.

——, 1973, Recent fault movement on Pahute Mesa, Nevada Test Site, from May 1970 through June 1973: U.S. Geological Survey Report USGS-474-137, 32 p.; available only from U.S. Department of Commerce National Technical Information Service, Springfield, Virginia 22161.

Sonder, L. J., England, P. C., and Hoseman, G. A., 1986, Continuum calculations of continental deformation in transcurrent environments: Journal of Geophysical Research, v. 91, no. B5, p. 4797–4810.

Spencer, J. E., and Chase, C. G., 1989, Role of crustal flexure in initiation of low-angle normal faults and implications for structural evolution of the Basin and Range Province: Journal of Geophysical Research, v. 94, p. 1765–1775.

Stauder, W., and Ryall, A. S., 1967, Spatial distribution and source mechanism of microearthquakes in central Nevada: Seismological Society of America Bulletin, v. 57, p. 1317–1345.

Steinbrugge, K. V., and Moran, D. F., 1956, Damage caused by the earthquakes of July 6 and August 23, 1954: Seismological Society of America Bulletin, v. 46, p. 15–33.

Stewart, J. H., 1978, Basin-Range structure in western North America; A review, in Smith, R. B., and Eaton, G. P., eds., Cenozoic tectonics and regional geophysics of the western Cordillera: Geological Society of America Memoir 152, p. 1–31.

——, 1980, Regional tilt patterns of late Cenozoic Basin-Range fault blocks, western United States: Geological Society of America Bulletin, v. 91, p. 460–464.

——, 1988, Tectonics of the Walker Lane belt, western Great Basin; Mesozoic and Cenozoic deformation in a zone of shear, in Ernst, W. G., ed., Metamorphism and cause of evolution of the western United States, Ruby volume 7: Englewood Cliffs, New Jersey, Prentice-Hall, p. 683–713.

Tarr, A. C., and Rogers, A. M., 1986, Analysis of earthquake data recorded by digital field seismic systems, Jackass Flats, Nevada: U.S. Geological Survey Open-File Report 86-420, 67 p.

Thenhaus, P. C., and Barnhard, T. P., 1989, Regional termination and segmentation of Quaternary fault belts in the Great Basin, Nevada and Utah: Seismological Society of America Bulletin, v. 79, p. 1426–1438.

Tocher, D., 1956, Movement on the Rainbow Mountain fault: Seismological Society of America Bulletin, v. 46, p. 10–14.

——, 1957, The Dixie Valley–Fairview Peak earthquakes of December 16, 1954; Introduction: Seismological Society of America Bulletin, v. 47, p. 299–300.

——, 1975, On crustal plates: Seismological Society of America Bulletin, v. 65, p. 1495–1500.

Townley, S. D., and Allen, M. W., 1939, Descriptive catalog of earthquakes of the Pacific coast of the United States 1769 to 1928: Seismological Society of America Bulletin, v. 29, p. 1–297.

VanWormer, J. D., and Ryall, A., 1980, Sierra Nevada–Great Basin boundary zone; Earthquake hazard related to structure, active tectonic processes, and anomalous patterns of earthquake occurrence: Seismological Society of America Bulletin, v. 70, p. 1557–1572.

Vetter, V. R., 1984, Focal mechanisms and crustal stress patterns in western Nevada and eastern California: Annales Geophysicae, v. 2, p. 699–710.

Vetter, V. R., and Ryall, A. S., 1983, Systematic change of focal mechanisms with depth in the western Great Basin: Journal of Geophysical Research, v. 88, p. 8237–8250.

Wallace, R. E., 1978, Patterns of faulting and seismic gaps in the Great Basin Province: U.S. Geological Survey Open-File Report 78-943, p. 857–868.

——, 1984a, Patterns and timing of late Quaternary faulting in the Great Basin Province and relation to some regional tectonic features: Journal of Geophysical Research, v. 89, p. 5763–5769.

——, 1984b, Fault scarps formed during the earthquakes of October 2, 1915, in Pleasant Valley, Nevada, and some tectonic implications in faulting related to the 1915 earthquakes in Pleasant Valley, Nevada: U.S. Geological Survey Professional Paper 1274-A, 33 p.

——, 1987, Grouping and migration of surface faulting and variations in slip rates on faults in the Great Basin Province: Seismological Society of America Bulletin, v. 77, no. 3, p. 868–876.

Wallace, R. E., and Whitney, R. A., 1984, Late Quaternary history of the Stillwater seismic gap, Nevada: Seismological Society of America Bulletin, v. 74, no. 1, p. 301–314.

Wallace, T. C., Helmberger, D. V., and Engen, G. R., 1983, Evidence of tectonic release from underground nuclear explosions in long-period P-waves: Seismological Society of America Bulletin, v. 73, p. 593–613.

——, 1985, Evidence of tectonic release from underground nuclear explosions in long-period S-waves: Seismological Society of America Bulletin, v. 75, p. 157–174.

Wallace, T. C., Helmberger, D. V., and Lay, J., 1986, Reply to Comments by A. Douglas, J. B. Yolung, and N. S. Lyman and a note on the revised moments for Pahute Mesa tectonic release: Seismological Society of America Bulletin v. 76, p. 313–318.

Wernicke, B., 1981, Low-angle normal faults in the Basin and Range Province; Nappe tectonics in an extending orogen: Nature, v. 291, p. 645–648.

Wernicke, B., Christiansen, R. L., England, P. C., and Sonder, L. J., 1987, Tectonomagmatic evolution of Cenozoic extension in the North American Cordillera, *in* Coward, M. P., and others, eds., Continental extensional tectonics: Geological Society of London Special Publication 28, p. 203–221.

Wernicke, B., Axen, G. J., and Snow, J. K., 1988, Basin and Range extensional tectonics at the latitude of Las Vegas, Nevada: Geological Society of America Bulletin, v. 100, p. 1738–1757.

Westphal, W. H., and Lange, A. L., 1967, Local seismic monitoring; Fairview Peak area, Nevada: Seismological Society of America Bulletin, v. 57, p. 1279–1298.

Whitten, C. A., 1957, Geodetic measurements in the Dixie Valley area: Seismological Society of America Bulletin, v. 47, p. 321–325.

Wright, L., 1976, Late Cenozoic fault patterns and stress fields in the Great Basin and westward displacement of the Sierra Nevada block: Geology, v. 4, p. 489–414.

Zoback, M. L., 1989, State of stress and modern deformation of the northern Basin and Range Province: Journal of Geophysical Research, v. 94, p. 7105–7128.

Zoback, M. L., and Beanland, S., 1986a, Stress and tectonism along the Walker Lane belt, western Great Basin: EOS Transactions of the America Geophysical Union, v. 67, no. 44, p. 1225.

—— , 1986b, Temporal variations in stress magnitude and style of faulting along the Sierran frontal fault system: Geological Society of America Abstracts with Programs, v. 18, p. 801.

Zoback, M. L., and Zoback, M. D., 1980, State of stress in the conterminous United States: Journal of Geophysical Research, v. 85, p. 6113–6156.

Zoback, M. L., Anderson, R. E., and Thompson, G. A., 1981, Cenozoic evolution of the state of stress and style of tectonism of the Basin and Range Province of the western United States: Royal Society of London Philosophical Transactions, series A, v. 300, p. 407–434.

Manuscript Accepted by the Society August 7, 1989

ACKNOWLEDGMENTS

The authors thank the anonymous reviewers for their helpful comments. Detailed reviews and comments were also provided by R. E. Anderson and W. B. Hamilton, whose help is gratefully acknowledged.

Chapter 11

Seismicity of the Intermountain Seismic Belt

Robert B. Smith and Walter J. Arabasz
Department of Geology and Geophysics, University of Utah, Salt Lake City, Utah 84112

INTRODUCTION

In this chapter we present an overview of the Intermountain seismic belt (ISB), a first-order feature of the Seismicity Map of North America (Engdahl and Rinehart, 1988). The ISB is a prominent northerly-trending zone of mostly shallow (<20 km) earthquakes, about 100 to 200 km wide, that extends in a curvilinear, branching pattern at least 1500 km from southern Nevada and northern Arizona to northwestern Montana (Fig. 1). Our study area, defined by the bounds of Figure 1, covers a sizable part of the western United States encompassing the ISB and is informally referred to herein as the Intermountain region.

Contemporary deformation in the ISB is dominated by intraplate extension. Forty-nine moderate to large earthquakes ($5.5 \leq M_S \leq 7.5$) since 1900 and spectacular late Quaternary faulting with a predominance of normal to oblique-normal slip make the Intermountain region a classic study area for intraplate extensional tectonics. Information from the Intermountain region, relating for example to paleoseismology (Schwartz, 1987), seismotectonic framework (Smith and others, 1989), contemporary deformation from geodetic measurements and seismic moments of earthquakes (Savage and others, 1985; Eddington and others, 1987), and strong ground motion in normal-faulting earthquakes (Westaway and Smith, 1989a) has added significantly to understanding extensional seismotectonics worldwide. Particularly valuable contributions have come from field and seismological observations of two large normal-faulting earthquakes in the Intermountain region—the 1959 Hebgen Lake, Montana, earthquake ($M_S = 7.5$) and the 1983 Borah Peak, Idaho, earthquake ($M_S = 7.3$)—both described herein. Our basic intent in this chapter is to provide an interpretive guide to the seismicity of the ISB. We also summarize and discuss observations from the Intermountain region that are relevant to general aspects of extensional intraplate tectonics.

The coherence of the ISB as a regional earthquake belt became apparent with evolving compilations of seismicity (Heck, 1938; Woolard, 1958; Ryall and others, 1966). The earthquake belt was well defined in Barazangi and Dorman's (1969) global seismicity map of shallow earthquakes and was first called the "Intermountain Seismic Belt" in joint abstracts by Sbar and Barazangi (1970) and Smith and Sbar (1970). Follow-up papers by Sbar and others (1972), and especially one by Smith and Sbar (1974), gave modern seismotectonic overviews (see also Smith, 1978; Arabasz and Smith, 1981; and Stickney and Bartholomew, 1987).

The ISB roughly follows the eastern margin of a broad region of late Cenozoic crustal extension in western North America. This seismically active boundary with more stable continental interior to the east has been interpreted as a subplate boundary (Smith and Sbar, 1974; Smith, 1978). It is well known that, on a regional scale, the ISB coincides with a persistent deformational belt in western North America that has been recurrently active since late Precambrian time (Levy and Christie-Blick, 1989; Anderson, 1989) and which is now characterized by pronounced lateral heterogeneities in crust-mantle structure across the ISB (e.g., Smith and others, 1989). Contemporary deformation in the region marks a continuation of late Cenozoic extension and volcanism (in the Yellowstone–Snake River Plain volcanic system and in southern Utah), whose various modern stages began roughly 10 to 15 m.y. ago (Anderson, 1989).

Regional-scale earthquake pattern

There is a general north-south regional continuity to the ISB (see the Seismicity Map of North America, Engdahl and Rinehart, 1988), but we can distinguish at least three parts—referred to herein as the southern, central, and northern ISB (Fig. 1)—for convenient reference. These subdivisions of the ISB may be arguable, but we believe distinctive features of the central ISB, as described in this paper, differentiate it from the ISB to the north and south. Referring to Figures 1 and 2 (see also Fig. 3 for additional features and place names), the southern ISB (36° to 42¼°N) coincides with a tectonic transition zone between the Basin and Range province on the west and the Colorado Plateau–Middle Rocky Mountain provinces on the east. In southwestern Utah at about 38°N there is a southward bifurcation of the ISB. A

Smith, R. B., and Arabasz, W. J., 1991, Seismicity of the Intermountain Seismic Belt, *in* Slemmons, D. B., Engdahl, E. R., Zoback, M. D., and Blackwell, D. D., eds., Neotectonics of North America: Boulder, Colorado, Geological Society of America, Decade Map Volume 1.

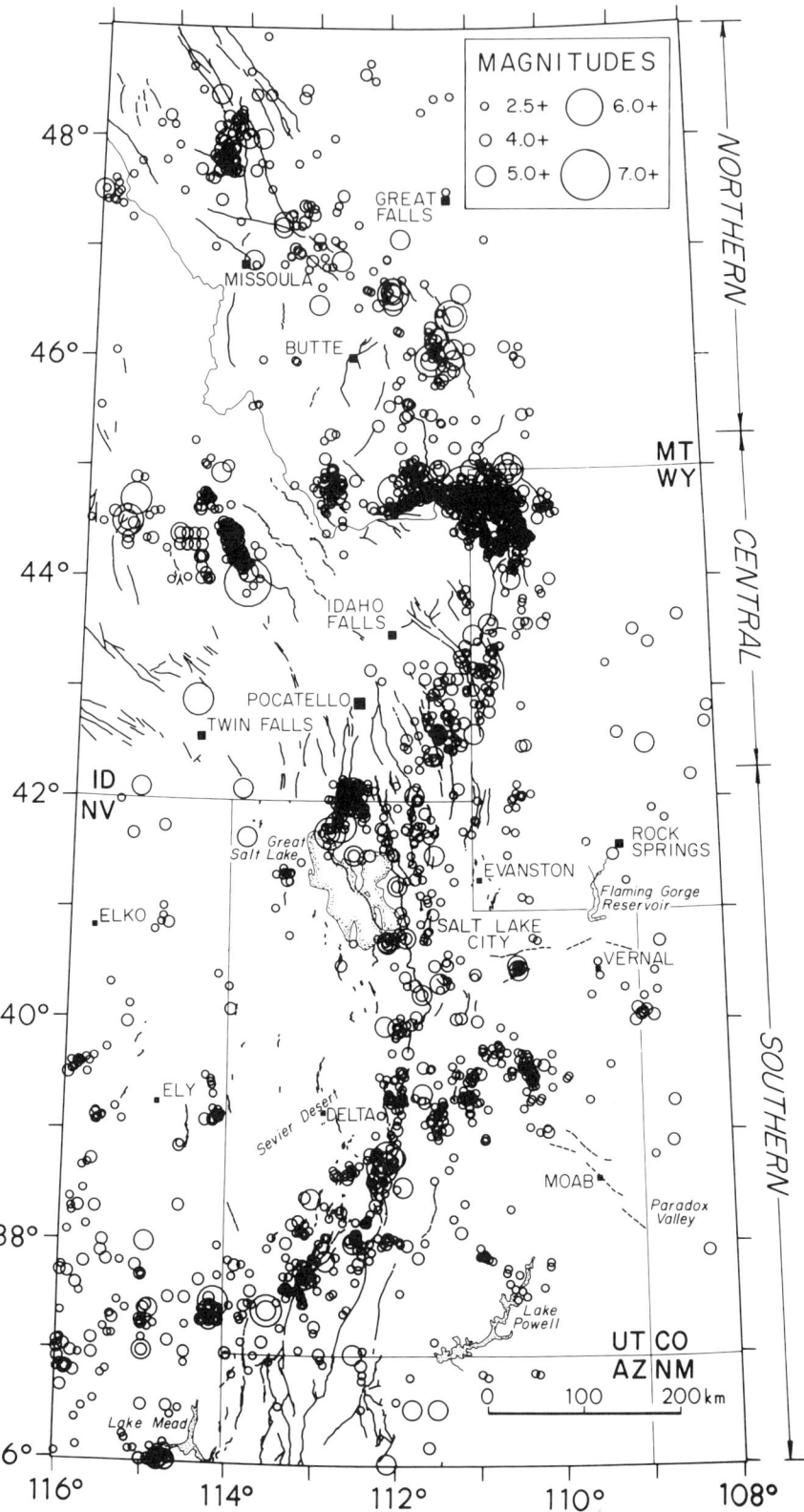

Figure 1. Earthquakes in the Intermountain region, 1900–1985, outlining the Intermountain seismic belt (ISB), together with selected Cenozoic faults identified in Figure 3. Earthquake data are from the compilation of Engdahl and Rinehart (1988; this volume). Northern, central, and southern parts of the ISB are delimited for reference.

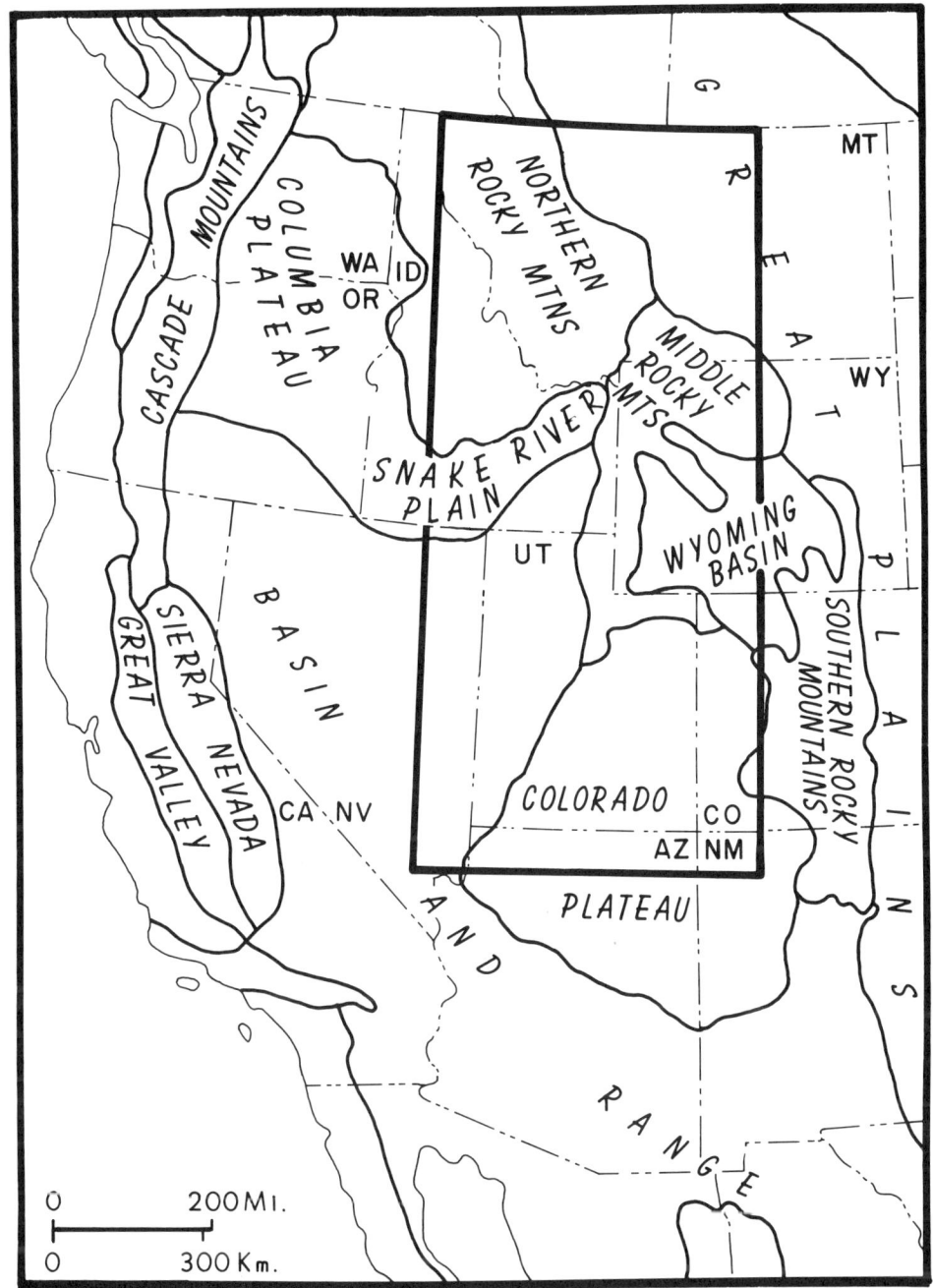

Figure 2. Map of the western United States showing physiographic provinces and location of study area (bold outline) shown in Figures 1, 3, 6, 8, 9, and 20.

distinct belt of seismicity continues southwestward some 200 km across southern Nevada, partly including induced earthquakes related to underground nuclear explosions at the Nevada Test Site in southern Nevada (Rogers and others, this volume); this belt transects the local north-south tectonic grain and coincides with the midpoint of a steep regional gravity gradient (Eaton and others, 1978) between the northern and southern sections of the Basin and Range province. The other part of the bifurcation is a weaker zone of scattered earthquakes that extends southward into central Arizona through a broad belt of Quaternary faulting (see Kruger-Kneupfer and others, 1985, Fig. 3).

In southwestern Utah (Fig. 1), earthquakes of the southern ISB follow a northeasterly structural trend to about 39°N, where both structure and the earthquake belt change to a northerly trend. Clustered earthquakes defining an inverted U-shaped pattern of epicenters in east-central Utah between 39° and 40°N are

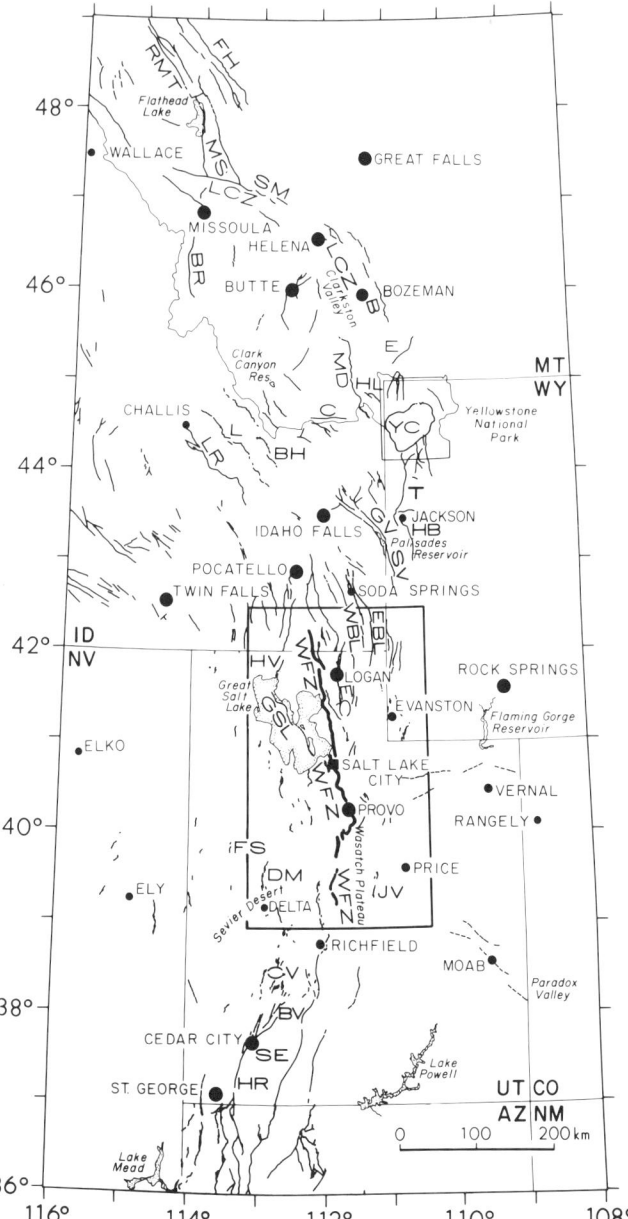

Figure 3. Location map of place names and selected late Cenozoic normal faults of the Intermountain region. Rectangle outlines Wasatch Front area shown in Figure 12, wherein other faults are identified. Abbreviations of faults are as follows: B = Bridger; BH = Beaverhead; BR = Bitterroot; BV = Beaver; C = Centennial; CV = Cove Fort; DM = Drum Mountains; E = Emigrant; EBL = East Bear Lake; EC = East Cache; FH = Flathead; FS = Fish Springs; GSL = Great Salt Lake; GV = Grand Valley; HB = Hoback; HL = Hebgen Lake; HR = Hurricane; JV = Joes Valley; L = Lemhi; LR = Lost River; MD = Madison; MS = Mission; SM = St. Mary's; SE = Sevier; SV = Star Valley; T = Teton; WBL = West Bear Lake. Other labeled tectonic features are: LCZ = Lewis and Clark Zone (see Fig. 18); RMT = Rocky Mountain trench; YC = Yellowstone Caldera.

mining-related (described below). Northward in Utah the ISB centers on the 380-km-long Wasatch fault, the preeminent normal fault zone of the eastern Basin and Range province, along which young mountain blocks have been uplifted to form a major west-facing physiographic scarp, called the Wasatch Front, with up to 2,300 m of relief.

The central ISB (42¼° to 45¼°N) follows, in part, the Basin and Range–Middle Rocky Mountain transition, but is complicated by having a westerly-trending branch; the result is an arcuate pattern that appears to "wrap around" the late Tertiary volcanic province of the eastern Snake River Plain (SRP; see Fig. 2). North of the Utah-Idaho border the ISB takes on a marked northeasterly trend, subparallel to the southeastern edge of the SRP and oblique to northwest-trending Quaternary normal faults in southeastern Idaho. The seismic belt continues northeasterly into western Wyoming to the vicinity of Jackson, immediately north of which there is a notable gap in seismicity coincident with the 70-km-long Teton fault. Intense seismicity occurs beneath the volcanically and hydrothermally active Yellowstone region and to its west in the Hebgen Lake region. A divergent belt of earthquake activity extends more than 400 km from Yellowstone Park in a west-southwest direction into central Idaho. This zone was originally described by Smith and Sbar (1974) as independent of the ISB and was termed the Idaho seismic zone (see also Smith, 1978). Stickney and Bartholomew (1987) similarly characterize it as an independent seismic zone, calling it the Centennial Tectonic Belt. This zone, however, forms part of an arcuate, parabolic pattern of seismicity flanking the SRP, with a vertex at Yellowstone Park that suggests causal influence by the Yellowstone-SRP (Y-SRP) volcanic system and related hot spot (Smith and others, 1985; Anders and others, 1989; Blackwell, 1989) and hence an integral relation with the main ISB.

The northern ISB (45¼° to 49°N) lies within the Northern Rocky Mountains province and extends more than 400 km in a northwest direction from Yellowstone Park to northwestern Montana, following a structural belt of Cenozoic basins bounded by Quaternary faulting of diverse trend. Earthquakes and west-northwest–striking faults between about 46½° and 48°N are interpreted by Stickney and Bartholomew (1987) to reflect, in part, an intraplate boundary called the Lewis and Clark Zone, which trends about N70°W through Missoula and Helena (Fig. 3).

SEISMOTECTONIC FRAMEWORK

Crustal structure and structural style

The geophysical framework of the Intermountain region has recently been summarized by Smith and others (1989). The crustal and upper-mantle velocity structure in the region is typified by the cross sections shown in Figures 4 and 5. The southern ISB coincides with a transition from thinner, extended crust and lithosphere on the west to thicker, more stable crust and

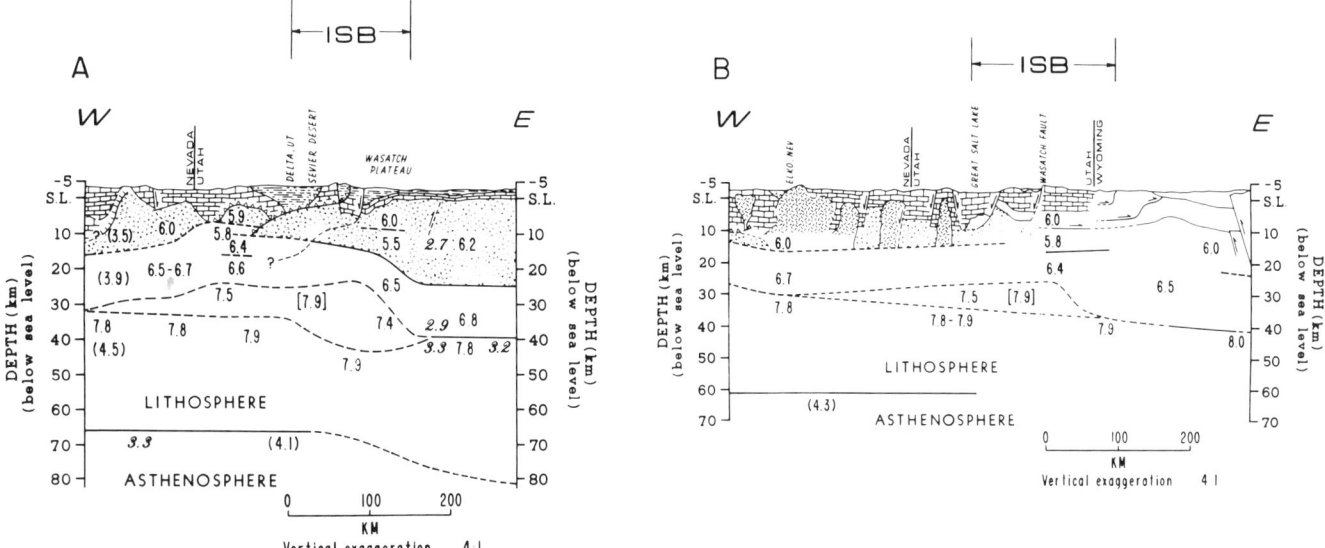

Figure 4. Cross sections of representative crustal and lithospheric structure across the southern Intermountain seismic belt (ISB), as summarized by Smith and others (1989). Profiles are roughly at latitude 38°N (A) and 41°-42°N (B); geographic reference features are shown in Figure 3. Superposed numbers indicate P-wave velocities (5.9-8.0 km/sec), S-wave velocities (3.5-4.5 km/sec, in parentheses), and densities (2.7-3.3 gm/cm^3, in italics); numbers in brackets are P_n-wave velocities corresponding to an alternative interpretation by Pakiser (1989) in which a mantle upwarp beneath the Basin and Range–Colorado Plateau transition has a uniform upper mantle velocity of 7.8 to 7.9 km/sec.

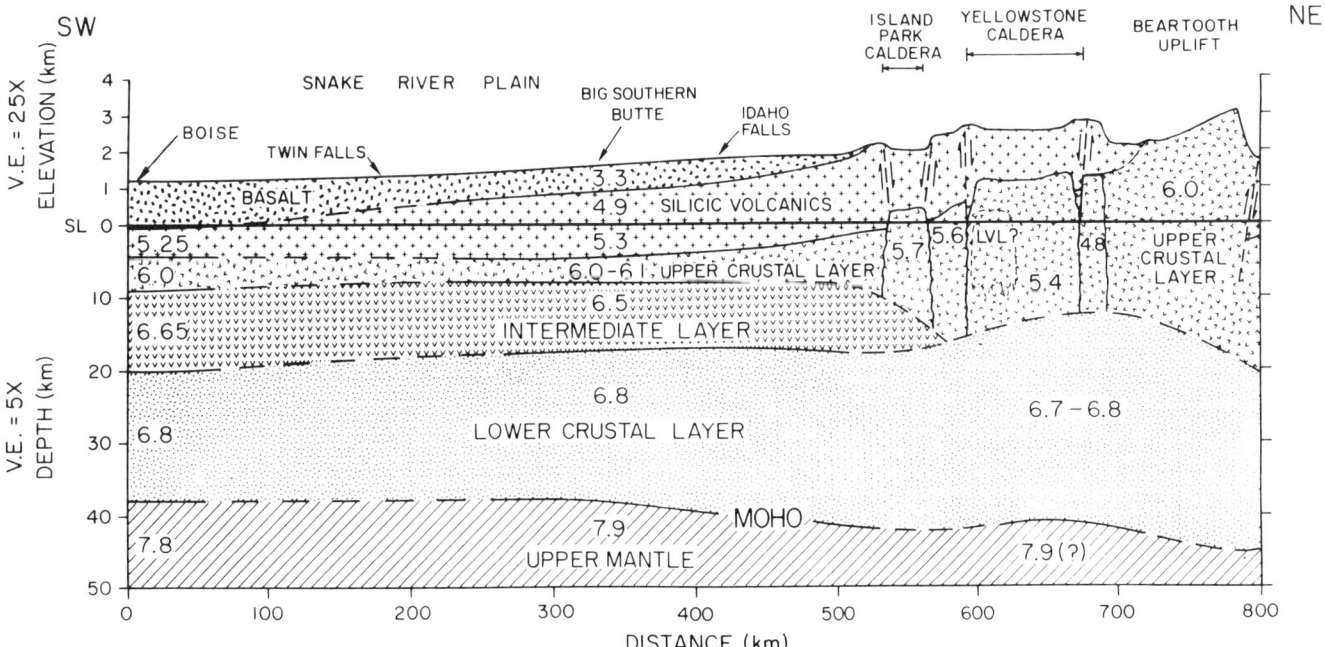

Figure 5. Cross section of crustal structure along the Snake River Plain in Idaho (representative of part of the central ISB) into the Beartooth uplift of Montana (representative of the northern ISB). Superposed numbers indicate P-wave velocities (km/sec). Data from 1978 and 1980 Yellowstone–Snake River Plain seismic-refraction experiment (Braile and others, 1982; Smith and others, 1982).

lithosphere on the east. The two profiles shown in Figure 4 extend from the central Basin and Range province in Nevada, where the Moho is about 30 km deep (below sea level) and the upper-mantle P_n velocity is 7.8 km/sec, to the Colorado Plateau–Middle Rocky Mountain provinces in Utah and Wyoming, where the Moho depth slightly exceeds 40 km and the P_n velocity reaches 8.0 km/sec. There is uncertainty about the depth of the true Moho beneath the transition region. If the top of a 7.8 to 7.9-km/sec layer marks the Moho beneath the Wasatch Front region of Utah (Fig. 4a), as identified by Loeb and Pechmann (1986), then the crust-mantle boundary is as deep as 45 km beneath the transition. Alternative interpretations imply the depth to the Moho may be as shallow as 25 km if observed velocities of 7.4 to 7.5 km/sec (Fig. 4) are the result of down-dip ray paths (see Smith and others, 1989, Fig. 4; Pakiser, 1989).

The P-wave velocity structure of the upper crust beneath the Y-SRP volcanic province, revealed by refraction/wide-angle reflection profiles, is more laterally heterogeneous than in the surrounding thermally undisturbed areas of the ISB. At Yellowstone a caldera-wide low-velocity body extends to depths of about 15 km and is thought to reflect a remnant magma reservoir with materials ranging from melts to hot, cooling granitic rocks (Smith and others, 1982). Along the SRP, the systematic decrease in elevation southwestward away from the caldera also reflects a systematic change in crustal structure. The near-surface basaltic layer thins northeastward from 2 km in southwestern Idaho, the suggested beginning of the trace of the Yellowstone hot spot, to zero thickness at Yellowstone; correspondingly, the deeper silicic layer thickens northeastward from zero thickness in southwestern Idaho to 2 km at the Yellowstone caldera, following the form of the surface topography (Fig. 5). A high-velocity, 6.5-km/sec layer cores the mid-crust of the eastern SRP and is interpreted as a solidified, mafic remnant of the crustal magma sources of the Yellowstone hot spot (Sparlin and others, 1982). This unusual high-velocity and high-density body may affect the overall strength and hence the seismic capability of the SRP. Note that neither the Moho nor the seismic velocity structure of the lower crust of the eastern SRP seem to have been altered by the youthful magmatism. The lower-crustal velocity structure beneath the SRP is the same as beneath adjacent thermally undisturbed regions.

We noted earlier that the ISB roughly follows the eastern margin of a broad domain of late Cenozoic extension in western North America (e.g., Eaton, 1982, Fig. 1). This margin also marks a thermal transition. The background heat flow is about 85 mW m^{-2} in the Basin and Range province and more than 100 mW m^{-2} in the SRP, contrasting with background values of less than 65 mW m^{-2} in the Colorado Plateau and in the area east of the Northern Rocky Mountains (Bodell and Chapman, 1982; Morgan and Gosnold, 1989). Locally, heat flow exceeds 1500 mW m^{-2} in the Yellowstone caldera (Blackwell, 1989). The ISB thus follows a structural and thermal transition to more stable continental interior with lower heat flow. The transition may be a locus of active lithospheric thinning (see Fig. 4A), occurring in general along the eastern margin of a regional-scale thermotectonic anomaly in the upper mantle. Effects of an active mantle hot spot associated with the Y-SRP system (discussed in a later section) are a special case.

Seismic-reflection profiling across Tertiary-Quaternary basins and active fault zones in the Basin and Range province has provided important information on upper-crustal structural style. The superposition of basin-range faulting upon pre-Neogene thrust belt structure, especially along the eastern margin of the Basin and Range province, is well known to be a fundamental and complicating factor (e.g., Smith and Bruhn, 1984). Anderson and others (1983), Allmendinger and others (1983), Smith and Bruhn (1984), and Smith and others (1989) have interpreted seismic-reflection data from throughout the northern Basin and Range province, identifying three characteristic styles of extensional basin development in the region: (1) relatively simple basins bounded by one or more planar normal faults dipping 45° to 60°; (2) asymmetric tilted basins displaced chiefly by a listric or planar low-angle normal fault; and (3) complex basins, typically with subbasins, associated with both planar and listric normal faults that sole into low-angle detachments.

Low-angle detachment faulting may have contributed substantially to Cenozoic crustal extension in at least part of the Intermountain region. The best example is the case of the Sevier Desert detachment (see sawteeth pattern beneath the Sevier Desert in Fig. 4A). High-quality seismic-reflection data (Allmendinger and others, 1983; Smith and Bruhn, 1984; Planke and Smith, 1991) show that the Sevier Desert detachment extends laterally with an average 10° to 15° westward dip at least 70 km from near the surface in central Utah to a depth of about 15 km beneath western Utah. The detachment has an estimated total surface area of 5,600 to 9,100 km^2 (Planke and Smith, 1991) and may have accommodated 30 to 60 km of Cenozoic extensional displacement (Allmendinger and others, 1983). In the Sevier Desert region, where the detachment lies only 3 to 5 km below the surface, prominent normal faults in the hanging wall do not cut the detachment but either abut or merge with it (Crone and Harding, 1984; Planke and Smith, 1991). The configuration of a high-angle normal fault, with Holocene(?) surface displacement, "directly connected" to the Sevier Desert detachment at relatively shallow depth, led Crone and Harding (1984) to raise concern about the detachment's seismogenic potential. Reviews of moderate to large normal-faulting earthquakes in diverse extensional regimes (Jackson and White, 1989; Doser and Smith, 1989) suggest that the low angle of dip of the Sevier Desert detachment make it an unlikely source of seismic slip, but the possibility of aseismic motion on the detachment cannot be ruled out.

There remain many uncertainties about the subsurface geometry of seismically active normal faults in the Intermountain region and whether seismic slip can occur on low-angle or listric normal faults known to be present. Moderately- to steeply-dipping planar geometries for the seismically active faults are typically inferred from aftershock locations and from the focal mechanisms of large earthquakes (e.g., Smith and others, 1985), a point we pursue later in this paper. Nevertheless, seismic-reflection evidence suggests that some segments of major normal faults with late Quaternary surface ruptures indeed have a listric subsurface geometry (Smith and Bruhn, 1984; Smith and others, 1989).

Throughout this paper we emphasize that the late Cenozoic structural style of the Intermountain region is dominated by normal faulting. There is growing awareness of the importance of strike-slip faulting as part of Neogene extensional deformation in the Basin and Range Province (Anderson, 1989), particularly in parts of the southern and western Great Basin (Rogers and others, this volume). In the Intermountain region, the best geologic evidence for Neogene strike-slip deformation is in the Sevier Valley area of south-central Utah (near Richfield, Fig. 3; Anderson and Barnhard, 1987). Focal mechanisms of background earthquakes in that same area also imply strike-slip faulting (Arabasz and Julander, 1986). Strike-slip focal mechanisms have also been observed for historical moderate-sized earthquakes in the northern ISB of Montana (Doser, 1989a), in the Hansel Valley area of northwestern Utah (near HV, Fig. 3; Doser, 1989b), and in southeastern Nevada (Smith and Sbar, 1974). Our present understanding of extension in the Intermountain region may underestimate the importance of strike-slip deformation.

Contemporary deformation

Regional stress field. Stress observations give important information about the pattern and mechanics of intraplate extension in the Intermountain region. Figure 6 shows the orientations of minimum horizontal compressive stress, S_{hmin}, variously deduced from the T-axes of focal mechanisms of moderate to large earthquakes, mapped fault-displacement vectors associated with Quaternary and Holocene slip events, orientations of volcanic dikes, borehole deformation, and in-situ hydrofracture stress measurements. The data are from a recent compilation by Zoback and Zoback (1989) for the continental United States.

Focal mechanisms and geologic indicators are the main sources of available stress information for the ISB. Given the ambiguity of selecting the correct fault plane in focal mechanisms, the principal stress axis directions were determined by assuming the standard Coulomb failure criterion with a coefficient of internal fraction of zero. This constraint places the maximum and minimum principal stress directions (corresponding to the P-axes, for maximum compression, and T-axes, for minimum compression) at 45° to the nodal planes. Because the dominant mode of contemporary deformation in the Intermountain region is extension, the T-axes give a general indication of relative motion within areas of coherent intraplate deformation and are consistent indicators of the directions of the minimum compressive stress for this region.

The overall stress field of the ISB (Fig. 6) is generally characterized by NE-trending S_{hmin} orientations in the northern ISB and the western part of the central ISB and ENE-to-ESE-trending S_{hmin} orientations in the southern ISB and the eastern part of the

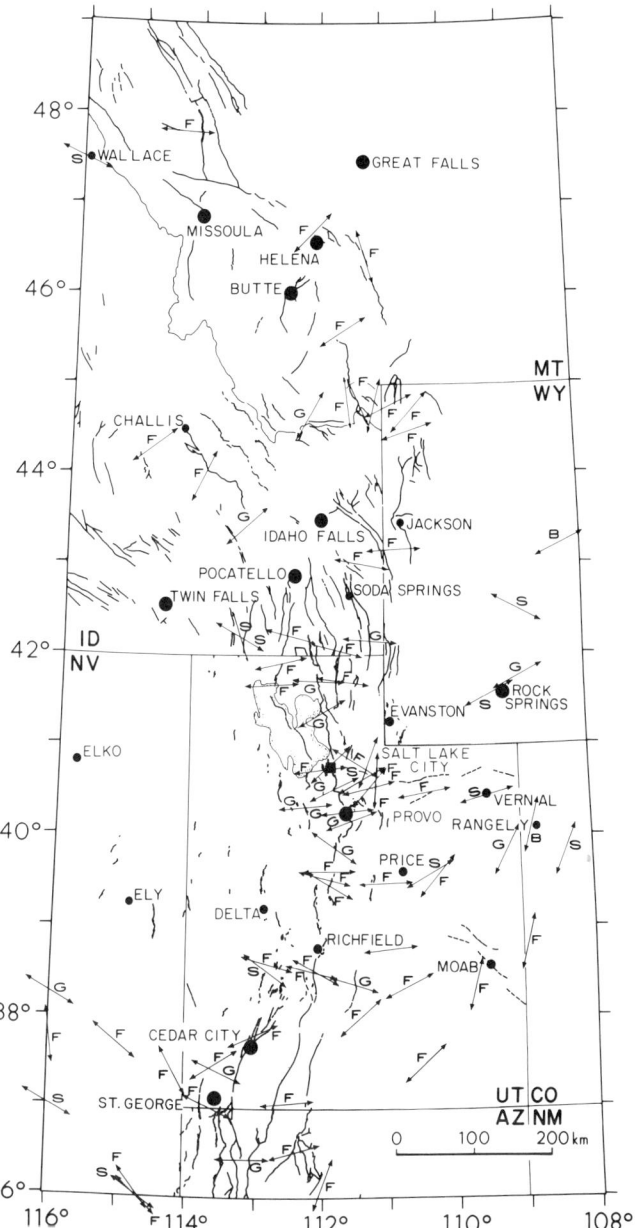

Figure 6. Map showing orientations of minimum horizontal compressive stress in the Intermountain region from a data compilation of Zoback and Zoback (1989), together with selected faults, as in Figure 3. Types of stress indicators are indicated as follows: B = borehole "breakout", F = focal mechanism, G = geologic, M = mixed, and S = in situ stress.

central ISB. The coherence of these orientations within the ISB and surrounding areas of the Rocky Mountains has led Zoback and Zoback (1989) to define a "Cordilleran extensional" province, larger than the traditional domain of the Basin and Range province and the Rio Grande rift. They distinguish the interior of the Colorado Plateau, however, as a distinct stress province with a WNW-trending orientation of maximum horizontal compressive stress S_{Hmax}. This is reflected in Figure 6 by the difference in

S_{hmin} orientations in eastern Utah and western Colorado from those along the main ISB to the west.

On the basis of focal-mechanism studies of small to large earthquakes in central Idaho and southern Montana, Smith and others (1977) and Stickney and Bartholomew (1987) distinguished a stress regime with N-trending S_{hmin} in the Hebgen Lake region, west of Yellowstone National Park, from the broader region of basin-range tectonism in the vicinity of the Montana-Idaho border where S_{hmin} is northeast-trending. Among other deviations from this pattern of northeast-trending S_{hmin} is a data point east of Helena for the 1925 Clarkston Valley earthquake of M_W = 6.6 (discussed below). This earthquake may have been more closely related to a stress field with northeast-southwest maximum horizontal compressive stress, characteristic of the more stable interior and thus transitional between the Basin and Range and the Great Plains (Doser, 1989a). We will comment on the stress implications of other sizable earthquakes in later sections.

An indicator of the relative magnitudes of the maximum (S_1), intermediate (S_2), and minimum (S_3) principal stresses has been defined by Bott (1959) as $\Phi = S_2 - S_3/S_1 - S_3$. Zoback (1989) and others have used this parameter to interpret regional stress variations within the Basin and Range province. Assuming that fault slip occurs in the direction of the maximum shear stress on a fault plane, then the orientation and sense of slip is governed by Φ and by the orientation of the principal stresses. For a normal faulting stress regime such as the ISB (S_1 vertical), if $\Phi = 0$, then the two horizontal stresses are equal and the predicted deformation is pure dip-slip for any fault orientation. For the maximum value of $\Phi = 1$, the vertical and maximum horizontal stresses are equal, and the predicted deformation is oblique-normal slip, transitional to strike-slip, depending on fault orientation.

Zoback (1989) and Bjarnason and Pechmann (1989) have recently summarized information on Φ-values for the southern ISB, variously from fault-slip data, focal mechanisms, in-situ stress measurements, and well-bore breakouts. In northern Utah, where the observed mode of faulting from fault-slip information and focal mechanisms is predominantly normal dip-slip and where S_{hmin} has an average east-west trend, Φ is highly variable. Averaged values of Φ (Zoback, 1989; Bjarnason and Pechmann, 1989) range from low values (0.0 to 0.3) for in-situ stress measurements and well-bore breakouts through intermediate values (0.3 to 0.7) for late Quaternary fault-slip measurements to slightly higher values (0.5 to 0.9) for focal mechanisms. In south-central Utah along the Basin and Range-Colorado Plateau transition, there is an observed mixture of dip-slip and strike-slip deformation, both in focal mechanisms (Arabasz and Julander, 1986) and in fault slip (Anderson and Barnhard, 1987). The observations can be explained by a high Φ-value and local changes in the relative magnitudes of (near-equal) vertical and maximum horizontal principal stresses under a relatively constant east-trending S_{hmin} (see Zoback, 1989). Bjarnason and Pechmann (1989) and Zoback (1989) have independently calculated an average Φ-value of 0.8 for grouped focal mechanisms in this area. There is little information on Φ-values for the central and northern ISB, except for a study of the main shock and aftershocks of the 1983 Borah Peak, Idaho, earthquake by Smith and others (in preparation), who find a Φ-value of 0.65, consistent with the observed oblique-normal faulting (described below).

Strain rates. Earthquake focal mechanisms and seismic moments have been used by Eddington and others (1987) to calculate regionalized strain rates and corresponding deformation rates produced by historical earthquakes in the western United States, following the method of Kostrov (1974). Available results for the central and southern ISB are shown in Figure 7. The figure shows 14 selected areas of inferred homogeneous stress in which the moment tensors of historical earthquakes have been summed and diagonalized to get the direction and magnitude of horizontal principal strain. Together with the S_{hmin} orientations in Figure 6,

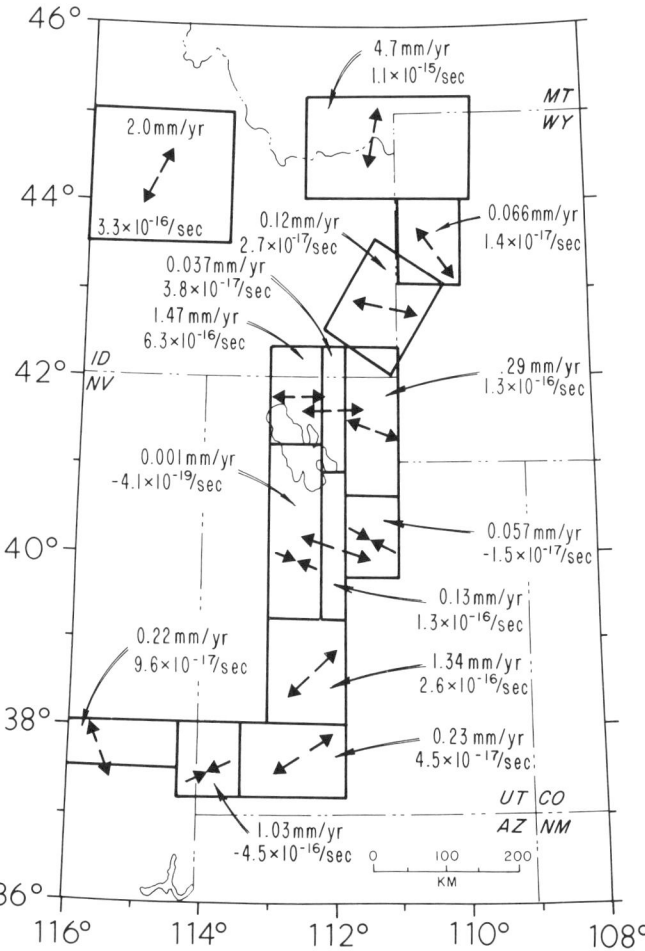

Figure 7. Strain and deformation rates in part of the Intermountain region based upon summations of seismic moments from historical earthquakes, after Eddington and others (1987). Earthquakes within areas of assumed homogeneous strain, shown by boxes, were used to determine the summed seismic moment. Information shown for each box includes: orientation of the horizontal principal strain axis (arrows); the deformation rate, in mm/yr (upper number); and the horizontal strain rate, in sec^{-1} (lower number).

the strain-rate and deformation data in Figure 7 provide an overall perspective of contemporary deformation in the central and southern ISB.

In the central ISB (Figure 7, top), historical seismicity in central Idaho implies NNE–SSW extensional strain and yields a strain rate of the order of 10^{-16}/sec. In the Hebgen Lake–Yellowstone Park region to the east, the most seismically active region of the entire U.S. Cordillera, the extensional strain rate of 1.1×10^{-15}/sec (4.7 mm/yr deformation rate) is more than three times greater, and the horizontal principal strain axis trends more northerly. For parts of the central ISB south of Yellowstone Park, much smaller extensional strain rates of the order of 10^{-17}/sec and deformation rates of 0.07 to 1.2 mm/yr were calculated. The northwest-southeast direction of horizontal principal strain in the Teton region is considered uncertain because of sparse data; in southeastern Idaho, a better-resolved extensional strain direction is nearly east-west.

In the southern ISB, general east-west extensional strain of the order of 10^{-16}/sec or smaller characterizes Utah's Wasatch Front area. Deformation rates range from 0.001 mm/yr in the western part of the Wasatch Front area to 1.5 mm/yr in the northwestern part. Regarding the latter subregion, we note that the depicted strain information would not be significantly affected by a recently revised focal mechanism (Doser, 1989b) implying predominantly strike slip rather than dip slip for the M = 6.6 Hansel Valley earthquake of 1934. Because the direction of the T-axis for the revised solution (Doser, 1989b) is nearly identical to that for the original dip-slip solution (Dewey and others, 1973) used by Eddington and others (1987), the averaged moment rate and direction of deformation given in Figure 7 would not be significantly changed. In southern Utah, extensional strain rates of 10^{-16}/sec to 10^{-17}/sec are comparable to those for most of the Wasatch Front area, but the strain direction is northeast-southwest. A possible change to northeast-southwest compression is suggested for one subregion in southwesternmost Utah. The significance of this local apparent change in mode of deformation is uncertain. The northwest-southeast extensional strain direction shown for a subregion in southeastern Nevada is consistent with that for other parts of southern and central Nevada analyzed by Eddington and others (1987).

Deformation rates intrinsically depend on the dimensions of the subregion being considered, so comparisons must be made with care. Nevertheless, data summarized by Eddington and others (1987) for the western United States make it evident that deformation rates for most of the intraplate ISB are one to two orders of magnitude lower than those along the western North American plate boundary. The ISB deformation rates can be evaluated in another way. Eddington and others (1987) summed earthquake deformation rates along east-west profiles across the northern and southern Great Basin, including the ISB, to obtain integrated extension rates of 8 to 10 and 3 to 4 mm/yr, respectively, with approximately 1 mm/yr taken up across the ISB (Doser and Smith, 1982). Estimates of Late Cenozoic total deformation rates of 1 to 20 mm/yr for the Great Basin determined

TABLE 1. RATES OF EXTENSION ACROSS THE GREAT BASIN *

Time	Deformation Rate (mm/yr)	Method
Late Cenozoic	3 - 20	Geological strain
Late Cenozoic	3 - 12	Heat flow
Holocene paleoseismicity	1 - 12	Fault-slip data
Historic seismicity	3.5 - 10	Historical earthquakes
Inferred Quaternary	<9	Intraplate models

* Adapted from Eddington and others (1987, Table 5); sources identified therein.

from other geologic and geophysical data (Table 1) are of the same magnitude as the contemporary earthquake-induced rates.

Quaternary faulting

To compare the regional seismicity of the ISB with active faulting, a generalized map was made of late Tertiary to Holocene normal faulting in the Intermountain region (Fig. 3) from the following sources. The locations of active fault traces compiled by Witkind (1975a, b, c) were digitized for Idaho (except for central Idaho), western Wyoming, and western Montana. Surface traces of late Quaternary faults of central Idaho and southwestern Montana, including the surface rupture of the 1983 Borah Peak earthquake (M_S = 7.3), were taken from compilations by Haller (1988) and Scott and others (1985). (A more detailed map of active faulting in the Y-SRP area is presented in Fig. 15, below.) Fault data for the Utah region are from a compilation of Arabasz and others (1987) for the Wasatch Front area (outlined in Fig. 3), supplemented elsewhere by data from Anderson and Miller's (1979) Quaternary fault map of Utah. Quaternary faults for northern Arizona are from the Arizona Quaternary fault map of Scarborough and others (1986). We did not attempt to make a complete compilation of fault data for eastern Nevada as this area was considered marginal to our discussion. Figure 3 thus serves as a fair representation of known or suspected active faults in late Cenozoic time in the Intermountain region, but we caution that the fault compilation is non-uniformly complete.

Age of normal faulting and correlation with topography. Dating the inception and evolution of normal faulting on individual faults in the Intermountain region is generally handicapped by a lack of suitable exposures and datable materials. There are some exceptions. Geochemical studies of altered fault rock in the exhumed footwall of the Wasatch fault (Parry and Bruhn, 1986, 1987) indicate an origin at 11 km depth 17.6 ± 0.7 m.y. ago. The Teton fault in Wyoming has a total vertical displacement of as much as 6 to 9 km that began 7 to 9 m.y. ago (Love and Reed, 1971). On a regional basis, there is evidence for two stages of Cenozoic extensional tectonism and normal faulting in many parts of the Basin and Range province. As reviewed by Levy and Christie-Blick (1989) and Anderson (1989), a first early stage of Cenozoic extension began about 37 Ma and was apparently restricted to a relatively narrow region of high strain in eastern Nevada, western Utah, and southern Idaho and was accompanied by calc-alkaline volcanism, detachment faulting, and core-complex formation. The second stage was the classic episode of Basin and Range extension and epeirogeny responsible for the present topography and the steeply dipping normal faults of the ISB. This modern stage generally began 15 to 10 Ma in the northern Basin and Range province, but earlier in the southern part of the province.

The distinctive north–northeast- to northwest-trending basin-range topography of the ISB involves sediment-filled basins and tilted range blocks bounded by large normal faults with significant Quaternary displacement. The observed coseismic deformation associated with normal faulting during large earthquakes and theoretical modeling suggest that both footwall uplift of the mountain blocks and hanging-wall subsidence of the adjacent asymmetric basins have been fundamental in developing basin-range topography (e.g., King and others, 1988; see also Fig. 24, below). For individual ranges, relative crest height along the range seems to correlate with the size and frequency of young fault displacements along the base of the range. For example, along the Teton fault in Wyoming, the maximum heights of Quaternary scarps ranging up to 50 m high correspond with the highest parts of the Teton Range (Smith and others, 1990a, b). The 1983 Borah Peak earthquake (M_S = 7.3) ruptured one of the most active central segments of the Lost River fault zone, which was coincidentally adjacent to the highest part of the Lost River Range (Scott and others, 1985), including Borah Peak, the highest point in Idaho. Crest heights along the Wasatch Range are greater along the active central segments of the Wasatch fault than along distal segments with lower late Pleistocene–Holocene slip rates (Schwartz and Coppersmith, 1984). Relative topographic relief serves as an indicative but insufficient guide to the location of segments of range-front normal faults likely to produce future surface-faulting earthquakes.

Threshold of surface faulting and maximum magnitude

Summaries of information and discussion about the minimum magnitude needed to produce coseismic surface faulting in the Intermountain region are given by Doser (1985a) and Arabasz and others (1987). The threshold magnitude appears to be in the range of $6.0 \leq M_L \leq 6.5$, based on the historical record of earthquakes in the ISB and in the Basin and Range province. Arabasz and others (1987) adopt $M_L = 6.3 \pm 0.2$ as an estimate of the threshold in the Utah region and argue that earthquakes up to this size can occur anywhere in the southern ISB, even where there is no geologic evidence for Quaternary surface faulting.

The M_S = 7.5 Hebgen Lake earthquake of 1959 (described in detail herein) is considered by some to represent the maximum earthquake size for the ISB (e.g., Doser, 1985a). This earthquake, however, was smaller than at least two other large earthquakes in the Basin and Range province. The 1872 Owens Valley, California, earthquake had an estimated moment magnitude of 7¾ to 8 and was associated with a predominantly strike-slip surface rupture up to 110 km long, with a maximum lateral offset of 7 m and a maximum vertical offset of 4.4 m; the 1915 Pleasant Valley, Nevada, earthquake of surface-wave magnitude 7.6 had a 60-km-long rupture and a maximum vertical displacement of 5.8 m (dePolo and others, 1989). The Hebgen Lake earthquake had a shorter rupture length, but had comparable displacement to these two earthquakes. Because there are adjoining fault segments in the ISB with potential rupture lengths exceeding that of the Hebgen Lake earthquake, and comparable to those of the Pleasant Valley and Owens Valley earthquakes, it is reasonable to consider maximum-magnitude earthquakes slightly greater than M_S = 7.5 for particular faults in the ISB (e.g., Arabasz and others, 1987, adopt a maximum magnitude of M_S = 7.5 to 7.7 for the Wasatch Front area).

Estimates of the maximum magnitude for earthquakes on any particular fault in the ISB have uncertainties relating not only to the prediction of future rupture characteristics but also to the conversion of those rupture characteristics to estimated magnitudes. This problem is particularly important for seismic hazard analysis. For example, in an evaluation of probabilistic ground shaking for the Wasatch Front, Youngs and others (1987) used fault length-magnitude relations of Bonilla and others (1984) to estimate maximum magnitudes of M_S = 7.2 to 7.5 for the longest segments of the Wasatch fault in northern Utah. In a seismotectonic study of the Jackson Lake dam in Wyoming, Gilbert and others (1983) selected a maximum credible earthquake of magnitude (unspecified scale) 7.5 based on comparisons of observed scarp lengths and fault length with rupture length-magnitude relations such as those of Slemmons (1977). Thenhaus and Wentworth (1982) suggested a maximum magnitude of 7½ for eastern Idaho and central and western Utah, and adopted a value of 7¾ specifically for the Wasatch fault. For a site at the Idaho National Engineering Laboratory near Idaho Falls, Idaho, Woodward-Clyde Consultants (1979) relied upon a global compilation of data for surface-faulting earthquakes to assign a peak ground acceleration, using the 1959, M_S = 7.5 Hebgen Lake earthquake as a maximum-magnitude event for that area.

THE EARTHQUAKE RECORD IN THE INTERMOUNTAIN REGION

The traditional earthquake record for the Intermountain region—consisting of historical seismicity (based on non-instrumental reports of felt earthquakes) and instrumental seismicity—is much less substantial than for other parts of the United States. This is because of the region's relatively late settlement, historically sparse population, and late seismographic coverage. In contrast, comparatively abundant information has been gathered in this part of the United States on the timing and character of prehistoric earthquakes from paleoseismology. In this section, we restrict attention to the traditional earthquake record.

Pre-instrumental information

Reports of historical seismicity in the Intermountain region date only from the mid 1800s when systematic modern settlement began. Diverse Indian cultures were well established in the region after A.D. 100 to 1300 but left no known record of specific felt earthquakes. During the 1820s and 1830s, a few hundred fur trappers made up most of the non-Indian population of the region. Mormon pioneers entered the Salt Lake Valley in 1847 and promptly started a regional colonizing program. Within a few decades, Mormon settlements extended throughout large portions of the Intermountain West (Wahlquist, 1981), with the exception of Montana, where the settlements began chiefly as mining camps after the 1850s. The first documented earthquakes in the Intermountain region date from 1850 in Utah (Arabasz and McKee, 1979), 1869 in Montana (Qamar and Stickney, 1983), 1871 in Wyoming (Hayden, 1872), and 1879 in Idaho (Townley and Allen, 1939). The highly non-uniform distribution of population before 1900 throughout the Intermountain region implies great variability in the threshold of detection and in location errors for pre-instrumental earthquakes.

The pre-instrumental earthquake record for the Intermountain region comes from multiple sources. Coffman and others (1982) cite many early reports, records, and compilations for earthquakes in the "Western Mountain Region" before 1928. For 1928 and later, annual reports published by the U.S. Department of Commerce under the title *United States Earthquakes* are key sources. Williams and Tapper (1953) made an important historical study of Utah earthquakes from 1850, the time of publication of the first newspaper in Utah, through 1949. Cook and Smith (1967) extended this record to 1965, including computer determinations of the first instrumental seismicity for the Utah region from systematic regional recording. Modern compilations of historical seismicity in the study area include those of Arabasz and McKee (1979) and Stover and others (1986), for the Utah region, and Qamar and Stickney (1983) for Montana.

Instrumental recording and seismic networks

Seismographic recording in the Intermountain region began with the installation of two modified Bosch-Omori pendulum seismographs on the University of Utah campus in Salt Lake City in 1907 (Arabasz, 1979). Figure 8 shows stages of subsequent instrumental coverage of the Intermountain region in 1948, 1968, and 1988. By 1948, electromagnetic seismographs were operating in at least eight locations in the region (dated circles, Fig. 8A). Systematic reporting of seismological data to the U.S. Coast and Geodetic Survey (USCGS) from Salt Lake City began in 1938, one year before the Bosch-Omoris were replaced by more modern instruments (Arabasz, 1979). For each of the other dated stations shown in Figure 8A, reporting to the USCGS began immediately after installation (see *United States Earthquakes*).

In 1968, there were at least 25 seismographic stations in the study area (triangles, Fig. 8A), virtually all with on-site recording. The changes from 1948 to 1968 almost exclusively reflect additions in the 1960s (see Poppe, 1980). These relate to stations in the western part of the area shown in Figure 8A installed with the motivation, in part, to record underground nuclear explosions from the Nevada Test Site, the development in Utah of a skeletal statewide network (Arabasz and others, 1979), local monitoring of mining-related seismicity in east-central Utah and in northern Idaho, dam-site monitoring at Flaming Gorge and Glen Canyon (Lake Powell) (Fig. 3), and added USGS coverage of the Hebgen Lake–Yellowstone area. The 1968 "snapshot" misses the presence of scattered Department of Defense, LRSM mobile seismic observatories operated on a temporary basis in the mid 1960s at a few dozen sites throughout the Intermountain region (see Poppe, 1980). The time frame does include, however, operation of a Department of Defense, VELA–Uniform array at the Uinta Basin Observatory (UBO) in northeastern Utah. Figure 8A also includes WWSSN stations installed in 1962 at Dugway (DUG), Utah, and in 1963 west of Bozeman (BOZ), Montana. The latter station operated until 1968 and was moved to Missoula, Montana, in 1973.

By the mid to late 1970s, short-period seismic telemetry networks had become well established in the Intermountain region. Seismographic coverage of the region shown for 1988 (Fig. 8B), with a total of nearly 150 stations, is chiefly a composite of three regional and five local networks (see caption for Fig. 8B). Representative reporting and details for some of the network monitoring are given by Stickney (1988) for Montana, Peyton and Smith (1990) for Yellowstone, King and others (1987) for eastern Idaho, Wood (1988) for western Wyoming–eastern Idaho, Nava and others (1990) for the Utah region, and Rogers and others (1987) for southern Nevada. (See also Wong and Humphrey, 1989, regarding local network monitoring within the Colorado Plateau of southeastern Utah between 1979 and 1987.)

Earthquake catalog

In this chapter we use the catalog through 1985 of Engdahl and Rinehart (1988; this volume), compiled for the 1988 Seismicity Map of North America and hereafter referred to as the DNAG catalog, in order to present an overview of the whole Intermountain region. Original pre-instrumental data are chiefly from sources already described in a preceding section. Sources of instrumental data vary with time, depending on the evolution of seismographic coverage in the region. Instrumental locations for sizable earthquakes in the Intermountain region were made in the

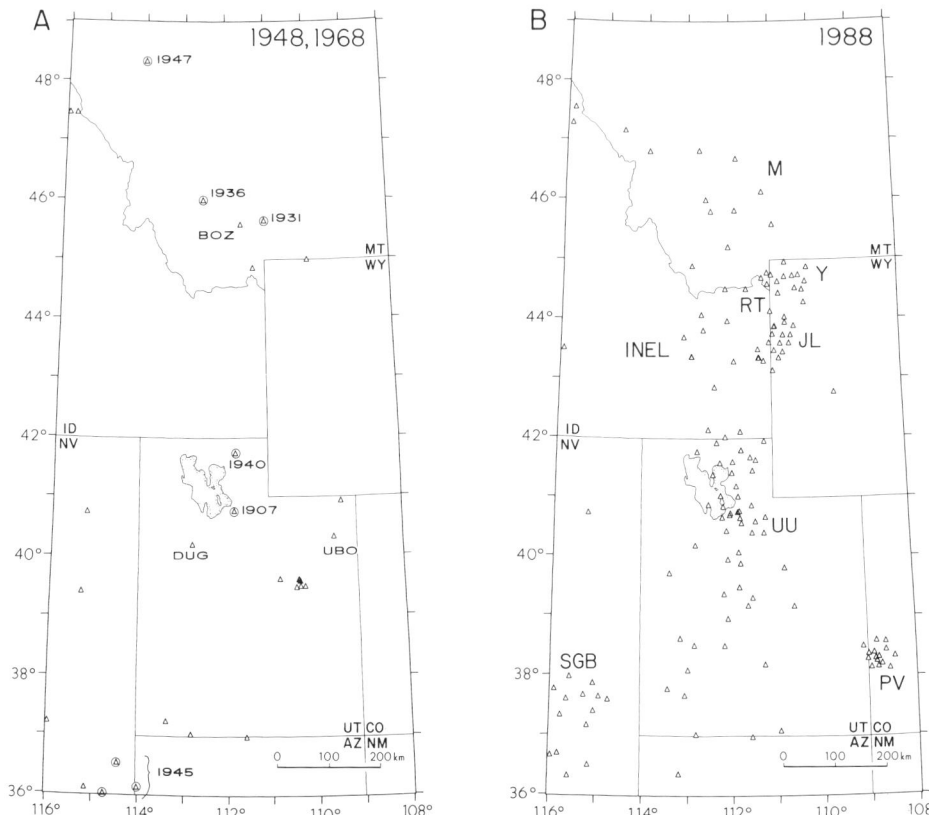

Figure 8. Maps showing representative distribution of seismographic stations in the Intermountain region at three selected times. A, Stations operating in 1948 (circles dated with year of installation) and in 1968 (triangles); circumscribed triangles, both 1948 and 1968. Key: 1907 = Salt Lake City; 1931 = Bozeman; 1936 = Butte; 1940 = Logan; 1945 = Boulder City, Overton, and Pierce Ferry (all surrounding Lake Mead). B, Stations operating in 1988, chiefly as parts of separate seismic networks. Key to seismic networks (from north to south): M = Montana (Montana Bureau of Mines and Geology, 12 sta., from ca. 1980); Y = Yellowstone (U.S. Geological Survey and University of Utah, 16 sta., from 1973); RT = Ricks-Teton (Ricks College, 5 sta., from 1972); JL = Jackson Lake (U.S. Bureau of Reclamation, 16 sta., from 1985); INEL = Idaho National Engineering Laboratory (6 stations, from ca. 1972; UU = University of Utah (57 sta. operated, 82 sta. recorded, from 1974); PV = Paradox Valley (U.S. Bureau of Reclamation, 15 sta., from 1983); SGB = Southern Great Basin (U.S. Geological Survey, 54 sta. [not all within figure], from 1978).

1920s to 1940s by the California Institute of Technology (Gutenberg and Richter, 1954) and routinely after the 1930s by the USCGS and later the U.S. Geological Survey (see *United States Earthquakes*). Figure 8A makes it evident that instrumental locations before the 1960s had to be based on recordings at widely spaced stations in the western U.S.

In mid-1962, the University of Utah began regional instrumental monitoring that later allowed compilation of an important catalog for the Utah region (36.75° to 42.50°N, 108.75° to 114.25°W), predating the installation of a modern (telemetered) regional network in 1974 (Arabasz and others, 1980). These data and subsequent, modern regional-network data make up primary sources of instrumental information in the DNAG catalog for the region (Engdahl and Rinehart, 1988). The latter include data from the University of Utah, the Montana Bureau of Mines and Geology, the Idaho National Engineering Laboratory, and the U.S. Geological Survey (Yellowstone National Park and southern Nevada). Other relevant data sources and compilations are described by Eddington and others (1987).

Time-varying thresholds of completeness since 1900 in the DNAG catalog are suggested by Engdahl and Rinehart (1988; this volume). For the study area, the overall record is best for the Utah region, where the threshold has been about magnitude 2.5 (3.0 in some distal areas) since 1962 and about magnitude 5¾ (Modified Mercalli Intensity VII) for perhaps the entire historic record (Rogers and others, 1976). For the region as a whole, our own subjective judgment indicates catalog completeness above magnitude 5¾ since 1900, above magnitude 5.0 since the 1920s to 1930s, and above magnitude 4.0 since the 1960s. Thresholds of completeness at and below magnitude 2.5 are associated with the modern regional network recording, but current completeness for the entire Intermountain region based on that recording (Fig. 8B) is at about magnitude 3.0.

The precision of instrumental earthquake locations in the region varies considerably in time and space (see original sources of data). Revised epicenters for instrumentally recorded earthquakes before the 1960s generally have uncertainties of tens of kilometers (e.g., Dewey and others, 1973; Qamar and Hawley, 1979). For earthquakes since the 1960s, epicentral precision reaches ±2 km or better within areas where seismographic spacing is of the order of a few tens of kilometers—as for some of the modern networks (Fig. 8B). Where the station spacing becomes greater, epicentral uncertainties are typically ±5 km, commonly increasing to ±10 km for events outside or in a distal part of the recording network. Reliable focal depths, requiring the presence of a recording station within roughly one focal depth of an earthquake's epicenter, are available for only a small fraction of the earthquakes in the DNAG catalog for the region.

Largest historical earthquakes

For the period from 1900 through 1985, the DNAG catalog contains 49 earthquakes in the Intermountain region with an indicated magnitude of 5.5 or greater, a selected threshold related to the potential for seriously damaging ground motions. These earthquakes are listed in Table 2 and plotted in Figure 9. The conventional magnitudes given in Table 2 are representative estimates, not necessarily the values listed in the DNAG catalog. Locations plotted in Figure 9 are directly from the DNAG catalog; refined locations, where available, are substituted in Table 2. In the following subsections we sequentially describe: (1) two known historical shocks of estimated magnitude 5.5 or greater from the pre-1900 period; (2) the four largest earthquakes in the Intermountain region's recorded history—all later than 1900, all with moment magnitudes greater than 6.5, and all but one with associated surface faulting (Table 2); and (3) other significant earthquakes after 1900 within each of the three main parts of the ISB. The descriptions are necessarily abbreviated, commonly citing one or more relevant summaries in place of original sources. For brevity, MMI signifies Modified Mercalli intensity.

Pre-1900 period. Two significant earthquakes occurred in the Intermountain region during the pre-1900 historical period. An earthquake on November 9, 1884, at 02:00 (local time) was felt strongly in Idaho, Utah, and Wyoming over at least 15,000 km^2 (Williams and Tapper, 1953). Descriptions of damage, MMI = VIII, and reports of at least six shocks felt at Paris, Idaho, in the Bear Lake Valley led Arabasz and McKee (1979) to assign an epicenter at 42.0°N, 111.3°W, arbitrarily on the Idaho-Utah border astride the active East Bear Lake fault, and to estimate a magnitude of 6.3, assuming a relation between MMI and magnitude from Gutenberg and Richter (Richter, 1958). A magnitude of at least 5½ seems likely. On November 4, 1897, at 02:29 (local time) a sizable earthquake occurred in southwestern Montana, with an assigned location of 45.0°N, 113°W, causing damage and resulting in MMI = VI at Dillon, Montana (Coffman and others, 1982). Estimating a felt area of about 500,000 km^2, Qamar and Stickney (1983) assigned a magnitude of 6.4, using an empirical relation between magnitude and felt area for Montana earthquakes.

1925 Clarkston Valley, Montana, earthquake. This 1925 earthquake (No. 11, Table 2), the second largest historical earthquake in Montana, occurred about 50 km northwest of Bozeman in the vicinity of Clarkston Valley, a late Cenozoic intermontane basin bounded on the east by the Clarkston Valley normal fault (Qamar and Hawley, 1979). Despite its significant size (M_{GR} = 6¾, M_W = 6.6, Table 2), the earthquake apparently produced no primary surface faulting, although ground cracks were observed at several localities (Pardee, 1926). The earthquake reached MMI = VIII, was felt over 800,000 km^2, caused major rockfalls, and resulted in considerable damage at nearby towns (Coffman and others, 1982; Qamar and Stickney, 1983). Seismologic data for the main shock (Doser, 1989a; Doser and Smith, 1989) indicate a mainshock focal depth of about 9 km, a subsurface rupture length of about 12 km, and oblique normal slip on a northwesterly-dipping plane with an orientation similar to that of the southern end of the Clarkston Valley fault. The main shock was preceded by at least one sizable foreshock and was followed by aftershocks as large as magnitude 4.8 (Doser, 1990).

1934 Hansel Valley, Utah, earthquake. The 1934 Hansel Valley (Kosmo) earthquake (No. 15, Table 2) remains the largest earthquake in the Utah region since 1850 and the only historical shock in the southern ISB known to have produced surface faulting. The earthquake occurred about 60 km west of the Wasatch fault in a sparsely populated, basin-range setting at the northern end of the Great Salt Lake (Fig. 3-2). The earthquake was felt over an area of 440,000 km^2 and reached MMI = VIII; it caused relatively minor damage (Coffman and others, 1982), although two deaths resulted, one direct and one indirect (Cook, 1972). Richter (1935) used the earthquake as an example in defining his magnitude scale, assigning a magnitude of 7.0 (probably overestimated because of an uncertain distance correction to Pasadena). The conventional magnitude of 6.6 assigned by Gutenberg and Richter (1954) is apparently a surface-wave magnitude and is identical to the earthquake's modern moment magnitude, M_W, of 6.6 calculated by Doser (1989b).

Shenon (1936) provided key documentation of geologic effects of the earthquake, including surface ruptures, rock slides, liquefaction, and other ground-water effects (see Arabasz, 1979, Doser, 1989b, and dePolo and others, 1989, for reference to other original sources). Shenon (1934) mapped four northerly-trending subparallel fractures displacing salt flats and unconsolidated late Quaternary sediments in the southwestern part of Hansel Valley over a zone about 6 km wide and 12 km long (see Doser, 1989b). Displacements were primarily vertical, up to a maximum of 50 cm, but a horizontal offset of 25 cm was also reported (dePolo and others, 1989). The relation of the 1934 surface rupturing to local geologic structure and neotectonics is of

TABLE 2. EARTHQUAKES IN THE INTERMOUNTAIN SEISMIC BELT
OF MAGNITUDE 5.5 AND GREATER, 1900 THROUGH 1985

No.	Date (GMT)	Time (GMT) hr mn	Lat. (°N)	Long. (°W)	Magnitude M_{conv}	M_w		Region	References
1.	1900 Aug 01	07:45	40.0	112.1	$(5\frac{1}{2}\pm)$	I/A	---	Eureka, Utah	1, 2, 3L
2.	1901 Nov 14	04:39	38.8	112.1	$(6\frac{1}{2}\pm)$	I/A	---	Southern Utah (Richfield)	1, 2, 3L
3.	1902 Nov 17	19:50	37.4	113.5	$(6\pm)$	I/A	---	Pine Valley, Utah	1, 2, 3L
4.	1905 Nov 11	21:26	42.9	114.5	$(5\frac{1}{2}\pm)$	I/A	---	South central Idaho (Shoshone)	2L, 4
5.	1909 Oct 06	02:50	41.8	112.7	$(6\pm)$	I/A	---	NW Utah (Hansel Valley)	1, 2, 3L
6.	1910 May 22	14:28	40.8	111.9	$(5\frac{1}{2}\pm)$	I/A	---	Salt Lake City, Utah	1, 2, 3L
7.	1912 Aug 18	21:12	36.5	111.5	$(5\frac{1}{2}\pm)$	I/A	---	NE of Williams, Ariz.	2L
8.	1914 May 13	17:15	41.2	112.0	$(5\frac{1}{2}\pm)$	I/A	---	Ogden, Utah	1, 2, 3L
9.	1921 Sep 29	14:12	38.7	112.2	$(6\pm)$	I/A	---	Elsinore, Utah	1, 2, 3L
10.	1921 Oct 01	15:32	38.7	112.2	$(6\pm)$	I/A	---	Elsinore, Utah, 2nd main shock	1, 2, 3L
11.	1925 Jun 28	01:21	46.00	111.50	$6\frac{3}{4}$	M_{GR}	6.6	Clarkston Valley, Mont.	(8, 12)L, (5, 6)S
12.	1928 Feb 29	22:38	46.6	112.0	$(5\frac{1}{2}\pm)?$	I/A	---	Helena, Mont.	7LS, 8S
13.	1929 Feb 16	03:00	46.1	111.3	5.6	I/A	---	Lombard, Mont.	8LS
14.	1930 Jun 12	09:15	42.6	111.0	5.8	REN	---	Grover, Wyo.	2L, 10S
15.	1934 Mar 12	15:05	41.77	112.67	6.6	M_{GR}	6.6	Hansel Valley, Utah	6S, 9LS
16.	1934 Mar 12	18:20	41.57	112.75	6	M_{GR}	5.9	Hansel Valley, Utah, aftershock	6S, 9LS
17.	1934 Apr 07	02:16	41.5	111.5 ?	5.5	REN	---	Hansel Valley, Utah, aftershock?	1, 7L, 10S
18.	1934 Apr 14	21:26	41.73	112.60	5.6	REN	---	Hansel Valley, Utah, aftershock	9L, 10S
19.	1934 May 06	08:09	41.96	112.82	$5\frac{1}{2}$	M_{GR}	---	Hansel Valley, Utah, aftershock	6S, 9L
20.	1935 Oct 12	07:50	46.62	111.97	5.9	I/A	---	Helena, Mont., swarm event	8LS, 11
21.	1935 Oct 19	04:48	46.80	112.00	$6\frac{1}{4}$	M_{GR}	6.2±	Helena, Mont., swarm event	5LS, 6S
22.	1935 Oct 31	18:37	46.62	111.97	6	M_{GR}	6.0±	Helena, Mont., swarm event	5LS, 6S
23.	1935 Nov 28	14:41	46.60	112.00	5.5	REN	---	Helena, Mont., swarm event	2L, 10S
24.	1936 May 13	14:06	46.60	112.00	5.7	?	---	Helena, Mont., swarm event	7LS, 8
25.	1936 May 22	02:19	46.60	112.00	5.7	?	---	Helena, Mont., swarm event	7LS, 8
26.	1944 Jul 12	19:30	44.41	115.06	6.1	PAS	---	Central Idaho (Seafoam)	2, 13LS
27.	1945 Feb 14	03:01	44.61	115.09	6.0	PAS	---	Central Idaho (Clayton ?)	2, 13LS
28.	1945 Sep 23	09:57	48.00	114.20	5.5	I/A	---	Flathead Lake, Mont.	2L, 8S
29.	1947 Nov 23	09:46	44.92	111.53	$6\frac{1}{4}$	M_{GR}	6.1	Virginia City, Mont.	6S, 5LS
30.	1952 Apr 01	00:37	48.00	113.80	5.5	I/A	---	Big Fork, Mont.	2L, 8S

TABLE 2. EARTHQUAKES IN THE INTERMOUNTAIN SEISMIC BELT
OF MAGNITUDE 5.5 AND GREATER, 1900 THROUGH 1985
(continued)

No.	Date (GMT)	Time (GMT) hr mn	Lat. (°N)	Long. (°W)	Magnitude M_{conv}	M_w	Region	References
31.	1959 Jul 21	17:39	37.00	112.50	5.7	PAS	---	Arizona-Utah border	2L, 7S
32a.	1959 Aug 18	06:37	44.88	111.11	6.3 m_b	6.3	Hebgen Lake, Mont., double event	14LS, 15S
32b.	**1959 Aug 18**	**06:37**	**44.84**	**111.03**	7.5 M_S	7.3	Hebgen Lake, Mont., double event	14L, (15,16)S
33.	1959 Aug 18	07:56	45.00	110.70	$6\frac{1}{2}$ BRK	---	Hebgen Lake, Mont., aftershock	15L, 20S
34.	1959 Aug 18	08:41	45.08	111.80	6 BRK	---	Hebgen Lake, Mont., aftershock	15L, 20S
35.	1959 Aug 18	11:03	44.94	111.80	$5\frac{1}{2}+$ BRK	---	Hebgen Lake, Mont., aftershock	15L, 20S
36.	1959 Aug 18	15:26	44.85	110.70	6.3 M_S	6.3	Hebgen Lake, Mont., aftershock	15LS
37.	1959 Aug 19	04:04	44.76	111.62	6 BRK	6.0	Hebgen Lake, Mont., aftershock	5L, (5, 20)S
38.	1962 Aug 30	13:35	41.92	111.63	5.7 M_S	5.6	Cache Valley (Logan), Utah	17LS
39.	1964 Oct 21	07:38	44.86	111.60	5.8 m_b	5.6	Hebgen Lake, Mont., aftershock	5LS, 7S
40.	1966 Aug 16	18:02	37.46	114.20	6.0 PAS	5.3	Southeast Nevada	7S, 15LS
41.	1966 Aug 18	10:09	37.30	114.20	5.6 m_b	---	Southeast Nevada, aftershock	7LS
42.	1975 Mar 28	02:31	42.06	112.53	6.0 M_S, GS	6.2	Pocatello Valley (Ida.-Utah border)	15S, 18LS
43.	1975 Jun 30	18:54	44.69	110.62	6.1 M_L, GS	---	Yellowstone Park, Wyo.	19LS
44.	1976 Dec 08	14:40	44.76	110.80	5.5 m_b, GS	---	Yellowstone Park, Wyo., aftershock	7LS
45.	**1983 Oct 28**	**14:06**	**43.97**	**113.92**	7.3 M_S, GS	6.9	Borah Peak, Idaho	13LS, 21S
46.	1983 Oct 28	19:51	44.05	113.92	5.8 M_L	5.4	Borah Peak, Idaho, aftershock	13L, (15, 21)S
47.	1983 Oct 29	23:29	44.24	114.06	5.8 M_L	5.5	Borah Peak, Idaho, aftershock	13L, (15, 21)S
48.	1983 Oct 29	23:39	44.24	114.11	5.5 m_b, GS	---	Borah Peak, Idaho, aftershock	13LS
49.	1984 Aug 22	09:46	44.37	114.06	5.8 M_L	5.6	Borah Peak, Idaho, aftershock	13L, (15, 22)S

Explanation:
 The local time (Mountain Standard Time) for the earthquakes in this table is found by subtracting seven hours from Greenwich Mean Time. In some cases (War Time, Daylight Savings Time), the difference is six hours. Non-instrumental and instrumental earthquake locations are listed with one- and two-decimal-point accuracy, respectively. **Earthquakes accompanied by surface faulting have their origin time and location in bold print.**
 Abbreviations for earthquake magnitude: M_{conv} = conventional magnitude, including M_{GR} (unified magnitude determined by Gutenberg and Richter), M_L (local magnitude), M_S (surface-wave magnitude), and m_b (body-wave magnitude); I/A = estimate of M_L based on intensity and/or felt area (values in parentheses based on authors' judgment here); REN = Reno, empirical estimate of M_{GR}; PAS = Pasadena, unspecified M_S or M_L (except 31 and 40, known to be M_L); BRK = Berkeley (all values here are M_L); GS = U.S. Geological Survey; M_w = moment magnitude (Hanks and Kanamori, 1979).
 Key to references, including source of location (L) and size (S): 1. Williams and Tapper (1953); 2. Coffman and others (1982); 3. Arabasz and McKee (1979); 4. Townley and Allen (1939); 5. Doser (1989a); 6. Gutenberg and Richter (1954); 7. Engdahl and Rinehart (this volume); 8. Qamar and Stickney (1983); 9. Doser (1989b); 10. Jones (1975); 11. Doser (1990); 12. Qamar and Hawley (1979); 13. Dewey (1987); 14. Doser (1985); 15. Doser and Smith (1989); 16. Abe (1981); 17. Westaway and Smith (1989b); 18. Arabasz and others (1981); 19. Pitt and others (1979); 20. R. Uhrhammer (personal communication, 1990); 21. Richins and others (1987); 22. Zollweg and Richins (1985).

continuing interest (McCalpin and others, 1987; dePolo and others, 1989), especially in light of recent seismic waveform modeling by Doser (1989b) that indicates a main-shock focal mechanism with nearly pure strike slip, rather than normal slip. The waveform modeling implies a main-shock focal depth of 8 to 10 km, left-lateral slip on a plane striking N38° to 48°E, and a subsurface rupture length of about 11 km (Doser, 1989b). Strong aftershocks were recorded at regional distances (Table 2).

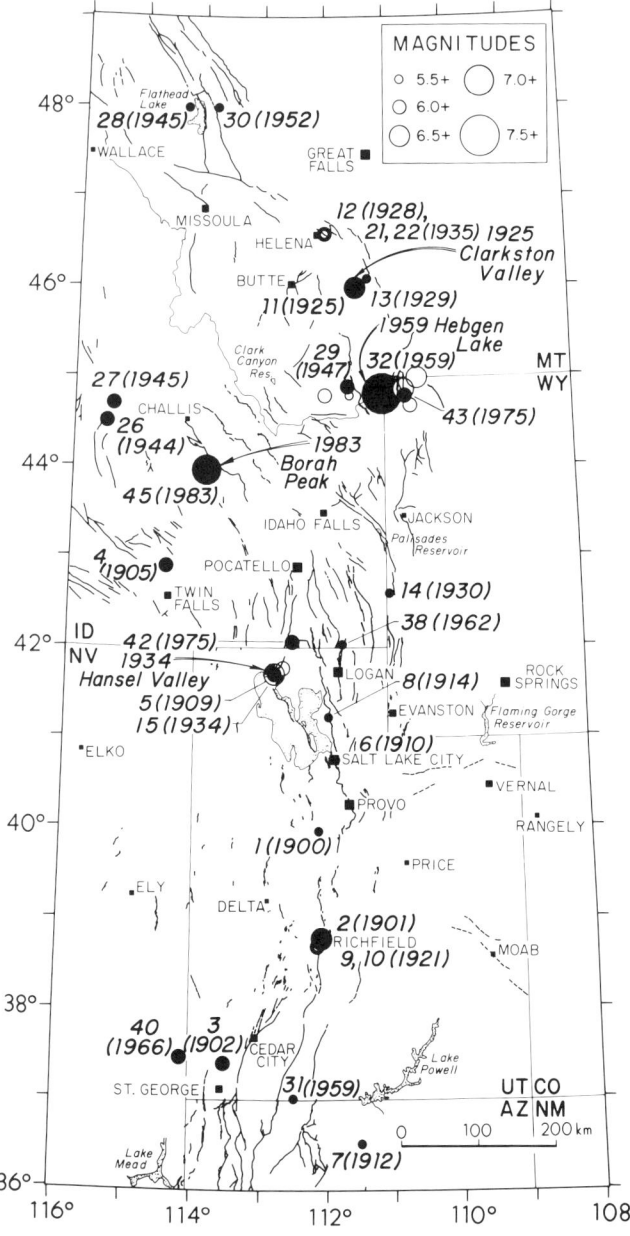

Figure 9. Map showing larger earthquakes in the Intermountain region, 1900–1985, together with selected faults, as in Figure 3. Plot includes main shocks of magnitude 5.5 and greater (solid circles) and aftershocks of magnitude 6.0 and greater (open circles). Numbers and dates for main shocks are keyed to Table 2. Four largest historical shocks known to exceed moment magnitude (M_W) 6.5 are indicated by name.

1959 Hebgen Lake, Montana, earthquake. The 1959 earthquake near Hebgen Lake in southwestern Montana (No. 32a, b, Table 2), directly west of Yellowstone National Park, was the first large normal-faulting earthquake in the Intermountain region in historical time and is distinguished as the largest recorded earthquake in the ISB (see special collection of papers in vol. 52, no. 1, of the *Bulletin of the Seismological Society of America,* 1962). The surface-wave magnitude (M_S) of 7.5 from Abe (1981) is a multistation estimate judged to be more accurate than the earthquake's previously estimated magnitude of 7.1, attributed to Pasadena by Tocher (1962). The occurrence of the $M_S = 7.3$ Borah Peak, Idaho, earthquake in 1983 (discussed below) prompted immediate comparison between the two large earthquakes. From comparisons of surface faulting (Hall and Sablock, 1985) and recorded seismograms (Bolt, 1984), the larger size of the Hebgen Lake earthquake was evident—confirmed by subsequent comparison of the two earthquakes' source parameters determined both seismically and geodetically (e.g., Barrientos and others, 1987). Restudy of Wood-Anderson seismograms at Berkeley and Pasadena for the 1959 earthquake and comparison with counterpart recordings for the 1983 earthquake led to a revised estimate of 7.7 to 7.8 for the local or Richter magnitude (M_L) of the 1959 earthquake and assignment of 7.2 (M_L) for the 1983 earthquake (Bolt, 1984). This relative size information is valuable. Given the large distances to Berkeley and Pasadena, however, the distance correlations—and hence the absolute values of M_L—are open to question.

The Hebgen Lake earthquake affected an area of 1.5 million km^2, reached MMI = X, caused 28 fatalities, and produced dramatic surficial geologic effects, including spectacular fault scarps, a catastrophic rockslide into the Madison River, basin subsidence of several meters, and hydrogeomorphic features associated with groundwater discharge (Witkind and others, 1962; Coffman and others, 1982). The earthquake produced a 26-km-long complex pattern of west- to northwest-trending normal faulting along the Hebgen and Red Canyon faults near the southern end of the Madison Range, where Laramide faults exert structural control (Witkind, 1964; Doser, 1985b). Trace lengths of surface faulting, including slip on the nearby north-south-trending Madison fault, sum to 61 km (Hall and Sablock, 1985). Maximum vertical displacement is variously cited as 6.7 m by dePolo and others (1989) and 5.5 ± 0.3 m by Bonilla and others (1984). Hall and Sablock (1985) use data from Witkind (1964) to estimate an average vertical displacement of 2.0 m on the Hebgen fault and 2.3 m on the Red Canyon fault.

Notable seismological details of the earthquake sequence (Doser, 1985b, 1989a) include: (1) the occurrence of the main shock as a multiple event, consisting of a shock of $m_b = 6.3$ at about 10 km depth followed 5 sec later by the principal $M_S = 7.3$ shock at 15 km depth; (2) nearly pure dip-slip motion during the main shock on one or more planes dipping 40° to 60°SW; and (3) the location of numerous strong aftershocks as large as $M_S = 6.3$ (Table 2) out to distances of 50 km from the main-shock epicenter. Barrientos and others (1987) recently reanalyzed the

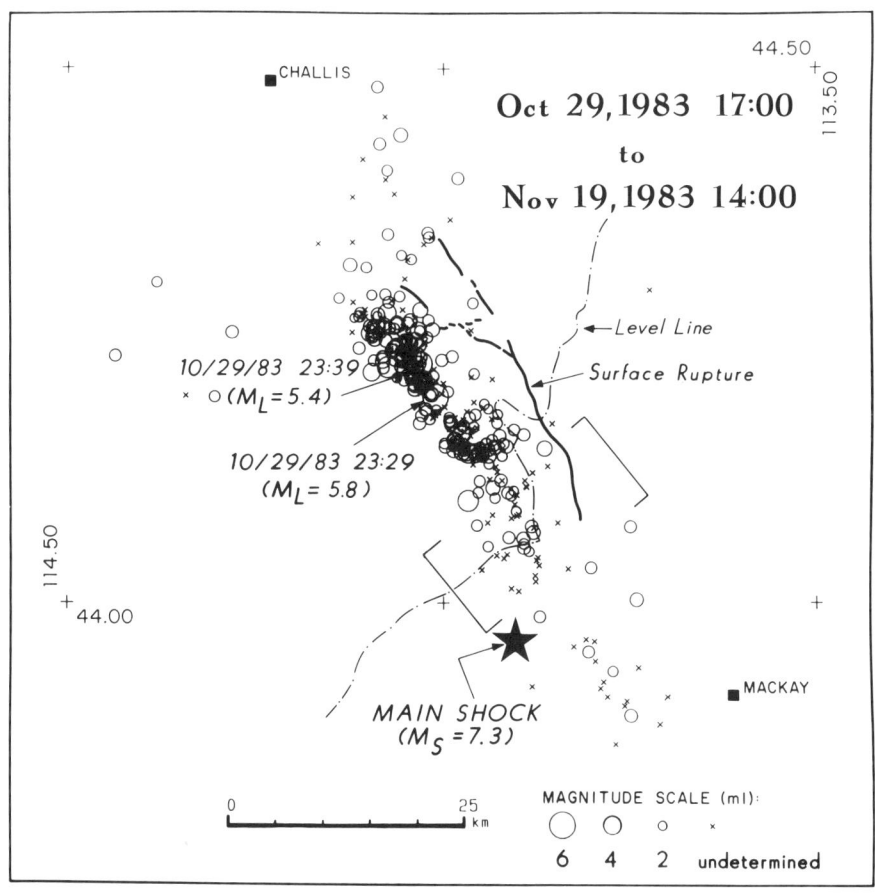

Figure 10. Map (from Richins and others, 1987) showing epicenters of the main shock and early aftershocks of the Borah Peak, Idaho, earthquake sequence, together with the trace of surface rupture. Brackets outline sample area for earthquakes shown in cross section of Figure 11 (middle); dot-dashed line, trace of geodetic level line from which data are displayed in Figure 11 (top).

static deformation field associated with the earthquake using a newly augmented geodetic data set. They interpret a complex source consisting of two en-echelon planes, 15 to 25 km long, which are coincident with the Hebgen and Red Canyon faults at the surface, extend to a depth of 10 to 15 km, dip 45° to 50°SW, and have coseismic dip slip of 7.0 and 7.8 m.

1983 Borah Peak, Idaho, earthquake. This 1983 surface-faulting earthquake (No. 45, Table 2) occurred in east-central Idaho, 60 km northwest of the Snake River Plain, in a sparsely settled area characterized by active late Quaternary basin-range faulting (Scott and others, 1985) but low historic seismicity (Dewey, 1987). The $M_S = 7.3$ earthquake is the second-largest historical earthquake in the Intermountain region. Importantly, the earthquake allowed abundant modern observations about the mechanics, subsurface rupture geometry, and seismic geology of a large normal-faulting earthquake (see special compendia of papers in U.S. Geological Survey Open-File Report 85-290, 1985, and vol. 77, no. 3, of the Bulletin of the Seismological Society of America, 1987).

The Borah Peak earthquake was felt over 670,000 km^2, reached MMI = VII at the nearby towns of Mackay and Challis (25 km and 65 km distant, respectively, from the main-shock epicenter), caused two deaths in Challis, and resulted in about $12.5 million of damage (Stover, 1985). The earthquake produced 36 km of surface faulting along the southwestern base of the Lost River Range, re-rupturing parts of the 140-km-long Lost River fault that had last broken about mid-Holocene time and a branch fault; vertical displacement along the new fault scarps reached a maximum of 2.7 m (0.8 m average), and net slip averaged 0.17 m of sinistral slip for every 1.00 m of dip slip (Crone and Machette, 1984; Crone and others, 1987). There is substantial information about the segmented behavior of the Lost River fault during the 1983 earthquake and about the fault's paleoseismology (see reviews by Crone and Haller, 1989, and dePolo and others, 1989).

Figure 10 (from Richins and others, 1987) illustrates the map distribution of aftershocks with respect to the main-shock epicenter and surface rupture. The earthquake sequence included sizable aftershocks (Table 2) but no foreshocks (Richins and others, 1987; Dewey, 1987). Seismic-waveform modeling by

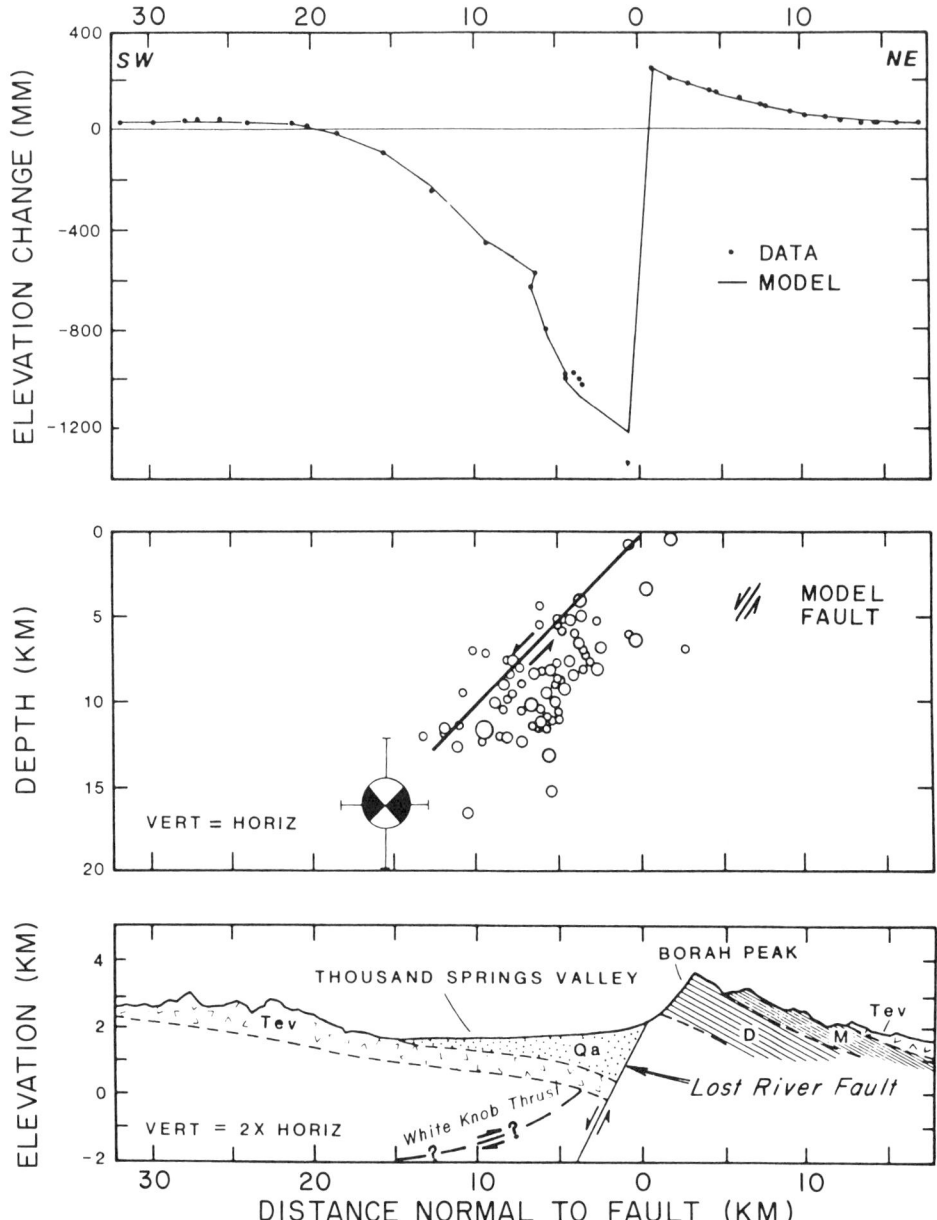

Figure 11. Diagram illustrating the subsurface geometry of faulting associated with the 1983 Borah Peak, Idaho, earthquake (after Stein and Barrientos, 1985). Top, plot of geodetically-observed coseismic elevation changes (dots) and the predicted elevation changes (line) of the coseismic dislocation model. Middle, cross section showing superposition of coseismic dislocation model, aftershock foci from the bracketed area shown in Figure 10, and the projection of the main-shock focus. Bottom, schematic geologic cross section from Bond (1978).

Doser and Smith (1985) indicates the main shock nucleated at a depth of about 16 km and propagated unilaterally northwestward toward the surface along a fault plane dipping 45° to 53° southwest (see Richins and others, 1987, Table 2, for a comparative tabulation of the main shock's source parameters). Figure 11, after Stein and Barrientos (1985), usefully illustrates some of the principal aspects of subsurface fault geometry and deformation. The data are displayed in transverse view to the Lost River fault (bottom) and are keyed to a northeast-southwest geodetic profile of coseismic elevation changes (top) along an irregular leveling route roughly transverse to the fault (Fig. 10). The middle panel shows a cross section of aftershock foci from Richins and others (1985), correlative with the bracketed sample area shown in Figure 10, together with the location and focal mechanism of the main shock and a planar dislocation model that matches the observed surface deformation. Focal mechanisms for 47 aftershocks suggest that most of the aftershock foci reflect complex fracturing on secondary structures adjacent to the main fault plane rather than seismic afterslip on a simple main-shock rupture plane (Richins and others, 1987).

Barrientos and others (1987) have refined the dislocation model represented in Figure 11 using supplementary geodetic data, but changes to the illustration would be slight. Their preferred model has a planar fault in the depicted section that dips 49° (instead of 47° as shown), extending to a depth of 14 km; the northern part of the fault is modeled with a dislocation extending only to 6 km depth. The modeling yields an average dip-slip displacement of 2.1 m on the southern dislocation and 1.4 m on the northern one. An important point is that the geodetic data do not permit a listric fault geometry (see also Stein and Barrientos, 1985). Source parameters determined by Barrientos and others (1987) from this geodetic modeling (slip = 1.4 to 2.1 m, static stress drop = 30 bars, moment = 2.9×10^{26} dyne-cm) are consistent with seismically-determined values for the earthquake (e.g., Doser and Smith, 1985: slip = 1.4 m, static stress drop = 17 bars, moment = 2.1×10^{26} dyne-cm).

Other significant earthquakes

Table 2 and Figure 9 give a succinct overview of other significant earthquakes in the Intermountain region besides the four largest just described. Using Table 2 for reference, to minimize repetition of information, we complete this section by mentioning the other main shocks of magnitude 6 and some notable smaller shocks.

Southern ISB. The 1901 Southern Utah (Richfield) earthquake (Event No. 2) appears to be the second largest historical shock in the southern ISB, although its equivalent magnitude and precise epicenter are not well known. The earthquake reached MMI = IX, was felt over 130,000 km², caused substantial damage at several towns and produced ground cracks (but no documented surface faulting), local liquefaction, and extensive rockslides (Williams and Tapper, 1953). In late 1921, following two and a half weeks of foreshock activity, the Elsinore area, 10 km southwest of Richfield, was struck by two damaging earthquakes of about magnitude 6¼ (Events No. 9 and 10), separated by 50 hours and with an intervening shock of magnitude 5¾ (Pack, 1921; Arabasz and Julander, 1986). Southwestern Utah was significantly affected in 1902 by a damaging earthquake (MMI = VIII) centered in Pine Valley (Event No. 3) and in 1966 by a sizable earthquake (m_b, USGS = 6.1; M_L, Univ. of Utah = 5.6) close to the Nevada-Utah border (Event No. 40) (Arabasz and others, 1979; Coffman and others, 1982). The 1966 event was notable for its strike-slip focal mechanism in a basin-range setting (Smith and Sbar, 1974; Rogers and others, this volume).

Two other magnitude 6 shocks in the southern ISB occurred in the Utah-Idaho border area. This includes a strong (MMI = IX) earthquake in 1909 (Event No. 5), assumed to have originated in the Hansel Valley area (Williams and Tapper, 1953), and the damaging M_L = 6.0 Pocatello Valley (Idaho-Utah border area) earthquake of 1975 (Event No. 42), notable for its discordant relation with the surface geology (Arabasz and others, 1981) and its space-time pattern of precursory seismicity (Arabasz and Smith, 1981). Another noteworthy earthquake is the M_L = 5.7 Cache Valley (Logan) earthquake of 1962 (Event No. 38), the most damaging yet in Utah's history (Cook, 1972; Rogers and others, 1976; Westaway and Smith, 1989b) and the only sizable earthquake in the Utah region for which good strong-motion recordings—one three-component set—currently exist (Smith and Lehman, 1979; Westaway and Smith, 1989a, b).

Central ISB. Four magnitude 6 main shocks have occurred in the Central ISB. Shocks of magnitude 6.1 (Event No. 26) and magnitude 6.0 (Event No. 27) occurred seven months apart in 1944 and 1945 in central Idaho along the eastern flank of the Idaho batholith, causing only minor damage in the remote mountainous setting (Coffman and others, 1982). Revised instrumental locations place both earthquakes relatively close to each other, suggesting a possible main-shock/large-aftershock relation (Dewey, 1987). The magnitude 6¼ Virginia City, Montana, earthquake of 1947 (Event No. 29) caused considerable damage in the Madison Valley (Coffman and others, 1982). The earthquake occurred about 50 km west-northwest of the 1959 Hebgen Lake earthquake and has been studied by Doser (1989a). Large aftershocks of the 1959 earthquake extended into the Yellowstone National Park region (Fig. 9) where an independent main shock occurred later in 1975. The M_L = 6.1 Yellowstone Park earthquake of 1975 (Event No. 43) caused moderate disruption in the national park (Coffman and others, 1982) and provided valuable information on the seismotectonics of the Yellowstone caldera (Pitt and others, 1979). A strong earthquake (MMI = VII), probably in the magnitude 5 range, which caused damage at Shoshone, Idaho, in 1905 (Event No. 4) (Coffman and others, 1982), has an uncertain epicenter but is significant for its possible association with the relatively aseismic Snake River Plain.

Northern ISB. The 1935–1936 Helena earthquakes (Event nos. 20 to 25) were part of a vigorous swarm of more than two thousand felt earthquakes that occurred within 10 to 25 km of Helena, between October 1935 and December 1936, within a poorly defined structural zone extending from Helena northwestward toward Marysville, an area of anomalously high heat flow (Smith and Sbar, 1974; Doser, 1989a). Earthquakes of magnitude 6¼ (Event No. 21) and magnitude 6 (Event No. 22) in October 1935 caused four deaths and severe damage in Helena (Coffman and others, 1982). Some important strong-ground-motion records were recorded locally (Westaway and Smith, 1989a). Other earthquakes in the northern ISB listed in Table 2, which caused only minor damage (Coffman and others, 1982; Qamar and Stickney, 1983), include: a 1928 shock near Helena (Event No. 12), perhaps a precursor to the 1935 swarm; a 1929 shock near Lombard (Event No. 13), 9 km north of Clarkston, possibly a late aftershock of the 1925 Clarkston Valley earthquake; and two shocks in the Flathead Lake region in 1945 (Event No. 28) and 1952 (Event No. 30).

DETAILED SEISMICITY

In earlier sections we gave an overview of the regional-scale patterns of earthquake activity in the ISB (Fig. 1) and described the largest earthquakes that have occurred historically (Fig. 9). In this section, we describe finer details of the spatial distribution of the seismicity portrayed in Figure 1, outlining what is known about the association of the seismicity with geologic structure. Observational data basically come from either long-term monitoring with telemetered seismic networks (Fig. 8B) or focused short-term monitoring using temporary arrays of portable seismographs. Some notable characteristics of observed seismicity throughout the ISB are: (1) the diffuse epicentral scattering of background earthquakes, with weak correlation to major active faults; (2) the conspicuous seismic quiescence of many major faults or fault segments that have been active in Holocene and late Quaternary time; (3) the predominance of focal depths shallower than 20 km; and (4) the prevalence of normal and oblique-normal seismic slip, but with local strike slip and reverse slip.

Southern ISB

Detailed summaries of the instrumental seismicity for major parts of the southern ISB have recently been given by Arabasz and others (1987) for the Wasatch Front area, by Arabasz and Julander (1986) for the Basin and Range–Colorado Plateau transition in central Utah, and by Wong and Humphrey (1989) for the Colorado Plateau. Here, we selectively adapt from and add to those summaries.

The recorded seismic history of the southern ISB has been distinctively characterized by abundant small- to moderate-sized earthquakes (magnitude ≤ 6.6) without a truly large surface-faulting earthquake, despite the widespread presence of late Pleistocene and Holocene fault scarps. The single instance of historical surface faulting in the southern ISB at Hansel Valley in 1934 (discussed earlier) was for an earthquake only slightly above the threshold of surface faulting. Thus, there is a lack of instrumental information relating to large-scale seismic slip on major faults in this region. Available earthquake observations may chiefly reflect seismic deformation on secondary structures.

The seismicity of the Wasatch Front area shown in Figure 12 displays no simple correlation between the distribution of background earthquakes and the traces of the numerous active faults, except for the general parallelism of this part of the ISB with the Wasatch fault. The depth above which 90 percent of the well-located earthquakes lie varies locally from about 11 to 17 km (Arabasz and others, 1987). Well-located shocks of magnitude 2.0 and greater in this area from 1962 to 1986 have a distinct peak in their depth distribution between 4.5 and 7.5 km (Bjarnason and Pechmann, 1989). An anomalously deep earthquake of $M_L = 3.8$ occurred at a depth of 90 km beneath northern Utah in 1979 (see Wong and Chapman, 1990).

The seismicity pattern of Figure 12B is representative of instrumental seismicity in the area since 1962 (see Arabasz and others, 1980); spatial clustering is due more to cumulative stationary activity than to isolated temporal bursts. An inverted Y-shaped pattern of clustered earthquakes on the Idaho-Utah border west of the Wasatch fault began to develop several months after the 1975 $M_L = 6.0$ Pocatello Valley earthquake (No. 42, Fig. 9) and persists to the present. Early aftershocks of the 1975 earthquake occurred mostly north of the Idaho-Utah border (Arabasz and others, 1981). Special studies of post-1975 earthquakes within the inverted Y-pattern (Jones, 1987; Chen, 1988) show that well-located foci are mostly shallower than 8 to 12 km deep, scatter beneath both horsts and grabens of the local basin-range structure, and display diverse seismic slip.

A linear north-south belt of seismicity about 15 to 40 km east of and parallel to the Wasatch fault (Fig. 12) is poorly understood but may be mechanically related to crustal flexure associated with the Wasatch fault and involving non-elastic mechanisms (Zandt and Owens, 1980; Owens, 1983). This seismicity is known to coincide in map view with the eastern leading edges of several Laramide thrust sheets (Smith and Bruhn, 1984). In cross-section view, however, the earthquake foci are diffusely scattered to about 20 km depth (Arabasz and others, 1987; Owens, 1983). In the northern part of the belt, the earthquakes lie east of the west-dipping East Cache and Wasatch faults, perhaps partly on a synthetic fault (Westaway and Smith, 1989b); south of 41°20′N, the epicentral belt follows a series of small, late Cenozoic structural basins within the Middle Rocky Mountains (Sullivan and others, 1988). In the lower right part of Figure 12, the prominent arcuate pattern of seismicity east of the Wasatch fault is mining related (see Induced seismicity).

What about earthquakes along the Wasatch fault itself? In addition to having no historic surface rupture, despite recurrent late Pleistocene and Holocene surface faulting on multiple segments (Machette and others, 1989; Schwartz and Coppersmith, 1984), the Wasatch fault has had little historical seismicity. As many as two, and perhaps no, earthquakes as large as magnitude 5 have occurred on the Wasatch fault in historical time (Arabasz and others, 1987). The most recent surface rupture occurred about 400 years ago (Machette and others, 1989) on a 40-km-long segment immediately north of Nephi (Fig. 12A). In terms of contemporary seismicity, Figure 12B shows a remarkable paucity of microseismicity along most of the Wasatch fault. There are a few local clusters of epicenters along the fault north of Brigham City, and more prominent clusters just west of the fault in the vicinity of Salt Lake City, at the northern end of Utah Valley (~40°20′N), in the vicinity of Goshen Valley (~40°00′N), and in a broadly scattered zone at the southern end of the Wasatch fault. For Goshen Valley and the southern Wasatch fault, hypocenters and corresponding focal mechanisms from portable-array studies show that background earthquakes west of the fault are not occurring on either a listric or a simple planar projection of the fault (Arabasz and Julander, 1986). Elsewhere along the Wasatch fault, cross sections suggest that very few well-located foci could be interpreted to lie on the fault—if one believes that the fault is a planar structure of moderate dip (Arabasz and others,

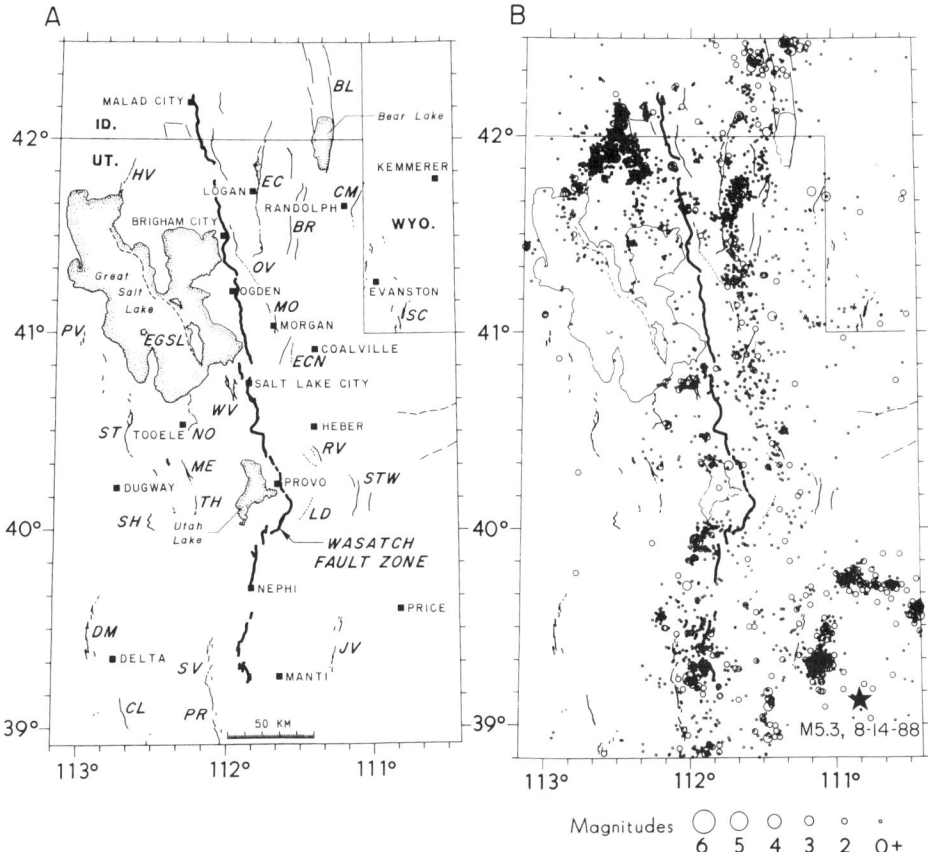

Figure 12. Active faulting and seismicity in the Wasatch Front area, outlined in Figure 3, from Arabasz and others (1987). A, Map showing traces of late Quaternary faulting, abbreviated as follows: BL = Bear Lake; BR = Bear River Range; CL = Clear Lake; CM = Crawford Mts.; DM = Drum Mts.; EC = East Cache; ECN = East Canyon; EGSL = East Great Salt Lake; HV = Hansel Valley; JV = Joes Valley; LD = Little Diamond Creek; ME = Mercur; MO = Morgan; NO = Northern Oquirrh; OV = Ogden Valley; PR = Pavant Range; PV = Puddle Valley; RV = Round Valley; SC = Sulphur Creek; SH = Sheeprock Mts.; ST = Stansbury Mts.; STW = Strawberry Valley; SV = Scipio Valley; TH = Topliff Hill; WV = West Valley. B, Epicenter map of all earthquakes located by the University of Utah Seismograph Stations in the Wasatch Front area, July 1, 1978 to December 31, 1986; star indicates location of 1988 San Rafael Swell, Utah, main shock described in Figure 14.

1987). But the same data are admittedly either inadequate or ambiguous (e.g., Pechmann and Thorbjarnardottir, 1984) for interpreting subsurface association with a listric projection of the fault.

Portable-seismograph studies throughout the transition zone between the Basin and Range and Colorado Plateau provinces in central and southwestern Utah reveal that low-angle structural discontinuities appear to play a fundamental role in separating locally intense upper-crustal seismicity above 6 to 8 km depth from less frequent background earthquakes at greater depth (Arabasz and Julander, 1986). This is illustrated in Figure 13A, where spatially discontinuous seismicity with depth beneath the Sevier Valley near Richfield, Utah, coincides with a low-angle detachment inferred from seismic-reflection data. The earthquakes beneath the northwestern side of the valley are background events recorded in 1981; those beneath the southeastern side are aftershocks of an $M_L = 4.0$ earthquake in May 1982.

Companion results from a portable-seismograph study of an earthquake swarm sequence ($M_L \leq 4.7$) in October 1982 near Soda Springs, Idaho (Fig. 13B), similarly show a depth distribution of upper-crustal earthquakes apparently influenced by pre-existing low-angle structures. Instead of being associated, as expected, with late Cenozoic basin-range faulting along the active Bear Lake fault, the seismicity is associated with secondary faults within a northwest-trending near-vertical zone in the hanging-wall block. Marked changes in the vertical distribution of foci coincide with pre-Neogene thrust faults. Focal mechanisms sampled from both above and below the Meade thrust indicate a predominance of strike slip on northwest-trending, steeply-dipping fault planes, with no evidence for seismic slip on a low-angle plane (see Arabasz and Julander, 1986).

East of the southern ISB, scattered seismicity within the Colorado Plateau (Figs. 1 and 2), described by Wong and Humphrey (1989), occasionally reaches the magnitude 6 range

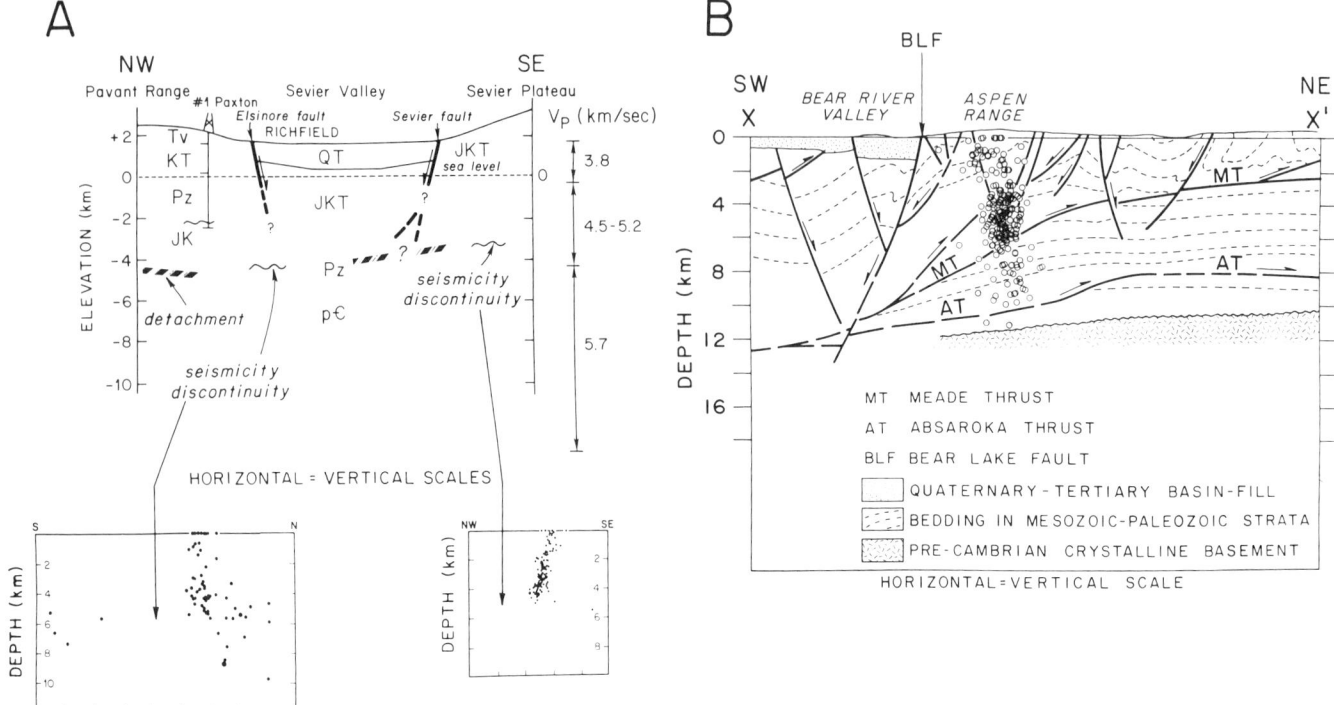

Figure 13. Results of two selected earthquake field studies carried out by the University of Utah illustrating the influence of low-angle structural discontinuities on seismicity, from Arabasz and Julander (1986). A, Schematic geologic cross section across the Sevier Valley near Richfield, Utah, showing spatially discontinuous seismicity with depth coincident with the location of a low-angle detachment inferred from seismic-reflection data. V_P = P-wave velocity; T_V = Tertiary volcanic rocks; other abbreviations, standard for geologic ages of rocks. B, Cross section showing the association of swarm seismicity in late 1982 near Soda Springs, Idaho, with geologic structure; earthquake data from Richins and others (1983) are superposed on a generalized geologic cross section from Dixon (1982) based on seismic-reflection profiling.

and appears chiefly to reflect normal to lateral seismic slip on buried Precambrian basement faults without evident surface expression. Focal depths predominate above 15-20 km, but the Colorado Plateau is distinctive in having observed seismicity in the lower crust, and locally 40 to 60 km deep in the uppermost mantle (Wong and Humphrey, 1989; Wong and Chapman, 1990). Figure 14 illustrates details of what may be a typical, moderate-sized crustal earthquake within the Colorado Plateau. The August 1988 San Rafael Swell earthquake (M_L = 5.3) involved oblique-normal slip on a buried Precambrian basement fault in an area of minimal historical seismicity where there are no active faults mapped in the overlying 3-km-thick sedimentary cover rocks of Mesozoic and Paleozoic age (Nava and others, 1988). Figure 14 also usefully illustrates some prerequisites for associating seismicity with geologic structure: (1) local seismographic control for hypocentral resolution, especially for precise focal depths; (2) sufficient seismicity for defining the spatial geometry of one or more active structures; and (3) a reliable focal mechanism for correlating with the geometry and sense of slip inferred from the seismicity.

Noteworthy microearthquake studies in the southern Utah–northern Arizona area have been reported by Johnson and Sbar (1987) and Kruger-Knuepfer and others (1985). Both studies present valuable focal-mechanism information relevant to regional stress orientations. Other key studies that summarize and discuss significant focal-mechanism information for the southern ISB include those by Bjarnason and Pechmann (1989), Wong and Humphrey (1989), Arabasz and Julander (1986), Zoback (1983), Arabasz and others (1980), Smith and Lindh (1978), and Smith and Sbar (1974). Besides implications for stress state (discussed earlier), the focal mechanisms also provide information on fault kinematics. Seismic slip predominates on fault segments of moderate (>30°) to steep dip. There is yet no convincing evidence, in the form of clustered earthquake foci and corroborating focal mechanisms, for seismic slip on either a downward-flattening or a low-angle normal fault in this region.

Central ISB

The central part of the ISB is distinctive in two regards. First, it has had two large surface-faulting earthquakes in historical time—the 1959 Hebgen Lake and 1983 Borah Peak earthquakes. Second, its seismicity reflects the apparent influence of the Y-SRP volcanic system. In map view this is illustrated by the arcuate pattern of seismicity described in the Introduction and shown in detail in Figure 15. This arcuate, parabolic pattern of earthquakes has been hypothesized to reflect lateral changes in deviatoric stresses in the wake of the relative passage of the Yellowstone hot spot (discussed in detail below), now centered beneath the Yellowstone caldera (Smith and others, 1985, in preparation; Anders and others, 1989; Blackwell, 1989). The seismogenic potential of the lithosphere axial to the SRP appears to be affected out to distances of more than 100 km. In Figure 15, for example, note the relative aseismicity of the SRP and the increase in seismicity away from the SRP along the trend of faults transverse to the plain.

Using Figure 15 as a guide, let us consider a counterclockwise circuit of the central ISB. Seismicity in the Utah-Idaho border region (already discussed) is near our defined juncture between the southern and central parts of the ISB. North of the Utah border there is an evident discordance between the northeast-trending seismicity belt and the northwest-trending late Cenozoic normal faulting. A refined compilation of earthquake locations verifies the regional seismicity pattern (Richins and Arabasz, 1985). On a local scale, the example of Figure 13B, from near Soda Springs in southeastern Idaho, illustrates the occurrence of small to moderate-sized background earthquakes near—but not on—one of the major active faults in this region. Portable-seismograph studies by the U.S. Bureau of Reclamation, in 1982–1983, of the region between Soda Springs and Jackson showed no obvious correlation between scattered microseismicity and local surficial faulting (Piety and others, 1986). Well-located earthquakes ($M_L \leq 3.0$) extended to 16 km depth, exhibited mostly normal slip, were apparently unaffected in their depth distribution by the presence of old thrustbelt structure, occurred abundantly within Precambrian basement below about 5 to 10 km depth, and appeared to be associated, in part, with buried normal faults having no surface expression (Piety and others, 1986).

Seismicity of the Idaho-Wyoming border area south of Yellowstone Park, encompassing the so-called Teton–Jackson Hole–southern Yellowstone region, has been summarized by Doser and Smith (1983) and by Smith and others (1990a, b) and is the target of network monitoring by the U.S. Bureau of Reclamation's Jackson Lake network (JL, Fig. 8B). Results described both by Doser and Smith (1983) from regional monitoring and

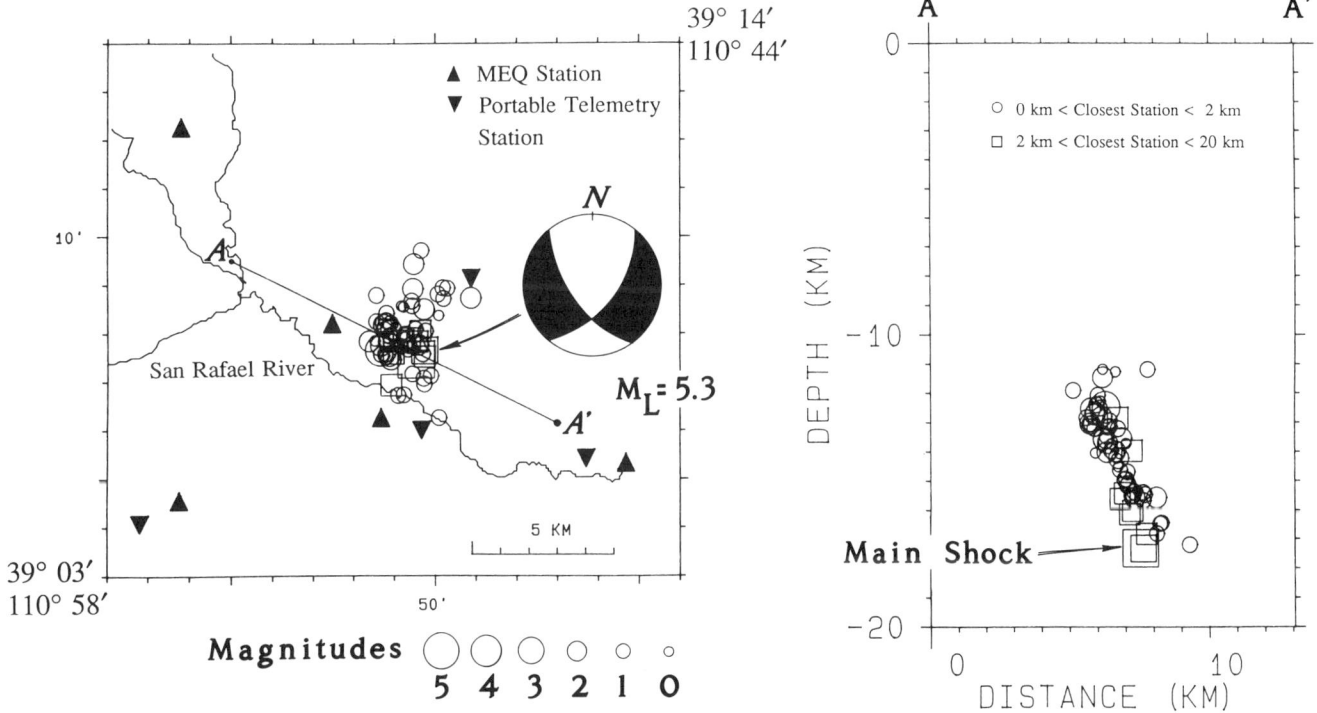

Figure 14. Details of the San Rafael Swell, Utah, earthquake sequence, August 14, 1988 to March 31, 1989, whose location is shown in Figure 12, adapted from Nava and others (1988) using revised, unpublished data from J. C. Pechmann and S. J. Nava, University of Utah. Left, map showing epicenters of the 66 best-located earthquakes in the sequence together with the main-shock focal mechanism (lower-hemisphere, compressional quadrant black). Right, cross section of earthquake foci projected onto the plane of line A-A' (left), the projection for which planar clustering was best defined; the earthquakes define a zone dipping 60° ± 5° SE, with a downdip extent of 6 km. The main-shock focal mechanism has a corresponding nodal plane striking N39°E and dipping 62°SE, with a slip-vector rake of 29° (down to the NE); uncertainties for that plane range from N20°E to N42°E in strike, 44° to 80° in dip to the SE, and 21° to 59° in slip-vector rake.

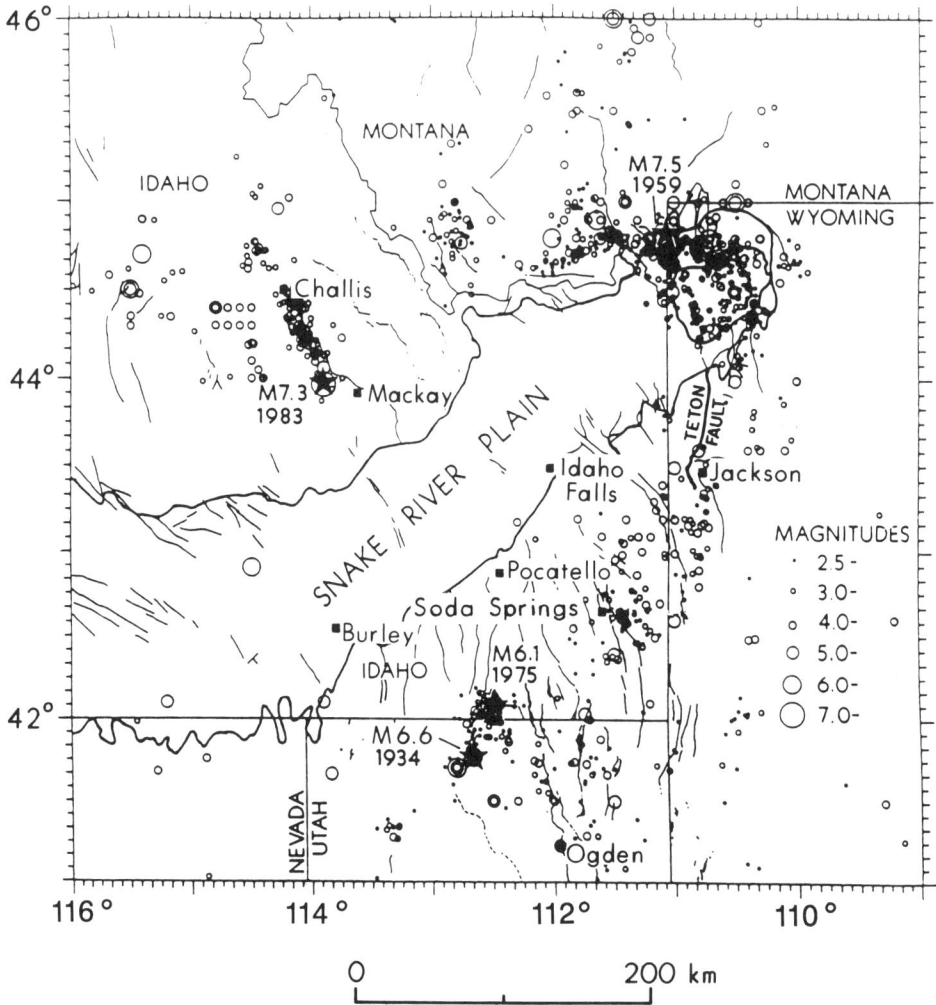

Figure 15. Map showing seismicity, 1900 to 1985, and selected Cenozoic faults of the central Intermountain seismic belt, from Smith and others (in preparation). Larger historical main shocks, as in Figure 9, shown for reference. Note the arcuate, parabolic pattern of seismicity flanking the Yellowstone–Snake River Plain basalt-rhyolite volcanic province.

local microearthquake studies and by Wood (1988) from the Jackson Lake network consistently depict: (1) diffusely scattered seismicity that correlates poorly with late Cenozoic normal faulting and, in an uncertain way, with relict thrustbelt structure in the subsurface; (2) seismic quiescence of the central part of the Teton fault, which had the greatest prehistoric surface displacements along the fault; (3) focal depths shallower than 15 km for most of the local earthquakes; and (4) fault-plane solutions with normal to strike slip reflecting general east-west extension. On a regional scale, perhaps the most noteworthy feature of the Teton region is its appearance as a distinct seismic gap in the ISB (Fig. 1), accentuated by the presence of the large and active Teton fault, which has had up to 50 m of late Quaternary displacement (Smith and others, 1990a, b).

Intense swarms of shallow earthquakes and occasional moderate-sized earthquakes as large as the $M_L = 6.1$ earthquake in 1975 near Norris Junction in Yellowstone Park (No. 43, Table 2) characterize the seismicity of the Yellowstone National Park region (Smith and others, 1977; Pitt, 1987; Nagy and Smith, 1988). The Yellowstone region is the site of one of the world's largest and most active hydrothermal-volcanic systems. During the past two million years, three catastrophic volcanic eruptions have expelled more than 3,500 km^3 of rhyolitic ashflow tuffs forming three calderas, and another 3,000 km^3 of similar material were extruded between explosive eruptions (Christiansen, 1984). A preponderance of geophysical evidence suggests that the seismicity of the Yellowstone region is directly influenced by the presence of magmas, partial melts, and hydrothermal activity at mid- to upper-crustal depths (Smith and others, 1974, 1977; Smith and Braile, 1984). Seismic slip on the boundaries of small upper-crustal blocks may reflect a combination of deformation caused by local transport of magma and hydrothermal fluid and by the regional tectonic stress field.

Figure 16 shows a representative view of background seismicity in the Yellowstone–Hebgen Lake region, together with an outline of the youngest, 600,000-year-old caldera. Earthquake clusters extend eastward from Hebgen Lake, Montana, along an east-west trend into Yellowstone National Park where they take on a northwest trend along distinct seismic zones about 25 km long that cross the caldera boundary. Within the caldera, earthquakes have not exceeded magnitude (M_L) 5.0 and generally have scattered epicenters; in the western part of the caldera, northwest-trending clusters of epicenters, together with aligned volcanic vents, may be related to buried, but still active, Quaternary faults (Christiansen, 1984). In several cases, there are good correlations between earthquake swarms and major changes in hydrothermal activity (Pitt and Hutchinson, 1982). Local faulting along the west side of Yellowstone Lake has Holocene displacements and appears to be seismically active. South of this area, seismicity has a general north-south trend as it extends southward into the Teton region. Older basin-range structure is inferred to have influenced the Quaternary tectonics of the Yellowstone region. Parts of the Gallatin and Teton normal fault systems, which generally have a northerly trend outside the Yellowstone region, presumably lie beneath the area now covered by the Quaternary volcanics of the Yellowstone Plateau.

Focal depths show conspicuous variations across the Yellowstone caldera (Fig. 17). Maximum focal depths outside the caldera are generally less than 15 to 20 km, and mostly less than 5 km beneath the inner caldera (Smith and others, 1977; Nagy and Smith, 1988). This pattern of earthquake shallowing suggests a thin layer of seismogenic brittle upper crust beneath the thermally active inner caldera. Rheologic considerations (e.g., Smith and Bruhn, 1984) imply that below about 5 km, the crust is in a quasi-plastic ductile state at temperatures in excess of 350°C, incapable of supporting large stresses. Note that the M_L = 6.1 earthquake in 1975 occurred along the caldera's northwest boundary.

Continuing our circuit of the central ISB in Figure 15, we next move westward from the Yellowstone region—following an

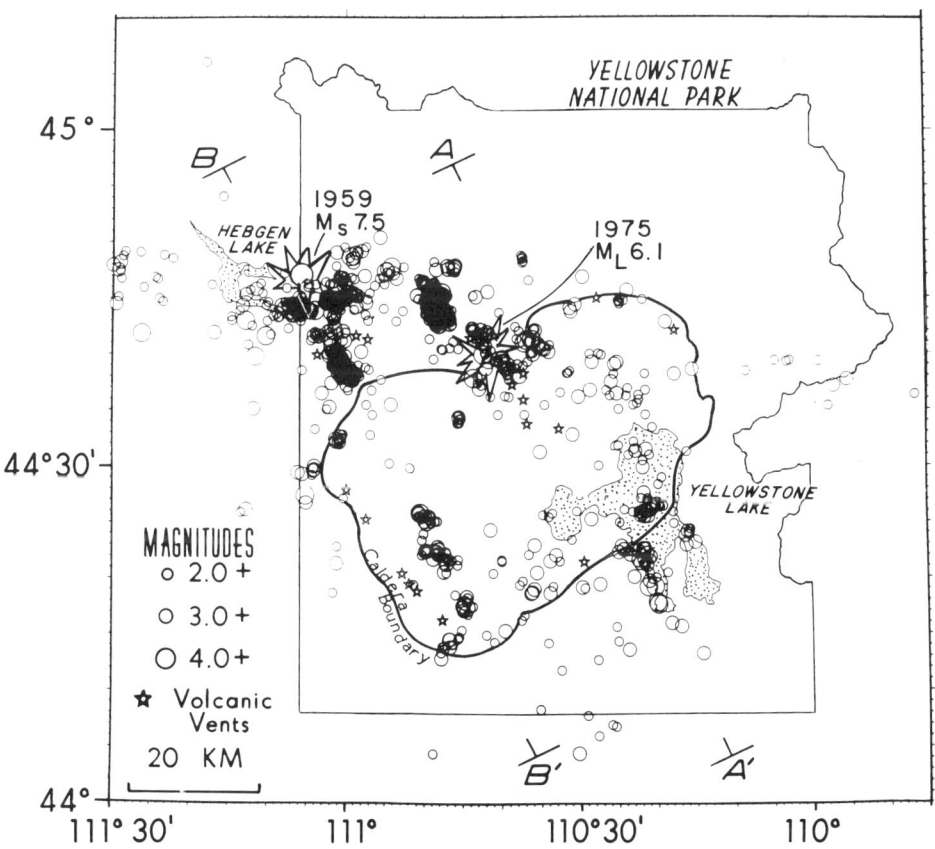

Figure 16. Map showing representative seismicity in the Yellowstone region of magnitude (M_L) 2.0 or greater, from compilations by the U.S. Geological Survey for 1973 to 1981 (Pitt, 1987) and by the University of Utah for 1984 to 1989 (e.g., Peyton and Smith, 1990). A-A' and B-B' define the locations of cross sections shown in Figure 17. Boundary of the Yellowstone caldera and epicenters of 1959 and 1975 main shocks (starbursts), as in Figure 9, shown for reference.

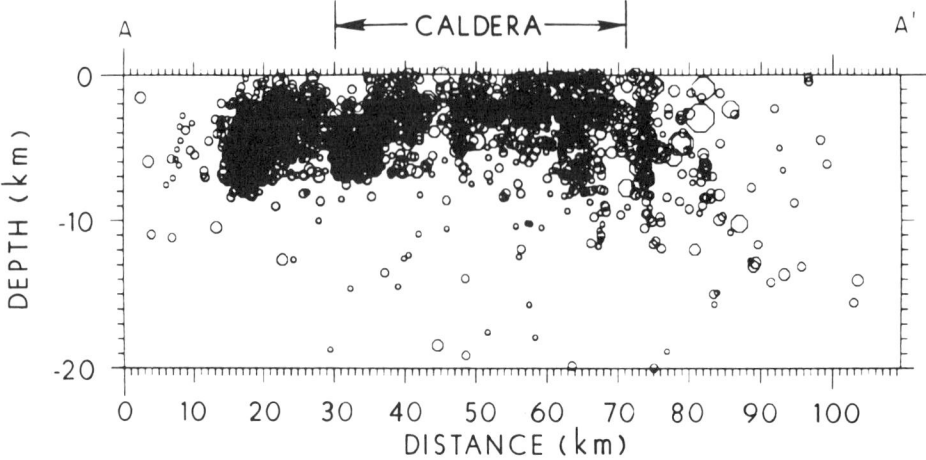

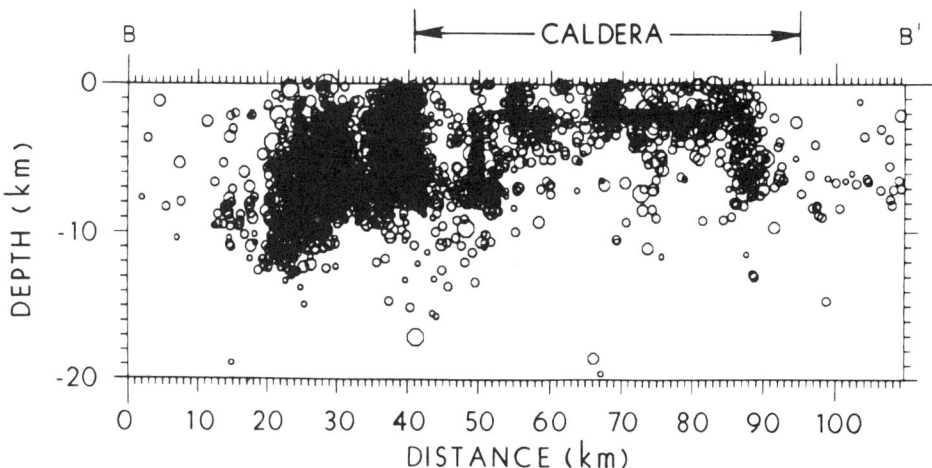

Figure 17. Cross sections of seismicity beneath the Yellowstone region, keyed to Figure 16. Data correspond to that described for Figure 16, but extended to all shocks of $M_L > 0$. Plots include only hypocenters within 10 km of the planes of section for which there were more than 7 recording stations, a map distance to the nearest recording station less than twice the focal depth, and an azimuthal gap in station coverage less than 180°.

east-west band of seismicity that passes through the Hebgen Lake region. The 1959 Hebgen lake main shock (discussed earlier) occurred within about 30 km of the Yellowstone caldera. The earthquake may have resulted from unusual lithospheric uplift and viscoelastic relaxation associated with the Yellowstone hot spot (Reilinger, 1986).

Along the northwest side of the SRP, Figure 15 shows a pronounced northwest alignment of epicenters between Mackay and Challis, which is aftershock activity of the 1983 Borah Peak earthquake on the Lost River fault. This pattern contrasts with the scatter of what we have called background seismicity elsewhere in the central ISB. The "turning on" of earthquakes on the Lost River fault emphasizes the relative seismic quiescence of the neighboring Lemhi and Beaverhead faults to the northeast. All three faults are part of a domain of active, latest Quaternary basin-range normal faulting northwest of the SRP (Scott and others, 1985). Hence, the paucity of earthquakes between the Lost River fault and the Idaho-Montana border marks another important seismic gap in the central ISB. Seismic surveillance of this region by the Idaho National Engineering Laboratory (King and others, 1987) shows small-magnitude earthquake activity near the central part of the Beaverhead fault, but very minimal microseismicity along the Lemhi fault. Central Idaho west of the Borah Peak earthquake zone (Fig. 15) is characterized by diffuse earthquake activity (Dewey, 1987; Smith and Sbar, 1974), earthquake swarms (Pennington and others, 1974; Smith, 1977), and extensive hot spring activity. Microearthquake studies reported by Smith (1977) suggest maximum focal depths of 10 to 15 km.

Northern ISB

The northern ISB, as we have delimited it north of the Hebgen Lake–Yellowstone region, lies entirely within western Montana (Fig. 1). The region has had no historical surface faulting, and the largest historical earthquake reached magnitude 6¾ (Fig. 9, Table 2). Here, we expand upon a recent summary of seismicity and faulting in western Montana–eastern Idaho by Stickney and Bartholomew (1987) to describe some detailed aspects of the northern ISB. For illustration, Figure 18 combines available information on Cenozoic faulting with a representative seven-year sample of instrumental seismicity from the Montana regional seismic network (M, Fig. 8B).

The area of Figure 18 contains at least three fundamentally different domains of Cenozoic basin-bounding extensional faults (see Fig. 3 for names): (1) large north-northwest-trending faults in northwestern Montana, (2) west-northwest-trending faults marking the Lewis and Clark Zone (LCZ), and (3) faults of variable trend south of the LCZ. Significantly, none of these domains include faulting of Holocene age; Holocene faulting in Montana is restricted to a belt along the northwest flank of the SRP (the Centennial Tectonic Belt of Stickney and Bartholomew, 1987), which we have described as part of the central ISB. Large faults in northwestern Montana have exerted strong control on Cenozoic topography, probably including significant Quaternary displacements (Pardee, 1950), but the current seismic potential of these faults is uncertain (Qamar and others, 1982). The LCZ is a pre-Cenozoic structural lineament about 400 km long, perhaps dating from the Proterozoic, that has been interpreted to reflect a fundamental intraplate boundary (see Stickney and Bartholomew, 1987). Major transcurrent shearing has been postulated for the LCZ, but the zone is not noted for any such slip in late Cenozoic time (Eardley, 1962). In its modern expression, the LCZ is considered to be a transitional zone, up to 50 km wide, that "divides a region of uniformly northwest-trending Laramide thrusts and folds and Tertiary normal faults to the north from a

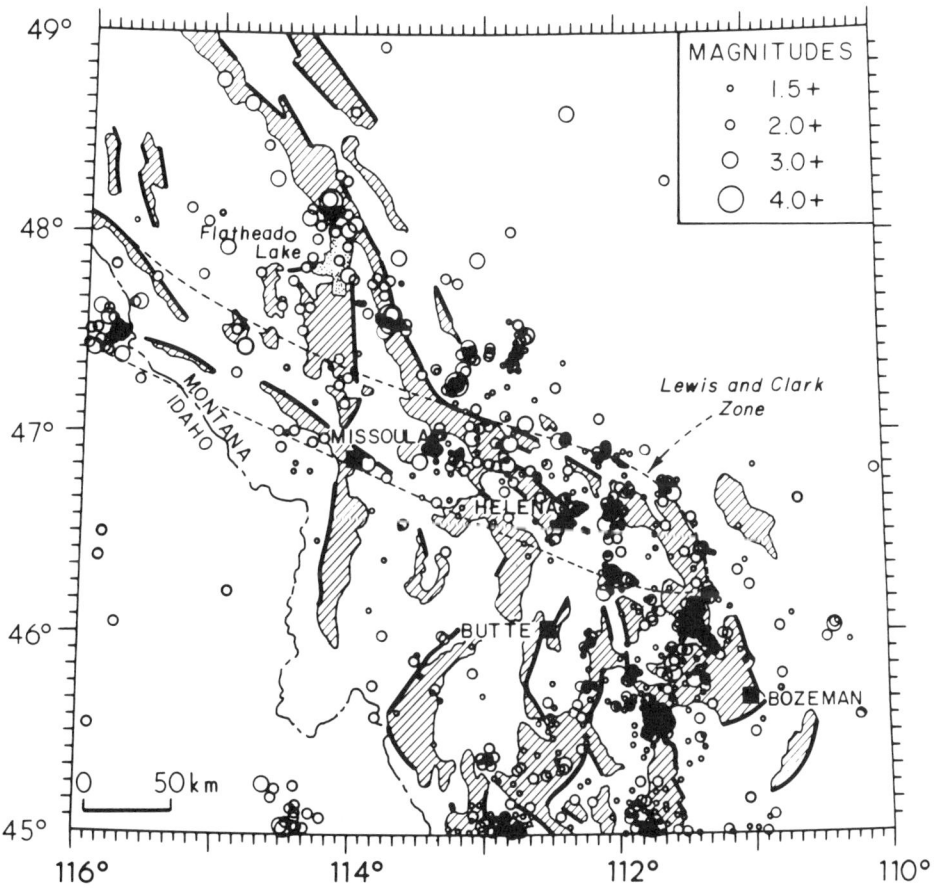

Figure 18. Seismicity map of the northern Intermountain seismic belt. Map shows all earthquakes of magnitude 1.5 and greater located by the Montana Bureau of Mines and Geology from 1982 through 1987 (M. C. Stickney, personal communication, 1990). Base map shows Cenozoic basins (hachured), basin-bounding extensional faults (heavy lines), and the outline of the Lewis and Clark Zone (short-dashed line) after Stickney and Bartholomew (1987).

region of batholithic intrusions and shorter Basin and Range structures with diverse trends to the south" (Stickney and Bartholomew, 1987).

The background seismicity shown in Figure 18 defines a belt about 200 km wide that follows the regional northwest trend of extensional basins in western Montana without definite association with the mapped faults. We describe features of this seismicity in each of the three domains of faulting outlined above. In northwestern Montana, historical seismicity has not exceeded magnitude 5.5 (Fig. 9) and has been dominant in the vicinity of Flathead Lake (Figs. 1, 9, 18), where significant earthquake swarms occurred in 1945, 1952, 1964, 1969, 1971, and 1975 (Qamar and others, 1982). A possible relation between seismicity and reservoir loading at Flathead Lake has been suggested by Dunphy (1972). Results from short-term seismographic recording near Flathead Lake in 1971 by Stevenson (1976; see also Sbar and others, 1972) showed clustering of small earthquakes on the west side of the lake, with focal depths extending from 5 to 12 km depth and foci defining a planar zone dipping 70° to the east-northeast. Diverse focal mechanisms in the vicinity of Flathead Lake appear mostly to reflect east-west to northwest-southeast extension (Qamar and others, 1982). Seismicity near Flathead Lake roughly marks the northernmost extent of the ISB as it dies out toward the Canadian border (Figs. 1 and 18). Coincidentally, Flathead Lake lies at the southern end of the Rocky Mountain trench, a large pre–basin-range graben that extends more than 800 km northwestward into Canada (Eardley, 1962; Pardee, 1950).

Stickney and Bartholomew (1987) attach great importance to the LCZ as the northern boundary of the active extensional regime of the Montana-Idaho part of the Basin and Range province. All identified late Quaternary faults and all historical earthquakes greater than magnitude 5.5 in the northern ISB are south of, or are included within, the LCZ. Seismicity shown in Figure 18 within the LCZ includes relatively dense clusters of earthquakes from east of Missoula to the southeastern end of the LCZ and a distinct cluster of events at the northwest end of the LCZ in the figure. The latter include both rockbursts related to deep mining near Wallace, Idaho, and local tectonic earthquakes (Stickney and Bartholomew, 1987). We earlier discussed the destructive swarm earthquakes of 1935–1936 near Helena. Earthquake studies reported by Friedline and others (1976) near Helena indicate earthquake clustering along a N50°W trend with focal-depth maxima of 15 to 20 km. Focal mechanisms for this area imply normal to lateral slip with consistent northeast-trending T-axes (Friedline and others, 1976; Stickney and Bartholomew, 1987; Doser, 1989a).

Background seismicity south of the LCZ in Figure 18 is concentrated in a northerly-trending zone about 100 km wide between the eastern part of the LCZ and the Hebgen Lake–Yellowstone region to the south. Recall that the Helena-Bozeman area has had the largest historical earthquakes in the northern ISB (Fig. 9). Densely clustered seismicity about 50 km northwest of Bozeman is in the Clarkston Valley area, one of the most persistently active earthquake zones in the northern ISB since the occurrence of the M = 6¾ Clarkston Valley earthquake in 1925 (Stickney and Bartholomew, 1987). Special studies of seismic activity in the Clarkston Valley area have been reported by Qamar and Hawley (1979), who document relatively shallow (<10 km) focal depths and both normal and strike-slip focal mechanisms with T-axes trending 55° to 95°—different from the trend of 127° for the 1925 main-shock mechanism (Doser, 1989a).

INDUCED SEISMICITY

Human activities have demonstrably triggered small to moderate earthquakes (M ≤ 5.0) in the Intermountain region, both within the main active zone of the ISB and in marginal, less seismically active areas. This includes documented cases of each of the three principal types of induced seismicity—associated, respectively, with reservoir impoundment, fluid injection, and mining. We proceed to summarize examples (see also Dewey and others, 1989, Table 1), referring to localities identified in Figure 3.

Reservoir-induced seismicity

The pre-eminent case of reservoir-induced seismicity (RIS) in the Intermountain region is that associated with Lake Mead along the Arizona-Nevada border—impounded by the 221-m-high Boulder Dam (now Hoover Dam) on the Colorado River in the mid-1930s. Locally felt earthquakes began soon after filling of what was then the world's largest reservoir, and a magnitude 5.0 shock occurred in 1939 about a year after the reservoir reached 80 percent capacity (Carder, 1970). Observations by Carder (1945) represent the first instance in which the phenomenon of RIS was recognized. By the mid-1970s more than 10,000 local earthquakes had been recorded near Lake Mead, with an irregular epicentral distribution inferred to be influenced chiefly by the variable permeability of sub-reservoir sedimentary rocks rather than by lateral differences in mass loading (Anderson and Laney, 1975). Rogers and Lee (1976; see also Rogers and others, this volume) describe results of detailed seismographic monitoring at Lake Mead during 1972–1973 and suggest observed RIS there is caused by small increases in pore pressure along existing faults.

Reviews of RIS on a worldwide basis (e.g., Simpson, 1976; Gupta and Rastogi, 1976; Gupta, 1985) typically mention five other likely or possible cases of RIS in the Intermountain region. These include: (1) Kerr Dam (Flathead Lake), Montana, one of at least 14 worldwide dams associated with earthquakes in the magnitude 4 range (Gupta, 1985; note that earthquakes of magnitude 4.6 and 4.9 occurred near Kerr Dam 6 and 13 years, respectively, after impoundment in 1958); (2) Glen Canyon (Lake Powell) and (3) Flaming Gorge dams in Utah, where there is uncertain evidence for impoundment-related *decreases* in seismicity (Simpson, 1976); and possible cases of RIS at (4) Clark Canyon Dam, Montana, and Palisades Dam, Idaho (Gupta,

1985; Simpson, 1976). The common occurrence of earthquake swarms throughout the region immediately surrounding Palisades Reservoir confounds arguments for causally relating such seismicity with the reservoir (see review by Piety and others, 1986). There, as elsewhere in the Intermountain region, adequacy of seismographic control is a critical issue in correlating small-magnitude earthquakes with reservoir impoundment. LaForge (1988), for example, rigorously tested for RIS in connection with 14 dams, 38 to 81 m high, built by the U.S. Bureau of Reclamation in north-central Utah and found no evidence for RIS. However, all but one of the reservoirs was filled before 1967 when seismographic control in the region was still relatively poor.

Fluid injection

After the U.S. Army accidentally triggered earthquakes by deep fluid injection near Denver, Colorado, in the early 1960s, the U.S. Geological Survey carried out a controlled experiment in the Rangely oil field in northwestern Colorado to study the effects of pore-pressure changes at depth on the triggering of small earthquakes (Raleigh and others, 1972). Water injection into a sandstone reservoir about 2 km deep began in late 1957 at Rangely for secondary oil recovery. The spatial coincidence of earthquakes (M ≤ 4.5) with the Rangely field can be shown at least as early as November 1962 when the UBO array (Fig. 8A), 50 to 80 km to the west-northwest, began operating (Munson, 1970). The detailed experiment carried out by the U.S. Geological Survey at Rangely in 1969–1970 successfully measured in situ stress state and showed how lowering and raising fluid pressures in a seismically active zone at depth could control the occurrence of earthquakes (Raleigh and others, 1972).

Arabasz (1984; see also Arabasz and Julander, 1986, Fig. 8) describes another possible case of seismicity related to fluid injection in the Intermountain region. In mid-1982 an "acid-breakdown hydrofrac" was made in a wellbore at a depth of about 5 km (Chevron U.S.A. #1 Chriss Canyon, total depth = 5,344 m) in the vicinity of the southern Wasatch fault. Hypocentral clustering of two earthquake swarms (M_L ≤ 2.1) two to three months later, within a few kilometers of the wellbore and with distance-delay times consistent with fluid diffusion, suggests triggering by the fluid injection. Geothermal steam production has been monitored for induced seismicity at two sites in the Intermountain region (Zandt and others, 1982): Roosevelt Hot Springs–Cove Fort, Utah (ca. 38°30′N, 112°45′W), and Raft River, Idaho (42°06′N, 113°23′W), but to our knowledge no significant induced seismicity has occurred.

Mining-related seismicity

Mining-related seismicity has been specially investigated in two parts of the Intermountain region—the northern and northwestern Colorado Plateau, chiefly in association with underground coal mining, and near Wallace, Idaho, in association with deep vein mines of the Couer d'Alene mining district. In the latter area, seismographic studies have focused on rockbursts (e.g., McLaughlin and others, 1976); however, sizable shocks up to magnitude 4 suggest tectonic stress release as well (Stickney and Bartholomew, 1987).

Seismicity in east-central Utah defines an inverted U-shaped pattern (Figs. 1 and 12B) coinciding with areas of extensive underground coal mining along an arcuate erosional escarpment of the eastern Wasatch Plateau and Book Cliffs. The association of both rockbursts and earthquakes (M_L ≤ 4.5) with sites of major coal extraction in this area has been evident since the late 1950s. Wong and Humphrey (1989) and Williams and Arabasz (1989; see also Smith and others, 1974) give good overviews of varied seismological investigations indicating that: (1) much of the seismicity appears to be mining induced, resulting from stress redistribution from both room-and-pillar and longwall coal extraction in mine workings down to 900 m below the surface; (2) abundant seismicity occurs *beneath* the mines to depths of 2 to 3 km; (3) time-varying rates of extraction at individual mines have influenced seismicity changes detected by regional seismic monitoring; and (4) source mechanisms appear variously to reflect extensional subsidence above mine workings and a mixture, at and below mine level, of seismic slip on prestressed reverse faults and possibly non-double-couple, implosional failures.

Figure 19A shows abundant submine events located in the Gentry Mountain area of the eastern Wasatch Plateau. Nearly all these events were recorded with ubiquitous dilatational first motions, inferred to be caused by implosional failure (Wong and others, 1989). Companion results from the nearby East Mountain area (Williams and Arabasz, 1989; Fig. 19B) also show reliably-located submine events, but mostly with reverse-faulting mechanisms (normal-faulting solutions 1 and 2 are for earthquakes west of the mining area beneath an active graben). About a third of the seismic events recorded in the East Mountain area were of the enigmatic type with all dilatational first motions. Given inadequate focal-depth resolution for many of these shallow events, Williams and Arabasz (1989) found that if these events were constrained to occur at mine level, their first-motion distributions were indeed incompatible with a double-couple source mechanism. However, the same first-motion observations could be fit with double-couple normal-faulting solutions if the sources were *above* mine level, perhaps reflecting overburden subsidence.

We refer the reader to Wong and Humphrey (1989) for a review of other varied work on mining-induced seismicity in the Colorado Plateau. This includes (1) studies by the U.S. Geological Survey of seismicity induced by underground coal mining near Somerset in western Colorado and (2) studies by Wong and coworkers of seismicity induced by potash mining in the north-central Colorado Plateau involving brine extraction from a previous room-and-pillar mine at 1 km depth.

PATTERNS OF EARTHQUAKE OCCURRENCE

Observations about patterns of earthquake occurrence in the ISB are fundamental both for scientific understanding of earthquake behavior in the region as well as for basic evaluations of

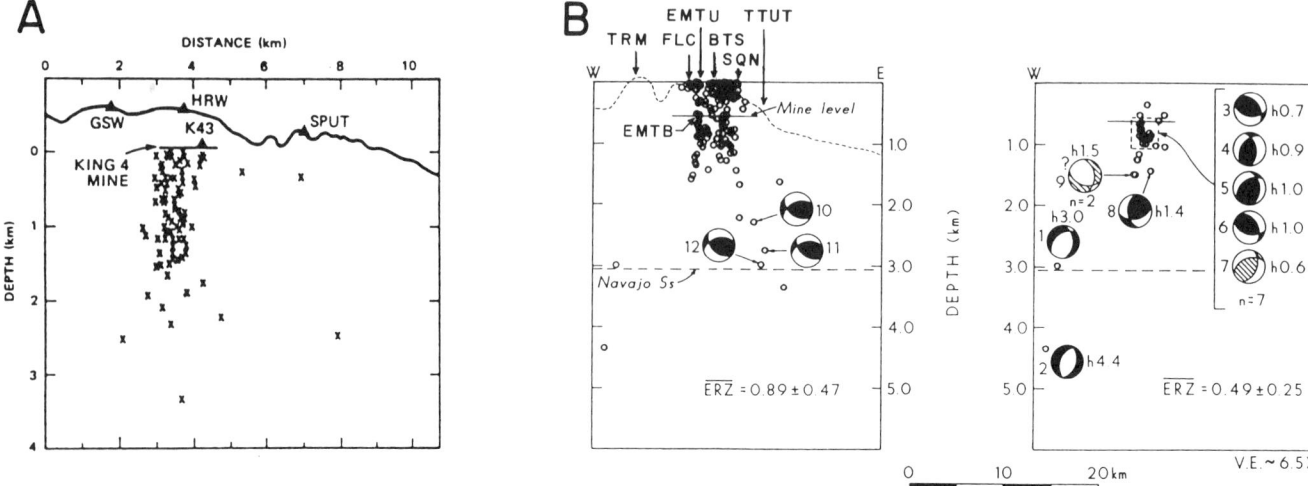

Figure 19. Mining-related seismicity in the eastern Wasatch Plateau, Utah. A, Cross section from a special study in the Gentry Mountain area (from Wong and others, 1989) showing ground surface (heavy line), seismic events (x's) within 500 m of the plane of section, and closest seismographs (triangles). B, Cross sections from the East Mountain area, about 20 km south of Gentry Mountain (from Williams and Arabasz, 1989) showing ground surface (short-dashed line), seismic events (small circles) within 5 km of the plane of section, seismograph locations (letter codes) within 1 km of the plane of section, and schematic equatorial-plane projection of 12 focal mechanisms—dilatational quadrants are white and compressional quadrants either black (for single-event solutions) or hachured (for composite solutions); the left panel of B includes events located with standard errors in focal depth (ERZ) less than 2.0 km; the right panel includes only the "best" located events meeting rigorous criteria for reliability of focal depth, h.

earthquake hazards and risk. In this section we briefly summarize relevant information on (1) earthquake swarms, (2) earthquake recurrence, and (3) the space-time distribution of earthquakes in the ISB.

Earthquake swarms

Earthquake swarm activity, the clustering of earthquakes of similar size in space and time without an outstanding main shock, is a common feature in parts of the ISB. Such earthquake swarms tend to occur in and near areas of Quaternary volcanism or high heat flow. Smith and Sbar (1974) present a good general discussion, describing notable areas of historical swarm activity in the ISB. These include: Flathead Lake and Helena, Montana, in the northern ISB; Yellowstone, central Idaho, and southeastern Idaho in the central ISB; and areas in western and southwestern Utah in the southern ISB. We have already described sources of information for most of these areas in preceding sections. Additional information on earthquake swarm activity is given by Arabasz and Julander (1986) for southwestern Utah and by Piety and others (1986) for southeastern Idaho. The destructive earthquake swarm near Helena, Montana, in 1935–1936 (described earlier) that included shocks of magnitude 6 and 6¼ (Table 2) is the most outstanding example of swarm seismicity in the ISB. Elsewhere in the ISB, the largest earthquakes in individual swarms have been in the upper magnitude 4 range or smaller.

Earthquake recurrence in the ISB

Earthquake recurrence specifies the distribution of earthquake sizes (e.g., magnitudes or intensities) and their frequency of occurrence on a fault or in a specified area. Here, we basically want to describe how often moderate to large earthquakes occur in the ISB and point the interested reader to more specific information for recurrence modeling. Modern methods for quantifying earthquake recurrence using information from observational seismology and from geologic studies of the age, frequency, and rupture characteristics of prehistoric earthquakes are summarized by McGuire and Arabasz (1990) and Schwartz and Coppersmith (1986) (see also Doser and Smith, 1982).

The most commonly used relations to describe the relative number of earthquakes as a function of size is the well-known Gutenberg-Richter relation (Richter, 1958): $\log_{10} N(m) = a - bm$, where $N(m)$ is the number of earthquakes of magnitude m or greater per unit time (and ideally per unit area) and a and b are constants. The average inter-event time or recurrence interval for earthquakes of a particular magnitude m or greater is given by $1/N(m)$. There is growing recognition that the Gutenberg-Richter relation may appropriately apply to a region but not necessarily to an individual fault (e.g., Schwartz and Coppersmith, 1986). It is also recognized that the historical and instrumental earthquake record cannot confidently be extrapolated to estimate the frequency of occurrence of large surface-faulting earthquakes in the

ISB and that information from late Quaternary faulting is therefore essential (Schwartz and Coppersmith, 1984; Arabasz and others, 1987). Thus, historical and instrumental records of seismicity in the ISB are important for modeling the recurrence of earthquake sizes up to the threshold of surface faulting, and paleoseismology is needed for estimating how often large surface-faulting earthquakes will occur.

A rough estimate of the recurrence interval of sizable earthquakes for the *whole* ISB can be made from Table 2. Neglecting aftershocks and secondary events, 27 independent main shocks of approximate magnitude 5.5 and greater occurred in the ISB from 1900 through 1985, which gives an average inter-event time of 3 years for such shocks somewhere in the ISB. Corresponding estimates from Table 2 for thresholds of approximate magnitude 6.0 and 6.5 are about 6 years and 17 years, respectively. For earthquakes of magnitude 7.0 or greater, we simply note that two such events occurred during the 85-year observation period, insufficient for meaningful modeling of inter-event times. Recurrence intervals vary, of course, with the particular region and the total area being considered. For example, for the 85,000 km^2 of the Wasatch Front area (Fig. 12), rigorous recurrence modeling of instrumental seismicity for 1962 through 1985 (Arabasz and others, 1987) yields average recurrence intervals of 24 years for $M_L \geq 5.5$, 54 years for $M_L \geq 6.0$, 120 years (extrapolated) for $M_L \geq 6.5$, and 280 years (extrapolated) for $M_L \geq 7.0$. For other examples of recurrence modeling in parts of the ISB, we refer the reader to Youngs and others (1987), Doser and Smith (1982), Piety and others (1986), Stickney and Bartholomew (1987), and Algermissen and others (1982). Comparison of seismicity parameters among these and other published reports for the ISB must be made with care, checking whether the parameters were determined with sufficient rigor, whether the parameters are intrinsically comparable, and whether the parameters describe the occurrence only of independent main shocks or of all earthquakes (see McGuire and Arabasz, 1990).

Average slip rates on normal faults in the ISB are one to two orders of magnitude lower than for those on major plate-boundary faults, typically being about 1 mm/yr for the most active faults like the Wasatch fault (Schwartz and Coppersmith, 1984; Machette and others, 1987) and the Teton fault (Byrd and Smith, 1990), and a few tenths of a millimeter or less on faults elsewhere in the region (e.g., Schwartz, 1987; Youngs and others, 1987; Scott and others, 1985; Stickney and Bartholomew, 1987). Corresponding recurrence intervals for surface rupture on an individual fault segment are about 2,000 yr on the most active parts of the Wasatch fault, are uncertain on the Teton fault, and are typically several thousand or tens of thousands of years on other faults. The issue of uniform versus time-varying recurrence (see Schwartz, 1988) has become an important consideration for estimating expected rates of occurrence of large earthquakes in the ISB. On the Wasatch fault, for example, the record of surface-faulting earthquakes during the past 6,000 yr leads to an average recurrence interval of 415 yr for a large surface-faulting earthquake somewhere on the fault, but an accelerated rate of faulting between about 400 and 1,500 yr ago implies such an earthquake once every 220 yr (Machette and others, 1989). Comparably detailed paleoseismological data do not exist for other faults in the ISB.

How do rates of intraplate faulting and earthquake activity in the ISB compare to those along the North American plate boundary in California? Recurrence intervals of thousands of years for surface rupture on individual fault segments in the ISB compare to much shorter intervals of hundreds of years for large earthquakes on the most active parts of the San Andreas fault system (Schwartz and Coppersmith, 1984). Maximum-size earthquakes of about magnitude 7¾ in the ISB compare to a value of about 8½ on the San Andreas fault. We remarked earlier that deformation rates for most of the intraplate ISB are one to two orders of magnitude lower than along the western North America plate boundary. Comparison between seismicity in the ISB and California can be made using seismicity rates determined in a uniform way by Algermissen and others (1982, Table 1) for source zones throughout the United States, normalizing those values per unit area. The mean rate of earthquakes equivalent in size to MMI = V (about magnitude 4) per year per 1,000 km^2 is approximately 8×10^{-2} for the 41 seismic source zones depicted by Algermissen and others (1982, Fig. 2) in the main seismically active part of California. The mean value of that same rate for their seismic zones making up the main ISB is approximately 2×10^{-2}. Thus, normalized seismicity in the ISB, on average, is lower by about a factor of 4 compared to that along the plate boundary in California.

Space-time patterns of seismicity in the ISB

To get an overview of variations of earthquake activity along the ISB as a function of space and time, as captured by the DNAG catalog, we have plotted those data in a conventional space-time format and show the results in Figure 20. The same DNAG catalog data used for Figure 1 were sorted for the four sample areas shown in Figure 20A, prescribing a magnitude threshold of 3.0 and the time period from 1930 through 1985. We chose 1930, judging that epicentral precision had become sufficient by that time to make meaningful spatial comparisons, and the date precedes the occurrence of sizable earthquakes in the ISB in the mid-1930s and 1940s. The sorted earthquakes are plotted in Figure 20B with latitude as the space coordinate, given the general north-south trend of the ISB, recognizing the limitation that the northern and southernmost parts of the ISB trend obliquely to the latitude ordinate. Data for central Idaho had to be excluded to prevent confusion in projection.

As usual, there are evident artifacts that must be accounted for in space-time plots of this type, the most obvious being those due to catalog incompleteness. Increases in numbers of earthquakes in the early 1960s and locally in the mid-1970s correspond to improvements in seismographic coverage that we described in the section on instrumental recording and seismic networks. Despite recognizable problems, Figure 20 reveals some

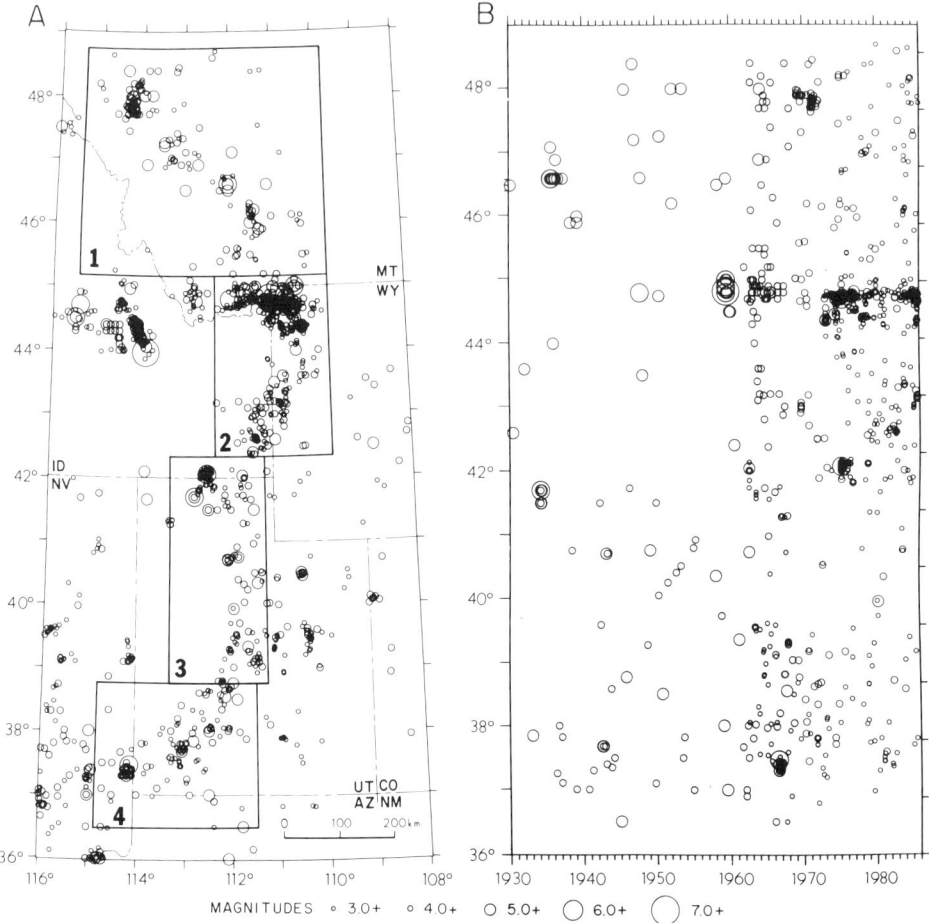

Figure 20. Space-time seismicity of the Intermountain seismic belt, 1930 to 1985. A, Map showing earthquakes from Figure 1 of magnitude 3.0 and greater since 1930, together with sample boxes used for the space-time plot in B. B, Space-time plot of earthquakes as a function of time (abscissa) and latitude (ordinate), keyed to A.

important observations that warrant attention. We proceed from north to south, using the boxes numbered from 1 to 4 for reference.

For the northern ISB, the space-time projection of earthquakes in Box 1 shows an understandable increase in smaller-magnitude earthquakes after about the mid-1960s, but there is a noticeable decrease in the number of larger earthquakes ($M \geqslant 5$) after that time, compared to earlier decades. Seismicity in the Hebgen Lake–Yellowstone region dominates the picture for the part of the central ISB bounded by Box 2. The projection of the Hebgen Lake–Yellowstone seismicity reveals the most intense earthquake activity in the ISB. There was a marked increase in the detection of small earthquakes following the installation of the Yellowstone seismic network in 1973, but similarly sensitive networks elsewhere in the ISB do not show the same level of seismicity. The $M_S = 7.5$ Hebgen Lake earthquake of 1959 is the largest shock at that latitude. Nearby prior occurrence of the $M = 6¼$ Virginia City earthquake 12 years earlier in 1947 suggests that earthquake may have been a preshock (Doser, 1989a). Both the map and space-time plot emphasize the existence of the Teton seismic gap, more than 50 km long, immediately south of the Yellowstone region. As earlier described, there is a similarly prominent gap northwest of the SRP between the Borah Peak earthquake zone and the Idaho-Montana border (Fig. 20A).

Box 3 essentially covers the Wasatch Front area. The $M_L = 6.0$ Pocatello Valley earthquake of 1975, at about 42°N, was preceded by seismic quiescence within 50 km that began 6.4 years before the main shock (Arabasz and Smith, 1981). One of the most striking features of the space-time projection of the earthquakes of Box 3 is the sparseness of earthquakes ($M \geqslant 3.0$) since the mid-1960s along a zone more than 200 km long between about 39.5°N and 41.5°N, corresponding to the most geologically active part of the Wasatch fault (Fig. 12). Note that there has been effective seismographic coverage of the Wasatch Front area since the early 1960s (Fig. 8B). Smith (1972) first noted the relative contemporary quiescence of the Wasatch Front area compared to neighboring segments of the ISB, and Arabasz and Smith (1981, Fig. 8; see also Griscom, 1980) portrayed the relative quiescence in space-time view. The apparent decrease in background seismicity in the Wasatch Front area beginning in the mid-1960s compared to prior decades was statistically tested by Arabasz (1984), who found that one could not reject at the 95-percent confidence level the hypothesis that the apparent decrease was simply random. The pattern of earthquake clustering in the Wasatch Front area, however, does have significant features. Based on a statistical analysis of the Utah earthquake catalog for 1962 through 1985 by Shimizu (1987), Veneziano and others (1987) observed that, compared to earthquake behavior in neighboring areas, there is a distinct paucity of secondary events, relative to the number of main events, in the vicinity of the Wasatch fault between 39.5°N and 41.5°N. We refer the reader to Arabasz and others (1987, Fig. 17) for a detailed space-time plot of microseismicity within a 30-km-wide zone along the Wasatch fault for 1962 through 1986.

The space-time plot of earthquakes in Box 4 at the southern end of the ISB shows a somewhat similar pattern to that for Box 1. Despite a marked improvement in seismographic coverage for the area of Box 4 in the mid-1970s, background seismicity has been noticeably lower since that time, compared to activity during the 1962 through 1975 period. Earthquake sizes in the Utah region have been measured instrumentally in a uniform way since 1962 (Arabasz and others, 1979). Although not rigorously analyzed here, the space-time overview of the entire ISB in Figure 20 points out general patterns that need to be addressed. The occurrence of the 1983 Borah Peak earthquake in an area of prior low seismicity and prehistoric surface faulting, which should have been recognized as a seismic gap, serves as a useful reminder.

CONCLUDING REMARKS

The mechanics and subsurface geometry of normal faulting are topics of great current interest and importance, especially for assessing ground deformation and peak ground motions associated with large normal-faulting earthquakes. The observation in parts of the Basin and Range province of steep planar faults (dips > 45°) associated with large normal-faulting earthquakes in the same structural setting where late Tertiary to Quaternary low-angle and listric normal faults are present in the subsurface poses a quandary. In this final section, we discuss observations and hypotheses relating to large normal-faulting earthquakes and to the diffuse background seismicity that dominates the earthquake record of the Intermountain region. We also consider implications of earthquake focal depths for rheology and the possible influence of the Yellowstone hot spot on the seismotectonics of the ISB, including its influence on Quaternary faulting of the SRP and surrounding region. Discussion of earthquakes in the Yellowstone region also provides insights into the relation between magma transport and seismicity.

Correlation of seismicity and geologic structure

A recurring observation in the ISB is the lack of distinct correlation between scattered background seismicity and mapped Cenozoic faulting. Numerous descriptions of background seismicity in the region published during the last decade include remarks to this effect. The problematic correlation has been discussed at length for the southern ISB by Arabasz and Julander (1986; see also Arabasz and Smith, 1981; Zoback, 1983; Smith and Bruhn, 1984). As outlined by Arabasz and Julander (1986), and generalized to the ISB as a whole, the basic problems include: (1) uncertain subsurface structure, which typically is more complex along the main seismic belt than is apparent from the surface geology, commonly because of the superposition of basin-range faulting upon older thrust-belt structure; (2) observations of discordance between surface fault patterns and seismic slip at depth; (3) limited opportunity to observe large-scale seismic slip because of only three cases of historic surface faulting; and (4) inadequate hypocentral resolution commonly resulting from regional seismic monitoring. In order to correlate seismicity with structure, there is a critical need (aptly illustrated by Fig. 14) for local seismographic control, especially for good focal-depth resolution, sufficient seismicity for defining the spatial geometry of active structures, and reliable focal mechanisms for correlating observed seismicity with fault geometry and the sense of slip. Focal mechanisms also allow assessment of the principal stress directions.

On the basis of special earthquake studies in mostly the southern ISB, a working hypothesis was offered by Arabasz (1984; see also Arabasz and Julander, 1986) to explain observations of diffuse background seismicity. Background seismicity, it was suggested, is fundamentally influenced by variable mechanical behavior and internal structure of individual plates within the seismogenic upper crust. Diffuse epicentral patterns appear to result from the superposition of relatively intense shallow seismicity within upper-crustal plates and less frequent background earthquakes at greater depth. Favorable conditions for block-interior rather than block-boundary microseismic slip may also contribute to the epicentral scatter. Some aspects of the working hypothesis are shown schematically in Figure 21A, which depicts (following Arabasz and Julander, 1986): (a) a predominance locally of seismicity within a lower plate; (b) nucleation of a large normal-faulting earthquake near the base of the seismogenic layer, hypothetically on an old thrust ramp, and with linkage to a shallow structure; (d) occurrence of a moderate-sized earthquake and aftershocks on a secondary fault where an underlying detachment restricts deformation to the upper plate; (e) diffuse block-interior microseismicity predominating within an upper plate—perhaps responding to extension enhanced by gravitational backsliding on an underlying detachment; and (f) diffuse block-interior microseismicity within a lower plate where frequency of occurrence is markedly lower than in the overlying plate.

While Figure 21A suitably illustrates many features of background seismicity in the southern ISB, particularly central Utah, it is not adequately general for the whole ISB. One shortcoming of the sketch in Figure 21A is that it does not explicitly depict the spatial relation of seismicity to Precambrian basement. In some parts of the ISB, Precambrian basement was faulted by

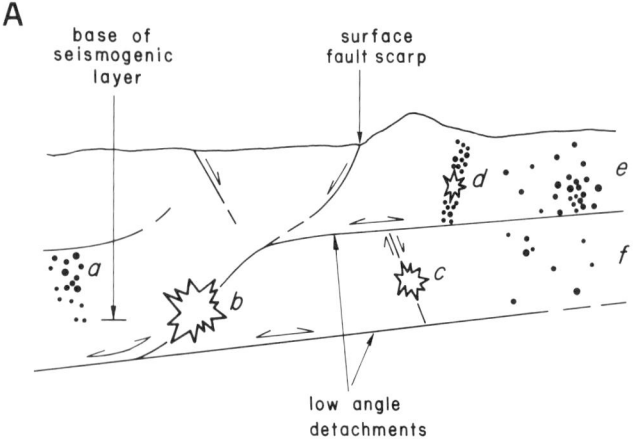

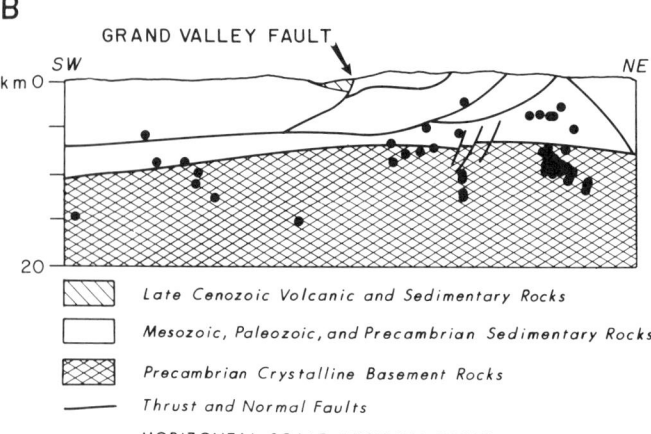

Figure 21. A, Schematic geologic cross section of the upper crust illustrating the inferred association of seismicity with geologic structure in the southern Intermountain seismic belt (from Arabasz and Julander, 1986). Starbursts indicate hypothesized foci of moderate-to-large earthquakes; small circles, microseismicity; lines in subsurface, faults; two-directional arrows, extensional backsliding on pre-existing low-angle faults possibly formed as thrust faults. Letters identify examples referred to in text. Base of seismogenic layer is approximately 10 to 15 km depth. B, Cross section (from Piety and others, 1986) showing distribution of well-located background earthquakes (circles), together with generalized geology from interpretations of seismic-reflection data by Dixon (1982), across the Grand Valley fault in the central Intermountain seismic belt. Line of section trends N60°E and crosses the northern end of Palisades Reservoir (Fig. 3).

The present thickness of sedimentary cover rocks, the extent of involvement of Precambrian basement in older compressional deformation and subsequent extension, and whether normal faulting penetrates the entire crust (Zandt and Owens, 1980) all are likely to be important local factors governing the pattern of background seismicity. The influence of these respective factors throughout the ISB is not yet understood, but there are accumulating observations, such as seismic slip on discrete Precambrian basement faults at depth (Fig. 14), the broad involvement of Precambrian basement in contemporary extension (Fig. 21B), and perhaps the regional, flexural(?) deformation of Precambrian basement on the eastern, footwall side of the Wasatch fault (Fig. 12B).

Earthquake focal depths and rheology

Since the mid-1970s, accurate hypocenter data have been acquired by regional and portable seismic networks in the Intermountain region that permit the construction of reliable focal-depth histograms. Using some of these data, Sibson (1982) and Smith and Bruhn (1984) hypothesized seismogenic models based on theoretical depths for peaks in maximum shear stress at the boundary between the brittle upper crust and a quasi-plastic layer. These models in a general way account for the maximum depths of nucleation of large normal faulting earthquakes and for the maximum depths of background seismicity, corresponding to the base of the seismogenic layer. The models involve a temperature-dependent, depth-varying power law for creep combined with a linear brittle-behavior criterion (see representative plot in Fig. 22). In Smith and Bruhn's (1984) model for extensional normal-faulting regimes, the maximum focal depths of large normal-faulting earthquakes correlate approximately with the 80th percentile of focal depths for smaller background earthquakes, similar to the findings of Sibson (1982) for the San Andreas fault system. Scholz (1990) predicts the thickness of the seismogenic layer, and hence the maximum focal depths of earthquakes, using both a similar temperature criterion as that described above and additional fault-velocity constraints.

Qualitative arguments of Sibson (1982) and Smith and Bruhn (1984) suggest that the theoretically derived transition depth from brittle to quasi-plastic flow for silica-rich rocks is controlled primarily by a critical temperature of approximately 350°C to 450°C and occurs at or near the depth of maximum shear stress (Figure 22). At this depth, short-term strain rates greater than 10^{-4}/sec are necessary to achieve brittle failure during earthquakes within the more ductile, intermediate-depth crustal material. In theory, this is the critical depth for nucleation of the largest magnitude earthquakes.

On the basis of the observed heat flow and extrapolated thermal gradients of the ISB, Smith and Bruhn (1984) inferred that this critical depth of earthquake nucleation would not exceed ~10 km, but large stress drops for magnitude-7 earthquakes could produce locally higher strain rates allowing their nucleation at mid-crustal depths of about 15 km ± 5 km. In the cooler

pre-Neogene thrusting and is allochthonous, as in parts of the Wasatch Front area (Smith and Bruhn, 1984); in others areas, such as that studied by Piety and others (1986) in the central ISB, Precambrian crystalline basement rocks lie beneath a basal detachment and are autochthonous. For the latter area we noted that relatively abundant background microseismicity was located by Piety and others (1986), as shown in Figure 21B, within the Precambrian crystalline basement, and they observed no marked change in the vertical distribution of foci that could be associated with the basal detachment.

lithosphere of the Colorado Plateau and Rocky Mountains east of the ISB, where background heat flow is less than 65 mW m^{-2}, maximum focal depths exceed 30 to 40 km and are attributed to deeper depths for the critical isotherms (Wong and Chapman, 1990).

The influence of shallow, high temperatures on earthquake depth distributions was described for the Yellowstone caldera (Fig. 17), which is characterized by an extremely high heat flow of 1500 mW m^{-2}. The observed lateral variation in focal-depth maxima reflects the combined influence of conductive and convective heat flow, hypothesized to produce the abrupt shallowing of the critical 350°C to 450°C isotherm beneath the inner caldera

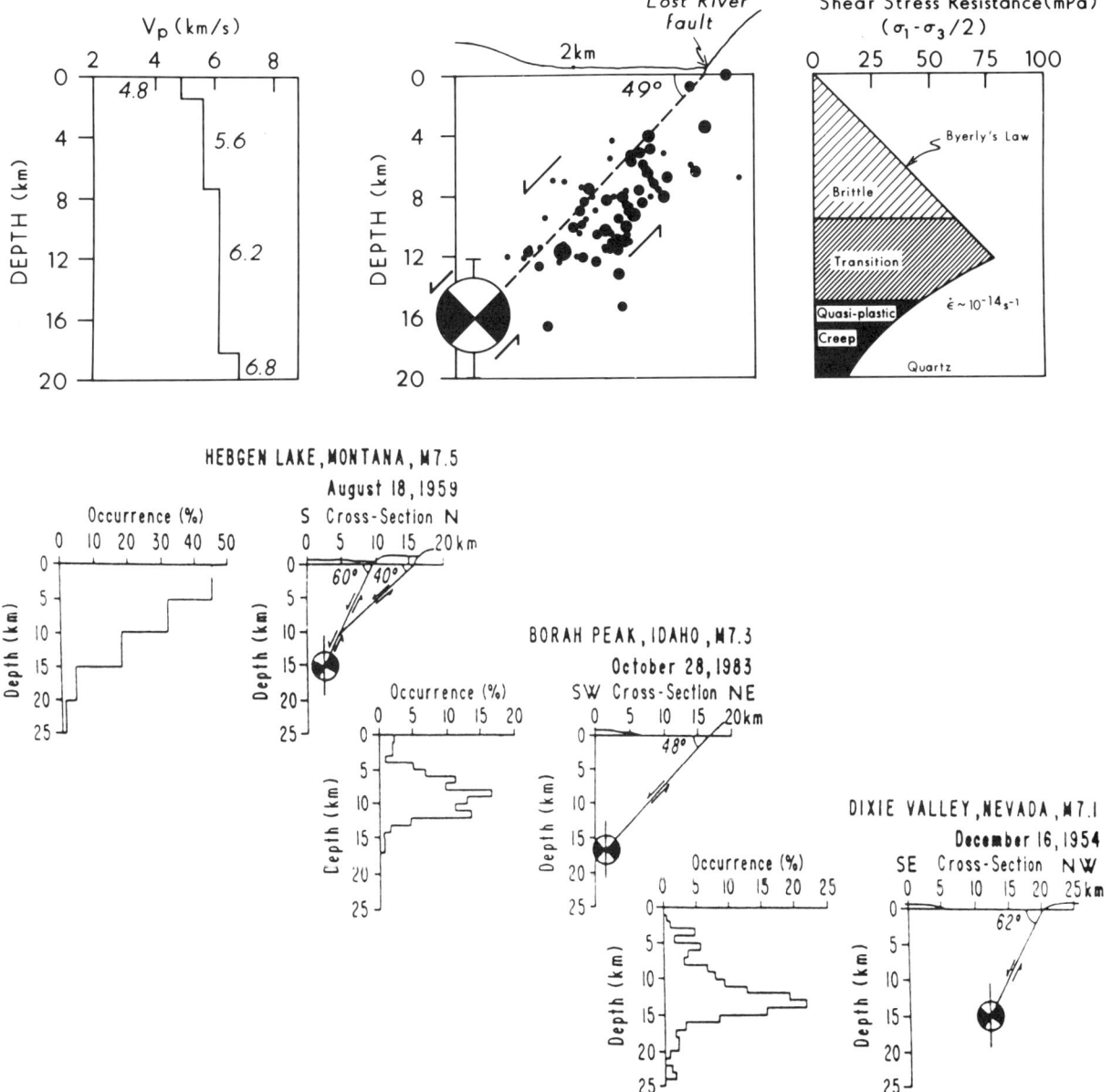

Figure 22. Hypothetical model for large (M >7.0) Basin and Range earthquakes, from Smith and others (1985). Upper, P-wave velocity model and subsurface fault geometry associated with the 1983, M_S = 7.3, Borah Peak, Idaho, earthquake (after Richins and others, 1985), together with a rheological model for the upper crust showing shear stress versus depth for a quartz rheology (after Smith and Bruhn, 1984). Lower, Fault-plane geometries and corresponding focal-depth histograms for three large historical earthquakes in the Basin and Range province; sources of data: Doser (1985b) and a compilation of microearthquake focal depths from University of Utah student theses, for the Hebgen Lake earthquake; Doser and Smith (1985) and Richins and others (1985), for the Borah Peak earthquake; and Okaya and Thompson (1985) and Doser (1986), for the Dixie Valley earthquake.

(Smith, 1989). Possibilities for the mechanisms controlling earthquake depth distributions include crystallization of rhyolite and basaltic melts, hydrothermal fluid flow into the shallow crust of the caldera, and the shallowing of superheated brines above a magma source (Pelton and Smith, 1982; Dzurisin and others, 1990).

Influence of the Yellowstone hot spot on the seismicity of the ISB

The occurrence of the 1983 Borah Peak earthquake on the flank of the SRP focused attention on the hypothesis that the Yellowstone hot spot and its track, the late Cenozoic Y-SRP volcanic province, are influencing the contemporary seismotectonics of the surrounding region (Smith and others, 1985; Scott and others, 1985; Anders and others, 1989). The unusual parabolic-shaped pattern of earthquakes surrounding the eastern SRP (Figs. 1 and 15) has been hypothesized to reflect the influence of lateral variations in deviatoric stresses, lithospheric subsidence, and high temperatures (Smith and others, 1985, in preparation; Blackwell, 1989), or an integrated loss of strength of upper-crustal volcanic material underlying the SRP (Anders and others, 1989). In all these hypothesized models, the thermomechanical response of the lithosphere may influence the seismogenic potential of the central ISB at map distances up to 100 km or more away from the SRP, thus altering the regional seismicity pattern.

The influence of the Yellowstone hot spot on the Quaternary tectonics of the ISB is further characterized in Fig. 23 (from Smith and others, 1990a), where late Cenozoic normal faults, calderas, and ages of plate and volcanic progression are plotted. A hypothesized "shoulder" of lithospheric subsidence flanking the SRP is also shown. Figure 23 shows that segments of normal faults with latest Quaternary displacements are systematically located away from the boundary of the SRP, whereas segments adjacent to the SRP are characterized by older Quaternary ruptures (Scott and others, 1985; Smith and others, 1985; Anders and others, 1989). Anders and others (1989) plot a double parabola with an apex at Yellowstone that envelops both the youngest faulting and background seismicity surrounding the relatively aseismic SRP. Their inner parabola follows approximately, but not exactly, the coarse dashed line sketched in Figure 23 marking the outer boundary of the subsidence shoulder around the SRP. The epicenter of the 1983 Borah Peak main shock is seen in Figure 23 to be on the outer boundary of the subsidence shoulder and, hence, along the inner margin of the SRP's flanking seismicity and along the boundary delimiting older and younger faulting.

Brott and others (1981) and Blackwell (1989) have shown that there is up to 700 m of systematic topographic subsidence southwestward along the SRP, away from the Yellowstone caldera along the 800-km-long track of the Yellowstone hot spot, that matches the cooling curve for the passage of a crustal heat source. These observations and the parabolic pattern of seismicity of the central ISB suggest that a thermal anomaly due to passage of the Yellowstone hot spot extends laterally beyond the SRP, systematically influencing regional seismicity in northern Utah, eastern Idaho, and western Wyoming.

Models for large Basin-and-Range-type normal-faulting earthquakes

In the past four decades, three of the largest earthquakes in the western United States were normal-faulting events that occurred in the Basin and Range province: the M_S = 6.8 Dixie Valley, Nevada, earthquake of 1954; the M_S = 7.5 Hebgen Lake, Montana, earthquake of 1959; and the M_S = 7.3 Borah Peak, Idaho, earthquake of 1983. Two of these earthquakes occurred in the ISB, and the other occurred in the central Basin and Range province. Studies of these earthquakes have provided new information on the geometry and mechanism of normal-faulting earthquakes, briefly summarized here. Some primary characteristics of these large earthquakes, illustrated in Figure 22, include rupture on planar normal faults, dipping 40° to 60°, and nucleation at mid-crustal depths of about 15 km, near the depth of the brittle/ductile transition.

Surface deformation for large normal-faulting earthquakes involves both coseismic deformation and slow deformation between the large earthquakes (King and others, 1988). Figure 24, from Smith and Richins (1984), shows a compilation of geodetic data representing basically coseismic ground deformation associated with the Dixie Valley, Hebgen Lake, and Borah Peak earthquakes. Hanging-wall subsidence dominates the observed surface deformation, which extends laterally up to 20 km from the surface fault trace and is a function of the causative fault's width, dip, and coseismic slip. For the Hebgen Lake earthquake, geodetic observations are not available for determining coseismic uplift of the footwall. To model the surface deformation of large normal-faulting earthquakes, King and others (1988) calculated the coseismic and long-term response due to a planar shear-dislocation within an elastic layer underlain by a viscous half-space. Their results show that, for coseismic deformation, hanging-wall subsidence is predicted to be roughly two to four times greater than footwall uplift for faults dipping 45° to 60°; long-term adjustments involving stress relaxation and deformation due to erosion and sediment-deposition tend to broaden the profile of surface deformation, thereby slightly raising both the hanging-wall and footwall blocks (King and others, 1988).

Low-angle and listric faults

The working model outlined in Figure 22 and described above for large normal-faulting earthquakes is not simply compatible with observations from seismic-reflection data and geologic mapping that document Quaternary listric and low-angle normal faulting in many subsurface locations in the central and eastern Basin and Range province (Royse and others, 1975; Anderson and others, 1983; Smith and Bruhn, 1984; Smith and others, 1989). These structures typically flatten at maximum depths of 4 to 6 km, at similar depths to much of the Sevier Desert detachment of western Utah, described in the section on seismotectonic framework.

The question of the seismogenic capability of shallow dipping faults in the ISB has been addressed by various workers by scrutinizing compilations of focal mechanisms (e.g., Zoback, 1983; Arabasz and Julander, 1986; Bjarnason and Pechmann, 1989; Doser and Smith, 1989), and all have similarly found the predominance of seismic slip on planes of moderate ($> 30°$) to steep dip, with mean dips in the range of 45° to 60°. Some fault-plane solutions can be found with one low-angle nodal plane; however, in those cases there is no corroborating evidence in the form of clustered earthquake foci on either a downward-flattening or planar low-angle normal fault to support selection of the low-angle nodal plane as the plane of seismic slip (Arabasz

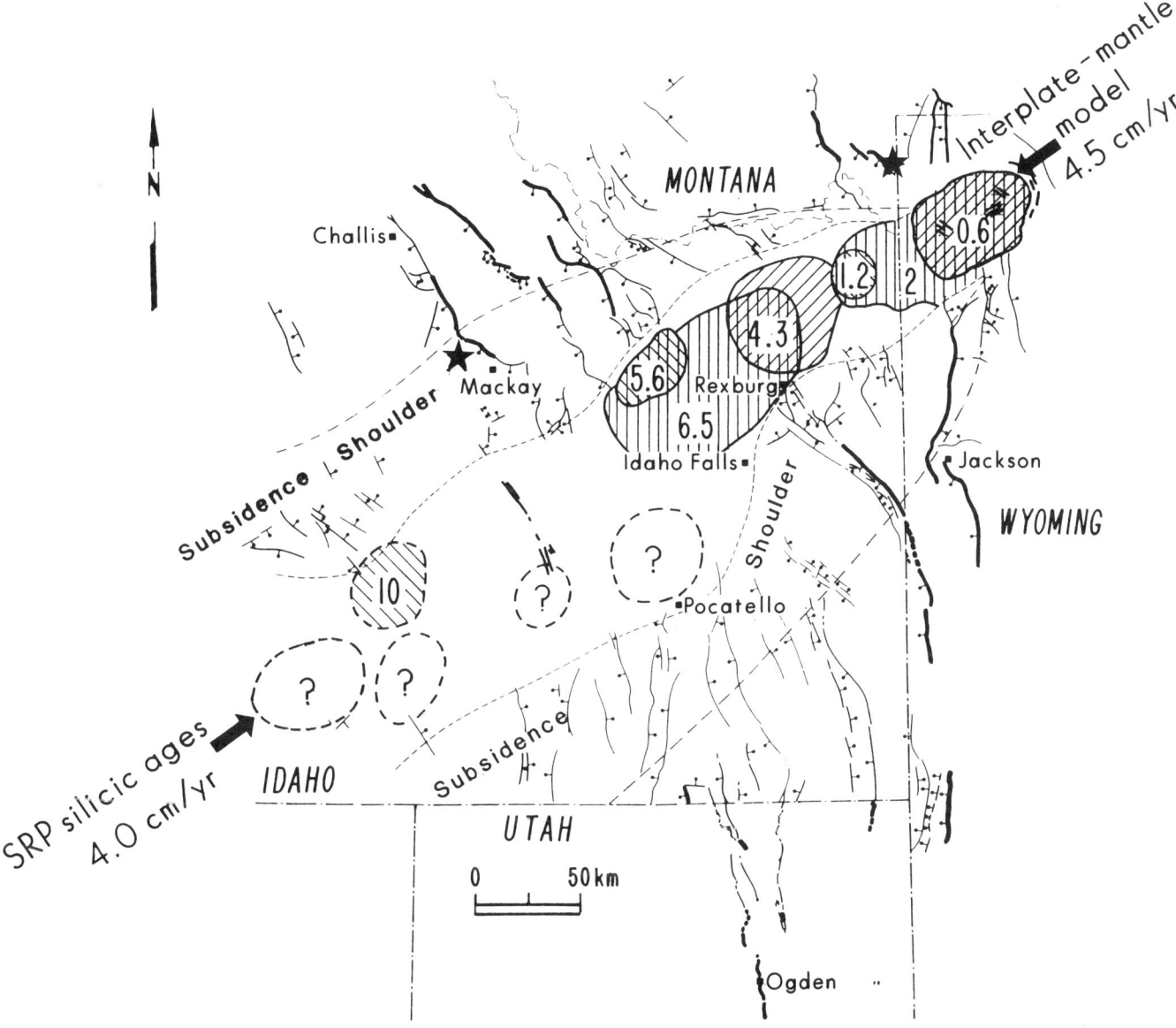

Figure 23. Seismotectonic framework of the Snake River Plain–Yellowstone region, from Smith and others (1990a). Map shows the northeastward decreasing age, in millions of years, of volcanic calderas (hachured) along the eastern Snake River Plain (ESRP), associated with passage of the Yellowstone hot spot, together with the boundary of an hypothesized shoulder of lithospheric subsidence (coarse dashed line), and late Quaternary normal faulting (marked by a heavy line weight where most recent displacement is Holocene). Arrow at Yellowstone (upper right) shows the direction of motion of the North American plate over the Yellowstone hot spot at a relative velocity of 4.5 cm/yr—in agreement with the northeastward space-time progression of calderas at a rate of 4.0 cm/yr. Fine dashed line indicates boundary of the Snake River Plain volcanic province; stars, the epicenters of the 1959 Hebgen Lake and 1983 Borah Peak earthquakes, as in Figure 9.

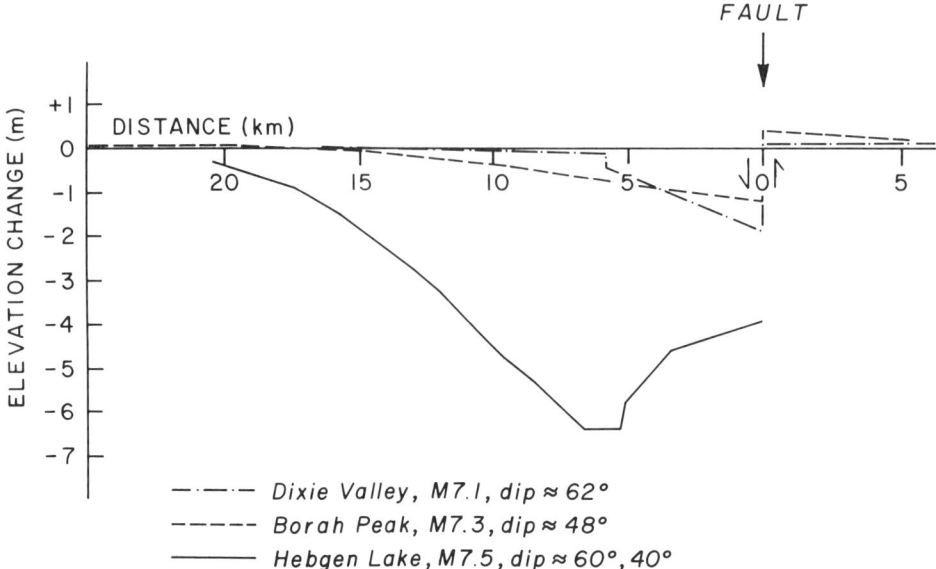

Figure 24. Coseismic vertical ground deformation, chiefly hanging-wall subsidence, produced by the three large scarp-forming, normal-faulting earthquakes identified in Figure 22, from Smith and Richins (1984). Sources of data: Savage and Hastie (1969) and Reil (1957), for the Dixie Valley earthquake; Stein and Barrientos (1985), for the Borah Peak earthquake; and Savage and Hastie (1966), for the Hebgen Lake earthquake.

and Julander, 1986). An investigation by Doser and Smith (1989) of 57 focal mechanisms for moderate to large normal and oblique-slip earthquakes throughout the U.S. Cordillera, including the ISB, revealed no evidence for nodal planes shallower than 30°. Jackson and White (1989) examined a global distribution of normal-faulting focal mechanisms with the same finding. Thus there appears to be no present evidence for the nucleation or slip of an historical earthquake in the ISB on a low-angle or listric normal fault.

Satisfactory answers to questions about the mechanical origin and seismic capability of low-angle normal faults are elusive. Melosh (1990) recently posed a mechanism for listric faulting in the Basin and Range that required a rotation of the maximum and minimum principal stresses, with the principal stress vertical at the surface and rotating to 45° at depths of a few kilometers. His model assumes an elastic upper crust and a viscoelastic lower crust, where the principal-stress rotation flattens the fault. Although this is an attractive hypothesis, the occurrence of large normal-faulting earthquakes at mid-crustal depths of about 15 km on planar faults dipping 45° to 60° argues for a homogeneous stress field to at least the maximum depth of crustal earthquakes in the Intermountain region.

ACKNOWLEDGMENTS

We are indebted to many colleagues who through the years have shared with us their data and ideas about the geology and geophysics of the Intermountain region. We thank E. R. Engdahl for providing the earthquake catalog, and for patiently encouraging this manuscript; S. J. Nava and D. Mason for special help in preparing the manuscript; M. L. Zoback for stress data used in Figure 6; M. C. Stickney for earthquake data used in Figure 18 and for comments on the manuscript; and J. O. Byrd for helpful discussions. J. C. Pechmann and an anonymous reviewer provided valuable comments and advice that improved the manuscript. R. S. Stein and I. G. Wong kindly gave permission to use figures from their papers. Support for preparation of this manuscript was provided by the U.S. Geological Survey, NEHRP Grant No. 14-08-0001-G1762, and by the state of Utah. These agencies, together with the National Science Foundation, the U.S. Bureau of Reclamation, and the National Park Service, have long provided support and cooperation for data acquisition and research that form the basis of this paper.

REFERENCES CITED

Abe, K., 1981, Magnitudes of large shallow earthquakes from 1904 to 1980: Physics of the Earth and Planetary Interiors, v. 27, p. 72–92.

Algermissen, S. T., Perkins, D. M., Thenhaus, P. C., Hanson, S. L., and Bender, B. L., 1982, Probabilistic estimates of maximum acceleration and velocity in the contiguous United States: U.S. Geological Survey Open-File Report 82-1033, 99 p.

Allmendinger, R. W., and 7 others, 1983, Cenozoic and Mesozoic structure of the eastern Basin and Range province, Utah from COCORP seismic reflection data: Geology, v. 11, p. 532–536.

Anders, M. H., Geissman, J. W., Piety, L. A., and Sullivan, J. T., 1989, Parabolic distribution of circumeastern Snake River Plain seismicity and latest Quaternary faulting: Migratory pattern and association with the Yellowstone hotspot: Journal of Geophysical Research, v. 94, p. 1589–1621.

Anderson, L. W., and Miller, D. G., 1979, Quaternary fault map of Utah, *in* Fugro, Inc., Consulting Engineers & Geologists Technical Report: Long Beach, California, 35 p., scale 1:500,000.

Anderson, R. E., 1989, Tectonic evolution of the Intermontane system; Basin and Range, Colorado Plateau, and High Lava Plains, *in* Pakiser, L. C., and Mooney, W. D., eds., Geophysical framework of the continental United States: Geological Society of America Memoir 172, p. 163–176.

Anderson, R. E., and Barnhard, T. P., 1987, Neotectonic framework of the central Sevier Valley area, Utah, and its relationship to seismicity, *in* Gori, P. L., and Hays, W. W., eds., Assessment of regional earthquake hazards and risk along the Wasatch Front, Utah, Vol. 1: U.S. Geological Survey Professional Paper 87-585, p. F1–134.

Anderson, R. E., and Laney, R. L., 1975, The influence of late Cenozoic stratigraphy on distribution of impoundment-related seismicity at Lake Mead, Nevada-Arizona: U.S. Geological Survey Journal of Research, v. 3, no. 3, p. 337–343.

Anderson, R. E., Zoback, M. L., and Thompson, G. A., 1983, Implications for selected subsurface data on the structural form and evolution of some basins in the northern Basin and Range province, Nevada and Utah: Geological Society of America Bulletin, v. 94, p. 1055–1072.

Arabasz, W. J., 1979, Historical review of earthquake-related studies and seismographic recording in Utah, *in* Arabasz, W. J., Smith, R. B., and Richins, W. D., eds., Earthquake studies in Utah 1850 to 1978: Salt Lake City, University of Utah Special Publication, p. 33–56.

Arabasz, W. J., 1984, Earthquake behavior in the Wasatch Front area: Association with geologic structure, space-time occurrence, and stress state, *in* Gori, P. L. and Hays, W. W., eds., Proceedings, Workshop XXVI, Evaluation of regional and urban earthquake hazards and risk in Utah: U.S. Geological Survey Open-File Report 84-763, p. 310–339.

Arabasz, W. J., and Julander, D. R., 1986, Geometry of seismically active faults and crustal deformation within the Basin and Range—Colorado Plateau transition in Utah, *in* Mayer, L., ed., Cenozoic tectonics of the Basin and Range province: A perspective on processes and kinematics of an extensional origin: Geological Society of America Special Paper 208, p. 43–74.

Arabasz, W. J., and McKee, M. E., 1979, Utah earthquake catalog 1850-June 1962, *in* Arabasz, W. J., Smith, R. B., and Richins, W. D., eds., Earthquake studies in Utah 1850 to 1978: Salt Lake City, University of Utah Special Publication, p. 119–121; 133–143.

Arabasz, W. J., and Smith, R. B., 1981, Earthquake prediction in the Intermountain seismic belt—An intraplate extensional regime, *in* Simpson, D. W., and Richards, P. G., eds., Earthquake prediction—An international review: American Geophysical Union, Maurice Ewing Series 4, p. 238–258.

Arabasz, W. J., Smith, R. B., and Richins, W. D., eds., 1979, Earthquake studies in Utah 1850 to 1978: Salt Lake City, University of Utah Seismograph Stations Special Publication, 552 p.

Arabasz, W. J., Smith, R. B., and Richins, W. D., 1980, Earthquake studies along the Wasatch Front, Utah: Network monitoring, seismicity, and seismic hazards: Bulletin of the Seismological Society of America, v. 70, p. 1479–1499.

Arabasz, W. J., Richins, W. D., and Langer, C. J., 1981, The Pocatello Valley (Idaho-Utah border) earthquake sequence of March to April 1975: Bulletin of the Seismological Society of America, v. 71, p. 803–826.

Arabasz, W. J., Pechmann, J. C., and Brown, E. D., 1987, Observational seismology and the evaluation of earthquake hazards and risk in the Wasatch Front area, Utah, *in* Gori, P. L., and Hays, W. W., eds., Assessment of regional earthquake hazards and risk along the Wasatch Front, Utah, Vol. 1: U.S. Geological Survey Professional Paper 87-585, p. D1–39 (revised, 1991, U.S. Geological Survey Professional Paper 1500-D [in press]).

Barazangi, M., and Dorman, J., 1969, World seismicity maps compiled from ESSA, Coast and Geodetic Survey, epicenter data, 1961–1967: Bulletin of the Seismological Society of America, v. 59, p. 369–380.

Barrientos, S. E., Stein, R. S., and Ward, S. N., 1987, Comparison of the 1959 Hebgen Lake, Montana and the 1983 Borah Peak, Idaho, earthquakes from geodetic observations: Bulletin of the Seismological Society of America, v. 77, p. 784–808.

Bjarnason, I. T., and Pechmann, J. C., 1989, Contemporary tectonics of the Wasatch front region, Utah, from earthquake focal mechanisms: Bulletin of the Seismological Society of America, v. 79, p. 731–755.

Blackwell, D. D., 1989, Regional implications of heat flow of the Snake River Plain, northwestern United States: Tectonophysics, v. 164, p. 323–343.

Bodell, J. M., and Chapman, D. S., 1982, Heat flow in the north-central Colorado Plateau: Journal of Geophysical Research, v. 87, p. 2869–2884.

Bolt, B. A., 1984, The magnitudes of the Hebgen Lake 1959 and Idaho 1983 earthquakes [abs.]: Earthquake Notes, v. 55, no. 1, p. 13.

Bond, J. G., 1978, Geologic map of Idaho, Idaho Bureau of Mines and Geology, Moscow, Idaho, scale 1:500,000.

Bonilla, M. G., Mark, R. K., and Lienkaemper, J. J., 1984, Statistical relations among earthquake magnitude, surface rupture length, and surface fault displacement: Bulletin of the Seismological Society of America, v. 74, p. 2379–2411.

Bott, M. H., 1959, The mechanism of oblique slip faulting: Geological Magazine, v. 96, p. 109–117.

Braile, L. W., and 9 others, 1982, The Yellowstone–Snake River Plain seismic profiling experiment: Crustal structure of the eastern Snake River Plain: Journal of Geophysical Research, v. 87, p. 2597–2609.

Brott, C. A., Blackwell, D. D., and Ziagos, D. P., 1981, Thermal and tectonic implications of heat flow in the eastern Snake River Plain, Idaho: Journal of Geophysical Research, v. 86, p. 11,709–11,734.

Byrd, J. O., and Smith, R. B., 1990, Dating recent faulting and estimates of slip rates for the southern segment of the Teton fault, Wyoming: Geological Society of America, Rocky Mountain Section Abstracts with Programs, v. 22, p. 4–5.

Carder, D. S., 1945, Seismic investigations in the Boulder Dam area, 1940–1944, and the influence of reservoir loading on local earthquake activity: Bulletin of the Seismological Society of America, v. 35, p. 175–192.

Carder, D. S., 1970, Reservoir loading and local earthquakes, *in* Adams, W. M., ed., Engineering seismology—The works of man: Geological Society of America, Engineering Geology Case Histories, no. 8, p. 51–61.

Chen, G. J., 1988, A study of seismicity and spectral source characteristics of small earthquakes: Hansel Valley, Utah and Pocatello Valley, Idaho, areas [M.S. thesis]: Salt Lake City, University of Utah, 119 p.

Christiansen, R. L., 1984, Yellowstone magmatic evolution: Its bearing on understanding large-volume explosive volcanism, *in* Explosive volcanism: Inception, evolution, and hazards: Studies in Geophysics: Washington, D.C., National Academy Press, p. 84–95.

Coffman, J. L., von Hake, C. A., and Stover, C. W., 1982, Earthquake history of the United States (revised edition): Washington, D.C., U.S. Government Printing Office, U.S. Department of Commerce and Department of the Interior Publication 41-1, 258 p.

Cook, K. L., 1972, Earthquakes along the Wasatch Front, Utah—The record and the outlook, *in* Environmental Geology of the Wasatch Front, 1971: Utah Geological Association Publication 1, p. H1–29.

Cook, K. L., and Smith, R. B., 1967, Seismicity in Utah, 1850 through June 1965: Bulletin of the Seismological Society of America, v. 57, p. 689–718.

Crone, A. J., and Haller, K. M., 1989, Segmentation of Basin-and-Range normal faults: Examples from east-central Idaho and southwestern Montana, *in* Schwartz, D. P., and Sibson, R. H., eds., Proceedings, Conference XLV, Fault segmentation and controls of rupture initiation and termination, Palm Springs: U.S. Geological Survey Open-File Report 89-315, p. 110–130.

Crone, A. J., and Harding, S. T., 1984, Relationship of late Quaternary fault scarps to subjacent faults, eastern Great Basin, Utah: Geology, v. 12, p. 292–295.

Crone, A. J., and Machette, M. N., 1984, Surface faulting accompanying the Borah Peak earthquake, central Idaho: Geology, v. 12, p. 664–667.

Crone, A. J., and others, 1987, Surface faulting accompanying the Borah Peak earthquake and segmentation of the Lost River fault, central Idaho: Bulletin of the Seismological Society of America, v. 77, p. 739-770.

dePolo, C. M., Clark, D. G., Slemmons, D. B., and Aymard, W. H., 1989, Historical Basin and Range province surface faulting and fault segmentation, *in* Schwartz, D. P., and Sibson, R. H., eds., Proceedings, Conference XLV, Fault segmentation and controls of rupture initiation and termination, Palm Springs: U.S. Geological Survey Open-File Report 89-315, p. 131-162.

Dewey, J. W., 1987, Instrumental seismicity of central Idaho: Bulletin of the Seismological Society of America, v. 77, p. 819-836.

Dewey, J. W., Dillinger, W. H., Taggart, J., and Algermissen, S. T., 1973, A technique for seismic zoning: Analysis of earthquake locations and mechanisms in northern Utah, Wyoming, Idaho, and Montana, *in* Harding, S. T., ed., Contributions to seismic zoning: NOAA Technical Report ERL 267-ESL 30, p. 29-48.

Dewey, J. W., Hill, D. P., Ellsworth, W. L., and Engdahl, E. R., 1989, Earthquakes, faults, and the seismotectonic framework of the contiguous United States, *in* Pakiser, L. C., and Mooney, W. D., eds., Geophysical framework of the continental United States: Geological Society of America Memoir 172, p. 541-575.

Dixon, J. S., 1982, Regional structural synthesis, Wyoming salient of western overthrust belt: American Association of Petroleum Geologists Bulletin, v. 66, p. 1560-1580.

Doser, D. I., 1985a, The 1983 Borah Peak, Idaho and 1959 Hebgen Lake Montana earthquakes: Models for normal fault earthquakes in the Intermountain seismic belt, *in* Stein, R. S., and Bucknam, R. C., eds., Proceedings, Workshop XXVIII, On the Borah Peak, Idaho, Earthquake, Vol. A: U.S. Geological Survey Open-File Report 85-290, p. 368-384.

Doser, D. I., 1985b, Source parameters and faulting processes of the 1959 Hebgen Lake, Montana, earthquake sequence: Journal of Geophysical Research, v. 90, p. 4537-4555.

Doser, D. I., 1986, Earthquake processes in the Rainbow Mountain-Fairview Peak-Dixie Valley, Nevada region, 1954-1959: Journal of Geophysical Research, v. 91, p. 12,572-12,586.

Doser, D. I., 1989a, Source parameters of Montana earthquakes (1925-1964) and tectonic deformation in the northern Intermountain seismic belt: Bulletin of the Seismological Society of America, v. 79, p. 31-50.

Doser, D. I., 1989b, Extensional tectonics in northern Utah-southern Idaho, U.S.A., and the 1934 Hansel Valley sequence: Physics of the Earth and Planetary Interiors, v. 54, p. 120-134.

Doser, D. I., 1990, Foreshocks and aftershocks of large (M ≥ 5.5) earthquakes within the western Cordillera of the United States: Bulletin of the Seismological Society of America, v. 80, p. 110-128.

Doser, D. I., and Smith, R. B., 1982, Seismic moment rates in the Utah region: Bulletin of the Seismological Society of America, v. 72, p. 525-551.

Doser, D. I., and Smith, R. B., 1983, Seismicity of the Teton-southern Yellowstone region, Wyoming: Bulletin of the Seismological Society of America, v. 73, p. 1369-1394.

Doser, D. I., and Smith, R. B., 1985, Source parameters of the 28 October 1983 Borah Peak, Idaho, earthquake from body wave analysis: Bulletin of the Seismological Society of America, v. 75, p. 1041-1051.

Doser, D. I., and Smith, R. B., 1989, An assessment of source parameters of earthquakes in the Cordillera of the western United States: Bulletin of the Seismological Society of America, v. 79, p. 1383-1409.

Dunphy, G. J., 1972, Seismic activity of the Kerr Dam—S.W. Flathead Lake area, Montana, *in* Earthquake research in NOAA: U.S. Department of Commerce, NOAA Technical Report ERL236-ESL121, p. 59-61.

Dzurisin, D., Savage, J. C., and Fournier, R. O., 1990, Recent crustal subsidence at Yellowstone caldera, Wyoming: Journal of Volcanology, v. 52, p. 247-270.

Eaton, G. P., 1982, The Basin and Range province: Origin and tectonic significance: Annual Review of Earth and Planetary Science, v. 10, p. 409-440A.

Eaton, G. P., Wahl, R. R., Prostka, H. J., Mabey, D. R., and Kleinkopf, M. D., 1978, Regional gravity and tectonic patterns: Their relation to late Cenozoic epeirogeny and lateral spreading in the western Cordillera, *in* Smith, R. B. and Eaton, G. P., eds., Cenozoic tectonics and regional geophysics of the western Cordillera: Geological Society of America Memoir 152, p. 51-92.

Eardley, A. J., 1962, Structural geology of North America (second edition): New York, Harper and Row, 743 p.

Eddington, P. K., Smith, R. B., and Renggli, C., 1987, Kinematics of Basin and Range intraplate extension, *in* Coward, M. P., Dewey, J. F., and Hancock, P. L., eds., Continental extensional tectonics: Geological Society of London Special Publication 28, p. 371-392.

Engdahl, E. R., and Rinehart, W. A., 1988, Seismicity map of North America, Continent-Scale Map-004, Decade of North American geology, Geological Society of America, scale 1:5,000,000, 4 sheets.

Freidline, R. A., Smith, R. B., and Blackwell, D. D., 1976, Seismicity and contemporary tectonics of the Helena, Montana area: Bulletin of the Seismological Society of America, v. 66, p. 81-95.

Gilbert, J. D., Ostenna, D., and Wood, C., 1983, Seismotectonic study of Jackson Lake Dam and Reservoir, Minidoka project, Idaho-Wyoming: U.S. Bureau of Reclamation, 123 p.

Griscom, M., 1980, Space-time seismicity patterns in the Utah region and an evaluation of local magnitude as the basis of a uniform earthquake catalog [M.S. thesis]: Salt Lake City, University of Utah, 134 p.

Gupta, H. K., 1985, The present status of reservoir induced seismicity investigations with special emphasis on Koyna earthquakes: Tectonophysics, v. 118, p. 257-279.

Gupta, H. K., and Rastogi, B. K., 1976, Dams and earthquakes: Amsterdam, Elsevier, 229 p.

Gutenberg, B., and Richter, C. F., 1954, Seismicity of the earth and associated phenomena (second edition): Princeton, New Jersey, Princeton University Press, 310 p.

Hall, W. B., and Sablock, P. E., 1985, Comparison of the geomorphic and surficial fracturing effects of the 1983 Borah Peak, Idaho earthquake with those of the 1959 Hebgen Lake, Montana earthquake, *in* Stein, R. S., and Bucknam, R. C., eds., Proceedings, Workshop XXVIII, The Borah Peak, Idaho, earthquake, Vol. A: U.S. Geological Survey Open-File Report 85-290, p. 141-152.

Haller, K. M., 1988, Segmentation of the Lemhi and Beaverhead faults, east-central Idaho, and Red Rock fault, southwest Montana, during the Late Quaternary [M.S. thesis]: University of Colorado, Boulder, 141 p.

Hayden, F. V., 1872, Preliminary report of the United States Geological Survey of Montana and portions of adjacent territories: U.S. Geological and Geographical Survey of the Territories Fifth Annual Report (for 1871), p. 82.

Heck, N. H., 1938, Earthquakes and the western mountain region: Bulletin of the Geological Society of America, v. 49, p. 1-22.

Jackson, J. A., and White, N. J., 1989, Normal faulting in the upper continental crust: Observations from regions of active extension: Journal of Structural Geology, v. 11, no. 1/2, p. 15-36.

Johnson, P. A., and Sbar, M. L., 1987, A microearthquake study of southwest Utah - northwest Arizona: Transition between the Basin and Range province and Intermountain seismic belt: Bulletin of the Seismological Society of America, v. 77, p. 579-596.

Jones, A. E., 1975, Recordings of earthquakes at Reno, 1916-1951: Bulletin of the Seismological Laboratory, Mackay School of Mines, University of Nevada, Reno, 199 p.

Jones, C. H., 1987, A geophysical and geological investigation of extensional structures, Great Basin, western United States [Ph.D. dissertation]: Cambridge, Massachusetts Institute of Technology, 226 p.

King, G. C., Stein, R. S., and Rundle, J. B., 1988, The growth of geological structures by repeated earthquakes, 1. Conceptual framework: Journal of Geophysical Research, v. 93, p. 13,307-13,318.

King, J. J., Doyle, T. E., and Jackson, S. M., 1987, Seismicity of the eastern Snake River plain region, Idaho, prior to the Borah Peak, Idaho, earthquake: October 1972-October 1983: Bulletin of the Seismological Society of America, v. 77, p. 809-818.

Kostrov, V. V., 1974, Seismic moment and energy of earthquakes, and seismic flow of rock: Izvestiya, Academy of Sciences, USSR, Physics of the Solid

Earth (English edition), No. 1, p. 13–21.

Kruger-Knuepfer, J. L., Sbar, M. L., and Richardson, R. M., 1985, Microseismicity of the Kaibab Plateau, northern Arizona and its tectonic implications: Bulletin of the Seismological Society of America, v. 75, p. 491–505.

LaForge, R. C., 1988, An analysis of reservoir-induced seismicity in the back valleys of the Wasatch Mountains, in Sullivan, J. T., Martin, R. A., and Foley, L. L., Seismotectonic study for Jordanelle Dam, Bonneville unit, Central Utah Project, Utah: U.S. Bureau of Reclamation, Seismotectonic Report no. 88-6, Appendix B.

Levy, M., and Christie-Blick, N., 1989, Pre-Mesozoic palinspastic reconstruction of the eastern Great Basin (western United States): Science, v. 245, p. 1454–1462.

Loeb, D. T., and Pechmann, J. C., 1986, The P-wave velocity structure of the crust-mantle boundary beneath Utah from network travel time measurements [abs.]: Earthquake Notes, v. 57, no. 1, p. 10.

Love, J. D., and Reed, J. C., Jr., 1971, Creation of the Teton landscape, the geological story of Grand Teton National Park: Jackson, Wyoming, Grand Teton Natural History Association, 120 p.

Machette, M. N., Personius, S. F., and Nelson, A. R., 1987, Quaternary geology along the Wasatch fault zone: segmentation, recent investigations, and preliminary conclusions in Gori, P. L., and Hays, W. W., eds., Assessment of regional earthquake hazards and risk along the Wasatch Front, Utah, Vol 1: U.S. Geological Survey Open-File Report 87-585, p. A1–72.

Machette, M. N., Personius, S. F., Nelson, A. R., Schwartz, D. P., and Lund, W. R., 1989, Segmentation models and Holocene movement history of the Wasatch fault zone, Utah, in Schwartz, D. P., Sibson, R. H., eds., Proceedings, Conference XLV, Fault segmentation and controls of rupture initiation and termination, Palm Springs: U.S. Geological Survey Open-File Report 89-315, p. 229–245.

McCalpin, J., Robison, R. M., and Garr, J. D., 1987, Neotectonics of the Hansel Valley-Pocatello Valley corridor, northern Utah and southern Idaho, in Gori, P. L., and Hays, W. W., eds., Assessment of regional earthquake hazards and risk along the Wasatch Front, Utah, Vol. 1: U.S. Geological Survey Open-File Report 87-585, p. G1–44.

McGuire, R. K., and Arabasz, W. J., 1990, An introduction to probabilistic seismic hazard analysis, in Ward, S. H., ed., Geotechnical and environmental geophysics, Volume 1: Review and tutorial: Society of Exploration Geophysics, Investigations in Geophysics No. 5, p. 333–353.

McLaughlin, W. C., Waddell, G. G., and McCaslin, J. G., 1976, Seismic equipment used in rock-burst control in the Coeur d'Alene mining district, Idaho: U.S. Bureau of Mines Report of Investigation 8138, 29 p.

Melosh, H. J., 1990, Mechanical basis for low-angle normal faulting in the Basin and Range province: Nature, v. 343, p. 331–335.

Morgan, P., and Gosnold, W. D., 1989, Heat flow and thermal regimes in the continental United States, in Pakiser, L. C., and Mooney, W. D., eds., Geophysical framework of the continental United States: Geological Society of America Memoir 172, p. 493–522.

Munson, R. C., 1970, Relationship of effect of waterflooding of the Rangely oil field on seismicity, in Adams, W. M., ed., Engineering seismology—The works of man: Geological Society of America, Engineering Geology Case Histories, no. 8, p. 39–49.

Nagy, W. C., and Smith, R. B., 1988, Seismotectonic implications of the 1985–86 Yellowstone earthquake swarm from velocity and stress inversion [abs.]: Seismological Research Letters, v. 59, no. 1, p. 20.

Nava, S. J., Pechmann, J. C., and Arabasz, W. J., 1988, A Swell earthquake in the Colorado Plateau [abs.]: Seismological Research Letters, v. 60, n. 1, p. 30.

Nava, S. J., and 7 others, 1990, Earthquake catalog for the Utah region: January 1, 1986 to December 31, 1988: Salt Lake City, University of Utah Seismograph Stations Special Publication, 96 p.

Okaya, D. A., and Thompson, G. A., 1985, Geometry of Cenozoic extensional faulting: Dixie Valley, Nevada: Tectonics, v. 4, p. 107–126.

Owens, T. J., 1983, Normal faulting and flexure in an elastic-perfectly plastic plate: Tectonophysics, v. 93, p. 129–150.

Pack, F. J., 1921, The Elsinore earthquakes in central Utah, September 29 and October 1, 1921: Bulletin of the Seismological Society of America, v. 11, p. 155–165.

Pakiser, L. C., 1989, Geophysics of the Intermontane system, in Pakiser and Mooney, in Pakiser, L. C., and Mooney, W. D., eds., Geophysical framework of the continental United States: Geological Society of America Memoir 172, p. 235–248.

Pardee, J. T., 1926, The Montana earthquake of June 27, 1925: U.S. Geological Survey Professional Paper 147-B, 17 p.

Pardee, J. T., 1950, Late Cenozoic block faulting in western Montana: Bulletin of the Geological Society of America, v. 51, p. 359–406.

Parry, W. T., and Bruhn, R. L., 1986, Pore fluid and seismogenic characteristics of fault rock at depth on the Wasatch Fault, Utah: Journal of Geophysical Research, v. 91, p. 730–744.

Parry, W. T., and Bruhn, R. L., 1987, Fluid inclusion evidence for minimum 11 km vertical offset on the Wasatch fault, Utah: Geology, v. 15, p. 67–70.

Pechmann, J. C., and Thorbjarnardottir, B., 1984, Investigation of an M_L 4.3 earthquake in the western Salt Lake Valley using digital seismic data, in Gori, P. L., and Hays, W. W., eds., Proceedings, Workshop XXVI, Evaluation of regional and urban earthquake hazards and risk in Utah: U.S. Geological Survey Open-File Report 84-763, p. 340–365.

Pelton, J. R., and Smith, R. B., 1982, Contemporary vertical surface displacements in Yellowstone National Park: Journal of Geophysical Research, v. 87, p. 2745–2761.

Pennington, W., Smith, R. B., and Trimble, A. B., 1974, A microearthquake survey of parts of the Snake River Plain and Central Idaho: Bulletin of the Seismological Society of America, v. 64, p. 307–312.

Peyton, S. L., and Smith, R. B., 1990, Earthquake catalog for the Yellowstone National Park region: January 1, 1989 through December 31, 1989: Salt Lake City, University of Utah Special Publication, 33 p.

Piety, L. A., Wood, C. K., Gilbert, J. D., Sullivan, J. T., and Anders, M. H., 1986, Seismotectonic study for Palisades Dam and Reservoir, Palisades project: U.S. Bureau of Reclamation, Engineering and Research Center, Seismotectonic Division, Denver, Colorado, and Pacific Northwest Region Geology Branch, Boise, Idaho, Seismotectonic Report no. 86-3, 198 p.

Pitt, A. M., 1987, Catalog of earthquakes in the Yellowstone National Park-Hebgen Lake region, Wyoming, Montana, and Idaho for the years 1973 to 1981: U.S. Geological Survey Open File Report, p. 87-61.

Pitt, A. M., and Hutchinson, R. A., 1982, Hydrothermal changes related to earthquake activity at Mud Volcano, Yellowstone National Park, Wyoming: Journal of Geophysical Research, v. 87, p. 2762–2766.

Pitt, A. M., Weaver, C. S., and Spence, W., 1979, The Yellowstone Park earthquake of June 30, 1975: Bulletin of the Seismological Society of America, v. 69, p. 187–205.

Planke, S., and Smith, R. B., 1991, Cenozoic extension and evolution of the Sevier Desert Basin, Utah, from seismic reflection, gravity, and well log data: Tectonics, v. 10, p. 345–365.

Poppe, B. B., 1980, Directory of world seismograph stations: U.S. Department of Commerce Report SE-25, Part 1, v. 1, 465 p.

Qamar, A., and Hawley, B., 1979, Seismic activity near the Three Forks Basin, Montana: Bulletin of the Seismological Society of America, v. 69, p. 1917–1929.

Qamar, A. I., and Stickney, M. C., 1983, Montana earthquakes, 1869–1979—Historical seismicity and earthquake hazard: Montana Bureau of Mines and Geology Memoir 51, 80 p.

Qamar, A., Kogan, J., and Stickney, M. C., 1982, Tectonics and recent seismicity near Flathead Lake, Montana: Bulletin of the Seismological Society of America, v. 72, p. 1591–1599.

Raleigh, C. B., Healy, J. H., and Bredehoeft, J. D., 1972, Faulting and crustal stress at Rangely, Colorado, in Flow and Fracture of Rocks, Monograph 16: American Geophysical Union, p. 275–284.

Reil, O. E., 1957, The Dixie Valley-Fairview Peak, Nevada, earthquakes of December 16, 1954: Damage to Nevada highways: Bulletin of the Seismological Society of America, v. 47, p. 349–352.

Reilinger, R., 1986, Evidence for postseismic viscoelastic relaxation following the 1959 M = 7.5 Hebgen Lake, Montana, earthquake: Journal of Geophysical Research, v. 91, p. 9488–9494.

Richins, W. D., and Arabasz, W. J., 1985, Seismicity of southeastern Idaho based on seismic monitoring using regional and temporary local networks [abs.]: Geological Society of America Rocky Mountain Section Meeting Abstracts with Programs, v. 17, n. 4, p. 261.

Richins, W. D., Arabasz, W. J., and Langer, C. J., 1983, Episodic earthquake swarms ($M_L \leq 4.7$) near Soda Springs, Idaho, 1981–82: Correlation with local structure and regional tectonics [abs.]: Earthquake Notes, v. 54, no. 1, p. 99.

Richins, W. D., Smith, R. B., Langer, C. J., Zollweg, J. E., King, J. J., and Pechmann, J. C., 1985, The 1983 Borah Peak, Idaho, earthquake: Relationship of aftershocks to the main shock, surface faulting, and regional tectonics, in Stein, R. S., and Bucknam, R. C., eds., Proceedings, Workshop XXVIII, On the Borah Peak, Idaho, earthquake, Vol. A: U.S. Geological Survey Open-File Report 85-290, p. 285–310.

Richins, W. D., and 6 others, 1987, The 1983 Borah Peak, Idaho, earthquake and its aftershocks: Bulletin of the Seismological Society of America, v. 77, p. 694–723.

Richter, C. F., 1935, An instrumental earthquake magnitude scale: Bulletin of the Seismological Society of America, v. 25, p. 1–32.

Richter, C. F., 1958, Elementary seismology: San Francisco, W. H. Freeman and Company, 768 p.

Rogers, A. M., and Lee, W. H., 1976, Seismic study of earthquakes in the Lake Mead, Nevada-Arizona region: Bulletin of the Seismological Society of America, v. 66, p. 1657–1681.

Rogers, A. M., Algermissen, S. T., Hays, W. W., and Perkins, D. M., 1976, A study of earthquake losses in the Salt Lake City, Utah, area: U.S. Geological Survey Open-File Report 76-89, 357 p.

Rogers, A. M., Harmsen, S. C., and Meremonte, M. E., 1987, Evaluation of the seismicity of the southern Great Basin and its relationship to the tectonic framework of the region: U.S. Geological Survey Open-File Report 87-408, 196 p.

Royse, F., Warner, M. A., and Reese, D. L., 1975, Thrust belt structural geometry and related stratigraphic problems, Wyoming–Idaho–Northern Utah, in Bolyare, D. W., ed., Deep drilling frontiers in the central Rocky Mountains: Rocky Mountain Association of Geologists, Denver, p. 41–54.

Ryall, A., Slemmons, D. B., and Gedney, L. S., 1966, Seismicity, tectonism, and surface faulting in the western United States during historic time: Bulletin of the Seismological Society of America, v. 56, p. 1105–1135.

Savage, J. C., and Hastie, L. M., 1966, Surface deformation associated with dip-slip faulting: Journal of Geophysical Research, v. 71, p. 4897–4904.

Savage, J. C., and Hastie, L. M., 1969, A dislocation model for the Fairview Peak, Nevada, earthquake: Bulletin of the Seismological Society of America, v. 59, p. 1937–1948.

Savage, J. C., Lisowski, M., and Prescott, W. H., 1985, Strain accumulation in the Rocky Mountain states: Journal of Geophysical Research, v. 90, p. 10,310–10,320.

Sbar, M. L., and Barazangi, M., 1970, Tectonics of the Intermountain seismic belt, western United States, Part I—Microearthquake seismicity and composite fault plane solutions: Geological Society of America Abstracts with Programs, v. 2, p. 675.

Sbar, M. L., Barazangi, M., Dorman, J., Scholz, C. H., and Smith, R. B., 1972, Tectonics of the Intermountain seismic belt, western United States: Microearthquake seismicity and composite fault plane solutions: Geological Society of America Bulletin, v. 83, p. 13–28.

Scarborough, R. B., Manges, C. M., and Pearthree, P. A., 1986, Map of Late Pliocene–Quaternary (post 4 m.y.) faults, folds, and volcanic rocks in Arizona: Tucson, Arizona Bureau of Geology and Mineral Technology, Geological Survey Branch, Map 22, scale 1:1,000,000.

Scholz, C. H., 1990, The mechanics of earthquakes and faulting, Cambridge University Press, 439 p.

Schwartz, D. P., 1987, Earthquakes of the Holocene: Reviews of Geophysics, v. 25, p. 1197–1202.

Schwartz, D. P., 1988, Geological characterization of seismic sources: Moving into the 1990s, in Von Thun, J. L., ed., Earthquake engineering and soil dynamics II—Recent advances in ground motion evaluation: American Society of Civil Engineers, Geotechnical Special Publication 20, p. 1–42.

Schwartz, D. P., and Coppersmith, K. J., 1984, Fault behavior and characteristic earthquakes: examples from the Wasatch and San Andreas fault zones: Journal of Geophysical Research, v. 89, p. 5681–5698.

Schwartz, D. P., and Coppersmith, K. J., 1986, Seismic hazards: New trends in analysis using geologic data, in Active tectonics: Washington, D.C., National Academy Press, p. 215–230.

Scott, W. E., Pierce, K. L., and Hait, M. H., Jr., 1985, Quaternary tectonic setting of the 1983 Borah Peak earthquake, central Idaho: Bulletin of the Seismological Society of America, v. 75, p. 1053–1066.

Shenon, P. J., 1934, Hansel Valley earthquake, March 12, 1934 (unpublished report): U.S. Geological Survey.

Shenon, P. J., 1936, The Utah earthquake of March 12, 1934 (extracts from unpublished report), in Neuman, F., United States earthquakes 1934: U.S. Department of Commerce, Coast and Geodetic Survey, ser. 593, p. 43–48.

Shimizu, Y., 1987, Earthquake clustering in the Utah region [M.S. thesis]: Cambridge, Massachusetts Institute of Technology, 149 p.

Sibson, R. H., 1982, Fault zone models, heat flow, and the depth distribution of earthquakes in the continental crust of the United States: Bulletin of the Seismological Society of America, v. 72, p. 151–163.

Simpson, D. W., 1976, Seismicity changes associated with reservoir loading: Engineering Geology, v. 10, p. 123–150.

Slemmons, D. B., 1977, State-of-the-art for assessing earthquake hazards in the United States, Report 6: Faults and earthquake magnitude: U.S. Army Engineer Waterways Experiment Station Miscellaneous Paper S-73-1, 129 p.

Smith, R. B., 1972, Contemporary seismicity, seismic gaps, and earthquake recurrences of the Wasatch Front; in Cook, K. L., ed., Environmental geology of the Wasatch Front, 1971: Utah Geological Association, Publication 1, p. I1–19.

Smith, R. B., 1977, Intraplate tectonics of the western North American Plate: Tectonophysics, v. 37, p. 323–336.

Smith, R. B., 1978, Seismicity, crustal structure, and intraplate tectonics of the interior of the western Cordillera, in Smith, R. B., and Eaton, G. P., eds., Cenozoic tectonics and regional geophysics of the western Cordillera: Geological Society of America Memoir 152, p. 111–144.

Smith, R. B., 1989, Kinematics and dynamics of the Yellowstone hotspot [abs.]: EOS American Geophysical Union Transactions, v. 70, p. 1356.

Smith, R. B., and Braile, L. W., 1984, Crustal structure and evolution of an explosive silicic volcanic system at Yellowstone National Park, in Explosive volcanism: Inception, evolution, and hazards: Studies in Geophysics: Washington, D.C., National Academy Press, p. 96–111.

Smith, R. B., and Bruhn, R. L., 1984, Intraplate extensional tectonics of the eastern Basin-Range: Inferences on structural style from seismic reflection data, regional tectonics, and thermal-mechanical models of brittle-ductile deformation: Journal of Geophysical Research, v. 89, p. 5733–5762.

Smith, R. B., and Lehman, J. A., 1979, Ground response spectra from M_L 5.7 Logan (Cache Valley) earthquake of 1962, in Arabasz, W. J., Smith, R. B., and Richins, W. D., eds., Earthquake Studies in Utah 1850 to 1978: Salt Lake City, University of Utah Seismograph Stations Special Publication, p. 487–495.

Smith, R. B., and Lindh, A. G., 1978, Fault-plane solutions of the western United States; A compilation, in Smith, R. B., and Eaton, G. P., eds., Cenozoic tectonics and regional geophysics of the western Cordillera: Geological Society of America Memoir 152, p. 107–109.

Smith, R. B., and Richins, W. D., 1984, Seismicity and earthquake hazards of Utah and the Wasatch Front: Paradigm and paradox, in Gori, P. L., and Hays, W. W., eds., Proceedings, Workshop XXVI, Evaluation of regional and urban earthquake hazards and risk in Utah: U.S. Geological Survey Open-File Report 84-763, p. 73–112.

Smith, R. B., and Sbar, M. L., 1970, Seismicity and tectonics of the Intermountain seismic belt, western United States, Part II, Focal mechanism of major earthquakes: Geological Society of America Abstracts with Programs, v. 2, p. 657.

Smith, R. B., and Sbar, M. L., 1974, Contemporary tectonics and seismicity of the western United States with emphasis on the Intermountain seismic belt: Geological Society of America Bulletin, v. 85, p. 1205–1218.

Smith, R. B., Winkler, P. L., Anderson, J. G., and Scholz, C. H., 1974, Source mechanisms of microearthquakes associated with underground mines in eastern Utah: Bulletin of the Seismological Society of America, v. 64, p. 1295–1317.

Smith, R. B., Shuey, R. T., Pelton, J. R., and Bailey, J. P., 1977, Yellowstone hot spot: Contemporary tectonics and crustal properties from earthquake and aeromagnetic data: Journal of Geophysical Research, v. 82, p. 3665–3676.

Smith, R. B., and 8 others, 1982, The Yellowstone–eastern Snake River Plain seismic profiling experiment: Crustal structure of Yellowstone: Journal of Geophysical Research, v. 84, p. 2583–2596.

Smith, R. B., Richins, W. D., and Doser, D. I., 1985, The 1983 Borah Peak, Idaho earthquake: Regional seismicity, kinematics of, and tectonic mechanism, *in* Stein, R. S., Bucknam, R. C., eds., Proceedings, Workshop XXVIII, The Borah Peak, Idaho earthquake, Vol. A.: U.S. Geological Survey Open-File Report 85-290, p. 236–263.

Smith, R. B., Nagy, W. C., Julander, D. R., Viveiros, J. J., Barker, C. A., and Gants, D. G., 1989, Geophysical and tectonic framework of the eastern Basin and Range–Colorado Plateau–Rocky Mountain transition, *in* Pakiser, L. C., and Mooney, W. D., eds., Geophysical framework of the continental United States: Geological Society of America Memoir 172, p. 205–233.

Smith, R. B., Byrd, J. O., and Susong, D. D., 1990a, Neotectonics and structural evolution of the Teton fault, *in* Roberts, S., ed., Geologic field tours of western Wyoming and parts of adjacent Idaho, Montana, and Utah, field trip no. 6: The Geological Survey of Wyoming, Public Information Circular No. 29, p. 126–138.

Smith, R. B., Byrd, J. O., Susong, D. D., Sylvester, A. G., Bruhn, R. L., and Geissman, J. W., 1990b, Three Year Progress Report, An evaluation of earthquake hazards of the Grand Teton National Park emphasizing the Teton fault (unpublished report to the University of Wyoming–National Park Service Research Center): Salt Lake City, University of Utah, 149 p.

Sparlin, M. A., Braile, L. W., and Smith, R. B., 1982, Crustal structure of the Eastern Snake River Plain determined from ray trace modeling of seismic refraction data: Journal of Geophysical Research, v. 87, p. 2619–2633.

Stein, R. S., and Barrientos, S. E., 1985, Planar high-angle faulting in the Basin and Range: Geodetic analysis of the 1983 Borah Peak, Idaho, earthquake: Journal of Geophysical Research, v. 90, p. 11,355–11,366.

Stevenson, P. R., 1976, Microearthquakes at Flathead Lake, Montana: A study using automatic earthquake processing: Bulletin of the Seismological Society of America, v. 66, p. 61–80.

Stickney, M. C., 1988, Montana seismicity, 1986: Butte, Montana Bureau of Mines and Geology Open-File Report 204, 39 p.

Stickney, M. C., and Bartholomew, M. J., 1987, Seismicity and late Quaternary faulting of the northern Basin and Range province, Montana and Idaho: Bulletin of the Seismological Society of America, v. 77, p. 1602–1625.

Stover, C. W., 1985, Isoseismal map and intensity distribution for the Borah Peak, Idaho, earthquake of October 28, 1983, *in* Stein, R. S., and Bucknam, R. C., eds., Proceedings, Workshop XXVIII, The Borah Peak, Idaho, earthquake, Vol. A: U.S. Geological Survey Open-File Report 85-290, p. 401–408.

Stover, C. W., Reagor, B. G., and Algermissen, S. T., 1986, Seismicity map of the State of Utah: U.S. Geological Survey Miscellaneous Field Studies Map MF-1856, scale 1:1,000,000.

Sullivan, J. T., Nelson, A. R., LaForge, R. C., Wood, C. K., and Hansen, R. A., 1988, Central Utah regional seismotectonic study for USBR dams in the Wasatch Mountains: U.S. Bureau of Reclamation Seismotectonic Report 88-5, 269 p.

Thenhaus, P. C., and Wentworth, C. M., 1982, Map showing zones of similar ages of surface faulting and estimated maximum earthquake size in the Basin and Range province and selected adjacent areas, U.S. Geological Survey Open-File Report 82-742.

Tocher, D., 1962, The Hebgen Lake, Montana, earthquake of August 17, 1959, MST,: Bulletin of the Seismological Society of America, v. 52, p. 153–162.

Townley, S. D., and Allen, M. W., 1939, Descriptive catalog of earthquakes of the Pacific Coast of the United States, 1769 to 1928: Bulletin of the Seismological Society of America, v. 29, p. 1–297.

Veneziano, D., Shimizu, Y., and Arabasz, W. J., 1987, Suppressed earthquake clustering in the Wasatch Front region, Utah [abs.]: EOS American Geophysical Union Transactions, v. 68, p. 1368.

Wahlquist, W. L., 1981, Atlas of Utah: Provo, Utah, Brigham Young University Press, Weber State College, 300 p.

Westaway, R., and Smith, R. B., 1989a, Strong ground motion in normal-faulting earthquakes: Geophysical Journal of the Royal Astronomical Society, v. 96, p. 529–559.

Westaway, R., and Smith, R. B., 1989b, Source parameters of the Cache Valley (Logan), Utah, earthquake of 30 August 1962: Bulletin of the Seismological Society of America, v. 79, p. 1410–1425.

Williams, D. J., and Arabasz, W. J., 1989, Mining-related and tectonic seismicity in the East Mountain area, Wasatch Plateau, Utah, U.S.A.: PAGEOPH, v. 129, p. 345–368.

Williams, J. S., and Tapper, M. L., 1953, Earthquake history of Utah, 1850–1949: Bulletin of the Seismological Society of America, v. 43, p. 191–218.

Witkind, I. J., 1964, Reactivated faults north of Hebgen Lake, in the Hebgen Lake, Montana, earthquake of August 17, 1959: U.S. Geological Survey Professional Paper 435, p. 37–50.

Witkind, I. J., 1975a, Preliminary map showing known and suspected active faults in Idaho: U.S. Geological Survey Open-File Report 75-278, scale 1:500,000.

Witkind, I. J., 1975b, Preliminary map showing known and suspected active faults in Wyoming: U.S. Geological Survey Open-File Report 75-279, scale 1:500,000.

Witkind, I. J., 1975c, Preliminary map showing known and suspected active faults in Montana: U.S. Geological Survey Open-File Report 75-285, scale 1:500,000.

Witkind, I. J., Myers, W. B., Hadley, J. B., Hamilton, W., and Fraser, G. D., 1962, Geologic features of the earthquake at Hebgen Lake, Montana, August 17, 1959: Bulletin of the Seismological Society of America, v. 52, p. 163–180.

Wong, I. G., and Chapman, D. S., 1990, Deep intraplate earthquakes in the western United States and their relationship to lithospheric temperatures: Bulletin of the Seismological Society of America, v. 80, p. 589–599.

Wong, I. G., and Humphrey, J. R., 1989, Contemporary seismicity, faulting, and the state of stress in the Colorado Plateau: Geological Society of America Bulletin, v. 101, p. 1127–1146.

Wong, I. G., Humphrey, J. R., Adams, J. A., and Silva, W. J., 1989, Observations of mine seismicity in the eastern Wasatch Plateau, Utah, U.S.A.: A possible case of implosional failure: PAGEOPH, v. 129, p. 369–405.

Wood, C., 1988, Earthquake Data—1986, Jackson Lake Seismograph Network, Jackson Lake Dam, Minidoka project, Wyoming: U.S. Bureau of Reclamation Seismotectonic Report 88-1, 41 p.

Woodward-Clyde Consultants, 1979, A seismic hazard study for the Treat Reactor Facility, at the INEL, Idaho (unpublished report prepared for Idaho National Engineering Laboratory, Idaho Falls, Idaho): San Francisco, California, Woodward-Clyde Consultants, 109 p.

Woolard, G. P., 1958, Areas of tectonic activity in the United States as indicated by earthquake epicenters: EOS (American Geophysical Union Transactions), v. 39, p. 1135–1150.

Youngs, R. R., Swan, F. H., Power, M. S., Schwartz, D. P., and Green, R. K., 1987, Probabilistic analysis of earthquake ground shaking hazard along the Wasatch Front, Utah, *in* Gori, P. L., and Hays, W. W., eds., Assessment of regional earthquake hazards and risk along the Wasatch Front, Utah, Vol. 2: U.S. Geological Survey Professional Paper 87-585, p. M1–110.

Zandt, G., and Owens, T. J., 1980, Crustal flexure associated with normal faulting and implications for seimicity along the Wasatch Front, Utah: Bulletin of the Seismological Society of America, v. 70, p. 1501–1520.

Zandt, G., McPherson, L., Schaff, S., and Olsen, S., 1982, Seismic baseline and inducation studies, Roosevelt Hot Springs, Utah and Raft River Idaho: Salt Lake City, University of Utah Research Institute, Earth Science Laboratory Technical Report, Department of Energy Contract No. DE-AS07-78ID01821, 58 p.

Zoback, M. L., 1983, Structure and Cenozoic tectonism along the Wasatch fault zone, Utah, in Miller, D. M., Todd, V. R., and Howard, K. A., eds., Tectonics and stratigraphy of the eastern Great Basin: Geological Society of America Memoir 157, p. 3–27.

Zoback, M. L., 1989, State of stress and modern deformation of the northern Basin and Range province: Journal of Geophysical Research, v. 94, p. 7105–7128.

Zoback, M. L., and Zoback, M. D., 1989, Tectonic stress field of the continental United States, in Pakiser, L. C., and Mooney, W. D., eds., Geophysical framework of the continental United States: Geological Society of America Memoir 172, p. 523–539.

Zollweg, J. E., and Richins, W. D., 1985, Late aftershocks of the 1983 Borah Peak, Idaho, earthquake and related activity in central Idaho, in Gori, P. L., and Hays, W. W., eds., Proceedings, Workshop XXVI, Evaluation of regional and urban earthquake hazards and risk in Utah: U.S. Geological Survey Open-File Report 84-763, p. 345–356.

MANUSCRIPT ACCEPTED BY THE SOCIETY FEBRUARY 25, 1991

Printed in U.S.A.

Chapter 12

Seismicity of the Rio Grande rift in New Mexico

Allan R. Sanford and Lawrence H. Jaksha
Geoscience Department and Geophysical Research Center, New Mexico Institute of Mining and Technology, Socorro, New Mexico 87801
Daniel J. Cash
Los Alamos National Laboratory, Los Alamos, New Mexico 87545

INTRODUCTION

Groups at the New Mexico Institute of Mining and Technology, Los Alamos National Laboratory, and the U.S. Geological Survey's Albuquerque Seismological Laboratory have engaged in instrumental studies of earthquakes in New Mexico since 1960, with particular emphasis on the Rio Grande rift. The three organizations have also collaborated on producing seismicity maps for the state of New Mexico.

The first section of this chapter discusses the distribution of earthquake activity throughout the state with respect to the Rio Grande rift, young fault movements, recent volcanism, and broad regional uplift. The second part summarizes results of specific studies that have produced fairly detailed knowledge of seismicity in several regions of the rift.

GENERAL FEATURES OF RIO GRANDE RIFT SEISMICITY

Background

Any discussion of the seismicity of the Rio Grande rift (RGR) in New Mexico depends on the assumed geographic limits of the structure, a subject of considerable debate for several decades. As originally defined by Bryan (1938) and later adopted by Kelley (1952), the RGR was a narrow chain of prominent structural depressions 50 to 100 km wide extending 550 km north-south through central New Mexico from the Colorado boundary to Mexico (Fig. 1). This definition was adopted by Sanford and others (1972) in an early paper on the seismicity of the rift. Included in that paper were instrumental data for the period 1960 through 1970. For the 11-yr period, only about 30 percent of all earthquakes in New Mexico with magnitude greater than 2.7 were located within the rift boundaries used by Bryan and Kelley.

In a second paper on the seismicity of the RGR, Sanford and others (1979) used broader boundaries (Fig. 1) proposed for the rift by Chapin (1971). A third paper on the seismicity of the

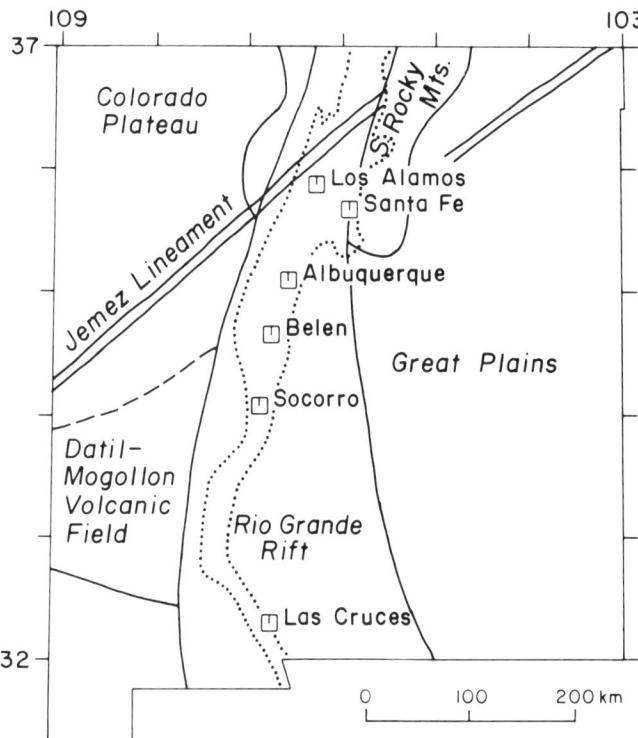

Figure 1. Boundaries proposed for the Rio Grande rift. Dotted lines indicate the approximate boundaries for the rift given by Bryan (1938) and later adopted by Kelley (1952). Their rift only included the narrow chain of prominent structural depressions through which the Rio Grande flows. Solid lines show the approximate boundaries of the rift proposed by Chapin (1971) in relation to other physiographic provinces in New Mexico. Chapin extended the margins of the rift to include structural basins adjacent to those through which the Rio Grande flows.

entire state of New Mexico (Sanford and others, 1981) used essentially the same boundaries. Despite the greater width of the RGR identified by Chapin (Fig. 1), most of the recorded activity reported in these two papers fell outside the rift. On the basis of instrumental data through 1977, the Great Plains and Colorado Plateau—two provinces considered to be stable on geologic

Sanford, A. R., Jaksha, L. H., and Cash, D. J., 1991, Seismicity of the Rio Grande rift in New Mexico, *in* Slemmons, D. B., Engdahl, E. R., Zoback, M. D., Blackwell, D. D., eds., Neotectonics of North America: Boulder, Colorado, Geological Society of America, Decade Map Volume 1.

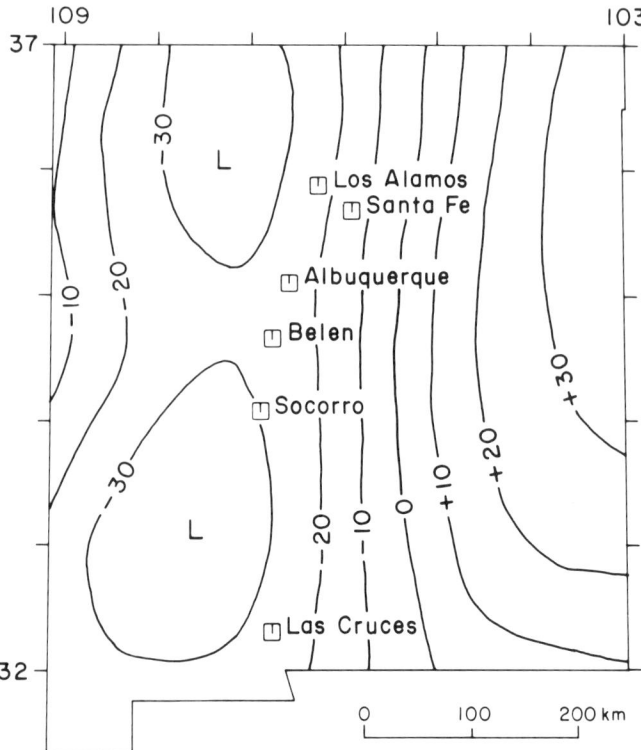

Figure 2. High-altitude magnetic map of New Mexico. The map is based on the United States composite magnetic anomaly map continued upward to an altitude of 160 km (Taylor and others, 1983). Contour interval is 10 n T.

anomaly has a maximum delay of 1.5 seconds and an east-west extent of approximately 340 km, of which about 60 km underlies the Colorado Plateau and about 70 km underlies the Great Plains.

Shown in Figure 2 is a magnetic map for New Mexico based on the composite magnetic anomaly map of the United States by Taylor and others (1983). The broad north-south–oriented low that appears to be associated with the RGR is one of the strongest and most clearly defined in the conterminous United States. The boundaries for this anomaly are somewhat subjective but clearly fall beneath the Colorado Plateau and the Great Plains and well beyond the maximum boundaries of the RGR in Figure 1.

Gable and Hatton (1983) presented geologic evidence for surface uplift during the past 10 m.y. over a broad region centered on the RGR (Fig. 3). The general character of the uplift extends uninterrupted north through Colorado and south-southeast through west Texas. The east-west extent and generally north-south orientation of the uplift suggests a relation between it and the asthenospheric-lithospheric structures underlying it.

On the basis of geophysical, topographical, and geological information, Eaton (1986) concluded that the Southern Rocky Mountains extend southward through the entire length of New Mexico. Accordingly, the broad geophysical anomalies and surface uplift described above are the signature of the Southern Rocky Mountains, and the RGR is a system of axial grabens along this major north-trending structure. In order to include the

evidence—were found to have a level of seismicity equivalent to or greater than that of the RGR.

Starting about 10 years ago, geophysical data began to accumulate that indicated that the RGR boundaries could be extended beyond those proposed by Chapin (1971) in Figure 1. Cordell (1978) summarized geophysical evidence for a broad asthenospheric-lithospheric structure centered beneath the chain of grabens shown in Figure 1. Cordell found that the main feature of east-west gravity profiles through New Mexico is a broad negative Bouguer anomaly centered roughly on the axis of the rift. According to Cordell (1978), the anomaly is a consequence of subcrustal structure related to a Rio Grande rift system, which includes a broad uplifted region as well as the obvious grabens and basins usually considered to be the RGR. The width of the anomaly appears to be at least 500 km along the New Mexico–Colorado border and broadens progressively to the south. Thus, the gravimetric signature of the rift system, and possibly the structure that produced it, extends beneath the presumably stable Colorado Plateau and Great Plains Provinces, well beyond Chapin's boundaries of the rift shown in Figure 1.

Other geophysical observations support the existence of a broad asthenospheric-lithospheric structure beneath the RGR. Davis and others (1984) found a traveltime residual anomaly for teleseismic P-wave arrivals along a profile crossing the RGR. The

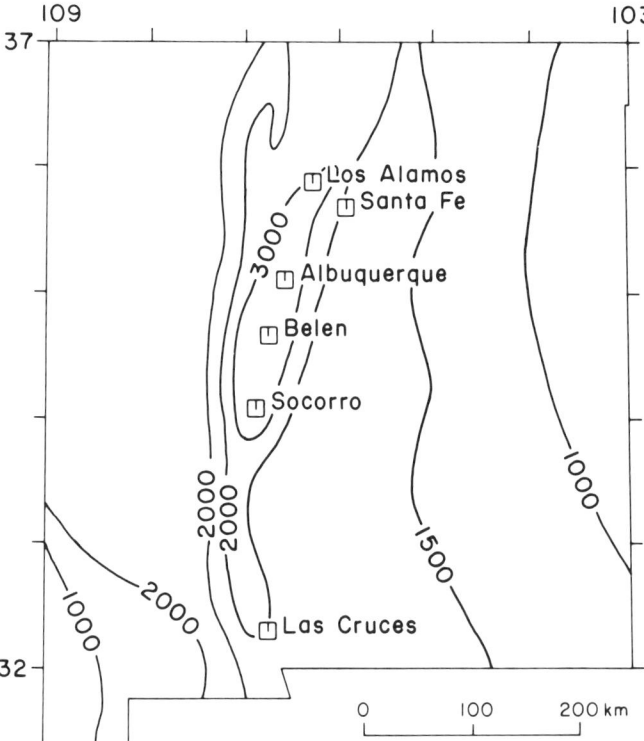

Figure 3. Surface uplift in meters during the past 10 m.y. Contours are approximate and based on the work of Gable and Hatton (1983). Maximum uplift has occurred in regions now occupied by some of the major structural depressions of the rift.

possibility that the RGR in New Mexico is simply a by-product of a more major structure, we examine the seismicity throughout the state without any preconceived idea where boundaries for the rift may lie.

Seismicity prior to 1962

Information on locations and strengths of earthquakes prior to 1962 is based mostly on noninstrumentally determined values of earthquake intensity. A major weakness in determining the strengths and locations of earthquakes from intensity observations is that the method depends on population density. In sparsely settled areas, shocks may go completely unreported or reported at low intensity values that do not indicate the true strengths of the earthquakes. Much of New Mexico is sparsely settled now, and it was sparser in the historic past. Even in areas where the population density was relatively high, such as some sections of the Rio Grande Valley, the point of maximum intensity or area of perceptibility may not have been defined adequately to determine the true strength of the event.

The reliability of the early reports of earthquakes must also be considered. Some of the intensities for strong earthquakes in New Mexico prior to 1900 were based on reports from local residents tens of years after the shocks (Bagg, 1904). Newspaper accounts of earthquakes have also been used to estimate earthquake intensity (Northrop, 1961, 1976). In most cases, this method has proved to be fairly reliable. However, in at least two instances the effects of earthquakes in the Rio Grande Valley at Socorro were exaggerated in reports appearing in newspapers in Albuquerque and El Paso (Sanford, 1963; Ashcroft, 1974).

Although settlement by the Spanish began in the early seventeenth century, little is known of seismic activity in New Mexico prior to its becoming a territory of the United States in 1848. No doubt, reports of earthquakes exist in Spanish and Mexican archives; such information, however, is difficult to extract, and to our knowledge no such attempt has been made. The earliest report of earthquakes after U.S. occupation is the description of a swarm of shocks in the Rio Grande rift at Socorro by a U.S. Army surgeon (Hammond, 1966). The swarm, which contained 22 felt shocks, commenced on 11 December 1849, and lasted until 8 February 1850. No shock in this swarm was reported felt at distances greater than 25 km. Similar sequences of shocks located away from population centers along the Rio Grande Valley or elsewhere in the state could easily have gone unreported before the start of instrumental studies.

For the period 1849 to 1961, Northrop (1961, 1976) cites evidence, primarily from old newspaper files, for over 600 felt earthquakes in New Mexico. Approximately 95 percent of these shocks occurred along a 150-km section of the RGR from Albuquerque to Socorro; the majority in the 75 km from Belen to Socorro (Fig. 1). The concentration of reported activity in the Belen-Socorro area cannot be attributed to population density because the population from Belen to Albuquerque has always exceeded the population from Belen to Socorro.

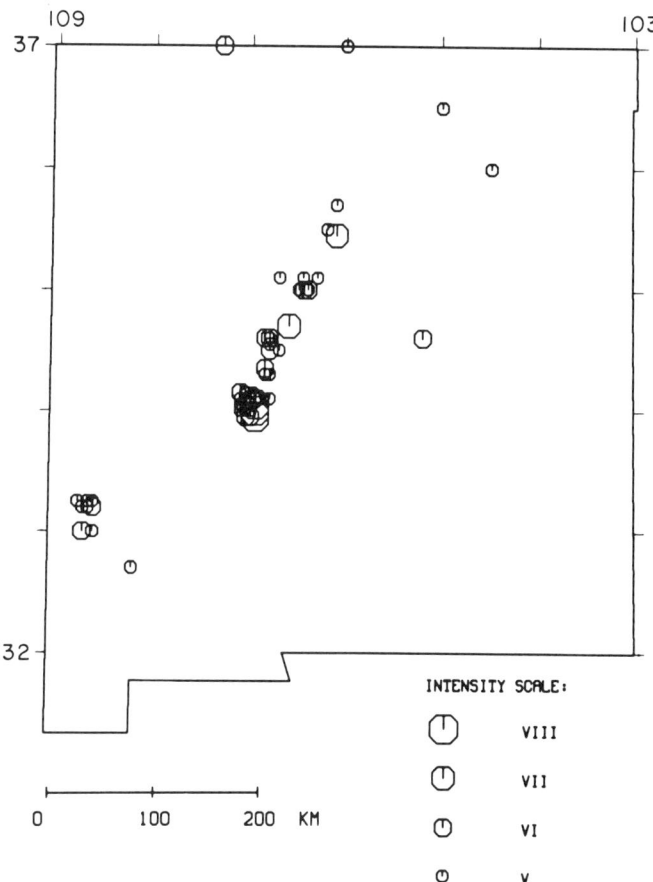

Figure 4. Earthquake activity in New Mexico prior to 1962. Plotted are the approximate epicenters for earthquakes with maximum reported intensities (Modified Mercalli) of V or greater. The information comes primarily from Coffman and van Hake (1973) and Sanford and others (1981). Locations of some events have been adjusted slightly to avoid overlapping of symbols.

However, when considering the entire state, Northrop's data are influenced by the distribution of population. Population density has been higher in the Albuquerque-to-Socorro section of the Rio Grande Valley than in most other sections of the state. To reduce bias arising from the distribution of population, only shocks with maximum reported intensities (Modified Mercalli) of V or greater are shown in Figure 4. The primary sources of data are Coffman and von Hake (1973) and Sanford and others (1981). Locations of events in Figure 4 were arbitrarily adjusted to avoid overlapping of symbols. With few exceptions, shocks for this time period can be placed only at the nearest population center. The earliest earthquake on the map occurred 28 April 1868 near Socorro; the latest on 3 July 1961, also near Socorro.

Of the 58 earthquake epicenters plotted in Figure 4, 39 are well within the RGR boundaries proposed by Chapin (Fig. 1), and 36 of these lie within the narrow confines of Bryan's rift (Fig. 1). This historical activity correlates well with the distribution of population in New Mexico, which was and is heavily concentrated along the Rio Grande. Therefore, bias toward population centers probably exists in this pre-1962 earthquake data, particularly for events with maximum intensities of V, which constitute about 60 percent of the epicenters in Figure 4. However, the concentration of earthquakes with maximum intensities of VII or VIII along the Rio Grande Valley from 1868 through 1961 is real because events of this strength would have been observed in all but the most isolated areas of the state. Included in this group are three earthquakes near Socorro (34.0°N and 107.0°W) that are the strongest earthquakes in New Mexico since 1868. The three events, felt over areas on the order of 200,000 to 250,000 km^2, were part of a swarm that commenced in early July 1906 and continued into January 1907 (Reid, 1911; Sanford, 1963). Although the evidence is not conclusive, the isoseismals for the three strong shocks in the swarm suggest hypocenters beneath the Socorro Mountain horst block, a relatively young north-south structural feature in the central part of the main rift graben (Chapin and Seager, 1975). The December 1935 swarm that was centered near Belen (34.6°N, 106.80°W) in the Albuquerque basin also seems to have originated near the axis of the central rift graben at that location. At Los Lunas, 18 km north of Belen, the shocks of the 1935 swarm were felt much more weakly than at Belen—an unlikely observation if the epicenters were on the rift margins, which are located approximately 30 km to the east and west of the two communities. Recent instrumental studies (Jaksha and Sanford, 1986) in the same region show considerable seismic activity within the main rift graben but little associated with its well-defined boundary faults (see later section).

Late Pliocene and Quaternary basalt flows, from north of Albuquerque to south of Socorro, are generally confined to the central part of the main rift graben (Kelley and Kudo, 1978; Bachman and Mehnert, 1978). This observation, together with the location of earthquake swarms before and after 1962, may indicate that magma is continuing to be injected along or near the axis of the main rift grabens, particularly in the vicinity of Socorro.

Besides the Socorro and Belen swarms, the best-documented earthquakes for the pre-1962 period were (1) a single shock near the axis of the RGR at 35.45°N and 106.1°W, and (2) a swarm centered in the southwestern corner of the state at 33.20°N and 108.70°W. The RGR event, known as the Cerrillos earthquake, occurred on 28 May 1918 and had a maximum Modified Mercalli intensity of VII to VIII (Olsen, 1979). The swarm in southwestern New Mexico occurred from September 1938 through early June 1939 and contained six shocks of maximum intensity V or VI (Taggart and Baldwin, 1982).

Instrumental data: 1962 to 1986

Instrumental epicenters for 89 New Mexico earthquakes with magnitudes equal to or greater than 2.5 are plotted in Figure 5. The map is believed to be an accurate representation of seismicity in New Mexico for the time period it covers, 1 January

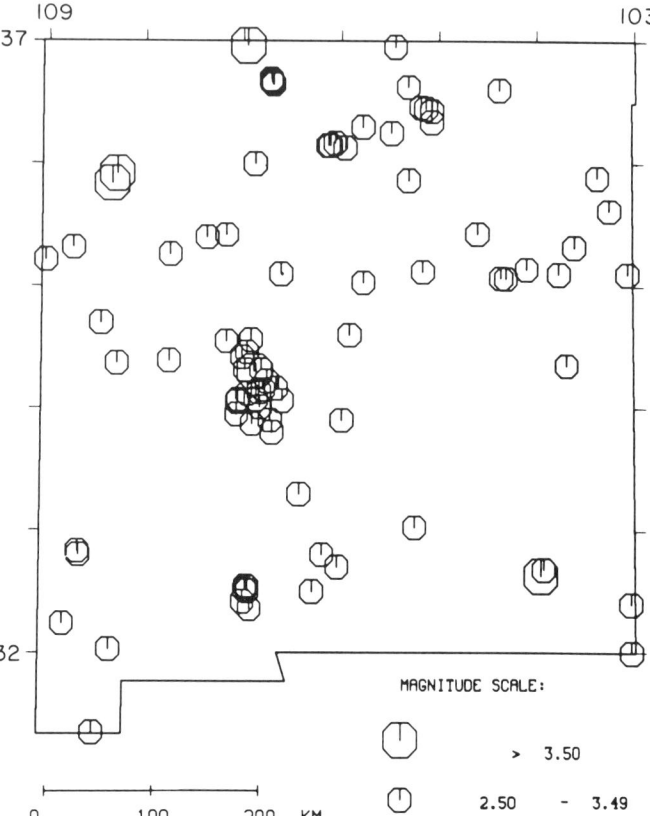

Figure 5. Earthquakes located in New Mexico from 1 January 1962 through 30 September 1986 with magnitudes equal to or greater than 2.5. The map is believed to accurately portray seismicity throughout the state for the study period.

1962 through 30 September 1986. A second map of instrumental epicenters (Fig. 6) with a minimum magnitude of 1.5 is incomplete both spatially and temporally. On this map, epicenters for small earthquakes are most numerous toward the central and north-central areas where most of the seismograph stations were located from 1962 through 1986. In addition, there is a progressive increase in weak shocks after 1962 because the number of seismograph stations operating in New Mexico increased with time. For most of the discussion that follows, we have used the second map because it gives the best indication of significant spatial variations in seismicity in the region of most interest, the western two-thirds of New Mexico.

Strongest earthquakes

Only six earthquakes on Figures 5 and 6 had magnitudes of 3.5 or more: three on the Colorado Plateau, two within Chapin's rift boundaries shown in Figure 1, and one on the Great Plains. The strongest earthquake during the 1 January 1962 through 30 September 1986 period was located on the Colorado Plateau at 37.0°N and 107.0°W. The quake, which occurred on 23 January

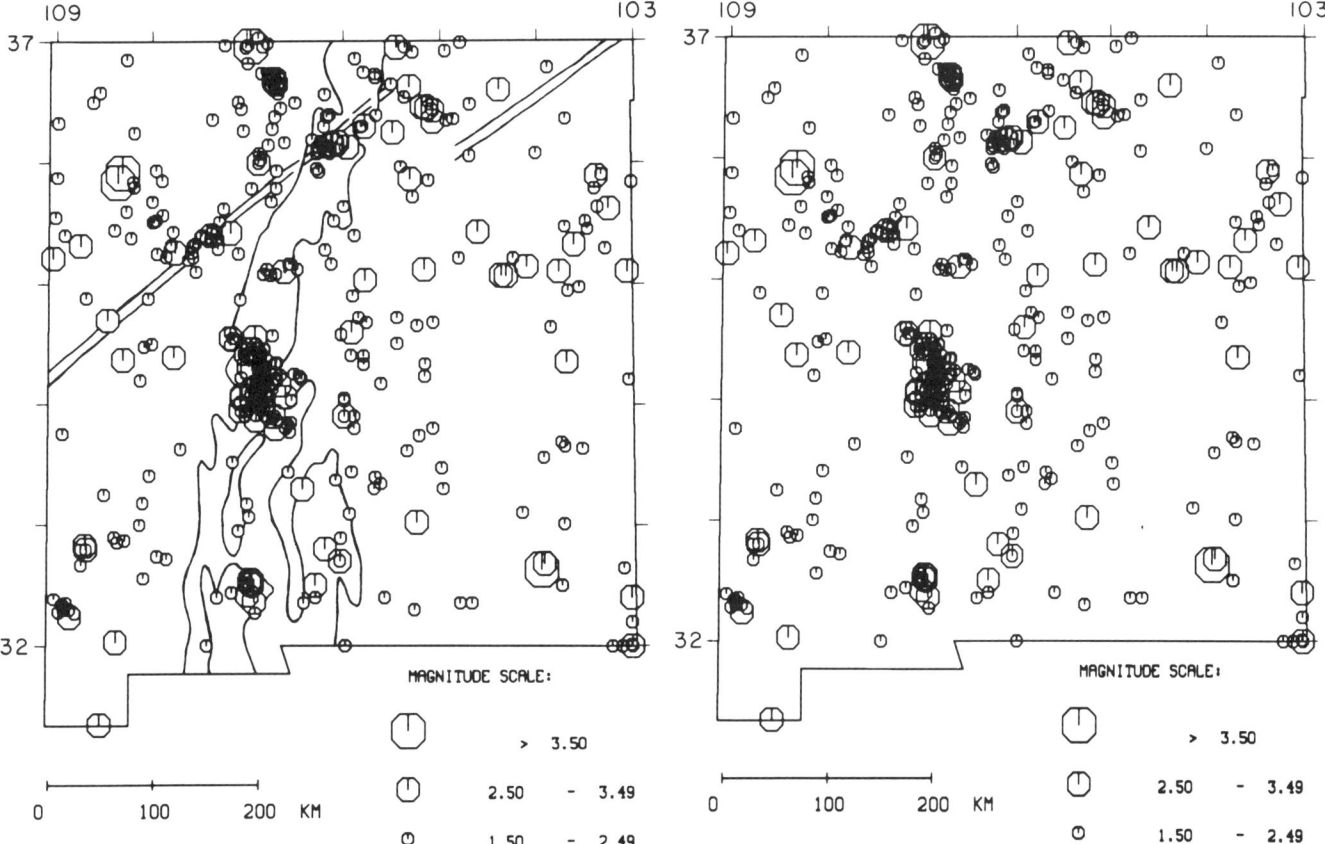

Figure 6. Earthquakes located in New Mexico from 1 January 1962 through 30 September 1986 with magnitudes equal to or greater than 1.5. Epicenters for earthquakes with magnitudes less than 2.5 are most numerous towards the central and north-central areas of the state where most of the seismograph stations were located from 1962 through 1986.

Figure 7. Location of the main Rio Grande rift grabens, the axis of the Jemez lineament, and New Mexico earthquake activity from 1 January 1962 through 30 September 1986. Position of the main rift grabens is from Chapin (1971) and the axis of the Jemez lineament is estimated from the position of young volcanic centers given by Luedke and Smith (1978).

1966, had M_s and m_b magnitudes of 4.6 to 4.9 and 5.1 to 5.5, respectively (Cash, 1971; Herrmann and others, 1980).

Seismicity, faulting, and volcanism

Main rift grabens and young surface fault movements. Figure 7 shows the distribution of earthquake activity relative to the location of the main grabens of the RGR. These fault-bounded depressions have vertical displacements many times greater than any structure within several hundred kilometers east or west of the rift axis. For the time period of the instrumental observations, earthquake activity along the main grabens of the RGR was not uniformly distributed and was not unusually high relative to the rest of the state.

The positions of faults that are known to have produced surface offsets within the past five million years (Machette, 1982; Callender and others, 1983; Hawley and Love, 1981; Seager, 1980) are plotted on the New Mexico seismicity map in Figure 8. The map is believed to show a reasonably complete inventory of all young faults in New Mexico (John Hawley, personal communication, 1987). Most of these faults are located within or border the principal grabens of the RGR, some close to plotted earthquake clusters, but many where little or no earthquake activity has been located. Striking examples of the latter are the regions of high fault density centered at (1) 35.5°N and 106.5°W, and (2) 33.2°N and 107.5°W. Low levels of seismicity along faults having geologic evidence of recent movement is a frequent observation in the western United States (Schwartz and Coppersmith, 1984).

Recent volcanism. The Jemez lineament is shown in Figure 9 cutting diagonally across the northern half of New Mexico. Most of the centers of volcanism in the state that are younger than 5 Ma cluster along this lineament. Other major centers are the Taos volcanic field (36.7°N, 105.9°W) in the northern rift valley and the Potrillo field (32.0°N, 107.2°W) in the southern rift valley. Some young volcanic areas appear to be seismically active: for example, (1) the Jemez lineament from 107.2° to 107.8°W; (2) at the Albuquerque volcanoes, 35.1°N, 106.8°W

(Jaksha and others, 1981); and (3) near the vent of a very recent eruption at 33.9°N and 106.0°W (Luedke and Smith, 1978). On the other hand, some sites of recent volcanism have experienced little earthquake activity in the past 25 years; the Jemez lineament in northeastern New Mexico is one example.

Discussion

Perhaps the most interesting and puzzling characteristic of the seismicity maps of New Mexico (Figs. 5 and 6) is the lack of well-defined seismic trends that correlate with young tectonic/volcanic features or boundaries between physiographic provinces. Given these maps alone, the RGR cannot be clearly identified using the boundaries of either Bryan or Chapin (Fig. 1). The other major young structure, the Jemez lineament and its young volcanics, is not evident on these maps either. In addition, the physiographic provinces shown in Figure 1 are not clearly defined by the seismicity; for example, the presumably stable Colorado Plateau and Great Plains provinces have had a level of seismicity since 1962 that is almost comparable to the RGR. On the other hand, the seismicity does appear to correlate fairly well with the broad gravity, P-wave traveltime, and magnetic anomalies described earlier in this chapter.

We do not know whether the diffuse pattern of seismicity seen in Figures 5 and 6 is representative of the long-term seismicity of the region. The absence of a clear correlation between seismicity and the main RGR grabens might suggest an inadequate record of earthquake activity. However, even if we assume that our sample under-represents activity along the main rift grabens, the problem of explaining the modest level of earthquake activity well away from these structures in all areas of the state remains. The possibility exists that the broad pattern of modern seismicity in New Mexico is related to the broad surface uplift that has occurred during the past 10 m.y. (Fig. 3).

SEISMIC STUDIES OF SPECIFIC AREAS

North-central New Mexico, 1973 to 1984

Introduction. The Los Alamos Seismograph Network (LASN) began operation in late 1973. For the period September 1973 through December 1984, approximately 2,000 earthquakes were located in northern New Mexico; the results have been reported in a series of seismicity catalogs published by Los Alamos National Laboratory (LANL). More than 20 of these events were felt, and the largest, near Crownpoint, New Mexico (35.9°N

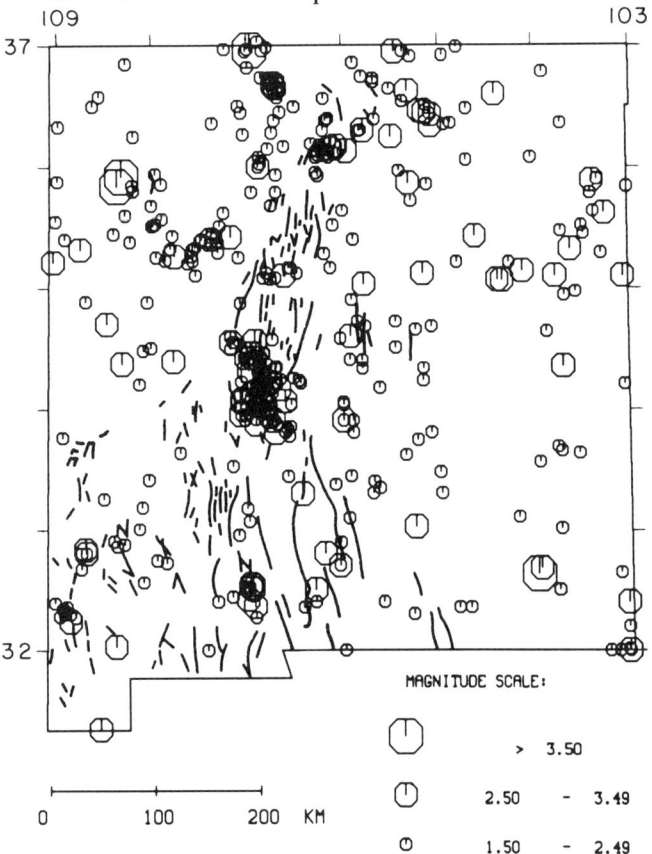

Figure 8. Location of faults with surface offsets within the past 5 m.y. and New Mexico earthquake activity from 1 January 1962 through 30 September 1986. Position of faults is from Machette (1982), Callender and others (1983), Hawley and Love (1981), and Seager (1980).

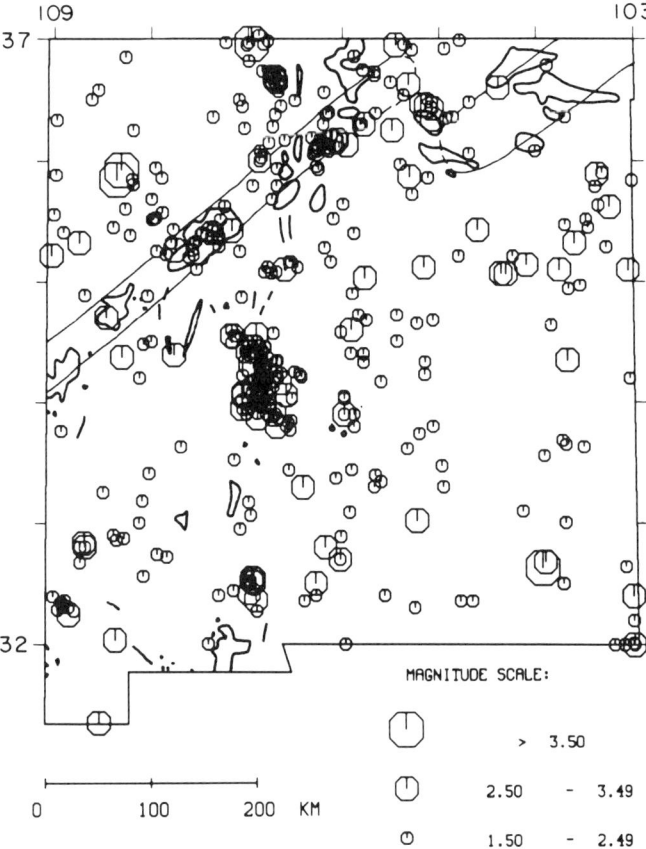

Figure 9. Location of centers of volcanism less than 5 Ma and New Mexico earthquake activity from 1 January 1962 through 30 September 1986. Position of volcanic centers is from Luedke and Smith (1978). The centers that fall within the outlined zones oriented northeast-southwest follow the trend of the Jemez lineament.

and 108.2°W) in 1976, had a magnitude M_L = 4.6 (m_b = 5.2; Wong and others, 1984).

Figure 10 is a map of stations of the LASN. (Not all stations were in operation all of the time, and only two still exist.) These stations, along with those of New Mexico Institute of Mining and Technology that operated in the Socorro area and those of the U.S. Geological Survey that operated throughout northern New Mexico, were used in calculating epicenters. Although the catalogs include all epicenters located in northern New Mexico and adjacent portions of Arizona and Colorado, most of the epicenters found in the easternmost and westernmost parts of northern New Mexico are very uncertain. For that reason the epicenter map, Figure 11, is restricted to an area bounded by 34.75° to 37.25° N and 104.75° to 107.75° W; this area was chosen because it is approximately centered on the LASN and the RGR, and has epicenters for which we are fairly confident.

Seismicity-structure relationships. Earthquake locations in Figure 11 are not defined well enough to associate individual events with specific faults. However, zones of seismicity appear to correlate with zones of faulting or other important structural features. Most notable of these are the Nacimiento uplift, the Gallina-Archuleta arch, the Jemez volcanic lineament, and a number of other features of the northern RGR. The low level of seismicity in the Jemez Mountains calderas and the San Felipe fault zone is also noteworthy.

Nacimiento uplift. The uplift was first formed during the east-west compressional tectonics of Laramide time. Reverse or high-angle thrust faulting, up to the west, occurred on a fault dipping to the east under the uplift. Because the current tectonics of the region involves east-west extension, the seismicity observed beneath the uplift may be occurring on the plane of weakness formed by the older Nacimiento fault reactivated as a normal fault.

Gallina-Archuleta arch. A Laramide east-west compressional structure continues northward from the Nacimiento uplift as the Gallina-Archuleta arch. A prominent belt of seismicity continues from the Nacimiento uplift along the arch and includes a concentration of about 150 earthquakes near 36.7° N, 106.7° W, about two-thirds of which occurred in 1982. This cluster is centered in the region of two reservoirs, El Vado and Heron, and may therefore be a case of reservoir-induced seismicity. In contrast, the Abiquiu Reservoir, lying just north of the Jemez Moun-

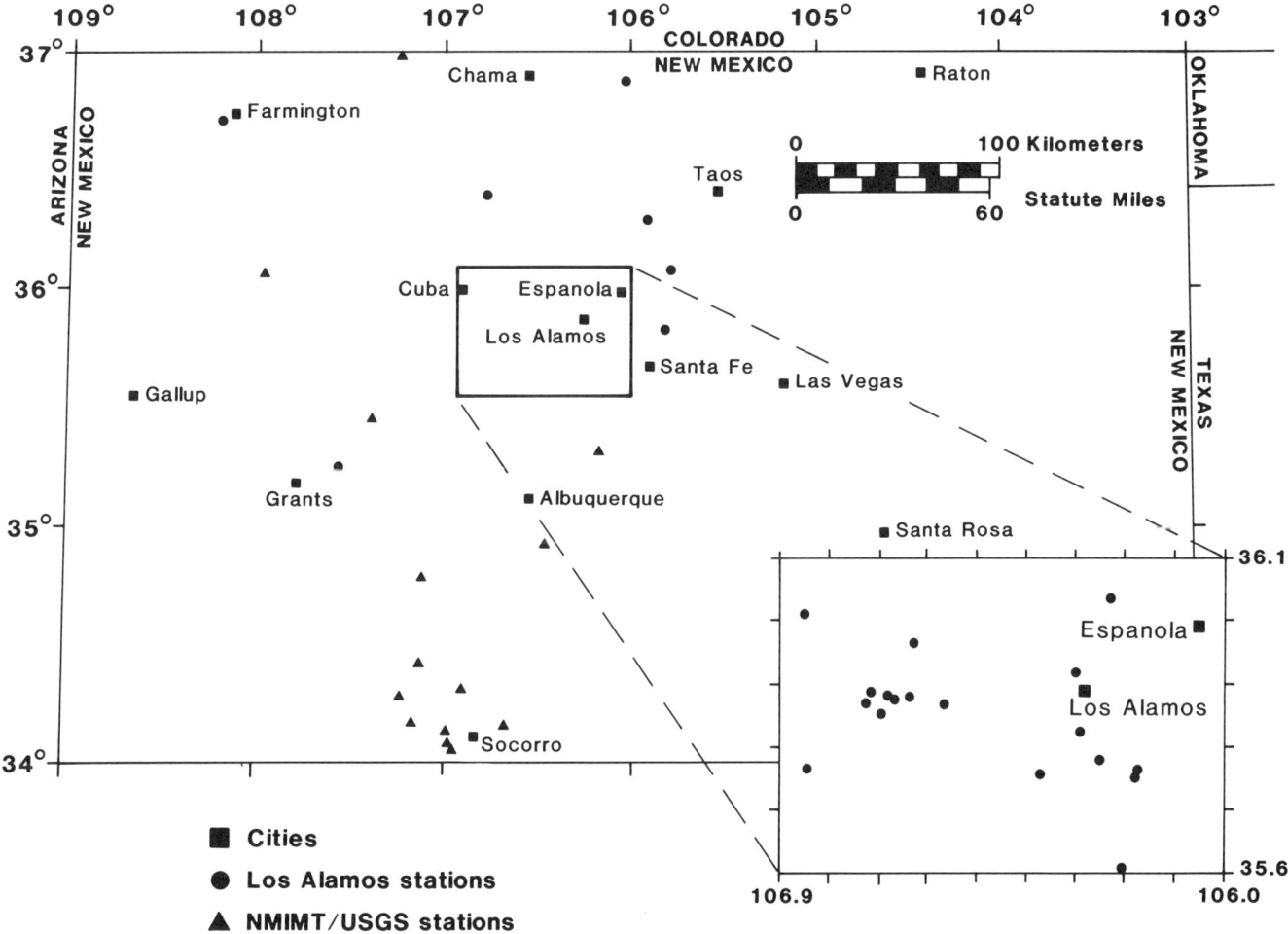

Figure 10. Location of seismograph stations in the Los Alamos Seismograph Network from September 1973 through December 1984. Not all of the stations shown were in operation for the entire period.

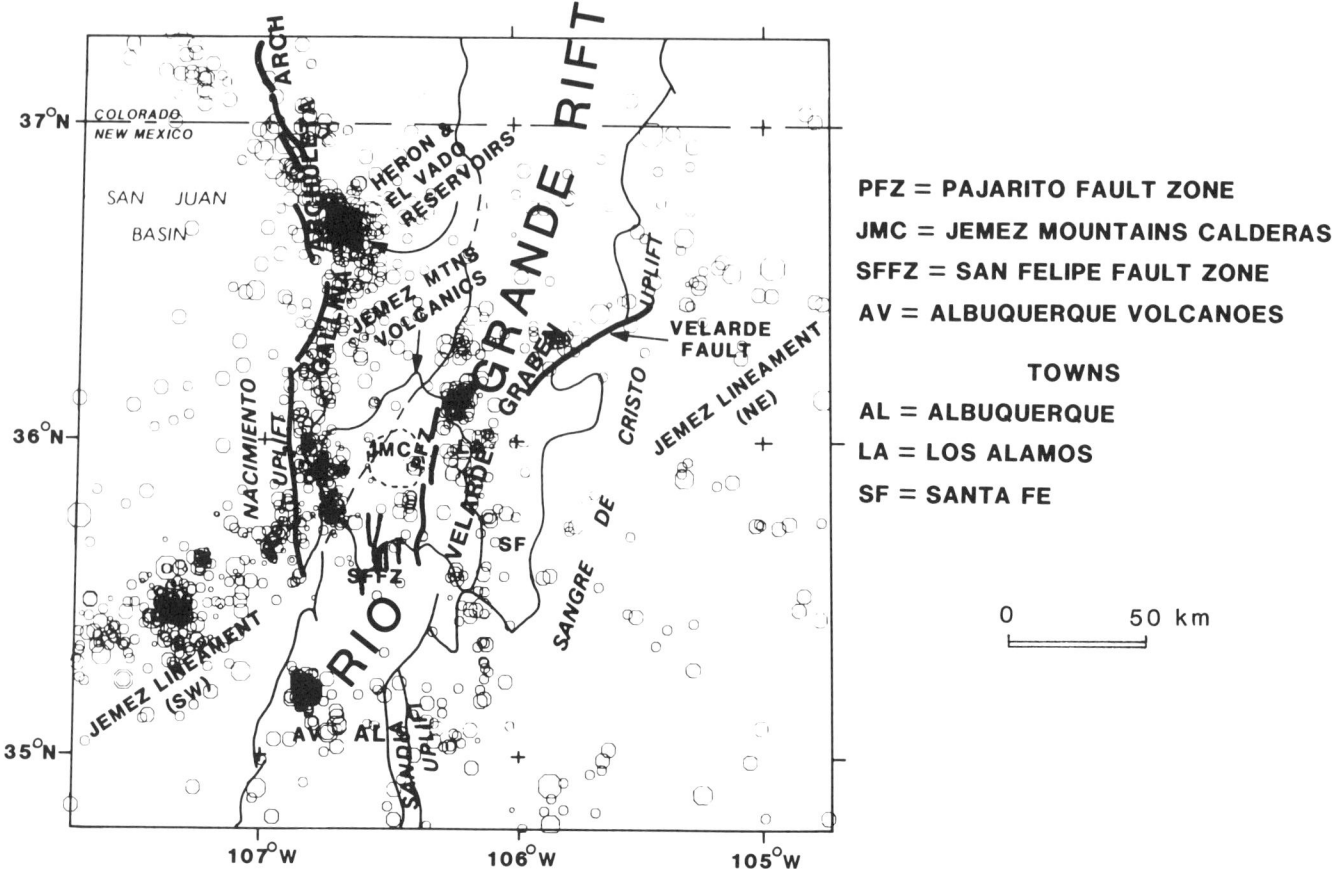

Figure 11. Earthquakes in north-central New Mexico for the period September 1973 through December 1984. Octagon symbols in five different sizes are used to specify the magnitudes. For the smallest to largest symbols, the range in magnitudes is (1) <0.0, (2) 0.0 to 0.9, (3) 1.0 to 1.9, (4) 2.0 to 2.9 and (5) >3.0.

tains and not more than 55 km S30°E from the El Vado and Heron Reservoirs, experiences very little seismicity.

The 1966 Dulce earthquake occurred near where the Gallina-Archuleta arch crosses the New Mexico–Colorado border. This earthquake had an epicenter at 36.98°N and 107.02°W, a magnitude $m_b = 5.5$, a maximum intensity of VII on the Modified Mercalli scale, and caused moderate damage (Cash, 1971; Herrmann and others, 1980).

Jemez volcanic lineament. Near the south end of the Nacimiento uplift, a southwest trend of epicenters coincident with a section of the Jemez volcanic lineament appears in the seismicity data. Mayo (1958) first described this southwest-northeast alignment of extrusive volcanics, which contains basalts as young as 1,000 years (Nichols, 1946). On the east side of the RGR, the lineament continues generally northeastward, but more broadly, and offset to the south. There is no clear relation between seismicity and the eastern segment of the Jemez lineament.

Central RGR. Several features of the seismicity of the central RGR deserve mention. The Pajarito fault, which has significant vertical offset (down to the east), occurs on the east side of the Jemez Mountains just west of Los Alamos. The fault offsets the 1.1 to 1.4 Ma Bandelier Tuff 100 to 150 m at Los Alamos (Bailey and Smith, 1978). Despite its relatively recent age and significant throw, the Pajarito fault was not very active near and south of Los Alamos in the 11-year period of coverage shown in Figure 11. North of Los Alamos the Pajarito fault zone was much more active. This activity is part of a band of seismicity that runs north-south along the center of the Velarde graben in the Espanola Basin and on southward to about 50 km east of Albuquerque.

Northeast of the Jemez Mountains, a band of seismicity strikes northeastward, possibly an extension of the seismic zone along the Jemez lineament to the southwest. Here it follows the trace of the Velarde fault that extends from the Velarde graben into the Sangre de Cristo uplift. There is a concentration of events where the projection of the Velarde fault intersects the Pajarito fault zone. This area of seismicity, which lies about 25 km north of Los Alamos, may have been the site of two of four earthquakes felt generally throughout the Los Alamos region since the town was established in the early 1940s. One occurred in 1952, two in 1971, and one in 1973. All took place before the establishment of LASN, and therefore, the epicenters are not precisely known.

The seismicity along the Velarde fault and that along the southwestern segment of the Jemez lineament are separated by the nearly aseismic region of the Jemez Mountains calderas (also known individually as the Valles and Toledo calderas). Very few earthquakes have occurred within the calderas, despite the presence of numerous faults associated with ring fractures and resurgent doming that followed the formation of the calderas about 1 Ma. The relative aseismicity may be attributed to high crustal temperatures or to the relief of stress on the surrounding active fault zones. The high heat flow of the caldera region is well known (Edwards and others, 1978); the concomitant high crustal temperatures may cause the crust to be ductile enough that it yields aseismically. On the other hand, the relative absence of seismicity may be attributed to stress being relieved on active fault zones immediately outside the calderas, most notably the north-south zones of activity along the Nacimiento uplift on the west and in the Velarde graben on the east.

Immediately to the south of the Jemez Mountains lies the San Felipe fault zone. This area exhibits characteristics similar to the Jemez calderas in that it contains a high density of normal faults, no older than Pleistocene, yet it is relatively aseismic. These faults are particularly visible in the Quaternary-Tertiary basalts of Santa Ana Mesa near the center of the aseismic area (Smith and others, 1970). No temperature or heat-flow measurements have been made in this area, but the young age of the volcanics could suggest high crustal temperatures here also.

Albuquerque area

Introduction. Earlier in this chapter we noted, on a large scale, some correlation between contemporary seismicity and broad, youthful uplift of the entire western two-thirds of New Mexico. The central depressions of the Rio Grande rift, as described by Bryan (1938; Fig. 1), are almost invisible on our seismicity map (Figs. 5 and 6). This is interesting because the Rio Grande depression is cut by numerous young fault scarps (Fig. 8), and most of the faults with suspected Holocene ground rupture in New Mexico occur within its margins (Seager, 1981; Machette, 1982; Machette and Colman, 1983). To study seismicity on a finer scale within the Rio Grande grabens and its bounding uplifts we now briefly reexamine a data set collected by the U.S. Geological Survey in 1976 to 1981 and reported by Jaksha and Sanford (1986).

Data. In the fall of 1975 the U.S. Geological Survey installed an eight-station seismic network near Albuquerque. By August 1977, through cooperative agreements with Los Alamos National Laboratory and the New Mexico Institute of Mining and Technology (NMIMT), the network had expanded to 13 stations. The station locations are shown in Figure 12, and the technical parameters of the instrumentation are given by Jaksha and others (1978).

Hypocenters were obtained from the computer program HYPO 71 (Lee and Lahr, 1975), using the plane-layered crustal model derived for the Rio Grande rift by Toppozada and Sanford (1976) and a Poisson's ratio of 0.25. An analysis of local mining explosions suggested that hypocenters within the seismic net were located to within about 2 km horizontally of their true positions. Where focal depth was well determined it was found that all such seismicity occurred within the upper crust and averaged about 9 km deep. Magnitudes were assigned using a duration equation deduced by Newton and others (1976). Magnitudes in the data set ranged from negative values up to 3.2.

Seismicity. All of the earthquake epicenters located during the 66-month study (~1,000 events) are plotted in Figure 12. Those events that occurred within or near the Albuquerque and Estancia basins (from 34.2°N to 35.7° N and 105.5° W to 107.5° W) are also plotted on the tectonic map of Woodward and others (1975) in Figure 13. In addition, Figure 13 shows composite fault plane solutions constructed during the study. The numbers 1 to 6 on Figure 12 represent features of particular interest in the data that are briefly discussed below.

Figure 12-1 is the relocated epicenter for both of the last two widely felt earthquakes in Albuquerque. Those two shocks, on 28 November 1970, and 4 January 1971, had magnitudes near 4.0; the 1971 event caused about $40,000 damage (Northrup, 1982).

Jaksha and others (1981) studied a swarm of small earthquakes that occurred in the Albuquerque volcano area (Fig. 12-2) in late 1978 and early 1979. The seismicity was located very close to the fissure through which the volcanic rocks erupted, but a one-to-one relationship could not be established. Focal depths for these shocks averaged about 9.5 km, and a composite fault-plane solution (Fig. 13) was interpreted to suggest down-to-the-west normal faulting on a plane striking N5°E.

Figure 12-3 denotes the seismicity associated with the Socorro magma body. This large and complex area exhibited the largest number of earthquakes and contained one of the two largest events in the data. A summary of results from ongoing seismic studies at Socorro is given later in this chapter.

A seismic zone extending northeastward from Grants, New Mexico, and associated geographically with the Jemez lineament is labeled 4 in Figure 12. This area exhibited the second highest number of located earthquakes during the study, the largest having magnitude 2.7. Like the Socorro area, earthquake swarms were commonly observed along this feature. A composite fault-plane solution for this section of the Jemez lineament (Fig. 13) suggests a preponderance of strike-slip faulting.

The most intense earthquake swarm observed during this study did not occur in the Rio Grande rift or along the Jemez lineament but south of Gran Quivera National Monument (Fig. 12-5). The activity spanned 11 months and contained six events with magnitudes greater than 1.5. This source zone was well outside the recording network, so the events are poorly located and an accurate count of the number involved was not obtained.

Seismicity associated with the Estancia basin east of Albuquerque (Fig. 12-6; Fig. 13) is diffuse. A number of shocks near station EST define a crudely north-striking zone subparallel to the Pedernal Hills uplift. The Tijeras fault zone, a long-lived area of crustal adjustment, exhibited a low level of seismicity during the study.

Among the conclusions that Jaksha and Sanford (1986) arrived at from studying this data were:

1. Central Rio Grande rift seismicity occurs mainly between Belen and Socorro and within metropolitan Albuquerque. Activity is largely intrabasin, and both normal and strike-slip faulting are exhibited. The average T-axis for the composite fault plane solutions in Figure 13 trends west-northwest to east-southeast.

2. During the study period, most of the young faults of the central rift valleys were seemingly inactive, and the central rift as a whole could not be delineated on the basis of seismicity. Areas of relatively high seismicity appear to correlate better with some form of magmatism than with tectonism.

Socorro area

Background. The Socorro area has long been recognized as a region of unusual seismic activity; the mode of occurrence is primarily in swarms; the intensity of earthquake activity, both in numbers and strengths, is the highest of any area along the RGR (Bagg, 1904; Reid, 1911; Northrop, 1945, 1947). Considerable effort has been devoted to documenting and understanding the seismicity of this region. Studies based primarily on reports of felt earthquakes, the only data available prior to 1960, have been published by Sanford (1963) and Northrop (1976). Papers emphasizing the results of instrumental studies are numerous, but those that best summarize the seismicity of the Socorro area and

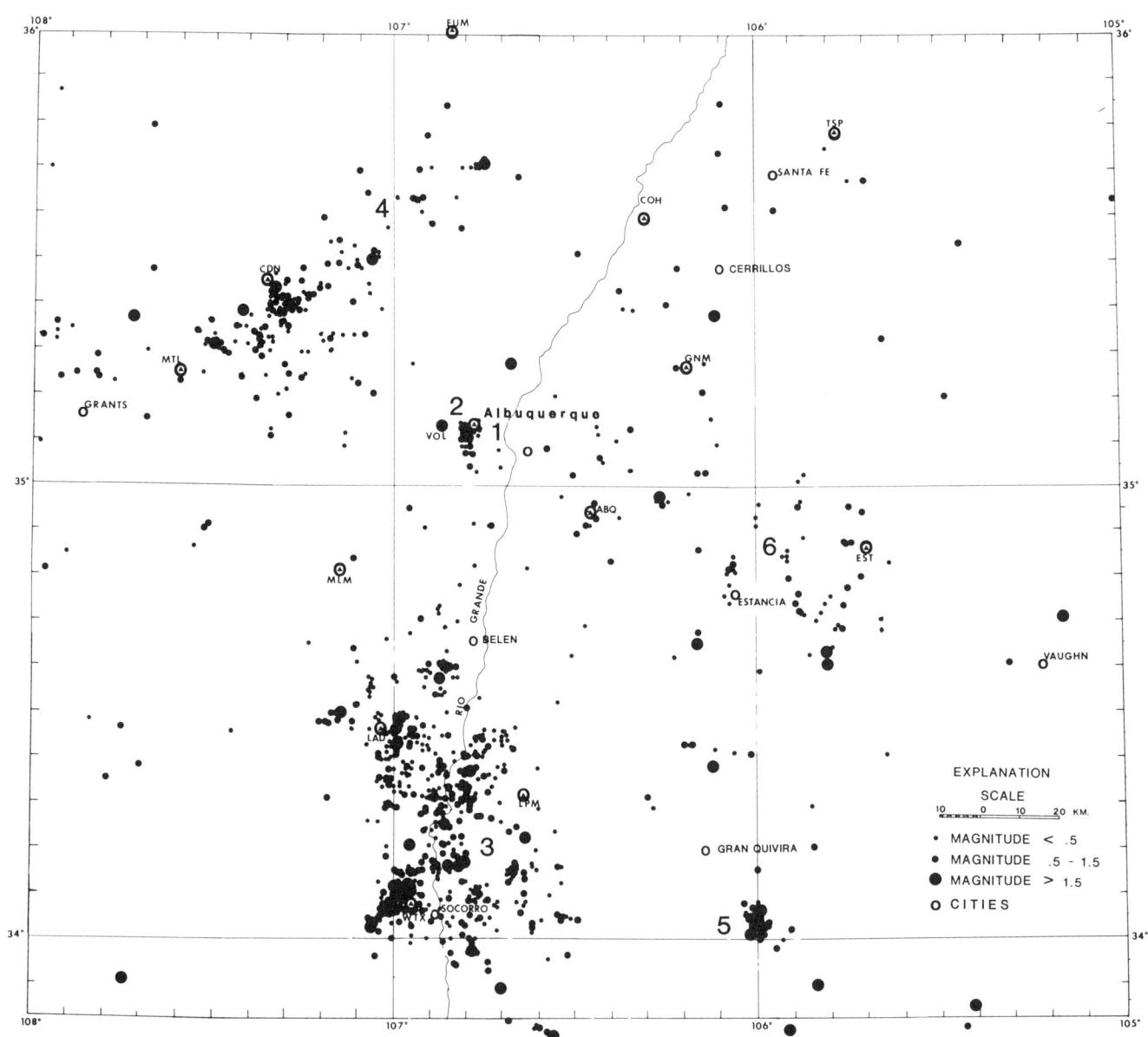

Figure 12. Seismicity map of the Albuquerque study area for a 66-month period between January 1976 and November 1981. Seismograph stations locations are designated by three letters and a symbol that consists of a circle surrounding a small triangle (from Jaksha and Sanford, 1986).

its relation to activity elsewhere along the rift are Sanford and others (1972) and Sanford and others (1979). The purpose of this section is to review results of recent studies and to discuss possible mechanisms for the unusual seismicity of the Socorro area.

Recent seismic activity. Figure 14 is a map of epicenters for 497 Socorro area earthquakes that occurred from 1 September 1982 through 31 December 1985. All events plotted fall within the magnitude range –0.5 to 4.0 and have HYPO71 (Lee and Lahr, 1975) quality *B* or better solutions. The crustal model used was a half-space with a velocity of 5.85 km/sec and a Poisson's ratio of 0.25. Station corrections were used to account for the highly variable Phanerozoic geology throughout the region. Magnitudes were calculated from durations using an empirical formula derived at LANL (Newton and others, 1976) and confirmed by studies at NMIMT (Ake and others, 1983).

The pattern of activity in Figure 14 is similar to that ob-

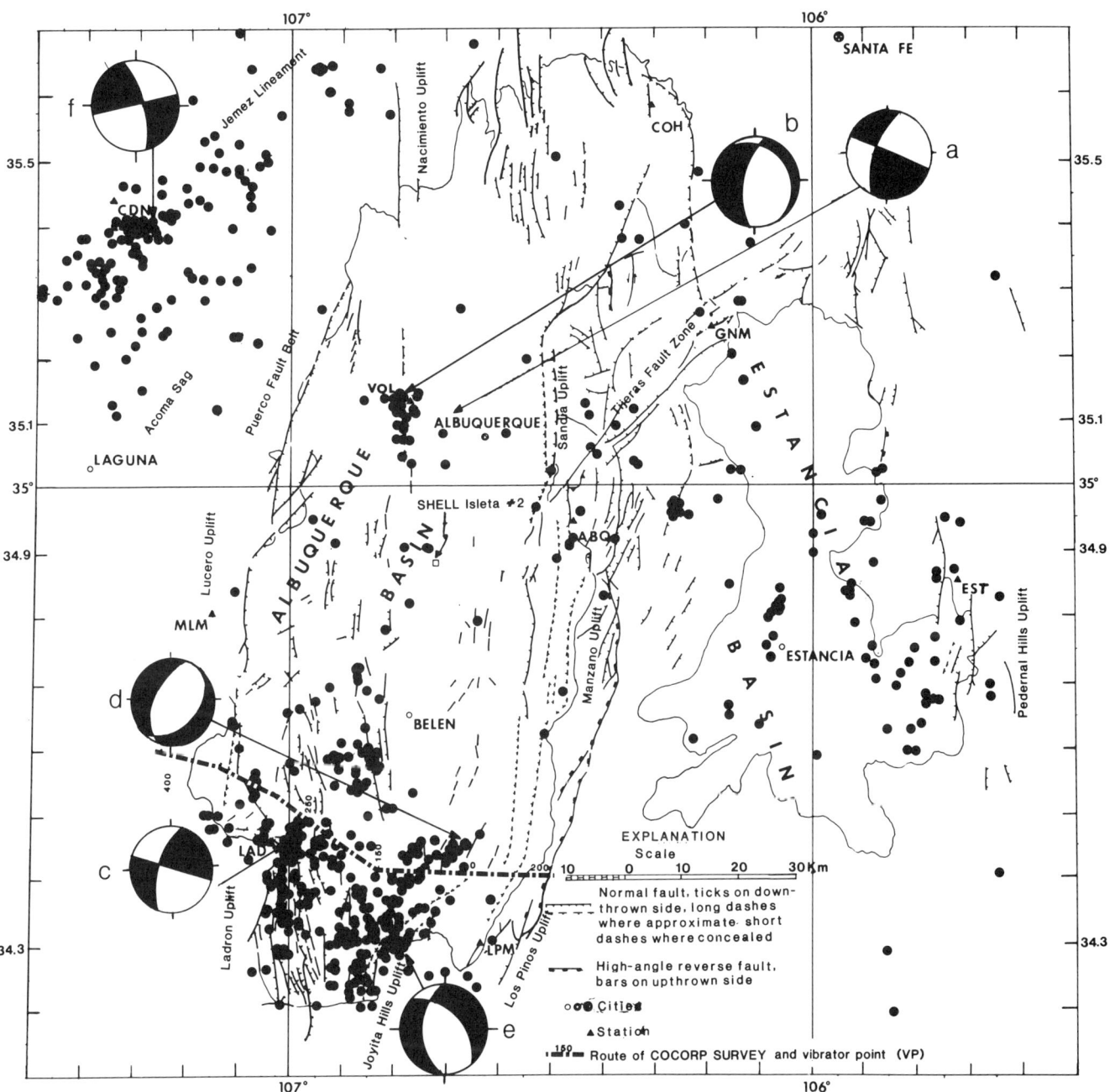

Figure 13. Tectonic map of the Albuquerque and Estancia Basins and surrounding features (modified from Woodward and others, 1975). The small solid circles are earthquake epicenters for the period January 1976 through November 1981, undifferentiated with respect to magnitude. The focal mechanism diagrams are lower hemisphere projections with compressional quadrants shaded. Data used in the construction of these fault-plane solutions is given in Jaksha and Sanford, 1986.

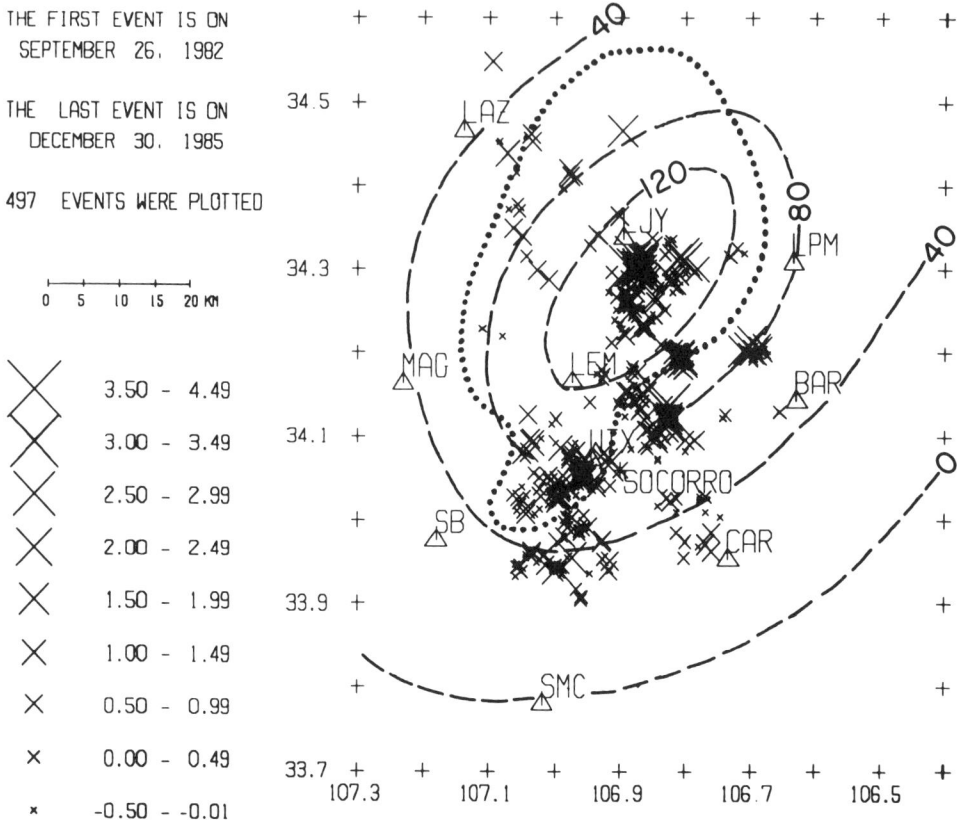

Figure 14. Earthquake activity in the Socorro area from September 1982 through December 1985. Superposed on the seismicity map are dashed contours of uplift in millimeters for the period 1911 through 1980 (Larsen and others, 1986) and the dotted outline of the midcrustal magma body (Rinehart and Sanford, 1981). Seismograph stations locations are designated by letters and a triangular symbol.

tained in all earlier studies (Sanford and others, 1972; Sanford and others, 1979; Sanford and others, 1983; Jaksha and Sanford, 1986), although the regions of most intense activity shift with time. The absence of epicenters around the perimeter of the network in Figure 14 is not the result of plotting good-quality solutions only. Maps of all epicenters, regardless of quality, show areas of very low seismicity south, east, and west of the region of relatively high seismicity in Figure 14. The same pattern for Socorro area seismicity is shown in Figure 12, a map based on a completely different data set than that used in Figure 14.

Depths of focus. A study of focal depths based on quality *A* HYPO 71 solutions from all studies in the Socorro area produced the histogram shown in Figure 15 (King, 1986). For the 513 earthquakes, 99 percent occur at depths from 4 to 12 km, 75 percent from 7 to 11 km. The lower boundary of the seismogenic zone is quite sharp, the upper boundary fairly diffuse. An analysis of the effect of random reading errors indicates that individual *A* solutions could have errors in depth on the order of ±1.6 km (2 standard deviations). Therefore, both lower and upper boundaries of the seismogenic zone could be sharper than indicated by the histogram.

Fault-plane solutions. From a study of first-motions of 534 well-located microearthquakes that occurred from May 1975

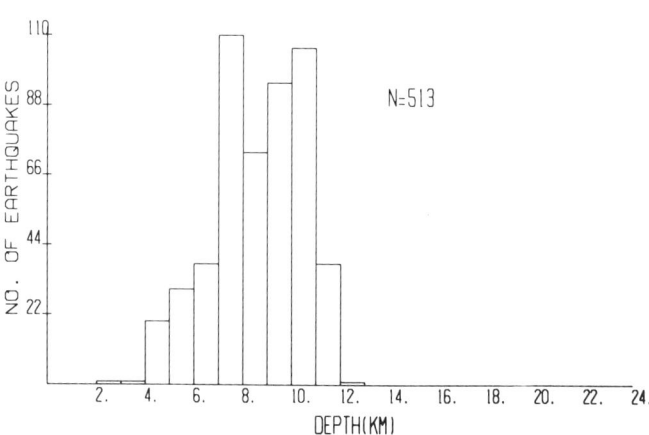

Figure 15. Distribution of focal depths in the Socorro area. All depths are for earthquakes with quality *A* HYP071 solutions. Data are from a study by King (1986).

to January 1978, Weider (1981) was able to obtain composite fault-plane solutions for seven regions in the Socorro area south of 34.35°N. Six of the seven solutions have nearly pure normal dip-slip movement, with their T-axes close to the same orientation, azimuth N75°E (±10°s.d.) and plunge 7.7° (±2.8°s.d.).

The dips of the nodal planes for these six solutions range from 36° to 60° and average 46°. The average focal depth for earthquakes used in the solutions is 9 km. If faulting at this depth is assumed to be listric on fault planes whose dips are the average of the observed nodal-plane dips (46°), then the average minimum depth at which the listric faults can become flat is 13 km.

The seventh composite fault-plane solution is for events near 34.3°N, 106.9°W. This solution has significant strike-slip motion as well as normal dip-slip motion; one nodal plane strikes N8°W with an unusually low dip of 30°, and the other strikes N68°W with an unusually high dip of 74°. The T axis for this solution has a N43°E azimuth and a 26° plunge. Later studies in the same area as this composite solution indicate additional tectonic complexities. Between 25 February 1983 and 16 March 1983, an earthquake swarm consisting of 300 recorded shocks occurred in a 3.2-km^2 region centered at 34.31°N and 106.87°W (Jarpe, 1984). The fault-plane solution for the strongest earthquake of the swarm, a magnitude 4.0 event on 2 March 1983, indicates a normal dip-slip mechanism with nodal plane strikes and dips of N7°W, 51°W and N7°E, 38°E. Although this solution agrees in type and strike with those obtained to the south, other shocks in this swarm do not. Some events have considerable strike-slip as well as normal movement on generally north-striking nodal planes; others have nearly pure normal dip-slip on east-west–striking nodal planes. In addition, more than half of the aftershocks have first-motion distributions and amplitudes that cannot be explained by the standard double-couple-without-moment mechanism (DCWM). In the most extreme cases, compressions are observed over the entire focal sphere except for one or two dilatations. Jarpe and others (1984) have proposed that the observed first-motions are the result of simultaneous shear along and opening of normal faults, the latter caused by injection of magma. This mechanism involves superposition of the DCWM and the linear vector dipole (LVD). Such a combination, with variable strengths for the DCWM and LVD, qualitatively satisfies the first-motion observations.

Young surface fault movements. On Figure 16, faults showing surface movement within the past 5 m.y. (Machette and McGimsey, 1982; Machette, unpublished data) have been superposed on a map of the Socorro area seismicity. All faults shown have normal dip-slip surface offsets, and their average strike is close to north-south. As discussed earlier, the contemporary earthquake activity is also occurring, for the most part, on normal faults with approximately north-south strikes. Because 99 percent of the earthquakes occur between 4 and 12 km deep, only shocks with epicenters located several kilometers downdip from a fault in Figure 16 have a chance of being associated with the subsurface projection of that fault. The only region where this occurs is the approximately 10-km-wide zone of epicenters extending from approximately 34.1°N to approximately 34.4°N at 106.9°W. Within this zone, most epicenters are located several kilometers downdip from normal faults dipping inward on both sides of the zone. Therefore, these earthquakes could have occurred on the subsurface projection of one or both of the bounding faults. The epicenters and focal depths are not known with sufficient precision to determine whether planar or curved (listric) projections best fit the data.

Away from the central zone, most earthquake activity cannot be associated with the young faults, and conversely, many young faults (some of the youngest) appear to be nearly or totally aseismic. A particularly good example is the La Jencia fault in Figure 16, which has had movement along some segments as recently as a few thousand years ago (Machette, 1986). The Borah Peak earthquake was generated on a large normal fault that showed no precursory activity on regional networks for a decade or two before the event (Dewey, 1987). The La Jencia fault, which is being monitored by a local network, demonstrates quiescence of geologically active faults down to magnitudes near zero.

Lineament and cauldrons. Also shown on Figure 16 are the locations of the Morenci lineament, a major transverse shear zone, and the boundaries of Tertiary cauldrons (Chapin, 1983). The path of the Morenci lineament intersects very few epicenters, fewer than perhaps any other diagonal path through the region. The absence of activity may indicate weakness in the crust and thus may indirectly confirm the existence of the shear zone. There is no apparent correlation between the presently observed seismicity and cauldron boundaries.

Uplift and extension. Shown in Figure 14 along with the earthquake epicenters are contours of surface uplift determined from releveling of elevation bench marks in the Socorro area (Larsen and others, 1986). The contours represent the deformation that accumulated from 1911 through approximately 1980. In the area of maximum uplift in Figure 14, geomorphic evidence indicates an average rate of uplift of 1.8 mm/yr during the past 20,000 years (Ouchi, 1983), which is in good agreement with the leveling results. Bachman and Mehnert (1978) found that sand deposits of the ancestral Rio Grande dear Socorro rise 85 m in elevation northward over a distance of 11 km. Assuming the ancestral Rio Grande had the existing river gradient of 1 m/km, the tectonic change in elevation is 74 m. Assuming an uplift rate of 1.8 mm/yr, the tilt of the sand deposits suggests commencement of uplift about 40,000 years ago.

Savage and others (1985) summarized the results of geodetic measurements made in the Socorro area from 1972 through 1984. The principal strain rates obtained for the region were 0.009 ± 0.016 μ strain/yr N84°E and -0.018 ± 0.017 μ strain/yr N6°W (extension taken as positive). Thus, the trilateration data across the RGR established that no significant strain had accumulated within the precision of the measurements during the 12-year period. However, the measured principal strain in the N84°E direction places an upper limit (at 2 standard deviations) of about 4 mm/yr of east-west spreading across the approximately 100-

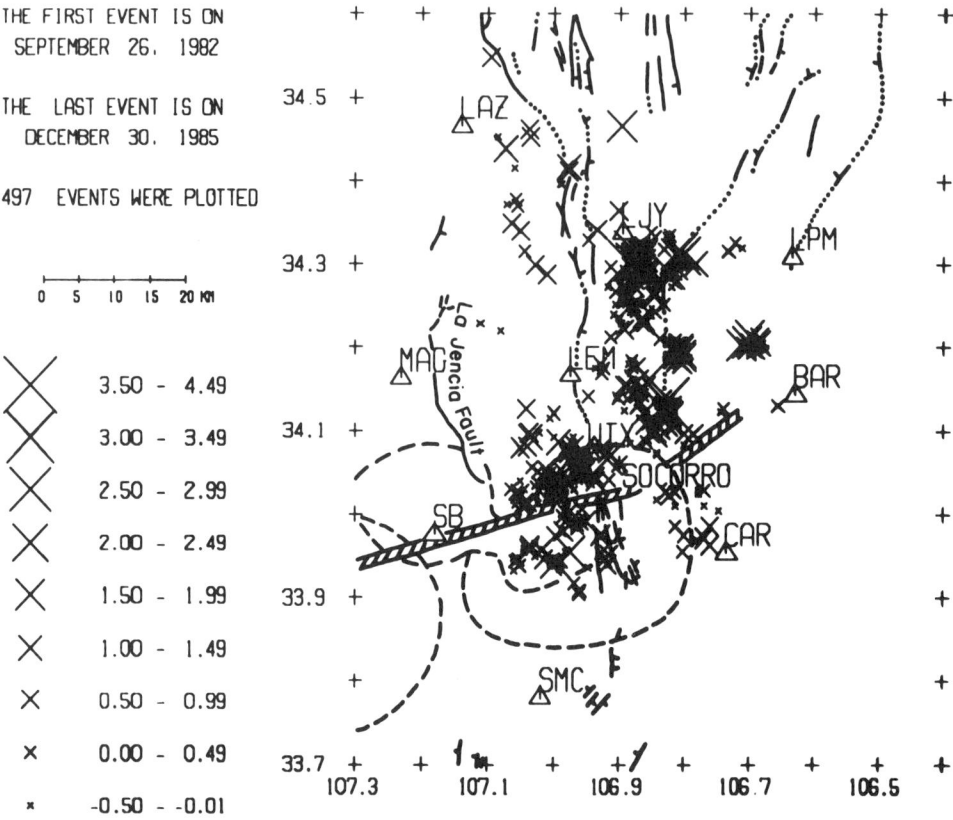

Figure 16. Location of faults with surface offsets within the past 5 m.y. and Socorro area seismicity from September 1982 through December 1985. Position of the faults is based on Machette and McGimsey (1982) and Machette (unpublished work). Also shown are the position of the Morenci lineament (hachered) and boundaries of Tertiary cauldrons (dashed) (from Chapin, 1983).

km breadth of surface uplift in Figure 14. Therefore, the trilateration results are consistent with the level-line data inasmuch as the horizontal movement accompanying 2 mm/yr of vertical uplift is much less than 4 mm/yr (Dieterich and Decker, 1975).

Midcrustal magma body. The dotted line on Figure 14 follows the margins of a thin midcrustal magma body that underlies the Socorro area at a depth of about 19 km. This extensive layer of magma was first detected in microearthquake studies at NMIMT (Sanford and others, 1973; Sanford and others, 1977; Rinehart and others, 1979) and later confirmed by a program of crustal reflection profiling directed by Cornell University (Brown and others, 1979; Brown and others, 1980; Brocher, 1981). The minimum lateral extent of the magma body is 1,700 km^2; its thickness is on the order of 130 to 190 m (Brocher, 1981; Ake and Sanford, 1988). Reilinger and Oliver (1976), Reilinger and others (1980), and Larsen and others (1986) have demonstrated that the observed surface uplift can be explained by inflation of the midcrustal magma body. Assuming an average rate of uplift of 1.8 mm/yr, the thickness of the magma body indicates an age of 70,000 to 105,000 years, which is considerably greater than the 40,000 years based on geologic evidence.

Discussion. Seismicity and surface uplift in the Socorro area are approximately centered on the position of the midcrustal magma body (Fig. 14). This suggests that the unusual concentration of earthquake activity at Socorro (Figs. 5 and 6) is related to inflation of the magma body. Considered here are two possible ways in which the magma body might induce seismicity. First, according to the models of Dieterich and Decker (1975), 160 m of inflation of a horizontal sill 20 km deep and 40 km wide would produce horizontal strains over the crest of the uplift on the order of $1.5 \cdot 10^{-3}$. This amount of stretching is probably adequate to produce movement on preexisting faults; therefore, some of the young faults and earthquakes in the central part of Figure 16 may be a consequence of inflation of the midcrustal magma body.

Secondly, most of the major and minor earthquakes in the Socorro area during the past 140 years have occurred in swarms. This mode of occurrence suggests that small quantities of magma might move upward from the midcrustal magma body into the brittle seismogenic zone and induce swarms of earthquakes. Two observations supporting the existence of small quantities of magma in the seismogenic zone are (1) low crustal velocities (Ward and others, 1981) and (2) low Q values in areas of high seismicity (Carpenter and Sanford, 1986). As mentioned earlier, the non-DCWM mechanisms observed for some earthquakes in the February to March 1983 swarm are also thought to be indicative of magmatic intrusion into the brittle crust.

REFERENCES CITED

Ake, J. P., and Sanford, A. R., 1988, New evidence for the existence and internal structure of a thin layer of magma at mid-crustal depths near Socorro, New Mexico: Bulletin of the Seismological Society of America, v. 78, p. 1335–1359.

Ake, J., Sanford, A., and Jarpe, S., 1983, A magnitude scale for central New Mexico based on signal duration: New Mexico Institute of Mining and Technology Geophysics Open-File Report 45, 26 p.

Ashcroft, B., 1974, The July 1906 earthquakes in Socorro: New Mexico Historical Review, v. 49, p. 325–330.

Bachman, G. O., and Mehnert, H. H., 1978, New K-Ar dates and late Pliocene to Holocene geomorphic history of the central Rio Grande region: Geological Society of America Bulletin, v. 89, p. 283–293.

Bagg, R. M., 1904, Earthquakes in Socorro, New Mexico: American Geologist, v. 34, p. 102–104.

Bailey, R. A., and Smith, R. L., 1978, Volcanic geology of the Jemez Mountains, New Mexico, in Hawley, J. W., compiler, Guidebook to Rio Grande rift in New Mexico and Colorado: New Mexico Bureau of Mines and Mineral Resources Circular 163, p. 184–196.

Brocher, T. M., 1981, Geometry and physical properties of the Socorro, New Mexico, magma bodies: Journal of Geophysical Research, v. 86, p. 9420–9432.

Brown, L. D., and 7 others, 1979, COCORP seismic reflection studies of the Rio Grande rift, in Riecker, R. E., ed., Rio Grande rift; Tectonics and magmatism: American Geophysical Union, p. 169–184.

Brown, L. D., Chapin, C. E., Sanford, A. R., Kaufman, S., and Oliver, J. E., 1980, Deep structure of the Rio Grande rift from seismic reflection profiling: Journal of Geophysical Research, v. 85, p. 4773–4800.

Bryan, K., 1938, Geology and ground-water conditions of the Rio Grande depression in Colorado and New Mexico, Regional planning, Part 4; Rio Grande joint investigation in the upper Rio Grande Basin: Washington, D.C., National Resource Committee, v. 1, pt. 2, sec. 1, p. 197–225.

Callender, J. F., Seager, W. R., and Swanberg, C. A., 1983, Late Tertiary and Quaternary tectonics and volcanism; Geothermal resources of New Mexico; National Oceanic and Atmospheric Administration National Geophysical Data Center, Scientific Map Series 1 sheet, scale 1:500,000.

Carpenter, P. J., and Sanford, A. R., 1985, Apparent Q for upper crustal rocks of the central Rio Grande rift: Journal of Geophysical Research, v. 90, p. 8661–8674.

Cash, D. J., 1971, The Dulce, New Mexico, earthquake of January 23, 1966; Location, focal mechanism, magnitude, and source parameters [Ph.D. thesis]: Socorro, New Mexico Institute of Mining and Technology, 89 p.

Chapin, C. E., 1971, The Rio Grande rift; 1, Modifications and additions: New Mexico Geological Society 22nd Annual Field Conference Guidebook, p. 191–201.

—— , 1983, Selected tectonic elements of the Socorro region: New Mexico Geological Society 34th Annual Field Conference Guidebook, p. 97.

Chapin, C. E., and Seager, W. R., 1975, Evolution of the Rio Grande rift in the Socorro and Las Cruces areas: New Mexico Geological Society 21st Annual Field Conference Guidebook, p. 297–321.

Coffman, J. L., and von Hake, C. A., 1973, Earthquake history of the United States: U.S. Department of Commerce, National Oceanic and Atmospheric Administration Publication 41-1, revised ed. (through 1970), p. 59–88.

Cordell, L., 1978, Regional geophysical setting of the Rio Grande rift: Geological Society of America Bulletin, v. 89, p. 1073–1090.

Davis, P. M., Parker, E. C., Evans, J. R., Iyer, H. M., and Olsen, K. H., 1984, Teleseismic deep sounding of the velocity structure beneath the Rio Grande rift: New Mexico Geological Society 35th Annual Field Conference Guidebook, p. 29–38.

Dewey, J. W., 1987, Instrumental seismicity of Central Idaho: Bulletin of the Seismological Society of America, v. 77, p. 819–836.

Dieterich, J. H., and Decker, R. W., 1975, Finite element modeling of surface deformation associated with volcanism: Journal of Geophysical Research, v. 80, p. 4094–4102.

Eaton, G. P., 1986, A tectonic redefinition of the Southern Rocky Mountains, in Johnson, B., and Bally, A. W., eds., Intraplate deformation; Characteristics, processes, and causes: Tectonophysics, v. 132, p. 163–194.

Edwards, C. L., Reiter, M., Shearer, C., and Young, W., 1978, Terrestrial heat flow and crustal radioactivity in northeastern New Mexico and southern Colorado: Geological Society of America Bulletin, v. 89, p. 1341–1350.

Gable, D. J., and Hatton, T., 1983, Maps of vertical crustal movements in the conterminous United States over the last 10 million years: U.S. Geological Survey Miscellaneous Investigations Series Map I-1315, scale 1:5,000,000, 1:10,000,000.

Hammond, J. F., 1966, A surgeon's report on Socorro, New Mexico, 1852: Santa Fe, New Mexico, Stagecoach Press, 47 p.

Hawley, J. W., and Love, D. W., 1981, Overview of geology related to environmental concerns in New Mexico, in Wells, S. G., and Lambert, P. W., eds., Environmental geology and hydrology in New Mexico: New Mexico Geological Society Special Publication 10, p. 1–10.

Herrmann, R. B., Dewey, J. W., and Park, S., 1980, The Dulce, New Mexico, earthquake of 23 January 1966: Bulletin of the Seismological Society of America, v. 70, p. 2171–2183.

Jaksha, L. H., and Sanford, A. R., 1986, Earthquakes near Albuquerque, New Mexico, 1976–1981: Journal of Geophysical Research, v. 91, p. 6293–6303.

Jaksha, L. H., Locke, J., Thompson, J. B., and Garcia, A., 1978, Reconnaissance seismology near Albuquerque, New Mexico: U.S. Geological Survey Open-File Report 78–339, 28 p.

Jaksha, L. H., Locke, J., and Gebhart, H. J., 1981, Microearthquakes near the Albuquerque volcanoes, New Mexico: Geological Society of America Bulletin, v. 92, p. 31–36.

Jarpe, S. J., 1984, Characteristics and possible cause of an earthquake swarm in the central Rio Grande rift 28 km north of Socorro, New Mexico; February and March, 1983: Socorro, New Mexico Institute of Mining and Technology Geophysics Open-File Report 50, 74 p.

Jarpe, S. P., Sanford, A. R., and Jaksha, L. H., 1984, Evidence for magmatic intrusion during an earthquake swarm in the central Rio Grande rift [abs.]: EOS Transactions of the American Geophysical Union, v. 65, p. 1001.

Kelley, V. C., 1952, Tectonics of the Rio Grande depression of central New Mexico: New Mexico Geological Society 3rd Annual Field Conference Guidebook, p. 93–105.

Kelley, V. C., and Kudo, A. M., 1978, Volcanoes and related basalts of Albuquerque Basin, New Mexico: New Mexico Bureau of Mines and Mineral Resources Circular 156, 30 p.

King, K. M., 1986, Investigation of the seismogenic zone in the vicinity of Socorro, New Mexico, from an analysis of focal depth distributions: Socorro, New Mexico Institute of Mining and Technology Geophysics Open-File Report 57, 124 p.

Larsen, S., Reilinger, R., and Brown, L., 1986, Evidence for ongoing crustal deformation related to magmatic activity near Socorro, New Mexico: Journal of Geophysical Research, v. 91, p. 6283–6292.

Lee, W.H.K., and Lahr, J. C., 1975, HYPO 71 (Revised); A computer program for determining hypocenter, magnitude, and first motion pattern of local earthquakes: U.S. Geological Survey Open-File Report 75–311, 113 p.

Luedke, R. G., and Smith, R. L., 1978, Map showing distribution, composition, and age of late Cenozoic volcanic centers in Arizona and New Mexico: U.S. Geological Survey Miscellaneous Investigations Series Map I-1091-A, scale 1:1,000,000.

Machette, M. N., 1982, Quaternary and Pliocene faults in the La Jencia and southern part of the Albuquerque-Belen Basins, New Mexico; Evidence of fault history from fault scarp morphology and Quaternary geology: New Mexico Geological Society 33rd Annual Field Conference Guidebook, p. 161–169.

——, 1986, History of Quaternary offset and paleoseismicity along the La Jencia fault, central Rio Grande rift, New Mexico: Bulletin of the Seismological Society of America, v. 76, p. 259–272.

Machette, M. N., and Colman, S. M., 1983, Age and distribution of Quaternary faults in the Rio Grande rift; Evidence from morphometric analysis of fault scarps: Geological Society of America Abstracts with Programs, v. 15, p. 320.

Machette, M. N., and McGimsey, R. G., 1982, Quaternary and Pliocene faults in the Socorro and western part of the Fort Sumner 1°×2° Quadrangles, New Mexico: U.S. Geological Survey Miscellaneous Field Studies Map MF–1465–A, pamphlet, 11 p.

Mayo, E. B., 1958, Lineament tectonics and some ore districts of the Southwest: American Institute of Mining, Metallurgical, and Petroleum Engineers Transactions, p. 1169–1175.

Newton, C. A., Cash, D. J., Olsen, K. H., and Homuth, E. F., 1976, LASL seismic programs in the vicinity of Los Alamos, New Mexico: Los Alamos Scientific Laboratory Informal Report LA–6406–MS, 42 p.

Nichols, R. L., 1946, McCartys basalt flow, Valencia County, New Mexico: Geological Society of America Bulletin, v. 57, p. 1049–1086.

Northrop, S. A., 1945, Earthquake history of central New Mexico [abs.]: Geological Society of America Bulletin, v. 56, p. 1185.

——, 1947, Seismology in New Mexico [abs.]: Geological Society of America Bulletin, v. 58, p. 1268.

——, 1961, Earthquakes of central New Mexico: New Mexico Geological Society 33rd Annual Field Conference Guidebook, p. 171–178.

——, 1976, New Mexico's earthquake history, 1849–1975: New Mexico Geological Society Special Publication 6, p. 77–87.

——, 1982, Earthquakes of Albuquerque country: New Mexico Geological Society 33rd Annual Field Conference Guidebook, p. 171–178.

Olsen, K. H., 1979, The seismicity of north-central New Mexico with particular reference to the Cerrillos earthquake of May 28, 1918: New Mexico Geological Society 30th Field Conference Guidebook, p. 65–75.

Ouchi, S., 1983, Effects of uplift on the Rio Grande over the Socorro magma body, New Mexico: New Mexico Geological Society 34th Annual Field Conference Guidebook, p. 54–56.

Reid, H. F., 1911, Remarkable earthquakes in central New Mexico in 1906 and 1907: Bulletin of the Seismological Society of America, v. 1, p. 10–16.

Reilinger, R. E., and Oliver, J. E., 1976, Modern uplift associated with a proposed magma body in the vicinity of Socorro, New Mexico: Geology, v. 4, p. 583–586.

Reilinger, R. E., Oliver, J. E., Brown, L. D., Sanford, A. R., and Balazs, E., 1980, New measurements of crustal doming over the Socorro magma body, New Mexico: Geology, v. 8, p. 291–295.

Rinehart, E. J. and Sanford, A. R., 1981, Upper crustal structure of the Rio Grande Rift near Socorro, New Mexico, from inversion of microearthquake S-wave reflections: Seismological Society of America Bulletin, v. 71, p. 437–450.

Rinehart, E. J., Sanford, A. R., and Ward, R. M., 1979, Geographic extent and shape of an extensive magma body at midcrustal depths in the Rio Grande rift near Socorro, New Mexico, in Riecker, R. E., ed., Rio Grande rift; Tectonics and magmatism: American Geophysical Union, p. 237–251.

Sanford, A. R., 1963, Seismic activity near Socorro: New Mexico Geological Society 14th Field Conference Guidebook, p. 146–154.

Sanford, A. R., and 5 others, 1972, Seismicity of the Rio Grande rift in New Mexico: New Mexico Bureau of Mines and Mineral Resources Circular 120, 19 p.

Sanford, A. R., Alptekin, O. S., and Toppozada, T. R., 1973, Use of reflection phases on microearthquake seismograms to map an unusual discontinuity beneath the Rio Grande rift: Bulletin of the Seismological Society of America, v. 63, p. 2021–2034.

Sanford, A. R., and 6 others, 1977, Geophysical evidence for a magma body in the crust in the vicinity of Socorro, New Mexico, in Heacock, J. G., ed., The earth's crust: American Geophysical Union Geophysical Monograph 20, p. 385–403.

Sanford, A. R., Olsen, K. H., and Jaksha, L. H., 1979, Seismicity of the Rio Grande rift, in Riecker, R. E., ed., Rio Grande rift; Tectonics and magmatism: American Geophysical Union, p. 145–168.

——, 1981, Earthquakes in New Mexico, 1849–1977: New Mexico Bureau of Mines and Mineral Resources Circular 171, 20 p.

Sanford, A. R., Jaksha, L. H., and Wieder, D. P., 1983, Seismicity of the Socorro area of the Rio Grande rift: New Mexico Geological Society 34th Annual Field Conference Guidebook, p. 127–131.

Savage, J. C., Lisowski, M., and Prescott, W. H., 1985, Strain accumulation in the Rocky Mountain states: Journal of Geophysical Research, v. 90, p. 10310–10320.

Seager, W. R., 1980, Quaternary fault system in the Tularosa and Hueco basins, southern New Mexico and West Texas: New Mexico Geological Society 31st Annual Field Conference Guidebook, p. 131–136.

——, 1981, Geology of the Organ Mountains and the southern San Andres Mountains, New Mexico: New Mexico Bureau of Mines and Mineral Resources Memoir 36, 97 p.

Smith, R. L., Bailey, R. A., and Ross, C. J., 1970, Geologic map of the Jemez Mountains, New Mexico: U.S. Geological Survey Miscellaneous Geologic Investigations Map I–571, scale 1:250,000.

Schwartz, D. P., and Coppersmith, K. J., 1984, Fault behavior and characteristic earthquakes; Examples from the Wasatch and San Andreas fault zones: Journal of Geophysical Research, v. 89, p. 5681–5698.

Taggart, J., and Baldwin, F., 1982, Earthquake sequence of 1938–1939 in Mogollon Mountains, New Mexico: New Mexico Geology, v. 4, no. 4, p. 49–52.

Taylor, P. T., and 5 others, 1983, Observing the terrestrial gravity and magnetic fields in the 1990's: EOS Transactions of the American Geophysical Union, v. 64, p. 609–611.

Toppozada, T. R., and Sanford, A. R., 1976, Crustal structure in central New Mexico interpreted from the GASBUGGY explosion: Bulletin of the Seismological Society of America, v. 66, p. 877–886.

Ward, R. M., Schlue, J. W., and Sanford, A. R., 1981, Three-dimensional velocity anomalies in the upper crust near Socorro, New Mexico: Geophysical Research Letters, v. 8, p. 553–556.

Wieder, D. P., 1981, Tectonic significance of microearthquake activity from composite fault-plane solutions in the Rio Grande rift near Socorro, New Mexico: Socorro, New Mexico Institute of Mining and Technology Geophysics Open-File Report 37, 159 p.

Wong, I. G., Cash D. J., and Jaksha, L. H., 1984, The Crownpoint, New Mexico, earthquakes of 1976 and 1977: Bulletin of the Seismological Society of America, v. 74, p. 2435–2449.

Woodward, L. A., Callender, J. F., and Zilinski, R. E., 1975, Tectonic map of the Rio Grande rift, New Mexico: Geological Society of America Map Chart Series MC–11, scale 1:500,000.

MANUSCRIPT ACCEPTED BY THE SOCIETY MARCH 20, 1989

ACKNOWLEDGMENTS

We are grateful to many students and co-workers at New Mexico Tech, Los Alamos National Laboratory, and the USGS Albuquerque Seismological Laboratory for their contributions through the years to the study of the Rio Grande rift and New Mexico seismicity. Without their assistance, this paper would not have been possible. We also wish to thank Walter Arabasz and James Dewey for thoughtful and thorough reviews of this chapter.

Chapter 13

Seismotectonics of the central United States

B. J. Mitchell, O. W. Nuttli*, R. B. Herrmann, and W. Stauder
Department of Earth and Atmospheric Sciences, Saint Louis University, St. Louis, Missouri 63103

INTRODUCTION

The seismicity rate in the central United States is, for the most part, low compared to that in more active regions such as California. Occasionally, however, severe earthquakes do occur, and certain regions, such as the New Madrid seismic zone, are characterized by a seismicity level well above that in surrounding areas. Earthquakes in midplate regions have been considered to be enigmatic since they cannot easily be associated with major deformation as can earthquakes near plate margins. The past decade has, however, seen a large increase in our understanding of midplate earthquakes, particularly those in concentrated zones where seismic activity has been well monitored.

For purposes of the present chapter we define the central United States as that region between the Appalachian and Rocky Mountains shown in Figure 1. The eastern boundary is taken to coincide with the eastern border of Ohio, to continue southward into West Virginia and to pass in a southwestward direction through a sliver of West Virginia, through eastern Kentucky and central Tennessee just west of the Appalachian Mountains. It continues in a southwestward direction through a corner of Alabama, intersects the state line of Mississippi at about 35.5° latitude, and coincides with the Alabama-Mississippi boundary southward to the Gulf Coast. The western boundary coincides with the western state lines of North Dakota, South Dakota, Nebraska, Kansas, Oklahoma, and the western boundary of the panhandle of Texas. It continues southward through west Texas to Mexico. The northern boundary is the international boundary between the United States and Canada, and the southern boundary coincides with the southern boundaries of Mississippi, Louisiana, and Texas. The total area encompassed by these borders is in excess of 3,000,000 km^2.

The level of seismicity in the central United States is compared to that of a 200,000 km^2 region of southern California in Figure 2. The curve for the central United States lies below that of southern California by a factor of more than 10 for magnitude 4 earthquakes and about 30 for magnitude 6 earthquakes. This low level of activity has characterized seismicity in the central United States for more than 150 years.

Three factors, however, make the study of earthquakes in the central United States important. First, at least one region (the New Madrid seismic zone) has been the site of great earthquakes in the past and is likely to be susceptible to future large earthquakes. Second, the rate of attenuation of seismic waves in the central United States is much lower than that in the more active regions of the western United States. This is true for both the Lg phase (Nuttli, 1973a) and for longer period surface waves (Mitchell, 1973, 1975). This low attenuation means that seismic waves will be felt, and cause damage, over a much larger region in the central United States than would waves from an earthquake of similar magnitude in regions such as California where attenuation is much greater. The lowest attenuation for Lg waves occurs in a north-south–trending band from southern Missouri to Lake Michigan with attenuation increasing both eastward and westward from that band (Singh and Herrmann, 1983). Mitchell and Hwang (1987) explained those increases as being due to increasing thicknesses of young sediments. A third factor that is unusual for earthquake occurrence in the central United States and makes it important to understand earthquakes and faulting there is related to the way midplate earthquakes scale. Nuttli (1983) showed that for midplate earthquakes, the seismic moment is proportional to the fourth power of corner period; for this reason, stress drop increases with moment, rather than remaining constant, as it would for plate-margin earthquakes. As a result, large-magnitude earthquakes in midplate regions can occur with relatively short rupture lengths. Thus, relatively small faults have the potential for producing large earthquakes and strong ground motion. Because of these factors, relatively small earthquakes might cause damage over a wide area in the central United States. Thus, earthquake risk can be comparatively high even though the recurrence intervals for major earthquakes are much longer than those in more tectonically active regions. For a discussion of earthquake risk in different tectonic settings, readers are referred to Nuttli (1981a).

In the following sections we will discuss several aspects of seismicity in the central United States. Because of particular problems with both historical and instrumentally recorded earthquakes in this region we will first discuss methods of magnitude

---

*Deceased February 9, 1988.

Mitchell, B. J., Nuttli, O. W., Herrmann, R. B., and Stauder, W., 1991, Seismotectonics of the central United States, *in* Slemmons, D. B., Engdahl, E. R., Zoback, M. D., and Blackwell, D. D., eds., Neotectonics of North America: Boulder, Colorado, Geological Society of America, Decade Map Volume 1.

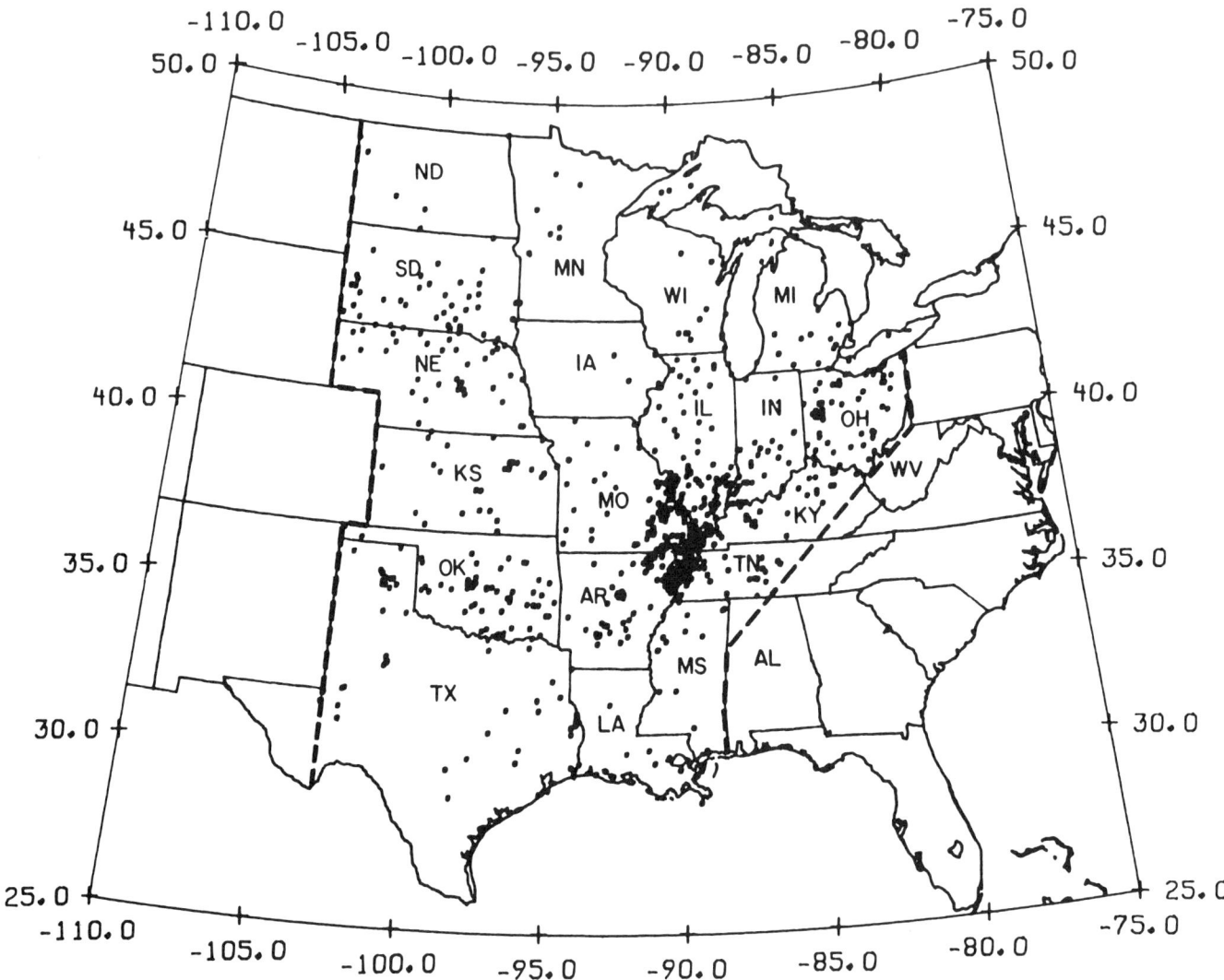

Figure 1. Map of earthquake epicenters with $m_b \geq 3.0$ in the central United States between 1811 and 1987. Heavy dashed lines indicate the eastern and western boundaries of the region considered in this study.

determination that have been used. We will then consider patterns of seismicity and propose a new classification of seismogenic zones for the central United States. Locations and fault-plane solutions of larger earthquakes will be related to geologic features and to the crustal stress field in each region. Finally, we will discuss the seismicity, structure, and tectonics of the New Madrid seismic zone, a region where extensive research efforts have yielded a wealth of information on intraplate earthquakes over the last decade.

MAGNITUDE DETERMINATIONS

Three problems arise in assigning magnitude values to earthquakes in the central United States. The first problem results from the lack of adequate instrumental data until relatively recent times. The second is a result of differences in P-wave amplitude attenuation at regional distances. The third arises from an inadequate distribution of seismograph stations, a problem that still exists. For these reasons, various methods have been employed to determine magnitudes over the time period covered by this study.

Instrumentally determined magnitudes began to be obtained in the 1930s and 1940s and were restricted to events large enough to be recorded teleseismically (more than 3,000 km). Seismologists employed methods developed by Gutenberg and Richter (1936) for surface waves, and by Gutenberg (1945a, b), as revised by the U.S. Coast and Geodetic Survey, for body waves. In the 1960s, St. Louis University established five new stations in the central United States, which included short-period instruments. These, together with a few other stations having short-period seismographs, provided a data set from which m_b could be determined using regional P_n amplitudes (Evernden, 1967) and regional Lg amplitudes (Nuttli, 1973a). These developments were

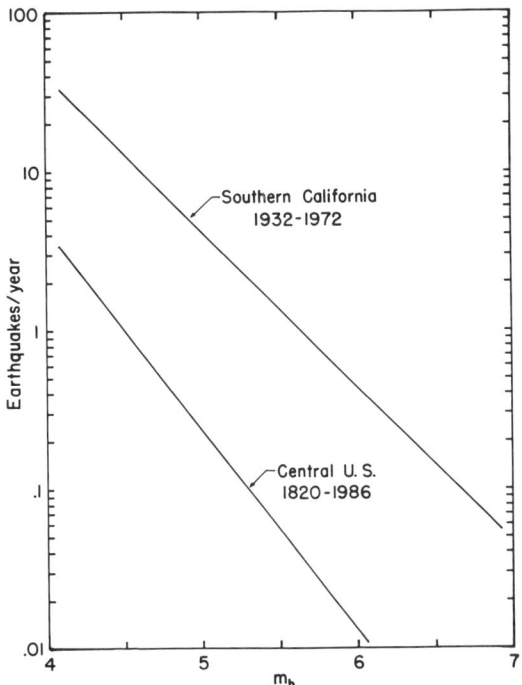

Figure 2. Comparison of yearly numbers of earthquakes in the central United States for the period 1820 to 1986 with those in southern California for the period 1932 to 1972.

important because most central United States earthquakes have m_b values less than 4.5 and are consequently too small to produce measurable 1-sec P-wave amplitudes at teleseismic distances.

Through the 1960s and early 1970s the U.S. Coast and Geodetic Survey, and later the U.S. Geological Survey, gave m_b values for central United States earthquakes obtained from P-wave amplitudes at regional distances (less than 3,000 km) using standard m_b calibration curves (e.g., Richter, 1958, p. 688). These are applicable at teleseismic distances, but are not generally appropriate at regional distances because of the heterogeneous character of the upper 600 km of the mantle, the region that affects the amplitudes of P-waves arriving at regional distances. The m_b values of central United States earthquakes obtained from regional data contained in the Preliminary Determination of Epicenters (PDE) reports of those agencies are therefore often unreliable.

In the mid-1970s several microearthquake networks were established in the central United States. Most had vertical-component seismographs only, with a frequency response peaking at 10 to 30 Hz. They include the central Mississippi Valley network of St. Louis University; the lower Mississippi Valley and eastern Tennessee networks of Memphis State University; the central and eastern Kentucky networks of the University of Kentucky; the Anna, Ohio, area network of the University of Michigan; the Oklahoma network of the University of Oklahoma; and the Kansas network of the Kansas Geological Survey and University of Kansas. These networks provide amplitude and magnitude data for the microearthquakes occurring within their areas of detection. Special magnitude formulas are needed because the recorded P and Lg waves have periods much less than 1 sec. Most of these formulas are based on the amplitudes of high-frequency Lg waves or the duration of the seismic coda.

Many earthquakes, some of which were major, occurred in the United States before seismographic recording was possible. Magnitude determinations of these historical earthquakes require nonstandard methods. Three methods have been employed. One relates magnitude to maximum intensity (Nuttli, 1974, 1981a), another relates magnitude to the area enclosed by a selected isoseismal or the felt area (Nuttli and Zollweg, 1974), and a third relates magnitude to the set of intensity versus epicentral distance values (Nuttli and others, 1979). The first of these methods is the least reliable since maximum intensity may be poorly determined either because there may be no town near the point at which ground shaking was greatest or because local site conditions may amplify ground shaking in the epicentral area. With good intensity data, magnitudes estimated by the second and third methods have an uncertainty (one standard deviation) of about 0.2 magnitude units, about the same as those for magnitudes determined from seismographic data, whereas uncertainties for the first method are estimated to be 0.3 to 0.5 magnitude units.

The magnitude threshold of historical earthquakes depends on the distribution of human population. Parts of the central United States were already settled by the latter part of the eighteenth century, whereas parts of the plains states were not settled until at least 100 years later. Because of the low attenuation of seismic-wave energy in the central United States, it is likely that all normal-depth earthquakes of epicentral intensity VIII or larger (m_b of 5.8 or larger) that occurred anywhere in the central United States since 1800 are known. All earthquakes of epicentral intensity VII or greater (m_b of 5.3 or larger), except for very shallow earthquakes, likely are known for the entire area back to about 1890, and for the eastern part to about 1800. For the period 1900 to 1960 the threshold of reported earthquakes is about V to VI (m_b of about 4.4) for the plains states, and IV to V (m_b of about 4.0) for the remainder of the central United States.

For the central United States as a whole, the m_b threshold for the period of the early 1960s to the mid-1970s is about 4.0. Certain areas, such as the New Madrid fault region, however, had a lower threshold; others, such as northern Alabama, may have had a higher threshold.

With the establishment of the microearthquake networks in the mid- and late-1970s, the m_b threshold dropped to as low as 1 or 2 for earthquakes occurring within the areas of the networks. Outside these areas the threshold is about 3.5.

PATTERNS OF SEISMICITY

The map in Figure 1 shows the locations of known earthquakes of $m_b \geqslant 3.0$ from 1811 through 1987 in the central United States (Nuttli, 1981b). The central Mississippi Valley (northeast Arkansas, southeast Missouri, southern Illinois, southwestern In-

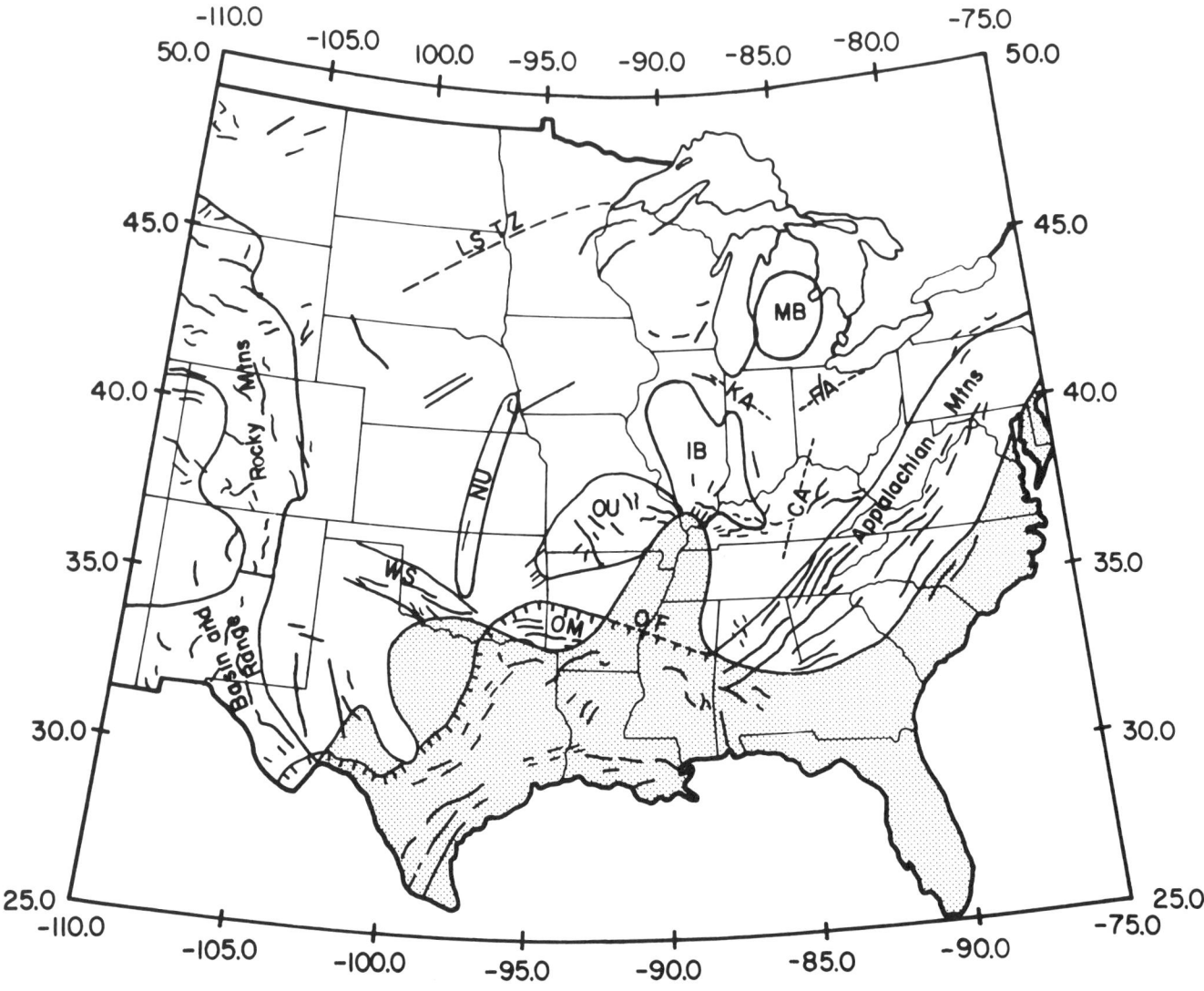

Figure 3. Simplified map of faults and selected tectonic features for the central United States and adjacent regions. LSTZ, Lake Superior tectonic zone; MB, Michigan basin; KA, Kankakee arch; FA, Findlay arch; CA, Cincinnati arch; IB, Illinois basin; OU, Ozark uplift; NU, Nemaha uplift; WS, Wichita system; OM, Ouachita system; and OF, Ouachita frontal thrust. Shading denotes sediments of the coastal provinces.

diana, western Kentucky and western Tennessee) stands out as the region of highest earthquake activity. The remainder of the map area shows a rather diffuse distribution of epicenters, although some states, such as Ohio and South Dakota, have more than others, such as Iowa, North Dakota, and Texas. Although some differences in epicenter density are related to population density, there is no one-to-one correlation between the two. The clusterings of epicenters in northern Ohio and near the boundary between Missouri and southern Illinois at about 38.5° latitude are undoubtedly due to the long-time operation of seismograph stations at the cities of Cleveland and St. Louis, respectively.

Small earthquakes, such as those of body-wave magnitude 3 to 4, have fault lengths of only 0.5 to 1 km (Nuttli and others, 1989). Since their epicenters can be mislocated by at least as much as several kilometers, it is impossible to unequivocally associate the epicenters with small geologic features. Some general conclusions can, however, be drawn by comparing the epicenters of Figure 1 with the fault locations of Figure 3. First, it is clear that many of the earthquakes cannot be associated with mapped faults. Faults are undoubtedly present in all cases, but are obscured by surface sediments or have not as yet been mapped. Second, many faults are conspicuous by the absence of significant earthquake activity near them. This lack of activity must be real, and the variable rates of seismicity along different faults and different portions of the same fault must be due to geologic factors such as the orientation of the stress field in the crust with respect to the fault orientation, or to physical properties of the fault zone itself. These factors will be discussed in later sections.

Association of seismicity with geologic features

Although earthquakes of $m_b \geq 3.0$ occur almost everywhere in the central United States (Fig. 1), more restricted patterns emerge if epicenters are plotted only for earthquakes of $m_b \geq 4.5$. The majority of the epicenters of those earthquakes appear to be associated with a few types of basement structures, most of which can be classified as rifts, uplifts, basins, or former plate boundaries. The boundaries drawn in Figure 4 surround several of these features and represent a new classification of seismic zones in the central United States. Some regions are identical to those of Nuttli and Brill (1981), but others have been added or modified to take into account knowledge about the tectonics and/or structural characteristics of each zone.

The zones indicated in Figure 4, in some cases at least, are somewhat arbitrary. Because of the long recurrence intervals and short historic record for earthquakes in the central United States, it is not possible to delineate seismogenic zones very precisely. When a much longer historical record is available it may become clear that earthquakes of $m_b \geq 4.5$ are concentrated in smaller regions. It is also possible that some new zones should be added or some existing zones extended. The delineation of active faults through geological or geophysical studies may also serve to define seismogenic zones more precisely. In the meantime, however, we have attempted to associate each zone with a distinct system of faults or with the largest tectonic feature with which earthquakes of $m_b \geq 4.5$ may be associated.

The New Madrid fault zone. The New Madrid fault zone extends southwestward from southern Illinois, through the Missouri boot heel and western Kentucky, into northeastern Arkansas (Fig. 5). Young sediments of the Mississippi Embayment cover the active portion of this zone and prevent surface observation of faulting. Surface faults are, however, observed north of the embayment in the Illinois-Kentucky mineral district. They are predominantly high-angle normal faults forming a complex horst and graben system, the faults of which have undergone vertical offsets as large as 900 m and have also undergone smaller horizontal offsets (Hook, 1974; Trace, 1974). These faults can be traced northward to the southern margin of the Shawneetown-Rough Creek fault zone. Although activity on these faults is minor, they have been included as part of the New Madrid fault zone since they are oriented in the same direction as trends of epicenters in the active portion of the zone and appear to be continuous with them.

Figure 4 indicates that all earthquakes of $m_b \geq 5.9$ in the central United States since 1811 have occurred in the New Madrid fault zone. The earthquake sequence that began there on December 11, 1811, included several strong earthquakes over a period of three months, and smaller aftershocks continued until at least 1817. These have been discussed extensively by Fuller (1912) and Nuttli (1973b). Those authors identified three principal earthquakes, one on December 16, 1811 ($m_b = 7.2$, $M_S = 8.6$); a second on January 23, 1812 ($m_b = 7.1$, $M_S = 8.4$); and a third on February 7, 1812 ($m_b = 7.3$, $M_S = 8.7$). Street (1982)

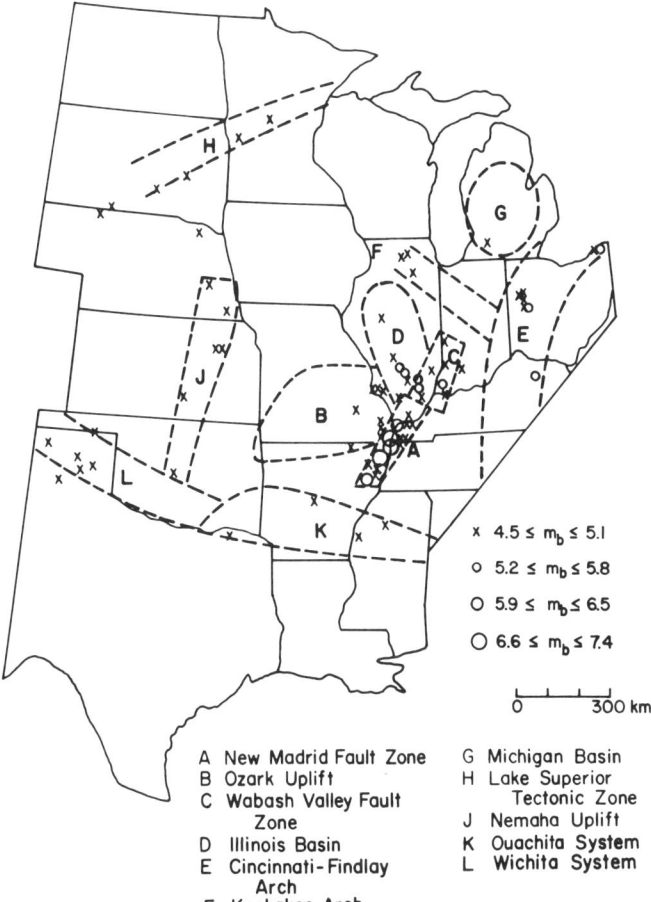

Figure 4. Map of proposed seismic zones and earthquake epicenters with $m_b \geq 4.5$ in the central United States between 1811 and 1987.

identified another major shock that occurred about six hours after the initial event, for which he estimated $m_b = 7.0$.

Figure 6 compares the cumulative number of earthquakes for three months following the initial event of December 16, 1811, with those that occurred in the entire central United States over a 167-year period between 1820 and 1986. The number of events for the three-month period exceeds that for the 167-year period by a factor of more than 5 at $m_b = 4.5$ and more than 10 at $m_b = 6.0$. This comparison shows that the 1811 to 1812 sequence dominates the earthquake history of the central United States.

Because the active portion of the New Madrid fault zone is overlain by a thick sequence of alluvial deposits, there is no surface evidence that indicates present-day fault movement. Microearthquake data have, however, allowed the delineation of the active portions of the fault zone. Figure 5 shows the pattern of seismicity determined by Stauder and others (1976). It indicates that the active faults form a complex pattern through portions of southern Missouri, northwestern Tennessee, and northeastern Arkansas. The most active branch of the fault zone extends in a north-northwest to south-southeast direction for about 70 km

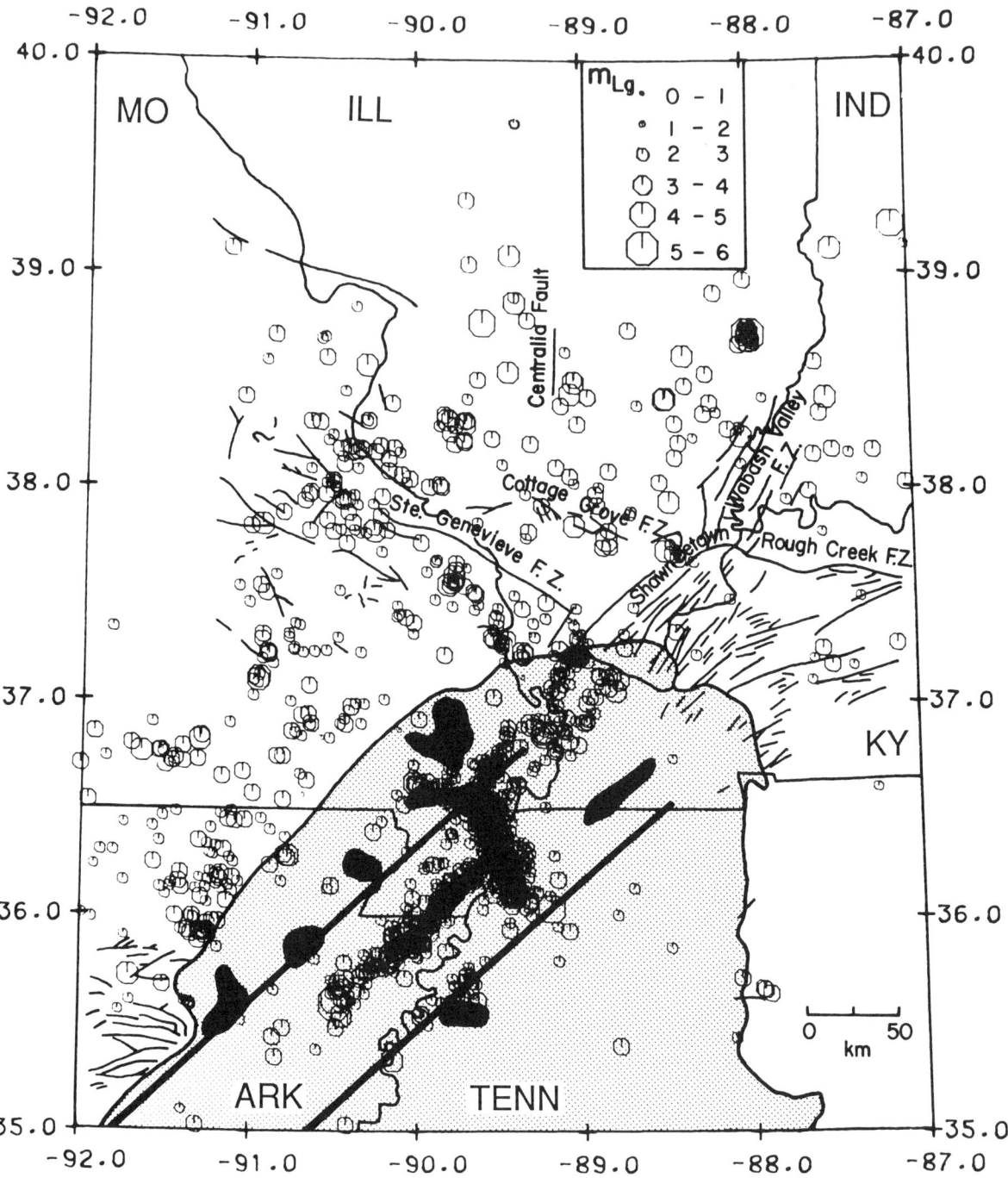

Figure 5. Map of earthquake epicenters (circles) and mapped faults in the northern Mississippi Embayment (shaded) and surrounding regions. Blackened areas indicate Mesozoic plutons and the heavy straight lines are approximate boundaries of a graben inferred from gravity and magnetic data.

between 36.6° and 36.0° latitude. A reflection survey (Zoback, 1979) identified many buried faults in this region with vertical offsets as large as 60 m. The longest branch extends for about 100 km in a northeasterly direction from Arkansas and intersects the most active branch about 30 km from its southernmost end. At the northern end of the most active trend the seismicity bifurcates into two branches, one extending northeastward for about 30 km and the other extending westward for about 35 km.

Nuttli (1973b) used intensity information to infer that the initial shock of December 16, 1811, occurred near the southernmost end of the active portion of the fault zone and that the last large shock (February 7, 1812) occurred near the bifurcation in the northern portion of the active zone. McKeown (1982) finds that the distribution of sand blows in the upper Mississippi Embayment supports the location inferred by Nuttli for the first shock but not the location inferred for the later shock.

A curious feature of the 1811 to 1812 earthquakes is that they seem to have been accompanied by a very small amount of surface faulting (Fuller, 1912; Nuttli and Herrmann, 1984). This small movement may be at least partially explained by the spectral scaling relations discussed in an earlier section for midplate earthquakes. Using these relations, Nuttli (1983) estimated that fault displacements for the three largest events in the New Madrid sequence to be 580, 470, and 850 cm.

The two largest earthquakes since 1812 occurred near the southern terminus (January 4, 1843; $m_b = 6.0$) and near the northern terminus (October 31, 1895; $m_b = 6.2$) of the active portions of the fault zone. The largest earthquake to have occurred since the start of operation of the St. Louis University/USGS/NRC network was of $m_b = 5.0$ and was also located near the southern terminus of the active region (Fig. 5). All earthquakes of $m_b \geq 3.0$ are similarly located near the ends of the main trends of seismic activity.

The great preponderance of earthquakes that occur in the New Madrid fault zone are at depths less than 10 km, but several are at greater depths, some in excess of 20 km. Most of the deeper earthquakes between 1974 and 1987 have been restricted to the same regions where the largest earthquakes have occurred. Most of the larger earthquakes in this zone are indeed deep, but many small events in the same regions also occur at considerable depth. It is noteworthy that the increases in depth and magnitude occur abruptly near the termini of active zones rather than gradually with distance along any portion of the zones.

Russ (1979) studied features in Holocene sediments exposed in an exploratory trench across an active portion of the fault zone in the northwestern Tennessee. He found no evidence for significant fault movement in that region during the 1811 to 1812 earthquakes. He found, however, evidence for two periods of faulting prior to 1800 and concluded that at least two earthquakes strong enough to liquify sediments and generate faulting occurred prior to the 1811 earthquakes. He deduced a recurrence interval of 600 years or less for large events, a figure that compares favorably with recurrence values inferred from historical seismicity (Nuttli, 1974).

Johnston and Nava (1985) determined recurrence rates and made probability estimates for earthquakes in the New Madrid seismic zone. They estimated that there is a 40 to 63 percent probability of an $m_b \geq 6.0$ (M_S 6.3) and a 1 percent probability of an $m_b \geq 7.0$ (M_S 8.3) event occurring by the year 2000. These probabilities increase to 86 to 97 percent for $m_b \geq 6.0$ (M_S 6.3) and 4 percent for $m_b \geq 7.0$ (M_S 8.3) by the year 2035.

From observed recurrence intervals in the New Madrid region, Nuttli (1983) estimated that movement along the New Madrid fault zone averages about 1 cm/yr. If that rate of movement were to have been continuous over a long time period, it would have tremendous implications for the geologic structure of the New Madrid region. For example, 1 m.y. of movement would produce displacements of 10 km. Since such deformation is not observed in the New Madrid region, recurrence intervals at some time in the past must have been much lower than they are at the present time. Russ (1979) finds evidence for major earthquake activity as early as 2,000 years ago. Thus, the change in rate of activity is likely to have occurred before that time. The present rate of fault movement must therefore have begun in the late Pleistocene, but earlier than 2,000 years ago.

Fault-plane solutions for the New Madrid seismic zone appear in the inset of Figure 7 (Herrmann, 1979; Andrews and others, 1985). These solutions indicate that right-lateral strike-slip

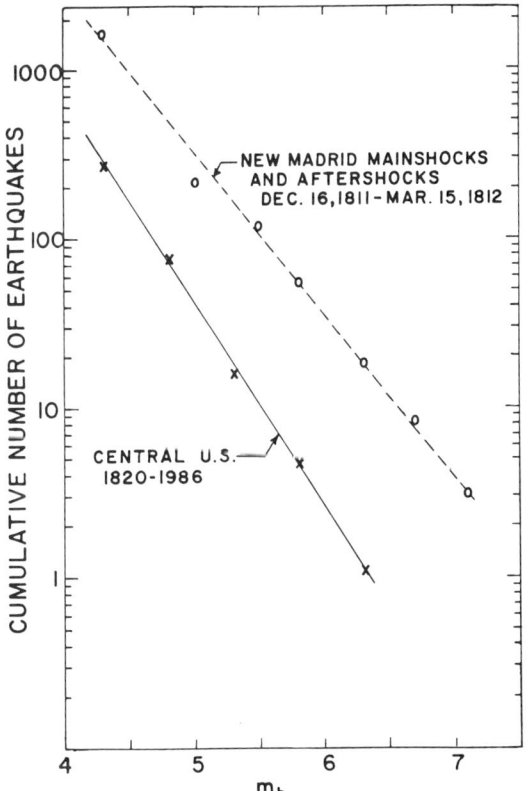

Figure 6. Comparison of the cumulative number of New Madrid mainshocks and aftershocks for the three-month period between December 16, 1811, and March 15, 1812, with the total cumulative number of earthquakes in the entire central United States for the 167-year period 1820 to 1986.

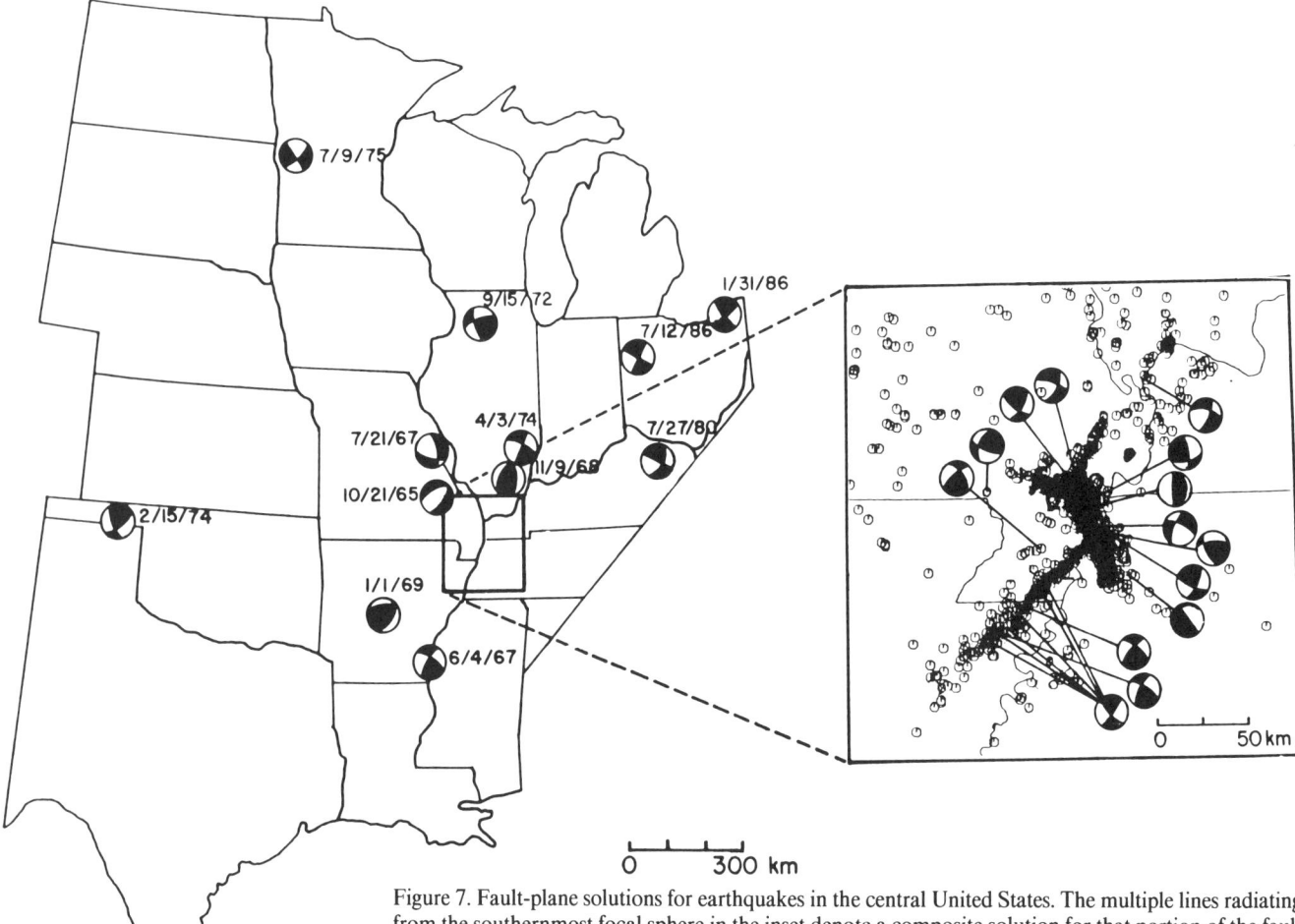

Figure 7. Fault-plane solutions for earthquakes in the central United States. The multiple lines radiating from the southernmost focal sphere in the inset denote a composite solution for that portion of the fault zone.

motion predominantly characterizes the northeast-southwest–trending portions of the fault zone both to the north and to the south of the north-northwest–trending segment. Fault-plane solutions on the north-northwest–trending segment are mixed, some indicating predominant strike-slip motion and some indicating predominant reverse motion. The inconsistency of the solutions in this region may mean that fault orientations do not correspond exactly with the main trend of this segment. Stauder and others (1976) suggested that several small offsets not aligned with the predominant trend might occur there. Composite fault-plane solutions for each portion of the fault zone (Herrmann and Canas, 1978; O'Connell and others, 1982) generally agree with the solutions for individual events. With only two exceptions, the solutions throughout the New Madrid fault zone are those that would be expected for a stress field in which the greatest principal stress is oriented in a direction between east-west and northeast-southwest.

The New Madrid seismic zone has been studied extensively to determine how seismicity there might be related to structural elements and subsurface features in the region. A summary of findings from several of those studies is presented in a later section.

The Illinois basin. The Illinois basin covers about two-thirds of the state of Illinois and about one-fourth of Indiana (Fig. 3). It is a spoon-shaped basin with the deepest part of the bowl in southeastern Illinois; the long axis is aligned slightly west of north. The LaSalle anticline separates the eastern flank of the basin from the deepest part. In considering the seismicity of this region, we have separated the western portion of the basin from the remaining portion on the basis of strike direction and densities of mapped faults. The area covered by the eastern portion of the basin approximately coincides with the Wabash Valley fault zone discussed in the following section; thus, the area referred to in this section is the western and larger portion of the Illinois basin.

Four events of $4.5 \leq m_b \leq 5.1$ and two events of $5.2 \leq m_b \leq 5.8$ have occurred in this zone since 1812 (Fig. 4). The few faults that have been mapped in the western part of the Illinois basin lie in the southern part of Illinois and are oriented in a north-south direction. Four of the events in Figure 4 are located near the Centralia fault (Fig. 5), one lies farther south—perhaps on the Cottage Grove fault zone—and one lies farther north where no faults are mapped. No fault-plane solutions are available for this zone.

The Wabash Valley fault zone. As indicated above, this

fault zone lies in the eastern part of the Illinois basin, but is considered to be a separate seismic zone on the basis of the orientations and densities of mapped faults. This zone lies in the region of a rift system postulated to extend northeastward from the rift known to lie beneath the New Madrid seismic zone (Braile and others, 1982). Faults of the Wabash Valley system extend north-northeastward for about 95 km from just north of the Shawneetown fault zone and span an area about 25 km wide (Bristol and Treworgy, 1979). They are predominantly parallel, high-angle, normal faults and bound horsts and grabens along which individual fault displacements reach 146 m. They are separated from the faults in the mineral district of the northern part of the New Madrid fault zone by the Cottage Grove–Shawneetown–Rough Creek fault zone. Because of that discontinuity and because their strike differs significantly from the strikes of faults in the New Madrid fault zone (Heyl and McKeown, 1978) and the Illinois basin, we consider them to be a distinct fault system.

Three earthquakes of $4.5 \leq m_b \leq 5.1$ and five of $5.2 \leq m_b \leq 5.8$ have occurred in this zone. Another event of $4.5 \leq m_b \leq 5.1$ lies just east of the zone as mapped in Figure 4 and might be considered to be associated with it. The earthquake of November 9, 1968, in this zone is the largest earthquake to have occurred so far in the twentieth century in the central United States. Its body-wave magnitude was 5.5 (Stauder and Nuttli, 1970).

Fault-plane solutions are available for two events in this zone (Fig. 7). The more northerly event (April 3, 1974) exhibits strike-slip motion (Herrmann, 1979), whereas the more southerly event is characterized by reverse faulting (Stauder and Nuttli, 1970; Herrmann, 1979). The greatest principal stress direction in this region inferred from the fault-plane solution for the northern event is east-northeast, and that for the southern event is east-southeast.

The Ozark uplift. The Ozark uplift is fractured in the form of a ruptured dome that has undergone repeated mild uplift beginning in Precambrian time. The dome lies predominantly in southern Missouri and northern Arkansas and is centered about 80 km northwest of the Mississippi Embayment (Fig. 3). Geologic evidence indicates that uplift occurred until at least post-Paleozoic time and may still be continuing (McCracken, 1971). Several faults have been mapped in this region; most trend in a northwest-southeast direction, but others trend approximately northeast-southwest and east-west.

Six earthquakes of $m_b \geq 4.5$ have occurred in the Ozark uplift, all of which are located in the eastern region (Fig. 4). Tikrity (1968) recognized three major tectonc elements of the Ozark uplift: the St. Francis dome that trends southwestward across the central part of the uplift, the Ozark arch that trends westward across southern Missouri, and the Decatursville dome in the northwestern part of the uplift. Earthquakes of $m_b \geq 4.5$ have occurred in the first two of these elements, but not the third. Only the St. Francis dome is thought to have been active in recent geologic time, although small events are distributed over most of the uplift (Fig. 1). Since the events in the eastern part of the uplift cannot be related to a single tectonic feature, we have chosen to include the entire Ozark uplift in our list of seismogenic regions at the present time.

Two fault-plane solutions (Fig. 7) have been determined in the Ozark uplift. The events of October 21, 1965 (Mitchell, 1973; Herrmann, 1979), and of July 21, 1967 (Herrmann, 1979), both indicate normal faulting with a nearly vertical greatest-principal-stress axis. The solutions thus suggest that uplift is occurring in the Ozarks at the present time.

The Cincinnati-Findlay arch. Although the Cincinnati arch and the Findlay arch appear as separate arches on tectonic maps, they are considered to form a single seismogenic zone in the present paper. The boundaries of this zone are somewhat arbitrary, but we define this zone as extending through the central lowlands from Tennessee to northern Ohio (Fig. 3). Mapped faults occur only in Kentucky and northern Ohio.

Three earthquakes of $5.2 \leq m_b \leq 5.8$ and four earthquakes of $4.5 \leq m_b \leq 5.1$ have occurred in this zone since 1811 (Fig. 4). It includes the Anna, Ohio, region where a number of moderately damaging earthquakes have occurred (Bradley and Bennett, 1965). The most recent of these occurred on July 12, 1986, with $m_b = 4.5$. Recent earthquakes of $m_b \geq 5.2$ occurred in northern Kentucky (July 27, 1980) and northeastern Ohio (January 31, 1986). Nicholson and others (1988) suggest that the latter event may have been triggered by injection of fluids in deep disposal wells operating within 15 km of the epicentral region.

Fault-plane solutions are available for three earthquakes in this region. The earthquakes of July 27, 1980 (Herrmann and others, 1982), January 31, 1986 (Herrmann and Nguyen, 1986), and July 12, 1986 (Schwartz and Christensen, 1988) all are characterized by strike-slip motion. The events in northern Kentucky and western Ohio give solutions that are consistent with a maximum principal stress axis oriented slightly north of east, and the event in northern Ohio gives a solution consistent with a maximum principal stress axis oriented slightly south of east.

The Kankakee arch. The Kankakee arch extends northwestward from the Cincinnati arch into northern Illinois (Fig. 3). Nuttli and Brill (1981) included this arch with the Cincinnati and Findlay arches to form a single seismogenic zone. We separate it in the present study because of the near absence of seismicity in the southeastern part of this zone (Fig. 1). That lack of seismicity may further suggest that this zone should be restricted to the vicinity of the Sandwich fault zone in northern Illinois. Kolata and others (1978) find, however, that the Sandwich fault zone has been aseismic in historic time; therefore, present-day seismicity must be associated with as-yet-identified subsurface faults in that region. Since we do not know anything about those faults, we have chosen, at least for the present, to designate the entire Kankakee arch region as the seismogenic zone in which earthquakes in this region occur.

Three earthquakes of $4.5 \leq m_b \leq 5.1$ occurred near the northwestern end of the Kankakee arch (Fig. 4). A fault-plane solution was obtained for the earthquake of September 15, 1972 (Herrmann, 1979). It indicates strike-slip faulting and a maximum principal stress direction: northeast-southwest. This direction is significantly more northerly than those inferred from earthquakes in the New Madrid fault zone, the Wabash Valley fault zone, and the Cincinnati-Findlay arch region. As discussed in a later section, however, this solution is considered less reliable than those for most of the other earthquakes.

The Lake Superior tectonic zone. The Lake Superior tectonic zone extends about 1,000 km southwestward from Lake Superior (Fig. 3) and is described by Sims and others (1980) as a zone of distinctive tectonism that coincides with the boundary between two Archean crustal segments, a greenstone-granite terrain to the north and an older gneiss terrain to the south. Although movement along this tectonic zone occurred primarily in the Precambrian, minor differential movements have been indicated by thinning of Cretaceous strata and by historic and recent earthquakes.

Four earthquakes of $4.5 \leqslant m_b \leqslant 5.1$ have been located in this tectonic zone: two in Minnesota and two in South Dakota (Fig. 4). Two additional earthquakes have been located to the southwest of the zone and appear to lie along an extension of it. Brill and Nuttli (1983) placed all six of these earthquakes in a seismogenic zone termed the Colorado lineament. The Colorado lineament was proposed by Warner (1978) as a continuous belt of Precambrian faults extending from northwestern Arizona to central Minnesota. Wong (1983) and Becker and Zeltinger (1983), however, question the continuity of that feature and suggest that segments of it should be considered to be separate regions of earthquake activity in the central United States. We consequently restrict this zone to that region of Minnesota and South Dakota shown in Figure 4. Further study will be required before extension of the Lake Superior tectonic zone to the southwest can be justified.

Only one fault-plane solution has been obtained for this zone (Herrmann, 1979). That event, in western Minnesota, indicates strike-slip motion and a maximum principal stress axis, which is oriented in a north-south direction. That direction contrasts with directions inferred for most of the strike-slip solutions discussed earlier, which range between east-west and northeast-southwest. Although one nodal plane for the fault-plane solution lines up well with the trend of the tectonic zone, this solution is again considered less reliable than most others for reasons discussed in a later section.

The Nemaha uplift. The Nemaha uplift extends for about 600 km from eastern Nebraska southward to the Wichita uplift in central Oklahoma (Fig. 3). It consists of a number of small crustal blocks, typically 5 to 8 km wide and 8 to 30 km long, which were raised sharply along the axis of the uplift (Luza and Lawson, 1979). They were uplifted and eroded during Late Mississippian and Early Pennsylvanian time and later covered by Pennsylvanian and Permian sediments. Faults are mapped near the northern end and near the southern end of the uplift.

Five earthquakes of $4.5 \leqslant m_b \leqslant 5.1$ are associated with this feature (Fig. 4). A sixth occurred near its southern end and may be associated with either the Nemaha uplift or the Wichita uplift. None of these earthquakes occurred in the last two decades; thus, no fault-plane solutions are available for this region.

The Wichita system. The Wichita system, shown on Figure 3, is a band of continental uplifts, which includes the Arbuckle Mountains in south-central Oklahoma, the Wichita uplift proper in southwestern Oklahoma, the Amarillo uplift in the Texas panhandle, and some adjacent troughs. This west-northwest–trending system is the result of forces that affected the region in later Paleozoic time. The basins are sites where great thicknesses of sediments have been deposited (King, 1977). The largest of these is the Anadarko basin, which lies to the north of all three uplifts.

The Wichita Mountains proper and the Anadarko basin are separated by a complex system of northwest-trending faults. Although few earthquakes presently occur there (Lawson, 1985), recent geological studies of the Meers fault in this system indicate that major movement occurred there in late Holocene time (Luza and others, 1987). The surface rupture on the Meers fault is at least 37 km in length, and movement at the surface reaches 5 m vertically and about 20 m horizontally. Using these dimensions and assuming an earthquake repeat time $>2,000$ years, Ramelli and others (1987) conclude that the Meers fault may be capable of producing an earthquake with a surface wave magnitude greater than 7.5.

All earthquakes of $m_b \geqslant 4.5$ have occurred in the westernmost portion of the system in the Amarillo uplift or in the Anadarko basin just to the north (Fig. 4). The uplift is a region of significant oil and gas production. Since all of these earthquakes occurred after 1925 and since oil production began in 1921 (Landes, 1951), it is possible that some seismicity in this region may be related to the withdrawal of oil and gas.

One fault-plane solution is available for the Wichita system (Herrmann, 1979). That solution (Fig. 7) is for an earthquake in the Anadarko basin and indicates predominant reverse faulting in a stress field in which the greatest principal stress axis is oriented approximately east-west.

The Ouachita system. Whereas the Wichita system was formed within the continent, the Ouachita system was formed outside the continental shelf. The frontal part of the exposed Ouachitas includes thrust faults that are directed toward the continental craton (Thomas, 1985). The boundary between the central uplift and frontal thrust faults has been interpreted by Viele (1979) to be part of a subduction complex. The Ouachita seismic zone, for purposes of this chapter, is taken to include the Ouachita Mountains exposed in southeastern Oklahoma (Fig. 3) as well as a region south of the Ouachita frontal thrust that extends eastward through Mississippi (Fig. 4). The portion of the Ouachita front extending through Texas has been devoid of earthquakes of $m_b \geqslant 4.5$ in historic and recent times. Four earthquakes of $4.5 \leqslant m_b \leqslant 5.1$ have occurred in this zone since 1811. Two of these occurred near the frontal thrust and two were far to the south of it.

Herrmann (1979) obtained two fault-plane solutions for this region. The event of June 4, 1967, located about 100 km south of the thrust, exhibits predominant strike-slip faulting. The maximum principal stress axis is oriented approximately east-west. The earthquake of January 1, 1969, is located very close to the frontal thrust in central Arkansas. It exhibits reverse faulting along a fault oriented in a northeasterly direction. The maximum principal stress axis inferred from this solution is oriented in a northwesterly direction about 60° from the normal direction to the thrust front. This event, however, is another case where the solution is considered less reliable than others.

The largest earthquake swarm ever recorded in the central or eastern United States occurred just north of the Ouachita frontal thrust beginning in January 1982 (Chiu and others, 1984). The Tennessee Earthquake Information Center recorded more than 30,000 earthquakes concentrated in a volume of only about 45 km^2. Composite focal mechanisms obtained by Chiu and others (1984) indicate predominantly strike-slip motion, with the maximum principal stress axis oriented northeast-southwest.

Summary of seismicity patterns. Patterns of earthquakes of $m_b \geqslant 3.0$ throughout the central United States include only one clear major concentration of activity (Fig. 1). That concentration lies in the New Madrid fault zone, adjacent regions of the Wabash Valley fault zone, and small regions of the Illinois basin and Ozark uplift. A few minor concentrations—such as those near Anna, Ohio, in the Texas panhandle, and in central Oklahoma—also occur, but these are difficult to distinguish from surrounding, more random activity.

Most earthquakes of $m_b \geqslant 4.5$ can be associated with 11 seismogenic regions (Fig. 4), which can be classified as rift zones, uplifts, basins, or former plate boundaries. Some of these are extensive linear features; we do not, however, see any compelling evidence for long throughgoing seismically active zones such as those postulated to extend from southeastern Missouri to the St. Lawrence Valley (Woollard, 1958).

It seems fair to say that earthquakes of m_b less than 4.5 might occur almost anywhere in the central United States, but that their frequency of occurrence will not be the same, even in the regions outside the so-called source zones. Inasmuch as earthquakes of m_b less than 4.5 seldom cause any damage, unless they are of very shallow depth and are situated immediately beneath a town, they are primarily of academic interest. However, a word of caution needs to be added. Professor Xie Yushou of the Institute of Geophysics of the State Seismological Bureau of China (personal communication, 1985) noted that the epicentral region of the great 1556 earthquake in Shanshi Province of China, which is reported to have killed more than 800,000 people, is at the present time seismically quiet, with only a few microearthquakes occurring there per year. Therefore, we have to question whether our 100- to 200-year history of earthquake activity in the central United States is adequate to identify the locations of all future large earthquakes.

Earthquake activity and crustal stresses in the central United States

Most of the earthquake focal mechanisms in Figure 7 are those that would be expected for a stress field where the maximum principal stress axis is oriented between east-west and northeast-southwest. The consistency of these directions and those from in situ stress measurements led Zoback and Zoback (1980) to identify this region as the Mid-continent Stress Province within which the crustal stress field is relatively uniform. Gough and others (1983) found evidence that it extends northward to the Arctic coast in Canada. The east-west direction for the maximum principal stress axis is consistent with observations of crustal shortening between Massachusetts and Texas observed using Very Long Baseline Interferometry (Carter and others, 1985).

Some mechanisms of Figure 7, however, are clearly inconsistent with the idea of a uniform stress field everywhere in the central United States. The two mechanisms in the Ozark uplift both have a vertically oriented maximum principal stress axis; their least principal stress axes are both horizontal, but differ by about 90°. This suggests that vertical compression is the dominant component of the stress field in the Ozarks. Thus, the fault-plane solutions in this region suggest that uplift is still going on in the Ozarks.

Other mechanisms for which the principal maximum stress directions differ from east-west or northeast-southwest are the event of July 9, 1975, in Minnesota, the event of September 15, 1972, in northern Illinois, and the event of January 1, 1969, in Arkansas. Herrmann (1979) indicates that the solutions for all three of these events provide poor agreement with portions of the data set that he used. These solutions are therefore thought to be more poorly constrained than most of the others, and the orientations of the stress fields in those regions may differ greatly from those inferred from the deduced fault-plane solutions shown in Figure 7.

Another region where a fault-plane solution is available is the Gulf coast of southern Louisiana. An earthquake there of m_b = 3.8 on October 16, 1983, yielded a solution that implies normal faulting on a northeast-trending fault (Stevenson and Agnew, 1988). This orientation agrees with that of mapped growth faults in the region and implies north-south extension and a vertically oriented maximum princpal stress axis. Thus, the orientation of the stress field in this part of the Gulf coast region appears to differ from that throughout most of the central United States.

Anderson (1986) estimated strain rates throughout the central and eastern United States from historical seismicity and from several seismicity models. He estimated typical rates to be on the order of 10^{-12} to 10^{-11} per year, but that higher rates occur near sites of historical earthquakes. The rate in the New Madrid region is estimated as being nearly 10^{-8} per year. Anderson suggests that denudation and deposition rates may control the rate of earthquake occurrence in some parts of the central and eastern United States by modifying the vertical stress due to load. If these mechanisms do control the rate of earthquake activity in some regions, the New Madrid fault zone is possibly such a region, since it lies in the Mississippi Embayment where sediments are accumulating.

EVOLUTION AND STRUCTURE OF THE NEW MADRID FAULT ZONE

In recent years, much has been learned about the structure and tectonic history of the New Madrid region. Burke and Dewey (1973) suggested that the upper Mississippi Embayment is the failed arm of a triple junction, and Ervin and McGinnis (1975) interpreted gravity data for the region to infer the existence of a rift structure, termed the Reelfoot rift, which correlates spatially with the zone of greatest earthquake activity. They suggested that it was formed in the late Precambrian–early Paleozoic and was reactivated in Mesozoic time. Ervin and McGinnis (1975) and Cordell (1977) interpreted a positive gravity anomaly as being caused by a mass excess in the lower crust of the New Madrid region, which is due to dense mateials intruded during rift formation. Using magnetic data, Hildenbrand and others (1977) delineated the edges of a graben within the rift region and found it to be approximately 70 km wide by 200 km long, trending in a northeast-southwest direction; it offsets the Precambrian surface by about 2 km. The most northeasterly segment of the active portion of the New Madrid fault complex lies close to a portion of the northwestern boundary of the graben (Fig. 5), and a few earthquakes lie along the southeastern boundary, but most of the earthquakes lie between the two boundaries.

Hildenbrand and others (1977) also identified several mafic or ultramafic intrusions in the New Madrid region that occurred along the edges of the New Madrid graben and inferred that they are of Mesozoic age. They are thought to be due to reactivation of the Reelfoot rift during a period of extension.

Zoback and others (1980) and Hamilton and Zoback (1982) reported the results of extensive seismic reflection profiling in the region of the New Madrid seismic zone. Vertical offset of Paleozoic rock along the seismic trend extending southwestward into Arkansas (Fig. 5) is about 1 km. The southern portion of the north-northwest–trending zone shows about 80 m of high-angle reverse faulting. Their work indicates, following the formation of the Reelfoot rift in late Precambian or Early Cambrian, that some of the faults in the New Madrid region were active through the Paleozoic. The main episode of deformation and fault activity occurred in the late Paleozoic and was a period of mostly extensional motion on faults north of the embayment. Since the middle Tertiary, faulting has occurred in an east-west–oriented compressional regime. Those authors conclude that modern seismicity results from east-west compression and occurs along fault zones and zones weakened by intrusive activity that originally formed as a result of rifting.

Braile and others (1982), on the basis of seismicity, gravity data, and magnetic data, postulated that the Reelfoot rift splits into three distinct arms near the southern tip of Illinois. One arm continues about 300 km northeastward into Indiana, one extends about 250 km northwestward to near St. Louis, and the other extends about 150 km eastward into Kentucky. Braile and others envision these arms to be outlined by igneous intrusions, which they infer from magnetic data. They conclude that the evolution of this rift system has been marked by several periods of extension and compression since the late Precambrian.

Al-Shukri and Mitchell (1987) analyzed traveltimes from several earthquakes at both teleseismic and regional distances in the region surrounding the New Madrid seismic zone. Using the method of Aki and others (1977), they inferred a three-dimensional velocity structure of the crust and upper mantle. Also, they verified earlier findings by Mitchell and others (1977) that seismic velocities in the New Madrid region are predominantly lower than average immediately below the region of greatest seismic activity and that anomalous velocity structure extends through the entire crust and upper mantle to depths of at least 150 km. Al-Shukri and Mitchell (1987) found several details in velocity structure that coincide with features inferred from gravity, magnetic, and geological studies. They found that slightly higher than average velocities coincide with the portion of the lower crust in which densities were found to be higher than average and also that somewhat higher than average velocities correspond with many of the mafic or ultramafic intrusions inferred in earlier studies. Low velocities occur in the upper crust, coinciding roughly in spatial extent with the graben mapped by Hildenbrand and others (1977).

In a more detailed study of velocities in the upper crust in and near the New Madrid fault complex, Al-Shukri and Mitchell (1988) found a striking correlation of low seismic velocities with the active portions of the New Madrid fault complex. They inverted 1,393 observations of traveltimes from 211 local earthquakes to infer seismic velocities in that region. Figure 8 shows variations of seismic velocities in the sedimentary section (taken to be 5 km thick) and in the upper crystalline crust to a depth of 14 km. The number of earthquakes shown in that figure is a subset of the total number of earthquakes shown in Figure 5 for which 10 stations or more were used for the epicenter and depth determinations. Both portions of the crust were divided into blocks of 14×14 km, and velocities were obtained using a modification of the method of Aki and Lee (1977) for local earthquakes. The maximum reductions in compressional wave velocity are at least 7 percent in the sedimentary layer and at least 4 percent in the crystalline crust. Al-Shukri and Mitchell (1988) interpreted the low seismic velocities as being due to fluid-filled cracks in the regions of most intense tectonic activity. They found that the reductions in velocity are consistent with those of laboratory measurements of Nur and Simmons (1969) if pore pressure at depth is about one-half or more of the total pressure. McKeown (1982) has suggested that volatile content in the New Madrid region may be enhanced because of characteristics associated with a rift. Such regions are highly faulted and are associated with an abundance of alkaline rocks with a high content of

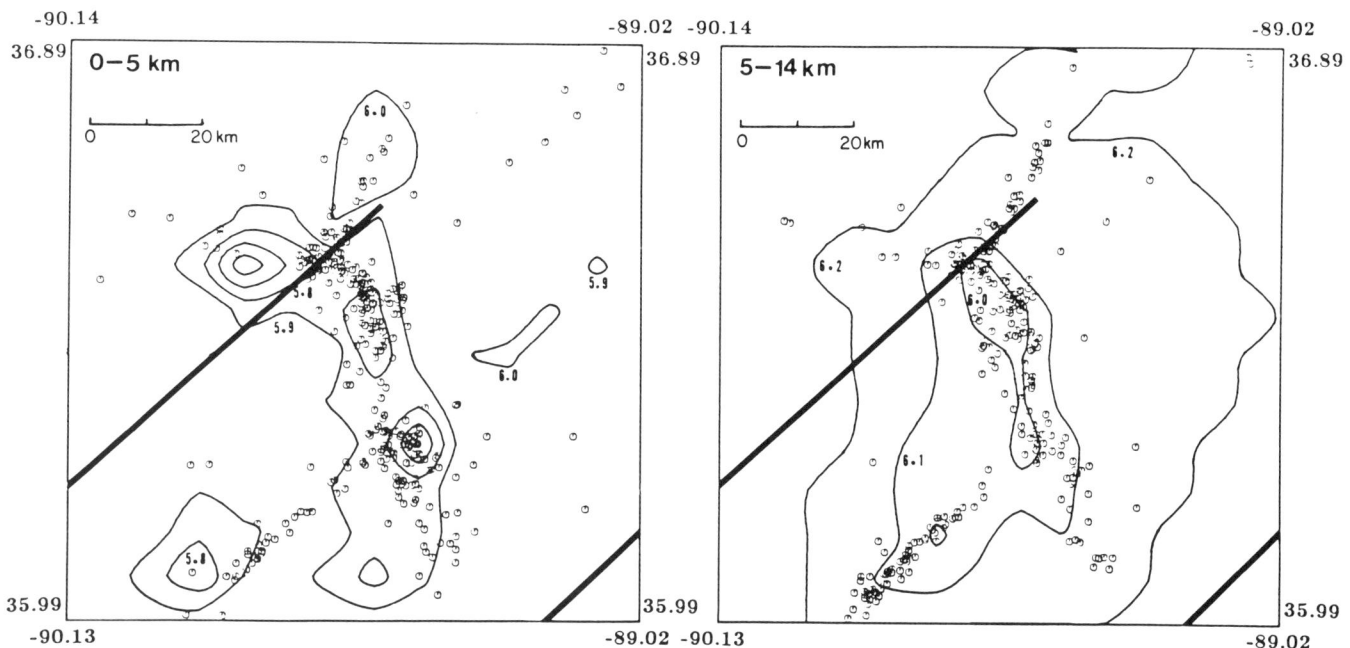

Figure 8. A three-dimensional model of the New Madrid seismic zone. Left, compressional-wave velocities in the upper 5 km of the crust. Right, compressional-wave velocities at depths between 5 and 14 km. The contour interval is 0.1 km/s. Circles denote earthquakes that occurred within each layer between 1974 and 1985 and were located using 10 or more stations. The heavy straight lines are approximate boundaries of a graben inferred from gravity and magnetic data (modified from Al-Shukri and Mitchell, 1988).

volatiles. He proposed that the enhanced volatile content results in higher porosity and pore pressure in the rift region than in adjacent regions. Thus, in addition to causing lower velocities, this situation is conducive for the generation of earthquakes.

Four heat-flow measurements and several hundred bottom-hole temperature measurements indicate a modest heat flow anomaly in the New Madrid region (Swanberg and others, 1982). Although the magnitude and spatial extent of this anomaly can be explained as being caused by a large intrusion of Eocene age (Mitchell and others, 1984), it might also be caused by convective flow of fluids in the fault zone or frictional heating on the fault.

The most widely accepted model for earthquake activity in the New Madrid seismic zone is one wherein earthquakes result from the release of accumulated stress along old lines of weakness in the crust. This mechanism was proposed by Sbar and Sykes (1973) and Sykes (1978) and was invoked to explain New Madrid earthquakes by Zoback and Zoback (1980) and Braille and others (1982). According to that mechanism, in order for earthquakes to occur, the fault zones should be appropriately oriented in the stress field in each region. Until the work of Al-Shukri and Mitchell (1988), a remaining question was why most of the earthquakes in the New Madrid region are restricted to a relatively small portion of the faulted area there. Several mapped faults north of the Mississippi Embayment are oriented in the same direction as the active New Madrid faults, but are nearly devoid of earthquake activity. The work of Al-Shukri and Mitchell (1988) suggests that the reason may be that the concentration of interstitial fluids in the active portions of the fault complex is greater than that in the inactive portions.

CONCLUSIONS

Minor earthquake activity occurs throughout the entire central United States. Much greater than average activity can, however, occur in more restricted regions. Seismogenic zones are associated with a variety of tectonic features including rifts, uplifts, basins, and former plate boundaries. The locations of earthquakes of $m_b \geqslant 4.5$ lead us to classify 11 zones in the central United States as seismogenic. These are the New Madrid fault zone, the Wabash Valley fault zone, the Ozark uplift, the Illinois basin, the Michigan basin, the Cincinnati-Findlay arch, the Kankakee arch, the Lake Superior tectonic zone, the Nemaha uplift, the Wichita seismic zone, and the Ouachita seismic zone. This classification may be modified as our record of earthquake activity becomes longer and as more detailed geological information is obtained. Activity in the New Madrid seismic zone dominates the earthquake history of the central United States. An extraordinary sequence of events that occurred there in the winter of 1811–1812 included four of the largest events ever to have been recognized in North America. The locations of historic and recent earthquakes that can be accurately obtained suggest larger earthquakes occur predominantly near the ends of branches of the active part of the New Madrid seismic zone.

Most earthquakes in the central United States have fault-plane solutions that would be expected if the stress field is oriented so that the greatest principal stress direction is between east-west and northeast-southwest. Exceptions occur in the Ozark uplift where the greatest principal stress is close to vertical, and along the Ouachita frontal thrust where it is oriented northwest-southeast. These directions may indicate that the Ozarks are still uplifting and that thrust motion is still occurring on the Ouachita front. Another exception is the Gulf Coast region where normal faulting occurs along growth faults and the greatest principal stress axis is oriented vertically.

Gravity, magnetic, and seismic reflection data have provided evidence that the New Madrid fault zone is a reactivated ancient rift that has undergone a complex history since Precambrian time. Extrapolation of present-day fault movement into the past predicts extensive surface deformation, which is not observed. Fault motion at the prsent time must therefore be much greater than that which occurred prior to the late Pleistocene. The most active portions of the New Madrid fault zone are characterized by low seismic velocities, which can be explained by the presence of interstitial fluids in cracks in the crust. The presence of such fluids may explain why some faults in intraplate regions are active while others are not.

REFERENCES CITED

Aki, K., and Lee, W.H.K., 1976, Determination of the three-dimensional velocity anomalies under a seismic array using first P arrival times from local earthquakes, 1. Homogeneous initial model: Journal of Geophysical research, v. 81, p. 4381–4399.

Aki, K., Christoffersson, K., and Husebye, E. S., 1977, Determination of the seismic structure of the lithosphere: Journal of Geophysical Research, v. 82, p. 277–269.

Al-Shukri, H. J., and Mitchell, B. J., 1987, Three-dimensional velocity variations and their relation to the structure and tectonic evolution of the New Madrid seismic zone: Journal of Geophysical Research, v. 92, p. 6377–6390.

—— , 1988, Reduced seismic velocities in the source zone of New Madrid earthquakes: Bulletin of the Seismological Society of America, v. 78, p. 1491–1509.

Anderson, J. G., 1986, Seismic strain rates in the central and eastern United States: Bulletin of the Seismological Society of America, v. 76, p. 273–290.

Andrews, M. C., Mooney, W. D., and Meyer, R. P., 1985, The relocation of microearthquakes in the northern Mississippi Embayment: Journal of Geophysical Research, v. 90, p. 10223–10236.

Becker, D. J., and Zeltinger, J. M., 1983, Comment on 'Seismicity of the Colorado Lineament': Geology, v. 11, p. 559–560.

Bradley, E. A., and Bennett, T. J., 1965, Earthquake history of Ohio: Bulletin of the Seismological Society of America, v. 55, p. 745–752.

Braile, L. W., Keller, G. R., Hinze, W. J., and Lidiak, E. G., 1982, An ancient rift complex and its relation to contemporary seismicity in the New Madrid seismic zone: Tectonics, v. 1, p. 225–237.

Brill, K. G., and Nuttli, O. W., 1983, Seismicity of the Colorado Lineament: Geology, v. 11, p. 20–24.

Bristol, H. M., and Treworgy, J. D., 1979, The Wabash Valley fault system in southeastern Illinois: Illinois State Geological Survey Circular 509, 19 p.

Burke, K., and Dewey, J. F., 1973, Plume-generated triple junction; Key indications in applying tectonics to old rocks: Journal of Geology, v. 81, p. 406–433.

Carter, W. E., Robertson, D. S., and MacKay, J. R., 1985, Geodetic radio interprometric surveying: Applications and results: Journal of Geophysical Research, v. 90, p. 4577–4587.

Chiu, J. M., Johnston, A. C., Metzger, A. G., Haar, L., and Fletcher, J., 1984, Analysis of analog and digital records of the 1982 Arkansas earthquake swarm: Bulletin of the Seismological Society of America, v. 74, p. 1721–1742.

Cordell, L., 1977, Regonal positive gravity anomaly over the Mississippi Embayment: Geophysical Research Letters, v. 4, p. 285–287.

Ervin, C. P., and McGinnis, L. D., 1975, Reelfoot rift-reactivated precursor to the Mississippi Embayment: Geological Society of America Bulletin, v. 86, p. 1287–1295.

Evernden, J. F., 1967, Magnitude determination at regional and near-regional distances in the United States: Bulletin of the Seismological Society of America, v. 57, p. 591–639.

Fuller, M. L., 1912, The New Madrid earthquake: U.S. Geological Survey Bulletin 494, 119 p.

Gough, D. I., Fordjor, C. K., and Bell, J. S., 1983, A stress province boundary and tractions on the North American plate: Nature, v. 305, p. 619–621.

Gutenberg, B., 1945a, Amplitudes of P, PP, and S and magnitudes of shallow earthquakes: Bulletin of the Seismological Society of America, v. 35, p. 57–69.

—— , 1945b, Magnitude determinations for deep-focus earthquakes: Bulletin of the Seismological Society of America, v. 35, p. 117–130.

Gutenberg, B., and Richter, C. F., 1936, On seismic waves, 3: Gerland's Beitraege zur Geophysik, v. 47, p. 73–131.

Hamilton, R. M., and Zoback, M. D., 1982, Tectonic features of the New Madrid seismic zone from seismic-reflection profiles, in McKeown, F. A., and Pakiser, L. C., eds., Investigations of the New Madrid earthquake region: U.S. Geological Survey Professional Paper 1236, p. 31–38.

Herrmann, R. B., 1979, Surface wave focal mechanisms for eastern North American earthquakes with tectonic implications: Journal of Geophysical Research, v. 84, p. 3547–3552.

Herrmann, R. B., and Canas, J., 1978, Focal mechanism studies in the New Madrid seismic zone: Bulletin of the Seismological Society of America, v. 68, p. 1095–1102.

Herrmann, R. B., and Nguyen, B. V., 1986, Focal mechanism studies of the January 31, 1986, Perry, Ohio, earthquake [abs.]: Earthquake Notes, v. 57, p. 107.

Herrmann, R. B., Langston, C. A., and Zollweg, J. E., 1982, The Sharpsburg, Kentucky, earthquake of 27 July 1980: Bulletin of the Seismological Society of America, v. 60, p. 973–981.

Heyl, A. V., and McKeown, F. A., 1978, Preliminary seismotectonic map of central Mississippi valley and environs: U.S. Geological Survey Miscellaneous Field Studies Map MF-1011, scale 1:500,000.

Hildenbrand, T. G., Kane, M. F., and Stauder, W., 1977, Magnetic and gravity anomalies in the northern Mississippi Embayment and their relation to seismicity: U.S. Geological Survey Miscellaneous Field Study Map MF-914, scale 1:1,000,000.

Hook, J. W., 1974, Structure of the fault systems in theIllinois–Kentucky fluorspar district, in Hutchinson, D. W., ed., A symposium on the geology of fluorspar; Proceedings of the ninth forum on geology of industrial minerals: Kentucky Geological Survey, Series X, Special Publication 22, p. 77–86.

Johnston, A. C., and Nava, S. J., 1985, Recurrence rates and probability estimates of the New Madrid seismic zone: Journal of Geophysical Research, v. 90, p. 6737–6753.

King, P. B., 1977, The evolution of North America (revised edition): Princeton, New Jersey, Princeton University Press, 197 p.

Kolata, D. R., Treworgy, J. D., and Buschbach, T. C., 1978, The Sandwich fault zone of northern Illinois: Illinois Geological Survey Circular 505, 26 p.

Landes, K. K., 1951, Petroleum geology: New York, Chapman and Hall, Ltd., 660 p.

Lawson, J. E., Jr., 1985, Seismicity at the Meers fault [abs.]: Earthquake Notes, v. 56, no. 1, p. 2.

Luza, K. V., and Lawson, J. E., 1979, Seismicity and tectonic relationships of the Nemaha uplift in Oklahoma, Part 2: Oklahoma Geological Survey Report NUREG/CR-0875, prepared for the U.S. Nuclear Regulatory Commission, 81 p.

Luza, K. V., Madole, R. F., and Crone, A. J., 1987, Investigations of the Meers fault, southwestern Oklahoma: Oklahoma Geological Survey Special Publication 87-1, 75 p.

McCracken, M. H., 1971, Structural features of Missouri: Missouri Geological Survey and Water Resources Report of Investigations 49, 99 p.

McKeown, F. A., 1982, Overview and discussion, in McKeown, F. A., and Pakiser, L. C., eds., Investigations of the New Madrid earthquake region: U.S. Geological Survey Professional Paper 1236, p. 1–14.

Mitchell, B. J., 1973, Radiation and attenuation of Rayleigh waves from the southeastern Missouri earthquake of October 21, 1965: Journal of Geophysical Research, v. 78, p. 886–899.

—— , 1975, Regional Rayleigh wave attenuation in North America: Journal of Geophysical Research, v. 80, p. 4904–4916.

Mitchell, B. J., and Hwang, H. J., 1987, Effect of low-Q sediments and crustal Q on Lg attenuation in the United States: Bulletin of the Seismological Society of America, v. 66, p. 1197–1210.

Mitchell, B. J., Cheng, C. C., and Stauder, W., 1977, Three-dimensional velocity model of the lithosphere beneath the New Madrid seismic zone: Bulletin of the Seismological Society of America, v. 67, p. 1061–1074.

Mitchell, B. J., Kohsmann, J., and Al-Shukri, H. J., 1984, Anomalous crust and upper mantle properties in the New Madrid area, in Gori, R. L., and Hays, W. W., eds., Proceedings of the Symposium on the New Madrid Seismic Zone: U.S. Geological Survey Open-File Report 84-770, p. 33–63.

Nicholson, C., Roeloffs, E., and Wesson, R. L., 1988, The northeastern Ohio earthquake of 31 January 1986; Was it induced?: Bulletin of the Seismological Society of America, v. 78, p. 188–217.

Nur, A., and Simmons, G., 1969, The effect of saturation on velocity in low porosity rocks: Earth and Planetary Science Letters, v. 7, p. 183–193.

Nuttli, O. W., 1973a, Seismic wave attenuation and magnitude relations for eastern North America: Journal of Geophysical Research, v. 78, p. 876–885.

—— , 1973b, The Mississippi Valley earthquakes of 1811–1812; Intensities, ground motion, and magnitudes: Bulletin of the Seismological Society of America, v. 63, p. 227–248.

—— , 1974, Magnitude recurrence relations for central Mississippi valley earthquakes: Bulletin of the Seismological Society of America, v. 64, p. 1189–1207.

—— , 1981a, Similarities and differences between western and eastern United States earthquakes and their consequences for earthquake engineering, in Beavers, J. G., ed., Earthquakes and earthquake engineering; Eastern United States, vol. 1: Ann Arbor, Michigan, Science Press, p. 25–52.

—— , 1981b, Catalog of central United States earthquakes since 1800 of m_b ≥3.0, Appendix B2, in Approach to seismic zonation for siting nuclear electric power generating facilities in the eastern United States: U.S. Nuclear Regulatory Commission Report NUREG/CR-1577, p. B2-1–B2-31.

—— , 1983, Average seismic source-parameter relations for mid-plate earthquakes: Bulletin of the Seismological Society of America, v. 73, p. 519–535.

Nuttli, O. W., and Brill, K. G., 1981, Earthquake source zones in the central United States determined from historical seismicity, in Barstow, N. L., Brill, K. G., Nuttli, O. W., and Pomeroy, P. W., eds., Approach to seismic zonation for siting nuclear electric power generating facilities in the eastern United States: U.S. Nuclear Regulatory Commission Report NUREG/CR-1577, p. 98–143.

Nuttli, O. W., and Herrmann, R. B., 1984, Source characteristics and strong ground motion of New Madrid earthquakes, in Gori, R. L., and Hays, W. W., eds., Proceedings of the Symposium on the New Madrid Seismic Zone: U.S. Geological Survey Open-File Report 84-770, p. 330–352.

Nuttli, O. W., and Zollweg, J. W., 1974, The relation between felt area and magnitude for the central United States: Bulletin of the Seismological Society of America, v. 64, p. 73–85.

Nuttli, O. W., Bollinger, G. A., and Griffiths, D. W., 1979, On the relation between modified Mercalli intensity and body-wave magnitude: Bulletin of the Seismological Society of America, v. 69, p. 893–909.

Nuttli, O. W., Jost, M., Herrmann, R. B., and Bollinger, G. A., 1989, Numerical modeling of the 1886 Charleston, South Carolina, earthquake: U.S. Geological Survey Bulletin 1586 (in press).

O'Connell, D., Bufe, C. G., and Zoback, M. D., 1982, Microearthquakes and faulting in the area of New Madrid, Missouri–Reelfoot Lake, Tennessee, in McKeown, F. A., and Pakiser, L. C., eds., Investigations of the New Madrid earthquake region: Geological Survey Professional Paper 1236, p. 31–38.

Ramelli, A. R., Slemmons, D. B., and Brocoum, S. J., 1987, The Meers fault: Tectonic activity in southwestern Oklahoma: U.S. Nuclear Regulatory Commission Report NUREG/CR-4852, 50 p.

Richter, C. F., 1958, Elementary seismology: San Francisco, California, W. H. Freeman and Company, 768 p.

Russ, D. P., 1979, Late Holocene faulting and earthquake recurrence in the Reelfoot Lake area, northwestern Tennessee: Geological Society of America Bulletin, part 1, v. 90, p. 1013–1018.

Sbar, M. L., and Sykes, L. R., 1973, Contemporary compressive stress and seismicity in eastern North America; An example of intraplate tectonics: Geological Society of America Bulletin, v. 84, p. 1861–1882.

Schwartz, S. Y., and Christensen, D. H., 1988, The 12 July 1986 St. Mary's, Ohio, earthquake and recent seismicity in the Anna, Ohio, seismogenic zone: Seismological Research Letters, v. 59, p. 57–62.

Sims, P. K., Card, K. D., Morey, G. B., and Peterman, Z. E., 1980, The Great Lakes tectonic zone; A major crustal structure in central North America: Geological Society of America Bulletin, part I, v. 91, p. 690–698.

Singh, S., and Herrmann, R. B., 1983, Regionalization of crustal coda Q in the continental United States: Journal of Geophysical Research, v. 88, p. 527–538.

Stauder, W., and Nuttli, O. W., 1970, Seismic studies; South-central Illinois earthquake of November 9, 1968: Bulletin of the Seismological Society of America, v. 60, p. 973–981.

Stauder, W., Kramer, M., Fischer, G., Schaefer, S., and Morrissey, S. T., 1976, Seismic characteristics of southeast Missouri as indicated by a regional telemetered microearthquake array: Bulletin of the Seismological Society of America, v. 66, p. 1953–1964.

Stevenson, D. A., and Agnew, J. D., 1988, Lake Charles, Louisiana, earthquake of 16 October 1983: Bulletin of the Seismological Society of America, v. 78, p. 1463–1474.

Street, R. L., 1982, A contribution to the documentation of the 1811–1812 Mississippi valley earthquake sequence: Earthquake Notes, v. 53, p. 39–52.

Swanberg, C. A., Mitchell, B. J., Lohse, R. L., and Blackwell, D. D., 1982, Heat flow in the upper Mississippi Embayment, in McKeown, F. A., and Pakiser, L. C., eds., Investigations of the New Madrid earthquake regions: U.S. Geological Survey Professional Paper 1236, p. 185–189.

Sykes, L. R., 1978, Intraplate seismicity, reactivation of pre-existing zones of weakness, alkaline magmatism, and other tectonism postdating continental fragmentation: Reviews of Geophysics and Space Physics, v. 16, p. 621–688.

Thomas, W. A., 1985, The Appalachian–Ouachita connection; Paleozoic orogenic belt at the southern margin of North America: Annual Reviews of Earth and Planetary Sciences, v. 13, p. 175–199.

Tikrity, S. S., 1968, Tectonic genesis of the Ozark uplift [Ph.D. thesis]: Seattle, Washington University, 196 p.

Trace, R. D., 1974, Illinois–Kentucky fluorspar district, in Hutcheson, D. W., ed., A symposium on the geology of fluorspar; Proceedings of the ninth forum on geology of industrial minerals: Kentucky Geological Survey, Series X, Special Publication 22, p. 58–76.

Viele, G. W., 1979, Geologic map and cross section, eastern Ouachita Mountains, Arkansas: Geological Society of America Map Chart Series MC-28F, 8 p., scale 1:250,000.

Warner, L. A., 1978, The Colorado lineament; A middle Precambrian wrench fault system: Geological Society of America Bulletin, v. 89, p. 161–171.

Wong, I. G., 1983, Comment *on* 'Seismicity of the Colorado Lineament': Geology, v. 11, p. 558–559.

Woollard, G. P., 1958, Areas of tectonic activity in the United States as indicated by earthquake epicenters: EOS Transactions of the American Geophysical Union, v. 49, p. 1135–1150.

Zoback, M. D., 1979, Recurrent faulting in the vicinity of Reelfoot Lake, northwestern Tennessee: Geological Society of America Bulletin, part 1, v. 90, p. 1019–1024.

Zoback, M. L., and Zoback, M., 1980, State of stress in the conterminous United States: Journal of Geophysical Research, v. 85, p. 6113–6156.

Zoback, M. L., and 5 others, 1980, Recurrent intraplate tectonism in the New Madrid seismic zone: Science, v. 209, p. 971–976.

MANUSCIPT ACCEPTED BY THE SOCIETY MARCH 20, 1989

ACKNOWLEDGMENTS

We thank Kenneth Brill of the Department of Earth and Atmospheric Sciences, St. Louis University, Paul Heigold and Dennis Kolata of the Illinois Geological Survey, Jerry Vineyard of the Missouri Division of Geology and Land Survey, and George Viele of the Department of Geology and Geophysics, University of Missouri, for helpful discussions on the geology of various regions of the central United States. This chapter benefitted from constructive reviews by Frank McKeown of the U.S. Geological Survey and by Arch Johnston of the Center for Earthquake Research and Information, Memphis State University. Mr. Haydar Al-Shukri assisted in preparing some of the figures.

This work was partially supported by the U.S. Geological Survey under Contract 14-08-0001-21262 and by the U.S. Nuclear Regulatory Commission under Contract NRC-04-86-121.

Printed in U.S.A.

Chapter 14

The seismicity and seismotectonics of eastern Canada

John Adams and Peter Basham
Geophysics Division, Geological Survey of Canada, 1 Observatory Crescent, Ottawa, Ontario K1A 0Y3, Canada

INTRODUCTION

Eastern Canada—Canada east of the Cordillera, and extending north from the United States border to the Arctic Ocean—contains about two-thirds of the stable craton of the North American plate. Much of this large area appears to be substantially aseismic, although it contains several zones of significant seismicity and a few other regions of lower-level seismicity. The seismicity of the southern part, together with the adjacent United States, was compiled comprehensively by Smith (1962, 1966), who collected earthquake reports compiled by others, analyzed original records where possible, and decided on the best location and magnitude for each earthquake. Smith's maps have been widely used by others (e.g., Yang and Aggarwal, 1981). Further earthquake analysis, including spatial distribution, recurrence rates, and relationship to geologic structure, was made by Basham and others (1979) and refined by Basham and others (1982a). The latter paper, although the most thorough seismicity compilation to date, was compiled for an engineering seismic-hazard study and is not widely known or circulated, although the conclusions were published (Basham and others, 1985). Other recent reviews are Hasegawa and others (1985), which deals with the crustal stresses driving the eastern Canadian seismicity, and Hasegawa (1986), which discusses the seismotectonics of southeastern Canada.

Within the southern part of the continental region, seismicity is clustered in four zones. In three of these zones—western Quebec, which includes a band of earthquakes along the Ottawa River and a second band north of the river; Charlevoix, a repetitive source of large earthquakes and a continuous source of small earthquakes; and the lower St. Lawrence, a diffuse zone of mostly small earthquakes—most of the earthquakes are thrust events occurring at depths of 5 to 25 km within the Grenville cratonic basement, apparently chiefly through the reactivation of a Paleozoic rift fault system along the St. Lawrence and Ottawa Rivers. The fourth zone, the northern Appalachians, includes the Miramichi earthquakes of 1982, which were due to shallow (<10 km) thrust faulting within a sheet of rocks that have been thrust over the older basement.

Along the eastern margin of the continent (North Atlantic Ocean, Labrador Sea, Baffin Bay), the seismicity includes the 1929 M7.2 Grand Banks and 1933 M7.3 Baffin Bay earthquakes. These and smaller earthquakes appear to be concentrated at the ocean-continent transition, perhaps by reactivation of the Mesozoic rift faults created when the North Atlantic was formed. In the Labrador Sea, earthquakes also occur on the extinct spreading ridge and its associated transform faults.

The seismicity of northeastern Canada was poorly known until the installation of the standard seismograph stations in the early 1960s. After about 15 years of lower magnitude seismicity had been recorded, Basham and others (1977) attempted the first comprehensive assessment of northern Canadian seismicity. Although a number of earthquakes in smaller sub-regions have since been described in more detail, as will be outlined below, much of our knowledge of the seismicity and seismotectonics of northern Canada has not advanced greatly beyond that available in 1977.

Within the Decade of North American Geology (DNAG) series, chapters by Zoback and others (1986) and Fujita and others (1989) discuss aspects of eastern Canadian seismicity; Rogers and Horner (this volume) discuss the seismicity of western Canada, and other chapters in this volume discuss the seismicity of adjacent regions.

In this chapter, we discuss briefly the regional seismicity and relate it to the seismotectonics as it is currently understood. Developments to our understanding have happened faster in the more accessible southern part of eastern Canada monitored by digital seismographs, so that description is more extended. We begin in southern Canada, discussing each cluster of seismicity in turn, and proceed from eastern Ontario, down the St. Lawrence River, to the earthquakes along the Atlantic margin, and the earthquakes of the Arctic, before summarizing our inferences about the causes of earthquakes in eastern Canada.

Some of the ideas we support have previously been suggested by others, e.g., Woollard (1969) and Kumarapeli and Saull (1966) for the activity of the St. Lawrence Valley, and Sykes (1978) for the trend between Boston and Ottawa. We believe that many reached approximately the correct conclusions, but that our improved knowledge of the seismicity, together with improvements to the understanding of the geologic history of eastern Canada, have allowed us to substantially refine the earlier ideas.

We now feel that all the earthquakes in eastern Canada

Adams, J., and Basham, P., 1991, The seismicity and seismotectonics of eastern Canada, *in* Slemmons, D. B., Engdahl, E. R., Zoback, M. D., and Blackwell, D. D., eds., Neotectonics of North America: Boulder, Colorado, Geological Society of America, Decade Map Volume 1.

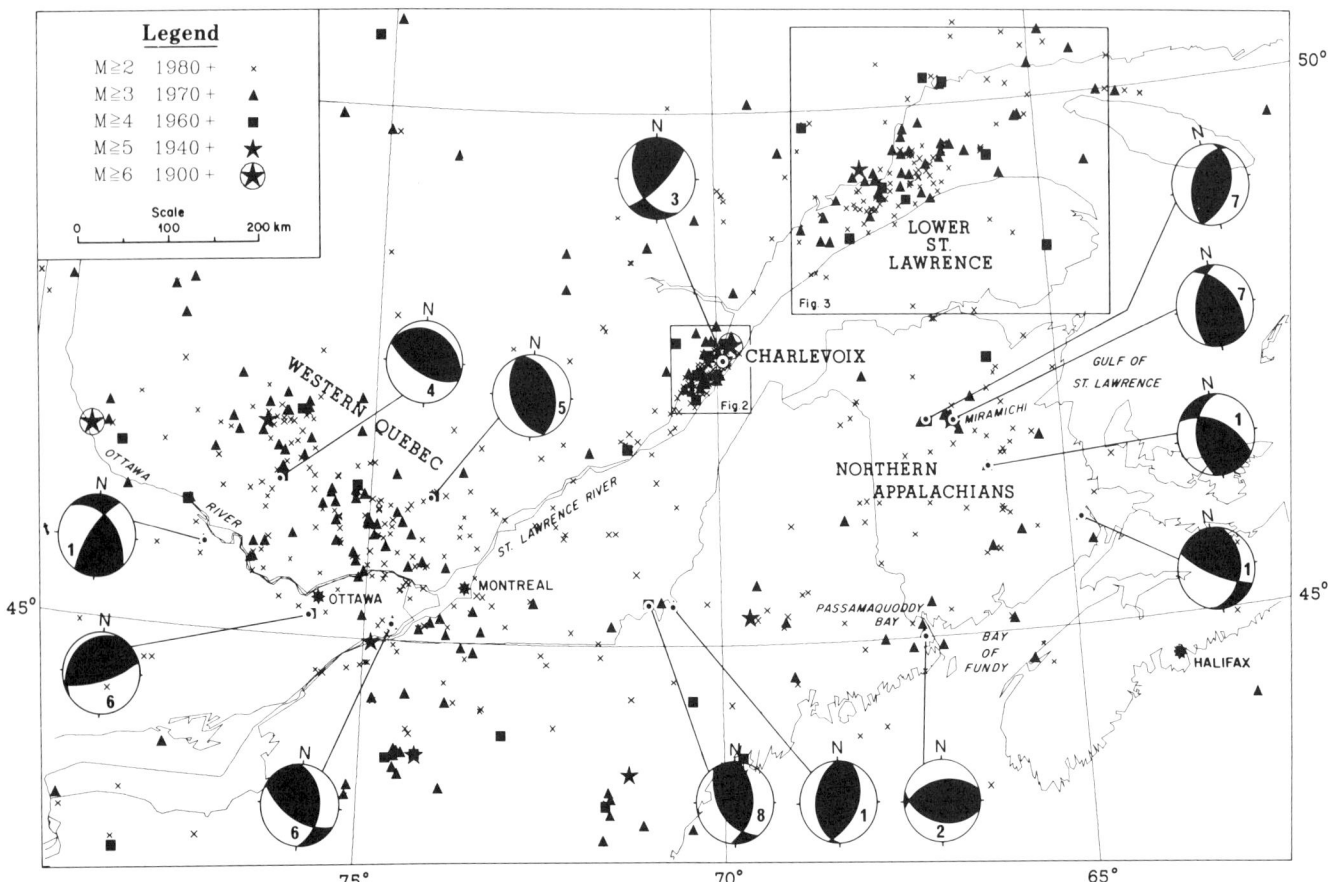

Figure 1. Seismicity of southeastern Canada identifying the four most active zones. The map shows a selection of earthquakes prior to July 1987, as detailed in the legend, chosen to eliminate many of the older, poorly located earthquakes. Focal mechanisms (lower hemisphere projections with compressional quadrants shaded) are from: 1, Adams and others, 1988; 2, Ebel, 1985, as modified by Adams; 3, Hasegawa and Wetmiller, 1980; 4, Horner and others, 1978; 5, Horner and others, 1979; 6, Wahlstrom, 1987b; 7, Wetmiller and others, 1984; and 8, Yang and Aggarwal, 1981.

appear to be occurring within a regional stress field dominated by northeast-to-east compression, and that most large earthquakes have occurred near Paleozoic or younger rift structures that surround or break the integrity of the North American craton.

SOUTHEASTERN CANADA

Western Quebec

A significant cluster of earthquakes occurs in the Grenville Province of the Canadian Shield, predominantly in western Quebec but extending into eastern Ontario across the Ottawa River. An earthquake with magnitude about 6 occurred at or near Montreal in 1732 (Leblanc, 1981). During this century, earthquakes of M6.2 occurred near Lake Timiskaming in 1935, and M5.6 near Cornwall, Ontario, in 1944.

For the last twenty years all earthquakes have been M4.3 or less, and most have been located north of the Ottawa River. Significant recent earthquakes, most studied by monitoring aftershocks, were: near Maniwaki in 1975, M4.2 (Horner and others, 1978); St. Donat in 1978, M4.1 (Horner and others, 1979); Cornwall in 1981, M3.3 (Schlesinger-Miller and others, 1983); Timiskaming in 1982, M4.3; and North Gower (near Ottawa) in 1983, M4.1 (Wahlstrom, 1987a). In an extension of the exposed Grenville Province into the Adirondacks of New York State, the largest recent earthquake was M5.2 at Goodnow in 1983.

In detail (Fig. 1), the seismicity appears to occur in two bands, probably sufficiently distinct to be considered separate zones. The first band, trending slightly west of northwest, lies along the Ottawa River from Lake Timiskaming (near circled star at left side of Fig. 1) to Ottawa and thence widens to extend southeast to Cornwall and east to Montreal. It includes the larger earthquakes near Timiskaming (1935, 1982), Rolphton (1963), North Gower (1983), Cornwall (1944, 1981), and Montreal (1732). The second band, containing more but smaller earthquakes, trends slightly north of northwest and extends from

Montreal to the Baskatong Reservoir, about 200 km north of Ottawa. The last decade of monitoring by the Eastern Canada Telemetered Network (ECTN) shows that the gap between the two bands is reasonably well defined at the northwestern end by an absence of even small earthquakes; however near the St. Lawrence River the two bands merge.

Field monitoring of aftershock sequences has provided good estimates for the focal depth of some earthquakes (Maniwaki, 17 km; St. Donat, 7 km; Cornwall, 16 km; North Gower, 12 km), and these together with some approximate depths computed from the digital ECTN suggest most earthquakes lie between 5 and 20 km, in the upper crust and within the Precambrian basement.

Focal mechanisms have been determined for about 40 earthquakes in the zone; the large number relative to the rest of eastern Canada reflecting the relatively dense network of digital seismographs. In addition to the four earthquakes just mentioned, Wahlstrom (1987b) and Adams and others (1988) have derived mechanisms for smaller events, and a selection is shown on Figure 1. Almost all mechanisms have near-horizontal P-axes, and represent mainly thrust earthquakes. This evidence for high horizontal compression is confirmed by other evidence for regional stresses in eastern Canada (Hasegawa and others, 1985; Adams, 1987, 1989). Although the regional stress field appears to have the compression axis in the northeast-to-east octant, the P-axes for some of the western Quebec earthquakes are distinctly different. In a region about 100 km northeast of Ottawa, a patch contains six earthquakes with south-southeast–trending P-axes, while outside the patch, six earthquakes have northeast-trending P-axes. It appears that within the patch the principal and secondary horizontal stresses have become reversed relative to the regional field, perhaps suggesting the two stress magnitudes are not very different (Adams, 1989).

Forsyth (1981) has shown that the earthquakes in the first band, including the larger historical earthquakes, may be associated with rift faults along the Ottawa River that were formed in the Paleozoic and offset Silurian strata. The rift faults are part of a large structure that extends from the Ottawa Valley and from rifts along Lake Champlain, down the St. Lawrence River (Kumarapeli and Saull, 1966; Kumarapeli, 1985), and through southern Labrador to the Labrador Sea (Gower and others, 1986). Related structures may extend from Timiskaming northwest to Kapuskasing, Ontario (Forsyth and others, 1983).

Although the last time of normal movement on these faults is not known, there is tenuous evidence that they might have been reactivated in the Mesozoic during the initial opening of the North Atlantic, as shown by Jurassic kimberlites in Ontario and New York State. However, if the rift faults underwent a major reactivation in the Mesozoic, it is surprising that so little hard evidence has been found.

Regardless of age, recent focal mechanisms of moderate earthquakes (Adams and others, 1988) and a study of seismicity near the 1935 Timiskaming earthquake suggest the Ottawa Valley rift faults are seismically active. All the Timiskaming earthquakes have occurred in a zone 50 km long by 10 km wide that trends northwest parallel to the rift faults, and they appear to have focal mechanisms consistent with thrust faulting on northwest-striking faults (Vonk and Adams, in preparation). These observations support the northwest-striking of the two possible thrust mechanisms derived for the main shock by Ebel and others (1986).

Assigning probable causative structures for the second band of seismicity is less easy. Forsyth (personal communication, 1987) sees no evidence for any well-defined northwest-trending features north of the Ottawa River, and earlier, Forsyth (1981) suggested that the earthquakes there might be occurring on or between structural boundaries within the northeast-trending Central Metasedimentary Belt. We suggest that the second band of seismicity is due to crustal fractures formed during the passage of North America over a hot spot between 140 and 120 Ma (Crough, 1981). The younger path of the hot spot is marked by the Monteregian Hills, a line of Early Cretaceous intrusions that lie mostly east-southeast of Montreal. In turn, these lie parallel to, but somewhat offset from, the still younger New England Seamount Chain (Sykes, 1978). Recent dating of the chain (Duncan, 1984) confirms Crough's drift of North America over the hot spot at about 25 mm/yr. The rate is consistent with the age of the Monteregian Hills, and of a northwest-trending kimberlite dyke at Kirkland Lake dated at 151 ± 8 Ma (Poole and others, 1970), to the northwest of the seismicity in western Quebec. Crough (1981) demonstrates that the passage of the hot spot caused a local uplift of the shield, resulting in erosion of at least 1 km at Montreal and perhaps 6 to 7 km in New England. Although the hot spot may have been less well developed when western Quebec passed over it, we suggest that the Precambrian crust in western Quebec may have been thermally stressed and fractured by differential uplift during the passage of the hot spot. This recent weakening of the North American craton has localized the release of seismic energy today. Under New England the plutonism may have been sufficient to heal any deep-crustal fractures. Although central Quebec and coastal Labrador are predicted to have passed over similar hot spots at about the same time as did western Quebec (Duncan, 1984, Fig. 3), there is no seismicity associated with their paths.

Charlevoix

The Charlevoix zone is historically the most active in eastern Canada, with at least five earthquakes of magnitude 6 or greater (1663, 1791, 1860, 1870, and 1925), which are described by Smith (1962, 1966). Magnitudes for these events were reassessed by Basham and others (1982a). Because of the historical seismicity, Charlevoix is thought to be the most probable site for a future large earthquake in eastern Canada, and so has been intensively studied (Buchbinder and others, 1983, 1988).

Hypocenters located by a 6-station local network since the mid-1970s demonstrate that most earthquakes are confined to a zone that is about 80 km long by 35 km wide (Fig. 2), mainly

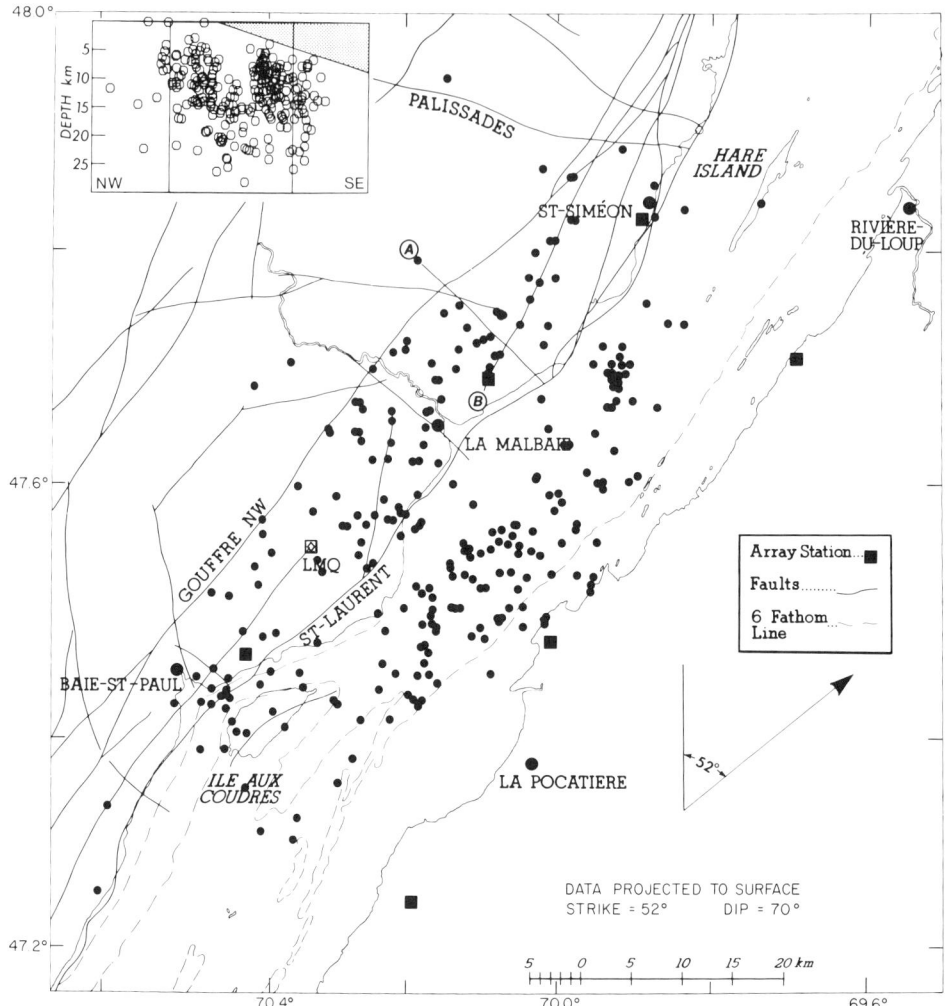

Figure 2. Microearthquakes (1977–1983) in the Charlevoix area (Anglin, 1984). In order to show their relation to the mapped surface faults, the epicenters have been moved to the surface up the dip of the regional rift faults. The inset is a northwest-southeast cross section of the hypocenters to show their depth distribution; all lie beneath the Paleozoic sedimentary wedge (shaded).

under the St. Lawrence River (Anglin, 1984). A relocation of some of the early instrumentally recorded earthquakes (Stevens, 1980) suggests that the larger earthquakes also occurred in this zone, perhaps initiating preferentially at the ends. The recent largest earthquakes are: 1939, M5.6; 1952, M5.2; 1979, M5.0; and 1989, M4.4.

Earthquake focal depths are well determined for the past decade of microearthquakes, and are mostly between 5 and 25 km (Fig. 2 inset). Paleozoic sediments that crop out on the south shore are only a few kilometers thick, so all the activity is occurring within the Precambrian basement (unshaded area on Fig. 2 inset). Stereo plots (Anglin, 1984) demonstrate that most of the microearthquakes are occurring on northeast-striking planes that dip to the southeast. A projection of the hypocenters to the surface along the postulated faults (Fig. 2) suggests the activity is confined between Paleozoic rift faults mapped on the north shore and a bathymetric feature near the river's south shore. Further,

the earthquakes do not extend downriver beyond the cross-cutting Saguenay graben faults (e.g. Palissades fault on Fig. 2; Basham and others, 1982b).

A number of earthquake focal mechanisms have been produced for Charlevoix. Field experiments in 1974 yielded six mechanisms (Leblanc and Buchbinder, 1977); Hasegawa and Wetmiller (1980) produced a mechanism for the 1979 M5.0 event (see Fig. 1); Lamontagne (1987) derived a suite of composite mechanisms; and Adams and others (1988) gave mechanisms of two recent events. No well-determined mechanism is available for the older large earthquakes, although Ebel and others (1986) suggest that both the 1925 and the 1939 earthquakes involved thrust faulting on either northeast- or northwest-striking planes.

The available mechanisms have not proved easy to interpret. In general, the derived P-axes lie in the east quadrant, and the mechanisms represent thrust or combination thrust/strike-slip faulting. No plane that might represent the rift faults is clearly identifiable in all of the mechanisms. At least some of the earthquakes appear to be occurring on northwest-trending transverse faults that offset the rift fault system (Fig. 2).

Charlevoix, the Ottawa River, and the lower St. Lawrence seismic zones all lie along the same rift system. However, at Charlevoix the rift structures have been complicated by a Late Devonian meteorite impact that caused ring faulting and distributed fracturing (between Baie-St-Paul and La Malbaie on Fig. 2). Lamontagne's (1987) composite mechanisms suggest that microearthquakes within the impact structure have more varied mechanisms than those outside it, due to the extra structural complexity caused by the impact. Considerable attention is being paid as to whether the earthquakes at Charlevoix occur because of the associated impact structure (in which case the seismicity could be considered to be localized and unlikely to occur elsewhere) or whether the earthquakes just happened to be coincident with the impact (in which case other parts of the rift fault system could become similarly active). Because other meteorite craters in Canada are not seismically active, because the earthquakes at Charlevoix extend downriver beyond the impact structure, and because the hypocentral trends suggest reactivation of the rift faults, current opinion is that the impact structures are not the controlling factor in the seismicity.

Lower St. Lawrence

Similar to Charlevoix, the Lower St. Lawrence earthquakes also occur mainly under the St. Lawrence River and may involve the old rift faults, although a correlation was previously made between the seismicity and gravitationally induced stresses (Goodacre and Hasegawa, 1980). The record of felt earthquakes extends back at most 100 years in this sparsely populated area, and none is likely to have been much larger than M5. The largest earthquake with a well-determined magnitude is M4.8 in 1944. Epicenters located by the ECTN in the last five years lie almost exclusively under the river, which at this point is an estuary 50 to 100 km wide. Despite the lack of known larger earthquakes, the zone has magnitude 3 and 4 earthquakes as often as the more confined Charlevoix zone.

Early epicenter maps (Smith, 1962, 1966) show a scattering of earthquakes extending onto the north and south shore. Relocation of some of these early epicenters (Adams and others, 1989) has affirmed that most actually occurred under the river (Fig. 3). Half of the relocated epicenters lie just offshore but parallel to the northern shoreline, and are inferred to be occurring on offshore faults that have controlled the shape of the coastline. The only confirmed significant activity on the north shore involves a M4.1 earthquake in 1975 and its aftershocks induced by the filling of the Manic 3 reservoir, and a cluster of M3 events in 1966 spatially associated with the Manic 2 reservoir (Leblanc and Anglin, 1978).

Reliable estimates of earthquake depth have been obtained for two earthquakes recorded in 1983 and 1984 by the Yellowknife array, 30 degrees to the northwest. Phases interpreted as pP and sP gave hypocentral depths of 17 and 19 km (Adams and others, 1988). Other earthquakes for which approximate depths have been computed from the ECTN network lie mostly between 10 and 20 km. Like the Charlevoix zone, this places them within the Grenville basement and well beneath the onlapping St. Lawrence platform and the overthrust Appalachian sedimentary rocks. The Manic earthquakes lie 50 to 100 km outside the rift system and, being induced earthquakes, are shallower than earthquakes under the St. Lawrence River.

Focal mechanisms have been derived for seven earthquakes (Adams and others, 1988); they have P-axis orientations only slightly less variable than at Charlevoix, and also indicate mostly thrust faulting in response to compression from the east quadrant. Significantly, five of the mechanisms have a common plane that strikes parallel to the river and dips to the southeast; the remaining two represent thrust faulting on northwest-trending planes. Taken together, the northeast-striking focal planes, the position of the larger earthquakes relative to the coastline, and the distribution of the smaller earthquake epicenters suggest that the Paleozoic rift faults are the chief active structure. Composite mechanisms of the Manic 3 induced earthquakes indicate thrust faulting on northwest-trending planes (Leblanc and Anglin, 1978; Adams and others, 1988), in agreement with the regional stress field.

Northern Appalachians

The northern Appalachian region, which includes most of New Brunswick and extends into New England, is a zone of relatively uniform seismicity. Significant historical earthquakes include those near Passamaquoddy Bay on the New Brunswick/

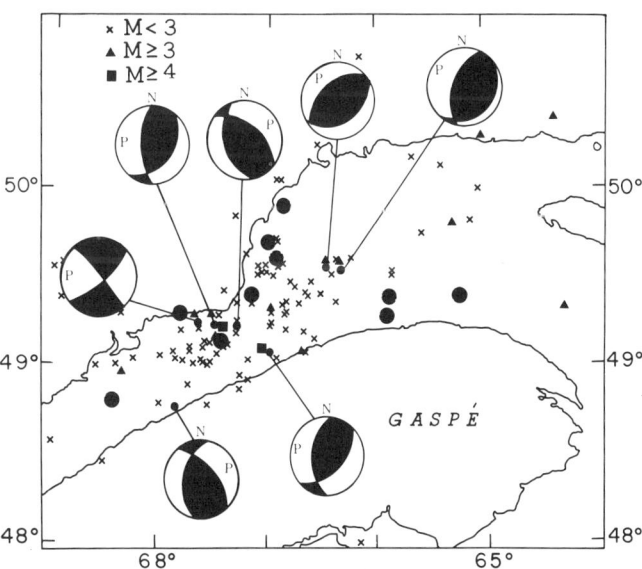

Figure 3. Lower St. Lawrence seismicity from 1981 to 1987. Large circles represent relocated epicenters of M3.5 to M4.8 earthquakes from the period 1944 to 1968 (Adams and others, 1989). Focal mechanisms shown are those derived by Adams and others (1988).

Maine border in 1817 (M4.8), 1869 (M5.2), and 1904 (M5.9), and near Moncton in 1855 (M5.2) (Leblanc and Burke, 1985).

Recent, well-located seismicity has occurred throughout New Brunswick (perhaps excluding the northwest corner), the Bay of Fundy, and southwestern Nova Scotia. Significant recent events in the region include: the Quebec-Maine border 1973, M4.8 (Wetmiller, 1975; Yang and Aggarwal, 1981); Miramichi 1982, M5.7, M5.1, M5.4, M5.0 (Wetmiller and others, 1984); Trousers Lake 1982, M4.7 (Wetmiller and others, 1984); and a series of six M3 earthquakes near Passamaquoddy Bay in 1983/84. Within New Brunswick, apparent concentrations of earthquakes occur in the Miramichi Highlands and near Passamaquoddy Bay. Outside of the northern Appalachians zone, the general lack of recent seismicity in the rest of Nova Scotia, in Prince Edward Island, and in the Gulf of St. Lawrence has been confirmed for the historical record by searching old newspapers for felt earthquake reports (Ruffman and Peterson, 1988). The lack of seismicity in the Gaspé, Quebec, also seems to be real. Likewise the rest of the Appalachian fold belt through the island of Newfoundland is substantially aseismic.

Focal mechanisms are available for the Miramichi main shock and several clusters of its aftershocks, and for the Trousers Lake earthquake (Wetmiller and others, 1984); for two earthquakes in east-central New Brunswick (Adams and others, 1988); for two earthquakes near the Quebec-Maine border (Wetmiller, 1975; Yang and Aggarwal, 1981; Adams and others, 1988); and for an earthquake at Passamaquoddy Bay (Ebel, 1985), as summarized on Fig. 1. All represent predominant thrust faulting, and all but the one at Passamaquoddy Bay were in response to northeast- to east-directed compression.

The Miramichi earthquakes, because of their size and numerous aftershocks, have been unusually well studied. Even today (1988), six years after the main-shock sequence, M3 aftershocks are occurring every few months; recent aftershocks have been as large as M4.1. Field aftershock surveys in 1982 (Wetmiller and others, 1984) established the general pattern of activity.

The 9 January 1982 main shock, M5.7, has been interpreted as a thrust with rupture up-dip on a west-dipping plane. A magnitude 5.1 aftershock occurred 3.5 hours later, probably on the lower northern portion of this plane. On 11 January a M5.4 aftershock ruptured (probably up-dip) a conjugate east-dipping plane and was followed by an intense aftershock sequence. Finally the 31 March M5.0 aftershock occurred as a repeat rupture on the upper northern portion of the west-dipping plane (Wetmiller and others, 1984; Basham and Adams, 1984; Basham and Kind, 1986). Microearthquake monitoring for short periods in 1983 and 1985 confirmed that the epicentral region has remained active and may have expanded slightly with time. Although considerable effort, including trenching, was made to locate a surface rupture, none was found, suggesting that the expected near-surface vertical offset of 0.3 to 0.5 m was distributed over a zone several hundred meters wide and not confined to a single plane.

Detailed geological and geophysical investigations have confirmed that the earthquakes all occurred within a single granodiorite pluton, but not on the obvious west-northwest–trending faults. No strong reason has emerged as to why the earthquakes occurred at the Miramichi site. If Miramichi-type earthquakes occurred regularly at the site, more surface evidence for thrusting, such as a degraded fault scarp, would be expected. Therefore, despite considerable effort, we do not understand why the earthquakes occurred where they did, and have no evidence that they occur often at the Miramichi site. Thus we must consider that Miramichi-sized earthquakes could occur anywhere in the northern Appalachians zone.

While aftershocks of the Miramichi and Trousers Lake earthquakes were shallow, all less than 9 km, earthquakes near Passamaquoddy Bay are deeper, perhaps 10 to 16 km, based on a 12-km depth from waveform modeling for a 1984 earthquake (Ebel, personal communication, 1988) and computed depths for other earthquakes. Earthquakes in the Appalachian part of Canada may thus be significantly shallower than in the Grenville basement at Charlevoix and western Quebec, perhaps reflecting a thinner brittle upper crust in the young Appalachian belt (Hasegawa, 1986, Fig. 9). In the Appalachians of New England, all earthquakes appear to be shallower than 13 km (Ebel, 1984). The shallow depths may indicate that the earthquakes are confined to the rocks above a shallow, sub-Appalachian detachment zone, although judging from the Quebec-Maine seismic reflection line, the detachment dips more steeply than in the southern United States and may not extend beneath southern New Brunswick (Spencer and others, 1989). Northern New Brunswick seismicity may be compared with the central Virginia seismic zone, where earthquakes occur above the detachment (Bollinger and others, this volume), and may follow the general style for Appalachian seismicity, with earthquakes above the detachment to the east and below the detachment to the west of the crystalline overthrust sheet (Seeber and Armbruster, 1988), perhaps because, as at Charlevoix, the sedimentary rocks above the detachment are insufficiently brittle. Inhibition of seismicity in the basement beneath the eastern part of the overthrust sheet might be due to thermal or other effects. If a sub-Appalachian detachment were important for controlling Appalachian seismicity, it is possible that the Miramichi earthquakes nucleated at the base of the overthrust sheet and ruptured upward to the surface, while the Trousers Lake main shock and aftershocks—all at essentially the same depth—may have occurred on the detachment itself.

Southeastern continental margin

Although poorly monitored and little studied, the seismicity of the southeastern continental margin of Canada is clearly higher than that at many comparable passive margins, and higher even than the same margin off the eastern United States. The description below is condensed from Adams (1986), which relies heavily on data as yet unpublished.

About half of the earthquakes off the southeastern margin occur in the Laurentian Slope seismic zone, site of the M7.2

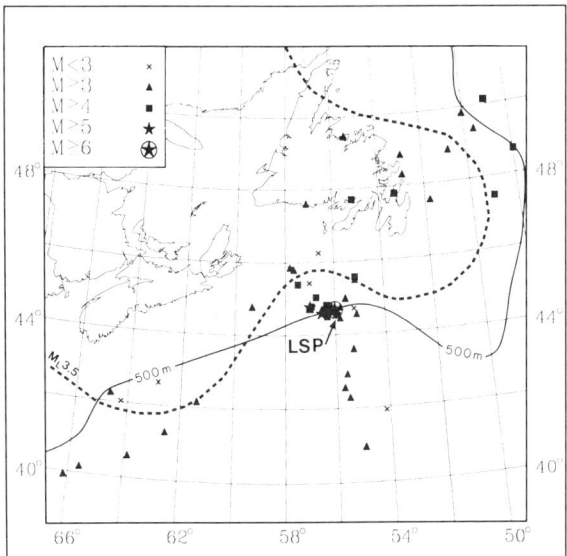

Figure 4. Seismicity of the southeastern margin (from unpublished work by Adams and Wahlstrom). Earthquakes in Nova Scotia and to the northwest are omitted. Inland of the striped line, M3.5 and larger earthquakes are thought to have been completely detected and located since 1983. Arrow marked "LSP" points to the many earthquakes near the epicenter of the 1929 "Grand Banks" earthquake. The 500-m isobath approximates the position of the ocean-continent transition.

"Grand Banks" earthquake of 1929 (Fig. 4). This earthquake caused a large submarine slump and consequent tsunami that killed 27 people on the south coast of Newfoundland. The larger historical earthquakes, and many recent smaller earthquakes, have been systematically relocated and found to lie within a 100 km (east-to-west) by 35 km (north-to-south) box at the mouth of the Laurentian Channel (Adams, 1986). In addition to the 1929 earthquake and its immediate aftershocks, there have been four M5 earthquakes since 1951, the most recent in 1975. The elongation of the seismic zone, and the location of the 1929 epicenter at the eastern end are consistent with Adams' 1986 hypothesis that current earthquakes represent belated aftershocks of the 1929 earthquake. If so, the earthquake appears to have ruptured westward along a fault about 70 km long.

Little is known about the source properties of these earthquakes. The 1975 earthquake appears to have occurred at a depth of 30 km, placing it in the upper mantle in this area of thinned continental crust, and appears to have involved thrusting on planes which, although not tightly constrained, probably strike roughly east to west, i.e., in response to compression from the north or northeast (Hasegawa and Hermmann, 1989). The depth and mechanism of the 1929 earthquake are uncertain, as the available seismograms are generally poor. Hasegawa and Kanamori (1987) have suggested that the long-period seismic source may have been the large submarine slump itself, although they admit that the slump may have been triggered by a "smaller" earthquake. Their explanation, if correct, has significant implications for the seismotectonic interpretations of large earthquakes on Canada's eastern margin, or indeed for other continental margins.

Our problems with the slump hypothesis are: (1) the only energy in the slump mass is its potential energy, and it is not clear that the mass can travel downslope fast enough for its energy to appear in the seismic coda; (2) the slump mass required by the Hasegawa and Kanamori model is 5 to 10 times larger than that computed for the initial slump by Piper and Aksu (1987); (3) the computed intermediate-period body-wave and long-period surface-wave magnitudes are similar (Hasegawa and Kanamori, 1987, Table 1), suggesting that the longer periods were not greatly augmented by low-frequency slump-generated energy relative to shorter periods; (4) the immediate aftershocks (up to magnitude 6) and the continuing seismicity are not easily explained; (5) the one earthquake of known depth (1975) is a mantle event; and (6) a magnitude 6 to 6.5 triggering earthquake is not ruled out by the data, particularly if its mechanism produces an S-wave node at Uccle and Kew.

Outside of the Laurentian Slope zone, scattered earthquakes occur northeast of Newfoundland and seaward of Nova Scotia. Some of the epicenters off northeast Newfoundland correlate with fracture zones (see Fig. 6 and section on Labrador Sea seismicity below), and perhaps with a structural trend extending northeast from eastern Newfoundland to the continental margin, but others cannot yet be related to structure. South from the Laurentian Slope zone, a trend of seismicity of unknown origin occurs along the Laurentian Fan.

Off Nova Scotia the earthquakes on Figure 4 suggest that the transition zone from oceanic to continental crust is active, although not currently at a very high level. None of these earthquakes has been larger than M4. Elsewhere, the transition is too poorly monitored to show if it is active at the low level we can detect off Nova Scotia.

The cause of earthquakes along the eastern margin, and thus a rationale for their distribution, are established in only a general way. Studies of stress directions from oil-well breakout data (Podrouzek and Bell, 1985) confirm that the margin is subject to the same northeast-directed compression as the rest of eastern North America (Adams, 1989). While some studies have indicated local areas of near-surface faulting (e.g., Orpheus Graben off Nova Scotia), many of the earthquakes along the margin probably occur near the ocean-continent transition on the deep-crustal rift faults formed during the opening of the Atlantic Ocean. Under the current northeast compression regime these normal faults would be reactivated as thrust or strike-slip faults.

As the numbers, location, and nature of the offshore earthquakes are poorly understood, Basham and Adams (1983) produced a speculative model that suggests rare large earthquakes can occur along the whole margin, using the pervasive Mesozoic rift faults as the causative structure, with a rate of about one M7 earthquake per thousand years per thousand kilometers of margin. The model implies that inactive parts of the margin are quiescent but potentially seismogenic, and that recent earthquakes in active regions like the Laurentian Slope represent belated aftershocks that will diminish over the next century without producing another M7 earthquake. That such large earthquakes may have occurred in the past is suggested by prehistoric submarine slumps elsewhere along the margin (Piper and others, 1985).

Southeastern background seismicity

Some scattered seismicity lies in less intense clusters outside the seismic zones discussed above with reference to Figures 1 and 4, and is shown only on Figure 7. Here, we briefly mention these areas, again moving from west to east. The shield areas of Ontario (Basham and Cajka, 1985; Wetmiller and Cajka, 1989) and Quebec show very low seismicity except for earthquakes near Cochrane and Iroquois Falls in northern Ontario, which may lie on an extension of the Ottawa-Timiskaming rift structure toward Kapuskasing (Forsyth and others, 1983). Farther north, the earthquakes around and in James Bay have an unknown origin.

A cluster of small earthquakes in the Burlington/Niagara Falls area of Ontario is poorly understood, but in part may represent very shallow stress release (Wetmiller, 1980), and in part may be related to a seismicity trend that extends from Lake Ontario and Lake Erie to Ohio and into the Mesozoic rift at New Madrid in the central United States. On Figure 7, we follow Woollard (1969) and speculatively extend the recent weakening of the craton by the St. Lawrence rift under lakes Ontario and Erie. Such weakening might account for the above trend in seismicity. Evidence for post-Proterozoic rifting in the north-central United States is beginning to emerge (e.g. Phipps, 1988), but has not yet been integrated into a full seismotectonic analysis.

Earthquakes near Quebec City may lie on the St. Lawrence rift system. The trend of earthquakes extending from Sept Iles across easternmost Quebec and southern Labrador may lie on an extension of the St. Lawrence rift system mapped by Gower and others (1986).

NORTHEASTERN CANADA

We continue our discussion of eastern Canadian seismicity and seismotectonics by proceeding northward along the continental margin of the eastern Arctic, inland to the northeastern portion of the Canadian Shield, and northward into the Arctic archipelago, completing the picture with a brief discussion of the Arctic Ocean margin (Fig. 5).

Labrador Sea

The seismicity of the Labrador Sea includes six earthquakes since 1934 in the magnitude 5.0 to 5.6 range. The most recent moderate earthquake was M4.7 in 1986, which although 200 km offshore, was felt in Nain, Labrador. There are older reports of

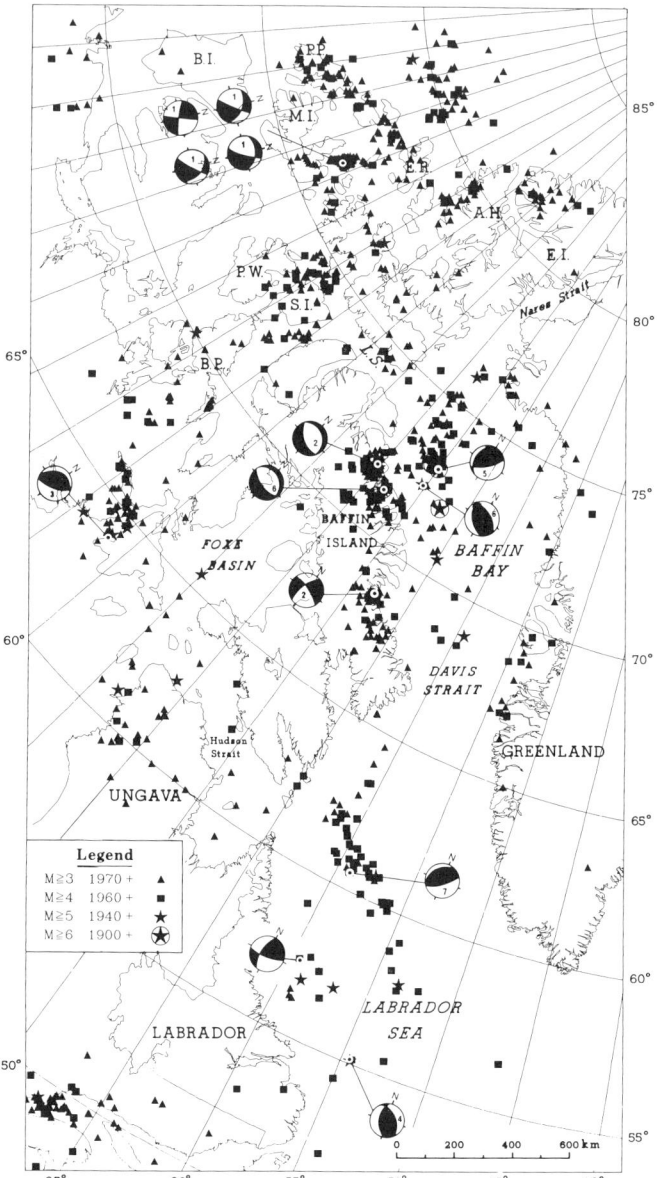

Figure 5. Seismicity of northeastern Canada showing a selection of earthquakes prior to July 1987, as detailed in the legend. Focal mechanisms (lower hemisphere projections with compressional quadrants shaded) are from: 1, Hasegawa, 1977; 2, Hashizume, 1973; 3, Hashizume, 1974; 4, Hashizume, 1977; 5, Sleep and others, 1989; 6, Stein and others, 1979; 7, Sykes and Sbar, 1974; and (blank), Adams, unpublished. Geographic features are identified by initials as follows: AH, Axel Heiberg Island; BI, Banks Island; BP, Boothia Peninsula; EI, Ellesmere Island; ER, Ellef Ringes Island; LS, Lancaster Sound; MI, Melville Island; PP, Prince Patrick Island; PW, Prince of Wales Island; SI, Somerset Island.

felt earthquakes from fishing villages along the Labrador coast as early as 1809 (Staveley and Adams, 1985). Currently, epicenters for these older events are assigned to the locations at which they were felt. However, 25 years of monitoring by the Canadian Network has provided no evidence that significant earthquakes are occurring onshore in this region. These older events likely occurred offshore, and may have been larger than hitherto thought (Basham and Adams, 1983).

The Labrador Sea was produced by rifting in the Middle Cretaceous and sea-floor spreading between 95 and 50 Ma. Srivastava and Tapscott (1986) have identified a central ridge and associated fracture zones from seismic and gravity profiles and linear magnetic anomalies. Adams and Simmons (1991) systematically relocated earthquake epicenters in the Labrador Sea and showed that seismicity can be associated with the extinct ridge and fracture zones (Fig. 6), even though the spreading ridge has been extinct for 50 m.y. The apparent decrease in activity to the southeast along the ridge is due to an increase in the earthquake detection threshold away from station FRB on Baffin Island.

No earthquakes are known in the Labrador Sea between the Labrador Ridge and Greenland to the northeast (perhaps also a problem with detection threshold), but on the Canadian side a separate trend of seismicity occurs at the continental margin off Labrador. In general terms, we associate the earthquakes with preexisting faults near the inactive ridge (a relatively young zone of weakness within the oceanic crust) and beneath the rifted continental margin. The best-studied ridge earthquake is the 1969 M5.0 (Sykes and Sbar, 1974), which represents compression along the ridge axis. On the slope, Hashizume (1977) determined a thrust mechanism for the 1971 magnitude 5.6 earthquake at a depth of 16 km, implying compression normal to the margin, and Adams (unpublished, 1987) determined a mechanism for the 1986, M4.7 earthquake, implying compression along the margin (Fig. 5).

North of the point where the ridge meets the continental margin, the Southeast Baffin shelf shows a trend of small earthquakes along the former transform continental margin toward Davis Strait (Fig. 5). Davis Strait itself appears to show little activity, at least at current detection levels (M about 3.5), until the southern portion of Baffin Bay is reached.

Baffin

The largest earthquake known to have occurred in northeastern Canada was the M7.3 event in Baffin Bay in 1933. This earthquake had aftershocks as large as M6.5, and M6 events have since occurred in the Bay in 1945, 1947, and 1957. Precise boundaries for the Baffin seismicity are difficult to define on the basis of geological and geophysical data, but there seems to be some evidence for a separation between the activity in the Bay (partly on oceanic and partly on transitional crust) and that on Baffin Island on true continental crust.

Jackson and others (1979) found evidence for sea-floor spreading and an extinct spreading center in the deep central region of Baffin Bay. Srivastava and Tapscott (1986) suggest rifting could have initiated in the latest Cretaceous and continued into the early Tertiary. However, there is now little or no seismic activity in the deep central bay, and the seismicity is confined almost exclusively to the landward side of the 2,000-m bathymetric contour in the northwestern segment of the bay, which delimits the thick sedimentary sequence (Basham and others, 1977). Sleep and others (1989) have established a focal depth of 10 km for the 1933 event (a correction to the 65-km depth estimated by Stein and others, 1979) and 24 km for a 1976 M5.4 event farther south in Baffin Bay.

Prior to 1960, only one earthquake is known to have occurred on Baffin Island, a magnitude 5.5 to 6.0 event in 1935, but with the development of the northern Canadian seismograph network in the 1960s the northeastern portion of the island was found to be highly active. The seismicity appears to be confined to the coastal region and does not occur much farther inland than the heads of the fjords. The earthquakes are concentrated in the regions of Buchan Gulf and Home Bay with a gap between these two regions, but because of the short history and the swarm-like nature of the seismicity, this gap may be a zone of temporary quiescence. In any case, the tectonic significance of such gaps is unclear. All available evidence on focal depths suggests the earthquakes are shallow. Hashizume (1973) determined depths of 6 to 9 km for the 1970 M4.4 and 1972 M5.1 events, Liu and Kanamori (1980) a depth of 7 km for the 1963 M6.1 event. A temporary deployment of ocean-bottom seismometers and land stations also established shallow focal depths on the island (Reid and Falconer, 1982).

Thrust faulting has been established for Baffin Bay earthquakes of 1933, M7.3 (Sleep and others, 1989) and 1976, M5.4 (Stein and others, 1979). In contrast, the island earthquakes show normal faulting, which is rare for eastern Canada: 1963, M6.2 (Stein and others, 1979) and 1972, M4.5 (Hashizume, 1973). On the basis of these focal mechanisms, Stein and others (1979) proposed a model for the entire passive margin of eastern Canada with thrust faulting seaward of the 1,000-m isobath and normal or strike-slip faulting for more landward events as due to Pleistocene deglaciation reactivating faults remaining from the rifting. Sleep and others (1989) developed this model further for the Baffin region, suggesting that the recent removal of surface loads by glacial erosion and by deglaciation is the major source of local stress that produces flexure of the lithosphere, and is comparable in magnitude to and superimposed on the compressional stress in the North American plate due to Mid-Atlantic Ridge spreading. The spreading stress enhances the shallow tension produced by flexure within the continental lithosphere and shallow compression produced by flexure in the oceanic lithosphere. Quinlan (1984) has argued, however, that postglacial rebound is capable of triggering earthquakes in prestressed regions but rarely capable of dictating the focal mechanism of these earthquakes.

The region surrounding the Baffin seismicity discussed above, and the continental margin of Greenland adjacent to Baf-

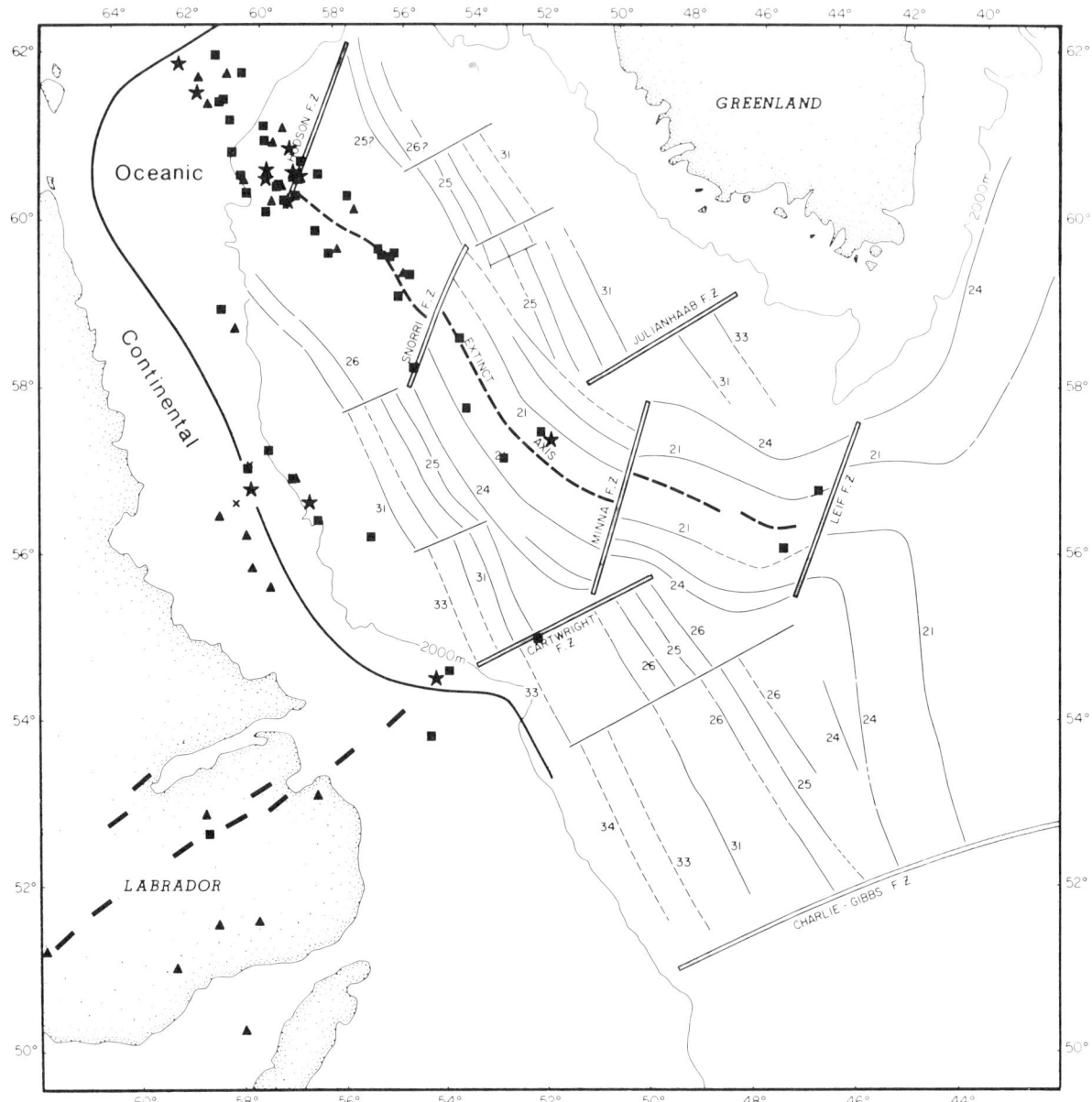

Figure 6. Seismicity of the Labrador Sea (Adams and Simmons, 1991) showing relation of earthquakes to the extinct spreading ridge and transform faults (as taken from Srivastava and Tapscott, 1986), to the boundary between continental and oceanic crust, and to a zone of crustal weakness that extends across southern Labrador.

fin Bay, have experienced low levels of both historical and recent seismicity. A trend of epicenters appears to extend westward from northern Baffin Bay into Lancaster Sound (Wetmiller and Forsyth, 1982), perhaps along the Cretaceous-Tertiary Lancaster Aulacogen (Kerr, 1980). Wetmiller and Forsyth (1982) have shown that Nares Strait, between Ellesmere Island and extreme northwestern Greenland, is currently aseismic. The region around Nares Strait marks the Tertiary collision of Greenland with North America, and deformation during that event may have sealed any old zones of weakness.

Boothia-Ungava

Basham and others (1977) recognized an arcuate band of seismicity that extends southward from the Boothia Peninsula, across northern Hudson Bay, the Ungava Peninsula, and eastward through Hudson Strait, connecting with the northern end of the Labrador Sea. With a center of postglacial uplift over Foxe Basin and a high differential rate of uplift on northeastern Baffin Island, they speculated that the Baffin Island–Foxe Basin block is responding independently to postglacial uplift (see also Hasegawa and Basham, 1989), and may be decoupled from the rest of the shield to the southwest along the arc of seismicity. The correlation to geology is best at the north end where the seismicity lies along the Boothia Uplift from Somerset and Prince of Wales Islands northward, meeting the Sverdrup Basin in the region of Grinnell Peninsula on the northwest tip of Devon Island (Wetmiller and Forsyth, 1982). The Boothia Uplift, which has been geologically active from the Paleozoic to the Cretaceous (Okulitch and others, 1986), was most recently active in a Cretaceous–Tertiary rifting episode in which north-south normal faults followed structural trends established earlier (Kerr, 1977, 1980). The uplift continues to be seismically active at present, although Hashizume's (1974) focal mechanism of a M5.0 at 21-km depth on Southampton Island (Fig. 5) represents thrust faulting in response to northeast compression rather than extension.

Sverdrup Basin

Most of the remaining seismicity in the Canadian Arctic archipelago can be spatially associated with the Sverdrup Basin, a 1,000-km-long northeast-southwest regional depression (between Melville and Ellef Ringnes Islands), in which the sedimentary rocks reach thicknesses of 10 km. The seismicity is characterized by intense low-magnitude swarms such as that which occurred on Prince Patrick Island in 1965 (Smith and others, 1968), by intense moderate-magnitude swarms such as that which started abruptly in 1972 in the Byam Martin Channel northeast of Melville Island (Basham and others, 1977; Hasegawa, 1977), and by single earthquakes with only small aftershocks such as the M5.2 event on western Axel Heiberg Island in 1975. Because of the swarm-like nature of the seismicity and the short instrumental observation period (25 years), it is unlikely that all potentially active regions of the basin have been identified.

A feature running through the basin that shows some geologic and geophysical correlation with the highest levels of seismicity is the Gustaf-Lougheed Arch (Forsyth and others, 1979; Basham and others, 1982a). This arch is a structurally significant feature, visible in the Bouguer anomaly contours, that divides the western Sverdrup Basin into two separate sub-basins. Superimposed on the arch is a series of northeasterly trending minor magnetic highs reflecting mineralized faults or intrusive dikes.

The focal mechanisms for the four largest earthquakes in the Byam Martin Channel swarm (M5.1 to 5.7), while dominantly strike-slip, show some deviatoric tension at depths from 9 km (near the base of the sedimentary rocks) to 31 km (Hasegawa, 1977). This suggests that the fractures or dikes may be the loci of current activity, and there may remain a small component of tension similar to that responsible for the opening of the Arctic Ocean Basin in the Early Cretaceous (Forsyth and others, 1979).

Arctic Ocean margin

The seismicity along the Arctic Ocean margin offshore of the Canadian Arctic archipelago is concentrated in distinct clusters in the Beaufort Sea and northwest of Ellef Ringnes Island, with only very scattered activity elsewhere (Basham and others, 1977; Wetmiller and Forsyth, 1978; Hasegawa and others, 1979; Wetmiller and Forsyth, 1982; Forsyth and others, 1988). The largest earthquake, M6.5, occurred in the Beaufort Sea in 1920 (see also the discussion of Beaufort Sea seismicity in Rogers and Horner, this volume). Elsewhere along the margin there are no known events larger than about M5.5, and no focal depths or focal mechanisms are yet available.

The rifted margin was formed in Early Cretaceous time, possibly when northern Alaska rotated counterclockwise away from Arctic Canada (Sweeney and others, 1978). The ocean-continent transition is characterized by a zone of negative magnetic anomalies that extends from the Beaufort Sea to north of Ellesmere Island. A series of elliptical free-air gravity anomalies lie over major sediment accumulations near the shelf-slope break. Forsyth and others (1988) have recently provided the following interpretation of the seismicity in terms of gravity and magnetic anomalies, bathymetry, and margin structures. The zone of rift faults separating continental and oceanic crust is inferred to lie immediately seaward of the magnetic lows. Although four major elliptical gravity anomalies lie along the margin, significant seismicity is associated with them only where the shelf break extends distinctly seaward of the magnetic low; i.e., where the sediments have prograded over more oceanic crust. This suggests that earthquakes occur on the rift-related structures chiefly where the oceanic or transitional crust is loaded by sediments. Seismicity is much lower where the sediment is loading continental crust (e.g., northwest of Banks Island). Perhaps for similar reasons, very little of the Beaufort seismicity extends landward of the gravity and magnetic anomalies, although the seismicity northwest of Ellef Ringnes Island extends onto the shelf and may connect with the seismicity of the Gustaf-Lougheed Arch discussed above.

SUMMARY

We summarize our conclusions regarding the causes of seismicity in eastern Canada in Figure 7. Almost all the significant earthquakes in the continental part of southeastern Canada can be spatially, and probably causally, associated with the Paleozoic rift system along the St. Lawrence, the chief exceptions being the northern band of seismicity in western Quebec (probably related to the hot-spot trace) and the Appalachian seismicity southeast of the rift system. To a first approximation, the continent-wide stress field is uniform, represents compression from the east-northeast octant, and causes thrust or thrust/strike-slip earthquakes.

Along the eastern margin of the continent, the seismicity includes the 1929 M7.2 Grand Banks and 1933 M7.3 Baffin Bay earthquakes. These and smaller earthquakes concentrated near the ocean–continent transition appear to be thrust events occurring through the reactivation of the Mesozoic rift faults formed during the opening of the Atlantic. In the mid Labrador Sea, earthquakes are associated with the extinct spreading ridge and its associated transform faults.

In Arctic Canada, continental earthquakes occur on Baffin Island, along an arcuate band between the Boothia and the Un-

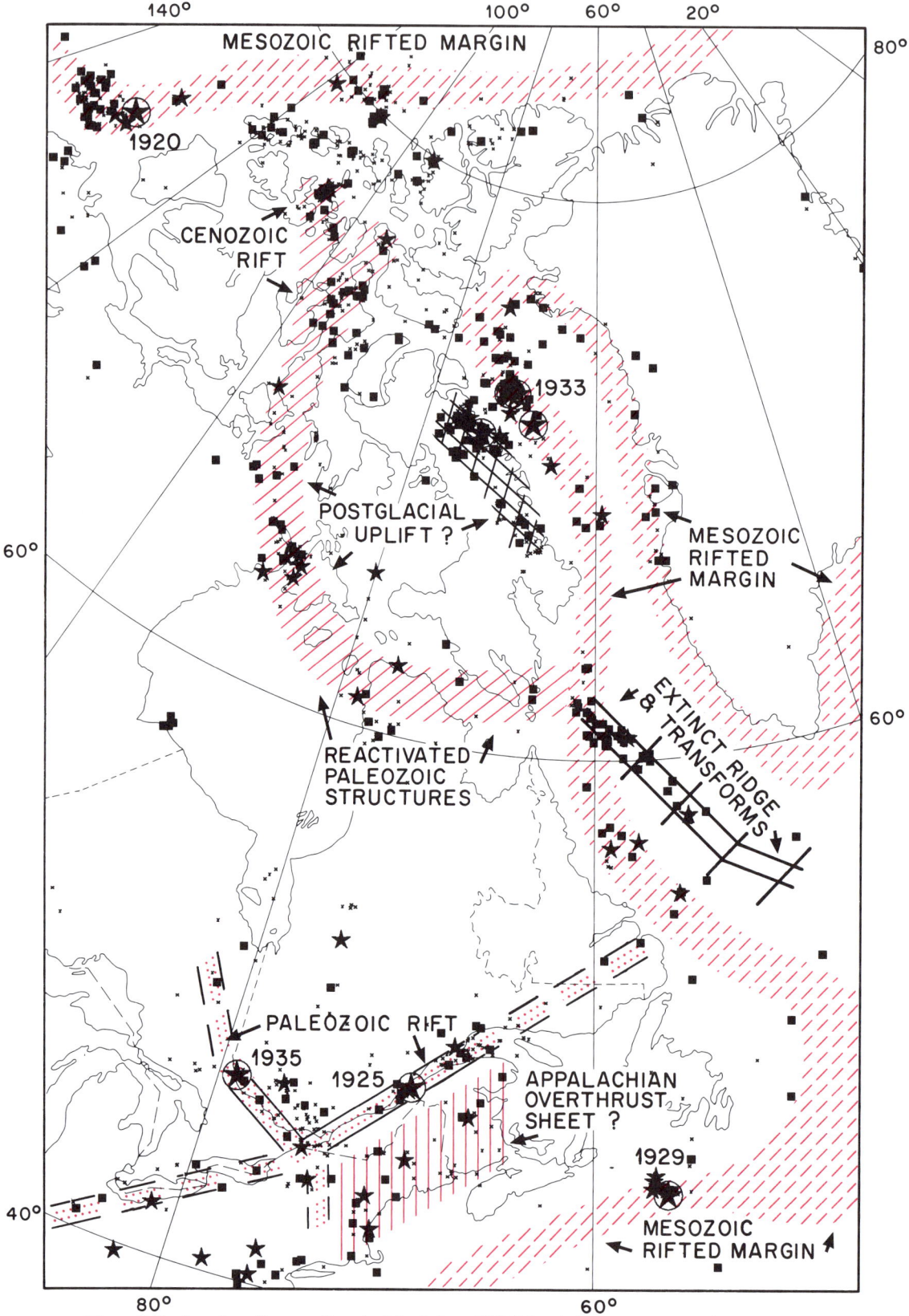

Figure 7. Earthquakes of eastern Canada (M ≥ 3 since 1970; M ≥ 4 since 1960; M ≥ 5 since 1940; M ≥ 6 since 1900) together with an interpretative framework for the cause of the seismicity. Note that the extension of the Paleozoic rift structures through Lakes Ontario and Erie is speculative (see text).

gava Peninsulas, and in the Sverdrup Basin. The Baffin and Boothia–Ungava earthquakes are spatially associated with steep gradients in the postglacial uplift rate, suggesting that they may occur because of differential uplift. The Baffin Island earthquakes are unique in eastern Canada in that they represent normal faulting. The Sverdrup earthquakes represent strike-slip deformation beneath a thick accumulation of sediments. The passive Arctic Ocean margin has a rifted ocean-continent transition comparable to the Atlantic margin, but it appears to be seismically active mainly where it has been recently loaded by thick sediments.

It is therefore clear from the above discussion and Figure 7 that most of the larger earthquakes can be associated with the Paleozoic or younger rift systems that surround or break the integrity of the North American craton. By contrast, the largest earthquakes in the unbroken craton (in Canada, but outside the seismic zones discussed above) are probably not much larger than M5. Coppersmith and others (1987) and Johnston (1989) have come to similar conclusions from a study of world-wide earthquakes in "stable continental interiors." They found that 71 percent of the seismicity of stable continental interiors was associated with imbedded continental rifts and continental passive margins (one-sided rifts). Further, all of their 17 M7 or larger earthquakes are strongly associated with the imbedded rifts or passive margins.

Both the Paleozoic and the Mesozoic rift systems in eastern Canada are continuous features, many thousands of kilometers long. Despite their continuity and the uniform stress field, the rift systems are only sporadically active, showing seismicity in clusters (e.g., Charlevoix and the lower St. Lawrence) or single large earthquakes (e.g., the "Grand Banks" earthquake).

For both rift systems there is a lack of geologic evidence for continual activity at present rates. We, like many others, have been confounded by the high levels of seismic activity at Charlevoix relative to other places in eastern Canada. At such high rates (M7 every few hundred years), the implied rates of geologic deformation would amount to kilometers over a million years. Clearly, kilometers of uplift (if due to thrusting), or even kilometers of strike-slip motion, would have been recognized at Charlevoix, had they occurred. That they have not, we infer, must be due to intermittent activity at Charlevoix and along the remaining rift system, perhaps with a time constant of thousands to tens of thousands of years.

For single large earthquakes such as the "Grand Banks," a similar argument can be made for intermittent activity on comparable time scales. This leads to the classic dilemma for hazard estimates that need to be made for low levels of probability: will the next large earthquake occur in a recognized seismic zone or not?

REFERENCES CITED

Adams, J., 1986, Changing assessment of seismic hazard along the southeastern Canadian margin: Proceedings, Third Canadian Conference on Marine Geotechnical Engineering, St. John's Newfoundland, 11–13 June, 1986, p. 41–53.
——, 1987, Canadian crustal stress database—a compilation to 1987: Geological Survey of Canada Open File 1622, 130 p.
——, 1989, Crustal stresses in eastern Canada, in Gregersen, S., and Basham, P. W., eds., Earthquakes at North Atlantic Passive Margins; Neotectonics and postglacial rebound: Dordrecht, Kluwer Academic Publishers, p. 289–297.
Adams, J., and Simmons, D. G., 1991, Relocation of earthquakes in the Labrador Sea and southern Labrador: Geological Survey of Canada Open File 2326, 103 p.
Adams, J., Sharp, J., and Stagg, M., 1988, New focal mechanisms for southeastern Canadian earthquakes: Geological Survey of Canada Open File 1892, 109 p.
Adams, J., Sharp, J., and Connors, K., 1989, Revised epicentres for earthquakes in the Lower St. Lawrence seismic zone, 1928–1968: Geological Survey of Canada Open File 2072, 82 p.
Anglin, F. M., 1984, Seismicity and faulting in the Charlevoix zone of the St. Lawrence Valley: Bulletin of the Seismological Society of America, v. 74, p. 595–603.
Basham, P. W., and Adams, J., 1983, Earthquakes on the continental margin of eastern Canada; Need future large events be confined to the locations of large historical events?: U.S. Geological Survey Open-File Report 83-843, p. 456–467.
——, 1984, The Miramichi, New Brunswick earthquakes—near-surface thrust faulting in the northern Appalachians: Geoscience Canada, v. 11, p. 115–121.
Basham, P. W., and Cajka, M. G., 1985, Contemporary seismicity in northwestern Ontario, in The geoscience program; Proceedings of the 17th information meeting of the nuclear fuel waste management program: Pinewa, Manitoba, Atomic Energy of Canada Limited Technical Report TR-299, v. 2, p. 367–374.
Basham, P. W., and Kind, R., 1986, GRF broad-band array analysis of the 1982 Miramichi, New Brunswick, earthquake sequence: Journal of Geophysics, v. 60, p. 120–128.
Basham, P. W., Forsyth, D. A., and Wetmiller, R. J., 1977, The seismicity of northern Canada: Canadian Journal of Earth Sciences, v. 14, p. 1646–1667.
Basham, P. W., Weichert, D. H., and Berry, M. J., 1979, Regional assessment of seismic hazard in eastern Canada: Bulletin of the Seismological Society of America, v. 69, p. 1567–1602.
Basham, P. W., Weichert, D. H., Anglin, F. M., and Berry, M. J., 1982a, New probabilistic strong seismic ground motion maps of Canada; A compilation of earthquake source zones, methods, and results: Earth Physics Branch Open File 82-33, 202 p.
Basham, P. W., Morel-a-l'Huissier, P., and Anglin, F. M., 1982b, Earthquake risk at Gros Cacouna, Quebec, and Melford Point, Nova Scotia: Earth Physics Branch Open File 82-2, 51 p.
Basham, P. W., Weichert, D. H., Anglin, F. M., and Berry, M. J., 1985, New probabilistic strong seismic ground motion maps of Canada: Bulletin of the Seismological Society of America, v. 75, p. 563–595.
Buchbinder, G.G.R., Kurtz, R. D., and Lambert, A., 1983, A review of time-dependent geophysical parameters in the Charlevoix region, Quebec: Earthquake Prediction Research, v. 2, p. 149–166.
Buchbinder, G.G.R., Lambert, A., Kurtz, R. D., Bower, D. R., Anglin, F. M., and Peters, J., 1988, Twelve years of geophysical research in the Charlevoix seismic zone: Tectonophysics, v. 156, p. 193–224.
Coppersmith, K. J., Johnston, A. C., and Arabasz, W. J., 1987, Methods for assessing maximum earthquakes in the central and eastern United States: Palo Alto, California, E.P.R.I. Research Project 2556-12, Working Report, Electric Power Research Institute, 108 p.

Crough, S. T., 1981, Mesozoic hotspot epeirogeny in eastern North America: Geology, v. 9, p. 2–6.

Duncan, R. A., 1984, Age progressive volcanism in the New England seamounts and the opening of the central Atlantic Ocean: Journal of Geophysical Research, v. 80, p. 9980–9990.

Ebel, J. E., 1984, Statistical aspects of New England Seismicity from 1975 to 1982 and implications for past and future earthquake activity: Bulletin of the Seismological Society of America, v. 74, p. 1311–1329.

—— , 1985, A study of seismicity and tectonics in New England: U.S. Nuclear Regulatory Commission, NUREG/CR-4354 RA, 88 p.

Ebel, J. E., Somerville, P. G., and McIver, J. D., 1986, A study of the source parameters of some large earthquakes in northeastern North America: Journal of Geophysical Research, v. 91, p. 8231–8247.

Forsyth, D. A., 1981, Characteristics of the western Quebec seismic zone: Canadian Journal of Earth Sciences, v. 18, p. 103–119.

Forsyth, D. A., Mair, J. A., and Fraser, I., 1979, Crustal structure of the central Sverdrup Basin: Canadian Journal of Earth Sciences, v. 16, p. 1581–1598.

Forsyth, D. A., and 9 others, 1983, Comparative study of the geophysical and geological information in the Timiskaming-Kapuskasing area: Atomic Energy of Canada Ltd., Technical Report TR-238, 52 p.

Forsyth, D. A., Broome, J., Embry, A. F., and Halpenny, J., 1988, Aeromagnetic, gravity, and earthquake features [of the Canadian Polar Margin]: GEOS, Energy Mines and Resources Canada, v. 17, p. 12–16.

Fujita, K., Cook, D. B., Hasegawa, H. S., Forsyth, D., and Wetmiller, R. J., 1989, Seismicity and focal mechanisms of the Arctic region and the North American plate boundary in Asia, in Grantz, A., Johnson, L., and Sweeney, J. F., eds., The Arctic Ocean region: Boulder, Colorado, Geological Society of America, The Geology of North America, v. L, p. 79–100.

Goodacre, A. K., and Hasegawa, H. S., 1980, Gravitationally induced stresses at structural boundaries: Canadian Journal of Earth Sciences, v. 17, p. 1286–1291.

Gower, C. F., Erdmer, P., and Wardle, R. J., 1986, The Double Mer Formation and the Lake Melville rift system, eastern Labrador: Canadian Journal of Earth Sciences, v. 23, p. 359–368.

Hasegawa, H. S., 1977, Focal parameters of four Sverdrup Basin, Arctic Canada, earthquakes in November and December of 1972: Canadian Journal of Earth Sciences, v. 11, p. 2481–2494.

—— , 1986, Seismotectonics in eastern Canada; An overview with emphasis on the Charlevoix and Miramichi regions: Earthquake Notes, v. 57, p. 83–94.

Hasegawa, H. S., and Basham, P. W., 1989, Spatial correlation between seismicity and postglacial rebound in eastern Canada, in Gregersen, S., and Basham, P. W., eds., Earthquakes at North Atlantic Passive Margins—Neotectonics and postglacial rebound: Dordrecht, Kluwer Academic Publishers, p. 483–500.

Haseagawa, H. S., and Herrmann, R. B., 1989, A comparison of the source mechanisms of the 1975 Laurentian Channel earthquake and the tsunamigenic 1929 Grand Banks event, in Gregersen, S., and Basham, P. W., eds., Earthquakes at North Atlantic Passive Margins; Neotectonics and postglacial rebound: Dordrecht, Kluwer Academic Publishers, p. 547–562.

Hasegawa, H. S., and Kanamori, H., 1987, Source mechanism of the magnitude 7.2 Grand Banks earthquake of November 1929—double couple or submarine landslide?: Bulletin of the Seismological Society of America, v. 77, p. 1984–2004.

Hasegawa, H. S., and Wetmiller, R. J., 1980, The Charlevoix earthquake of 19 August 1979 and its seismo-tectonic environment: Earthquake Notes, v. 51, p. 23–37.

Hasegawa, H. S., Chou, C. W., and Basham, P. W., 1979, Seismotectonics of the Beaufort Sea: Canadian Journal of Earth Sciences, v. 16, p. 816–830.

Hasegawa, H. S., Adams, J., and Yamazaki, K., 1985, Upper crustal stresses and vertical stress migration in eastern Canada: Journal of Geophysical Research, v. 90, p. 3637–3648.

Hashizume, M., 1973, Two earthquakes on Baffin Island and their tectonic implications: Journal of Geophysical Research, v. 78, p. 6069–6081.

—— , 1974, Surface wave study of earthquakes near northwestern Hudson Bay, Canada: Journal of Geophysical Research, v. 79, p. 5458–5468.

—— , 1977, Surface-wave study of the Labrador Sea earthquake, 1971 December: Geophysical Journal of the Royal Astronomical Society, v. 51, p. 149–168.

Horner, R. B., Stevens, A. E., Hasegawa, H. S., and Leblanc, G., 1978, Focal parameters of the July 12, 1975, Maniwaki, Quebec, earthquake—an example of intraplate seismicity in eastern Canada: Bulletin of the Seismological Society of America, v. 68, p. 619–640.

Horner, R. B., Wetmiller, R. J., and Hasegawa, H. S., 1979, The St. Donat, Quebec, earthquake sequence of February 18–23, 1978: Canadian Journal of Earth Sciences, v. 16, p. 1892–1898.

Jackson, H. R., Keen, C. E., Falconer, R.K.H., and Appleton, K. P., 1979, New geophysical evidence for sea-floor spreading in Baffin Bay: Canadian Journal of Earth Sciences, v. 16, p. 2122–2135.

Johnston, A. C., 1989, The seismicity of 'Stable Continental Interiors,' in Gregersen, S., and Basham, P. W., eds., Earthquakes at North Atlantic Passive Margins; Neotectonics and postglacial rebound: Dororecht, Kluwer Academic Publishers, p. 299–327.

Kerr, J. W., 1977, Cornwallis Fold Belt and the mechanism of basement uplift: Canadian Journal of Earth Sciences, v. 14, p. 1374–1401.

—— , 1980, Structural framework of Lancaster Aulacogen, Arctic Canada: Geological Survey of Canada Bulletin 319, 24 p.

Kumarapeli, P. S., 1985, Vestiges of Iapetan rifting in the west of the northern Appalachians: Geoscience Canada, v. 12, p. 55–59.

Kumarapeli, P. S., and Saull, V. A., 1966, The St. Lawrence Valley system—a North American equivalent of the east African rift valley system: Canadian Journal of Earth Sciences, v. 3, p. 639–658.

Lamontagne, M., 1987, Composite P-nodal solution analysis of earthquakes from the Charlevoix seismic zone: Canadian Journal of Earth Sciences, v. 24, p. 2118–2129.

Leblanc, G., 1981, A closer look at the September 16, 1732, Montreal earthquake: Canadian Journal of Earth Sciences, v. 18, p. 539–550.

Leblanc, G., and Anglin, F. M., 1978, Induced seismicity at the Manic 3 reservoir, Quebec: Bulletin of the Seismological Society of America, v. 68, p. 1469–1485.

Leblanc, G., and Buchbinder, G., 1977, Second microearthquake survey of the St. Lawrence Valley near La Malbaie, Quebec: Canadian Journal of Earth Sciences, v. 14, p. 2778–2789.

Leblanc, G., and Burke, K.B.S., 1985, Re-evaluation of the 1817, 1855, 1869, and 1904 Maine-New Brunswick area earthquakes: Earthquake Notes, v. 56, p. 107–123.

Liu, H.-L., and Kanamori, H., 1980, Determination of source parameters of midplate earthquakes from the waveforms of body waves: Bulletin of the Seismological Society of America, v. 70, p. 1989–2004.

Okulitch, A. V., Packard, J. J., and Zolani, A. I., 1986, Evolution of the Boothia Uplift, Arctic Canada: Canadian Journal of Earth Sciences, v. 23, p. 350–358.

Phipps, S. P., 1988, Deep rifts as sources for alkaline intraplate magmatism in eastern North America: Nature, v. 334, p. 27–31.

Piper, D.J.W., and Aksu, A. E., 1987, The source and origin of the 1929 Grand Banks turbidity current inferred from sediment budgets: Geomarine Letters, v. 56, p. 177–182.

Piper, D.J.W., Farre, J. A., and Shor, A., 1985, Late Quaternary slumps and debris flows on the Scotian Shelf: Bulletin of the Geological Society of America, v. 96, p. 1508–1517.

Podrouzek, A. J., and Bell, J. S., 1985, Stress orientations from wellbore breakouts on the Scotian Shelf, eastern Canada: Geological Survey of Canada Paper 85-1B, p. 59–62.

Poole, W. H., Sanford, B. V., Williams, H., and Kelley, D. G., 1970, Geology of southeastern Canada, in Douglas, R.J.W., ed., Geology and economic minerals of Canada: Geological Survey of Canada Economic Geology Report 1, p. 228–304.

Quinlan, G., 1984, Postglacial rebound and the focal mechanisms of eastern Canadian earthquakes: Canadian Journal of Earth Sciences, v. 19, p. 1018–1023.

Reid, I., and Falconer, R.K.H., 1982, A seismicity study in northern Baffin Bay: Canadian Journal of Earth Sciences, v. 19, p. 1518–1531.

Ruffman, A., and Peterson, J., 1988, Pre-confederation historical seismicity of Nova Scotia with an examination of selected later events: Geological Survey of Canada Open File 1917, 900 p.

Schlesinger-Miller, E. A., Barstow, N. L., and Kafka, A. L., 1983, The July 1981 earthquake sequence near Cornwall, Ontario, and Massena, New York : Earthquake Notes, v. 54, p. 11–26.

Seeber, L., and Armbruster, J. G., 1988, Seismicity along the Atlantic seaboard of the U.S.: Intraplate neotectonics and earthquake hazard, in Sheridan, R. E., and Grow, J. A., eds., The Atlantic Continental Margin, U.S.: Boulder, Colorado, Geological Society of America, The Geology of North Ameria, v. I-2, p. 565–582.

Sleep, N. H., Kroeger, G., and Stein, S., 1989, Canadian passive margin stress field inferred from seismicity: Journal of Geophysical Research (in press).

Smith, W.E.T., 1962, Earthquakes of eastern Canada and adjacent areas 1534–1927: Publications of the Dominion Observatory, v. 26, p. 271–301.

— , 1966, Earthquakes of eastern Canada and adjacent areas 1928–1959: Publications of the Dominion Observatory, v. 32, p. 87–121.

Smith, W.E.T., Whitham, K., and Piché, W. T., 1968, A microearthquake swarm in 1965 near Mould Bay, Northwest Territories, Canada: Bulletin of the Seismological Society of America, v. 58, p. 1991–2011.

Spencer, C., and 6 others, 1989, Allochthonous units in the northern Appalachians—results from the Quebec-Maine seismic and refraction surveys: Tectonics, v. 8, p. 677–696.

Srivastava, S. P., and Tapscott, C. R., 1986, Plate kinematics of the North Atlantic, in Vogt, P. R., and Tucholke, B. E., eds., The western North Atlantic region: Boulder, Colorado, Geological Society of America, The Geology of North America, v. M, p. 379–404.

Staveley, M., and Adams, J., 1985, Historical seismicity of Newfoundland: Earth Physics Branch Open File 85-22, 73 p.

Stein, S., Sleep, N. H., Geller, R. J., Wang, S. C., and Kroeger, G. C., 1979, Earthquakes along the passive margin of eastern Canada: Geophysical Research Letters, v. 6, p. 537–540.

Stevens, A., 1980, Reexamination of some larger La Malbaie, Quebec, earthquakes (1924–1978): Bulletin of the Seismological Society of America, v. 70, p. 529–557.

Sweeney, J. F., Irving, E., and Geuer, J. W., 1978, Evolution of the Arctic Basin, in Sweeney, J. F., ed., Arctic Geophysical Review: Canadian Earth Physics Branch Publication, v. 45, p. 91–100.

Sykes, L. R., 1978, Intraplate seismicity, reactivation of preexisting zones of weakness, alkaline magmatism, and other tectonism postdating continental fragmentation: Reviews of Geophysics and Space Physics, v 16, p. 621–688.

Sykes, L. R., and Sbar, M. L., 1974, Focal mechanism solutions of intraplate earthquakes and stresses in the lithosphere, in Kristjansson, ed., Geodynamics of Iceland and the North Atlantic area: Dordrecht, Netherlands, D. Reidel, p. 207–224.

Wahlstrom, R., 1987a, The North Gower, Ontario, earthquake of 11 October, 1983—Focal mechanism and aftershocks: Seismological Research Letters, v. 58(3), p. 65–72.

— , 1987b, Focal mechanisms of earthquakes in southern Quebec, southeastern Ontario, and northeastern New York with implications for regional seismotectonics and stress field characteristics: Bulletin of the Seismological Society of America, v. 77, p. 891–924.

Wetmiller, R. J., 1975, The Quebec–Maine border earthquake, 15 June 1973: Canadian Journal of Earth Sciences, v. 12, p. 1917–1928.

— , 1980, Investigation of earthquakes in Burlington, Ontario: Seismological Service of Canada Internal Report 80–5. 18 p.

Wetmiller, R. J., and Cajka, M. G., 1989, Tectonic implications of the seismic activity recorded by the northern Ontario seismograph network: Canadian Journal of Earth Sciences, v. 26, p. 376–386.

Wetmiller, R. J., and Forsyth, D. A., 1978, Seismicity of the Arctic, 1908–1975, in Sweeney, J. F., ed., Arctic Geophysical Review: Canadian Earth Physics Branch Publication, v. 45, p. 15–24.

— , 1982, Review of seismicity and other geophysical data near Nares Strait, in Dawes, P. R., and Kerr, J. W., eds., Nares Strait and the drift of Greenland; A conflict in plate tectonics: Meddelelser om Grønland Geoscience, v. 8, p. 261–274.

Wetmiller, R. J., Adams, J., Anglin, F. M., Hasegawa, H. S., and Stevens, A. E., 1984, Aftershock sequences of the 1982 Miramichi, New Brunswick, earthquakes: Bulletin of the Seismological Society of America, v. 74, p. 621–653.

Woollard, G. P., 1969, Tectonic activity in North America as indicated by earthquakes, in Hart, P. J., ed., The Earth's crust and upper mantle: American Geophysical Union Geophysical Monograph 13, p. 125–133.

Yang, J. P., and Aggarwal, Y. P., 1981, Seismotectonics of northeastern United States and adjacent Canada: Journal of Geophysical Research, v. 86, p. 4981–4998.

Zoback, M., Nishenko, S. P., Richardson, R. M., Hasegawa, H. S., and Zoback, M. D., 1986, Mid-plate stress, deformation, and seismicity, in Vogt, P. R., and Tucholke, B. E., eds., The western North Atlantic region: Boulder, Colorado, Geological Society of America, The Geology of North America, v. M, p. 297–312.

Manuscript Accepted by the Society November 16, 1988

ACKNOWLEDGMENTS

We thank M. J. Berry, H. S. Hasegawa, and R. J. Wetmiller for valuable critical comments on drafts of this manuscript. We also thank the participants at various workshops whose discussions over the past year have enabled us to clarify our thoughts on the seismotectonics of eastern Canada. Geological Survey of Canada Contribution Number 36987.

NOTES ADDED IN PROOF

The original manuscript for this paper was submitted to the editors in October 1988. In the subsequent three years there have been some important developments related to the seismicity and seismotectonics of eastern Canada, and we take this opportunity offered by the editors to bring the readership up to date.

Saguenay Earthquake, 25 November 1988. This M (Lg) 6.5 earthquake, the largest to take place in eastern North America in more than 50 years, occurred in the Grenville Province of the Canadian Shield, 35 km south of Chicoutimi, Quebec, in a hitherto aseismic region ~40 km northwest of the northwest corner of the Charlevoix box in Figure 1. The earthquake was unusual in that it occurred at a depth of 29 km (North and others, 1989), and important in that it produced the first set of high-quality strong motion data recorded for a strong earthquake in eastern North America (Munro and Weichert, 1988). The earthquake was preceded by an M4.8 foreshock on November 23. Three field seismographs, which had been installed within 10 km of the epicenter when the mainshock occurred, provided important data for accurate hypocenter location. The P-nodal focal mechanisms of the foreshock, main shock, and largest aftershock all exhibit dominantly thrust faulting, consistent with the northeast to east regional compression. The distribution of the small number of well-located aftershocks does not suggest an obvious rupture surface. Research in progress by R.A.W. Haddon suggests that the rupture occurred on a long, narrow fault and had significant southward directivity, contributing stronger than normal ground motions in the direction of most of the strong-motion instruments. High-frequency ground motions from the Saguenay earthquake considerably exceeded theoretical predictions (e.g., Atkinson, 1990), and there is an unresolved debate as to whether the earthquake was typical or atypical of eastern earthquakes. Lamontagne and others (1990) have shown that a major lineament, in our view the southern boundary of the Saguenay graben, passes very close to the epicenter. We infer the graben, a feature similar in age and origin to the Ottawa graben, is probably the causative feature. We did not identify it as a part of the Paleozoic rift on Figure 7 because of its (then) very low seismicity.

Ungava Earthquake, 25 December 1989. This M6.3 earthquake occurred near the center of the Ungava Peninsula in northern Quebec, just to the east of the cluster of smaller earthquakes shown in Figure 5. It produced the first surface faulting from a historical intraplate earthquake in North America (Adams and others, 1991). Field surveys during the summers of 1990 and 1991 have mapped the surface rupture, mainly on the basis of deformed lake shorelines and torn and buckled muskeg, which was frozen at the time of the earthquake. The arcuate scarp is 8.5 km long and the fault dips steeply to the southeast. The maximum throw is 1.8 m, with the southeast side upthrown, together with a significant amount of left-lateral strike slip. Accurately located aftershocks are shallow, suggesting the rupture does not extend to depths greater than 5 km. The geologically determined value of seismic moment is similar to the value determined from seismological observations. It is judged that the preservation potential of this evidence for surface faulting is not great; i.e., such earthquakes will not leave a strong geological record of faulting. Hence prehistoric surface ruptures of earthquakes elsewhere in the Canadian Shield may be difficult to find.

Iapetus Rift Faults. Our terms "Paleozoic Rift System" and "St. Lawrence Rift System" in the chapter were adopted for geographical and historical reasons arising from Kumarapeli's work. We used them for the zone of normal faults along the St. Lawrence River that weakened the craton in post-Ordovician times and are now reactivated by the present compressional stress field. R. Wheeler (personal communication, 1991) suggests that "Iapetan passive continental margin" better describes the genesis of these faults and should be applied to a very large geological province, with presumed similar earthquake potential, extending from Labrador to Alabama. Wheeler's idea of a large geological province is in accord with our own hypothesis of the large-scale causes of earthquakes in eastern Canada (Figure 7) and with the worldwide study of Johnston (1989). Hence, we will probably use the term "Iapetan rifted margin" in future papers.

In reviews of this chapter and through other publications (e.g., Adams and Basham, 1989) we have received some criticism for the speculative extension of the "Paleozoic Rift System" through Lakes Ontario and Erie in Figure 7. It has the implication that the largest earthquakes associated with the rift features (M7) are likely to occur near critical facilities around the lakes. Our reason for the speculative extension was simply to point out that the higher rate of moderate-magnitude earthquakes in and near Lakes Ontario and Erie than in adjacent regions is essentially unexplained. Work in progress by the Geological Survey of Canada in cooperation with Ontario Hydro on interpreting seismic reflection data from these lakes shows no evidence that the basement contains a through-going rift fault system similar to that in the St. Lawrence valley, although post-Paleozoic extensional faulting is well known near Picton on the north shore and near Attica on the south shore of Lake Ontario. There remains the essentially continuous zone of seismicity, through Lakes Ontario and Erie, that extends from the St. Lawrence valley to the New Madrid zone in northern Arkansas, first noticed by Wollard (1969). An important addition to this trend of seismicity is the recent discovery by Obermeier and others (1991) of liquefaction features in the Wabash valley bordering Indiana and Illinois, interpreted to have been produced by earthquakes as large as M6.7, significantly larger than any in the historical record.

ADDITIONAL REFERENCES

Adams, J., and Basham, P. W., 1989, The seismicity and seismotectonics of Canada east of the Cordillera: Geoscience Canada, v. 16, p. 3–16.

Adams, J., Wetmiller, R. J., Hasegawa, H. S., and Drysdale, J., 1991, The first surface faulting from a historic intraplate earthquake in North America, Nature, v. 352, p. 617–619.

Atkinson, G. M., 1990, A comparison of eastern North American ground motion observations with theoretical predictions: Seismological Research Letters, v. 61, p. 171–180.

Lamontagne, M., Wetmiller, R. J., and Du Berger, R., 1990, Some results from the 25 November, 1988 Saguenay, Quebec, earthquake: *in* Current Research, Part B, Geological Survey of Canada, Paper 90-1B, p. 115–121.

Munro, P. S., and Weichert, D. H., 1988, The Saguenay earthquake of November 25, 1988; Processed strong motion records: Geological Survey of Canada Open File Report No. 1996, 145 p.

North, R. G., and 8 others, 1989, Preliminary results from the November 25, 1988 Saguenay (Quebec) earthquake: Seismological Research Letters, v. 60, p. 89–93.

Obermeier, S. F., and 9 others, 1991, Evidence of strong earthquake shaking in the Lower Wabash Valley from prehistoric liquefaction features: Science, v. 251, p. 1061–1063.

Printed in U.S.A.

The Geology of North America
Decade Map Volume 1
1991

Chapter 15

Earthquake activity in the northeastern United States

John E. Ebel and Alan L. Kafka
Weston Observatory, Department of Geology and Geophysics, Boston College, Weston, Massachusetts 02193

INTRODUCTION

The northeastern United States has one of the longest records of reported earthquake activity in North America (Fig. 1). Earthquake activity in this region was noticed by the early European settlers within the first few decades after their arrival. Since the early colonial days, the increasing population density throughout the Northeast led to more frequent reports of usually minor, but sometimes not-so-minor and occasionally even damaging earthquakes. Instrumental seismic monitoring began in the early 1900s, and routine reporting of the earthquake activity in the region began in 1938 with the initiation of the Northeastern Seismic Association (NESA) bulletins. The first telemetered regional seismic network was operated in northern New England by Weston Observatory from 1962 to 1968, but it was not until the early 1970s that the present regional networks were established. The number of seismic stations in the northeastern United States steadily increased between 1970 and 1974. By 1975, a number of institutions operating seismic networks in the region formed a cooperative group known as the Northeastern United States Seismic Network (NEUSSN; Fig. 2). The data recorded by the NEUSSN has enhanced our ability to study the regional seismic activity, and analysis of these data provides insight into the possible causes of earthquakes in this region.

Since at present the northeastern United States is not located on or near a plate boundary, there is no obvious plate-tectonic interpretation of why earthquakes occur here. The present-day tectonic setting of the region is a passive continental margin. The nearest plate boundaries are the Mid-Atlantic Ridge to the east, the subduction/transform system along the western margin of North America to the west, and the northern boundary of the Caribbean plate to the south. All of these plate boundaries are located several thousand kilometers from the northeastern United States. The last major tectonic activity in this region postdates the Triassic separation of North America from Africa and occurred in the latter part of the Mesozoic. This activity is delineated by igneous rocks found in the White Mountain magma series of New England, in the Montegerian Hills of southern Quebec, and along the New England seamount chain (Sykes, 1978). In the more distant geologic past, the area experienced at least two continental collision and rifting episodes, of which the Appalachian mountain system is a remnant expression (e.g., Bird and Dewey, 1970). While the present-day seismicity is quite low compared to that of most plate boundaries, its persistence and its potential for producing damaging earthquakes make it the subject of both scientific inquiry and public concern.

Three fundamental, and as yet unanswered, questions about the regional seismicity provide the underlying framework of this chapter: (1) What are the forces that cause earthquakes in this area? (2) What geologic or tectonic features are seismically active? (3) What is the potential for future large earthquakes?

After summarizing the current understanding of earthquake activity in the northeastern United States, we discuss the extent to which the current state of knowledge helps to address these fundamental questions. First, we discuss the spatial distribution of earthquake activity. Next, we summarize the available data on the state of stress in the crust and on the relationship between earthquakes and mapped faults. Thirdly, we discuss the temporal distribution of the earthquake activity. Finally, we present a map of seismic zones in the Northeast.

SPATIAL DISTRIBUTION OF EARTHQUAKE ACTIVITY

Epicenters of earthquakes recorded in the northeastern United States and adjacent Canada since the beginning of the colonial period are shown in Figure 1. Locations of the older epicenters are based upon felt reports, while the locations of earthquakes since the 1930s have been determined primarily from instrumental data. The earthquakes shown in Figure 1 indicate that much of the area is seismically active but that the distribution of earthquake activity is not spatially uniform. The most active zones in the Northeast are: (1) eastern, coastal, and western Maine; (2) central and southern New Hampshire; (3) eastern Massachusetts; (4) eastern Rhode Island; (5) central and southern Connecticut; (6) southern New York; (7) northeastern, central, and southern New Jersey; (8) eastern Pennsylvania; (9) western New York; (10) northeastern New York; and (11) northwestern Vermont. All of these zones have been seismically active throughout both historic and instrumental time, and the largest earthquakes in the

Ebel, J. E., and Kafka, A. L., 1991, Earthquake activity in the northeastern United States, *in* Slemmons, D. B., Engdahl, E. R., Zoback, M. D., and Blackwell, D. D., eds., Neotectonics of North America: Boulder, Colorado, Geological Society of America, Decade Map Volume 1.

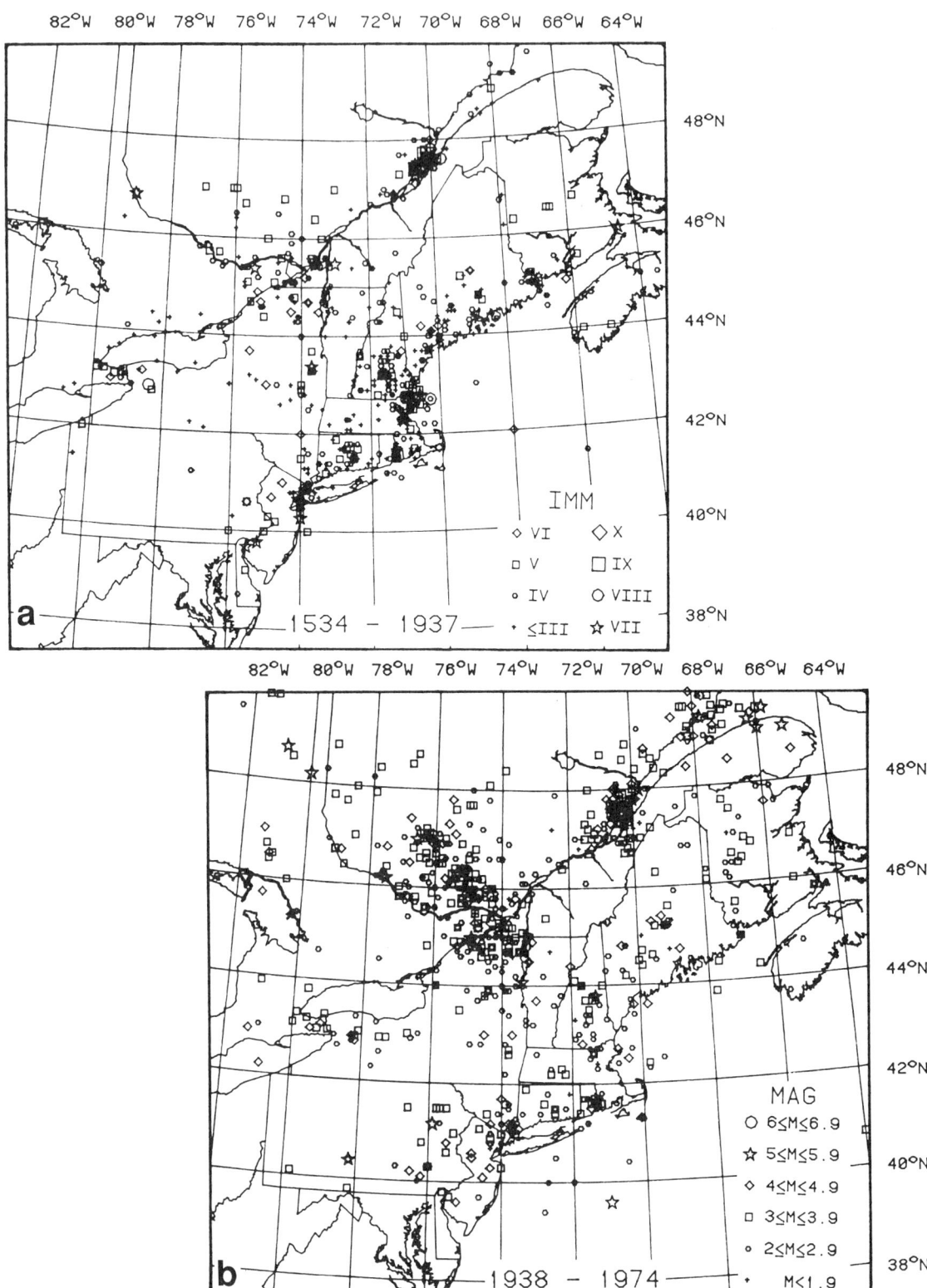

Figure 1. The northeastern United States (including the states of Maine, New Hampshire, Vermont, Massachusetts, Rhode Island, Connecticut, New York, New Jersey, Pennsylvania, and Delaware) showing (a) historical (1534 to 1937), (b) early instrumental (1938 to 1974), and (c) network (1975 to 1985) seismicity. The seismicity in western Quebec and nearby Ontario is not complete in these plots.

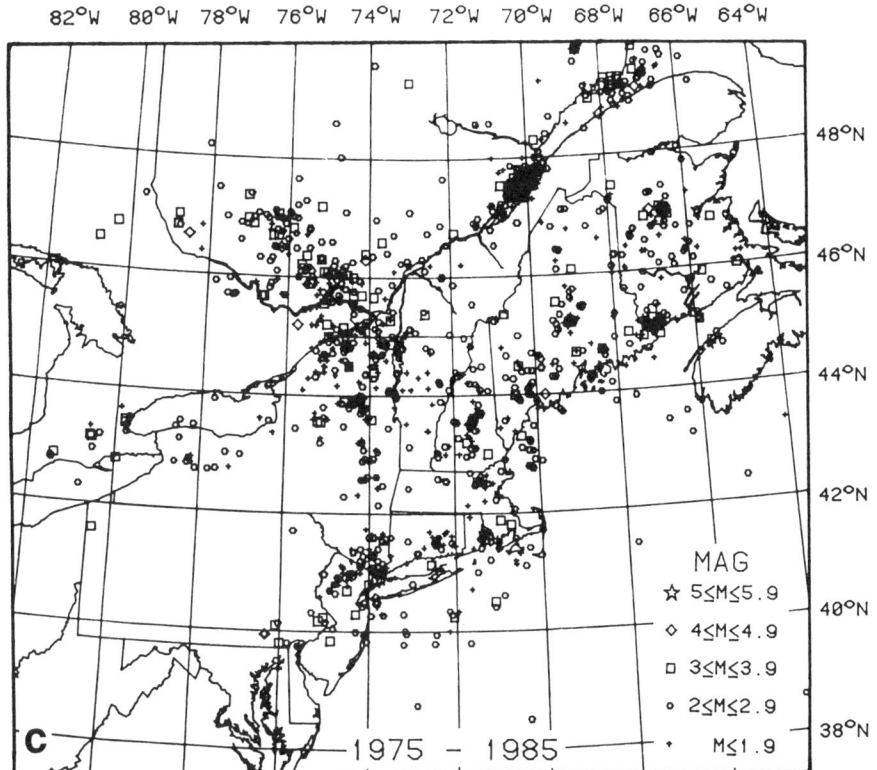

Northeast have all occurred in one or another of these active areas (Fig. 1).

The largest earthquakes recorded in the northeastern United States are: the 1727 and 1755 earthquakes located off the northeastern coast of Massachusetts; the 1884 earthquake near New York City; the 1904 earthquake in eastern Maine; the 1929 earthquake in western New York; the 1940 earthquakes in central New Hampshire; the 1944 earthquake at the northern New York/Canadian border; and the 1983 earthquake in north-central New York. The magnitudes of all these earthquakes, except perhaps the 1884 event, probably exceeded m_{bLg} of 5.0 (Street and Turcotte, 1977; Street and Lacroix, 1979), and all of them caused at least minor damage.

There is a general correlation between the spatial distribution of epicenters determined from the network data and that of the historical seismicity. This correlation between network and historical seismicity has been discussed by Yang and Aggarwal (1981) and Ebel (1984). In the section of this chapter on Seismic zones in the northeastern United States, we demonstrate that the general features of the spatial distribution of seismicity have been fairly stable since about the mid-1500s. Apparently, whatever the process is that causes earthquakes in this region, that process has been spatially stable for the past several hundred years.

Although the relatively low density of local seismic stations (Fig. 2) limits the resolution of focal depth for many of the events, the station density is sufficient to conclude that earthquakes in the area generally occur in the upper half of the crust. The deepest hypocenters located by the NEUSSN stations generally occur at about 20 km, and the shallowest earthquakes are about 1 km deep (Pomeroy and others, 1976; Ebel and others, 1982; Anglin, 1984; Filipkowski, 1986). Events deeper than 20 km tend to occur more frequently in adjacent Canada rather than in the northeastern United States. Earthquakes as deep as 33 km have been reported in the Charlevoix seismic zone—a cluster of activity centered at about 47.5°N, 70.3°W (Fig. 1; also see Anglin, 1984; and Adams and Basham, this volume).

In an effort to summarize the depth distribution of earthquakes recorded in the Northeast, we analyzed the depths reported for a sample of earthquakes from the NEUSSN bulletins. This sample consisted of all events recorded between 1975 and 1983, and from those events we extracted all earthquakes for which there was (a) at least one station located within a distance of twice the depth; and (b) a total of at least four stations recording the event. A histogram of those data is shown in Figure 3. For this set of data, the mean depth is 9 km and the standard deviation is 5 km. Because of the criterion of requiring at least one nearby station, the depths shown in the histogram are probably biased toward deeper events (which would be more likely to be recorded by a station within twice the depth, given the NEUSSN station spacing). From Figure 3 it appears that the earthquake activity tends to cluster in a zone between about 5 and 15 km beneath the earth's surface (although events shallower than 5 km could be more numerous than Fig. 3 implies since they were systematically excluded from these statistics).

An independent measure of the depth range where the larger earthquakes tend to occur in this region can be obtained from the results of waveform modeling of larger earthquakes that were recorded teleseismically. Also shown in Figure 3 is the range of

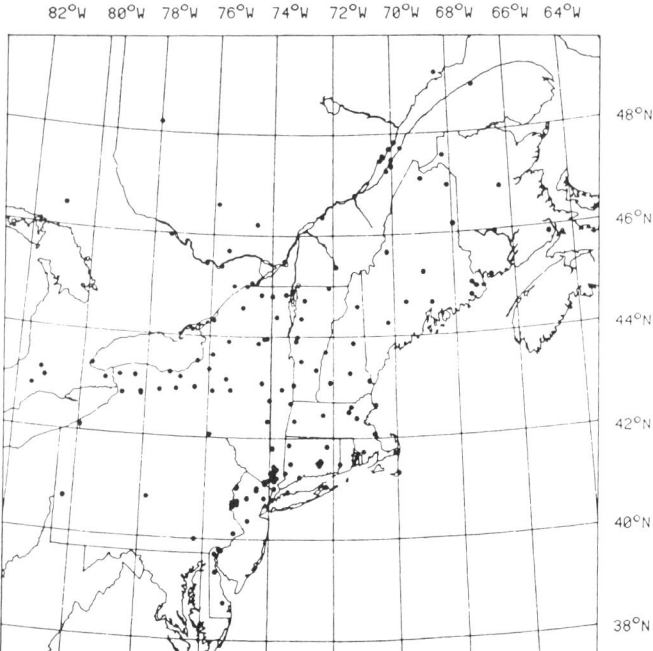

Figure 2. Stations of the Northeastern United States Seismic Network (NEUSSN). These stations are operated by Lamont-Doherty Geological Observatory of Columbia University, Massachusetts Institute of Technology, Weston Observatory of Boston College, Woodward-Clyde Consultants, State University of New York at Stony Brook, Pennsylvania State University, and Delaware Geological Survey.

depths that Ebel and others (1986) found from teleseismic waveform modeling of the largest earthquakes of this century in northeastern North America. Those events were found to occur at depths rangings from 8 to 10 km. Also, Nabelek (1984) and Basham and Kind (1986) modeled teleseismic waveforms from the January 9, 1982, earthquake in New Brunswick, Canada (m_b=5.7), and they concluded that the depth of that event was 6 km. Similar depth ranges were observed from aftershock surveys of larger earthquakes. For example, Wetmiller and others (1984) found that aftershocks of the 1982 New Brunswick earthquakes occurred at depths ranging from very near the surface to about 7 km, and Seeber and others (1984) found that aftershocks of the 1983 earthquake at Goodnow, New York (m_b=5.1), were confined to a depth range of 7 to 8.5 km.

All of these depth estimates suggest that in the northeastern United States, earthquakes occur in the upper 20 km of the crust. This depth range is consistent with the idea that only the upper half of the crust behaves in a brittle fashion (Kirby, 1980).

EARTHQUAKES, STRESS FIELD, AND MAPPED FAULTS

The most commonly accepted explanation for the earthquakes in the Northeast is that ancient zones of weakness are being reactivated in the present-day stress field (e.g., Sykes, 1978). In this model, preexisting faults and zones of weakness from earlier orogenic episodes persist in the intraplate crust, and by way of analogy with plate-boundary seismicity, earthquakes occur when the present-day stress is released along these zones of weakness. Much of the recent research on the cause of earthquakes in the northeastern United States has, therefore, involved attempts to identify preexisting faults and determine whether they are favorably oriented so as to be reactivated by the present-day stress field. While this concept of reactivation of old zones of weakness is commonly assumed, the identification of individual active geologic features has proven to be quite difficult; unlike the situation along plate boundaries, it is not at all clear whether faults mapped at the earth's surface in the Northeast are the same faults along which the earthquakes are occurring.

State of stress

The intraplate stress field in the northeastern United States is generally assumed to be the result of a combination of forces generated by plate-tectonic processes. Two large-scale sources of stress that have been considered in a number of studies are aesthenospheric drag and ridge push (e.g., Richardson and others, 1979). Another source of stress that could be significant in plate interiors is aesthenospheric counterflow (e.g., Chase, 1979; Hager and O'Connell, 1979). These stress models are discussed more extensively elsewhere in this volume. Within the context of the ancient zones of weakness hypothesis, it is important to characterize accurately the observed modern stress field to determine whether preexisting faults are favorably oriented to be reactivated.

Zoback and Zoback (1985, and this volume) analyzed the present-day stress field throughout the interior of the North American plate, and Gephart and Forsyth (1985) discussed the relationship between the present-day stress field and earthquake focal mechanisms in New England. In an effort to summarize the present state of knowledge regarding the stress field in the Northeast, we have reviewed data from focal mechanism studies in the

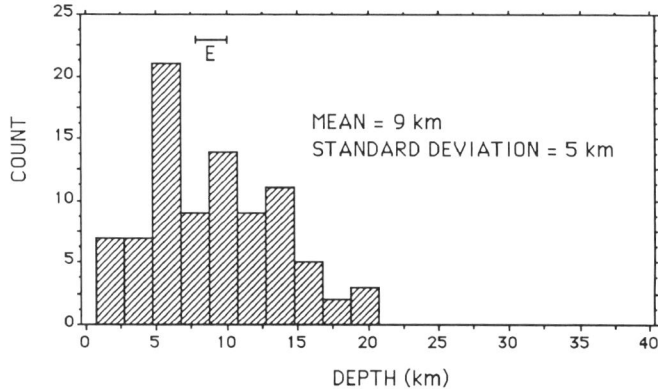

Figure 3. Histogram of depths of earthquakes recorded by the NEUSSN in the northeastern United States from 1975 to 1983. Bar labelled E indicates the range of depths that Ebel and others (1986) found from teleseismic waveform modelling of some of the largest earthquakes in North America.

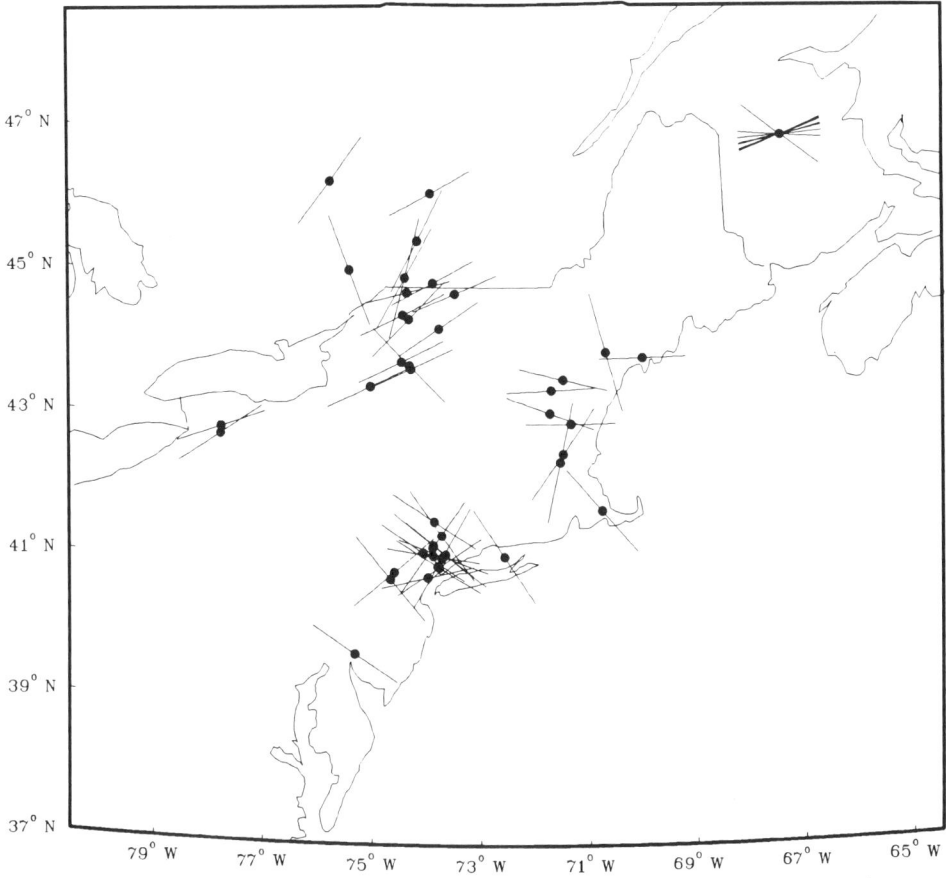

Figure 4. Map showing azimuths of P axes from focal mechanisms of earthquakes in the northeastern United States. The data shown were taken from references listed in caption of Figure 5.

region and adjacent parts of Canada (Figs. 4 and 5). Figure 4 shows the azimuths of the horizontal component of P axes from numerous studies superposed on a map of the region. Figure 5 shows equal-area stereographic plots of the P axes and the T axes from the various studies. A total of 53 earthquakes make up our data set, including ten focal mechanisms from the 1982 Miramichi, New Brunswick, earthquake sequence. There are a number of cases where the same earthquake was published in two or more papers with radically different results (most often with regard to the azimuths of P axes), and we excluded those events from this analysis. We computed the average P axis and T axis for this data set by vectorially averaging all of the axes that intercept a given focal hemisphere. Because the P axes tend to cluster in the east and west directions (Fig. 5), we performed the calculation in the western hemisphere and found the average P axis to have an azimuth of 266° and a plunge of 1°. The standard deviation of the P axes relative to this direction is 38°. The T-axis calculation was done for a lower hemisphere projection, and the average T axis was determined to have an azimuth of 105° and a plunge of 88°. The standard deviation of the observations around this value is 32°. To reduce the influence of the Miramichi events on the average stress directions, we recalculated them using only three events from this sequence, the largest events from January, April, and June, 1982. For this set of 46 focal mechanisms, the average P axis has an azimuth of 86°, a plunge of 1°, and a standard deviation of 40°; the average T axis has an azimuth of 2°, a plunge of 89°, and a standard deviation of 33°.

Thus, on average for the entire northeastern United States and parts of adjacent Canada, the maximum compressive stress inferred from focal mechanism data is essentially horizontal and trends approximately east-west. Also, the minimum stress, as inferred from the average T axis, is nearly vertical. There is, of course, no reason to assume that the stress field doesn't vary across the region studied, and in the section on Seismic zones in the northeastern United States (see below in this chapter), we discuss the variation of average orientations of P and T axes for several subregions of the northeastern United States and adjacent Canada.

Seismicity and mapped faults

If the stress field in the upper crust can be accurately characterized, and if reactivation of ancient faults is the cause of earthquakes in the northeastern United States, then seismic zones could be delineated by identifying the faults and determining their

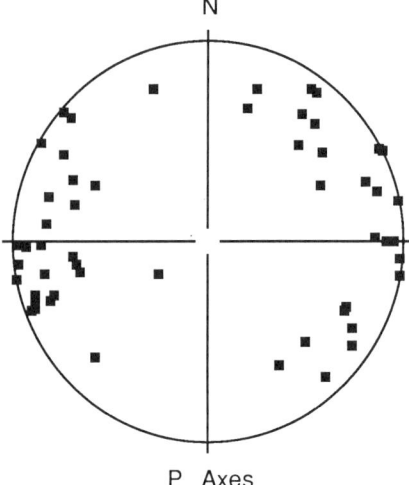

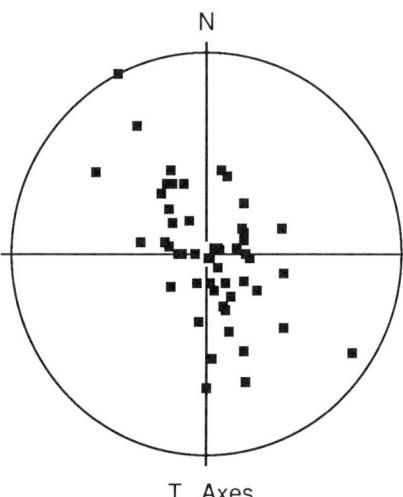

Figure 5. Lower hemisphere stereographic plots of (a) P axes and (b) T axes. The data shown were taken from the following studies: Sbar and others (1970, 1972, 1975), Herrmann (1979), Horner and others (1978), Graham and Chiburis (1980), Yang and Aggarwal (1981), Pulli and Toksöz (1981), Kafka (1982), Kafka and others (1982), Schlesinger-Miller and others (1983), Wetmiller and others (1984), Wahlstrom (1985), Quittmeyer and others (1985), and Filipkowski (1986). If angles were not published in numerical form, angles were measured from published stereographic projections.

orientation relative to the stress field. The situation is apparently more complex, however, since the regional seismicity does not reveal any obvious alignments of epicenters along mapped faults or other known structural boundaries. It is thus quite difficult to prove the existence of any seismically active faults on a regional basis. Detailed studies of seismicity on a more local scale, especially results of aftershock monitoring of larger events, also yield ambiguous results with regard to the identification of active features. In this section, we discuss several examples of such detailed studies, and we conclude that an unequivocal correlation between mapped faults and earthquake activity is, at best, the exception rather than the rule.

One of the larger earthquakes that occurred since the installation of the NEUSSN stations was the magnitude 4.7, Gaza, New Hampshire, earthquake of 1982. The location of that earthquake and its aftershocks are shown in Figure 6a. Both the Gaza, New Hampshire, earthquake and the 1983 Dixfield, Maine, earthquake (magnitude 4.4; Fig. 6b) occurred in locations where field mapping has failed to show surface faults in the vicinity of the epicenters (Ebel and McCaffrey, 1984; Brown and Ebel, 1985). Microearthquake swarms near Moodus, Connecticut, have occurred near a locality where a fault has been inferred but never confirmed (Ebel, 1985). In apparent contrast, aftershocks of the magnitude-4.0 earthquake near Bath, Maine, in 1979 occurred on or very near the Cape Elizabeth fault (Ebel, 1983), and this may be evidence that, at least in some cases, earthquakes in the Northeast are associated with mapped faults. However, the mere spatial association of earthquakes with ancient faults that are mapped near the surface does not necessarily imply that the earthquakes are occurring on those faults. The faults could, for example, be the nucleation points for the earthquakes, and yet the earthquake movements themselves may not have actually occurred on the preexisting fault surfaces.

Another event of interest is the magnitude 5.1 Goodnow, New York, earthquake of 1983 that occurred in the central Adirondack Mountains. That event was large enough to be felt throughout the northeastern United States and into southeastern Canada, and aftershock surveys provided a well-constrained location for the source region. Seeber and others (1984) and Seeber (1987) argued that the aftershocks of the Goodnow earthquake occurred on a steeply dipping fault striking N 15° ± 10°W and dipping 60° ± 10° beneath a surface lineament striking north-northwest, although this lineament had not been previously confirmed to be a fault. The focal mechanism for this earthquake supports the trend seen in the aftershocks, but no surface fault breakage was found from the event.

Several other associations between smaller earthquakes and surface lineaments striking north-northwest to northwest in the Adirondack Mountains were proposed by Yang and Aggarwal (1981). Based on the orientations of faults inferred from earthquake focal mechanisms and on the spatial distribution of microearthquake activity, they argued that such basement lineaments could be the surface expression of presently active faults. While this could be an example of a possible correlation between lineations mapped on the surface and the occurrence of earthquakes, it doesn't confirm that those lineations are in fact expressions of active faults.

Another possible candidate for present-day seismic activity that results from reactivation of an ancient zone of weakness is the Ramapo fault system in northern New Jersey and southeastern New York. The Ramapo fault system forms the northwestern margin of the Triassic-Jurassic Newark basin. This fault system is about 120 km long and appears to have been active at various times throughout geologic history, including the Precambrian,

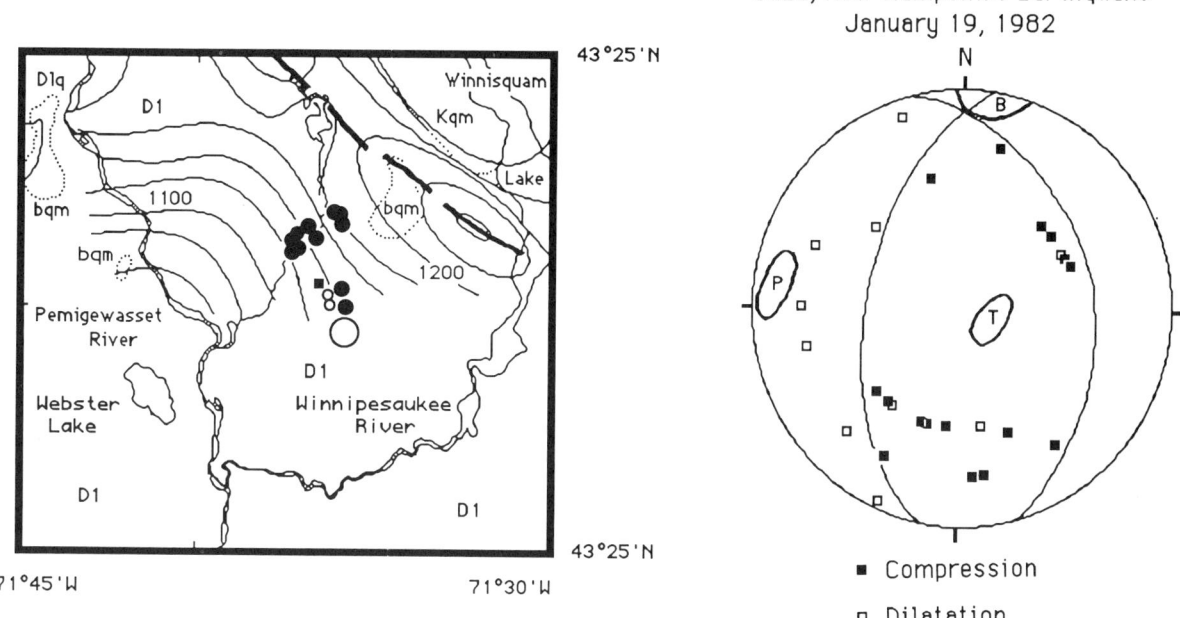

Figure 6a. Epicentral area and fault-plane solution of the 1982, Gaza, New Hampshire, earthquake. Large open circle indicates relative location of mainshock, and small open circles indicate relative locations of immediate aftershocks. Location of master event is shown by the closed square. Closed circles indicate locations of other aftershocks. Aeromagnetic contours are also shown. Heavy dashed line indicates the trend of the Laconia aeromagnetic anomaly.

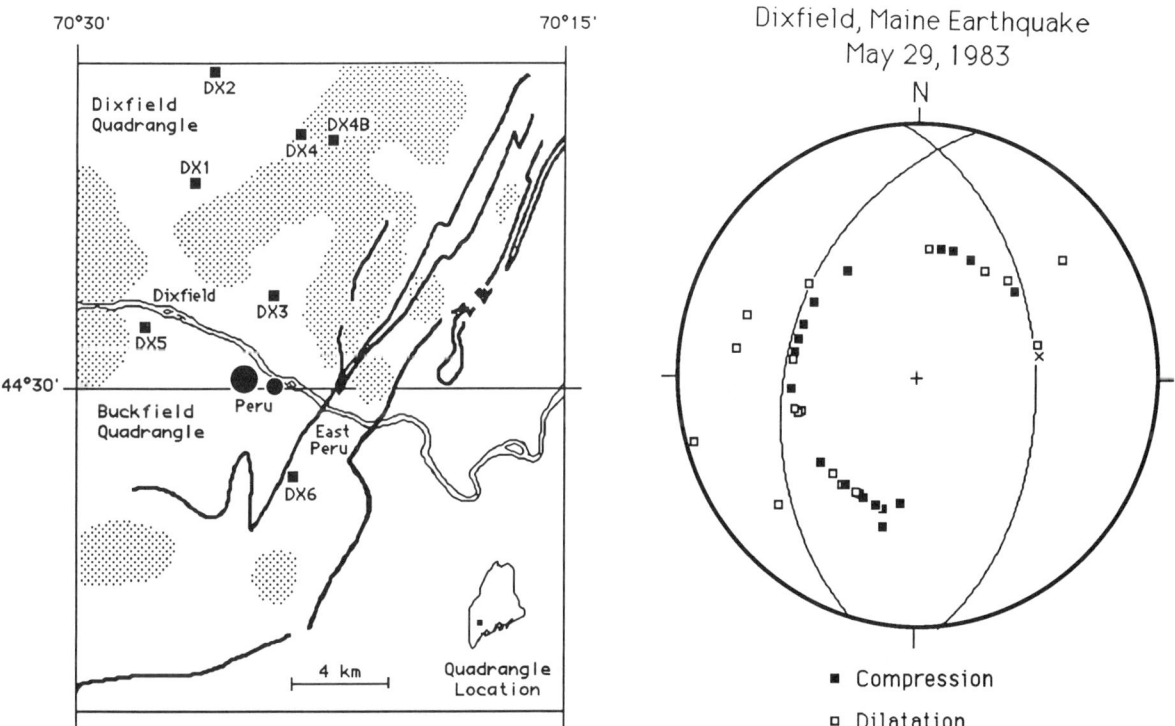

Figure 6b. Epicentral area and fault plane solution of the 1983, Dixfield, Maine, earthquake. Large closed circle indicates the location of the mainshock, and small closed circle indicates the location of a well located aftershock. Closed squares indicate the locations of portable seismic stations. Thick solid lines are mapped faults.

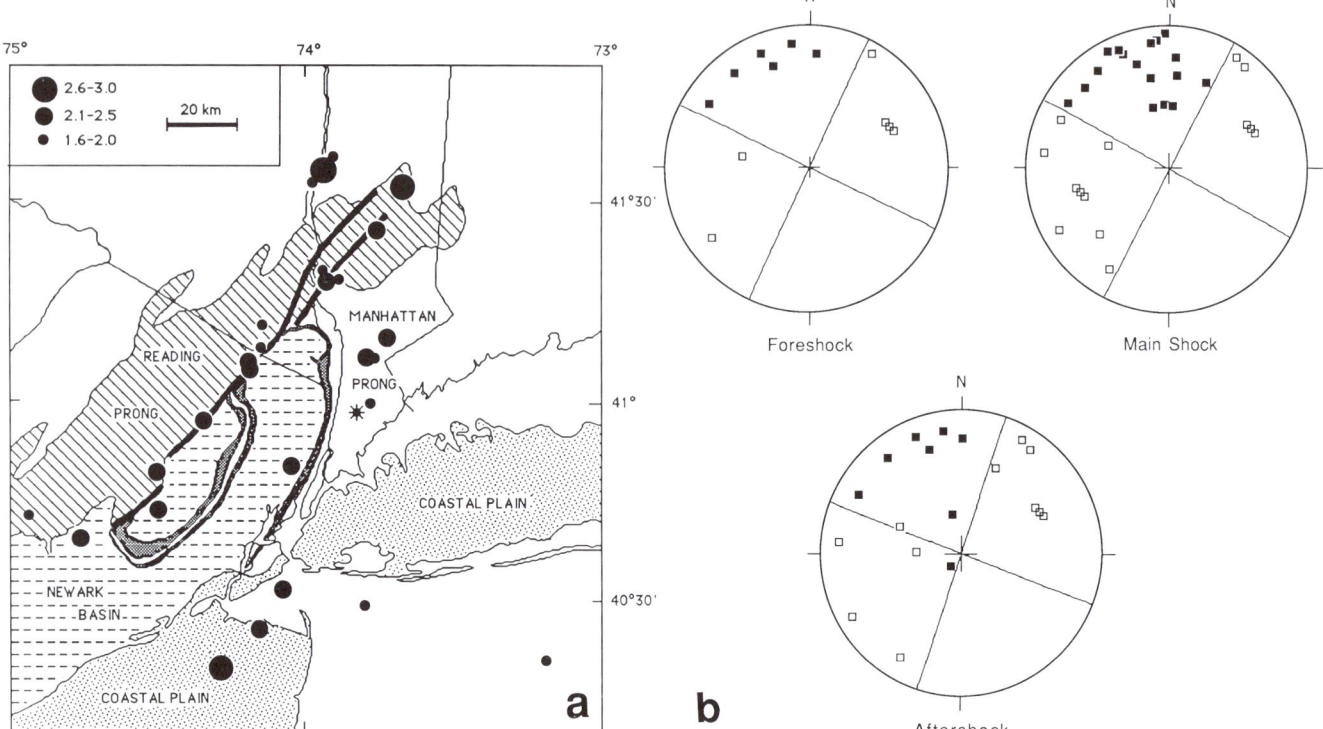

Figure 7. (a) Earthquakes recorded by stations of the NEUSSN in the greater New York City area. Closed circles are m_{bLg} magnitudes estimated from signal duration as described by Kafka and others (1985). Thick solid lines represent the surface trace of the Ramapo fault. Star indicates location of the Ardsley, New York, earthquake sequence of October 1985. (b) Fault-plane solution for the Ardsley, New York, earthquake sequence (adapted from Filipkowski, 1986).

Paleozoic, Triassic, and Jurassic (Ratcliffe, 1971). Thus, the Ramapo fault could be an example of a major throughgoing fault that is being reactivated by the present-day stress field in the northeastern United States.

Aggarwal and Sykes (1978) studied locations, depths, and focal mechanims of earthquakes in the greater New York City area, and they concluded that seismic activity in that region is concentrated along several northeast-trending faults, of which the Ramapo fault appears to be the most active. More recent studies of earthquakes in the New York City area, however, have produced results that suggest a more complicated relationship between earthquakes and mapped faults in that region. For example, Seborowski and others (1982) argued that focal mechanisms of three earthquakes that occurred on or very near the Ramapo fault have fault planes transverse to the mapped trace of the fault. Moreover, they argued that the microearthquake seismicity near the northern end of the Ramapo fault trends northwest, transverse to the trend of major geologic structures mapped at the surface. In addition, based on the distribution of earthquakes recorded by the NEUSSN, it is not clear that the Ramapo fault is more active than other parts of the New York City area. Kafka and others (1985) argued that although about half the earthquakes in the New York City area occur within 10 km of the Ramapo fault, earthquakes larger than those recorded near the Ramapo fault are located as far as 50 km from that fault in a variety of geologic structures that surround the northern part of the Newark basin (Fig. 7). Indeed, the largest earthquake recorded by the modern NEUSSN stations in the New York City area occurred near Ardsley, New York (October 19, 1985; magnitude 3.8), about 25 km from the mapped trace of the Ramapo fault (Fig. 7). Fault-plane solutions for the Ardsley, New York, earthquake and a foreshock and aftershock are shown in Figure 7b. The nodal planes are fairly well constrained, with one possible fault plane trending northeast and the other plane trending northwest. Thus, in the New York City area, as in other parts of the Northeast, it is not clear which mapped faults, if any, are the faults that are currently active.

Some situations seem more clear-cut than others. For example, in the case of an earthquake near Lancaster, Pennsylvania (magnitude 4.1, April 23, 1984), the locations of the main event, aftershocks, and previous events in that region correlate well with mapped geologic lineaments and also correlate with one fault plane of the focal mechanisms found for the 1984 events (Stockar and Alexander, 1986; Armbruster and Seeber, 1987). While such studies have indicated the possible orientations and positions of active faults, no microseismic monitoring has yet confirmed the activity of surface-mapped faults.

Geological studies of recent movement on faults mapped in the northeastern United States have also failed to verify that surface faults are correlated with earthquake occurrence. When

Anderson and others (1984) examined evidence for crustal warping in coastal Maine, they concluded that significant downwarping of about 9 mm/yr is taking place in easternmost Maine. However, Reilinger and others (1986) contradicted Anderson and others (1984) by arguing that easternmost Maine can be subsiding by no more than 1 to mm/yr. This locality is associated with a relatively high rate of seismicity, but there is no apparent relationship between that seismicity and mapped bedrock faults. Oliver and others (1970) discussed a number of localities in the Northeast where post-glacial faulting had been observed. They concluded that the correlation between the pattern of observed faults and that of recent seismicity is not very convincing, but they note that this lack of correlation could be due to inadequacies in either or both sets of data.

Interestingly, there are some areas in the region that appear to be experiencing deformation at the present time, and these areas are associated with earthquake activity. Isachsen (1975) argued that the Adirondack Mountain region is presently being uplifted, and this area is one of the most seismically active parts of the Northeast. Also, Hutchinson and others (1981) showed reflection seismic evidence of post-Pleistocene displacements on faults beneath Lake George in New York. This region has also been the site of persistent earthquake activity.

TEMPORAL ASPECTS OF SEISMICITY

Based on historical and early instrumental catalogues of earthquakes (Smith, 1962, 1966; Chiburis, 1981; and Nottis, 1983), as well as the record of network seismicity since 1975, the history of the earthquake activity in the northeastern United States and southeastern Canada is documented for the past several hundred years (Fig. 1). While there are certainly fluctuations in the number of earthquakes in any given time period, the overall level of activity appears to have been fairly constant throughout recorded time (Chiburis, 1981; Shakal and Toksöz, 1977; Ebel, 1984). Ebel (1984) argued that seismicity in New England has behaved in a temporally random manner in recent years, and that there are only vague hints of spatial migrations of seismicity with time.

Examples of aftershocks, multiple main shocks, and earthquake swarms appear in the seismic history of the Northeast. While aftershocks are not abundant in most cases, they have been routinely observed after the larger earthquakes. There are cases where not one but two or more main shocks of comparable size have taken place within a few hours or a few days of each other. Striking examples of this phenomenon are the shocks that occurred on December 20 and 24, 1940, near Ossipee, New Hampshire, each with magnitude about 5.4 (Ebel and others, 1986), and the three earthquakes above magnitude 5.0 that occurred at Miramichi, New Brunswick, in January 1982 (Wetmiller and others, 1984).

Earthquake swarms have been observed in the northeastern United States. For example, the Blue Mountain Lake area in the Adirondack Mountains of northern New York and an area just north of Moodus, Connecticut, have both been the sites of a number of microearthquake swarms in recent years (Yang and Aggarwal, 1981; Ebel and others, 1982). In addition, a swarm of earthquakes with magnitudes approaching 4.0 took place near Augusta, Maine, in July, 1967.

The establishment of the NEUSSN in 1975 has provided the first reliable and complete catalogue of earthquakes for the northeastern United States. The catalogue for the region is probably complete down to about magnitude 2.0 from 1975 to the present. Using data from that catalogue, Ebel (1987) determined a cumulative recurrence relation for the northeastern United States. This recurrence relation, based on data between magnitudes 2.0 and 4.5 for the time period from October 1975 through June 1986, is:

$$\text{Log N } (/\text{yr}/10{,}000 \text{ km}^2) = 1.72(\pm 0.06) - 0.93(\pm 0.02) \text{ M} \qquad (1)$$

where N is the number of earthquakes/year/10,000 km^2 with magnitude greater than or equal to M.

In equation (1) and in all other recurrence relations discussed in this paper, M is the magnitude scale adopted for reporting magnitudes in the NEUSSN bulletins. Those magnitudes are determined by measuring the largest sustained amplitude of the Lg wave, dividing that amplitude by the period of the signal (A/T), and then substituting that value of A/T into the Nuttli (1973) formula for m_{bLg}. However, the Nuttli (1973) formula was intended to be applied to Lg waves of 0.8 to 1.3 sec periods, and such long-period Lg waves are not generally observed on NEUSSN records. This practice of using higher frequency Lg waves in the Nuttli (1973) formula when calculating magnitudes of northeastern United States earthquakes is discussed by Ebel (1982), Herrmann and Kijko (1983), and Kafka and others (1985).

Ebel (1984) published a recurrence relation for New England (Maine, New Hampshire, Vermont, Massachusetts, Connecticut, and Rhode Island) based on data from October 1, 1975, to November 30, 1982. The b-value he found for that study area was 0.894, smaller than that of 0.93 for the larger northeastern United States area represented by (1) above. Normalizing Ebel's (1984) a-value to 10,000 km^2, his relation for New England is:

$$\text{Log N } (/\text{yr}/10{,}000 \text{ km}^2) = 1.55(\pm 0.09) - 0.84(\pm 0.03) \text{ M}, \qquad (2)$$

which was found from data between magnitudes 2.0 and 4.5. Recent work on earthquake magnitudes for events recorded at Weston Observatory in Massachusetts since 1938 (Ebel, unpublished data, 1987) has led to the calculation of a recurrence relation for a significantly longer time period than those for equations (1) or (2). In order to compare the sizes of events prior to 1975 with those reported by the modern NEUSSN, magnitudes were calculated for all events recorded at Weston Observatory before 1975 using the same procedure that is used for calculating magnitudes in the NEUSSN bulletins. For the time period from January 1, 1938, to June 30, 1986, the recurrence relation for the northeastern United States found from events between magnitudes 3.0 and 5.0 is*:

$$\text{Log } N \,(/\text{yr}/10{,}000 \text{ km}^2) = 1.42 \,(\pm 0.04) - 0.83 \,(\pm 0.02)\, M \qquad (3).$$

In comparing equations (1) and (3), the a-value and b-value of 1.72 and 0.93 for equation (1) are greater than the corresponding values (1.42 and 0.83) for equation (3). The a-value represents the rate of occurrence of small-magnitude events, and the higher value of equation (1) may be indicative of a more complete sampling of the small-magnitude seismicity since 1975 compared to that since 1938. The b-value is related to the proportion of the number of large events to the number of small events in the data set, with a smaller b-value meaning a larger such ratio. The smaller b-value for equation (3) could be due to the larger time period used in computing the relation. Because larger magnitude earthquakes occur less frequently than smaller ones, the larger time period represented in equation (3) has a better chance of meaningfully sampling the rate of activity at larger magnitudes than does the shorter time period represented by equation (1). Thus, the b-value of equation (3) may be due to the long-term rate of occurrence of larger magnitude events combined with a lower a-value.

In comparing equations (2) and (3), it is seen that the b-values for these two relations are virtually identical, while the a-value from equation (2) is higher than that for (3). Ebel (1984, 1985) noted variations in the rate of earthquake occurrence with time in New England, and the higher a-value in equation (2) than in (3) may reflect the fact that the seismic activity in New England from 1975 to 1982 was higher than the longer term average.

Linearly extrapolating to larger earthquakes, the recurrence relation determined from the recent network data, equation (1), predicts that in the northeastern United States there should be an average of one earthquake of at least magnitude 5, 6, and 7 every 14, 122, and 1,039 years, respectively. In contrast, return times for the northeastern United States computed in the same way from equation (3) for magnitudes 5, 6, and 7 are 9, 64, and 436 years, respectively. Since the year 1900, there have been seven earthquakes of magnitude 5.0 or more in the northeastern United States (including the pair of main shocks at Ossipee, New Hampshire, in 1940). Only one earthquake of magnitude 6.0 or more is known in the region from the historic record since 1700 (the 1755 Cape Ann, Massachusetts, earthquake). Equation (1) predicts six earthquakes of magnitude 5.0 or more in 86 years and two earthquakes of magnitude 6.0 or more in 286 years. In contrast, equation (3) predicts ten and four earthquakes, respectively. Ebel (1984) noted that equation (2) overestimates the number of larger earthquakes expected in the historic record. Thus, equation (1) seems to be more consistent with the long-term seismicity than either equations (2) or (3).

Another important point is that the existence of a maximum possible magnitude for all earthquakes in a region would necessitate that the recurrence relation roll off from its linear trend at high magnitudes (e.g., Berrill and Davis, 1980; Main and Burton, 1984). No such roll-off is seen in the data used to compute equations (1), (2), and (3), but if there is a maximum possible earthquake in the northeastern United States, it would mean that the above return times would be underestimates of the actual values.

SEISMIC ZONES IN THE NORTHEASTERN UNITED STATES

We now turn to the problem of developing a seismic zonation map for the northeastern United States. While other zonation maps for the region have been put forth previously (e.g., Hadley and Devine, 1974; Yang and Aggarwal, 1981; Bernreuter and others, 1985), the additional data accumulated in this study have caused us to reexamine this problem. The inability to identify particular geologic faults that are seismically active, as discussed above in the section on state of stress, has important consequences regarding seismic zonation of this region. It means that seismic zonation maps must be based primarily on the known seismic history and that geological input into the selection of seismic zones cannot be given as much weight as the observed occurrence of earthquakes. Furthermore, the accumulation of new earthquake data can lead to the repositioning of zone boundaries, especially when earthquakes occur in localities of little or no previously known activity. The relatively low rate of earthquake occurrence in general and the widespread scatter of epicenters means that some of the possible sites of larger, potentially damaging earthquakes have probably not yet been recognized. Only the continued accumulation of seismic data will clarify this issue. Furthermore, the inability to identify seismically active faults means that geological arguments cannot be used to constrain the largest possible earthquake magnitude that can be experienced in the region. Thus, we cannot rule out the possibility that an earthquake as large as the 1886 earthquake near Charleston, South Carolina (m_b = 6.6 to 6.9; Nuttli and others, 1979), or the 1929 earthquake near Grand Banks, Newfoundland (M = 7.2), could occur in the northeastern United States.

Our characterization of seismic zones is based on the following criteria, listed in order of importance: (1) rate of seismicity from the NEUSSN catalogue (1975 to 1985); (2) rate of seismicity from the historical catalogue (1534 to 1937); (3) rate of seismicity from the early instrumental catalogue (1938 to 1975); (4) occurrence of moderate and/or large (m_b > 5.0) earthquakes; and (5) distinctive geologic or geophysical characteristics.

Since our seismic zonation of the Northeast relies heavily on the observed seismicity, it is clear that the primary assumption underlying that map is that earthquake processes in northeastern North America are stationary, at least over a period of time on the order of about 50 to 100 years. While that assumption cannot be verified, the historical seismicity (1534 to 1974) and network seismicity (1975 to 1985) are discussed below, and the pattern of

---

*Note: equations (1) and (3) correct errors in Ebel, 1987.

seismicity does appear to be relatively stable over the past several hundred years, as is shown in the discussion of the historical (1534 to 1974) and the NEUSSN seismicity (1975 to 1985) below.

To evaluate the stability of the pattern of seismicity in the northeastern United States and southeastern Canada, we divided the entire region into a grid of 1° × 1° cells, and we determined the rate of seismicity for each cell. The number of earthquakes per 10,000 km^2 was determined for each cell for both the historical record of earthquakes and for the network data. For the historical catalogue, a lower magnitude cut-off was chosen at 3.4, and a lower intensity cut-off was chosen at IV. For the network data, a lower cut-off was chosen at 2.0. While these values do not reflect the inhomogeneities in both space and time inherent in the historic and recent catalogues, we feel that there are sufficient statistics to establish the stability of seismicity patterns. The results of the seismicity-rate analysis are shown in Figure 8. Note that the general features of the pattern of earthquake activity are, for the most part, quite similar in the two data sets. In both cases, the most active cell corresponds to the Charlevoix zone (cell centered at 47.5°N, 70.5°W). That cell has 114 events/10,000 km^2 in the historical catalogue and 127 events/10,000 km^2 in the network catalogue. The Charlevoix zone also has a high rate of seismicity in the early instrumental period of time (Fig. 1) and has some of the deepest earthquakes in northeastern North America.

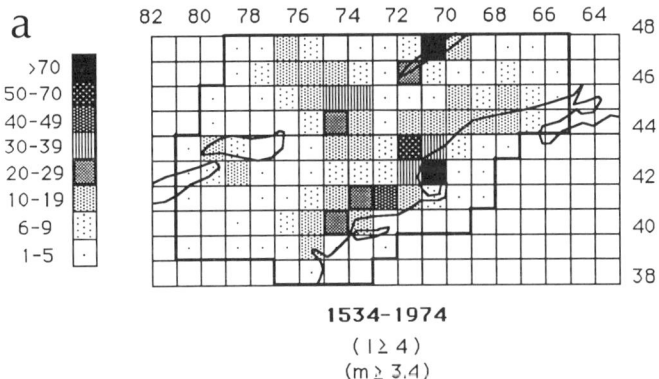

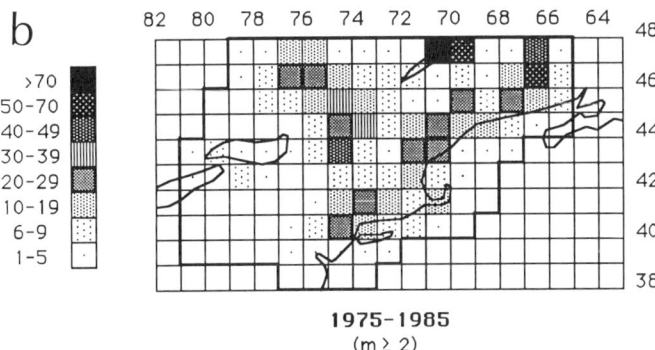

Figure 8. Number of earthquakes/10,000 km^2 in each 1° × 1° cell for (a) the historical and early instrumental catalogue, and (b) the network catalogue.

In addition to the cluster of activity associated with the Charlevoix zone, there are two distinct trends of activity in both the historical and network data. One trend parallels the Appalachian Mountains from New Jersey (cell centered at 40.5°N, 74.5°W) through the New England states to the cell centered at 45.5°N, 67.5°W. This northeast trend of seismic activity has been discussed by a number of investigators; Yang and Aggarwal (1981) refer to it as part of the Appalachian Province, which extends from northern Virginia to New Brunswick, Canada. It is clear from the distribution of seismicity rates shown in Figure 8 that the area surrounding the 1982, New Brunswick, Canada, earthquakes (two cells centered at 46.5°N, 66.5°W, and 47.5°N, 66.5°W) has a significantly lower rate of activity in the historical data than in the network data. That lower rate of activity is probably an artifact of the effect of low population density on the completeness of the historical catalogue, since the New Brunswick region has only 100 years of known activity, but it is rather active in both the early instrumental and the network periods of time (Fig. 1).

Another trend that is apparent in the historical and network data is a northwest-trending zone extending from about 43°N, 73°W to 48°N, 77°W (Fig. 8). That zone has also been identified by several investigators, and Yang and Aggarwal (1981) referred to it as the Adirondack–Western Quebec province. Adams and Basham (this volume) discuss the Canadian part of this zone.

The western New York State area has only a moderate rate of activity characterized by the four cells centered at 43°N, 79°W. However, that area has been the site of a number of fairly large earthquakes, including an intensity VIII earthquake in 1929 and a magnitude 4.7 earthquake in 1966, both of which were centered near Attica, New York.

One significant difference between the pattern of seismicity rates from the historical data and that from the network data is in the area near the cell centered at 42.5°N, 70.5°W. That cell, which corresponds to the vicinity of Cape Ann, Massachusetts, is very active in the historical catalogue (99 events/10,000 km^2), but has only 9 events/10,000 km^2 in the network catalogue. This difference could be due to a relatively higher density of population in coastal areas of Massachusetts (particularly during the earlier part of the historical record), or to a long period of aftershocks of the 1755 Cape Ann, Massachusetts, earthquake, the largest event recorded historically in the northeastern United States, or perhaps to both of these effects.

Using the above criteria, we can define the following seismic zones in the northeastern United States and adjacent Canada: (1) the Adirondack–western Quebec zone; (2) the New England–New Brunswick Appalachians zone; (3) the southern New York–eastern Pennsylvania–New Jersey zone; and (4) the western New York State zone.

These zones are illustrated in Figure 9. Basham and others (1982, 1985) published zones for Canada, and we show their zones as dashed lines on the Canadian side of the international border in Figure 9. While we have selected the Adirondack–western Quebec seismicity to be a single zone, we have subdi-

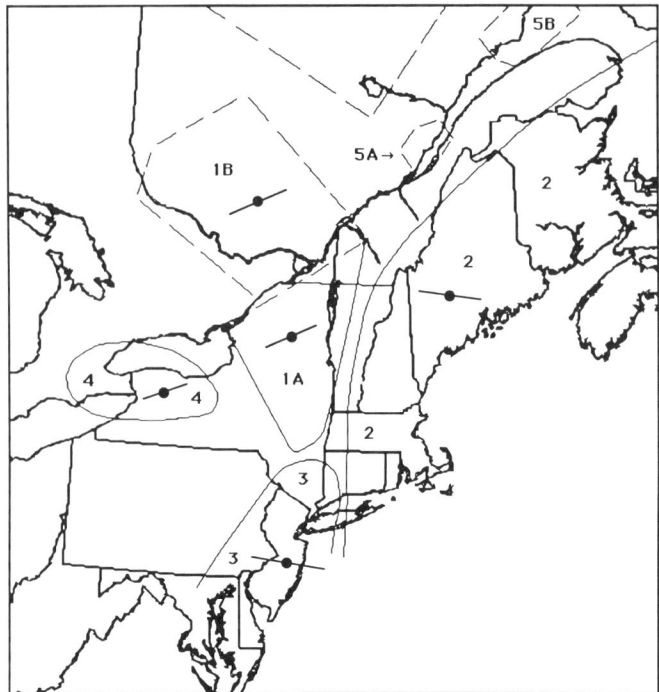

Figure 9. Map of seismic zones in the northeastern United States and adjacent areas. The zones in Canada are after Basham and others (1982, 1985). The average P-axis azimuth in a given zone is shown by a line drawn through a closed circle.

vided it into two subzones (1A and 1B) to be consistent with Basham and others' (1982, 1985) interpretation of the seismic zones there. For each of the zones, we have estimated the orientation of the stress field by calculating the average P and T axes using the same methodology that we applied to the entire NEUSSN region. Only one zone (western New York State) had an average T-axis plunge that differed by more than 9° from vertical. That region had only two focal mechanisms, both of which were very similar, and the average T-axis plunge for those two events was 40° ± 42°. The average P-axis plunges ranged from 5° to 8° for all four zones analyzed. The P-axis azimuths were 67° ± 35°, 97° ± 37°, 99° ± 40°, and 68° ± 8° for zones 1, 2, 3, and 4 respectively. The average P-axis azimuths for zones 1A and 1B are 67° ± 30° and 67° ± 46°, respectively, and the T axes for each zone have plunges of 85° ± 30°.

There are many similarities between the seismic zones we have outlined here and those published by other investigators, but there are also some important differences. Our zone 1, the Adirondack–western Quebec zone, is similar to that described by Yang and Aggarwal (1981). It was the existence of Grenville crust in the Adirondacks and the broad-scale scatter of epicenters from the Adirondacks into western Quebec that led us to define this zone. However, Basham and others (1982, 1985) and Adams and Basham (this volume) separate the western Quebec seismicity from that in the Adirondacks, and Adams and Basham (this volume) further subdivide the western Quebec zone based on their observation of separate bands of seismicity. Our zone 3, the southern New York–eastern Pennsylvania–New Jersey zone, includes the seismicity around the Newark Triassic basin, as recognized by Kafka and others (1985), as well as the active Lancaster, Pennsylvania, area. In their analysis, Yang and Aggarwal (1981) had lumped this area with their equivalent of our zone 2 (New England–New Brunswick Appalachian zone) into a central and northern Appalachian zone.

SUMMARY AND CONCLUSIONS

1. The northeastern United States has been characterized by moderate seismicity throughout its recorded history. Although the cause of this earthquake activity is unclear, there is evidence that the forces causing these earthquakes are, at least in part, related to a combination of stresses caused by plate-tectonic processes. Other sources of stress (e.g., topography, sediment loading along the coastal margin, or post-glacial rebound) may also be involved but to a lesser extent. Also, the fact that the earthquakes tend to occur in some places more than others probably is an indication that there are some structures being preferentially activated or reactivated in the present stress field, even if those structures are not yet identified.

2. Earthquake depths are primarily within the upper half of the crust. The mean depth from the network data is 9 km. The deepest events in the northeastern United States are no deeper than 20 km, although in eastern Canada, events as deep as 33 km have been reported. Depths of the largest events in the northeastern United States and adjacent Canada are about 6 to 10 km.

3. The regional maximum compressive stress is essentially horizontal and trends approximately east-west. The regional minimum stress is very close to vertical. The average P axis trend is 86° ± 40°. For the various seismic regions in the Northeast, the trend of the average P axes ranges from about 67° to 99°.

3. There is no obvious correlation between the earthquakes and mapped faults on a regional scale, and such correlations on a local scale for individual events seem to be the exception rather than the rule.

5. The seismicity rate over the past 50 years seems to have been fairly constant, although there may have been a local increase in that rate beginning in 1982. Extrapolations of recurrence curves calculated from the instrumental data overestimate the number of large earthquakes observed in the historic catalogue. This could be due either to natural fluctuations in the occurrences of the large earthquakes or to a roll-off of the recurrence curve at large magnitudes due to the existence of a maximum possible earthquake magnitude for the region.

REFERENCES CITED

Aggarwal, Y. P., and Sykes, L. R., 1978, Earthquakes, faults, and nuclear power plants in southern New York and northern New Jersey: Science, v. 200, p. 425–529.

Anderson, W. A., and 10 others, 1984, Crustal warping in coastal Maine: Geology, v. 12, p. 677–680.

Anglin, F. M., 1984, Seismicity and faulting in the Charlevoix zone of the St. Lawrence Valley: Bulletin of the Seismological Society of America, v. 74, p. 595–603.

Armbruster, J. G., and Seeber, L., 1987, The 23 April 1984 Martic earthquake and the Lancaster seismic zone in eastern Pennsylvania: Bulletin of the Seismological Society of America, v. 77, p. 877–890.

Basham, P. W., and Kind, R., 1986, GRF broad-band array analysis of the 1982 Miramichi, New Brunswick, earthquake sequence: Journal of Geophysical Research, v. 60, p. 120–128.

Basham, P. W., Weichert, D. H., Anglin, F. M., and Berry, M. J., 1982, New probabilistic strong seismic ground maps of Canada; A compilation of earthquake source zones, methods, and results: Earth Physics Branch Open-File 82-33, 202 p.

— , 1985, New probabilistic strong seismic ground motion maps of Canada: Bulletin of the Seismological Society of America, v. 75, p. 563–595.

Bernreuter, D. L., Savy, J. B., Mensing, R. W., Chen, J. C., and Davis, B. C., 1985, Seismic hazard characterization of the eastern United States: Lawrence Livermore National Laboratory Report UCID-20421, 621 p.

Berrill, J. B., and Davis, R. O., 1980, Maximum entropy and the magnitude distribution: Bulletin of the Seismological Society of America, v. 70, p. 1823–1831.

Bird, J. M., and Dewey, J. F., 1970, Lithosphere plate continental margin tectonics and the evolution of the Appalachian orogen: Geological Society of America Bulletin, v. 81, p. 1031–1060.

Brown, E. J., and Ebel, J. E., 1985, An investigation of the January 1982 Gaza, New Hampshire, aftershock sequence: Earthquake Notes, v. 56, no. 4, p. 125–133.

Chase, C. G., 1979, Asthenospheric counterflow; A kinematic model: Geophysical Journal of the Royal Astronomical Society, v. 56, p. 1–18.

Chiburis, E. F., 1981, Seismicity, recurrence rates, and regionalization of the northeastern United States and adjacent southeastern Canada: U.S. Nuclear Regulatory Commission Report NUREG/CR-2309, 76 p.

Ebel, J. E., 1982, ML measurements for northeastern United States earthquakes: Bulletin of the Seismological Society of America, v. 72, p. 1367–1378.

— , 1983, A detailed study of the aftershocks of the 1979 earthquake near Bath, Maine: Earthquake Notes, v. 54, no. 4, p. 27–40.

— , 1984, Statistical aspects of New England seismicity from 1975 to 1982 and implications for past and future earthquake activity: Bulletin of the Seismological Society of America, v. 74, p. 1311–1329.

— , 1985, A study of seismicity and tectonics in New England: U.S. Nuclear Regulatory Commission Report NUREG/CR-4354, 87 p.

— , 1987, The seismicity of the northeastern United States, in Seismic hazards, ground motions, soil-liquification, and engineering practice in eastern North America: National Center for Earthquake Engineering Research Technical Report NCEER-87-0025, p. 178–188.

Ebel, J. E., and McCaffrey, J. P., 1984, The Mc = 4.4 earthquake near Dixfield, Maine: Earthquake Notes, v. 55, no. 3, p. 21–24.

Ebel, J. E., Vudler, V., and Celata, M., 1982, The 1981 microearthquake swarm near Moodus, Connecticut: Geophysical Research Letters, v. 9, p. 397–400.

Ebel, J. E., Somerville, P. G., and McIver, J. D., 1986, A study of the source parameters of some large earthquakes of northeastern North America: Journal of Geophysical Research, v. 91, p. 8231–8247.

Filipkowski, F., 1986, The use of short-period Rg waves as a depth discriminant for seismic events in the upper crust of the northeastern United States [M.S. thesis]: Chestnut Hill, Massachusetts, Boston College, 81 p.

Gephart, J. W., and Forsyth, D. D., 1985, On the state of stress in New England as determined from focal mechanisms: Geology, v. 13, p. 70–72.

Graham, T., and Chiburis, E. F., 1980, Fault plane solutions and state of stress in New England: Earthquake Notes, v. 51, p. 3–12.

Hadley, J. B., and Devine, J. F., 1974, Seismotectonic map of the eastern United States: U.S. Geological Survey Miscellaneous Field Study Map MF-620, scale 1:5,000,000.

Hager, B. H., and O'Connell, R. J., 1979, Kinematic models of large-scale flow in the earth's mantle: Journal of Geophysical Research, v. 84, p. 1031–1048.

Herrmann, R. B., 1979, Surface wave focal mechanisms for eastern North American earthquakes with tectonic implications: Journal of Geophysical Research, v. 84(B7), p. 3543–3552.

Herrmann, R. B., and Kijko, A., 1983, Short-period Lg magnitudes; Instrument, attenuation, and source effects: Bulletin of the Seismological Society of America, v. 73, p. 1835–1850.

Horner, R. B., Stevens, A. E., Hasegawa, H. S., and Leblanc, G., 1978, Focal parameters of the July 12, 1975, Maniwaki, Quebec, earthquake; An example of intraplate seismicity in eastern Canada: Bulletin of the Seismological Society of America, v. 68, p. 619–640.

Hutchinson, D. R., Ferrebee, W. M., Knebel, H. J., Wold, R. J., and Isachsen, Y. W., 1981, Sedimentary framework of the southern basin of Lake George, New York: Journal of Quaternary Research, v. 15, p. 44–61.

Isachsen, Y. W., 1975, Possible evidence for contemporary doming of the Adirondack Mountains, New York, and suggested implications for regional tectonics and seismicity: Tectonophysics, v. 29, p. 169–181.

Kafka, A. L., 1982, Seismicity and geologic structures in the Manhattan prong; Similarities and contrasts with the Hudson Highlands [abs.]: Earthquake Notes, v. 53, no. 3, p. 20.

Kafka, A. F., Schlesinger-Miller, E. A., and Barstow, N. L., 1982, The Cheesequake, New Jersey, earthquake of January 30, 1979; An inquiry into seismic activity in the Atlantic Coastal Plain Province of New Jersey [abs.]: Earthquake Notes, v. 53, no. 3, p. 21.

Kafka, A. F., Schlesinger-Miller, E. A., and Barstow, N. L., 1985, Earthquake activity in the greater New York City area; Magnitudes, seismicity, and geologic structures: Bulletin of the Seismological Society of America, v. 75, p. 1285–1300.

Kirby, S. H., 1980, Tectonic stresses in the lithosphere; Constraints provided by the experimental deformation of rocks: Journal of Geophysical Research, v. 85, p. 6353–6363.

Main, I. G., and Burton, P. W., 1984, Information theory and the earthquake frequency-magnitude distribution: Bulletin of the Seismological Society of America, v. 74, p. 1409–1426.

Nabelek, J. L., 1984, Determination of earthquake source parameters from inversion of body waves [Ph.D. thesis]: Cambridge, Massachusetts Institute of Technology, 361 p.

Nottis, G., 1983, Epicenters of northeastern United States and southeastern Canada, onshore and offshore; Time period 1534–1980: New York State Geological Survey Map and Chart Series no. 38, scale 1:1,000,000.

Nuttli, O. W., 1973, Seismic wave attenuation and magnitude relations for eastern North America: Journal of Geophysical Research, v. 78, p. 876–885.

Nuttli, O. W., Bollinger, G. A., and Griffiths, D. W., 1979, On the relation between Modified Mercalli Intensity and body-wave magnitude: Bulletin of the Seismological Society of America, v. 69, p. 893–909.

Oliver, J., Johnson, T., and Dorman, J., 1970, Postglacial faulting and seismicity in New York and Quebec: Canadian Journal of Earth Science, v. 7, p. 579–590.

Pomeroy, P. W., Simpson, D. W., and Sbar, M. L., 1976, Earthquakes triggered by surface quarrying—the Wappingers Falls, New York, sequence of June 1974: Seismological Society of America Bulletin, v. 66, p. 685–700.

Pulli, J. J., and Toksöz, M. N., 1981, Fault plane solutions for northeastern United States earthquakes: Bulletin of the Seismological Society of America, v. 71, p. 1875–1882.

Quittmeyer, R. C., Statton, C. T., Mrotek, K. A., and Houlday, M., 1985, Possible implications of recent microearthquakes in southern New York State: Earth-

quake Notes, v. 56, p. 35–42.
Ratcliffe, N. M., 1971, The Ramapo fault system in New York and adjacent northern New Jersey; A case of tectonic heredity: Geological Society of America Bulletin, v. 82, p. 125–142.
Reilinger, R., Larsen, S., and Krumhansl, P., 1986, Reanalysis of contemporary crustal downwarping of coastal Maine [abs.]: Earthquake Notes, v. 57, no. 4, p. 113.
Richardson, R. M., Solomon, S. C., and Sleep, N. H., 1979, Tectonic stress in the plates: Reviews of Geophysics and Space Physics, v. 17, p. 981–1019.
Sbar, M. L., Rynn, J.M.W., Gumper, F. J., and Lahr, J. C., 1970, An earthquake sequence and focal mechanism solution, Lake Hapatcong, northern New Jersey: Bulletin of the Seismological Society of America, v. 60, p. 1231–1243.
Sbar, M. L., Armbruster, J., and Aggarwal, Y. P., 1972, The Adirondack, New York, earthquake swarm of 1971 and tectonic implications: Bulletin of the Seismological Society of America, v. 62, p. 1303–1317.
Sbar, M. L., Jordan, R. J., Stephens, C. D., Hickett, T. E., Woodruff, K. D., and Sammis, C. G., 1975, The Delaware–New Jersey earthquake of February 28, 1973: Bulletin of the Seismological Society of America, v. 65, p. 85–92.
Schlesinger-Miller, E., Barstow, N. L., and Kafka, A. L., 1983, The July 1981 earthquake sequence near Cornwall, Ontario, and Massena, New York: Earthquake Notes, v. 54, p. 11–26.
Seborowski, K. D., Williams, G., Kelleher, J. A., and Statton, C. T., 1982, Tectonic implications of recent earthquakes near Annsville, New York: Bulletin of the Seismological Society of America, v. 72, p. 1601–1609.
Seeber, L., 1987, Problems in intraplate seismogenesis and earthquake hazard, in Jacob, K. H., ed., Proceedings of the Symposium on Earthquake Hazards, Ground Motions, Soil Liquefaction, and Engineering Practice in Eastern North America: National Center for Earthquake Engineering Research Technical Report NCEER-87-0025, p. 143–162.
Seeber, L. E., Cranswick, E., Armbruster, J., and Barstow, N., 1984, The Oct. 1983 Goodnow aftershock sequence; Regional seismicity and structural features in the Adirondacks [abs.]: EOS American Geophysical Union Transactions, v. 65, p. 240.
Shakal, A. F., and Toksöz, M. N., 1977, Earthquake hazard in New England: Science, v. 195, p. 171–173.
Smith, W.E.T., 1962, Earthquakes of eastern Canada and adjacent areas; 1535–1927: Ottawa, Canada, Publication of the Dominion Observatory, v. 25, p. 271–301.

———, 1966, Earthquakes of eastern Canada and adjacent areas; 1928–1959; Ottawa, Canada, Publication of the Dominion Observatory, v. 32, p. 87–121.
Stockar, D. V., and Alexander, S. S., 1986, The Lancasater seismic zone; Geologic and geophysical evidence for a north-south trending, presently active zone of weakness through Lancaster County, Pennsylvania [abs.]: EOS Transactions of the American Geophysical Union, v. 67, p. 314.
Street, R., and Lacroix, A., 1979, An empirical study of New England seismicity; 1727–1977: Bulletin of the Seismological Society of America, v. 69, p. 159–175.
Street, R. L., and Turcotte, F. T., 1977, A study of northeastern North American spectral moments, magnitudes, and intensities: Bulletin of the Seismological Society of America, v. 67, p. 599–614.
Sykes, L. R., 1978, Intraplate seismicity, reactivation of preexisting zones of weakness, alkaline magmatism, and other tectonism postdating continental fragmentation: Reviews of Geophysics and Space Physics, v. 16, p. 621–688.
Wahlstrom, R., 1985, The North Gower, Ontario, earthquake of 11 October 1983; Focal mechanism and aftershocks: Earthquake Notes, v. 56, no. 4, p. 35–143.
Wetmiller, R. J., Adams, J., Anglin, F. M., Hasegawa, H. S., and Stevens, A. E., 1984, Aftershock sequences of the 1982, Miramichi, New Brunswick, earthquakes: Bulletin of the Seismological Society of America, v. 74, p. 621–653.
Yang, J. P., and Aggarwal, Y. P., 1981, Seismotectonics of northeastern United States and adjacent Canada: Journal of Geophysical Research, v. 86, p. 4981–4998.
Zoback, M. L., and Zoback, M. D., 1985, Uniform NE to ENE maximum horizontal compressive stress throughout mid-plate North America [abs.]: EOS Transactions of the American Geophysical Union, v. 66, p. 1056.

Manuscript Accepted by the Society May 16, 1989

ACKNOWLEDGMENTS

We thank John Adams, John Armbruster, and Bob Engdahl for their comments and suggestions about an earlier version of this paper. We also acknowledge the efforts of Bruce Bouck and Zayne Zakimi who helped assemble the data that we used in this paper. This work was supported by the U.S. Nuclear Regulatory Commission under contract no. 04-085-113-01.

NOTES ADDED IN PROOF

Subsequent to the completion of the text of this chapter, Ebel and Bouck (1988) published new analyses of the focal mechanisms of 15 earthquakes in the northeastern United States. These focal mechanisms were included in the principle stress direction averages given in section 3.1. Ebel (1990) reported on the focal mechanisms for three additional earthquakes which occurred in northern New England (near Berlin, New Hampshire, MN = 4.0; Dixfield, Maine, Mc = 3.3; and Albion, Maine, MN = 4.0) in late 1988. The P axes for these events were oriented northeast-southwest, north-northeast–south-southwest, and northwest-southeast. More recently, an earthquake of MN = 4.5 took place at Summit, New York, on June 17, 1991, and was a thrust earthquake with a northeast-southwest P axis. These four earthquakes show deviations from the average P-axis direction for the region. The cause of these deviations is not understood at present.

Seeber and Dawers (1989) reported strong geologic evidence for a northwest-trending fault zone (the Dobbs Ferry fault) in the surface geology above the 5-km deep hypocenter of the 1985 Ardsley, New York, earthquake. There is, however, no direct evidence that the seismicity associated with the Ardsley earthquake extends upward to the mapped trace of the Dobbs Ferry fault. Based on an assumed relationship between the Ardsley earthquake and the Dobbs Ferry fault, Seeber and Dawers (1989) argue that minor faults with little or no accumulated slip can be the sources of significant earthquakes in northeastern North America, although the maximum earthquake on such features may be limited by segmentation along the faults. In contrast, Ebel and Kafka (1990) reported new seismic evidence showing that the hypocentral depth of the January 19, 1982, Gaza, New Hampshire, earthquake (Fig. 6) was only 1 km. Even with this extremely shallow depth of focus, no geologic structure has yet been found on the surface over the hypocenter to explain why this earthquake took place where it did.

ADDITIONAL REFERENCES

Ebel, J. E., 1990, Deviations of the regional stress field in northeastern North America at the times of moderate and large earthquakes: Seismological Research Letters, v. 61, p. 149.
Ebel, J. E., and Bouck, B. R., 1988, New focal mechanisms for the New England region; Constraints for the regional stress regime: Seismological Research Letters, v. 59, p. 183–187.
Ebel, J. E., and Kafka, A. L., 1990, A Reanalysis of the mb = 4.4 Gaza, N.H. earthquake of 1982; Source depth, seismic moment, stress drop and strong ground motions: EOS, Transactions of the American Geophysical Union, v. 71, p. 1440.
Seeber, L., and Dawers, N., 1989, Characterization of an intraplate seismogenic fault in the Manhattan Prong, Westchester, Co., N.Y.: Seismological Research Letters, v. 60, p. 71–78.

Printed in U.S.A.

Chapter 16

Seismicity of the southeastern United States; 1698 to 1986

G. A. Bollinger
Seismological Observatory, Virginia Polytechnic Institute and State University, Blacksburg, Virginia 24061
Arch C. Johnston
Center for Earthquake Research and Information, Memphis State University, Memphis, Tennessee 38152
Pradeep Talwani
Department of Geology, University of South Carolina, Columbia, South Carolina 29208
Leland T. Long
School of Geophysical Sciences, Georgia Institute of Technology, Atlanta, Georgia 30332
Kaye M. Shedlock
Branch of Geological Risk Assessment, U.S. Geological Survey, Box 25046, Denver Federal Center, Denver, Colorado 80225
Matthew S. Sibol and Martin C. Chapman
Seismological Observatory, Virginia Polytechnic Institute and State University, Blacksburg, Virginia 24061

INTRODUCTION

Prior to 1970 there were fewer than 10 seismographs sited in the southeastern United States—a study region defined herein as all of the states of West Virginia, Virginia, North Carolina, South Carolina, Georgia, and Alabama and the eastern portions of Kentucky and Tennessee (see Sibol and others, 1986, for a detailed description). Multistation seismic networks began to be installed in 1974, and by the end of 1977 there were 53 operational seismograph stations. As of January 1986, the Southeastern U.S. Seismic Network (SEUSSN) Seismicity Bulletin (Sibol and others, 1986) listed 147 stations in the region (Fig. 1). Thus, several years of instrumentally monitored seismicity data for the region have been acquired to supplement the historical data base that extends back to as early as 1698.

The purpose of this chapter is to review the seismicity of the southeastern United States as documented by the studies to date of the historical and network data bases. Data acquisition and analyses are derived in large measure from the collective and individual efforts of the members of the SEUSSN coalition—a group of some 10 member institutions that operate seismographs in the region (Sibol and others, 1986).

We divide consideration of the data base into a Historical Seismicity period (pre-network: 1698 to June 30, 1977), and a recent period of Instrumental Seismicity (post-network: July 1, 1977 to January 1, 1986). The reasons for that division are the differences in the levels of completeness and accuracy with respect to earthquake size and location. Prior to the mid-1970s, the earthquake catalog consists primarily of earthquakes large enough to be felt by people, whereas in recent years, many smaller earthquakes have been located instrumentally. Accuracy of epicenter locations for the pre-network shocks averages in the 10s of kilometers, and depths of epicenters are undetermined. Much higher epicentral accuracy (a few kilometers or less) and

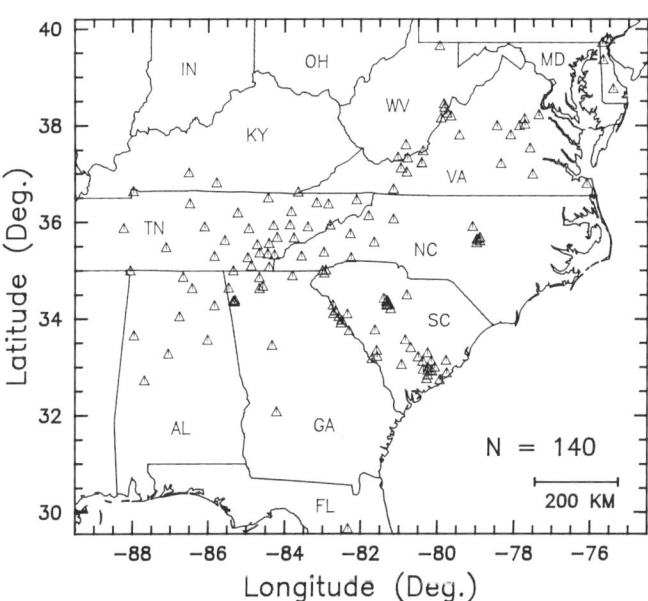

Figure 1. Southeastern U.S. Seismic Network (SEUSSN). Configuration as of December 31, 1985; N = 140 = number of operational seismic stations shown as triangles (Sibol and others, 1986). Very closely spaced stations shown as a single triangle. The network seismicity maps shown herein (Figs. 3–5, 7–10) are based on the data from this network beginning on July 1, 1977 with N = 53, and ending on December 31, 1985.

Bollinger, G. A., Johnston, A. C., Talwani, P., Long, L. T., Shedlock, K. M, Sibol, M. S., and Chapman, M. C., 1991, Seismicity of the southeastern United States; 1698 to 1986, *in* Slemmons, D. B., Engdahl, E. R., Zoback, M. D., and Blackwell, D. D., eds., Neotectonics of North America: Boulder, Colorado, Geological Society of America, Decade Map Volume 1.

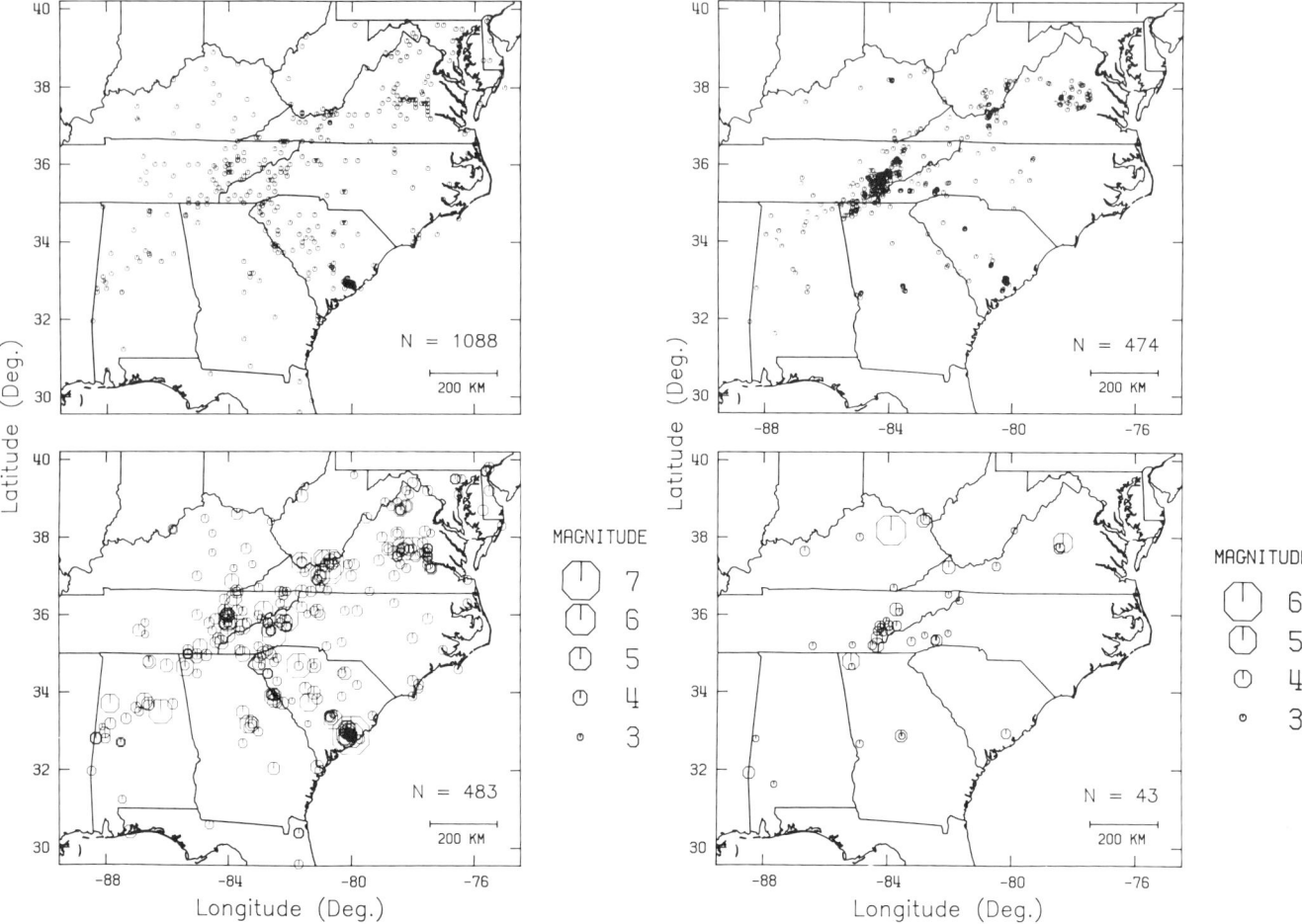

Figure 2. Historical southeastern U.S. Seismicity: 1698 to 1977. Upper: Epicenters for earthquakes with $m_b \geq 0.0$ shown as small circles. Lower: Epicenter map with epicenters shown as octagons whose size is scaled to magnitude for $m_b \geq 3$; Intensity/Felt Area conversion to m_b after Sibol and others, 1987. N equals the number of earthquakes plotted.

Figure 3. Network southeastern U.S. Seismicity: 1977 to 1985. Upper: Epicenters for earthquakes with $M_D \geq 0.0$ shown as small circles (M_D = duration magnitude). Lower: Epicenter map with epicenters shown as octagons whose size is scaled to magnitude for $M_D \geq 3$. N equals the number of earthquakes plotted.

some reliable focal depths (less than 5-km error estimates) have been obtained recently with the network data.

We begin with a section on Historical Seismicity, followed by a general overview of earthquake focal depths and mechanisms in a section on Instrumental Seismicity. Then, a detailed consideration of Network Seismicity on a state-by-state basis is presented. Next, a Regional Summary synthesizes the historical, instrumental, and network seismicity results with respect to the regional host geologic/physiographic provinces. Finally, we conclude with an enumeration of topical areas for which there is currently a lack of consensus among the co-authors of this paper.

HISTORICAL SEISMICITY

The SEUSSN earthquake catalog (see SEUSSN contributors, 1985, for a general description) lists 1,088 shocks (483 with $M > 3$; $M = m_b, m_b(L_g), m_D$ and M_L) for the pre-network period, 1698 to 1977 (Fig. 2) and 474 events (43 with $M > 3$) for the network period, 1977 through 1985 (Fig. 3). Our evaluations of catalog completeness as a function of time are conservative and are presented in conjunction with the *Seismicity Map of North America* (this volume). Briefly, we judge the recent network data to be complete to $m_b = 2.5$, the historical period between 1930 and 1977 complete to $m_b = 4.5$, and the historical period between 1870 and 1930 complete at the $m_b = 5.7$ level.

The historical seismicity of the region has been discussed in some detail by Bollinger (1973a, b) and by Hadley and Devine (1974). All of those studies noted the decidedly nonrandom spatial distribution of epicenters, with patterns that are parallel to, as well as oblique to, the northeasterly tectonic fabric of the host region. Specifically, the Appalachian Highlands (Valley and Ridge and Blue Ridge provinces) displays seismic activity throughout its extent (south of latitude 40° north), whereas the Piedmont and Coastal Plain provinces do not exhibit such ubiquitous release of seismic energy. Both the Piedmont and the Coastal Plain have appreciable seismicity only in Virginia, South Carolina, and Georgia (Figs. 2, 3). It is important to note that the epicenters located by network monitoring (Fig. 3) display the same general spatial pattern as do the historical, noninstrumental epicenters (Fig. 2). Even though the network events are generally smaller than the historical shocks, the overall patterns of active/inactive areas appear to extend down to those lower energy levels. However, important exceptions in detail exist in this historical versus modern epicenter correspondence. For example, relative decreases in modern seismic activity are seen in the northern Virginia Appalachians and the South Carolina Piedmont, while relative increases in seismicity have occurred in the northeastern Kentucky Plateau and in the southeastern Tennessee Appalachians compared to the historical seismicity.

The frequency of earthquake occurrence for the entire region was determined by Bollinger (1973a) as,

$$\log N = 3.01 - 0.59 I_o, \qquad V \leq I_o \leq VIII,$$

where N = number of earthquakes/year of intensity I_o, and I_o equals the maximum Modified Mercalli Intensity (MMI). The above recurrence equation is based entirely on historical seismicity. That is, it does not include the last several years of network data, which are complete to lower magnitude levels than the older data set. Preliminary results for the region based on all of the data sets are:

$$\log N_c = 2.88 - 0.80 \, m_b(L_g), \qquad 2 \leq m_b(L_g) \leq 6,$$

where N_c = cumulative number of earthquakes/yr $\geq m_b(L_g)$ (F. Davison, 1986, personal communication).

INSTRUMENTAL SEISMICITY

A major benefit of seismic network monitoring is the determination of accurate focal depths. Focal depths are, in general, difficult to determine because of their linear dependence on time of origin. At least one seismograph whose epicentral distance is less than the focal depth is generally required to offset that dependence. In the Southeast, the networks are not dense everywhere. That results in only a fraction (308/474 = 65 percent) of the monitored seismic events having their focal depths determined with modest accuracy (±5 km or less; see Fig. 4). The 90

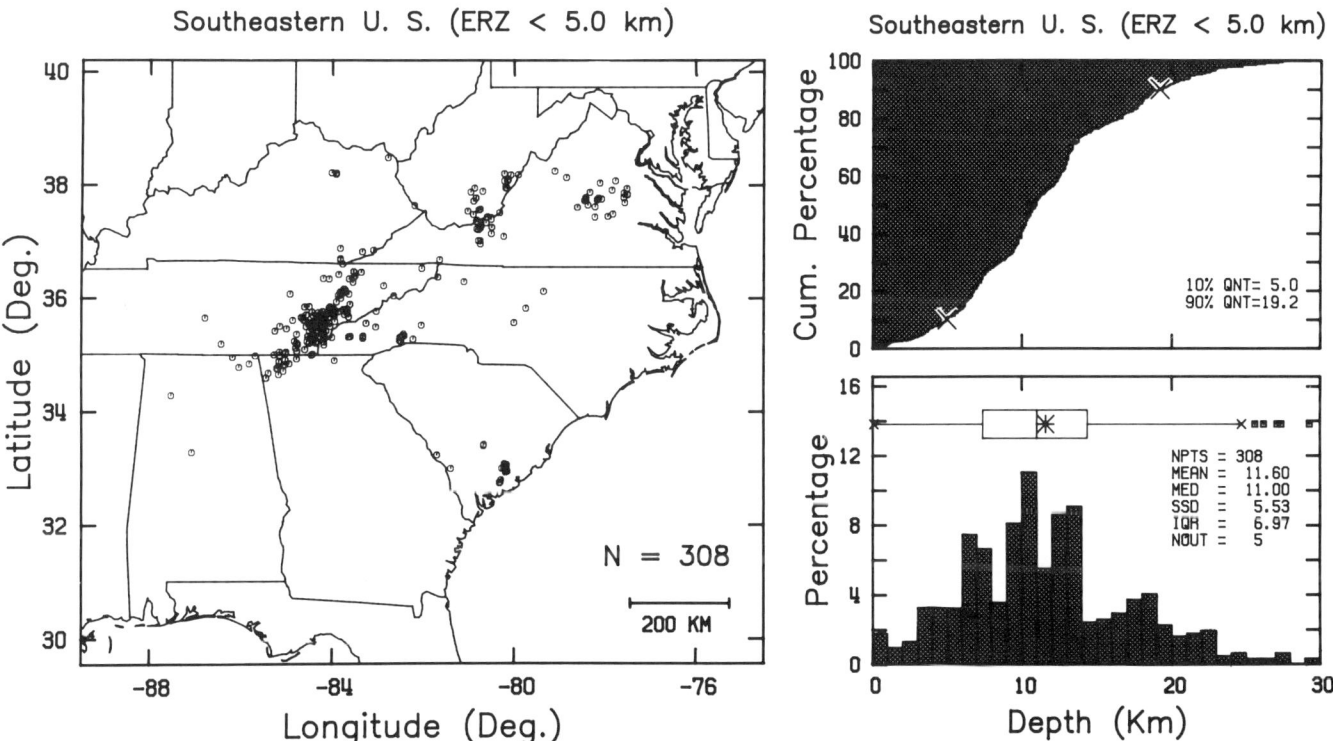

Figure 4. Distribution of earthquake focal depths in the southeastern U.S. From network seismicity (1977 to 1985) shocks whose errors in depth determination (ERZ; see Lahr, 1980) were estimated to be equal to or less than ±5 km. Left: Epicenter map. Right: (top) Cumulative distribution with 10 percent and 90 percent quantiles (QNT) indicated by X's and (bottom) histogram at 1-km depth intervals. Above the histogram is a statistical box plot (see Ott, 1984, p. 43–47) displaying the mean (star), the 25-percent and 75-percent quantiles (box outline), the median (center bar), the upper and lower adjacent values (X's), and outliers (squares). NPTS = number of depths used; MEAN = average depth; MED = median depth; SSD = sample standard deviation; IQR = inner quantile range; and NOUT = number of outliers present in the data.

percent depth for those shocks (i.e., the depth above which 90 percent of all the foci lie) is 19 km. That depth can be used as an estimate of the total thickness of the brittle seismogenic crust available for strain accumulation (Sibson, 1982). A worldwide survey of focal depths versus lithospheric thermal and mechanical properties concluded that earthquakes near passive margins tend to occur in the top 20 km of the continental crust (Chen and Molnar, 1983). The *peak* in the focal depth distribution for the southeastern United States occurs at 10 to 11 km. From their study of the depth distributions of earthquake foci and maximum stresses, Meissner and Strehlau (1982) correlate such seismicity peaks with the location of a high-strength region of the crust wherein larger shocks tend to nucleate.

The above depths are actually characteristic only of the Appalachian Highlands (Fig. 5); comparable depth characteristics are much shallower for the relatively fewer Piedmont and Coastal Plain earthquakes: 13 km (90 percent value) and 7 to 8 km (peak value). The focal depth data in those latter areas are sparse, 24 percent of the total for the region, but they are probably representative. If such is indeed the case, then there is either a significant (30 percent) difference in the thicknesses of the seismogenic crust between the adjacent provinces or a significant difference in the earthquake mechanisms. An example of this difference would be tectonic versus reservoir-induced shocks. These characteristics are discussed in greater detail in Bollinger and others (1985).

Of the 44 focal-mechanism solutions derived thus far from network data in the region (Fig. 6), 37 are for single earthquakes (SEFM = single earthquake focal mechanism) and seven are composited from multiple, adjacent earthquakes (CEFM = composited earthquakes focal mechanism). SEFMs are generally considered more representative of fault geometry in that they do not carry the additional assumption of the multiple events having occurred on the same fault plane or on planes of similar orientations in a uniform stress field as is required by the CEFM. Trends of the P-axes, which are estimates of the axes of pressure (maximum compressive stresses or sigma-one) in a nonfriction environment (i.e., they are at 45 degrees to the nodal [fault] planes) are also plotted in Figure 6. Obviously, seismic failure may occur in nature on preexisting zones of weakness, which may in turn bear no simple geometric relation to the principal stress directions. However, stress-direction estimates derived from studies of *families of focal-mechanism solutions* (Gephart and Forsyth, 1984; Vasseur and others, 1983) have been shown to be consistent with similar estimates obtained by independent means (e.g., slickenside measurements).

The family of focal-mechanism solutions for the Southeast exhibits a decided preference for strike-slip faulting on steeply dipping planes—right-lateral on northerly trending planes or left-lateral on the companion easterly planes. Throughout the region the average orientation of the associated P-axes is predominantly subhorizontal with northeasterly trends. Exceptions to these generalizations are found in central Virginia (Piedmont) and at Newberry (Piedmont) and Charleston (Coastal Plain), South Carolina. In the central Virginia Piedmont, some dip-slip mechanisms (reverse faulting) are seen, but the most striking characteristic is a depth dependence in the trends of the P-axes. That is, the shallow foci (3 to 7 km) have the regional northeasterly orientation, but the deeper foci (8 to 13 km) are mostly *northwesterly* (4 cases) with some north-northeasterly (two cases); all are subhorizontal (Munsey and Bollinger, 1985). The depth of this 90° change in P-axis orientation is near that estimated for the Appalachian detachment (Glover and others, 1982), possibly indicating some type of mechanical decoupling in the vicinity of that structural discontinuity. Another possibility is that the two primarily horizontal principal stresses are of approximately the same magnitude, and thus, a 90° shift is rather easily accommodated. Note that this change is in a vertical, not a lateral, direction. Nelson and Talwani (1985) did not require a 90° change in the direction of the P-axes for CEFM of the same central Virginia earthquakes studied with both SEFM and CEFMs by Munsey and Bollinger (1985).

Newberry, South Carolina, focal mechanisms are predominantly reverse movements on northwesterly striking planes that dip northeast or southwest, whereas the Charleston CEFMs exhibit subhorizontal and subvertical nodal planes in addition to a regional strike-slip faulting. Talwani (1982) interprets a change of faulting style with depth at Charleston. Specifically, he finds reverse faulting on a steeply dipping plane for earthquakes in the 4 to 8 km depth interval, and strike-slip faulting (similar to others elsewhere in the region) for events in the 9 to 13 km depth range. His results, along with those by Tarr and others (1981), all have northeasterly, subhorizontal P-axes. Strike-slip faulting was exhibited by an SEFM at the Savannah River Plant in the Coastal Plain of South Carolina, and thrust faulting was observed for three earthquakes in the Piedmont of South Carolina near the community of Newberry.

CEFMs for reservoir-induced seismicity were strike-slip at Lake Jocassee and thrust faulting at Monticello reservoir in the South Carolina Piedmont.

NETWORK SEISMICITY

Seismicity of Virginia, West Virginia, and eastern Kentucky

The historical record of earthquakes in Virginia extends back to 1774 (Reagor and others, 1980a) and depicts a persistent, low-level release of strain energy in the central and southwestern portions of that commonwealth (see Figs. 2, 3), with diffuse extensions into southern West Virginia and eastern Kentucky. To characterize that spatial distribution, Bollinger (1973a, b) defined the Central Virginia Seismic Zone (CVSZ) and the Southern Appalachian Seismic Zone (SASZ). The 1875 Goochland County earthquake (MMI = VII, m_b = 5) in the central portion of the state (Oaks and Bollinger, 1986) and the 1897 Giles County earthquake (MMI = VIII, m_b = 5.8) near the Virginia–West Virginia border (Bollinger and Hopper, 1971) are the two largest known shocks in the state. Temporally, earthquake occurrence has varied, with the largest shocks occurring just before the turn

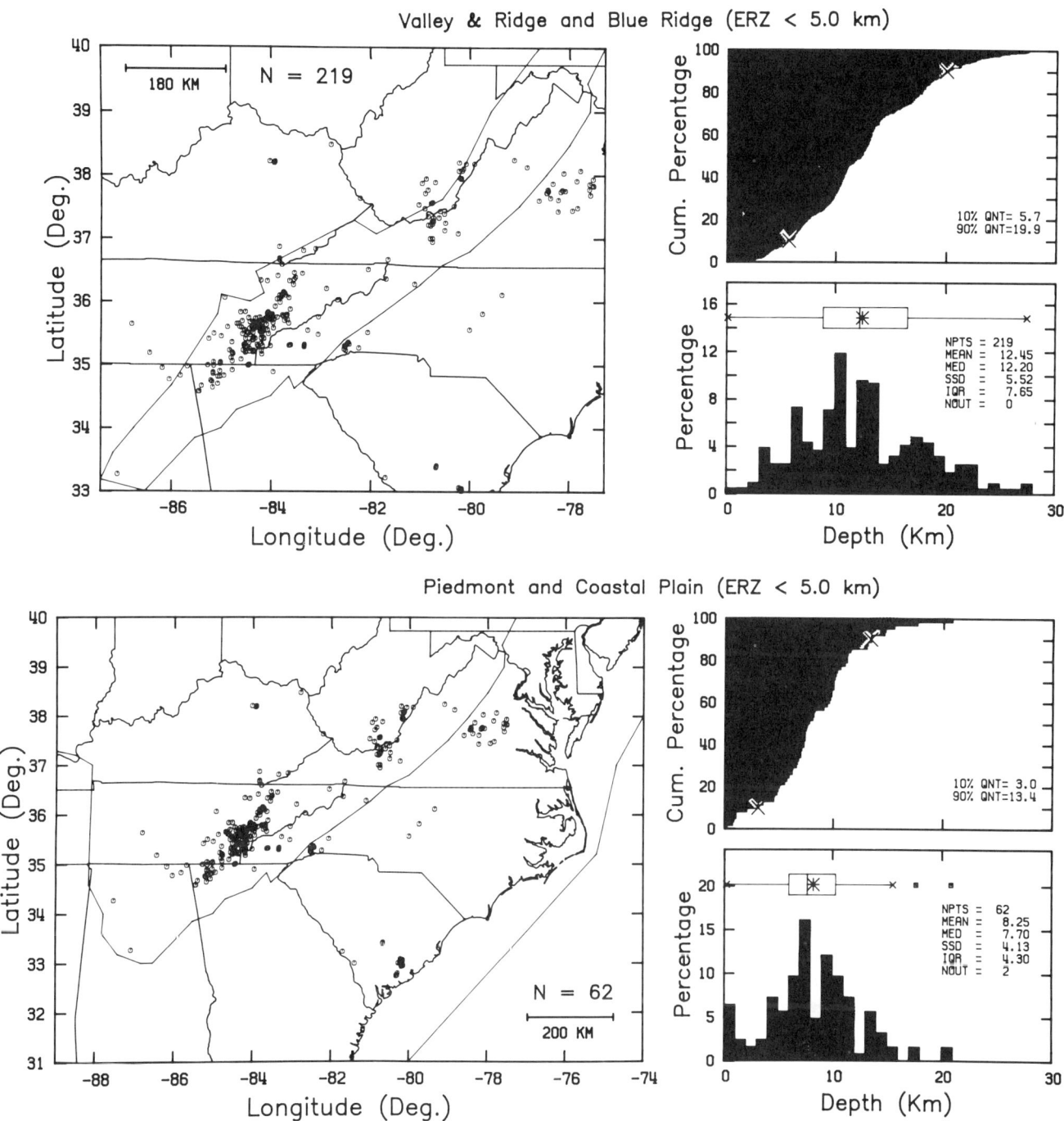

Figure 5. Distribution of earthquake focal depths in the southeastern U.S. by geologic province. Left: (upper) Epicenter map with combined Valley and Ridge and Blue Ridge provinces indicated between northeasterly oriented, connected straight-line segments; (lower) Epicenter map with combined Piedmont and Coastal Plain provinces indicated between northeasterly trending, connected straight-line segments. Right: Cumulative distribution and histogram plots (see caption, Figure 3, for explanation).

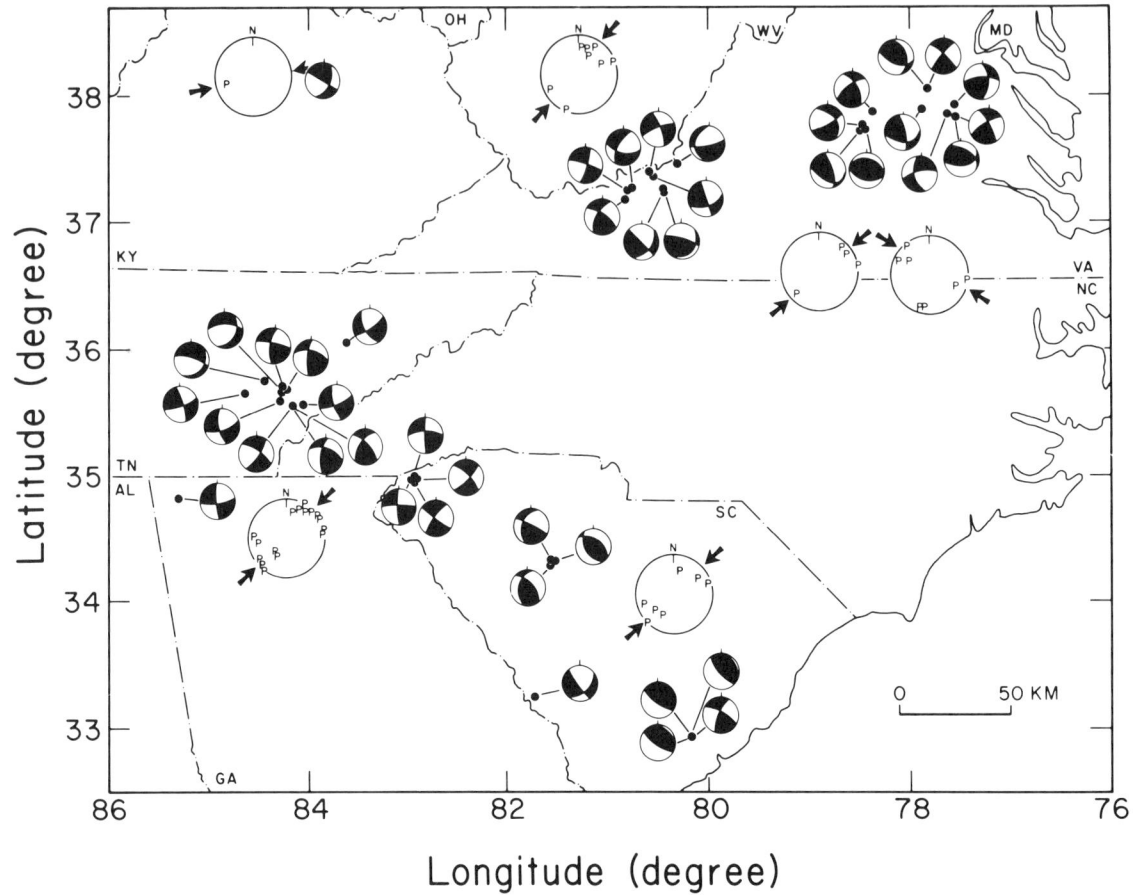

Figure 6. Focal mechanism solutions for southeastern U.S. earthquakes. Shown are lower hemisphere, equal-area plots. Smaller circles are for the focal mechanism solutions (orthogonal nodal planes with compressional quadrants shaded). Larger circles are plots of P-axes (maximum compressive stress), one axis plot (P) per focal-mechanism solution. The two P-axis diagrams for the central Virginia earthquakes are for different depths: the left diagram is for shallow shocks at 3 to 7 km depths; the right diagram is for deeper shocks at 8 to 13 km. Heavy arrows indicate average P-axis direction. Sources and solution quality (A = excellent, B = good, C = fair): Kentucky—Herrmann and others, 1982 (A); Virginia—Munsey and Bollinger, 1985 (GCVSZ = A-B; CVSZ = B-C); Tennessee—Teague and others, 1986 (A); Georgia—Long and Liow, 1986 (A); South Carolina (southeastern)—left; Tarr, 1977 (top = C), Tarr and others, 1981 (lower = C), and right: Talwani, 1982 (A; B). All of the focal mechanisms shown are for individual earthquakes except those by Talwani, 1982, and by Tarr and others, 1981. The Talwani focal mechanisms are for different depth intervals: upper for 4 to 8 km and lower for 9 to 13 km; southwestern (Savannah River Plant): Talwani and others, 1985 (C); North-Central (Newberry, SC): Rawlins, 1986 (C); Northwestern (Lake Jocassee, SC): Rastogi and Talwani, 1984 (A-B). See Figures 7–10 for details of place locations.

of the century and no damaging, $m_b \geqslant 5$, events since that time except for the single m_b = 5.2, MMI = VII, Kentucky earthquake in 1980.

Network monitoring in Virginia was initiated during 1977 and 1978. Since that time, studies based on data derived from a 20-station network (Fig. 1) have shed light on the nature of the seismic characteristics of the state and its environs. Important benchmark studies include those of the Giles County Seismic Zone by Bollinger and Wheeler (1982, 1983), of the CVSZ by Bollinger and Sibol (1985); and of both zones by Munsey and Bollinger (1985) and Bollinger and others (1986). The major characteristics of the two zones are very different even though they are separated by only some 200 km.

Giles County, Virginia, Seismic Zone (GCVSZ). The GCVSZ, a well-defined zone that strikes northeasterly and dips nearly vertically, is about 40 km long, 10 km wide, and from 5 to 25 km deep. The orientation of the zone is about 20° counterclockwise to the east-northeasterly trend of the overlying, detached southern Appalachian structures (Valley and Ridge province) and subparallel to the northeasterly trend of the central Appalachian structures in the northern part of the state. Additionally, the GCVSZ is beneath the rocks detached by thrusting.

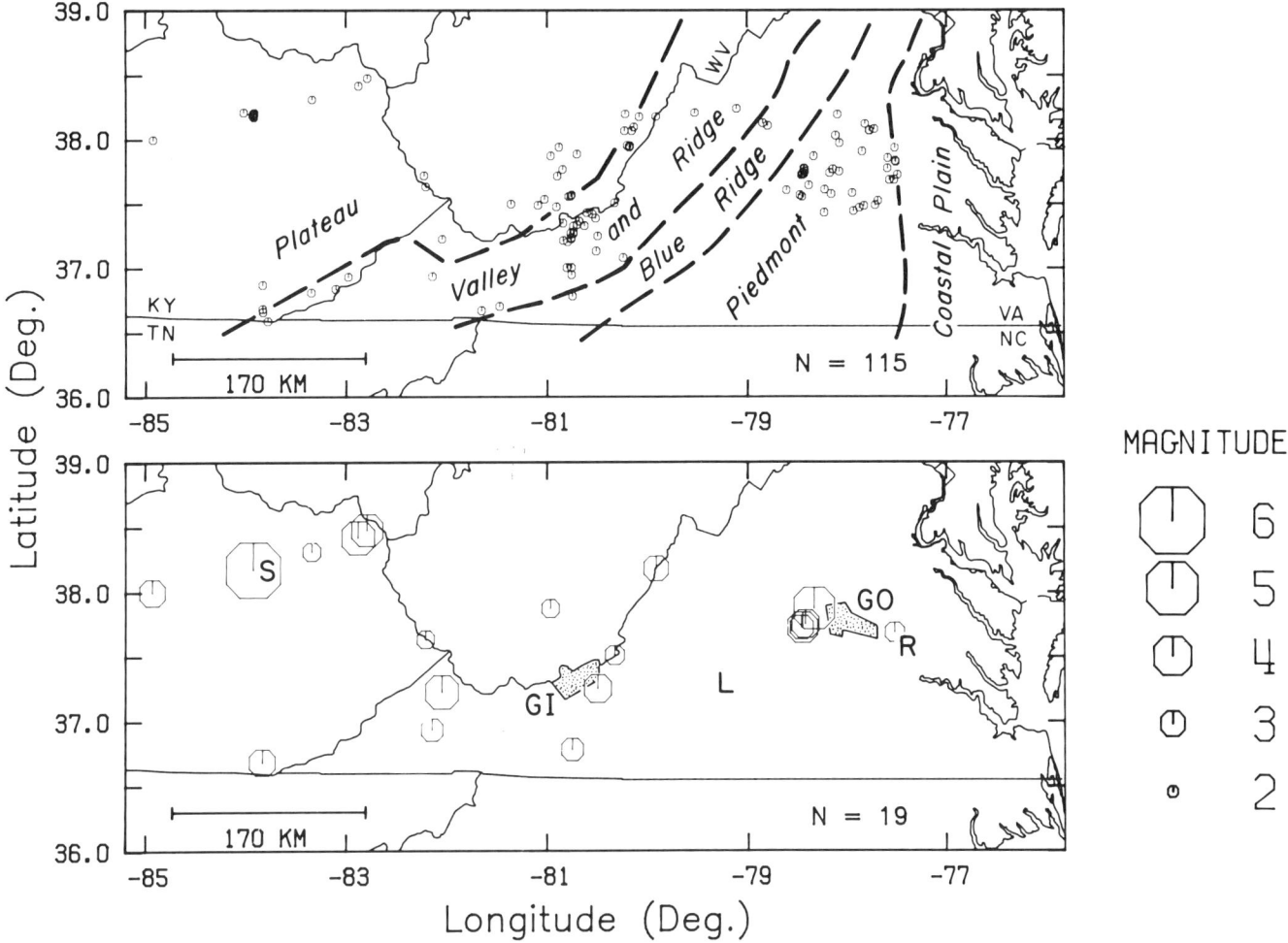

Figure 7. Network seismicity maps for Virginia and environs—1977 through 1985. Upper: Epicenters for earthquakes with $M_D \geq 0.0$ shown by small circles. Light dashed lines separate geologic provinces as labeled. Lower: Epicenters for earthquakes with $M_D \geq 2.5$ shown as octagons whose diameters are scaled to M_D. N equals the number of earthquakes plotted. Locations: GI (shaded area) = Giles County, VA; GO (shaded area) = Goochland County, VA; L = Lynchburg, VA; R = Richmond, VA; S = Sharpsburg, KY.

The detachment separation is especially important because it demonstrated, for the first time, that a correspondence between bedrock geologic structures observed at the surface, e.g., faults, and the seismogenic structures in the region might be completely lacking.

It is likely that the release of seismic energy within the GCVSZ is the result of reactivation of one or more faults formed initially by extensional stresses during Eocambrian time. Present-day focal mechanisms (Fig. 6) exhibit mainly strike-slip motions on steeply dipping (>70°) planes that are right-lateral on the northerly striking nodal planes or left-lateral on the easterly striking nodal planes. There are also some nodal plane pairs that are nearly vertical and subhorizontal in orientation. But the P-axes (maximum compressive stresses) estimates are uniformly of northeasterly (NNE to ENE) strike with subhorizontal trends, and they are the same as P-axes estimates elsewhere in the region (Fig. 6).

Seismicity adjacent to the Giles County zone. The GCVSZ is not an isolated seismogenic structure. Network monitoring has defined some seismic activity to the north in West Virginia and to the south in Virginia toward the North Carolina state line (Fig. 7). The relation, if any, of this outlying seismicity to the main, more active zone is unknown at this time. Additionally, there is a spatially separate scattering of small shocks in westernmost Virginia. All of this adjacent or "halo" seismicity is outside the seismograph networks and is, therefore, less well located and much less apt to yield any reliable focal mechanisms.

Central Virginia Seismic Zone (CVSZ). This zone is an area of persistent, low-level seismicity in the Piedmont province of Virginia. Its north-south dimension is about 120 km, and it extends some 150 km in an east-west direction from Richmond to Lynchburg. Seventy-five percent of the foci are in the upper third of the crust, at an average depth of 8 km (see Fig. 5), which places them *above* the Appalachian detachment. This spatial distribution of earthquake foci is diffuse, both horizontally and vertically. Such a pattern of earthquake foci favors multiple, rather than singular, seismogenic structures. The focal-mechanism solutions for the CVSZ corroborate this source multiplicity (Fig. 6) by exhibiting a mixture of reverse and strike-slip faulting on planes with variable strike and generally steep dip (>60°). There is a tendency for strike-slip motion to dominate over dip-slip as the mode of faulting for deeper earthquakes in central Virginia, a result consistent with the increase of lithostatic pressure with depth causing an interchange of the relative sizes of the three principal stresses. (Talwani [1982] proposes the same effect for earthquakes near Charleston, South Carolina.)

There is, however, a distinct separation of the P-axes with focal depth: For shallow shocks located 3 to 7 km deep, the trend of these maximum compressive stress estimates is northeasterly, whereas the deeper earthquakes at 8 to 13 km have northwesterly trending stresses. The correspondence of that separation to the depth of the Appalachian detachment (8 to 10 km) indicates that some mechanical decoupling may exist. This is another important geologic characteristic to be detected and documented for the first time in the region by the careful analysis of seismic network data. It is, however, disputed by Nelson and Talwani (1985), who interpret the data there as not requiring such a change. (Additionally, Long [personal communication, 1986] reports that he has not observed a similar P-axis shift in Georgia or South Carolina).

Seismicity in eastern Kentucky. Both historically and recently, easternmost Kentucky has maintained a low level of very diffuse earthquake activity (see Fig. 7). Interestingly, however, an m_b = 5.2 shock occurred in 1980 in the north central portion of the state near the small community of Sharpsburg (Herrmann and others, 1982; see Figs. 6 and 7 herein).

Summary. Eight years of monitoring with a 20-station regional seismic network has produced epicenters ($0 \leq m_b \leq 4$), focal depths, and mechanisms of adequate number and quality to reveal considerable differences between the two seismically active portions of Virginia. Despite their spatial proximity (about 200 km apart), the GCVSZ and CVSZ exhibit remarkable differences in their geologic and mechanical characteristics. In Giles County, seismic energy is released by predominantly strike-slip faulting within a nearly vertical, tabular zone (40 km long) that is *below* the Appalachian décollement. In central Virginia, the seismicity is derived from mixed strike-slip and dip-slip faulting in a large coin-shaped volume (120 × 150 km; ≈ 13 km vertical thickness) *above* the major detachment faulting. Compressive stress estimates, as derived from focal-mechanism solutions, are northeasterly and subhorizontal except for the deeper (8 to 13 km) shocks in central Virginia where a northwesterly trend may be present.

The causes for the observed variability are unknown. The two zones are in different geologic provinces, but because the Giles County activity is subdetachment, that aspect does not appear to be a controlling factor. The zones may also be in different tectonostratigraphic (suspect) terranes, and that difference could be relevant. Clearly, favorably oriented and/or weak faults are being reactivated under the current in situ stress regime. The recently proposed hydroseismicity model (Costain and others, 1986) ascribes such observed seismicity variations in Virginia and throughout the Southeast to differences in drainage basin hydrological characteristics and in upper-crustal fracturing, and to the chemical effects of hydrolytic weakening of rocks along pre-existing fractures.

Seismicity of North Carolina and Tennessee

Most of the historical seismicity in North Carolina and Tennessee is contained within the Blue Ridge and Valley and Ridge physiographic provinces—part of the Southern Appalachian Seismic Zone defined by Bollinger (1973a, b) (Fig. 8). Reinbold and Johnston (1987) list 108 historical earthquakes in North Carolina and Tennessee; the first reported earthquake is in North Carolina in November 1776. There are nearly twice as many felt events (69 compared to 39) reported from eastern Tennessee as there are from western North Carolina. However, the only shock with an estimated $m_b \geq 5$ occurred near Waynesville, North Carolina, on February 21, 1916. The largest known event in eastern Tennessee is the 1973 m_b = 4.6 earthquake in the Alcoa-Maryville area just south of Knoxville (Bollinger and others, 1976).

In east Tennessee, nearly half the seismic events (33) originate within 25 km of Knoxville. The earthquakes in western North Carolina appear to run somewhat parallel to bedrock structure, and MacCarthy (1956) suggests an alignment of 11 events paralleling the Blue Ridge front. The only clustering of events is in the Asheville-Hendersonville area, where a circle of 25 km radius includes 15 of the 39 events. A very interesting swarm of more than 75 earthquakes in two months occurred at Stone Mountain in 1874. Earthquakes were also reported there in 1848, 1876, 1880, and 1884. Field inspection by Ferguson and Stewart (1975) showed "huge, relatively-recent looking splits in the mountains there that could have been due to this swarm." (See Fig. 8 for the locations discussed above.)

The remainder of North Carolina has experienced a relatively low level of historical seismicity. There is a clustering of about 11 epicenters near Winston-Salem in the Carolina Piedmont. These events are first recorded in the diaries of the Moravians who settled that area and apparently end with the last of the diaries around 1830.

The SEUSSN instrumental data base for the region is shown in Figure 8 for the time period 1977 to 1986. Some larger regional events occurring between 1928 and 1977 have been instrumentally relocated by Dewey and Gordon (1984), but true local or even regional seismicity monitoring was not available during that period. The quality of instrumental coverage of this portion of the southern Appalachian zone increased significantly during the period 1980 to 1983, with the installation of the 18-station Southern Appalachian Regional Seismic Network (SARSN) by the Center for Earthquake Research and Information (CERI) at Memphis State University, a 7-station southeast-

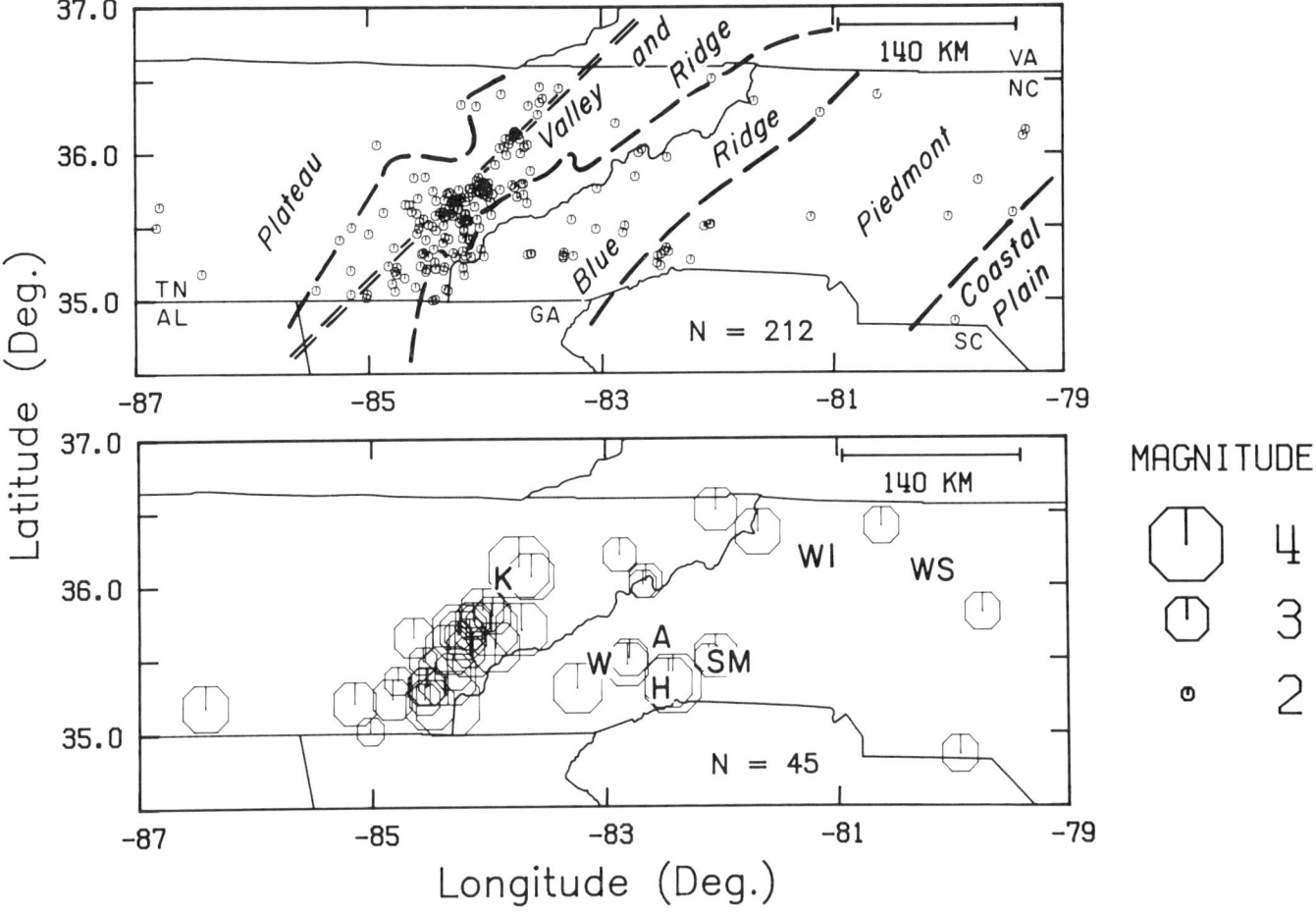

Figure 8. Network seismicity maps for the Tennessee–North Carolina border area—1977 through 1985. Upper: Epicenters for earthquakes with $M_D \geq 0.0$ shown by small circles. Heavy dashed line = New York–Alabama lineament (King and Zeitz, 1978). Light dashed lines separate geologic provinces as labeled. Lower: Epicenters for earthquakes with $M_D \geq 2.5$ shown as octogons whose diameters are scaled to M_D. N equals the number of earthquakes plotted. A = Asheville, NC; H = Hendersonville, NC; K = Knoxville, TN; W = Waynesville, NC; WI = Wilkesboro, NC; SM = Stone Mtn, NC; WS = Winston-Salem, NC.

ern Tennessee network operated by the Georgia Institute of Technology, and a 20-station network operated by the Tennessee Valley Authority (for a station location map, see Fig. 1).

A conservative estimate of the level of completeness for detection/location of earthquakes throughout the two-state region is M = 2.5, but this regional level is affected by the relatively poor coverage of eastern North Carolina. In the most active portion of the region (the southern Appalachians), Johnston and others (1985) estimated a detection completeness threshold below magnitude 1 and a locational completeness threshold of about M = 1.8.

Eastern Tennessee. The detailed instrumental seismicity of the Tennessee/North Carolina Appalachians compiled over the last 10 years (Fig. 8) closely emulates the spatial pattern of historical seismicity of the last two centuries (Fig. 2). Eastern Tennessee, principally within the Valley and Ridge physiographic province, exhibits the highest level of seismic activity. This zone has recently merited numerous special studies for two reasons: (1) it has been the most seismically active region in the Southeast over the past 10 years, even though the largest known shock of the region, the 1973 M = 4.6 Alcoa-Maryville earthquake (Bollinger and others, 1976), is smaller than other known events in the Southeast; and (2) the region contains a high concentration of critical facilities, principally nuclear power generating plants, so that an accurate characterization of its seismic hazard takes on added importance.

Johnston and others (1985) and Long and Liow (1986) discussed the seismotectonics of eastern Tennessee in terms of recent results from seismic networks and potential field studies. Teague and others (1986) studied the regional stress regime and style of faulting, utilizing both P-wave first motions and P/S-wave amplitude ratios in single-event and composite focal mechanisms. As previously mentioned, Reinbold and Johnston (1987) have recently completed a new compilation of the historical seismicity of the southern Appalachians, consulting previous regional catalogs and over 100 regional newspapers as primary sources.

From these investigations and others in progress, a reasonably well-constrained seismotectonic model for eastern Tennessee has emerged that has many similarities to the GCVSZ already discussed herein. Active faulting is occurring on steeply dipping fault planes, the majority of which are located beneath the master Appalachian décollement. Focal mechanisms of the region's larger events are remarkably similar, yielding estimates of right-lateral strike-slip faulting on north-south nodal planes or left-lateral slip on east-west planes. Some variation from this faulting style for smaller events or composite mechanisms is evident in Figure 6, but all yield a consistent P-axis orientation of northeast-southwest.

Johnston and others (1985) have argued that a spatial control of eastern Tennessee seismicity is exerted by major structural features in the basement crust at seismogenic depths (about 7 to 25 km) beneath the Appalachian detachment. In map view, between 80 and 90 percent of the Tennessee–North Carolina seismicity lies between deep-seated linear structures or boundaries that have been identified by their magnetic field signature (King and Zietz, 1978; Nelson and Zietz, 1983). The most extensive such feature, the New York–Alabama (NY-AL) lineament (Fig. 8), is apparently not seismogenic itself, as its strike of about N30°E is subparallel to the tectonic fabric of the overthrust Appalachians and is not consistent with the northerly and easterly focal mechanism nodal planes. However, the average depth of seismicity occurring in Tennessee inboard (northwest) of the NY-AL lineament is 16.5 km ($\sigma = 6.6$ km). In contrast, the average depth of Tennessee/North Carolina seismicity outboard (southeast) of the NY-AL lineament is 12.8 km ($\sigma = 4.6$ km).

Western North Carolina. Recently Williams and Talwani (1986) obtained the first focal mechanism in western North Carolina for a 1983 earthquake very near the Brevard zone that separates the Blue Ridge and Piedmont provinces. Nodal plane constraint was poor, so that either the strike-slip solution typical of eastern Tennessee or atypical low-angle reverse faulting was possible. However, both yielded a northeast-southwest, nearly horizontal P-axis, implying a consistent regional stress regime throughout the southern Appalachians.

Western Tennessee and eastern North Carolina. Western Tennessee, as part of the Mississippi embayment/New Madrid seismic zone, is covered elsewhere in this volume and will not be included here. Moreover, eastern North Carolina, comprising the Coastal Plain and eastern portions of the Carolina Piedmont, has been virtually aseismic both historically and in recent times.

Summary. The above findings suggest that the NY-AL lineament may be a major crustal boundary, separating cratonic North American crust of Grenville age from a more heterogeneous, perhaps allochthonous, basement crust underlying most of the Valley and Ridge and Blue Ridge overthrust provinces. The principal difficulty with this interpretation is the presence of obducted Grenville-age rocks to the east in the Blue Ridge and Piedmont provinces, which would support the interpretation of the NY-AL lineament as a structural feature *within* Grenville basement rather than as a boundary between crustal blocks of different composition and/or age.

Aside from the major question posed by the nature of the NY-AL lineament, there are several other important but unresolved seismic problems in this region.

1. *Maximum earthquake potential.* Unlike the GCVSZ to the north, the hypocentral pattern in Tennessee–North Carolina is diffuse, with no obvious alignments. Nodal planes do not align with the major basement structures so far identified, suggesting the seismicity is occurring on a collection of small-scale (reactivated Eocambrian?) faults rather than on a relatively few, larger scale, throughgoing features. If this interpretation proves correct, it might explain the lack of large (M > 5.5) earthquakes in the region.

2. *Temporal variability.* Close examination of Figures 2, 3, and 8 reveals some smaller scale deviations within the overall consistency of the pattern of seismic energy release between the historical and instrumental data. Western North Carolina (Blue Ridge province) had a significantly higher seismicity level historically than has been measured thus far with network instrumentation. Historically, the Coastal Plain of eastern North Carolina was not totally aseismic as it has been since 1977. The importance of these temporal variations in activity is difficult to assess, but they do suggest that exceptions exist to the key assumption in seismic hazard analysis that seismic energy release is stationary in time (and thus also spatially).

3. The Coastal Plain and Piedmont provinces of North Carolina (Fig. 5) are regions that have been neglected in seismic studies. Why they are apparently nearly aseismic when the Virginia Piedmont to the north and the South Carolina Coastal Plain to the south are active is unknown and introduces a major element of uncertainty into seismic hazard assessments of the Atlantic seaboard.

Seismicity of South Carolina

The most well-known South Carolina earthquake occurred in the Coastal Plain near Charleston on August 31, 1886 (MMI = X, $m_b = 6.7$, $M_s = 7.7$; Nuttli and others, 1986). The earliest known earthquake in historic times also occurred near Charleston in 1698 (Bollinger and Visvanathan, 1977). The largest historical earthquake in the Piedmont portion of the state was the MMI = VII–VIII, 1913, Union County earthquake (Taber, 1913; Reagor and others, 1980a).

The initial studies of seismicity in this area were by Dutton (1889) and Taber (1913, 1914). More recently, Bollinger wrote a series of papers describing the seismicity of South Carolina (1972) and the southeastern U.S. (1973a, b), compiled an early catalog of earthquakes in the southeastern U.S. (1975), and described the pre-1886 earthquakes in the Charleston area (Bollinger and Visvanathan, 1977). He described the diffuse spatial

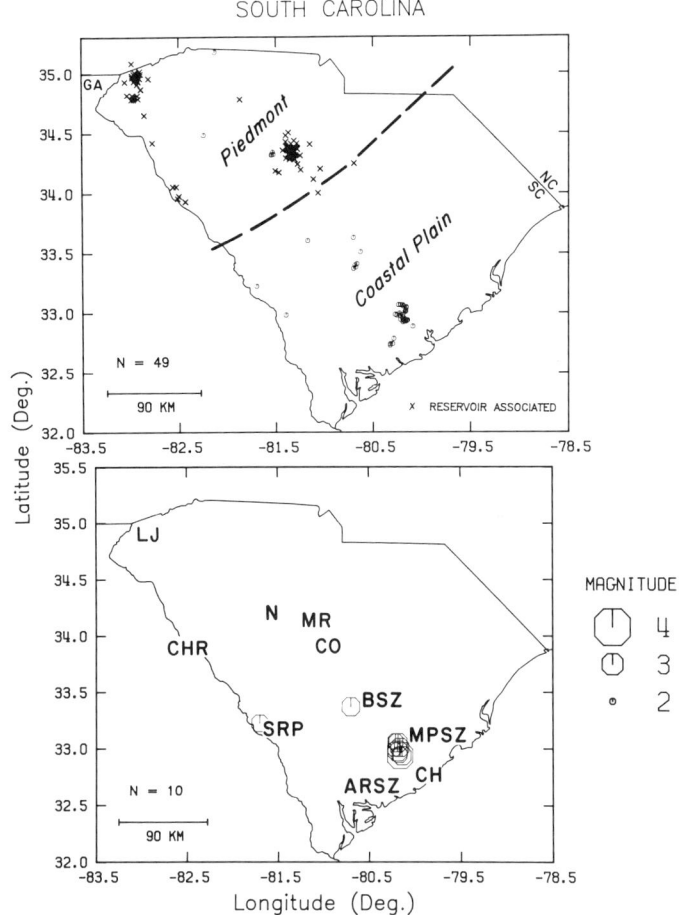

Figure 9. Network seismicity maps for South Carolina—1977 through 1985. Upper: Epicenters for earthquakes with $M_D \geq 0.8$ shown by small circles. Light dashed line separates the Piedmont and Coastal Plain provinces as labeled. Lower: Epicenters for earthquakes with $M_D > 2.5$ shown as octagons whose diameters are scaled to M_D. N equals the number of earthquakes plotted. X indicates an epicenter of a reservoir-associated earthquake; shown are a total of 484 such shocks. Locations: ARSZ = Adams Run, SC; BSZ = Bowman, SC; CH = Charleston, SC; CO = Columbia, SC; LJ = Lake Jocassee; MR = Monticello Reservoir; MPSZ = Middleton Place—Summerville, SC locale; N = Newberry, SC; SRP = Savannah River Plant; CHR = Clarks Hill Reservoir.

pattern of seismicity in the state as being a part of the northwest-trending South Carolina–Georgia seismic zone (SCGSZ). Visvanathan (1980) incorporated earlier catalogs to publish a list of felt earthquakes in South Carolina for the period 1698 to 1975. A newspaper search for historical earthquakes not previously listed, or erroneously listed in earlier catalogs, led Seeber and others (1982) to suggest the occurrence of a spatial "donut" pattern of epicenters preceding the 1886 Charleston shock. Seeber and Armbruster (1983) also discovered additional earthquakes in the area between 1886 and 1889 not listed in earlier catalogs.

Seismic network monitoring began in South Carolina in March 1973 with a reconnaissance field survey in the meizoseismal zone of the 1886 earthquake (Tarr and King, 1974). This survey was followed by the establishment of a 10-station permanent statewide network in May 1974 (Tarr, 1977). That network has developed into the present-day South Carolina Network (SCNET) of 19 stations and has been supplemented by three local mininetworks, two in the vicinity of reservoirs and the third in the vicinity of the Savannah River Plant near the South Carolina–Georgia border (Fig. 1).

Reservoir-induced seismicity (RIS) has been monitored since its inception in October 1975 at Lake Jocassee in the extreme northwestern corner of the state (Fig. 9). That seismicity was initially monitored by a network of portable seismographs through 1980 and then by a permanent three-station network (Talwani and others, 1980). In November 1977, South Carolina Electric and Gas Company installed a four-station network in the vicinity of Monticello Reservoir, near Columbia, prior to the impoundment of the reservoir (Fig. 9). Following the subsequent inception of RIS there, the network was expanded to ten stations in May and June 1978 (Talwani and others, 1980). In 1976, the E. I. duPont de Nemours Company established a three-station network at the Savannah River Plant.

The nature of seismicity near Charleston was the subject of papers by Tarr (1977), Talwani (1982), and Tarr and Rhea (1983). An early result of these studies was the identification of two different seismic regions (Tarr and others, 1981). The first region, in the Coastal Plain province, consists of distinct clusters of seismicity. A diffuse pattern of seismicity characterizes the second regime in the Piedmont and upper Coastal Plain, except in the vicinity of reservoirs.

Seismic zones in the Coastal Plain. Tarr and others (1981) noted that the seismicity in the Coastal Plain province occurred in three distinct zones: the Middleton Place–Summerville seismic zone (MPSSZ), some 20 km northwest of Charleston; the Bowman seismic zone (BSZ), about 60 km northwest of the MPSSZ; and the Adams Run seismic zone (ARSZ; actually a cluster of four earthquakes with $M < 2.5$), about 30 km southwest of MPSSZ (Fig. 9). In addition, they noted other spatially isolated cases of individual earthquakes that were large enough to be felt by residents in the Coastal Plain.

The Middleton Place–Summerville Seismic Zone (MPSSZ). Coincident with the meizoseismal area of the 1886 Charleston earthquake, the MPSSZ is the most active seismic zone in the Coastal Plain province. In the last decade, it has been the focus of an intense multidisciplinary effort aimed at understanding the seismotectonics of the area (Rankin, 1977; Gohn, 1983; and Hays and Gori, 1983). Reviews of those recent studies (Talwani, 1985) and hypotheses (Dewey, 1985) point to a lack of agreement on the cause and nature of seismicity in this host area for the important 1886 shock. However, several significant results have emerged.

Talwani (1982) and Rhea (1987) separately used different velocity model revision procedures to relocate the instrumentally recorded earthquakes in the MPSSZ. Talwani inferred the existence of two intersecting buried faults by the alignment of relocated hypocenters. A shallow, northwest-trending fault, inferred from hypocenters lying between 4 and 8 km and located along the Ashley River, was labeled the Ashley River fault (the ARF is at ARSZ in Fig. 9). It is postulated to intersect a deeper, north-northeasterly–trending Woodstock fault (WF; extends from ARSZ to MPSZ in Fig. 9) inferred from the planar distribution of hypocenters between 9 and 13 km. Conversely, Rhea (1987) reports that the MPSZ is a north-south–trending, diffuse zone of hypocenters, about 28 km long and 14 km wide. Calculated hypocentral depths in her study lie between about 2 and 10 km.

Composite focal-mechanism solutions of the events used to define the northwest-striking ARF (Tarr and others, 1981; Talwani, 1982), and that for a single magnitude-3.8 event on November 22, 1974 (Tarr, 1977), indicate that the seismicity may be occurring on a steep, southwesterly dipping reverse fault, with the southwest side being upthrown, or on a very shallowly dipping thrust fault with the northeast side up. Both nodal planes have a northwesterly strike (Fig. 6). Composite fault-plane solutions of events used to define the WF suggest strike-slip motion (Talwani, 1982). A common feature of the various focal-mechanism solutions is that the P-axes are nearly horizontal and oriented east-northeasterly, consistent with other stress estimates elsewhere in the region. All focal mechanism solutions are accompanied by a number of inconsistent P-wave first motions; see the caption for Figure 6 for evaluations of focal-mechanism solution quality.

A search for preserved evidence of prehistoric earthquakes in surficial sediments (paleoseismology) in the MPSSZ has revealed the presence of at least two pre-1886 earthquakes (Talwani and Cox, 1985; Obermeier and others, 1985). Using dateable organic matter and its crosscutting relationships with extruded sands interpreted to be caused by prehistoric earthquakes, Talwani and Cox estimated a recurrence time of 1,500 to 1,800 yr for destructive earthquakes in the MPSSZ. Obermeier and others (1986) discovered multiple generations of prehistoric Holocene earthquake-induced sandblows at widespread sites in coastal South Carolina. These sandblows are more widely distributed spatially than those associated with the 1886 shock, suggesting multiple shocks at various locations, or at least one shock, stronger than that in 1886 but with the same general epicentral location.

A synthesis of geological, geophysical, and seismological data has led Talwani (1986) to present an explanation of the seismicity in the MPSSZ as the result of intersecting structures.

Bowman Seismic Zone (BSZ). In an area for which there were no other known earthquakes in historical times, a series of felt ($3 < M_L < 4$) events occurred near Bowman between 1971 and 1974 (Tarr and King, 1974; Tarr, 1977). The BSZ (Fig. 9) has continued to experience sporadic seismicity since 1974. Between November 1977 and early 1984, four additional stations were installed in the BSZ, and the subsequent hypocentral locations suggested that most of the earthquakes were 6 km deep or shallower and apparently related to the border faults of a buried Triassic basin. No fault plane solutions have thus far been obtained.

Adams Run Zone (ARSZ). This zone (Fig. 9) was identified by Tarr and others (1981) on the basis of four earthquakes with magnitudes less than 2.5, three of which occurred in a two-day period in December 1977. In spite of increased instrumentation, no events have been detected here since October 1979. No fault plane solutions have been obtained for these events. Until there is considerably more seismic activity at this locale, it should be referred to as a "cluster" rather than a "zone." The seismicity here has been interpreted by Talwani (1982) to occur at the southwest terminus of a proposed Woodstock fault.

Other earthquakes in the Coastal Plain. The most significant post-1979 event in the Coastal Plain, outside of the three zones described above, was the magnitude-2.6 earthquake of June 29, 1985. This event, which occurred within the Savannah River Plant (Talwani and others, 1985), was about 1 km deep and was located at the intersection of an inferred shallow northwest-trending feature and a border fault on the northeast-trending Dunbarton Triassic basin. A focal-mechanism solution suggests mostly left-lateral strike-slip motion on the northeast-oriented border fault dipping about 46° to the southeast, or mostly right-lateral strike-slip motion on a northwesterly striking plane with subvertical dip. The inferred direction of the P-axis was north-northeasterly.

Piedmont province. Most of the seismicity in the Piedmont province since 1979 has been spatially associated with reservoirs. Major exceptions were the two swarms of earthquakes in the vicinity of Newberry (Fig. 9) that occurred in July and August 1982 and in April 1983 (Rawlins, 1986). Five earthquakes with magnitudes greater than 2.0 were felt there. The earthquakes were all found to be shallow, in the top 4 km, and occurred along the flanks of a large granitic pluton. Three focal-mechanism solutions were obtained for the larger events (Fig. 6). They all suggest reverse movement on northwest-striking faults that dip northeast or northwest, with the P-axes subhorizontal and oriented in the regional east-northeasterly direction.

Reservoir-induced seismicity (RIS) at Lake Jocassee. Reservoir-induced seismicity has been monitored since its inception at Lake Jocassee in October 1975 (Talwani, 1981). Low-level seismicity continues there and is being monitored on a three-station network operated by Duke Power Company around the lake (Figs. 1, 9). The largest induced earthquake, a magnitude-3.9 shock, occurred there on August 25, 1979. The seismicity is confined to the top 5 km, and four composite focal-mechanism solutions exhibit right-lateral or left-lateral strike-slip faulting, with the P-axes oriented around N60°E (Fig. 6; Rastogi and Talwani, 1984), in excellent agreement with stress measurements obtained there from hydrofracture and overcoring (Talwani, 1984).

RIS at Monticello Reservoir. Monticello Reservoir, in addition to being the subject of RIS monitoring since its inception in December 1977, has also been the location of intensive geophysical and borehole investigations. The seismicity through 1985 was confined to the vicinity of the reservoir, was shallower than about

6 km, and was spatially associated with fractured rocks surrounding plutons. Comparison of 22 CEFMs (not shown in Fig. 6) with borehole data suggests that seismicity is occurring by thrust faulting on preexisting fractures, with the P-axes oriented northeasterly (Talwani, 1984).

Seismicity of Alabama and Georgia

The historical record of earthquakes in Alabama and Georgia begins in the late 1800s (Stover and others, 1978; Long, 1982; Steigert, 1984) and depicts a low-level release of energy in central Georgia and a moderate to low-level release of energy along the southwestern extension of the SASZ (Bollinger, 1973a, b) into Alabama and northwestern Georgia. The October 18, 1916, earthquake near Irondale (a Birmingham suburb) in central Alabama (MMI = VII) and the March 5, 1914, earthquake near Covington and Mansfield in central Georgia (MMI = VII, m_b = 4.9; Allison, 1980) are the two largest known shocks in the area.

Network monitoring in northern Georgia was initiated in 1975, and in Alabama in 1981 (Fig. 1). Earlier regional instrumental coverage from the early 1960s was available from Worldwide Standard Seismograph Network (WWSSN) stations at Atlanta, Georgia (ATL), Spring Hill, Alabama (SHA), and Oxford, Mississippi (OXF) and from temporary networks or isolated independent stations. Since that time, studies based on data from 1- to 5-station networks near reservoirs, and a regional network of about 10 stations in northern Alabama and Georgia, have helped define the seismic characteristics of the Appalachian Highlands (Valley and Ridge and Blue Ridge provinces), the crystalline Piedmont, and the Coastal Plain (Fig. 10). Significant differences in the major characteristics of the seismicity of the various geologic provinces are summarized in Steigert, 1984; Long, 1982; Marion and Long, 1980; and Allison, 1980.

Earthquakes in the Coastal Plain. The Coastal Plain sediments cover the southern half of Alabama and Georgia and conceal remnant crustal features of extension that occurred during the early opening of the Atlantic Ocean. The extensional thinning of the crust and the centers of mafic volcanism distinguish the crust of the Coastal Plain from the thicker crust of the Piedmont and Appalachian Highlands to the north and northwest. Fewer than 10 small earthquakes or event swarms are known or suspected to have occurred in the Alabama and Georgia Coastal Plain since the beginning of instrumental monitoring there. The best-documented event occurred December 27, 1976, in Toombs County (Fig. 10; Lance and others, 1977) and was felt with MMI = V. Such Coastal Plain events are distinguished from the significant Charleston earthquake of 1886 only by their smaller magnitudes. Network data have not been available for the Coastal Plain, but depths of focus are believed to be at typical crustal depths of 5 to 15 km, similar to depths observed in the Charleston, South Carolina, epicentral zone.

One exception to the mid-crustal origin for Coastal Plain seismicity is the sequence of events that occurred near the border

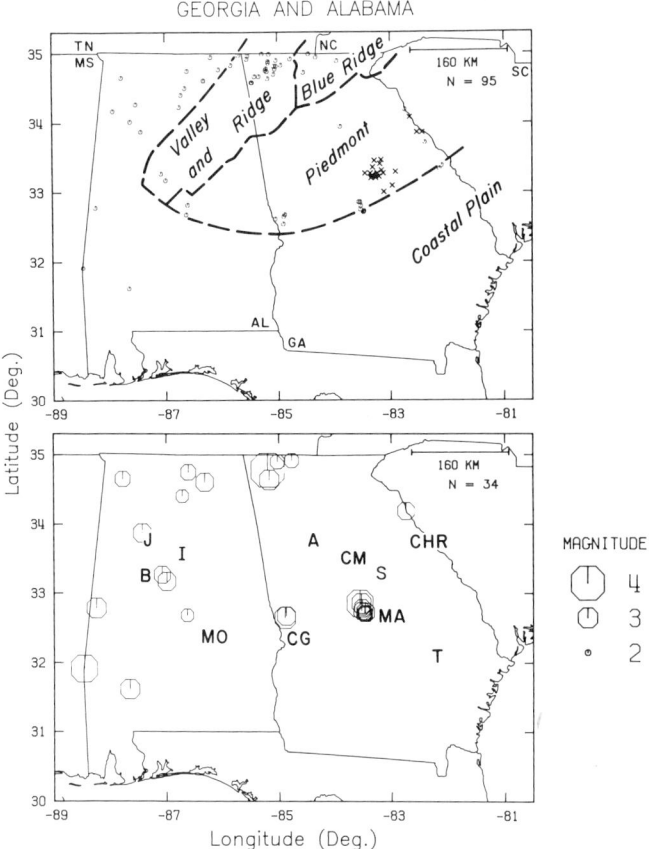

Figure 10. Network seismicity maps for Georgia and Alabama—1977 through 1985. Upper: Epicenters for earthquakes with $M_D \geq 0.6$ shown by small circles. Light dashed lines separate geologic provinces as labeled. Lower: Epicenters for earthquakes with $M_D > 2.5$ shown as octogons whose diameters are scaled to M_D. N equals the number of earthquakes plotted. X indicates an epicenter of a reservoir-associated earthquake; shown are a total of 93 such shocks. Locations: A = Atlanta, GA; B = Brookwood, AL; CG = Columbus, GA; CHR = Clarks Hill Reservoir; CM = Covington-Mansfield, GA locale; I = Irondale, AL (Birmingham suburbs); J = Jasper, AL; MA = Macon, GA; MO = Montgomery, AL; T = Toombs County, GA; S = Lake Sinclair.

of Mississippi and Alabama (32N,88.5W) between 1977 and 1981. These m_b = 3.0 to 3.6 events were felt strongly (MMI = V) in a small region near their epicenters. They are believed to have had very shallow hypocenters, suggesting a possible causal association with fluid injection in nearby secondary oil recovery wells.

Earthquakes in the Piedmont. The seismicity of the Piedmont follows generally the central portion of that province usually identified with the Charlotte and Carolina slate belts. In those belts in Georgia, two historical earthquake epicentral zones are notable. The first is in central Georgia; the second is near Georgia's border with South Carolina (Figs. 2, 3, and 10). In recent years, other seismic events have occurred near Macon and Columbus, Georgia (Jones and others, 1985). The Alabama

Piedmont has not experienced notable earthquake activity. The most active Piedmont area is in central Georgia, with eight felt events with MMI = III to VII. Two events of MMI = VI (1875 and 1974) have been felt in the Clarks Hill Reservoir area (Long, 1982; Reagor and others, 1979).

In the Piedmont, reliable hypocentral depths have been computed only in aftershock surveys and other studies using data from densely spaced stations. The typical depth is less than 2 km. Hence, a distinction can be made here between natural events and reservoir-induced events only on the basis of temporal and spatial association with reservoir impoundment. The observations for all Piedmont events are consistent with these earthquakes being caused by a release of stress along existing planes of weakness, such as near-surface fractures—i.e., focal mechanisms often correspond to the directions of prominent joint sets. Such strain release could be facilitated by strength deterioration through weathering processes or fluid-pressure changes, caused for example, by reservoir impoundment. A general model for the Piedmont earthquakes has been developed from seismic monitoring data and through geophysical studies in the reservoir areas (Guinn, 1980; Marion and Long, 1980; Jones and others, 1985) in which the fault area is limited by the depth penetration of joints or other planes of weakness, typically on the order of 4 km.

The two major centers of Piedmont seismicity are near reservoirs (Fig. 10): Lake Sinclair in central Georgia (33.3N,83.3W) and the Clarks Hill Reservoir on Georgia's border with South Carolina (34.2N,82.5W). Although the recent activity in these zones can be attributed to reservoir inducement, the historical seismicity predates any significant impoundment of the rivers in the Piedmont. Not all reservoirs in the Piedmont have induced measurable seismicity, suggesting that variations in the in situ stresses and/or host-rock mechanics may be the determining factor(s) in the occurrence of Piedmont earthquakes.

Earthquakes in the Appalachian Highlands. The northeast-southwest trend of the seismicity of Alabama and northwest Georgia parallels the structure of the Southern Appalachian Mountains and lies subparallel to the NY-AL lineament (see Tennessee and North Carolina section herein). These earthquakes populate the same depth ranges and exhibit the same general characteristics as other earthquakes of the SASZ to the northeast in southeastern Tennessee, but at a lower level of activity. The recurrence relation for Alabama is $\log N_c = 1.3 - 0.6\, m_b$ (L_g), where N_c = number of earthquakes/year $\geq m_b$ (L_g) (Steigert, 1984). The largest shock occurred east of Birmingham on October 18, 1916, and was felt with MMI = VII over an area that included most of Alabama, northern Georgia, and central Tennessee.

A possible transverse north-south seismic trend (Steigert, 1984) may bisect Tennessee and northern Alabama. A comparison of gravity data with earthquakes occurring in this Alabama-Tennessee Seismic Zone suggested to Steigert (1984) that a correlation may exist between that possible zone and a north-south–trending change in crustal structure.

The analysis of seismic network data for Alabama has been made difficult by the numerous large (m_b = 2 to 3) mining explosions and occasionally larger mine collapses or rockbursts (mb = 2 to 4), a series of which occurred from 1974 to 1980 near Jasper, Alabama. On May 7, 1986, near Brookwood, Alabama, a mine failure occurred that was felt locally at MMI = IV and registered m_b = 4.0 instrumentally.

REGIONAL SUMMARY

The seismicity in the southeastern United States occurs in three physiographic provinces: the Appalachian Highlands (combined Valley and Ridge and Blue Ridge provinces), the Appalachian Piedmont, and the Atlantic Coastal Plain (Fig. 11). Earthquakes in all three provinces are located in rather poorly defined zones, rather than being distributed randomly throughout each province. These zones are composed of lineations or clusters, either embedded within the sparser regional distribution or surrounded by a "halo" of seismic energy release of a level lower than that in the zone proper. The seismicity, in general, is low but persistent in each of the zones. The maximum horizontal stress, as inferred from the P-axes of focal mechanism solutions, is compressive, northeasterly in trend, and subhorizontal, although there is some localized variation, particularly with respect to depth, in all three provinces. Earthquake focal depths are some 25 percent deeper in the Appalachian Highlands than in the other two provinces, indicating a greater thickness of seismogenic crust there. However, it is the Atlantic Coastal Plain province that was host to the region's largest known shock by approximately a full magnitude unit.

Appalachian Highlands

A generally diffuse distribution of earthquakes occurs from the Virginia–West Virginia border area southwestward into central Alabama, with concentrated, northeasterly trending lineations in eastern Tennessee and in Giles County, Virginia. In vertical section, the majority of foci in the province are at depths of 5 to 25 km, *below* the Appalachian décollement. The average focal depth is 12 km, and the 90 percent depth is 20 km. Approximately 76 percent of the earthquakes in the southeastern United States occur in these provinces. The release of seismic energy is generally on steeply dipping planes (>60°), with a decided preference for strike-slip motions: dextral on northerly trending planes and sinistral on easterly striking planes. Those movements are believed to be reactivation of faults formed initially by an extensional stress regime during Eocambrian time. The Giles County, Virginia, seismic zone was the location of the region's second largest shock (m_b = 5.8, MMI = VIII) in 1897.

Appalachian Piedmont province

This province accounts for about 13 percent of the regional seismicity. The most extensive and persistent seismic activity is in the Central Virginia Seismic Zone (CVSZ). The foci in that zone

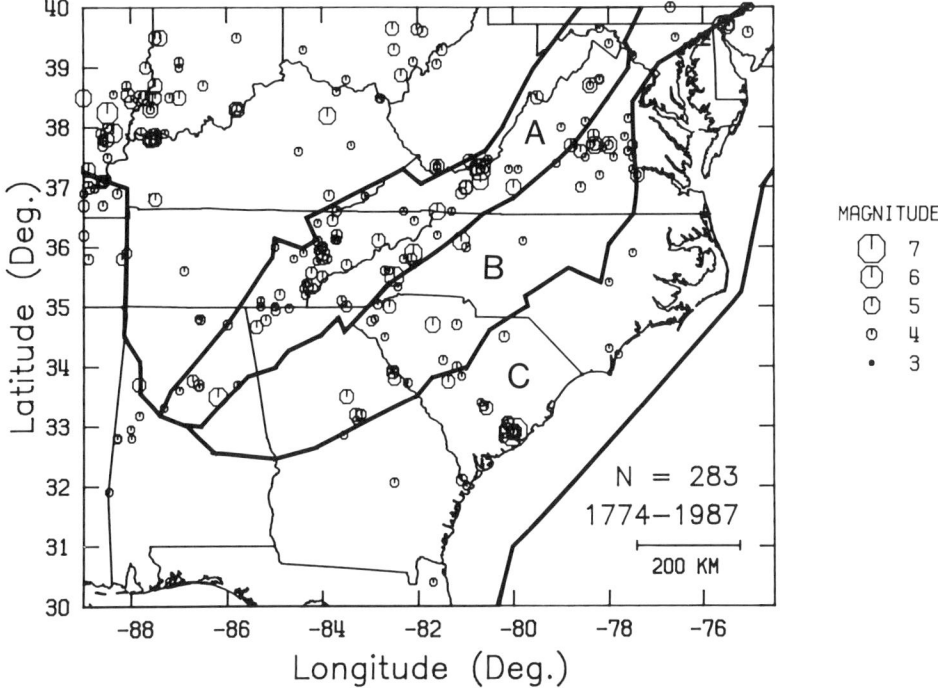

Figure 11. Southeastern U.S. seismicity (1774 to 1987) and geologic provinces. Epicenters for earthquakes with M ≥ 3.5 shown by small circles; provinces enclosed by solid straight line segments: A = Valley and Ridge and Blue Ridge; B = Piedmont; C = Coastal Plain.

are at an average depth of 9 km, and their 90 percent depth is 15 km. Again, focal-mechanism nodal planes are steeply dipping (>60°), with average orientations of north-northwesterly (dextral strike slip) and east-northeasterly (sinistral strike slip). The earthquake activity during the recent decade—m_b (L_g) ≤ 4.2—has occurred across a band 120 km (N-S) by 150 km (E-W) and primarily (approximately 75 percent) *above* the Appalachian décollement. The extended spatial distribution of the CVSZ, together with its primarily low-magnitude shocks (maximum historical: m_b = 5.0, MMI = VII in 1875), suggests multiple source structures. The maximum horizontal stress, as inferred from focal-mechanism solutions of earthquakes, is compressive, northeasterly, and subhorizontal, although there is some question as to whether the deeper CVSZ mechanisms indicate a northwesterly trend (Munsey and Bollinger, 1985; Nelson and Talwani, 1985).

Additional historical earthquakes have occurred in the South Carolina Piedmont and, to a lesser extent, in the Georgia Piedmont. However, that historical seismicity is not especially well documented. A recent earthquake swarm near the town of Newberry was composed of shocks that were small (M_L ≤ 2.5), shallow (≤ 4 km deep), and spatially associated with a buried granitic body (Rawlins, 1986).

The very low level of seismicity in the Piedmont sections of North Carolina and Georgia, compared to the more active adjacent areas in Virginia and South Carolina, is puzzling indeed.

The South Carolina Piedmont has been the location for rather extensive reservoir-induced seismicity (RIS) during the recent decade. This RIS is shallow (≤ 5 km) and is composed of small, tightly clustered earthquakes (M_L ≤ 4) and, where focal mechanisms have been determined, either of thrust or strike-slip types. This RIS in the South Carolina Piedmont stands in contrast to the lack of such activity in the adjoining states of North Carolina and Georgia.

Atlantic Coastal Plain province

The largest earthquake in the southeastern United States during historical times occurred in this province in 1886 (m_b = 6.7, M_s = 7.7, MMI = X). The epicenter was probably about 20 to 30 km northwest of Charleston in the Middleton Place–Summerville seismic zone (MPSSZ). The province as a whole accounts for some 11 percent of the regional seismicity. Other than a few small earthquakes in Georgia and Alabama, Atlantic Coastal Plain seismicity has been concentrated in three seismic zones in South Carolina. Nearly 70 percent of the instrumentally recorded Coastal Plain earthquakes have occurred in the largest of these three zones, the MPSSZ. The activity in the MPSSZ, coupled with the small Bowman seismic zone (BSZ) about 60 km to the northwest, represents most of the seismicity in the northwest-trending regional South Carolina–Georgia seismic zone of Bollinger (1973a, b). The hypocenters of Coastal Plain shocks are distributed throughout the upper crust (≤ 13 km deep). The focal-mechanism solutions as presented herein (Fig. 6), mostly CEFMs, indicate a north-northeast maximum compressive stress.

An intriguing feature of Coastal Plain seismicity is the apparent northwest trend, perpendicular to the northeasterly tectonic fabric of the adjoining Piedmont and Appalachian Highlands and their associated seismicity. Although various interpretations have been put forward, the faults associated with Coastal Plain earthquakes have not been unambiguously identified. Relocation of hypocenters in combination with analyses of related geological and potential field data led Talwani (1982, 1986) to identify two possible causative structures. Similar studies by Rhea (1987) produced a single zone interpretation.

As noted previously, in Georgia and Alabama, the Coastal Plain earthquakes are spatially and temporally sparse. The earthquakes of the Georgia Piedmont have been interpreted to represent failure on planes of weakness or joints within 4 km of the surface. Their mechanism is the same as that proposed for reservoir-induced earthquakes.

FINAL COMMENTS

As documented throughout this chapter, the instrumental data base for the southeastern United States has developed to a stage in which important advances in understanding the seismotectonics of the region have been made. The data set, however, is not so extensive as to be conclusive for several important topics. Consequently, differences in interpretation exist between the workers in the region. We have tried to call attention to these differences throughout this chapter, but space limitations prevent a full discussion of all of the arguments by opposing viewpoints. For emphasis we note the following areas of disagreement.

Focal depths of Piedmont earthquakes

G. Bollinger and co-workers report depths down to and below about 10 to 12 km, whereas P. Talwani locates foci only down to some 6 km and L. Long argues for the existence of only very shallow focal depths—2 km or less. Given the intrinsic difficulty in focal-depth determination, this is not a simple problem; only time and special studies, e.g., aftershock surveys *throughout* the province, will resolve it.

Azimuth of the principal horizontal stresses in the Virginia Piedmont

G. Bollinger and his students find a 90° change in azimuth (northeast to northwest) with increasing depth for focal mechanism P-axes in this seismic zone. P. Talwani and his students, using the same input data, conclude that such a change is not required. Clearly, more data are also needed for resolution here.

Seismotectonic model(s) for the Charleston, South Carolina, Seismicity

This is clearly the most controversial topic in the seismicity of the region. One school of thought, headed by P. Talwani, contends that the problem is essentially solved and that two intersecting faults compose the seismogenic structure. Talwani's model is a reasonable interpretation of the hypocentral locations. However, U.S. Geological Survey researchers interpret the spatial distribution of hypocenters, all associated with earthquakes having $M_L \leq 3.8$, to be an indication of small-scale faulting distributed over an appreciable volume. Again, more data and careful analyses are in order.

The above differences in interpretation reflect an important aspect of our current level of understanding. Different workers are persuaded to varying levels of confidence by the data currently available. It will indeed be interesting to watch the picture sharpen into focus as the accuracy and adequacy of the data base improves with time.

REFERENCES CITED

Allison, J. D., 1980, Seismicity of the Central Georgia Seismic Zone [Masters thesis]: Atlanta, Georgia Institute of Technology, 204 p.

Bollinger, G. A., 1972, Historical and recent seismic activity in South Carolina: Bulletin of the Seismological Society of America, v. 62, p. 851–864.

——, 1973a, Seismicity of the southeastern United States: Bulletin of the Seismological Society of America, v. 63, p. 1785–1808. Also see: 1974, Errata, v. 64, p. 733–734.

——, 1973b, Seismicity and crustal uplift in the southeastern United States: American Journal of Science, v. 273-A, p. 396–408.

——, 1975, A catalog of southeastern United States earthquakes, 1754 through 1974: Blacksburg, Virginia Polytechnic Institute and State University Research Division Bulletin 101, 68 p.

Bollinger, G. A., and Hopper, M. G., 1971, Virginia's two largest earthquakes—December 22, 1875, and May 31, 1897: Bulletin of the Seismological Society of America, v. 61, p. 1033–1039.

Bollinger, G. A., and Sibol, M. S., 1985, Seismicity, seismic reflection studies, gravity, and geology of the central Virginia seismic zone; Part I, Seismicity: Geological Society of America Bulletin, v. 96, p. 49–57.

Bollinger, G. A., and Visvanathan, T. R., 1977, The seismicity of South Carolina prior to 1886, *in* Rankin, D. W., ed., Studies related to the Charleston, South Carolina, earthquake of 1886; A preliminary report: U.S. Geological Survey Professional Paper 1028, p. 33–42.

Bollinger, G. A., and Wheeler, R. L., 1982, The Giles County, Virginia, seismogenic zone; Seismological results and geological interpretations: U.S. Geological Survey Open-File Report 82–585, 95 p.

——, 1983, The Giles County, Virginia, seismic zone: Science, v. 219, p. 1063–1065.

Bollinger, G. A., Langer, C. J., and Harding, S. T., 1976, The eastern Tennessee earthquake sequence of October through December, 1973: Bulletin of the Seismological Society of America, v. 66, p. 525–547.

Bollinger, G. A., Chapman, M. C., Sibol, M. S., and Costain, J. K., 1985, An analysis of earthquake focal depths in the southeastern U.S.: Geophysical Research Letters, v. 12, p. 785–788.

Bollinger, G. A., Snoke, J. A., Sibol, M. S., and Chapman, M. C., 1986, Virginia regional seismic network; Final report (1977–1985): Washington, D.C., U.S. Nuclear Regulatory Commission Report NUREG/CR–4502, 57 p.

Chen, W-P., and Molnar, P., 1983, Focal depths of intracontinental and intraplate earthquakes and their implications for the thermal and mechanical properties of the lithosphere: Journal of Geophysical Research, v. 88, p. 4183–4214.

Costain, J. K., Bollinger, G. A., and Speer, J. A., 1986, Hydroseismicity—A hypothesis for the role of water in the generation of intraplate seismicity: Seismological Research Letters, v. 58, p. 41–64.

Dewey, J. W., 1985, A review of recent research on the seismotectonics of the southeastern seaboard and an evaluation of hypotheses on the source of the 1886 Charleston, South Carolina, earthquake: U.S. Nuclear Regulatory Commission Report NUREG/CR-4339, 44 p.

Dewey, J. W., and Gordon, D. W., 1984, Map showing recomputed hypocenters of earthquakes in the eastern and central United States and adjacent Canada: U.S. Geological Survey Miscellaneous Field Studies Map MF-1699.

Dutton, C. E., 1889, The Charleston earthquake of August 31, 1886: U.S. Geological Survey Ninth Annual Report, 1887–1888, p. 203–528.

Ferguson, J. F., and Stewart, D. M., 1975, Summary of North Carolina seismicity in the 18th and 19th centuries: Earthquake Notes, v. 46, p. 27–36.

Gephart, J. W., and Forsyth, D. W., 1984, An improved method for determining the regional stress tensor using earthquake focal mechanism data; Application to the San Fernando earthquake sequence: Journal of Geophysical Research, v. 89, p. 9305–9320.

Glover, L. III, Costain, J. K., and Coruh, C., 1982, Vibroseis reflection seismic structure along the Blue Ridge and Piedmont James River profile, north central Virginia: Geological Society of America Abstracts with Programs, v. 14, p. 497.

Gohn, G. S., ed., 1983, Studies related to the Charleston, South Carolina, earthquake of 1886; Tectonics and seismicity: U.S. Geological Survey Professional Paper 1313, 375 p.

Guinn, S. A., 1980, Earthquake focal mechanisms in the southeastern United States: Washington, D.C., U.S. Nuclear Regulatory Commission Report NUREG/CR-1503, 150 p.

Hadley, J. B., and Devine, J. F., 1974, Seismotectonic map of the eastern United States: U.S. Geological Survey Miscellaneous Field Studies Map MF-620, scale 1:5,000,000.

Hays, W. W., and Gori, P. L., eds., 1983, A workshop on "The Charleston, South Carolina, earthquake and its implications for today," in Proceedings of Conference 20: U.S. Geological Survey Open-File Report 83-843, 502 p.

Herrmann, R. B., Langston, C. A., and Zollweg, J. E., 1982, The Sharpsburg, Kentucky, earthquake of 27 July 1980: Bulletin of the Seismological Society of America, v. 72, p. 1219–1239.

Johnston, A. C., Reinbold, D. J., and Brewer, S. I., 1985, Seismotectonics of the southern Appalachians: Bulletin of the Seismological Society of America, v. 75, p. 291–312.

Jones, F. B., Long, L. T., Chapman, M. C., and Zelt, K.-H., 1985, Columbus, Georgia, earthquakes of October 31, 1982: Earthquake Notes, v. 56, p. 55–61.

King, E. R., and Zietz, I., 1978, The New York–Alabama lineament; Geophysical evidence for a major crustal break in the basement beneath the Appalachian basin: Geology, v. 6, p. 312–318.

Lahr, J. C., 1980, HYPOELLIPSE/MULTICS—A computer program for determining local earthquake parameters, magnitude and first-motion pattern: U.S. Geological Survey Open-File Report 80-59, revised April, 1980, 59 p.

Lance, R. J., Fogle, G. H., and Long, L. T., 1977, Report on the earthquake of December 27, 1976, in southern Georgia: Earthquake Notes, v. 48, p. 51–56.

Long, L. T., 1982, Seismicity in Georgia, in Hardin, D., Beck, B. F., and Morrow, E., eds., Proceedings, Symposium on the Geology of the Southeastern Coastal Plain, 2nd, Americus, Georgia, March 1979: Atlanta, Georgia, Geological Survey Information Circular 53, p. 202–210.

Long, L. T., and Liow, J. S., 1986, A study of seismicity and earthquake hazard in northern Alabama and adjacent parts of Tennessee and Georgia: Washington, D.C., U.S. Nuclear Regulatory Commission Report NUREG/CR-4706, 60 p.

MacCarthy, G. R., 1956, A marked alignment of earthquake epicenters in western North Carolina and its tectonic implications: Elisha Mitchell Scientific Society Journal, v. 72, p. 274–276.

Marion, G. E., and Long, L. T., 1980, Microearthquake spectra in the southeastern United States: Bulletin of the Seismological Society of America, v. 70, p. 1037–1054.

Meissner, R., and Strehlau, J., 1982, Limits of stresses in continental crusts and their relation to the depth-frequency distribution of shallow earthquakes: Tectonics, v. 1, p. 73–89.

Munsey, J. W., and Bollinger, G. A., 1985, Focal mechanism analyses for Virginia earthquakes (1978–1984): Bulletin of the Seismological Society of America, v. 75, p. 1613–1636.

Nelson, A. E., and Zietz, I., 1983, The Clingman lineament, other aeromagnetic features, and major lithotectonic units in part of the southern Appalachian Mountains: Southeastern Geology, v. 24, p. 147–157.

Nelson, K., and Talwani, T., 1985, Reanalysis of focal mechanism data for the central Virginia seismogenic zone [abs.]: Earthquake Notes, v. 56, p. 76.

Nuttli, O. W., Bollinger, G. A., and Herrmann, R. B., 1986, The 1886 Charleston, South Carolina, earthquake; A 1986 perspective: U.S. Geological Survey Circular 985, 52 p.

Oaks, S. D., and Bollinger, G. A., 1986, The epicenter of the m_b 5, December 22, 1875, Virginia earthquake; New findings from documentary sources: Earthquake Notes, v. 57, p. 65–75.

Obermeier, S. F., Gohn, G. S., Weems, R. E., Gelinas, R. L., and Rubin, M., 1985, Geologic evidence for recurrent moderate to large earthquakes near Charleston, South Carolina: Science, v. 227, p. 408–411.

Obermeier, S. F., Jacobson, R. B., Weems, R. E., and Gohn, G. S., 1986, Holocene and Late Pleistocene(?) earthquake-induced sandblows in coastal South Carolina [abs.]: Earthquake Notes, v. 57, p. 17.

Ott, L., 1984, An Introduction to statistical methods and data analysis: Boston, Duxbury Press, 775 p.

Rankin, D. W., ed., 1977, Studies related to the Charleston, South Carolina, earthquake of 1886; A preliminary report: U.S. Geological Survey Professional Paper 1028, 204 p.

Rastogi, B. K., and Talwani, P., 1984, Reservoir-induced seismicity at Lake Jocassee in South Carolina, USA, in Gowd, T. N., and others, eds., Proceedings of Indo-German workshop on rock mechanics: Hyderabad, India, National Geophysical Research Institute, p. 225–232.

Rawlins, J., 1986, Seismotectonics of the Newberry, South Carolina, earthquakes [M.S. thesis]: Columbia, University of South Carolina, 68 p.

Reagor, B. G., Stover, C. W., and Algermissen, S. T., 1979, Seismicity map of the state of South Carolina: U.S. Geological Survey Miscellaneous Field Studies Map MF-1225, scale 1:1,000,000.

—— , 1980a, Seismicity map of the state of South Carolina: U.S. Geological Survey Miscellaneous Field Studies Map MF-1060, scale 1:1,000,000.

Reinbold, D. J., and Johnston, A. C., 1987, Historical seismicity in the southern Appalachian seismic zone: U.S. Geological Survey Open-File Report 87-433, 859 p.

Rhea, S., 1987, Wave conversions from earthquakes and a new velocity model for the South Carolina Coastal Plain: Bulletin of the Seismological Society of America, v. 77, p. 2143–2151.

Seeber, L., and Armbruster, J. G., 1983, Large strain effects of the 1886 South Carolina earthquake, in Hays, W. W., and Gori, P. L., eds., A workshop on "The 1886 Charleston, South Carolina, earthquake and its implications for today," Proceedings of Conference 20: U.S. Geological Survey Open-File Report 83-843, p. 142–149.

Seeber, L., Armbruster, J. G., and Bollinger, G. A., 1982, Large-scale patterns of seismicity before and after the 1886 South Carolina earthquake: Geology, v. 10, p. 382–386.

SEUSSN Contributors, 1985, Availability of a six-year (1977–1983) earthquake catalog for the southeastern United States derived from network monitoring: Bulletin of the Seismological Society of America, v. 75, p. 629–633.

Sibol, M. S., Bollinger, G. A., and Mathena, E. C., eds., 1986, Seismicity of the southeastern United States, July 1, 1985–December 31, 1985: Blacksburg, Virginia Polytechnic Institute and State University Seismological Observatory Bulletin 17, 78 p.

Sibol, M. S., Bollinger, G. A., and Birch, J. B., 1987, Estimation of magnitudes in central and eastern North America using intensity and felt area: Bulletin of the Seismological Society of America, v. 77, p. 1635–1654.

Sibson, R. H., 1982, Fault zone models, heat flow, and the depth distribution of earthquakes in the continental crust of the United States: Bulletin of the Seismological Society of America, v. 72, p. 151–163.

Steigert, W. F., 1984, Seismicity of the Southern Appalachian Seismic Zone in Alabama: Geological Survey of Alabama Circular 119, 106 p.

Stover, C. W., Reagor, B. G., Algermissen, S. T., and Long, L. T., 1979, Seismicity map of the state of Georgia: U.S. Geological Survey Miscellaneous Field Studies Map MF–1060, scale 1:1,000,000.

Taber, S., 1913, The South Carolina earthquake of January 1, 1913: Bulletin of the Seismological Society of America, v. 3, p. 6–13.

——, 1914, Seismic activity in the Atlantic Coastal Plain near Charleston, South Carolina: Bulletin of the Seismological Society of America, v. 4, p. 108–160.

Talwani, P., 1981, Earthquake prediction studies in South Carolina, *in* Simpson, D. W., and Richards, P. G., eds., Earthquake prediction; An international review, M. Ewing series 4: American Geophysical Union, p. 381–393.

——, 1982, Internally consistent pattern of seismicity near Charleston, South Carolina: Geology, v. 10, p. 654–658.

——, 1984, Orientation of stress in southeastern U.S. [abs.]: Geological Society of America Abstracts with Programs, v. 16, 202 p.

——, 1985, Current thoughts on the cause of the Charleston, South Carolina, earthquakes: South Carolina Geology, v. 29, no. 2, p. 19–38.

——, 1986, Seismotectonics of the Charleston region, *in* Proceedings, Third U.S. National Conference on Earthquake Engineering, v. 2: El Cerrito, California, National Geophysical Research Institute, p. 15–24.

Talwani, P., and Cox, J., 1985, Paleoseismic evidence for recurrence of earthquakes near Charleston, South Carolina: Science, v. 229, p. 379–381.

Talwani, P., Rastogi, B. K., and Stevenson, D., 1980, Induced seismicity and earthquake prediction studies in South Carolina: U.S. Geological Survey Tenth Technical Report, Contract No. 14–08–0001–17670, 211 p.

Talwani, P., Rawlins, J., and Stephenson, D., 1985, The Savannah River Plant, South Carolina, earthquake of June 9, 1985, and its tectonic setting: Earthquake Notes, v. 56, no. 4, p. 101–106.

Tarr, A., 1977, Recent seismicity near Charleston, South Carolina, and its relationship to the August 31, 1886, earthquake: U.S. Geological Survey Professional Paper 1028–D, p. 43–57.

Tarr, A. C., and King, K. W., 1974, South Carolina seismic program: U.S. Geological Survey Open-File Report 74–58, 15 p.

Tarr, A. C., and Rhea, S., 1983, Seismicity near Charleston, South Carolina, March 1973 to December 1979, *in* Gohn, G. S., ed., Studies related to the Charleston, South Carolina, earthquake of 1886; Tectonics and seismicity: U.S. Geological Survey Professional Paper 1313, p. R1–R17.

Tarr, A., Talwani, P., Rhea, S., Carver, D., and Amick, D., 1981, Results of recent South Carolina seismological studies: Bulletin of the Seismological Society of America, v. 71, p. 1883–1902.

Teague, A. G., Bollinger, G. A., and Johnston, A. C., 1986, Focal mechanism analyses for eastern Tennessee earthquakes (1981–1983): Bulletin of the Seismological Society of America, v. 76, p. 95–109.

Vasseur, G., Etchecopar, A., and Philip, H., 1983, State of stress inferred from multiple focal mechanisms: Annales Geophysicae, v. 1, p. 291–297.

Visvanathan, T. R., 1980, Earthquakes in South Carolina 1698–1975: South Carolina Geological Survey Bulletin 40, 61 p.

Williams, R. T., and Talwani, P., 1986, Seismicity and focal mechanisms in the Great Smoky Mountains [abs.]: Earthquake notes, v. 57, p. 106.

MANUSCRIPT ACCEPTED BY THE SOCIETY OCTOBER 14, 1988

ACKNOWLEDGMENTS

Our appreciation is extended to Brian Mitchell and an anonymous referee for their critical reading of this manuscript. Research support was provided by the Nuclear Regulatory Commission under contracts NRC-04-79-214 and NRC-04-85-107 (ACJ), NRC-04-77-134 and NRC-04-85-121 (GAB), and NRC-04-85-122 (LTL) and by the Tennessee Valley Authority Contract TV-58767A (ACJ). D. J. Reinbold assisted with the historical seismicity of Tennessee and North Carolina. K. M. Shedlock is supported by a Nuclear Regulatory Commission grant to the U.S. Geological Survey. S. Rhea assisted with the seismicity of South Carolina. Reservoir-induced seismicity studies in South Carolina were supported by grants from the U.S. Geological Survey, Nos. 14-08-0001-19252, 21229, and 22010 (PT). Continued support from the South Carolina Electric and Gas Company for monitoring at the Monticello Reservoir is also acknowledged (PT). Studies in Georgia and Alabama were also funded by the Geological Survey of Alabama, No. GSA-80-3053, the Georgia Power Company, and the U.S. Army Corps of Engineers (LTL).

The Geology of North America
Decade Map Volume 1
1991

Chapter 17

Seismotectonics of Middle America

James W. Dewey
National Earthquake Information Center, U.S. Geological Survey, MS 967, Box 25046, Denver Federal Center, Denver, Colorado 80225

Gerardo Suárez
Instituto de Geofísica, UNAM, Ciudad Universitaria, México 04510, D.F.

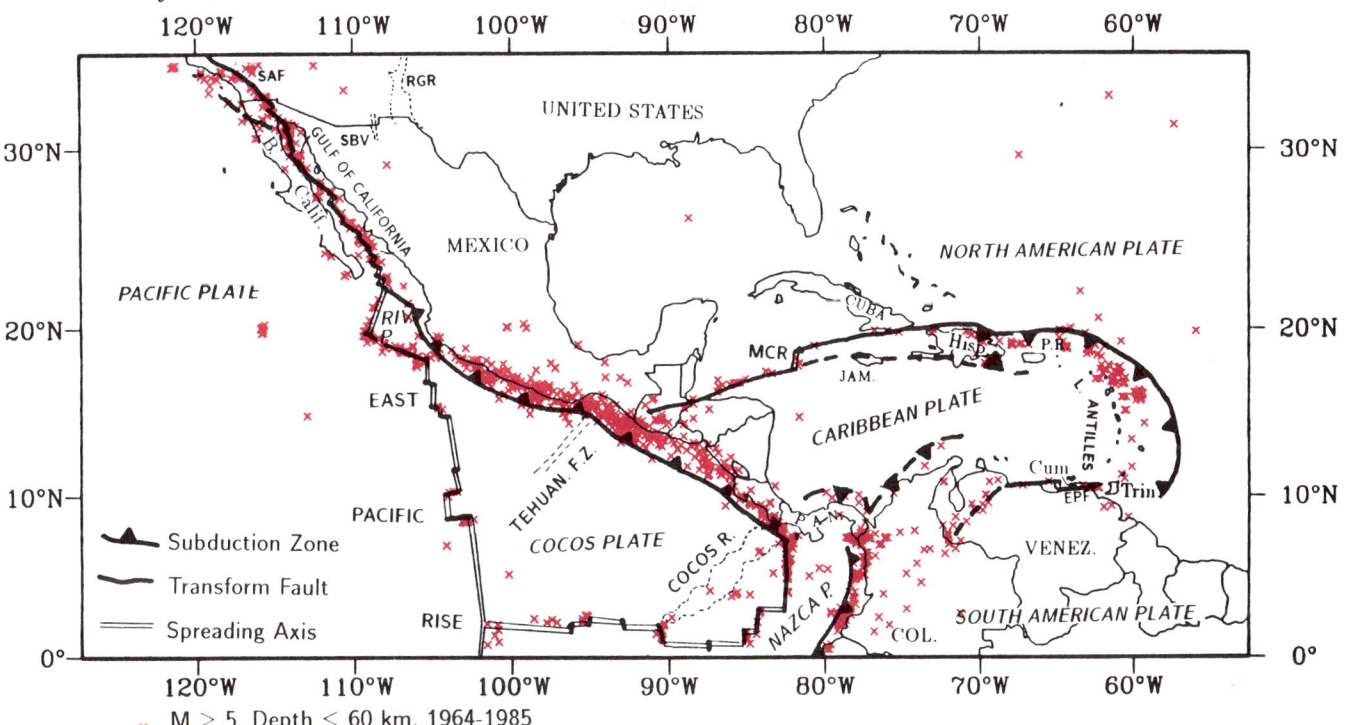

Figure 1. Shallow earthquakes of Middle America. Primary plate boundaries are shown solid. Dashed boundaries are those commonly considered to be microplate boundaries or secondary loci of deformation within a broad plate boundary zone. Breaks in the network of plate boundaries correspond to regions in which seismicity data do not strongly support any one of several proposed plate boundaries. Epicenters are plotted from the data base used to create the Seismicity Map of North America (Engdahl and Rinehart, 1988; this volume). Cum, Cumana; EPF, El Pilar fault; MCR, Mid-Cayman Rise; RGR, Rio Grande Rift; SAF, San Andreas fault; SBV, San Bernardino Valley.

INTRODUCTION

The seismotectonics of the Middle America region—Mexico, Central America, the Antilles, and northwestern South America—are dominated by the interaction of six major tectonic plates (Fig. 1). Middle America contains the three main types of plate boundaries—boundaries of lithospheric divergence, boundaries of lithospheric convergence, and transform-fault boundaries. Seismicity is concentrated in the plate-boundary regions. Many crustal earthquakes occur as slip on the principal interfaces between plates and have slip vectors that are parallel to directions of relative plate motion. In addition to shocks occurring on the plate interfaces, substantial crustal seismicity occurs in plate interiors within 100 to 200 km of some interfaces. Mantle earthquakes define subducted lithosphere to depths of more than 150 km in the major Middle American subduction zones (Fig. 2). As is also observed elsewhere in the world, some important plate-boundary elements appear to be intrinsically aseismic.

Dewey, J. W., and Suárez, G., 1991, Seismotectonics of Middle America, *in* Slemmons, D. B., Engdahl, E. R., Zoback, M. D., and Blackwell, D. D., eds., Neotectonics of North America: Boulder, Colorado, Geological Society of America, Decade Map Volume 1.

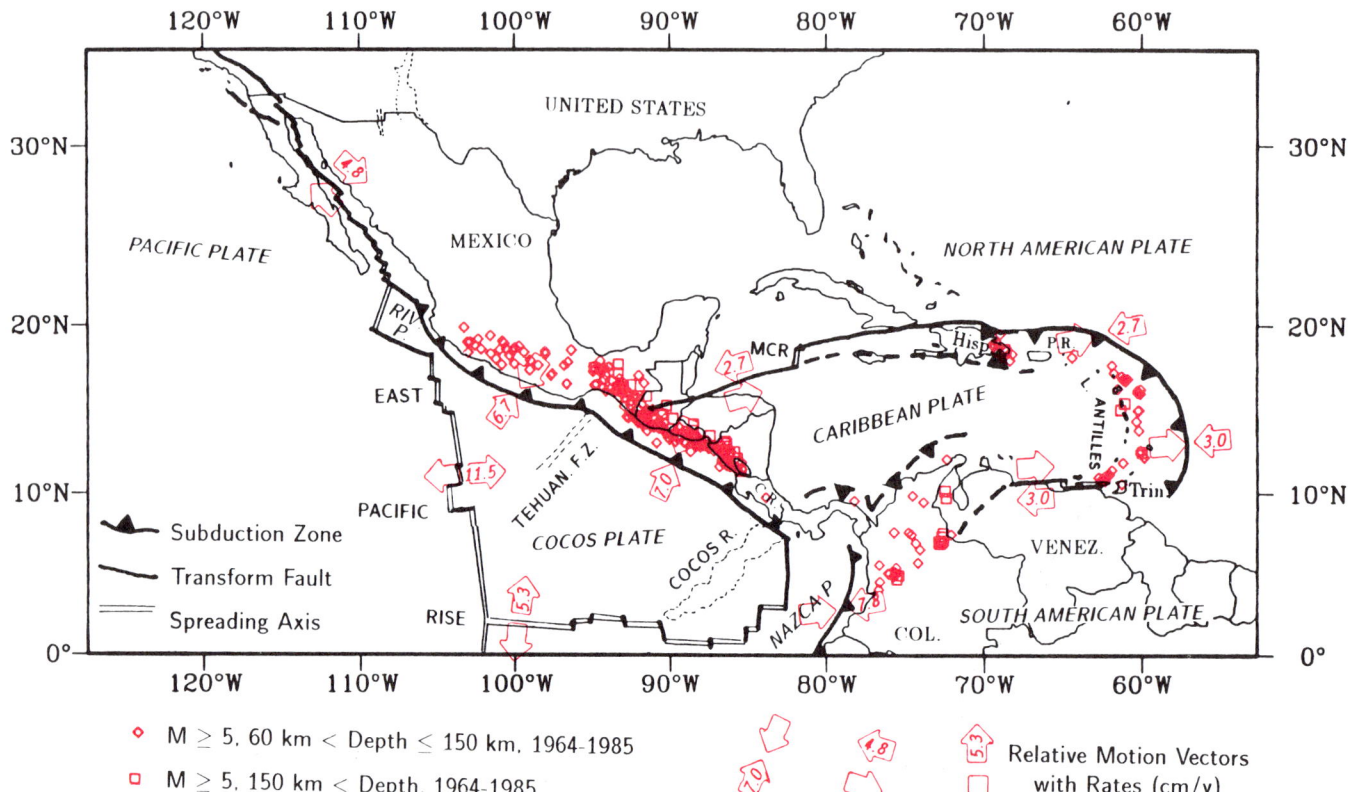

Figure 2. Plate velocities and zones of mantle earthquakes in Middle America. Most motion vectors are determined with a variation of model AM1-2 of Minster and Jordan (1978) in which the Caribbean plate has been assigned zero absolute velocity in order to achieve a compromise with the Caribbean model of Sykes and others (1982). The relative velocity between the North American and Pacific plates in northwest Mexico is based on NUVEL-1 of Demets and others (1987). Mantle earthquakes are from the data base used to create CSM-4 (Engdahl and Rinehart, this volume). The deepest plotted hypocenter has a depth of 236 km. MCR, Mid-Caymen Rise.

We will elaborate on the map that was prepared in conjunction with this volume, "Continental Scale Map 4, Seismicity Map of North America" (CSM-4; Engdahl and Rinehart, 1988). Tectonic models that have been proposed to account for the seismicity will be reviewed and we will discuss some important characteristics of the seismicity that are not obvious in the map. The data base itself is treated in detail by Engdahl and Rinehart (this volume). Magnitudes (M) used in this chapter are those used by Engdahl and Rinehart in plotting CSM-4.

NORTHERN MEXICO

Most earthquakes in Mexico north of 23°N occur on or near the predominantly transform plate boundary across which the Pacific plate is moving northwest with respect to the North American plate at a velocity of about 5 cm/y (Fig. 2).

The principal boundary, beneath the Gulf of California, consists of a series of transform faults offset by small spreading centers or pull-apart basins (Rusnak and others, 1964). The plate boundary underlying the northern third of the Gulf has seismicity characteristics intermediate between those of the continental San Andreas system of California and the oceanic transform-fault system to the south: strike-slip earthquakes occur on the transform faults, and both normal-faulting and strike-slip earthquakes occur in diffuse right-stepping offsets of the transform faults (Thatcher and Brune, 1971; Reichle and Reid, 1977; Goff and others, 1987). Seismicity of the southern Gulf of California is similar to that associated with rapidly spreading oceanic rises and their transform faults worldwide, with virtually all teleseismically recorded shocks resulting from strike-slip motion on the transform faults and with the spreading centers producing little activity (Sykes, 1968; Isacks and others, 1968; Reichle and Reid, 1977). The distribution of sonobuoy-located microearthquakes, the surface-wave spectra of large earthquakes, modeling of body-waveforms of large earthquakes, and the long-term rate of moment release suggest that seismic rupture on the Gulf transform faults is confined to the uppermost 10 to 15 km of the Earth's crust (Reichle and others, 1976; Munguía and others, 1977; Munguía and Brune, 1984; Goff and others, 1987).

Instrumental seismicity in northern Baja California, west of the plate boundary, has been concentrated on right-lateral strike-slip faults trending west-northwest to northwest (Shor and Roberts, 1958; Allen and others, 1960; Reyes and others, 1975; Johnson and others, 1976; Rebollar and Reichle, 1987). To the northwest, these faults trend into a system of faults along and off the coast of southern California (Allen and others, 1960). Goff and others (1987) suggest that the faults of northern Baja may accommodate 30 percent (about 15 mm/y) of the total relative motion between the Pacific and North American plates.

The destructive Sonora earthquake of 3 June 1887 (M = 7, 31°N, 109°W; CSM-4) was associated with 76 km of normal faulting along the east side of the north-trending San Bernardino Valley (Herd and McMasters, 1982). The source region of the 1887 shock continues to be active, producing small shocks with normal faulting and strike-slip faulting source mechanisms (Natali and Sbar, 1982). The San Bernardino Valley is similar to the Rio Grande Rift of New Mexico in being a locus of relatively high extensional tectonism and late Neogene volcanism within the southern Basin and Range (Seager and Morgan, 1979).

EAST PACIFIC RISE

The seismicity of the East Pacific Rise south of Mexico is typical of a fast-spreading oceanic rise system (Isacks and others, 1968). Microearthquake studies using temporary arrays of hydrophones and ocean-bottom seismographs (Prothero and Reid, 1982; Tréhu and Solomon, 1983) are consistent with the pattern of seismicity implied by the distribution of teleseismically recorded earthquakes (CSM-4, Fig. 1): seismicity is heavily concentrated in the transform fault zones, with only a few microearthquakes possibly emanating from the spreading centers. The transform fault-zones themselves appear to be somewhat complex and may include small spreading centers and pull-apart basins (Prothero and Reid, 1982; Tréhu and Solomon, 1983). The microearthquake studies imply a seismogenic crust that extends only a few kilometers below the ocean floor.

The seismicity near 20°N, 116°W, though situated in the interior of the Pacific plate, is probably best viewed as a consequence of tectonic stress caused by the proximity of a spreading center. Worldwide, oceanic intraplate seismicity tends to be higher in young, near-ridge lithosphere than in old, midplate lithosphere (Bergman and Solomon, 1984; Wiens and Stein, 1984). Near-ridge intraplate earthquakes may be due to high thermoelastic stresses resulting from rapid cooling of the lithosphere as it moves away from the ridge axis (Bergman and Solomon, 1984; Wiens and Stein, 1984).

SUBDUCTION OF THE RIVERA PLATE

The rotation pole for motion of the Rivera plate with respect to North America apparently lies within the Rivera plate, and the character of displacement along the boundary of the two plates changes from strike-slip in the north to underthrusting in the south (Fig. 1). The rotation vector of Minster and Jordan (1979) implies a 1.8 cm/yr rate of subduction at 19°N 106°W, a value that is rather low compared to that of most subducting plates worldwide. The Rivera lithosphere being consumed at the trench is of late Miocene age and is therefore among the youngest lithosphere being subducted anywhere in the world (Klitgord and Mammerickx, 1982). Although the slow rate of convergence and the high temperatures of a young oceanic plate might have been thought conducive to aseismic subduction, the largest thrust-fault earthquake along the Middle American trench in this century, the earthquake of 3 June 1932 (M = 8.4), had its nucleus near the diffuse Rivera–Cocos–North American triple junction (Eissler and McNally, 1984) and ruptured predominantly along the Rivera–North American boundary (Singh and others, 1985b).

SUBDUCTION OF THE COCOS PLATE BENEATH THE NORTH AMERICAN PLATE IN CENTRAL MEXICO

The tectonic regime of central Mexico is dominated by the ongoing subduction of the Cocos plate beneath the western margin of the North American plate (Fig. 1, Fig. 3). The subducted Cocos plate west of about 96°W dips shallowly (~15°) and is seismogenic to a depth of about 100 km (Jiménez and Ponce, 1978; Burbach and others, 1984). The dip is among the most shallow worldwide and probably reflects the relatively young age of the subducting oceanic lithosphere (Bevis and Isacks, 1984).

Thrust-fault earthquakes occurring on the gently dipping interface between the Cocos and North American plates (Fig. 3) have repeatedly damaged coastal cities and towns and, farther inland, Mexico City and other large population centers. Many segments of the subduction zone have recurrence times of 30 to 40 years for thrust-fault earthquakes of magnitude 7.5 or larger; these times are relatively short compared with those in other subduction zones worldwide (Kelleher and others, 1973; Singh and others, 1981; McNally and Minster, 1981). The seismically active thrust interface appears to extend to a depth of about 30 km: the P-waveforms of recent large shocks imply that their nuclei were at depths of shallower than 30 km (Chael and Stewart, 1982; Beroza and others, 1984; Astiz and others, 1987), and locally recorded aftershocks whose hypocenters are consistent with their occurrence on a shallowly dipping thrust interface are concentrated at depths less than 30 km (Stewart and others, 1981; Stolte and others, 1986; UNAM Seismology Group, 1986).

Since 1973, the great thrust-fault earthquakes in the Mexican subduction zone have occurred in previously identified "seismic gaps." A seismic gap is a plate-boundary segment in which enough time has elapsed since the preceding great earthquake that elastic strain energy sufficient to produce another great earthquake may have accumulated across the plate boundary (Kelleher and others, 1973; McCann and others, 1979; Nishenko and Singh, 1987). The great Mexican earthquakes that have occurred in previously identified gaps (Fig. 3) are those of 30 January 1973 (M = 7.5; Reyes and others, 1979), 29 November 1978 (M = 7.8; Singh and others, 1979; Stewart and others, 1981; Ohtake and others, 1981), 14 March 1979 (M = 7.6; Valdés and others, 1982), and 19 September 1985 (M = 8.1; UNAM Seismology Group, 1986). The 1985 earthquake, however, also demonstrated limitations in current understanding of seismic gaps

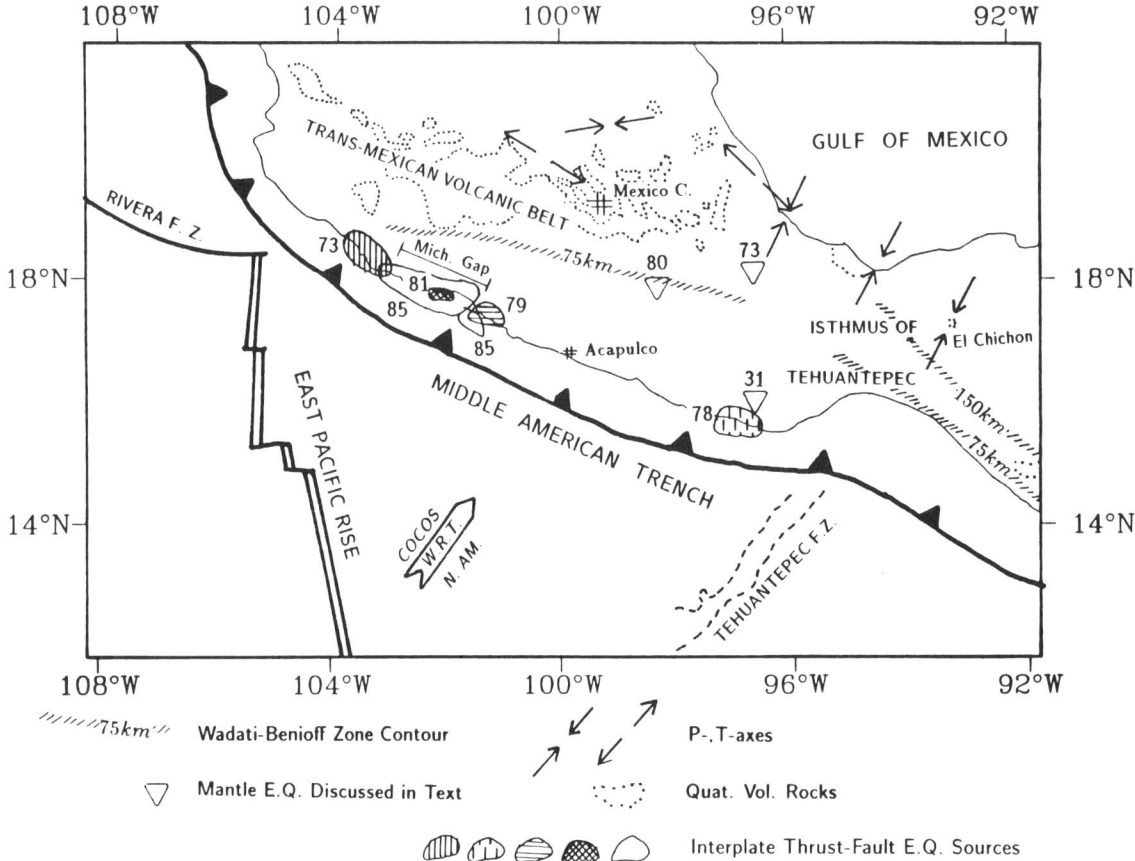

Figure 3. Mexican subduction zone. Sources of recent great interplate earthquakes are shown as defined by locally recorded aftershocks (Singh and others, 1979; UNAM Seismology Group, 1986) and are labeled by the year of the earthquake. Mantle earthquakes discussed in the text are also labeled by their year of occurrence. In the region west of 96°W the downdip edge of the Wadati-Benioff zone is close to the 75-km contour. P- and T-axes are for shallow-focus earthquakes in the North American plate (Havskov and others, 1983a; Suárez and Ponce, 1986; Suárez, Assumpcao, and Ponce, personal communication, 1988). P-axes (converging arrows) are plotted for earthquakes having a reverse sense of dip-slip motion; T-axes (diverging arrows) are plotted for earthquakes having a normal-sense of dip-slip motion.

and in the data base to which the gap methodology is applied. The "Michoacan gap" (Fig. 3), in which the earthquake occurred, was at one time thought to be possibly aseismic (McNally and Minster, 1981; Singh and others, 1981). The occurrence of an M = 7.3 earthquake on 25 October 1981 demonstrated that the thrust interface was not totally aseismic (Havskov and others, 1983b; LeFevre and McNally, 1985). The source region of the 1981 shock appeared to subdivide the Michoacan gap into three moderate-sized segments that would not be large enough to individually produce great earthquakes. The rupture of the great earthquake of 19 September 1985 was not confined to a single segment, however, but propagated from the northern segment, through or around the source region of the 1981 earthquake, to conclude with substantial moment release in the southern part of the gap (UNAM Seismology Group, 1986, and other papers in the v. 13, no. 6, of *Geophysical Research Letters*; Anderson and others, 1986). At this writing, the thrust interface in central Guerrero (100°W to 101°W, last ruptured in several large earthquakes in 1899 to 1911) is widely viewed as a potentially dangerous seismic gap (Nishenko and Singh, 1987).

Focal mechanisms of Mexican subduction-zone earthquakes reliably located at depths greater than 40 km imply that these shocks occur within the subducted Cocos plate, rather than at the plate interface. Most earthquakes within the subducted Cocos plate have normal-fault focal mechanisms, with axes of maximum extensional strain (T-axes) that are approximately parallel to the dip of the sinking lithosphere (Jiménez and Ponce, 1978; LeFevre and McNally, 1985; McNally and others, 1986; González-Ruiz and McNally, 1988). Downdip orientation of T-axes may reflect the importance of slab-pull forces transmitted along the lithospheric stress guide (Isacks and Molnar, 1971). Alternatively, downdip orientations of T-axes may reflect in-plate stresses due to bending about a horizontal axis (Isacks and Barazangi, 1977; Fujita and Kanamori, 1981).

The Oaxaca earthquake of 15 January 1931 (M = 8.0; Fig. 3) is an example of a destructive normal-fault shock that occurred within the subducted Cocos plate immediately downdip from the seismogenic thrust-fault interface between the plates (Singh and others, 1985a). The shock's nucleus was at a depth of about 40 km, and the size of the earthquake suggests that rupture may have broken through the subducted lithosphere from top to bottom (Singh and others, 1985a). Examples of destructive normal-fault shocks that occurred near the downdip edge of the seismically active part of the subducted slab (Fig. 3) are the earthquakes of 28 August 1973 (Depth = 80 km; M = 7.3; Meehan, 1973; Singh and Wyss, 1976) and 24 October 1980 (Depth = 65 km; M = 7.0; Yamamoto and others, 1984).

A belt of shallow-focus earthquakes in continental Mexico extends several hundred kilometers inland from the Middle America trench (Fig. 1). Earthquakes occurring in the Isthmus of Tehuantepec have been characterized by reverse faulting, with axes of maximum compressional strain (P-axes) oriented approximately parallel to the direction of relative plate motions at the trench (Fig. 3). These orientations of the P-axes are consistent with strong coupling across the plate interface due to the impingement of the Tehuantepec Fracture Zone at the subduction zone (Suárez and Ponce, 1986). West of 96°W along the Trans-Mexican Volcanic Belt, ongoing work of Suárez, Assumpcao, and Ponce (personal communication, 1988) shows normal faulting and reverse faulting mechanisms for shallow intraplate shocks that are quite close to each other (Fig. 3). The proximity of the two different modes of faulting may imply that horizontal compressive stresses transmitted from the Middle American thrust interface are approximately balanced by stresses arising from the action of gravity on the high topography of the Trans-Mexican Volcanic Belt. The stress relief necessary to maintain the balance is accomplished through the different types of faulting. This mechanism has been proposed to explain the proximity of reverse and normal faults in the central Andes of Peru (Dalmayrac and Molnar, 1981; Suárez and others, 1983). Because of their shallow focal depths, shocks within the overriding North American plate, such as the 1912 and 1920 earthquakes in central Mexico, have been extremely destructive (e.g., Urbina and Camacho, 1913; Camacho and others, 1922).

CARIBBEAN PLATE VELOCITY

Central America, northernmost South America, and the West Indies are on the margins of the Caribbean plate (Figs. 1 and 2), and the seismotectonics of these regions are therefore determined by the velocity of the Caribbean plate. There are presently several alternative models for motions of the Caribbean plate (Minster and Jordan, 1978; Sykes and others, 1982; Stein and others, 1988). The models are generally consistent in implying that the Caribbean plate has a low absolute velocity in a hotspot reference frame, that it is moving south-southwest with respect to the Cocos plate, and that it is moving predominantly eastward with respect to the American plates. The models differ, however, in ways that are tectonically significant and important for seismic hazard evaluations. Along the northern, eastern, and southern boundaries of the Caribbean plate, for example, the relative plate velocities of Sykes and others (1982) are almost twice those of Minster and Jordan (1978). The models also differ on the senses and magnitudes of the components of plate-motion that are normal to the trends of the north and south boundaries of the Caribbean plate, and on where each model introduces microplates or broad zones of deformation between the Caribbean and surrounding plates. Circum-Caribbean relative-velocity vectors in Figure 2 were computed by setting the absolute velocity of the Caribbean plate to zero in the hotspot-frame model (AM1-2) of Minster and Jordan (1978). The resulting relative motion vectors are close to the means of the vectors predicted by the models of Minster and Jordan and of Sykes and others (1982).

CENTRAL AMERICAN SUBDUCTION ZONE

A zone of intermediate-depth earthquakes, continuous to within the precision of hypocenter locations, extends from southeasternmost Mexico to northern Costa Rica (Bevis and Isacks, 1984; Burbach and others, 1984; Fig. 2). This Central American Wadati-Benioff zone has generally much steeper dips than the Mexican Wadati-Benioff zone, and seismicity in many segments extends to depths of approximately 200 km (Dewey and Algermissen, 1974; Carr, 1976; Burbach and others, 1984). The steeper dip of most of the Central American Wadati-Benioff zone relative to that of the Mexican Wadati-Benioff zone is probably in part a consequence of the greater age, and hence greater density, of the oceanic lithosphere south of the Tehuantepec Fracture Zone (Couch and Woodcock, 1981). The dip of the Central American Wadati-Benioff zone becomes shallower at its southern end, and the zone disappears several hundred kilometers north of the Cocos-Nazca-Caribbean triple junction (Burbach and others, 1984; Guendel and McNally, 1986; Fig. 2). The shoaling and disappearance of the Wadati-Benioff zone to the south occur where the age of subducted oceanic crust again becomes much younger (Lonsdale and Klitgord, 1978), where the buoyant Cocos Ridge enters the trench (Fig. 2), and where the current episode of subduction began only within the last few million years (Gardner and others, 1987). The southward disappearance of the intermediate-depth earthquake zone may plausibly be attributed to one or more of these characteristics of the southernmost plate boundary.

The thrust interface in Central America between the Cocos and Caribbean plate is seismically active (Molnar and Sykes, 1969; Dean and Drake, 1978), but the percentage of plate-motion accommodated by seismic slip appears to be substantially lower than along the Cocos–North American interface in Mexico (McNally and Minster, 1981). This is consistent with a worldwide tendency for oceanic plates that are attached to steeply dipping subducted slabs to be less strongly coupled to the overriding plate than oceanic plates attached to shallowly dipping subducted slabs (e.g., Lay and others, 1982). Although the southern end of the Central American Wadati-Benioff zone lies well north of the Cocos-Nazca-Caribbean triple junction, the shallow thrust interface is active to the triple junction (Fig. 1; Adamek and others, 1987). The largest instrumentally recorded earthquakes on or near the Central American thrust interface have had magnitudes of about 8 (McNally and Minster, 1981).

In much of Central America, the most destructive historical earthquakes have been shallow intraplate shocks of moderate magnitude (5 to 6.5) that occur in the densely populated regions surrounding the Quaternary volcanoes (White and Harlow, this volume). The earthquakes typically occur as strike-slip faulting. Both left-lateral faults, striking at a large angle to the trend of the Middle American arc, and right-lateral faults, approximately parallel to the arc, have been implicated as sources (Carr and Stoiber, 1977; White and others, 1987). Recurring earthquakes affecting the same city do not necessarily represent separate slip-episodes on the same fault segment but, instead, slip on different members of a family of faults. The catastrophic Managua earthquakes of 1931 and 1972, for example, appear to have involved rupture on different, parallel fault strands within a 5-km-wide system of faults (Fig. 4). There is no consensus on the regional tectonic process that is responsible for these earthquakes; the shocks have been attributed to back-arc extension above the Middle American subduction zone, to oblique convergence at the plate margin to the west that cannot be entirely accommodated by slip on the plate interface, and to unstable triple junctions at the north and south ends of the Central American arc (Mann and others, 1989; White and Harlow, this volume).

PANAMA

Relative motion between the Nazca and Caribbean plates must be largely accommodated in the region of the Isthmus of Panama. The geometrically simplest boundary between rigid Nazca and Caribbean plates would be an east-northeast–striking, left-lateral transform fault (Jordan, 1975). Instrumentally recorded, shallow-focus earthquakes have occurred over a wide area both north and south of the Isthmus, however, and many have involved reverse faulting rather than strike-slip faulting (Pennington, 1981; Wolters, 1986; Adamek and others, 1988). South-directed underthrusting of the Caribbean lithosphere along the northern coast of the Isthmus is implied by a deformed accretionary wedge north of the Isthmus (Bowin, 1976) and by the occurrence of mantle earthquakes landward of the accretionary wedge (Wolters, 1986; Adamek and others, 1988). These observations suggest that the Nazca and Caribbean plates are separated by microplates or by a zone of distributed deformation rather than by a simple boundary (Bowin, 1976; Pennington, 1981; Adamek and others, 1988).

NORTHERN SOUTH AMERICA

The seismicity of northernmost South America is a consequence of the predominantly eastward motion of the Nazca and Caribbean plates with respect to South America (Molnar and Sykes, 1969; Case and others, 1971). West of about 68°W, the undeformed interiors of the Nazca and Caribbean plates are separated from the undeformed interior of the South American plate by a broad region of Cenozoic deformation and seismicity that is sometimes viewed as comprising microplates or tectonic flakes (Silver and others, 1975; Bowin, 1976; Pennington, 1981; Kafka and Weidner, 1981; Mann and Burke, 1984; Dewey and Pindell, 1985). For the coast of northern Venezuela between 68°W and 63°W, an east-west transform boundary between the two plates accounts for the historical and instrumental seismicity (Molnar and Sykes, 1969). The intersection of the east-west transform system with the north-south Antilles subduction zone is not sharply defined by the distribution of shallow-focus earthquakes.

The presence of an easterly or southeasterly dipping Wadati-Benioff zone in northern Colombia and western Venezuela (Dewey, 1972; Pennington, 1981; Schneider and others, 1987) is consistent with the hypothesis that subduction occurs along the Caribbean coast of Colombia (Krause, 1971; Kellogg and Bonini, 1982). The lack of a great earthquake in the four centuries of recorded history for the Caribbean coast of Colombia (CSM-4; Ramírez, 1969) is not definite evidence against an active subduction zone, because such active arc segments as the southern Lesser Antilles subduction zone (to be discussed below) have also experienced no great underthrust earthquake in the historical period.

The Boconó fault zone (Rod, 1956; Schubert, 1982; Fig. 5) is commonly postulated to accommodate a significant fraction of transform displacement between the undeformed interiors of the Caribbean and South American plates. Slip-rates implied by various measurements of offsets of Pleistocene moraines are from 0.3 cm/y to 1.4 cm/y (Schubert, 1982). The Boconó fault, like California's San Andreas fault (Allen, 1981) probably has sections that rupture infrequently (recurrence times of centuries) in very large earthquakes but that are quiescent at both high- and low-magnitude levels for most of the time between very large earthquakes. The last great rupture in the Boconó fault zone was probably during the earthquake of 26 March 1812, whose mei-

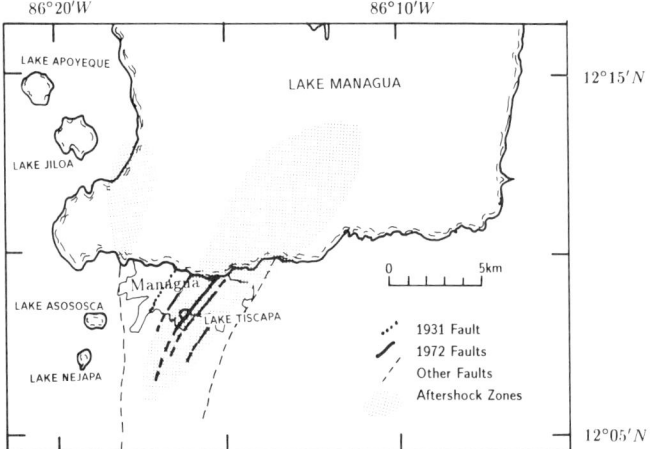

Figure 4. Managua, Nicaragua, showing the aftershock zones observed after the 1972 earthquake and faults that produced surface breakage in 1931 and 1972 (Brown and others, 1973; Langer and others, 1974; Ward and others, 1974).

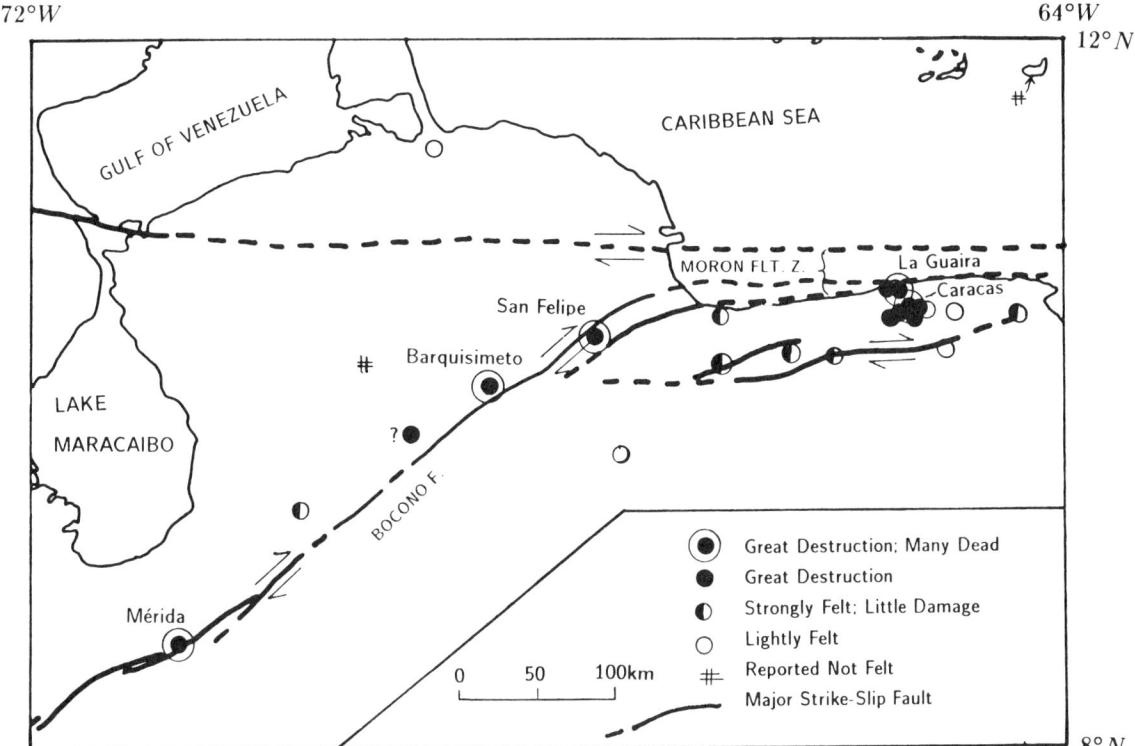

Figure 5. Damage distribution of Venezuelan earthquake of 26 March 1812. Only sites specifically mentioned in compilations of von Humboldt (1889) and Centeno-Grau (1940) are plotted. Faults are those of Schubert (1982, 1984) and Schubert and Krause (1984).

zoseismal zone follows the fault zone quite closely (Fig. 5). Although the Boconó fault zone has not produced a great earthquake in the instrumental era of seismology, small and moderate shocks have originated in some sections of the zone (Dewey, 1972).

Strike-slip faulting along the north-central coast of Venezuela accommodates much of the eastward motion of Caribbean lithosphere with respect to the South American craton (Dengo, 1953; Schubert and Krause, 1984). Damage in and near Caracas caused by the earthquake of 1812 suggests that, in addition to slip on the Boconó fault, the earthquake involved rupture of an east-west fault in the Morón fault zone (Fig. 5). Two epicenters are plotted in CSM-4 for the 1812 earthquake in order to emphasize that both the Boconó and Morón faults seem to have been involved and to acknowledge that along-strike variations in maximum intensity have been interpreted in terms of several discrete sites of maximum energy-release along the Boconó-Morón trend (Fiedler, 1961). The Caracas earthquake of 29 July 1967 (Hanson and Degenkolb, 1969; Espinosa and Algermissen, 1972) probably also occurred on an east-west strike-slip fault along the coast (Molnar and Sykes, 1969). A north-south fault for the source of the 1967 earthquake was proposed on the basis of modeling P and S waveforms (Rial, 1978), but an east-west fault matches better with the P and S waveforms and agrees with the distribution of damage in the earthquake and the location of aftershocks (Suárez and Nabelek, 1989). Geologically recent, east-west, transcurrent faulting (Smith, 1953; Schubert, 1984) also occurs south of the principal plate boundary, in the interior ranges of north-central Venezuela (Fig. 5).

The transform boundary between the Caribbean and South American plates in northeastern Venezuela follows the El Pilar fault between 64°W and 62.8°W (Pérez and Aggarwal, 1981; Vierbuchen, 1984; Fig. 1). The magnitude 6.9 earthquake of 17 January 1929 was associated with surface faulting on the El Pilar system near Cumana (Paige, 1930). Earthquakes in 1766 and 1853 also produced intense damage at Cumana and were felt at locations distant from Cumana with intensities that were similar (in 1853) or larger (in 1766) than the intensities with which the 1929 shock was felt (Centeno-Grau, 1940). The damage reports for the shocks of 1766 and 1853 do not permit an unambiguous association of the earthquakes with the El Pilar fault, but the intensity distributions are at least consistent with such an association.

Between 62.8°W and about 60°W, the boundary between the South American and Caribbean plates changes from an east-west transform to the north-south Lesser Antilles subduction zone. The historic and instrumental seismicity of this region has been too low to determine how the change in boundary orientation is accomplished at shallow depth. The transform boundary may extend on the trend of the El Pilar fault east of 62.8°W across northern Trinidad and thence to a hinge fault at the south end of the Antilles arc (e.g., Vierbuchen, 1984; Mann and others,

1989). If so, most of the transform boundary has not slipped seismically in historic times; no shock in the history of northern Trinidad (Robson, 1964) has produced the level of destruction expected with complete seismic rupture of the postulated boundary. Likewise, there is no seismicity east of Trinidad that would correspond to shallow-focus tearing of the American plate lithosphere and hinge faulting at the east end of the transform boundary. Alternate tectonic models in which the locus of maximum deformation extends southeast from the El Pilar fault at about 62.8°W (Pérez and Aggarwal, 1981; Speed, 1985) face generally similar problems, although Speed (1985) envisions a distributed deformation east of 62.8°W that need not require shallow-focus hinge faulting and that might occur largely aseismically. At upper-mantle depths, earthquakes near the south end of the Antilles Wadati-Benioff zone at 10.9°N, 62.8°W (Fig. 2) exhibit focal mechanisms consistent with the subducting American lithosphere detaching from the surface lithosphere along an east-west hinge fault (Molnar and Sykes, 1969; Tomblin, 1975).

THE ANTILLES ARC

The Antilles arc, from Trinidad to eastern Hispaniola, reflects the subduction of the relatively old oceanic lithosphere of one or both of the American plates. Wadati-Benioff zones extend to depths of about 200 km beneath the Antilles arc (Molnar and Sykes, 1969; Wadge and Shepherd, 1984; Dorel 1981; Fig. 2). Presumably, the slow (several mm/yr) motion of the North American with respect to the South American plate is accommodated along a boundary (possibly diffuse) that intersects the Antilles arc between 10°N and 20°N (Minster and Jordan, 1978).

At shallow depths within the Lesser Antilles arc, seismicity has been substantially higher north of 14°N than to the south. This has been true in both instrumental and historical eras. Earthquakes of magnitude about 8 occurred in the northern Lesser Antilles in 1843 and, probably, in 1690 (Robson, 1964; Dorel, 1981). South of 14°N, the largest historical shallow-focus earthquakes have had magnitudes of about 7 (Dorel, 1981). During the instrumental era, most moderate and large shocks have been shallow intraplate shocks near the plate boundary, occurring in both the overriding Caribbean and the subducting American plates (Robson and others, 1962; McCann and others, 1982; Stein and others, 1982; Morgan and others, 1988). Stein and others (1982) propose that much of the underthrusting of the American plates is accomplished by aseismic slip on the plate interface, because the ocean floor is old, relatively dense, and therefore poorly coupled to the overriding Caribbean plate. Stein and others (1982) postulate that the great historic earthquakes of the northern Lesser Antilles were intraplate earthquakes, like many of the instrumentally recorded shocks of recent decades. McCann and Sykes (1984), on the contrary, hypothesize that coupling between the American and Caribbean plates is rather high in the northern Lesser Antilles, due to the presence of high topography on the subducting sea floor. From the observed distribution of intensities in 1843, McCann and Sykes argue that the earthquake was most likely an interface-thrust event; the current scarcity of interface-thrust earthquakes would reflect a quiescent phase of the seismic cycle on the thrust interface. McCann and Sykes (1984) suggest that the inactivity of the thrust interface in the southern Lesser Antilles may indicate aseismic slip on the plate interface due to high pore pressure at the base of the thick sedimentary wedge in that section of the arc.

Although the trend of the plate boundary north of Puerto Rico is nearly parallel to the direction of relative plate motion, a Wadati-Benioff zone dips south from the Puerto Rico trench (Schell and Tarr, 1978; Fisher and McCann, 1984). Focal mechanisms of instrumentally recorded small- and moderate-magnitude earthquakes suggest that the approximately easterly motion of the Caribbean with respect to North America is accommodated by strike-slip motion on the gently south-dipping interface between the Caribbean plate and the subducting North American plate (Molnar and Sykes, 1969; Frankel, 1982).

Eastern Hispaniola has subduction zones along both its northern and southern coasts. At the north, the North American plate is being subducted toward the south. The northern subduction zone has a mantle seismic zone and has experienced great earthquakes in the instrumental era, the locations and aftershock distributions of which are consistent with thrust faulting on the shallow plate interface (McCann and Sykes, 1984). Northward subduction from the south is suggested by a sedimentary wedge offshore of southeastern Hispaniola that is deformed in a manner similar to sedimentary wedges bordering active trenches on the margin of the Pacific (Ladd and others, 1977). Although subduction from the south does not produce as many earthquakes as subduction from the north coast of Hispaniola, a moderate-magnitude shock in southeastern Hispaniola had a focal mechanism consistent with thrusting on a gently north-dipping interface, and some upper-mantle shocks have hypocenters that would put them in a north-dipping Wadati-Benioff zone (Byrne and others, 1985).

WESTERN HISPANIOLA TO THE MIDDLE AMERICA TRENCH

Between Hispaniola and the Middle America Trench, the plate boundary is predominantly a transform boundary. The distribution of historical seismicity (Shepherd and Aspinal, 1980; Sykes and others, 1982) and Neogene tectonism (Burke and others, 1980) suggest that the boundary between Hispaniola and the Mid-Cayman Rise comprises two branches—one, probably the principal branch, passing through northern Hispaniola and just south of Cuba, and the other passing through southern Hispaniola and near Jamaica (Fig. 1). The southern branch has been quiescent in recent decades (Shepherd and Aspinal, 1980) and is not well marked by the seismicity in CSM-4. Nevertheless, several historical earthquakes on this branch must have had magnitudes of about 7.5 or larger (Sykes and others, 1982). For example, the earthquake that in 1692 caused the foundering of the town of

Port Royal, on Jamaica's south coast, was destructive across the entire island and produced a large tsunami on the island's north coast (Taber, 1920).

The North American-Caribbean plate boundary in Guatemala includes the Polochic and Motagua faults (Schwartz and others, 1979; Burkart, 1983). The Guatemala earthquake of 4 February 1976 (M = 7.5) involved 270 km of strike-slip rupture on the Motagua fault (Plafker and others, 1976; Kanamori and Stewart, 1978; Langer and Bollinger, 1978; Fig. 6). White (1985) places the Guatemalan earthquake of 22 July 1816 (M ~ 7½ to 7¾) on the western end of the Polochic fault. The trends of the great strike-slip faults of the plate boundary are parallel to predicted plate-motion vectors in eastern Guatemala, but depart substantially from the predicted east-northeast trend of the relative-motion vectors in central and western Guatemala (Fig. 6), implying substantial "intraplate" deformation adjacent to the principal boundary in central and western Guatemala. This deformation apparently occurs as strike-slip and normal faulting south of the Motagua fault (Dengo and others, 1970; Plafker, 1976; Burkart and Self, 1985; White and Harlow, this volume). Some of the normal faults were reactivated by seismic slip on the Motagua fault in 1976 (Plafker and others, 1976; Langer and Bollinger, 1978; Fig. 6).

CONCLUDING REMARKS

In our treatment of individual plate boundaries in Middle America, we have frequently referred to earthquakes whose locations and sizes have been estimated from intensity observations rather than from instrumental data. We noted, for example, that the coast of northern Venezuela, though not having experienced large numbers of moderate-magnitude epicenters in recent decades, has an earlier history of major earthquakes that is consistent with its being the principal boundary between the Caribbean and South American plate. There is probably much more to be gained in the study of pre-instrumental seismicity. There are many plate-boundary segments for which pre-instrumental earthquakes still need to be systematically catalogued or for which magnitudes of pre-instrumental earthquakes need to be estimated using regionally derived intensity-attenuation laws (Singh and others, 1980).

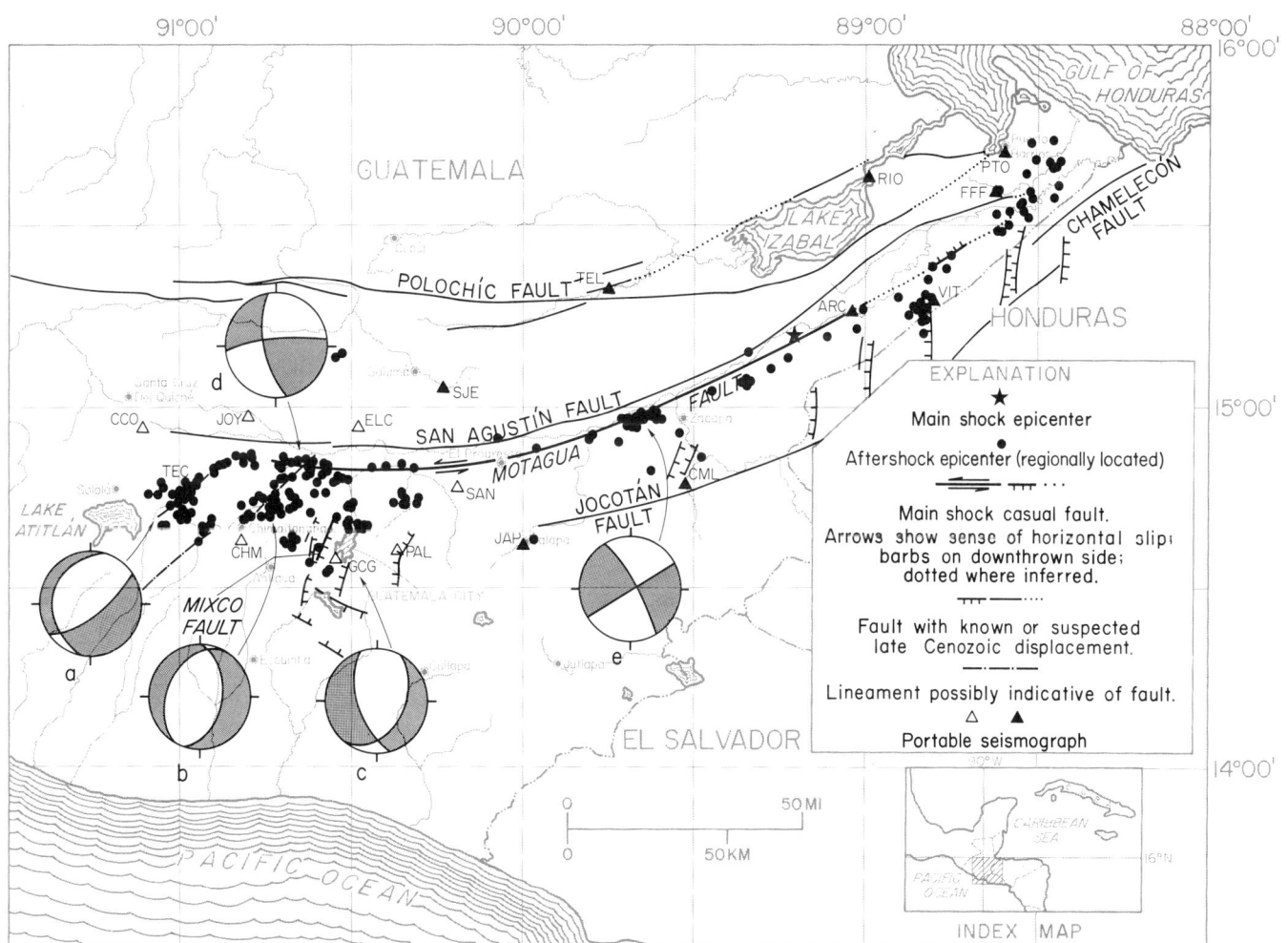

Figure 6. Aftershocks and focal mechanisms of the Guatemala (Motagua fault) earthquake of 4 February 1976 (from Langer and Bollinger, 1979).

The number of permanent seismographic networks capable of locating local earthquakes to within 5 or 10 km continues to increase in Middle America (White and Harlow, this volume), and improvements in the global seismographic network permit more reliable determination of focal mechanisms for smaller earthquakes than in the past. In several regions of North America surveyed by other papers in this volume, decades of precisely located hypocenters and well-constrained focal mechanisms have revealed seismotectonic processes that would have been invisible with the type of data upon which our review of Middle American seismicity is largely based. As high-quality seismicity data accumulate in Middle America, we are likely to see clarification of the seismotectonics in such currently recognized puzzle spots as the Central American triple junctions and northwestern South America, and we may see our models unravel in some areas that now seem simple with the current sparse data sets.

REFERENCES CITED

Adamek, S., Tajima, F., and Wiens, D. A., 1987, Seismic rupture associated with subduction of the Cocos plate: Tectonics, v. 6, p. 757–774.

Adamek, S., Frohlich, C., and Pennington, W. D., 1988, Seismicity of the Caribbean–Nazca boundary: Constraints on microplate tectonics of the Panama region: Journal of Geophysical Research, v. 93, p. 2053–2075.

Allen, C. R., 1981, The modern San Andreas fault, in Ernst, W. G., ed., The geotectonic development of California: Englewood Cliffs, New Jersey, Prentice-Hall Inc., p. 511–534.

Allen, C. R., Silver, L. T., and Stehli, F. G., 1960, Agua–Blanca fault; A major transverse structure of northern Baja California, Mexico: Geological Society of America Bulletin, v. 71, p. 457–482.

Anderson, J. G., and 6 others, 1986, Strong ground-motion from the Michoacan, Mexico, earthquake: Science, v. 233, p. 1043–1049.

Astiz, L., Kanamori, H., and Eissler, H., 1987, Source characteristics of earthquakes in the Michoacan seismic gap in Mexico: Bulletin of the Seismological Society of America, v. 77, p. 1326–1346.

Bergman, E. A., and Solomon, S. C., 1984, Source mechanisms of earthquakes near mid-ocean ridges from body waveform inversion; Implications for the early evolution of oceanic lithosphere: Journal of Geophysical Research, v. 89, p. 11415–11441.

Beroza, G., Rial, J. A., and McNally, K. C., 1984, Source mechanisms of the June 7, 1982, Ometepec, Mexico, earthquake: Geophysical Research Letters, v. 11, p. 689–692.

Bevis, M., and Isacks, B. L., 1984, Hypocentral trend surface analysis; Probing the geometry of Benioff zones: Journal of Geophysical Research, v. 89, p. 6153–6170.

Bowin, C., 1976, The Caribbean; Gravity field and plate tectonics: Geological Society of America Special Paper 169, 79 p.

Brown, R. D., Ward, P. L., and Plafker, G., 1973, Geologic and seismologic aspects of the Managua, Nicaragua, earthquakes of December 23, 1972: U.S. Geological Survey Professional Paper 838, 34 p.

Burbach, G. V., Frohlich, C., Pennington, W. D., and Matumoto, T., 1984, Seismicity and tectonics of the subducted Cocos plate: Journal of Geophysical Research, v. 89, p. 7719–7735.

Burkart, B., 1983, Neogene North American–Caribbean plate boundary across northern Central America; Offset along the Polochic fault: Tectonophysics, v. 99, p. 251–270.

Burkart, B., and Self, S., 1985, Extension and rotation of crustal blocks in northern Central America and effect on the volcanic arc: Geology, v. 13, p. 22–26.

Burke, K., Grippi, J., and Sengor, A. M., 1980, Neogene structures in Jamaica and the tectonic style of the northern Caribbean plate boundary zone: Journal of Geology, v. 88, p. 375–386.

Byrne, D. B., Suárez, G., and McCann, W. R., 1985, Muertos Trough subduction; Microplate tectonics in the northern Caribbean?: London, Nature, v. 317, p. 420–421.

Camacho, H., Flores, T., and Salazar, L., 1922, Terremoto Mexicano del 3 de Enero de 1920: Boletín 38 Instituto Geologico de México, México D.F., 107 p.

Carr, M. J., and Stoiber, R. E., 1977, Geologic setting of some destructive earthquakes in Central America: Geological Society of America Bulletin, v. 88, p. 151–156.

Case, J. E., Durán S.L.G., López, R. A., and Moore, W. R., 1971, Tectonic investigations in western Colombia and eastern Panama: Geological Society of America Bulletin, v. 82, p. 2685–2712.

Centeno-Grau, M., 1940, Estudios Sismológicos: Caracas, Litografía del Comercio, 555 p.

Chael, E. P., and Stewart, G. S., 1982, Recent large earthquakes along the Middle American trench and their implications for the subduction process: Journal of Geophysical Research, v. 87, p. 329–338.

Couch, R., and Woodcock, S., 1981, Gravity and structure of the continental margins of southwestern Mexico and northwestern Guatemala: Journal of Geophysical Research, v. 86, p. 1829–1840.

Dalmayrac, B., and Molnar, P., 1981, Parallel thrust and normal faulting in Peru and constraints on the state of stress: Earth Planetary Science Letters, v. 55, p. 473–481.

Dean, B. W., and Drake, C. L., 1978, Focal mechanism solutions and tectonics of the Middle American arc: Journal of Geology, v. 86, p. 111–128.

Demets, C., Gordon, R. G., Stein, S., and Argus, D. F., 1987, A revised estimate of Pacific–North American motion and implications for western North America plate boundary zone tectonics: Geophysical Research Letters, v. 14, p. 911–914.

Dengo, G., 1953, Geology of the Caracas region, Venezuela: Geological Society of America Bulletin, v. 64, p. 7–40.

Dengo, G., Bohnenberger, O., and Bonis, S., 1970, Tectonics and volcanism along the Pacific marginal zone of Central America: Geologische Rundschau, v. 59, p. 1215–1230.

Dewey, J. F., and Pindell, J. L., 1985, Neogene block tectonics of eastern Turkey and northern South America; Continental applications of the finite difference method: Tectonics, v. 4, p. 71–83.

Dewey, J. W., 1972, Seismicity and tectonics of western Venezuela: Bulletin of the Seismological Society of America, v. 62, p. 1711–1751.

Dewey, J. W., and Algermissen, S. T., 1974, Seismicity of the Middle America arc-trench system near Managua, Nicaragua: Bulletin of the Seismological Society of America, v. 64, p. 1033–1048.

Dorel, J., 1981, Seismicity and seismic gaps in the Lesser Antilles arc and earthquake hazard in Guadeloupe: Geophysical Journal of the Royal Astronomical Society, v. 67, p. 679–695.

Eissler, H. K., and McNally, K. C., 1984, Seismicity and tectonics of the Rivera plate and implications for the 1932 Jalisco, Mexico, earthquake: Journal of Geophysical Research, v. 89, p. 4520–4530.

Engdahl, E. R., compiler, and Rinehart, W. A., preparer, 1988, Seismicity map of North America: Boulder, Colorado, Geological Society of America, Geology of North America Continent-Scale Map-004, scale 1:5,000,000, 4 sheets.

Espinosa, A. F., and Algermissen, S. T., 1972, A study of soil amplification factors in earthquake damage areas, Caracas, Venezuela: National Oceanic and Atmospheric Administration Technical Report ERL280–ESL31, 201 p.

Fiedler, G., 1961, Areas afectadas por terremotos en Venezuela: Boletín Geología, Memoria III, Congreso Geológia Venezolano IV, p. 1791–1801.

Fisher, K. M., and McCann, W. R., 1984, Velocity modeling and earthquake relocation in the northeast Caribbean: Bulletin of the Seismological Society of America, v. 74, p. 1249–1262.

Frankel, A., 1982, A composite focal mechanism for microearthquakes along the northeastern border of the Caribbean plate: Geophysical Research Letters, v. 9, p. 511–514.

Fujita, K., and Kanamori, H., 1981, Double seismic zones and stresses of intermediate depth earthquakes: Geophysical Journal of the Royal Astronomical Society, v. 66, p. 131–156.

Gardner, T. W., Hare, P. W., Pazzaglia, F. J., and Sasowsky, I. D., 1987, Evolution of drainage systems along a convergent plate margin, Pacific Coast, Costa Rica, in Graf, W. L., ed., Geomorphic systems of North America: Boulder, Colorado, Geological Society of America, Centennial Special Volume 2, p. 357–372.

Goff, J. A., Bergman, E. A., and Solomon, S. C., 1987, Earthquake source mechanisms and transform fault tectonics in the Gulf of California: Journal of Geophysical Research, v. 92, p. 10485–10510.

González-Ruiz, J. R., and McNally, K. C., 1988, Stress accumulation and release since 1882 in Ometepec, Guerrero, Mexico; Implications for failure mechanisms and risk assessments of a seismic gap: Journal of Geophysical Research, v. 93, p. 6297–6317.

Guendel, F., and McNally, K. C., 1986, High resolution evidence of smooth Benioff zone gradations approaching the southern terminus of the Middle America trench [abs.]: EOS Transactions of the American Geophysical Union, v. 67, p. 1114.

Hanson, R. D., and Degenkolb, H. J., 1969, The Venezuelan earthquake July 29, 1967: New York, American Iron and Steel Institute, 176 p.

Havskov, J., de la Cruz-Reyna, S., Singh, S. K., and others, 1983a, Seismic activity related to the March–April, 1982, eruptions of El Chichon volcano, Chiapas, Mexico: Geophysical Research Letters, v. 10, p. 293–296.

Havskov, J., Singh, S. K., Nava, E., Domínguez, T., and Rodríguez, M., 1983b, Playa Azul, Michoacan, Mexico earthquake of 25 October, 1981 ($M_s = 7.3$): Bulletin of the Seismological Society of America, v. 73, p. 449–458.

Herd, D. G., and McMasters, C. R., 1982, Surface faulting in the Sonora, Mexico, earthquake of 1887: Geological Society of America Abstracts with Programs, v. 14, p. 172.

Isacks, B. L., and Barazangi, M., 1977, Geometry of Benioff zones; Lateral segmentation and downwards bending of the subducted lithosphere, in Talwani, M., and Pitman, W. C., eds., Island arcs, deep sea trenches, and back-arc basins: American Geophysical Union Maurice Ewing Series 1, p. 99–114.

Isacks, B., and Molnar, P., 1971, Distribution of stresses in the descending lithosphere from a global survey of focal-mechanism solutions of mantle earthquakes: Reviews of Geophysics and Space Physics, v. 9, p. 103–174.

Isacks, B., Oliver, J., and Sykes, L. R., 1968, Seismology and the new global tectonics: Journal of Geophysical Research, v. 73, p. 5855–5899.

Jiménez, Z., and Ponce, L., 1978, Focal mechanism of six large earthquakes in northern Oaxaca, Mexico, for the period 1928–1973: Geofisica Internacional, v. 17, p. 379–386.

Johnson, T. L., Madrid, J., and Koczynski, T., 1976, A study of microseismicity in northern Baja California, Mexico: Bulletin of the Seismological Society of America, v. 66, p. 1921–1929.

Jordan, T. H., 1975, The present-day motions of the Caribbean plate: Journal of Geophysical Research, v. 80, p. 4433–4439.

Kafka, A. L., and Weidner, D. J., 1981, Earthquake focal mechanisms and tectonic processes along the southern boundary of the Caribbean plate: Journal of Geophysical Research, v. 86, p. 2877–2888.

Kanamori, H., and Stewart, G. S., 1978, Seismological aspects of the Guatemala earthquake of February 4, 1976: Journal of Geophysical Research, v. 83, p. 3427–3434.

Kelleher, J., Sykes, L., and Oliver, J., 1973, Possible criteria for predicting earthquake locations and their application to major plate boundaries of the Pacific and Caribbean: Journal of Geophysical Research, v. 78, p. 2547–2585.

Kellogg, J. N., and Bonini, W. E., 1982, Subduction of the Caribbean plate and basement uplifts in the overriding South American plate: Tectonics, v. 1, p. 251–276.

Klitgord, K. D., and Mammerickx, J., 1982, Northern East Pacific Rise; Magnetic anomaly and bathymetric framework: Journal of Geophysical Research, v. 87, p. 6725–6750.

Krause, D. C., 1971, Bathymetry, geomagnetism, and tectonics of the Caribbean Sea north of Colombia, in Donnelly, T. W., ed., Caribbean geophysics, tectonics, and petrologic studies: Geological Society of America Memoir 130, p. 35–54.

Ladd, J. W., Shih, T.-C., and Tsai, C. J., 1977, Cenozoic tectonics of central Hispaniola and adjacent Caribbean Sea: American Association of Petroleum Geologists Bulletin, v. 65, p. 466–489.

Langer, C. J., and Bollinger, G. A., 1979, Secondary faulting near the terminus of a seismogenic strike-slip fault; Aftershocks of the 1976 Guatemala earthquake: Bulletin of the Seismological Society of America, v. 69, p. 427–444.

Langer, C. J., Hopper, M. G., Algermissen, S. T., and Dewey, J. W., 1974, Aftershocks of the Managua, Nicaragua, earthquake of December 23, 1972: Bulletin of the Seismological Society of America, v. 64, p. 1005–1016.

Lay, T., Kanamori, H., and Ruff, L., 1982, The asperity model and the nature of large subduction zone earthquakes: Earthquake Prediction Research, v. 1, p. 3–71.

LeFevre, L. V., and McNally, K. C., 1985, Stress distribution and subduction of aseismic ridges in the Middle America subduction zone: Journal of Geophysical Research, v. 90, p. 4495–4510.

Lonsdale, P., and Klitgord, K. D., 1978, Structure and tectonic history of the eastern Panama Basin: Geological Society of America Bulletin, v. 98, p. 981–999.

Mann, P., and Burke, K., 1984, Neotectonics of the Caribbean: Reviews of Geophysics and Space Physics, v. 22, p. 309–362.

Mann, P., Schubert, C., and Burke, K., 1989, Review of Caribbean neotectonics, in Dengo, G., and Case, J. E., eds., The Caribbean region: Boulder, Colorado, The Geology of North America, v. H (in press).

McCann, W. R., and Sykes, L. R., 1984, Subduction of aseismic ridges beneath the Caribbean plate; Implications for the tectonics and seismic potential of the northeastern Caribbean: Journal of Geophysical Research, v. 89, p. 4493–4519.

McCann, W. R., Nishenko, S. P., Sykes, L. R., and Krause, J., 1979, Seismic gaps and plate tectonics; Seismic potential for major boundaries: Pure and Applied Geophysics, v. 117, p. 1082–1147.

McCann, W. R., Dewey, J. W., Murphy, A. J., and Harding, S. T., 1982, A large normal-fault earthquake in the overriding wedge of the Lesser Antilles subduction zone; The earthquake of 8 October, 1974: Bulletin of the Seismological Society of America, v. 72, p. 2267–2283.

McNally, K. C., and Minster, J. B., 1981, Nonuniform seismic slip rates along the Middle America trench: Journal of Geophysical Research, v. 86, p. 4949–4959.

McNally, K. C., González-Ruiz, J. R., and Stolte, C., 1986, Seismogenesis of the 1985 great ($M_s = 8.1$) Michoacan, Mexico, earthquake: Geophysical Research Letters, v. 13, p. 585–588.

Meehan, J. F., 1973, Reconnaissance report of the Veracruz, Mexico, earthquake of August 28, 1973: Bulletin of the Seismological Society of America, v. 64, p. 2011–2025.

Minster, J. B., and Jordan, T. H., 1978, Present-day plate motions: Journal of Geophysical Research, v. 83, p. 5331–5354.

—— , 1979, Rotation vectors for the Philippine and Rivera plates [abs.]: EOS Transactions of the American Geophysical Union, v. 60, p. 958.

Molnar, P., and Sykes, L. R., 1969, Tectonics of the Caribbean and Middle American regions from focal mechanisms and seismicity: Geological Society of America Bulletin, v. 80, p. 1639–1684.

Morgan, F. D., and 5 others, 1988, The earthquake hazard alert of September 1982 in southern Tobago: Bulletin of the Seismological Society of America, v. 78, p. 1550–1562.

Munguía, L., and Brune, J. N., 1984, High stress drop events in the Victoria, Baja California, earthquake swarm of 1978 March: Geophysical Journal of the Royal Astronomical Society, v. 76, p. 725–752.

Munguía, L., Reichle, M., Reyes, A., Simons, R., and Brune, J., 1977, Aftershocks of the 8 July, 1975, Canal de los Ballenas, Gulf of California, earthquake: Geophysical Research Letters, v. 4, p. 507–509.

Natali, S. G., and Sbar, M. L., 1982, Seismicity in the epicentral region of the

1887 northeastern Sonora earthquake, Mexico: Bulletin of the Seismological Society of America, v. 72, p. 181–196.

Nishenko, S. P., and Singh, S. K., 1987, Conditional probabilities for the recurrence of large and great interplate earthquakes along the Mexican subduction zone; 1986–2006: Bulletin of the Seismological Society of America, v. 77, p. 2094–2114.

Ohtake, M., Matumoto, T., and Latham, G. V., 1981, Evaluation of the forecast of the 1978 Oaxaca, southern Mexico, earthquake based on a precursory seismic quiescence, *in* Simpson, D. W., and Richards, P. G., eds., Earthquake prediction; An international review: American Geophysical Union Maurice Ewing Series 4, p. 53–61.

Paige, S., 1930, The earthquake at Cumana, Venezuela, January 17, 1929: Bulletin of the Seismological Society of America, v. 20, p. 1–10.

Pennington, W. D., 1981, The subduction of the eastern Panama Basin and the seismotectonics of northwestern South America: Journal of Geophysical Research, v. 86, p. 10753–10770.

Pérez, O. J., and Aggarwal, Y. P., 1981, Present-day tectonics of the southeastern Caribbean and northeastern Venezuela: Journal of Geophysical Research, v. 86, p. 10791–10804.

Plafker, G., 1976, Tectonic aspects of the Guatemala earthquake of 4 February 1976: Science, v. 93, p. 1201–1208.

Plafker, G., Bonilla, M. G., and Bonis, S. B., 1976, Geologic effects, *in* Espinosa, A. F., ed., The Guatemalan earthquake of February 4, 1976; A preliminary report: U.S. Geological Survey Professional Paper 1002, p. 38–51.

Prothero, W. A., and Reid, I. D., 1982, Microearthquakes on the East Pacific Rise at 21°N and the Rivera Fracture Zone: Journal of Geophysical Research, v. 87, p. 8509–8518.

Ramírez, J. E., 1969, Historia de los terremotos en Colombia: Bogotá, Editorial Argra, 218 p.

Rebollar, C. J., and Reichle, M. S., 1987, Analysis of the seismicity detected in 1982–1984 in the northern Peninsular Ranges of Baja California: Bulletin of the Seismological Society of America, v. 77, p. 173–183.

Reichle, M., and Reid, I., 1977, Detailed study of earthquake swarms from the Gulf of California: Bulletin of the Seismological Society of America, v. 67, p. 159–171.

Reichle, M. S., Sharman, G. F., and Brune, J. N., 1976, Sonobuoy and teleseismic study of Gulf of California transform fault earthquake sequences: Bulletin of the Seismological Society of America, v. 66, p. 1623–1641.

Reyes, A., and 6 others, 1975, A microearthquake survey of the San Miguel fault zone, Baja California, Mexico: Geophysical Research Letters, v. 2, p. 56–59.

Reyes, A., Brune, J. N., and Lomnitz, C., 1979, Source mechanism and aftershock study of the Colima, Mexico, earthquake of January 10, 1973: Bulletin of the Seismological Society of America, v. 69, p. 1819–1840.

Rial, J. A., 1978, The Caracas, Venezuela, earthquake of July 1967; A multiple source event: Journal of Geophysical Research, v. 83, p. 5405–5415.

Robson, G. R., 1964, An earthquake catalogue for the eastern Caribbean: Bulletin of the Seismological Society of America, v. 54, p. 785–832.

Robson, G. R., Barr, K. G., and Smith, G. W., 1962, Earthquake series in St. Kitts–Nevis 1961–1962: Nature, v. 195, p. 972–974.

Rod, E., 1956, Strike-slip faults of northern Venezuela: American Association of Petroleum Geologists Bulletin, v. 40, p. 457–476.

Rusnak, G. A., Fisher, R. L., and Shepard, F. P., 1964, Bathymetry and faults of Gulf of California, *in* von Andel, T., and Shor, G. G., eds., Marine geology of the Gulf of California: American Association of Petroleum Geologists Memoir 3, p. 59–75.

Schell, B. A., and Tarr, A. C., 1978, Plate tectonics of the northeastern Caribbean Sea region: Geologie en Mijnbouw, v. 57, p. 319–324.

Schneider, J. F., Pennington, W. D., and Meyer, R. P., 1987, Microseismicity and focal mechanisms of the intermediate-depth Bucaramanga nest: Journal of Geophysical Research, v. 92, p. 13913–13926.

Schubert, C., 1982, Neotectonics of Boconó fault, western Venezuela: Tectonophysics, 85, p. 205–220.

—— , 1984, Basin formation along the Boconó–Morón–El Pilar fault system, Venezuela: Journal of Geophysical Research, v. 89, p. 5711–5718.

Schubert, C., and Krause, F. F., 1984, Morón fault zone, north-central Venezuela borderland; Identification, definition, and neotectonic character: Marine Geophysics Researches, v. 6, p. 257–273.

Schwartz, D. P., Cluff, L. S., and Donnelly, T. W., 1979, Quaternary faulting along the Caribbean–North American plate boundary in Central America: Tectonophysics, v. 52, p. 431–445.

Seager, W., and Morgan, P., 1979, Rio Grande rift in southern New Mexico, west Texas, and northern Chihuahua, *in* Reicker, R., ed., Rio Grande rift; Tectonics and magmatism: American Geophysical Union, p. 87–106.

Shepherd, J. B., and Aspinall, W. P., 1980, Seismicity and seismic intensities in Jamaica, West Indies; A problem in risk assessment: Earthquake Engineering and Structural Dynamics, v. 8, p. 315–335.

Shor, G. G., Jr., and Roberts, E., 1958, San Miguel, Baja California Norte, earthquakes of February 1956; A field report: Bulletin of the Seismological Society of America, v. 48, p. 101–116.

Silver, E. A., Case, J. E., and MacGillivary, H. J., 1975, Geophysical study of the Venezuelan borderland: Geological Society of America Bulletin, v. 86, p. 213–226.

Singh, S. K., and Wyss, M., 1976, Source parameters of the Orizaba earthquake of August 28, 1973: Geofísica Internacional, v. 16, p. 165–194.

Singh, S. K., and 5 others, 1979, The Oaxaca, Mexico, earthquake of 29 November 1978; A preliminary report on aftershocks: Science, v. 207, p. 1211–1213.

Singh, S. K., Reichle, M., and Havskov, J., 1980, Magnitude and epicenter estimations of Mexican earthquakes from isoseismic maps: Geofísica Internacional, v. 19, p. 269–284.

Singh, S. K., Astiz, L., and Havskov, J., 1981, Seismic gaps and recurrence periods of large earthquakes along the Mexican subduction zone; A reexamination: Bulletin of the Seismological Society of America, v. 71, p. 827–843.

Singh, S. K., Suárez, G., and Domínguez, T., 1985a, The Oaxaca, Mexico, earthquake of 1931; Lithospheric normal faulting in the subducted Cocos plate: London, Nature, v. 317, p. 56–56.

Singh, S. K., Ponce, L., and Nishenko, S. P., 1985b, The great Jalisco, Mexico, earthquakes of 1932; Subduction of the Rivera plate: Bulletin of the Seismological Society of America, v. 75, p. 1301–1313.

Smith, R. J., 1953, Geology of the Los Teques–Cua region, Venezuela: Geological Society of America Bulletin, v. 64, p. 41–64.

Speed, R. C., 1985, Cenozoic collision of the Lesser Antilles arc and continental South America and the origin of the El Pilar fault: Tectonics, v. 4, p. 41–69.

Stein, S., Engeln, J. F., Wiens, D. A., Fujita, K., and Speed, R. C., 1982, Subduction seismicity and tectonics in the Lesser Antilles arc: Journal of Geophysical Research, v. 87, p. 8642–8664.

Stein, S., and 9 others, 1988, A test of alternative Caribbean plate relative motion models: Journal of Geophysical Research, v. 93, p. 3041–3050.

Stewart, G. S., Chael, E. P., and McNally, K. C., 1981, The 1978 November 29, Oaxaca, Mexico, earthquake; A large simple event: Journal of Geophysical Research, v. 86, p. 5053–5060.

Stolte, C., and 7 others, 1986, Fine structure of a postfailure Wadati–Benioff zone: Geophysical Research Letters, v. 13, p. 577–580.

Suárez, G., and Nabelek, J., 1989, The 1967 Caracas earthquake; Fault geometry, direction of rupture propagation, and seismotectonic implications: Journal of Geophysical Research (in press).

Suárez, G., and Ponce, L., 1986, Intraplate seismicity and crustal deformation in central Mexico [abs.]: EOS Transactions of the American Geophysical Union, v. 67, p. 1114.

Suárez, G., Molnar, P., and Burchfiel, B. C., 1983, Seismicity, fault plane solutions, depth of faulting, and active tectonics of the Andes of Peru, Ecuador, and southern Colombia: Journal of Geophysical Research, v. 88, p. 10403–10428.

Sykes, L. R., 1968, Seismological evidence of transform faults, sea-floor spreading, and continental drift, *in* Phinney, R. A., ed., History of the earth's crust: Princeton, New Jersey, Princeton University Press, p. 120–150.

Sykes, L. R., McCann, W. R., and Kafka, A. L., 1982, Motion of Caribbean plate during last 7 million years and implications for earlier Cenozoic movements:

Journal of Geophysical Research, v. 87, p. 10656–10676.

Taber, S., 1920, Jamaica earthquakes and the Bartlett trough: Bulletin of the Seismological Society of America, v. 10, p. 55–89.

Thatcher, W., and Brune, J. N., 1971, Seismic study of an oceanic ridge earthquake swarm in the Gulf of California: Geophysical Journal of the Royal Astronomical Society, v. 22, p. 473–489.

Tomblin, J. F., 1975, The Lesser Antilles and Aves Ridge, in Nairn, A.E.M., and Stehli, F. G., eds., The Gulf of Mexico and the Caribbean; The ocean basins and margins, v. 3: New York, Plenum, p. 467–500.

Tréhu, A. M., and Solomon, S. C., 1983, Earthquakes in the Orozco transform zone; Seismicity, source mechanisms, and tectonics: Journal of Geophysical Research, v. 88, p. 8203–8225.

UNAM Seismology Group, 1986, The September 1985 Michoacan earthquakes; Aftershock distribution and history of rupture: Geophysical Research Letters, v. 13, p. 573–576.

Urbina, F., and Camacho, H., 1913, La zona megaseismica de Acambay-Tixmadeje: Boletín 32 Instituto Geologico de México, México D.F., 135 p.

Valdés, C., Meyer, R., Zuñiga, R., Havskov, J., and Singh, S., 1982, Analysis of the Petatlan aftershocks; Numbers, energy release, and asperities: Journal of Geophysical Research, v. 87, p. 8519–8529.

Vierbuchen, R. C., 1984, The geology of the El Pilar fault zone and adjacent areas in northeastern Venezuela, in Bonini, W. E., Hargraves, R. B., and Shagam, R., eds., The Caribbean–South American plate boundary and regional tectonics: Geological Society of America Memoir 162, p. 189–212.

von Humboldt, A., 1889, Personal narrative of travels to the equinoctal regions of America during the years 1799–1804 by Alexander von Humboldt and Aimé Bonpland (translated and edited by Thomasina Ross, v. 1): London, George Bell and Sons, 505 p.

Wadge, G., and Shepherd, J. B., 1984, Segmentation of the Lesser Antilles subduction zone: Earth and Planetary Science Letters, v. 71, p. 297–304.

Ward, P. L., Gibbs, J., Harlow, D., and Aburto Q., A., 1974, Aftershocks of the Managua, Nicaragua, earthquake and the tectonic significance of the Tiscapa fault: Bulletin of the Seismological Society of America, v. 64, p. 1017–1030.

White, R. A., 1985, The Guatemala earthquake of 1816 on the Chixoy–Polochic fault: Bulletin of the Seismological Society of America, v. 75, p. 455–473.

White, R. A., Harlow, D. H., and Alvarez, S., 1987, The San Salvador earthquake of October 10, 1986; Seismological aspects and other recent local seismicity: Earthquake Spectra, v. 3, p. 419–434.

Wiens, D. A., and Stein, S., 1984, Intraplate seismicity and stresses in young oceanic lithosphere: Journal of Geophysical Research, v. 89, p. 11442–11464.

Wolters, B., 1986, Seismicity and tectonics of southern Central America and adjacent regions with special attention to the surroundings of Panama: Tectonophysics, v. 128, p. 21–46.

Yamamoto, J., Jiménez, Z., and Mota, R., 1984, El temblor de Huajuapan de León, Oaxaca, México, del 24 de Octubre de 1980: Geofísica Internacional, v. 23, p. 83–110.

MANUSCRIPT ACCEPTED BY THE SOCIETY MARCH 21, 1989

ACKNOWLEDGMENTS

We thank Macelo Assumpcao and Lautaro Ponce for providing results of their studies of Trans-Mexican Volcanic Belt focal-mechanisms prior to publication. Stuart Nishenko and Randall White gave helpful reviews.

Printed in U.S.A.

The Geology of North America
Decade Map Volume 1
1991

Chapter 18

Tectonic implications of upper-crustal seismicity in Central America

Randall A. White
Office of Earthquakes, Volcanoes, and Engineering, U.S. Geological Survey, 345 Middlefield Road, Menlo Park, California 94025

INTRODUCTION

The western boundary of the Caribbean Plate is delineated by a wide band of seismicity associated with the Middle American Trench (Continental Scale Map 4, *Seismicity Map of North America,* this volume). Most of this seismicity occurs along the interface between the Caribbean Plate and the subducting Cocos Plate, at and beneath the Middle American Trench; this seismicity is dealt with by Dewey and Suarez (this volume). In this chapter I concentrate on seismicity within Central America that occurs both within the Caribbean Plate and along the transcurrent Caribbean–North American plate boundary (hereafter called the "CARB-NOAM boundary"); within the Caribbean Plate, seismicity occurs principally along the volcanic front (hereafter called the "volcanic zone"); other seismicity also occurs within the wedge-shaped region between the volcanic zone and the CARB-NOAM boundary, and within the complicated and poorly understood region of southern Costa Rica. Except possibly within the latter region, all such seismicity is probably confined to the upper latter 20 km of the crust (hereafter called "upper-crustal" seismicity).

Unlike seismicity along the subduction zone, upper-crustal seismicity within Central America is not conspicuous in international earthquake catalogs or on global seismicity maps. This is due to the fact that, at any particular magnitude level, upper-crustal earthquakes are infrequent compared with those along the subduction zone. However, because they are very shallow and often near heavily populated areas, upper-crustal earthquakes with magnitudes as small as M 5 can produce significant damage in Central America, whereas subduction-zone earthquakes generally must have magnitudes of at least M 7 to produce such damage. Upper-crustal earthquakes actually produce significant damage much more frequently than do subduction zone events. During this century alone, upper-crustal earthquakes have ravaged each of the capital cities of Guatemala, El Salvador, and Nicaragua at least twice and have killed a total of about 40,000 people (White and Harlow, in preparation).

With the installation of high-gain seismograph networks in Central America, it has become possible to study the upper-crustal seismicity of the region. High-gain networks were temporarily deployed, following the disastrous December 23, 1972, Managua earthquake (M_s 6.2), which killed 11,000 people; these networks quickly demonstrated their value in determining the location and orientation of active faults (Ward and others, 1974; Langer and others, 1974). By mid-1975, three permanent high-gain seismograph networks were installed and operating, in Costa Rica, Guatemala, and Nicaragua. The first two of these have since expanded and additional networks have begun operation in Costa Rica, El Salvador, and Guatemala.

In this chapter I shall (1) briefly review the history and current status of each permanent high-gain seismograph network in Central America; (2) present examples of upper-crustal microseismicity data produced by these networks; (3) review published earthquake focal mechanisms for events along the volcanic zone and CARB-NOAM boundary; (4) describe results of historical earthquake studies that constrain magnitudes and occurrence rates of upper-crustal earthquakes; and (5) show that a simple fore-arc sliver model, first described for Central America by Harlow and White (1985), is compatible with the volcanic zone observations and can explain apparent correlations between

White, R. A., 1991, Tectonic implications of upper-crustal seismicity in Central America, *in* Slemmons, D. B., Engdahl, E. R., Zoback, M. D., and Blackwell, D. D., eds., Neotectonics of North America: Boulder, Colorado, Geological Society of America, Decade Map Volume 1.

major ruptures at the western end of the CARB-NOAM boundary and activity, both seismic and volcanic, along the volcanic zone.

SEISMOTECTONIC SETTING

By far the most seismically active tectonic feature of the region is the Central American subduction zone, along the Pacific coast (Fig. 1). There the Cocos Plate collides with, and dips beneath, the Caribbean and North American Plates (Dewey and Suarez, this volume). Above the subduction zone lie the principal volcanoes of the Central American chain, which extend from central Costa Rica, offshore from which the Cocos Ridge enters the trench and the subduction zone shoals and disappears, northwestward to the Mexico-Guatemala border, where the chain terminates against the Chixoy-Polochic fault of the CARB-NOAM plate boundary.

The volcanic chain so dominates the landscape that it is featured on the flags of several Central American countries. The volcanoes stretch about 1,060 km along the Pacific coast and lie about 160 to 175 km landward from trench. The chain is among the most active in the world, with about ten volcanoes active during the past decade and five active during an average year (Simkin and others, 1981). The volcanoes are among the most closely spaced in the world, with an average of 25 km between the 42 volcanoes with evidence of Holocene activity; however, at the ends of the chain, Holocene volcanic centers are more widely spaced, from 18 to 30 km, and summit elevations are higher, averaging more than 3,300 m, compared with spacings of 12 to 18 km and average summit elevations of less than 1,000 m near the center of the chain (Simkin and others, 1981; Carr and others, 1982). Ben-Avraham and Nur (1980) point out that edifice heights are nearly constant over the entire length of the chain, however. Stoiber and Carr (1973) note that the principal volcanoes of the chain are grouped into seven linear segments, with each segment offset in a right-lateral sense from adjoining segments.

The transcurrent plate boundary between the Caribbean and North American Plates trends east-west through central Guatemala. In the Gulf of Honduras to the east, the plate boundary is expressed principally as the Swan Fracture Zone (Fig. 1); within Guatemala however, the plate boundary is expressed primarily as two subparallel, left-lateral strike-slip faults: the Chixoy-Polochic fault to the north, and the Motagua fault to the south (Fig. 2). The faults have created dramatic parallel river valleys 40 to 50 km apart, each with vertical relief of more than 1,500 m. To the west, these faults approach, and disappear beneath, the Quaternary deposits of the volcanic chain; if and where they intersect the trench is unknown. Burkart (1978) has estimated a total offset of 132 km along the Chixoy-Polochic fault since the middle Miocene. The total offset along the Motagua fault is unknown but probably similar (Schwartz and others, 1979). Thrust and reverse faults subparallel with, and adjacent to, the Motagua fault appear not to have been significantly active during the Quaternary (Schwartz and others, 1979). On either side of the Motagua and Chixoy-Polochic faults lie additional faults with major topographic expression: the Jocotan-Chamelecon to the southeast, and unnamed faults to the north (Fig. 2); though geologic evidence of recent strike-slip faulting has not been reported along them, they may form minor but active parts of the transcurrent plate boundary (Schwartz and others, 1979).

Within the wedge-shaped area between the Motagua fault and the volcanic chain, there is a broad area of crustal extension that is bounded to the east approximately by the Honduras Depression (Fig. 2). Within this area and west of Guatemala City, faults splay to the southwest from the Motagua fault and terminate at the volcanic chain; from Guatemala City eastward, this area is dominated by normal faults, including faults of the Mixco and Ipala grabens, which trend north-south to northwest-southeast.

Prior to 1973, very little was known about the seismicity, either modern or historical, of any of the faults of the volcanic zone, the transcurrent zone, or of the wedge-shaped area of crustal extensional.

CENTRAL AMERICAN SEISMOGRAPH AND STRONG-MOTION NETWORKS

Table 1 presents basic data on permanent high-gain seismograph networks that have apertures of at least 50 km and have

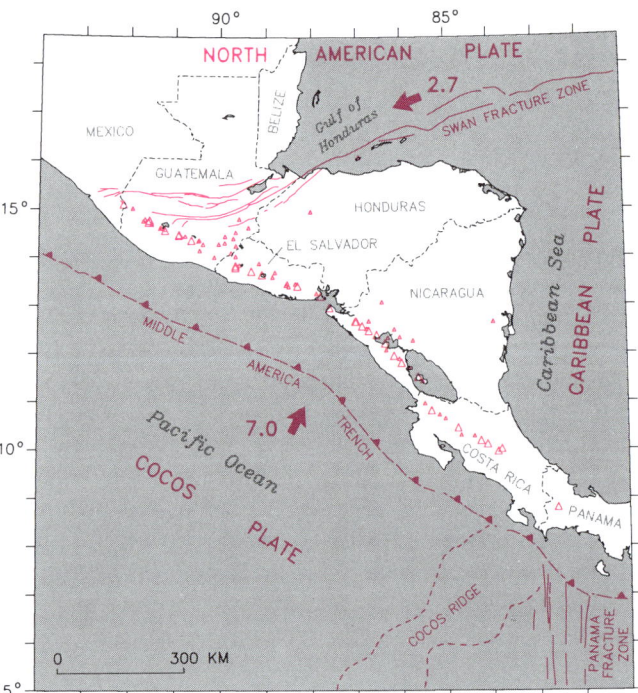

Figure 1. General tectonic setting of the Central American region. Large and small triangles indicate volcanoes with historical and Holocene volcanic activity, respectively (Simkin and others, 1981). Faults are from Case and Holcombe (1980). Plate-motion vectors are relative to the Caribbean Plate, from Dewey and Suarez (this volume).

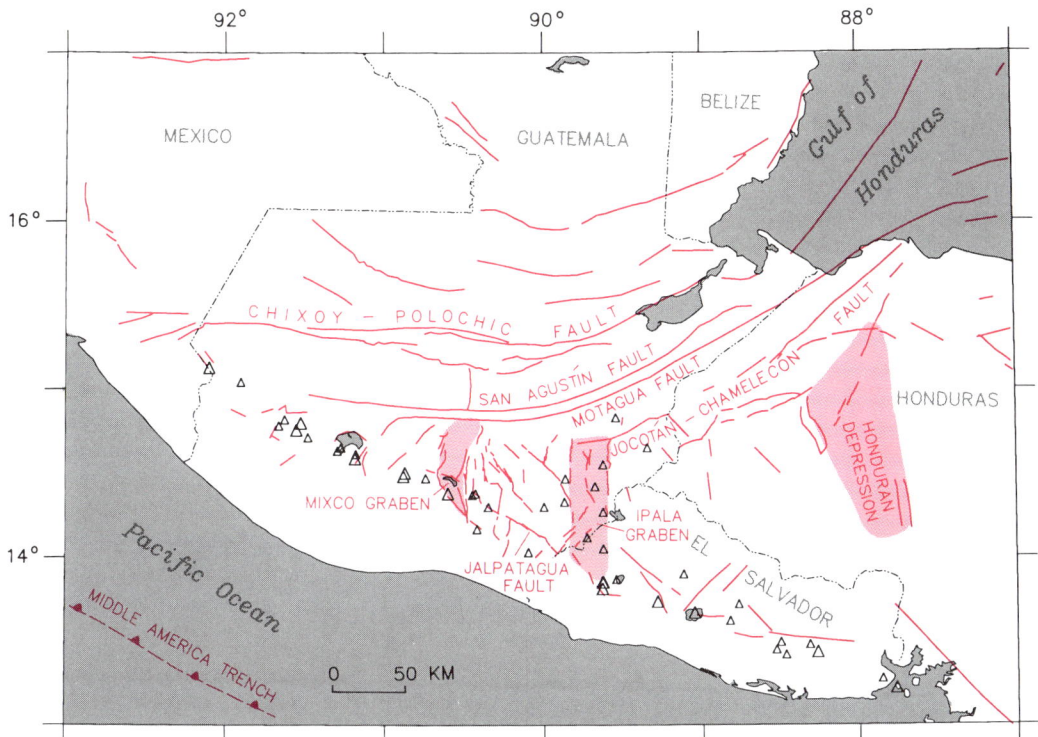

Figure 2. Map of volcanoes and faults in northwestern Central America. Large and small triangles indicate volcanoes with historical and Holocene volcanic activity, respectively (Simkin and others, 1981). Solid lines show the principal Quaternary faults of (1) the transcurrent CARB-NOAM plate boundary (from north to south: unnamed faults, the Chixoy-Polochic fault, the San Agustin fault, Motagua fault, and Jocotan-Chamelecon fault); (2) the volcanic chain; (3) the Honduras Depression; and (4) the wedge-shaped area of extension bounded by the Motagua fault, the volcanic chain, and the Honduras Depression (modified from Case and Holcombe, 1980; faults in southeastern Guatemala are from Carr, 1976). Pink areas indicate the Mixco and Ipala grabens and the Honduran Depression.

operated for at least four years. The table shows agencies responsible for daily operation of each network, network apertures at inception and at maximum, yearly estimates of the average number of sites with operating high-gain vertical-component seismographs, and the title and frequency of publication of microseismicity catalogs for each network. The yearly average number of operating seismographs was estimated by averaging the maximum number of P-wave arrival times listed in the network catalogs for local events of M $\geq$3.0 (this number may underestimate the actual number if distant stations are routinely ignored for hypocentral solutions). Figure 3 shows the general areas covered by each network during the years indicated and shows the M $\geq$3 earthquakes detected by each network during those years.

Guatemalan seismograph networks

In a cooperative program with the Instituto Nacional de Sismologia, Vulcanologia, Meteorologia, e Hidrologia (INSIVUMEH) of Guatemala, the U.S. Geological Survey installed a six-station seismograph network around Guatemala City in early 1975. The network recorded several thousand aftershocks of the M_w 7.5, February 4, 1976, Guatemala earthquake (see White and Harlow, 1979, 1980). INSIVUMEH expanded the network to more than 20 total stations, including one long-period seismograph. Two strong-motion accelerographs have operated intermittently since 1976.

The Instituto Nacional de Electrificación (INDE), the national electric power company, has operated an 11-station network, including one 3-component station, intermittently since 1982. The network monitors seismicity near and beneath a potential dam site near the Chixoy-Polochic fault.

The El Salvador national seismograph and strong-motion network

In a cooperative program between the Centro de Investigaciones Geotecnicas (CIG) of El Salvador and the U.S. Geological Survey, ten seismograph stations were installed in western El Salvador in early 1984, and ten strong-motion accelerographs were installed throughout the city of San Salvador in 1985. At least eight stations of each network functioned well during the M_s 5.4 October 10, 1986, San Salvador earthquake (See Shakal

TABLE 1. AVERAGE NUMBER OF SEISMOGRAPH STATIONS OPERATING BY YEAR

Network*	Start Date†	Aperture§ Start	Aperture§ Max	1975	1976	1977	1978	1979	1980	1981	1982	1983	1984	1985	1986	1987	1988	1989	1990
INSIVUMEH	3/1975	40	200	6	6	8	14	16	16	16	16	14	14	12	10	10	12	10	9
INDE	1/1982	35	50								8	9	0	0	8	7	?	0	0
CIG	2/1984	150	150										10	10	8	7	8	8	9
IIS/INETER	3/1975	200	200	13	15	16	17	17	16	15	12	8?	4?	1?	1?	1?	1?	1?	2?
ICE	3/1975	60	120	6	8	7	5	5	5	5	7	9	9	8	8	8	8	8	8
UCR	?/1977	60	60			5	5	5	5	5	5	5	5	5	5	4	4	4	6
UNA	?/1984	80	300										7	11	12	13	14	14	14

Note: This table is meant to give a general idea of the *average* number of vertical short-period high-gain seismograph stations operating at one time during any given year; this number was estimated by averaging the maximum number of P-phase arrival time readings for m >3 events as listed in the network catalogs; for large networks this number may underestimate the actual average number of stations in operation (see text).

*INSIVUMEH = Instituto Nacional de Sismologia, Vulcanologia, Meterologia e Hidrologia (federal agency, part of Ministerio de Comunicaciones, Transporte y Obras Publicas); catalog, *Boletín Sismologico* (annual); address, INSIVUMEH, Sección de Sismología, 7a Avenida 14-57 Zona 13, Guatemala, C.A.

INDE = Instituto Nacional de Electrificacion (autonomous federal agency); catalog, *Boletín Sismico* (quarterly/semiannual); address, Proyecto Hidroelectrico Chulac, INDE, Guatemala, C.A.

CIG = Centro de Investigaciones Geotecnicas (federal agency, part of Ministerio de Obras Publicas); catalog, *Informe Sismico Mensual* (monthly); address, Departamento Investigaciones Sismologicas, Apartado 109, San Salvador, El Salvador, C.A.

IIS = Instituto de Investigaciones Sismicas, absorbed in 1979 by INETER = Instituto Nacional de Estudias Territoriales (federal agency, part of the Ministerio de Planificación); catalog, *Catalogo de Temblores de Nicaragua* (annual), address (as of 1979), INETER, Direccion de Geología y Geofísica, Apartado #1761, Managua, Nicaragua.

ICE = Instituto Costarricense de Electricidad (autonomous federal agency); catalog (joint with UCR), *Boletín Sismico Nacional* (quarterly); address, Sección Sismologia e Ingeniería Sismica, Departmento de Geología, Instituto Costarricense de Electricidad, Apartado 10032, 1000 San Jose, Costa Rica.

UCR = Universidad de Costa Rica; catalog (joint with ICE), *Boletín Sismico Nacional* (quarterly); address, Sección de Sismología, Escuela Centroamericana de Geología, Cuidad Universitaria "Rodrigo Facio," Apartado 35, San Jose, Costa Rica.

UNA = Universidad Nacional; catalog, *Catalogo de Temblores* (monthly); address, Observatorio Vulcanologico y Sismologico de Costa Rica, Campus Omar Dengo, Universidad Nacional, Apartado 86, Heredia 3000, Costa Rica.

†Start = initial month and year network began recording seismograms.

§Aperture = aperture for the network (km) for the first year of operation (start) and for the year with maximum number of stations operating (max).

and others, 1987; White and others, 1987; Harlow and others, in preparation).

The Nicaraguan national seismograph and strong-motion network

In 1974, under a cooperative program between the Instituto de Investigaciones Sismicas (IIS) of Nicaragua and the U.S. Geological Survey, 17 seismograph stations, including three 3-component stations, were installed throughout western Nicaragua in early 1975. Twenty strong-motion accelerographs were also installed in the six largest cities. In 1981, IIS became a branch of the Instituto Nacional de Estudias Territoriales (INETER).

Costa Rican seismograph networks

Six seismograph stations were installed in Costa Rica in early 1974 under a cooperative program between the Instituto Costarricense de Electricidad (ICE) and the University of Texas, Austin. The network detected more than 5,000 earthquakes through 1978 (Matumoto, 1978), mostly aftershocks of the April 14, 1973, Tilaran earthquake (M_s 6.5). The network has expanded and, in 1982, merged with the network of the University of Costa Rica (UCR) at San Jose, and now spans most of western Costa Rica. The UCR network began regular operation of five stations in 1977 in a cooperative program with the Organization of American States. Since 1982, the ICE and UCR networks publish a joint catalog; the joint network is called the Red Sismologica Nacional (RSN, or national seismograph network).

In early 1984, a joint program was begun between the Universidad Nacional (UNA) at Heredia, Costa Rica, and the University of California at Santa Cruz. The network began operation with eight stations, including one 3-component station, and currently is composed of about 16 stations covering most of the country.

LOCAL NETWORK DATA

A major contribution of these local networks has been the cataloging of upper-crustal earthquakes of Central America, the vast majority of which are below the detection threshold of regional and teleseismic stations. I have collected and studied copies of most Central American network catalogs through 1987. The catalogs contain hypocentral parameters for local earthquakes and some also contain arrival times for moderate regional events. Hypocentral determinations are based principally on P-phase arrival times supplemented with a few S-phase arrival times. I estimate that about one-half the hypocenters in the network catalogs meet these following criteria for A- and B-quality solutions: (1) at least six P-phase and two S-phase arrival times are used, (2) the root-mean-square of travel-time residuals is less than 0.5 sec, and (3) the azimuthal separation between any two stations used, as viewed from the epicenter is less than 180° (i.e., the epicenter lies within the perimeter of the network); hypocenters meeting these criteria generally have computed standard errors of less than 3 km in the horizontal dimension and less than 30 percent of the estimated depth in the vertical dimension. Study of the A- and B-quality hypocentral solutions indicates that epicenters of upper-crustal shocks fall primarily within two zones: (1) the volcanic zone, a 20-km-wide belt along the volcanic chain; and (2) the CARB-NOAM boundary, a 120-km-wide zone along the transcurrent Caribbean–North American plate boundary through Guatemala. Other upper-crustal seismicity has also been detected from the wedge-shaped area between these two zones and from southern Costa Rica. The volcanic zone seismicity has especially serious implications for Central America because 60 to 70 percent of the population lives within the volcanic zone.

TYPES OF RECORDED SEISMIC SIGNALS

Along the volcanic zone, both volcanic and tectonic earthquakes occur. Volcanic earthquakes are classified into four main types by Minakami (1974): (1) A type, (2) B type, (3) explosion quakes, and (4) volcanic tremor. The latter three types originate

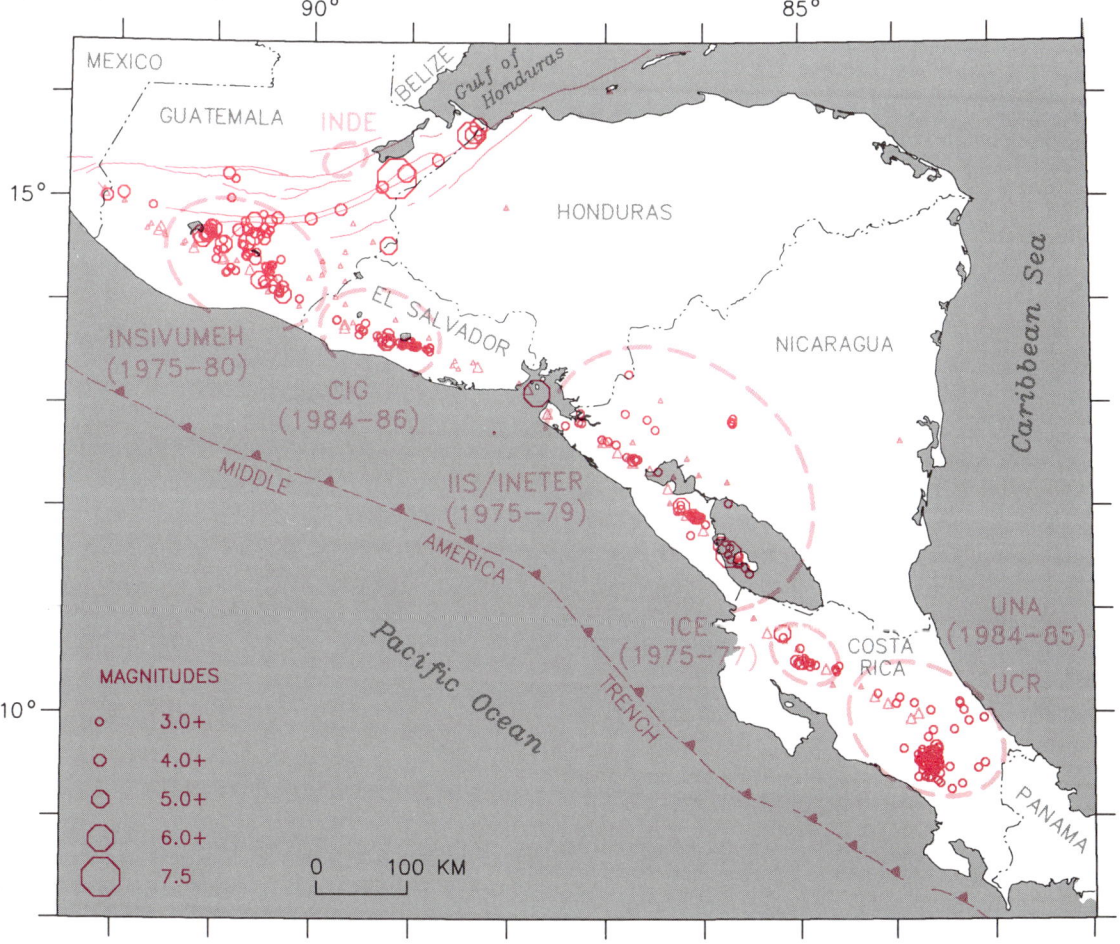

Figure 3. Sample of upper-crustal epicenters within Central America. Shown are A- and B-quality epicenters of M ≥3 from five different networks during the years indicated (see text for explanation of quality). Dashed ovals indicate approximate aperture for each network during the years shown. Note that the vast majority of the seismicity occurred within 10 km of the volcanic axis (Fig. 1); the absence of seismicity between networks is undoubtedly a result of poor network coverage in those areas. See Table 1 for definition of abbreviations.

within the upper-most 1 km beneath active craters, while A-type events originate from 1 to 20 km beneath volcanoes. A-type earthquakes have high-frequency waveforms that appear similar to shallow tectonic quakes; magnitudes range up to six, and the temporal pattern of accompanying seismicity is "swarm type" (largest shock embedded in sequence and not much larger than the largest fore- and aftershocks). B-type earthquakes have very small magnitudes and have waveforms with periods from 0.2 to 1.0 sec with unclear S-phase onset; the temporal pattern of accompanying seismicity is also swarm type. Explosion quakes accompany explosive eruptions, have longer dominant periods than the above, and usually contain a high-frequency signal, corresponding to the air-phase, embedded within the waveform. Volcanic tremor is longer period, similar to explosion-quakes, and is more or less continuous over minutes to hours.

During 1973 and 1974, Harlow (McNutt and Harlow, 1983) operated high-gain seismographs at several active volcanoes along the Central American volcanic chain. All four types of volcanic earthquakes were observed, the last three were weak signals detectable only in the immediate vicinity of erupting volcanoes. Explosion earthquakes were recorded during small explosive eruptions of Fuego volcano. Volcanic tremor with a dominant frequency of about 1 Hz was recorded during the eruptive stages of volcanoes in Guatemala and Nicaragua. McNutt and Harlow (1987) found that B-type earthquakes, with dominant frequencies of 2.5 to 4.5 Hz, occurred more frequently as volcanic activity increased. For B-type earthquakes, the slope of the magnitude-frequency relation (i.e., the "b-value", ranged from 1.7 to 2.9). Some B-type events with unusually long durations were recorded at Fuego and may represent an intermediate stage between B-type events and tremor. More recently, these same types of volcanic earthquakes have been reported at Arenal volcano by Alvarado and Barquero (1987).

A-type volcanic earthquake waveforms appear identical with those from tectonic events, but a comparison by Okada (1983) shows significant differences: (1) A-type seismicity tends to have a magnitude-frequency relation with higher b-values, from 1.3 to 1.8, that are less log-linear; and (2) the onset of seismicity tends to be more swarmlike than for tectonic seismicity, which generally occurs as main shock-aftershock sequences and has b-values from 0.6 to 1.2. If both A-type volcanic earthquakes and tectonic earthquakes result from rock fracture, the existence of subsurface magma in the vicinity of A-type earthquakes may explain the difference in accompanying seismic parameters. Yuan and others (1984) recognized that small swarms of A-type earthquakes accompanied several eruptions of Fuego volcano in 1975. A larger swarm of A-type events occurred during January 1977, however, when no volcanic activity was observed. During this swarm, seismicity levels were highest at regularly spaced intervals, b-values varied systematically between periods of high and low seismicity, and a large simultaneous tilt event was recorded that was too large to be due entirely to the seismic energy released during the swarm. From this, Yuan and others (1984) conclude that, although no eruption occurred, the swarm and tilt event represents a brittle response to subsurface magma movement.

Tectonic earthquakes are generally of larger amplitude and are much more widely distributed than volcanic earthquakes. For this reason, tectonic events are the most common type of earthquake recorded by Central American seismograph networks, and although a few A-type volcanic earthquakes are occasionally large enough to locate, the vast majority of hypocentral locations published in local microseismicity catalogs are for tectonic earthquakes.

MICROSEISMICITY AND FOCAL MECHANISMS

Microseismicity epicenters along Central America originate primarily from within two zones: (1) within the upper crust along and behind the volcanic arc (Fig. 3), and (2) along the Central American subduction zone. For earthquakes along and behind the volcanic arc, including earthquakes along the CARB-NOAM boundary, all high-quality hypocentral solutions (i.e., events well within network perimeters and that have at least eight azimuthally well distributed P-phase and four S-phase arrival times, each with a travel-time residual less than 0.4 sec) have depths of less than 20 km; therefore, all originated within the upper crust.

Between the megathrust zone and the volcanic zone, some seismic activity may occur along the coast of Costa Rica, where the megathrust zone is somewhat shallow dipping. Along the coast of Nicaragua, however, where the megathrust zone is more steeply inclined and therefore less tightly coupled, no events have yet been detected between the megathrust interface and the volcanic zone (Fig. 4).

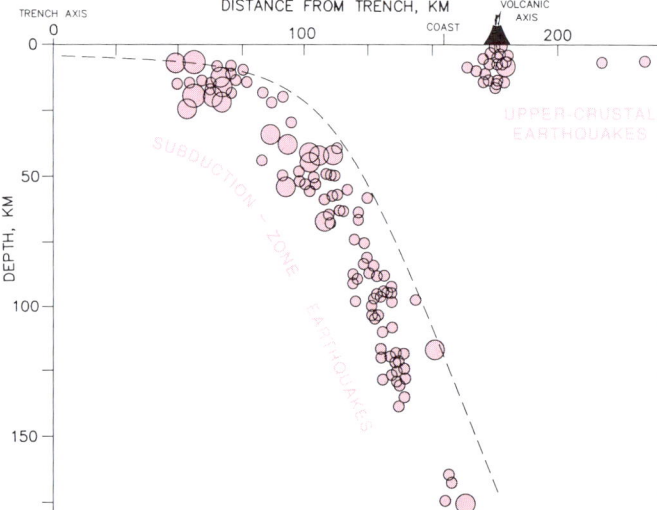

Figure 4. Cross section through Nicaragua of seismicity of the Central American subduction zone and upper crust (modified from Aburto, 1975). The volcano symbol indicates the axis of the volcanic front. The slab dip implied by locally recorded data may be exaggerated by refraction within the slab; the dashed line denotes the dip implied by teleseismically located shocks (Dewey and Algermissen, 1974; Burbach and others, 1984). Note that the volcanic axis lies approximately 150 km vertically above the upper surface of the down-going slab, about 50 km greater for most other arcs. Also note the total absence of seismicity between the down-going slab and the volcanic axis.

Approximately 10,000 local earthquakes of M ≥2 were recorded by the Nicaraguan IIS network from 1975 through 1982. Of these, approximately ten times as many earthquakes originated from the subduction zone as originated within the upper crust. From data in the catalog of the International Seismological Centre, approximately the same ratio applies along the entire length of Central America back to at least 1963 for earthquakes of M ≥4.5.

Costa Rica

The first relatively long-term study (mid-1976 through 1979) of microseismicity along the volcanic chain was the work of Montero and Dewey (1982). They investigated the seismicity around the southeastern limit of the volcanic arc using data from the UCR network. They show that the seismicity was essentially confined to the upper 20 km of crust, tended to occur near Quaternary volcanoes, and tended to occur in swarms.

Figure 5 shows examples of M ≥3 upper-crustal epicenters within Costa Rica. Epicenters west of 84.5° West longitude are from ICE network data from 1974 through 1976, while the epicenters to the southeast are from the UNA network during 1984 and 1985. The small cluster of events at 10.5°N, 85°W are aftershocks of the April 14, 1973, Tilarian (M_s 6.5) earthquake. The large cluster of epicenters to the southeast are aftershocks of the July 3, 1983 San Isidro (M_s 6.2) earthquake. The seismicity accompanying both the 1973 and 1983 events was typical for main shock-aftershock sequences.

Published focal mechanisms for Costa Rica (Table 1 and Fig. 5) all have generally east-west to southeast-northwest tension axes. Right-lateral slip along a northwest-trending vertical plane is indicated for mechanism C4 from the trend for aftershock epicenters and mapped faults in the area (W. Montero, personal communication, 1988). For composite mechanism C6, dip-slip motion along a nearly vertical northeast-trending plane is indicated from the strike of associated seismicity and from mapped faults at the surface, which essentially connect the two volcanoes Irazu, to the southwest, and Turrialba, to the northeast (Guendel, 1985). The preferred plane is unknown for the other four mechanisms.

Nicaragua

Temporary networks installed immediately after the December 23, 1972, M_s 6.2 Managua earthquake (Brown and others, 1973; Langer and others, 1974; Ward and others, 1974) show that the aftershocks occurred less than 8 km deep and delineate the main shock rupture zone along the Tiscapa fault. Left-lateral surface slip was observed along this fault, which strikes northeast and essentially spans a right-stepping offset in the chain. At least three adjacent and parallel faults also showed minor left-lateral slip at that time. A fourth showed features consistent with left-lateral slip during a similar earthquake in March 31, 1931 (Brown and others, 1973). Aftershock epicenters also delineate a north-striking secondary fault just west of the city.

Figure 6 shows epicentral determinations for events with M ≥2, from the permanent Nicaragua IIS network catalogs from 1975 through 1980. The upper-crustal seismicity is concentrated

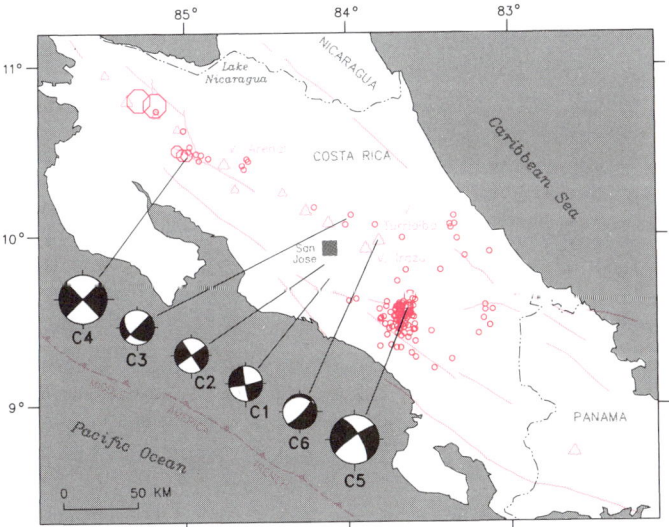

Figure 5. Upper-crustal microseismicity and focal mechanisms of Costa Rica. Epicenters east of 84.5°W are from UNA network catalogs for 1984 and 1985; epicenters west of 84.5°W are from the ICE network (Matumoto, 1978) for 1974 through 1976. Only quality A and B events of M ≥3 are shown (see text for explanation of quality). Focal mechanisms are equal-area lower-hemisphere projections; compressional quadrants are black. Mechanisms and preferred planes, if any, are discussed in the text. Large and small triangles indicate volcanoes with historical and Holocene volcanic activity, respectively (Simkin and others, 1981).

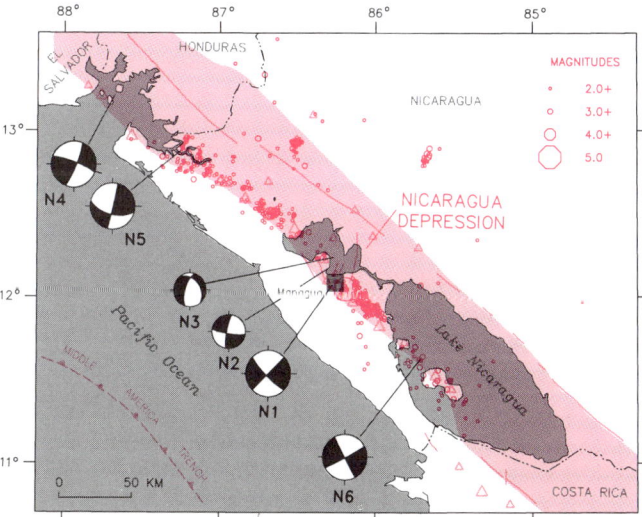

Figure 6. Upper-crustal microseismicity and focal mechanisms of Nicaragua. Epicenters were relocated by Harlow from data supplied by IIS, Nicaragua, for the period 1975 to 1980. Only quality A and B events of M ≥2 are shown (see text for explanation of quality). Focal mechanisms are equal-area lower-hemisphere projections; compressional quadrants are black. Mechanisms and preferred planes, if any, are discussed in the text. Large and small triangles indicate volcanoes with historical and Holocene volcanic activity, respectively (Simkin and others, 1981). Pink area indicates the Nicaragua Depression.

in a belt 10 to 20 km wide along the volcanic chain (the apparent spreading of seismicity near the Honduras and Costa Rica borders is probably due to larger epicentral errors at the periphery of the network). None of the seismic events illustrated in the Figure 6 coincided with volcanic eruptions. Although a few minor ash eruptions occurred at four different volcanoes during this time, volcanic seismicity was always too small to locate (i.e., of $M < 1$; A. Aburto Q., personal communication, 1988). Events shown in Figure 6 had depths between 0 and 20 km, with the vast majority between 5 and 15 km. These events occurred both singly and in swarms lasting a few days and consisting of tens to a few hundred events. Very few events located seaward of the volcanic chain. Landward of the chain, locatable seismicity was also minor and makes up mainly two clusters, at 13°N, 86°30'W and 12°50'N, 85°40'W, plus a few additional events along the eastern edge of the Nicaragua Depression.

Published focal mechanisms for Nicaragua (Table 1 and Fig. 6) have tension axes oriented from east-northeast to southeast. The interpretation for mechanism N1, for the December 23, 1972, Managua earthquake main shock (M_s 6.2), is left-lateral strike-slip motion along the northeast-trending Tiscapa fault, known from observed surface faulting and alignment of aftershock epicenters (Brown and others, 1973). Composite mechanisms for aftershocks by Langer and others (1974) and Brown and others (1973) are very similar to mechanism N1. Composite mechanisms N2 and N3 are for a north-striking alignment of epicenters; thus the more northerly striking planes are preferred. The preferred plane is unknown for mechanisms N4 to N6.

El Salvador

Recent microseismicity was located exclusively within a 15-km-wide band along the volcanic zone in El Salvador (Fig. 7). The M_s 5.4, October 10, 1986, San Salvador, earthquake (the largest event within El Salvador on Fig. 7) was followed by an aftershock sequence typical of main shock–aftershock sequences; the aftershocks occurred principally along a near-surface, vertically dipping, NNE-trending fault located directly beneath the city of San Salvador (White and others, 1987; Harlow and others, in preparation). Several aftershocks and possible surface faulting indicate a secondary fault zone just west of the city, striking northwest, subparallel with the volcanic zone (Harlow and others, in preparation). Data from the CIG strong-motion stations show that the main-shock rupture lasted about three seconds and produced anomalously large ground motions for the magnitude (Harlow and others, in preparation).

White and others (1987) examined local low-gain seismograph data since 1952 and concluded that upper-crustal seismicity within El Salvador has occurred almost exclusively within the volcanic zone back to at least 1952. Most of the seismicity since 1952 occurred in 20 swarms, each comprising ten to a few hundred events occurring over a few days, at eight different sites along the chain. Only one swarm, at San Miguel volcano in 1987, was related to observed eruptive activity.

Published focal mechanisms for El Salvador (Table 2 and Fig. 8) all show virtually pure strike slip, with one plane parallel with the volcanic arc. Mechanism S1 implies almost pure strike-slip motion with one nodal plane parallel to the volcanic arc. Mapped high intensity contours for the May 3, 1965, San Salvador earthquake (M_s 6.25) show a strong west-northwest elongation (Rosenbleuth and Prince, 1966) and an similar apparent alignment for locally recorded foreshocks (White and others, 1987), best support the west-northwest–striking plane as the preferred plane for mechanism S1. For mechanism S2, the northeast-striking plane is preferred from the alignment of aftershocks (Harlow and others, in preparation). For mechanism S3, the west-northwest–striking plane is the probable fault plane, based on the alignment of aftershocks and the orientation of mapped surface faults.

Guatemala

Except from 1976 through 1978, when aftershocks of the February 4, 1976, Guatemala earthquake dominated local seismicity (discussed further below), upper-crustal seismicity within Guatemala was concentrated primarily within 15 km of the volcanic chain southeast of Guatemala City (Fig. 7); swarms occurred at six different locations along the 110-km-long stretch from the El Salvador border to the center of the volcanic zone within Guatemala. White and others (1980) studied one particularly intense swarm that occurred during 1979 and 1980 at 14°N08', 90°W15' near the El Salvador border. The swarm included about 100,000 events of $M > 1$, with the largest earthquake (M 5.0) occurring about a month into the sequence, after which the seismicity level fluctuated for about a year before quieting down. High-quality epicenters define a fault zone that is north trending at the volcanic axis but curves and merges with the Jalpatagua fault, a right-lateral strike-slip fault (Carr, 1976) parallel with, and 15 km north of, the volcanic axis.

Compared with the rest of the volcanic front in Central America, seismic activity has been unusually low near the westernmost Guatemalan volcanoes. Only at volcano Tacana, at the Mexican border, has notable seismicity been reported (Quevec and Molina, 1986). There, main shock–aftershock sequences occurred in December 1985 and February 1986 (main shocks were of M 5.0) about 10 km to the east-southeast of the volcano; seismicity then migrated toward the summit crater and increased dramatically until a small phreatic explosion occurred, then decayed as for a normal main shock-aftershock sequence.

The February 4, 1976, Guatemala earthquake (M_w 7.5) was, by far, the largest upper-crustal earthquake within Central America this century, rupturing the Motagua fault for more than 240 km from near the Caribbean coast to west of Guatemala City and killing about 23,000 people (Espinosa, 1976). White and others (1984) analyzed locally recorded seismograms for the ten months prior to the main shock: they found less than 40 microearthquakes ($2 < M < 4.5$), which formed an oval pattern encompassing the impending rupture zone, plus a few events along the fault, including the two largest ($4 < M < 4.5$) within

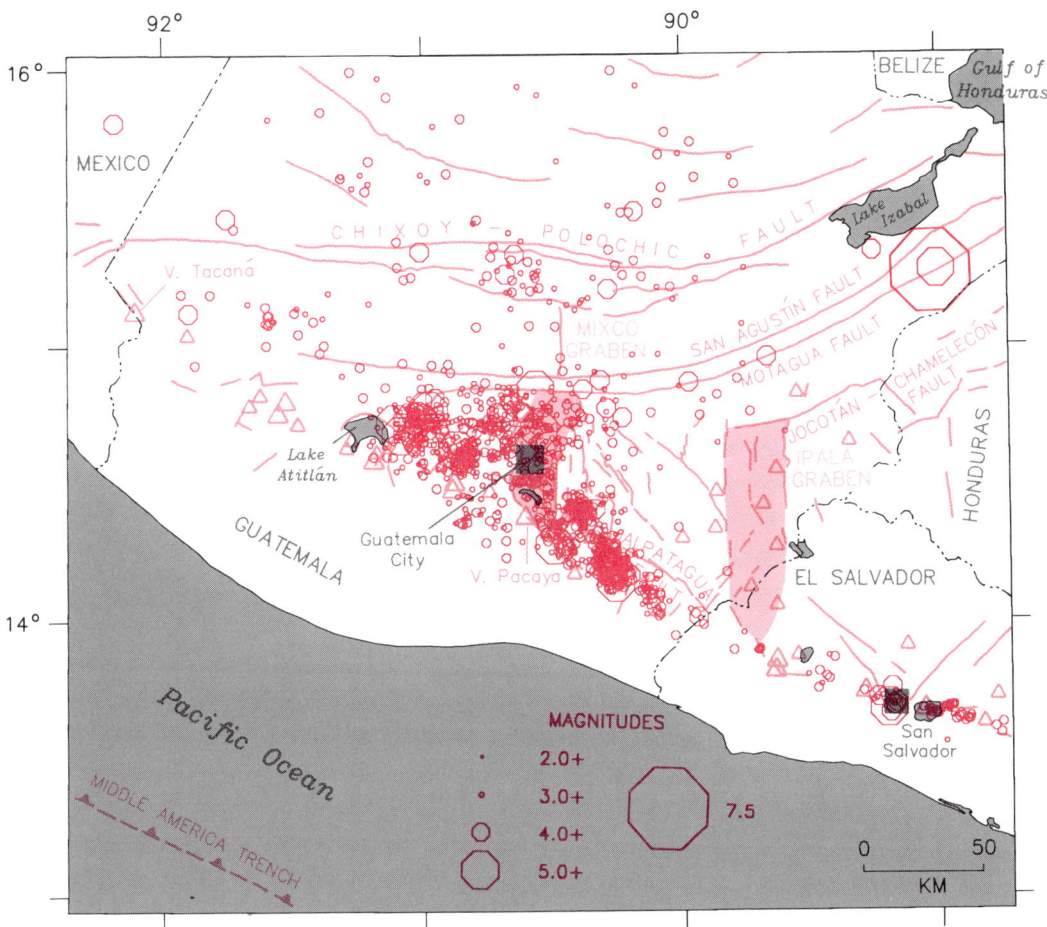

Figure 7. Upper-crustal microseismicity of El Salvador and southern Guatemala. Epicenters for El Salvador are from the CIG catalogs for 1984 through 1986; epicenters for Guatemala are for 1975 through 1980. Data are from Harlow and White (1980), White and Harlow (1979, 1980), and INSIVUMEH annual catalogs. Only quality A and B events of M ≥2.5 are shown (see text for explanation of quality). Large and small triangles indicate volcanoes with historical and Holocene volcanic activity, respectively (Simkin and others, 1981). Faults are from Carr (1976) and Case and Holcombe (1980). Pink areas indicate the Mixco and Ipala grabens.

10 km of the main shock epicenter. White and others (1984) found nothing in the data, however, that would strongly have suggested that a destructive earthquake was imminent.

The 1976 main shock produced 3 m of pure left-lateral surface faulting (Plafker, 1976). The rupture apparently initiated near Lake Izabal and propagated principally westward. Kanamori and Stewart (1978) show that the main shock was complex, essentially composed of several smaller events, some possibly along splay faults to the south. Thousands of aftershocks recorded by the INSIVUMEH network were analyzed by White and Harlow (1979, 1980). The epicenters are distributed along several eastward-dipping faults that splay to the southwest from the western end of the Motagua fault at and west of Guatemala City. The epicenters and faults terminate abruptly against the volcanic chaing (Fig. 7). Data in Langer and Bollinger (1979) suggest that the aftershocks were concentrated within the western half of the rupture zone and show that the splay pattern of the faults there is predicted from the theoretical stress trajectories at the terminus of strike-slip faults.

Although aftershocks of the 1976 earthquake decreased normally with time, they dominated the upper-crustal seismicity of Guatemala through 1978. Destructive, somewhat late, aftershocks occurred on July 29, 1978 (M_s 5.0), August 9, 1980 (M_s 6.5), and September 29, 1982 (M_s 5.4): these aftershocks occurred at the western and eastern ends of the rupture zone and off its center, respectively, at locations where aftershocks would be expected to concentrate due to shear-stress increase after a main shock, according to Chinnery (1963) and Das and Scholz (1981).

Seismicity east of Guatemala City, within the wedge area between the volcanic zone and the Motagua fault, has generally been very low. There is no clear evidence for seismicity along the Jocotan-Chamelecon fault, for example. Although Harlow

TABLE 2. FOCAL MECHANISM PARAMETERS

Id*	Year	Month	Day	Hour	Minute	Latitude North	Longitude West	Deep§ (km)	Mag†	Dir	Dp	Rak	Reference‡
C1	77	01	00	C		9°43.50	84°06.65	10.0	C	236	90	180	1
C2	77	12	00	C		9°48.00	84°06.65	10.0	C	260	86	-174	1
C3	78	01	00	C		10°02.40	84°00.00	10.0	C	226	45	180	1
C4	73	04	14	C		10°27.00	84°54.20	10.0	6.5	226	90	-180	2
C5	83	07	03	17	14	9°39.00	83°40.00	10.0	6.2	325	80	-20	2
C6	82	09	00	C		10°00.00	83°49.80	5.0	C	136	78	-105	2
N1	72	12	23	06	30	12°09.00	86°16.20	5.0	6.2	135	90	-10	3
N2	73	01	00	C		12°12.00	86°16.00	5.0	C	280	80	-170	4
N3	72	12	28	C		12°12.00	86°18.00	5.0	C	66	66	-120	5
N4	82	01	12	05	48	13°07.36	87°38.34	3.0	6.0	21	81	171	6
N5	84	08	31	04	42	12°50.80	87°06.20	5.0	5.4	103	88	-30	6
N6	85	12	16	02	44	11°43.80	85°50.40	5.0	6.0	152	87	-2	7
S1	65	05	03	10	01	13°43.20	89°07.20	10.0	6.25	25	80	170	8
S2	86	10	10	17	49	13°08.00	89°20.40	11.0	5.4	302	78	8	9
S3	86	10	13	22	24	13°15.00	89°40.00	5.0	4.6	25	80	180	9
G1	79	10	09	07	49	14°08.05	90°15.06	8.5	5.0	261	65	-84	10
G2	76	02	04	09	01	15°09.50	89°11.00	10.0	7.5	155	89	-9	11
G3	76	02	20	C		14°57.00	89°39.00	10.0	C	152	90	10	12
G4	76	02	20	C		14°50.00	90°42.00	10.0	C	354	80	-20	12
G5	76	02	20	C		14°48.00	90°58.00	10.0	C	142	70	-75	12
G6	76	02	20	C		14°40.00	90°47.00	10.0	C	100	50	-104	12
G7	76	02	20	C		14°30.00	90°37.00	10.0	C	117	38	-54	12
G8	76	02	06	18	19	14°45.00	90°33.00	10.0	5.7	313	67	-26	13
G9	78	08	16	16	35	15°16.80	90°16.20	18.0	2.9	170	90	0	14
G10	78	09	15	07	58	15°18.00	89°52.80	6.0	1.7	170	90	0	14
G11	78	09	20	C		15°55.80	90°48.00	4.0	C	126	68	-90	14
G12	71	10	12	09	44	15°42.60	91°04.60	10.0	5.7	14	86	11	15

*Id = identification number of events discussed in text or on Figures 5, 6, and 8.
†C in Time and Mag columns indicates that this is a composite focal mechanism for several events on or about the date shown; Mag = the larger of M_s or m_b for events of Mag ≥5.0; for smaller events it may be either local amplitude or duration magnitude.
§Deep = depth in km (approximate for composite events).
**Focal plane = three parameters for either focal plane as follows: Dir = down-dip direction (i.e., perpendicular to the strike direction); Dp = dip angle of the plane; Rak = the angle between the strike direction and the slip vector, where 0 = left-lateral faulting, ±180 = right-lateral faulting, 90 = reverse faulting, and -90 = normal faulting. Events used for composite mechanisms generally have magnitudes less than 3 and, because first motions are less certain for events of this size, composite mechanisms should be considered less certain.
‡1 = Montero and Dewey (1982); 2 = Guendel (1988); 3 = Algermissen and others (1974); 4 = Brown and others (1973); 5 = Langer and others (1974); 6 = PDE's (CMT solution); 7 = Needham (1986); 8 = Molnar and Sykes (1969); 9 = Harlow and others (in preparation); 10 = White and others (1980); 11 = Dewey and Julian (1976); 12 = Langer and Bollinger (1979); 13 = Burbach and others (1984); 14 = Woodward-Clyde (1979); 15 = Dean and Drake (1978).

(1976) recorded seismicity from the immediate vicinity of the Jocotan-Chamelecon fault (near and parallel to the Honduras-Guatemala border), the seismicity may equally well have originated along north-trending normal faults there, which show evidence of recent faulting (Schwartz, personal communication, 1988).

Just 3 to 4 km north of the Montagua fault lies the San Agustin fault. Microseismicity has not yet been clearly associated with this fault. Epicenters in INSIVUMEH catalogs that appear to lie along the San Agustin fault may have originated along the Motagua fault.

Harlow (1976) first recorded microseismicity along the Chixoy-Polochic fault in 1973. Later, Woodward-Clyde Consultants (1979) operated two small-aperture networks just north of the fault and recorded minor seismicity along and north of the fault. The largest well-recorded events along the fault have been of about magnitude 5. Earthquakes with magnitudes between 4 and 5 have been detected along faults located between the Motagua and Chixoy-Polochic faults and also along unnamed faults north of the latter (INSIVUMEH catalogs), indicating that these faults are also active to some degree.

Fault plane solution G1 is the only published mechanism for

near the CARB-NOAM boundary. With but one exception, the largest earthquakes at 20 separate locations along the volcanic zone were all of M_s 6 to 6.5. The lone exception occurred in southeastern Guatemala on July 14, 1930, and measured M_s 6.9. This event occurred within the longest stretch (120 km) of the volcanic chain with no evidence of historical eruptive activity (White and Harlow, in preparation). White and Harlow (in preparation) find that maximum magnitude of volcanic-zone events at particular locations may be related to the distance between adjacent centers of historical eruption.

White and Harlow (in preparation) also compiled a list of the dead and injured for destructive volcanic-chain events in Central America. Since 1900, volcanic-zone earthquakes of M_s 5.4 to 6.9 have killed a total of about 17,000 people. Even though subduction-zone events of that size are roughly an order of magnitude more frequent along the Pacific coast of Central America, and the largest events there have magnitudes approaching M_s 8, only about 200 people are known to have died from subduction-zone events since 1900 (Rinehart and others, 1982). Although the above casualty figures may be skewed by the paucity of recent large subduction-zone events (none of $M_s > 7.4$ since 1950), the severity of the seismic hazard from moderate upper-crustal volcanic zone earthquakes is nevertheless readily apparent.

Recurrence intervals for volcanic-chain earthquakes are difficult to determine because historical data from prior to the twentieth century are meager for all but the largest population centers. Furthermore, specific causal faults are difficult to identify. Quaternary volcanic deposits are thick along most of the chain, and very few events rupture the surface. When earthquakes ruptured the surface at Managua during 1931 and 1972, active faults were found to be very closely spaced—five faults were found within 4 km (Brown and others, 1973).

For one location along the volcanic zone, the city of San Salvador, historic documents indicate that nine upper-crustal earthquakes ($5.6 \leq M \leq 6.5$), from at least four sources within or adjacent to the city, have badly damaged San Salvador since 1700 (Harlow and others, in preparation); this is an average rate of about once every 30 years. Historical data are moderately complete for three additional locations along the volcanic zone: Amatitlan, Guatemala (Diaz, 1931); San Vicente, El Salvador (Larde, 1960); and Managua, Nicaragua (Leeds, 1974); these data indicate that serious damage from upper-crustal earthquakes occurs at these localities every 30 to 60 years.

The CARB-NOAM plate boundary

Destructive historical earthquakes have occurred over an area about 120 km wide along the plate boundary through Guatemala and northwestern Honduras and illustrate the diffuse nature of the plate boundary there (White, 1982). White (1984) catalogued 25 destructive historical earthquakes, most previously unknown, along the CARB-NOAM boundary since 1530 (Fig. 9 shows the largest of these). The largest ($M \geq 7.25$) earthquakes occurred along the Chixoy-Polochic and Motagua faults, with

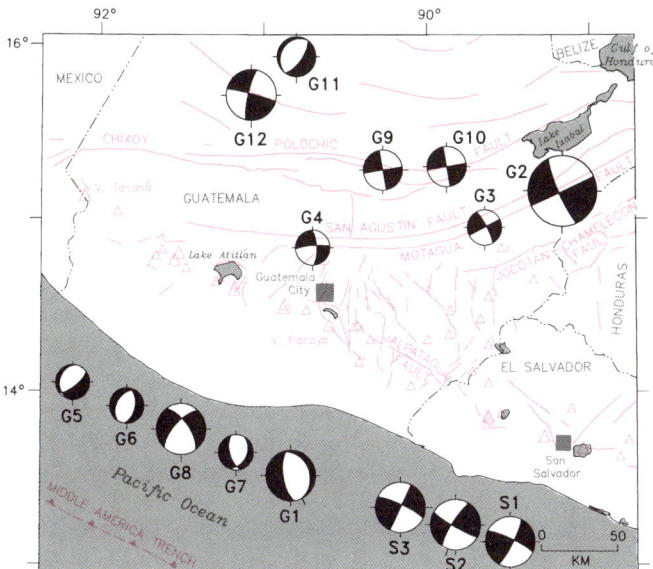

Figure 8. Focal mechanisms for El Salvador and Guatemala. Focal mechanisms are equal-area lower-hemisphere projections; compressional quadrants are black. Mechanisms and preferred planes are discussed in the text. Large and small triangles indicate volcanoes with historical and Holocene volcanic activity, respectively (Simkin and others, 1981). Faults are from Carr (1976) and Case and Holcombe (1980). Pink areas indicate the Mixco and Ipala grabens.

the volcanic zone in Guatemala (Table 2 and Fig. 8); from the accompanying seismicity, the steeply westward-dipping plane is preferred. From observed surface rupture at the location of solutions G2 to G4, and from mapped faults at G9, G10, and G12, the preferred mechanism for each is almost pure left-lateral strike-slip faulting along the more easterly striking plane. Solutions G5 to G8 are for secondary faulting after the February 4, 1976, earthquake. The eastward-dipping plane is preferred for mechanism G5, from a mapped fault at that location. For mechanism G6, the preferred plane is unknown. The preferred fault planes for G7 and G8 are the east-dipping and the northeast-striking planes, respectively, based on the orientation of the mapped Mixco fault located there. Solutions G9 and G10 are for events along the east-striking Chixoy-Polochic fault, indicating pure left-lateral slip along the fault. Solution G11 indicates normal faulting along a northeast-trending fault, but whether the fault plane dips to the northwest or southeast is unknown.

HISTORICAL UPPER-CRUSTAL EARTHQUAKES OF CENTRAL AMERICA

The volcanic zone

The spatial pattern of destructive upper-crustal earthquakes since 1900 (Fig. 9) is virtually identical with the pattern of recent microseismicity (Fig. 3): most events occurred within 10 km of the volcanic axis, and most of the remainder occurred along or

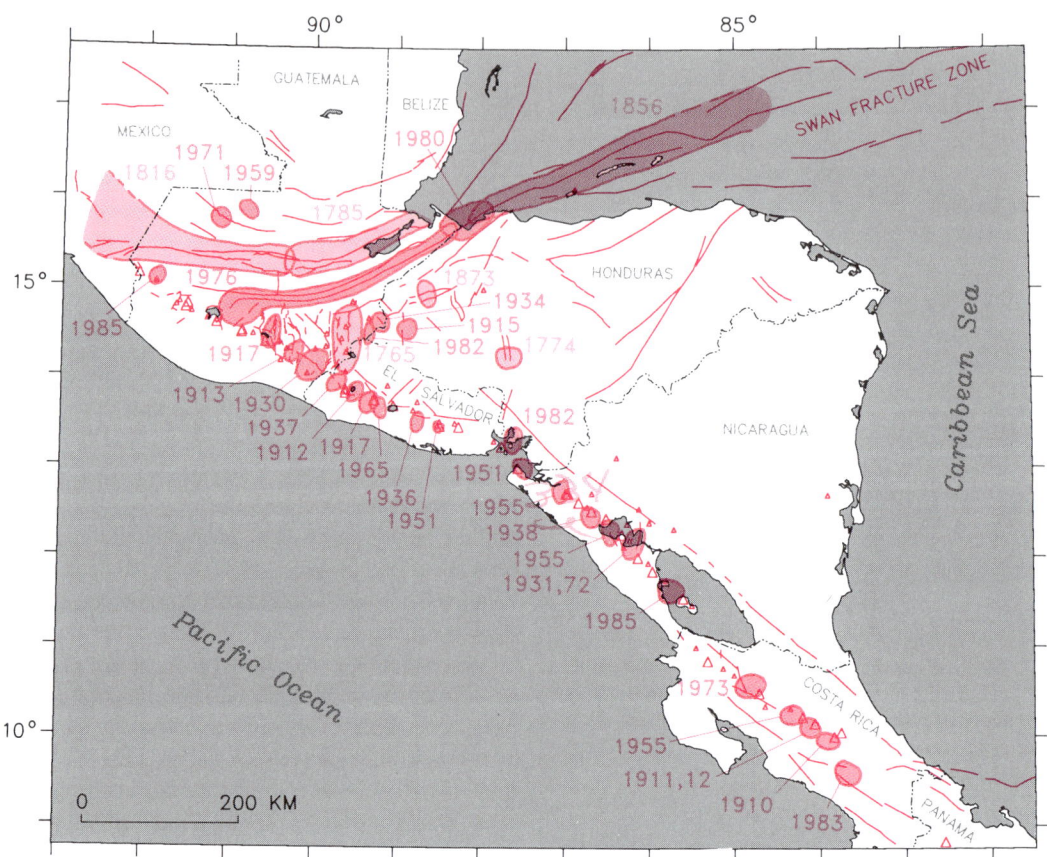

Figure 9. Destructive historical earthquakes (M ≥5.7) of Central America. Enclosed areas indicate the approximate areas of Modified Mercalli intensity VII and greater damage from earthquakes (a few intensity contours are extended over water to indicate probable extent of strong shaking). Red areas indicate earthquakes of this century (from White and Harlow, in preparation). Pink areas indicate larger historical earthquakes from 1700 to 1900 (from White, 1984).

smaller (6 ≤ M ≤ 6.2) earthquakes occurring along faults just north of the Chixoy-Polochic fault and between it and the Motagua fault (White, 1984). The catalog begins with what may be a major earthquake along the Chixoy-Polochic fault in 1538, followed by apparent quiescence along the zone through the 1600's. The late eighteenth and early nineteenth centuries represent an active period of at least 14 damaging earthquakes that culminated in the two-stage rupture of the entire Chixoy-Polochic fault, with the eastern half rupturing on January 6, 1785 (M 7.5), and the western half on July 22, 1816 (M 7.5 to 7.75). After five years of damaging aftershocks, quiescence set in, lasting until about 1930. An active period with three damaging earthquakes on August 10, 1945, February 20, 1959, and October 12, 1971, may have culminated with the February 4, 1976, rupture (M_s 7.5) of the Motagua fault. This active period may have ended after damaging aftershocks on July 29, 1978, August 9, 1980, and September 29, 1982.

The historical record shows no major historical rupture of the Motagua fault prior to 1976. Schwartz (1985) found geologic evidence, from trenching along the Motagua fault, for a major rupture between A.D. 1250 and 1550. From trenching along the Chixoy-Polochic fault, he found supporting evidence for a major earthquake near 1538. From this and the data in the catalog of White (1984), the average recurrence interval between major through-going ruptures along the CARB-NOAM boundary is about 225 ± 50a, whether calculated from 1250 or from 1538.

The area between the Motagua fault and the volcanic chain

Historical earthquakes have occurred along several faults within the wedge-shaped area within Guatemala (Fig. 9). In December 1917 and January 1918, a series of five earthquakes with magnitudes between 5.5 and 6.25 probably ruptured the entire length of the north-trending Santa Catarina fault just east of Guatemala City, from the volcanic chain to the Motagua fault (White and Harlow, in preparation). Farther east, large earthquakes apparently ruptured north-trending faults of the Ipala graben in May 1733 and on June 2, 1765. Moment magnitudes, estimated from the areal extent of damage by White (1984), are M 7.25 to M 7.5 for the 1765 earthquake, which apparently ruptured from the Jocotan-Chamelecon fault south to the El Salvador border, and M 7.25 ± 0.25 for the 1733 earthquake (located within the damage area for the 1765 earthquake in Fig. 9).

Along the Honduras-Guatemala border, earthquakes of M_S 6.25 and 5.4 ruptured northwest-trending faults on February 3, 1934, and September 29, 1982, respectively (White and Harlow, in preparation). Although no focal mechanism could be produced for the 1982 event, 10 cm of vertical slip was observed for about 9 km along an unnamed north-northwest–trending fault (E. Sanchez B., personal communication, 1982).

East of this, earthquakes may have ruptured other northwest-trending faults on March 9, 1733, during October 1873 to May 1874, and on December 29, 1915 (M_S 6.3; Grases, 1975; White, 1984). A fault or faults along the north-northwest–trending valley near Comayagua apparently ruptured on March 22, 1610, October 14, 1774, and in early 1809 (Grases, 1975). East of the Honduras Depression, both microseismicity and historical seismicity levels are low.

Historical versus predicted slip rates along CARB-NOAM boundary

Assuming that slip per event along the Motagua and Chixoy-Polochic faults is about 3 m (both the estimated slip for the July 22, 1816 earthquake [White, 1985], and the maximum slip observed for the February 4, 1976, earthquake [Plafker, 1976]), the estimated historical slip rate for the two faults combined is about 13 mm/yr. Though no large historical events are known along faults subparallel with the Motagua and Chixoy-Polochic faults, destructive moderate events such as the February 20, 1971, earthquake indicate that these faults may contribute another few mm/yr of slip over the long term, between the Caribbean and North American Plates. Within the wedge-shaped area of crustal extension between the volcanic zone and the CARB-NOAM boundary, the historical earthquakes during May 1733 (M 7.25) and June 2, 1765, (M 7.25 to 7.5) probably contribute about 4 to 8 mm/yr of additional slip between the plates there. The net historical slip rate is therefore estimated to be about 17 to 21 mm/yr.

Consider four models for the relative motion between the Caribbean and North American Plates, by ascending rate: model NUVEL-1 of Demets and others (1990; 12 mm/yr), model RM2 of Minster and Jordan (1978; 20 mm/yr), the model of Figure 1 (27 mm/yr), and the model of Sykes and others (1982; 34 mm/yr). The historic rate is closest to the value predicted by model RM2 of Minster and Jordan (1978). Though the historic rate does not rule out any of the models, it probably argues for a model with a rate of at least 20 mm/yr due to the distributed nature of the plate boundary in this area. In northern California, for example, where the transcurrent Pacific–North American plate boundary likewise approaches a triple junction, the San Andreas fault alone accounts for only 50 to 75 percent of the observed relative plate motion, with considerable relative motion occurring hundreds of kilometers inland (Lisowski and others, 1991).

DISCUSSION OF VOLCANIC-ZONE SEISMICITY

Although volcanic-type earthquakes indicative of magma movement have been reported along the Central American volcanic zone (McNutt and Harlow, 1983; Yuan and others, 1984; Alvarado and Barquero, 1987), tectonic-type main shock–aftershock sequences with main shocks as large as M_S 6.5 have also occurred recently (White and Harlow, in preparation). Those events that have been well studied were clearly tectonic in nature (Brown and others, 1973; Ward and others, 1974; White and others, 1987; Harlow and others, in preparation). Okada (1983) finds that the largest observed and expected earthquakes produced by the movement of magma are no greater than about magnitude 5.5, which suggests that many events, including the largest, along the Central American volcanic zone result not from the movement of magma, but from regional tectonic stresses.

In fact two distinct types of temporal energy release patterns accompany large earthquake sequences along the volcanic chain: main shock–aftershock sequences, and a type that seems to be hybrid between these and volcanic swarm–type sequences. Main shock–aftershock sequences are characterized by having a main shock that is more the one magnitude unit larger than any fore- or aftershock and which occurs at or near the beginning of the sequence. Such main shock–aftershock sequences were those that were associated with the M_S 6.2 December 23, 1972, Managua earthquake (Algermissen and others, 1974); the M_S 6.5, April 14, 1973, Tilaran earthqake (Plafker, 1973); and the M_S 5.4, October 10, 1986, San Salvador earthquake (White and others, 1987). Hybrid sequences appear similar to main shock–aftershock sequences but have extensive foreshocks and more slowly decaying aftershock sequences. Such hybrid sequences accompanied the M_S 6.0 May 3, 1965, San Salvador earthquake (Lomnitz and Schulz, 1966) and also were recorded from southeastern Guatemala in 1979 (White and others, 1980). From the local catalogs, smaller hybrid sequences, each with at least 50 earthquakes of M >2 recorded in a three-day period from within a 3-km radius, apparently have occurred at the rate of roughly 1 sequence/yr per 100 km of volcanic arc. Local low-gain seismograph data indicate that a similar rate applies in El Salvador back to at least 1952 (White and others, 1987). The hybrid pattern suggests some interplay between tectonic forces and the special physical properties usually present along volcanic zones, such as the highly fractured and heterogeneous medium, the nonuniform stress field, high heat flow, and the presence or proximity of magma.

Of the focal mechanisms along the volcanic arc, all have tension axes that are essentially horizontal and strike more or less east-west. Tension axes strike more to the east-northeast from Guatemala into northern Nicaragua, strike more directly to the east from Central Nicaragua through northern Costa Rica, and show a mix of easterly and southeasterly strikes in central Costa Rica. Preferred fault planes for solutions N1 and S2 indicate left-lateral strike-slip faulting along faults perpendicular with the volcanic arc, while those for C4, S3, and possibly S1 indicate right-lateral stirke-slip faulting along faults subparallel with the arc. Solutions N3 and G1 indicate normal faulting along north-striking faults.

The fact that all of the focal mechanisms along the volcanic zone show horizontal tension axes is typical for volcanic arcs above convergent margins (Nakamura and Uyeda, 1980). North-striking tensional geologic features, such as rifts and alignments of cinder cones have been noted in Nicaragua by McBirney and Williams (1965) and demonstrate that east-west tension probably persisted throughout the Holocene. McBirney and Williams also note that these geologic features suggest strike-slip faulting along the volcanic zone. Indeed, although the only well-observed surface faulting along the volcanic zone, for the Managua earthquakes of March 31, 1931, and December 23, 1972, was oriented perpendicular to the zone, all of the focal mechanisms along the volcanic zone imply a stress field that is also compatible with transcurrent faulting along the zone.

THE FORE-ARC SLIVER MODEL

Harlow and White (1985) suggest that focal mechanisms and geologic features along the volcanic zone are compatible with the interpretation of the volcanic zone as a leaky right-lateral shear zone. They suggest that the true relative plate motion vector between the Caribbean and Cocos Plates is slighly oblique and physically decoupled into two components: a large, well-known normal component, manifested as convergence along the Middle America Trench; and a small, previously unrecognized transverse component, manifested as right-lateral shear along the volcanic zone. Thus, the portion of the over-riding plate between the trench and the volcanic chain might be viewed as a microplate. A model for such decoupling during oblique convergence was first proposed by Fitch (1972) for southeast Asia and the western Pacific. Later, Jarrard (1986) termed the microplate portion of the overriding plate a "fore-arc sliver" and recognized them at several other convergent plate margins.

The volcanic zone as leaky shear zone helps explain apparent decoupling across the volcanic axis in Guatemala as evidenced by: (1) major grabens of the wedge area are bounded on the south by the volcanic chain (Plafker, 1976); and (2) epicenters of some 3,000 aftershocks of the February 3, 1976, earthquake, which were located near and west of Guatemala City (White and Harlow, 1979, 1980), are sharply bounded on the south by the volcanic chain.

White and others (1989) find evidence suggesting that the level of eruptive activity at any particular volcano along the zone varies with the cycle of very large to great earthquakes along the megathrust zone offshore. They propose a simple model wherein compressive stress across the fore-arc sliver, and therefore the volcanic zone, decreases immediately after a great megathrust earthquake, which leads to generally increased eruptive activity. As the compressive stress gradually increases during the rest of the cycle, eruptive activity decreases. For example, Santa Maria Volcano produced its first and largest historical eruption (Simkin and others, 1981) about six months after the largest megathrust event (M_s 7.9) in more than 220 years occurred beneath the area on April 18, 1902 (White and Cifuentes, in preparation).

By similar reasoning, an episode of significant left-lateral slip along the western end of the CARB-NOAM plate boundary should also be expected to temporarily decrease compressive stress across the volcanic zone and lead to increased eruptive activity locally. The following suggests that this is, in fact, the case: (1) since the February 4, 1976, rupture of the Motagua fault, Pacaya Volcano, located just south of Guatemala City, has become one of the most active volcanoes in the world (McClelland and others, 1989); and (2) shortly after the only documented historical rupture of the western Chixoy-Polochic fault in 1816 (White, 1985), the first historical eruptions of volcanoes Tajumulco and Atitlan were reported (Simkin and others, 1981).

CONCLUSIONS

1. All reliable hypocenters within Central America that originate within the Caribbean Plate or along the CARB-NOAM boundary have depths of less than about 20 km.

2. A dominant feature of the upper-crustal microseismicity pattern is the concentration of seismicity in a nearly continuous belt less than 20 km wide along the axis of principal Quaternary volcanoes; that is, along the volcanic front.

3. Two distinct patterns of temporal tectonic energy release are found along the volcanic front: main shock–aftershock sequences, and sequences that are hybrid between main shock–aftershock sequences and volcanic swarm–type sequences.

4. Along the volcanic front, all fault-plane solutions show approximately east-west extension. There is evidence for left-lateral strike-slip faulting along faults striking perpendicular to the volcanic axis and for right-lateral slip along arc-parallel faults.

5. Historical earthquakes of $M_s \geq 6$ have occurred at 20 distinct locations along the volcanic front since 1900; only three can be said to have been accompanied by observed eruptive activity at the nearest volcano. These earthquakes are all larger than the largest earthquakes expected from magma movement, thus they probably result from tectonic rather than from volcanic stresses.

6. The largest known historical earthquakes along the volcanic front have magnitudes of M_s 6 to 6.5 everywhere except southeastern Guatemala, where magnitudes reach M_s 6.9. The maximum magnitude at any particular location along the volcanic chain is apparently related to the distance between adjacent, historically active volcanoes.

7. At the transcurrent CARB-NOAM plate boundary, instrumentally recorded seismicity of magnitude 4 or greater has been detected along the Motagua and Chixoy-Polochic faults, along faults located between them, and along unnamed faults just north of the Chixoy-Polochic fault. Historical earthquakes of magnitude 6 or greater have also occurred in the vicinity of these faults. Seismicity has yet to be clearly associated with the San Agustin fault and Jocotan-Chamelecon faults.

8. Most of the slip along the transcurrent CARB-NOAM plate boundary in Central America is accommodated along the

Motagua and Chixoy-Polochic faults. Since 1610, major ruptures have occurred along each of these faults only once: the Motagua fault in 1976 (M_w 7.5); and the Chixoy-Polochic in two parts, the eastern portion in 1785 (M 7.25 to 7.5), and the western portion in 1816 (M ~7.5).

9. Within the wedge-shaped area of extensional tectonics, bounded by the volcanic zone on the southwest, the Motagua fault on the north, and the Honduras Depression on the east, damaging earthquakes have occurred infrequently along several different normal faults. Magnitudes for the two largest historical events, near the junction of the Guatemala, El Salvador, and Honduras borders, are about 7.25 (1733) and 7.5 (1765). Elsewhere within the extensional area, including all of Honduras, reliable magnitudes of historical earthquakes are no larger than about magnitude 6.5.

10. Estimates of historical coseismic slip since 1538 along the Motagua and Chixoy-Polochic faults account for a relative slip rate between the Caribbean and North American Plates of about 13 mm/yr. Slip along other faults subparallel with the CARB-NOAM boundary may contribute another few mm/yr of slip between the plates. Extensional faulting within the wedge-shaped area between the volcanic zone and the CARB-NOAM plate boundary probably contributes another 4 to 8 mm/yr of slip between the plates. The estimated net historical slip is about 17 to 22 mm/yr across the plate boundary in this region.

11. Along the volcanic front, spatial and temporal seismicity patterns, focal mechanisms, and geology are compatible with the interpretation of the volcanic zone as a leaky right-lateral shear zone, driven by a small but non-negligible oblique component of convergence between the Caribbean and Cocos Plates. Modelling the region between the volcanic zone and the subduction zone as a microplate or "fore-arc sliver" can explain apparent modulation of volcanic and seismic activity along the volcanic zone by cycles of M ≥7.5 earthquakes along adjacent sections of both the megathrust zone and the CARB-NOAM plate boundary.

REFERENCES CITED

Aburto Q., A., 1975, Reporte de los temblores ocurridos en Nicaragua: Nanagua, Nicaragua, Boletín del Instituto de Investigaciones Sismicas, v. 1, 51 p.

Algermissen, S. T., Dewey, J W., Dillinger, W. H., and Langer, C. J., 1974, The Managua, Nicaragua, earthquake of December 13, 1972; Location, focal mechanism, and intensity distribution: Bulletin of the Seismological Society of America, v. 64, p. 993–1003.

Alvarado, G., and Barquero, R., 1987, Las senales sismicas del Volcán Arenal (Costa Rica) y su relación con las fases eruptivas (1968–1986): Ciencia y Technología, v. 11, p. 15–35.

Ben-Avraham, Z., and Nur, A., 1980, The elevation of volcanoes and their edifice heights at subduction zones: Journal of Geophysical Research, v. 85, p. 4325–4335.

Brown, R. D., Jr., Plafker, G., and Ward, P. L., 1973, Geologic and seismologic aspects of the Managua, Nicaragua, earthquakes of December 23, 1972: U.S. Geological Survey Professional Paper 838, 34 p.

Burbach, G. V., Frohlich, C., Pennington, W. D., and Matumoto, T., 1984, Seismicity and tectonics of the subducted Cocos Plate: Journal of Geophysical Research, v. 89, p. 7719–7735.

Burkart, B., 1978, Offset across the Polochic fault of Guatemala and Chiapas, Mexico: Geology, v. 6, p. 328–332.

Carr, M. J., 1976, Underthrusting and Quaternary faulting in northern Central America: Geological Society of America Bulletin, v. 87, p. 825–829.

Carr, M. J., Rose, W. I., and Stoiber, R. E., 1982, Regional distribution and character of active andesitic volcanism, Central America, in Thorpe, R. S., ed., Andesites; Orogenic andesites and related rocks: New York, John Wiley, p. 149–166.

Case, J. E., and Holcombe, T. L., 1980, Geologic-tectonic map of the Caribbean region: U.S. Geological Survey Miscellaneous Investigations Map I-1100, 2 sheets, scale 1:2,500,000.

Chinnery, M. A., 1963, The stress changes that accompany strike-slip faulting: Bulletin of the Seismological Society of America, v. 53, p. 921–932.

Das, S., and Scholz, C. H., 1981, Off-fault aftershock clusters caused by shear stress increase?: Bulletin of the Seismological Society of America, v. 71, p. 1669–1675.

Dean, B. W., and Drake, C. L., 1978, Focal mechanism solutions and tectonics of the Middle America Arc: Journal of Geology, v. 86, p. 111–128.

DeMets, C., Gordon, R. G., Argus, D. F., and Stein, S., 1990, Current plate motions: Geophysical Journal International, v. 101, p. 425–478.

Dewey, J. W., and Algermissen, S. T., 1974, Seismicity of the Middle America arc-trench system near Managua, Nicaragua: Bulletin of the Seismological Society of America, v. 64, p. 1033–1048.

Dewey, J. W., and Julian, B. R., 1976, Main event source parameters from teleseismic data, in Espinossa, A. F., ed., The Guatemala earthquake of February 4, 1976; A preliminary report: U.S. Geological Survey Professional Paper 1002, 90 p.

Diaz, V. M., 1931, Conmociones terrestres en la America Central 1469–1930: Guatemala City, El Santuario, 268 p.

Espinosa, A. F., ed., 1976, The Guatemala earthquake of February 4, 1976; A preliminary report: U.S. Geological Survey Professional Paper 1002, 90 p.

Fitch, T. J., 1972, Plate convergence, transcurrent faulting, and internal deformation adjacent to southeast Asia and the western Pacific: Journal of Geophysical Research, v. 77, p. 4432–4460.

Grases, J., 1975, Sismicidad de la region asociada a la cadena volcanica Centroamericana del cuaternario, v. 2: Caracas, Universidad Central de Venezuela and the Organization of American States, 253 p.

Guendel, F., 1985, Earthquake swarms at Irazu volcano, in Catalogo de Temblores 1984: Heredia, Costa Rica, Observatorio Vulcanologico y Sismologico, Universidad Nacional, p. 100–104.

—— , 1988, Seismotectonics of Costa Rica; An analytical view of the southern terraines of the Middle America Trench [Ph.D. thesis]: Santa Cruz, University of California, 157 p.

Harlow, D. H., 1976, Instrumentally recorded seismic activity prior to main event, in Espinosa, A. F., ed., The Guatemala earthquake of February 4, 1976; A preliminary report: U.S. Geological Survey Professional Paper 1002, p. 12–16.

Harlow, D. H., and White, R. A., 1980, Preliminary catalog of seismicity prior to the Guatemala earthquake of February 4, 1976: U.S. Geological Survey Open-File Report 80–60, 24 p.

—— , 1985, Shallow earthquakes along the volcanic chain of Central America; Evidence for Oblique subduction [abs.]: Earthquake Notes, v. 55, p. 28.

Jarrard, R. D., 1986, Relationships among subduction parameters: Reviews of Geophysics, v. 24, p. 217–284.

Kanamori, H., and Stewart, G. S., 1978, Seismological aspects of the Guatemala earthquake of February 4, 1976: Journal of Geophysical Research, v. 83, p. 3427–3434.

Langer, C. J., and Bollinger, G. A., 1979, Secondary faulting near the terminus of a seismogenic strike-slip fault; Aftershocks of the 1976 Guatemala earthquake: Bulletin of the Seismological Society of America, v. 69, p. 427–444.

Langer, C. J., Algermissen, S. T., Dewey, J. W., and Hopper, M. G., 1974, Aftershocks of the Managua, Nicaragua, earthquake of December 23, 1972: Bulletin of the Seismological Society of America, v. 64, p. 1005–1016.

Larde, J., 1960, Historia seismica y erupcio-volcanica de El Salvador; 1, Obras completas: El Salvador, Ministerio de Cultura, Departamento Editorial, 576 p.

Leeds, D. J., 1974, Catalog of Nicaraguan earthquakes: Bulletin of the Seismological Society of America, v. 64, p. 1132–1158.

Lisowski, M., Savage, J. C., and Prescott, W. H., 1991, The velocity field along the San Andreas fault in central and southern California: Journal of Geophysical Research (in press).

Lomnitz, C., and Schulz, R., 1966, The San Salvador earthquake of May 3, 1965: Bulletin of the Seismological Society of America, v. 56, p. 561–575.

Matumoto, T., 1978, Report to the Instituto Costarricense de Electricidad, 50 p. (unpublished internal report).

McBirney, A. R., and Williams, H., 1965, Volcanic history of Nicaragua: University of California Publications in Geological Sciences, v. 55, 65 p.

McClelland, L., Simkin, T., Summers, M., Nielsen, E., and Stein, T. C., eds., 1989, Global volcanism 1975–1985: Smithsonian Institution Scientific Event Alert Network, p. 489–492.

McNutt, S. R., and Harlow, D. H., 1983, Seismicity at Fuego, Izalco, and San Cristobal volcanoes, Central America, 1973–1974: Bulletin Volcanologique, v. 46-3, p. 283–297.

Minakami, T., 1974, Seismology of volcanoes in Japan, in Civetta, L., Gasparini, P., Luongo, G., and Rapolla, A., eds., Developments in solid earth geophysics, v. 6: Amsterdam, Elsevier, p. 1–28.

Minster, J. B., and Jordan, T. H., 1978, Present-day plate motions: Journal of Geophysical Research, v. 83, p. 5331–5354.

Molnar, P., and Sykes, L. R., 1969, Tectonics of the Caribbean and Middle America regions from focal mechanisms and seismicity: Geological Society of America Bulletin, v. 84, p. 1651–1658.

Montero, P., W., and Dewey, J. W., 1982, Shallow-focus seismicity, composite focal mechanisms, and tectonics of the Valle Central of Costa Rica: Bulletin of the Seismological Society of America, v. 72, p. 1611–1626.

Nakamura, K., and Uyeda, S., 1980, Stress gradient in the arc–back arc regions and plate subduction: Journal of Geophysical Research, v. 85, p. 6419–6428.

Needham, R. E., 1986, Catalog of first motion focal mechanisms, 1984–1985, v. 1: U.S. Geological Survey Open-File Report 86–520A, p. 63–84.

Okada, H., 1983, Comparative study of earthquake swarms associated with major volcanic activities, in Shimozuru, D., and Yokoyama, I., eds., Arc volcanism; Physics and tectonics: Terra Scientific Publishing Co., p. 43–61.

Plafker, G., 1973, Field reconnaissance of the effects of the earthquake of April 13, 1973, near Laguna de Arenal, Costa Rica: Science, v. 63, p. 1847–1856.

—— , 1976, Tectonic aspects of the Guatemalan earthquake of 4 February 1976: Science, v. 193, p. 1201–1208.

Quevec, R., E. R., and Molina, C., E., 1986, Actividad sismica en los alrededores del Volcan Tacana: Report of the Instituto Nacional de Sismologia, Vulcanologia, Meterologia, e Hidrologia, 41 p.

Rinehart, W., and 5 others, compilers, 1982, Seismicity of Middle America: U.S. Department of Commerce, National Oceanic and Atmospheric Administration and U.S. Geological Survey, National Earthquake Information Service, scale 1:8,000,000.

Rosenbleuth, E., and Prince, J., 1966, El temblor de San Salvador, 3 Mayo 1965; Ingenieria sismica: Revista de la Sociedad Mexicana de Ingenieria, A.C., p. 33–60.

Schwartz, D. P., 1985, The Caribbean–North American plate boundary in Central America; New data on Quaternary tectonics: Earthquake Notes, v. 55, p. 28.

Schwartz, D. P., Cluff, L. S., and Donnelly, T. W., 1979, Quaternary faulting along the Caribbean–North American plate boundary in Central America: Tectonophysics, v. 52, p. 431–445.

Shakal, A. F., Huang, M., and Linares, R., 1987, Processed strong motion data: Earthquake Spectra, v. 3, p. 456–482.

Simkin, T., and 5 others, 1981, Volcanoes of the World; A regional directory, gazetteer, and chronology of volcanism during the last 10,000 years: Stroudsburg, Pennsylvania, Hutchinson Ross Publishing Co., 232 p.

Stoiber, R. E., and Carr, M. J., 1973, Quaternary volcanic and tectonic segmentation in Central America: Bulletin Volcanologique, v. 37, p. 304–325.

Sykes, L. R., McCann, W. R., and Kafka, A. L., 1982, Motion of the Caribbean plate during the last 7 million years and implications for earlier Cenozoic movements: Journal of Geophysical Research, v. 87, p. 10,656–10,676.

Ward, P. L., Gibbs, J., Harlow, D. H., and Aburto, Q., A., 1974, Aftershocks of the Managua, Nicaragua, earthquake and the tectonic significance of the Tiscapa fault: Bulletin of the Seismological Society of America, v. 64, p. 1017–1029.

White, R. A., 1982, The diffuse nature of the CARB-NOAM boundary in Guatemala [abs.]: EOS Transactions of the American Geophysical Union, v. 63, p. 81.

—— , 1984, Catalog of historic seismicity in the vicinity of the Chixoy-Polochic and Motagua faults, Guatemala: U.S. Geological Survey Open-File Report 84–88, 34 p.

—— , 1985, The Guatemala earthquake of 1816 on the Chixoy-Polochic fault: Bulletin of the Seismological Society of America, v. 75, p. 455–473.

White, R. A., and Harlow, D. H., 1979, Preliminary catalog of aftershocks of the Guatemala earthquake of February 4, 1976, from the area between Guatemala City and Lake Atitlan: U.S. Geological Survey Open-File Report 79–864, 63 p.

—— , 1980, Preliminary catalog of seismicity from south-central Guatemala, July 1 through December 31, 1976: U.S. Geological Survey Open-File Report 80–83, 39 p.

White, R. A., Cifuentes, I. L., Harlow, D. H., and Sanchez, B., E., 1980, Preliminary report to the government of Guatemala on the on-going earthquake swarm in the Department of Santa Rosa, Guatemala: U.S. Geological Survey Open-File Report 80–800, 24 p.

White, R. A., Harlow, D. H., and Quevec, E. R., 1984, Microseismicity of the western Motagua fault prior to the 4 February 1976 Guatemala earthquake [abs.]: EOS Transactions of the American Geophysical Union, v. 65, p. 1109.

White, R. A.. Alvarez, S., and Harlow, D. H., 1987, The San Salvador earthquake of October 10, 1986; Seismological aspects and recent local seismicity: Earthquake Spectra, v. 3, p. 419–434.

White, R. A., Carr, M. J., and Harlow, D. H., 1989, An episodic model for volcanic arc activity in Central America [abs.]: International Association of Volcanology and Chemistry of the Earth's Interior General Assembly on Continental Magmatism, Santa Fe, New Mexico, June 25–July 1, 1989, p. 292.

Woodward-Clyde Consultants, 1979, Microearthquake investigation for the Chulac and Xalala hydroelectric projects: Final report for the Instituto Nacional de Electrificación, Guatemala, 49 p.

Yuan, A.T.E., Harlow, D. H., and McNutt, S. R., 1984, Seismicity and the eruptive activity at Fuego Volcano, Guatemala, February 1975–January 1977; Journal of Volcanology and Geothermal Research, v. 21, p. 277–296.

MANUSCRIPT ACCEPTED BY THE SOCIETY DECEMBER 11, 1990

ACKNOWLEDGMENTS

I wish to express my thanks and appreciation to Frederico Guendel of the UNA (Costa Rica) network, Walter Montero of the RSN (Costa Rica) network, Antonio Gonzales, Roberto Linares, and Salvador Alvarez of the CIG (El Salvador) network (from 1984 to 1986), Eddy Sanchez and Edgar Quevec of the INSIVUMEH (Guatemala) network, and Arturo Aburto Q. of the IIS (Nicaragua) network (from 1975 to 1980), for their help with supplying tape versions of their network catalogues and other materials, some published and some unpublished, that were used in this review but are not widely available. I especially wish to thank Dave Harlow of the U.S. Geological Survey (USGS) in Menlo Park, California, for help and guidance working in Central America over the years. I thank Dave Harlow and Chris Stevens of the USGS in Menlo Park and James Dewey and an anonymous reviewer of the USGS in Golden, Colorado, for valuable criticism and improvements to the manuscript.

Printed in U.S.A.

Chapter 19

Tectonic stress field of North America and relative plate motions

Mark D. Zoback
Department of Geophysics, Standford University, Stanford, California 94305
Mary Lou Zoback
U.S. Geological Survey, Menlo Park, California 94025

INTRODUCTION

The tectonic stress field is both the cause and result of active geologic processes. At the largest scale, stresses in the earth's lithosphere arise from such processes as mantle convection and lithospheric density imbalances. The forces generated by these processes are ultimately responsible for driving the plates. Examples include the ridge-push force at spreading centers and the trench-pull force associated with negative buoyancy of subducting plates (utilizing the terms of Forsyth and Uyeda, 1975). A variety of important processes act at more regional scales. These include stresses generated by relative plate motion, magmatic and thermal processes, regional crust and lithospheric thickness and density variations, and topography (e.g., Jeffreys, 1976; Bott and Dean, 1972; Artyushkov, 1973; Fleitout and Froideveaux, 1983), and lithospheric flexure (e.g., McNutt, 1984). Improved knowledge of the tectonic stress field is intimately tied to improved understanding of the mechanisms that drive plate motion, the dynamics of faulting along both plate boundaries and in intraplate areas, and the overall mechanical, thermal, and rheological constraints on volcanism, mountain building, basin formation, and other active geologic processes.

We have been accumulating data on the orientation and relative magnitude of the tectonic stress field for over a decade. The study described here is based upon integration of our previous analyses of regional stress patterns in the conterminous United States (Zoback and Zoback, 1980, 1981, 1989; Zoback and others, 1986, 1987) and the studies of Canada (Adams and Bell, this volume), Mexico and Central America (Suter, this volume), and Alaska (Eastabrook and Jacob, this volume). These studies made use of previous compilations of data on the tectonic stress field in parts of the United Staes and Canada (e.g., Sbar and Sykes, 1973; Haimson, 1977; Bell and Gough, 1979; Gough and Bell, 1981; Sbar, 1982; Gough and others, 1983; Hasagawa and others, 1985; Mount and Suppe, 1987; Plumb and Cox, 1987; Evans and others, 1989).

The first objective of this paper is to thoroughly document the ways in which in situ stress data have been used to compile the North American stress data base. As discussed below, various types of data have been used to define the tectonic stress field. We present a detailed "Quality Table" (Table 1) for establishing a qualitative measure of the reliability of the data and the degree to which we believe the different stress indicators represent the tectonic stress field. Although this table is somewhat subjective, it has evolved over a decade and represents our opinion of the "state of the art" of mapping the tectonic stress field. This table was previously presented by Zoback and Zoback (1989) without detailed discussion. The criteria in the table were used to prepare a compilation of in-situ stress data in the conterminous United States (Zoback and Zoback, 1989) and a preliminary version of a global stress map (Zoback and others, 1989). These criteria also form the basis for the other compilations of stress data in this volume. Data on the state of stress in North America are synthesized in a 1:5,000,000 map of stress data in North America (M. L. Zoback and others, 1991) and released on CD Rom (National Geophysical Data Center, 1990) without discussion or synthesis. The Canadian data set was published in tabular form by Adams (1987).

The second objective of this paper is to briefly overview the platewide stress distribution throughout North America and to define and discuss provinces in which the orientation and relative magnitude of the tectonic stresses are approximately constant. We discuss these provinces in the context of current tectonics and the sources of tectonic stress acting on the North American Plate. As the states of stress in these provinces are comprehensively discussed in this volume by the authors cited above, it is not necessary to repeat their arguments. Instead, we only summarize key points, point out similarities and differences between stress

---

Zoback, D., and Zoback, M. L., 1991, Tectonic stress field of North America and relative plate motions, *in* Slemmons, D. B., Engdahl, E. R., Zoback, M. D., and Blackwell, D. D., eds., Neotectonics of North America: Boulder, Colorado, Geological Society of America, Decade Map Volume 1.

TABLE 1. QUALITY-RANKING SYSTEM FOR STRESS ORIENTATIONS

	A	B	C	D
Focal Mechanism	Average P-axis or formal inverstion of four or more single-event solutions in close geographic proximity (at least one event M ≥ 4.0, other events M ≥ 3.0)	Well-constrained single-event solution (M ≥ 4.5) or average of two well-constrained single-event solutions (M ≥ 3.5) determined from first motions and other methods (e.g., moment tensor wave-form modeling, or inversion)	Single-event solution (constrained by first motions only, often based on author's quality assignment) (M ≥ 2.5) Average of several well-constrained composites (M ≥ 2.0)	Single composite solution Poorly constrained single-event solution Single-event solution for M < 2.5 event
Wellbore Breakout	Ten or more distinct breakout zones in a single well with S.D. ≤ 12° and/or combined length > 300 m Average of breakouts in two or more wells in close geographic proximity with combined length > 300 m and S.D. ≤ 12°	At least six distinct breakout zones in a single well with S.D. ≤ 20° and/or combined length > 100 m	At least four distinct breakouts with S.D. < 25° and/or combined length > 30 m	Less than four consistently oriented breakouts or > 30 m combined length in a single well Breakouts in a single well with S.D. ≥ 25°
Hydraulic Fracture	Four or more hydrostatic orientations in a single well with S.D. ≤ 12°, depth > 300 m Average of hydrofrac orientations for two or more wells in close geographic proximity, S.D. ≤ 12°	Three or more hydrofrac orientations in a single well with S.D. < 20° Hydrofrac orientations in a single well with 20° < S.D. < 25°	Hydrofrac orientations in a single well with 20° < S.D. < 25° Distinct hydrofrac orientation change with depth, deepest measurements assumed valid One or two hydrofrac orientations in a single well	Single hydrofrac measurements at < 100 m depth
Petal Centerline Fracture			Mean orientation of fractures in a single well with S.D. < 20°	
Overcore	Average of consistent (S.D. ≤ 12°) measurements in two or more boreholes extending more than two excavation radii from the excavation wall, and far from any known local disturbances, depth > 300 m	Multiple consistent (S.D. < 20°) measurements in one or more boreholes extending more than two excavation radii from excavation well, depth > 100 m	Average of multiple measurements made near surface (depth > 5 to 10 m) at two or more localities in close proximity with S.D. ≤ 25° Multiple measurements at depth > 100 m with 20° < S.D. < 25°	All near surface measurements with S.D. > 15°, depth < 5 m All single measurements at depth Multiple measurements at depth with S.D. > 25°
Fault Slip	Inversion of fault-slip data for best-fitting mean deviatoric stress tensor using Quaternary age faults	Slip direction on fault plane, based on mean fault attitude and multiple observations of the slip vector. Inferred maximum stress at 30° to fault	Attitude of fault and primary sense of slip known, no actual slip vector	Offset coreholes Quarry pop-ups Post-glacial surface fault offsets
Volcanic Vent Alignment*	Five or more Quaternary vent alignments or parallel dikes with S.D. < 12°	Three or more Quaternary vent alignments or parallel dikes with S.D. < 20°	Single well-exposed Quaternary dike Single alignment with at least five vents	Volcanic alignment inferred from less than five vents

S.D. = standard deviation.
*Volcanic alignments must be based, in general, on five or more vents or cinder cones. Dikes must not be intruding a subparallel regional joint set.

provinces from the perspective of the state of stress within the entire plate, and, in effect, attempt to integrate our previous work on the state of stress within the conterminous United States with the studies for Canada, Mexico, and Central America and Alaska.

The final objective of this paper is to take a detailed look at the state of stress along the western margin of the North American Plate to examine the manner in which relative plate motion afects the stress field along the plate boundary. As noted along the San Andreas fault by Zoback and others (1987) and Mount and Suppe (1987), the state of stress along the plate margin in western California results from the interaction of relative plate motion and the frictional strength of San Andreas fault. We examine the state of stress along the western margin of North America in detail to examine the relationship between relative plate motions and the state of stress in areas of transform faulting (the San Andreas and Queen Charlotte systems), subduction (the Aleutians, Juan de Fuca, and Mid-America systems), and combined spreading and transform motion (the Gulf of California).

TECTONIC STRESS INDICATORS

In both this and our previous compilations of in situ stress data, we define criteria that identify in situ stress indicators indicative of the tectonic stress field. By this we mean stresses associated with large-scale processes capable of causing relatively wide scale faulting and crustal deformation. For practical purposes we assume that any source of stress is a tectonic stress if it affects areas with dimensions of several times the elastic plate thickness of the lithosphere. Other investigators have suggested more specific criteria to distinguish "tectonic" from "nontectonic" stresses. For example, Bott and Kusznir (1984) argued that for any source of stress to be tectonic it must be renewable. Consistent with both of these definitions, sources of stress that are clearly tectonic are those that arise from absolute and relative plate motion, lithospheric flexure, density inhomogeneities in the crust and lithosphere, and magmatism.

We have used four general categories of stress indicators that we believe have the potential for recording the tectonic stress field. These are well-constrained earthquake focal mechanisms, stress-induced wellbore breakouts, in-situ stress measurements at depth, and young geologic data such as volcanic alignments and fault-slip events. Each type of stress indicator is discussed below. As noted in our previous studies (and illustrated below), the correlation between the tectonic stress orientations deduced from these techniques is quite good. In fact, the general lack of significant variations of stress orientation with depth is fundamentally important to the ability to map the tectonic stress field. Perhaps it is fair to say that is both significant and surprising that the different stress indicators correlate so well when they sample such markedly different crustal depths. Several areas where this does not seem to be the case are mentioned below. We also assume that principal stresses in the upper few kilometers of the earth's crust generally act in the vertical direction (corresponding to the weight of the overburden, S_v) and in two orthogonal horizontal directions (corresponding to the least and greatest principal stresses, S_{hmin} and S_{Hmax}, respectively). The validity of this assumption is borne out by the very small number of crustal earthquake focal mechanisms in which either the P or T axis is not observed to be within 10 to 15° of horizontal or vertical (e.g., Zoback and others, 1989).

Each type of stress indicator has distinct advantages and disadvantages. A quality ranking system has been developed (Table 1) and is intended to indicate both the overall reliability of a given type of data as a stress indicator and the degree to which it records the tectonic stress field. Table 1 was originally presented by Zoback and Zoback (1989) for their study of the state of stress in the conterminous United States. We have added additional clarifying comments to their data, but the number of observations and standard deviations associated with each quality has not changed.

An important goal in establishing the quality rankings for the tectonic stress indicators in Table 1 was to be sufficiently conservative. We would rather disregard some possibly reliable indicators of tectonic stress than to include with the quality criteria indicators that are potentially affected by nontectonic sources of stress. The criteria have generally become more conservative since our original 1980 paper.

While the specific entries in this table are discussed at length below, an issue affecting the quality of all potential stress indicators is that of depth. Because tectonic stresses are usually so small at shallow depth (because of the negligible frictional strength of fractured rock in the very near-surface), the surficial stress field (i.e., that in the upper 10 m) is often dominated by the effects of open fractures and weathering, thermoelastic stresses due to diurnal and seasonal temperature changes, topographically induced stresses, and rock anisotropy (see discussion by Engelder and Sbar, 1984). For these reasons, all stress measurements at depths less than 10 m are considered to be of D quality (Table 1), which means that they are retained in the data base but are not presented on maps. This issue is most critical for the abundant overcoring, or stress relief, stress measurements that have been made at depths of only a few meters. Similarly, surficial geological features such a glacial "pop-ups" and offset coreholes in road cuts are not currently considered to be reliable tectonic stress indicators becuase of their shallow depth, although we did use some of these types of data in our original compilation of stress data in the conterminous United States (Zoback and Zoback, 1980). Although surface observations of fault slip and volcanic vent alignments could also be affected by nontectonic surficial processes, in general, we think these features record the near-surface expression of deeper processes and thus frequently contain extremely valuable information about the tectonic stress field.

The exact depths associated with specific quality-rankings are clearly subjective. Stress measurements from depths greater than 10 m are given progressively higher qualities (depending on their internal consistency, Table 1) because the magnitudes of the tectonic stresses are expected to increase with depth, due to the increase in the frictional strength of the crust; and non-tectonic sources of stress, related to weathering, thermal stresses, and topography, decrease rapidly with depth. Thus, internally consistent stress measurements from depths greater than 300 m are given an A quality, from depths greater than 100 m are given a B quality, and so on, unless there is some other reason to suspect that the measurements might be influenced by nontectonic sources of stress such as pronounced topography or a near-by subsurface excavation.

Earthquake focal plane mechanisms

The advantages of utilizing well-constrained earthquake focal plane mechanisms to map the stress field are fairly obvious: earthquakes record stress-induced deformation at mid-crustal depths and sample relatively large volumes of rock, and focal-plane mechanisms provide information on both the orientation and relative magnitude of the in-situ stress field. More, and better constrained, focal mechanisms for mapping the stress field are available now than ever before due to the continued improvement of regional and global networks and the fact that it is now routine for moment tensor inversions to be computed for larger events (e.g., Dziewonski an Woodhouse, 1983).

An important disadvantage of utilizing earthquake focal plane mechanisms as stress indicators is that focal-plane mechanisms record deformation and not stress. Body-wave earthquake focal plane mechanisms define two possible orthogonal fault planes and slip vectors. The P and T axes are, by definition, the bisectors of the dilatational and compressional quadrants of the focal mechanism, respectively. Thus, they are not the maximum and minimum principal stress directions (as is often assumed) but are the compressional and extensional strain directions for the two possible faults. As most crustal earthquakes appear to occur on preexisting faults (rather than resulting from new fault breaks), the slip vector is a function both of the orientation of the fault and the orientation and relative magnitude of the principal stresses, and the P and T axes of the focal plane mechanism do not correlate directly with principal stress directions. In an attempt to rectify this problem, Raleigh and others (1972) showed that if the nodal plane of the focal mechanism corresponding to the fault is known, it is preferable not to use the P axes of the focal-plane mechanism but instead to assume an angle between the maximum horizontal stress and the fault plane defined by the coefficient of friction of the rock. Because the coefficient of friction of many rocks is often in the range of 0.6 to 0.8 (Byerlee, 1978), Raleigh and others suggested that the expected angle between the fault plane and the direction of maximum principal stress would be expected to be about 30°.

Unfortunately, for the majority of intraplate events, we often do not know which focal plane corresponds to the fault plane and we certainly do not know the exact coefficient of friction of the rock. What makes this matter worse is that if the coefficient of friction of the fault is quite low, the direction of maximum compression can be anywhere in the dilatational quadrant and the P axis could differ from the true maximum stress direction by as much as 45° (MacKenzie, 1969). In fact, Zoback and others (1987) excluded as tectonic stress indicators right-lateral strike-slip focal plane mechanisms right on the San Andreas. They argued that if the frictional strength of the San Andreas was quite low, as appreciable data indicates (see below), the P axes of these focal-plane mechanisms simply reflect the geometry of the fault and slip vector, and not the orientation of the maximum horizontal stress. In most intraplate areas, however, P axes do seem to represent good approximations of the maximum horizontal stress direction, apparently because intraplate earthquakes do not seem to occur on faults with extremely low friction (Zoback and Zoback, 1981, 1989; Zoback, 1991). Nevertheless, because earthquake focal plane mechanisms basically represent strain and not stress, no single earthquake focal mechanism ranks higher than a B-quality rating (Table 1), even when it constrained by both body waves and surface waves. Single events constrained by body waves only are given a C rating and composite mechanisms are given D ratings.

To optimize the use of focal plane mechanism data for determining stress orientations it is necessary to consider multiple events in a given region and use either the average P-axis direction as the maximum horizontal stress direction or to formally invert a group of focal-plane mechanisms to determine the orientation and relative magnitude of the principal stress tensor (see, e.g., Angelier, 1979, 1984; Gephart and Forsyth, 1984; Michael, 1984, 1987; Reches, 1987).

Stress-induced wellbore breakouts

Wellbore breakouts result naturally from compressive failure of rock around a wellbore due to the elastic stress concentration around the hole. As illustrated in Figure 1a (modified from Zoback and Healy, 1991), breakouts are centered at the azimuth of minimum horizontal compression, S_{hmin}, where the elastic compressive stress concentration is greatest (Bell and Gough, 1979, 1983; Cox, 1983; Zoback and others, 1985). Conversely, induced hydraulic fractures, which form in tension, initiate at the azimuth of maximum principal stress, S_{Hmax}, where the stress concentration is least compressive. As discussed below, hydraulic fractures can be induced intentionally to measure in-situ stress magnitude or they can occur inadvertently during drilling under certain in-situ stress and borehole conditions (e.g., Moos and Zoback, 1990). Figure 1b shows a three-dimensional perspective view of a section of the Cajon Pass scientific drillhole where an intentionally induced hydraulic fracture and a small stress-induced breakout are at right angles to each other. This image was constructed through special processing of data from a device known as an ultrasonic borehole televiewer (Zemanek and others, 1970) as discussed by Barton and others (1991). The orthogonality of breakouts and hydraulic fractures has now been demonstrated in a number of boreholes in a variety of tectonic settings (e.g., Hickman and others, 1985, in the Appalachian Plateau of western New York; Stock and others, 1985, in the Basin and Range province in southern Nevada; Paillet and Kim,

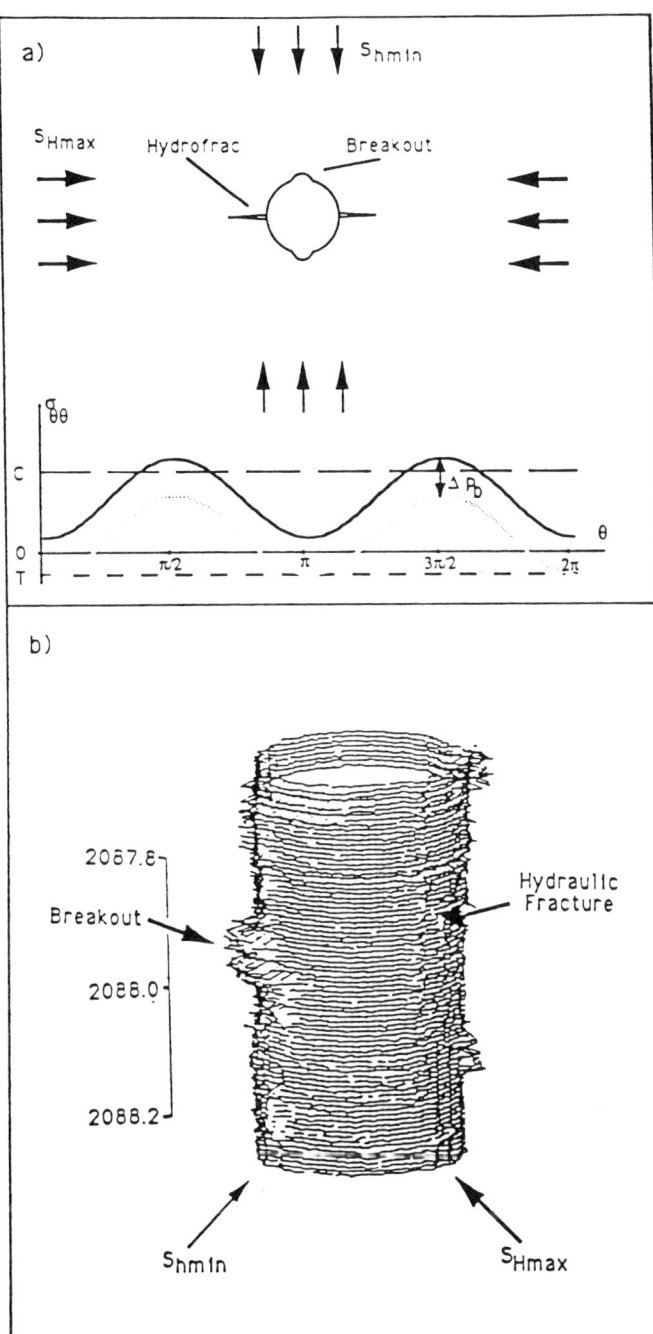

Figure 1. a) Elastic stress concentration around a circular hole (after Kirsch, 1898) demonstrating that breakouts form at the azimuth of S_{hmin} where the concentraton of circumferential stress is most compressive and hydraulic fractures (upon pressurization of the borehole by an amount ΔP_b) at the azimuth of S_{Hmax} where the original stress concentration is least compressive. b) The shape of a section of the Cajon Pass borehole where an induced hydraulic fracture is oriented 90° from a small, naturally occurring wellbore breakout.

1985, in the Columbia Plateau of southern Washington; Shamir and others, 1988, in a deep borehole along the San Andreas fault).

Unfortunately, the majority of stress-induced breakout data do not come from borehole televiewer surveys but from magnetically oriented four-arm caliper logs, which are part of the dipmeter logging tool commonly used in the petroleum industry. Despite the relatively low sensitivity of this technique, breakouts analyzed by both methods generally compare quite well when the breakouts are well developed (see, e.g., Plumb and Hickman, 1985; and Shamir and others, 1988). Numerous studies have shown that caliper-determined breakout orientations are remarkably consistent in a given well and a given oil-field (Bell and Gough, 1979; Gough and Bell, 1981; Klein and Barr, 1986; Plumb and Cox, 1987; Mount and Suppe, 1987), and breakouts have proven to be a very reliable measure of stress orientation in many areas (see compilation of various types of in-situ stress data in Europe by Müller and others, 1991; and worldwide stress compilation of Zoback and others, 1989). The depth range of breakouts, typically 1 to 4 km, provide an important bridge between the depths at which most crustal earthquakes occurs (3 to 15 km) and the depths of most in situ stress measurements (<2 km) and geologic observations at the surface.

The assignment of A, B, C, and D qualities to wellbore breakout data is based on the frequency, overall length, and consistency of occurrence of breakouts in a given well. The standard deviations associated with the various qualities shown in Table 1 were determined from our experience working with breakout data in a number of boreholes. High standard deviations ($>\sim 25°$) represent either scattered, or bimodal distributions due to factors such as the following. (1) Data quality is poor because either the rock is strong compared to the stress concentration and breakouts are poorly developed, or the rock might be so weak that failure is pervasive and it is difficult to distinguish between stress-induced breakouts and more pervasive "wash-outs" (Plumb and Hickman, 1985). (2) Both wellbore breakouts and drilling-induced hydraulic fractures occur in a given well (see Stock and others, 1985; Moos and Zoback, 1990). In such cases it is common to observe bimodal distributions of the directions of wellbore enlargements as the borehole tends to elongate in orthogonal directions. In such cases, it may not be possible to determine which direction corresponds to the hydrofrac (and S_{Hmax}) and which corresponds to the breakout (and S_{hmin}). (3) Significant variations in stress orientation can occur with depth (see, e.g., Shamir and others, 1988; Shamir and Zoback, 1991), although this case seems to be rare. In Ohio, Evans (1989) has documented unusual changes in stress magnitude with depth, and he suggests that the shallow stress field may be somewhat decoupled from the deeper stress field by thick salt deposits in the area. A similar phenomenon has been called upon to explain unusual stress orientation at relatively shallow depth in the Jura Mountains of Switzerland (Becker and others, 1987).

In situ stress measurements

Many hydraulic fracturing and overcoring stress measurements have been made to measure the tectonic stress field. Hydraulic fracturing stress measurements have been widely discussed in the literature and several compilations of papers related to this techique have appeared in recent years (see papers in compilations edited by Zoback and Haimson, 1983; Stephannson, 1986; Haimson, 1989). Hydraulic fracturing involves pressurizing an isolated section of a borehole until a tensile fracture is induced at the point of least compressive stress around the hole and extends away from the borehole. It is an extremely effective technique for measuring the magnitude of the least compressive stress because the fracture propagates in a plane perpendicular to the direction of minimum compression. Hence, the pressure required to propagate this fracture at low flow rates and the "shut-in" pressure (measured immediately after pumping is stopped and pressure in the pipe is "shut-in" by closing a valve), is essentially equal to S_{hmin}.

Under ideal circumstances, the magnitude and orientation of the maximum horizontal stress can also be determined. This requires that the well or borehole be drilled parallel to one of the principal stresses (typically a vertical well is drilled that is parallel to the vertical principal stress) and that the rock behavior is essentially linearly elastic. The most common problem in utilizing hydraulic fracturing data is that the difficulty and expense associated with the measurements generally results in relatively few measurements in a given borehole. Utilizing magnetically oriented impression packers to determine the azimuth of the induced hydrofrac is both a time-consuming and difficult process, but under ideal circumstances it can yield excellent results.

Overcoring, or stress-relief stress measurements, are quite common in the mining and civil-engineering applications. While capable of measuring the complete stress tensor (see an overview of these techniques in McGarr and Gay, 1978), obvious problems with overcoring measurements include the difficulty of making the measurements at distances more than a few tens of meters from a free surface, the small volume of rock in which the stress field is measured (typically a few cm^3) and the need to compute stress from observed strain, which requires knowledge of the rocks' elastic properties (this is especially difficult in anisotropic rock).

The rationale for the assignments of A, B, C, or D quality for hydraulic fracturing data is similar to that for the breakouts alluded to above. As the deep overcoring data generally come from mines, we add to the quality criteria the obvious requirement that the boreholes in which the measurements are made are not affected by excavations or any other known perturbations to the tectonic stress field. To exclude the possibility that rock anisotropy may affect the overcoring measurements we add the requirement for A-quality data that data come from at least two different boreholes at about the same depth.

Fault slip

The techniques for mathematically inverting earthquake focal mechanism data to derive a "best-fitting" stress tensor mentioned above were derived from similar techniques originally developed to infer stress orientation and relative magnitude from measured slip directions on fault planes (striae or slickensides) exposed at the surface (Angelier, 1979, 1984; Carey, 1979; Reches, 1987; Carey-Gailhardis and Mercier, 1987). For this technique to work well it is necessary to document similar-age slip directions on faults of varied orientation at a given local. One significant drawback is the assumption that the observed deformation results from a spatially uniform stress tensor without any complexities due to fault interaction. A second is assuring that the slip on the different faults occurred over a period of time during which the stress field was uniform. In addition, in some cases the slip that is observed at the surface might only represent the near-surface response to deeper slip. An example of this is the occurrence of superficial normal faulting in the 1979 El Asnam earthquake that occurred as a flexural response to deeper reverse faulting (King and Vita-Finzi, 1981). Despite these potential problems, this method has been successfully used in many places to determine the orientation and relative magnitude of the contemporary stress field as well as stress fields that existed in the past.

The highest quality rating for fault-slip data are assigned to well-constrained inversions of fault-slip observations of Quaternary age. In cases when multiple tectonic events are recorded at a given site, only the youngest event is used. B-quality data involves using multiple observations of slip vectors within a major fault zone to define a mean fault attitude and slip vector. These data are treated as a paleo-focal-plane mechanism, and it is assumed that the maximum principal stress is 30° from the fault in the plane of the slip vector (Raleigh and others, 1972). C-quality data are taken from faults with Quaternary-age slip (with the exception of the eastern United States, see Zoback and Zoback, 1989) in which the primary sense of slip is known but no actual slip vector has been determined. A variety of data fall into the D category. These include offset coreholes along road cuts that do not move in a down-slope direction and surficial postglacial "pop-ups" and fault offsets in quarries. As mentioned above, some of these data were given higher rankings in our 1980 compilation.

Volcanic vent alignments

Dikes and sills are natural analogs of hydraulic fractures and propagate in a plane perpendicular to the least principal stress (Anderson, 1951). Nakamura and others (1977) first demonstrated in Alaska and Japan that the trend of flank volcanos and vents radiating from a central volcano can be used to trace stress trajectories correlative with the direction of plate convergence.

The flank volcanos are apparently fed by an underground dike radiating out from the volcano with an azimuth controlled by the regional stress field. Many others have used individual dike zones and cinder-cone alignments within volcanic fields to define the orientation of both the contemporary and past stress fields (see, e.g., Thompson and Zoback, 1978; Aldrich and Laughlin, 1984). Where available, such features appear to be reliable indicators of tectonic-stress orientation.

For individual alignments within volcanic fields that lack a single central volcano, we have assigned quality on the basis of the consistency of the trend and the number and/or length of the volcanic alignments. Five or more parallel dikes or cinder cones of Quaternary age are given an A quality if the standard deviation of the orientation is <12°. In the case of fewer (3 or less), or less parallel (S.D. < 20°), features, a B-quality rating is given. C-quality data are for single dikes or a single vent alignment of at least five vents. If there are fewer than five events the data point is given a D quality. For Nakamura's data based on the mean elongation direction for a population of flank eruptions, we utilized his quality rankings.

TECTONIC STRESSES IN NORTH AMERICA

Figure 2 is an integrated map of tectonic stress indicators in North America utilizing the criteria cited above. Each data point indicates the direction of maximum horizontal compression. The

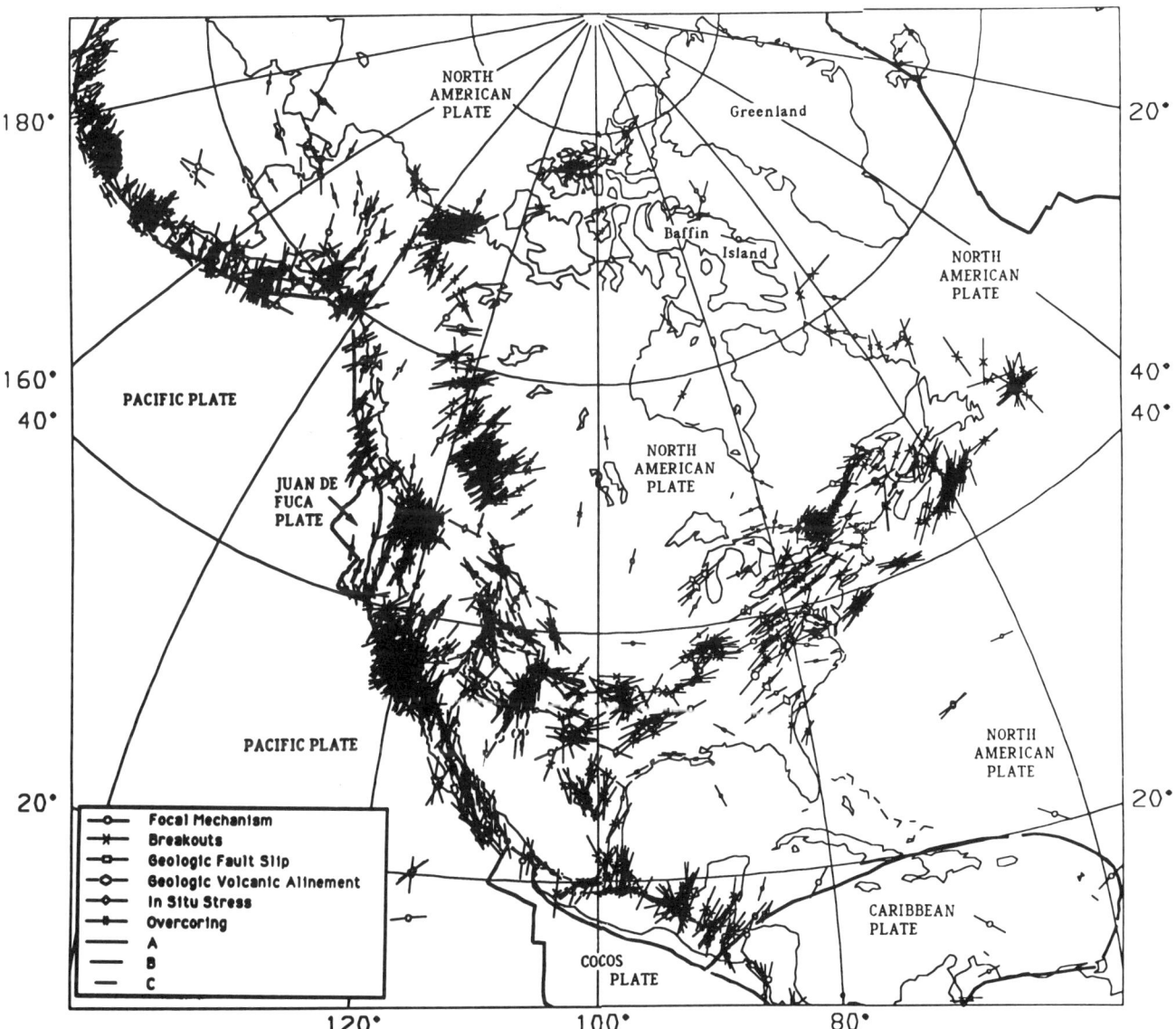

Figure 2. Stress map of North America showing the directions of maximum horizontal compressive stress determined from various types of indicators. The types of indicators and quality are indicated in the legend and explained in the text. Areas of particularly dense data quality are shown in more detail in subsequent figures.

symbol at the center of each data point indicates the type of stress indicator (see legend). The length of the symbols indicates the quality of the individual data points as defined in Table 1. Only A-, B-, and C-quality data are shown. In areas of active crustal extension (normal faulting or combined normal and strike-slip faulting), the direction of maximum horizontal compression corresponds to the intermediate principal stress as the maximum principal stress is vertical.

Perhaps one of the most striking features of Figure 2 is the extreme variability of data density. In many large portions of the plate there is extremely little data because of the absence of both earthquakes and oil wells from which to obtain breakout data. This is especially true in the middle of the plate. In other place, like central California, the extremely high data density makes the individual data points difficult to see in Figure 2. These areas are shown in more detail in subsequent figures.

As discussed above, each type of stress indicator has distinct advantages and disadvantages and samples different depths. The ability to use these different types of stress indicators is clearly critical to our ability to map the tectonic stress field. To assess the degree to which the different stress indicators yield similar results, we compare maximum horizontal stress directions from earthquake focal plane mechanisms with the other types of indicators for the portion of the eastern United States and southeasternmost Canada in Figure 3a. We selected this region to in order to consider a large area with numerous stress indicators of all types and a relatively constant stress field. By comparing earthquake focal plane mechanisms and all other types of data, we can directly compare the influence of depth on stress orientation. As shown in Figure 3b, the mean direction of maximum horizontal compression indicated by the relatively shallow data (breakouts, stress measurements, and geologic indicators are shown by the solid bars) is N69°E + 27°. For the earthquake focal-mechanism data (the open bars) it is virtually the same (N67°E + 29°). We take this correlation as a clear demonstration that over very large regions, there are generally no systematic differences between maximum principal stress directions indicated by either relatively shallow or midcrustal stress indicators. A possible exception to this is the zone of localized seismicity in the La Malbaie area of southern Canada (Adams, 1987) where thrust earthquakes occur along west-northwest striking planes and P axes are uniformly oriented east-northeast above 9 km. Deeper events, however, show thrusting on planes with a variety of strikes perhaps indicating that below 9 km the two horizontal principal stresses are both much greater than the vertical stress and have approximately the same magnitude. The relatively numerous east-west maximum principal stress values that differ noticeably from the overall trend come mostly from relatively shallow breakouts in the Illinois basin (Dart, 1985), focal-plane mechanisms from the New Madrid seismic zone, and several events located throughout the region shown in Figure 3a.

Figure 4 is a generalized version of Figure 2 in which stress provinces are defined and both average stress directions and relative stress magnitudes are indicated. The primary criteria we have used in defining a stress province is that it is a region of relatively consistent horizontal principal stress orientation and style of faulting (i.e., consistent relative stress magnitudes). A summary of the stress provinces shown in Figure 4 is presented in Table 2. In Figure 4, inward-pointed arrows indicate a compressive stress regime (reverse faulting and strike-slip faulting) and the arrows are shown in the direction of maximum horizontal compression. Outward-pointed arrows indicate extensional provinces (normal faulting and strike-slip faulting) and the arrow is shown in the direction of minimum horizontal compressive stress. As discussed previously, stress provinces are often correlative with regions of relatively constant average crustal thickness, elevation, heat flow, and overall tectonic activity (Zoback and Zoback, 1989). The stress provinces summarized in Table 2 and shown in Figure 4 are discussed in detail below. Within the Midplate and Gulf Coast provinces we also show in Figure 4 the direction of absolute plate motion of North America (as determined by Minster and Jordan, 1978, using a mantle hotspot reference frame) at the center of each region defined by the 20 by 20° latitude and longitude grid.

Several general observations can be made about the state of stress throughout North America based on Figures 2 and 4. First, almost everywhere that there are multiple data points, there is a well-defined orientation of maximum horizontal stress—implying that the two horizontal principal stresses are not, in general, of comparable magnitude. Because there are no detectable changes in the orientation of maximum horizontal principal stress with depth as discussed above, the maps of horizontal stress orientation that we present are thought to be generally representative of the stress orientation throughout the upper (seismogenic) part of the crust. The second general observation about the state of stress in North America is that it is compressional throughout most of the plate. The primary evidence for this is the fact that earthquake focal mechanisms range between strike-slip and reverse faulting throughout the interior of North America. With the exception of the Texas/Louisiana Gulf Coast and several intraplate basins discussed below, extensional states of stress are restricted to the western Cordillera (the Cordilleran Extension, Colorado Plateau, Southern Great Plains, and Trans-Mexico Volcanic Belt provinces) and a poorly documented region of back-arc extension north of the Aleutian arc. Compilation of in-situ stress data from around the world (Zoback and others, 1989) also indicate that these generalizations are true on a global scale. That is, well-defined and relatively uniform stress fields can be documented over large regions, no systematic stress changes are observed with depth, and most midplate regions are in compression—extension, being generally limited to thermally uplifted regions.

It is much more difficult to generalize about the absolute magnitude of crustal stresses as extremely few in-situ stress measurements are available for depths greater than 1 to 2 km. An additional problem is that it is clear that in some places that high horizontal stresses are observed at shallow depth that are not representative of deeper processes. Although the most obvious examples of this are glacial pop-ups and the buckling of quarry floors, in some areas, highly compressional stresses are observed

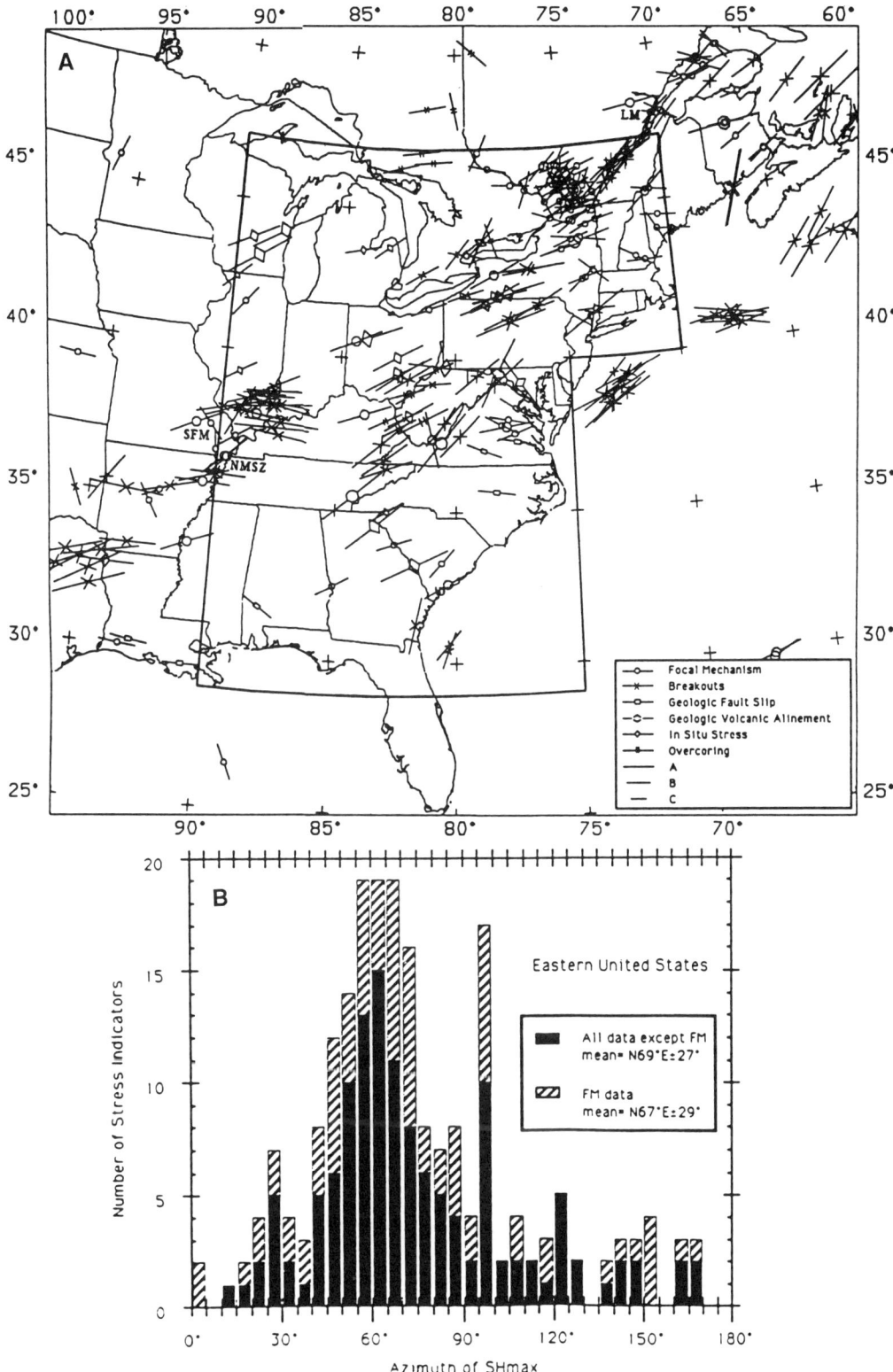

Figure 3. a, Stress map of eastern United States and southeasternmost Canada showing the directions of maximum horizontal compressive stress. NMSZ, SFM, and LM stand for the New Madrid seismic zone, St. Francois Mountains, and LaMalbaie, respectively. Data points within area enclosed by the polygon on the stress map are shown in the histogram (b) below comparing relatively shallow in situ stress indicators (wellbore breakouts, stress measurements, etc.) with earthquake focal plane mechanisms. Note that the mean and standard deviation of both sets of data are essentially identical.

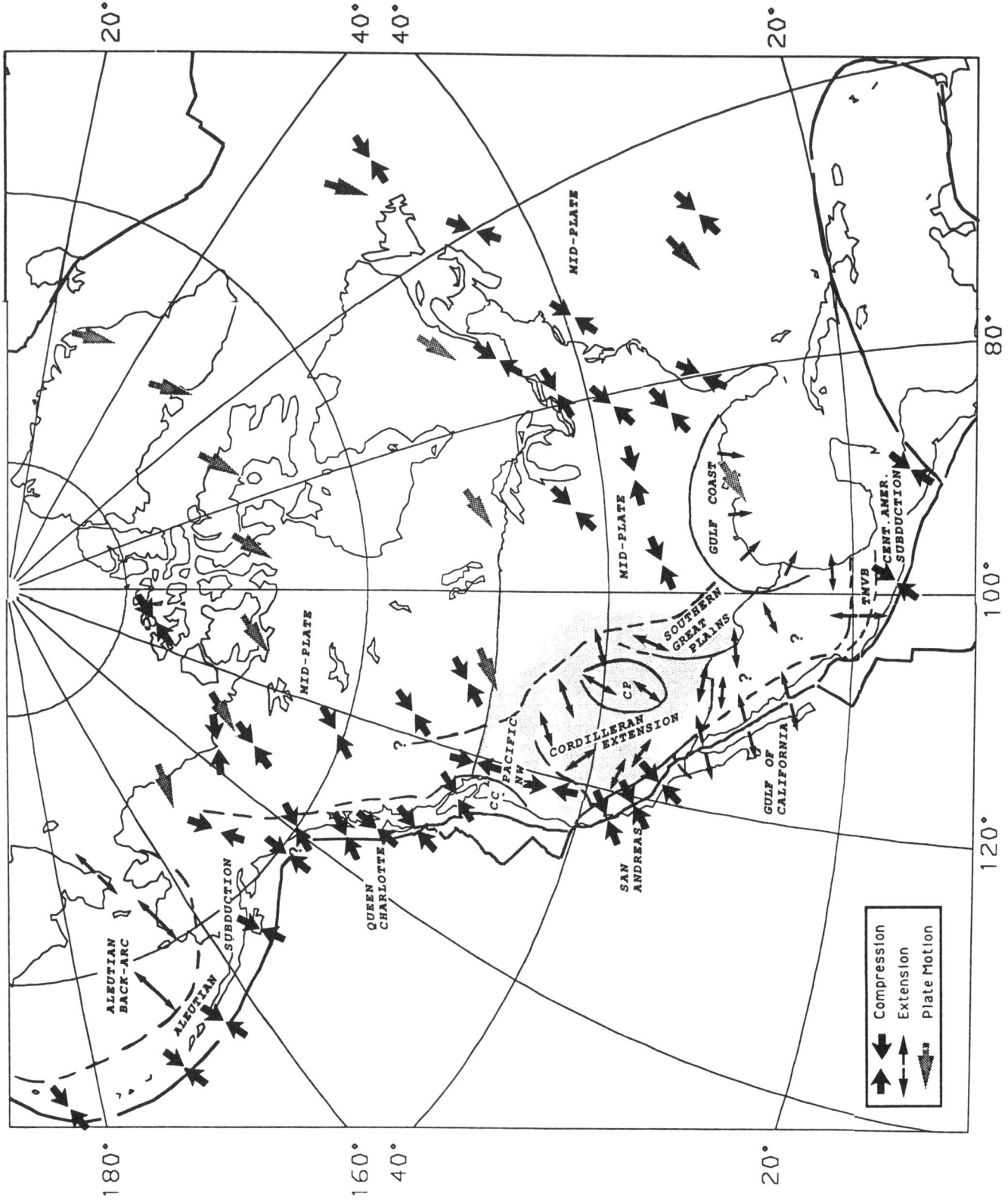

in the upper 1-km that are often sufficiently large to cause shallow reverse faulting and small earthquakes but do not seem to extrapolate to greater depth (see, e.g., McGarr and Gay, 1978; Zoback and Hickman, 1982; Zoback and Healy, 1984; Rummel and Mahring-Erdmann, 1986; Stephansson and others, 1986). In these cases, even though the orientation of maximum principal stress seems to remain constant with depth, there is a change of relative stress magnitude with depth because the vertical stress increases more rapidly than the horizontal stresses. Two somewhat extreme examples of this phenomena come from the deep mines of South Africa and the region of the 1975 Oroville earthquake in California. In both places a highly compressional stress field exists in the upper ~ 0.5 km, although at greater depth the stress field is clearly extensional (McGarr and Gay, 1978; Anderson and Zoback, 1989). In general, there is insufficient data to define whether significant variations of relative stress magnitude with depth are common.

STRESS PROVINCES OF NORTH AMERICA

In this section we discuss the individual stress provinces shown in Figure 4 and summarized in Table 2. As mentioned above, the states of stress in these provinces are discussed elsewhere in this issue and by Zoback and Zoback (1989), and we only summarize key points from the perspective of the state of stress within the entire plate. For clarity, we divide this discussion into four sections: plate interior provinces, western cordilleran provinces, subduction provinces, and transform provinces. As the state of stress along the western margin of North America is dominated by its interaction with the Pacific, Juan de Fuca, and Cocos Plates, we investigate the state of stress along the western boundary of North America in the context of relative plate motions in some detail.

Plate interior provinces

The stable interior of North America is characterized by the Midplate stress province that dominates most of Canada, the central and eastern United States, and most of the western Atlantic (see Zoback and others, 1986) and several intraplate sedimentary basins (e.g., the Texas-Louisiana Gulf Coast, the Scotian Shelf, and Grand Banks) that are discussed separately below.

Midplate province. As shown in Figures 2 and 4, the principal characteristics of the Midplate stress province are its relatively uniform east-northeast direction of maximum horizontal compression (over an extremely large area) and the consistent compressional state of stress (strike-slip and reverse faulting) within it. The United States portion of this province has been recently discussed by Zoback and Zoback (1989) building on previous work by Sbar and Sykes (1973), Haimson (1977), and Zoback and Zoback (1980). Adams and Bell (this volume) discuss this province and concentrate on Canada. Zoback and others (1986) discussed this province from the perspective of the North Atlantic. Throughout the United States, the principal style of faulting is strike slip and in southern Canada the faulting is principally reverse faulting. Thus, in general, the relative magnitudes of the principal stresses are $S_{Hmax} > S_{hmin} > S_v$ in southeastern Canada and $S_{Hmax} > S_v > S_{hmin}$ in the central eastern United States. While a number of localized variations of principal stress directions can be seen in Figure 2, the only known exceptions to the overall compressional stress state within the Midplate province are occurrences of normal faulting earthquakes in Baffin Island, Canada, and the St. Francois Mountains of Missouri.

As discussed previously by Zoback and Zoback (1981, 1989) and Zoback and others (1989), a remarkable correlation can be observed in Figure 4 between the direction of maximum horizontal compressive stress and the direction of absolute plate motion in the Midplate province of North America. From the northernmost islands of Arctic Canada, to the southeastern United States and western North Atlantic, observed directions of maximum horizontal compression generally correlate well with the absolute motion direction. Unfortunately, one cannot simply interpret the generally good correlation between the S_{Hmax} and absolute motion directions in North America as proof that the state of stress results from drag forces as the plate moves over a relatively stable asthenosphere (see Zoback and Zoback, 1989; Zoback and others, 1989). One reason for this is that it is essentially impossible to distinguish the plate-motion direction from the direction of ridge push in North America; thus, it is not possible to say whether a "push" from the ridge or a "drag" from the base is responsible for the state of stress in the Midplate province (see also the discussion by Richardson, 1991). Nevertheless, some sort of causal relation is apparent between the forces that drive the North American Plate and the state of stress within it, although this simple relationship is not observed for other large continental plates (Zoback and others, 1989).

Several localized exceptions to this general pattern are apparent in Figures 2 and 4. One interesting variation of stress

Figure 4. Generalized stress map of North America. Interpretative stress provinces shown in the figure are summarized in Table 2 and discussed in detail in the text. As explained in legend, inward-pointed arrows indicate directions of maximum horizontal compression in areas of reverse and strike-slip faulting and thinner, outward-pointed arrows indicate the direction of least horizontal compression in areas of normal faulting and strike-slip faulting. Single, large shaded arrows indicate direction of absolute plate motion. Abbreviations: CC, Cascade convergence; CP, Colorado Plateau; TMVB, Trans-Mexican volcanic belt. Shaded region in the western United States indicates the area with elevations over 1,100 m based on 1° average elevations and is correlative with the most active part of the Cordilleran extension province.

TABLE 2. STRESS PROVINCES OF NORTH AMERICA

	Stress Regime	Principal Stress Direction Maximum	Principal Stress Direction Minimum	Remarks
PLATE-INTERIOR PROVINCES				
Midplate	SS/TF*	ENE	NNW/Vertical	Encompasses most of intraplate North America east of the U.S. Cordillera including the western Atlantic basin. Earthquake focal plane mechanisms in southeastern Canada are predominately thrust and those in the United States are predominately strike-slip. Excellent correlation between direction of maximum horizontal stress with absolute plate-motion direction as well as the direction of ridge push from the Mid-Atlantic ridge.
Gulf Coast	NF	Vertical	SSE	Extensive gulfward extension and growth faulting within Coastal Plain sediments. State of stress in the underlaying basement is not known.
Cordilleran Extension	NF/SS	Vertical	WNW (WSW-WNW)†	Broad region of variable magnitude extension including the Basin and Range, Rio Grande rift, Northern Rocky Mountain, and Snake River Plain regions of western United States. Extent into Canada and Mexico is uncertain. Correlative with zone of high heat flow, elevation, and thin crust. Pronounced strike-slip deformation (with constant S_{hmin} direction) along western boundary of province (Walker Lane). Predominant S_{hmin} direction within the province is WNW but direction varies between WSW and WNW.
Colorado Plateau/ Southern Great Plains	NF	Vertical	NNE	Unique extension direction, thicker crust, and very low rate of crustal deformation distinguishes the Colorado plateau and Southern Great Plains from the surrounding Cordilleran extension province.
SUBDUCTION-RELATED PROVINCES				
Aleutian Subduction	TF/SS	NNW	Vertical/ENE	NNW S_{Hmax} direction observed throughout most of Alaska south of the Brooks Range is consistent with the NNW direction of relative motion of the Pacific Plate with respect to North America. Thrust faulting along shallow, low-angle subduction zone and within accretionary prism. Strike-slip faulting in volcanic arc. Pronounced oblique subduction in western Aleutians.
Aleutian Back-Arc	NF	Vertical	NNW?	While poorly documented, this province may encompass most of the Bering Sea, Seward Peninsula, and parts of northeasternmost Siberia. Direction of minimum horizontal compression is approximately perpendicular to the arc.
Central American Subduction	TF/SS	NE	Vertical/NW	S_{Hmax} direction parallel to direction of relative motion of Cocos Plate with respect to North America.
Trans-Mexican Volcanic Belt	NF	Vertical	N (NNE-N)	Pronounced volcanic alignments and grabens define this stress province. Calc-alkalic composition and location of volcanic belt consistent with this zone as a volcanic arc of the Central American subduction zone. Unlike most volcanic arcs, however, the S_{Hmax} direction is normal to the convergence direction rather than parallel to it.
Cascade Convergence	TF	NE	Vertical	Poorly defined, this province is based on ~10 earthquakes in Vancouver Island and sparse breakout data in western Washington and Oregon. The NE S_{Hmax} direction distinguishes the province from the Pacific Northwest province to the east. Apparent S_{Hmax} direction is ~30° oblique to convergence direction of the Juan de Fuca Plate, suggesting a weak coupling between the state of stress in the North American Plate and subduction of the Juan de Fuca Plate.

TABLE 2. STRESS PROVINCES OF NORTH AMERICA (continued)

	Stress Regime	Principal Stress Direction Maximum	Principal Stress Direction Minimum	Remarks
SUBDUCTION-RELATED PROVINCES (continued)				
Pacific Northwest	SS/TF	N	E/Vertical	N–S compression throughout this province is accompanied by SS faulting in the Cascades and active folding and thrust faulting in Puget Sound and Columbia Plateau. Source of this N–S compression may be movement of the Pacific Plate with respect to North America, perhaps suggesting a weak coupling between the Juan de Fuca Plate and North American Plates.
TRANSFORM-RELATED PROVINCES				
San Andreas	SS//TF	NE	NW/Vertical	This province includes the San Andreas fault zone and a broad zone of crustal shortening adding to it in which S_{Hmax} is oriented at a very high angle to the strike of the fault (~85° in central California and ~70° in southern California). This is especially surprising as the direction of relative plate motion between the Pacific and North American Plates in central California is only slightly convergent. The source of the near fault-normal compression direction is believed to be the result of the San Andreas, and its principal branches, having a markedly lower stress than the surrounding crust.
Queen Charlotte	SS//TF	NE	NW/Vertical	Similar to the San Andreas, the S_{Hmax} direction adjacent to the Queen Charlotte is at a high to the strike of the fault (~60 to 70°). However, the data set is quite sparse and based primarily on earthquake focal plane mechanisms. In addition, the Queen Charlotte is obliquely oriented to the direction of relative plate motion and thus must accommodate appreciable shortening across it.
Gulf of California	SS/NF	N-NW/Vertical	E/NE	Although generally considered as a zone of mid-ocean spreading, most of the length of the Gulf of California is a transform plate boundary with spreading limited to a number of rather short segments. Geologic indicators and earthquake focal plane mechanisms indicate strike-slip and normal faulting in Baja California with a S_{hmin} direction that ranges between NE (fault-normal extension) to E–W, a direction more consistent with the expected from conventional faulting.

*NF = normal faulting; SS = strike-slip faulting; TF = thrust faulting.
†Directions in italics refer to range of stress directions observed throughout a province.

within the Midplate stress province is associated with continental margin of the United States and eastern Canada. From Georgia (~26°N latitude) to Newfoundland (~42°N latitude) the direction of maximum horizontal compression inferred from breakouts in offshore petroleum exploration wells seems to be suparallel the topographic contours of the continental slope. A possible explanation of this phenomena is that the superposition of topographic or crustal density effects along the continental slope are tending to rotate the direction of S_{Hmax} either northward or eastward depending on the strike of the slope (Dart and Zoback, 1987; Zoback and Zoback, 1989). Perhaps the reason this pattern is not seen farther north is that the continental slope strikes at such a large angle to the direction of S_{Hmax} that the superimposed stresses do not cause the tectonic stresses to rotate.

Another deviation from the general east-northeast direction of maximum horizontal stress in the Midplate province is the generally east-west–oriented S_{Hmax} directions observed in the region of the New Madrid seismic zone and Illinois basin. The origin of this apparent stress rotation is not known. It is also clear in Figure 2 that the state of stress along the Labrador and Baffin Shelves is quite complex and cannot be characterized by uniform stress directions (see Adams and Bell, this volume). While it has been suggested that the large Baffin Island earthquake of 1933 was at least partially induced by deglaciation (Stein and others, 1979), no single mechanism would seem to be able to explain all of the variability observed in this region. A local variation of S_{Hmax} orientations is also observed in eastern Virginia and Maryland where reverse fault offsets of Miocene and younger age faults appear to indicate northwest-southeast compression. While some of these offsets may indicate a stress field considerably older than that indicated by the other data in this compilation, these features are nonetheless quite unusual.

Intraplate basins. The seaward extension observed in the Texas-Louisiana Gulf Coast stress province is quite disimilar to the compressional state of stress in the Midplate. We now interpret the size of this province to be somewhat larger than in our previous studies based on recently acquired data on stress magnitudes in wells in northeastern Texas that indicate an extensional state of stress. While it is clear that the pervasive normal faulting in this province is primarily restricted to the sedimentary section in this province (and the degree to which basement rocks are involved is unknown), the large extent of this province and the extreme thickness of the Gulf Coast sediments (about 10 km) warrant definition of a distinct stress province. A magnitude-5.0 thrust faulting earthquake in the center of the Gulf of Mexico occurred in 1978 (Frohlich, 1982). It is been suggested that the compression associated with this earthquake may result from flexure induced by the load of sediments in the Gulf (Frohlich, 1982; Nunn, 1985). Regardless of the exact cause of this earthquake, its style of faulting seems to demonstrate that the state of stress within the sedimentary basin cannot be extrapolated to the deeper crust with confidence.

In-situ stress and fluid pressure data from the Gulf Coast can be used to show that the state of stress within the sedimentary basin is controlled by the frictional strength of the ubiquitous normal faults both onshore and offshore. In areas of normal faulting, the minimum principal stress S_3 is equivalent to S_{hmin} and the maximum principal stress S_1 is equivalent to S_v (Anderson, 1951). We can extend Anderson's faulting theory to predict stress magnitudes at depth through utilization of simplified two-dimensional Mohr-Coulomb failure theory and the concept of effective stress. As suggested by Sibson (1974) and Brace and Kohlstedt (1980), we assume that the maximum shear stress in the crust cannot exceed that required to cause motion on preexisting faults that are optimally oriented to the principal stress field. Expressed in terms of the principal stresses, the limiting stress ratio can be written (after Jaeger and Cook, 1979) as:

$$\sigma_1/\sigma_3 = (S_1 - P)/(S_3 - P) = \left[(1+ \mu^2)^{1/2} + \mu\right]^2, \qquad (1)$$

where μ is the coefficient of friction and P is the pore pressure. It is common to refer to σ_i as a component of effective stress and S_i as a component of the total stress. A large number of in-situ stress measurements in areas of active faulting have shown this to be generally correct (e.g., Raleigh and others, 1972; McGarr and Gay, 1978; Brace and Kohlstedt, 1980; Zoback and Hickman, 1982; Pine and others, 1983; Zoback and Healy, 1984, 1991; Stock and others, 1985; Baumgärtner and Zoback, 1989; Baumgärtner and others, 1990) for coefficients of friction in the range consistent with laboratory-determined values of 0.6 to 1.0 (Byerlee, 1978). In the case of normal faulting, faulting is controlled by the difference between the effective vertical principal stress ($S_v - P$) and the effective least horizontal principal stress ($S_{hmin} - P$), which correspond to σ_1 and σ_3, respectively, in equation 1.

Data on the magnitude of the least horizontal principal stress, S_{hmin}, are available from hydraulic fracturing and leak-off tests in oil wells in the region (Breckels and Van Eekelen, 1981). These data show that at depths less than about 1.5 km, S_{hmin} is about 60 percent of the vertical stress. S_{hmin} increases rapidly at greater depth until, at depths greater than about 4 km, S_{hmin} is only slightly less than the vertical stress. Unfortunately, to test the applicability of equation 1 we need to know the value of pore pressure and published values of pore pressure in the Gulf Coast are quite rare. Bruce (1984) argues that pore pressure at depth in the Gulf generally corresponds to a hydrostatic fluid pressure gradient to a depth of about 1.5 km, and then increases with depth to a nearly lithostatic fluid pressures at depths greater than about 3 to 4 km. If this is the case, it is straightforward to show that a pore pressure increase with depth of this type is consistent with the general increase of S_{hmin} with depth reported by Breckels and Van Eekelen (1981). Equation 1 shows that for reasonable coefficients of friction, S_{hmin} should, in fact, be about 60 percent of S_v when hydrostatic pore pressure exists and 96 percent of S_v when pore pressure is about 95 percent of lithostatic. Thus overall, the generalized increase of S_{hmin} with depth reported by Breckels and Van Eekelen (1981) suggests that the magnitude of the least principal stress results from the equilibrium between the stress state and the frictional strength of the ubiquitous normal faults in the region.

S_{hmin} data as a function of depth have also been compiled for several intraplate basins in Canada (Bell and Babcock, 1986; Bell, 1989). As discussed by Adams and Bell (this volume), the least principal stress in the Scotian Shelf and Grand Banks basins increase with depth in a manner generally similar to that in the Gulf Coast and it is possible that an extensional stress state may also exist within these basins. Unfortunately, we cannot be confident of this interpretation for several reasons. First, we do not have the necessary pore pressure information to evaluate the validity of equation 1. Second, the available S_{hmin} data are limited in depth and may only reflect the state of stress in the upper several kilometers and not the crust as a whole. Nevertheless, it is interesting that within the region of North America generally characterized by a compressional stress field, limited regions of extensional stress exist within sedimentary basins.

Western Cordilleran provinces

This group of stress provinces comprises the most tectonically active parts of interior North American Plate and includes the Cordilleran extension province of the western United States and northern Mexico, the Southern Great Plains province, and the Colorado Plateau province of the western United States (Fig. 4). Although the tectonics of much of the thermally uplifted western United States and Mexico are primarily extensional, the distinct stress provinces within the Cordillera are defined on the basis of different directions of extension, crustal thickness, heat flow, and rates of tectonic activity. Only three earthquake focal mechanisms are available for the uplifted Canadian Cordillera, all of which indicate thrust faulting. Thus, although the data coverage is far too sparse to characterize the state of stress in the Canadian Cordillera, it is clear that the active extensional tectonism of the United States Cordillera does not seem to extend into Canada. Also note that a large region of central Mexican Cordilleran region is uncharacterized in Figure 4 due to the complete absence of data. Two additional stress provinces, the Pacific Northwest and the Trans-Mexican Volcanic Belt, lie geographically within the western North American Cordilleran region. As the state of stress in these provinces is probably related to relative plate motions between the Pacific and Cocos plates and the North American plate, these provinces are discussed in the next section on subduction-related provinces.

Cordilleran extension province. This stress province coincides generally with the broad thermally elevated region of the western United States (e.g., Eaton, 1979) and extends southward into northern Mexico. It incorporates the classic Basin and Range subprovince of extensive ~north-south-trending range-bounding normal faults, as well as the regions surrounding the Colorado Plateau: the Rio Grande Rift, and the middle Rocky Mountains and Denver Basin regions (Fig. 4). All these areas are characterized by extensional tectonics with a S_{hmin} orientation varying between west-northwest to east-northeast. Although seismicity occurs throughout the region (Engdahl and Rinehart, this volume), the rate of tectonic activity in southern parts of this province (northern Mexico, Arizona, New Mexico) and northeastern parts (middle Rocky Mountains and Denver Basin) it is considerably less than the part of the province coincident with the "Great Basin" of Nevada and Utah. As the state of stress, deformation and tectonics of much of the Cordilleran extension province have been recently discussed in detail (Zoback, 1989; Zoback and Zoback, 1989), we only briefly summarize key points in here.

Distinct subprovinces of uniform S_{hmin} orientation can be defined within the Cordilleran Extension province: (1) the northern Basin and Range and Rio Grande rift regions are characterized by a west-northwest S_{hmin}, (2) in southern Arizona, New Mexico, and northern Mexico, Quaternary faulting patterns indicate that the S_{hmin} direction is approximately east-west (Muehlberger and others, 1978), and (3) north of the Snake River Plain and in the Denver Basin and middle Rocky Mountains region (Colorado and Wyoming) the S_{hmin} orientation averages about east-northeast. Note that along the eastern boundary of the Cordilleran stress province the S_{hmin} directions are generally perpendicular to the topographic bulge suggesting a causal relationship. In contrast, strike-slip faulting is common along the western boundary of this province (suggesting that $S_v \sim S_{Hmax} >> S_{hmin}$), particularly in regions within the Walker Lane belt of western Nevada (Stewart, 1988), such as Owens Valley historic faulting at the eastern edge of the Sierra Nevada (Beanland and Clark, 1989), and the Furnace Creek fault zone near Death Valley (Burchfiel and others, 1987). The occurrence of strike-slip faulting along the western margin of the province indicates a component of superimposed shear related to the San Andreas transform system. The percentage of Pacific–North American Plate motion being accommodated by deformation within the northern Basin and Range is currently a topic of debate (Minster and Jordan, 1984; DeMets and others, 1987).

Colorado Plateau and Southern Great Plains Provinces. As illustrated in Figures 2 and 4, these two regions are unique within the western United States Cordillera in showing generally north-northeast, rather than approximately east-west extension. The Colorado Plateau, a relatively stable tectonic block throughout much of Phanerozoic time, is surrounded by active west-northwest extension on both its western and eastern border. Deformation within the Colorado Plateau interior is dominated by small-magnitude normal faulting earthquakes (Wong and Humphrey, 1989). Similarly, the southern Great Plains stress province, which is defined largely on the basis of trends of Quaternary basaltic volcanic vent alignments, is bordered on its western side by the west-northwest extension in the Rio Grande rift. It is difficult to define the northeastern boundary of the Southern Great Plains Province using currently available data. However, the extensional state of stress in this province is markedly different than the compression observed in the Midplate province. Southeast of this province a change in extension direction occurs as southeastward (gulfward) extension is observed in the Gulf Coast province.

Both the Colorado Plateau and Southern Great Plains have

354

crustal thickness 10 to 15 km greater than the 30- to 35-km-thick crust in the surrounding actively extending areas (Braile, 1989). Available geophysical data including seismic refraction, heat flow, gravity, deep electrical sounding, and teleseismic P-residuals suggest a dramatic thinning and heating of the upper mantle beneath the actively extending areas surrounding the Colorado Plateau interior and the Southern Great Plains. Thompson and Zoback (1978) first suggested that a "ridge push-type force" acting on the colder, denser upper mantle lids beneath the Colorado Plateau and Southern Great Plains may be responsible for the apparent nearly 90° rotation S_{hmin} orientations within these provinces relative to the actively extending areas.

Subduction-related plate boundary provinces

Nakamura and Uyeda (1980) attempted to synthesize the states of stress often found in continental plates that are overriding subduction zones. As shown in Figure 5 (from Nakamura and Uyeda, 1980) the state of stress can be divided into three distinct regimes: (1) in the accretionary wedge and forearc region the state of stress is highly compressional (thrust faulting) with the direction of S_{Hmax} parallel to the direction of relative plate motion; (2) in the volcanic arc the state of stress is strike-slip, again with the direction of S_{Hmax} parallel to the direction of relative plate motion; and (3) in the back-arc region the state of stress is extensional, with the direction of extension in the roughly parallel with direction of convergence. As pointed out by Nakamura and Uyeda, it is not necessary for all three of the stress domains illustrated in Figure 5 to be present in a given plate overriding a subduction zone; however, the general pattern shown in Figure 5 is widely observed. Thus, Figure 5 provides a useful framework discussion of the state of stress associated with the subduction zones along the western boundary of North America.

In an attempt to better understand the state of stress and tectonism in the regions adjacent to the western North American Plate boundary we have prepared a series of detailed stress maps in an oblique mercator projection about the pole of relative motion between the North American and the appropriate adjacent oceanic plate. In this type of projection, plate boundaries that are small circles on a globe become straight lines as does the direction of relative motion between the plates.

Aleutian subduction province and Aleutian back-arc province. As discussed in detail by Nakamura and others (1977) and Estabrook and Jacob (this volume) the state of stress in this province is dominated by the relative motion between the Pacific Plate and North America. All of the stress domains described in the Nakamura and Uyeda model can be found in the Aleutian subduction province although the region of back-arc extension is poorly documented. As shown by the oblique Mercator projection in Figure 6 (for a pole of relative motion between North America and the Pacific Plate after Minster and Jordan, 1978), we define this stress province to include the majority of southern Alaska, the Aleutian island arc, and the Bering Sea. The Aleutian back-arc province shown in Figure 6 is not discussed in detail because it is documented only by a scatter of normal faulting earthquakes and volcanic alignments within an extremely large area (see Estabrook and Jacob, this volume, for more details).

In the Aleutian subduction province, compression subparallel to the direction of relative plate motion is seen in the subduction zone, volcanic arc, and throughout an immense region behind the arc in Alaska (Fig. 6). Definition of the northern boundary of this province is difficult to define. There is little data in northwestern Alaska and a complex state of stress is observed in the area of the McKenzie Delta and the Yukon (Adams and Bell, this volume). It does appear, however, that the compressive state of stress associated with subduction may be limited to the region south of the Brooks Range and the Seward Peninsula (Estabrook and Jacobs, this volume).

Although most of the earthquakes concentrated along the plate boundary indicate compression directions that are generally in the direction of relative plate motion, some systematic differences are observed that indicate a partitioning of slip into down-dip and strike-slip components (see Fitch, 1972; Ekström and Engdahl, 1989). In the cases in which the relative motion direction is quite oblique (such as in the western Aleutians), there is an appreciable oblique component to subduction. As first de-

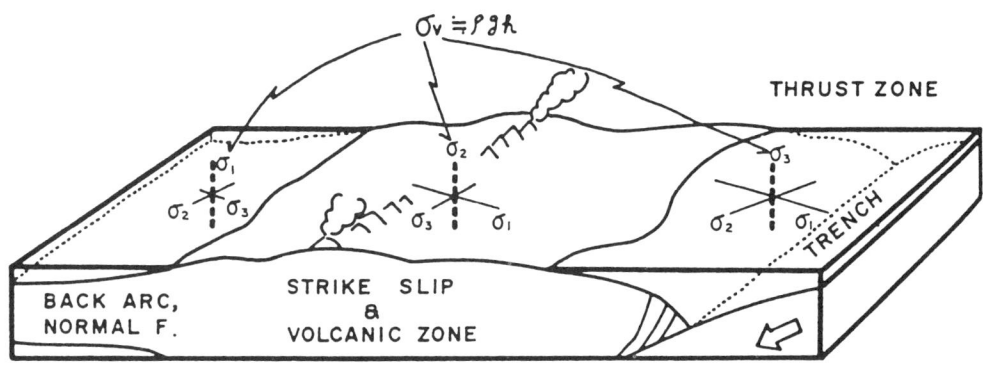

Figure 5. Generalized states of stress in shallow subduction zone, are, and back-arc regions (from Nakamura and Uyeda, 1980).

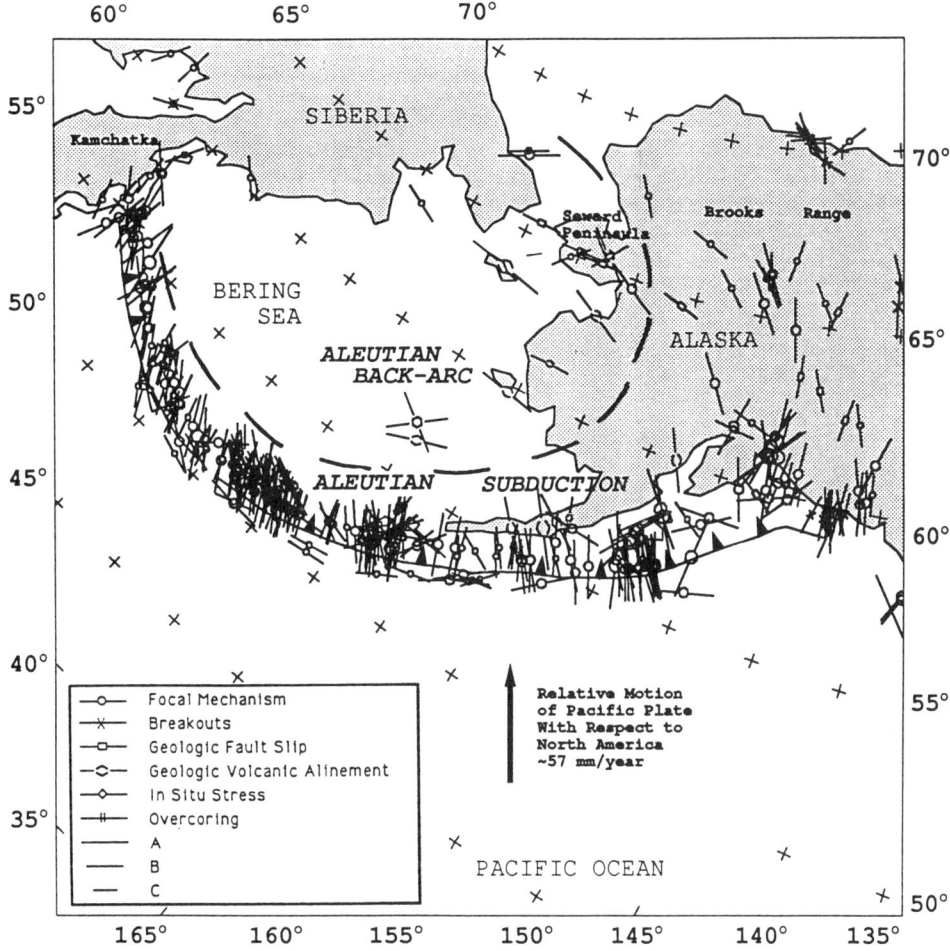

Figure 6. Stress map of Alaska and the Bering Sea. Oblique Mercator projection for a pole of relative motion between the Pacific Plate and North America after Minster and Jordan (1978). In this projection the Pacific Plate is moving toward North America along straight lines parallel to the arrow shown. The rate of convergence of the Pacific Plate is after DeMets and others (1990).

scribed by Fitch (1972), in areas of oblique subduction, slip partitions into dip-slip movement in the subduction zone and strike-slip movement behind the trench. As illustrated by Geist and others (1988), Ekström and Engdahl (1989), Ryan and Scholl (1989), and Estabrook and Jacobs (this volume), however, there is also evidence for both dip-slip and strike-slip motion accompanying subduction in several places along the Aleutian arc where the relative motion is just slightly oblique. In this case, purely horizontal slip may be occurring along the strike-slip faults paralleling the arc because of their low frictional resistence (see the discussion below for the San Andreas fault).

Figures 2 and 6 also show the normal-faulting earthquakes along the outer rise in the Pacific Plate that result from plate flexure (Chapple and Forsyth, 1979). These earthquakes involve normal faulting perpendicular to the subduction zone and are shown in the figures by the data points with a direction of maximum horizontal compression parallel to the subduction zone on the seaward side of the plate. Although it would have been possible to define this as a separate stress province, it did not seem warranted as the state of stress and crustal deformation there is so closely tied to the subduction process.

Central American subduction province. Figure 7 presents the state of stress in Mexico and Central America in an oblique Mercator projection about the pole of relative rotation between the Cocos Plate and North America (from Minster and Jordan, 1978). As shown in this figure, the Central American subduction province covers the southernmost portion of the North American Plate. It is separated from the Caribbean Plate by the ~east-west–trending Motagua transform system and its southwestern boundary is the Middle America trench (Fig. 7). Due to deficiencies in the stress catalog, few earthquakes are associated with the thrusting in the shallow subduction zone. However, thrust-faulting earthquakes associated with subduction of the Cocos Plate beneath North America are well known (Lefevre and McNally, 1985) and folding and crustal shortening in the direction of relative plate motion is well documented in the

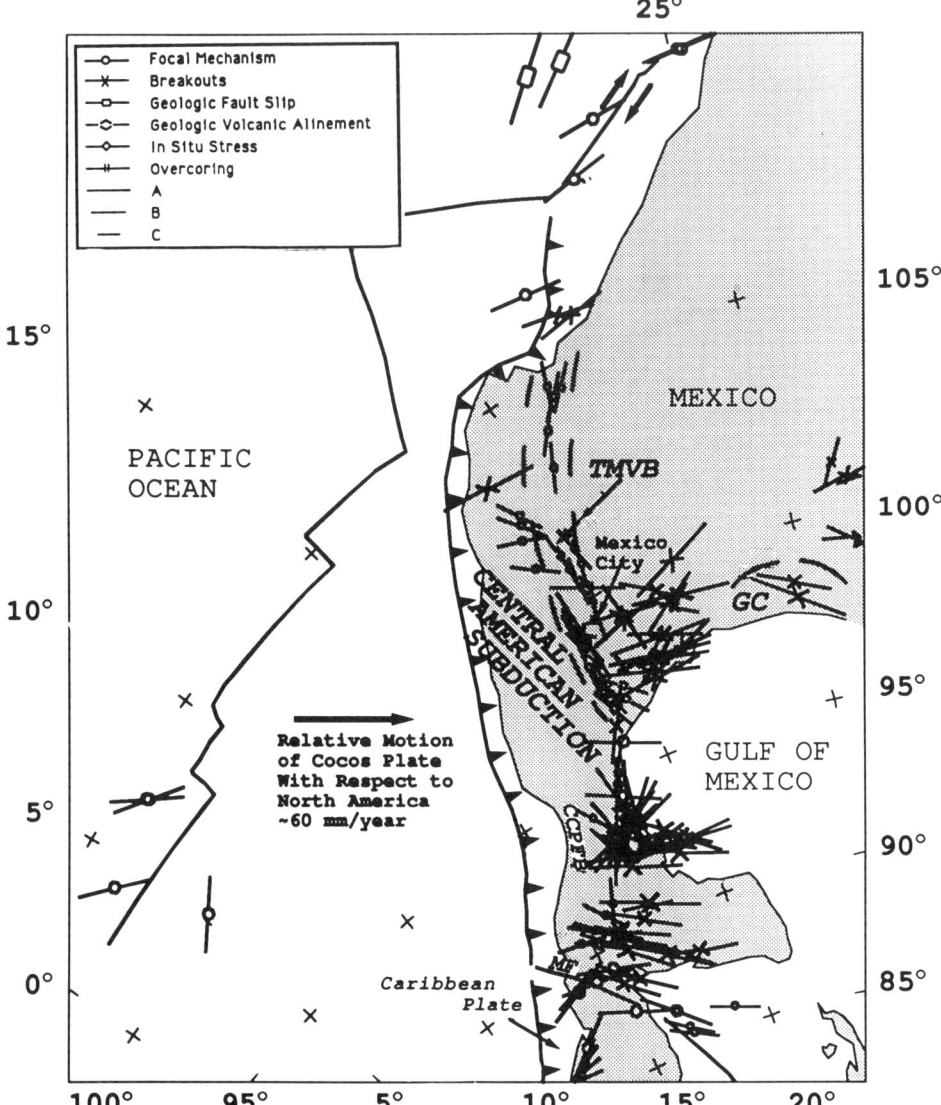

Figure 7. Stress map of Central America. Oblique Mercator projection for a pole of relative motion between the Cocos Plate and North America after Minster and Jordan (1978). The rate of convergence of the Cocos Plate is after DeMets and others (1990). and stand for the TMVB, Trans-Mexican volcanic belt; CCPFB, Compeche-Chiapas-Peten fold belt; MF, Motagua fault; GC, Gulf Coast.

accretionary prism by marine geophysical data (Moore and others, 1982; Moore and Shipley, 1988). As illustrated in Figure 7, the average S_{Hmax} direction in southernmost Mexico is generally consistent with that expected from relative motion of the Cocos Plate, although the state of stress in the region near 18°N, 92°W is clearly quite complex. As discussed by Suter (this volume), there is appreciable strike-slip deformation along the Chiapas fold belt, a series of strike-slip faults that are subparallel to the trench but well inland in southern Mexico. Suter refers to this region as the Compeche-Chipas-Peten fold belt stress province. We, however, simply include it in the Central American subduction province (Fig. 7) on the basis of similar stress orientations.

In the context of the states of stress characterized by Nakamura and Uyeda (1980), we see that the state of stress observed in the Central American subduction province is generally consistent with the region of high horizontal compression often found closest to the trench. There is, however, a marked transition from this compression to active extension observed in the Trans-Mexican volcanic belt (TMBV in Fig. 7).

Trans-Mexican volcanic belt. As defined by Suter (this volume), this narrow (~100 km wide) arcuate zone of active volcanism and approximately north-south extension extends across southern Mexico at a latitude of about 19°N. This province is distinguished from the Cordilleran extension province, which may lie immediately to the northeast, by its distinctly different extension direction. S_{hmin} orientations in the Trans-

Mexican volcanic belt are generally north-south in the central and eastern parts of the province, but somewhat more northeast-southwest in the western part. With the exception of a single focal-plane mechanism (which does not fit the overall stress pattern and shows ~east-west extension), all of the data on in-situ stress direction in this province comes from the alignment of volcanic cones. As pointed out by Suter (this volume), the trends of the volcanic cones are remarkably parallel to the number of active grabens within the province.

Nixon and others (1987) relate the extension and volcanism of this province to subduction and consider this region as a calc-alkalic volcanic belt. As is evident in Figure 7, the trend and location of the Trans-Mexican volcanic belt is approximately that which would normally be expected for a volcanic arc considering the geometry of the trench and the direction of relative movement between the Cocos Plate and North America. Note, however, that S_{Hmax} orientations within this belt are perpendicular to the convergence direction, not parallel to it as is observed in most island arcs on continental plates (Fig. 5; Nakamura and Uyeda, 1980). The extension direction in the Trans-Mexican volcanic belt, generally perpendicular to the trench and perpendicular to the convergence direction, is more similar to the stress pattern Nakamura and Uyeda describe for back-arc basins. The discontinuous nature of the volcanic arc in southern Mexico is interesting and not understood, as farther south in Central America in Panama and Costa Rica there is a well-developed volcanic arc and significant strike-slip faulting occurs there between the trench and volcanic arc along the Soná-Azuero and the Celmira-Ballena faults (Mann and Corrigan, 1990), which may be a continuation of the Compeche-Chiapas-Peten fold belt.

Cascade convergence province. To the north of the Mendocino triple junction, the western boundary of the North American Plate is a subduction boundary with the Juan de Fuca Plates subducting beneath northernmost California, Oregon, Washington, and southern British Columbia (Fig. 8). Figure 8 is an oblique Mercator projection about the pole of relative motion between the Juan De Fuca Plate and North America (after DeMets and others, 1987). While there is not a well-developed trench or Benioff zone associated with subduction of the Juan de Fuca Plate, availalbe marine geophysical evidence indicates a rate of relative convergence of approximately 41 mm/yr (DeMets and others, 1990). The state of stress in the region (Magee and Zoback, 1989) is indicated by earthquake focal plane mechanisms in western Washington and on Vancouver Island as well as several breakouts in wells along the Columbia River in western Washington. The sparse S_{Hmax} orientations within the continental portion of this province direction are generally ~45° to the direction of relative plate motion (Fig. 7), quite unlike the Aleutian or Central American subduction zones. This may suggest rather weak coupling between the subducting Juan de Fuca Plate and overriding North America. However, there is evidence of strain accumulation in the upper plate that has been interpreted to be the result of aseismic slip at depth along the megathrust (Savage and others, 1981). As such strain would accumulate only if the upper portion of the subduction was locked, this interpretation of the geodetic strain data suggest that the two plates are coupled at shallow depth. If this interpretation is correct, it is not clear why this coupling is not reflected in the stress field.

Pacific Northwest province. The Pacific Northwest stress province is shown in Figures 4 and 8. This province includes most of the North American lithosphere overlying the subducted Juan de Fuca Plate including the Cascade volcanic arc. It is characterized by ~north-south compression associated with strike-slip faulting earthquakes along the Cascades and folding and reverse faulting earthquakes farther east. While all but the western boundary of this province is very poorly defined, the state of stress in this region is distinctly different from the extension observed to the southeast in the Cordilleran extension province and the more northeasterly compression observed along the coast in the Cascade convergence province. The origin of the state of stress in the Pacific Northwest province does not appear directly related to the convergence between the Juan de Fuca and North American Plates. However, while a strike-slip stress regime within the Cascade volcanic arcs is consistent with the general Nakamura and Uyeda model (Fig. 5), the roughly north-south S_{Hmax} orientations differ markedly from it as S_{Hmax} is not parallel to the convergence direction but is at an angle of about 60 to 70°. In fact, there are abundant indicators of ~north-south compression both to the east and west of the Cascades. As suggested previously by Sbar (1982) and Zoback and Zoback (1989), the north-south compression throughout the Pacific Northwest province may be related to a broad zone of northwest shear arising from Pacific–North American relative plate motion.

Transform-related plate-boundary provinces

This category of stress provinces includes the San Andreas and Queen Charlotte transform systems and the mixed transform-/oceanic rifting plate boundary stress province of the Gulf of California.

San Andreas province. Figure 9 presents a map of S_{Hmax} orientation in the San Andreas stress province plotted in an oblique Mercator projection using the pole of relative motion between the Pacific and North American Plates after Minster and Jordan (1978). Because the data density is so high in central California, where the majority of data come from earthquake focal-plane mechanisms and wellbore breakouts, only A- and B-quality data are shown on this map.

It is immediately clear in Figure 9 that, although the strike of the San Andreas fault in central California is almost parallel to the direction of relative plate motion, the direction of maximum horizontal compression is almost perpendicular to the relative plate motion direction. While quite surprising, this stress orientation is consistent with abundant geologic data indicating crustal shortening (folding and thrust faulting) nearly perpendicular to the trend of the San Andreas throughout central California (see, e.g., Page, 1981). Zoback and others (1987) and Mount and Suppe (1987) discuss the state of stress along the San Andreas

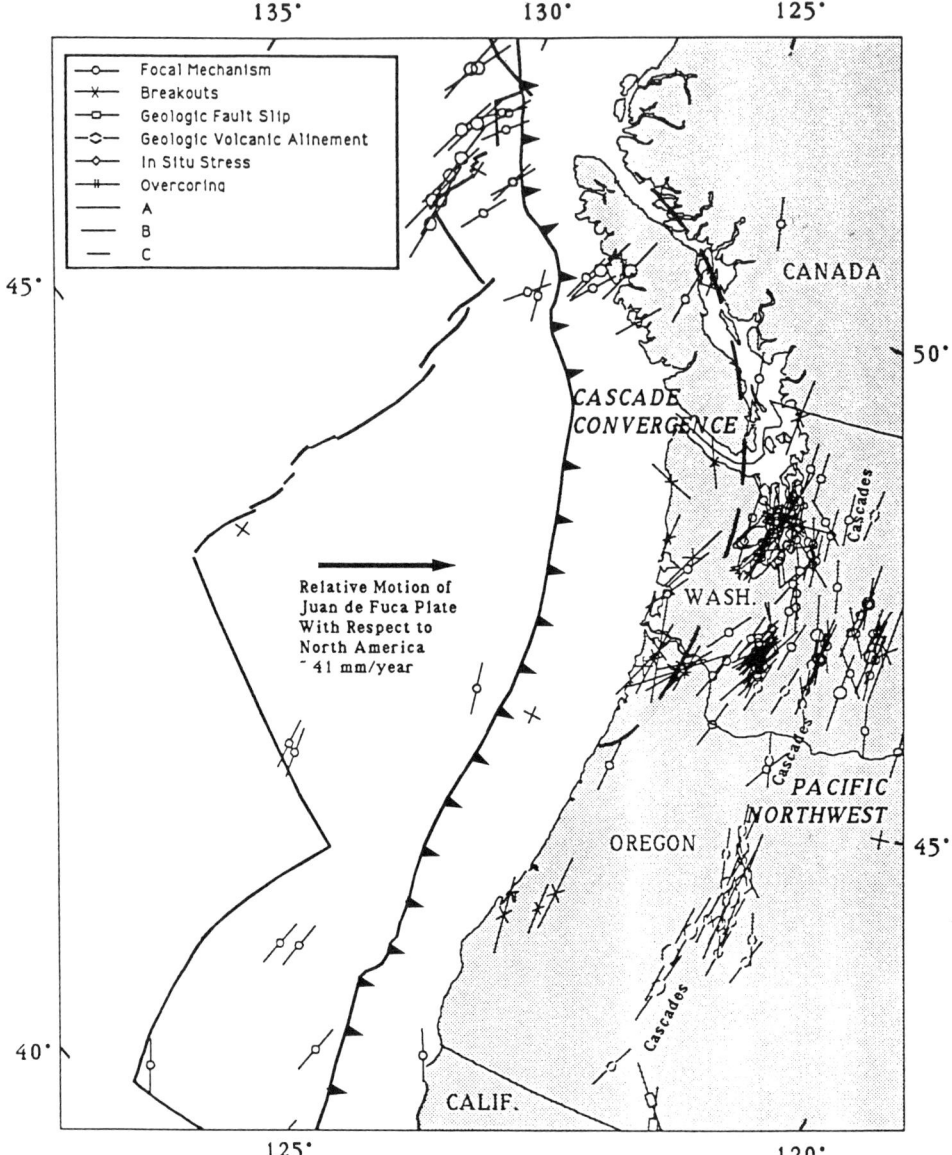

Figure 8. Stress map of western Oregon, Washington, and southwestern British Columbia. Oblique Mercator projection for a pole of relative motion between the Pacific Plate and North America after DeMets and others (1990). The rate of convergence of the Juan de Fuca Plate is also after DeMets and others (1990). The Cascade convergence province is poorly documented but appears to indicate compression about 45° from the direction of relative plate motion.

system in detail, especially with respect to the implications of the near fault-normal compression for the low frictional strength of the fault. There is substantial data from conductive heat flow measurements in shallow boreholes along the San Andreas (typically ~300 m deep), which also suggests that the San Andreas moves at extremely low levels of shear stress (see recent reviews of these data by Lachenbruch and Sass, 1988, 1991). Because of this, P axes determined from the focal plane mechanisms of right-lateral strike-slip earthquakes on the San Andreas may simply reflect the geometry of fault movement and not the orientation of the maximum horizontal stress responsible for that motion (MacKenzie, 1969); therefore there were not included in Zoback and others (1987). A number of these events are included in Figure 9 and show the expected pattern of P axes oriented 45° to the strike of the San Andreas for right-lateral strike-slip earthquakes located directly on the San Andreas and its principal branches.

In addition to the heat-flow data along the San Andreas and pattern of crustal stresses shown in Figure 9, considerable other information indicates that the San Andreas moves at extremely

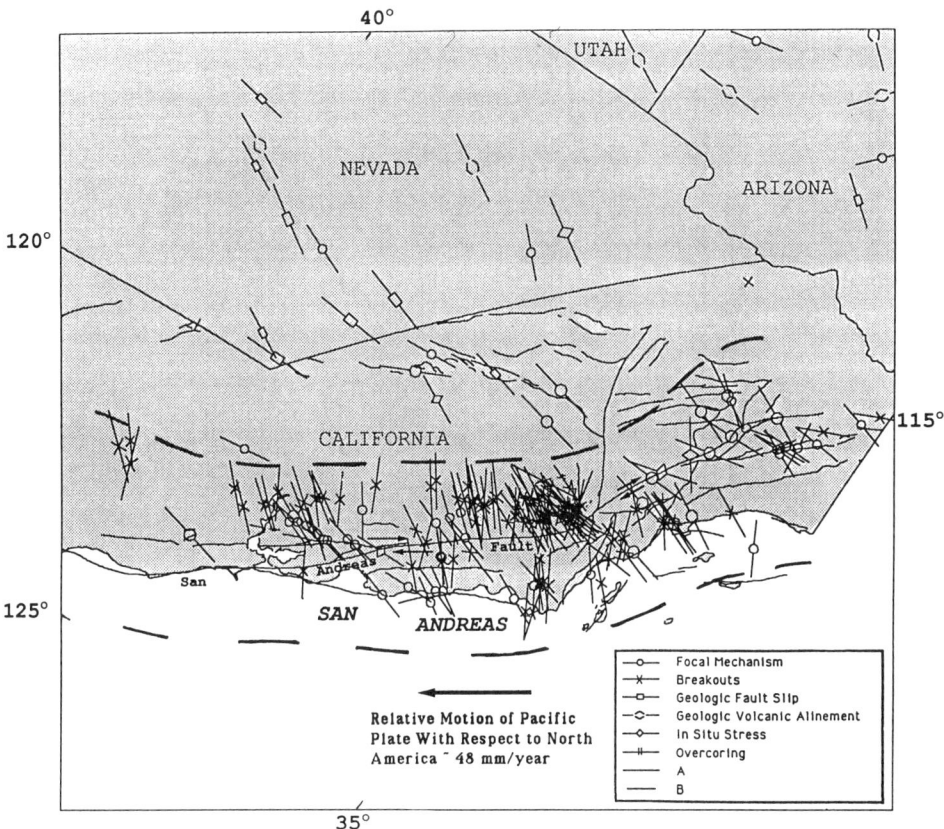

Figure 9. Stress map of California and western Nevada. Oblique Mercator projection for a pole of relative motion between the Pacific Plate and North America. The rate of motion of the Pacific Plate is after DeMets and others (1990). The San Andreas stress province comprises a broad zone adjacent to the San Andreas where the direction of maximum horizontal compression is at a very high angle to the strike of the fault.

low shear stress. It was proposed almost 20 years ago that the orthogonal relationship between oceanic transform faults and spreading centers implies that the transforms slip at extremely low levels of shear stress (Lachenbruch and Thompson, 1972; Oldenburg and Brune, 1972, 1975). More recently, a complete lack of shear stress on planes parallel to the San Andreas was found in a 3.5-km-deep borehole drilled adjacent to the fault at Cajon Pass, California (Zoback and Healy, 1991), a site presumably quite late in the earthquake cycle. In this same borehole, no evidence of fault-generated frictional heat was found nor was any evidence found of heat redistribution due to thermal convection, thereby substantiating the measurements in the shallow boreholes along the length of the fault (Lachenbruch and Sass, 1991). In light of such an impressive set of data, it is worth considering the relationship between the state of stress and crustal deformation along transform plate boundaries if the frictional strength of the transform is quite low.

Figure 10 a and b is a representation of two possible states of stress along weak transform fault zones based on an extension of the two-dimensional Mohr failure analysis of strike-slip faulting described previously by Zoback and others (1987). The basis of this simple two-dimensional stress analysis is the assumption that the San Andreas fault zone represents a "weak fault inbedded in a strong crust." If the average shear strength of the San Andreas fault is quite low (~10 MPa), compared to the average strength of the upper crust adjacent to the fault (~100 MPa), then the horizontal principal stresses near the fault have to be oriented approximately parallel and perpendicular to the fault so as to minimize shear stress in the plane of the weak fault. Either fault-normal compression or fault-normal extension would result depending on whether the relative plate motion is convergent, as has been the case for the central San Andreas for the past 4 m.y. (left panel of Fig. 10a), or divergent, as was the case 10 to 4 Ma (left panel of Fig. 10b). As illustrated in Figure 11, a remarkably different relationship between stress and crustal deformation occurs along weak transform margins than predicted on the basis of conventional faulting theory. On the left side of Figure 11, in a and b, we illustrate that a conventional way to account for compression of extensional deformation along transform boundaries is in the context of stepovers, or bends, in fault strike. While localized compressional and extensional features certainly exist along discontinuities of major transforms (see, e.g., Bilham and King, 1989), a pervasive pattern of fault-normal compressional stress and crustal shortening (right side of Fig. 11a) such as cur-

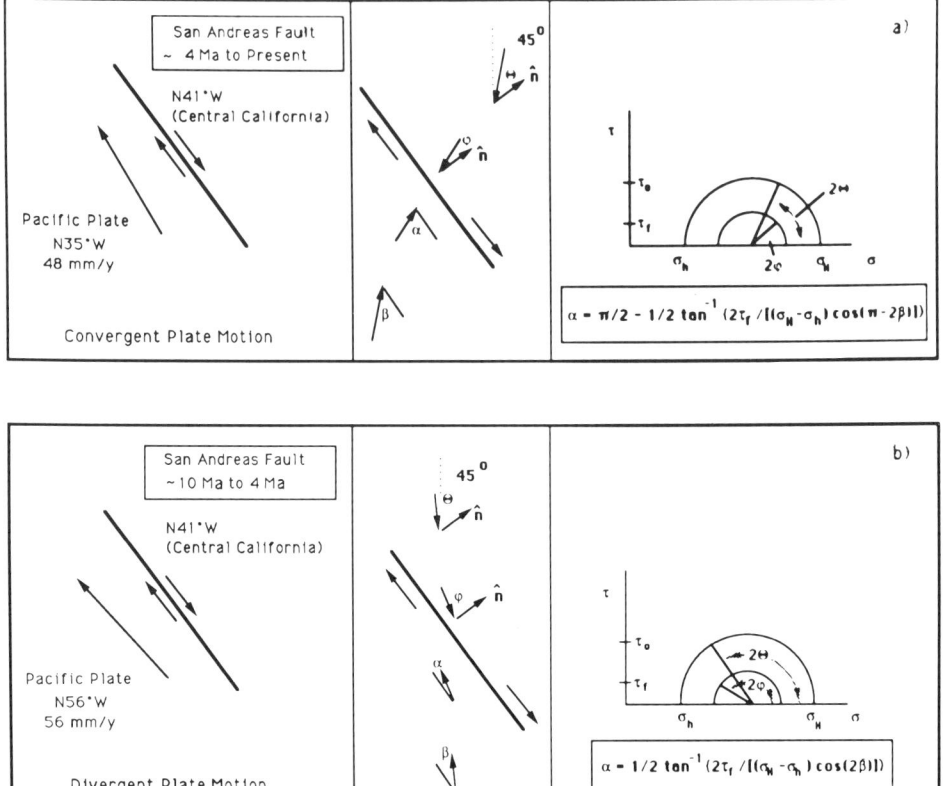

Figure 10. A simple two-dimensional frictional model (after Zoback and others, 1987) to predict the change in stress orientation in the vicinity of a weak transform fault as illustrated in the right panels. The left sides of the panels indicate the relative plate motion directions for the past 10 m.y. in central California along the San Andreas fault system. For about the last 4 m.y. the relative plate motion has been convergent (upper left), and from about 10 Ma to 4 Ma it was divergent (lower left). Using the simple model, if the plate motion is relatively convergent (a), the direction of maximum principal stress near the fault is either near-perpendicular (middle panel), whereas if the plate motion is relatively divergent (b), the direction of least horizontal compresson is perpendicular to the strike of the fault (middle panel). The angles between the strike of the fault and the maximum principal stress in the vicinity of the fault is defined by the angle α and that far from the fault is defined the angle β (see also Fig. 12).

rently observed in central California, or fault-normal extension (right side of Fig. 11a) such as currently observed along the Dead Sea transform (Eyal and Reches, 1983), are a logical consequence of weak faults imbedded in a strong crust.

Using equations originally derived by Zoback and others (1987; which are shown in the right panels of Fig. 10 with a typographical error that appeared in the original corrected here), the orientation of the maximum principal stress near the fault can be computed as a function of the far-field stress orientation. Figure 12 presents the values of the far-field and near-field angles for different values of average fault strength. Note that except when the far-field stress orientation is almost exactly 45° to the fault, a rotation of the near-field maximum stress direction is predicted and the amount of this rotation depends on the assumed strength of the fault. The sense of rotation (resulting in either fault-normal compression or fault-normal extension) depends on whether the far-field stress orientation relative to the fault is a result of convergent or divergent plate motion. In the former case it is assumed that the far-field direction of maximum horizontal compression is more than 45° to the San Andreas and in the latter case it is assumed that it is less than 45°. Although this model is quite simple, it provides a conceptual framework for interpreting fault-normal extension or fault-normal compression near major, lithospheric-scale transform faults. In addition, this model helps understand rapid transitions from one mode to another in response to changes in relative plate motion. For example, throughout much of middle to late Miocene time, tectonism adjacent to the San Andreas was dominated by basin formation and fault-normal extension; from 4 to 5 Ma the modern compressional deformation was initiated along much of the length of the San Andreas creating folds and reverse faults striking subparallel to the San Andreas and closing the previously formed basins.

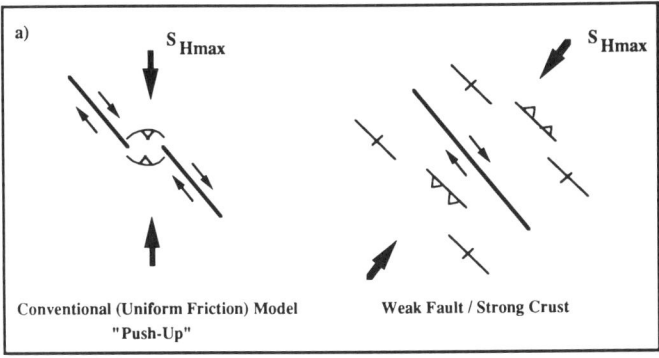

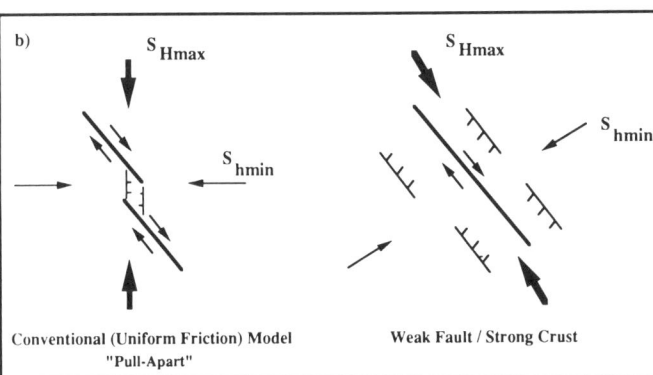

Figure 11. The occurrence of compressional and extensional features along strike-slip faults is usually explained in the context of offsets in fault strike and step-overs (left side). However, for the case of a weak fault inbedded in a strong crust the relationship between the state of stress and patterns of deformaton are as illustrated on the right side of the figure and either fault-normal compression (a) or fault-normal extension (b) is expected.

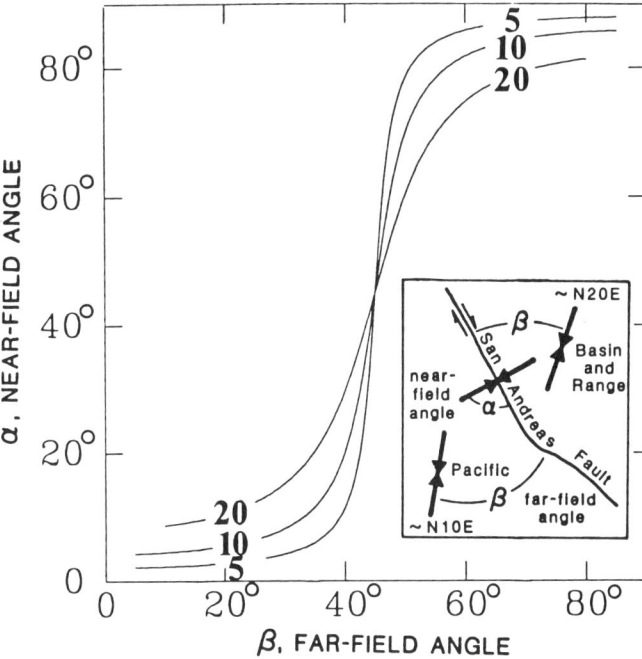

Figure 12. Utilizing the model illustrated in Figure 10, the relationship between the far-field and near-field direction of maximum principal stress can be determined for values of average fault strength of 5, 10, and 20 MPa. When the far-field stress direction is more than 45° (corresponding to convergent plate motion) near fault-normal compression occurs in the near field. In the case of divergent plate motion (a farfield stress direction of less than 45°), fault-normal extension occurs. Angles α and β are the same as in Figure 10.

Page and Engebretsen (1984) and Cox and Engebretson (1985) suggested that this major change in style of tectonism along the San Andreas fault, which occurred approximately 4 to 5 Ma, corresponds to a change in absolute motion of the Pacific Plate. This resulted in relative motion vectors that changed from slightly divergent in Miocene time to slightly convergent. Zoback and others (1987) argued that the fault-normal compression currently observed along the San Andreas began due to this change in relative plate motion. Similarly, Eyal and Reches (1983) and Letouzey (1986) discuss a transition from fault-normal compression to fault-normal extension in the vicinity of the Dead Sea Rift, which also can be related to far-field changes in relative plate motion vectors.

South of the "big bend" of the San Andreas in California, the pattern of stress is somewhat more complex than in central California (Fig. 9). The direction of maximum horizontal compression near the San Andreas fault (Jones, 1988) and in the Los Angeles basin (Hauksson, 1990) make a 60 to 70 angle with the overall strike of the San Andreas. Nevertheless, there seem to be an appreciable amount of data along the Pacific Coast and in the eastern Transverse Ranges that indicates compression at a very high angle to the strike of the San Andreas suggesting that perhaps the state of stress in the Los Angeles basin is somewhat anomalous to the overall pattern, perhaps related to anomalously dense material in the lower crust beneath the western Transverse Ranges (Sonder, 1991).

In the southernmost part of this province, the Salton Trough-Cerro Prieto region, there are very few data (other than the focal-plane mechanisms associated with a complex pattern of crustal extension and right-lateral strike-slip faulting) with which to try to define the stress field. Nevertheless, there are two wellbore breakouts that indicate fault-normal extension in this region. While it is difficult to draw much of a conclusion on the basis of only two data points, it is interesting that the evidence for fault-normal extension comes from a place where the strike of the San Andreas is more northerly than elsewhere. In fact, if one considers the southern extension of San Andreas system to be the Imperial/Cerro Prieto fault system in southernmost California and Mexico, a component of divergent plate motion would be expected in this region (e.g., Lachenbruch and others, 1985).

Queen Charlotte province. The Queen Charlotte fault is a major transform fault system connecting the Juan de Fuca Ridge with the Aleutian subduction zone. As shown in the oblique

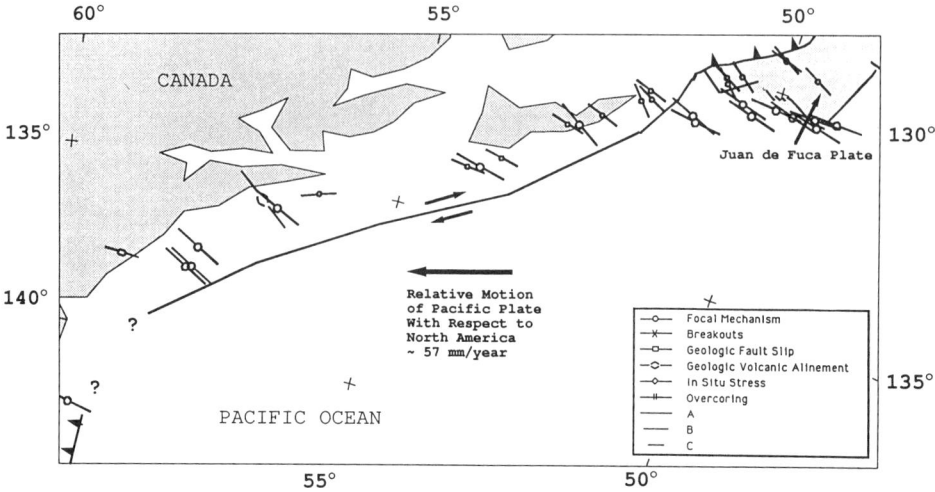

Figure 13. Stress map of the Queen Charlotte fault system. Oblique Mercator projecton for a pole of relative motion between the Pacific Plate and North America after Minster and Jordan (1978). The rate of motion of the Pacific Plate is after DeMets and others (1990). Complexity of the northern end of plate boundary near the intersection of the Queen Charlotte and Aleutian subduction provinces is ignored.

Mercator projection stress map of this province (Fig. 13), all of the stress information in this region comes from earthquake focal plane mechanisms. Since the fault is offshore for much of its length, and detection and accurate location of relatively small earthquakes (M < 4) is difficult, and there are relatively few earthquake focal plane mechanisms to indicate the state of stress. Nevertheless, the available focal-plane mechanisms shown in Figure 13 (which include both strike-slip and thrust-faulting events) are clearly not directly on the trace of the fault, and remarkably, like the San Andreas, there is a large (60 to 70°) angle between the inferred S_{Hmax} orientations and the strike of the fault. Significantly, the trace of the Queen Charlotte, as represented on Figure 13, is far from being a small circle about the pole of relative motion between the North American and Pacific Plates. This oblique trend of the Queen Charlotte relative to the relative velocity vector would also require a significant component of compression across the region adjacent to the fault. Thus, while the Queen Charlotte data set is sparse, the available data do suggest that the maximum horizontal stress is at a high angle to the fault. In the context of the discussion above for the San Andreas, the observed pattern of stress may be the result from the weakness of this plate boundary, but clearly the available data are much too sparse to draw this conclusion with certainty.

Gulf of California province. The boundary between the Pacific and North American Plates in the Gulf of California is principally a transform fault with sea-floor spreading restricted to the mouth of the Gulf (Lonsdale and Lawver, 1980). To the north, the Gulf of California evolves into the San Andreas stress province via the Cerro Prieto/Imperial fault systems in the Salton Trough. As shown in oblique Mercator projection in Figure 14, the available data on the state of stress in the Gulf of California consists of a number of geologic stress indicators along the Baja Peninsula and a scattering of earthquake focal plane mechanisms elsewhere (see Suter, this volume). The shape of the plate boundary as shown in Figure 14 is simplified after Lonsdale and Lawver (1980). Northwest-trending transform faulting and normal faulting dominates the tectonics of Baja California (Allen and others, 1960; Lomnitz and others, 1970; Angelier and others, 1981; Engdahl and Rinehart, this volume) as well as the San Benito and Tosco Abreojos faults to the west of the Baja Peninsula.

Utilization of the weak fault model discussed above for explanation of the stress data in the Gulf of California stress province is problematic. As along the San Andreas, we should probably ignore the focal plane mechanism data going up the center of the Gulf, which could presumably be located on the plate boundary and thus yielding no useful information aobut stress direction. If we do this, we are basically left with the stress data along Baja California that seem to be almost bimodally distributed. About half the data show a ~northeast-southwest S_{hmin} direction indicating transform-normal extension basically consistent with the pattern expected for low-friction faults. This is especially true along the southern tip of the peninsula. The other half of the data, however, indicate an approximately east-west direction of least horizontal stress, a direction more consistent with that expected from conventional faulting theory. Thus, it is not clear at this time whether the state of stress and crustal deformation in Baja is supportive of the weak transform hypothesis or not.

CONCLUSIONS

With suitable care, a variety of types of data can be used to map the crustal stress field. Despite their somewhat subjective nature, the criteria and quality-ranking system outlined in Table 1

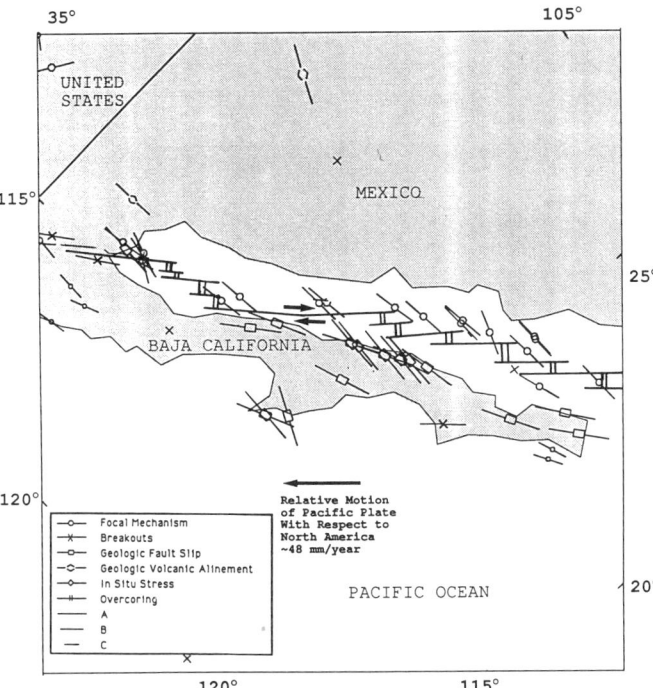

Figure 14. Stress map of the Gulf of California region. Spreading centers and transform faults generalized from Lonsdale and Lawver (1980). Oblique Mercator projection for a pole of relative motion between the Pacific Plate and North America after Minster and Jordan (1978). The rate of motion of the Pacific Plate is after DeMets and others (1990).

seems to work fairly well in defining the direction of horizontal principal stresses within the interior of the North American Plate. The general invariance of horizontal principal stress orientations with depth in the crust seen in previous studies is confirmed here. As summarized in Table 2, regionally uniform stress orientations are observed throughout broad regions that are generally correlative with physiographic provinces of North America. In many places the boundaries between these provinces are uncertain because of deficiencies of the data set.

Overall, the tectonic stress field can be seen to correlate with large-scale plate-tectonic processes such as plate-driving forces in the Midplate stress province, relative plate motion along the western plate boundary, and regional extension in the Cordilleran extension province. In many places the first-order uniformity can be seen to be modulated by interaction with more local-scale crustal processes such as the effects we attribute to low-friction transform faults along the San Andreas, Queen Charlotte, and Gulf of California transform systems, and flexural forces related to sediment loading on the continental shelf of the eastern seaboard of the United States and in the Gulf of Mexico. As more detailed data become available, it should be possible to quantitatively model these processes, more precisely estimate crustal stress magnitudes, and gain a better understanding of the mechanisms responsible for the crustal faulting and deformation.

ACKNOWLEDGMENTS

We thank Marian Magee for her help in maintaining the North American in-situ stress data base and furthering development of the plotting programs used in this study; and to Susanne Kothe for helping us to prepare the figures. We also appreciate the beneficial reviews of David Hill and Walter Mooney. This manuscript was written while MDZ was on sabbatical leave at the University of Karlsruhe, Germany. The financial support of the Alexander von Humboldt Stiftung is gratefully acknowledged.

REFERENCES CITED

Adams, J., 1987, Canadian crustal stress database; A compilation to 1987; Geological Survey of Canada Open-File 1622, 130 p.

Aldrich, M. J., Jr., and Laughlin, A. W., 1984, A model for the tectonic development of the southeastern Colorado Plateau boundary: Journal of Geophysical Research, v. 89, p. 10207–10218.

Allen, C. R., Silver, L. T., and Stehli, F. G., 1960, Agua Blanca fault; A major transverse structure of northern Baja California, Mexico: Geological Society of America Bulletin, v. 71, p. 457–482.

Anderson, E. M., 1951, The Dynamics of faulting and dyke formation with applications to Britain (second edition): Edinburgh, Oliver and Boyd, 206 p.

Anderson, R., and Zoback, M. D., 1989, The Oroville, California scientific drillhole; Did surface compression arrest propagation of rupture during the Oroville 1975 normal-faulting earthquake?: Scientific Drilling, v. 1, p. 82–89.

Angelier, J., 1979, Determination of the mean principal directions of stresses for a given fault population: Tectonophysics, v. 56, p. T17–T26.

—— , 1984, Tectonic analysis of fault slip data sets: Journal of Geophysical Research, v. 89, p. 5835–5848.

Angelier, J., Colletta, B., Chorowicz, L., Ortlieb, L., and Rangin, C., 1981, Fault tectonics of the Baja California Peninsula and the opening of the Sea of Cortez, Mexico: Journal of Structural Geology, v. 3, p. 347–357.

Artyushkov, E. V., 1973, Stresses in the lithosphere caused by crustal thickness inhomogeneities: Journal of Geophysical Research, v. 78, p. 7675–7708.

Barton, C. A., Tessler, L., and Zoback, M. D., 1991, Interactive analysis of borehole televiewer data, in Palaz, I., and Sengupta, S., eds., Automated pattern analysis in petroleum exploration: New York, Springer Verlag, xxx p.

Baumgärtner, J., and Zoback, M. D., 1989, Interpretation of hydraulic fracturing pressure-time records using interactive analysis methods (abs.): International Journal of Rock Mechanics and Mining Sciences and Geomechanics, v. 26, p. 461–1470.

Baumgärtner, J., Rummel, F., and Zoback, M. D., 1990, Hydraulic fracturing in situ stress measurements to 3 km depth in the KTB pilot hole VB; A summary of preliminary data evaluation: KTB-Report, v. 90-6a, p. 353–400.

Beanland, S., and Clark, M. M., 1989, The Owens Valley fault zone, eastern California, and surface rupture associated with the 1872 earthquake (abs.): Seismological Research Letters, v. 58, p. 32.

Becker, A., Blümling, P., and Müller, W. H., 1987, Recent stress field and neotectonics in the eastern Jura Mountains, Switzerland: Tectonophysics, v. 135, p. 277–288.

Bell, J. S., 1989, Investigating stress regimes in sedimentary basins using information from oil industry wireline logs and drilling records, in Hurst, A., ed., Geological applications of wireline logs: Geological Society of London Special Publication 48, p. 305–325.

Bell, J. S., and Babcock, E. A., 1986, The stress regime of the Western Canadian Basin and implications for hydrocarbon production: Bulletin of Canadian Petroleum Geology, v. 34, p. 364–378.

Bell, J. S., and Gough, D. I., 1979, Northeast-southwest compressive stress in Alberta; Evidence from oil wells: Earth Planetary Science Letters, v. 45, p. 475–482.

——, 1983, The use of borehole breakouts in the study of crustal stress, *in* Zoback, M. D., and Haimson, B. C., eds., Hydraulic fracturing stress measurements: Washington, D.C., National Academy Press, p. 201–209.

Bilharn, R., and King, G., 1989, The morphology of strike-slip faults; Examples from the San Andreas fault, California: Journal of Geophysical Research, v. 94, p. 10204–10214.

Bott, M.H.P., and Dean, D. S., 1972, Stress systems at young continental margins: Nature Physical Science, v. 235, p. 23–25.

Bott, M.H.P., and Kusznir, N., 1984, The origin of tectonic stress in the lithosphere: Tectonophysics, v. 105, p. 1–13.

Brace, W. F., and Kohlstedt, D. L., 1980, Limits on lithospoheric stress imposed by laboratory experiments: Journal of Geophysical Research, v. 85, p. 6248–6252.

Braile, L. W., 1989, Crustal structure of the continental interior: Geological Society of America Memoir 172, p. 285–315.

Breckels, I. M., and Van Eekelen, H.A.M., 1981, Relationship between horizontal stress and depth in sedimentary basins, *in* 56th Annual Fall Technical Conference, Society of Petroleum Engineers of AIME, San Antonio, Texas, October 5–7, 1981: Society of Petroleum Engineers, Paper SPE 10336, 7 p.

Bruce, C. H., 1984, Smectite dehydration; its relation to structural development and hydrocarbon accumulation in northern Gulf of Mexico basin: American Association of Petroleum Geologists Bulletin, v. 68, p. 673–683.

Burchfiel, B. C., Hodges, K. V., and Royden, L. H., 1987, Geology of Panamint Valley–Saline Valley pull-apart system, California; Palinspastic evidence for low-angle geometry of a Neogene range-bounding fault: Journal of Geophysical research, v. 92, p. 10422–10426.

Byerlee, J. D., 1978, Friction of rock: Pure Applied Geophysics, v. 116, p. 615–626.

Carey, E., 1979, Recerche des directions principales de contraintes associees au jeu d'une population de failes: Revue de Geologie Dynamique et Geographie Physique, v. 21, p. 57–66.

Carey-Gailhardis, E., and Mercier, J. L., 1987, A numerical method for determining the state of stress using focal mechanisms of earthquake populations; Application to Tibetal teleseisms and microseismicity of southern Peru: Earth Planetary Science Letters, v. 82, p. 165–179.

Chapple, W., and Forsythe, D., 1979, Earthquakes and bending of plates at trenches: Journal of Geophysical Research, v. 84, p. 6729–6749.

Cox, A., and Engebretson, D., 1985, Change in motion of Pacific Plate at 5 Myr BP: Nature, v. 313, p. 472–474.

Cox, J. W., 1983, Long axis orientation in elongated boreholes and its correlation with rock stress data, *in* Proceedings, 24th Annual Logging Symposium, Society of Professional Well Log Analysts: Society of Professional Well Log Analysts, 17 p.

Dart, R., 1985, Horizontal stress directions in the Denver and Illinois Basins from the orientations of borehole breakouts: U.S. Geological Survey Open-File Report 85-733, 41 p.

Dart, R., and Zoback, M. L., 1987, Principal stress orientations on the Atlantic continental shelf inferred from the orientations of borehole elongations: U.S. Geological Survey Open File Report 87-283, 43 p.

DeMets, C., Gordon, R. G., Stein, S., and Argus, D. F., 1987, A revised estimate of Pacific–North American motion and implications for western North American Plate boundary zone tectonics: Geophysical Research Letters, v. 14, p. 911–914.

DeMets, C., Gordon, R. G., Argus, D. F., and Stein, S., 1990, Current plate motions: Geophysical Journal International, v. 101, p. 425–478.

Dziewonski, A., and Woodhouse, J., 1983, An experiment in systematic study of global seismicity: Centroid-moment tensor solutions for 201 moderate and large earthquakes of 1981: Journal of Geophysical Research, v. 88, p. 3247–3271.

Eaton, G. P., 1979, Regional geophysics, Cenozoic tectonics, and geologic resources of the Basin and Range Province and adjoining regions, *in* Newman, G. W., and Goode, H. D., eds., 1979 Basin and Range Symposium: Denver, Rocky Mountain Association of Geologists and Utah Geological Association, p. 11–39.

Ekström, G., and Engdahl, R., 1989, Earthquake source parameters and stress distribution in the Adak Island region of the central Aleutian Islands, Alaska: Journal of Geophysical Research, v. 94, p. 15499–15519.

Engelder, T., and Sbar, M. L., 1984, Near-surface in situ stress; Introduction: Journal of Geophysical Research, v. 89, p. 9321–9322.

Evans, K., 1989, Appalachian stress study 3; Regional scale stress variations and their relation to structure and contemporary tectonics: Journal of Geophysical Research, v. 94, p. 17619–17645.

Evans, K., Engelder, T., and Plumb, T. A., 1989, Appalachian stress study 1; A detailed description of in situ stress variations in Devonian shales of the Appalachian Plateau: Journal of Geophysical Research, v. 94, p. 7129–7154.

Eyal, Y., and Reches, Z., 1983, Tectonic analysis of the Dead Sea rift region since the Late Cretaceous based on megastructures: Tectonics, v. 2, p. 167–185.

Fitch, T. J., 1972, Plate convergence, transcurrent faults, and internal deformation adjacent to Southeast Asia and the western Pacific: Journal of Geophysical Research, v. 77, p. 4432–4460.

Fleitout, L., and Froidevaux, C., 1983, Tectonics and topography of a lithosphere containing density heterogeneities: Tectonics, v. 2, p. 315–324.

Forsyth, D., and Uyeda, S., 1975, On the relative importance of the forces of plate motion: Geophysical Journal of the Royal Astronomical Society, v. 43, p. 163–200.

Frohlich, C., 1982, Seismicity of the central Gulf of Mexico: Geology, v. 10, p. 103–106.

Geist, E., Childs, J. R., and Scholl, D. W., 1988, The origin of summit basins of the Aleutian Ridge; Implications for block rotation of an arc massif: Tectonics, v. 7, p. 327–341.

Gephart, J. W., and Forsythe, D., 1984, An improved method for determining the regional stress tensor using earthquake focal plane mechanism data; Application to the San Fernando earthquake sequence: Journal of Geophysical Research, v. 89, p. 9305–9320.

Gough, D. I., and Bell, J. S., 1981, Stress orientations from oil well fractures in Alberta and Texas: Canadian Journal of Earth Sciences, v. 18, p. 1358–1370.

Gough, D. I., Fordjor, C. K., and Bell, J. S., 1983, A stress province boundary and tractions on the North American Plate: Nature, v. 305, p. 619–621.

Haimson, B. C., 1977, Crustal stress in the continental United States as derived from hydrofracturing tests, *in* Heacock, J. C., ed., The Earth's crust: American Geophysical Union Monograph Series 20, p. 575–592.

——, ed., 1989, Hydraulic fracturing stress measurements, Special issue: International Journal of Rock Mechanics and Mining Sciences and Geomechanics, v. 26, 685 p.

Hasegawa, H. S., Adams, J., and Yamazaki, K., 1985, Upper crustal stresses and vertical stress migration in eastern Canada: Journal of Geophysical Research, v. 90, p. 3637–3648.

Hauksson, E., 1990, Earthquakes faulting and stress in the Los Angeles Basin: Journal of Geophysical Research, v. 95, p. 15365–15394.

Hickman, S. H., Healy, J. H., and Zoback, M. D., 1985, In situ stress, natural fracture distribution, and borehole elongation in the Auburn geothermal well, Auburn, New York: Journal of Geophysical Research, v. 90, p. 5497–5512.

Jaeger, J. C., and Cook, N.G.W., 1979, Fundamentals of Rock Mechanics (third edition): New York, Chapman and Hall, 593 p.

Jeffreys, H., 1976, The Earth: Cambridge, England, Cambridge University Press, 574 p.

Jones, L., 1988, Focal mechanisms and the state of stress on the San Andreas fault in southern California: Journal of Geophysical Research, v. 93, p. 8869–8891.

King, G.C.P., and Vita-Finzi, C., 1981, Active folding in the Algerian earthquake of 10 October 1980: Nature, v. 292, p. 22–26.

Kirsch, G., 1898, Die Theorie der Elastizitat und die Bedurfnisse der Festigkeitslehre: Zeitschrift des Vereines Deutscher Ingenieure, v. 42, p. 797–807.

Klein, R. J., and Barr, M. V., 1986, Regional state of stress in western Europe *in* Proceedings, International Symposium on Rock Stress and Rock Stress Measurements, Stockholm, 1–3 September: Lulea, Sweden, Centek Publishers, 694 p.

Lachenbruch, A. H., and Sass, J. H., 1988, The stress heat-flow paradox and preliminary thermal results from Cajon Pass: Geophysical Research Letters, v. 15, p. 981–984.

—— , 1991, Heat flow from Cajon Pass, fault strength and tectonic implications: Journal of Geophysical Research (in press).

Lachenbruch, A. H., and Thompson, G. A., 1972, Oceanic ridges and transform faults; Their intersection angles and resistance to place motion: Earth Planetary Science Letters, v. 15, p. 116–122.

Lefevre, L. V., and McNally, K. C., 1985, Stress distribution and subduction of aseismic ridges; In the Middle America subduction zone: Journal of Geophysical Research, v. 90, p. 4495–4510.

Letouzey, J., 1986, Cenozoic paleo-stress pattern in the Alpine foreland and structural interpretation in a platform basin: Tectonophysics, v. 132, p. 215–231.

Lomnitz, C., Mooser, F., Allen, C. R., Brune, J. N., and Thatcher, W., 1970, Seismicity and tectonics of the northern Gulf of California region, Mexico: Geofisica Internacional, v. 10, p. 37–48.

Lonsdale, P., and Lawver, L., 1980, Immature plate boundary zones studied with a submersible in the Gulf of California: Bulletin Geological Society of America, v. 91, p. 555–569.

MacKenzie, D. P., 1969, The relationship between fault plane solutions for earthquakes and the directions of the principal stresses: Seismological Society of America Bulletin, v. 59, p. 591–601.

Magee, M. and Zoback, M. L., 1989, Present state of stress in the Pacific northwest: EOS, v. 40, p. 1332–1333.

Mann, P., and Corrigan, J., 1990, Model for late Neogene deformation in Panama: Geology, v. 18, p. 558–562.

McGarr, A., and Gay, N. C., 1978, State of stress in the earth's crust: Annual Review of Earth Planetary Science, v. 6, p. 405–436.

McNutt, M., 1984, Lithospheric flexure and thermal anomalies: Journal of Geophysical Research, v. 89, p. 11180–11194.

Michael, A., 1984, Determination of stress from slip data; Faults and folds: Journal of Geophysical Research, v. 89, p. 11517–11526.

—— , 1987, Use of focal mechanisms to determine stress; A control study: Journal of Geophysical Research, v. 92, p. 357–368.

Minster, J. B., and Jordan, T. H., 1978, Present-day plate motions: Journal of Geophysical Research, v. 83, p. 5331–5354.

—— , 1984, Vector constraints on Quaternary deformation of the western United States east and west of the San Andreas fault, *in* Crouch, J. K., and Bachman, S. B., eds., Tectonics and sedimentation along the California Margin: Los Angeles, S.E.P.M., Pacific Section, Fieldtrip Guidebook 38, p. 1–16.

Moore, G. F., and Shipley, T. H., 1988, Mechanisms of sediment accretion in the Middle America Trench of Mexico: Journal of Geophysical Research, v. 93, p. 8911–8927.

Moore, J. C., and 5 others, 1982, Geology and tectonic evolution of a juvenile accretionary terrane along a truncated convergent margin; Synthesis of results from Leg 66 of the Deep Sea Drilling Project, southern Mexico: Geological Society of America Bulletin, v. 93, p. 847–861.

Moos, D., and Zoback, M. D., 1990, Utilization of observations of well bore failure to constrain the orientation and magnitude of crustal stresses; Application to conitnental Deep Sea Drilling Project and Ocean Drilling Project boreholes: Journal of Geophysical Research, v. 95, p. 9305–9328.

Mount, V. S., and Suppe, J., 1987, State of stress near the San Andreas fault; Implications for wrench tectonics: Geology, v. 15, p. 1143–1146.

Muehlberger, W. R., Belcher, R. C., and Goetz, L. K., 1978, Quaternary faulting in Trans-Pecos, Texas: Geology, v. 6, p. 337–340.

Müller, B., and 6 others, 1991, Regional patterns of stress in Europe: Journal of Geophysical Research (in press).

Nakamura, K., and Uyeda, S., 1980, Stress gradient in arc-back arc regions and plate subduction: Journal of Geophysical Research, v. 85, p. 6419–6428.

Nakamura, K.K, Jacob, K. H., and Davies, J. N., 1977, Volcanoes as possible indicators of tectonic stress orientation; Aleutians and Alaska: Pure and Applied Geophysics, v. 115, p. 87–112.

National Geographic Data Center, 1990, Geophysics of North America: Boulder, Colorado, NOAA, National Geophysical Data Center, product no. 975-B27-001, CD ROM disk.

Nixon, G. T., Demant, A., Armstrong, R. L., and Harakal, J. E., 1987, K-Ar and geologic data bearing on the age and evolution of the trans-Mexican volcanic belt: Geofisica Internacional, v. 26, p. 109–158.

Nunn, J., 1985, State of stress in the northern Gulf Coast: Geology, v. 13, p. 429–432.

Oldenburg, D. W., and Brune, J. N., 1972, Ridge transform fault spreading pattern in freezing wax: Science, v. 178, p. 301–304.

—— , 1975, An explanation for the orthogonality of ocean ridges and transform faults: Journal of Geophysical Research, v. 80, p. 2575–2585.

Page, B., 1981, The southern Coast Ranges *in* Ernst, W. G., ed., The geotectonic development of California: Englewood Cliffs, New Jersey, Prentice-Hall, p. 329–417.

Page, B., and Engebretson, D. C., 1984, Correlation between the geologic record and computed plate motions for central California: Tectonics, v. 3, p. 133–155.

Paillet, F., and Kim, K., 1985, Character and distribution of borehole breakouts and their relationship to in situ stresses in deep Columbia River basalts: Journal of Geophysical Research, v. 92, p. 6223–6234.

Pine, R. J., Ledingham, P., and Merrifield, C. M., 1983, In situ stress measurements in the Carmenellis granite, II; Hydrofracture tests at Rosemanowes Quarry to depths of 2,000 m: International Journal of Rock Mechanics, Mining Sciences, and Geomechanics, Abstracts, v. 20, p. 63–72.

Plumb, R. A., and Hickman, S. H., 1985, Stress-induced borehole elongation; A comparison between the four-arm dipmeter and the borehole televiewer in the Auburn geothermal well: Journal of Geophysical Research, v. 90, p. 5513–5521.

Plumb, R. A., and Cox, J. W., 1987, Stress directions in eastern North America determined to 4.5 km from borehole elongation measurements: Journal of Geophysical Research, v. 92, p. 4805–4816.

Raleigh, C. B., Healy, J. H., and Bredehoeft, J. D., 1972, Faulting and crustal stress at Rangely, Colorado, *in* Heard, J. C., and others, eds., Flow and fracture of rocks: Washington, D.C., American Geophysical Union, Geophysics Monograph Series, p. 275–284.

Reches, Z., 1987, Determination of the tectonic stress tensor from slip along faults that obey the coulomb yield criterion: Tectonics, v. 6, p. 849–861.

Richardson, R. M., 1991, Ridge forces, absolute plate motions, and the intraplate stress field: Journal of Geophysical Research (in press).

Rummel, F., and Mohring-Erdmann, G. G., 1986, Stress constraints and hydrofracturing stress data for the continental crust: Pure and Applied Geophysics, v. 124, p. 875–895.

Ryan, H. F., and Scholl, D. W., 1989, The evolution of fore-arc structures along an oblique convergent margin, central Aleutian arc: Tectonics, v. 8, p. 497–516.

Savage, J. C., Lisowski, M., and Prescott, W. H., 1981, Geodetic strain measurements in Washington: Journal of Geophysical Research, v. 86, p. 4929–4940.

Sbar, M. L., 1982, Delineation and interpretation of seismotectonic domains in western North America: Journal of Geophysical Research, v. 87, p. 3919–3928.

Sbar, M. L., and Sykes, L. R., 1973, Contemporary compressive stress and seismicity in eastern North America; An example of intraplate tectonics: Geological Society of America Bulletin, v. 84, p. 1861–1882.

Shamir, G., and Zoback, M. D., 1991, A crustal stress orientation profile to 3.5 km depth near the San Andreas fault at Cajon Pass, California: Journal of Geophysical Research (in press).

Shamir, G., Zoback, M. D., and Barton, C. B., 1988, In situ stress orientation near the San Andreas fault; Preliminary results to 2.1 km depth from the Cajon

Pass scientific drillhole: Geophysical Research Letters, v. 15, p. 989–992.

Shamir, G., and Zoback, M. D., 1991, A crustal stress orientation profile to 3.5 km depth near the San Andreas fault at Cajon Pass, California: Journal of Geophysical Research (in press).

Sibson, R. H., 1974, Frictional constraints on thrust, wrench and normal faults: Nature, v. 249, p. 542–544.

Sonder, L., 1991, Effects of density contrasts on the orientation of stresses in the lithosphere; Relation to principal stress directions in the Transverse Ranges, California: Tectonics, v. 9, p. 761–777.

Stein, S., Sleep, N. H., Geller, R. J., Wang, S. C., and Kroeger, G. C., 1979, Earthquakes along the passive margin of eastern Canada: Geophysical Research Letters, v. 6, p. 537–540.

Stephannson, O., ed., 1986, Proceedings, International Symposium on Rock Stress and Rock Stress Measurements, Stockholm, 1-3 September: Lulea, Sweden, Centek Publishers, 694 p.

Stewart, J. H., 1988, Tectonics of the Walker Lane belt, western Great Basin; Mesozoic and Cenozoic deformation in a zone of shear, in Metamorphism and crustal evolution of the western United States: Englewood Cliffs, New Jersey, Prentice-Hall, Rubey Volume VII, p. 683–713.

Stock, J. M., Healy, J., and Svitek, J., 1985, Hydraulic fracturing stress measurements at Yucca Mountain, Nevada, and relationship to the regional stress field: Journal of Geophysical Research, v. 90, p. 8691–8706.

Thompson, G. A., and Zoback, M. L., 1978, Regional geophysics of the Colorado Plateau: Tectonophysics, v. 61, p. 149–181.

Wong, I. G., and Humphrey, J. R., 1989, Contemporary seismicity, faulting, and the state of stress in the Colorado Plateau: Geological Society of America Bulletin, v. 101, p. 1127–1146.

Zemanek, J., Glenn, E. E., Norton, L. J., and Caldwell, R. L., 1970, Formation evaluation by inspection with the borehole televiewer: Geophysics, v. 35, p. 254–269.

Zoback, M. D., and Haimson, B. C., eds., 1983, Workshop on Hydraulic Fracturing Stress Measurements, U.S. National Committee for Rock Mechanics: National Academy Press, p. 44–54.

Zoback, M. D., and Healy, J. H., 1984, Friction, faulting and in situ stress: Annales Geophysicae, v. 2, p. 689–698.

——, 1991, In situ stress measurements to 3.5 km depth in the Cajon Pass scientific research borehole; Implications for the mechanics of crustal faulting: Journal of Geophysical Research (in press).

Zoback, M. D., and Hickman, S., 1982, In situ study of the physical mechanisms controlling induced seismicity at Monticello Reservoir, South Carolina: Journal of Geophysical Research, v. 87, p. 6959–6974.

Zoback, M. D., and Zoback, M. L., 1981, State of stress and intraplate earthquakes in the central and eastern United States: Science, v. 213, p. 96–109.

Zoback, M. D., Moos, D., Mastin, L., and Anderson, R. N., 1985, Wellbore breakouts and in situ stress: Journal of Geophysical Research, v. 90, p. 5523–5530.

Zoback, M. D., and 12 others, 1987, New evidence on the state of stress of the San Andreas fault system: Science, v. 238, p. 1105–1111.

Zoback, M. L., 1989, State of stress and modern deformation of the northern Basin and Range province: Journal of Geophysical Research, v. 94, p. 7105–7128.

——, 1991, First and second-order patterns of stress in the lithosphere; World Stress Map Project: Journal of Geophysical Research (in press).

Zoback, M. L., and Zoback, M. D., 1980, State of stress of the conterminous United States: Journal of Geophysical Research, v. 85, p. 6113–6156.

——, 1989, Tectonic stress field of the conterminous United States: Boulder, Colorado, Geological Society of America Memoir, v. 172, p. 523–539.

Zoback, M. L., Nishenko, S. P., Richardson, R. M., Hasegawa, H. S., and Zoback, M. D., 1986, Mid-plate stress, deformation, and seismicity, in Vogt, P. R., and Tucholke, B. E., eds., The Geology of North America, vol. M.; The Western North Atlantic Region: Boulder, Colorado, Geological Society of America, p. 297–312.

Zoback, M. L., and 28 others, 1989, Global patterns of intraplate stress; A status report on the world stress map project of the International Lithosphere Program: Nature, v. 341, p. 291–298.

Zoback, M. L., and 8 others, 1991, Stress map of North America: Boulder, Colorado, Geological Society of America, Continent Scale Map CSM-5, scale 1:5,000,000.

MANUSCRIPT ACCEPTED BY THE SOCIETY MAY 3, 1991

The Geology of North America
Decade Map Volume 1
1991

Chapter 20

Crustal stresses in Canada

John Adams
Geophysics Division, Geological Survey of Canada, 1 Observatory Crescent, Ottawa, Ontario K1A 0Y3, Canada
J. Sebastian Bell
Institute of Sedimentary and Petroleum Geology, Geological Survey of Canada, Calgary, Alberta T2L 2A7, Canada

INTRODUCTION

Although there has been much interest in crustal stresses in Canada in previous decades, notably related to deep mining (e.g., Herget, 1980), it is only in the last decade that the rate of data collection has accelerated. Because the amount of the data began to exceed the space available in tables (e.g., Bell and Babcock, 1986; Franklin and Hungr, 1978; Hasegawa and Adams, 1981; Hasegawa and others, 1985), the Geological Survey of Canada began a computer data base of crustal stress data in 1985 (Adams, 1987). This was prompted largely by the need for Canadian input to the Decade of North American Geology (DNAG) Stress Map of North America and later to the International Lithosphere Program World Stress Map. The timing was fortuitous in that a great deal of oil-well breakout data was beginning to be published (e.g., Bell and Gough, 1981), and a renewed emphasis was being placed on earthquake focal mechanisms (e.g., Wahlstrom, 1987; Adams and others, 1988).

In this chapter we discuss the sources of in situ stress information for Canada, the nature of the information provided by the various methods, the recognition and definition of stress provinces, regions with anomalous stress signatures, the effects of glaciation, and finally, possible causal mechanisms. All localities, including sedimentary basins mentioned in the text, are shown on Figure 1. We conclude that Canada east of the Cordillera is being compressed in a northeast-southwest direction in what is an extensive and relatively uniform stress province, which also includes the eastern and midcontinental United States. The stress signature is probably dominated by the forces driving the tectonic plates. Although deglaciation severely stressed the North American plate about 10,000 years ago, these stresses no longer dominate the direction of the stress field, and apparently make only a small contribution to the contemporary stress field.

STRESS DATA SOURCES FOR CANADA

The information summarized in this article is contained in the Canadian Crustal Stress Database (Adams, 1987; and updates since then), which contains 1,221 entries as of October 1989 when the maps in this chapter were plotted. The stress data have been compiled from the literature and from pre-publication personal communications. The database was intended to be a complete data source so that entries for individual overcoring measurements made in the same hole, early attempts at focal mechanisms, and data that may pertain to earlier (noncontemporary) stress fields are included, even though their value may be questionable. The data (Fig. 2) are assigned a subjective quality designation, and are identified by specific fields in the database so that any type of data can be identified, extracted, or ignored if deemed appropriate.

The full Canadian database of 1,221 entries on 31 October 1989 comprised:

• 1,221 entries, of which 113 pertain to postglacial stress regimes as discussed later in this chapter;

• 1,108 contemporary stress entries, of which 204 are subsidiary entries to an average entry that is deemed to be more representative (146 of these are individual overcoring results), and 39 are entries deemed to be superceded by later work;

• 865 representative entries compose the "good data set" mapped in this chapter, of which 281 entries are non-Canadian data. Most of these are overcoring results from New England and earthquake focal mechanisms from Alaska, Washington, and New England. They are substantially the same as the measurements given by Zoback and Zoback (this volume) and they are included to allow interpretation of the Canadian data set.

• 584 Canadian entries, which form the core data set. Of this data set, the number of entries for each quality level are: A = 227, B = 224, C = 116, D = 14, and E = 3 (Fig. 2). Thus, 77 percent are considered to be good entries.

TYPES OF DATA

The measurements have been made by a number of different techniques of varying applicability and accuracy, as are classified and summarized in Adams (1987). From the 584-entry Canadian data set, 60 percent represent breakout measurements, 28 percent earthquake focal mechanisms, 6.5 percent overcoring measurements, and 5.5 percent other methods (Fig. 2).

Because Zoback and Zoback (this volume) give a fuller de-

Adams, J., and Bell, J. S., 1991, Crustal stresses in Canada, *in* Slemmons, D. B., Engdahl, E. R., Zoback, M. D., and Blackwell, D. D., eds., Neotectonics of North America: Boulder, Colorado, Geological Society of America, Decade Map Volume 1.

scription of the methods involved, here we only summarize those that have been important in Canada and give typical Canadian examples. The number in brackets after the first mention of each method is the number of entries in the Canadian data set.

Earthquake focal mechanisms (162)

Focal mechanisms have provided deep stress data in areas distant from most human endeavors; the method provides the sense of faulting, the relative magnitudes of the three principal stresses, and (in an approximate way) their orientation. The method works well for the largest earthquakes, for which the P, B, and T axes are good estimators of the stress orientations (e.g., along the plate boundary on Canada's west coast), but is also useful for smaller earthquakes within a plate, particularly in southeastern Canada (e.g., Wahlstrom, 1987; Adams and others, 1988). It is the only method that provides stress information at mid-crustal or greater depths.

When deriving regional stress data from individual earthquake mechanisms, there naturally is some question as to whether the deviatoric compression (P) axis of the earthquake focal mechanism is a good representation of SHmax. For double-couple focal mechanisms, the data are used to constrain a pair of orthogonal nodal planes. The P-axis is constrained to bisect the dilational quadrants so defined, and therefore lies at an angle of 45° to the planes. Mechanically, other failure angles are possible, or even more likely, but their assignment requires that the fault plane first be identified as one of the two equivalent nodal planes, and in general this cannot be done. Variability of P-axis orientation has sometimes been ascribed to the preferential failure of preexisting planes of weakness in a uniform stress field. In areas with a strong structural grain, the preponderance of failure planes with a single orientation may even lead to a systematic rotation of the P-axis relative to the regional stress field driving the earthquakes. Despite these provisos, earthquake focal mechanisms provide useful indications of the nature of faulting at depth within the Canadian lithosphere.

Breakouts (350)

Borehole breakouts have been identified and analyzed in some 350 exploration wells in Canadian onshore and offshore

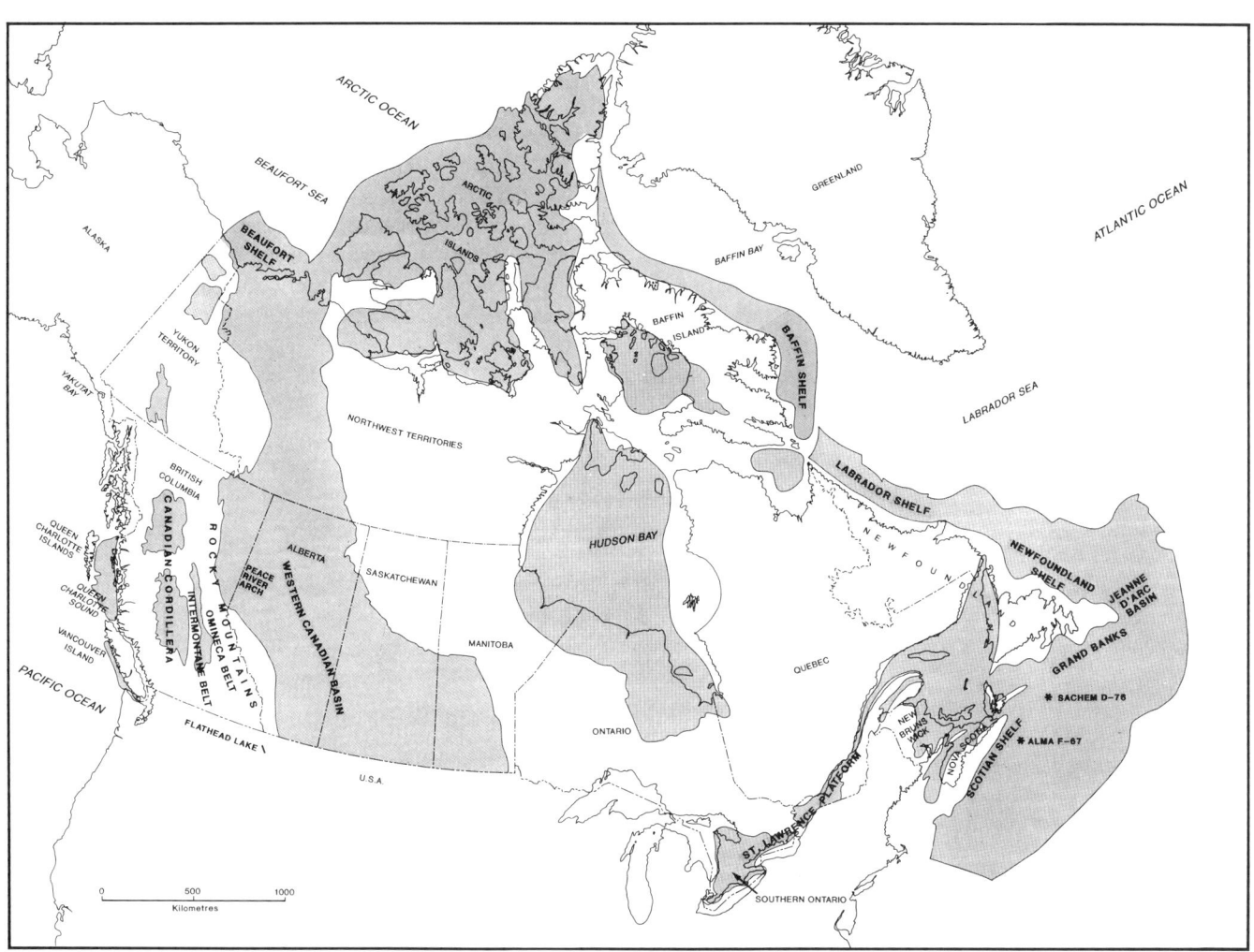

Figure 1. Map of Canada showing localities mentioned in the text. Phanerozoic sedimentary basins are stippled. Stars denote positions of two exploration wells drilled offshore of eastern Canada.

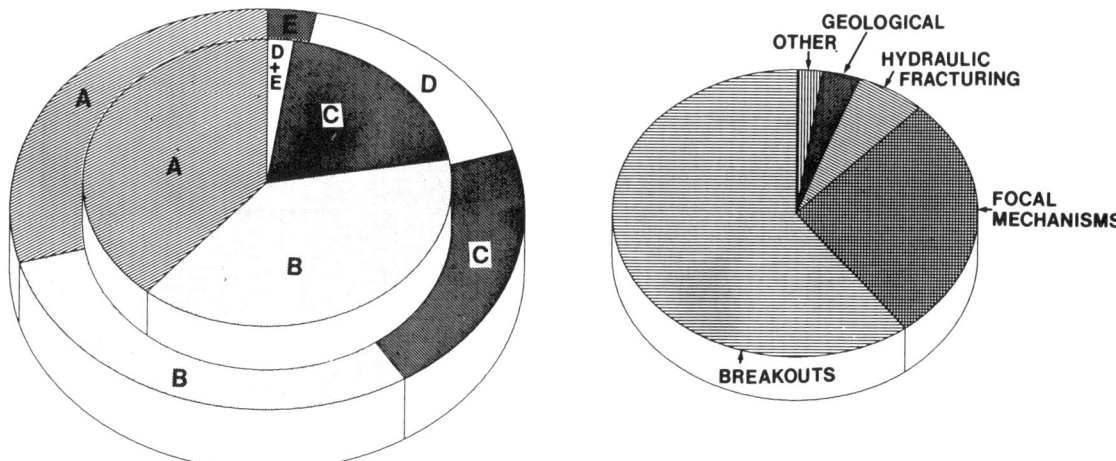

Figure 2. Pie charts showing: Left, the composition of the full database by quality (A through E), with a raised center showing the composition of the restricted "Canadian" data set; right, the types of stress indicators used for the "Canadian' data set.

sedimentary basins. Breakout orientation is recorded by 4 arm dipmeters, borehole televiewers, or equivalent tools, so that information can be obtained only from wells logged with such instruments.

In most wells there is strong directional consistency from top to bottom of the logged interval, implying little change in stress orientation with depth, rock type, or rock age. Breakouts identify horizontal stress anisotropy, and their long axes parallel SHmin (Bell and Gough, 1979), so they can be used to map horizontal principal stress trajectories and identify the relative horizontal stress magnitudes. They do not provide absolute magnitude data (Zoback and others, 1985) and they do not, on their own, indicate the prevalent fault regime, since vertical stress information is also needed. Nor do breakouts distinguish between compressive stress provinces and extensional stress provinces. In certain situations, breakouts may be aligned perpendicular to SHmin due to inadvertent hydraulic fracturing during drilling (Dart and Zoback, 1989) or possibly because of large stress magnitude contrasts between SV, the vertical stress, and SHmin, the smallest horizontal stress (Woodland, 1987). Borehole conditions and rock fabric effects may also give rise to breakouts that do not form parallel to SHmin (Babcock, 1978; Cox, 1983; Bell and Babcock, 1986).

Breakouts have played a major role in defining stress provinces in Canada and in documenting the remarkably consistent horizontal stress orientations in parts of Canada. Information on stress orientations from breakouts has been obtained between subsurface depths of 100 m to 6 km and is particularly abundant in the federally administered offshore exploration areas of Canada.

Hydraulic fracturing (6)

Hydraulic fracturing has provided a small number of stress magnitude measurements from relatively deep holes in Canada (e.g., Kry and Gronseth, 1982). The method relies on estimating the pressure at which induced fractures close from pressure-time records (Haimson and Fairhurst, 1970), and there has been much discussion (e.g., Nolte, 1988a; 1988b; McLennan and Roegiers, 1982) of the correct interpretation procedures. What is obtained is the magnitude of the least principal stress (usually horizontal) and, if an impression packer or televiewer tool is used, orientation information. Vertical stress magnitudes rely on estimates of overburden load, and the larger horizontal stress magnitude is inferred from pressure data obtained during the hydraulic fracturing program (Hubbert and Willis, 1957; Bredehoeft and others, 1976; Cornet and Valette, 1984). Stress magnitudes and orientations have been obtained from controlled hydraulic fracturing investigations at several sites in eastern Canada and from a few oil wells in western Canada (McLellan, 1988; Sarda and others, 1988). There are also many records of commercial fracture programs undertaken as part of enhanced production feasibility studies or as reservoir fracture programs designed to increase flow rates of hydrocarbons. Analysis of the records of these "frac jobs" can, in some cases, yield estimates of the magnitude of the least principal stress; such data has been obtained from some 140 wells in the Western Canadian Basin (Woodland and Bell, 1989; McLellan, 1988; Bell and Price, 1989). The stress magnitude data set has been augmented by analysis of oil-well fracture pressures, which provide a large body of less precise estimates (Breckels and Van Eekelen, 1981; Ervine and Bell, 1987). Fracture pressure indications of least principal stress magnitudes have been obtained from wells drilled offshore of eastern Canada and on the Beaufort Shelf.

Overcoring (38)

Overcoring methods employing strain-relief measurement have been widely used in southeastern Canada and have given numerous stress orientation and magnitude values in the upper

200 m (e.g., Lo, 1978) and some deeper values from mines (e.g., Herget, 1980). The problem with these measurements is that they have to be made close to free surfaces (mine openings), they sample very small volumes of strained rock (typically 1 to 10 dm^3), and so often represent local distortions of the regional stress field. In the near-surface data particularly, wide variations in magnitude and orientation among closely spaced measurements are common. However, at this time, the only measurements of vertical principal stress magnitudes that have been made in Canada are those derived from overcoring.

Indirect techniques

Less direct stress measurements come from *shear-wave splitting* (2), which can measure the preferred orientation of open fluid-filled cracks (Crampin, 1987). This approach to mapping stress trajectories has not yet been widely applied in Canada, though preliminary results (Buchbinder, 1989; Silver and Chan, 1988) are encouraging. In addition, induced *permeability trends* (2) in oil fields that can be attributed to induced hydraulic fracture have been used to infer SHmax (largest horizontal stress) directions in the Western Canadian Basin (Dusseault, 1977; Hassan, 1982).

Finally, there are the geologic indications of contemporary lithospheric stress in Canada. These include: *quarry-floor buckles* (10), which occur after removal of the rock load and strike normal to the principal horizontal stress (Adams, 1982); *pop-ups* (1), which occur in open county (White and Russell, 1982); and *offset boreholes* (2) in road cuts, which represent the release of confining stress (Bell, 1985). Another stress-release phenomenon is the *"squeeze"* (8) that occurs in narrow slot-shaped excavations such as canals (e.g., Lo, 1978) or tunnels. Such phenomena are common in the flat-lying limestones of southern Ontario, and have been observed elsewhere in eastern Canada, but have usually been ignored unless they had engineering consequences (e.g., Lo, 1978). If due care is taken to eliminate measurements where topography exerts a strong control, the resultant directions agree well with regional stress indicators, and indeed may give stress-orientation information quickly and cheaply.

Deep geologic indicators such as the strike of *active folds* (1), the alignment of volcanic vents, and the direction of slip on surface faults have not yet been widely used, although they may have considerable potential in western Canada. Triangulation and trilateration, which measure *geodetic strain* (2), have been used in western Canada to estimate plate-margin deformation and may also be used to infer the consequent stress directions (Dragert, 1986; Dragert and Rogers, 1988).

ORIENTATION OF PRINCIPAL HORIZONTAL STRESS

As others have shown, the stress tensor is usually aligned with one component near-vertical, and for almost all of Canada this vertical component does not have the largest magnitude.

These conditions describe a thrust or a strike-slip faulting regime, which can conveniently be represented by the azimuth of principal compression, being the horizontal projection of the (often gently dipping) SHmax. Most "stress maps" are maps of this largest horizontal component, and they show in a reasonable way the "flow" of stress directions across a plate. We plot, for example: the azimuth of largest horizontal compression measured by overcoring, the normal to the breakout, pop-up and buckle strikes, the direction of vertical hydraulic fractures, and the azimuth of the P-axis for thrust and strike-slip earthquakes. We could have chosen to plot the B-axis for normal faulting earthquakes to represent the largest horizontal stress direction (as was done by Zoback and others, 1989, in their color map), but instead we plot outward-pointing arrows in the T-axis direction to show the deviatoric extension on our single-color maps. For our representations, we have made the bar lengths proportional to data quality, so that the eye is not biased by poor data. Zoback and Zoback (this volume) have addressed stress data quality and have introduced a grading scheme for the World Stress Map Project. For the maps in this chapter, the data quality criteria are somewhat different and were evolved earlier (Adams, 1987) in response to Canadian needs. Both approaches, however, identify the best data.

As shown on the accompanying maps (Figs. 3, 4, and 5), we find that the regional pattern of stress orientations for most of Canada is relatively simple. Detailed maps for Alberta–British Columbia (Fig. 4) and southeastern Canada (Fig. 5) show the amount and general consistency of the data in these well-studied regions. The trend of the stress field changes across the country with ENE- to NE-directed stresses in the southeast and NE- to NNE-directed stresses in the northwest. As we will discuss later, this difference is consistent with the changing azimuth of the plate motion across North America.

STRESS MAGNITUDES

Direct measurements

Stress magnitudes have been measured by overcoring and hydraulic fracturing in Canada (Adams, 1987). A large number of overcoring measurements have been made in crystalline rocks of the Canadian Shield from the surface to depths of 2 km (Fig. 6). Although there is considerable variation in the stress magnitudes obtained, it is clear that high horizontal stresses are widely present and that both SHmax and SHmin often exceed overburden pressures at the depth of measurement (e.g., Haimson and others, 1986). The difference between SHmax and SHmin also appears to increase with depth in the Canadian Shield (Fig. 7). The same picture emerges from stress magnitudes measured in the overlying Paleozoic rocks of the St. Lawrence Platform (Fig. 8). Most of these data come from very shallow overcoring measurements and are highly scattered, but nearly all record SHmax and SHmin magnitudes above overburden pressure levels. This is also observed with hydraulic fracture stress magnitudes to 300 m depth (Fig. 8).

The causes of these anomalously high horizontal stresses are not totally understood. In Canada they have been attributed to retention of residual stresses (Eisbacher and Bielenstein, 1971) and to variations in elastic moduli (Herget, 1973). Elsewhere, variations in seasonal temperatures are believed to have induced near-surface horizontal thermal stresses that argumented the regional stress field (Sbar and others, 1984). It is also possible that the method of stress measurement is a factor. Doe and others (1983) found that overcoring yielded greater stress magnitudes than those obtained by hydraulic fracturing in the same rock masses.

Overcoring and hydraulic fracture measurements have also been made at several sites in Western Canada (Fig. 9). All the measurements were made in Mesozoic sediments and, except locally, point to a stress regime where SHmin is less than SV, the overburden load.

Indirect measurements

Records of oil-well drilling and commercial hydraulic fracturing have been used to provide indications of stress magnitude (Breckels and Van Eekelen, 1981; Ervine and Bell, 1987; Woodland and Bell, 1989; McLellan, 1988; Bell, 1989a, 1990). A practical handbook is Bell (1989b).

Commercial mini-frac and microfrac programs aimed at enhancing hydrocarbon production provide pressure records suitable for obtaining SHmin magnitudes (Woodland and Bell, 1989; McLellan, 1988). Figure 10 illustrates such data obtained from

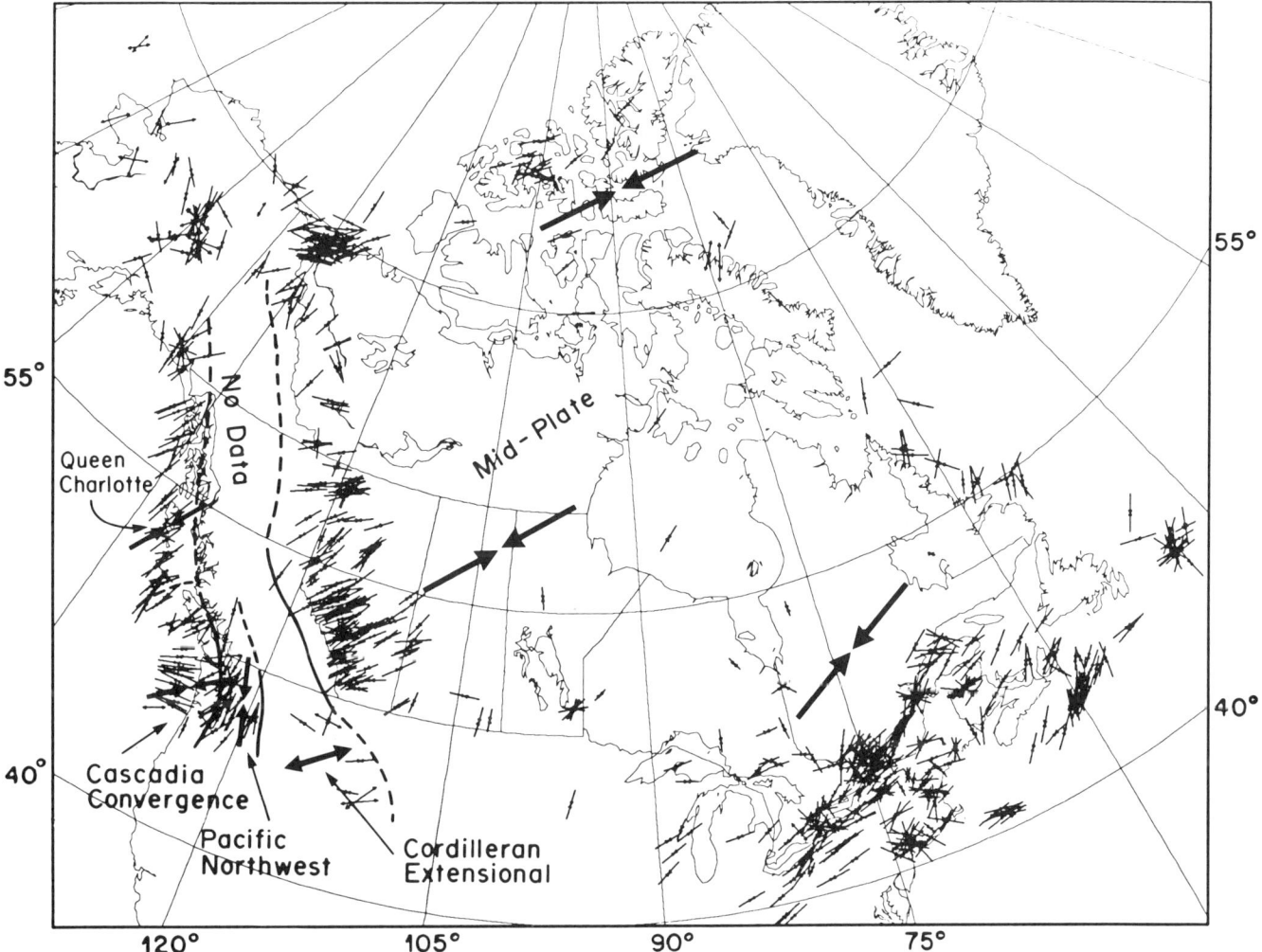

Figure 3. Stress map of Canada showing all data believed to relate to the contemporary stress field. Data in this and subsequent figures not otherwise attributed is taken from the current version of the Canadian Stress Database (i.e., Adams, 1987, modified by additional Canadian data). Bar length is proportional to data quality, with inward-pointing arrows showing the direction of maximum horizontal compression, and the few outward-pointing arrows the deviatoric extension or minimum horizontal compression direction. Dashed lines indicate the approximate boundaries of stress provinces for the northern part of North America, and pairs of bold arrows show the regional trend in each province. Data type for each measurement can be found on the DNAG stress map of North America.

fracture records of hydrocarbon wells in the Western Canadian Basin. Most of the values apply to Mesozoic rocks, and nearly all were obtained from hydrocarbon-bearing sandstone. There is a degree of scatter. Some of this is probably real, since nonlinear increases in SHmin are to be expected and have been observed where multiple measurements have been made in single wells (Kry and Gronseth, 1982; McLellan, 1988). Part of this scatter, however, may be due to different operators' practices in estimating fracture-closure pressures. There is a suggestion that absolute SHmin magnitudes increase toward the overthrust Rocky Mountains, but an increase in stress gradients with depth could also account for the variation. More data are needed to clarify the situation.

Leak-off tests, or pressure integrity tests, are often run below cemented casing in hydrocarbon exploration wells; they involve initiating a small hydraulic fracture in the wall of the well. First-cycle tests estimate formation breakdown pressures, whereas second-cycle and subsequent tests measure fracture opening and fracture propagation pressures. Usually only the pressure at which the drilling mud begins to leak off into the formation is recorded, but in some recent tests, pressure decline records have been released (Bell, 1990). From these one can establish fracture closure pressures and, hence, the smallest principal stress. In most cases, though, leak-off pressures can only be used to constrain the magnitude of the smallest principal stress, which for all the cases discussed here is SHmin.

On the Scotian Shelf, offshore of Nova Scotia, leak-off occurred in every test at pressures less than the corresponding overburden load. This clearly indicates that SHmin magnitudes are less than SV at all depths above 6 km (Fig. 11). Below that depth, as a consequence of overpressured pore fluids, SHmin may actually become greater than SV (Bell, 1990). In other East Coast

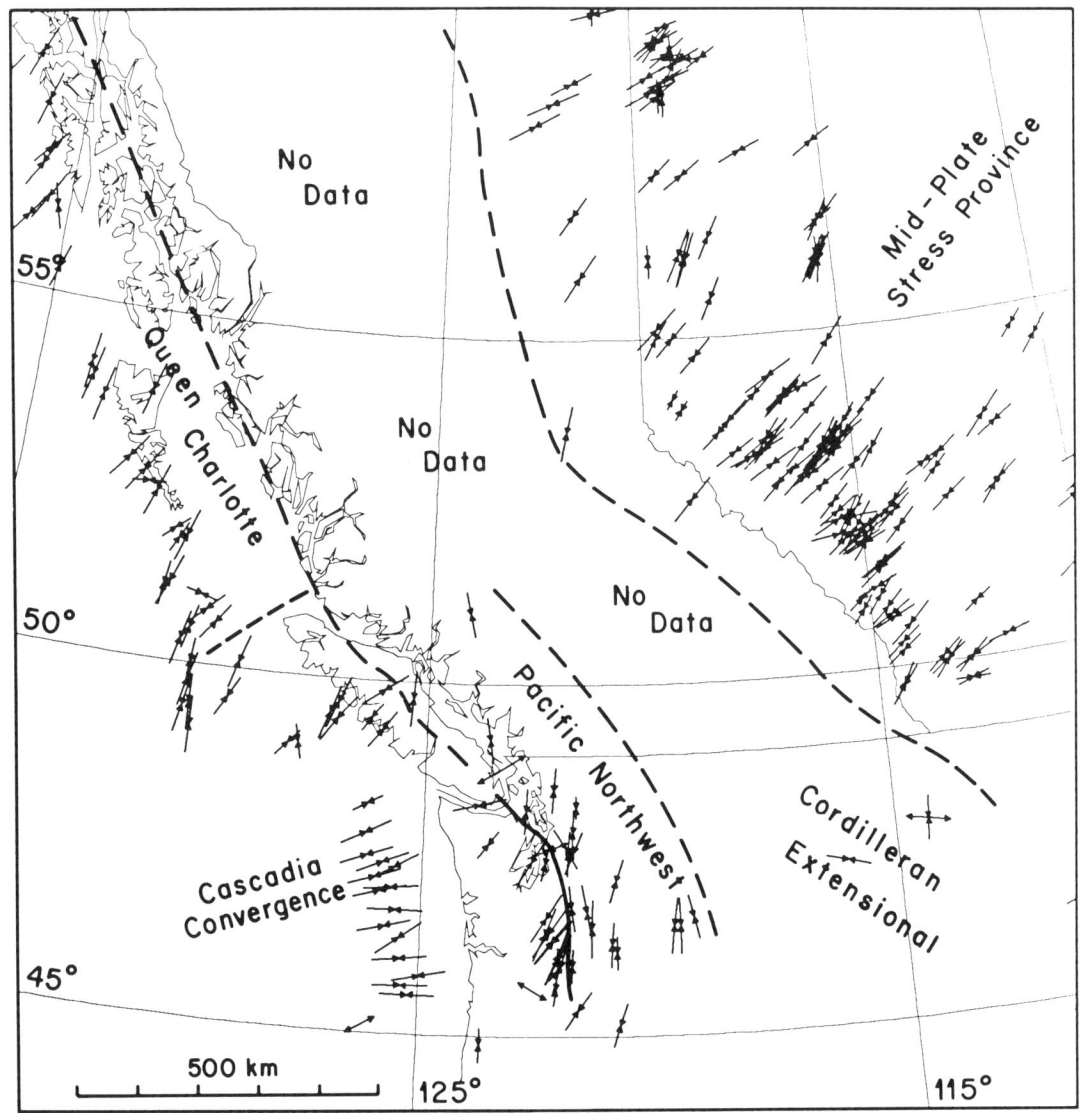

Figure 4. Detail of Figure 3 for southwestern Canada showing the consistency of the breakout directions in Alberta (data from a few other methods are represented) and stress directions along the Pacific margin.

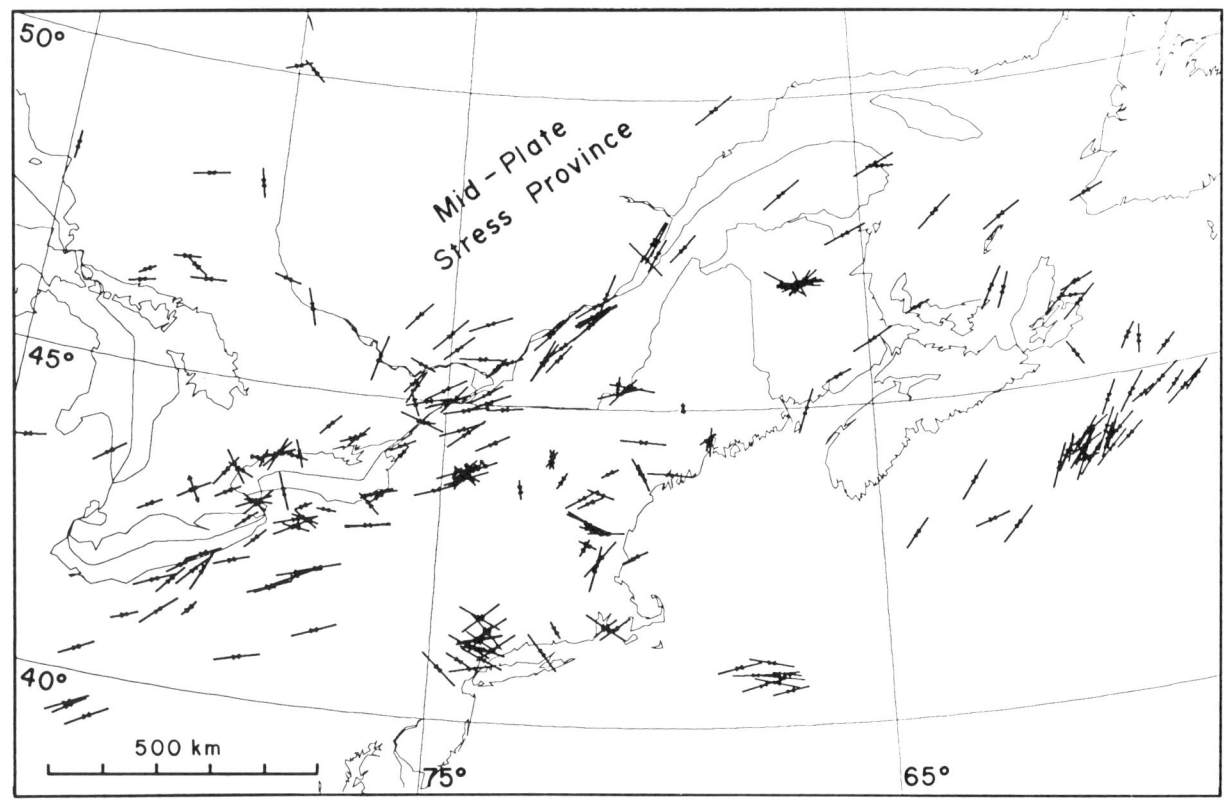

Figure 5. Detail of Figure 3 for southeastern Canada and adjacent United States showing the consistency of data from a variety of stress indicators. Stress directions inferred from earthquakes deeper than 9 km are not plotted for reasons discussed in the text.

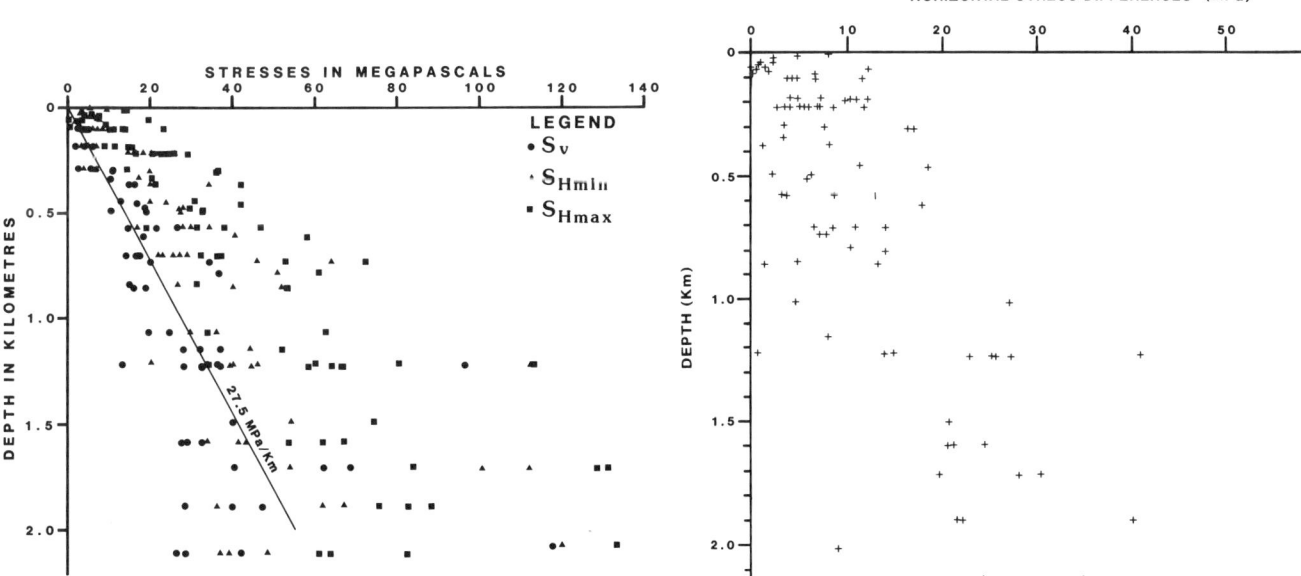

Figure 6. Principal-stress magnitudes obtained by overcoring measurements made in the crystalline rocks of the Canadian Shield to depths of 2 km. Note that SHmin exceeds SV at many sites. The data plotted here are listed in Adams (1987).

Figure 7. Horizontal principal-stress differences for rocks of the Canadian Shield plotted against depth. The difference in stress magnitude appears to increase with depth.

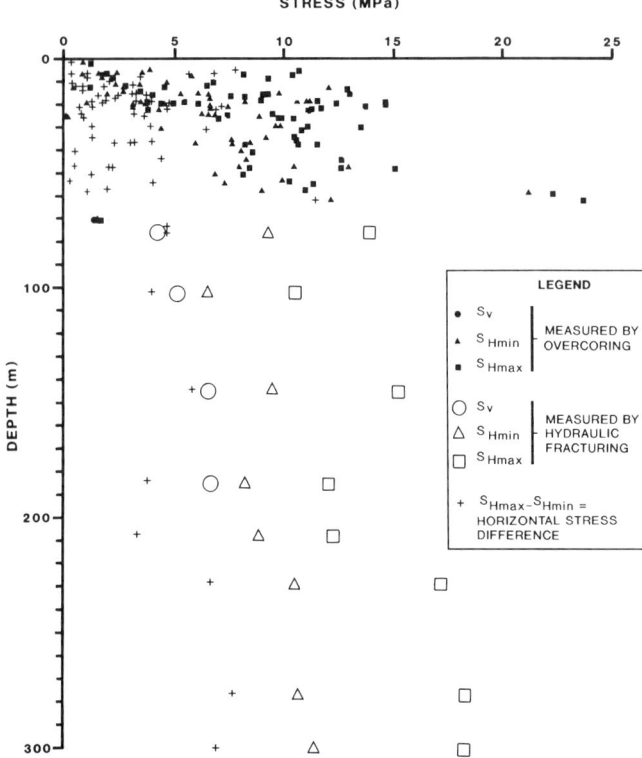

Figure 8. Stress magnitudes measured in Paleozoic sedimentary rocks of the St. Lawrence Platform, southeastern Canada, to depths of 300 m. The scatter of values obtained by overcoring measurements at shallow depths changes to a more uniform increase with depth as measured by hydraulic fracturing. Note that the difference in horizontal principal-stress magnitudes appears to increase with depth.

offshore areas, such as the Grand Banks, Northeast Newfoundland Shelf, Labrador Shelf, and southeastern Baffin Shelf, leak-off data are available to depths of 4 to 5 km and uniformly indicate that SHmin magnitudes are less than SV. Leak-off tests give no indication of SHmax values, but estimates can be obtained from Ervine and Bell's (1987) approximation:

SHmax = 2 (leak-off pressure) − pore pressure

If repeated leak-off tests have been run, and pressure decline records are available, more accurate estimates of SHmax can be made (Bell, 1990). At shallow depths in basins offshore of eastern Canada it appears likely that SV often exceeds, SHmax. At greater depths, and especially within overpressured intervals, leak-off pressures approach SV, so SHmax probably becomes the largest principal stress at these levels. It should be borne in mind that most of the leak-off tests were conducted in shales, so the results apply largely to this lithology. Fracture tests of selected sandstones have been undertaken locally following drillstem testing (Ervine and Bell, 1987), and in many cases the sandstones have fractured at higher pressures than the adjacent shales. Therefore, the stress-magnitude/depth profiles are probably less smooth than Figure 11 might suggest.

In the Beaufort Sea, unusually high leak-off pressures were recorded in wells drilled on shale-cored anticlines. Many were close to SV in magnitude, and below 3 km, leak-off occurred in some wells at pressures greater than the overburden load (Fig. 12). Preliminary interpretation suggests that some Mesozoic and Cenozoic clastic rocks at such depths beneath the Beaufort Shelf may be subject to a stress regime where SHmax > SHmin > SV, and may be undergoing contemporaneous compressive deformation.

In whichever way one interprets the stress magnitude data now available in Canada, it is clear that there are significant regional variations in the absolute magnitudes of the horizontal principal stresses (Fig. 13). We have not yet investigated what this may mean, or how the variations might have arisen.

Relative magnitudes

Earthquake focal mechanisms provide the sense of faulting and the relative magnitudes of the stress components. A significant component of **normal faulting** is confirmed only for two earthquake focal mechanisms on Baffin Island: 1963, M6.2 (Stein and others, 1979) and 1972, M4.5 (Hashizume, 1973). The seismicity appears to be confined to the coastal region, under and seaward of the 2-km elevations of the island, and all available evidence suggests that the earthquakes are relatively shallow. Stein and others (1979) proposed that the stresses resulting from Pleistocene deglaciation were reactivating Mesozoic rift faults, with thrust faulting occurring seaward of the 1,000-m isobath and normal or strike-slip faulting for more landward events. Stein and others (1989) developed this model further, suggesting that the recent removal of surface loads by glacial erosion and by deglaciation (and not addition of loads by sediment accumulation) was the major source of local stress that produced flexure of the lithosphere. By their computations, this added stress is comparable in magnitude to, and is superimposed upon, the compressional stress induced in the North American plate due to Mid-Atlantic Ridge spreading. Quinlan (1984) has argued, however, that while postglacial rebound is capable of triggering earthquakes in pre-stressed regions, it is rarely capable of dictating the focal mechanism of these earthquakes. Elsewhere in eastern Canada the few other normal faulting mechanisms are associated with small earthquakes, many being aftershocks of larger events.

Earthquake fault mechanisms with dominant **strike-slip** motion are found near the Queen Charlotte Transform (Pacific–North America plate boundary) and within the craton in the Arctic Archipelago. Among others, the focal mechanisms for the four largest earthquakes in the Byam Martin Channel swarm (M5.1 to M5.7), while dominantly strike-slip, show some deviatoric tension at depths from 9 km (near the base of sedimentary rocks) to 31 km (Hasegawa, 1977).

Thrust faulting dominates in the rest of Canada, although many focal mechanisms have small to moderate components of strike-slip displacement.

Thus, with the exception of those on Baffin Island, the

earthquake focal mechanisms represent mainly thrust, strike-slip, or combined thrust/strike-slip motion, thus confirming that the greatest stress component Smax is seldom vertical; i.e., that most of the Canadian crust is subject to high horizontal compressive stresses.

Complete stress tensor

The complete stress tensor has been measured only by overcoring in mines or engineering excavations in Canada. Most of the measurements were made in eastern Canada, mostly at sites in the Canadian Shield. As discussed earlier, the results have not always given clear indications of the regional stress regime. In sedimentary basins it has been necessary to assume that one principal stress is vertical and two are horizontal, without making confirmatory measurements. Hydraulic fracture programs do not give independent measurements of SHmax magnitudes because the values are obtained by calculation and depend on the assumption that the pressures were measured in circular well bores surrounded by perfectly isotropic rocks. For these reasons, we are not yet at the stage when we can describe the regional stress signatures in Canada in terms of tightly constrained stress tensors.

SPECIAL TOPICS

Effects of glaciation

Deglaciation severely stressed the North American plate about 10,000 years ago. Post-glacial near-surface stresses were dominated by the radial flexural ("fiber" or longitudinal) stresses induced near the retreating ice margin. Because the ice cap was roughly circular in plan, and was superposed on a continent with a fairly uniform northeast-directed compression field, the unloading stresses were often orthogonal (i.e., N-S or NW-SE) to the contemporary regional stresses in southeastern Canada. Figure 14 shows the near-surface postglacial direction of maximum horizontal compression, taken to be normal to the strike of small postglacial thrust faults that are inferred to have formed in the time period soon after deglaciation. As can be seen by comparison with Figure 3, which shows the contemporary stress-field orientation, the stress field revealed by the postglacial features is more consistent with flexural stresses radial to the ice margin than with the contemporary stress field (see Adams, 1989, for a fuller discussion).

That the postglacial stresses no longer dominate the direction of the stress field in southeastern Canada suggests in this region that residual, glaciation-induced stresses make only a small contribution to the magnitude of the contemporary stress field. However, during the writing of this manuscript, Adams became aware that new earthquake focal mechanisms in the northern Quebec–Labrador Sea–Baffin Island region had P-axes closer to north-south than east-west. When taken with the normal faulting mechanisms on Baffin Island, which give east-west deviatoric extension, they suggest that over a broad area the regional northeast-directed field may have been modified, perhaps by an extensional component radial to the ice sheet or normal to the

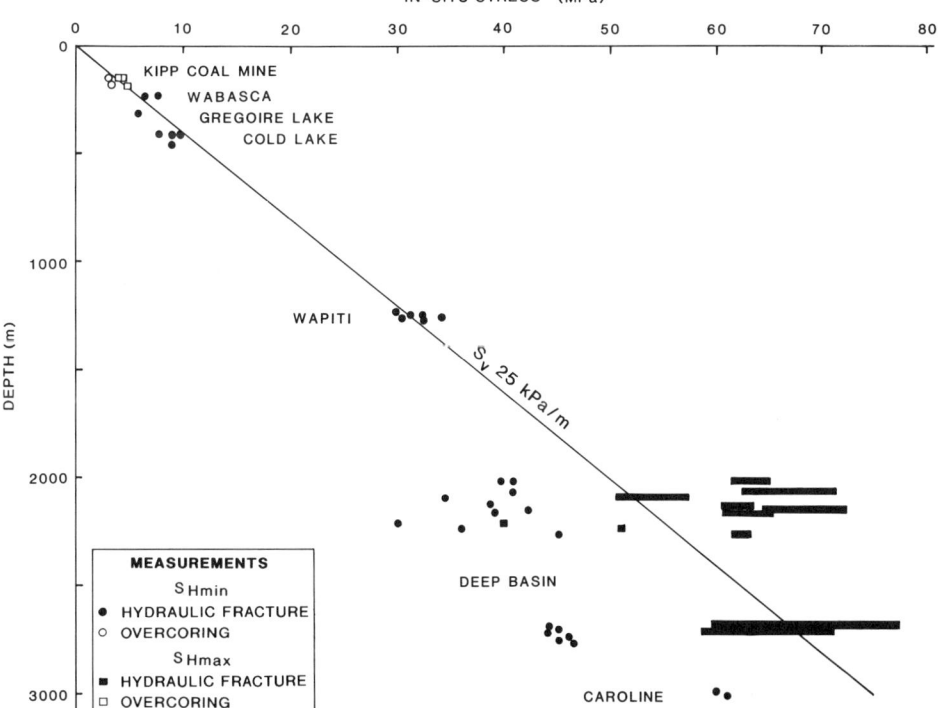

Figure 9. Measurements of stress magnitudes made in the Western Canadian Sedimentary Basin. Overcoring was used at the Kipp Mine; the other values were obtained by hydraulic fracturing. All of the data are summarized in Bell and Babcock (1986) except those from Wapiti, which are derived from McLellan (1988).

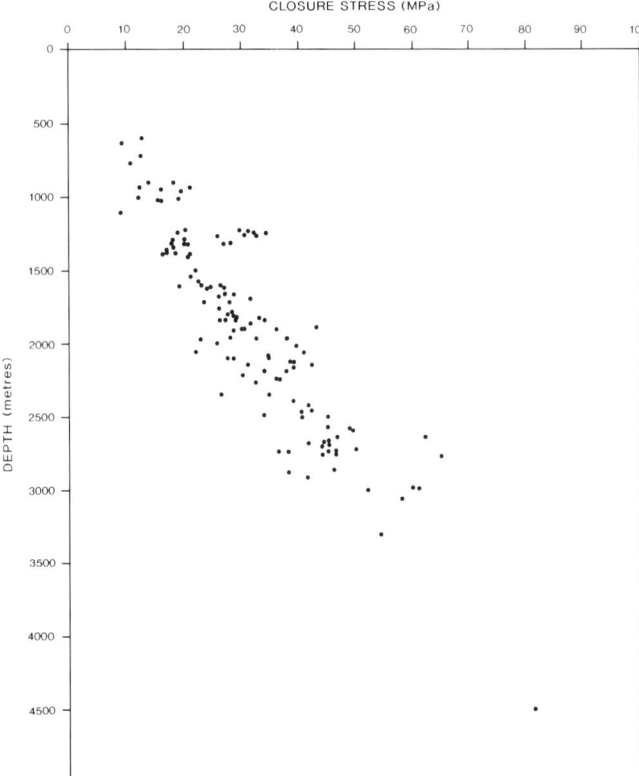

Figure 10. Magnitudes of SHmin estimated from industrial mini-frac and microfrac programs run in 132 wells in the Western Canadian Basin. The rocks fractured were mainly Mesozoic sandstones that were deemed capable of enhanced hydrocarbon production.

continental margin. These findings may have significant implications for the origin and nature of seismicity in the glaciated craton of eastern Canada (see Adams and Basham, 1989; Hasegawa and Basham, 1989).

Refraction of stresses

The large amount of stress-orientation data now allows recognition of regional stress-orientation signatures and, as a corollary, identification of regions where the horizontal principal stress trajectories are anomalous. Such a situation occurs above the Peace River Arch in western Canada. In an area dominated by the regional NE-SW compression direction, SHmax is deflected to a NNE-SSW direction (Fig. 15; see also Hansen and Mount, 1990). It is suspected that this is a case of stress refraction caused by stress transmission through a rock mass with laterally variable elastic properties. Bell and Lloyd (1989) were able to mimic the observed stress trajectories with a simple two-dimensional finite-element model.

Stress refraction of this type does not appear likely to deflect stress trajectories more than 10 to 20° in azimuth. Larger directional anomalies are likely to have other causes.

Stress trajectory reorientation near free surfaces

It is well known (e.g., Jaeger and Cook, 1976) that one of the three principal stresses at a point must be oriented perpendicular to any adjacent free surface and that, in the vicinity of a free surface, stress trajectories will bend to accomplish this. This occurs at the surface of the Earth and permits the approximation that one principal stress is vertical in topographically simple settings (McGarr and Gay, 1978). A classic example of stress trajectory bending occurs near the San Andreas fault, which is apparently acting today as "an almost free surface" (Mount and Suppe, 1987). Similar phenomena have been identified near the Viking graben in the North Sea (Clauss and others, 1989), in offshore eastern Canada near the Alma F-67 well on the Scotian Shelf (Bell, 1989c), and in the Jeanne d'Arc Basin on the Grand Banks of Newfoundland (Bell, 1989a). In the latter area (Fig. 16), the regional orientation of SHmax is NE-SW (Adams, 1987), but several wells that were drilled on downthrown blocks adjacent to the Murre and Mercury faults exhibit E-W SHmax orientations based on breakout azimuths. It appears that these two faults could be acting as free surfaces and deflecting stress trajectories toward them. If this proves to be the case, there are a number of implications. Preliminary studies imply that only a few faults are potential free surfaces, and these are faults that extend upward close to the sea floor and downward into highly overpressured zones. They appear to be possible conduits for upward fluid migration as well as surfaces prone to movement, causing or, in response to, earthquakes. In other words, the stress trajectory changes may diagnose free surfaces that correspond to faults that are active in both the hydraulic and tectonic sense.

Deriving stress directions from disparate focal mechanisms

Two seismic zones in southeastern Canada—Charlevoix and the Lower St. Lawrence zones—lie along the northeast-trending Paleozoic St. Lawrence Rift fault system (Adams and Basham, 1989) and are believed to involve reactivation of the old rift faults or of orthogonal transverse faults. For each zone, a number of earthquake mechanisms have been derived (e.g., Adams and others, 1988). They display a wide range of P-axis trend (Charlevoix 029 to 130°; Lower St. Lawrence 052 to 145°), and further analysis is clearly needed to understand why this is so.

Figure 17 shows the result of applying the method of Angelier and Mechler (1977) to the seven Charlevoix and eleven Lower St. Lawrence mechanisms. The nodal planes are drawn for each mechanism successively, while keeping track of sectors of the focal sphere that have compressional or dilatational first arrivals for all mechanisms. As can be seen from the two plots, only small sectors of consistent first motion are common to all mechanisms: for dilatation they lie toward the outside of the focal sphere near the east and west axes; for compression they are near the vertical. If we relax the requirement that the P-axis must lie at 45° to the nodal planes, but still must lie in the dilational quandrants, the common dilatational areas on each plot imply that all the earthquakes could have occurred in response to E–W–directed horizontal compression (common dilational field trending east-west and plunging gently). Thus, this method seems to recover regional stress data from a suite of varying earthquake mechanisms, as was also found for the southeastern United States by Teague and others (1986), and for the United Kingdom by Marrow and Walker (1988). The Canadian data may even show a slight systematic clockwise rotation of the regional NE- to ENE-directed stress field because of the NE-trending rift faults.

The implications of constraining the P-axes to the small dilatational areas common to all mechanisms are not yet fully appreciated. For some mechanisms it implies that the fault may have been driven by stresses acting at angles very diferent from the 45° initially assumed.

Change in the stress field with depth

Within the Canadian Shield there are stress anomalies in the basement rocks that imply that local conditions can significantly modify the regional field. Most focal mechanisms for earthquakes in southeastern Canada represent thrust or thrust/strike-slip motion and have gently dipping P-axes. Such earthquake data, not identified by depth, were often used to suggest that the stress field in the adjacent United States was very uniform (Yang and Ag-

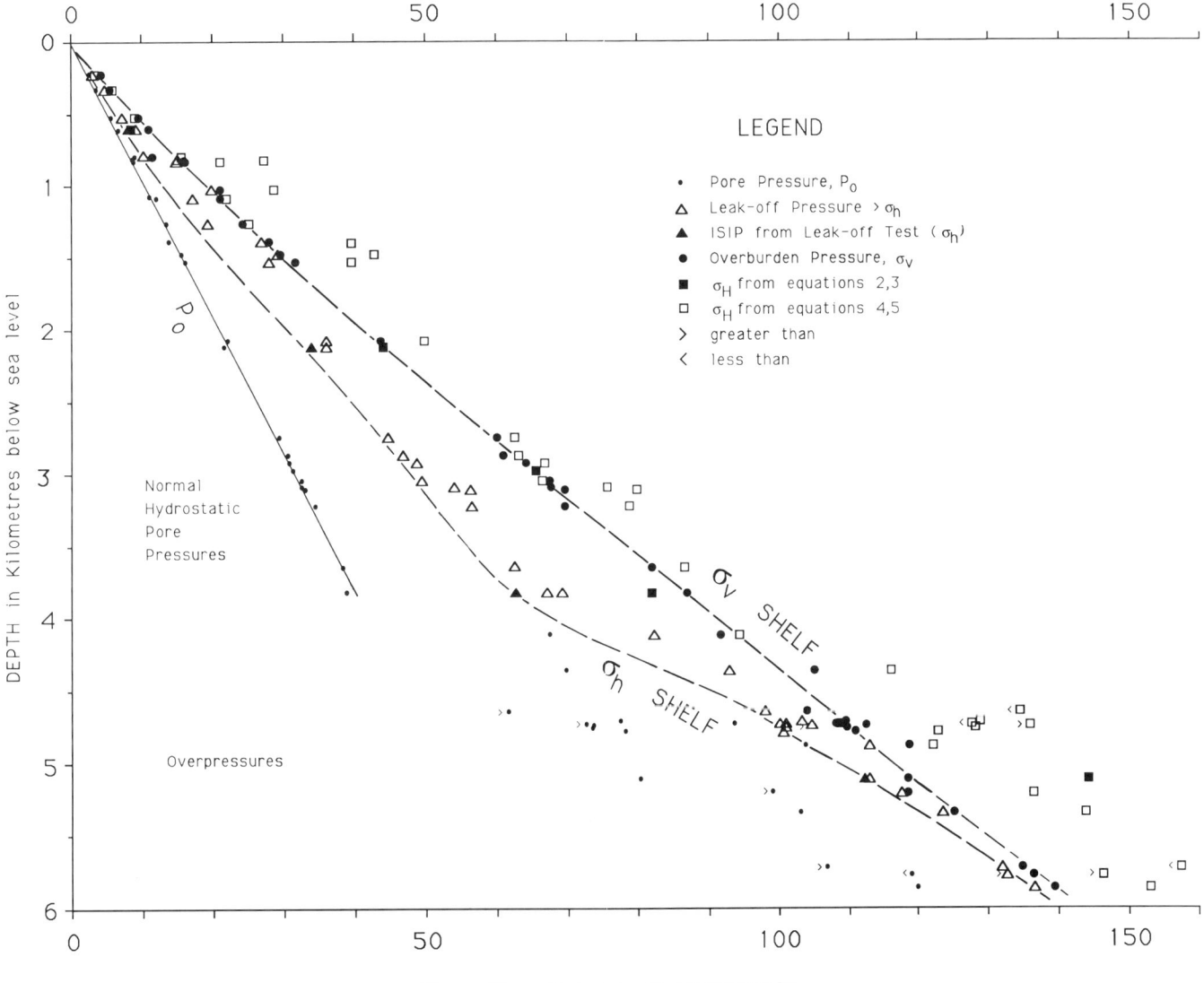

Figure 11. Stress magnitudes from the central part of the Scotian Shelf. Leak-off tests have been used to constrain SHmin, as explained in the text. SV was obtained by integrating density logs and SHmax is calculated (and probably overestimated) from a generalized relation. The increase in SHmin at the top of the overpressured sequence is real, as is the reduction in the pressure difference between the pore pressures and leak-off pressures. Within the overpressured zone, SHmax is greater than SV; above it, SV is probably the largest principal stress.

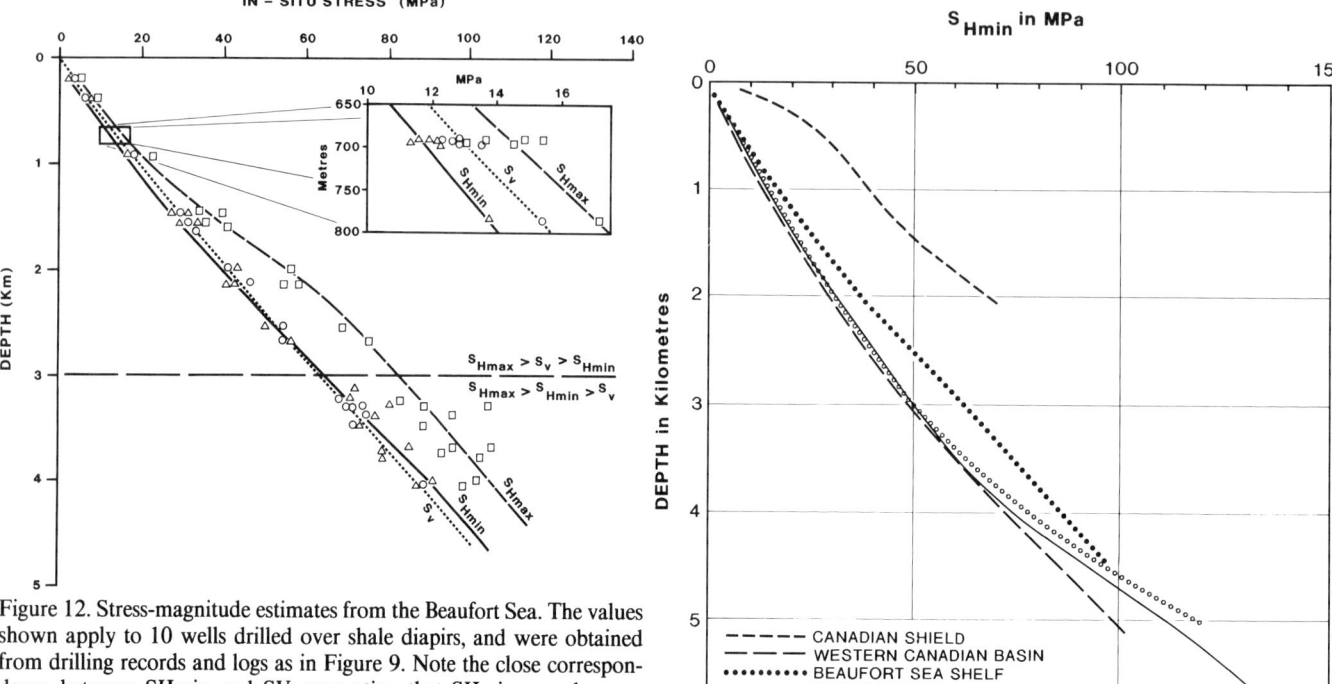

Figure 12. Stress-magnitude estimates from the Beaufort Sea. The values shown apply to 10 wells drilled over shale diapirs, and were obtained from drilling records and logs as in Figure 9. Note the close correspondence between SHmin and SV, suggesting that SHmin may become greater than SV below 3 km.

Figure 13. Magnitude/depth profiles of SHmin for various geologic provinces in Canada. The curves give mean values and suggest that stress magnitude gradients vary laterally.

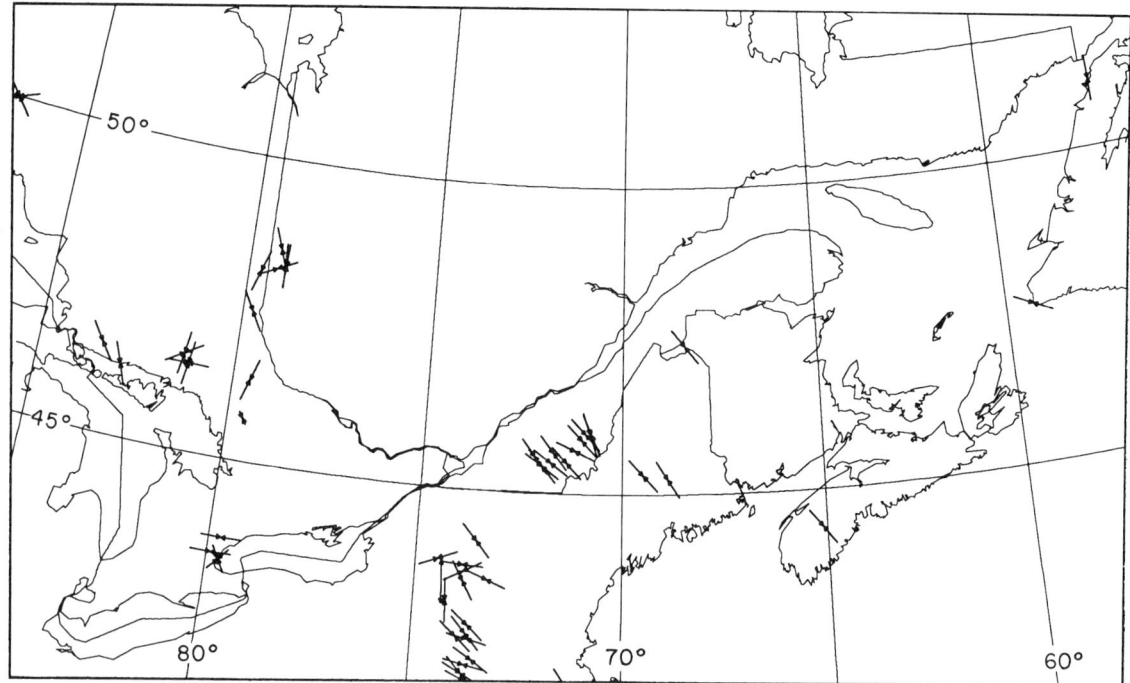

Figure 14. Map of stress orientations for southeastern Canada as inferred from small thrust faults thought to relate to a transient stress field during the deglaciation.

garwal, 1981), while the Canadian data, which was sparse in earlier years, was thought to be of comparatively poorer quality. However, analysis of P-axes orientation as a function of depth reveals a change (discontinuity) in the consistency of the deviatoric compression direction. While the change appears to occur in other places, such as the Lower St. Lawrence seismic zone (Basham and Adams, this volume), it is most easily seen in the region around Ottawa and Montreal, where there are many focal mechanisms, and the earthquakes extend deeper than 20 km (Fig. 18). For earthquakes shallower than 9 km (Fig. 18A), the gently dipping P-axes are in the northeast quadrant (065 ± 025°) for both Canadian and United States data. For deeper earthquakes (Fig. 18B), the P-axes appear to have no consistent orientation, and a paucity of deeper earthquakes in the United States is apparent. Other types of stress measurements (Fig. 18C; all from depths <5 km) confirm the uniform trend of the shallow-earthquake P-axes, suggesting that it is the deeper earthquakes that are anomalous.

A plausible explanation is that the horizontal components of the regional stress field change smoothly with depth so that the stress field becomes almost isotropic below 10 km, and the direction of principal compression is poorly defined, and hence, easily perturbed by local features.

Anomalous stress orientations in seismically active regions

Depending on the nature of the perturbing features discussed above, the deep stress field may appear to be chaotic or to be organized on a local scale. One such "organized" anomalous patch lies northeast of Ottawa (Fig. 18B). Earthquakes in a 70 × 50 km area (all deeper than 10 km) have P-axes consistently orthogonal to the first-order regional field. One intriguing interpretation of this patch is that it may represent the "ghost" of a large (M7) earthquake that occurred in prehistoric times. The expected stress drop (about 10 MPa) during such an earthquake is similar to the expected difference SHmax – SHmin (e.g., see Figs. 11 and 12), so that the secondary horizontal stress SHmin

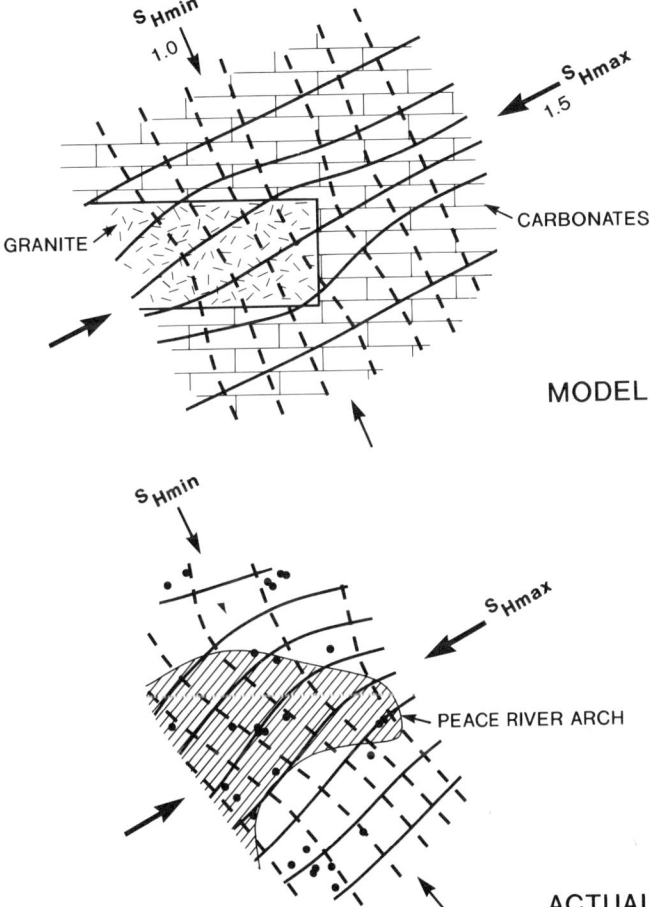

Figure 15. Comparison between the actual horizontal principal-stress trajectories in the Peace River Arch area of western Canada and those predicted by a two-dimensional finite-element model (Bell and Lloyd, 1989). The model used the following parameters: carbonates—Young's Modulus 65 GPa, Poisson's Ratio 0.3; granite—Young's Modulus 40 GPa, Poisson's Ratio 0.25. The correspondence suggests that the observed variation in stress orientation may be due to stress refraction caused by lateral variations in the elastic properties of the rocks being compressed.

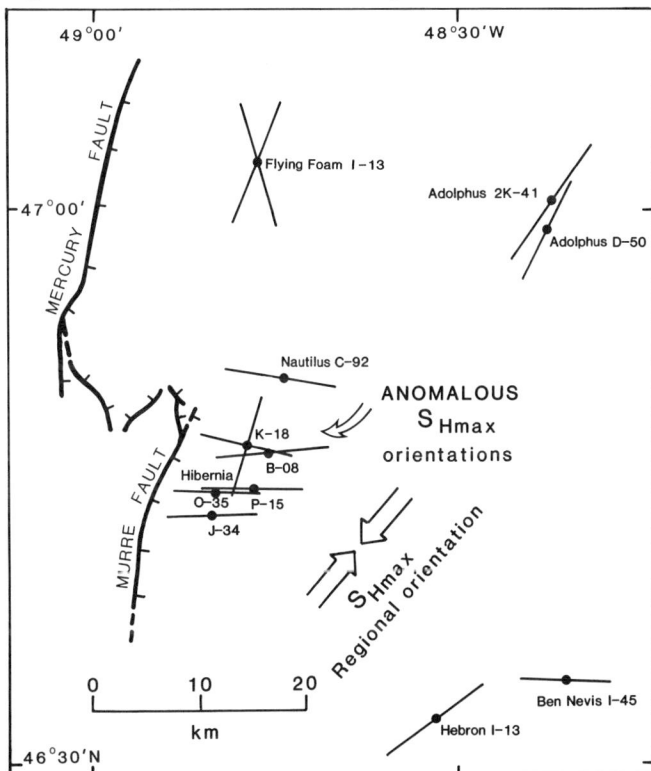

Figure 16. Breakout-derived mean SHmax orientations for wells in the Jeanne d'Arc Basin, offshore eastern Canada. The regional SHmax direction in this area is NE-SW. The E-W stress orientations at Nautilus, Hibernia, and perhaps Ben Nevis are believed to be due to the deflection of SHmax trajectories toward N-S–trending faults that are acting as "free surfaces." One such structure is the Murre fault, a listric normal structure active in latest Tertiary time. Its fault plane extends to within 500 m of the sea floor and, at depth, bends toward the Hibernia wells. The Mercury fault is a similar structure, and there are numerous other deeper faults (not shown) that offset Lower Cretaceous and older rocks.

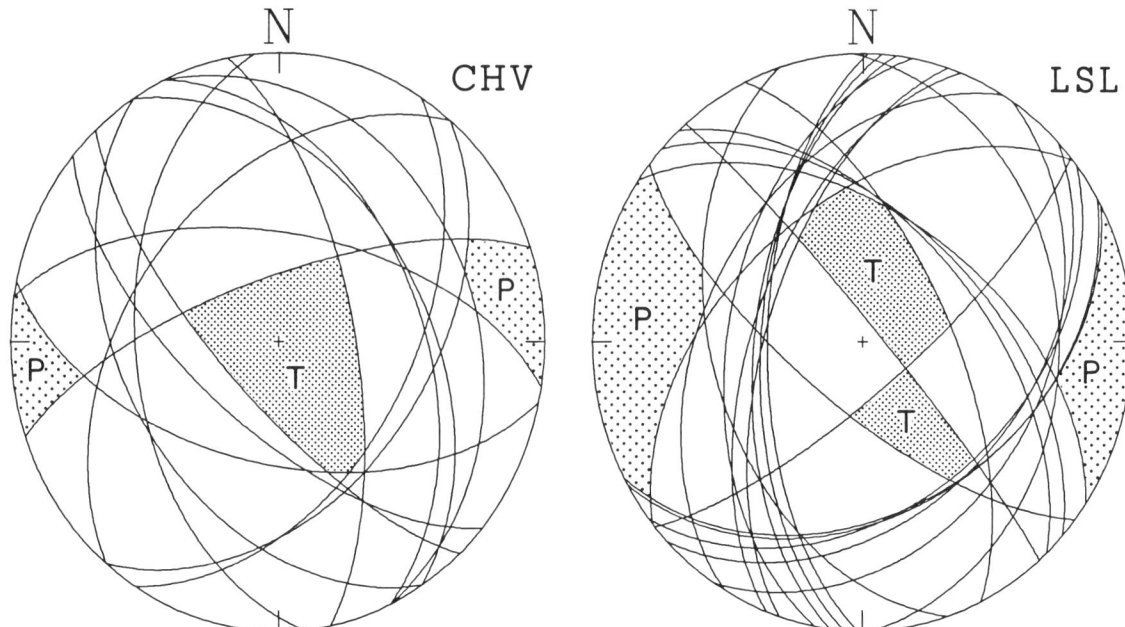

Figure 17. Combined plots of 7 Charlevoix (left) and 11 Lower St. Lawrence (right) earthquakes showing the common dilatational sectors (stippled and labeled P) and the common compressional sectors (stippled and labeled T) for each.

might become SHmax, and so the orientation of the stress field would flip 90°.

Understanding these second-order effects—whether due to local structure or to a change in the applied plate tectonic field with depth—will be important for explaining other stress anomalies and may prove valuable for improving seismic-hazard assessment. For example, it is no longer possible to conclude that a northeast-trending surface fault is unlikely to generate large earthquakes just because it is poorly oriented relative to the regional stress field. Past large (M>6) earthquakes in eastern Canada mostly nucleated at depths between 10 and 20 km (e.g., Ebel and others, 1986), and so may be caused by stresses that differ from those seen at shallow depths.

DISCUSSION

Stress provinces in Canada

Implicit in much of the above review and discussion is the assumption that stress provinces can be defined and mapped in Canada. Thanks to the large increase in stress-direction data gathered over the last decade and information bearing on whether the stress regime is compressional or extensional, Canada can be subdivided presently into five stress provinces. A selection of the best horizontal stress orientation data is shown on Figure 3 together with our preliminary interpretation of stress provinces for the northern part of North America.

Almost everywhere in Canada the observations indicate a compressive stress field with the largest principal stress σ_1 being horizontal, although on Baffin Island the earthquake focal mechanisms show deviatoric extension and normal faulting (i.e., σ_1 is vertical), as discussed earlier. The compiled data support the now widely accepted suggestions of Gough, Bell, and others that the whole of the Canadian lithosphere east of the Cordillera is being compressed in a generally northeast-southwest direction. Earlier workers had considerably fewer data available and sometimes concluded that what we now identify as regional stresses were dominated by residual or local stresses. However, we now find that the Canadian data support the existence of an extensive and relatively uniform *Mid-Plate* stress province (MPSP) of North America, which also includes the eastern and midcontinent United States (Zoback and Zoback, 1988). The eastern and northern boundaries of this province lie beyond the Canadian continental margin, and may be taken to extend to near the Mid-Atlantic Ridge. The western boundary is more problematic. The province clearly includes the eastern part of the Canadian Rockies (Cordillera) from western Alberta to the Beaufort Sea. However, in the central Cordillera there is an almost complete lack of stress data, precluding a definitive statement.

To the west, from northern Vancouver Island to Yakutat, the transform boundary between the Pacific and North American plates is well defined, and the direction of slip between the plates is at about 45° to the stresses in the MPSP. Therefore, here it is likely that the stress field of the MPSP blends smoothly into the *Queen Charlotte* stress province. Farther north, the Yukon-Alaska region is affected by the subduction of the Pacific plate beneath southern Alaska, and we see evidence for rotation of the SHmax direction to the north so as to be normal to the surface trace of the plate boundary (perhaps due to the Pacific plate acting as an indentor).

Although earthquakes in western Alaska show deviatoric extension in a roughly north-south direction, earthquakes in central Alaska show mostly strike-slip mechanisms (Biswas and others, 1986), and so it is unlikely that the western Alaska extensional province extends into Canada. The northern boundary between tensional and compressional regimes drawn by Biswas and others (1986) passes into the Beaufort Sea on the Alaskan side of the Canada–U.S. border. Two earthquakes in the Beaufort Sea (1968, 1975) have predominantly strike-slip mechanisms, though 1975 at a focal depth of 40 km has a moderate component of normal faulting (Hasegawa and others, 1979). In the shallow crust, magnitudes estimated from oil-well records suggest a thrust or strike-slip faulting regime (Fig. 12). Thus, at this time the placement of the stress province boundary offshore is uncertain.

Off Vancouver Island (and off the Oregon-Washington margin of the United States) compressional folds at the base of the continental slope indicate active subduction of the Juan de Fuca plate and NE-directed compression in a direction similar to that of the Queen Charlotte stress province to the north. The compression direction is supported by the few earthquake focal mechanisms on western Vancouver Island, and has been termed the *Cascadia Convergence* stress province in the adjacent United States. For western Washington, Weaver and Smith (1983) have suggested an abrupt change in SHmax orientation from the NE-SW compression in the Cascadia Convergence stress province to N-S compression (essentially parallel to the margin) inland in the North American crust. Our data for British Columbia are incomplete, but there is evidence from crustal earthquakes that this N-S-directed *Pacific Northwest* stress province extends into the Coastal Range of British Columbia. As shown on Figure 3, it is problematic how far toward the *Mid-Plate* stress province the Pacific Northwest stress province extends, because of the lack of stress orientation data in the Cordillera.

South of the Alberta–U.S. border lies the *Cordilleran Extensional* stress province, which includes the Basin and Range area (Zoback and Zoback, 1980, 1989). Although normal faulting earthquake mechanisms occur at Flathead Lake, Montana, 70 km south of Canada (Smith and Sbar, 1974), no normal faulting mechanisms have yet been documented in adjacent areas in Canada. However, Gough (1986) has assembled considerable evidence favoring contemporary mantle upflow beneath southeastern British Columbia and proposed that an extensional stress province overlies a crustal region of high electrical conductivity in the Intermontane and Ominec Belts. In addition to the high electrical conductance, Gough (1986) cites the high heat flux, low density, low seismic velocity, thin lithosphere, Tertiary volcanism, and late Tertiary normal faulting as evidence for partial melting at high crustal levels promoted by mantle upflow. He concludes that contemporary crustal extension about an E-W axis is probable. We have no stress measurements from this region, but have identified it as a possible extensional stress prov-

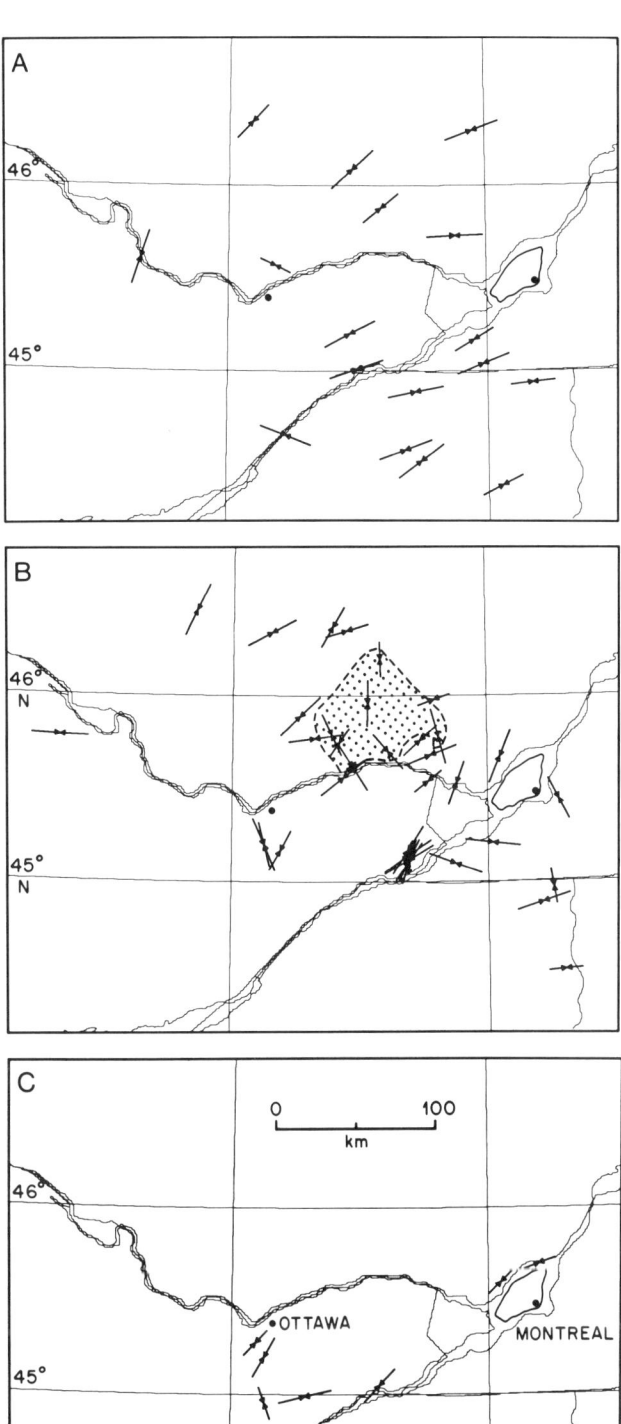

Figure 18. Stress directions for the Ottawa-Montreal region, compiled by Adams (1987) and including focal mechanisms derived by Wahlstrom (1987). A: P-axes of earthquakes shallower than 9 km. B: P-axes of earthquakes deeper than 9 km. C: stress directions derived from stress indicators other than earthquakes. Stippled region north of Ottawa shows the anomalous patch that has a uniform stress direction orthogonal to the regional field.

ince. It is shown (on Fig. 3) extending as far north as the mapped limit of the Canadian Cordilleran regional conductor (Gough, 1986), but could continue farther northwestward.

Age of the Canadian stress regimes

For the westernmost stress provinces in Canada, off Vancouver Island and the Queen Charlotte Islands, most of the stress data are provided by earthquake focal mechanisms, so the stress regime is clearly contemporary. Elsewhere, in areas where most directional data are provided by hydraulic fracturing and breakouts, one can question whether relict stresses might be contributing to the overall stress signature. Breakout orientations may supply part of the answer for the MPSP. They are uniformly oriented in sedimentary rocks of all ages, except where local structural conditions distort the regional stress field. Uniform, roughly NW-SE, long-axis orientations have been measured in more than 350 wells in rocks ranging in age from Lower Paleozoic to Upper Miocene. The uniform orientation of the breakouts (e.g., Fig. 19) implies that the stress regime they diagnose need be no older than the youngest rocks in which they occur. If pre–Late Miocene stress regimes in the MPSP had had significantly different SHmax orientations, and had the rocks retained a memory of them, we should expect to see systematic differences in breakout azimuth in the older rocks. This is not the case, so we must conclude that the observed stress regime need be no older than Late Miocene. Rocks have to become somewhat indurated before they can develop breakouts, so an even more recent age is implied, and the data are perfectly compatible with the stress regime being contemporary.

We see two implications. Firstly, while older, *similarly* oriented stress regimes might have existed in pre-Miocene time (and probably did in eastern Canada since the modern Atlantic Ocean was established), their trace is not distinguishable from the contemporary field. Secondly, where older, *differently* oriented stress regimes existed during Phanerozoic time (e.g., that documented for the Late Paleozoic by Craddock and van der Pluijm, 1989), they cannot have left any macroscopic mechanical memory in the rocks, for otherwise there would be a distinctive influence on the breakouts that form today.

Similar inferences can be made about the stresses induced by the repeated glaciation/deglaciation cycles in southeastern Canada. As discussed above, observations in southern Canada (Fig. 14) suggest that the near-surface rocks experienced NW-SE–directed radial stresses following the deglaciation. As these mechanical stresses no longer control the orientation of the present stress field, which is NE-directed, they clearly have not been "locked in" the rocks, but have been released, perhaps by partial unflexing, or otherwise dissipated over the past 10,000 years. In this respect, the observed postglacial faults can be considered a stress-relief phenomenon.

Thus, we consider that it is the fate of all ancient applied stresses to fade away relative to any contemporary stress field. The time-scale on which the stresses fade may be scale-dependent, so that small blocks of very strong rock, isolated from other rock masses by joints, faults, and shear zones, may contain relict stresses that have been locked in for a very long period of time, whereas the same relict stresses in the rock mass as a whole have been completely dissipated, whether by relaxation of the rock mass as a whole or by tiny movements on the joints of the upper crust. Thus, the scale of the jointing, the scale of the sampling technique, and the strength of the rock to long-term viscoelastic deformation, may control whether relict stresses are measured. In this context, the near-surface sampling of strong, granitic rock by a small-sample-volume method such as overcoring is more likely to observe relict stresses, while breakouts in weak sedimentary rock measured over a large length of wellbore are more likely to report only the contemporary field. In part, we think this explains the greater variability of near-surface overcoring results relative to breakout orientations.

The need for a stress-renewal mechanism

Earthquakes and stress-relief features are continually releasing the deviatoric stresses in the brittle part of the Earth's crust. The source of the stress is immaterial—they merely release that

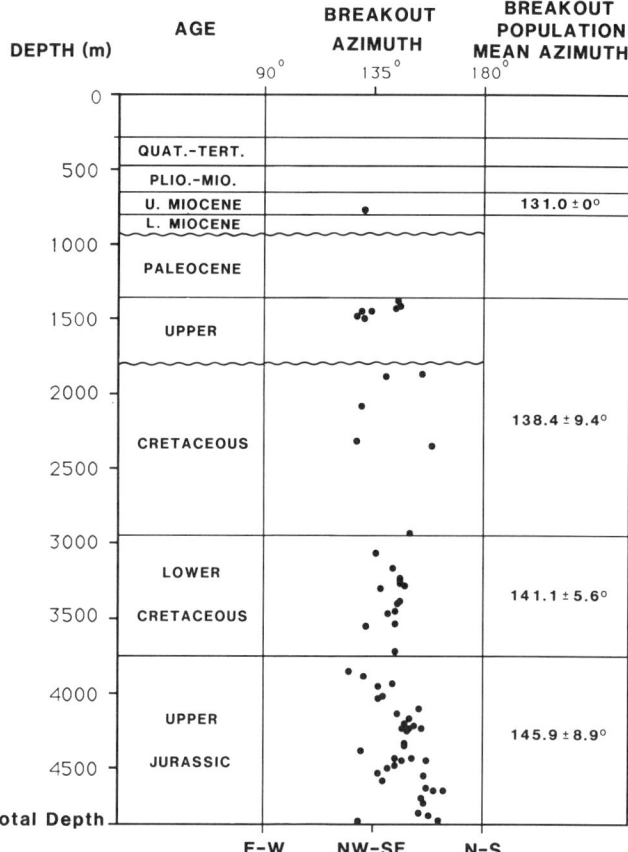

Figure 19. Mean azimuths of 64 breakouts measured in the Sachem D-76 well, Scotian Shelf, offshore eastern Canada. Note the consistency in breakout orientation for rocks that range in age from Late Jurassic to late Miocene. This implies that the contemporary stress regime need be no older than late Miocene.

which is available. Over millions of years, all deviatoric stresses would have been thus released, were it not for a stress-renewal mechanism. The stress-renewal mechanism ensures that the stresses from the contemporary applied stress are renewed while stresses relating to older stress fields are dissipated without being renewed. This allows the contemporary field to dominate, as we observe.

Our best candidate for a renewal mechanism is ridge push/ mantle drag on the asthenosphere, together with a stress-migration mechanism similar to that modelled by Kusznir and Bott (1977), by which the load on the ductile lower crust and upper mantle is transferred to the brittle upper crust (Hasegawa and others, 1985), where it is released by earthquakes and by aseismic stress release within the uppermost, fractured part of the crust.

Causes of Canadian crustal stresses

The crustal stress data presented here can be accounted for in several ways. There is clearly a burial effect whereby all principal stresses increase in magnitude with depth, which can be documented to 6 km. It is also clear that principal stress anisotropy is almost totally pervasive. Breakouts are ubiquitous in wells, testifying to unequal magnitudes of horizontal principal stress magnitudes and all the overcoring and hydraulic fracturing data point to unequal principal stress magnitudes. Only locally in overpressured sediments on the continental shelves do the data suggest stress magnitude isotropy, and this is not clearcut by any means. Absolute magnitudes of horizontal stress with respect to depth appear to vary between geologic provinces.

What is most striking is the similarity of horizontal principal stress orientation—the geometry of relative magnitude—over most of the Canadian landmass. The NE-SW SHmax (in detail, ENE in southeast Canada and NNE in Arctic) signature of the Mid-Plate stress province is a global phenomenon on the scale of a lithospheric plate that requires a causal mechanism of similar scale. We believe that it is most logically explained as a manifestation of the forces driving the plates and it may be significant that the absolute plate motion vectors derived by Minster and Jordan (1978) coincide closely with mean SHmax azimuths measured in the Mid-Plate stress province (Zoback and others, 1989). At least for this stress province, a case can be made for basal drag on the lithosphere imparting the stress signature. The drag could be exerted by the lithosphere sliding over the asthenosphere (Zoback and Zoback, 1980) or by mantle convection propelling the plate (Gough, 1984). We favor the latter mechanism.

As we have seen, crustal stresses in Canada are predominantly compressive. On Baffin Island, and locally in southern British Columbia, there may be regions of extension, and in the upper, hydrostatically pressured sediments on the Scotian Shelf it appears likely that SV is the largest principal stress (Bell, 1990). The common orientation of the horizontal principal stresses over crustal plate-sized areas of the lithosphere calls for a generating mechanism of similar magnitude. From what we can observe with our present state of knowledge, horizontal stress magnitudes vary from place to place in Canada within the Mid-Plate stress province. They are highest in the Canadian Shield and lowest in the upper part of the Scotian Shelf (Fig. 13). Intermediate values are found in western Canada and in the Beaufort Sea Shelf. It is as though the lithosphere were receiving an "external regional signal" that is being modified (in terms of strength, but not direction) by "local circumstances." The "external regional signal" is believed to be basal shear applied to lithosphere by a convecting mantle, and the "local circumstances" may be a combination of lithospheric thickness, local structure, rock properties, and possibly, structural history. For example, the extensional behavior of the Scotian Shelf may be due to its being oriented approximately parallel to SHmax, whereas basins on the Grand Banks and Labrador Shelf trend almost at right angles, and might thus be subject to greater compression. The high horizontal stresses measured in the Canadian Shield may be a result of greater lithospheric thickness, and possibly, the rocks themselves are capable of transmitting more compression than the softer clastics in the offshore eastern Canadian basins.

CONCLUSIONS

We have summarized here information bearing on the contemporaneous state of stress in Canada. Stress orientation information is sufficiently widespread and consistent to allow us to recognize four stress provinces, which are characterized by directional signatures of SHmax that form small circles on the surface of the Earth. Within these stress provinces, we suspect there are lateral gradients in stress magnitudes, but our data base is not yet adequate to define them confidently. We can, however, identify some areas with anomalous horizontal principal stress orientations and have sufficient supporting data to suggest causes such as stress refraction (Fig. 15), active faults (Fig. 16), and (more speculatively) past earthquakes (Fig. 18B). This is encouraging, particularly as it relates to assessing earthquake hazard and other applications discussed above.

Nevertheless, we still need to address deficiencies in the stress database, and data remain to be compiled in several areas. Overcoring results in mines are considered proprietary information that can be obtained only after significant personal lobbying; past large earthquakes have not yet been analyzed for focal mechanisms; there are more breakouts to be identified and studied in wells; and there is a huge source of stress magnitude estimates in the records of hydraulic fracture treatments and leak-off tests run in western Canadian wells. We also need to identify and understand additional indirect indicators of the contemporary stress regime (like pop-ups) so that we are better able to use simple observational criteria.

Even with all these additional data, there are many large regions of Canada where there are no obvious sources of stress information, and these parts of the stress map may remain blank for some time to come. Therefore, we must seek to understand the regional patterns in the available data, to test theories that will account for the nature of stress field, and to devise models that will allow extrapolation into those areas we cannot sample.

REFERENCES CITED

Adams, J., 1982, Stress-relief buckles in McFarland Quarry, Ottawa: Canadian Journal of Earth Sciences, v. 19, p. 1883–1887.
—— , 1987, Canadian crustal stress database; A compilation to 1987: Geological Survey of Canada Open File 1622, 130 p.
—— , 1989, Postglacial faulting in eastern Canada; Nature, origin and seismic hazard implications: Tectonophysics, v. 163, p. 323–331.
Adams, J., and Basham, P. W., 1989, Seismicity and seismotectonics of Canada east of the Cordillera: Geoscience Canada, v. 16, p. 3–16.
Adams, J., Sharp, J., and Stagg, M., 1988, New earthquake focal mechanisms for eastern Canada: Geological Survey of Canada Open File 1892, 109 p.
Angelier, J., and Mechler, P., 1977, Sur une méthode graphique de recherche des contraintes principales également utilisable en tectonique et en séismologie; La méthode des dièdres droits: Bulletin de la Société Géologique de France, v. 19, p. 1309–1318.
Babcock, E. A., 1978, Measurement of subsurface fractures from dipmeter logs: American Association of Petroleum Geologists Bulletin, v. 62, p. 1111–1126.
Bell, J. S., 1985, Offset boreholes in the Rocky Mountains of Alberta, Canada: Geology, v. 13, p. 734–737.
—— , 1989a, Investigating stress regimes in sedimentary basins using information from oil industry wireline logs and drilling records, in Hurst, A., ed., Geological applications of wireline logs: Geological Society of London Special Publication 48, p. 305–325.
—— , 1989b, Stress in sedimentary basins seminar: Geological Survey of Canada Open File 2140, 96 p.
—— , 1989c, Vertical migration of hydrocarbons at Alma, offshore eastern Canada: Bulletin of Canadian Petroleum Geology, v. 37, p. 358–364.
—— , 1990, The stress regime of the Scotian Shelf, offshore eastern Canada, to 6 kilometers depth and the implications for rock mechanics and hydrocarbon migration: in Maury, V., and Fourmaintraux, D., eds., Rock at great depth, v. 3: Rotterdam, A. A. Balkema, p. 1243–1265.
Bell, J. S., and Babcock, E. A., 1986, The stress regime of the Western Canadian Basin and implications for hydrocarbon production: Bulletin of Canadian Petroleum Geology, v. 34, p. 364–378.
Bell, J. S., and Gough, D. I., 1979, Northeast–southwest compressive stress in Alberta; Evidence from oil wells: Earth and Planetary Science Letters, v. 45, p. 475–482.
—— , 1981, Intraplate stress orientations from Alberta oil wells, in Evolution of the Earth: American Geological Union Geodynamics Series, v. 5, p. 96–104.
Bell, J. S., and Lloyd, P. F., 1989, Modelling of stress refraction in sediments around the Peace River Arch, western Canada, in Current research, Part D: Geological Survey of Canada Paper 89-1D, p. 49–54.
Bell, J. S., and Price, P. R., 1989, In-situ stresses in Western Canada; Implications for the oil industry: Poster presented at Forum '89, Oil and Gas Activities in Canada, Geological Survey of Canada, February 27–28, Calgary.
Biswas, N. N., Aki, K., Pulpan, H., and Tytgat, G., 1986, Characteristics of regional stresses in Alaska and neighboring areas: Geophysical Research Letters, v. 13, p. 177–180.
Breckels, I. M., and Van Eekelen, H.A.M., 1981, Relationship between horizontal stress and depth in sedimentary basins, in 56th Annual Fall Technical Conference, San Antonio, Texas, October 5–7, 1981: Society of Petroleum Engineers, Paper SPE10336, 7 p.
Bredehoeft, J. D., Wolff, R. G., Keys, W. S., and Shuter, E., 1976, Hydraulic fracturing to determine the regional in-situ stress field, Piceance Basin, Colorado: Geological Society of America Bulletin, v. 87, p. 250–280.
Buchbinder, G.G.R., 1989, Azimuthal variations in P-wave travel times and shear-wave splitting in the Charlevoix seismic zone: Tectonophysics, v. 165, p. 293–302.
Clauss, B., Marquert, G., and Fuchs, K., 1989, Stress orientations in the North Sea and Fennoscandia, a comparison to the central European stress field, in Gregersen, S., and Basham, P. W., eds., Earthquakes at North Atlantic passive margins; Neotectonics and postglacial rebound: Dordrecht, Kluwer Academic Publishers, p. 277–287.
Cornet, F. H., and Valette, B., 1984, In situ stress determination from hydraulic injection test data: Journal of Geophysical Research, v. 89, p. 11527–11537.
Cox, J. W., 1983, Long axis orientation in elongated boreholes and its correlation with rock stress data, in Proceedings, 24th Annual Logging Symposium, Calgary, June 27–30, 1983: Society of Professional Well Log Analysts, 17 p.
Craddock, J. P., and van der Pluijm, B. A., 1989, Late Paleozoic deformation of the cratonic carbonate cover of eastern North America: Geology, v. 17, p. 416–419.
Crampin, S., 1987, Geological and industrial implications of extensive-dilatancy anisotropy: Nature, v. 328, p. 491–496.
Dart, R. L., and Zoback, M. L., 1989, Wellbore breakout stress analysis within the central and eastern continental United States: Log Analyst, January–February 1989, p. 12–23.
Doe, T. W., Hustrulid, W. A., Leijon, B., Ingvald, K., and Strindell, L., 1983, Determination of the state of stress at the Stripa Mine, Sweden, in Zoback, M. D., and Haimson, B. C., eds., Hydraulic fracturing stress measurements: Washington, D.C., National Academy Press, p. 119–129.
Dragert, H., 1986, A summary of recent geodetic measurements of surface deformation on Vancouver Island: Royal Society of New Zealand Bulletin, v. 24, p. 29–37.
Dragert, H. and Rogers, G. C., 1988, Could a megathrust earthquake strike southwestern British Columbia?: Geos, v. 17, no. 3, p. 5–8.
Dusseault, M. B., 1977, State of stress and hydraulic fracturing in the Athabasca oil sands: Journal of Canadian Petroleum Technology, v. 16, p. 19–27.
Ebel, J. E., Somerville, P. G., and McIver, J. D., 1986, A study of the source parameters of some large earthquakes in northeastern North America: Journal of Geophysical Research, v. 91, p. 8231–8247.
Eisbacher, G. H., and Bielenstein, H. V., 1971, Elastic strain recovery in Proterozoic rocks near Elliott Lake, Ontario: Journal of Geophysical Research, v. 76, p. 2012–2021.
Ervine, W. B., and Bell, J. S., 1987, Subsurface in-situ stress magnitudes from oil well drilling records; An example from the Venture area, offshore Eastern Canada: Canadian Journal of Earth Sciences, v. 24, p. 1748–1759.
Franklin, J. A., and Hungr, O., 1978, Rock stresses in Canada and their relevance to engineering projects: Rock Mechanics, Supplement 6, p. 25–46.
Gough, D. I., 1984, Mantle upflow under North America and plate dynamics: Nature, v. 311, p. 428–433.
—— , 1986, Mantle upflow tectonics in the Canadian Cordillera: Journal of Geophysical Research, v. 91, p. 1909–1919.
Haimson, B. C., and Fairhurst, C., 1970, In-situ stress determination at great depth by means of hydraulic fracturing, in Somerton, W., ed., Rock mechanics; Theory and practice; Proceedings of the 11th Symposium on Rock Mechanics: New York, American Institute of Mining Engineers, p. 559–584.
Haimson, B. C., Lee, C. F., and Huang, J.H.S., 1986, High horizontal stresses at Niagara Falls, their measurement, and the design of a new hydroelectric plant, in Stephansson, O., ed., Proceedings of the International Symposium on Rock Stress and Rock Stress Measurements: Stockholm, 1–3 September, 1986: Centek, p. 615–624.
Hansen, K. M., and Mount, V. S., 1990, Smoothing and extrapolation of crustal stress orientation measurements: Journal of Geophysical Research, v. 95, p. 1155–1165.
Hasegawa, H. S., 1977, Focal parameters of four Sverdrup Basin Arctic Canada earthquakes in November and December of 1972: Canadian Journal of Earth Sciences, v. 14, p. 2481–2494.
Hasegawa, H. S., and Adams, J., 1981, Crustal stresses and seismotectonics in eastern Canada: Earth Physics Branch Open File 81-12, 42 p.
Hasegawa, H. S., and Basham, P. W., 1989, Spatial correlation between seismicity and postglacial rebound in eastern Canada, in Gregersen, S., and Basham,

P. W., eds., Earthquakes at North Atlantic passive margins; Neotectonics and postglacial rebound: Dordrecht, Kluwer Academic Publishers, p. 483–500.

Hasegawa, H. S., Chou, C. W., and Basham, P. W., 1979, Seismotectonics of the Beaufort Sea: Canadian Journal of Earth Sciences, v. 16, p. 816–830.

Hasegawa, H. S., Adams, J., and Yamazaki, K., 1985, Upper crustal stresses and vertical stress migration in eastern Canada: Journal of Geophysical Research, v. 90, p. 3637–3648.

Hashizume, M., 1973, Two earthquakes on Baffin Island and their implications: Journal of Geophysical Research, v. 78, p. 6069–6081.

Hassan, D., 1982, A method for predicting hydraulic fracture azimuth and the implication thereof to improve hydrocarbon recovery, in 33rd Annual Technical Meeting of the Petroleum Society, Calgary, June 6–9, 1982: Canadian Institute of Mining and Metallurgy, paper 82-33.

Herget, G., 1973, Variation of rock stresses with depth at a Canadian iron ore mine: International Journal of Rock Mechanics and Mining Science, v. 10, p. 37–51.

—— , 1980, Regional stresses in the Canadian Shield, in Underground rock engineering: Canadian Institute of Mining and Metallurgy Special Volume 22, p. 9–16.

Hubbert, M. K., and Willis, D. G., 1957, Mechanics of hydraulic fracturing: Transactions of the Society of Petroleum Engineers of AIME, v. 210, p. 153–166.

Jaeger, J. C., and Cook, N.G.W., 1976, Fundamentals of rock mechanics, 2nd ed.: London, Chapman and Hall, 585 p.

Kry, P. R., and Gronseth, J. M., 1982, In-situ stresses and hydraulic fracturing in the Deep Basin, in 33rd Annual Technical Meeting of the Petroleum Society, Calgary, June 6–9, 1982: Canadian Institute of Mining and Metallurgy, paper 82-33-21.

Kusznir, N. J., and Bott, M.H.P., 1977, Stress concentration in the upper lithosphere caused by underlying visco-elastic creep: Tectonophysics, v. 43, p. 247–256.

Lo, K. Y., 1978, Regional distribution of in-situ horizontal stresses in rocks in southern Ontario: Canadian Geotechnical Journal, v. 15, p. 371–381.

Marrow, P. C., and Walker, A. B., 1988, Lleyn earthquake of 1984 July 19; Aftershock sequence and focal mechanism: Geophysical Journal, v. 92, p. 487–493.

McGarr, A., and Gay, N. C., 1978, State of stress in the Earth's crust: Annual Review of Earth and Planetary Sciences, v. 6, p. 405–436.

McLellan, P., 1988, In-situ stress magnitudes from hydraulic fracturing treatment records; A feasibility study: Geological Survey of Canada Open File Report 1947, 30 p.

McLennan, J. D., and Roegiers, J.-C., 1982, How instantaneous are instantaneous shut-in pressures?, in 57th Annual Technical Conference and Exhibition, New Orleans, Louisiana: Society of Petroleum Engineers, paper 11064, 10 p.

Minster, J. B., and Jordan, T. H., 1978, Present day plate motions: Journal of Geophysical Research, v. 83, p. 5331–5354.

Mount, V. S., and Suppe, J., 1987, State of stress near the San Andreas fault; Implications for wrench tectonics: Geology, v. 15, p. 1143–1146.

Nolte, K. G., 1988a, Principles for fracture design based on pressure analysis: Production Engineering, Society of Petroleum Engineers, February 1988, p. 22–30.

Nolte, K. G., 1988b, Application of fracture design based on pressure analysis: Production Engineering, Society of Petroleum Engineers, February 1988, p. 31–42.

Podrouzek, A. J., and Bell, J. S., 1985, Stress orientations from wellbore breakouts on the Scotian Shelf, eastern Canada: Geological Survey of Canada Paper 85-1B, p. 59–62.

Quinlan, G., 1984, Postglacial rebound and the focal mechanisms of eastern Canadian earthquakes: Canadian Journal of Earth Sciences, v. 19, p. 1018–1023.

Sarda, J.-P., Perreau, P. J., and Deflandre, J.-P., 1988, Acoustic emission interpretation for estimating hydraulic fracture extent; Laboratory and field studies, in 63rd Annual Technical Conference and Exhibition, Houston Texas, October 2–5, 1988: Society of Petroleum Engineers, paper 18192.

Sbar, M. L., Richardson, R. M., and Flaccus, C., 1984, Near-surface in-situ stress; 1, Strain relaxation measurements along the San Andreas fault in southern California: Journal of Geophysical Research, v. 89, p. 9323–9332.

Silver, P. G., and Chan, W. W., 1988, Implications for continental structure and evolution from seismic anisotropy: Nature, v. 335, p. 34–39.

Smith, R. B., and Sbar, M. L., 1974, Contemporary tectonics and seismicity of the western United States with emphasis on the intermountain seismic belt: Geological Society of America Bulletin, v. 85, p. 1205–1218.

Stein, S., Sleep, N. H., Geller, R. J., Wang, S. C., and Kroeger, G. C., 1979, Earthquakes along the passive margin of eastern Canada: Geophysical Research Letters, v. 6, p. 537–540.

Stein, S., Cloetingh, S., Sleep, N. H., and Wortel, R., 1989, Passive margin earthquakes, stresses, and rheology, in Gregersen, S., and Basham, P. W., eds., Earthquakes at North Atlantic passive margins; Neotectonics and postglacial rebound: Dordrecht, Kluwer Academic Publishers, p. 231–259.

Teague, A. G., Bollinger, G. A., and Johnston, A. C., 1986, Focal mechanism analysis for eastern Tennessee earthquakes (1981–1983); Bulletin of the Seismological Society of America, v. 76, p. 95–109.

Wahlstrom, R., 1987, Focal mechanisms of earthquakes in southern Quebec, southeastern Ontario, and northeastern New York, with implications for regional seismotectonics and stress field characteristics: Bulletin of Seismological Society of America, v. 77, p. 891–824.

Weaver, C., and Smith, S., 1983, Regional tectonic and earthquake hazard implications of a crustal fault zone in southwestern Washington: Journal of Geophysical Research, v. 88, p. 10371–10383.

White, O. L., and Russell, D. J., 1982, High horizontal stresses in southern Ontario; Their orientation and their origin: Proceedings, IV Congress, International Association of Engineering Geology, New Delhi, v. V, p. V39–V54.

Woodland, D. C., 1987, Borehole stability in western Canada: Poster presented at the 38th Annual Technical Meeting, Petroleum Society of CIM, Calgary, June 7–10, 1987.

Woodland, D. C., and Bell, J. S., 1989, In-situ stress magnitudes from mini-frac records in western Canada: Journal of Canadian Petroleum Technology, v. 28, p. 22–31.

Yang, J. P., and Aggarwal, Y. P., 1981, Seismotectonics of northeastern United States and adjacent Canada: Journal of Geophysical Research, v. 86, p. 4981–4998.

Zoback, M. L., and Zoback, M., 1980, State of stress in the conterminous United States: Journal of Geophysical Research, v. 85, p. 6113–6156.

—— , 1988, Tectonic stress field of the continental United States, in Pakiser, L., and Mooney, W., eds., Geophysical framework of the continental United States: Geological Society of America Memoir 172, p. 523–540.

Zoback, M. D., Moos, D., Mastin, L., and Anderson, R. N., 1985, Wellbore breakouts and in-situ stress: Journal of Geophysical Research, v. 90, p. 5523–5530.

Zoback, M. L., and 28 others, 1989, Global patterns of tectonic stress: Nature, v. 341, p. 291–298.

MANUSCRIPT ACCEPTED BY THE SOCIETY AUGUST 1, 1990

ACKNOWLEDGMENTS

The stress database used in this chapter was established by a number of undergraduate university students working under Adams' supervision and funded by the Office of Energy Research and Development of Energy, Mines, and Resources Canada as part of the Geological Survey of Canada's research into seismic hazards to conventional energy systems. Henry Hasegawa sparked Adams' interest in crustal stresses, and Frank Anglin provided the mapping routines. Bell's work in crustal stress studies dates from his involvement in formulating the breakout hypothesis with D. I. Gough, and evolved in subsequent investigations with Gough and many other colleagues. He is particularly grateful to the Geological Survey of Canada for the support and encouragement provided for stress compilation studies. We thank Henry Hasegawa and two anonymous reviewers for comments on this manuscript. This chapter is also Geological Survey of Canada contribution 14390.

NOTES ADDED IN PROOF

Our ideas involving the regional extent of the stress fields and the age of the stress field are further developed in Adams and Bell (1990) and Bell and Adams (1990). We have also considered further the anomalous breakouts found in the Alma well on the Scotian shelf, and suggested that the rotations of the breakout orientation sometimes seen in wells are due to open fractures crossing or passing near to the well (Bell and others, 1991).

ADDITIONAL REFERENCES

Adams, J., and Bell, J. S., 1990, Mapping regional stress provinces in Canada—a progress report: *in* Proceedings, Stresses in underground structures, CANMET specialty conference: Ottawa, p. 3–12.

Bell, J. S., and Adams, J., 1990, Do the rocks remember? How contemporary are regional stresses in Canada? *in* Proceedings, Stresses in underground structures, CANMET specialty conference: Ottawa, p. 13–22.

Bell, J. S., Caillet, G., and Adams, J., 1991, Attempts to detect open fractures and non-sealing faults with dipmeter logs: *in* Proceedings, Geological applications of wireline logs II: Geological Society of London, in press.

Chapter 21

Stress indicators in Alaska

Charles H. Estabrook* and Klaus H. Jacob
Lamont-Doherty Geological Observatory of Columbia University, Palisades, New York 10964

INTRODUCTION

The interaction of tectonic plates has been used to predict the direction of tectonic stresses well away from plate boundaries. These studies have used individual stress indicators to infer trajectories of tectonic stress (maximum horizontal compression or MHC; e.g., Nakamura and others, 1977, 1980; Nakamura and Uyeda, 1980; Zoback and Zoback, 1980). The indicators are often derived from earthquake focal mechanisms (slip vectors and/or pressure, tension axes). Taking such data from a wide variety of mechanisms allows one to infer the orientation of the causative stress field (e.g., Angelier, 1979; Gephart and Forsyth, 1984) despite the fact that individual mechanisms are solely geometrical strain indicators that describe the orientation of the fault on which the earthquake occurred (McKenzie, 1969). Differences of up to 45° have been observed between the in situ stress indicators (borehole hydrofractures and breakouts) and the geometrical strain indicators (faults, earthquake focal mechanisms; Mount and Suppe, 1987; Zoback and others, 1987). With these preconditions in mind, we infer the directions for maximum horizontal compressive stress (MHC) from individual indicators for Alaska and its surroundings.

Several studies have used stress and strain indicators to infer stress directions across the Alaska region. Nakamura and others (1977, 1980) used volcanic stress and geologic fault strain indicators, and Davies (1983), Biswas and others (1986), and Estabrook and others (1988) used earthquake focal mechanism strain indicators to infer stress trajectories across interior Alaska. Nakamura and others (1977, 1980), Nakamura and Uyeda (1980), and Biswas and others (1986) have proposed a transition from a compressional environment near the Pacific/North American (PAC/NAM) Plate boundary to a tensional environment in western and northeast Alaska. Nakamura and Uyeda (1980) proposed a mantle source for the broad regional extension. Nakamura and others (1977, 1980), in their study of the Aleutian arc, found that compressive stress extends at least into the arc. Many authors have determined earthquake focal mechanisms in studies of the plate interaction along the PAC/NAM (Boyd and Nábělek, 1988; Bhattacharya and Biswas, 1979; Biswas and others, 1986; Chandra, 1974; Cormier, 1975; Ekström and Engdahl, 1989; Frohlich and others, 1982; Gedney, 1970; Hong and Fujita, 1981; House and Jacob, 1983; Jacob, 1987; LaForge and Engdahl, 1979; Lahr and others, 1988; Newberry, 1983; Page and others, 1989; Pérez and Jacob, 1980; Schell and Ruff, 1989; Spence, 1977; Stauder, 1968a, b; Stauder and Bollinger, 1964, 1966a, b; Stauder and Udias, 1963; Stephens and others, 1980; Wu and Kanamori, 1973). In our discussion of the MHC indicators, we rely on many ideas from the previously mentioned papers, though much of the interpretation is strictly our own.

We have compiled a set of all available stress indicators from Alaska and adjacent areas to produce maps that display the direction of MHC. The data include earthquake focal mechanisms, volcanic stress indicators, geologically mapped fault data, and borehole breakout data. The locations of the stress indicators are between 160°E to 125°W and 48°N to 72°N, an area including all of Alaska, the western part of the Yukon Territory, the Aleutian Islands, the Bering Sea, and part of eastern Siberia. Since stress is a physical quantity that cannot be directly observed, but generally can be inferred from strain, slip, breakage, or other deformational indicators, we develop rules that we apply to obtain the MHC direction. A total of 621 stress indicators (Fig. 1; Estabrook and Jacob, 1991) were collected and categorized in four quality groups A through D, from most to least reliable. We then made maps of MHC for various regions and found that, in general, the direction of MHC strongly reflects the interaction between the Pacific and North American Plates along the plate boundary that varies from pure strike-slip to underthrusting.

DATA ANALYSIS

Earthquakes

We restrict earthquake focal mechanisms (549 individual solutions) to have hypocenter depth ≤ 65 km in an attempt to depict only the stresses near the Earth's surface (i.e., only in the upper plate when one plate subducts beneath another, or to depict the stresses near the plate interface) while attempting to eliminate imaging of stresses in the lower plate. Though this depth may be somewhat deeper than the PAC/NAM plate interface in

---

*Also at Department of Geological Sciences of Columbia University, Palisades, New York 10964.

Estabrook, C. H., and Jacob, K. H., 1991, Stress indicators in Alaska, *in* Slemmons, D. B., Engdahl, E. R., Zoback, M. D., and Blackwell, D. D., eds., Neotectonics of North America: Boulder, Colorado, Geological Society of America, Decade Map Volume 1.

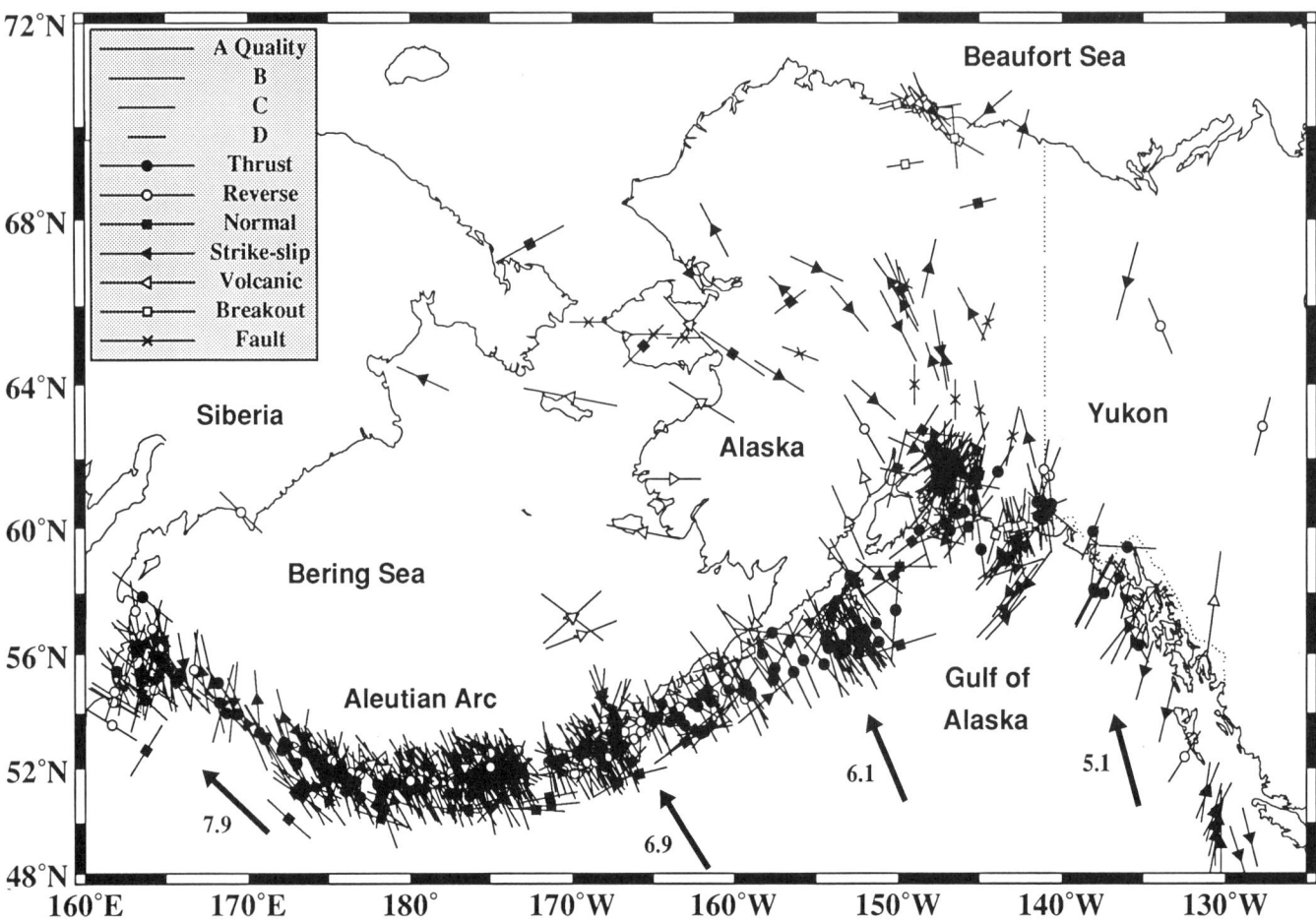

Figure 1. Maps (Mercator projection) showing the direction of maximum horizontal compression (MHC) from a variety of sources, including seismic focal mechanisms, volcanic, geologic fault, and borehole breakout data. Black arrows are NUVEL1 (DeMets and others, 1990) Pacific/North American relative plate-motion vectors, with velocities in centimeters/year. See text for quality descriptions. Figures 2 through 7 provide enlarged views of pertinent areas of the Aleutian arc and interior Alaska.

the Alaska-Aleutian subduction zone, errors in earthquake hypocenters require us to overestimate the average depth of the interface to avoid missing events. It is likely that some earthquakes in the downgoing Pacific Plate are shallower than 65 km depth; we may, therefore, include in some instances stress indicators in the shallowest portions of the lower plate. Earthquakes that occur seaward of the Aleutian Trench are included in our data set because they occur where the Pacific Plate is the plate seen at the surface, even though farther north the Pacific Plate becomes the downgoing plate. Earthquakes that occurred prior to 1960 are included, but because station distribution, uniformity, and instrument polarity are unpredictable, they are assigned our lowest quality rating. When several authors determined focal mechanisms for the same earthquakes, solutions obtained using body-wave or surface-wave analysis are chosen over first-motion mechanisms.

The earthquake focal mechanisms used here as stress indicators were determined in other studies. They include published solutions, several theses, and unpublished sources. All mechanisms are converted to pressure, tension, and null axes (P, T, and B) from other descriptions. It is assumed that the principal stress orientations are parallel to these axes, with maximum compression parallel to the P-axis and minimum compression parallel to the T-axis. The P, T, B-axes, if published, are used as is; otherwise, they are determined graphically or computed from the strike, dip, and rake description using formulas from Ben-Menahem and Singh (1981, p. 194). While the MHC description is most applicable to compressional stress regimes, we apply it to a geographic area in a heterogeneous stress state. For example, the bending of the subducting plate seaward of the trench axis describes an extensional system with the MHC direction orthogonal to that arcward of the trench. We define β_i as the plunge angle of a given axis (where i = P, T, B for the P, T, or B axes) with respect to the horizontal plane, and Ω_i as the azimuth of the trend direction of the P, T, or B-axis or of the slip vector S, measured clockwise from north in the horizontal plane. We separate the mechanisms into four faulting categories (normal, strike-slip, reverse, and thrust faulting), and assign Ω according to the following criteria:

Normal Faulting	$\beta_P > (\beta_B$ and $\beta_T)$	$\rightarrow$	Ω_B = MHC
Strike-Slip Faulting	$\beta_B > (\beta_P$ and $\beta_T)$	$\rightarrow$	Ω_P = MHC
Reverse Faulting	$\beta_T \geqslant (\beta_P$ and $\beta_B)$ and $\beta_P \leqslant 20°$	$\rightarrow$	Ω_P = MHC
Thrust Faulting	$\beta_T \geqslant (\beta_P$ and $\beta_B)$ and $\beta_P > 20°$	$\rightarrow$	Ω_S = MHC

The reason we use Ω_S instead of Ω_P for thrust events is that the slip vector has a shallower dip than the P-axis; therefore it is more likely to be parallel to the causative stress system (horizontal) along the plate boundary (Minster and Jordan, 1978; DeMets and others, 1990). If the frictional strength of the dipping plate boundary were quite low; it might distort the stress field in a manner analogous to a mature strike-slip fault (i.e., San Andreas) in which the maximum compressive stress direction rotates to being nearly normal to the fault very close to the fault (Mount and Suppe, 1987; Zoback and others, 1987). In this case, MHC derived from the slip vector direction would be very different from the stress direction close to the plate interface.

The earthquake data are each assessed for the quality of the solution and sorted into four categories. The "A" quality describes events with good depth control and two well-constrained nodal planes, using body-wave inversion techniques. The "B" quality contains events whose depths are uncertain while having constrained nodal planes. This includes body-wave inversion mechanisms, first-motion solutions using long-period P and S data, the largest short-period solutions, and those events from the Harvard centroid moment tensor catalog (CMT) with $M_w \geqslant 6.0$ (references to 1977 to 1988 CMT events are contained in Dziewonski and others, 1989). The "C" quality set contains long-period solutions with only one well-constrained nodal plane, the smaller CMT solutions ($M_w < 6.0$), and short-period solutions with one well-constrained plane. The "D" quality set contains all pre-1960 solutions regardless of quality, poorly constrained long- and short-period events (local network data), and composite solutions.

Volcanic indicators

Nakamura and others (1977, 1980) have demonstrated that elongations of volcanic edifices and alignments of parasitic cones of an active volcano are oriented perpendicular to the direction of minimum horizontal compression (i.e., magma-filled dikes are preferentially injected along tension cracks). The azimuths of the features are presumed to be parallel to the direction of the MHC. The volcanic indicators represent an average stress system over a geologically significant time period; hence they indicate the stress field at the time of formation. The sources of the data are Nakamura and others (1977, 1980) and one new point from this study for a total of 32 indicators. We follow the convention described by Nakamura and others (1977, 1980) using as MHC the direction of the elongation of height contours and of alignments of parasitic cones on polygenetic volcanoes in predominantly compressional regimes; and the direction of fissure eruptions along tension cracks in monogenetic volcanic, tensional (rifting) environments. The quality of the indicator is taken from Nakamura and others (1980) and is based on the crater distribution, flank craters on one or both sides of the main crater, elongation of the main edifice, and the occurrence of normal faults parallel to the crater distribution. Indicators rated by Nakamura and others (1980) as excellent, good, poor, and very poor become our quality A, B, C, and D.

Geologic fault indicators

Geologic faults can give an approximation of the principal stresses acting at the time of faulting (Zoback and Zoback, 1980). The strike and the sense of motion give the direction of maximum horizontal shortening. We assume that the instantaneous direction of maximum horizontal shortening approximates the MHC direction, a condition that is only absolutely true in a homogeneous medium. However, since MHC fault indicators often parallel estimates obtained from other sources, we feel the approximation is justified. For regimes with reverse and thrust faults we use the direction normal to the strike of the fault as MHC. For strike-slip regimes, we use the direction of maximum shortening, assumed at 45° to the strike of faults, as MHC. For normal faulting we use the direction parallel to the faults as MHC. The fault data and assigned quality come from Nakamura and others (1977, 1980) and King (1969) and include 14 indicators. Following Nakamura and others (1980), we assign quality based on lengths of Late Quaternary faults: "A" for faults longer than 200 km, "B" between 200 and 10 km, "C" for less than 10 km, "D" quality for those faults whose Late Quaternary history is suspect.

Oil well breakouts

Oil well breakouts provide an in situ stress indicator (Bell and Gough, 1979). MHC is assumed to be the direction normal to the azimuth of borehole elongation from breakouts. Borehole elongation is attributed in most cases to spalling from shear-stress concentrations at the bore wall surface. The spalling elongates the borehole generally in the direction of the least horizontal compression. However, the data can be ambiguous because borehole elongations can vary with depth in the hole, because hydrofracturing in the direction of maximum horizontal compression competes with the breakout spalling, and because preexisting material strength anisotropies (e.g., schistosity, steeply dipping bedding, or joint patterns) may cause spurious borehole elongations and "washouts." Unlike fault and earthquake data, however, these observations generally are direct products of the regional stress rather than strain fields.

Data sources for the breakouts are 26 indicators from wells at Alaska's North Slope (Jacob, 1986) and the Gulf of Alaska (Suppe, personal communication, 1986). Four-arm dip meter logs were obtained from the North Slope for 60 wells, of which

20 were analyzed for well elongations. Of these, several were rejected due to large scatter and/or rotation with depth. Averages were computed for the entire depth of each hole; however, two peaks in the distribution of trend directions of MHC occur at ~90° and ~135°. In the Gulf of Alaska there is a consistent 90° rotation of some of the elongations with depth, which may be attributable to hydrofracturing. Breakout data are assigned quality "C" for higher quality and "D" for unstable holes from the North Slope.

RESULTS AND INTERPRETATIONS

Plate-boundary stresses

Figure 1 depicts the MHC patterns for different quality data. Quality A data alone do not contain sufficient data to convey a coherent picture. Quality A and B combined convey a relatively simple picture. While inclusion of quality C and D data increases the density and coverage, they also introduce scatter and local apparent inconsistencies. Figures 2 through 7 show enlarged views of key areas that are pertinent to discussion of individual indicators. In this discussion of the MHC indicators, we rely on the ideas of many authors, though much of the interpretation is our own.

In the eastern Gulf of Alaska, southeast Alaska, and adjacent Canada (east of 140°W) a predominant southwest-northeast compression prevails (Figs. 1 and 2). The consistency hardly deteriorates even when quality C and D solutions are included. It is based on well-established right-lateral strike-slip motion along the Queen Charlotte–Fairweather fault system (Plafker and others, 1978) on the major plate boundary between the Pacific and North American Plates, and perhaps newly incipient strike-slip faulting on a conjugate set of faults within the Pacific Plate in the offshore area in the northern Gulf of Alaska (south of 59.5°N, 142 to 145°W; Lahr and others, 1988). A few stress indicators inland in Canada, within the North American Plate, east of 140°W, corroborate this consistent southwest-northeast trend of the MHC (see Adams and Bell, this volume, for a more complete discussion of the Canadian data). The apparently inconsistent quality C indicator in southeast Alaska at 59.5°N, 136°W has a thrust mechanism (M_w 5.0) with an east-west–trending MHC that preceded the first of the 1987 Gulf of Alaska earthquakes (Lahr and others, 1988) by three days. If the MHC direction is

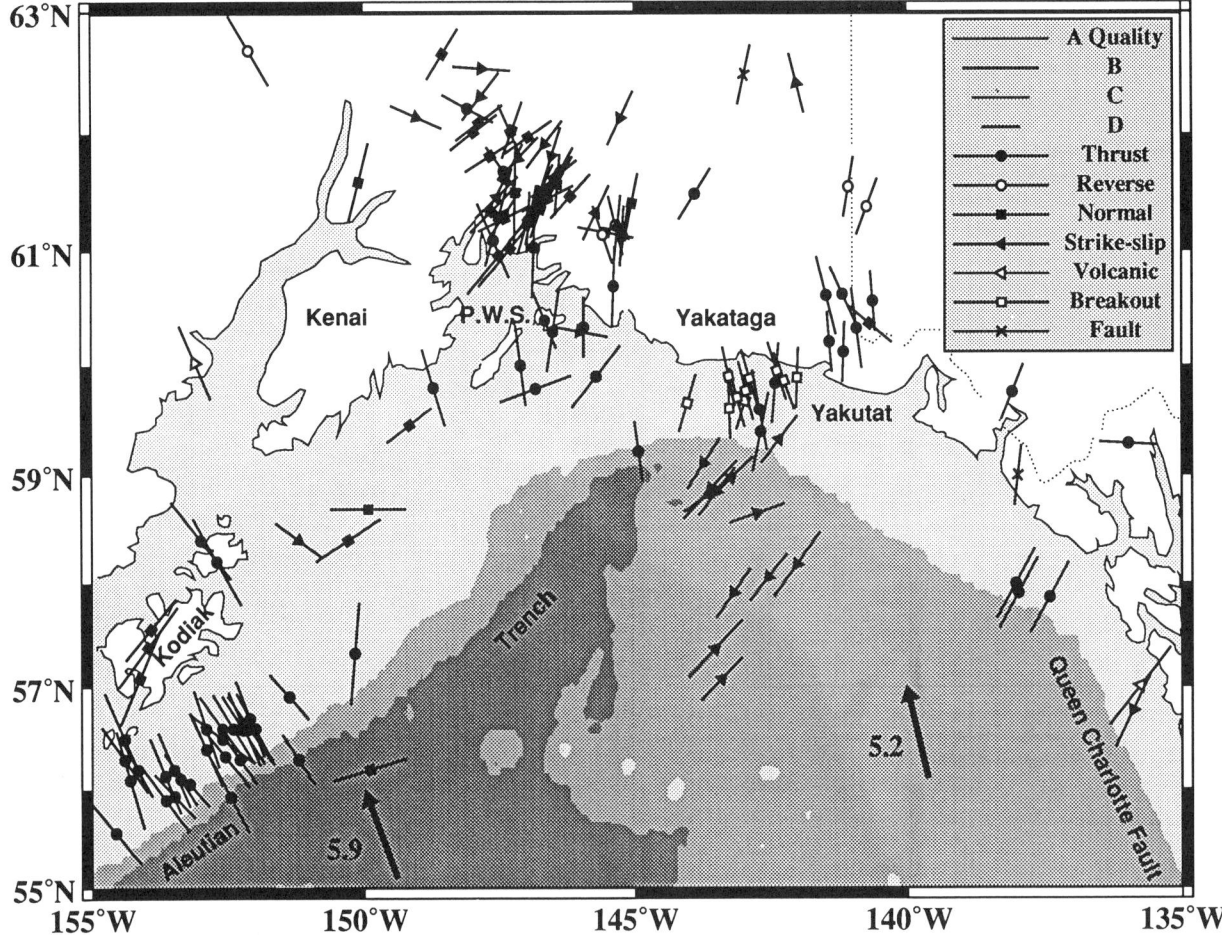

Figure 2. Map of MHC indicators for the Gulf of Alaska region. Bathymetry (from DBDB5 data base) shows shading changes at 2,000-m bathymetric contour intervals. Shown are quality A, B, and C. See Figure 1 for description of plate-motion vectors.

valid (this size event is about the smallest for which CMT inversions are performed, Dziewonski and others, 1981), then there may have been a temporal change in the direction of the regional stress field prior to the large Gulf of Alaska earthquakes.

In the Yakutat-Yakataga strike-slip-to-underthrusting transition zone (Fig. 2, 59 to 61°N, 140 to 146°W), some scatter occurs with generally north-trending MHC (parallel to plate vector) from mostly thrust earthquake indicators onshore and borehole breakouts along the Pamplona zone (the proposed continental shelf extension of the Aleutian Trench, e.g., Bruns, 1983). Within the Pacific Plate, the indicators—mostly from strike-slip solutions—trend northeast. This change in direction of MHC in less than 100 km may indicate a decoupling of the stresses between the Pacific Plate and the overlying Yakutat block (Pérez and Jacob, 1980), or that the strike-slip and breakout indicators are recording different things (e.g., stresses at different depth levels in the two separate plate or tectonic units, or distortion of the strain field by preexisting faults). The MHC indicators in the eastern Gulf of Alaska are not parallel to the Pacific/North America Plate convergence direction (southeast-northwest; DeMets and others, 1990). These MHC directions are in contrast to MHC from normal faulting solutions beneath and north of Prince William Sound (north of 61°N, 145 to 150°W). The latter may signal complex bending stresses in the descending or even overriding plate as subduction occurs beneath the concave north-dipping plate boundary (Suárez, unpublished data; Page and others, 1989). These normal faulting earthquake indicators are, however, probably in the lower plate, which in general, is not the subject of this chapter (the depths for these indicators are between 25 and 45 km). Note that the indicators north of Prince William Sound parallel the strike-slip indicators in the Gulf of Alaska.

West of 150°W (Fig. 3), the PAC/NAM plate boundary is strongly dominated by MHC directions parallel to the southeast-northwest–directed relative plate-motion vector. These data are mostly interplate thrust earthquakes just north of the Aleutian Trench, while north of the underthrusting region, a few less consistent volcanic stress indicators imply that similar directions of

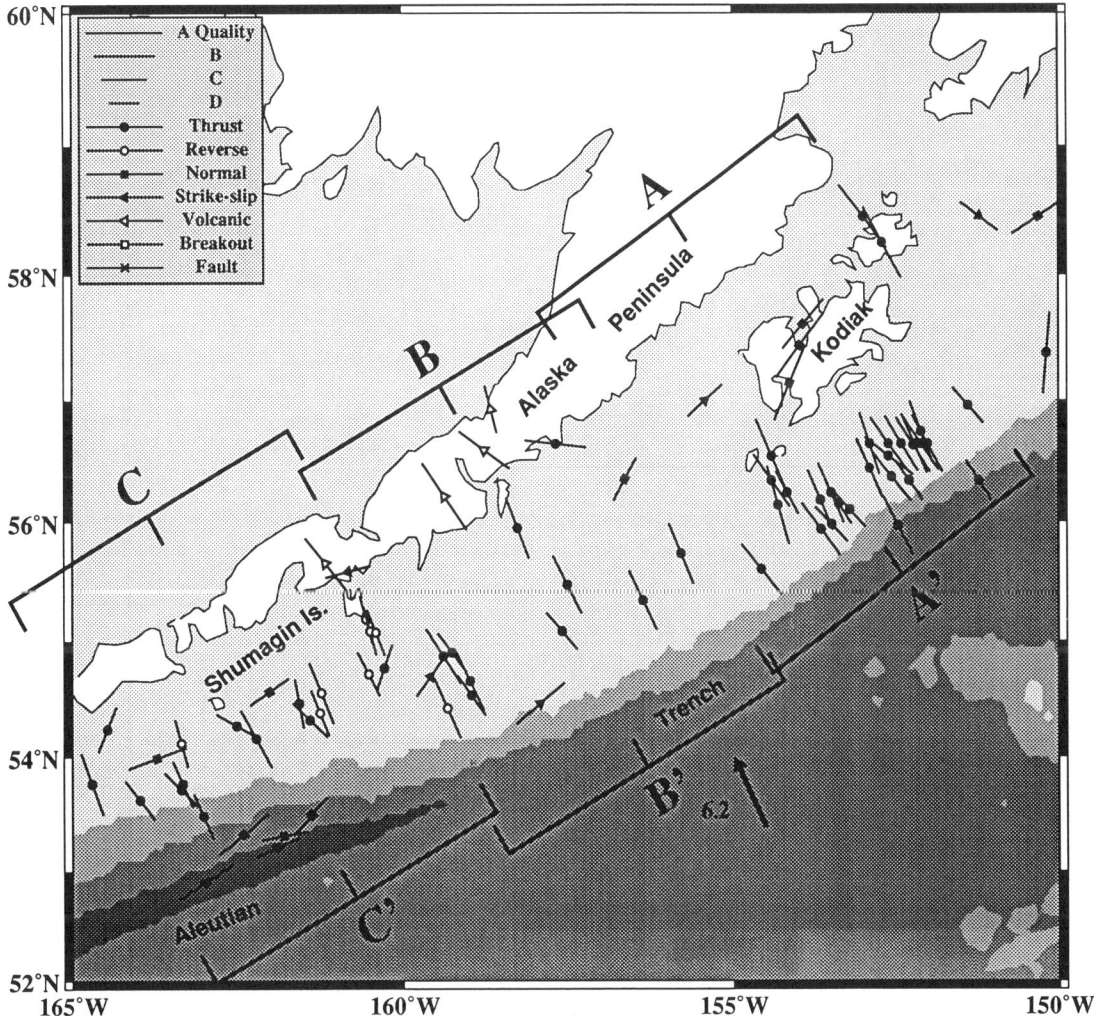

Figure 3. Map of MHC indicators for the eastern Aleutians and location of cross sections shown in Figure 4. Shown are quality A, B, and C. See Figure 1 for description of plate-motion vectors.

compressive stress possibly extend as far north as the chain of Aleutian volcanos (Nakamura and others, 1980). A few exceptional observations (near Kodiak Island) signify local tension possibly in the lower plate (Jacob, 1986; see discussion of vertical cross sections below). However, nonvolcanic data do not exist that corroborate the arc-normal compressive stress in the volcanic arc.

Near the trench, several consistently trench-parallel MHC indicators originate from normal faulting apparently due to bending in the Pacific Plate as the plate approaches the trench (e.g., Chapple and Forsyth, 1979; Forsyth, 1982; Frohlich and others, 1982; Ward, 1983). Thrust indicators are roughly parallel to the plate convergence direction, while the trench-parallel normal faulting indicators do not reflect the convergence direction but are due purely to the static geometry of subduction. A unique situation occurs in the western Aleutians where the normal indicators due to bending seaward of the trench become nearly parallel to the thrusting indicators along the plate interface due to the oblique convergence of the Pacific Plate (Cormier, 1975; Spence, 1977).

The variation in depth of stress indicators due to the subduction process is shown in cross sections across the eastern Aleutian arc (Figs. 3 and 4). P-axes are plotted for these events in contrast to MHC for the maps. These three sections all show seaward-plunging P-axes for the underthrusting events, and no indicators in the upper plate. The vertical extent of the underthrusting indicators is probably greater than in reality due to poor depth control. In the Kodiak and Shumagin regions (sections A and C of Fig. 4), the locus of underthrusting (shallow seaward plunging P-axes) starts at about 70 km north of the trench, while in the 1938 rupture zone (section B; 155 to 158°W) the thrusting events are slightly farther from the trench. These distances are similar to those determined from local network seismicity by Kienle and others (1983) for the Kodiak region and Hudnut and Taber (1987) and Taber and others (this volume) for the Shumagins. The width of the locus of indicators measured in the downdip direction varies from 120 km near Kodiak through 90 km in the Shumagins to about 50 km in the 1938 zone; however, the narrow width in the 1938 zone may be an artifact of the small number of events. The deepest extent of these indicators may define the downdip extent of the great underthrusting events that occur in these areas (e.g., McCann and others, 1979). In the Shumagin region (section C), reverse fault indicators occur north of and at greater depth than the thrust indicators; this may mean that the dip of the main thrust zone is not constant (section A) but increases to the north, as seen in local network seismicity (Hudnut and Taber, 1987; Taber and others, this volume). Normal faulting at depths greater than 30 km, seen as the high-angle indicators in section A, occurs beneath and to the southwest of Kodiak Island and is perhaps related to tensional stresses in the upper part of the downgoing slab due to an increase in slab dip toward the northwest (Jacob, 1986). No bending events occur near the trench in the Kodiak region (section A) nor in the 1938 zone (section B). In the case of the 1938 zone, this may be a

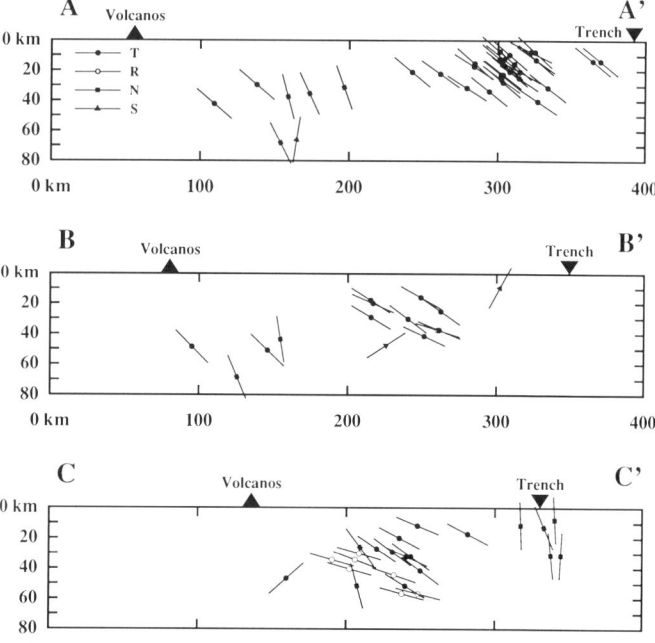

Figure 4. (A) Cross section across Kodiak Island showing P-axes of focal mechanisms. See Figure 3 for location of sections. Shown are quality A, B, and C (see explanation in Fig. 2). Volcanic, borehole breakout, and geologic fault indicators are not shown. (B) Cross section across the rupture zone of the 1938 earthquake. (C) Cross section across the Shumagin Islands region.

temporal effect since normal faulting seaward of the trench often occurs soon after a major underthrusting earthquake (e.g., Christensen and Ruff, 1988; Lay and others, 1989) and it has been over 50 years since the earthquake. In section C, bending events occur in two groups: one near the trench, and a second 100 to 130 km north of the trench. These may represent two separate loci for downward flexure of the Pacific Plate.

The consistency in direction of the MHC indicators in the great earthquake rupture zones is not uniform. In the 1964 rupture zone (145 to 155°W; Fig. 2), excluding the Kodiak Island and Prince William Sound regions, only one thrusting earthquake has occurred since 1964, although some normal faults have occurred south of the Kenai Peninsula (59°N, 150°W). This is in contrast to the 1938 rupture zone (155 to 158°W; Fig. 3) where there are numerous thrusts. The rupture zone of the great 1957 earthquake (166 to 179°W; Fig. 5) has a very large number of underthrusting indicators. The number of thrusting indicators dramatically increases when the overriding plate changes from continental North America to the oceanic Bering Sea (about 168°W): west of this point there is a greater abundance of magnitude 5 and greater seismicity (Taber and others, this volume). The 1964 earthquake is thought to have involved two large asperities (Christensen and Beck, 1989), while the 1957 event may have been a complex rupture (Hartzell and Heaton, 1985). Thus this may be related to differences in strength distribution and down-

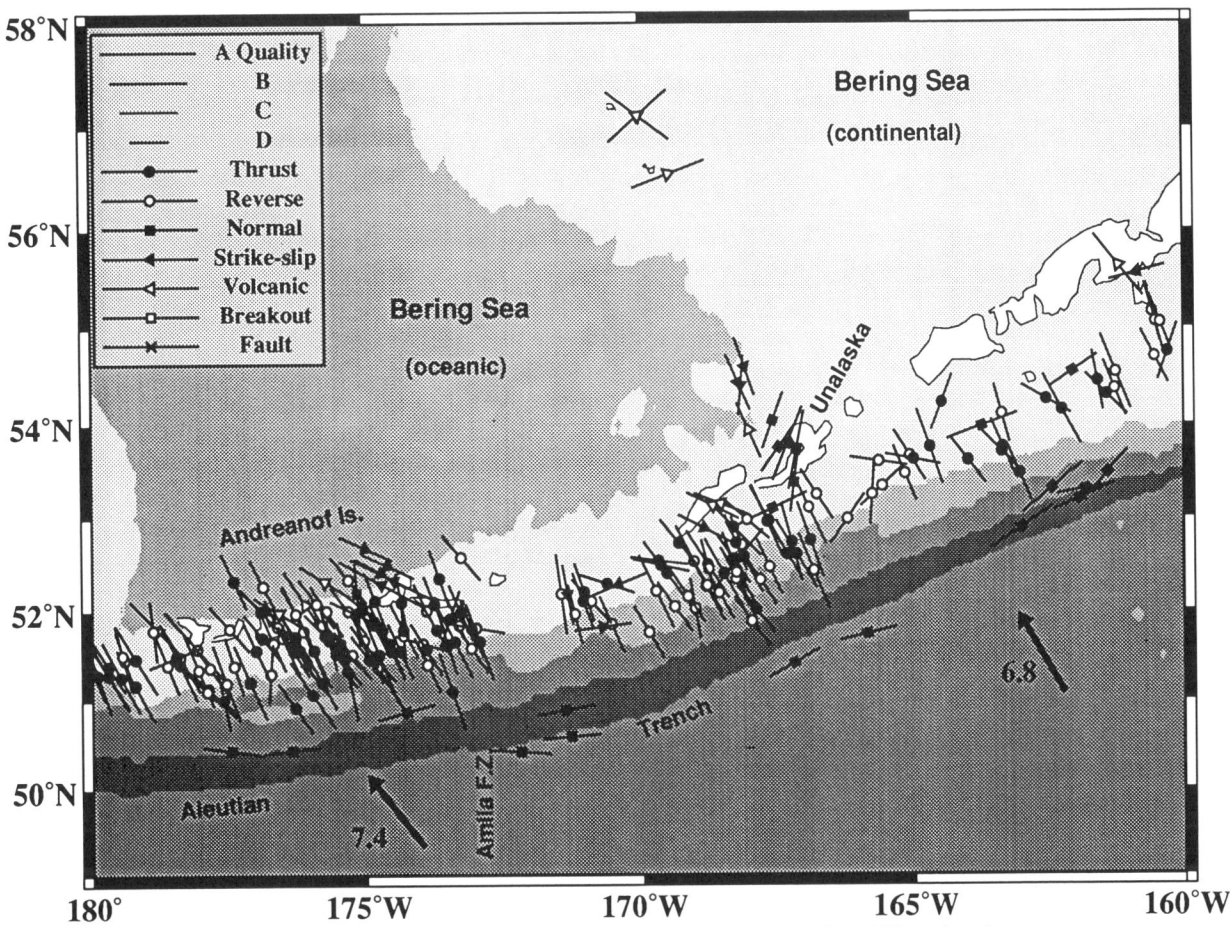

Figure 5. Map of MHC indicators for the central Aleutians, principally the 1957 earthquake rupture zone. Shown are quality A, B, and C. See Figure 1 for description of plate-motion vectors.

dip width of the locked portion of the plate boundary. The 1965 rupture zone (179°W to 170°E; Fig. 6) also contains a large number of underthrusting indicators, which also may relate to the complexity of the rupture (Ruff and Kanamori, 1983). Variations along arc in the amount of sediment filling the trench (Scholl and Marlow, 1974; Jacob and others, 1977) may also be a factor that influences the roughness and asperity distribution on the plate interface (Ruff, 1989).

The bending-related, normal-faulting MHC indicators in the Pacific Plate of the central and eastern Aleutians are located near the trench and trend strictly trench-parallel. They are more numerous in the central and western Aleutians (Figs. 5 and 6) than in the eastern Aleutians–Alaska Peninsula region (Figs. 2 and 3). This pattern correlates with increasing slab-dip angle (see Taber and others, this volume, for review) and decreasing volcanic arctrench distance (Jacob and others, 1977) toward the western part of the arc, implying that bending-related earthquakes occur near the trench when bending stresses due to the geometry of subduction are above a minimum value (Chapple and Forsyth, 1979; Forsyth, 1982; Ward, 1983). The trench-parallel bending stress indicators occur in clusters that seem to increase in number and trench normal extent in those areas that appear deficient in interplate activity. It may reflect either spatial-temporal variations of bending stresses related to the earthquake cycle of great earthquakes at each plate-boundary segment, or it may be a stationary feature related to permanent differences along the arc in the degree of coupling at the plate interface (Spence, 1977; Dmowska and others, 1988; Christensen and Ruff, 1988; Lay and others, 1989).

At 173°W (Fig. 5) there is a paucity of thrust indicators (and no PDE [Preliminary Determination of Epicenters] seismicity), although there are two bending indicators. The great 1957 earthquake ruptured across this gap, but it corresponds to the eastern edge of the 1986 Andreanof Islands earthquake (Engdahl and others, 1989; Boyd and Nábělek, 1988). Note that there is a ~20° rotation of thrust indicators from one side of the gap to the other. House and Jacob (1983) showed that this gap, coinciding with the location of the Amlia fracture zone, corresponds to a bend or tear in the downgoing Pacific Plate with the eastern slab segment descending more steeply than the western segment. The clockwise rotation of MHC just east of the gap is consistent with an edge effect along the bend or tear. The large number of reverse fault plate boundary indicators between 167 and 170°W (mostly CMT solutions) supports the idea that the PAC/NAM plate interface dips more steeply than along adjacent sections of the arc.

There is an unusual group of back-arc events north of Unalaska Island (near 168°W; Fig. 5), consisting mostly of strike-slip events with north-trending MHC (implying northeast- or northwest-striking planes) roughly parallel to the continental margin of the Bering Sea shelf that separates the oceanic from the continental portions of the Bering Sea and normal faulting events closer to Unalaska. These events indicate that northerly compression persists north of the island arc, while east-west to northwest-southeast extension (north- to northeast-trending MHC indicators) are found in the arc itself. Note that about 400 km north of the arc, MHC rotates to nearly east-west.

In the western Aleutians, west of 180° (Fig. 6), and up to the intersection with Kamchatka and the Kuril Trench, the MHC is generated by increasingly arc-parallel thrusting with northwest-directed slip in the fore-arc gap between trench and islands (53°N, 171°E). Here, MHC rotates ~20° clockwise from the nearly arc-parallel PAC/NAM vector, which may be caused by partial decoupling of the thrust motion (which usually trends arc-normal) from arc-parallel right-lateral strike-slip motion (165 to 170°E) within the back-arc region and in the volcanic arc itself (Cormier, 1975; Newberry, 1983; Ekström and Engdahl, 1989). To explain the systematic misfit between the observed and predicted slip vectors in the presence of oblique convergence, Ekström and Engdahl (1989) propose a model in which arc-parallel relative motion is accommodated along a weak strike-slip fault zone near the volcanic arc. This strike-slip motion produces a distinct set of MHC indicators pointing in a north to north-northwest direction (near 54°N, 171°E), rotated clockwise from the thrust-related MHC by at least 30°. This is probably an effect of preexisting zones of weakness determining the preferred fault plane rather than a change in the stress field. Slip vectors for the thrust and strike-slip solutions are nearly identical (Cormier, 1975). The occurrence of strike-slip faulting within and at the northern edge of the arc has been linked to the movement of tectonic blocks independent of the North American Plate (Spence, 1977; Geist and others, 1988; Ekström and Engdahl, 1989). Engdahl and others (1977) showed that the western Aleutian slip vectors may be biased due to the presence of unmod-

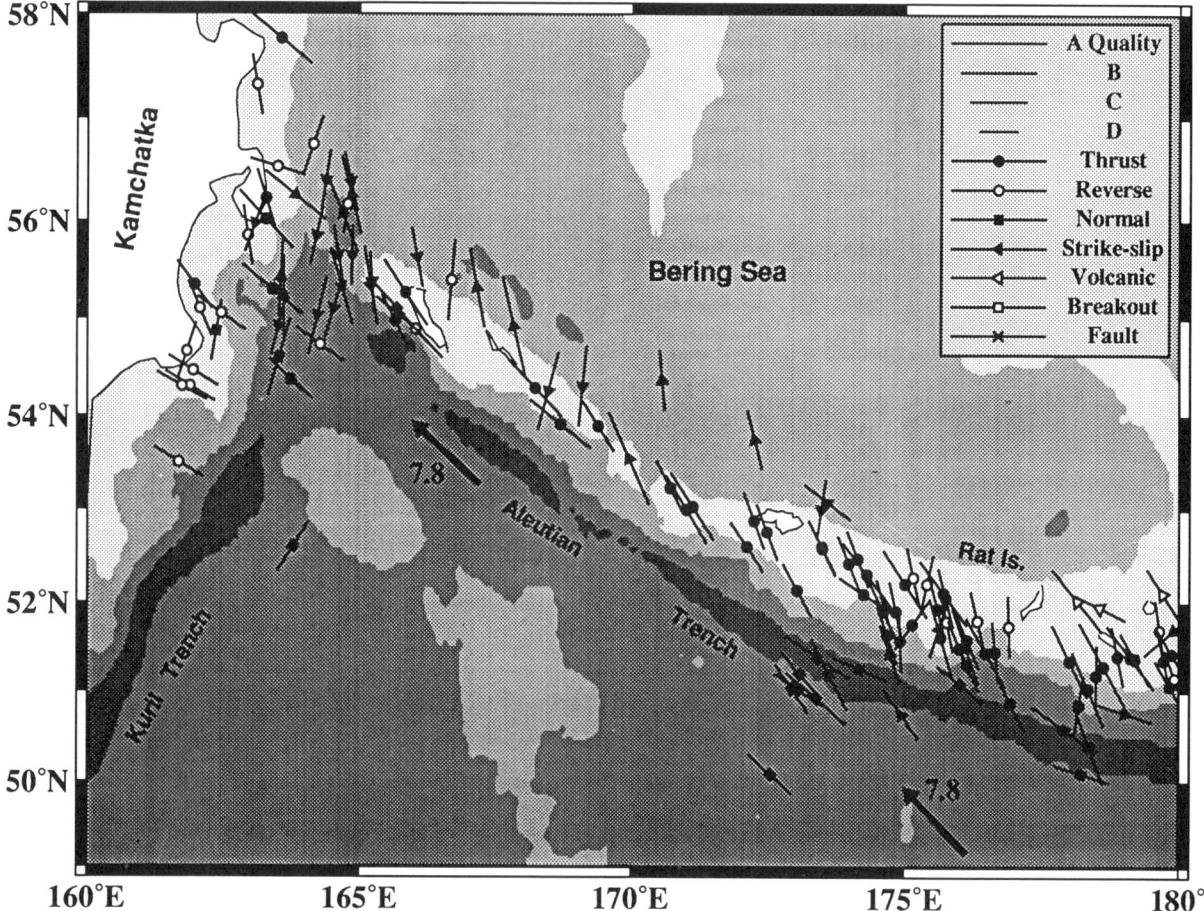

Figure 6. Map of MHC indicators for the western Aleutians and Kamchatka, including the site of the 1965 Rat Islands earthquake. Shown are quality A, B, and C. See Figure 1 for description of plate-motion vectors.

eled lateral heterogeneities (i.e., the subducted slab), which would affect ray paths to European seismic stations. DeMets and others (1990) suggest that the misfit may be due to a product of the two effects.

As one moves from east to west across the cusp of the junction of the western Aleutian and northern Kuril Trenches near Kamchatka, there is a counterclockwise rotation of interplate MHC thrust indicators. Note, however, that there is a large number of north- to north-northeast–trending strike-slip MHC indicators that occur right at the cusp. The indicators from these events are parallel to the strike of the Kamchatka Trench and the B-axes of several arc-parallel normal faults. The sense of motion on the strike-slip solutions implies right-lateral motion on a northwest-striking plane (Aleutian Trench; Cormier, 1975; Newberry, 1983). A northeast-striking plane would imply left-lateral shearing of the Aleutian arc as it collides with the Kamchatka Trench (Newberry, 1983). Obviously, more work needs to be done on this complex corner.

Plate-interior stress directions

The observations of MHC away from the PAC/NAM plate boundary, in the continental interior of Alaska and adjacent portions of Canada, are much sparser and thus less stable because seismicity and magnitude levels are lower (Figs. 1 and 7; Biswas and others, 1986; Estabrook and others, 1988). The general observation is that there is north-south compression through the continent near 145°W, just west of the Alaska-Canada boundary, from the Gulf of Alaska coast into central Alaska. It is not clear whether it extends northward into the Brooks Range or the Beaufort Sea (Fig. 7); although along the Canadian margin, MHC has a strong northeast orientation (Adams and Bell, this volume). The

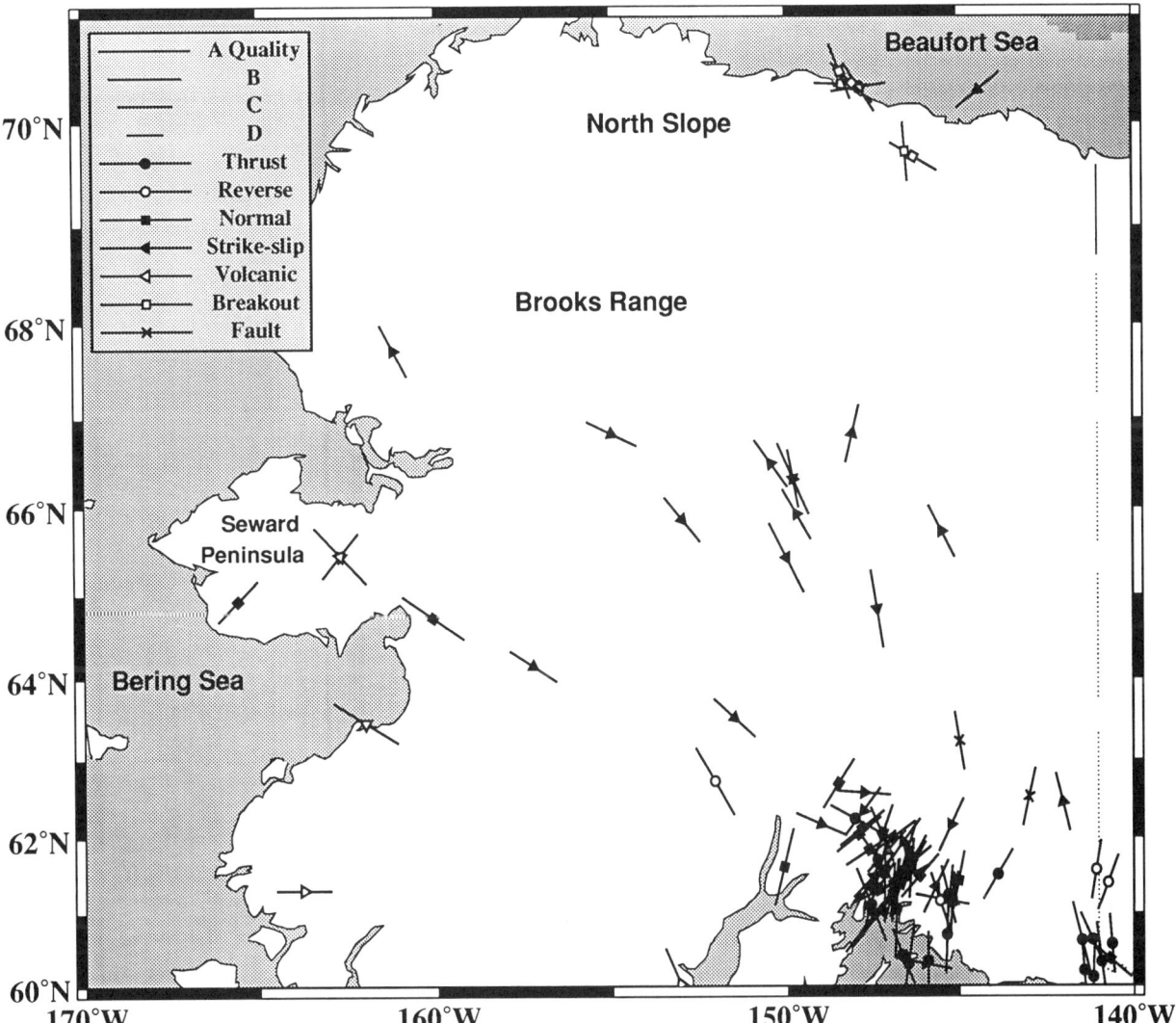

Figure 7. Map of MHC indicators for continental Alaska and the Bering shelf. Shown are quality A, B, and C. See Figure 1 for description of plate-motion vectors.

MHC orientation gradually turns from north in central Alaska to west in western Alaska where it may be associated with tensional rift volcanism on the Seward Peninsula (65°N, 165°W; Turner and others, 1981; Biswas and others, 1983). Unfortunately, no stress indicators are available in the oceanic, abyssal portions of the Bering Sea (except the few events just north of Unalaska discussed earlier and along the Siberian coast). Therefore, we do not know whether this north-south extension actually continues into the northern and central regions of the Bering Sea portion of the North American Plate, or how far the north-south compression extends north from the Aleutian arc.

The general pattern of MHC directions is fan shaped, emerging from the northernmost apex of the Gulf of Alaska (Fig. 8). This pattern is consistent with a pattern of slip lines one would expect if the Pacific Plate were the rigid indentor into a plastically deforming North American continent (Nakamura and others, 1980; Davies, 1983; Estabrook and others, 1988). When the lowest-quality indicators are included (Figs. 1 and 8), the MHC direction gradually rotates counterclockwise, from north-south in the Yakataga area to northwest trending in central Alaska. This rotation of MHC directions (although obscured by the near absence of indicators in southwest Alaska and on the Bering Sea shelf) supports the indentor hypothesis. Deformation in central Alaska occurs predominantly along preexisting zones of weakness as strike-slip faults (Estabrook and others, 1988) and along smaller, more northerly striking faults (Huang and Biswas, 1983; Estabrook and others, 1988). The large scatter of the MHC directions on the Alaskan Beaufort Sea coast may indicate that the two highest principal stresses are quite close in magnitude, and hence, small fluctuations in their respective magnitudes may "flip" the MHC direction by almost 90°. The stress trajectory across the Canadian Beaufort Sea margin is based on breakout directions from Adams and Bell (this volume) that are rather uniform in direction.

It is unclear whether the stress patterns in the interior of Alaska can be attributed to widespread active mantle upwelling in the fashion of general back-arc extension (Nakamura and Uyeda, 1980; Biswas and others, 1986), or whether they should be seen as a by-product of the collision of the PAC/NAM Plates in the apex of the Gulf of Alaska, gradually giving way to a

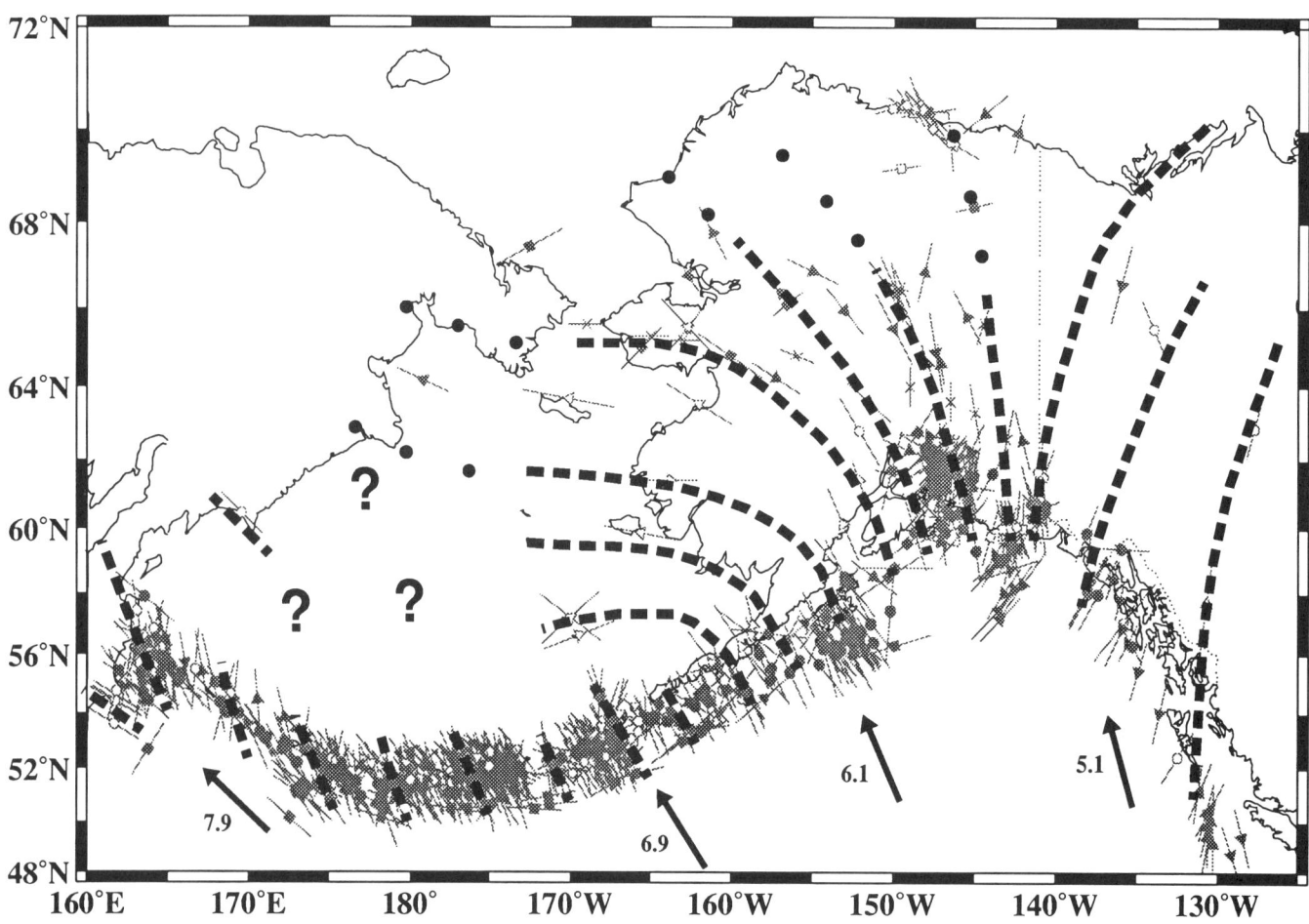

Figure 8. Map of MHC-determined stress trajectories for Alaska, the Aleutians, and the Bering Sea. Trajectories are dotted where they are the least well known. This map is based on the maps of Nakamura and others (1980) and Estabrook and others (1988). Shown in gray are the MHC indicators from Figure 1. See text for discussion.

strike-slip–dominated regime in central Alaska, and finally grading into a tensional regime toward the Bering Sea margin (Davies, 1983; Estabrook and others, 1988). Alternative explanations exist. One invokes the existence of a Bering Sea Plate separate from the North American Plate (Stone, 1983), although Cook and others (1986) argue that the North American/Eurasia Plate boundary is in eastern Siberia.

Another hypothesis presented by Jacob (1984) is that a mantle source may explain the simultaneous occurrence of compression in southern, eastern, and northeastern Alaska, the extension on the Seward Peninsula, and a diffuse zone of alkali volcanism in western and southwest Alaska (Beikman, 1980; Kienle and Swanson, 1983). Mantle excess pressure may build up beneath the Bering Sea margin and western central Alaska as the North American Plate moves at 1 cm/yr in a southwest direction relative to a fixed hot-spot–anchored reference system (Minster and Jordan, 1978; Jacob, 1987). In doing so, the mantle material may get confined and squeezed between the northward-dipping Pacific Plate in the south and the root system of the Brooks Range in the north. A perhaps oversimplified but more descriptive analogy might be a fish swimming with an open mouth, the jaws being pushed apart. The apex of the jaws, or their joint, would be in eastern central Alaska, the opening pointing to the Bering Sea, resulting in crustal extension in central Alaska, while retaining compression all around: in the south at the convergent PAC/NAM plate margin, north of and in the Brooks Range, and in the eastern central Alaskan highlands. The latter, however, may be more dominated by the PAC/NAM indentation than from the fluid-dynamic mantle confinement as described. To unravel these hypotheses, one would require not only the pattern of MHC directions, but also accurate knowledge of the magnitudes of all three principal stresses as a function of depth, and of the recent strain history. This information on the spatial variations of the components of the stress and strain tensors is not available at present. Future satellite-supported geodetic measurements (GPS) may help to unravel the kinematic deformation patterns, though not the underlying stress system itself.

CONCLUSIONS

The large-scale MHC pattern that emerges is generally consistent with a compressively coupled PAC/NAM plate boundary with right-lateral slip dominating in both southeast Alaska and the western Aleutians, and arc-normal compression from the Gulf of Alaska towards the central Aleutians. Bending stresses in the Pacific Plate resulting in plate-boundary-parallel MHC directions are very consistent. Some along arc variation is seen in the numbers of arc normal indicators, perhaps temporally related to the occurrence of great earthquakes along the Alaska-Aleutian arc. Away from the convergent margin, MHC indicators, defined mostly by strike-slip and normal faults in central and western Alaska, change direction from north-northwest trending in the east to west-trending in the tensional environment of the Seward Peninsula. Stress indicators are not consistent in direction across the Brooks Range to the Beaufort Sea coast, implying that north-trending compression due to plate convergence in southern Alaska may have diminished to the point of not being the dominant MHC indicator.

REFERENCES CITED

References marked with an asterisk are sources for data in figures and are not otherwise cited in the text.

Angelier, J., 1979, Determination of the mean principal directions of stresses for a given fault: Tectonophysics, v. 56, p. T17–T26.
Beikman, H. M., 1980, Geologic map of Alaska: U.S. Geological Survey State Geologic Map, scale 1:2,500,000.
Bell, J. S., and Gough, D. I., 1979, Northeast-southwest compressive stress in Alberta; Evidence from oil wells: Earth and Planetary Science Letters, v. 45, p. 475–482.
Ben-Menahem, A., and Singh, S. J., 1981, Seismic waves and sources: New York, Springer-Verlag, 1108 p.
Bhattacharya, B., and Biswas, N. N., 1979, Implications of North Pacific plate tectonics in central Alaska; Focal mechanisms of earthquakes: Tectonophysics, v. 53, p. 99–130.
Biswas, N. N., Pujol, J., Tytgat, G., and Dean, K., 1983, Synthesis of seismicity studies for western Alaska: Anchorage, Alaska, National Oceanic and Atmospheric Administration Report NA81-RAC00112, 69 p.
Biswas, N. N., Aki, K., Pulpan, H., and Tytgat, G., 1986, Characteristics of regional stresses in Alaska and neighboring areas: Geophysical Research Letters, v. 13, p. 177–180.
Boyd, T. M., and Nábělek, J. L., 1988, Rupture process of the Andreanof Islands earthquake of May 7, 1986: Bulletin of the Seismological Society of America, v. 78, p. 1653–1673.
Bruns, T. R., 1983, Model for the origin of the Yakutat block; An accreting terrane in the northern Gulf of Alaska: Geology, v. 11, p. 718–721.
Chandra, U., 1974, Seismicity, earthquake mechanisms, and tectonics along the western coast of North America, from 42°N to 61°N: Bulletin of the Seismological Society of America, v. 64, p. 1529–1549.
Chapple, W. M., and Forsyth, D. W., 1979, Earthquakes and bending of plates at trenches: Journal of Geophysical Research, v. 84, p. 6729–6749.
Christensen, D., and Beck, S., 1989, Rupture process of the great 1964 Alaska earthquake [abs.]: EOS Transactions of the American Geophysical Union, v. 70, p. 1225.
Christensen, D. H., and Ruff, L. J., 1988, Seismic coupling and outer rise earthquakes: Journal of Geophysical Research, v. 93, p. 13421–13444.
Cook, D. B., Fujita, K., and McMullen, C. A., 1986, Present-day plate interactions in northeast Asia; North American, Eurasian, and Okhotsk Plates: Journal of Geodynamics, v. 6, p. 33–51.
Cormier, V. F., 1975, Tectonics near the junction of the Aleutian and Kuril-Kamchatka Arcs and a mechanism for middle Tertiary magmatism in the Kamchatka Basin: Geological Society of America Bulletin, v. 86, p. 443–453.
Davies, J. N., 1983, Seismicity of the interior of Alaska; A direct result of Pacific–North American Plate convergence? [abs.]: EOS Transactions of the American Geophysical Union, v. 64, p. 90.
DeMets, C., Gordon, R. G., Argus, D. F., and Stein, S., 1990, Current plate motions: Geophysics Journal International, v. 101, p. 425–478.

Dmowska, R., Rice, J. R., Lovison, L. C., and Josell, D., 1988, Stress transfer and seismic phenomena in coupled subduction zones during the earthquake cycle: Journal of Geophysical Research, v. 93, p. 7869–7884.

Dziewonski, A. M., Chou, T.-A., and Woodhouse, J. H., 1981, Determination of earthquake source parameters from waveform data for studies of global and regional seismicity: Journal of Geophysical Research, v. 86, p. 2825–2852.

Dziewonski, A. M., Ekstrom, G., Woodhouse, J. H., and Zwart, G., 1989, Centoid-moment tensor solutions October-December 1988: Physics of the Earth and Planetary Interior, v. 57, p. 179–191.

Ekström, G., and Engdahl, E. R., 1989, Earthquake source parameters and stress distribution in the Adak Island region of the central Aleutian Islands, Alaska: Journal of Geophysical Research, v. 94, p. 15499–15519.

Engdahl, E. R., Sleep, N. H., and Lin, M-T, 1977, Plate effects in north Pacific subduction zones: Tectonophysics, v. 37, p. 95–116.

Engdahl, E. R., Billington, S., and Kisslinger, C., 1989, Teleseismically recorded seismicity before and after the May 7, 1986, Andreanof Islands, Alaska, earthquake: Journal of Geophysical Research, v. 94, p. 15481–15498.

Estabrook, C. H., Davies, J. N., and Stone, D. B., 1988, Seismotectonics of northern Alaska: Journal of Geophysical Research, v. 93, p. 12026–12040.

Estabrook, C. H., and Jacob, K. H., 1991, Table of stress indicators in Alaska: Palisades, N.Y., Lamont-Doherty Geological Observatory of Columbia University, Publications Office, 12 p.

Forsyth, D. W., 1982, Determination of focal depths of earthquakes associated with the bending of oceanic plates at trenches: Physics of Earth and Planetary Interiors, v. 28, p. 141–160.

Frohlich, C., Billington, S., Engdahl, E. R., and Malahoff, A., 1982, Detection and location of earthquakes in the central Aleutian subduction zone using land and ocean bottom seismograph stations: Journal of Geophysical Research, v. 87, p. 6853–6864.

Gedney, L., 1970, Tectonic stresses in southern Alaska in relationship to regional seismicity and the new global tectonics: Bulletin of the Seismological Society of America, v. 60, p. 1789–1802.

*—— , 1985, Stress trajectories across the northeast Alaska Range: Bulletin of the Seismological Society of America, v. 75, p. 1125–1134.

Geist, E. L., Childs, J. R., and Scholl, D. W., 1988, The origin of summit basins of the Aleutian ridge; Implications for block rotation of an arc massif: Tectonics, v. 7, p. 327–341.

Gephart, J. W., and Forsyth, D. W., 1984, An improved method for determining the regional stress tensor using earthquake focal mechanism data; Application to the San Fernando earthquake sequence: Journal of Geophysical Research, v. 89, p. 9305–9320.

Hartzell, S. H., and Heaton, T. H., 1985, Teleseismic time functions for large, shallow subduction zone earthquakes: Bulletin of the Seismological Society of America, v. 75, p. 965–1004.

Hong, T., and Fujita, K., 1981, Modelling of depth phases and source processes of some central Aleutian earthquakes: Earth and Planetary Science Letters, v. 53, p. 333–342.

House, L. S., and Jacob, K. H., 1983, Earthquakes, plate subduction, and stress reversals in the eastern Aleutian Arc: Journal of Geophysical Research, v. 88, p. 9347–9373.

Huang, P. Y., and Biswas, N. N., 1983, Rampart seismic zone of central Alaska: Bulletin of the Seismological Society of America, v. 73, p. 813–829.

Hudnut, K. W., and Taber, J. J., 1987, Transition from double to single Wadati-Benioff seismic zone in the Shumagin Islands: Geophysical Research Letters, v. 14, p. 143–146.

Jacob, K. H., 1984, Lithospheric stress indicators for Alaska [abs.]: EOS Transactions of the American Geophysical Union, v. 65, p. 1119.

—— , 1986, A pilot study of borehole breakouts as tectonic stress indicators in Alaska; Final report: National Science Foundation, 32 p.

—— , 1987, Seismicity, tectonics, and geohazards of the Gulf of Alaska regions, in Hood, D. W., and Zimmerman, S. T., eds., The Gulf of Alaska; Physical environment and biological resources: U.S. Department of Commerce, p. 145–184.

Jacob, K. H., Nakamura, K., and Davies, J. H., 1977, Trench-volcano gap along the Alaska-Aleutian arc; Facts and speculations on the role of terrigeneous sediments for subduction, in Talwani, M., and Pitman, W., eds., Island arcs, deep sea trenches, and back arc basins: American Geophysical Union Ewing v. 1, p. 243–258.

*Jordan, J., Dunphy, G., and Harding, S., 1968, The Fairbanks, Alaska, earthquakes of June 21, 1967; Preliminary seismological report: U.S. Coast and Geodetic Survey, 60 p.

Kienle, J., and Swanson, S. E., 1983, Volcanism in the eastern Aleutian Arc; Late Quaternary and Holocene centers, tectonic setting, and petrology: Journal of Volcanism and Geothermal Research, v. 17, p. 393–432.

Kienle, J., Swanson, S. E., and Pulpan, H., 1983, Magmatism and subduction in the eastern Aleutian Arc, in Shimozuru, D., and Yokoyama, I., eds., Physics and Tectonics: Tokyo, Terra Scientific Publishing Company, p. 191–224.

King, P. B., 1969, Tectonic map of North America: U.S. Geological Survey Geologic Maps of North America, scale 1:5,000,000.

LaForge, R., and Engdahl, E. R., 1979, Tectonic implications of seismicity in the Adak Canyon region, central Aleutians: Bulletin of the Seismological Society of America, v. 69, p. 1515–1532.

Lahr, J. C., Page, R. A., Stephens, C. D., and Christensen, D. H., 1988, Unusual earthquakes in the Gulf of Alaska and fragmentation of the Pacific Plate: Geophysical Research Letters, v. 15, p. 1483–1486.

Lay, T., Astiz, L., Kanamori, H., and Christensen, D. H., 1989, Temporal variation of large intraplate earthquakes in coupled subduction zones: Physics of the Earth and Planetary Interiors, v. 54, p. 258–312.

*Liu, H., and Kanamori, H., 1980, Determination of source parameters of midplate earthquakes from the waveforms of body waves: Bulletin of the Seismological Society of America, v. 70, p. 1989–2004.

*LeBlanc, G., and Wetmiller, R. J., 1974, An evaluation of seismological data available for the Yukon Territory and the Mackenzie Valley: Canadian Journal of Earth Sciences, v. 11, p. 1435–1454.

McCann, W. R., Nishenko, S. P., Sykes, L. R., and Krause, J., 1979, Seismic gaps and plate tectonics: Seismic potential for major boundaries: Pure and Applied Geophysics, v. 117, p. 1082–1147.

McKenzie, D. P., 1969, The relationship between fault plane solutions for earthquakes and the directions of the principal stresses: Bulletin of the Seismological Societ of America, v. 59, p. 591–601.

Minster, J. B., and Jordan, T. H., 1978, Present-day plate motions: Journal of Geophysical Research, v. 83, p. 5331–5354.

Mount, V. S., and Suppe, J., 1987, State of stress near the San Andreas fault; Implications for wrench tectonics: Geology, v. 15, p. 1143–1146.

Nakamura, K., and Uyeda, S., 1980, Stress gradient in arc–back arc regions and plate subduction: Journal of Geophysical Research, v. 85, p. 6419–6428.

Nakamura, K., Jacob, K. H., and Davies, J. N., 1977, Volcanoes as possible indicators of tectonic stress orientations; Aleutians and Alaska: Pageoph, v. 115, p. 87–112.

Nakamura, K., Plafker, G., Jacob, K. H., and Davies, J. N., 1980, A tectonic stress trajectory map of Alaska using information from volcanoes and faults: Bulletin of the Earth Research Institute, v. 55, p. 89–100.

Newberry, J. T., 1983, Seismicity and tectonics of the far western Aleutian Islands [M.S. thesis]: East Lansing, Michigan State University, 93 p.

Page, R. A., Stephens, C. D., and Lahr, J. C., 1989, Seismicity of the Wrangell and Aleutian Wadati-Benioff zones and the North American Plate along the Trans-Alaska Crustal Transect, Chugach Mountains, and Copper River basin, southern Alaska: Journal of Geophysical Research, v. 94, p. 16059–16082.

Pérez, O. J., and Jacob, K. H., 1980, Tectonic model and seismic potential of the eastern Gulf of Alaska and Yakataga seismic gap: Journal of Geophysical Research, v. 85, p. 7132–7150.

Plafker, G., Hudson, T., Bruns, T., and Rubin, M., 1978, Late Quaternary offsets along the Fairweather fault and crustal plate interactions in southern Alaska: Canadian Journal of Earth Sciences, v. 15, p. 805–816.

Ruff, L. J., 1989, Do trench sediments affect great earthquake occurrence in subduction zones?: Pageoph, v. 129, p. 263–282.

Ruff, L., and Kanamori, H., 1983, The rupture process and asperity distribution of three great earthquakes from long-period diffracted P-waves: Physics of Earth and Planetary Interiors, v. 31, p. 202–230.

Schell, M. M., and Ruff, L. J., 1989, Rupture of a seismic gap in southeastern Alaska; The 1982 Sitka earthquake (M_s 7.6): Physics of Earth and Planetary Interiors, v. 54, p. 241–257.

Scholl, D. W., and Marlow, M. S., 1974, Sedimentary sequence in modern Pacific trenches and the deformed circum-Pacific eugeosyncline, in Dott, R. H., and Shaver, R. H., eds., Modern and ancient geosynclinal sedimentation: Society of Economic Paleontologists and Mineralogists Special Publication 19, p. 193–211.

Spence, W., 1977, The Aleutian arc; Tectonic blocks, episodic subduction, strain diffusion, and magma generation: Journal of Geophysical Research, v. 82, p. 213–230.

Stauder, W., 1968a, Mechanism of the Rat Island earthquake sequence of February 4, 1965, with relation to island arcs and sea-floor spreading: Journal of Geophysical Research, v. 73, p. 3847–3858.

——, 1968b, Tensional character of earthquake foci beneath the Aleutian trench with relation to sea-floor spreading: Journal of Geophysical Research, v. 73, p. 7693–7701.

Stauder, W., and Bollinger, G. A., 1964, The S wave project for focal mechanism studies; Earthquakes of 1962: Bulletin of the Seismological Society of America, v. 54, p. 2199–2208.

——, 1966a, The focal mechanism of the Alaska earthquake of March 28, 1964, and its aftershock sequence: Journal of Geophysical Research, v. 71, p. 5283–5295.

——, 1966b, The S wave project for focal mechanism studies; Earthquakes of 1963: Bulletin of the Seismological Society of America, v. 56, p. 1363–1371.

Stauder, W., and Udias, A., 1963, S-wave studies of earthquake of the North Pacific; Part 2, Aleutian Islands: Bulletin of the Seismological Society of America, v. 53, p. 59–77.

Stephens, C. D., Lahr, J. C., Fogleman, K. A., and Horner, R. B., 1980, The St. Elias, Alaska, earthquake of February 28, 1979; Regional recording of aftershocks and short-term, pre-earthquake seismicity: Bulletin of the Seismological Society of America, v. 70, p. 1607–1633.

Stone, D. B., 1983, Present-day plate boundaries in Alaska and the Arctic, in Proceedings of the 1982 Symposium 3: Journal of the Alaska Geological Society, p. 1–14.

*Sykes, L. R., and Sbar, M. L., 1974, Focal mechanism solutions of intraplate earthquakes and stresses in the lithosphere, in Kristiansson, L., ed., Geodynamics of Iceland and the North Atlantic area: Boston, D. Reidel Publishing Company, p. 207–224.

Turner, D. L., Swanson, S. E., and Wescott, E., 1981, Continental rifting; A new tectonic model for geothermal exploration of the central Seward Peninsula, Alaska: Transactions of the Geothermal Resource Council, v. 5, p. 213–216.

Ward, S. H., 1983, Body wave inversion; Moment tensors and depths of oceanic intraplate bending earthquakes: Journal of Geophysical Research, v. 88, p. 9315–9330.

Wu, F. T., and Kanamori, H., 1973, Source mechanism of February 4, 1965, Rat Island earthquake: Journal of Geophysical Research, v. 78, p. 6082–6092.

Zoback, M. L., and Zoback, M., 1980, State of stress in the conterminous United States: Journal of Geophysical Research, v. 85, p. 6113–6156.

Zoback, M. D., and 11 others, 1987, New evidence on the state of stress of the San Andreas fault system: Science, v. 238, p. 1105–1111.

Manuscript Accepted by the Society November 15, 1990

ACKNOWLEDGMENTS

We thank Tom Boyd and Gerardo Suárez for allowing us to use unpublished focal mechanism data, and to John Suppe for supplying unpublished breakout data. Roger Davis, Javier Pacheco, Walter Smith, and Pål Wessel wrote the Lamont plotting packages. Geoff Abers, John Adams, John Beavan, Göran Ekström, Bob Engdahl, and Mark Zoback provided thoughtful reviews of the manuscript. This research was supported by the Department of Energy grant DE-FG02-84ER13221E (C.H.E.). The study of borehole breakouts in northern Alaska was supported by the National Science Foundation grant EAR-83-09658 (K.H.J.). This is Lamont-Doherty contribution number 4722.

Printed in U.S.A.

The Geology of North America
Decade Map Volume 1
1991

Chapter 22

State of stress and active deformation in Mexico and western Central America

Max Suter
Institute of Geology, National University of Mexico, Apartado 70-296, México DF 04510

INTRODUCTION

We present a tectonic interpretation of the stress-orientation data for Mexico and western Central America shown on the Decade of North American Geology (DNAG) continental-scale stress map of North America (Zoback and others, 1991). Geographically, the study area comprises Mexico, northern and western Guatemala, and Belize (Fig. 1). The plate-tectonic setting of the region is the southernmost part of the North American Plate and a part of the Pacific Plate in the form of the Baja California Peninsula (Fig. 1). The stress-orientation data are displayed in their neotectonic framework (Figs. 3 to 6). They are listed for some of the acquisition techniques (Tables 1 and 2); additionally the entire dataset will be deposited with the U.S. National Geophysical Data Center in Boulder, Colorado. In the text we follow geographic regions and review, for each region, first the state of active deformation and then the state of stress.

Our stress-orientation maps contain, to our knowledge, all available intraplate measurements as well as interplate measurements in the transtensional Gulf of California and transpressional Motagua-Polochic plate boundary zones between the North American and Pacific, and the North American and Caribbean Plates, respectively (Fig. 1). However, we did not compile focal mechanisms of seismic events located in the Wadati-Benioff zone between the North American, and the Cocos and Rivera Plates at the Pacific margin of Mexico and Guatemala (Fig. 1). For this subduction zone, the stress distribution is presently much better resolved (e.g., Lefevre and McNally, 1985) than for the area analyzed in this chapter.

DATA ACQUISITION AND PROCESSING

Our stress-orientation data are derived from drillhole elongations (Fig. 2), focal mechanisms, volcanic alignments, and fault-slip analysis.

Drillhole elongations

The directions of maximum horizontal stress, S_H, resulting from drillhole elongations (Table 1) are based on the work of Suter (1987) for northeastern Mexico (43 sites), and 46 new measurements from wells located in northwestern, central, and southern Mexico, Guatemala, and Belize. The applied data acquisition and processing techniques were presented in Suter (1987). The depth intervals (Table 1) are indicated with reference to the surface elevation, with the exception of offshore sites where depth is with respect to sea level.

The data taken from Suter (1987) are modified in the following ways (Table 1). Instead of rounding the elongation-direction mean values to sectors of 5°, the unrounded mean value and one standard deviation are now indicated. Additionally, some of the data have been reprocessed with different filters and reinterpreted. Furthermore, the quality-ranking criteria currently in use for the World Stress Map of the International Lithosphere Program (Zoback and others, 1989; Mastin and others, 1989) have partly been adopted. There are four qualities, A >B >C >D. The ranking is based on the amount of distinct breakout zones and the dispersion of the breakout-direction data. In breakout zones >5 m, each 5-m interval was considered a distinct zone. Quality A is assigned to records of a single well with ≥10 distinct breakout zones and a standard deviation SD ≤12°, and to averaged records from two or more wells of close geographic proximity (sites TB-05 to TB-07). Quality B is assigned to records of a single well with ≥6 distinct breakout zones and 12° <SD <20°. Quality C is assigned to records of a single well with ≥4 distinct breakout zones and 20° ≤SD ≤25°. Quality D is assigned to records of a single well with <4 consistently oriented breakouts.

Focal mechanisms

The focal mechanism solutions and S_H directions were compiled by G. Suárez. The database includes entries from intraplate earthquakes, as well as from interplate events at the Gulf of California and Motagua-Polochic transform plate boundaries. For reverse and strike-slip earthquakes, the azimuth of the P axis was taken as the S_H direction, whereas for normal faulting events, where the P axis is subvertical, the S_H direction was assumed to

---

Suter, M., 1991, State of stress and active deformation in Mexico and western Central America, *in* Slemmons, D. B., Engdahl, E. R., Zoback, M. D., and Blackwell, D. D., eds., Neotectonics of North America: Boulder, Colorado, Geological Society of America, Decade Map Volume 1.

be oriented at 90° from the T axis. The quality ranking is based on Zoback and others (1989). Additionally, the quality of earthquakes showing several inconsistent first-motion polarities was reduced by one grade.

Since S_H directions deduced from focal mechanisms are assumed to be at 45° to the orientation of the fault plane, they may be in error as much as ±45°. The S_H directions derived from thrust and normal fault focal mechanisms are probably much better constrained than S_H directions derived from strike-slip focal mechanisms. For the former two fault types, the variability of the angle between the maximum and minimum principal stresses and the fault plane occurs in a vertical plane and will therefore not affect the direction of S_H.

Cinder-cone alignments

S_H directions deduced from vent alignments of monogenetic volcanoes are based on the hypothesis that the vents will form normal to the least horizontal far-field stress, S_h (Nakamura, 1977), because the magma will preferentially migrate in fracture pore-space that is statistically oriented perpendicular to the least principal stress, or will induce new extension fractures perpendicular to the least principal stress. The hypothesis was confirmed by comparing, on a regional scale, data resulting from this technique with independent stress measurements (Nakamura and others, 1977).

The data sources based on this technique are seven meas-

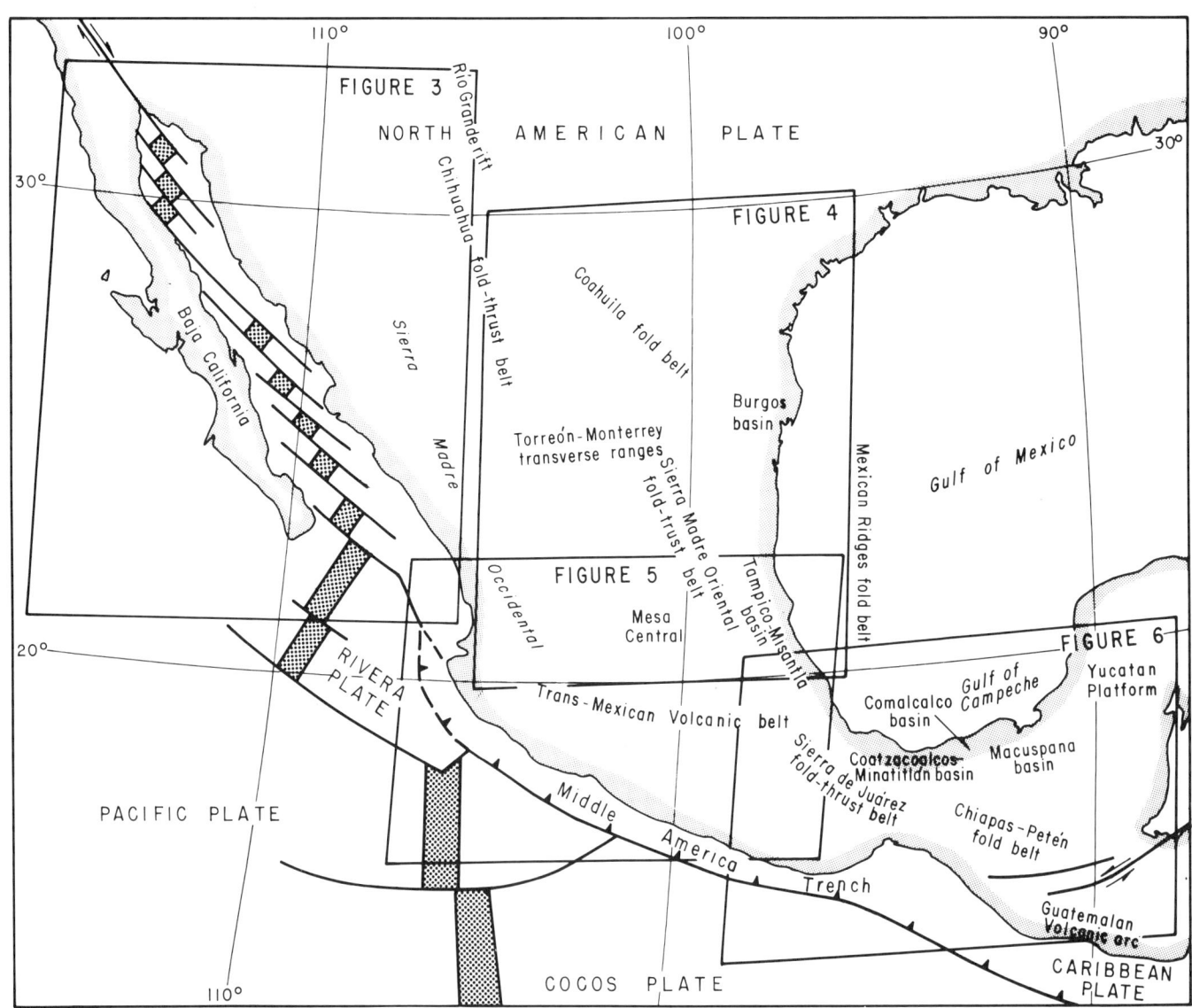

Figure 1. Outline map of the study area with the locations of the tectonic features and physiographic regions mentioned in the text. Insets: Figures 3 to 6.

urements from the central part of the trans-Mexican volcanic belt (Suter and others, 1990) and 40 new measurements, which were determined in a systematic search for alignments of Quaternary cinder cones documented in the literature of the trans-Mexican volcanic belt and from Quaternary volcanic fields north of the trans-Mexican volcanic belt in the states of Baja California Norte, Chihuahua, Durango, and San Luis Potosí (Table 2). Each datum location corresponds to the midpoint of a volcanic alignment. The geometry of the alignment is defined by its midpoint, orientation, and length. The orientation was determined by a linear least squares fit.

These data are grouped in four qualities, A >B >C >D (Table 2). The ranking is based on the amount of vents. A penalty is applied, which is based on the standard deviation of the least

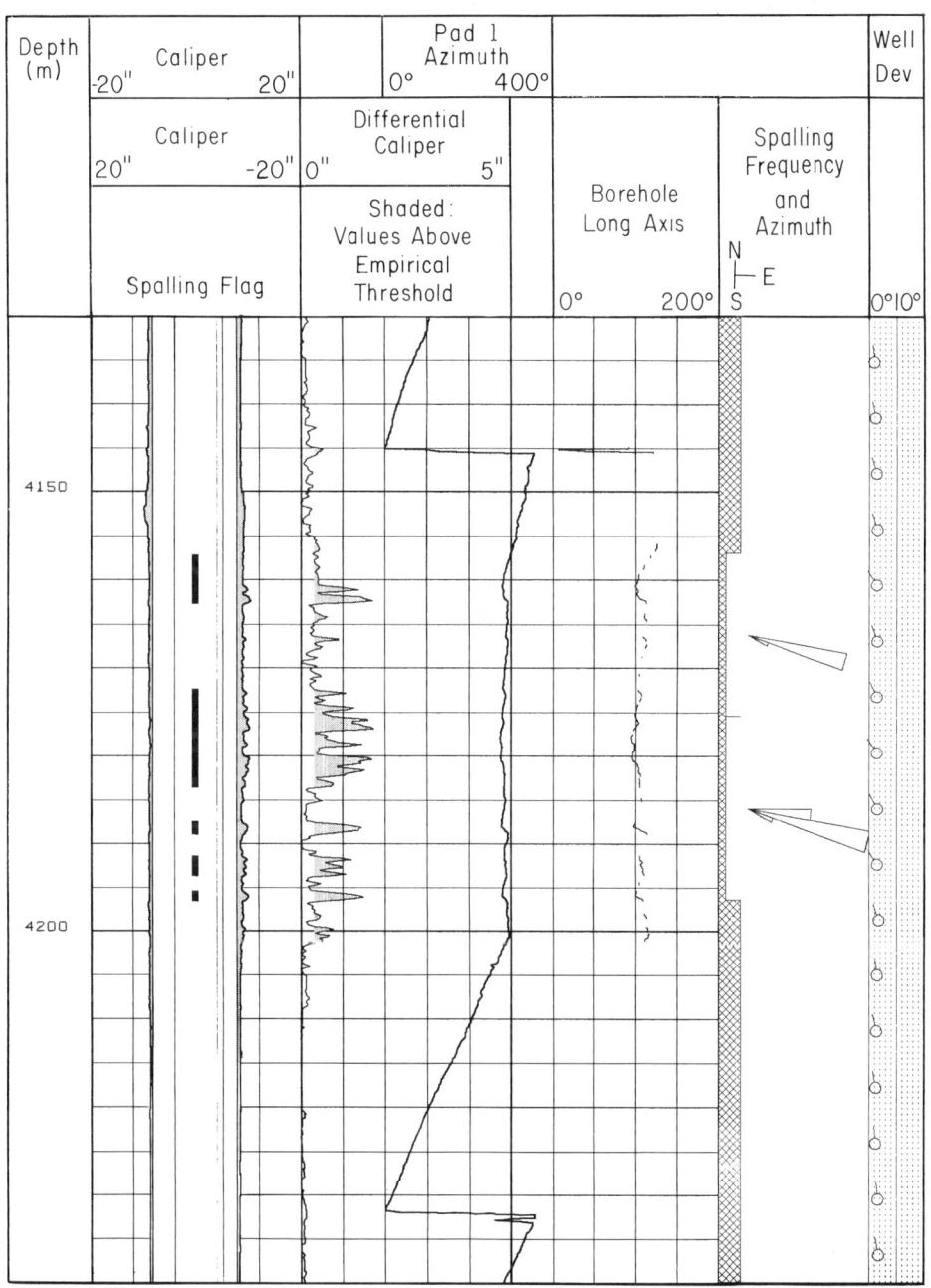

Figure 2. Determination of the direction of the least horizontal stress based on wellbore elongations. Site TB-08. The processing algorithm is described in Suter (1987). A clear correlation can be observed between the increased differential caliper and the sonde rotation-velocity anomaly in the spalled depth interval 4,160 to 4,200 m. The preferential elongation direction is 105° and does not coincide with the well-inclination azimuth, which precludes a control of the elongation direction by the well deviation.

TABLE 1. S_H DIRECTION FROM DRILLHOLE ELONGATIONS

	Lat. (°)	Long. (°)	Depth (m)	S_H (°)	Variance	St. Dev. (°)	Events	Quality	Ref.*
				BELIZE					
BE-01	17.25	-88.79	66- 236	35.5	0.067	10.6	11	A	N
BE-02	17.97	-88.39	122- 275	39.6	0.047	8.8	4	C	N
BE-03	18.44	-88.39	335- 389	12.7	0.014	4.8	6	B	N
				GUATEMALA					
GL-01	16.04	-90.14	399- 583	37.5	0.003	2.1	4	C	N
GL-02	17.08	-90.34	799-1188	34.4	0.008	3.7	7	B	N
GL-03	15.99	-90.36	366- 718	23.8	0.002	1.8	4	C	N
GL-04	17.38	-90.70	831-1094	12.7	0.005	2.9	5	C	N
GL-05	16.14	-90.36	432- 768	31.6	0.079	11.6	17	A	N
GL-06	17.20	-90.75	185- 557	30.3	0.003	2.4	3	D	N
GL-07	16.06	-89.59	453- 751	45.2	0.011	4.3	63	A	N
GL-08	15.92	-90.23	978-1351	37.9	0.030	7.0	11	A	N
GL-09	17.51	-90.76	562-1414	25.6	0.028	6.8	27	A	N
GL-10	16.05	-90.32	418- 496	41.4	0.009	3.8	4	C	N
GL-11	16.06	-90.10	474-1348	43.8	0.034	7.5	24	A	N
				MEXICO					
Baja California									
BC-19	25.06	-111.95	1180-2549	129.7	0.059	10.0	7	B	N
BC-20	31.87	-114.97	2495-4373	149.3	0.036	7.7	7	B	N
Campeche									
CA-01	19.08	-91.92	2443-4575	22.5	0.013	4.7	12	A	N
CA-02	19.49	-92.28	1690-2600	41.1	0.050	9.2	6	B	N
CA-03	19.15	-92.28	3995-4161	13.2			1	D	N
CA-04	18.94	-92.83	2638-3920	37.5	0.022	6.0	20	A	N
CA-05	19.19	-92.73	4692-4810	3.1			1	D	N
CA-06	18.79	-92.60	4245-4416	167.8	0.033	7.4	5	C	N
CA-07	19.39	-92.27	3513-4357	7.2	0.001	1.4	11	A	N
CA-08	19.39	-92.27	2775-3480	179.1	0.011	4.2	10	A	N
Chiapas									
CS-01	17.51	-93.40	3002-3283	117.0	0.037	7.8	16	A	N
CS-02	17.29	-93.00	1920-4302	58.5	0.053	9.5	4	C	N
CS-03	17.53	-93.47	3778-5540	76.6	0.030	7.1	6	B	N
CS-04	17.49	-92.19	3471-5386	24.4	0.030	7.1	54	A	N
CS-05	17.30	-92.63	1801-4597	17.2	0.032	7.3	59	A	N
CS-06	17.55	-93.24	3710-4063	163.6	0.079	11.6	19	A	N
CS-07	17.80	-93.37	5154-5960	151.9	0.025	6.4	19	A	N
Chihuahua									
CH-01	29.54	-104.42	846-4296	21.2	0.210	19.7	13	B	1
CH-02	28.85	-104.67	646-3563	4.4	0.003	2.0	3	D	1
CH-03	26.47	-104.51	3366-4912	178.2	0.011	4.3	3	D	1
Coahuila									
CO-01	28.50	-100.51	1796-2279	24.1	0.011	4.3	41	A	1
CO-02	26.94	-101.42	2203-4614	147.2	0.011	4.3	6	B	1
CO-03	26.94	-101.36	4036-4256	178.8	0.019	5.6	6	B	1
CO-04	27.82	-103.18	2451-3356	161.1	0.145	16.0	16	B	1
CO-05	28.94	-100.77	1496-1626	7.1	0.042	8.4	6	B	1
CO-06	27.84	-102.62	2474-2846	174.7	0.041	8.3	6	B	1
CO-07	28.96	-100.65	1391-1856	19.1	0.011	4.3	13	A	1
CO-08	27.08	-101.43	2360-2856	172.9	0.034	7.6	9	B	1
CO-09	26.10	-101.21	933-1426	114.5	0.010	4.1	4	C	1
CO-10	27.11	-102.01	891-2849	155.4	0.003	2.2	22	A	1
CO-11	27.60	-102.04	397-1466	148.0	0.062	10.3	20	A	1
CO-12	26.88	-101.01	384-2897	154.9	0.071	11.0	20	A	1
CO-13	27.07	-101.05	1250-3454	158.9	0.182	18.2	23	B	1
CO-14	27.75	-101.70	546-3348	126.2	0.263	22.4	3	D	1

TABLE 1. S_H DIRECTION FROM DRILLHOLE ELONGATIONS (continued)

	Lat. (°)	Long. (°)	Depth (m)	S_H (°)	Variance	St. Dev. (°)	Events	Quality	Ref.*
				MEXICO (CONTINUED)					
Colima									
CL-01	19.08	-103.57	2385-2791	4.8	0.048	9.0	16	A	N
Federal District									
DF-02	19.34	-99.16	796-2196	83.8	0.048	9.0	18	A	N
Hidalgo									
HI-01	21.04	-98.25	764-1736	2.0	0.002	1.6	12	A	1
HI-02	21.14	-98.41	1329-1793	164.1	0.001	1.3	2	D	1
HI-03	20.40	-99.03	1118-2986	162.7	0.116	14.2	6	B	1
HI-04	20.36	-99.26	264-2586	7.6	0.050	9.2	9	B	1
Nuevo León									
NL-01	26.39	-100.85	1701-3798	161.4	0.012	4.5	6	B	1
NL-02	26.67	-100.77	4180-4451	168.6	0.013	4.7	5	C	1
Querétaro									
QU-01	21.45	-99.43	286-993	31.8	0.150	16.3	6	B	1
San Luis Potosí									
SL-01	21.71	-98.85	1899-2742	31.1	0.242	21.4	7	C	1
SL-02	21.98	-99.11	2896-3933	6.6	0.030	7.1	32	A	1
SL-03	21.62	-99.00	2337-3189	145.0	0.161	17.0	8	B	1
SL-04	22.07	-100.01	1979-3222	153.6	0.072	11.1	17	A	1
Sinaloa									
SI-01	22.18	-105.85	496-896	165.8	0.097	12.9	22	B	N
SI-02	21.97	-105.97	1190-3160	7.9	0.032	7.3	8	B	N
Sonora									
SO-01	31.27	-114.25	407-4239	14.1	0.106	13.6	16	B	N
Tabasco									
TB-01	18.15	-93.96	2513-4153	112.6	0.027	6.7	17	A	N
TB-02	18.34	-93.76	830-1309	101.2	0.026	6.6	9	B	N
TB-03	18.10	-93.98	2598-3646	81.7	0.002	1.8	21	A	N
TB-04	18.20	-92.82	4427-5250	31.7	0.257	22.1	5	C	N
TB-05	18.09	-93.45	5022-5709	144.5	0.032	7.3	6	A	N
TB-06	18.12	-93.25	4398-5267	159.7	0.061	10.2	21	A	N
TB-07	18.10	-93.45	5225-5983	133.7	0.060	10.1	12	A	N
TB-08	18.29	-92.72	5561-6370	31.7	0.015	4.9	27	A	N
Tamaulipas									
TA-01	27.27	-99.94	1881-2781	20.4	0.041	8.3	10	A	1
TA-02	24.71	-98.49	316-2594	23.6	0.123	14.7	16	B	1
TA-03	27.00	-99.48	943-2286	32.5	0.064	10.4	12	A	1
TA-04	26.43	-99.28	1862-2250	4.9	0.090	12.5	8	B	1
TA-05	26.45	-99.12	3261-3691	32.7	0.006	3.1	8	B	1
TA-06	24.72	-98.08	2201-3406	33.6	0.019	5.6	11	A	1
TA-07	26.91	-99.43	161-1092	0.8	0.055	9.6	22	A	1
TA-08	23.12	-99.93	2661-2720	136.3	0.003	2.0	3	D	N
Veracruz									
VE-01	20.92	-97.67	699-2196	1.7	0.131	15.2	4	C	1
VE-02	21.03	-97.79	834-1281	0.0	0.002	1.6	12	A	1
VE-03	20.96	-97.80	1097-1636	8.9	0.002	1.8	14	A	1
VE-04	20.85	-97.72	2248-2911	13.9	0.011	4.3	27	A	1
VE-05	20.53	-97.23	3286-3801	175.4	0.008	3.7	22	A	1
VE-06	20.47	-97.47	1046-2052	17.6	0.005	3.0	27	A	1
VE-07	21.24	-98.27	439-1720	21.0	0.063	10.4	8	B	1
VE-08	21.55	-97.78	301-2742	155.0	0.001	1.2	6	B	1
VE-13	19.01	-96.46	2151-4500	121.7	0.047	8.9	8	B	N

*References: N = new; 1 = Suter (1987)

TABLE 2. S_H DIRECTION FROM VOLCANIC ALIGNMENTS

	Lat. (°)	Long (°)	S_H (°)	Length (km)	St. Dev. (m)	Vents	Quality	Ref.*
			Baja California					
BC-37	30.47	-116.00	136.0	6.5	70	6	A	N
BC-38	29.36	-114.48	154.8	3.6	109	4	B	N
BC-39	29.32	-114.36	133.6	15.2	479	7	A	N
BC-40	28.70	-113.86	143.6	4.4	733	6	A	N
			Chihuahua					
CH-04	27.89	-104.09	160.5	3.3	230	5	A	N
			Durango					
DU-01	24.36	-104.64	135.7	14.9	971	12	B	N
DU-02	24.29	-104.34	2.0	12.7	343	11	A	N
			Federal District					
DF-01	19.33	-98.98	75.3	16.6	297	11	A	N
			Guanajuato					
GU-01	20.11	-101.27	75.9	4.8	429	7	A	N
GU-02	20.09	-100.88	99.6	4.6	121	6	A	N
			Hidalgo					
HI-06	19.79	-98.47	61.4	17.4	715	12	A	N
HI-07	19.97	-98.63	55.7	8.2	323	5	A	N
HI-08	19.73	-98.35	51.7	8.0	258	7	A	N
			Jalisco					
JA-01	20.54	-103.21	112.7	44.6	547	9	A	N
JA-02	19.37	-102.64	57.2	11.7	575	7	A	N
JA-03	20.79	-103.99	125.8	17.0	1726	10	C	N
			México					
ME-01	19.19	-99.41	89.2	8.9	153	9	A	N
ME-02	19.17	-98.81	72.3	2.6	190	5	A	N
			Michoacán					
MI-01	19.88	-100.23	81.5	7.8	409	4	B	1
MI-02	19.84	-100.34	89.6	5.2	198	4	B	1
MI-03	19.85	-100.52	92.5	8.2	281	9	A	N
MI-04	19.88	-100.77	90.0	12.4	548	4	B	N
MI-05	19.78	-100.90	88.1	5.4	308	8	A	N
MI-06	19.74	-101.29	81.2	7.5	562	5	A	N
MI-07	19.28	-102.42	50.8	20.4	1059	10	B	N
MI-08	19.13	-102.08	18.8	10.9	1290	6	B	N
MI-09	19.06	-101.32	36.0	28.4	1987	23	C	N
MI-11	19.82	-100.33	40.1	3.8	98	4	B	1
MI-12	19.83	-100.33	80.1	2.6	399	4	B	1
MI-13	19.77	-100.44	77.0	3.8	611	5	A	1
MI-14	19.91	-100.12	116.5	9.8	1094	4	C	1
MI-15	19.80	-100.35	76.5	5.3	62	4	B	1
			Morelos					
MO-01	19.10	-99.27	89.0	9.4	512	12	A	N
			Nayarit					
NA-01	21.21	-104.57	131.5	11.9	825	9	B	N
NA-02	21.42	-104.66	130.0	34.1	441	20	A	N
NA-03	21.19	-104.81	106.2	3.0	250	7	A	N

TABLE 2. S$_H$ DIRECTION FROM VOLCANIC ALIGNMENTS (continued)

	Lat. (°)	Long (°)	S$_H$ (°)	Length (km)	St. Dev. (m)	Vents	Quality	Ref.*
				Puebla				
PU-01	19.12	-97.54	90.5	17.3	440	7	A	N
PU-02	19.26	-97.40	76.2	24.3	692	7	A	N
PU-03	19.32	-97.65	82.4	13.2	40	5	A	N
PU-04	19.06	-98.33	92.8	41.0	1227	9	B	N
PU-05	18.93	-97.35	44.1	2.9	77	4	B	N
PU-06	19.68	-97.44	42.3	5.3	753	9	B	N
				San Luis Potosi				
SL-05	22.41	-100.78	168.2	11.5	1285	5	B	N
				Veracruz				
VE-09	18.44	-95.16	124.5	36.0	454	13	A	N
VE-10	18.29	-94.86	114.2	30.0	712	5	A	N
VE-11	19.59	-97.05	52.2	4.7	663	6	A	N
VE-12	19.17	-96.96	67.2	6.8	751	8	B	N

*References: N = new; 1 = Suter and others (1990)

squares fit. Quality A is assigned to alignments with ≥5 vents and quality B to alignments with four vents. No alignments were determined for <4 vents. No penalty is applied to alignments with a standard deviation SD ≤750 m. A penalty of one rank (A → B, B → C) is applied to alignments with 750 m <SD <1,500 m. A penalty of two ranks (A → C, B → D) is applied to alignments with 1,500 m ≤SD ≤2,250 m. A penalty of three ranks (A → D, B → out) is applied to alignments with SD >2,250 m.

Fault-slip data

Directions of maximum horizontal stress, S$_H$, determined from the inversion of fault-slip data are based on Angelier and others (1981) for Baja California, Colletta and Ortlieb (1984) for northwestern Sonora (Fig. 3), and Suter and others (1990) for the central part of the trans-Mexican volcanic belt (see Fig. 5). Also here, we applied the quality-ranking criteria currently in use for the World Stress Map of the International Lithosphere Program (Zoback and others, 1989; Mastin and others, 1989), following which all our data are ranked as being of A quality, which is assigned to the best-fitting mean deviatoric stress tensor resulting from an inversion of fault-slip data.

RESULTS

Northwestern Mexico

Northwestern Mexico, including the Gulf of California, the Baja California Peninsula, and the part of mainland Mexico that borders the Gulf of California (Sonora, Sinaloa, and Nayarit States), is characterized by active right-lateral strike-slip and normal faulting (Fig. 3). The displacement is concentrated at the Gulf of California plate boundary, which is being deformed by right-lateral transform faults connecting short spreading centers. The Gulf of California plate boundary is linked through the Cerro Prieto and Imperial faults with the San Andreas fault system. Active deformation occurs also along right-lateral faults that cut northwesterly across Baja California (Allen and others, 1960; Lomnitz and others, 1970; Stock and Hodges, 1990; Suárez and others, 1990, Fig. 3). These faults show up as linear trends on the map of recent seismicity by Stover (1986), whereas their trend is somewhat blurred in Engdahl and Rinehart (1988) due to a larger number of data points and enclosure of older data. These faults most probably cause a partial displacement transfer from the northernmost Gulf of California transform segments to the California continental borderland faults (Wong and others, 1987; Legg and others, 1989). Additionally, right-lateral intraplate strike-slip faulting along the western margin of peninsular California (Fig. 3) is evidenced by seismicity (Molnar, 1973, events 15 and 16; Rebollar and others, 1982; Rebollar and Reichle, 1987; Engdahl, 1988) and sea-floor topography (Spencer and Normark, 1979), as well as from outcrops in the Vizcaino Peninsula (Angelier and others, 1981).

The slip-vector azimuth from 15 focal-plane solutions of strike-slip events (centroid depth <20 km) in the Gulf of California is 130 ± 4° (Goff and others, 1987). This agrees with geodetic measurements of the same parameter in the central part of the Gulf (134°; Kasser and others, 1987), and with the general strike of the transform segments of the Gulf of California plate boundary, which is 134° in the southern Gulf adjacent to the Mazatlán ridge, 130° in the central part of the Gulf, and 134° for the Cerro Prieto fault (Fig. 3; Bischoff and Niemitz, 1980).

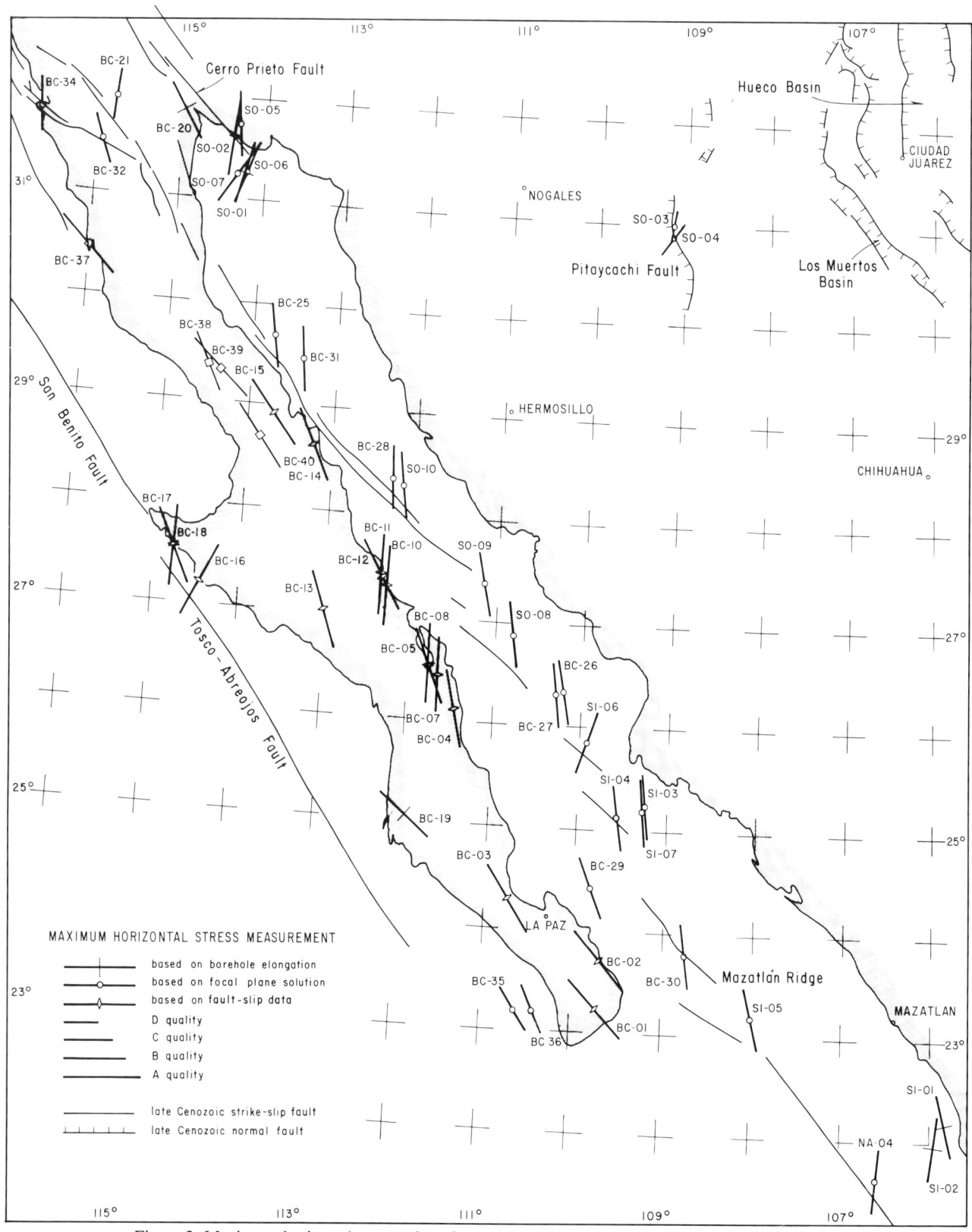

Figure 3. Maximum horizontal stress orientations and late Cenozoic faults in northwestern Mexico. Except in Sonora, only data of A to C quality are graphed. The location of the transform faults in the Gulf of California is based on the scarps in the bathymetric maps by Bischoff and Niemitz (1980).

The average amount of slip for the last 3 Ma along the Gulf of California transform plate-boundary was estimated as 48 mm/a by DeMets and others (1987) from the spreading rate directly north of the Tamayo fracture zone, in the southern Gulf of California, based on magnetic-anomaly data. Following a kinematic model by Goff and others (1987), the slip is partitioned in the northern Gulf, from where presumably a component equivalent to the slip rate along the San Andreas fault (34 ± 3 mm/a; Minster and Jordan, 1987) is transferred along the Cerro Prieto fault and other faults nearby and parallel to it (Fig. 3), whereas a remaining 15 mm/a are transmitted along faults crossing the Baja California Peninsula (Fig. 3) to the California continental borderland faults, bounding a relatively coherent block in between. A kinematic model for southern California by Weldon and Humphreys (1986) also requires that the San Andreas system (including the San Andreas and San Jacinto faults) accounts for just under two-thirds of the plate motion, and the coastal system (Newport-Inglewood fault) for about one-third.

Stress measurements available for northwestern Mexico (Fig. 3) are from four sources. Most are based on the inversion of fault striation data from the Baja California Peninsula (Angelier and others, 1981, Fig. 16b) and northwestern Sonora (Colletta and Ortlieb, 1984). The S_H azimuth compiled for 19 measurement stations of these authors has a mean of 166°, a circular variance of 0.054, and a standard deviation of 10°. Other data include four cinder cone alignments from the Baja California Peninsula (BC-37 to BC-40, Fig. 3 and Table 2), compiled focal mechanisms (Fig. 3), and new breakout data from five wells (depth <4.2 km) in Baja California and the Gulf of California (Table 1; S_H azimuth: 166° ± 13°, circular variance: 0.091). The stress data based on the inversion of fault striations and the breakouts indicate independently that the plate boundary is oriented at an angle of 35° to the maximum horizontal far-field stress (Fig. 3) (Suter, 1990a). The results are somewhat different if we include the directional stress data derived from focal mechanisms. In that case, the S_H direction is 180 ± 10° for the northern Gulf, 170 ± 7° for the central Gulf, and 163 ± 10° for the southern Gulf of California (Table 3). Most of the sites of the breakout and fault striation data are located on either side at distances >50 km from the plate boundary with the exception of sites BC-14, BC-15, BC-20, and SO-01. The drillhole elongation measurements BC-20 and SO-01 are located at <10 km from the plate boundary near the Cerro Prieto fault and in the Wagner basin, respectively (Fig. 3).

We now compare the slip and stress directions in the Gulf of California region with the values of the same parameters farther north along the Pacific–North America Plate boundary, in southern and central California. The directions of slip are practically the same in the two areas; 139 ± 2° along the San Andreas fault (Minster and Jordan, 1987), and 134° along the Cerro Prieto transform segment in the northern Gulf of California. The far-field stress directions, however, are different: In the Gulf of California, S_H is—based on breakout and striation analysis—oriented 167° (i.e., 32° clockwise with respect to the slip direction; Fig. 3), which is in agreement with the S_H direction expected from the empirical Coulomb-Mohr expression (Jaeger and Cook, 1976) and experimental results (Patterson, 1978) for right-lateral horizontal shear on a vertical plane. In southern California, S_H directions determined by an inversion of focal mechanism data are generally the same as in the Gulf of California, but the angle between S_H and the San Andreas fault is approximately 65° (Jones, 1988). In central California, on the other hand, S_H is apparently oriented 44 ± 2°, or at 85° clockwise to the San Andreas fault within a zone of approximately 100 km on either side of the fault (Mount and Suppe, 1987; Zoback and others, 1987). We would therefore expect a higher resolved shear stress along the transform segments of the Gulf of California plate boundary than along the San Andreas fault, if the magnitudes of the far-field stresses are equal at the two positions. Furthermore, this higher resolved shear stress may be caused by a higher coefficient of friction at the Gulf of California plate boundary as compared to the San Andreas fault.

Based on the stress data in Figure 3, we define the "Gulf of California stress province" (Table 3; see Fig. 7), which is characterized by a strike-slip stress regime with the principal stress directions $S_{SSE} > S_v > S_{ENE}$. Local perturbations in this regional stress field can be expected in the area of the small transtensional basins located at the Gulf of California plate boundary (Bischoff and Niemitz, 1980). The Gulf of California stress province has an approximate extension of 300 km in the west-east direction and 1,000 km in the north-south direction (see Fig. 7). It can be delimited in the north from the San Andreas fault stress province (Zoback and Zoback, 1980, 1989), which is represented in central California by a stress regime with $S_{NE} > S_v > S_{SE}$ (Mount and Suppe, 1987). Both stress provinces are characterized by a strike-slip stress regime, but the directions of S_H and S_h differ 58° between the two provinces. However, the S_H direction in the Gulf of California stress province is comparable to the results of the stress measurements for southern California by Jones (1988), and to breakouts in a well at the southwestern margin of the Salton Sea (Zoback and others, 1987, Fig. 1). In the east, the Gulf of California stress province can be delimited in the state of Sonora from the southern Basin and Range stress province (Fig. 7), which is there characterized by a stress field with $S_v > S_{NNE} > S_{ESE}$ (see below). In the southeast, the Gulf of California stress province can be delimited from the stress province of the trans-Mexican volcanic belt at the northwestern end of the Tépic-Zacoalco graben (see Figs. 5 and 7). In the south, its limit probably lies in the area of the Tamayo fracture zone (Fig. 1), where a thrust-fault focal mechanism with west-east–striking nodal planes was determined by Goff and others (1987; Fig. 5).

Northeastern Mexico

Northeastern Mexico is an intracratonic area with few indications of neotectonic activity (Fig. 4). An exception is the Rio Grande Rift which is volcanically and tectonically active, and the southern continuation of which into Chihuahua is formed by the

TABLE 3. STRESS PROVINCES

Stress Province	Sites	Amount of Sites		Mean	S_H 1 St. Dev. Circular Variance		Relative Magnitude of Principal Stresses
Gulf of California a. Northern part (north of lat. 30°)	BC-20 to BC-24, BC-32 to BC-34, BC-37, and SO-01 to SO-07.	16	Quality-weighted Unweighted	177.1° 177.4°	0.084 0.068	12.0° 10.7°	$S_N > S_v > S_E$
b. Central part (30° > lat. > 26°)	BC-04 to BC-18, BC-25 to BC-28, BC-31, BC-38 to BC-40, and SO-08 to SO-10.	26	Quality-weighted Unweighted	166.5° 166.9°	0.040 0.037	8.2° 7.9°	$S_{SSE} > S_v > S_{ENE}$
c. Southern part (south of lat. 26°)	BC-01 to BC-03, BC-19, BC-29, BC-30, BC-35, BC-36, NA-04, and SI-01 to SI-07	16	Quality-weighted Unweighted	162.0° 162.9°	0.059 0.056	10.0° 9.7°	$S_{SSE} > S_v > S_{ENE}$
Southern Basin and Range– Rio Grande Rift		46	Quality-weighted Unweighted	0.3° 179.6°	0.088 0.079	12.3° 11.7°	$S_v > S_N > S_E$
a. Northeastern Mexico (Chihuahua, Coahuila, Durango, Neuvo León, and Tamaulipas States; Chihuahua and Coahuila fold-thrust belts, Burgos Basin)	CH-01 to CH-04, CO-01, CO-03 to CO-08, CO-10 to CO-13, DU-01, DU-02, NL-01, NL-02, and TA-01 to TA-08	27	Quality-weighted Unweighted	180.0° 178.5°	0.079 0.077	11.6° 11.5°	$S_v > S_N >> S_E$
b. East-central Mexico (Hidalgo, San Luis Potosí, Querétaro, and part of Veracruz State; Sierra Madre Oriental fold-thrust belt and Tampico-Misantla Basin)	HI-01 to HI-05, SL-01 to SL-05, QU-01, and VE-01 to VE-08	19	Quality-weighted Unweighted	1.7° 1.3°	0.076 0.082	11.4° 11.9°	$S_v > S_N > S_E$
Southern Great Plains (part of Coahuila State)	CO-02, CO-09, and CO-14.	3	Quality-weighted Unweighted	132.9° 129.1°	0.035 0.028	7.6° 6.9°	$S_v > S_{SE} > S_{NE}$
Trans-Mexican Volcanic Belt a. Eastern part (parts of the States of Hidalgo, Puebla, and Veracruz)	HI-06 to HI-08, PU-01 to PU-03, PU-05 PU-06, VE-11, VE-12, VE-16, and VE-17	12	Quality-weighted Unweighted	62.6° 62.7°	0.036 0.036	7.7° 7.8°	$S_v > S_{ENE} > S_{SSE}$
b. Central part (Federal District, México, Morelos and Guanajuato States, parts of the States of Puebla and Michoacán)	DF-01, DF-02, ME-01, ME-02, MO-01, GU-01, GU-02, MI-01 to MI-06, MI-10 to MI-15, and PU-04	20	Quality-weighted Unweighted	83.0° 82.9°	0.026 0.033	6.6° 7.4°	$S_v > S_E > S_N$
c. Southwestern Michoacán and southeastern Jalisco States	MI-07 to MI-09, and JA-02.	4	Quality-weighted Unweighted	42.9° 41.0°	0.036 0.034	7.8° 7.5°	$S_v > S_{NE} > S_{SE}$

TABLE 3. STRESS PROVINCES (continued)

Stress Province	Sites	Amount of Sites		S_H Mean / Circular Variance / 1 St. Dev.			Relative Magnitude of Principal Stresses
d. Colima Graben (Colima State)	CL-01	1		4.8°			$S_v > S_N > S_E$
e. Tépic–Zacoalco Graben (Jalisco and Nayarit States)	JA-01, JA-03, and NA-01 to NA-03.	5	Quality-weighted Unweighted	120.1° 121.3°	0.017 0.015	5.2° 5.0°	$S_v > S_{ESE} > S_{NNE}$
Gulf Coast stress province in southern Veracruz and western Tabasco States (without lower crust focal-mechanism-derived stress data)	VE-09, VE-10, VE-13, and TB-01 to TB-03.	6	Quality-weighted Unweighted	109.8° 109.9°	0.033 0.031	7.4° 7.2°	$S_v > S_{ESE} > S_{NNE}$
Chiapas-Tabasco (southeastern Mexico and western Central America) a. Tabasco and Campeche States (Macuspana Basin and Gulf of Campeche)	CA-01 to CA-08, TB-04, and TB-08.	10	Quality-weighted Unweighted	20.1° 17.7°	0.044 0.046	8.6° 8.7°	$S_{NNE} > S_{ESE} > S_v$
b. Oaxaca, Chiapas, and western Guatemala (Chiapas and Petén fold belts)	GL-01, GL-03, GL-05 to GL-08, GL-10 to GL-15, GL-20 to GL-22, GL-25, CS-02 to CS-05, CS-08, CS-09, and OX-01.	23	Quality-weighted Unweighted	40.8° 39.6°	0.060 0.063	10.0° 10.3°	$S_{NNE} > S_v > S_{ESE}$
c. Northern Guatemala and Belize	BE-01 to BE-03, GL-02 GL-04, and GL-09.	6	Quality-weighted Unweighted	27.3° 26.8°	0.015 0.018	5.0° 5.4°	$S_{NNE} > S_{ESE} > S_v$
Ipala (between the Motagua-Polochic transform system and the Guatemalan volcanic arc)	GL-17 to GL-19	3	All data are of the same quality	20.6°	0.037	7.9°	$S_v > S_{NNE} > S_{ESE}$

Hueco and Los Muertos grabens (Figs. 3 and 4; Seager and Morgan, 1979). Farther east and parallel to the Rio Grande rift zone, the Salt basin and Presidio grabens, documented in Texas by Muehlberger and others (1978), trend southward across the Rio Grande (Rio Bravo) River into northeastern Chihuahua and northwestern Coahuila (Fig. 4; Charleston, 1981, p. 32 and Plate A). The Salt basin graben is partly seismically active (Muehlberger and others, 1978). Southwest of the Rio Grande rift, the 1887 Sonoran earthquake caused an average 3-m vertical displacement along an 80-km segment of the Pitaycachi fault (Natali and Sbar, 1982; Pearthree and others, 1990) in northeastern Sonora (Fig. 3). Furthermore, there are several Cenozoic grabens on the Mesa Central (Fig. 4; Tristan, 1986; Aranda and others, 1990) that may have been partly active during Quaternary time. Ongoing deformation also occurs on the slope of the Gulf of Mexico continental margin (Fig. 4) with the formation of the Mexican Ridges fold belt and concurrent normal faulting directly west of the fold belt and parallel to its trend (Buffler and others, 1979).

Published stress measurements are limited to focal mechanisms of microseismicity along the Pitaycachi fault in Sonora (Natali and Sbar, 1982) and an event north of Mexico City (HI-05, Fig. 4), as well as from cinder cone alignments from Quaternary volcanic fields in Chihuahua (CH-04) and Durango (DU-01 and DU-02) States, from Quaternary maar alignments in San Luis Potosí State (SL-05), and borehole elongations from 46

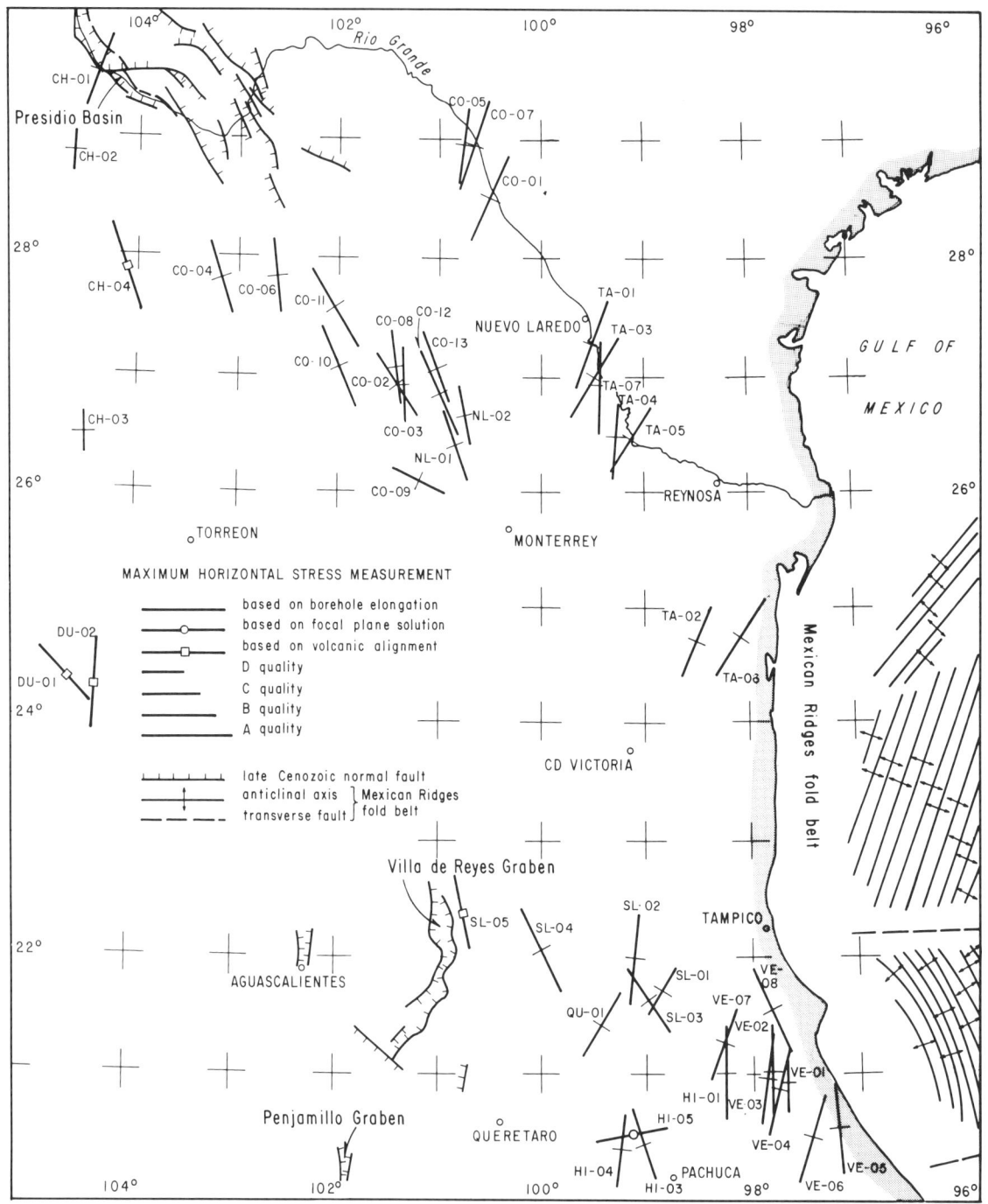

Figure 4. Maximum horizontal stress orientations and late Cenozoic faults in northeastern Mexico. Except in Chihuahua, only data of A to C quality are graphed; for the remaining data of D quality see Tables 1 to 3.

sites with a depth <4.9 km (Suter, 1987), all of which indicate, in general, a north-south–trending S_H direction (Fig. 4). As mentioned above, some of the breakout data have been reprocessed with different filters and reinterpreted. Furthermore, instead of rounding the elongation-direction mean values to sectors of 5°, we indicate now the unrounded mean and one standard deviation (Table 1).

The results of the stress measurements in Chihuahua and Sonora (Fig. 4) are comparable with the state of stress in southern New Mexico (Aldrich and others, 1986). The S_H direction is parallel to the trend of the Pitaycachi fault, which indicates a normal fault-type stress field with $S_v > S_{NNE} > S_{ESE}$.

The S_H direction determined in the states of Coahuila and Nuevo León in the area of the Laramide Coahuila fold belt is parallel to the Rio Grande rift and the Quaternary faults in western Texas and northeastern Chihuahua (Muehlberger and others, 1978; Seager and Morgan, 1979), which suggests a stress field with $S_v > S_{NNW} > S_{ENE}$ (Figs. 4 and 7, Table 3).

Four sites in Coahuila (CO-2, CO-09, CO-13, and CO-14; Figure 4) apparently belong to the southern Great Plains stress province (see Fig. 7). The extent of this stress province is defined by Zoback and Zoback (1980, 1989) and Aldrich and others (1986) inside the United States, where it is characterized by extensional tectonism (basaltic volcanism and normal fault focal mechanisms) and a stress field with $S_v > S_{ESE} > S_{NNE}$. This corresponds to a 90° rotation of S_H with respect to the Basin and Range–Rio Grande rift stress province. The boundary between the Basin and Range and southern Great Plains stress province is shown in Figure 7. It forms an outlier into Coahuila State and is farther southeast parallel to the Rio Grande (Rio Bravo) River, where its location is constrained by our data and stress measurements by Schafer (1980) in Zavala and Frio Counties (Suter, 1987, Fig. 3).

Following W. Muehlberger (written communication, 1987), the La Babia and San Marcos faults, which cross Coahuila in an ESE to SE direction (Charleston, 1981, Plate A; McKee and others, 1984, 1990) in the area of the southern Great Plains stress province, may have been reactivated during the Quaternary by dilatant right-lateral slip. Site CO-09 is located very close to the San Marcos fault, and site CO-14 near the La Babia fault. The orientation of S_H at these two sites with respect to the San Marcos and La Babia faults would favor such an extensional reactivation.

In the Cenozoic Burgos basin in the states of Coahuila and Tamaulipas, we previously defined the Burgos basin stress province (Suter, 1987), which is characterized by a SSW-NNE trend of S_H (Fig. 4). In analogy with the situation in the area of the Coahuila fold belt, the Burgos basin is most likely characterized by a normal-fault-type stress field with $S_v > S_{NNE} > S_{ESE}$. The change in the S_H direction from NNW to NNE between the sites in the Coahuila fold belt and those in the Burgos basin is gradual (Fig. 4) and perpendicular to the topographic contours of the Monterrey-Torreón transverse ranges, a west-east–trending salient of the Cordilleran fold-thrust belt with an average altitude difference >1,000 m with respect to the Coahuila fold belt and the Burgos basin. The altitude contours of the mountain front can be noted on the DNAG continent-scale stress map of North America (Zoback and others, 1991). We hypothesize that the N-S–trending far-field component of S_H is locally deflected into a NNW to NNE direction by this gravity load. This would allow calculation of the magnitude of the far-field stress components, if the gravity load could be quantified (Hafner, 1951; Savage and Swolfs, 1986).

In the Cenozoic Tampico-Misantla basin, the direction of S_H is NNE and shows a very low dispersion (Fig. 4). By extrapolating the situation farther NNW, we think a stress field with $S_v > S_{NNE} > S_{ESE}$ is most likely. This hypothesis is supported by the existence of a N-S–trending province of alkaline to hyperalkaline magmatism of Oligocene to Quaternary age (Cantagrel and Robin, 1979), which crosses the Tampico-Misantla basin and is typical of continental rifts. Cenozoic normal faults and igneous dikes documented in the western part of this stress province (Suter, 1990b) also trend NNE, but their age is not known in more detail. Further support for the proposed stress field in the Tampico-Misantla basin is the existence of NNE-trending Cenozoic extension structures such as the Villa de Reyes graben (Tristan, 1986; Aranda and others, 1990) in the Mesa Central, west of the Tampico-Misantla basin (Fig. 4).

In the Tampico-Misantla basin, as well as in the Burgos basin, S_H is not parallel to the continental margin (Fig. 4). This suggests that the two basins are controlled by the same stress field as the craton farther west, whereas farther north, in Texas and Louisiana, the Gulf coastal plain is apparently part of the gravity-controlled Gulf coast stress province (Zoback and Zoback, 1980, 1989). We speculate that in northeastern Mexico the Gulf coast stress province is located offshore, on the continental slope east of the Burgos and Tampico-Misantla basins (Fig. 7), where numerous active normal faults have been documented (Buffler and others, 1979; Pew, 1982).

Our stress measurements and the neotectonic deformation suggest that northeastern Mexico is under extension with $S_v > S_N > S_E$ (Table 3), and forms the southeastern continuation into Mexico of the southern Basin and Range–Rio Grande rift stress province (see Fig. 7). Inside Mexico, this stress province has an approximate expanse of 1,000 km in a west-east direction in northern Mexico, and 1,000 km in a north-south direction. It can be delimited in the west from the Gulf of California stress province, in the northeast from the Great Plains stress province, in the east from the Gulf Coast stress province, and in the south from the stress province of the trans-Mexican volcanic belt (Fig. 7 and Table 3). The existence of NNW-trending late Cenozoic normal faults within (Querétaro-Taxco lineament, Nixon and others, 1987) and south (Oaxaca fault, Centeno, 1988; Centeno and others, 1990) of the trans-Mexican volcanic belt suggests that the southern Basin and Range stress province extended, before the development of the younger trans-Mexican volcanic belt, farther south than at present.

An important conclusion derived from our stress measurements for northeastern Mexico is that the deformation of the Mexican Ridges fold belt on the slope of the Gulf of Mexico continental margin cannot be caused by compressional far-field stresses, since S_H at far-field distance west of the belt is nearly parallel to the fold axial trend of the belt (Figs. 4 and 7).

Central Mexico

Central Mexico is tectonically characterized by an active WNW-ESE–trending calc-alkaline volcanic arc, the trans-Mexican volcanic belt (TMVB), which is related to subduction along the Middle America trench (Nixon and others, 1987).

The western part of the TMVB is affected by continental rifting in form of the W-E-trending Chapala, the N-S-trending Colima, and the NW-SE-trending Tépic-Zacoalco grabens (Fig. 5) which intersect at a ridge-ridge-ridge triple junction southwest of Lake Chapala (Allan, 1986; Allan and others, 1991).

The central part of the TMVB is affected by W-E-trending Quaternary normal faults (Fig. 5; Mooser, 1969; Bloomfield, 1975; Fries and others, 1977; Johnson, 1987; Suter and others, 1990). Historical seismicity is known from several of the faults (Urbina and Camacho, 1913; Astiz, 1980; Suter and others, 1990), whereas no major active faults have been documented from the eastern part of the TMVB. However, the neotectonics of the latter region remains fairly undocumented.

Our stress measurements for central Mexico include compiled focal mechanisms for shallow events (HI-05 and MI-10), two new drillhole elongations (DF-02 and CL-01, Table 1), two stress measurements based on fault-slip analysis (MI-16 and MI-17; Suter and others, 1990), and 39 stress measurements based on alignments of Quaternary cinder cones (Table 2). Due to space problems, not all of the S_H data based on volcanic alignments could be presented on Figure 5. The fault striations, as well as the focal mechanism MI-10, indicate a normal-fault-type stress regime with a minor left-lateral strike-slip component.

In the central part of the TMVB, the S_H direction derived from our stress measurements is WSW-ENE and shows a very low dispersion (Fig. 5). In the western part of the belt, S_H is oriented parallel to the trends of the Colima and Tépic-Zacoalco grabens (Fig. 5), whereas in the eastern limit of the belt a case can be made for a gradual transition of S_H from its WSW-ENE direction in the TMVB to the WNW-ESE direction typical in the Tampico-Misantla basin (Fig. 5).

The above-presented data allow us to define the "stress province of the trans-Mexican volcanic belt" (Fig. 7), which is characterized by a stress field with $S_v > S_{ENE} > S_{NNW}$ in its central and eastern parts, whereas west and south of the ridge-ridge-ridge triple junction, a normal-fault-type stress field, with S_H directed parallel to the Colima and Tépic-Zacoalco grabens, is

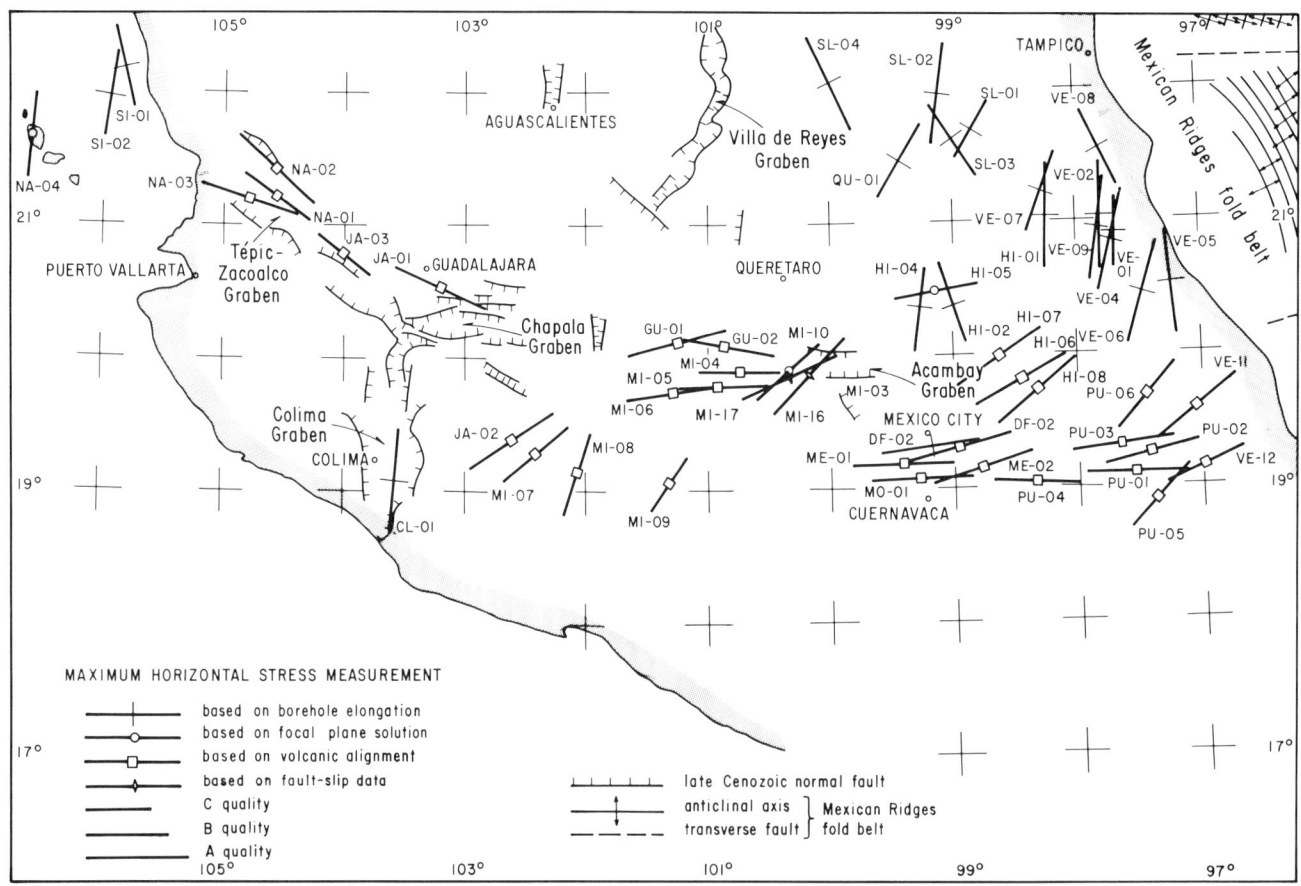

Figure 5. Maximum horizontal stress orientations and late Cenozoic faults in central Mexico. Only data of A to C quality are graphed; for data of D quality see Tables 1 to 3.

most likely (Fig. 5). This stress province can be delimited in the north from the southern Basin and Range stress province, and in the northwest from the Gulf of California stress province (Fig. 7). The southern extension of the stress province of the trans-Mexican volcanic belt remains undefined, since no stress data are known at a shallow crustal level in the states of Guerrero and Oaxaca (Figs. 5 and 6).

Southern Mexico and western Central America

At the Caribbean–North American Plate boundary, Quaternary deformation has occurred in Guatemala along the Motagua and Polochic faults (Fig. 6; Schwartz and others, 1979) in the form of left-lateral strike-slip. A continuation of the Polochic fault into southern Mexico has been documented by Burkart and others (1987). However, there is no alignment of seismic events along the probable extension of the Polochic fault to the west of its known surface trace (Guzmán, 1985; Guzmán and others, 1989), and geophysical and topographic sea-floor data of the continental margin (Couch and Woodcock, 1981; Sánchez, 1981) suggest that the fault does not reach the Middle America trench. Apparently there is no discrete trench–transform fault–trench triple junction between the Caribbean, North American, and Cocos Plates.

In Mexico, at least part of the relative movement between the two plates seems to occur along a system of left-lateral strike-slip faults documented in the Chiapas fold belt by Sánchez (1979) and Meneses (1985). Guzmán and others (1989) have documented an alignment of shallow and intermediate-depth earthquakes in the Chiapas fold belt, with fault plane solutions indicating left-lateral strike-slip displacement. It seems therefore that the southernmost part of the boundary between the Caribbean and North American Plates consists on a large scale of an en échelon array of three left-lateral strike-slip segments, with the westernmost segment being composed of the seismically active strike-slip faults in the Chiapas fold belt (Fig. 6). The late Quaternary slip rate along the plate boundary, which is estimated at 0.45 to 1.8 cm/a by Schwartz and others (1979) based on offsets of stream terraces by the Motagua fault, and at 1.9 cm/a in the model by Minster and Jordan (1978), seems to be distributed in the Chiapas fold belt over four or more subparallel, segmented left-lateral strike-slip faults of 100 to 200 km length (Figure 6).

South of the Motagua fault, there are several Cenozoic grabens with their axes perpendicular to the trace of the Motagua fault (Fig. 6; Plafker, 1976). Some of them, such as the Guatemala City and Ipala grabens, were activated during aftershocks of the 1976 Motagua earthquake (Langer and Bollinger, 1979). Mann and Burke (1984) hypothesized that this intraplate exten-

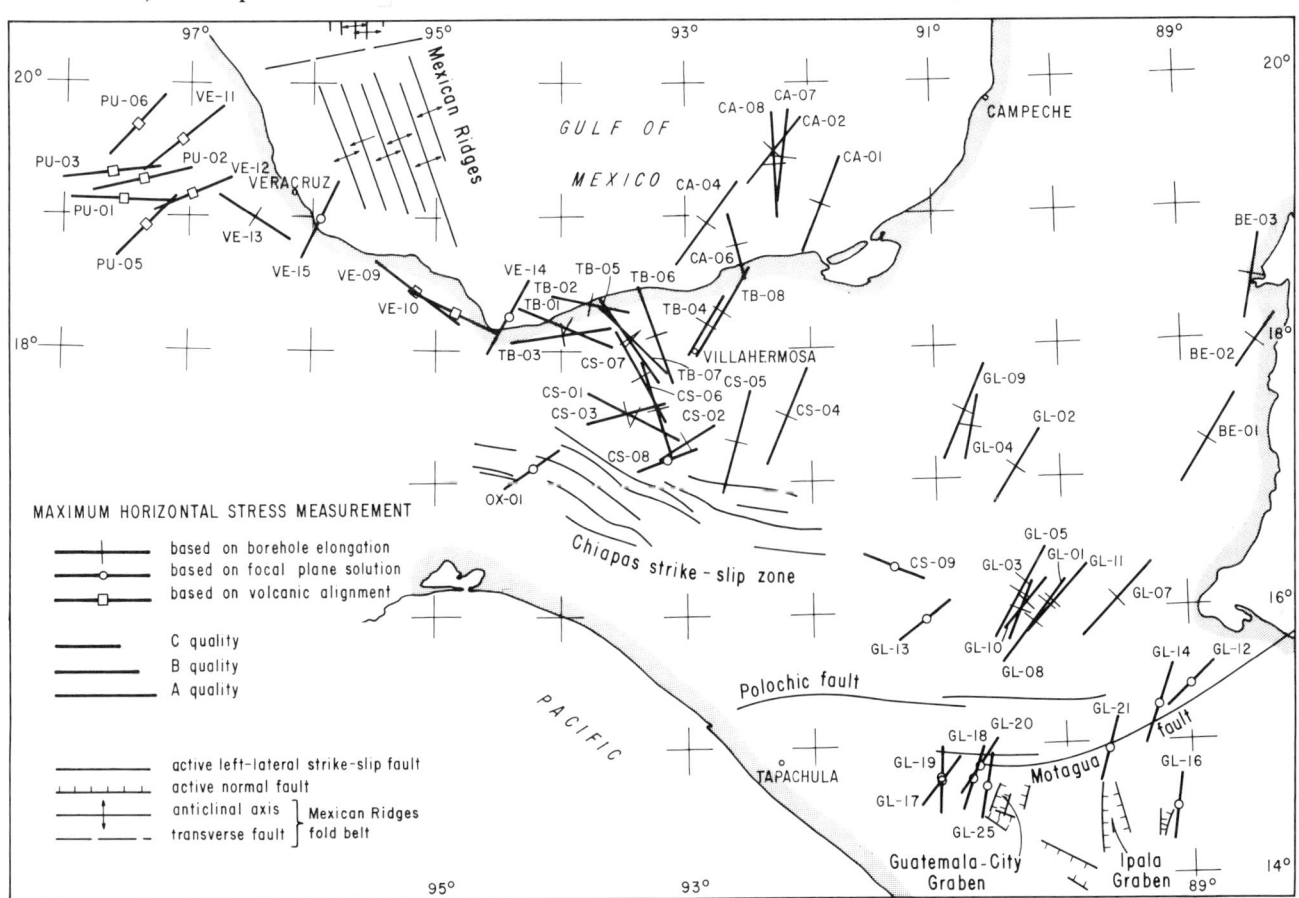

Figure 6. Maximum horizontal stress orientations and late Cenozoic faults in southern Mexico and western Central America. Only data of A to C quality are graphed; for data of D quality see Tables 1 to 3.

sion is due to the arcuate shape of the transform plate boundary (Fig. 6), similar to the extension occurring in the outer arc of a fold hinge.

Our stress data for southern Mexico and western Central America (Fig. 6) can be grouped into two clusters, each of which shows a well-defined trend of S_H.

1. The S_H azimuth from six near-surface measurements in the coastal plain of the Gulf of Mexico, in the Coatzacoalcos-Minatitlán basin of southern Veracruz and western Tabasco States (Fig. 6), based on volcanic alignments and borehole elongations, is $110 \pm 7°$ and subperpendicular to the continental margin. Active normal faulting along the upper continental slope (Pew, 1982, and unpublished Pemex seismic sections) suggests a stress field with $S_v > S_{NNE} > S_{ESE}$. The measured S_H direction is different from the direction of the same parameter in the trans-Mexican volcanic belt, as well as in the Gulf coastal plain north of the trans-Mexican volcanic belt (Figs. 4 and 5). We speculate that this area belongs to the Gulf Coast stress province (Zoback and Zoback, 1980, 1989), which would imply that there is an offshore continuity, along the continental slope, between the onshore areas of this stress province in Texas and in southern Veracruz (Fig. 7).

2. Thrust-fault focal mechanisms from the same area in southern Veracruz and western Tabasco at a depth of about 20 km, on the other hand, indicate a stress field with $S_{NNE} > S_{ESE} > S_v$ (Fig. 6), which implies polarity changes of the principal stresses with depth. The S_H direction in the lower crust is similar to the one in the Campeche-Chiapas-Petén stress province farther southeast (Figs. 6 and 7).

Borehole elongations from 23 wells (depth <6.4 km) in the seismically active Chiapas-Petén fold belt and in the adjacent foreland indicate a WNW-ESE– to NW-SE–trending S_H azimuth of $28 \pm 7°$ (Table 1 and Fig. 6). A subset of 12 wells closest to the left-lateral Motagua-Polochic transform system has an S_H azimuth of $33 \pm 4°$, which is at an angle of 30 to 60° counterclockwise from this plate boundary (Suter, 1990c). Based on the left-lateral strike-slip focal mechanisms in the Chiapas-Petén fold belt and along the Motagua fault, the stress field of this region, which we denominate "Campeche-Chiapas-Petén stress province" (Fig. 7), is most likely characterized by $S_{NE} > S_v > S_{SE}$ near

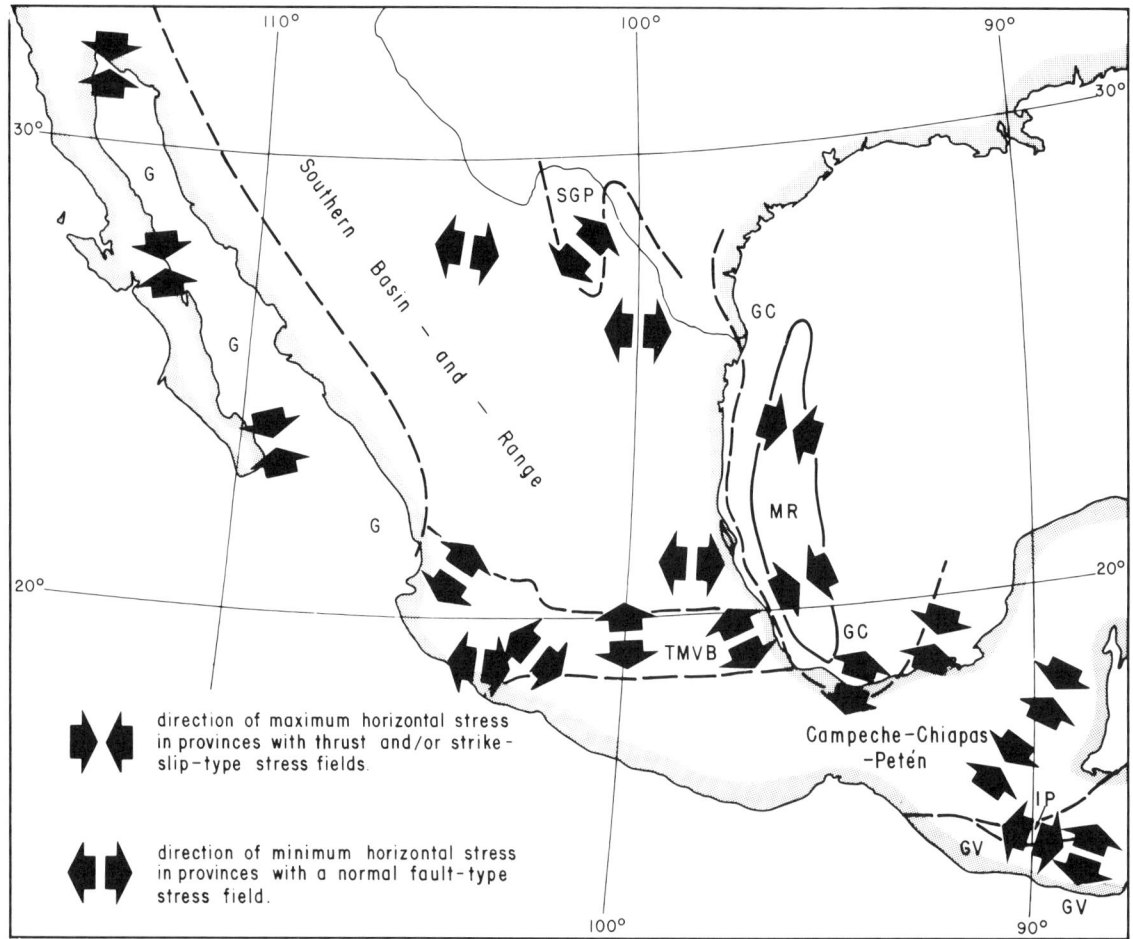

Figure 7. Stress provinces defined for Mexico and western Central America. G, Gulf of California stress province; GC, Gulf Coast stress province; GVA, stress province of the Guatemalan volcanic arc; IP, Ipala stress province; SGP, Southern Great Plains stress province; TMVB, stress province of the trans-Mexican volcanic belt.

the transform system on its northern side, and $S_{NE} > S_{SE} > S_v$ in the North American Plate interior. The stress province extends approximately 500 km east-west and 400 km north-south. It comprises the Chiapas and Petén fold belts, the Cenozoic Comalcalco and Macuspana basins northwest of the Chiapas fold belt, the Gulf of Campeche, and probably most of the Yucatan Peninsula (Figs. 6 and 7). In the southeast, this stress province is delimited by the Motagua-Polochic transform system, whereas in the southwest, its extension remains undefined due to the lack of stress measurements at a shallow crustal depth. In the northwest, a gradual transition in the direction of S_H can be observed, in Tabasco and northwestern Chiapas, between this stress province and the above-mentioned cluster of homogeneous upper-crust S_H directions along the Gulf of Mexico continental margin, in southern Veracruz and northwestern Tabasco States (Fig. 6). On the other hand, the S_H direction remains approximately the same between the two areas if we compare the data from Chiapas-Tabasco with the focal mechanism–derived stress data from the lower crust in southern Veracruz State.

Directly south of the Motagua-Polochic transform system, the small "Ipala stress province" (Fig. 7 and Table 3) is defined for the area of the seismically active Guatemala City and Ipala grabens (Fig. 6), which trend perpendicular to the trace of the Motagua fault. South of these grabens, a different stress field can be expected in the area of the Guatemalan volcanic arc, which has in its crest normal faults trending parallel to the axis of the arc (Fig. 6; Plafker, 1976, Fig. 1). This is similar to the situation in the central part of the trans-Mexican volcanic belt. The normal faults inside the volcanic arc are perpendicular to the Guatemala City graben. The Guatemalan volcanic arc, as well as the adjacent fore arc and trench are not included in this study.

A lack of crustal stress measurements exists in our data set for a large area in the States of Oaxaca and Guerrero (Fig. 6); this area corresponds tectonically to the overriding plate above the subduction zone between the Cocos and North American Plates. The stress field at the depth of the subduction zone, on the other hand, is well constrained from focal-mechanism solutions. A segment of the subduction zone, approximately 100 km wide and parallel to the trench, is characterized by thrust faulting, whereas its deeper and more landward part is affected by normal faulting in the lower plate (Lefevre and McNally, 1985; Suárez and others, 1990).

The accretionary wedge adjacent to the Middle America trench is being shortened (Moore and others, 1982; Moore and Shipley, 1988); the maximum principal stress in that area is most likely perpendicular to the Middle America trench.

SOURCES OF STRESS

Intraplate stress regimes

The intraplate stress regimes of our study area are characterized by horizontal deviatoric tension (in the sense of Turcotte and Schubert, 1982); they include the Basin and Range, the trans-Mexican volcanic belt, the Gulf Coast stress province in southern Veracruz and Tabasco States, and the region of the Guatemala City and Ipala grabens south of the Motagua fault, as well as the Guatemalan volcanic arc (Fig. 7). We believe that this horizontal deviatoric tension is caused in all our stress regimes, with the exception of the southern Basin and Range–Rio Grande stress province(?), by flexural stresses.

These flexural or bending stresses have various origins. In the case of the trans-Mexican and Guatemalan volcanic arcs, which are characterized by high relief and crestal normal faults striking parallel to the axes of the arcs, the bending stresses are due to the volcanic topographic loads, and bending occurs around a subhorizontal axis. In the case of the Ipala and Guatemala City grabens, bending occurs around a vertical axis and is a geometrical adjustment to the horizontal relative movement along an arcuate transform fault. In the case of the extension perpendicular to the continental margin of the Gulf of Mexico at a shallow crustal level in western Tabasco and southern Veracruz, the flexural stresses are most probably due to sedimentary surface loads acting on the continental shelf.

The observed stress and deformation in the central and eastern part of the trans-Mexican volcanic belt can be explained by the superposition of two sources of stress. Local intraplate bending stresses related to the high elevation of the trans-Mexican volcanic belt probably cause the observed crestal normal faults, which strike parallel to the axis of the volcanic arc. The observed left-lateral strike-slip component, on the other hand, can be explained by far-field stresses caused by loads applied at the boundary between the Cocos and North American Plates. The orientation of the normal faults with respect to the direction of the vector of relative plate motion, as well as the coincidence of the latter with the direction of S_H in the central and eastern part of the trans-Mexican volcanic belt favor such an interpretation.

An understanding of the horizontal stress components of deviatoric tension in the southern Basin and Range–Rio Grande rift stress province requires a physical model of the Basin and Range tectonics. The stresses are related to distributed extension of the lithosphere in an area characterized by high regional elevation and heat flow, and are most likely controlled by vertical boundary forces at the base of the lithospheric plate (Lachenbruch and Sass, 1978; Eaton, 1982). Following Sonder and others (1987) and Wernicke and others (1987), the extension is a result of gravitationally driven thinning of a lithosphere previously thickened in late Mesozoic and earliest Tertiary times. In other models, the crust is not thickened, but only thermally elevated.

As noted before (Suter, 1987), there is no correlation between the direction of S_H and the direction of absolute plate motion in the above-mentioned stress provinces. This is further evidence that the stress distribution in the intraplate stress provinces of our study area is not controlled by plate-driving forces.

Interplate stress regimes

The interplate stress regimes of our study area include the Gulf of California and the Campeche-Chiapas-Petén stress provinces. Both are of strike-slip type, with a tensional and a compressional horizontal deviatoric stress component, and are related to transform plate boundaries. However, these two stress provinces cover not only the plate boundary, but also the adjacent intraplate area: in the case of the Gulf of California, on either side of the plate boundary; and in the case of the Motagua-Polochic transform zone, only on its northern side.

The orientations of S_H with respect to the Motagua-Polochic and Gulf of California plate boundaries are 30 to 60° counterclockwise and 35° clockwise, respectively. These are favorable orientations for left-lateral and right-lateral strike-slip, respectively; experimental results show that a minimum in strength occurs when the maximum principal compressive stress is inclined at 30°, or somewhat more, to a plane of weakness (Patterson, 1978). It seems therefore that the stress field of these two stress provinces is controlled by displacement loads at the plate boundary. Furthermore, our results suggest that the configuration at the San Andreas fault in central California, where S_H is nearly normal to the fault (Mount and Suppe, 1987; Zoback and others, 1987), is exceptional.

SUMMARY AND OUTLOOK

We have presented and interpreted new and compiled stress measurements that constrain the state of stress in the uppermost 30 km of the lithosphere in Mexico and western Central America. We also have reviewed the present state of knowledge of active deformation in this area.

Northwestern Mexico, including the Gulf of California, the Baja California Peninsula, and the part of mainland Mexico that borders the Gulf of California, is characterized by active normal and right-lateral strike-slip faulting. The displacement is concentrated at the Gulf of California plate boundary, which is being deformed by right-lateral transform faults connecting short spreading centers. Stress measurements are derived from focal mechanisms, the inversion of fault striation data (azimuth of maximum horizontal far-field stress S_H: 165 ± 10°), and new breakout data from 5 wells (depth <4.2 km, S_H azimuth: 166 ± 13°). These stress data indicate independently that S_H is oriented 35° clockwise from the direction of relative plate motion, which is constrained by focal-plane solutions of strike-slip events (centroid depth <20 km) in the Gulf of California (131 ± 8°), geodetic measurements in the central part of the Gulf (134°), and the general strike of transform segments of the Gulf of California plate boundary (130 ± 5°).

In northeastern Mexico, available stress measurements are limited to focal mechanisms of microseismicity along the Pitaycachi fault in Sonora and borehole elongations (depth <4.9 km), which have in general a W-E to NW-SE trend. These measurements, as well as the existence of N- to NNE-trending active normal faults, suggest that this low-seismicity intraplate area is under extension with $S_v > S_N > S_E$ and forms part of the southern Basin and Range stress province.

For central Mexico we define the trans-Mexican volcanic belt stress province, which is characterized by seismically active W-E-trending normal faults. Focal-plane solutions, the alignment of Quaternary cinder cones, and striation analysis of active faults indicate a stress field with $S_v > S_{ENE} > S_{NNW}$.

In southern Mexico and western Central America, we delimit two stress provinces: (1) The S_H azimuth from six measurements in the coastal plain of the Gulf of Mexico in southern Veracruz and western Tabasco States, based on volcanic alignments and borehole elongations, is 110 ± 7° and parallel to the continental margin. Active normal faulting along the continental slope suggests a stress field with $S_v > S_{NNE} > S_{ESE}$. Thrust-fault focal mechanisms from the same area at a depth of about 20 km, on the other hand, indicate a stress field with $S_{NNE} > S_{ESE} > S_v$, which implies vertical polarity changes of the principal stresses. (2) Borehole elongations from 23 wells (depth <6.4 km) in the seismically active Chiapas-Petén fold belt and in the adjacent foreland indicate an S_H azimuth of 28 ± 7° directed perpendicularly to the fold axes of the belt. A subset of 12 wells closest to the left-lateral Motagua-Polochic transform system has an S_H azimuth of 33 ± 4°, which is oriented at an angle of 30 to 60° counterclockwise from the plate boundary. These borehole elongations, in conjunction with left-lateral strike-slip focal mechanisms, suggest a stress field with $S_{NE} > S_v > S_{SE}$.

The intraplate stress regimes of our study area are characterized by horizontal deviatoric tension; they include the Basin and Range, the trans-Mexican volcanic belt, the Gulf Coast stress province in southern Veracruz and Tabasco States, and the region of the Guatemala City and Ipala grabens south of the Motagua fault, as well as the Guatemalan volcanic arc. We believe that this horizontal deviatoric tension is caused in all our stress regimes, with exception of the southern Basin and Range stress province, by local flexural stresses of various origins. In the case of the trans-Mexican and Guatemalan volcanic arcs, which are characterized by high relief and crestal normal faults trending parallel to the axes of the arcs, the bending stresses are due to the volcanic topographic loads, and bending occurs around a subhorizontal axis. In the case of the Ipala and Guatemala City grabens, bending occurs around a vertical axis and is a geometrical adjustment to the horizontal relative movement along an arcuate transform fault. In the case of the extension perpendicular to the continental margin of the Gulf of Mexico at a shallow crustal level in western Tabasco and southern Veracruz, the flexural stresses are most probably caused by sedimentary surface loads acting on the continental shelf.

The interplate stress regimes of our study area include the Gulf of California and the Campeche-Chiapas-Petén stress provinces, the stress distributions of which seem to be controlled by displacement loads at the Gulf of California and Motagua-Polochic plate boundaries, respectively. The orientation of S_H are 30 to 60° counterclockwise from the left-lateral Motagua-Polochic, and 35° clockwise from the right-lateral Gulf of California transform systems. The two stress provinces affect not only the plate boundary, but also the adjacent intraplate area; in the case of the Gulf of California, on either side of the plate boundary; but in the case of the Motagua-Polochic transform zone, only on its northern side.

Until three years ago, there were no published stress data for the intraplate area of Mexico and western Central America, with the exception of some focal mechanisms. Today, we have a database including close to 200 sites, based on various measurement techniques. Where do we go from here? We see three ways to proceed. (1) The homogeneity and density of the site distribution can certainly still be improved. (2) Little has been done to analyze stress measurements on a regional scale, and some algorithms may have to be developed specifically for that purpose: Stress provinces are clusters formed by two linearly dependent variables and one circular independent variable and could be defined algorithmically. Stress trajectories can be modeled as lines of a smooth unit vector field either statistically (Mendoza, 1986; Hansen and Mount, 1990) or by fitting a spline to the stress directions given at the measurement sites (Westerveld, 1989). We may also try to be more quantitative about the determination of volcanic alignments; for example, by searching for the least variance in the output of directional filters (Kass and Witkin, 1985) applied to a given point cluster, or by applying other algorithms developed in image analysis for the determination of linear point arrays (i.e., Wadge and Cross, 1988). (3) It will be interesting to compare our set of real data with synthetic data resulting from the application of numerical modeling techniques; a match between the real and synthetic data will show us possible load configurations responsible for the stress distribution.

REFERENCES CITED

Aldrich, M. J., Jr., Chapin, C. E., and Laughlin, A. W., 1986, Stress history and tectonic development of the Rio Grande rift, New Mexico: Journal of Geophysical Research, v. 91, p. 6199–6211.

Allan, J. F., 1986, Geology of the northern Colima and Zacoalco grabens, southwest Mexico; Late Cenozoic rifting in the Mexican volcanic belt: Geological Society of America Bulletin, v. 97, p. 473–485.

Allan, J. F., Nelson, S. A., Luhr, J. F., Carmichael, I.S.E., and Wopat, M., 1991, Pliocene-Recent rifting in SW Mexico and associated alkaline volcanism, in Dauphin, J. P., and Simoneit, B.R.T., eds., The Gulf and Peninsular province of the Californias: American Association of Petroleum Geologists Memoir 47, p. 425–446.

Allen, C. R., Silver, L. T., and Stehli, F. G., 1960, Agua Blanca fault; A major transverse structure of northern Baja California, Mexico: Geological Society of America Bulletin, v. 71, p. 457–482.

Angelier, J., Colletta, B., Chorowicz, L., Ortlieb, L., and Rangin, C., 1981, Fault tectonics of the Baja California Peninsula and the opening of the Sea of Cortez, Mexico: Journal of Structural Geology, v. 3, p. 347–357.

Aranda, J. J., Aranda, J. M., and Nieto, A. F., 1990, Consideraciones acerca de la evolución tectónica durante el Cenozoico de la Sierra de Guanajuato y la porción meridional de la meseta central, México: Universidad Nacional Autónoma de México, Instituto de Geología, Revista, v. 8, p. 33–46.

Astiz, L., 1980, Sismicidad en Acambay, Estado de México; El temblor del 22 de febrero de 1979 [B.Sc. thesis]: México, D.F., Universidad Nacional Autónoma de México, 130 p.

Bischoff, J. L., and Niemitz, J. W., 1980, Bathymetric maps of the Gulf of California: U.S. Geological Survey Miscellaneous Investigation Series Map I-1244, scale 1:298,880.

Bloomfield, K., 1975, A late Quaternary monogenetic volcano field in central Mexico: Geologische Rundschau, v. 64, p. 476–497.

Buffler, R. T., Shaub, F. J., Watkins, J. S., and Worzel, J. L., 1979, Anatomy of the Mexican ridges, southwestern Gulf of Mexico: American Association of Petroleum Geologists Memoir 29, p. 319–327.

Burkart, B., Deaton, B. C., Dengo, C., and Moreno, G., 1987, Tectonic wedges and offset Laramide structures along the Polochic fault of Guatemala and Chiapas, Mexico; Reaffirmation of large Neogene displacement: Tectonics, v. 6, p. 411–422.

Cantagrel, J.-M., and Robin, C., 1979, K-Ar dating on eastern Mexican volcanic rocks; Relations between the andesitic and the alkaline provinces: Journal of Volcanology and Geothermal Research, v. 5, p. 99–114.

Centeno García, E., 1988, Evolución estructural de la falla Oaxaca durante el Cenozoico [M.Sc. thesis]: México, D.F., Universidad Nacional Autónoma de México, 156 p.

Centeno García, E., Ortega Gutiérrez, F., and Corona Esquivel, R., 1990, Oaxaca fault, Cenozoic reactivation of the suture between the Zapoteco and Cuicateco terranes, southern Mexico: Geological Society of America Abstracts with Programs, v. 22, p. 13.

Charleston, S., 1981, A summary of the structural geology and tectonics of the state of Coahuila, Mexico, in Lower Cretaceous stratigraphy and structure, northern Mexico: West Texas Geological Society Publication 81-74, p. 38–36.

Colletta, B., and Ortlieb, L., 1984, Deformations of middle and late Pleistocene deltaic deposits at the mouth of the Rio Colorado, northwestern Gulf of California, in Malpica-Cruz, V., and others, eds., Neotectonics and sea level variations in the Gulf of California area: Universidad Nacional Autónoma de México, Instituto de Geología, Symposium Proceedings, p. 31–53.

Couch, R., and Woodcock, S., 1981, Gravity and structure of the continental margins of southwestern Mexico and northwestern Guatemala: Journal of Geophysical Research, v. 86, p. 1829–1840.

DeMets, C., Gordon, R. G., Stein, S., and Argus, D. F., 1987, A revised estimate of Pacific–North America motion and implications for western North America plate boundary zone tectonics: Geophysical Research Letters, v. 14, p. 911–914.

Eaton, G. P., 1982, The Basin and Range province; Origin and tectonic significance: Annual Review of Earth and Planetary Sciences, v. 10, p. 409–440.

Engdahl, R. E., and Rinehart, W. A., 1988, Seismicity map of North America: Boulder, Colorado, Geological Society of America, Decade of North American Geology CSM–004, 4 sheets, scale 1:5,000,000.

Fries, C., Jr., Ross, C. S., and Obregón, A., 1977, Mezcla de vidrios en los derrames cineríticos Las Americas de la región de El Oro-Tlapujahua, Estados de México y Michoacán, parte centro-meridional de México: Universidad Nacional Autónoma de México, Instituto de Geología Boletín 70, 84 p.

Goff, J. A., Bergman, E. A., and Solomon, S. C., 1987, Earthquake source mechanisms and transform fault tectonics in the Gulf of California: Journal of Geophysical Research, v. 92, p. 10485–10510.

Guzmán Speziale, M., 1985, The triple junction of the North America, Cocos, and Caribbean plates; Seismicity and tectonics [M.A. thesis]: Austin, University of Texas, 66 p.

Guzmán Speziale, M., Pennington, W. D., and Matumoto, T., 1989, The triple junction of the North America, Cocos, and Caribbean plates; Seismicity and tectonics: Tectonics, v. 8, p. 981–997.

Hafner, W., 1951, Stress distributions and faulting: Geological Society of America Bulletin, v. 62, p. 373–398.

Hansen, K. M. and Mount, V. S., 1990, Smoothing and extrapolation of crustal stress orientation measurements: Journal of Geophysical Research, v. 95, p. 1155–1165.

Jaeger, J. C., and Cook, N.G.W., 1976, Fundamentals of rock mechanics, 2nd ed.: London, Chapman and Hall, 585 p.

Johnson, C. A., 1987, A study of neotectonics in central Mexico from Landsat thematic mapper imagery [M.S. thesis]: Coral Gables, Florida, University of Miami, 112 p.

Jones, L. M., 1988, Focal mechanisms and the state of stress on the San Andreas Fault in southern California: Journal of Geophysical Research, v. 93, p. 8869–8891.

Kass, M., and Witkin, A., 1985, Analyzing oriented patterns; Paper presented at International Joint Conference for Artificial Intelligence: University of California at Los Angeles, 9 p.

Kasser, M., and 7 others, 1987, Geodetic measurements of plate motions across the central Gulf of California, 1982–1986: Geophysical Research Letters, v. 74, p. 5–8.

Lachenbruch, A. H., and Sass, J. H., 1978, Models of extending lithosphere and heat flow in the Basin and Range province, in Smith, R. B., and Eaton, G. P., eds., Cenozoic tectonics and regional geophysics of the western Cordillera: Geological Society of America Memoir 152, p. 209–250.

Langer, C. J., and Bollinger, G. A., 1979, Secondary faulting near the terminus of a seismogenic strike-slip fault; Aftershocks of the 1976 Guatemala earthquake: Seismological Society of America Bulletin, v. 69, p. 427–444.

Lefevre, L. V., and McNally, K. C., 1985, Stress distribution and subduction of aseismic ridges in the Middle America subduction zone: Journal of Geophysical Research, v. 90, p. 4495–4510.

Legg, M. R., Luyendyk, B. P., Mammerickx, J., de Moustier, C., and Tyce, R. C., 1989, Sea beam survey of an active strike-slip fault; The San Clemente fault in the California continental borderland: Journal of Geophysical Research, v. 94, p. 1727–1744.

Lomnitz, C., Mooser, F., Allen, C. R., Brune, J. N., and Thatcher, W., 1970, Seismicity and tectonics of the northern Gulf of California region, Mexico: Geofísica Internacional, v. 10, p. 37–48.

Mann, P., and Burke, K., 1984, Cenozoic rift formation in the northern Caribbean: Geology, v. 12, p. 732–736.

Mastin, L., Muller, B., and Zoback, M. L., 1989, World stress map: EOS, American Geophysical Union Transactions, v. 70, p. 1520–1521.

McKee, J. W., Jones, N. W., and Long, L. E., 1984, History of recurrent activity along a major fault in northeastern Mexico: Geology, v. 12, p. 103–107.

—— , 1990, Stratigraphy and provenance of strata along the San Marcos fault, central Coahuila, Mexico: Geological Society of America Bulletin, v. 102, p. 593–614.

Mendoza, C. E., 1986, Smoothing unit vector fields: Mathematical Geology, v. 18, p. 307–322.

Meneses Rocha, J., 1985, Tectonic evolution of the strike-slip province of Chiapas, Mexico [M.Sc. thesis]: Austin, University of Texas, 315 p.

Minster, J. B., and Jordan, T. H., 1978, Present-day plate motions: Journal of Geophysical Research, v. 83, p. 5331–5354.

—— , 1987, Vector constraints on western U.S. deformation from space geodesy, neotectonics, and plate motions: Journal of Geophysical Research, v. 92, p. 4798–4804.

Molnar, P., 1973, Fault plane solutions of earthquakes and direction of motion in the Gulf of California and on the Rivera fracture zone: Geological Society of America Bulletin, v. 84, p. 1651–1658.

Mooser, F., 1969, The Mexican volcanic belt; Structure and development, in Proceedings, Pan-American Symposium on the Upper Mantle, Group 2, Upper Mantle, Petrology, and Tectonics, v. 2: México, D.F., Instituto de Geofísica, UNAM, p. 15–22.

Moore, G. F., and Shipley, T. H., 1988, Mechanisms of sediment accretion in the Middle America trench off Mexico: Journal of Geophysical Research, v. 93, p. 8911–8927.

Moore, J. C., and 5 others, 1982, Geology and tectonic evolution of a juvenile accretionary terrane along a truncated convergent margin; Synthesis of results from Leg 66 of the Deep Sea Drilling Project, southern Mexico: Geological Society of America Bulletin, v. 93, p. 847–861.

Mount, V. S., and Suppe, J., 1987, State of stress near the San Andreas fault; Implications for wrench tectonics: Geology, v. 15, p. 1143–1146.

Muehlberger, W. R., Belcher, R. C., and Goetz, L. K., 1978, Quaternary faulting in trans-Pecos Texas: Geology, v. 6, p. 337–340.

Nakamura, K., 1977, Volcanoes as possible indicators of tectonic stress orientation; Principle and proposal: Journal of Volcanology and Geothermal Research, v. 2, p. 1–16.

Nakamura, K., Jacob, K. H., and Davies, J. N., 1977, Volcanoes as possible indicators of tectonic stress orientation; Aleutians and Alaska: Pure and Applied Geophysics, v. 115, p. 87–112.

Natali, S. G., and Sbar, M. L., 1982, Seismicity in the epicentral region of the 1887 northeastern Sonoran earthquake, Mexico: Seismological Society of America Bulletin, v. 72, p. 181–196.

Nixon, G. T., Demant, A., Armstrong, R. L., and Harakal, J. E., 1987, K-Ar and geologic data bearing on the age and evolution of the trans-Mexican volcanic belt: Geofísica Internacional, v. 26, p. 109–158.

Patterson, M. S., 1978, Experimental faulting: New York, Springer-Verlag, 254 p.

Pearthree, P. A., Bull, W. B., and Wallace, T. C., 1990, Geomorphology and Quaternary geology of the Pitaycachi fault, northeastern Sonora, Mexico: Arizona Geological Survey Special Paper 7, p. 124–135.

Pew, E., 1982, Seismic structural analysis of deformation in the southern Mexican Ridges [M.S. thesis]: Austin, University of Texas, 102 p.

Plafker, G., 1976, Tectonic aspects of the Guatemala earthquake of 4 February 1976: Science, v. 193, p. 1201–1208.

Rebollar, C. J., and Reichle, M. S., 1987, Analysis of the seismicity detected in 1982–1984 in the northern Peninsular Ranges of Baja California: Seismological Society of America Bulletin, v. 77, p. 173–183.

Rebollar, C. J., Reyes, A., and Reichle, M. S., 1982, Estudio del enjambre de San Quintín, Baja California, México, ocurrido durante 1975: Geofísica Internacional v. 21, p. 331–358.

Sánchez Barreda, L. A., 1981, Geologic evolution of the continental margin of the Gulf of Tehuantepec in southeastern Mexico [Ph.D. thesis]: Austin, University of Texas, 191 p.

Sanchez Montes de Oca, R., 1979, Geología petrolera de la Sierra de Chiapas: Asociación Mexicana de Geólogos Petroleros Boletín, v. 31, p. 67–97.

Savage, W. Z., and Swolfs, H. S., 1986, Tectonic and gravitational stress in long symmetric ridges and valleys: Journal of Geophysical Research, v. 91, p. 3677–3685.

Schafer, J. N., 1980, A practical method of well evaluation and acreage development for the naturally fractured Austin Chalk Formation: The Log Analyst, v. 21, p. 10–23.

Schwartz, D. P., Cluff, L. S., and Donnelly, T. W., 1979, Quaternary faulting along the Caribbean–North American plate boundary in Central America: Tectonophysics, v. 52, p. 431–445.

Seager, W. R., and Morgan, P., 1979, Rio Grande rift in southern New Mexico, west Texas, and northern Chihuahua, in Riecker, R. E., ed., Rio Grande rift, tectonics and magmatism: American Geophysical Union, p. 87–106.

Sonder, L. J., England, P. C., Wernicke, B. P., and Christiansen, R. L., 1987, A phyical model for Cenozoic extension of western North America: Geological Society of London Special Publication 28, p. 187–201.

Spencer, J. E., and Normark, W. R., 1979, Tosco-Abreojos fault zone; A Neogene transform plate boundary within the Pacific margin of southern Baja California, Mexico: Geology, v. 7, p. 554–557.

Stock, J. M., and Hodges, K. V., 1990, Miocene to Recent structural development of an extensional accommodation zone, northeastern Baja California, Mexico: Journal of Structural Geology v. 12, p. 315–328.

Stover, C. W., 1986, Seismicity map of the conterminous United States and adjacent areas, 1975–1984: U.S. Geological Survey Map GP-984, scale 1:5,000,000.

Suárez Reynoso, G., and Ponce, L., 1986, Intraplate seismicity and crustal deformation in central Mexico [abs.]: EOS, American Geophysical Union Transactions, v. 67, p. 1114.

Suárez, R. G., Monfret, T., Wittlinger, G., and David, C., 1990, Geometry of subduction and depth of the seismogenic zone in the Guerrero gap, Mexico: Nature, v. 345, p. 336–338.

Suárez Vidal, F., Armijo, R., Morgan, G., Bodin, P., and Gastil, R. G., 1990, Framework of recent and active faulting in northern Baja California: American Association of Petroleum Geologists Memoir (in press).

Suter, M., 1987, Orientational data on the state of stress in northeastern Mexico as inferred from stress-induced borehole elongations: Journal of Geophysical Research, v. 92, p. 2617–2626 (correction in v. 93, p. 8085).

—— , 1990a, State of stress in the Gulf of California and along the San Andreas fault; a comparison: Geological Society of America Abstracts with Programs, v. 22, p. 87–88.

—— , 1990b, Hoja Tamazunchale 14Q-e(5) con Geología de la Hoja Tamazunchale, Estados de Hidalgo, Querétaro y San Luis Potosí: Universidad Nacional Autónoma de México, Instituto de Geología, Carta geológica de México, serie de 1:100,000 (geologic map sheet, structural sections, and explanations with extended English abstract).

—— , 1990c, State of stress in southeastern Mexico and western Central America (abs.): EOS, American Geophysical Union Transactions, v. 71, p. 461.

Suter, M., Quintero, O., and Johnson, C., 1990, Active faults, and state of stress in the central part of the trans-Mexican volcanic belt; Part 1, The Venta de Bravo fault: Journal of Geophysical Research (in revision).

Tristan, M., 1986, Estratigrafía y tectónica del graben de Villa de Reyes en los Estados de San Luis Potosí y Guanajuato, México: Universidad Autónoma de San Luis Potosí, Instituto de Geología, Folleto técnico 107, 91 p.

Turcotte, D. L., and Schubert, G., 1982, Geodynamics; Applications of continuum physics to geological problems: New York, John Wiley and Sons, 450 p.

Urbina, F., and Camacho, H., 1913, La zona megasísmica Acamby-Tixmadejé, Estado de México, conmovida el 19 de noviembre de 1912: Instituto Geológico de México Boletín 32, 125 p.

Wadge, G., and Cross, A., 1988, Quantitative methods for detecting aligned points; An application to the volcanic vents of the Michoacán-Guanajuato volcanic field, Mexico: Geology, v. 16, p. 815–818.

Weldon, R., and Humphreys, E., 1986, A kinematic model of southern California: Tectonics, v. 5, p. 33–48.

Wernicke, B. P., Christiansen, R. L., England, P. C., and Sonder, L. J., 1987, Tectonomagmatic evolution of Cenozoic extension in the North American Cordillera: Geological Society of London Special Publication 28, p. 203–221.

Westerveld, W. B., 1989, Plotting of closed field lines with an application to coherently excited atoms: Computers in Physics, v. 3, no. 5, p. 74–77.

Wong, V. O., Legg, M. R., and Suárez, F., 1987, Sismicidad y tectónica de la margen continental del sur de California y Baja California norte: Geofísica Internacional, v. 26, p. 459–478.

Zoback, M. D., and 12 others, 1987, New evidence on the state of stress of the San Andreas fault system: Science, v. 238, p. 1105–1111.

Zoback, M. L., and Zoback, M. D., 1980, State of stress in the conterminous United States: Journal of Geophysical Research, v. 85, p. 6113–6156.

—— , 1989, Tectonic stress field of the continental U.S., *in* Pakiser, L., and Mooney, W., eds., Geophysical framework of the continental United States: Geological Society of America Memoir 172, p. 523–539.

Zoback, M. L., and 7 others, 1991, Stress map of North America: Boulder, Colorado, Geological Society of America CSM-005, 4 sheets, scale 1:5,000,000.

Zoback, M. L., and others, 1989, Global patterns of tectonic stress: Nature, v. 341, p. 291–298.

MANUSCRIPT ACCEPTED BY THE SOCIETY APRIL 3, 1990

ACKNOWLEDGMENTS

Financial and logistical support by the National University of Mexico, grant SNI 881161 from the Mexican Secretariat of Education (SEP), and grant P221 CCON 892316 from the Mexican National Council for Science and Technology (CONACYT) are acknowledged. The presentation of this paper at a symposium about the World Stress Map Project of the International Lithosphere Program, held in 1989 at the 28th International Geological Congress in Washington, D.C., was financially supported by the 28th International Geological Congress and UNESCO.

Gerardo Suárez compiled the focal-mechanism solutions. Bill Muehlberger called my attention to the possible relation between the stress field at the sites CO-09 and CO-14, and the La Babia and San Marcos faults. Jorge Aranda, Francisco Suárez, and Joann Stock provided copies of manuscripts prior to publication, and Jorge Aranda also provided information concerning late Cenozoic volcanic fields to the north of the trans-Mexican volcanic belt. I also thank Mark Zoback for his invitation to write this contribution and John Oldow for his review.

Printed in U.S.A.

The Geology of North America
Decade Map Volume 1
1991

Chapter 23

Heat-flow patterns of the North American continent; A discussion of the Geothermal Map of North America

David D. Blackwell, John L. Steele, and Larry S. Carter
Department of Geological Sciences, Southern Methodist University, Dallas, Texas 75275

INTRODUCTION

The objective in creating a Geothermal Map of North America (Blackwell and Steele, 1991) was to show as accurately and completely as possible the state of knowledge of the geothermal field of the continent in all its variations. As a consequence, the types of information shown are combined in a way that is different from existing maps. The only other continent-wide map representations of aspects of temperature, geothermal gradient, or heat-flow data are the AAPG/USGS *Subsurface Temperature Map of North America* (1976b) and the *Geothermal Gradient Map of North America* (1976a) both at a 1:5,000,000 scale.

Geothermal gradient maps of the United States have been prepared by Kron and Stix (1982) and Nathenson and Guffanti (1988). However, because both temperature and interval gradient can be calculated if the heat flow and lithology are known, the basic quantity that is needed is heat flow. Therefore, these gradient maps have limited usefulness outside the range of direct observation. Numerous countrywide heat-flow maps have been described for North America. However, the knowledge of the thermal regime of North America has increased many fold in the last 10 to 15 years because of academic studies and because of the exploitation of geothermal energy and associated resource studies. Thus, it seemed to be an appropriate undertaking at this time to compile a continent-wide representation of the thermal field using this old and new knowledge. The preparation and publication of other continent-wide maps on a new digital base at the 1:5,000,000 scale is also part of the Decade of North American Geology (DNAG) program. The common scale will allow direct areal comparison of several types of geological and geophysical information; this capability was also an important consideration in the decision to compile the geothermal map.

The basic background types of data shown on the new map are the heat-flow sites with coded heat-flow values, and a color pattern based on the heat flow contoured at 10-mWm^{-2} intervals for the ocean and continental areas. The areas of the continental shelves are generally not included in the contouring because there are few heat-flow data for those regions. In addition to the heat-flow sites and color pattern, the locations of Quaternary volcanoes and major geothermal systems are indicated. Geothermal areas are shown with four different symbols. The areas are first subdivided based on whether the maximum temperature is above or below 150°C, on the basis of temperatures either measured or inferred from geochemical studies. The second category is whether or not there is information in the literature documenting heat-flow/geothermal gradient distributions based on the results of shallow exploration drilling and/or deep production drilling. A third major type of information shown involves areas with peculiar thermal characteristics, such as major areal effects on the conductive heat flow from ground-water flow. Also included in this group are contour and shade indicating the temperature on the major thermal aquifer in the north-central midcontinent region of the United States (the Dakota Sandstone), and the depth to the top of the geopressured zone in the Gulf Coast region.

References to the major sources of the various data are given in the map legend, and more detailed information is given in this paper. Compilers of the heat-flow data for various areas are also listed. The object of this chapter is to give some of the background information on the map, to describe some of the decisions made in the compilation and contouring, and to summarize in a brief form some conclusions that can be drawn from the resulting map. More detailed discussions of Canada (Jessop, this volume), the Canadian Cordillera (Lewis, this volume), the southwestern United States (Reiter and others, this volume), and the midcontinent region of the United States (Gosnold, this volume) can be found in the accompanying chapters in this volume.

The basic heat flow and ancillary data are available in compiled form. The point data base for the continent is available as one of the data sets (Blackwell and others, 1989) on the *Geophysics of North America* CD-ROM published by the National Oceanographic and Atmospheric Administration. The data for eastern Canada as compiled by A. M. Jessop, for the Canadian

---

Blackwell, D. D., Steele, J. L., and Carter, L. S., 1991, Heat-flow patterns of the North American continent; A discussion of the *Geothermal Map of North America*, in Slemmons, D. B., Engdahl, E. R., Zoback, M. D., and Blackwell, D. D., eds., Neotectonics of North America: Boulder, Colorado, Geological Society of America, Decade Map Volume 1.

423

Cordillera as compiled by Trevor Lewis, and for the remainder of the continent as compiled by the authors of this paper are included on that disk. The data included in the compilation for the United States are quite inclusive for individual heat-flow sites and include some data not in the original publications. The ideal data-base content is shown in Table 1. There is extensive information listed in addition to heat-flow value and location. A complete reference list for the United States heat-flow data base is contained in Blackwell and others (1988); a copy of the list can be obtained from the authors on request. A subset of the information for each site is also contained in a worldwide heat-flow compilation by Pollack and others (1991). The Pollack and others (1991) compilation also includes updated heat flow for the oceans. The compilation of oceanic heat-flow values by Jessop and others (1976), as updated by D. S. Chapman (personal communication, 1980), was used as the basic data base for the oceans and was updated by subsequent published data, most notably the addition of more continent-like data based on heat-flow measurements in the holes drilled by the Deep Sea Drilling Program (Hyndman and others, 1987).

A very simplified index map for the continent-wide data is shown in Figure 1. The heat-flow regions are taken from the 1:5,000,000-scale map, with some modifications as described below. The basis for the heat-flow contours in the oceans, not the values, is also shown on Figure 1. Reference to this map should be made as the discussion proceeds. The maps in this paper do not show the individual data points. These are shown on the 1:5,000,000 scale map. The most recent summaries of the heat flow in each area show most of the points (Sass and others, 1981, for Alaska; Jessop, this volume, for Canada; Morgan and Gosnold, 1989, for the United States; Ziagos and others, 1985, for Mexico).

OCEAN HEAT FLOW

Pacific Ocean

Because of the general correspondence of oceanic heat flow to the age of the oceanic lithosphere, and because of a general lack of a large heat-flow data set for the eastern part of the Pacific Ocean, the heat-flow contours shown in that area are based on the age of the lithosphere inferred from the magnetic anomaly lineations (see Fig. 1 for a division of the oceanic areas according to the contouring basis). The positions of magnetic anomalies of appropriate ages were taken from the *Plate-tectonic Map of the circum-Pacific Basin* (Drummond, 1981) and from the *Magnetic Anomaly Map of North America* (Committee . . . , 1987). The oceanic lithosphere age was determined at heat-flow intervals of 10 mWm^{-2} from 60 to 100 mWm^{-2} and at 120 and 150 mWm^{-2} from the curve of Sclater and others (1980, Fig. 4) based on a summary of the relation of heat flow to lithospheric age for the oceans.

Near the ridge crests, convection is the dominant heat-transfer mechanism, so within some distance of the ridge crest, the observed heat flow is usually significantly below the value predicted by the age-versus-heat-flow relation. Thus, for much of the area of the Pacific Ocean shown on the map, the actual measured heat-flow values are lower than indicated by the contouring. This situation seems dominant in the Cocos Plate where the measured heat-flow values over almost all the area shown on the map are significantly below those predicted by the age-versus-heat-flow relation. The general pattern of low heat flow in this area has recently been documented by new data discussed by Prol-Ledesma and others (1989). Along the ridge crests, of course, there are numerous high- and intermediate-temperature geothermal systems. These have proved to be so numerous that it did not seem worthwhile at this time to indicate individual areas where the systems have been located. The complexity of heat transfer near the ridge crests has been demonstrated in the map area for East Pacific Rise and the Gulf of California as far north as the Guaymas Basin (e.g., Lonsdale and Becker, 1985) and for the Juan de Fuca Ridge (Davis and Lister, 1977).

The area of the Juan de Fuca Plate is unique in that it is covered by sediment almost as quickly as the lithosphere is generated; therefore, hot fluid outflow may not dominate the heat-loss mechanism as much as it dominates in other areas of young lithosphere. Moran and Lister (1987) presented a detailed study of heat flow versus age for the Juan de Fuca Plate. This area is unusual because of the thick layer of sediment very near the ridge axis. Their locations are multiple penetration sites that have been carefully studied. Their results show heat-flow values for sites only 150 km or more from the ridge crest that are close to the predicted heat flow for the age of the crust after corrections for sedimentation and other effects. This combination of high heat flow and thick sediment cover results in high temperatures for the top of the ocean basement. The high temperature of top of the basement as it is subducted beneath the North American Plate is the primary reason for the low seismicity of the subducting lithosphere, not the thinness of the plate (Blackwell and others, 1982).

Data on the continental shelves are not common, but where present have been contoured with the land measurements, so the contours are based on observed data. Therefore, there is generally a gap between the land/shelf data and the data in the oceans. There are a number of important scientific questions, such as the vertical distribution of heat production in the lithosphere, that could be addressed by data from the continental shelves. Perhaps such studies will be carried out in the future.

Atlantic Ocean

In the Atlantic Ocean, as in the Pacific Ocean, the heat-flow data have been contoured based on the age of the lithosphere. As a result, most of the Atlantic Ocean off the eastern United States is shown to have a heat flow between 40 and 50 mWm^{-2}, while further to the east, the theoretical values are between 50 and 60 mWm^{-2}. The age corresponding to the 50-mWm^{-2} heat-flow contour is about 63 Ma. The location of this age of lithosphere was taken from a map by Vogt and Einwich (1979) and the

Magnetic Anomaly Map of North America (Committee . . . , 1987).

In general, in the area of the Atlantic Ocean southeast of the Labrador Sea shown on the map, the observed data on this relatively old, thickly sedimented lithosphere are in close agreement with the predicted heat flow. A similar agreement in such settings has been illustrated by Sclater and others (1980) to be typical of all ocean basins.

An anomaly may be associated with the Bermuda Rise, which may be part of a hot-spot track. Detailed studies in that area by Detrick and others (1986) document a slightly higher background heat flow in the vicinity of the Bermuda Rise. This area is characterized by a region of over 50-mWm^{-2} heat-flow values in an area of below 50-mWm^{-2} heat flow based on the age-versus-heat-flow relation.

In the North Atlantic the heat-flow data have also been contoured using the age inferred from interpretation of magnetic anomalies. The magnetic pattern is quite complicated and has been discussed in detail by Bott (1983) and by Nunns (1983). Their interpretations, as well as the magnetic anomalies shown on

TABLE 1. DESCRIPTION OF HEAT-FLOW DATA BASE IN BLACKWELL AND OTHERS (1989)

A. Description of 80-column "card image" format.

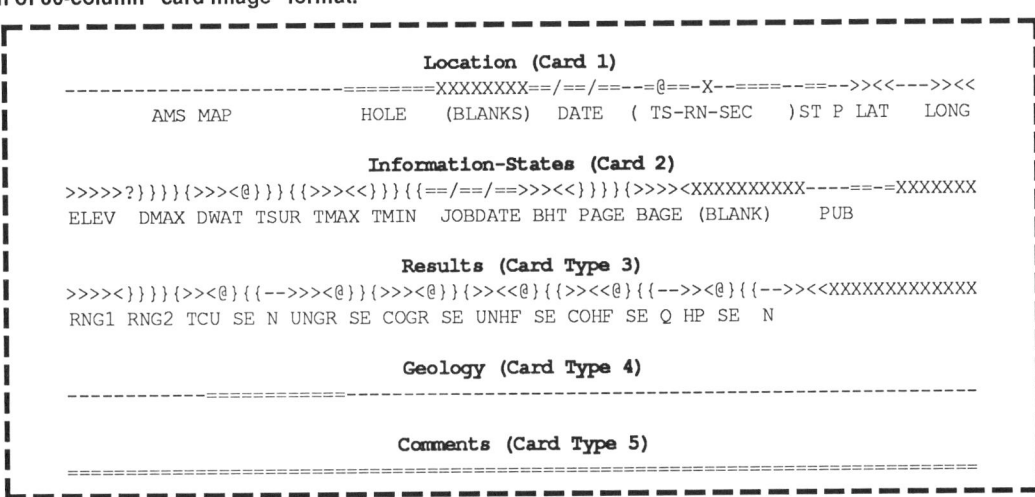

B. Explanation of fields

Card 1

AMS MAP	Name of 1:250,000-scale map covering site
HOLE	Hole name or identifier
DATE	Date hole was logged
TS-RN-SEC	Township location
ST, P	Codes identifing state and physiographic province
LAT, LONG	Latitude and longitude location
ELEV	Elevation of hole collar

Card 2

DMAX, DWAT	Maximum depth logged and depth to water table
TSUR	Temperature at surface (extrapolated)
TMAX, TMIN, BHT	Maximum, minimum, and bottom hole temperatures
JOBDATE	Date hole was completed
PAGE, BAGE	Age of province and basement
PUB	Reference to source of data

Card 3

RNG1, RNG2	Upper and lower depth interval for heat flow
TCU	Average thermal conductivity in interval
SE	Standard error of preceeding average quantity
N	Number of thermal conductivity samples
UNGR, COGR	Uncorrected and corrected gradient in interval
UNHF, COHF	Uncorrected and corrected heat flow in interval
Q	Code for quality, or error, of measurement
HP	Heat production

the *Magnetic Anomaly Map of North America* (Committee..., 1987), were used to reconstruct the theoretical position of the heat-flow contours within this region. Measured heat-flow data are quite sparse except just south of Iceland. The heat flow in the Laborador Sea is more problematic because there has been argument about the age of its opening (LePichon and others, 1971; Kristoffersen and Talwani, 1977; Talwani and Eldholm, 1977). Within this area there are some heat-flow observations, however, and the data agree in general with the contours shown, which are in turn consistent with an age of opening in the center of the Laborador Sea between 46 and 63 Ma. Roest and Srivastava (1989) propose a similar age range for the opening, although their orientation of the spreading is slightly different.

There have been extensive and intensive studies of the heat-

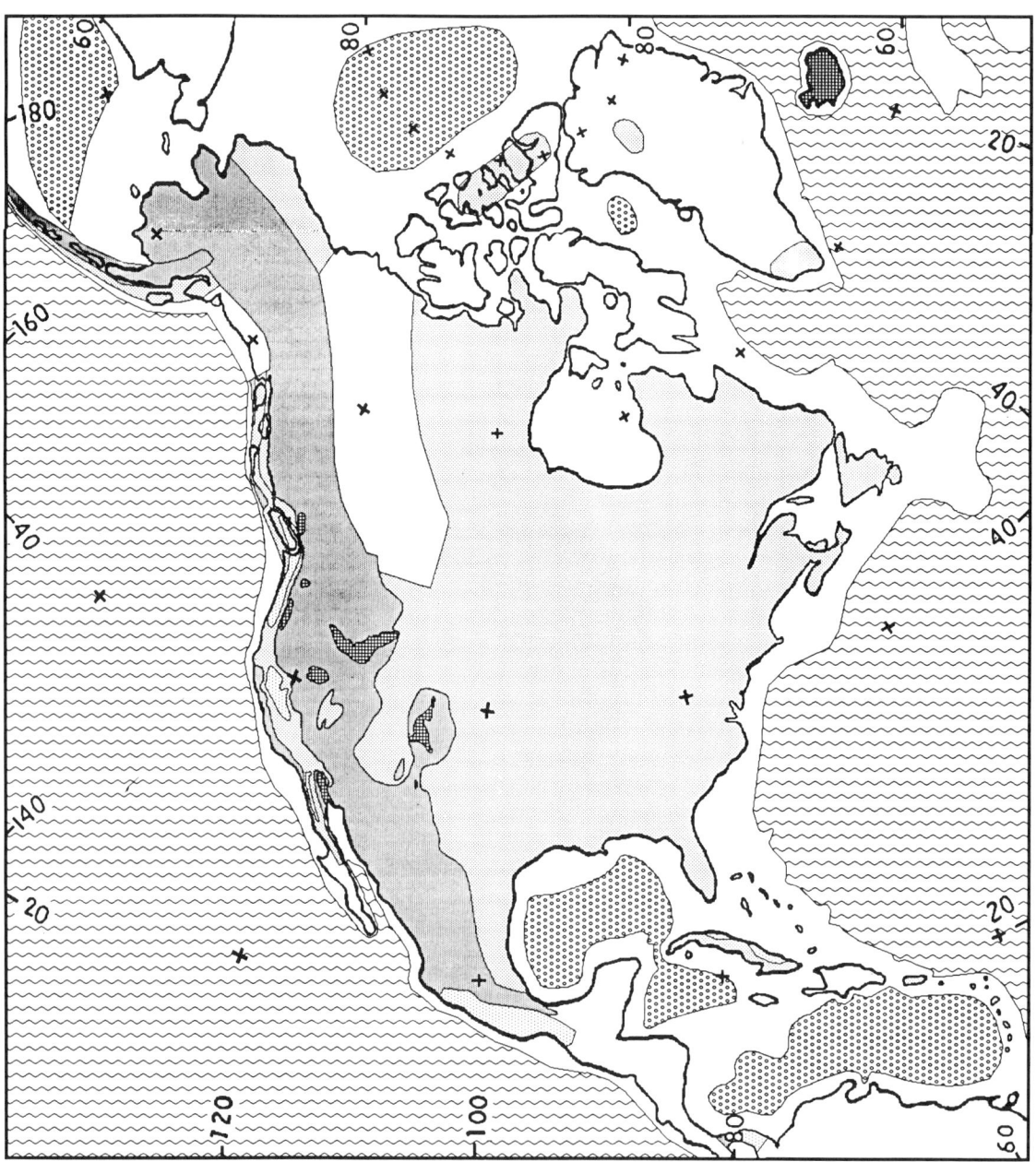

Figure 1. Index map for the DNAG *Geothermal Map of North America*. Areas of oceans with contours based on age are shown by the wave pattern. Areas of oceans with contours based on observed heat flow are shown by the open-dot pattern. Areas of continent with normal heat flow (Q_s of 50 mWm^{-2}, Q_r of 30 mWm^{-2}) are shown by the intermediate-density dot pattern. Areas of heat flow equal to or below normal in tectonic areas are shown by the low-density dot pattern. Areas of heat flow typical of the CTAZ (Q_s of 80 mWm^{-2}, Q_r of 60 mWm^{-2}) are shown by the high-density dot pattern. Areas with very high heat flow (generally greater than 100 mWm^{-2}) are shown by the cross-hatched pattern.

flow distribution in the Barbados accretionary sediment prism in the last few years. In some areas the heat flow is typical for the age of the ocean crust being subducted. In others there is evidence for fluid flow in the wedge (e.g., Langseth and others, 1988). As in the case of the ridges and land geothermal systems, there is no attempt to show the details of such settings on the map.

Gulf of Mexico and Caribbean Sea

Heat-flow contours in the Gulf of Mexico and the Caribbean Sea are based on observed data. The patterns that were observed cannot be closely related to the age of the sea floor as in the case of the Atlantic and Pacific Oceans because diagnostic magnetic anomaly patterns have not been recognized in these areas to identify the age of the oceanic crust. Thus these areas were contoured based on the observed heat-flow values.

There is a large area of moderate heat flow in the Caribbean Sea between the West Indies and Central America. This area is probably part of a back-arc heat high-flow zone related to the subduction of the Atlantic lithosphere beneath the Caribbean Plate.

Within the Gulf of Mexico, heat-flow values are generally low, in large part because of the high sedimentation rates, and particularly low heat-flow values are observed off the Mississippi Delta. If valid, these low values are probably related to the extremely high rate of sedimentation at the mouth of the delta. Some high values were found in the area of the Sigsbee Knolls, perhaps associated with the salt domes that underlie the uplifts.

Quite high heat-flow values are observed in the Cayman Trough, the plate boundary between the North American Plate and the Caribbean Plate. These high values are consistent with the idea that a limited segment of ridge crest exists in this area. The age and characteristics of spreading at this ridge have been summarized by Rosencrantz and others (1988).

Arctic Ocean and Bering Sea

Contours in the Arctic Ocean are based on observed heat-flow values. The Arctic Ocean data have recently been summarized on a DNAG map (Wetmiller and others, 1990). The data shown in the Arctic Ocean are taken primarily from that map, and references to the original data sources can be found in that publication. Heat flow in the part of the Arctic Ocean shown on the map is generally within the normal range and so is probably associated with lithosphere ages in excess of 50 to 60 Ma.

The contours in the Bering Sea are based on observed data. The heat flow is generally higher than the global average heat flow. The elevated heat flow is consistent with the back-arc setting and the heat flow in the parts of Alaska along strike of the subduction zone.

Iceland

While not part of an ocean basin, the characteristics of heat flow and geothermal activity in Iceland have some similarities to the mid-ocean ridges. The Mid-Atlantic ridge crosses the island so that it is the site of active extension, volcanism, and geothermal activity. The heat flow is generally high. The heat-flow values for the *Geothermal Map of North America* were calculated from a contoured geothermal gradient map of the island by Flovenz (1985), and the average thermal conductivity measured for the 1.9-km-deep Reydarfjordur drill hole (Oxburgh and Agrell, 1982) was used in combination with the contoured gradient to calculate heat flow. The resulting heat-flow map is certainly generalized, but interestingly shows some similarity to the inferred heat-flow-versus-age values in the adjoining Atlantic Ocean. Major geothermal systems locations were taken from Flovenz (1985) and a review of high and low temperature geothermal fields by Arnorsson (1986). The locations of Quaternary volcanoes are from Simkin and others (1981).

CONTINENTAL DATA

Heat Flow

Heat-flow data available on the various land masses are plotted with a code that divides the data into 10-mWm^{-2} intervals. Where there were multiple holes for a single reported site or multiple sites that would overlap at the scale of the map, the data were averaged and only a single value shown. Where data are of sufficient density, they are contoured at a 10-mWm^{-2} interval. Over much of the map there are insufficient data to constrain fully the contouring at that interval. However, a coarser interval would lose detail in the areas of high data density and is not fine enough to show some significant features. Therefore, for consistency, the 10-mWm^{-2} interval was used for the whole map. The data sites are shown so that the user can evaluate the reliability of the contours in a particular area of interest as needed. In any event, the contouring must be viewed as preliminary and primarily as an attempt to aid in outlining areas of heat-flow variation. Details of the contour pattern, unless based on data shown on the map, must be treated as inference. In some cases, contours follow known or presumed tectonic/physiographic/thermal trends in the absence of constraining heat-flow data. Whether or not these interpretations are valid remains to be investigated by collection of additional data.

Many heat-flow measurements have been made in Lake Superior using an oceanic approach with corrections for the effect of annual variations. There is significant variation in heat flow with position in the lake, but the size of the variations has not been corroborated with land data. There are no major conflicts with the land data, but the land data are sparse near the lake. For this reason the contours in the lake have not been carried onto the land, and the lake data stand by themselves.

The contoured areas within the continents range from less than 20 mWm^{-2} (in the Sierra Nevada Mountains) to greater than 120 mWm^{-2} (in such areas as Yellowstone, the Salton Trough, etc). However, only small areas of these extremes are shown. The cause of the variations is generally volcano/tectonic disturbances in the lithosphere and variations in the radioactive heat generation of the crust. In a few areas, more surficial factors affect the heat flow from the crust and mantle. The most common effect is rapid ground-water flow.

Fluid Flow

In an attempt to delineate areas where large-scale hydrologic disturbances of, or influences on, heat-flow data are common, some areas of the map have an overprint. Where there is documentation that the heat flow by conduction from the Earth's interior is disturbed by ground-water flow over large areas, there is a shaded overprint. This overprint is specialized in the case of a large area of the midcontinent region of the United States. In this area, calculated temperatures on a specific aquifer are contoured at 10°C intervals. This aquifer is the Dakota Sandstone of Cretaceous age. The contouring is based on depth to the aquifer, combined with heat-flow values logged from wells in the region, and geographic information on the thermal conductivity of the section above the aquifer. The details are described by Gosnold (1990, and this volume). In large areas east of the Rocky Mountains, ground-water flow in the aquifer has a significant basinwide effect on the heat flow. For example, the high values shown in northeastern Nebraska and eastern South Dakota (the area of heat-flow values above 60 mWm^{-2}; the area labeled DA on Fig. 2) probably are related to eastward flow (up dip) of warm water recharged at the west edge of the basin at high-elevation exposures of the Dakota aquifer along the edges of the Black Hills and the Front Range (Gosnold, 1990, and this volume).

In the case of the Snake River Plain aquifer in Idaho (Brott and others, 1981) and the various aquifers in the Prairies Basin in Alberta (Majorowic and others, 1984, 1985), the possibility of fluid-flow influence on observed heat flow is indicated by a single-density gray overprint on the Geothermal Map. More details of the effect of the fluid flow on the heat flow can be found in these references. The heat-flow estimates on which the contouring in Alberta is based are given by Beach and others (1987).

A somewhat different set of data is illustrated in the Gulf Coast in Texas and Louisiana by a shaded overprint with contours on the Geothermal Map. In this case the depth to the top of the occurrence of geopressure is shown (Bebout and others, 1983). This depth is of interest from a geothermal point of view because the fluids from the geopressure zone may be produced, and their thermal, mechanical, and hydrocarbon energy extracted (Wallace and others, 1979). Research and testing is underway to develop the technology to use this energy economically (Garg and others, 1981). In addition, the top of the geopressure zone is important as a zone of geothermal gradient change. Temperature gradients in the Gulf Coast generally range from 20 to 30°C/km above the geopressure zone, while at the top of the zone the gradients often increase to 35 to 45°C/km (e.g., see Bebout and others, 1979). Where deep thermal data are available, gradients at depth eventually return to lower values of 20 to 25°C/km. For this reason, and because of the potential geothermal applications, the depth to the top of the geopressure zone is shown on the map. The con-

Figure 2. Heat-flow map for the eastern United States. The contours are from the DNAG *Geothermal Map of North America* and are shown at 10-mWm^{-2} intervals. Each band of heat flow is shown by a different pattern. Abbreviations are explained in the text.

tours of the top of the zone are from Bebout and others (1983), interpolated to contour intervals of 500 m.

The origin of this change in gradient is controversial. It has been related to active fluid flow through the geopressure zone and/or to a low thermal conductivity for the shale that makes up the fluid seal. We favor the second explanation (Blackwell and Steele, 1989) in general. However, there is evidence for fluid flow both on the local (Wallace and others, 1977) and the large scale (Bodner and Sharp, 1988). A massive number of bottom-hole temperature data exist for the Gulf Coast (e.g., Jam and others, 1969; Kron and Stix, 1982), but there are no published heat-flow values for the area. We have estimated a value for a well near Houston on the *Geothermal Map of North America*. The temperature-depth curve for this well is shown in Backwell and Steele (1989).

Hot springs, geothermal systems, and volcanic centers

In addition to the areas of aquifer effects on heat flow, information is shown on the location of major hot springs systems. Unfortunately, simple plotting of all hot springs is not possible because no data base of uniform quality is available for all areas of North America. We have attempted to select from the literature major geothermal systems as indicated by their size and/or temperature. Only the larger/hotter systems were selected for display in an attempt to produce a display that will be more uniform across the continent.

In the United States the geothermal systems were plotted from the compilation of Brook and others (1979). Major Quaternary volcanoes were also taken from the same publication (Smith and Shaw, 1979). The major hot springs and their reservoir temperatures in Canada were taken from data supplied by Jessop (personal communication, 1988) and the studies of Souther (1975) and Souther and Halstead (1973). Locations of Quaternary volcanic centers were taken from Mathews (1986). In Mexico the major geothermal systems plotted were selected on the basis of data in Prol-Ledsma and Juarez (1986). The location and state of development of major geothermal systems in Central America is described by Dipippo (1986). Quaternary volcanoes were taken from Simkin and others (1981).

In the past ten years there has been extensive drilling in the vicinity of many geothermal systems to evaluate their potential for commercial exploitation of geothermal energy. Many short papers (or extended abstracts) have been published in the *Transactions* of the Geothermal Resources Council of Davis, California. A summary of references to temperature-depth data from geothermal areas is included in Blackwell and others (1988). In general the holes in the vicinity of geothermal systems are too closely spaced to be individually plotted on the 1:5,000,000 map scale. Geothermal systems with gradient and heat-flow data in the public domain are indicated by a different symbol than the systems without data in the public domain.

Heat production

A major component of the thermal characteristics of a continental area is the heat production of the upper crust from radioactive isotopes of uranium, thorium, and potassium. In plutonic terrains, there is generally a linear relation between the average radioactivity of the surface rocks and the heat flow in typical-depth drill holes (Birch and others, 1968; Lachenbruch, 1968; Roy and others, 1968). The implications of this relation have been thoroughly discussed in the literature, and the recent heat-flow publications cited in this paper have many references to these discussions. The relation is important in this discussion because the intercepts of the straight lines, the heat flow at zero heat production (Q_r, the "reduced" heat flow), determined in different areas is in some way a measure of the deep heat flow (i.e., the heat flow from the lower crust and mantle). Because this quantity varies much less than the measured surface heat flow (Q_s), it is used to define as thermal provinces contiguous areas that have similar values of "reduced" heat flow (Roy and others, 1972). For example, all of North America east of the Cordillera appears to have a similar Q_r of 30 ± 5 mWm^{-2}.

The heat production and the reduced heat flow are not shown on the *Geothermal Map of North America* to simplify the map presentation. However, the reduced heat flow for various areas will be used to discriminate different thermal provinces and mentioned as a characteristic parameter throughout this discussion. The index map in Figure 1 may be viewed in some sense as a simplified reduced heat-flow map because thermal provinces are depicted on it.

HEAT-FLOW DISCUSSION: UNITED STATES

Introduction

Blackwell (1971), Roy and others (1972), Lachenbruch and Sass (1977), and Sass and others (1981) have given general summaries of the heat flow in the United States. Swanberg and Morgan (1980) used the silica geothermometer in an attempt to characterize areas with no heat-flow data. General discussions focusing on the western United States have been given by Lachenbruch and Sass (1977) and Blackwell (1978). Because of these and many more area-specific papers, such as the chapters in this volume by Gosnold and by Reiter and others, most of the larger-scale thermal features are relatively well known.

Morgan and Gosnold (1989) have recently given a fairly complete summary of the heat flow in various regions of the United States as developed in the last ten years. The first object of this section is to supplement their discussion as it related specifically to features shown on the geothermal map and not to repeat their summary. Thus, Morgan and Gosnold (1989) should be read in conjunction with this section for a complete picture of United States heat flow. The second object of this discussion is to describe several features of the thermal pattern that are not well known, some of which are more easily recognized because of the detailed contour interval of the geothermal map.

Eastern United States

A simplified heat-flow map for the eastern United States is shown in Figure 2, based on the contours from the Geothermal Map. The contours are based on a total of 390 individual sites that are included on the large-scale heat-flow map. Physiographic provinces are also shown for reference: the Atlantic Coastal Plain (ALP), Interior Lowlands (IL), Ozark Plateau (OP), and Ouachita Mountains (OM). Heat-flow values are generally less than 60 mWm^{-2} east of the Cordillera. The major exception to this generalization is in the Appalachian Mountains (AM) where higher than average heat-flow values are associated with areas of high crustal radioactivity, such as in the White Mountains of New England (Birch and others, 1968; WM on Fig. 2). Similar relatively high heat-flow values in association with high crustal radioactivity are found in the southern Appalachians (Costain and others, 1986) and in the Maritime Provinces of Canada (Jessop, this volume).

There is a band of lower than average heat flow (generally less than 40 mWm^{-2}) that is west of and roughly parallel to the area of generally high heat flow in the Appalachians. It is not as easy, however, to associate this region with the cause, or causes, of the low heat flow. In New York the low heat flow is associated with the low heat generation in the Proterozoic Grenville-age anorthosite in the Adirondacks (Birch and others, 1968). The low heat flow in the central Appalachians might be associated with low crustal heat production or with regional redistribution of heat through large-scale ground-water flow that is not measured by the available heat-flow measurements (Smith and others, 1981). The data are not available to differentiate between these two hypotheses at the present time. The low heat flow values in Florida are probably related to aquifer effects because the measurements have been made in the extensive Floridan carbonate aquifer (Smith and Fuller, 1977).

In eastern Canada there is also low heat flow in the Grenville terrain according to recent heat-flow determinations described by Mareschal and others (1989). Unfortunately, all the measurements are in shallow holes in an area where climatic effects may be important to depths of several hundred meters. If the data are taken at face value the inference might be made that at least part of the low heat flow is associated with a crust that is generally low in heat production, possibly due to deep erosion and removal of upper crustal radioactive heat sources after the Precambrian collision of the Grenville terrain with North America.

In the Canadian Shield (Jessop, this volume) and in the central stable region of the United States the heat flow is generally between 40 and 50 mWm^{-2}, and local heat flow depends on the local value of crustal heat production. The only anomalous area is in Arkansas and Louisiana where high heat-flow values are found with no obvious explanation (Smith and Dees, 1982). Smith and Dees (1982) have proposed that structure (salt domes) might be part of the reason for the high heat flow.

The heat flow in the Great Plains (GP) in both the United States and Canada is complex. There are major perturbations to the crustal heat flow due to the flow of ground water in the major regional aquifers such as the Cretaceous Dakota Sandstone (DA in Fig. 2) in Colorado, Nebraska, and the Dakotas, as mentioned above. The situation is also complex in the Prairies Basin area of Canada where there is much discussion of the role of fluid flow in the observed heat-flow pattern (Majorowicz and others, 1984, 1985; Jessop and Vigass, 1989; Bachu, 1988; Jessop, 1990). Additional discussion and references are given by Jessop (this volume).

In spite of the many thousands of wells drilled in the Gulf Coastal Plain (GCP), as described above, there are few quantitative thermal data available. Discussions of bottom-hole temperatures have been presented for some areas, but thermal conductivity data are rare. On a regional scale, gradient decreases toward the Mississippi Delta in southern Louisiana. This pattern also appears (weakly) in the available heat-flow data. The most probable cause of this variation, if it is verified by further studies, is the cooling effect of the rapid sedimentation in the Gulf Coast Basin. A similar effect is seen in the data in the Gulf of Mexico. However, the thermal properties of the rocks in the basin are not known well enough to determine whether mechanisms such as sedimentation, regional changes in lithology, or basement structure are responsible for the effect. The second generalization that can be made concerning the Gulf Coast is that, in the areas of geopressure, the gradient typically varies in relation to the top of the geopressured zone as described in the general map discussion above. The contours to the top of the geopressure zone are shown on the 1:5,000,000 scale map.

Western United States

In contrast to the eastern United States, the factors that affect the heat flow in the western states are of varying types, are complex, and overlap. While increased heat flow is often associated with tectonic activity, as in the oceans, in the western United States, both high and low heat-flow anomalies are associated with tectonism (see Morgan and Gosnold, 1989). Because of the added complexity, the regional heat-flow patterns have been studied extensively. A total of 1,428 separate sites, many representing more than one hole (and not counting the many multiple-hole studies in geothermal systems), are shown on the map. A small-scale section of the *Geothermal Map of North America* for the western United States, with heat-flow contours only, is shown in Figure 3. The heat-flow contours in 10-mWm^{-2} intervals from the map are shown and the contours are patterned at 20-mWm^{-2} intervals for clarity. The major areas of geothermal significance are indicated by abbreviations; the key is in the figure caption. The areas, either individually or in combination, can be classified into thermal provinces in most cases.

For example, the largest province in the western United States includes the Northern Rocky Mountains (NR), Columbia Basin (CB), and Basin and Range Province (B&R) where the volcanism is older than about 17 Ma (see Blackwell, 1978). All these areas have a similar heat flow from below the upper crustal radiogenic layer (60 ± 10 mWm^{-2}); thus, these areas make up a thermal province in the sense of Roy and others (1972).

The Rio Grande rift (Reiter and others, this volume) and the Southern Rocky Mountains (Decker and others, 1988) have equal or higher heat flow from below the radiogenic layer. This area is geographically distinct and is discussed in more detail below.

Blackwell (1969) referred to the area of high heat flow encompassing the Basin and Range Province and the Northern Rocky Mountains of the United States Pacific Northwest as the Cordilleran Thermal Anomaly Zone (CTAZ). In a subsequent analysis of the heat flow/heat production in the Basin and Range Province, Blackwell (1978) concluded that the 60±10-mWm^{-2} intercept value was characteristic of areas with regional volcanism older than 17 Ma. More recent studies in Canada and Mexico, mentioned below, have found remarkably similar heat-flow characteristics for the areas to the north and south along the Cordillera.

Thermal subprovinces exist within this large area of the western United States. Notable among these are the Eureka low (EL) and the Battle Mountain high (BM) in the Basin and Range Province. Local and regional variation may be caused by fluid flow, local structural effects, heat production differences, presence of young volcanism, and other types of effects.

Other areas that have heat flow equal to or in most cases higher than the CTAZ are the Salton Trough (ST), the Snake

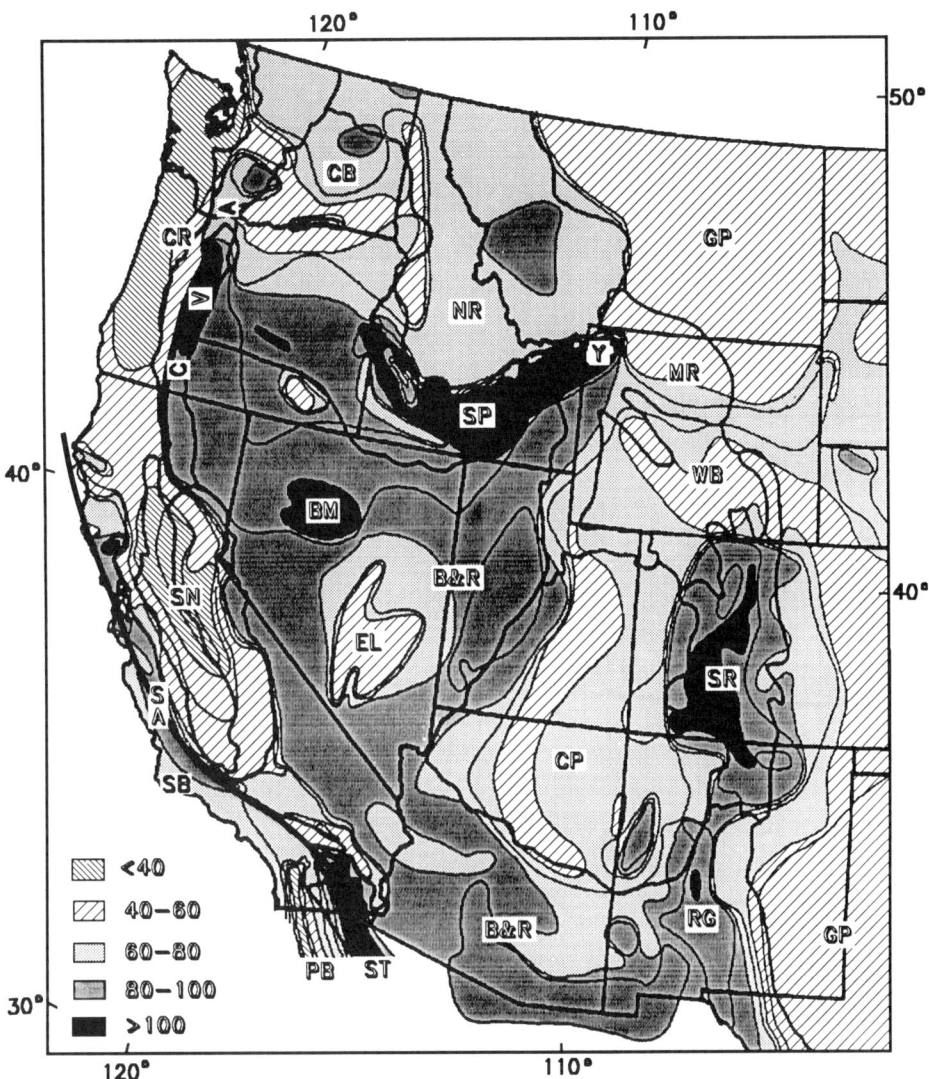

Figure 3. Heat-flow map for the western United States. The contours are from the DNAG *Geothermal Map of North America* and are shown at 10-mWm^{-2} intervals. Each pattern covers a 20-mWm^{-2} interval. Abbreviations refer to areas discussed in the text.

River Plain/Yellowstone region (SP, Y), and the Cascade volcanic arc (CVA). The heat flow in the Salton Trough is high due to the extremely rapid extension (Lachenbruch and others, 1985) associated with the formation of an ocean ridge in the Gulf of California. The heat flow is high in the Snake River Plain/Yellowstone region due to the effects of the Yellowstone hot spot (see discussion below). The heat flow is variable, but generally very high, in the Cascade volcanic arc due to active midcrustal intrusion there (Lewis and others, 1988; T. J. Lewis, this volume; Blackwell and others, 1990a, b). The High Cascade Range of Oregon, the east half of the southern Cascade Range in Washington (CR), and the part of the Cascade Range in British Columbia have very high heat flow and crustal temperatures.

West of these areas of high heat flow there is an almost continuous band of heat flow that is less than or equal to the heat flow in the stable eastern United States. This area includes the Western Cascade Range subprovince in Oregon, the coastal provinces in Oregon and Washington (CR), the Klamath Mountains, Sierra Nevada, and the Peninsular Ranges in southern California. The Sierra Nevada, as an area, has the lowest Q_r (only 20 mWm^{-2}) over a large region of a continent. It is even lower than the heat flow from below the upper crustal heat-production layer in the Stable Interior (30 ± 5 mWm^{-2}).m

The low heat flow is due to the fact that a subducting oceanic plate absorbs heat from the overlying wedge as it descends into the mantle. The heat flow is low from the trench to the edge of the volcanic arc. In contrast to the volcanic and back-arc regions, where penetrative convection of magma and extension are intracrustal sources, in the outer arc the thermal regime in the crust is conductive. Therefore, heat-conduction theory can be utilized to calculate the thermal recovery after subduction stops in an area such as the Sierra Nevada (SN) due to triple-point migration along the coast (see Atwater, 1970). The heat flow in the Sierra Nevada and Peninsular Ranges (PB) in southern California and in northern Baja California is consistent with heating from the bottom and sides of the cold block after the termination of subduction. The Salinian block (SB) does not fit this pattern because the heat flow is high in this Mesozoic plutonic terrain west of the San Andreas fault.

In fact, all along the gap in the volcanic arc that follows the west coast of the conterminous United States the heat flow rises again to the west of the region of low heat flow that is the remnant of the early and mid-Cenozoic subduction pattern. This area is coincident with the extent of the San Andreas fault zone (SA). The high heat flow east of the San Andreas and west of the Sierra Nevada Mountains is attributed to the effects of the hole in the plate (see the model described by Lachenbruch and Sass, 1980; and Zandt and Furlong, 1982). The Clear Lake volcanic center, site of The Geysers geothermal field, is the late Cenozoic location of the migrating volcanism along the east side of the San Andreas that may coincide with the migrating "hole."

Surprisingly the heat flow is also high along the west side of the fault zone in the granitic terrain of the Salinian block (SB) as well. Because this block was probably in an outer arc position analogous to the Peninsular Ranges and Sierra Nevada as recently as 5 Ma, the heat flow was low before the position of the San Andreas fault jumped inland at about 5 Ma. Yet its present-day surface and reduced heat flow are similar to the Basin and Range Province. The block is wide enough that it should not have heated up from the sides, and judging from the example of the other two blocks, it should not yet have heated up from the bottom. A possible resolution of this inconsistency is that the cold block is thin (and thus heating faster) because the cold-mantle part of the Salinian block is delaminated from the crustal block as it moves northward and is deformed around the bend in the fault zone at the Transverse Ranges. A deep, high-velocity, high-density, "root" in the upper mantle beneath the Transverse Ranges has been inferred on the basis of seismic and gravity data (Hadley and Kanamori, 1977; Sheffels and McNutt, 1986). If the anomaly is due to cold upper mantle that is being stripped from the base of the Salinian block as it goes around the bend in the San Andreas, then the crust alone could heat up in a few millions of years.

Whatever the explanation, the heat flow is high on both sides of the San Andreas fault, and there is no heat flow peak over the fault zone as should occur if there is significant frictional loss of heat in the seismogenic zone of the fault (see Lachenbruch and Sass, 1980, 1991). This lack of an anomaly has led to the hypothesis that the fault zone is weak (Zoback and others, 1987). The situation is of enough interest to spawn the drilling of a 4-km well to test the properties and heat flow of the fault zone at depth. Preliminary results of the drilling have been presented by Zoback and others (1988), and the heat flow from the well has been described by Sass and others (1991).

A third more or less north-trending area of high heat flow in the western United States occurs in the Rio Grande rift subprovince (RG) of the Basin and Range Province and in the Southern Rocky Mountains (SR). This north-south–trending zone is bounded on the east by the Great Plains (GP) Province with generally normal continental heat flow except in areas disturbed by aquifer flow as described above. On the west, the bounding province is the Colorado Plateau (CP), which has normal heat flow in spite of its high average elevation. This area has surface and reduced heat flow that is equal to or greater than the values for the CTAZ. The detailed characteristics of the two areas are described in Reiter and others (this volume) and Decker and others (1988). Of particular interest is evidence for contemporary magma chambers in both areas.

The Rio Grande rift is the most difficult heat-flow province to fit into a subduction-related category. The high heat flow east of the normal heat flow in the Colorado Plateau makes a simple back-arc setting doubtful. One possible scenario is a very low angle of subduction in the Oligocene from the coast to the east edge of the Colorado Plateau followed by a steep angle to form the San Juan volcanics as a volcanic arc. Reactivation of the still warm region by late Cenozoic extension related to triple-point effects (Atwater, 1970) could explain the high heat flow in New Mexico. However, the reduced heat flow is as high or higher in Colorado where extension is minor to nonexistent.

The north edge of the high heat flow in the Southern Rocky Mountains is quite abrupt with a half-width of 50 km (Decker and others, 1988). The heat flow in the Wyoming Basin (WB) and the middle Rocky Mountains (MR) to the north (and east of the CTAZ) is similar to that in the craton, with a surface heat flow of 50 mWm^{-2} and a reduced heat flow of 30 $\pm$ 5 mWm^{-2}. The heat-flow boundary almost coincides with the Archean/Proterozoic boundary between the Wyoming craton and the Southern Rocky Mountains Proterozoic terrain (about 1,420 to 1,500 Ma zircon ages; Bickford and others, 1986). Morgan (1985) proposed that Archean cratons have survived without experiencing major tectonism because the heat production is lower for these old cratons, and hence the lithosphere is thicker than for younger crusts. In Africa, Ballard and others (1987) found evidence for a thermal effect associated with the Archean terrains and inferred fundamental heat-flow differences between Archean and younger terrains. It is interesting that the Archean Wyoming craton has had a completely different tectonic history than any of the other areas in the Cordillera. While crustal compression has affected the area, volcanism and extension have not. The explanation may be related to the peculiarities of the plate-tectonic interactions, but it may also be related to the nature of Archean cratons. Further investigation of this question would seem to be a useful line of research.

The Snake River Plain/Yellowstone area (SP, Y) is a major thermal and tectonic province of the western United States. Its original geologic and geophysical characteristics have been completely overprinted in the late Cenozoic by a sequence of thermally dominated features related to passage of a hot spot (currently situated under the Yellowstone caldera) as described by Brott and others (1981). Of particular interest is the development, due to lithospheric thermal contraction, of a regional topographic profile that is very similar to the ocean ridge topographic profile. The heat flow in this area is generally above 100 mWm^{-2} (Blackwell, 1989).

HEAT-FLOW DISCUSSION: NORTH AMERICA

There are several overall patterns that appear when the heat-flow data for the whole continent are plotted and contoured. The general differences in heat flow between the stable eastern and the tectonically active western part of the continent are obvious. General continent-wide relations of heat flow to age have been extensively explored (Sclater and others, 1980; Vitorello and Pollack, 1980; Morgan and Sass, 1984). As pointed out by Morgan and Sass (1984), these average relations become more complicated in actual applications to specific North American terrains.

A very generalized heat-flow map of North America is shown in Figure 1. At this scale the major patterns that stand out are the normal heat flow in the eastern part of the continent and the variable heat flow in the western part of the continent associated with the Cordillera. General aspects of the heat flow in Canada have been described by Jessop (this volume), and general aspects of the heat flow in the Canadian Cordillera have been discussed by Lewis (this volume). The heat flow in Mexico has been discussed by Ziagos and others (1985) and by Prol-Ledsma and others (1989). A sketch of the heat-flow pattern in Alaska has been given by Sass and others (1981). In all areas there are broad regions characterized by high heat flow bounded on the west by regions of low heat flow. The heat flow in the part of Siberia shown on the map is normal based on the measurements described by Duchkov (1985). Only four heat-flow regimes are delineated on Figure 1. These are regions with heat flow less than 40 mWm^{-2}, 40 to 70 mWm^{-2}, 70 to 90 mWm^{-2} and more than 90 mWm^{-2}. Since the areas outlined are based on conductive heat transfer from the crust and upper mantle, areas disturbed by large-scale fluid flow are not shown as they are on the 1:5,000,000-scale map. Furthermore, because of the small scale, some areas of no data are included in the characterized areas. In a general way these areas can also be categorized on the basis of their reduced heat flow (Q_r). The four areas are characterized by Q_r values of <25, 30 ± 5, 60 ± 10, and >60 mWm^{-2}.

At this scale the eastern and central parts of North America are characterized by uniform and normal heat flow. Variations in heat flow are shown on the 1:5,000,000-scale map but not on Figure 1, because the dominant cause of heat-flow variation is differences in radiogenic heat production in the upper crust, and the whole area is thought to have the same upper mantle and lower crustal heat flow (about 25 to 30 mWm^{-2}). However, the areas within the Cordillera of western North America are quite different. All values of regional heat flow, from the lowest to the highest, are found there. About 75 percent of the area of the Cordillera is characterized by heat-flow values in the range of 70 to 90 mWm^{-2}. The heat-flow distribution is well established for the United States, for Mexico north of the TransMexico Volcanic Belt, and for the southern Canadian Cordillera. The patterns shown for the northern Canadian Cordillera, and for interior Alaska (Sass and others, 1981) are based on sparse data.

Where radioactivity data exist, the heat flow/heat production lines are characterized by intercept values at zero heat production of 60±10 mWm^{-2}. Such areas include the Sierra Madre Oriental and Occidental in Mexico (Smith and others, 1979); much of the Basin and Range Province in the United States (Roy and others, 1972; Blackwell, 1978); the Northern Rocky Mountains, Columbia Basin, and Blue Mountains in the northwestern United States; and the southern Canadian Cordillera in British Columbia (Lewis and others, 1985).

As noted above, Blackwell (1969) referred to the area of high heat flow as then known (the Basin and Range Province and the northern Rocky Mountains of the United States Pacific Northwest) as the Cordilleran Thermal Anomaly Zone (CTAZ). In a subsequent analysis of the heat flow/heat production in the Basin and Range Province, Blackwell (1978) concluded that the 60 ± 10-mWm^{-2} intercept value was characteristic of areas with regional volcanism older than 17 Ma. More recent studies in Canada and Mexico have found remarkably similar heat-flow characteristics for the areas to the north and south along the Cordillera. Furthermore, the correspondence of the eastern part of the region of high heat flow to the area characterized by decollement-type supracrustal deformation in the latter half of the Mesozoic is striking (see Drewes, 1978). The eastern border of this region, with minor exceptions, also defines the areas of Cenozoic volcanic activity in the Cordillera.

The area of high heat flow is bordered along its west side by areas of generally higher heat flow in the volcanic arcs that exist where subduction is still occurring off the west coast. West of the volcanic arcs, and west of the CTAZ where there are no contemporary volcanic arcs such as the Sierra Nevada, there is a more or less continuous band of low heat flow. This band of low heat flow is the outer arc of the subduction zone. Because heat flow is conductive in the crust and mantle above the slip zone, the heat flow takes several millions of years to recover when subduction stops, unless an active heat source is present. The low heat flow in the areas of western North America where no subduction is occurring today are remnants of the previous time when subduction was continuous along the whole west coast (Atwater, 1970).

Areas of higher heat flow within this larger area include the Battle Mountain region in Nevada, parts of the Cascade volcanic arc, parts of the Southern Rocky Mountains, and the Snake River Plain/Yellowstone region. As discussed above, the high heat flow of the volcanic arc and the Snake River Plain/Yellowstone area is related to arc volcanism and hot-spot activity, respectively. An area of similar heat flow includes parts of California in the vicinity of the San Andreas fault. This area is also not related to the CTAZ in the sense that the high heat flow there is not back-arc related.

In order to illustrate the categories of origins for the heat-flow anomalies in the western Cordillera the various areas discussed are shown in Table 2. The various provinces shown in

Figures 2 and 3, and equivalent areas in Canada, Alaska, and Mexico, are listed in the table as one axis, and the various types of anomalous areas are listed on the second axis. Most of the areas fall into one of the three categories related to subduction listed in Table 2. In the case of some areas in Mexico and Alaska, the association is by type of geologic province rather than measured heat flow because of lack of data in those areas.

The continent-wide view emphasizes the somewhat similar thermal characteristics of the Cordillera along strike and the extent to which these characteristics are related to the subduction process that has dominated the tectonics of the region for the last 100 m.y. or more. Much of the interior region of the Cordillera apparently has had high heat flow for at least 70 to 100 m.y., and the broad patterns that persist today, especially the eastern border of the high heat-flow back-arc zone, may have been established at that time. The major modifications of the pattern up to about 30 Ma may have been the variations of the width of the low heat flow, outer arc block as the subduction angle of the slab changed. After 30 Ma, changes were associated with the fragmentation of the outer arc zone associated with triple-point movements, but the basic thermal pattern has been remarkably stable.

TABLE 2. GENERALIZED HEAT FLOW SETTING OF VARIOUS AREAS OF CHARACTERIZED HEAT FLOW IN WESTERN NORTH AMERICA*

Province or area	Heat Flow				
	Normal	Low (OA)*	High (BA)*	Vol. Arc	Other
East of Cordillera	X				
Southern Rocky Mountains/RGR					X?
Wyoming Basin/ Middle RM	X				
Colorado Plateau	X				
Basin and Range			X		
Northern Rocky Mountains (US)			X		
Northern Rocky Mountains (Canada)			X		
Brooks Range (?)			X		
Columbia Basin			X		
High Lava Plains/ Blue Mountains			X		
Interior Plateaus (Canada)			X		
Yukon Plateaus			X		
Sierra Madre Oriental†			X		
Sierra Madre Occidental			X		
Sierra Nevada/Great Valley			X		
U.S. Northwest Coast		X			
Vancouver Island, Fjords		X			
Peninsular Ranges of Southern California		X			
Northern Baja California		X			
Sierra Madre del Sur (Mexico)		X			
Cascade Range				X	
Aleutian Range				X	
TransMexico Volcanic Belt				X	
Salton Trough					Extension
Snake River Plain/ Yellowstone					Hot Spot
San Andreas (east side)					Slab Hole(?)
(west side: Salinia)					Delamination(?)

*The abbreviations OA and BA refer to outer arc and back arc respectively.
†North of the TransMexico volcanic belt.

REFERENCES CITED

American Association of Petroleum Geologists and U.S. Geological Survey, 1976a, Geothermal gradient map of North America: U.S. Geological Survey, scale 1:5,000,000.

American Association of Petroleum Geologists and U.S. Geological Survey, 1976b, Subsurface temperature map of North America: U.S. Geological Survey, scale 1:5,000,000.

Arnorsson, S., 1986, Chemistry of gases associated with geothermal activity and volcanism in Iceland; A review: Journal of Geophysical Research, v. 91, p. 12261–12268.

Atwater, T., 1970, Implications of plate tectonics for the Cenozoic tectonic evolution of western North America: Geological Society of America Bulletin, v. 81, p. 3513–3536.

Bachu, S., 1988, Analysis of heat transfer processes and geothermal pattern in the Alberta Basin, Canada: Journal of Geophysical Research, v. 93, p. 7767–7781.

Ballard, S., Pollack, H. N., Skinner, N. J., 1987, Terrestrial heat flow in Botswana and Namibia: Journal of Geophysical Research, v. 92, p. 6291–6300.

Beach, R.D.W., Jones, F. W., and Majorowicz, J. A., 1987, Heat flow and heat generation estimates for the Churchill basement of the western Canadian basin in Alberta, Canada: Geothermics, v. 16, p. 1–16.

Bebout, D. G., Weise, B. D., Gregory, A. R., and Edwards, M. B., 1979, Wilcox sandstones in the deep subsurface along the Texas Gulf Coast, their potential for production of geopressured energy: Texas Bureau Economic Geology, Department of Energy Report ET28461, 219 p.

Bebout, D. G., and 5 others, 1983, Technical support for geopressured-geothermal well activities in Louisiana: Annual Report DOE/NV/10174-2, 454 p.

Bickford, M. E., Van Schmus, W. R., and Zietz, I., 1986, Proterozoic history of the midcontinent region of North America: Geology, v. 14, p. 492–496.

Birch, F., Roy, R. F., and Decker, E. R., 1968, Heat flow and thermal history of New York and New England, in Zen, E., White, W. S., Hadley, J. B., and Thompson, J. B., Jr., eds., Studies of Appalachian geology; Northern and maritime: New York, Interscience, p. 437–451.

Blackwell, D. D., 1969, Heat flow in the northwestern United States: Journal Geophysical Research, v. 74, p. 992–1007.

—— , 1971, The thermal structure of the continental crust, in Heacock, J. G., ed., The structure and physical properties of the Earth's crust: American Geophysical Union Monograph 14, p. 169–184.

—— , 1978, Heat flow and energy loss in the western United States, in Smith, R. B., and Eaton, G. P., eds., Cenozoic tectonics and regional geophysics of the western Cordillera: Geological Society of America Memoir 152, p. 175–208.

—— , 1989, Regional implications of heat flow of the Snake River Plain, northwestern United States: Tectonophysics, v. 164, p. 323–343.

Blackwell, D. D., and Steele, J. L., 1989, Thermal conductivity and sedimentary rocks: Measurement and significance, in Naeser, N. D., and McCulloh, T. H., eds., Thermal history of sedimentary basins; Methods and case histories: New York, Springer-Verlag, p. 13–36.

——— , eds., 1991, Geothermal map of North America: Boulder, Colorado, Geological Society of America, map CSM-007, 4 sheets, scale 1:5,000,000.

Blackwell, D. D., Bowen, R. G., Hull, D. A., Riccio, J., and Steele, J. L., 1982, Heat flow, arc volcanism, and subduction in northern Oregon: Journal of Geophysical Research, v. 87, p. 8769–8754.

Blackwell, D. D., Steele, J. L., and Carter, L. S., 1988, Geothermal database and Oregon Cascade thermal studies: Dallas, Texas, Southern Methodist University, Department of Energy Final Report DOE/ID/12623-1, 90 p.

Blackwell, D. D., Steele, J. L., and Carter, L. S., 1989, Heat flow data base for the United States, in Hittleman, A. M., Kinsfather, J. O., and Meyers, H., eds., Geophysics of North America: Boulder, Colorado, National Oceanographic and Atmospheric Administration, National Geophysical Data Center, CD-ROM.

Blackwell, D. D., Steele, J. L., Kelley, S. A., and Korosec, M., 1990a, Heat flow in the state of Washington and thermal conditions in the Cascade Range: Journal of Geophysical Research, v. 95, p. 19495–19516.

Blackwell, D. D., and 5 others, 1990b, Heat flow in the Oregon Cascade Range and its correlation with regional gravity, Curie point depths, and geology: Journal of Geophysical Research, v. 95, p. 19475–19493.

Bodner, D. P., and Sharp, J. M., Jr., 1988, Temperature variations in south Texas subsurface: American Association of Petroleum Geologists Bulletin, v. 72, p. 21–32.

Bott, M.H.P., 1983, Deep structure and geodynamics of the Greenland-Scotland Ridge; An introductory view, in Bott, M.H.P., Saxon, S., Talwani, M., and Thiede, J., eds., Structure and development of the Greenland-Scotland Ridge: New York, Plenum Press, p. 3–10.

Brook, C. A., and 5 others, 1979, Hydrothermal convection systems with reservoir temperatures ≥90°C, in Muffler, L.J.P., ed., Assessment of geothermal resources of the United States—1978: U.S. Geological Survey Circular 790, p. 18–85.

Brott, C. A., Blackwell, D. D., and Ziagos, J. P., 1981, Thermal and tectonic implications of heat flow in the eastern Snake River Plain, Idaho: Journal of Geophysical Research, v. 89, p. 11709–11734.

Committee for the Magnetic Anomaly Map of North America, 1987, Magnetic Anomaly Map of North America: Boulder, Colorado, Geological Society of America, map CSM-003, 4 sheets, scale 1:5,000,000.

Costain, J. K., and 5 others, 1986, Heat flow in the Piedmont and Atlantic Coastal Plain of the southeastern United States: Journal of Geophysical Research, v. 91, p. 2123–2135.

Davis, E. E., and Lister, C.R.B., 1977, Heat flow measured over the Juan De Fuca ridge; Evidence for widespread hydrothermal circulation in a highly heat transportive crust: Journal of Geophysical Research, v. 82, p. 4845–4860.

Decker, E. R., Heasler, H. P., Buelow, K. L., Baker, K. H., and Hallin, J. S., 1988, Significance of past and recent radioactivity studies in the southern Rocky Mountains: Geological Society of America Bulletin, v. 100, p. 1851–1885.

Detrick, R. S., VonHerzen, R. P., Parsons, B., Sanwell, D., and Dougherty, M., 1986, Heat flow observations on the Bermuda Rise and thermal models of midplate swells: Journal of Geophysical Research, v. 91, p. 3701–3723.

Dipippo, R., 1986, Geothermal energy developments in Central America: Geothermal Resources Council Bulletin, v. 15, no. 10, p. 3–13.

Drewes, H., 1978, The Cordilleran orogenic belt between Nevada and Chihuahua: Geological Society of America Bulletin, v. 89, p. 641–657.

Drummond, K. J., chairman, 1981, Plate-tectonic map of the circum-Pacific region; Northeast quadrant: American Association of Petroleum Geologists, scale 1:10,000,000.

Duchkov, A. D., ed., 1985, Heat flow data catalogue for Siberia (1966–1984): Novosibirsk, Institute of Geology and Geophysics, Siberian Branch, Academy of Sciences U.S.S.R., 82 p. (in Russian).

Florenz, O. G., 1985, Application of subsurface temperature measurements in geothermal prospecting in Iceland: Journal of Geodynamics, v. 4, p. 331–340.

Garg, S. K., Riney, T. D., and Fwu, J. M., 1981, Analysis of phase I flow data from Pleasant Bayou No. 2 geopressured well: Transactions of the Geothermal Resources Council, v. 5, p. 693–696.

Gosnold, W. D., Jr., 1990, Heat flow in the Great Plains of the United States: Journal of Geophysical Research, v. 95, p. 353–374.

Hadley, D. M., and Kanamori, H., 1977, Seismic structure of the Transverse Ranges, California: Journal of Geophysical Research, v. 88, p. 1469–1478.

Hyndman, R. D., Langseth, M. G., and VonHerzen, R. P., 1987, Deep sea drilling project geothermal measurements; A review: Reviews of Geophysics, v. 25, p. 1563–1582.

Jam, P. L., Dickey, P. A., and Tryggvason, E., 1969, Subsurface temperature in south Louisiana: American Association of Petroleum Geologists Bulletin, v. 53, p. 2141–2149.

Jessop, A. M., 1990, Comparison of industrial and high resolution thermal data in a sedimentary basin: PAGEOPH, v. 133, p. 251–267.

Jessop, A. M., and Vigass, L. W., 1989, Geothermal measurements in a deep well at Regina, Saskatchewan: Journal of Volcanology and Geothermal Research, v. 37, p. 151–166.

Jessop, A. M., Hobart, M. A., and Sclater, J. G., 1976, The world heat flow data collection-1975: Energy, Mines, and Resources Canada Geothermal Series 5, 125 p.

Klitgord, K. D., and Schouten, H., 1986, Plate kinematics of the central Atlantic, in Tucholke, B. E., and Vogt, P. R., eds., The western North Atlantic region: Boulder, Colorado, Geological Society of America, The Geology of North America, v. M, p. 351–378.

Kristoffersen, Y., and Talwani, M., 1977, Extinct triple junction south of Greenland and the Tertiary motion of Greenland relative to North America: Geological Society of America Bulletin, v. 88, p. 1037–1049.

Kron, A., and Stix, J., 1982, Geothermal gradient map of the United States exclusive of Alaska and Hawaii: Boulder, Colorado, National Oceanic and Atmospheric Administration National Geophysics Data Center, scale 1:2,500,000.

Lachenbruch, A. H., 1968, Preliminary model of the Sierra Nevada: Journal of Geophysical Research, v. 75, p. 3291–3300.

Lachenbruch, A. H., and Sass, J. H., 1977, Heat flow in the United States and the thermal regime of the crust, in Heacock, J. G., ed., The nature and physical properties of the Earth's crust: American Geophysical Union Geophysical Monograph 20, p. 626–675.

——— , 1980, Heat flow and energetics of the San Andreas fault zone: Journal of Geophysical Research, v. 90, p. 6185–6222.

——— , 1991, Heat flow from Cajon Pass, fault strength, and tectonic implications: Journal of Geophysical Research (in press).

Lachenbruch, A. H., Sass, J. H., and Galanis, S. P., Jr., 1985, Heat flow in southernmost California and the origin of the Salton trough: Journal of Geophysical Research, v. 90, p. 6709–6736.

Langseth, M. G., Westbrook, G. K., and Hobart, M. A., 1988, Geophysical survey of a mud volcano seaward of the Barbados Ridge accretionary complex: Journal of Geophysical Research, v. 93, p. 1049–1061.

LePichon, X., Hyndman, R. D., and Pautot, G., 1971, Geophysical study of the opening of the Labrador Sea: Journal of Geophysical Research, v. 26, p. 4724–4743.

Lewis, T. J., Jessop, A. M., and Judge, A. S., 1985, Heat flux measurements in southwestern British Columbia; The thermal consequences of plate tectonics: Canadian Journal of Earth Sciences, v. 22, p. 1262–1273.

Lewis, T. J., and 5 others, 1988, Subduction of the Juan de Fuca Plate; Thermal consequences: Journal of Geophysical Research, v. 93, p. 15207–15225.

Lonsdale, P., and Becker, K., 1985, Hydrothermal plumes, hot springs, and conductive heat flow in the southern trough of the Guaymas Basin: Earth and Planetary Science Letters, v. 73, p. 211–225.

Majorowicz, J. A., Jones, F. W., Lam, H. L., and Jessop, A. M., 1984, The variability of heat flow both regional and with depth in southern Alberta, Canada; Effect of ground-water flow: Tectonophysics, v. 106, p. 1–24.

Majorowicz, J. A., Jones, F. W., Lam, H. L., and Jessop, A. M., 1985, Terrestrial heat flow and geothermal gradients in relation to hydrodynamics in the Alberta Basin; Canada: Journal of Geodynamics, v. 4, p. 265–283.

Mareschal, J. C., and 7 others, 1989, New heat flow density and radiogenic heat production data in the Canadian Shield and the Quebec Appalachians: Canadian Journal of Earth Sciences, v. 26, p. 845–852.

Mathews, W. H., compiler, 1986, Physiographic map of the Canadian Cordillera: Geological Survey of Canada Map 1701A, scale 1:5,000,000.

Moran, J. E., and Lister, C.R.B., 1987, Heat flow across Cascadia Basin near 47°N, 128°W: Journal of Geophysical Research, v. 92, p. 11416–11432.

Morgan, P., 1985, Crustal radiogenic heat production and the selective survival of ancient continental crust: Journal of Geophysical Research, v. 90, supplement, p. C561–C570.

Morgan, P., and Gosnold, W. D., 1989, Heat flow and thermal regimes in the continental United States, in Pakiser, L. C., and Mooney, W. D., eds., Geophysical framework of the continental United States: Boulder, Colorado, Geological Society of America Memoir 172, p. 493–522.

Morgan, P., and Sass, J. H., 1984, Thermal regime of the continental lithosphere: Journal of Geodynamics, v. 1, p. 143–166.

Nathenson, M., and Guffanti, M., 1988, Geothermal gradients in the conterminous United States: Journal of Geophysical Research, v. 93, p. 6437–6450.

Nunns, A. G., 1983, Plate tectonic evolution of the Greenland-Scotland Ridge and surrounding regions, in Bott, M.H.P., Saxon, S., Talwani, M., and Thiede, J., eds., Structure and development of the Greenland-Scotland Ridge: New York, Plenum Press, p. 11–30.

Oxburgh, E. R., and Agrell, S. O., 1982, Thermal conductivity and temperature structure of the Reydarfjordur borehole: Journal of Geophysical Research, v. 87, p. 6423–6428.

Pollack, N., Johnson, J. R., and Harter, S. J., 1991, Global heat flow data compilation: Boulder, Colorado, National Oceanographic and Atmospheric Administration, National Geophysical Data Center (in press).

Prol-Ledesma, R. M., and Juarez, M. G., 1986, Geothermal map of Mexico: Journal of Volcanology and Geothermal Research, v. 28, p. 351–362.

Prol-Ledesma, R. M., and 9 others, 1989, Heat flow variations along the Middle America Trench: Marine Geophysical Research, v. 11, p. 69–76.

Roest, W. R., and Srivastava, S. P., 1989, Sea-floor spreading in the Laborador Sea; A new reconstruction: Geology, v. 17, p. 1000–1003.

Rosencrantz, E., Ross, M. I., and Sclater, J. G., 1988, Age and spreading history of the Cayman Trough as determined from depth, heat flow, and magnetic anomalies: Journal of Geophysical Research, v. 93, p. 2141–2147.

Roy, R. F., Decker, E. R., Blackwell, D. D., and Birch, F., 1968, Heat generation of plutonic rocks and continental heat flow provinces: Earth and Planetary Science Letters, v. 5, p. 1–12.

Roy, R. F., Blackwell, D. D., and Decker, E. R., 1972, Continental heat flow, in Robertson, E. C., ed., The nature of the solid earth: New York, McGraw-Hill, p. 506–543.

Sass, J. H., and 6 others, 1981, Heat flow from the crust of the United States, in Touloukian, T. S., Judd, W. R., and Roy, R. F., eds., Physical properties of rocks and minerals: New York, McGraw-Hill, CINDUS Data Series on Material Properties, v. II-2, p. 503–548.

Sass, J. H., Lachenbruch, A. H., Moses, T. H., Jr., and Morgan, P., 1991, Heat flow from a scientific research well at Cajon Pass, California: Journal of Geophysical Research (in press).

Sclater, J. G., Jaupart, C., and Galson, D., 1980, The heat flow through oceanic and continental crust and the heat loss of the earth: Reviews in Geophysics and Space Physics, v. 18, p. 269–311.

Sheffels, B., and McNutt, M., 1986, Role of subsurface loads and regional compensation in the isostatic balance of the Transverse Ranges, California; Evidence for intercontinental subduction: Journal of Geophysical Research, v. 91, p. 6419–6431.

Simkin, T., and 5 others, 1981, Volcanoes of the world; A regional directory, gazetteer, and chronology of volcanism during the last 10,000 years: Stroudsburg, Pennsylvania, Hutchinson Ross Publishing Co., 232 p.

Smith, D. L., and Dees, W. T., 1982, Heat flow in the Gulf Coastal Plain: Journal of Geophysical Research, v. 87, p. 7687–7693.

Smith, D. L., and Fuller, W. R., 1977, Terrestrial heat flow values in Florida and the effects of the aquifer system, in The geothermal nature of the Floridan Plateau: Florida Bureau of Geology Special Publication, p. 91–130.

Smith, D. L., Gregory, R. G., and Emhof, J. W., 1981, Geothermal measurements in the southern Appalachian Mountains and southeastern Coastal Plain: American Journal of Science, v. 328, p. 282–298.

Smith, D. L., Nuckels, C. E., Jones, R. L., and Cook, G. A., 1979, Distribution of heat flow and radioactive heat generation in northern Mexico: Journal of Geophysical Research, v. 84, p. 2371–2379.

Smith, R. L., and Shaw, H. R., 1979, Igneous-related geothermal systems, in Muffler, L.J.P., ed., Assessment of geothermal resources of the United States—1978: U.S. Geological Survey Circular 790, p. 12–18.

Souther, J. G., 1975, Geothermal potential of western Canada, in Second, U.N. Symposium on the Development and Use of Geothermal Resources: Washington, D.C., U.S. Government Printing Office, v. 1, p. 259–267.

Souther, J. G., and Halstead, E. C., 1973, Mineral and thermal waters of Canada: Canada Department of Energy, Mines and Resources Paper 73-18, p. 225–256.

Swanberg, C. A., and Morgan, P., 1980, The silica heat flow technique; Assumptions and applications: Journal of Geophysical Research, v. 85, p. 7206–7214.

Talwani, M., and Eldholm, O., 1977, Evolution of the Norwegian-Greenland Sea: Geological Society of America Bulletin, v. 88, p. 969–999.

Vitorello, I., and Pollack, H. N., 1980, On the variation of continental heat flow with age and the thermal evolution of continents: Journal of Geophysical Research, v. 85, p. 983–995.

Vogt, P. R., and Einwich, A. M., 1979, Magnetic anomalies and sea-floor spreading in the western Atlantic, and a revised calibration of the Keathly (M) geomagnetic reversal chronology, in Tucholke, B. E., and Vogt, P. R., eds., Initial reports of the Deep Sea Drilling Project: Washington, D.C., U.S. Government Printing Office, v. 43, p. 857–867.

Wallace, R. H., Jr., Kraemer, T. F., Taylor, R. E., and Wessleman, J. B., 1979, Assessment of geopressured-geothermal resources in the northern Gulf of Mexico basin, in Muffler, L.J.P., ed., Assessment of geothermal resources of the United States—1978: U.S. Geological Survey Circular 790, p. 132–155.

Wallace, R. H., Jr., Taylor, R. E., and Wesselman, J. B., 1977, Use of hydrogeologic mapping techniques in defining potential geopressured-geothermal reservoirs in the lower Rio Grande embayment, Texas, in Meriwether, J., ed., Proceedings, Third geopressured-geothermal energy conference: Lafayette, Louisiana, University of Southwestern Louisiana Center for Energy Studies, p. GI1–88.

Wetmiller, R. J., Langseth, M. G., Marshall, D. V., and Lachenbruch, A. H., 1990, Seismicity and heat flow of the Arctic, in Grantz, A., Johnson, L., and Sweeney, J. F., eds., The Arctic Ocean region: Boulder, Colorado, Geological Society of America, The Geology of North America, v. L, Plate 2.

Zandt, G., and Furlong, K. P., 1982, Evolution and thickness of the lithosphere beneath coastal California: Geology, v. 10, p. 376–381.

Zoback, M. D., and 12 others, 1987, New evidence on the state of stress of the San Andreas fault system: Science, v. 238, p. 1105–1111.

Zoback, M. D., Silver, L. T., Henyey, T., and Thatcher, W., 1988, The Cajon Pass scientific drilling experiment; Overview of phase 1: Geophysical Research Letters, v. 15, p. 933–936.

Ziagos, J. P., Blackwell, D. D., and Mooser, F., 1985, Heat flow and subduction in southern Mexico: Journal of Geophysical Research, v. 90, p. 5410–5420.

MANUSCRIPT ACCEPTED BY THE SOCIETY SEPTEMBER 26, 1990

ACKNOWLEDGMENTS

The map compilation was supported by the U.S. Department of Energy through contract ID/12623. Support for some aspects of the compilation also came from Southern Methodist University. Part of the support for the preparation of this paper came from the National Science Foundation under grant EAR-8817176. Sidney Baker assisted in the data compilation, and Kris Eckhardt helped in the preparation of this manuscript. Paul Morgan and Shari Kelley made helpful comments on the manuscript.

Printed in U.S.A.

The Geology of North America
Decade Map Volume 1
1991

Chapter 24

Terrestrial heat flow in Canada

Alan M. Jessop
Institute of Sedimentary and Petroleum Geology, Geological Survey of Canada, Energy, Mines and Resources, 3303 33rd Street N.W., Calgary, Alberta T2L 2A7, Canada

INTRODUCTION

The first heat-flow data in Canada were published in 1951 (Misener and others). This was twelve years after the earliest publication of completely measured heat flow (Bullard, 1939; Benfield, 1939), a period that included the Second World War. Canada was the third country, after the United Kingdom and the United States, to produce heat-flow data, and was followed by Australia, Poland, South Africa, and Iceland.

At present, heat-flow values have been published or submitted for publication from 298 sites in Canada, including the continental shelves. A list of these data is to be found in Jessop (1989). A "site" may include several boreholes if they are close together, the limiting distance being about 10 km. However, this definition cannot be adhered to strictly. A line of boreholes spaced at less than 10 km cannot be reconciled with this definition, and closely spaced measurements that reflect different geothermal settings cannot provide useful averages. In general, the concept of "site" is intended to permit the averaging of closely spaced measurements, for example, in several boreholes at a mining property. The criteria for combination are: (1) the geological setting is uniform, (2) the results are similar, and (3) the quality of the data is sufficiently low to indicate combination into one result.

Some of the continental data have been acquired by oceanic probe techniques in shallow lakes. A summary of all continental data by style of site and type of measurement is given in Table 1. Table 2 shows mean and distribution of the data by physiographic and tectonic setting, as shown in Figure 1. An additional 59 measurements have been made by single probe casts from ice in the Arctic Ocean or ships in Atlantic Ocean, Labrador Sea, and Baffin Bay, and numerous measurements have been made in the Pacific Ocean by "pogo-stick" techniques.

TABLE 1. SUMMARY OF HEAT-FLOW DATA IN AND AROUND CANADA*

	Single	Multiple	Total
Borehole on land	150	26	176
Lakes	2	7	9
Shelves or lakes	33	1	34
Shelves by oil-well	78	1	79
Total	263	35	298

*These numbers include data from the Canadian Cordillera, described in a separate chapter.

THE TECTONIC SETTING

The Canadian continental area, with the United States to the south, has the structure of a craton surrounded by later additions, as shown in Figure 1. The interior of the country is a large Precambrian craton, part of which is exposed to form the Canadian Shield, and the remainder is covered by sediments or shallow seas to form the platforms of the Great Plains, the southern Arctic islands and Hudson Bay. The Precambrian craton has a central core, the Superior Province of Archean age, surrounded by other provinces of Proterozoic age. The central craton is almost surrounded by younger additions. On the west is a complex cordilleran belt, which is described in a companion chapter (Lewis, this volume). In the north is the Sverdrup Basin, originating in late Paleozoic time. In the east is the Atlantic margin, a result of Paleozoic orogenic events. Only to the northeast, the shore of Baffin Bay and the Labrador Sea, does the central craton

Jessop, A. M., 1991, Terrestrial heat flow in Canada, *in* Slemmons, D. B., Engdahl, E. R., Zoback, M. D., and Blackwell, D. D., eds., Neotectonics of North America: Boulder, Colorado, Geological Society of America, Decade Map Volume 1.

TABLE 2. SUMMARY OF HEAT FLOW BY TECTONIC SETTING

Region	Number	Mean	Standard Deviation	Standard Error
		mW/m^2		
Nova Scotia Shelf	39	73.4	11.5	1.8
Newfoundland Shelf	28	78.1	15.1	2.8
Total eastern shelves	67	75.4	13.3	1.6
Atlantic margin	40	56.8	11.6	1.8
Superior Province	27	42.6	8.6	1.7
Churchill Province	7	43.0	11.2	4.2
Grenville Province	4	47.2	9.5	4.7
Total shield (excluding lakes)	40	43.3	9.3	1.5
Grenville platforms	17	40.2	5.6	1.4
Western platform	12	72.3	12.1	3.5
In Mesozoic section*		69.5	13.7	
In Paleozoic section		76.5	19.2	
Cordillera†	64	71.5	22.7	2.8

*Derived from maps of heat flow estimated from industrial data.
†These data are described in detail in a separate chapter.

meet the ocean where Cretaceous rifting has separated Greenland from Canada.

DISTRIBUTION OF HEAT-FLOW DATA

The distribution of data in terms of major physiographic units is summarized in Table 2. Not all units have sufficient data to merit the calculation of means and deviations from the mean. The Altantic margin, parts of the Canadian Shield, and parts of the Cordillera are well covered, but there are still significant parts of the country with few published data. Figure 2 shows histograms of heat flow in some regions of Canada, the details of which are reviewed in the following sections.

Most Canadian data have been corrected for effects of the recent glaciations, which vary widely from south to north. Although data on times of change and effective temperature of the present surface are not exact, the correction is believed to improve the measurement. Most corrections have been applied by means of a model based on Birch (1948) and reviewed in the Canadian setting by Jessop (1971).

The eastern continental shelf

On the eastern continental shelf heat flow has been estimated from a combination of industrial temperature data and integrated thermal resistance derived from net-rock analysis and conductivity by rock type. These estimates are not as reliable as

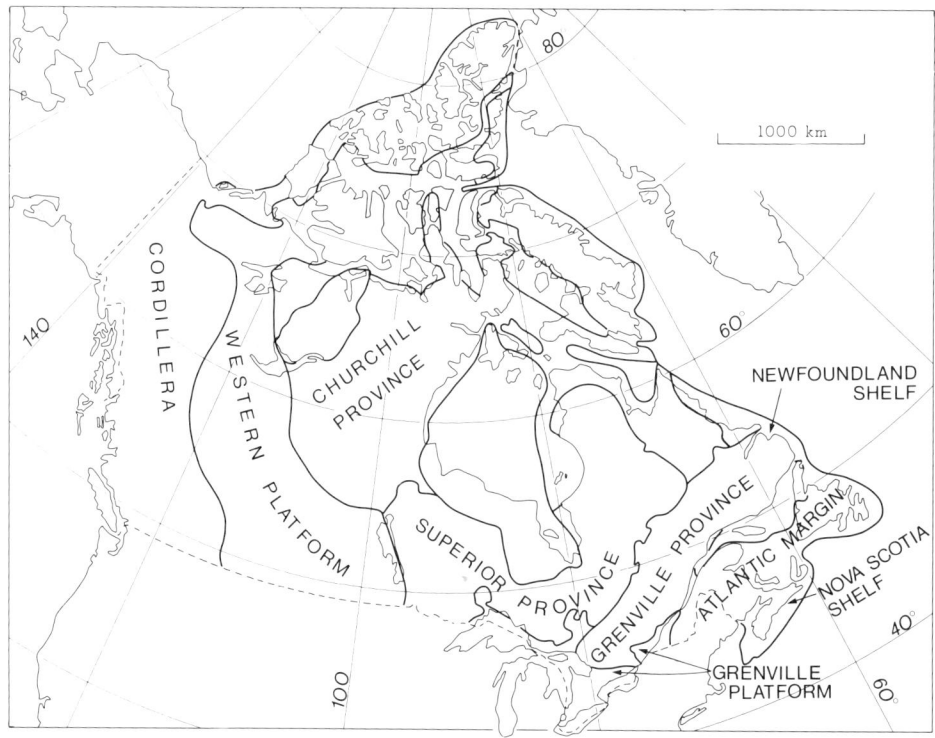

Figure 1. The physiographic regions of Canada in which heat-flow data are sufficiently numerous for the calculation of means in Table 2.

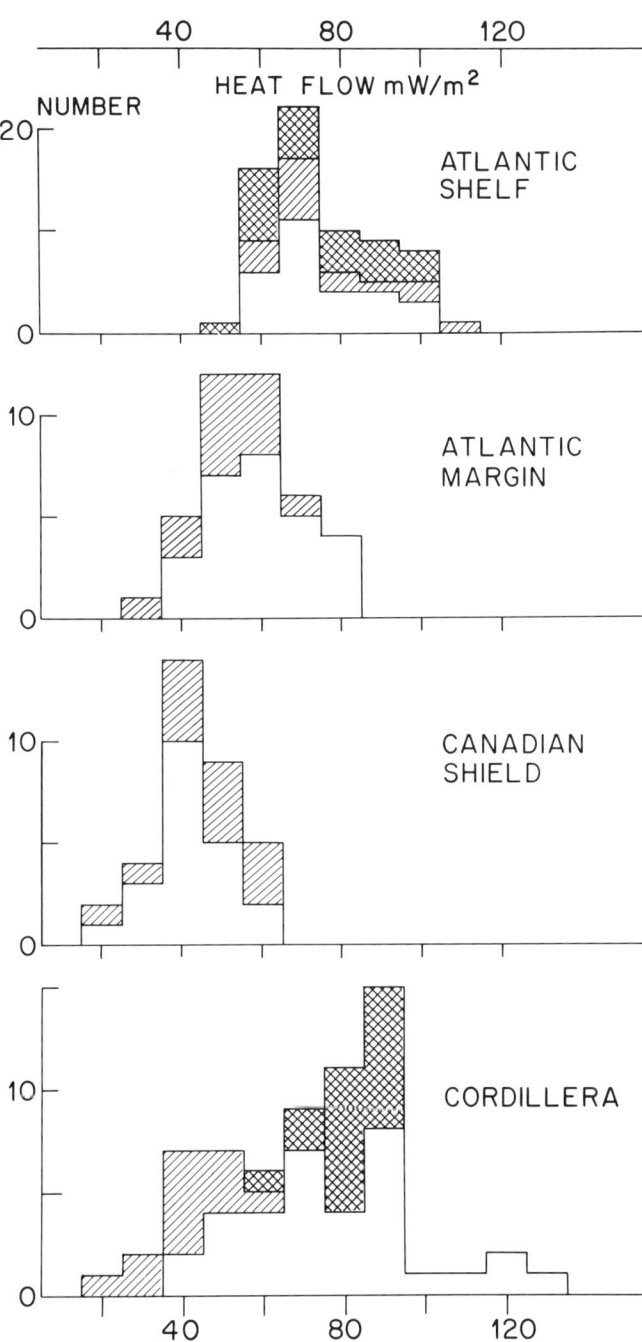

Figure 2. Histograms of heat flow in four of the main structural provinces of Canada. In the individual histograms the unshaded parts show the reliable data from deep boreholes, free from recognizable disturbances, while the lined and squared shadings, respectively, show extra data of significance as follows: Atlantic shelf, data of categories B and C (Reiter and Jessop, 1985); Atlantic margin, less reliable data; Canadian Shield, data from areas other than the Superior Province; Cordillera, data from coastal fjords and from shallow boreholes.

data derived from multiple measurements of temperature and conductivity, but they are the best that can be achieved at a reasonable cost on the continental shelf. The greatest potential problem in these data is the estimation of conductivity. It is virtually impossible to obtain samples in a state reasonably close to their condition before drilling, and most samples are no more than a small phial of broken fragments. Thus thermal resistance has been estimated from a combination of conductivities from the literature for each type of rock and an analysis from the lithological logs of the fraction of each type of rock in each section of the well. However, the error in the heat-flow estimates should be within the estimated 25 percent, and the patterns of heat flow revealed by these estimates should be valid, even though the whole set may be biased by erroneous conductivities for the more common rock types, particularly shale.

These data were divided by the authors into three categories of quality, based on internal vertical consistency (Reiter and Jessop, 1985). Figure 2 shows a histogram of these estimates, divided into the three categories from A (the best) to C (the worst). These data do not include the four points located in the Gulf of St. Lawrence and the Bay of Fundy. Although no definite interpretation could be given, it was suggested that hydrological redistribution of heat flow in the sediments is a probable cause of the observed patterns and some vertical variations in heat flow. Thus the rather high average heat flow may not be representative of the basement below the sediments, but merely a reflection of the enhanced transfer of heat by convection.

The Atlantic margin

Drury and others (1987) have shown that heat flow, based on good measurements of both temperature and conductivity, from the Appalachian region of Canada, but excluding Newfoundland, has an average value of 61 mW/m^2. Heat flow in the Carboniferous sediments of the Magdalen Basin is generally low, at about 40 to 50 mW/m^2, with zones of higher heat flow on the northwestern and southeastern flanks. However, the Fredericton graben, the southwestern corner of the basin, is part of an area of high heat flow, from 70 to 80 mW/m^2. The Fredericton graben is thought to be underlain by a granitic basement of radiogenic content similar to that of the batholiths on either side, where similarly high heat flow is observed.

Four of five measurements in Newfoundland have an average of 47 mW/m^2, rather lower than the average for the whole Atlantic margin, but these are from the "Dunnage Zone," an area of oceanic origin that does not extend to the mainland. The fifth site is in a granitic batholith and shows a high heat flow. The whole Atlantic Margin has been regarded as a "heat-flow province" (see Blackwell and others, this volume, for a definition) with an average heat flow of 57 mW/m^2, a reduced heat flow of 40 mW/m^2, and a depth constant of 10 km (Wright and others, 1980). However, some of the sites in granitic batholiths show a heat generation too high for the measured heat flow, suggesting that the layer of high heat generation is of limited thickness and not representative of the crust below a depth of about 3 km (Drury and others, 1987).

A total of 40 data points in the whole Atlantic margin, including four measurements from offshore wells in the Gulf of St. Lawrence and the Bay of Fundy, have an average of 57 mW/m^2. These values are illustrated in the histogram of Figure 2, where they are divided by quality, but the inclusion of the poorer data, indicated by shading, does not significantly modify the distribution shown by the reliable data.

The Precambrian craton

Heat flow in the Superior Province of the Canadian Shield is very uniform, with an average of 43 mW/m^2 and a standard deviation of 9 mW/m^2. There is a tendency for the heat flow at sites nearest to Hudson Bay to be the lowest, as for example at Nielsen Island, Chibougamou, and Otoskwin River, where the average is 27 mW/m^2, but these data are too few and too widely spaced to confirm a definite trend.

The relation of heat flow to heat generation in the Superior Province has been examined at length (Cermak and Jessop, 1971; Rao and Jessop, 1976; Jessop and Lewis, 1978; Drury and Taylor, 1987). Since erosion of the surface may have removed a thickness of crust comparable to the normal magnitude of the depth factor provided by various heat-flow provinces, the interpretation of such a relationship is not necessarily the same as for areas of crust of younger age. However, the data provide a reduced heat flow of 28 mW/m^2 and a depth factor of 14 km, which suggests a less advanced differentiation of the radioactive elements than is found in younger areas.

The histogram of Figure 2 reflects the uniformity of the measured heat flow, particularly in the Superior Province, which is shown by the unshaded space. The data from the other provinces reinforce the peak and tend to favour the higher side of the histogram. As shown in Table 2, the small number of data from the youngest province, the Grenville, average 47 mW/m^2. The differences between averages for individual provinces are not statistically significant.

The platform regions

Extensive analysis of a large set of temperature data, collected from records of industrial well logs and drill-stem tests, has shown that heat flow varies both laterally and vertically within the sedimentary strata of the Great Plains of Canada (Majorowicz and others, 1985, 1986). Heat flow has been estimated by statistical analysis of temperature data within squares of 10 × 10 or 30 × 30 km, combined with estimates of thermal resistance based on net-rock analysis of lithological logs of a network of wells, both above and below the Paleozoic-Mesozoic unconformity. Each heat-flow estimate thus represents an average heat flow within the chosen strata over an area of 100 or 900 km^2, depending on the density of the temperature data, in contrast with the normal measured heat flow in a single borehole at a single location. When plotted on a map, these estimates show that heat flow in the Mesozoic strata is low in the southwest and higher in the northeast, whereas heat flow in the Paleozoic strata shows the opposite trend. On this regional trend are superimposed variations of more local extent and lesser depth. This complex pattern has been interpreted as the result of water migration, driven on the regional scale by the recharge area of the Rocky Mountains toward the northeast, on intermediate scales by relative uplands within the plains, and on the local scale by small features such as escarpments and river valleys (Hitchon, 1984).

Figure 3 shows histograms of the heat flow, above and below the Paleozoic-Mesozoic unconformity. These histograms have been derived from the maps supplied by the University of Alberta (Jones and Majorowicz, personal communication, 1987), on a scale of 1:2,000,000. Only areas where heat-flow data are available in both sections were used, amounting to a total area of about 530,000 km^2, or about 45 percent of the whole platform area between latitudes 49 and 60°N and longitudes 96 and 120°W. Heat-flow distribution in the Mesozoic strata is restricted within a fairly narrow band of 50 to 80 mW/m^2, with an average and standard deviation of 70 and 14 mW/m^2. Heat flow in the Paleozoic strata is more widely spread, with an average and standard deviation of 76 and 19 mW/m^2. Given that less than half of the platform area is covered, the difference in average heat flow between the two levels is not significant. The parts of the platform not covered are generally in the northeast, where the sedimentary section is shallow. Since this part tends to have heat

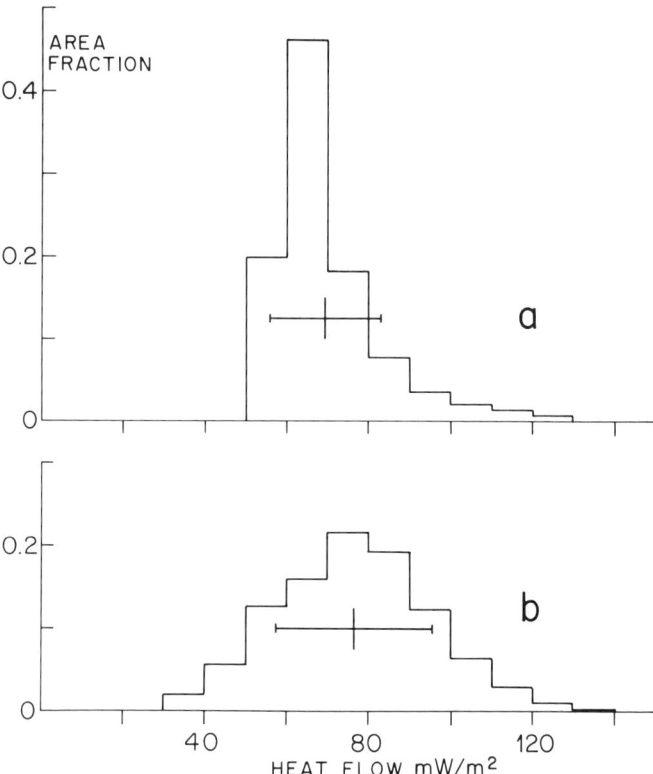

Figure 3. Histograms of heat flow in the Western Canada Sedimentary Basin, a, in the Mesozoic section, and b, in the Paleozoic section, estimated from maps. Means and standard deviations are shown by vertical and horizontal bars, respectively.

flow higher in the Mesozoic strata than in the Paleozoic, data from this area would probably eliminate the difference in average heat flow. Thus the different patterns of heat flow in the two levels seem to reflect a lateral redistribution of heat, rather than a system of sources and sinks, adding support to the theory of hydrodynamic distortion.

It has not been conclusively demonstrated that water flow is the only cause of the anomalies, either here or on the eastern continental shelf. Bachu (1988) has argued that convection is not capable of causing the observed basinwide thermal anomaly, and that convective heat transport is important on the local scale only. Bachu proposes heat flow from the mantle as the cause of the lateral variations, but, while deep sources can cause lateral variations in heat flow, they cannot account for the differences between the patterns in the Paleozoic and Mesozoic sections. Information on the distribution of heat flow from the Precambrian basement is sparse and poorly distributed, but there is no reason to suggest any greater variation than is observed in the exposed shield to the east. No other cause capable of the observed widespread redistribution of heat flow is recognized at present.

Experience with one deep well has shown that accurate measurement in a single deep borehole does not give a uniform vertical heat flow, and that even on the scale of a single hole, variation may be clearly seen (Jessop and Vigrass, 1989). Although problems of determination of conductivity of shales remain to be resolved, it is concluded that it is not possible to measure in the sediments a heat flow that is known to be representative of the impermeable basement. Wells deepened into the basement are necessary for the measurement of both heat flow and radioactive heat generation that may be associated with the crust as a whole.

The Cordillera

Heat flow in the Cordillera show a wide variation, as reflected by the histogram of Figure 2. These data are described in detail in an accompanying chapter (Lewis, this volume), and are included here only for comparison with other regions of Canada.

The oceans

Oceanic measurements have been made in the seas adjacent to Canada. Most of these have been made in the Pacific Ocean, in studies of the plate margins near the coast of British Columbia, and are described in a companion chapter. Small groups of data are available from the continental rise of the east coast (Lewis and Hyndman, 1976), Baffin Bay and the Labrador Sea (Pye and Hyndman, 1972), and the Arctic Ocean (Taylor and others, 1987).

One of the profiles of Lewis and Hyndman (1976) extends the estimates of Reiter and Jessop (1985) beyond the edge of the continental shelf, and seems to show the seaward size of a zone of high heat flow that coincides roughly with the edge of the shelf.

In Baffin Bay and the Labrador Sea, heat-flow data show no definite evidence of anomalies that might be associated with postulated spreading centres. It was concluded there is no active spreading at present in these areas, although slow spreading or residual effects of ancient spreading could not be ruled out.

Heat-flow measurements on the Alpha Rise were taken by using the ice as an observation platform. These data are in general agreement with a much larger set of data obtained from ice-island T-3 (Lachenbruch and Marshall, 1969; Langseth and others, 1990), and suggest an oceanic origin for the Alpha Rise (Taylor and others, 1986).

HEAT-FLOW PROFILES

Figure 4 shows a profile of heat flow on a great circle from Vancouver to Halifax. All data within 200 km of this line are projected on to it, and maximum and minimum envelopes have been drawn to include most of the data. This profile is similar to, but updated from, Figure 5 of Jessop and others (1984a). The

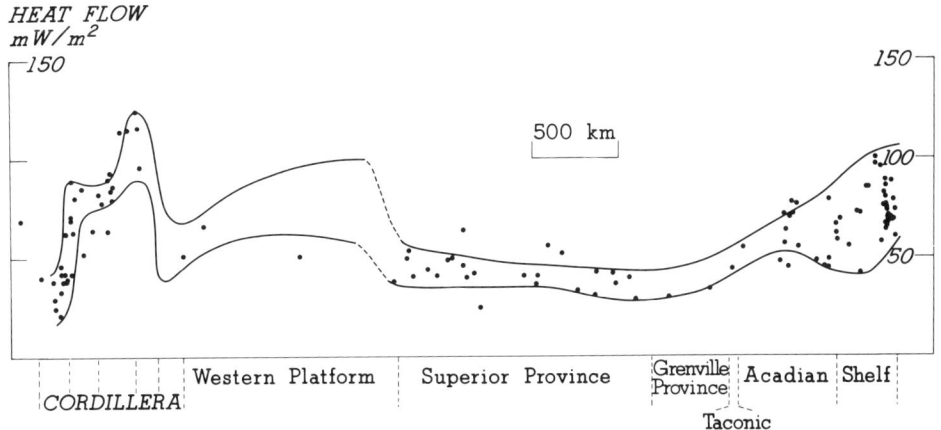

Figure 4. Profile of heat flow across southern Canada from Vancouver to Halifax. Data shown are within 200 km of the great circle profile. Envelopes over the Western Platform are estimated from the maps of Majorowicz and others (1985, 1986).

profile shows, from west to east, a steep rise across the coastal zone to the line of the Garibaldi Volcanics, high values in the Cordillera rising to a maximum in the Purcell Mountains, and a rapid drop at the Rocky Mountain Trench. In the Prairies envelopes are drawn from the maps of Majorowicz and others (1985, 1986), since there are few data from individual sites. The value of 51 mW/m² at Regina, taken as the best value to be derived from detailed measurements in the sedimentary section of a single well, falls outside the envelope. The connection between the apparent high heat flow in the sediments and the typical shield value, as measured at Winnipeg is obscure, and the envelopes are shown only as dashed lines. Continuing eastward, the Canadian Shield shows uniform heat flow typical of Precambrian regions, and the eastern margin shows a general rise with some high values on the Scotia Shelf.

Since the great-circle profile does not cross the margins at right angles, shorter profiles are shown in Figure 5, the western margin, and Figure 6, the eastern margin. Figure 5 shows heat flow rising from a low level over Vancouver Island to high background levels inland. Several abnormally high values have been observed in the vicinity of the Lillooet River, associated with the hydrothermal systems of the northern Cascade volcanic centers. This area is described in more detail in the accompanying paper by Lewis (this volume).

Figure 6 shows somewhat variable but normal heat flow in New Brunswick and Nova Scotia, with the highest values associated with large granitic intrusions. Offshore values reach a peak at the edge of the shelf and fall away rapidly on the continental rise. As discussed above, heat-flow estimates on the shelf are subject to uncertainties derived from convective redistribution and errors in estimation of conductivity.

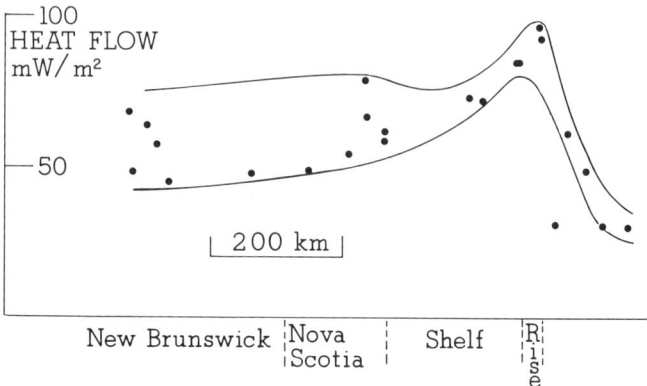

Figure 6. Profile of heat flow across the eastern continental margin from New Brunswick to the continental rise.

CONCLUSIONS

Many interpretations of the data summarized here have been presented in the papers in which the data were first published. The single main conclusion to be derived from the assembled data is that the Canadian terrain, and all other terrains, can be divided into two categories: one where the assumption of purely conductive heat flow is usually valid and its measurement may be achieved by a combination of measurements of temperature and thermal conductivity; and the other where the assumption of purely conductive heat flow is not valid, the measurable conductive component varies both laterally and vertically, and any one measured data point does not necessarily represent the conductive heat flow in the basement. The two categories can only be distinguished with certainty by direct observation.

The regions presenting the greatest distortion to traditionally determined conductive heat flow are the sedimentary basins, and much remains to be learned about the operative mechanisms of heat transfer, the details of the individual basins, and the variation of heat flow and temperature with time. Since heat flow is observed to vary both laterally and vertically, and is normally determined by the statistical analysis of low-quality data over specified areas, the results are difficult to present in a comprehensible form. For studies of the crust and upper mantle, direct measurements of heat flow in the basement would be preferable to observations in the sediments. However, the subject of geothermics is of profound importance in the assessment of thermal history of basins and in theories of generation and accumulation of hydrocarbons, which present a major economic incentive for the expansion of geothermal research.

In areas where conventional heat-flow measurement is possible, the observed data generally follow the trend of lower heat flow and lower variation of heat flow with greater tectonic age. The Superior Province of the Precambrian Shield is a particularly good example of uniform and moderate heat flow, in contrast with the Cordillera, where heat flow is higher and more varied. However, there are still gaps to be filled as opportunities arise.

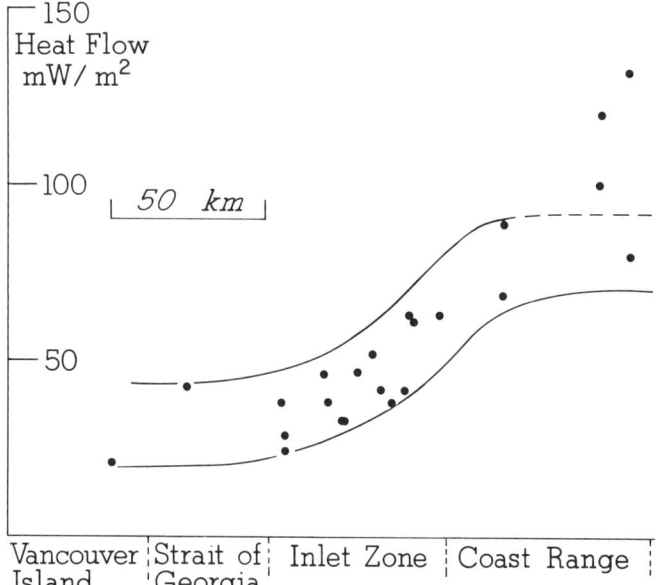

Figure 5. Profile of heat flow across the western continental margin from Vancouver Island to the Garibaldi Volcanic Belt.

REFERENCES CITED*

*Includes sources of information for the Canadian data on the Geothermal Map of North America.

Allis, R. G., and Garland, G. D., 1976, Geothermal measurements in five small lakes of northwest Ontario: Canadian Journal of Earth Sciences, v. 13, p. 987–992.

—— , 1979, Heat flow measurements under some lakes in the Superior Province of the Canadian Shield: Canadian Journal of Earth Sciences, v. 16, p. 1951–1964.

Bachu, S., 1988, Analysis of heat transfer processes and geothermal pattern in the Alberta Basin, Canada: Journal of Geophysical Research, v. 93, p. 7767–7781.

Beck, A. E., 1962, Terrestrial flow of heat near Flin Flon, Manitoba: Nature, v. 195, p. 368–369.

Beck, A. E., and Logis, Z., 1963, Terrestrial flow of heat in the Brent crater: Nature, v. 201, p. 383.

Beck, A. E., and Sass, J. H., 1966, A preliminary value of heat flow at the Muskox Intrusion near Coppermine, N.W.T., Canada: Earth and Planetary Science Letters, v. 1, p. 123–129.

Benfield, A. E., 1939, Terrestrial heat flow in Great Britain: Proceedings of the Royal Society of London, Series A, v. 173, p. 428–450.

Birch, F., 1948, The effects of Pleistocene climatic variation upon geothermal gradients: American Journal of Science, v. 246, p. 729–760.

Bullard, E. C., 1939, Heat flow in South Africa: Proceedings of the Royal Society of London, Series A, v. 173, p. 474–502.

Cermak, V., and Jessop, A. M., 1971, Heat flow, heat generation and crustal temperature in the Kapuskasing area of the Canadian Shield: Tectonophysics, v. 11, p. 287–303.

Davis, E. E., and Lewis, T. J., 1984, Heat flow in a back-arc environment; Intermontane and Omineca Crystalline Belts, southern Canadian Cordillera: Canadian Journal of Earth Sciences, v. 21, p. 715–726.

Drury, M. J., 1985, Heat flow and heat generation in the Churchill Province of the Canadian Shield, and their palaeotectonic significance: Tectonophysics, v. 115, p. 25–44.

Drury, M. J., and Lewis, T. J., 1983, Water movement within Lac du Bonnet Batholith as revealed by detailed thermal studies of three closely-spaced boreholes: Tectonophysics, v. 95, p. 337–351.

Drury, M. J., and Taylor, A. E., 1987, Some new measurements of heat flow in the Superior Province of the Canadian Shield: Canadian Journal of Earth Sciences, v. 24, p. 1486–1489.

Drury, M. J., Jessop, A. M., and Lewis, T. J., 1987, The thermal nature of the Canadian Appalachian crust: Tectonophysics, v. 133, p. 1–14.

Fou, J.T.K., 1969, Thermal conductivity and heat flow at St. Jerome, Quebec [M.Sc. thesis]: Montreal, McGill University, 79 p.

Garland, G. D., and Lennox, D. H., 1962, Heat flow in western Canada: Geophysical Journal of the Royal Astronomical Society, v. 6, p. 245–262.

Hitchon, B., 1984, Geothermal gradients, hydrodynamics, and hydrocarbon occurrences, Alberta, Canada: American Association of Petroleum Geologists Bulletin, v. 68, p. 713–743.

Hyndman, R. D., 1976, Heat flow measurements in the inlets of southwestern British Columbia: Journal of Geophysical Research, v. 81, p. 337–349.

Hyndman, R. D., Jessop, A. M., Judge, A. S., and Rankin, D. S., 1979, Heat flow in the Maritime Provinces of Canada: Canadian Journal of Earth Sciences, v. 16, p. 1154–1165.

Hyndman, R. D., and 5 others, 1982, Queen Charlotte fault zone; Heat flow measurements: Canadian Journal of Earth Sciences, v. 19, p. 1657–1669.

Jessop, A. M., 1968, Three measurements of heat flow in eastern Canada: Canadian Journal of Earth Sciences, v. 5, p. 61–68.

—— , 1971, The distribution of glacial perturbation of heat flow in Canada: Canadian Journal of Earth Sciences, v. 8, p. 162–166.

—— , 1986, Geothermal regimes and geothermal resources of sedimentary basins, in Hydrogeology of sedimentary basins; Application to exploration and exploitation; Proceedings of the Canadian/American Conference on Hydrogeology: U.S. National Water Well Association, p. 127–139.

—— , 1989, Heat flow data base for Canada, in Hittelman, A. M., Kinsfather, J. O., and Meyers, H., eds., Geophysics of North America: Boulder, Colorado, National Oceanographic and Atmospheric Administration, National Geophysical Data Center, CD-ROM No. 975-B27-001.

Jessop, A. M., and Judge, A. S., 1971, Five measurements of heat flow in southern Canada: Canadian Journal of Earth Sciences, v. 8, p. 771–716.

Jessop, A. M., and Lewis, T. J., 1978, Heat flow and heat generation in the Superior Province of the Canadian Shield: Tectonophysics, v. 50, p. 55–77.

Jessop, A. M., and Vigrass, L. W., 1989, Geothermal measurements in a deep well at Regina, Saskatchewan: Journal of Volcanology and Geothermal Research, v. 37, p. 151–166.

Jessop, A. M., Hobart, M. A., and Sclater, J. G., 1976, The world heat flow data collection, 1975: Ottawa, Energy, Mines and Resources, Earth Physics Branch Geothermal Series, v. 5, 125 p.

Jessop, A. M., Lewis, T. J., Judge, A. S., Taylor, A. E., and Drury, M. J., 1984a, Terrestrial heat flow in Canada: Tectonophysics, v. 103, p. 239–261.

Jessop, A. M., Souther, J. G., Lewis, T. J., and Judge, A. S., 1984b, Geothermal measurements in northern British Columbia and southern Yukon Territory: Canadian Journal of Earth Sciences, v. 21, p. 599–608.

Judge, A. S., and Beck, A. E., 1973, Analysis of heat flow data—Several boreholes in a Sedimentary Basin: Canadian Journal of Earth Sciences, v. 10, p. 1494–1507.

Lachenbruch, A. H., and Marshall, B. V., 1969, Heat flow in the Arctic: Arctic, v. 22, p. 300–311.

Langseth, M. G., Lachenbruch, A. H., and Marshall, B. V., 1990, Geothermal observations in the Arctic region, in Grantz, A., Johnson, L., and Sweeney, J. F., eds., The Arctic Ocean region: Boulder, Colorado, Geological Society of America, The Geology of North America, v. L, p. 133–152.

Law, L. K., Paterson, W.S.B., and Whitham, K., 1965, Heat flow determinations in the Canadian arctic archipelago: Canadian Journal of Earth Sciences, v. 2, p. 59–71.

Lewis, J. F., and Hyndman, R. D., 1976, Oceanic heat flow measurements over the continental margins of eastern Canada: Canadian Journal of Earth Sciences, v. 13, p. 1031–1038.

Lewis, J. F., and Jessop, A. M., 1981, Heat flow in the Garibaldi volcanic belt, a possible Canadian geothermal resource area: Canadian Journal of Earth Sciences, v. 18, p. 366–375.

Lewis, T. J., 1969, Terrestrial heat flow at Eldorado, Saskatchewan: Canadian Journal of Earth Sciences, v. 6, p. 1191–1197.

—— , 1984, Geothermal energy from Penticton Tertiary outlier, British Columbia: Canadian Journal of Earth Sciences, v. 21, p. 181–188.

Lewis, T. J., and Beck, A. E., 1977, Analysis of heat-flow data—detailed observations in many holes in a small area: Tectonophysics, v. 41, p. 41–59.

Lewis, T. J., and Souther, J. G., 1978, Meager Mountain, B.C.—A possible geothermal energy resource: Ottawa, Energy, Mines and Resources, Earth Physics Branch Geothermal Series, v. 9, 17 p.

Lewis, T. J., Jessop, A. M., and Judge, A. S., 1985, Heat flux measurements in southwestern British Columbia; The thermal consequences of plate tectonics: Canadian Journal of Earth Sciences, v. 22, p. 1262–1273.

Majorowicz, J. A., and Jessop, A. M., 1981, Regional heat flow patterns in the Western Canadian Sedimentary Basin: Tectonophysics, v. 74, p. 209–238.

Majorowicz, J. A., Jones, F. W., Lam, H. L., and Jessop, A. M., 1985, Terrestrial heat flow and geothermal gradients in relation to hydrodynamics in the Alberta Basin, Canada: Journal of Geodynamics, v. 4, p. 265–283.

Majorowicz, J. A., Jones, F. W., and Jessop, A. M., 1986, Geothermics of the Williston Basin in Canada in relation to hydrodynamics and hydrocarbon occurrences: Geophysics, v. 51, p. 767–779.

Mathews, W. H., 1972, Geothermal data from the Granduc area, northern Coast Mountains of British Columbia: Canadian Journal of Earth Sciences, v. 9, p. 1333–1337.

Misener, A. D., 1955, Heat flow and depth of permafrost at Resolute Bay, Cornwallis Island, N.W.T., Canada: Transactions of the American Geophysical Union, v. 36, p. 1055–1060.

Misener, A. D., Thompson, L.G.D., and Uffen, R. J., 1951, Terrestrial heat flow in Ontario and Quebec: Transactions of the American Geophysical Union, v. 32, p. 729–738.

Paterson, W.S.B., and Law, L. K., 1966, Additional heat flow determinations in the area of Mould Bay, Arctic Canada: Canadian Journal of Earth Sciences, v. 3, p. 237–246.

Pye, G. D., and Hyndman, R. D., 1972, Heat-flow measurements in Baffin Bay and the Labrador Sea: Journal of Geophysical Research, v. 77, p. 938–944.

Rankin, D. S., and Hyndman, R. D., 1971, Shallow water heat flow measurements in Bras d'Or Lake, Nova Scotia: Canadian Journal of Earth Sciences, v. 8, p. 96–101.

Rao, R.U.M., and Jessop, A. M., 1976, A comparison of the thermal characters of shields: Canadian Journal of Earth Sciences, v. 12, p. 347–360.

Reader, J. F., and Fairbanks, B. D., 1983, Heat flow in the vicinity of the Meager Volcanic Complex, southwestern British Columbia: Transactions of the Geothermal Resources Council, v. 7, p. 535–539.

Reiter, M. A., and Jessop, A. M., 1985, Estimates of terrestrial heat flow in offshore eastern Canada: Canadian Journal of Earth Sciences, v. 22, p. 1503–1517.

Sass, J. H., Killeen, P. G., and Mustonen, E. D., 1968, Heat flow and surface radioactivity in the Quirke Lake Syncline near Elliot Lake, Ontario, Canada: Canadian Journal of Earth Sciences, v. 5, p. 1417–1428.

Sass, J. H., Lachenbruch, A. H., and Jessop, A. M., 1971, Uniform heat flow in a deep hole in the Canadian Shield and its paleoclimatic implications: Journal of Geophysical Research, v. 76, p. 8586–8596.

Saull, V. A., Clark, T. H., Doig, R. P., and Butler, R. B., 1962, Terrestrial heat flow in the St. Lawrence Lowland of Quebec: Canadian Mining and Metallurgy Bulletin, v. 65, p. 63–66.

Taylor, A. E., and Judge, A. S., 1979, Permafrost studies in northern Quebec: Géographie physique et quaternaire, v. 33, p. 245–251.

Taylor, A. E., Judge, A. S., and Allen, V. S., 1986, Terrestrial heat flow from Project Cesar, Alpha Ridge, Arctic Ocean: Journal of Geodynamics, v. 6, p. 137–176.

Wright, J. A., Jessop, A. M., Judge, A. S., and Lewis, T. J., 1980, Geothermal measurements in Newfoundland: Canadian Journal of Earth Sciences, v. 17, p. 1370–1376.

Yorath, C. J., and Hyndman, R. D., 1983, Subsidence and thermal history of Queen Charlotte Basin: Canadian Journal of Earth Sciences, v. 20, p. 135–159.

ACKNOWLEDGMENTS

The author acknowledges the efforts of many scientific and technical colleagues who have contributed to the data summarized in this chapter. Scientists are represented in the bibliography, but the names of technical colleagues are not so listed. The work of V. S. Allen, Jean Bisson, and Roger Blais, and a series of student assistants is thus specifically acknowledged. The willing cooperation of many oil companies and mining companies in making wells and boreholes available for measurement is also gratefully acknowledged.

MANUSCRIPT ACCEPTED BY THE SOCIETY NOVEMBER 30, 1990

Chapter 25

Heat flux in the Canadian Cordillera

Trevor Lewis
Geological Survey of Canada, Pacific Geoscience Centre, Box 6000, Sidney, British Columbia V8L 4B2, Canada

INTRODUCTION

The rugged Canadian Cordillera has been and is being created by various dynamic processes that are dominated by past and present plate-tectonic motions and interactions. A summary of the tectonic models proposed for the Cordillera, together with the geologic record and geophysical data that support them, is given by Gabrielse and Yorath (DNAG volume: Geology of Canada Series: Cordilleran Orogen, Canada, in preparation). In this active area the heat flux is determined by various regimes and processes, and the reduced heat flow is a diagnostic indicator of the state of the lithosphere. In general, the heat flux is decreased by subducting plates, increased by uplift and erosion, and increased and decreased, respectively, by exothermic and endothermic reactions occurring within the crust. It is quite variable over small areas where crustal intrusions are advectively cooling. Relatively high heat flux occurs in regions that are undergoing crustal extension. The magnitude of these variations can be used to evaluate the rates at which these processes are occurring and/or to infer other details concerning the complex evolution of this region.

Current plate interactions along the western margin of the Canadian Cordillera (Riddihough and Hyndman, in preparation) are indicated in Figure 1. South of the triple junction, the Juan de Fuca plate is subducting under southern Vancouver Island and the coastal mainland; the smaller Explorer plate broke away and has now ceased to subduct. North of the triple junction, there is right-lateral transform movement along the Queen Charlotte fault with a small amount of oblique subduction. Farther to the north, relative movement is along a series of faults that splay out through southeastern Alaska into northern British Columbia and the Yukon. The related, large uplift rates in the St. Elias Mountains occur where the fault traces curve westward.

The interior of the southern Canadian Cordillera and the Foreland Belt, although not presently involved directly in the plate interactions, have directly and indirectly participated in a variety of such tectonic processes during the last 1,440 m.y. (Gabrielse and Yorath, in preparation). The most recent major event was an intense and widespread phase of volcanism and extension in the Eocene, which was accompanied by widespread granitic magmatism along the eastern boundary of the Coast Plutonic Complex. Although the crust affected by the Eocene heating has had time enough to cool, there has been no major subsidence. Also, all geophysical and geological studies indicate a thin, warm crust at present, similar to the Basin and Range province to the south in the U.S. Unlike in the Basin and Range, however, the amount of crustal extension here is small, and each of the numerous volcanic events since Eocene time has produced only a small volume of basic lava.

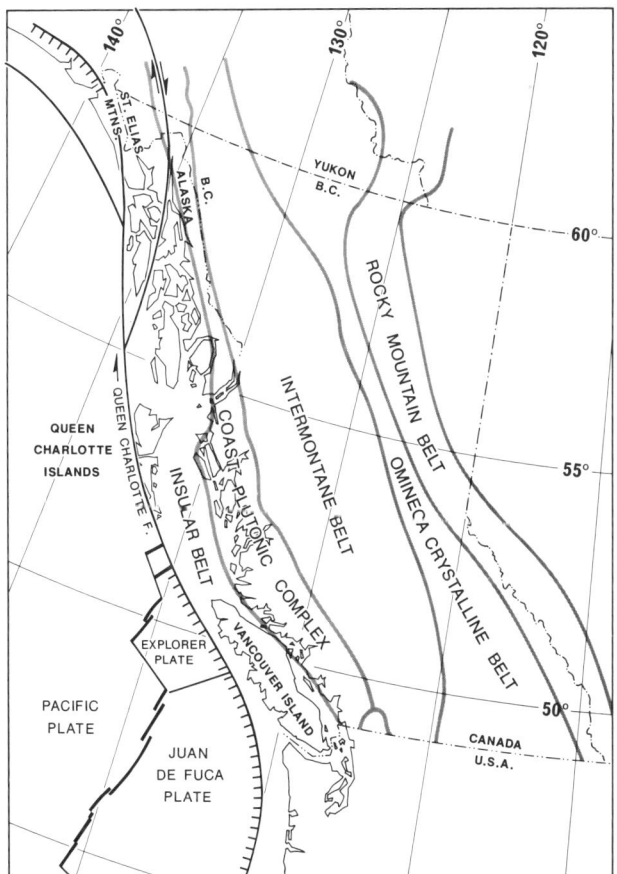

Figure 1. Map showing the plate boundaries and continental tectonic/physiographic belts.

Lewis, T., 1991, Heat flux in the Canadian Cordillera, *in* Slemmons, D. B., Engdahl, E. R., Zoback, M. D., and Blackwell, D. D., eds., Neotectonics of North America: Boulder, Colorado, Geological Society of America, Decade Map Volume 1.

This compilation includes a brief description of the geothermal measurements and corrections, the resulting heat flux and heat generation determinations, and a discussion of the relations between the heat flux and regional tectonics. Table 1 includes all of the published heat-flow data with their associated references. The data have been included on the *Geophysics of North America* CD-ROM as well (see Lewis, 1989). No references are included on the CD-ROM, but all original sources are referenced in Table 1. The data in the south form a unique profile across a subduction zone as well as a large inland region of extension that is a high heat-flow province. Near the Queen Charlotte transform fault, high heat flux from the young oceanic lithosphere is just 30 km seaward of more normal continental values of heat flux.

MEASUREMENTS AND CORRECTIONS

The heat-flow data come from various types of sites, including wells for hydrocarbon exploration and slim holes for stratigraphic tests, short holes in a tunnel, and surface and underground boreholes drilled for mineral exploration. Heat flux also was determined in shallow and deep boreholes drilled for geothermal energy assessment and exploration, as well as a few boreholes drilled specifically for heat-flow determinations. Shallow holes (100 m) yielded good results in moderate topography (Davis and Lewis, 1984; Bentkowski and Lewis, 1984, 1985), but results in rugged topography were inconsistent (Bentkowski and Lewis, 1983).

For the vast majority of boreholes, thermal conductivity measurements were made on an adequate number of diamond drilled core samples from the logged hole to accurately characterize the thermal conductivity value (Table 1, column labeled n). Most measurements were made on solid disks in a divided bar, for which the absolute accuracy of individual measurements is 5 percent. If whole cores were not available, chips or cuttings were measured by the cell method (Sass and others, 1971) or by a pulsed-needle probe technique (Lewis and others, in preparation). The latter two methods have an accuracy for individual measurements of 10 to 15 percent. For the overpressured offshore wells near Vancouver Island, the uncertainty in the value of the in situ porosity caused large uncertainties in the calculated formation thermal conductivity values. The conductivity of the rock matrix and the porosity were measured on cuttings using the pulsed-needle probe method. Average porosities obtained from the electrical logs and measured porosities were in good agreement. For wells near the Queen Charlotte Islands with normal hydrostatic pressures, porosities were obtainable from sonic logs, but conductivities of the rock matrix were estimated.

Oceanic probes were successfully used in some of the fjords and in Powell Lake (Hyndman, 1976; Lewis and others, 1988) to obtain a better distribution of data. Thermal conductivity measurements on early cruises were made using needle probe measurements on cores. Later measurements employed an in situ pulsed linear source method (Hyndman and others, 1979), which gave very reliable results at discrete depths. Large aperiodic bottom-water changes in some of the fjords along the British Columbian Coast (Lewis, 1983; Lewis and others, 1988) disturb the heat flux in the top few meters of the bottom sediments. Repeat measurements over a period of a few years using probes up to 12 m in length yielded enough data to obtain equilibrium thermal gradients from comparison with conductive models. Where borehole and marine results were measured at nearby sites, the corrected heat-flow values agreed to within 5 percent.

The desired value of equilibrium heat flux reflects the amount of heat that is flowing up through the upper crust to the surface. The observed conductive heat flux is often influenced by the local terrain, advective heat flow, and/or changes in the past climate. If the heat flux was modified or thought to be modified by such disturbances, corrections were applied, except for glacial corrections (e.g., Jessop, 1971). These are omitted from all results herein in order to facilitate valid comparisons with most of the other heat-flow values that are reported in this volume without glacial corrections. (Most glacial corrections to the heat flux were less than 10 percent). The reliability, Q in Table 1, of each tabulated heat flux is given as a percent of the equilibrium value. Some heat-flow data were not reported because of very large uncertainties.

The best reliability assigned, 5 percent, is for borehole sites in which deep temperature measurements were made and for which the thermal conductivity of a large number of disks from the same borehole were measured on a divided bar. Also the vertical heat flux was constant, and there was no evidence for large or uncertain corrections. For the best 12 sites, the average number of thermal conductivity measurements per hole was 44, and the average maximum depth was 489 m. Since the value of the reliability is somewhat subjective and varies immensely from investigator to investigator, these figures are given to indicate the quality of reliabilities in Table 1. For example, Sass and others (1985) give a reliability of 3 percent for hole NA1, which has a maximum depth of 112 m and for which 9 thermal conductivities were measured. The higher reliability values (lower Q values as percentages), given by Sass and others (1985) and Yorath and Hyndman (1983), have been adjusted for Table 1.

Topographic corrections for most results published before 1985 and for areas of subdued topography were made according to the method of Jeffreys (1940), except for the Granduc site where two-dimensional analog models were used to estimate the effects of topography and the overlying glacier. For more recently measured sites in rugged terrains and sites with sediment-filled valleys or fjords, corrections were calculated using a finite-element, two-dimensional model (Lee and Henyey, 1974; Finckh, 1981; Lewis and others, 1985). Very large corrections, as much as 30 percent, were recalculated for stations in the most severe topography. In fjords, the combination of a low-conductivity sediment prism, high mountains, and the warm rim effect produced large opposing corrections within the sediment wedge. The positioning of the site within the prism critically affected the magnitude and sign of this correction (Lewis and others, 1988).

Very small water flow through rock can transport relatively

TABLE 1. HEAT-FLOW DATA FOR THE CANADIAN CORDILLERA AND SOUTHEAST ALASKA*

Site	Latitude (deg N)	Longitude (deg W)	Elevation (m)	Depth (m)	n	Q (%)	q (mW/m^2)	A (µW/m^3)	References[†]
451	50°09.74'	123°23.66'	335	100	8	20[§]	58	0.89	2
454	50°07.61'	123°27.06'	290	100	8	20[§]	71	0.66	2
456	50°12.20'	123°32.19'	290	100	7	20[§]	63	0.70	2
457	50°11.61'	123°31.91'	290	100	4	20[§]	56	0.54	2
458	50°10.62'	123°31.60'	274	100	8	20[§]	39	0.72	2
461	50°27.45'	123°07.53'	762	100	9	20[§]	53	0.97	2
462	50°27.38'	123°10.36'	823	100	8	20[§]	56	0.66	2
463	50°27.20'	123°12.19'	914	100	8	20[§]	69	0.91	2
311	49°49.8'	120°49.7'	1,300	56	6	10	77	1.29	3
312	49°49.5'	120°47.7'	1,275	90	10	10	81	0.93	3
313	49°49.2'	120°42.7'	1,000	69	7	10	70	1.04	3
314	49°50.3'	120°36.5'	1,050	99	11	10	76	0.50	3
316	49°42.3'	120°26.5'	1,475	46	5	15	79	1.60	3
317	49°44.4'	120°24.4'	1,500	88	9	10	68	0.64	3
318	49°47.0'	120°29.3'	1,250	80	9	10	68	0.63	3
322	49°48.5'	119°59.8'	1,300	86	9	10	80	1.08	3
326	49°39.9'	119°15.0'	1,460	98	11	10	67	2.22	3
327	49°41.8'	119°12.5'	1,220	47	5	15	74	2.12	3
328	49°43.7'	119°09.8'	1,280	84	9	10	69	3.98	3
330	49°47.1'	119°06.2'	1,220	52	5	15	81	2.55	3
333	50°02.9'	118°27.5'	1,250	52	5	15	74	1.54	3
335	50°01.5'	118°33.3'	1,150	94	10	10	87	0.79	3
336	49°58.9'	118°37.7'	1,100	79	8	10	54	1.55	3
337	49°57.2'	118°40.7'	1,100	85	9	10	77	0.99	3
340	49°47.5'	118°43.5'	950	82	9	10	74	1.75	3
310-1	49°57.5'	119°47.2'	1,311	370	44	5	81	1.70	1, 3
310-2	49°34.2'	119°39.1'	389	475	60	5	73	2.00	1, 3
Bute	50°49.8'	124°53.5'	-360	3	5	35	46	0.8	4, 9, 11
Desolat.	50°05.5'	124°48.8'	-527	6	5	35	38	0.8	4, 9, 11
Powell	49°56.3'	124°32.0'	-350	2	5	55	25	0.8	4, 9, 11
Powell	49°56.1'	124°32.0'	-350	2	5	65	29	0.8	4, 9, 11
Howe Sd	49°25.0'	123°24.1'	-249	4	6	45	38	0.8	4, 9, 11
Indian	49°24.7'	122°52.2'	-220	4	5	20	63	0.8	4, 9, 11
Indian	49°23.5'	122°52.2'	-217	4	5	35	42	0.8	4, 9, 11
68-408	52°45.3'	132°02.9'	384	237	**	20	39		5
68-413	52°45.4'	132°02.9'	372	222	**	20	53		5
68-425	52°45.4'	132°02.9'	366	244	**	20	44		5
68-435	52°45.5'	132°03.1'	229	170	**	20	49		5
68-446	52°45.5'	132°03.1'	201	207	**	20	53		5
68-427	52°45.3'	132°03.0'	366	259	**	20	53		5
68-423	52°45.3'	132°03.0'	351	265	**	20	46		5
68-431	52°45.4'	132°02.9'	338	190	**	20	45		5
68-439	52°45.4'	132°03.0'	341	245	**	20	54		5
68-451	52°45.4'	132°03.0'	323	267	**	20	47		5
68-441	52°45.3'	132°03.0'	329	267	**	20	42		5
Q.C. Ter	52°32.3'	132°18.7'	-1,780	2		10	74		5
Q.C. Ter	52°32.5'	132°18.7'	-1,760	2		20	69		5
Q.C. Ter	52°32.9'	132°17.9'	-1,774	2		10	56		5
Q.C. Ter	52°34.7'	132°16.2'	-1,750	2		10	67		5
Q.C. Ter	52°36.8'	132°09.8'	-1,925	2		15	66		5
Q.C. Ter	52°24.3'	131°56.1'	-2,040	2		10	77		5
Q.C. Ter	52°10.4'	131°43.4'	-2,103	2		10	66		5
Q.C. Ter	52°09.5'	131°46.4'	-2,048	2		10	80		5
Q.C. Off	52°09.4'	132°01.7'	-2,846	2		10	247		5

TABLE 1. HEAT-FLOW DATA FOR THE CANADIAN CORDILLERA AND SOUTHEAST ALASKA* (continued)

Site	Latitude (deg N)	Longitude (deg W)	Elevation (m)	Depth (m)	n	Q (%)	q (mW/m^2)	A (μW/m^3)	References†
Q.C. Off	51°57.9'	131°31.8'	-2,040	2		10	104		5
Q.C. Off	51°57.0'	131°32.8'	-2,095	2		10	128		5
Q.C. Off	51°48.0'	131°15.2'	-1,886	2		20	98		5
Q.C. Off	52°13.2'	132°08.7'	-2,850	2		20	270		5
Q.C. Off	52°13.2'	132°08.2'	-2,850	2		15	265		5
Q.C. Off	52°16.0'	132°05.9'	-2,825	2		10	144		5
Q.C. Off	52°16.1'	132°05.6'	-2,825	2		10	134		5
Q.C. Off	52°16.1'	132°05.4'	-2,825	2		10	139		5
Q.C. Ter	52°20.5'	131°59.1'	-1,900	2		15	77		5
Q.C. Ter	52°20.4'	131°59.2'	-1,900	2		15	84		5
Q.C. Ter	52°20.0'	131°59.3'	-1,900	2		15	111		5
Q.C. Sd	51°34.6'	130°17.3'	-1,104	2		20	84		5
Q.C. Sd	51°30.3'	130°20.1'	-1,190	2		20	95		5
Q.C. Sd	51°34.5'	130°21.1'	-945	2		20	75		5
Q.C. Sd	51°32.6'	130°17.4'	-903	2		20	91		5
Q.C. Sd	51°31.9'	130°16.2'	-1,021	2		20	85		5
Q.C. Sd	51°33.1'	130°11.1'	-822	2		20	84		5
Q.C. Sd	51°25.2'	130°13.1'	-1,190	2		20	108		5
Q.C. Sd	51°23.0'	130°07.0'	-860	2		20	83		5
Q.C. Off	51°26.2'	130°26.0'	-1,227	2		20	71		5
Q.C. Off	51°30.2'	130°36.8'	-1,338	2		20	159		5
3	49°19.8'	119°37.6'	552	660	118	5	67		6
122-3	60°45.0'	135°11.0'	873	388	23	5	61	1.79	7
188-1	59°42.8'	133°24.1'	1,559	350	‡	10	103	6.53	7
188-2	59°42.8'	133°24.1'	1,452	305	‡	10	92	6.53	7
8	58°38.9'	130°00.6'	829	292	30	10	84	1.10	7
160-2	58°15.3'	129°49.0'		143	6	10	72	1.07	7
160-3	58°15.3'	129°49.0'		150	6	10	73	1.07	7
9	58°09.6'	129°51.9'	1,170	450	46	13	58	2.59	7
10	57°53.6'	130°51.3'	899	427	45	5	69	0.65	7
75	56°25.1'	129°09.2'	1,100	913	13	5	90		7
303-1	50°37.8'	123°23.8'	580	213	30	15	153	0.53	8, 9, 11, 13
303-2	50°34.8'	123°16.8'	350	196	17	15	93	1.10	8, 9, 11, 13
304-1	50°07.3'	123°23.6'	170	213	15	15	67	0.83	8, 9, 11, 13
304-2	50°06.4'	123°22.0'	190	244	29	15	93	0.99	8, 9, 11, 13
352-1	49°21.5'	119°41.6'	670	320	22	10	74		10
352-2	49°22.1'	119°46.1'	1,190	271	‡	15	109		10
352-5	49°21.9'	119°45.4'	1,070	150	‡	15	76		10
352-6	49°22.2'	119°38.8'	655	440	65	15	69		10
309-1	50°05.1'	123°19.1'	450	432	64	10	144		12
309-2	50°04.3'	123°19.8'	274	444	60	5	112		12
343	50°05.0'	123°12.3'	1,113	438	64	10	96		12
344	50°06.5'	123°16.0'	1,560	455	17	15	339	0.45	12
Jervis	50°03.4'	123°49.05'	-547	7	6	15	77	0.8	12, 9
Jervis	50°02.5'	123°51.00'	-543	7	6	15	51	0.8	12, 9
Jervis	50°02.23'	123°51.8'	-543	7	6	15	42	0.8	12, 9
Jervis	50°00.65'	123°56.6'	-555	7	6	15	44	0.8	12, 9
Jervis	50°00.1'	123°58.45'	-563	7	6	15	31	0.8	12, 9
Jervis	50°01.0'	123°54.20'	-512	7	6	15	45	0.8	12, 9
Jervis	50°04.8'	123°46.6'		9	‡	20	53	0.8	12, 9
Jervis	50°10.65'	123°56.1'	-305	9	‡	20	63	0.8	12, 9
Jervis	50°09.6'	123°53.15'	-329	7	6	15	85	0.8	12, 9
Jervis	50°08.28'	123°50.2'	-356	7	6	15	81	0.8	12, 9
Jervis	49°51.0'	123°54.1	-680	7	6	15	28	0.8	12, 9

TABLE 1. HEAT-FLOW DATA FOR THE CANADIAN CORDILLERA AND SOUTHEAST ALASKA* (continued)

Site	Latitude (deg N)	Longitude (deg W)	Elevation (m)	Depth (m)	n	Q (%)	q (mW/m^2)	A (μW/m^3)	References†
Jervis	49°51.0'	123°53.9'	-680	7	6	15	28	0.8	12, 9
Jervis	49°48.7'	123°55.97'	-680	7	6	15	42	0.8	12, 9
Jervis	49°47.5'	124°02.2'	-680	7	6	25	22	0.8	12, 9
Jervis	49°47.5'	124°02.2'	-680	12	‡	25	22	0.8	12, 9
Desolat	50°05.5'	124°48.8'	-497	7	6	15	32	0.8	12, 9
Homfray	50°11.7'	124°38.2'	-727	7	6	25	43	0.8	12, 9
Homfray	50°17.2'	124°39.3'	-718	7	6	25	40	0.8	12, 9
Homfray	50°14.6'	124°37.6'	-740	7	6	25	39	0.8	12, 9
Homfray	50°16.3'	124°38.4'	-733	7	6	25	35	0.8	12, 9
Saanich	48°38.1'	123°30.1'	-200	7	6	15	33	0.6	12, 11
354-2	48°51.95'	123°47.43'	510	118	15	10	31		12
354-4	48°52.51'	123°47.12'	450	204	6	15	30		12
354-5	48°52.47'	123°47.00'	510	208	6	15	43		12
354-6	48°52.52'	123°46.96'	490	182	7	15	34		12
358-1	49°48.34'	125°09.12'	108	426	15	10	38		12
358-2	49°23.11'	124°57.10'	121	470	24	10	43		12
358-3	49°06.05'	123°51.65'	15	374	12	15	35		12
321	48°39.0'	123°26.9'	10	445	46	5	35	0.83	12
339	50°43.3'	123°40.21'	808	399	37	5	120	0.60	12
I-87	48°56.6'	125°57.0'		3,726	18	25	50		12
I-65	48°54.6'	126°09.1'		3,042	9	25	55		12
D-14	48°53.0'	126°03.0'		2,433	14	25	38		12
J-14	49°23.6'	127°02.0		3,100	10	25	45		12
J-100	48°19.7'	125°44.0'		2,460	10	25	43		12
H-68	48°37.3'	125°39.1'		2,335	10	25	45		12
80-1	52°31.1'	122°17.4'	1,085	293	20	7	67	0.50	13
80-2	52°31.2'	122°17.6'	1,104	232	16	7	60	0.50	13
81-1	52°33.0'	121°38.3'	1,158	330	19	7	70		13
81-2	52°33.1'	121°38.5'	1,148	225	13	7	62		13
101	50°46'	122°48'	1,212	1,520	4	15	80		13
119	49°30.7'	121°17.3'	1,208	490	20	15	63		13
125-1	51°37.7'	118°25.0'	666	280	19	15	104	3.7	13
125-3	51°37.8'	118°25.0'	665	260	18	15	115	3.7	13
159	49°19.4'	121°55.5'	732	438	18	5	50	0.8	13, 9
161	50°28.5'	120°58.3'	1,195	460	106	10	72	0.8	13
162-1	51°31.2'	119°48.5'	1,708	275	17	7	101	4.8	13
162-2	51°31.1'	119°48.5'	1,683	250	21	7	119	4.8	13
163-1	49°34.6'	125°36.6'	392	780	90	10	36	0.6	13
163-2	49°34.3'	125°35.2'	312	650	67	10	41	0.6	13
107-1	49°36.7'	123°09.6'	-389	305§§	19	15	68	1.0	13
107-2	49°36.7'	123°09.5'	-389	107§§	7	20	56	1.0	13
107-3	49°36.6'	123°09.1'	-393	105§§	8	20	84	1.0	13
107-4	49°36.8'	123°10.1'	-386	229§§	‡	20	73	1.0	13
107-5	49°36.8'	123°10.1'	-386	114§§	‡	20	74	1.0	13
118	50°39.1'	120°28.5'	765	260	18	15	85	0.7	13
182	50°39.7'	120°30.6'	669	265	17	15	84	0.7	13
183	50°38.2'	120°27.5'	945	135	12	15	55	0.7	13
184-1	50°54.4'	121°24.6'	512	200	9	10	81	0.9	13
184-2	50°54.1'	121°24.5'	512	170	8	10	67	0.9	13
302	50°25.6'	122°50.2'	1,295	90	9	20	41	0.9	13
305-1	49°26.2'	118°25.4'	720	460	33	5	113	4.9	13, 19
305-2	49°25.1'	118°31.1'	750	450	34	5	111	4.9	13, 19
353	50°36.3'	127°28.6'	244	760	40	10	65	0.9	13
351-1	49°30.7'	123°22.0'	137	280	19	15	31	0.8	13, 9

TABLE 1. HEAT-FLOW DATA FOR THE CANADIAN CORDILLERA AND SOUTHEAST ALASKA* (continued)

Site	Latitude (deg N)	Longitude (deg W)	Elevation (m)	Depth (m)	n	Q (%)	q (mW/m²)	A (μW/m³)	References†
301-2	50°34.6'	123°27.6'	575	118	2	15	120		14
301-3	50°34.1'	123°28.3'	640	310	21	10	100		14
301-4	50°34.3'	123°29.1'	777	90	3	20	290		14
301-5	50°34.0'	123°30.2'	770	90	4	20	930		14
301-6	50°34.3'	123°30.3'	808	56	3	30	450		14
Granduc	56°12'	130°20'	800	1,500***	12	15	73		15
M6-79D	50°34.4'	123°29.9'	885	321	†††		1,660		16
M7-79D	50°34.1'	123°30.9'	900	367	†††		2,530		16
M8-79D	50°34.2'	123°32.3'	875	497	†††		309		16
M9-80D	50°31.2'	123°30.1'	765	1,142	†††		120		16
M11-80D	50°34.1'	123°28.8'	791	560	†††		185		16
M12-80D	50°32.6'	123°28.7'	792	604	†††		148		16
M14-81D	50°33.5'	123°27.5'	861	578	†††		146		16
L1-78D	50°41.1'	123°30.8'	760	603	†††		680		16
L2-80D	50°40.4'	123°32.3'	896	595	†††		262		16
L3-80D	50°40.5'	123°31.4'	972	1,010	†††		322		16
L4-81D	50°39.6'	123°32.0'	1,097	1,297	†††		207		16
L5-81D	50°40.5'	123°28.8'	774	660	†††		272		16
L6-81D	50°37.6'	123°25.0'	535	579	†††		284		16
L7-82D	50°38.9'	123°33.5'	1,808	421	†††		227		16
L8-82D	50°40.5'	123°33.5'	960	48	†††		176		16
MB1	54°47.0'	132°03.0	275	208	0	15	46		17
BC6	54°54.8'	132°08.0	335	108	1	15	69	4.2	17
BC7	54°54.6'	132°08.0'	274	165	1	15	95	4.2	17
NIB	55°04.0'	132°08.8'	24	129	6	25	51		17
NA1	55°09.8'	131°46.8'	62	112	9	10	68	0.8	17
QO1	55°24.4'	130°28.7'	458	945	7	10	115	2.8	17
SUK	55°37.1'	132°35.4'	153	212	0	20	40		17
ASC1	56°50.3'	133°18.9'	18	129	11	10	69		17
ASC2	56°50.4'	133°18.8'	25	211	5	10	66		17
WSB	57°05.5'	134°48.0'	125	168	4	10	150	1.2	17
GC8	58°04.6'	134°37.2'	549	102	3	20	46		17
GC9	58°04.5'	134°36.9'	462	189	9	20	65		17
SCOHO	53°33.6'	131°25.8'	-21	2,781	0	15	60		18
TYEE	53°18.9'	131°20.4'	-27	3,460	0	10	69		18
SOCKEYE	52°47'	130°58'	-31	4,773	0	10	78		18
MURRELT	52°24.7'	130°47.6'	-111	2,920	0	12	59		18
AUKLET	52°20.3'	130°36.6'	-169	2,371	0	16	50		18
HARLEQN	51°55.1'	129°58.2'	-140	3,242	0	16	45		18
OSPREY	51°35.1'	129°20.8'	-59	2,531	0	17	47		18
467	50°32.3'	127°20.1'	137	100	8	12	60	0.6	20, 23
468	50°33.2'	127°24.2'	61	100	8	12	62	0.6	20, 23
469	50°31.7'	127°23.0'	91	100	8	12	70	0.6	20, 23
471	50°29.3'	127°22.4'	213	100	8	12	67	0.6	20, 23
494-1	51°46.95'	120°22.17'	1,405	235	25	7	113	4.79	21
494-2	51°49.0'	120°28.04'	1,356	244	26	7	100	3.13	21
494-3	51°44.14'	120°09.04'	1,417	244	30	7	92	5.74	21
494-4	51°45.09'	120°04.78'	1,311	244	24	7	107	4.98	21
479	51°58.38'	121°10.00'	950	100	6	15	63	1.30	22
480	51°57.49'	121°09.83'	925	100	6	15	64	1.22	22
481	51°50.00'	121°08.91'	985	100	5	15	71	1.02	22
482	51°49.22'	121°08.96'	900	100	6	15	58	1.46	22
483	51°49.32'	121°08.30'	925	100	6	15	60	0.92	22
484	51°59.76'	120°56.87'	1,300	100	6	15	62	0.83	22

TABLE 1. HEAT-FLOW DATA FOR THE CANADIAN CORDILLERA AND SOUTHEAST ALASKA* (continued)

Site	Latitude (deg N)	Longitude (deg W)	Elevation (m)	Depth (m)	n	Q (%)	q (mW/m^2)	A (µW/m^3)	References[†]
485	51°59.73'	120°56.61'	1,300	100	6	15	62	0.81	22
486	51°59.89'	120°57.78'	1,250	100	6	15	50	0.61	22
487	51°59.81'	120°48.26'	1,240	97	6	15	63	0.78	22
488	52°00.02'	120°48.75'	1,240	100	5	15	65	0.71	22
489	52°00.49'	120°49.57'	1,255	100	6	15	57	0.84	22
355-1	51°09.0'	119°49.21'	1,502	219	17	10	57		23
355-2	51°08.65'	119°48.17'	1,524	671	23	7	73		23
355-3	51°08.97'	119°49.15'	1,495	425	18	7	84		23
356	55°28.4'	129°49.0'	655	292	22	20	85	2.53	23
357-1	53°21.77'	130°07.48'	33	218	14	10	61	0.81	23
357-2	53°21.89'	130°07.34'	33	389	29	10	76	0.81	23
359-1	54°13.81'	126°14.00'	1,225	245	20	7	82	1.80	23
359-2	54°13.86'	126°14.32'	1,225	209	10	10	78	1.80	23
491	49°08.7'	122°01.37'	10	274	20	20	68		24

*No climatic corrections have been made.
[†]1. Lewis and Werner, 1982; 2. Bentkowski and Lewis, 1983; 3. Davis and Lewis, 1984; 4. Hyndman, 1976; 5. Hyndman and others, 1982; 6. Jessop and Judge, 1971; 7. Jessop and others, 1984; 8. Lewis and Jessop, 1981; 9. Lewis, 1976; 10. Lewis, 1984; 11. Lewis and others, 1984; 12. Lewis and others, 1988; 13. Lewis and others, 1985; 14. Lewis and Souther, 1978; 15. Mathews, 1972; 16. Reader and Fairbank, 1983; 17. Sass and others, 1985; 18. Yorath and Hyndman, 1983; 19. Lewis and others, 1979; 20. Bentkowski and Lewis, 1984; 21. Bentkowski and Lewis, 1986; 22. Bentkowski and Lewis, 1985; 23. Bentkowski and Lewis, 1989; 24. Bentkowski and Lewis, 1988.
§The quality of the measured value is 20 percent, but large variations between nearby sites indicates the probable indirect influences of water flows.
**Conductivities were measured on 30 equivalent core samples from nearby with an equivalent average geological log.
‡Conductivities from samples from nearby boreholes or cores.
§§Collared underground.
***Maximum rock thickness above tunnel.
†††Measurements were made, but the number of samples is not given.

large amounts of heat, and in the process may perturb the conductive heat flux. In this rugged region, boreholes have intersected dipping horizons along which water was flowing before drilling commenced (e.g., Lewis and others, 1988). In such cases the deepest measured heat flux is taken to best represent the crustal heat flux. The possibility of one or more of these zones existing, undetected, beneath the bottom of any hole causes the largest uncertainty in our determinations of crustal heat flux in this region. Agreement between closely spaced holes or constant heat flux in a deep hole gives some confidence that such a situation does not exist at that particular site. The measured heat flux in shallow holes in the Garibaldi Volcanic Belt (Lewis and others, 1988) is somewhat inconsistent, and water flows are suspected (see Table 1, sites 451–463). Similarly, water flows associated with cooling intrusives within the Garibaldi Volcanic Belt cause highly variable heat flux such as that within the Meager Mountain geothermal area (Table 1, 301-2 through 301-6 and M6-79D through L8-82D). These particular results are tabulated without any corrections.

The concentrations of naturally occurring radioactive elements in many samples representing the crust in this area were determined using gamma-ray spectrometry (Lewis, 1974). The heat generation, A, calculated from these concentrations, is listed in Table 1 for samples associated with heat-flow determinations. In addition to borehole samples, large numbers of fresh-looking outcrop samples have been measured (Lewis, 1976; Lewis and others, 1985, unpublished data), and average values determined from the larger areas are shown in Figures 2 and 3.

HEAT FLUX AND HEAT GENERATION

The heat flux within the Canadian Cordillera varies from minimum values of 25 mW/m^2 above the subducting Juan de Fuca plate to values of several hundred mW/m^2 in the Meager Mountain geothermal area. Excluding geothermal areas associated with young volcanic belts, the highest values of heat flux on land, 120 mW/m^2, are associated with young acidic plutons. The average heat flux and standard deviation for each area are given in Table 2, as is the number, n, of independent sites that were averaged for each area. Except for the geothermal areas, boreholes within 5 km of each other are considered to compose a single site. The very large, local variations in heat flux in the geothermal area come from advective cooling of intrusive bodies. Elsewhere on land the larger standard deviations reflect the ranges of heat flux that are due to associated lateral variations of crustal heat generation (e.g., Omineca Crystalline Belt) or a varying heat flux from the lower crust (e.g., Transition Zone of the Coast Plutonic Complex). A large scatter in the heat flux from the young oceanic crust is expected, due to hydrothermal circulation and sedimentation effects, etc. As shown in Figures 2 and 3, only two regions have enough data to contour the heat flux and heat generation.

The heat generation varies from low values of 0.6 $\mu m/m^3$ in very basic plutonic rocks to values higher than 5.0 $\mu m/m^3$ in several acidic plutons. The variation generally reflects the mineralogy of the rock, and the more hydrous phases have the higher values (e.g., Webb and others, 1985). Within regions where enough data are available, the relatively younger intrusive bodies generally have a relatively higher heat generation (Lewis and others, 1985; Lewis and others, 1988). Values of heat generation from fresh-looking surficial samples were found to be in good agreement with results from cores of nearby boreholes.

The surficial heat flux is composed of heat generated in the upper crust and heat flowing upward from the lower crust. A heat-flow province is defined where there is a linear relation between q and A throughout the area. A simple interpretation of this linear relation equates a constant heat flux from the lower crust throughout the province to the component of q unrelated to A. Such a linear relation is unlikely to occur above a subduction zone because the heat that flows upward into the upper crust is quite variable. This result is illustrated in Figure 4 where the crosses represent heat-flow values in the subduction zone in southwestern British Columbia (Lewis and others, 1988). Since the heat generation in the Insular Belt and the Coast Plutonic Complex is very low, these values plot in a vertical band, indicating the irrelevance of the q-A relation in this area.

Lewis and others (1985) concluded that the southern interior of the Canadian Cordillera, composed of the Intermontane and Omineca Crystalline Belts, formed a heat-flow province with a very high reduced heat flux. This value is 58 mW/m^2 (64 mW/m^2 with glacial corrections). This area is probably part of a larger heat-flow province since Washington (Blackwell, 1974), and western Montana (Blackwell and Robertson, 1973) to the south have the same reduced heat flux and similar scale depths. The 90 mW/m^2 heat-flow contour (Fig. 2) within the Canadian portion of this province bounds the area of higher heat production plutons to the northeast, and may define the position of the ancient North American plate boundary. This high heat-flow province extends at least as far north as the Raft and Takomkane Batholiths near 52°N (see Fig. 2). The few heat-flow data north of 52°N within these two tectonic belts indicate a lower reduced heat flux there.

In most parts of the southern heat-flow province, upper crustal temperatures are now controlled by conductive heat flow. Even in areas where young, basic lavas are moving to the surface very quickly from great depths (Hickson, 1986), the crust generally is not being heated by such events (Bentkowski and Lewis, 1986). Figure 5 indicates temperatures that are considered to be valid in the upper crust within this region. The large variation in temperature is caused by the variation in heat generation, from 1 to 5 $\mu m/m^3$. It is possible to obtain the approximate heat flux and crustal temperatures at locations within the area where the heat flux has not been measured, given the heat generation representing the crust is sufficient, since the heat flux has been shown to vary linearly with the average upper crustal heat generation based on surficial samples.

Other geophysical parameters in southern British Columbia also indicate that the crust under this high heat-flow province is warm and thin. Seismic studies (Wickens, 1971; Berry and Forsyth, 1975) show that a low-velocity layer is very close to the base of the crust, and geomagnetic soundings (e.g., Caner and others, 1971) reveal an electrically conductive (low Z) region under the entire area. A magnetotelluric survey in the White Lake basin indicates high electrical conductivities rising to a minimum depth of 8 km (J. Delaurier, personal communication, 1988). Magnetic anomalies are quite subdued in comparison to the

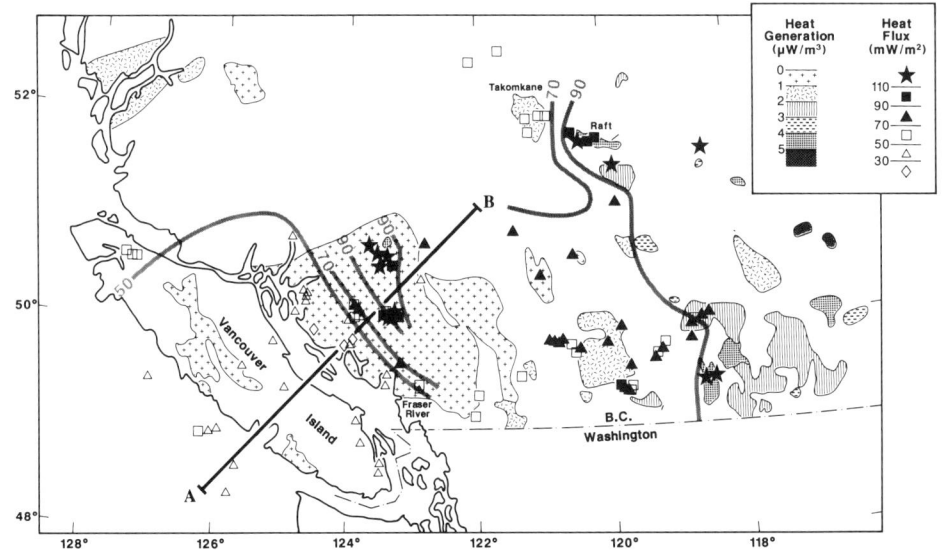

Figure 2. Heat flux and heat generation in the southern Canadian Cordillera.

Coast Plutonic Complex to the west or to the area east of the Rocky Mountain Trench (Haines and others, 1971). Beneath the small Tertiary basins (e.g., Church, 1982) that are filled with relatively low-thermal-conductivity volcanoclastic sediments, temperatures may be high as well (Lewis, 1984).

The high reduced head flux presents an interpretation problem. Conductive cooling of a hot lithosphere over the past 50 m.y. might produce the present observed heat flux, but the associated subsidence has not taken place. In some areas there has been significant amounts of erosion between 50 and 15 Ma (Ewing, 1981). Davis and Lewis (1984) suggested that the pres-

TABLE 2. SUMMARY OF HEAT FLUX BY REGION

Region	n	q (mW/m^2)	S.D. (mW/m^2)
Southern British Columbia			
Vancouver Island Shelf	6	46.0	5.9
S. Vancouver Island	7	36.7	3.6
N. Vancouver Island	5	64.9	4.6
Western CPC	6	29.5	8
Coastal CPC north of 50°	7	39.0	4.7
Transition Zone, CPC	11	50.6	15
Geothermal Zone, CPC	20	471	607
Eastern CPC	7	80.3	12
Intermontane	17	72.6	6
Omineca Crystalline Belt	22	83.2	19
Central British Columbia			
Oceanic crust, west of the Queen Charlotte Island	11	159.9	69
Queen Charlotte Island Terrace	11	75.2	14
Queen Charlotte Island	1	47.7	....
Queen Charlotte South	15	74.1	15
Southeastern Alaska, Northern B.C., and Southern Yukon			
Coastal	13	73.3	31
Inland	9	75.1	16

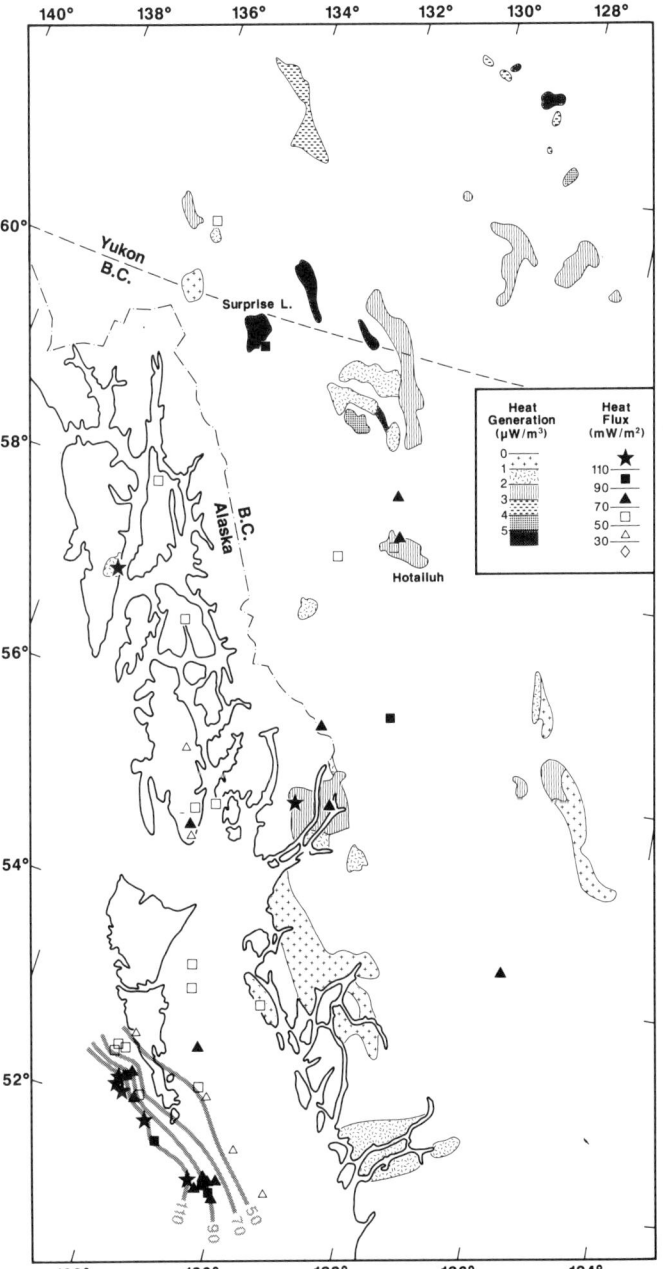

Figure 3. Heat flux and heat generation in southeastern Alaska, northern British Columbia, and southern Yukon.

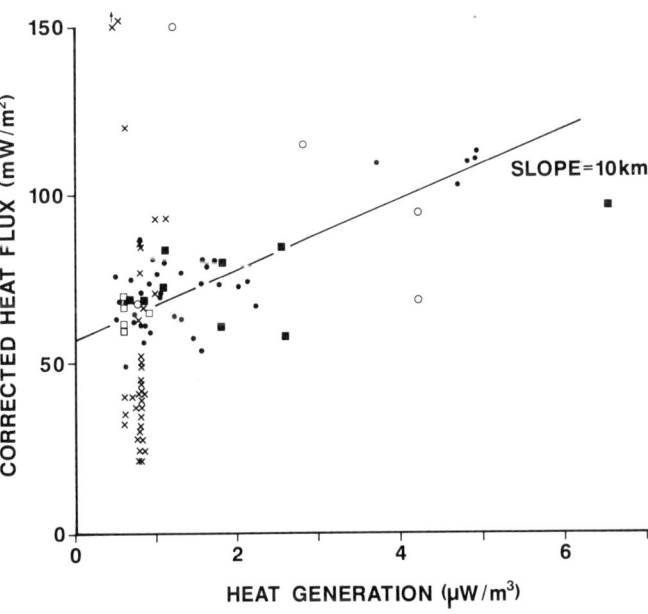

Figure 4. Heat flux versus heat generation. Data from above the subducting Juan de Fuca plate are shown as Xs; data from the southern heat-flow province to which the line is fitted are shown as solid circles; data from northern Vancouver Island are shown as open squares; data from southeastern Alaska are shown as open circles; all other data north of 52°N are shown as solid squares.

ent heat flux has been supported by asthenosphere flow toward the subducting Juan de Fuca plate. Although crustal extension is occurring at present, extension rates are very small. Alternatively, exothermic reactions at depth may be supplying the extra heat. These would be the result of rehydration of the crust after it was dehydrated during the large Eocene heating event. This process could prolong the cooling of the crust, maintain high temperatures and high electrical conductivities in the lower crust, and prevent large amounts of subsidence initially. However, since the region is all one heat-flow province the unlikely condition is required that the heat flux from these reactions be nearly constant over the entire area.

Large-scale thrust faults could have displaced the tops of batholiths from their deeper roots, producing a step-like distribution in the crustal heat generation distribution, as has been suggested for the southeastern United States (Costain and others, 1986). The scale depth of 10 km (Fig. 4) may be the average depth of such terrain-bounding faults, a depth that agrees well with the depths shown by Monger and others (1985) along crustal transects that cross this region. Sites where the heat flux is greater than predicted by the linear relation shown in Figure 4 (e.g., 125-Goldstream) may be locations where the fault plane is deeper than 10 km. As a result of such a model, the northern and southern boundaries of this heat-flow province are either zones of vertical displacement of the main thrust zone, blocking any further perpendicular component of thrusting, or (more likely) the heat flux coming from the lower crust is relatively lower, indicating a different "basement" or less asthenospheric flow. Between 52 and 58°N, there is such a boundary or a transition in the thermal structure of the lithosphere. The most likely position coincides with the young Anahim Volcanic Belt–Wells Gray volcanic lineament. The fixed position of the triple point for the last 10 m.y. (Riddihough, 1977) may indicate a much slower flow of the asthenosphere to the north.

HEAT FLUX AND PRESENT TECTONICS

The heat flux along the active western margin of the Cordillera is related to the present tectonics, not crustal heat generation (Fig. 4). In the south, a multi-disciplinary study across Vancouver Island, Lithoprobe I, investigated the subducting Juan de Fuca plate (Hyndman and others, in preparation). The associated heat-flow profile (Lewis and others, 1988) was continued inland past the end of the Lithoprobe I profile, and showed a large increase from 25 to 70 mW/m^2 over a distance of 20 km, 30 km seaward of the Garibaldi Volcanic Belt. This pattern is similar to a large transition observed in northern Oregon (Blackwell and others, 1982), and requires an apparent source of heat within 10 km of the surface. Differential uplift and erosion can cause large heat-flow contrasts and does contribute to this anomaly, but the geologic evidence for a difference in uplift rate over this 20-km distance of 10 km per 10 m.y., enough to produce a doubling of the heat flux, is lacking.

The heat-flow values, and crustal isotherms calculated from

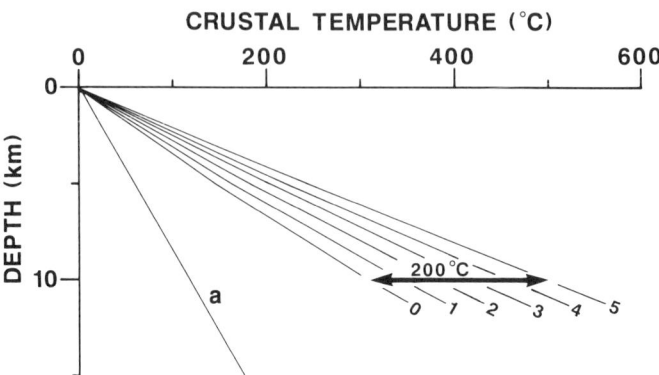

Figure 5. Estimated crustal temperatures. The minimum gradient curve above the subducting Juan de Fuca plate is labelled "a." For the southern B.C heat-flow province, the surficial heat generation (in $\mu m/m^3$) is specified on each of the curves.

them assuming conductive heat flux, are plotted in Figure 6. Asthenospheric flow, presumed responsible for the high reduced heat flux inland (Davis and Lewis, 1984), may continue southwestward to this transition zone where it starts to flow downward and northeastward. Lava generated at depths of 100 km where this downward flow is in contact with subducted oceanic crust is the probable source for the overlying Garibaldi Volcanic Belt and the high heat flux (Lewis and others, 1988).

As the oceanic plate is subducted, it is a heat sink, the heat being used to warm and dehydrate the crust. Lewis and others (1988) claim that the young oceanic lithosphere and underlying asthenosphere supply the necessary heat, and that the surface heat flux along this profile is the resulting reduced amount of heat that reaches the surface. Dehydration would occur as a series of reactions over a small range of temperatures near 450°C. Beneath central Vancouver Island, the 450°C isotherm, the top of a high electrical conductivity zone (Kurtz and others, 1986), and the top of the lower zone of seismic reflectors (E1 of Clowes and others, 1987) are all at the same depth, to within the accuracy of model calculations. The entire lower zone of seismic reflectors lies above the top of the subducting oceanic plate, as defined by seismicity (Rogers, personal communication, 1988), and within the underplated region as defined by Clowes and others (1987).

A large cold prism under the western Coast Plutonic Complex and the eastern edge of the Insular Belt (Lewis and others, 1988) can account for the large gravity (Riddihough, 1979) and magnetic (Coles and Currie, 1977) anomalies. In the Fraser Valley and western Washington (see Fig. 2), just south of this profile, the heat flux, gravity field, magnetic field, and crustal electrical resistivity (Law, personal communication, 1988) are not consistent with such a cold crust, although in Oregon the gravity, heat flux, and electrical resistivity data do resemble those along this profile.

The northern edge of the presently subducting material (Riddihough, 1984) forms a boundary that crosses the middle of Vancouver Island: subduction beneath northern Vancouver Is-

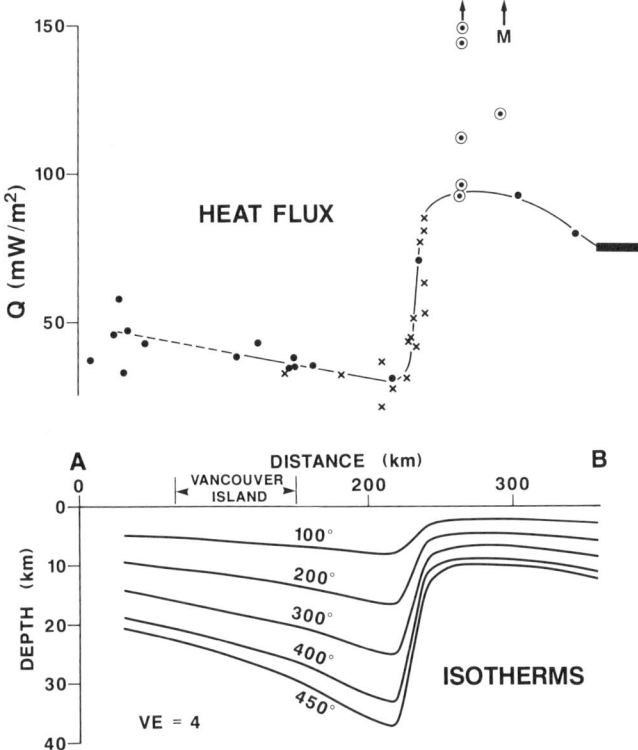

Figure 6. Heat flux and computed crustal temperatures along profile AB (Fig. 2), which follows the LITHOPROBE PHASE I profile across Vancouver Island. Marine data are shown as Xs and circled dots represent values from near young volcanics. The many values near Meager Mountain are indicated by an M.

land ceased at 4 Ma, and the much higher heat flux reflects this history (see Table 2). The gradual heating of the previously subducted crust should produce uplift, but the topography here is lower and more subdued than that of southern and central Vancouver Island. Continuing subduction that produces strains of 1 to 2 cm/yr (Lisowski and others, 1987; Dragert, 1987b) probably causes uplift in the south, and probably uplifted northern Vancouver Island until 4 Ma. The present observed relative uplift to the northeast between these two areas (Dragert, 1987a) may be related to continuing uplift to the south or to local distribution of stresses opposing the subducting Juan de Fuca plate.

The Queen Charlotte transform fault that extends northwest of the triple point forms the boundary between the American and Pacific plates. A small component of oblique subduction, the result of relative plate motions and geometry of the fault, occurs at the seaward side of Queen Charlotte Sound and Moresby Island, the southern Queen Charlotte Island (Hyndman and others, 1982). The high, average heat flux for the 7-Ma oceanic crust seaward of Moresby Island agrees well with the expected theoretical value of 190 mW/m^2. The decreasing heat flux landward to a value of 47 mW/m^2 on Moresby Island is consistent with present oblique underthrusting. A model proposed by Yorath and Hyndman (1983) includes rifting of Queen Charlotte Sound prior to 17 Ma and agrees with the higher heat flux measured on the edge of Queen Charlotte Sound as well as the paleo-heat flux determined for wells in the sound and Hecate Strait.

SUMMARY

The heat flux in the Canadian Cordillera varies from 25 to several hundred mW/m^2. In the south where most of the measurements were made, the variation is directly related to present tectonics or to the distribution of crustal heat generation. Southeastern British Columbia is a heat-flow province with a reduced heat flow of 58 mW/m^2 (uncorrected for glacial effects) and a scale depth of 10 km. The subducting Juan de Fuca plate beneath southwestern British Columbia causes very low heat flux until it reaches a depth of 100 km beneath the Coast Plutonic Complex, above which there is an abrupt transition to higher heat-flow values of 70 mW/m^2. Within the Garibaldi Volcanic Belt, cooling magmas produce large local variations in the heat flux. The geothermal pattern, the result of present and past tectonics, is imprinted across many different terrains and physiographic belts.

REFERENCES CITED

Bentkowski, W. H., and Lewis, T. J., 1983, Preliminary results of shallow drilling in the Garibaldi Volcanic Belt, British Columbia: Ottawa, Energy, Mines, and Resources, Earth Physics Branch Open-File Report 83-23, 38 p.

——, 1984, Preliminary results from shallow drilling in the Alert Bay Volcanic Belt, 1982: Ottawa, Energy, Mines, and Resources, Earth Physics Branch Open-File Report 84-2, 16 p.

——, 1985, Preliminary results from shallow drilling in the Anahim Volcanic Belt, B.C.: Ottawa, Energy, Mines, and Resources, Earth Physics Branch Open-File Report 85-9, 19 p.

——, 1986, Heat flux in the Raft batholith, central British Columbia: Ottawa, Energy, Mines, and Resources, Earth Physics Branch Open-File Report 1337, 24 p.

——, 1988, Thermal measurements in the Fraser Valley, British Columbia: Ottawa, Energy, Mines, and Resources, Geological Survey of Canada Open-File Report 1743, 19 p.

——, 1989, Thermal measurements in Cordilleran boreholes of opportunity: Ottawa, Energy, Mines, and Resources, Geological Survey of Canada Open-File Report 2048, 30 p.

Berry, M. J., and Forsyth, D. A., 1975, Structure of the Canadian Cordillera from seismic refraction and other data: Canadian Journal of Earth Sciences, v. 12, p. 182-208.

Blackwell, D. D., 1974, Terrestrial heat flow and its implications on the locations of geothermal reservoirs in Washington: State of Washington Department of Natural Resources Information Circular 50, p. 21-33.

Blackwell, D. D., and Robertson, E. C., 1973, Thermal studies of the Boulder batholith and vicinity, Montana: Society of Economic Geologists Guide Book for the Butte Field Meeting, 1973, p. D-1-D-6.

Blackwell, D. D., Bowen, R. G., Hull, D. A., Riccio, J., and Steele, J. L., 1982, Heat flow, arc volcanism, and subduction in northern Oregon: Journal of Geophysical Research, v. 87, p. 8735-8754.

Caner, B., Auld, D. R., Dragert, H., and Camfield, P. A., 1971, Geomagnetic depth sounding and crustal structure in western Canada: Journal of Geophysical Research, v. 76, p. 7181-7201.

Church, B. N., 1982, Notes on the Penticton Group; A progress report on a new stratigraphic subdivision of the Tertiary, south-central British Columbia: British Columbia Ministry of Energy, Mines, and Petroleum Resources Paper 1982-1, p. 12-16.

Clowes, R. M., and 6 others, 1987, Lithoprobe—southern Vancouver Island;

Cenozoic subduction complex imaged by deep seismic reflections: Canadian Journal of Earth Sciences, v. 24, p. 31–51.

Coles, R. L., and Currie, R. G., 1977, Magnetic anomalies and rock magnetizations in the southern Coast Mountains, British Columbia; Possible relation to subduction: Canadian Journal of Earth Sciences, v. 14, p. 1753–1770.

Costain, J. K., and 5 others, 1986, Heat flow in the Piedmont and Atlantic Coastal Plain of southeastern United States: Journal of Geophysical Research, v. 91, p. 2123–2135.

Davis, E. E., and Lewis, T. J., 1984, Heat flow in a back-arc environment, Intermontane Belt, southern Canadian Cordillera: Canadian Journal of Earth Sciences, v. 21, p. 715–726.

Dragert, H., 1987a, The fall (and rise) of central Vancouver Island; 1930–1985: Canadian Journal of Earth Sciences, v. 24, p. 689–697.

———, 1987b, Contemporary crustal deformation measurements on central Vancouver Island [abs.]: Pacific Section, Geological Association of Canada Symposium on Recent Crustal Movements in the Pacific Northwest, p. 15–16.

Ewing, T. E., 1981, Regional stratigraphy and structural setting of the Kamloops Group, south-central British Columbia: Canadian Journal of Earth Sciences, v. 18, p. 1464–1477.

Finkh, P., 1981, Heat flow measurements in 17 perialpine lakes: Geological Society of America Bulletin, v. 92, p. 452–514.

Haines, G. V., Hannaford, W., and Riddihough, R. P., 1971, Magnetic anomalies over British Columbia and the adjacent Pacific Ocean: Canadian Journal of Earth Sciences, v. 8, p. 387–391.

Hickson, C. J., 1986, Quaternary volcanism in the Wells Gray–Clearwater area, east-central British Columbia [Ph.D. thesis]: Vancouver, University of British Columbia, 307 p.

Hyndman, R. D., 1976, Heat flow measurements in the inlets of southwestern British Columbia: Journal of Geophysical Research, v. 81, p. 337–349.

Hyndman, R. D., Davis, E. E., and Wright, J. A., 1979, The measurement of marine geothermal heat flow by a multipenetration probe with digital acoustic telemetry and in-situ thermal conductivity: Marine Geophysical Research, v. 4, p. 181–205.

Hyndman, R. D., and 5 others, 1982, Queen Charlotte fault zone; Heat flow measurements: Canadian Journal of Earth Sciences, v. 19, p. 1657–1669.

Jeffreys, H., 1940, The disturbance of the temperature gradient in the Earth's crust by inequalities of height: Monthly Notices of the Royal Astronomical Society, Geophysical Supplement, v. 4, p. 309–312.

Jessop, A. M., 1971, The distribution of glacial perturbation of heat flow in Canada: Canadian Journal of Earth Sciences, v. 8, p. 162–166.

Jessop, A. M., and Judge, A. S., 1971, Five measurements of heat flow in southern Canada: Canadian Journal of Earth Sciences, v. 8, p. 711–716.

Jessop, A. M., Souther, J. G., Lewis, T. J., and Judge, A. S., 1984, Geothermal measurements in northern British Columbia and the southern Yukon Territory: Canadian Journal of Earth Sciences, v. 21, p. 599–608.

Kurtz, R. D., DeLaurier, J. M., and Gupta, J. C., 1986, A magnetotelluric sounding across the subducting Juan de Fuca plate: Nature, v. 321, p. 596–599.

Lee, T.-C., and Henyey, T. L., 1974, Heat-flow refraction across dissimilar media: Geophysical Journal of the Royal Astronomical Society, v. 39, p. 319–333.

Lewis, J. F., and Jessop, A. M., 1981, Heat flow in the Garibaldi volcanic belt, a possible Canadian geothermal energy resource area: Canadian Journal of Earth Sciences, v. 18, p. 366–375.

Lewis, T. J., 1974, Heat production measurement in rocks using a gamma-ray spectrometer with a solid state detector: Canadian Journal of Earth Sciences, v. 11, p. 526–532.

———, 1976, Heat generation in the Coast Range Complex and other areas of British Columbia: Canadian Journal of Earth Sciences, v. 13, p. 1634–1642.

———, 1983, Bottom water temperature variations as observed and recorded in the bottom sediments, Alice Arm and Douglas Channel, British Columbia, in Macdonald, R. W., ed., Proceedings, Workshop on the Kitimat Marine Environment: Canadian Technical Report of Hydrography and Ocean Sciences 18, p. 138–161.

———, 1984, Geothermal energy from the Penticton Tertiary outlier, British Columbia; An initial assessment: Canadian Journal of Earth Sciences, v. 21, p. 181–188.

———, 1989, Heat flow data base for the Canadian Cordillera, in Hittleman, A. M., Kinsfather, J. O., and Myers, H., eds., Geophysics of North America: Boulder, Colorado, National Oceanic and Atmospheric Administration, National Geophysical Data Center, CD-ROM No. 975-B27-001.

Lewis, T. J., and Souther, J. G., 1978, Meager Mountain, British Columbia; A possible geothermal energy resource: Ottawa, Energy, Mines, and Resources, Earth Physics Branch Geothermal Series 9, 17 p.

Lewis, T. J., and Werner, L., 1982, Geothermal gradients on the west side of Okanagan Lake, British Columbia: Ottawa, Energy, Mines, and Resources, Earth Physics Branch Open-File Report 82-6, 58 p.

Lewis, T. J., Allen, V. S., Taylor, A. E., and Jessop, A. M., 1979, Temperature observations during drilling of two 400-m wells in the Coryell intrusives north of Grand Forks, British Columbia: Ottawa, Energy, Mines, and Resources, Earth Physics Branch Open-File Report 79-4, 24 p.

Lewis, T. J., Bennets, H., Allen, V. S., and Chan, F., 1984, Uranium, thorium, and potassium concentrations and heat generated in samples of crustal rocks; A data file: Ottawa, Energy, Mines, and Resourcs, Earth Physics Branch Open-File Report 84-15, 100 p.

Lewis, T. J., Jessop, A. M., and Judge, A. S., 1985, Heat flux measurements in southwestern British Columbia; The thermal consequences of plate tectonics: Canadian Journal of Earth Sciences, v. 22, p. 1262–1273.

Lewis, T. J., and 5 others, 1988, Subduction of the Juan de Fuca plate; Thermal consequences: Journal of Geophysical Research, v. 93, p. 15207–15225.

Lisowski, M., Savage, J. C., Prescott, W. H., and Dragert, H., 1987, Strain accumulation across the Strait of Juan de Fuca and in the Olympic Mountains, Washington [abs.]: Pacific Section, Geological Association of Canada Symposium on Recent Crustal Movements in the Pacific Northwest, p. 17.

Matthews, W. H., 1972, Geothermal data from the Granduc area, northern Coast Mountains of British Columbia: Canadian Journal of Earth Sciences, v. 9, p. 1333–1337.

Monger, J.W.H., and others, 1985, Juan de Fuca plate to Alberta Plains, B-2: Geological Society of America Continent/Ocean Transect 7, scale 1:500,000.

Reader, J. F., and Fairbank, B. D., 1983, Heat flow in the vicinity of the Meager Volcanic Complex, southwestern British Columbia: Geothermal Resources Council Transactions, v. 7, p. 535–539.

Riddihough, R. P., 1977, A model for recent plate interactions off Canada's west coast: Canadian Journal of Earth Sciences, v. 14, p. 384–396.

———, 1979, Gravity and structure of an active margin; British Columbia and Washington: Canadian Journal of Earth Sciences, v. 16, p. 350–363.

———, 1984, Recent movements of the Juan de Fuca Plate system: Journal of Geophysical Research, v. 89, p. 6980–6994.

Sass, J. H., Lachenbruch, A. H., and Munroe, R. J., 1971, Thermal conductivity of rocks from measurements on fragments and its application to heat-flow determinations: Journal of Geophysical Research, v. 76, p. 3391–3401.

Sass, J. H., Lawver, L. A., and Munroe, R. J., 1985, A heat-flow reconnaissance of southeastern Alaska: Canadian Journal of Earth Sciences, v. 22, p. 416–421.

Webb, P. C., Tindle, A. G., Barritt, S. D., Brown, G. C., and Miller, J. F., 1985, Radiothermal granites of the United Kingdom; Comparison of fractionation patterns and variation of heat production for selected granites, in High heat production (HHP) granites, hydrothermal circulation, and ore genesis: London, Institute of Mining and Metallurgy, p. 409–424 p.

Wickens, A. J., 1971, Variations in lithospheric thickness in Canada: Canadian Journal of Earth Sciences, v. 8, p. 1154–1162.

Yorath, C. J., and Hyndman, R. D., 1983, Subsidence and thermal history of Queen Charlotte Basin: Canadian Journal of Earth Sciences, v. 20, p. 135–159.

MANUSCRIPT ACCEPTED BY THE SOCIETY AUGUST 1, 1990
GEOLOGICAL SURVEY OF CANADA CONTRIBUTION 42687, SEPTEMBER 15, 1988

Printed in U.S.A.

Chapter 26

An overview of heat flow in southwestern United States and northern Chihuahua, Mexico

Marshall Reiter, Margaret W. Barroll, and Jeffrie Minier*
New Mexico Bureau of Mines and Mineral Resources and Geoscience Department, New Mexico Institute of Mining and Technology, Socorro, New Mexico 87801

INTRODUCTION

Over the past two decades a great many heat-flow measurements and estimates have been made in the southwestern United States and neighboring areas of northern Chihuahua, Mexico. Although these data have enhanced our understanding of the geothermal regime in the southwestern United States, many questions have developed while analyzing the data. In this chapter we present some of the conclusions that can be made from the heat-flow data in the study areas. In addition, we discuss some of the important unanswered questions concerning the geothermal regime in various geologic provinces of the southwestern United States and northern Chihuahua, Mexico. We consider below the heat-flow data in the various geologic provinces of the study area, and when appropriate, interrelate observations from different geologic regions. Articles by Roy and others (1968), Lachenbruch and Sass (1977), and Blackwell (1978) review principles and concepts concerning heat flow in the western United States.

DISCUSSION OF HEAT-FLOW DATA

Colorado Plateau, interior region

The Colorado Plateau is a roughly circular region, about 700 km in diameter, that encompasses large areas in the four-corner states of Arizona, Colorado, New Mexico, and Utah (Fig. 1); it is a relatively stable region within the tectonically and volcanically active western United States. The region has been uplifted 1,000 to 1,500 m since Eocene time, but in general has maintained more lithologic and structural continuity than other areas of the western United States (King, 1959; Eardley, 1962). In keeping with its relatively stable tectonic setting, the heat flow in the Colorado Plateau is intermediate between higher values characteristic of neighboring geologically active regions, and lower values characteristic of geologically stable regions or regions inactive in Cenozoic times.

Within the Colorado Plateau there appears to be a large interior region where the average heat flow is intermediate, approximately 58 to 64 mW m^{-2} (Roy and others, 1972; Lachenbruch and Sass, 1977; Bodell and Chapman, 1982). Heat-flow data obtained from both relatively shallow and deep geothermal logs ($\gtrsim$750 m and $\lesssim$750 m, respectively) seem to support this conclusion (Reiter and Clarkson, 1983a). There does, however, remain a question concerning the uniformity of heat flow in the interior of the Colorado Plateau. Whereas the heat-flow data calculated from the shallow temperature logs suggest regions of relatively high and low heat flow; the data calculated from the deeper temperature logs suggest a more uniform heat flow (Fig. 2). The degree of uniformity in the heat flow of the Colorado Plateau interior is an important parameter in understanding the tectonics of the province; it relates to the geographic distribution of possible heat sources and heating processes that may be involved with the Colorado Plateau uplift. For example, a uniform heat flow in the Colorado Plateau interior may suggest rather broad, uniform, relatively deep heating of the Colorado Plateau during Cenozoic uplift (e.g., lithospheric heating and thinning). Alternatively, variable heat flow in the central plateau would suggest variations in crustal radioactivity, or variations in lower crustal and mantle temperatures, which might relate to differential elevation.

Although the difference between heat-flow means calculated from deep and shallow temperature logs of about 5 mW m^{-2} is rather small from an experimental point of view, such small differences in average heat flow for a region can significantly affect potential times estimated for the initiation of processes like lithospheric heating (see Reiter and Clarkson, 1983a). For example, a 400-K temperature change at 80 km depth will produce a heat-flow anomaly of 7 mW m^{-2}, or 12 mW m^{-2} after 20 or 30 m.y., respectively. Consequently, many heat-flow models may be quite sensitive to small heat-flow differences. Therefore, if the uncertainties in thermal models are to be at least partially resolved, a number of heat-flow measurements are required, based on deeper temperature logs at more evenly distributed sites in the Colorado Plateau interior.

---

*Present address: Daniel B. Stephens & Associates, Inc., 4415 Hawkins NE, Albuquerque, New Mexico 87109.

Reiter, M., Barroll, M. W., and Minier, J., 1991, An overview of heat flow in southwestern United States and northern Chihuahua, Mexico, *in* Slemmons, D. B., Engdahl, E. R., Zoback, M. D., and Blackwell, D. D., eds., Neotectonics of North America: Boulder, Colorado, Geological Society of America, Decade Map Volume 1.

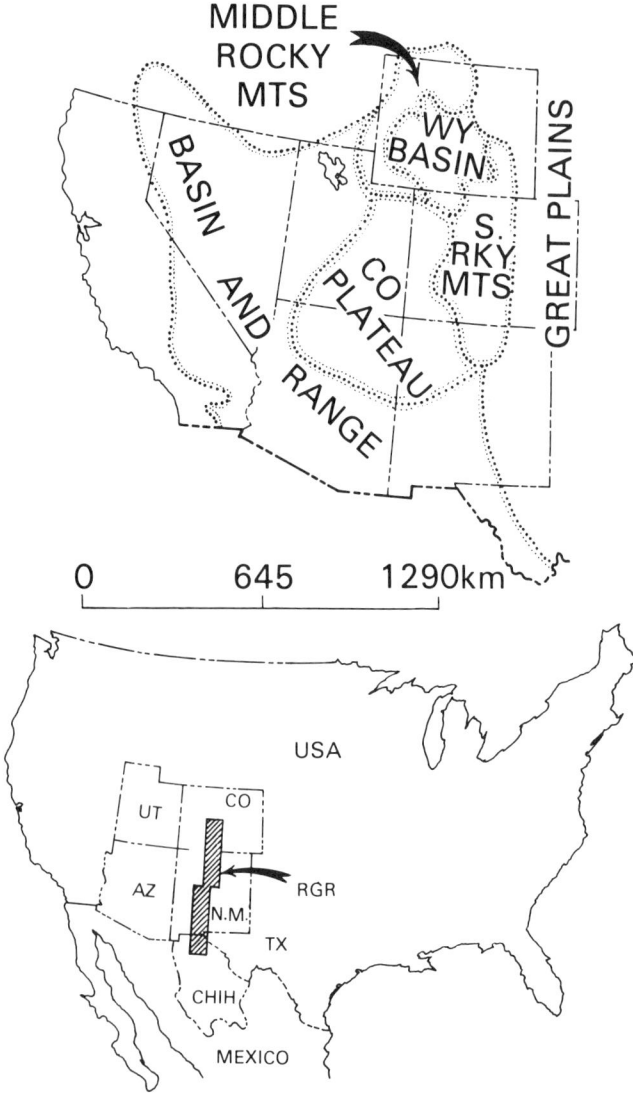

Figure 1. Location maps for this chapter. RGR, Rio Grande rift.

Colorado Plateau peripheral regions

Heat flow in the peripheral regions of the Colorado Plateau is often influenced by thermal processes in neighboring areas whose geologic environments are quite different. For example, deep heat-flow measurements in the San Juan Basin show a uniform increase in heat flow, from south to north, as the San Juan volcanic field is approached (see Fig. 1 in Reiter and Mansure, 1983). Thermal sources associated with the San Juan volcanic field have influenced the heat-flow distribution at appreciable distances into the San Juan Basin within the eastern part of the Colorado Plateau. This pattern of increasing heat flow, and the high heat-flow measurements in the San Juan volcanic field, have been considered in terms of heat-flow anomalies produced by subsurface thermal sources (Reiter and Clarkson, 1983b). The least-squares fitting of theoretical heat flow to observed heat flow allows the prediction of a large thermal source associated with the San Juan volcanic field. The diameter of the source is estimated to be about 100 km, the depth of the source about 30 km on the top to $\gtrsim 100$ km at the bottom. Because of the relatively good fit between observed data and data generated from a steady-state isothermal step, Clarkson (1984) suggests that the thermal source has been replenished since the major volcanism ceased at about 25 Ma.

High heat-flow estimates have also been calculated along the northwestern periphery of the Colorado Plateau where it borders the Basin and Range Province (Eggleston and Reiter, 1984). These heat-flow estimates are made by using bottom-hole temperatures recorded in deep petroleum tests to calculate geothermal gradients. Lithologic logs are used to characterize the rocks over appropriate depth intervals so that reasonable estimates of thermal conductivity can be calculated. The heat-flow estimates, along with some of the conventional heat-flow measurements in the region (Bodell and Chapman, 1982), suggest that the peripheral region of the Colorado Plateau, near the transition of the Colorado Plateau to the Basin and Range Province, has an elevated heat flow typical of the Basin and Range Province. If this suggestion is correct, then the thermal processes operating in the crust and upper mantle of the Colorado Plateau transition zone in this area may be quite similar to thermal processes occurring in the Basin and Range Province. High heat-flow estimates in the Salt Lake Valley, as well as in the Colorado Plateau transition, qualitatively agree with the energy-flux map suggested by Blackwell (1978).

From the limited data available, it would appear that heat-flow values in the southern regions of the Middle Rocky Mountains and Wyoming Basin are similar to the intermediate heat flows for the interior and northern parts of the Colorado Plateau (see Eggleston and Reiter, 1984, Fig. 1). As such, there does not appear to be a geothermal transition at the northern boundary of the plateau. In other peripheral regions of the Colorado Plateau it has been difficult to obtain reliable heat-flow values. For example, many of the heat-flow values in and near the San Francisco volcanic field are low (see Eggleston and Reiter, 1984, Fig. 1). The low values probably result because downward ground-water movement, due to the large recharge in the area, lowers the conductive geothermal gradient. We have noticed unusually low gradients in a number of wells from the area. In other areas, such as the southern and southwestern parts of the San Juan Basin, the relatively large variability of values may be due to the ground-water redistribution of heat. In fact, the large increase in heat flow across the San Juan Basin in the direction of the San Juan volcanic field went unnoticed until deep-temperature logs (1 to 2 km depth) were made for heat-flow measurements. It is believed that these deep-temperature data are below most of the disturbing effects of ground-water movement, and therefore provide a reasonably accurate picture of the conductive heat flow in the region. Similarly, deep-temperature data in other peripheral areas of the Colorado Plateau may be required to reasonably estimate conductive heat flow.

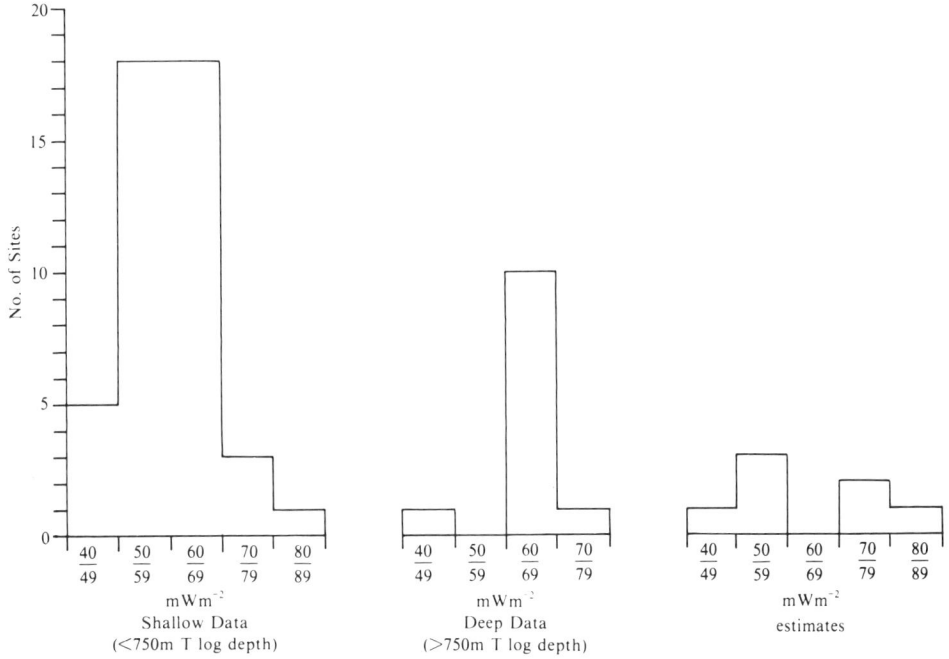

Figure 2. Histogram of heat-flow data in the Colorado Plateau interior.

Southern Basin and Range Province

For the present discussion, areas included in the southern Basin and Range Province are southern and western Arizona and southwestern New Mexico (Fig. 1; Rio Grande rift is discussed separately). The southern Basin and Range Province is a region of considerable Cenozoic volcanic and extensional tectonic activity. The crust is generally thinner than in the Colorado Plateau or Southern Rocky Mountains (25 to 35 km as compared to about 35 to 50 km; Pakiser, 1963; Thompson and Burke, 1974; Stewart, 1978). Upper mantle velocities in the southern Basin and Range Province are generally $\gtrsim 7.9$ km s^{-1} (Stewart, 1978). Modern topography in the southern Basin and Range Province results largely from high-angle faulting that occurred principally from about 13 to 10 Ma in a stress field where the least compressive stress was approximately west-southwest to east-northeast (Zoback and others, 1981). Much of the earlier extension seems to have occurred along low-angle faults or detachments (Zoback and others, 1981; Wernicke, 1981). Volcanics younger than 17 Ma are predominantly basalt or bimodal basalt and rhyolite; volcanics older than 17 Ma are predominantly calc-alkalic (Christiansen and Lipman, 1972).

The heat flow in the southern Basin and Range Province is typical for continental areas of considerable Cenozoic volcanic activity (e.g., see Stacy, 1969, p. 240). The mean heat flow for the southern Basin and Range Province in Arizona is very close to the mean heat flow for the entire Basin and Range Province (86 ± 19 and 89 ± 31 mW m^{-2}, respectively). The mean of 15 heat-flow values in southwestern New Mexico is slightly higher (94 ± 21 mW m^{-2}; see Fig. 1 in Shearer and Reiter, 1981). A histogram of the data in the southern Basin and Range Province of Arizona and southwest New Mexico is presented in Figure 3. A relatively large variability in the shallow data is illustrated.

Shearer and Reiter (1981) measured the radiogenic heat production of rocks, representing the upper crust, at ten heat-flow sites in the southern Basin and Range Province (Fig. 4). The data indicate a general trend of increasing heat flow with increasing heat production (as suggested for several geologic provinces by Roy and others, 1968; Lachenbruch, 1970). A least mean squares straight-line fit through the heat-flow–heat-production data pairs is presented by Shearer and Reiter (1981; see also Fig. 4). This linear relation has somewhat different values for the slope and intercept constants than the relation obtained by Roy and others (1968) using 12 of 15 data pairs over the entire Basin and Range Province. (The intercept and slope values are 72 mW m^{-2} and 6.2 km for the southern Basin and Range Province, and 59 mW m^{-2} and 9.4 km for the Basin and Range Province.) Indeed the least mean squares linear fit of both data sets is sensitive to the choice of data excluded. Therefore, consideration of data quality becomes very important if one wishes to accurately describe heat-flow increase as a function of crustal radioactivity.

Of the ten heat-flow–heat-production measurement sites in the southern Basin and Range Province, five are associated with heat-flow measurements where deep-temperature logs were made ($\gtrsim 650$ m). The heat-flow–heat-production data plot (Fig. 4) indicates that the deeper heat-flow sites show a bit less scatter from the least mean squares straight-line fit to the data than do the other sites. Much of the scatter in the complete data set results may be because of relatively near-surface perturbations to temperature. The deeper heat-flow measurements are scattered

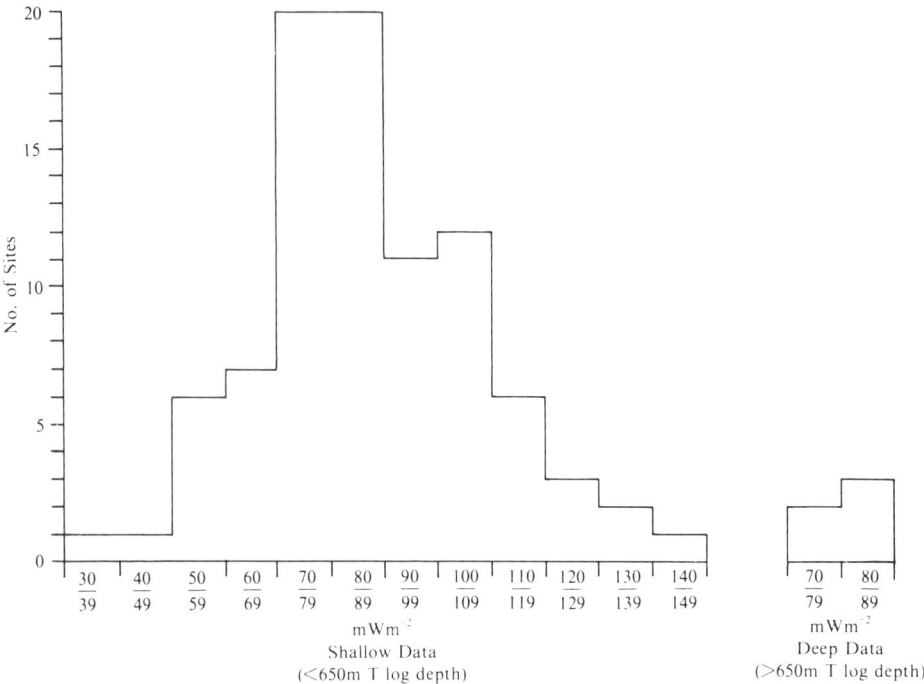

Figure 3. Histogram of heat-flow data in the southern Basin and Range Province.

over a large area. Because the heat-flow values are similar (Figs. 3 and 4), the deep conductive heat flux for the area might be rather uniform, modified slightly by variations in crustal radioactivity. Based on the results illustrated in Figure 2 of Shearer and Reiter (1981), we suggest that variations in crustal radioactivity may contribute to changes in heat flow of ~ 20 mW m^{-2} (i.e., crustal radiogenic changes could cause heat flows to vary from about 75 to 95 mW m^{-2}; also see Fig. 4). Higher quality data are required to obtain more certain estimates of the geothermal parameters associated with the slope and intercept of the least mean squares straight-line fit to the heat-flow–heat-production data.

Geothermal trends associated with nonradiogenic causes are not immediately obvious from the geothermal data in the southern Basin and Range Province (see Fig. 1 in Shearer and Reiter, 1981). Values for three large data groups in central Pinal County, and in western and eastern Pima County, vary little. However, the large variability of closely spaced sites in Graham County, Arizona, and in most of southwestern New Mexico, suggests ground-water movement. A group of five high heat-flow values in west-central Yuma County may indicate the presence of a thermal anomaly; however, heat-production data are necessary to preclude high crustal radioactivity. The data further suggest that a region of relatively subdued Basin and Range Province heat flow (subdued crustal radiogenic heat?) exists in central and southwestern Mohave County and west-central Yavapai County.

Southern Rocky Mountains

The Southern Rocky Mountains are a series of elongated ranges extending from north-central New Mexico through central Colorado to the Wyoming border (Fig. 1). Cutting through this region from north to south is the Rio Grande rift, which is discussed separately. The Front, Park, Sawatch, and Sangre de Cristo Ranges, and the Wet Mountains, are the principal high-relief structures; they expose Precambrian basement in a number of places at elevations of more than 4,000 m. Most of the present relief occurred during the past 10 m.y. along late Paleozoic and Laramide uplifts (Scott, 1975; Taylor, 1975; Tweto, 1975; Axelrod and Bailey, 1976). In southwestern Colorado, 60,000 km^3 of volcanics were erupted during Oligocene time, the remnants of which are the San Juan Mountains (Lipman and others, 1978). Crustal thickness for the Southern Rocky Mountains is estimated

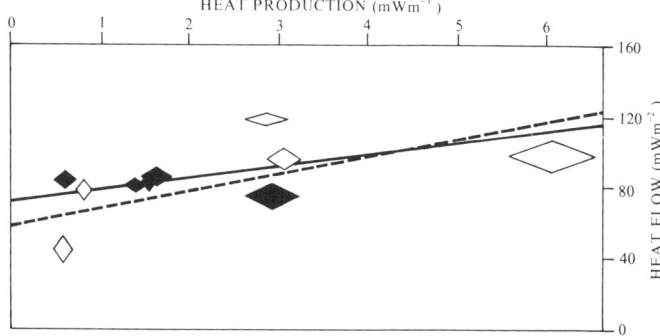

Figure 4. Heat flow versus heat production for ten sites in the southern Basin and Range Province (after Shearer and Reiter, 1981). Solid figures, sites where T logs exceed 650 m; open figures, T-log sites less than 650 m deep. Width and height of figure represent errors on the horizontal and vertical axes. Solid line is least mean squares fit of ten data points shown; dashed line is least mean squares fit of 12 data points over the entire Basin and Range Province (Roy and others, 1968).

to be about 50 km across the Front Range, similar to the crustal thickness in the Great Plains but greater than in the Colorado Plateau (Jackson and Pakiser, 1965). Seismic data also indicate a relatively shallow, crustal low-velocity layer under the San Juan volcanic field (Prodehl and Pakiser, 1980). Mantle material with high electrical conductivity seems to underlie the Southern Rocky Mountains; in northern New Mexico a highly conductive mantle seems to extend east of the Southern Rocky Mountains–Great Plains transition (Caner and others, 1967; Gough, 1974). From geomagnetic data, Schmucker (1970) suggested increased subsurface temperatures throughout the Southern Rocky Mountains region. From heat-flow data, Decker (1969) also suggested a high heat flow for the region.

The ranges of the Southern Rocky Mountains generally have high heat flow; the heat-flow means are about the same as the mean for the southern Basin and Range Province. The average heat flow for the eastern ranges (the Front, Wet, and Sangre de Cristo) is about 83 mW m^{-2}; and for the western ranges (the Elk Mountains and the Sawatch Range) the average heat flow is about 93 mW m^{-2} (see Fig. 5 for a histogram of the data). The average heat flow for the San Juan volcanic field is greater, about 135 mW m^{-2} (Fig. 5). As in other geologic regions, the data in the Southern Rocky Mountains are variable and geographically sparse, so that trends may be difficult to observe. Data in the San Juan volcanic field are uniformly higher than 84 mW m^{-2}. In most other regions of the Southern Rocky Mountains, values above and below 84 mW m^{-2} are found (see Figs. 2 and 3 in Reiter and others, 1979; and Fig. 5, this chapter). The basins along the transition from the eastern ranges to the Great Plains (i.e., the Las Vegas, Raton, and Denver basins; Edwards and others, 1978; see Fig. 6) seem to have high heat flow (the mean value for the basins is about 106 mW m^{-2}).

The high heat flow in the basins along the transition between the Southern Rocky Mountains and the Great Plains may be associated with subsurface thermal sources, or alternatively, hydrologic flow that is moving heat from the Southern Rocky Mountains. Majorowicz and others (1985, 1986) demonstrated the influence of ground-water recharge and discharge on the heat-flow pattern in the Williston Basin, Alberta, Canada. In the Denver Basin, northern Colorado, heat-flow values calculated from deep-temperature logs (about 2,100 m depth) are quite high (96 to 100 mW m^{-2}; Reiter and others, 1979). The sites are about 20 to 30 km east of the Front Range. If the temperature data are deep enough to avoid ground-water advection of heat, then it would appear that the high heat flow in the Denver Basin is caused by a geothermal source probably associated in some way with the Front Range. This phenomenon may also occur in the other transitional basins along the eastern front of the Southern Rocky Mountains. Data from depths greater than the marine shales (in which temperatures were logged in the Denver Basin) would be most useful in providing thermal conductivity values potentially less influenced by anisotropy (i.e., data below the thick Pierre Shale may provide more reliable conductivity values as well as better temperature information).

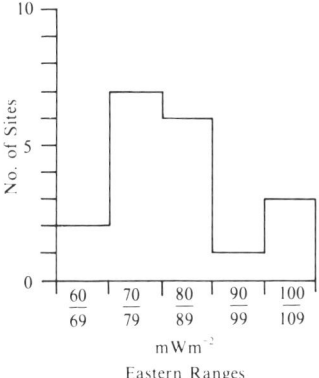
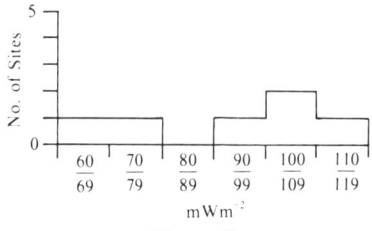
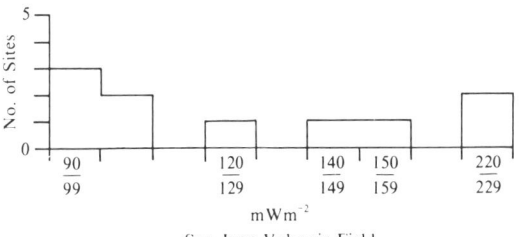

Figure 5. Histograms of heat-flow data in the Southern Rocky Mountains.

From the two high heat-flow values in the Denver Basin (discussed above) and two additional intermediate values in southern Wyoming (temperature logs to about 1,500 m and 2,100 m depth; Reiter and others, 1979), we suggest a rather abrupt north-south termination of high heat flow associated with the Front Range of the Southern Rocky Mountains. From a number of data in New Mexico, it would also appear that the high heat flow in the Sangre de Cristo Mountains changes to lower values just south of Santa Fe. These rather abrupt heat-flow changes at the northern and southern boundaries of the eastern ranges suggest mantle and lower-crustal thermal anomalies associated with the Southern Rocky Mountains. Detailed heat-flow profiles will be required to further estimate thermal source characteristics such as depth and timing.

The large amount of ground-water recharge in the ranges of the Southern Rocky Mountains may greatly lower the heat flow measured at many drill sites. As suggested before, downward-moving ground water can considerably lower near-surface temperature gradients. Considering the large recharge in the Southern Rocky Mountains, and the high values for the deep heat-flow data in the Denver Basin, the average heat flow measured in the

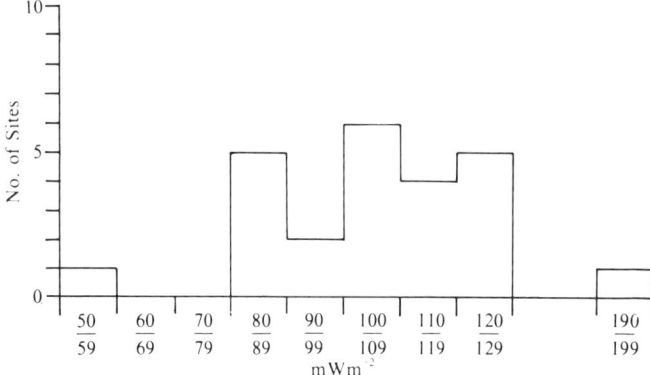

Figure 6. Histograms of heat-flow data in basins just east of the Southern Rocky Mountains.

Southern Rocky Mountains may be a low estimate of the conductive heat flux for the region (i.e., the heat flux that would relate to mid- and lower-crustal and upper mantle temperatures, and crustal radioactivity); we may be significantly underestimating the average or regional heat flow and therefore not fully appreciating the magnitude of the associated geothermal anomaly. Once again, deep-temperature logs (>1 to 2 km) may be required to obtain heat-flow values representative of crustal and mantle thermal conditions.

The Rio Grande rift and northern Chihuahua, Mexico

The Rio Grande rift is a series of deeply filled, en echelon grabens trending north-south from central Colorado through New Mexico and possibly into west Texas and northern Chihuahua (Fig. 1; Chapin, 1971, 1979; Gries, 1979; Seager and Morgan, 1979). It is a late Cenozoic, volcanically and tectonically active zone. The geologic activity along the rift is associated with a thinned crust, upwarped mantle, and subsurface magmas (e.g., Sanford and others, 1973; Cordell, 1978; Keller and others, 1979; Lipman and Mehnert, 1979; Hermance and Pedersen, 1980). Extension across the Rio Grande rift seems to increase from north to south (Woodward, 1977; Cordell, 1982; Golombeck and others, 1983). Cenozoic cauldrons, and extensive middle and late Tertiary and Quaternary volcanics, are found along the Rio Grande rift (Woodward and others, 1978).

An envelope of high heat flow seems to lie along the Rio Grande rift, although low heat-flow values are present as well (Fig. 7). In southern and central Colorado there are only two heat-flow measurements within the grabens that may be considered part of the Rio Grande rift: both of these measurements are greater than 84 mW m^{-2} (Reiter and others, 1979). In the Española Basin (north-central New Mexico), there are a number of high heat-flow measurements (Fig. 7, top; see also Plate 1 in Reiter and others, 1986). Some of these high values are probably associated with the young volcanics in the Española Basin. High and low values of heat flow were measured close to the Albuquerque Basin and in the basins of the Rio Grande rift in southern New Mexico (Fig. 7, middle and bottom; see also Plate 1 in Reiter and others, 1986).

Very high heat flows (more than about 200 mW m^{-2}) often occur in close proximity to large normal faults and/or caldera boundaries (Reiter and others, 1978). These features may provide avenues for the movement of hydrothermal fluids (magmas and heated ground water). Generally, however, it is difficult to attribute high heat-flow values to a specific cause (i.e., to either magmatic sources or ground-water advection). Morgan and others (1981) suggest that high heat flow at appropriate sites along the Rio Grande rift may be caused by ground water forced to move upward at the hydrologically constricted ends of grabens; this suggestion seems reasonable for a number of sites along the Rio Grande rift. Alternatively, Sanford and others (1973), Sanford and Einarsson (1982), and Schlue and others (1986) predict from seismic studies that magmas are present at intermediate (about 19 km) and shallow (about 2 to 6 km) crustal depths. Hydrologic and magmatic processes may well be operating in concert to produce the observed high heat flows.

New heat-flow estimates in the Albuquerque-Belen basin, in the basins of southern New Mexico, and in northern Chihuahua may provide additional insight into the geothermal regime of the Rio Grande rift (Reiter and others, 1986). These new heat-flow estimates are calculated by using gradients derived from bore-hole temperature data taken in deep oil tests, and thermal conductivities compiled for standard rock types found in lithologic logs (Reiter and others, 1986). Because of the generally decreasing porosity and permeability with depth, it is thought that the bore-hole temperature data for new heat-flow estimates are below many of the perturbing effects of ground-water advection. The estimates do, however, suffer from numerous potential errors.

A number of observations can be made about the heat-flow measurements and the heat-flow estimates along the Rio Grande rift (see Plate 1 in Reiter and others, 1986). The estimates in the Albuquerque-Belen basin are at locations some distance from conventional heat-flow measurements. The estimates show a northeast-southwest zone of high heat flow about midway north-south in the basin (see Plate 1 in Reiter and others, 1986). This zone of high heat flow is in proximity to a near-surface magma body suggested by Schlue and others (1986) from surface-wave studies. Data are not present in the southern part of the Albuquerque-Belen basin, and hence, it is not possible to estimate the extent of the high heat-flow region. Estimates in the northern part of the basin and along the western boundary show that the high heat flow in the zone across the middle of the basin decreases to intermediate values elsewhere. Two heat-flow estimates and two deep heat-flow measurements along the western edge of the basin have intermediate values (72 to 82 mW m^{-2}). Therefore, previous shallow measurements of high heat flow at other sites along the western boundary of the basin are suspect. Of course, the high heat flows along the western edge of the basin could

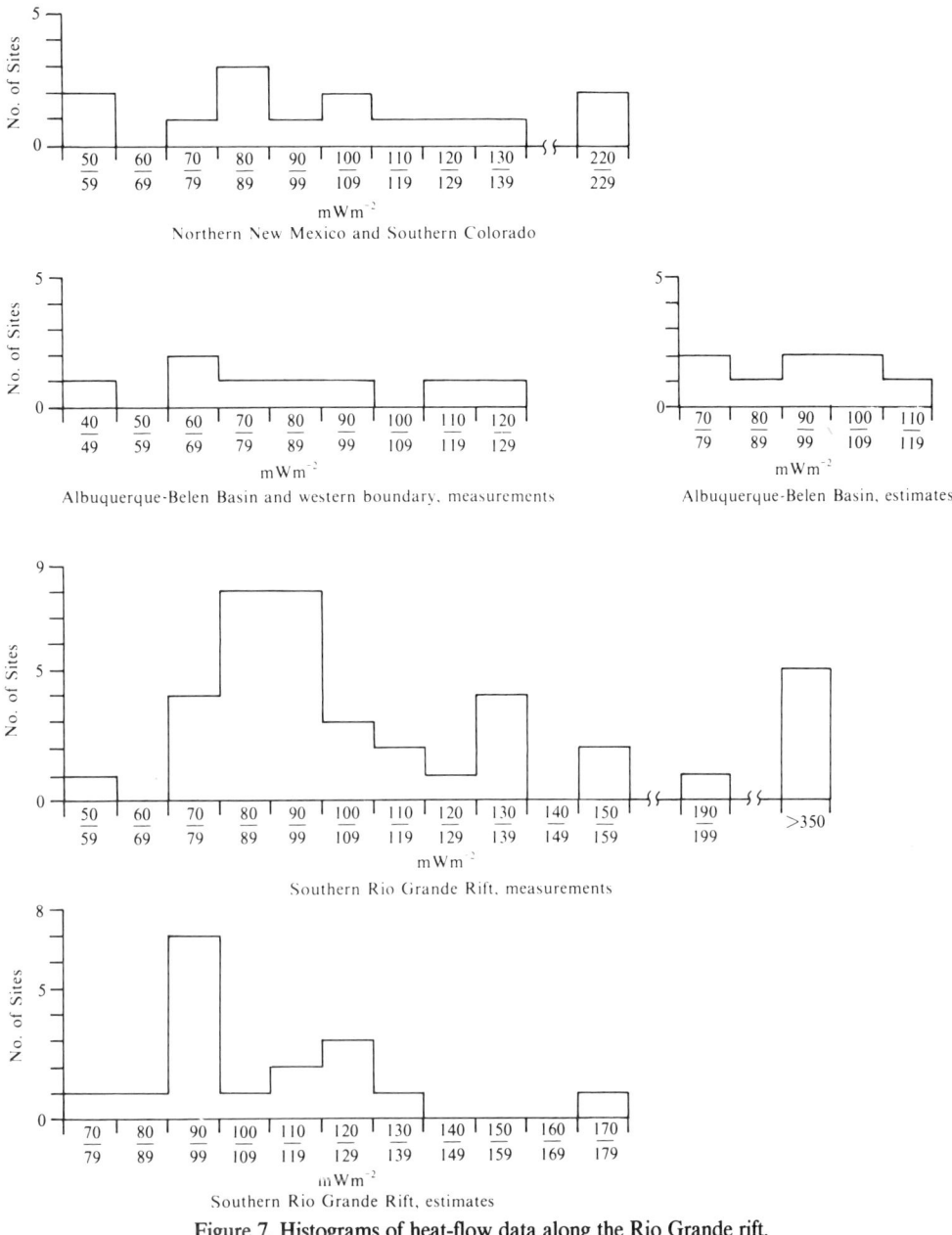

Figure 7. Histograms of heat-flow data along the Rio Grande rift.

result from a local thermal source, but in any case, the influence of ground-water advection is suspected.

In the southern part of the Rio Grande rift there are a few heat-flow measurements and estimates that are intermediate in value (about 70 to 80 mW m^{-2}); most of the data are above the average heat flow (89 mWm^{-2}) for the entire Basin and Range Province (Fig. 7, bottom; see also Plate 1 in Reiter and others, 1986). Between Socorro and Las Cruces there are eight sites where the heat-flow estimates are about 90 to 100 mW m^{-2}. Four of the sites are located in the northern part of the Jornada del Muerto, north of the Jornada Basin. The other four sites are scattered widely. Higher heat-flow estimates (117 to 135 mW m^{-2}) are also made at sites scattered over the region from Socorro to Las Cruces. From these data it is suggested that the regional or background heat flow for the rift area between Socorro and Las Cruces is about 95 mW m^{-2}; a value that closely agrees with the interpretation of previous data by Decker and Smithson (1975). It is also interesting to note that the mean of 15 measurements in extreme southwest New Mexico is 94 ± 21 mW m^{-2}. Some of the sites with higher heat-flow estimates in the southern Rio Grande rift (117 to 135 mW m^{-2}) are near large normal faults that may act as conduits for hydrothermal circulation to enhance geothermal gradients; but again the actual cause of the elevated geothermal gradients is generally ambiguous and difficult to quantify (e.g., the relative contributions of magma and hydrothermal circulation, crustal thinning, crustal radioactivity, etc.).

In the Jornada del Muerto a most interesting heat-flow pattern is emerging (Reiter, 1985; Reiter and Barroll, in preparation). At sites in the northern and central regions of the Jornada del Muerto, heat-flow values are characteristic of the southern Rio Grande rift (90 to 98 mW m^{-2}). The heat flow may increase slightly (about 9 mW m^{-2}) from the north to the central area. This pattern changes abruptly at sites in the southern Jornada del Muerto. Over a 5 to 10 km distance, heat-flow estimates change about 25 to 30 mW m^{-2} (from 98 mW m^{-2} to about 125 mW m^{-2}). This abrupt change suggests a heat source in the upper crust within the southern Jornada del Muerto. Possible causes may include intrusive activity, mantle degassing, and enhanced crustal radioactivity.

The mean of heat-flow estimates in the northern Jornada del Muerto (about 94 mW m^{-2}) is slightly greater than the mean heat flow for the Basin and Range Province. A relatively thin crust, determined from seismic and geothermal data (about 30 to 31 km), is consistent with high heat flow (Warren, 1978; Cook and others, 1979; Keller and others, 1979; Jaksha, 1982). Nevertheless, the Jornada del Muerto demonstrates little if any syn-rift subsidence or deposition (New Mexico Geological Society, 1982), and the primarily synclinal structure is likely to be a remnant feature of Laramide compression (Kelley, 1955). It would therefore appear that syn-rift, deviatoric, crustal tension was absent in the Jornada del Muerto. Consequently, this may be a region where crustal thinning has not involved upper-crustal extension, but rather some type of mantle–lower-crustal process with continued high upper-mantle temperatures.

The heat-flow measurements and estimates from Las Cruces south to the international border with Mexico are few and bimodal (i.e., intermediate–moderately high values along with very high values; see Plate 1 in Reiter and others, 1986). The intermediate–moderately high heat flows estimated in the Mesilla Basin are in qualitative agreement with estimates in northern Chihuahua, in the region between the Los Muertos and Hueco basins. Very high heat flows in west Texas, and on the eastern boundary of the Mimbres Basin, are thought to be caused largely by ground-water advection of heat along major faults; although magnetotelluric data have been interpreted as indicating partial melts at about 21 to 28 km depth east-southeast of El Paso (Hermance and Pedersen, 1980). Surface-wave data also suggest a thin crust (about 30 km) and low Sn velocities (about 4.4 km s^{-1}) extending well into northern Chihuahua along a zone of undetermined width (Keller and others, 1979).

Heat-flow estimates in northeastern Chihuahua, east of about 106°50′, are typically intermediate (about 61 to 83 mW m^{-2}; four values average about 73 mW m^{-2}; Reiter and Tovar, 1982). The estimates in north-central Chihuahua are higher (from about 106°50′ to about 107°30′; three values average about 85 mW m^{-2}). West of about 107°30′, heat flow increases; five estimates average about 95 mW m^{-2}. The high heat flow associated with the southern Rio Grande rift, perhaps interrupted by the Mesilla Basin, may continue in northern Chihuahua, shifted to the west of the Los Muertos Basin. Smith and Jones (1979) suggest the possibility of a southeast-trending, but discontinuous, incipient rift across the Central Plateau of northern Mexico. Data in the Hueco Basin would also provide useful information with regard to the possible continuation of the Rio Grande rift geothermal anomaly.

CONCLUSIONS

From presently available data a number of observations can be made concerning the geothermal regime in the southwestern United States; however, there also remain a great many unanswered questions. If regional subsurface temperatures are to be better appreciated and less ambiguously related to models of regional tectonism and volcanism, then data less perturbed by mechanisms such as ground-water advection will be necessary. Many of the deep data will be taken in sedimentary basins. Therefore, data below the commonly occurring thick marine shales (where thermal conductivity is profoundly dependent on anisotropy, porosity, sand content, etc.) will provide more reliable heat-flow values than are now available.

Some of the most interesting questions concerning geothermal studies in the southwestern United States involve the uniformity of heat flow within a geologic province. Deep data in the Colorado Plateau interior suggest a remarkably constant heat flow over large areas. A somewhat similar situation seems to exist with the deeper data in the southern Basin and Range Province (although more variability is present in the deeper data of the southern Basin and Range Province as compared with the deeper data from the Colorado Plateau interior). The uniformity of deep heat-flow values relates to the distribution of crustal radiogenic elements and the thermal processes operating in the lower crust and upper mantle. A uniform heat flow throughout the Colorado Plateau interior would be consistent with, and perhaps suggestive of, a uniform crustal radiogenic concentration that evolved over a large region within a given time period. A rather uniform heat flow in the southern Basin and Range Province would be consistent with less tectonic activity than elsewhere in the Basin and Range Province during the past 10 m.y.

Estimates of heat flow from deep bore-hole temperatures in the Albuquerque-Belen basin suggest a rather uniform, intermediate heat flow of about 72 to 82 mW m^{-2}. These data also indicate a northeast to southwest zone of high heat flow crossing the basin at about a north-south midpoint. Although these estimates are subject to considerable error, they have enabled us to better understand the basic geothermal picture of the Albuquerque-Belen basin by removing much of the variability (noise) commonly associated with near-surface problems in conventional data that are taken in geologically and hydrologically complex areas. Similarly, in the southern Rio Grande rift, most of the heat-flow estimates are more than about 95 mW m^{-2}, suggesting a large area of higher heat flow than in the Albuquerque-Belen basin. This conclusion is supported by conventional data. In the southern Rio Grande rift the high heat-flow estimates (more than about 130 mW m^{-2}) near large normal faults may be indicating that deep (more than several km) hydrothermal activity is occurring. Data in northern Chihuahua suggest an increase in heat flow as the Sierra Madre Occidental are approached.

Deep heat-flow data in the Southern Rocky Mountains will be important if we are to fully appreciate the magnitude of the associated geothermal anomaly. It is possible with the large snowfall and rainfall in the Southern Rocky Mountains that temperature gradients shallower than about 1,000 m are lowered by recharge. The basins just east of the Southern Rocky Mountains seem to have a higher heat flow than in the mountains. Does this data imply a different thermal regime in the basins, or a different hydrologic regime, or a larger and hotter anomaly associated with the Southern Rocky Mountains? High heat-flow measurements approaching the San Juan volcanic field and high heat-flow estimates approaching the Sierra Madre Occidental imply hotter subsurface conditions (heat replenishment) than one would expect from the age of the volcanics. Newer and better data will give us information concerning heat replenishment in the crust and upper mantle that may be quite different from what we would anticipate by surface geologic observations.

REFERENCES CITED

Axelrod, D. I., and Bailey, H. P., 1976, Tertiary vegetation, climate, and altitude of the Rio Grande depression, New Mexico–Colorado: Paleobiology, v. 2, p. 235–254.

Blackwell, D. D., 1978, Heat flow and energy loss in the western United States, in Smith, R. B., and Eaton, G. P., eds., Cenozoic tectonics and regional geophysics of the western Cordillera: Geological Society of America Memoir 152, p. 175–209.

Bodell, J. M., and Chapman, D. S., 1982, Heat flow in the north-central Colorado Plateau: Journal Geophysical Research, v. 87, p. 2869–2884.

Caner, B., Cannon, W. H., and Livingstone, C. E., 1967, Geomagnetic depth sounding and upper mantle structure in the Cordillera region of western North America: Journal of Geophysical Research, v. 72, p. 6335–6340.

Chapin, C. E., 1971, The Rio Grande rift; 1, Modifications and additions, in James, H. L., ed., Guidebook of the San Luis Basin: New Mexico Geological Society 22nd Field Conference, p. 191–201.

—— , 1979, Evolution of the Rio Grande rift; A summary, in Riecker, R. E., ed., Rio Grande rift; Tectonics and magmatism: American Geophysical Union, p. 1–5.

Christiansen, R. L., and Lipman, P. W., 1972, Cenozoic volcanism and plate tectonic evolution of western United States; 2, Late Cenozoic: Philosophical Transactions of the Royal Society of London, series A., v. 271, p. 249–284.

Clarkson, G., 1984, Implications for thermal histories of the San Juan Basin and San Juan Mountains since Late Cretaceous time [Ph.D. thesis]: Soccoro, New Mexico Institute of Mining and Technology, 107 p.

Cook, F. A., McCullar, D. W., Decker, E. R., and Smithson, S. B., 1979, Crustal structure and evolution of the southern Rio Grande rift, in Riecker, R. E., ed., Rio Grande rift; Tectonics and magmatism: American Geophysical Union, p. 195–208.

Cordell, L., 1978, Regional geophysical setting of the Rio Grande rift: Geological Society of America Bulletin, v. 89, p. 1073–1090.

—— , 1982, Extension in the Rio Grande rift: Journal of Geophysical Research, v. 87, p. 8561–8569.

Decker, E. R., 1969, Heat flow in Colorado and New Mexico: Journal of Geophysical Research, v. 75, p. 550–559.

Decker, E. R., and Smithson, S. B., 1975, Heat flow and gravity interpretation across the Rio Grande rift in southern New Mexico and west Texas: Journal of Geophysical Research, v. 80, p. 2542–2552.

Eardley, A. J., 1962, Structural geology of North America: New York, Harper and Row, 743 p.

Edwards, C. L., Reiter, M., Shearer, C., and Young, W., 1978, Terrestrial heat flow and crustal radioactivity in northeastern New Mexico and southeastern Colorado: Geological Society of America Bulletin, v. 89, 1341–1350.

Eggleston, R. E., and Reiter, M., 1984, Terrestrial heat-flow estimates from petroleum bottom-hole temperature data in the Colorado Plateau and the eastern Basin and Range Province: Geological Society of America Bulletin, v. 95, p. 1027–1034.

Golombek, M. P., McGill, G. E., and Brown, L., 1983, Tectonic and geologic evolution of the Española basin, Rio Grande rift; Structure, rate of extension, and relation to the state of stress in teh western United States: Tectonophysics, v. 94, p. 483–507.

Gough, D. I., 1974, Electrical conductivity under North America in relation to heat flow, seismology, and structure: Journal of Geomagnetism and Geoelectricity, v. 26, p. 105–123.

Gries, J. C., 1979, Problems of delineation of the Rio Grande rift into the Chihuahua tectonic belt of northern Mexico, in Riecker, R. E., ed., Rio Grnade rift; Tectonics and magmatism: American Geophysical Union, p. 107–113.

Hermance, J. F., and Pedersen, J., 1980, Deep structure of the Rio Grande rift; A magnetotelluric interpretation: Journal of Geophysical Research, v. 85, p. 3899–3912.

Jackson, W. H., and Pakiser, L. C., 1965, Seismic study of crustal structure in the Southern Rocky Mountains: U.S. Geological Survey Professional Paper 525-D, D85–D92.

Jaksha, L. H., 1982, Reconnaissance seismic refraction-reflection surveys in southwestern New Mexico: Geological Society of America Bulletin, v. 93, p. 1030–1037.

Keller, G. R., Braile, L. W., and Schlue, J. W., 1979, Regional crustal structure of the Rio Grande rift from surface wave dispersion measurements, in Riecker, R. E., ed., Rio Grande rift; Tectonics and magmatism: American Geophysical Union, p. 115–126.

Kelley, V. C., 1955, Regional tectonics of south-central New Mexico, in Fitzsimmons, J. P., and Krusekoph, H. H., eds., South-central New Mexico: New Mexico Geological Society, p. 96–104.

King, B. P., 1959, The evolution of North America: Princeton, New Jersey, Princeton University Press, 189 p.

Lachenbruch, A. H., 1970, Crustal temperature and heat production; Implications of the linear heat flow relation: Journal of Geophysical Research, v. 75, p. 3291–3300.

Lachenbruch, A. H., and Sass, J. H., 1977, Heat flow in the United States and the thermal regime of the crust, in Heacock, J. G., ed., The Earth's crust: American Geophysical Union Geophysical Monograph 20, p. 626–675.

Lipman, P. W., and Mehnert, H. H., 1979, The Taos Plateau volcanic field, northern Rio Grande rift, New Mexico, in Riecker, R. E., ed., Rio Grande rift; Tectonics and magmatism: American Geophysical Union, p. 289–311.

Lipman, P. W., Bruce, R. B., Hedge, C. E., and Steven, J. A., 1978, Petrologic evolution of the San Juan volcanic field, southwestern Colorado: Geological Society of America Bulletin, v. 89, p. 59–92.

Majorowicz, J. A., Jones, F. W., Lam, H. L., and Jessop, A. M., 1985, Regional variations of heat-flow differences with depth in Alberta, Canada: Geophysical Journal of the Royal Astronomical Society, v. 81, p. 479–487.

Majorowicz, J. A., Jones, F. W., and Jessop, A. M., 1986, Geothermics of the Williston Basin in Canada in relation to hydrodynamics and hydrocarbon occurrence: Geophysics, v. 51, p. 767–779.

Morgan, P., Harder, V., Swanberg, C. A., and Daggett, P. H., 1981, A groundwater convection model for Rio Grande rift geothermal resources: Transactions of the Geothermal Resource Council, v. 4, p. 193–196.

New Mexico Geological Society, 1982, New Mexico highway geologic map: New Mexico Geological Society in cooperation with New Mexico Bureau of Mines and Mineral Resources, Socorro, scale 1:1,000,000.

Pakiser, L., 1963, Structure of the crust and upper mantle in the western United States: Journal of Geophysical Research, v. 68, p. 5747–5756.

Prodehl, C., and Pakiser, L. C., 1980, Crustal structure of the Southern Rocky Mountains from seismic measurements: Geological Society of America Bulletin, v. 91, p. 253–269.

Reiter, M., 1985, Heat flow and thermal processes in the Jornada del Muerto, New Mexico, *in* Papers Presented to the Conference on Heat and Detachment in Crustal Extension on Continents and Planets: Houston, Texas, Lunar and Planetary Institute, p. 123–127.

Reiter, M., and Clarkson, G., 1983a, A note on terrestrial heat flow in the Colorado Plateau: Geophysical Research Letters, v. 10, p. 929–932.

——, 1983b, Geothermal studies in the San Juan Basin and the Four Corners area of the Colorado Plateau; 2, Steady-state models of the thermal sources of the San Juan volcanic field: Tectonophysics, v. 92, p. 253–269.

Reiter, M., and Tovar, J. C., 1982, Estimates of terrestrial heat flow in northern Chihuahua, Mexico, based upon petroleum bottom-hole temperatures: Geological Society of America Bulletin, v. 93, p. 613–624.

Reiter, M., and Mansure, A. J., 1983, Geothermal studies in the San Juan Basin and the Four Corners area of the Colorado Plateau; 1, Terrestrial heat flow measurements: Tectonophysics, v. 92, p. 233–251.

Reiter, M., Shearer, C., and Edwards, C. L., 1978, Geothermal anomalies along the Rio Grande rift in New Mexico: Geology, v. 6, p. 85–88.

Reiter, M., Mansure, A. J., and Shearer, C., 1979, Geothermal characteristics of the Rio Grande rift within the Southern Rocky Mountain complex, *in* Riecker, R. E., ed., Rio Grande rift; Tectonics and magmatism: American Geophysical Union, p. 253–267.

Reiter, M., Eggleston, R. E., Broadwell, B. R., and Minier, J., 1986, Estimates of heat flow along the Rio Grande rift in central and southern New Mexico: Journal of Geophysical Research, v. 91, p. 6225–6245.

Roy, R. F., Decker, E. R., Blackwell, D. D., and Birch, F., 1968, Heat flow in the United States: Journal of Geophysical Research, v. 73, p. 5207–5221.

Roy, R. F., Blackwell, D. D., and Decker, E. R., 1972, Continental heat flow, *in* Robertson, E. C., ed., The nature of the solid earth: New York, McGraw-Hill, p. 506–544.

Sanford, A. R., and Einarrsson, P., 1982, Magma chambers in rifts, *in* Palmason, G., ed., Continental and oceanic rifts: American Geophysical Union Geodynamic Series, v. 8, p. 147–168.

Sanford, A. R., Alptekin, O., and Toppozada, T. R., 1973, Use of reflection phases on microearthquake seismograms to map an unusual discontinuity beneath the Rio Grande rift: Bulletin of the Seismological Society of America, v. 63, p. 2021–2034.

Schlue, J. W., Singer, P. J., and Edwards, C. L., 1986, Shear wave structure of the upper crust of the Albuquerque-Belen basin from Rayleigh wave phase velocities: Journal of Geophysical Research, v. 91, p. 6277–6282.

Schmucker, U., 1970, Anomalies of geomagnetic variations in the southwestern United States: Bulletin of Scripps Institute of Oceanography, v. 13, p. 1–165.

Scott, G. R., 1975, Cenozoic surfaces and deposits in the southern Rocky Mountains, *in* Curtis, B. M., ed., Cenozoic history of the Southern Rocky Mountains: Geological Society of America Memoir 144, p. 227–248.

Seager, W. R., and Morgan, P., 1979, Rio Grande rift in southern New Mexico, west Texas, and northern Chihuahua, *in* Riecker, R. E., ed., Rio Grande rift; Tectonics and magmatism: American Geophysical Union, p. 87–106.

Shearer, C., and Reiter, M., 1981, Terrestrial heat flow in Arizona: Journal of Geophysical Research, v. 86, p. 6249–6259.

Smith, D. L., and Jones, R. L., 1979, Thermal anomaly in northern Mexico; An extension of the Rio Grande rift, *in* Riecker, R. E., ed., Rio Grande rift; Tectonics and magmatism: American Geophysical Union, p. 269–278.

Stacey, F. D., 1969, Physics of the Earth: New York, John Wiley and Sons, Inc., 324 p.

Stewart, J. H., 1978, Basin-range structure in western North America; A review, *in* Smith, P. B., and Eaton, B. P., eds., Cenozoic tectonics and regional geophysics of the western Cordillera: Geological Society of America Memoir 152, p. 1–31.

Taylor, R. B., 1975, Neogene tectonism in south-central Colorado, *in* Curtis, B. M., ed., Cenozoic history of the Southern Rocky Mountains: Geological Society of America Memoir 144, p. 211–226.

Thompson, G. A., and Burke, D. B., 1974, Regional geophysics of the Basin and Range Province: Annual Review of Earth and Planetary Sciences, p. 213–238.

Tweto, O., 1975, Laramide (Late Cretaceous-early Tertiary) orogeny in the Southern Rocky Mountains, *in* Curtis, B. M., ed., Cenozoic history of the Southern Rocky Mountains: Geological Society of America Memoir 144, p. 1–44.

Warren, R. G., 1978, Characterization of the lower crust-upper mantle of the Engle Basin, Rio Grande rift, from a petrochemical and field geologic study of basalts and their inclusions [M.S. thesis]: Albuquerque, University of New Mexico, 156 p.

Wernicke, B., 1981, Insights from Basin and Range surface geology for the process of large-scale divergence of the continental lithosphere; Paper presented to the Conference on the Processes of Planetary Rifting: Houston, Texas, Lunar and Planetary Institute, p. 90–92.

Woodward, L. A., 1977, Rate of crustal extension across the Rio Grande rift near Albuquerque, New Mexico: Geology, v. 5, p. 269–272.

Woodward, L. A., Callender, J. F., Seager, W. R., Chapin, C. E., Gries, J. C., Shaffer, W. L., and Zilinski, R. E., 1978, Tectonic map of the Rio Grande rift region in New Mexico, Chihuahua, and Texas, *in* Hawley, J. W., compiler, Guidebook to the Rio Grande rift in New Mexico and Colorado: New Mexico Bureau of Mines and Mineral Resources, sheet 2, scale 1:1,000,000.

Zoback, M. L., Anderson, R. E., and Thompson, G. A., 1981, Cainozoic evolution of the state of stress and the style of tectonism of the Basin and Range Province of the western United States: Philosophical Transactions of the Royal Society of London series A, v. 300, p. 407–434.

MANUSCRIPT ACCEPTED BY THE SOCIETY MAY 16, 1989

The Geology of North America
Decade Map Volume 1
1991

Chapter 27

Subsurface temperatures in the northern Great Plains

William D. Gosnold, Jr.
Department of Geology and Geological Engineering, Box 8068, University of North Dakota, Grand Forks, North Dakota 58202

INTRODUCTION

Geothermal resources within the Great Plains occur as strata-bound waters in regional aquifers. Generally, all aquifers in the Great Plains including and underlying the Inyan Kara and Newcastle Formations of the Dakota Group (Cretaceous) contain waters warm enough for development as geothermal resources (Table 1). Many of these aquifers occur throughout the Great Plains, but the formation names may differ across the province. The Dakota aquifers are the most extensive and the best mapped in the subsurface; thus, a contour map of temperatures on top of the Dakota Group provides a large-scale view of the minimum geothermal potential of the Great Plains. This temperature contour map is the principal contribution of this study to the geothermal resource map of North America (Blackwell and others, 1991).

HEAT FLOW AND TEMPERATURE DATA

Data on subsurface temperatures and geothermal aquifers in the Great Plains have been published primarily in technical reports for the U.S. Department of Energy (Sorey and others, 1983). Primary sources for data are reports for North Dakota (Harris and others, 1982; Gosnold, 1984), Nebraska (Gosnold and Eversoll, 1982), Kansas (Blackwell and Steele, 1989), and South Dakota (Gosnold, 1987). Analyses of these data are found in geothermal literature that are primarily conference proceedings (Gosnold, 1984, 1987; Gosnold and Eversoll, 1981, 1982) and in reports of the U.S. Geological Survey (Sammel, 1979; Sorey and others, 1983).

Downey (1986) synthesized data on the subsurface extent and hydrologic characteristics of several Paleozoic and Mesozoic aquifers in the northern Great Plains. Detailed information on depth, thickness, subsurface extent, and permeability of all aquifers for the western half of South Dakota is given by Rahn (1981). The results of Rahn's (1981) and Downey's (1986) work contribute significantly to analysis of the hydrogeologic conditions that affect subsurface temperatures in the northern Great Plains. Overall, these studies provide the stratigraphic, lithologic, hydrologic, and part of the temperature data base for this study.

Additional temperature data were taken from a number of sources, including Schoon and McGregor (1973), Combs and Simmons (1973), Sass and Galanis (1983), and Gosnold (1990).

ANALYSIS OF SUBSURFACE TEMPERATURES

Quantitative analysis of the thermal structure of the subsurface requires determination of several thermal properties of the strata as well as accurate determination of local heat flow. Assuming a conductive thermal regime, subsurface temperatures may be determined by Fourier's law of heat conduction:

$$Q = \frac{dT}{dZ} k, \quad (1)$$

where Q is heat flow in mW m^{-2}, k is thermal conductivity in W m^{-1} K^{-1}, and dT is the increment of temperature change in a vertical length, dZ. In a sedimentary basin where the vertical variation in thermal conductivity is controlled by the lithologies of the strata, temperature at any depth, Z, can be calculated by

$$T(Z) = T_o + \sum_{i=1}^{n} \frac{Z_i Q}{k_i}, \quad (2)$$

where T_o is surface temperature, Q is the conductive heat flow, Z_i is the thickness of the ith stratum, and k_i is the conductivity of the ith stratum.

The geologic section in the northern Great Plains consists in part of a thick (1 to 2 km) blanket of shale of Upper Cretaceous age with an effective thermal conductivity of about 1.2 W m^{-1} K^{-1} (Gosnold, 1990). Thus, most areas of the Great Plains are characterized by relatively high geothermal gradients within the shale units. The Paleozoic formations underlying the Cretaceous shales in the Great Plains are typically carbonates (see Table 1), which have an effective thermal conductivity of about 3.0 W m^{-1} K^{-1}. The net effect is a temperature profile in which temperature gradients vary significantly from the upper to the lower sections (Fig. 1). Regional heat flow values average about 60 mW m^{-2}; thus, a gradient of about 50 K km^{-1} characterizes the clastic rocks, and a gradient of about 20 K km^{-1} characterizes the carbonates.

It can be seen in Figure 1 that determining the temperature

Gosnold, W. D., Jr., 1991, Subsurface temperatures in the northern Great Plains, 1991, *in* Slemmons, D. B., Engdahl, E. R., Zoback, M. D., and Blackwell, D. D., eds., Neotectonics of North America: Boulder, Colorado, Geological Society of America, Decade Map Volume 1.

TABLE 1. GENERALIZED STRATIGRAPHY OF THE NORTH-CENTRAL GREAT PLAINS*

System	Rock Units	Lithology	Maximum Thickness
Quaternary	Coleharbor + unnamed units	Clay, silt, sand, gravel	510
Tertiary	White River	Siltstone, clay, sand	75
	Golden Valley	Clay, siltstone, lignite	65
	Fort Union	Silt, clay, sand	600
Cretaceous	Hell Creek	Sand	200
	Fox Hills	Silt, shale, sandstone	120
	Pierre	Shale	700
	Niobrara	Shale	75
	Carlile	Shale	120
	Greenhorn	Shale, shaly limestone	45
	Belle Fourche	Shale	105
	Mowry	Shale	55
	Newcastle	Sandstone, shale	45
	Skull Creek	Shale	40
	Inyan Kara	Sandstone	135
Jurassic	Morrison	Shale, siltstone	80
	Swift	Shale	150
	Rierdon	Shale	30
	Piper	Limestone, anhydrite, shale	190
Triassic	Spearfish	Siltstone, shale	225
Permian	Minnekahta	Limestone	12
	Opeche	Shale; dolomitic and silty	120
	Broom Creek	Sandstone, dolomite	100
Pennsylvanian	Amsden	Dolomite, sandstone	135
	Tyler	Shale, limestone	80
	Otter	Shale	60
Mississippian	Kibbey	Sandstone, limestone	75
	Madison	Limestone	600
Devonian	Bakken	Shale	35
	Three Forks	Siltstone, shale	75
	Birdbear	Limestone, dolomite	40
	Duperow	Limestone, dolomite	140
	Souris River	Dolomite, limestone	105
	Dawson Bay	Limestone, dolomite	55
	Prairie	Evaporites	200
	Winnipegosis	Dolomite, limestone	120
Silurian	Interlake	Dolomite, limestone	335
	Stonewall	Dolomite, limestone	35
Ordovician	Stony Mountain	Dolomite, limestone	60
	Red River	Limestone, dolomite	215
	Winnipeg Group	Siltstone, sandstone, shale	125
Cambrian	Deadwood	Limestone, sandstone, shale	300

*Stratigraphy generalized from Blumle and others (1986).

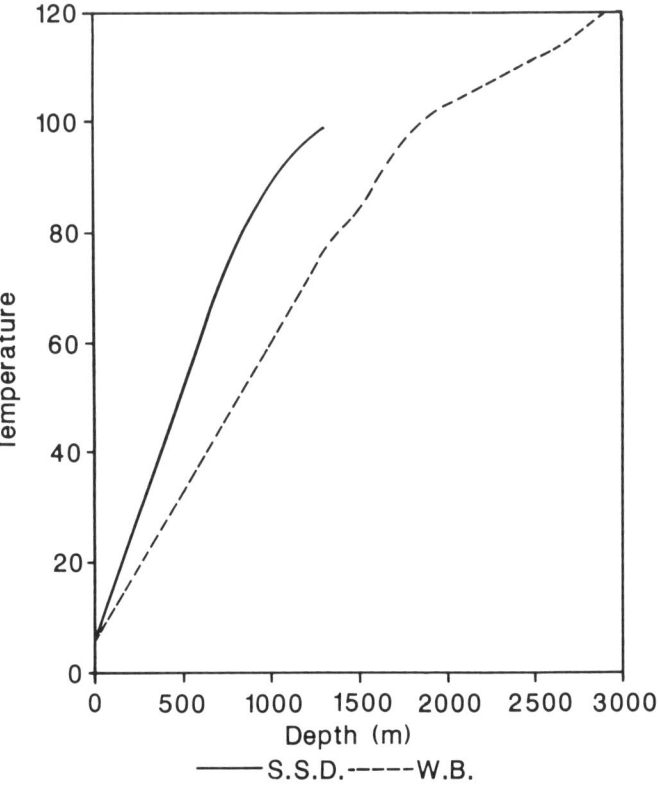

Figure 1. Temperature versus depth profiles for the Williston Basin (W.B.) and for southern South Dakota. The points indicated by symbols correspond to temperatures on formation tops and are taken from the data in Tables 2 and 3.

for one stratum using a linear temperature gradient based on a bottom-hole temperature measurement in another stratum can lead to large errors even if the bottom-hole temperature measurement is accurate. This observation can be applied generally to all sedimentary basins because the thermal conductivities of different stratigraphic units may be different. However, given knowledge of the thermal conductivities of stratigraphic units, heat flow, and that the heat flow is conductive, prediction of subsurface temperatures is relatively straight forward. For example, Table 2 gives subsurface temperatures for the principal stratigraphic formations in the Williston Basin assuming a constant heat flow value and using the appropriate thermal conductivity for each formation. The data in Table 2 were used to generate the example temperature profile for the Williston Basin in Figure 1.

This approach to predicting subsurface temperature may be applied generally to most sedimentary basins. However, if regional or local ground-water flow occurs within the basin, heat advection in flowing ground water can significantly alter the thermal structure of the subsurface from the purely conductive condition. Interestingly, regional ground water flow is the norm rather than the exception in the Great Plains, and significantly influences subsurface temperatures in most of South Dakota and parts of Nebraska and North Dakota (Gosnold, 1985, 1988, 1990).

TABLE 2. THERMOSTRATIGRAPHY OF A GENERALIZED SECTION OF THE WILLISTON BASIN

Stratigraphic Horizon	Temperature (°C)	Thickness (m)	Depth (m)	Thermal Conductivity (W m^{-1} K^{-1})	Gradient (K km^{-1})
Fort Union	6.0	152.4	0.0	1.2	50.0
Fox Hills	13.6	36.6	152.4	1.2	50.0
Pierre	15.4	1,127.8	189.0	1.1	54.5
Inyan Kara	77.0	42.7	1,316.8	1.6	37.5
Jurassic (undiv.)	78.6	61.0	1,359.4	1.4	42.9
Sundance	81.2	167.6	1,420.4	1.4	42.9
Spearfish	88.4	167.6	1,588.0	1.3	46.1
Minnekahata	96.1	12.2	1,755.7	3.0	20.0
Opeche	96.3	44.2	1,767.9	1.4	42.9
Minnelusa	98.2	137.2	1,812.1	2.2	27.3
Madison	102.0	390.1	1,949.2	3.5	17.1
Devonian (undiv.)	108.7	368.8	2,339.4	3.5	17.1
Cambrian (undiv.)	115.0	207.3	2,708.2	2.4	25.0
Precambrian	120.2		2,915.4		

Note: Heat flow = 60 mW m^{-2}. Surface temperature taken from NOAA climate data. Thermal conductivities from Gosnold (1987).

Although heat advection can significantly alter the thermal profile of a basin, the degree of thermal disturbance can be determined and included in calculations of subsurface temperatures. Hydrologic data in the Great Plains by Downey (1986), Rahn (1981), and Bredehoeft and others (1983) provide a general framework for the ground water flow systems. Downey (1986) has determined that regional ground water flow is eastward and northeastward at velocities ranging from about 10^{-8} m s^{-1} to 10^{-7} m s^{-1} in the Paleozoic and Mesozoic aquifers in the northern Great Plains. In addition, Bredehoeft and others (1983) and Neuzil and others (1984) suggest that vertical ground water flow through fractures in the Cretaceous shales in South Dakota occurs predominantly in the broad stream valleys at velocities of the order of 10^{-9} m s^{-1}. Both of these flow systems affect subsurface temperatures, and the general ground-water flow patterns can be combined with thermostratigraphic data to develop models for calculating the temperature disturbance. For example, a regional ground-water flow system in a confined aquifer is depicted crossing a vertical recharge system in Figure 2. This physical model applies to an east-west profile from the Black Hills to eastern South Dakota and was used to develop a numerical model that produced a heat-flow profile that compares favorably with the observed heat flow (Gosnold, 1988, 1990).

The numerical modeling approach is best applied to regional systems, but calculations at a single point can be accomplished analytically in several ways. For example, equation 2 can be modified to treat the effects of advective heat flow as vertical variation in conductive heat flow:

$$T = T_o + \sum_{i=1}^{n} \frac{Z_i - Q_i}{k_i} \quad (3)$$

where Q_i is the heat flow in the ith stratum. An example is given in Table 3, which is a point model that generally applies to south-central South Dakota and north-central Nebraska. The data in Table 3 were used to generate the temperature profile labelled S.S.D. (southern South Dakota) in Figure 1.

A different analytical model using a relation between heat flow, water velocity, and distance of flow for a vertical one-dimensional system (Lachenbruch and Sass, 1977) can be applied to vertical ground-water flow as follows:

$$\ln\left(\frac{Q_2}{Q_1}\right) = \frac{VD \, \sigma \, C_p}{k}, \quad (4)$$

where Q_1 is heat flow at the base of the zone of flow, Q_2 is heat flow at the top of the zone of flow, V is the Darcy velocity in m s^{-1}, D is the length of the zone of flow in meters, σ is density of

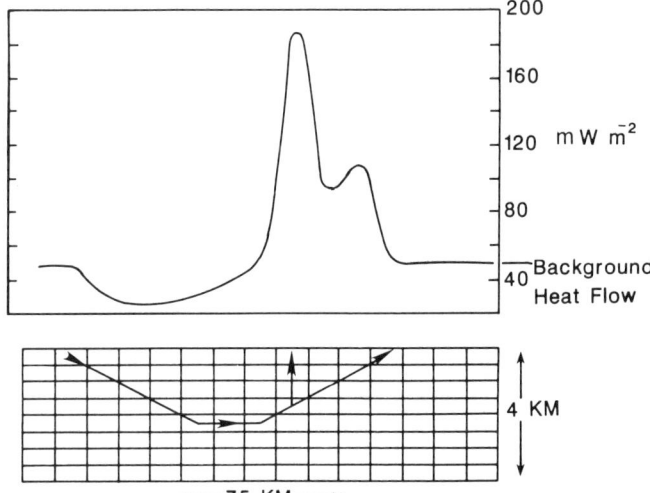

Figure 2. Physical model of the effects of regional and localized ground-water flow on heat flow. Ground-water velocity used in the model for both systems was 1.6×10^{-10} m s^{-1}.

TABLE 3. THERMOSTRATIGRAPHY OF A GENERALIZED SECTION OF SOUTHERN SOUTH DAKOTA

Stratigraphic Horizon	Temperature (°C)	Thickness (m)	Depth (m)	Thermal Conductivity (W m^{-1} K^{-1})	Gradient (K km^{-1})
Oligocene	6.0	57.9	0.0	1.2	83.3
Pierre	10.8	701.0	57.9	1.1	90.0
Newcastle	74.6	71.6	759.0	1.4	57.1
Dakota	78.7	79.2	830.6	1.6	50.0
Morrison	82.6	33.5	909.8	1.4	57.1
Sundance	84.5	56.4	943.4	1.4	57.1
Spearfish	87.8	41.1	999.8	1.3	61.5
Minnekahata	90.3	6.1	1,040.9	3.0	26.7
Opeche	90.4	45.7	1,047.0	1.4	57.1
Minnelusa	93.1	222.5	1,092.7	2.2	27.3
Cambrian (undiv.)	99.1	21.3	1,315.2	2.4	25.0
Precmabrian	99.7		1,336.6		

Note: Heat flow at surface = 100 mW m^{-2}, heat flow below Dakota = 80 mW m^{-2}, heat flow below Minnelusa = 60 mW m^{-2}. Surface temperature from NOAA climate data. Thermal conductivities from Gosnold (1987).

the fluid in Kg m^{-3}, C$_p$ is heat capacity of the fluid in W s Kg^{-1}, and k is thermal conductivity in W m^{-1} K^{-1}.

This model is best applied to regions having homogeneous permeability characteristics. However, it also provides reasonable estimates of the temperature disturbance in stream valleys in southern South Dakota where Bredehoeft and others (1983) suggest that upward ground-water flow occurs in fractures in the Pierre Shale. Relevant parameters for applying this model to south-central South Dakota include depths to the Dakota aquifer, which range from 1,000 m in the west to about 500 m in the east, and surface heat flow, which ranges from 80 mW m^{-2} in the west to 130 mW m^{-2} in the east. For an assumed heat flow of 60 mW m^{-2} below the advection zone (i.e., below the Dakota), equation 4 predicts that a vertical water flow velocity of the order of 10^{-7} m s^{-1} could account for the entire thermal anomaly. It is important to note that this flow velocity is about two orders of magnitude greater than the flow velocities given by Bredehoeft and others (1983). However, temperature profiles in stream valleys in the region show curves characteristic of upward flow velocities of the required magnitude (Gosnold, in preparation).

HEAT FLOW

Fifty new heat-flow values, 12 revised heat-flow values, and 28 published heat-flow values (Gosnold, 1990) make up the heat-flow database in the northern and central Great Plains. A large heat-flow anomaly in central South Dakota (Fig. 3) is largely due to ground-water flow (Gosnold, 1988, 1990). An additional data base of several hundred temperature-gradient measurements exists for South Dakota (Schoon and McGregor, 1973) and Nebraska (Gosnold and Eversoll, 1982). These temperature-gradient measurements provided crucial data for this study and were particularly important in delineating the high heat-flow zone in central South Dakota.

Schoon and McGregor (1973) measured several temperature gradients as high as 130 K km^{-1} west of the Missouri River in Gregory County, South Dakota. These temperature gradients are anomalously high partly because they were measured in the Pierre Shale (Cretaceous), which has an effective thermal conductivity of about 1.1 W m^{-1} K^{-1}. The thermal conductivity of the Pierre Shale was previously reported to be about 1.7 W m^{-1} K^{-1} (Sass and others, 1971; Combs and Simmons, 1973). However, recent measurements on shales indicate that the thermal conductivity is on the order of 1.1 to 1.2 W m^{-1} K^{-1} (Blackwell and Steele, 1989; Sass and Galanis, 1983). Thus, a normal continental heat flow of about 60 mW m^{-2} would cause geothermal gradients of 50 to 55 K km^{-1} in the Pierre Shale.

SUBSURFACE TEMPERATURE MAP

Temperatures on top of the basal Cretaceous aquifer (i.e., the Inyan Kara Formation in North Dakota), the Dakota Sandstone in eastern South Dakota and Nebraska, and the Lakota Formation in western Nebraska were projected using the heat-flow data from Gosnold (1990) and either equation 2 or equation 3, depending on absence or presence of an advective heat-flow component. Temperatures on top of the Dakota Sandstone in Nebraska were projected for 103 points where stratigraphic data were available from well logs (Gosnold and Eversoll, 1982). Temperatures on the Dakota Sandstone in South Dakota were projected to depth using a 14 by 14 square grid (Gosnold, 1987) on which stratigraphic data were deduced from structure contour maps prepared by Rahn (1981). Temperatures on the Inyan Kara in North Dakota were projected using a 9 by 10 square grid (Gosnold, 1984) on which stratigraphic data were deduced from structure contour maps prepared by Harris and others (1982). Temperatures on the Inyan Kara Formation in eastern Montana, eastern Wyoming, and eastern Colorado were

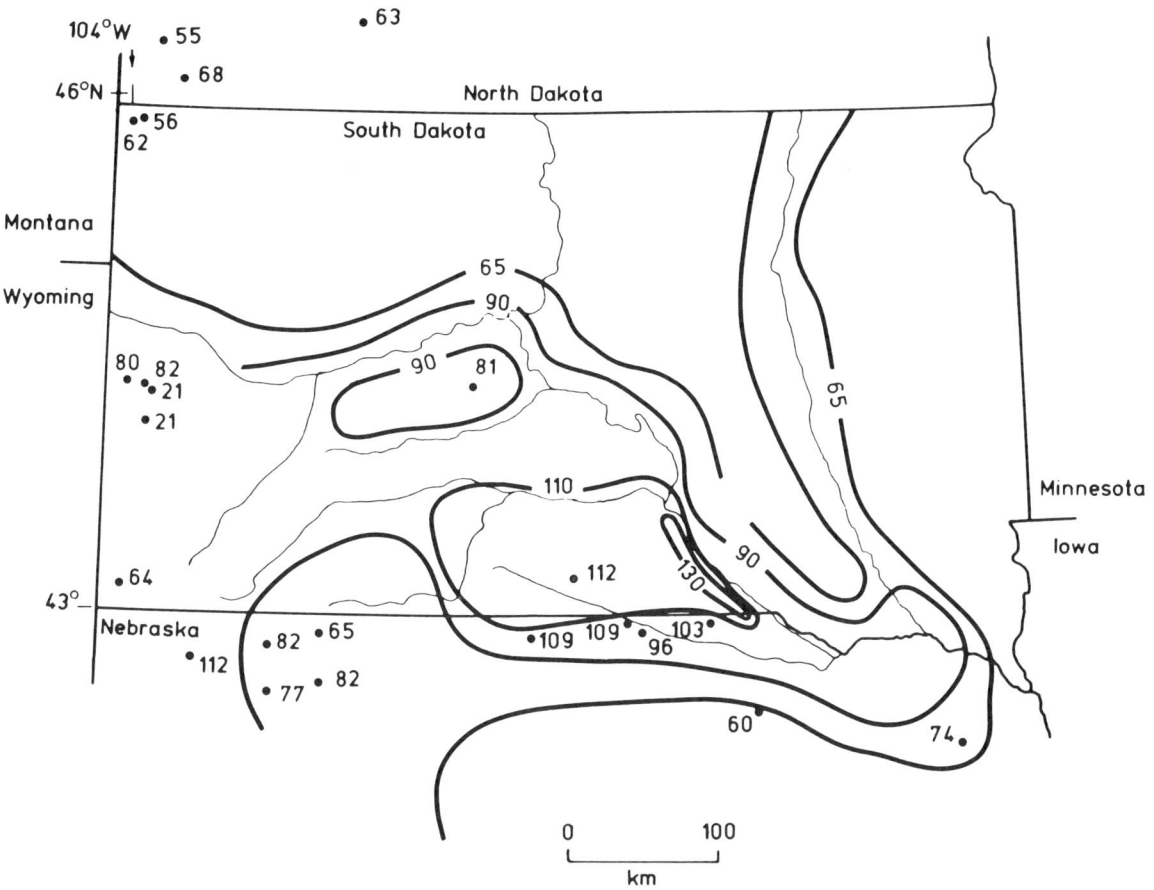

Figure 3. Heat-flow map of South Dakota. High heat flow areas are due to advection of heat in ground water.

predicted on the assumption that the stratigraphy at locations in North Dakota, South Dakota, and Nebraska provides a reasonable approximation for stratigraphy at nearby sites. Additional stratigraphic data were taken from the Geologic Atlas of the Rocky Mountain Region (Mallory, 1972). The depth to the Inyan Kara Formation in these regions was taken from King (1977) and Volk (1972).

REFERENCES CITED

Blackwell, D. D., and Steele, J. A., 1989, Heat flow and geothermal potential of Kansas, *in* Geophysics in Kansas: Kansas Geological Survey Bulletin 226, p. 267–295.

Blackwell, D. D., Steele, J. L., and Carter, L. C., 1991, Geothermal Map of North America: Boulder, Colorado, Geological Society of America, CSM-007, 4 sheets, scale 1:5,000,000.

Bluemle, J. P., Anderson, S. B., Andrew, J. A., Fischer, D. W., and LeFever, J. A., 1986, North Dakota stratigraphic column: North Dakota Geological Survey Miscellaneous Series 66, 1 page.

Bredehoeft, J. D., Neuzil, C. E., and Milley, P.C.D., 1983, Regional flow in the Dakota aquifer; A study of the role of confining layers: U.S. Geological Survey Water Supply Paper 2237, 45 p.

Combs, J., and Simmons, G., 1973, Terrestrial heat flow determinations in the north-central United States, 1973: Journal of Geophysical Research, v. 78, p. 441–461.

Downey, J. S., 1986, Geohydrology of bedrock aquifers in the northern Great Plains in parts of Montana, North Dakota, South Dakota, and Wyoming: U.S. Geological Survey Professional Paper 1402-E, 87 p.

Gosnold, W. D. Jr., 1984, Geothermal resources of the Williston Basin, North Dakota: Geothermal Resources Council Transactions, v. 8, p. 431–436.

——— , 1985, Heat flow and groundwater flow in the Great Plains of the United States: Journal of Geodynamics, v. 4, p. 247–264.

——— , 1987, Geothermal resource assessment of South Dakota: Transactions of the Geothermal Resources Council, v. 11, p. 355–360.

——— , 1988, Analysis of heat flow and groundwater flow in the South Dakota geothermal anomaly: Transactions of the Geothermal Resources Council, v. 12, p. 251–255.

——— , 1990, Heat flow in the Great Plains of the United States: Journal of Geophysical Research, v. 95, p. 353–374.

Gosnold, W. D., Jr., and Eversoll, D. A., 1981, Usefulness of heat flow data in regional assessment of low-temperature resources with special reference to Nebraska: Geothermal Resources Council Transactions, v. 5, p. 79–82.

——— , 1982, Geothermal resources of Nebraska: Boulder, Colorado, National Oceanic and Atmospheric Administration map, scale 1:500,000.

Harris, K. L., Howell, F. L., Wartman, B. L., and Anderson, S. B., 1982, An evaluation of hydrothermal resources of North Dakota: University of North Dakota Engineering Experimental Station Bulletin 82-08-EES-01, 210 p.

King, P. B., 1977, The evolution of North America, revised ed.: Princeton, New Jersey, Princeton University Press, 197 p.

Lachenbruch, A. H. and Sass, J. A., 1977, Heat flow and the thermal regime of the crust, *in* Heacock, J. G., ed., The Earth's Crust; Its nature and physical properties: American Geophysical Union Geophysical Monograph 20,

p. 626–675.

Mallory, W. W., 1972, Regional synthesis of the Paleozoic eratherm, *in* Mallory, W. W., ed., Geologic atlas of the Rocky Mountain region: Rocky Mountain Association of Geologists, p. 111–129.

Neuzil, C. E., Bredehoeft, J. D., and Wolff, R. G., 1984, Leakage and fracture permeability in the Cretaceous shales confining the Dakota aquifer in South Dakota, *in* Jorgensen, D. G., and Signor, D. C., eds., Geohydrology of the Dakota aquifer: Worthington, Ohio, National Water Well Association, p. 113–120.

Rahn, P. H., 1981, Groundwater stored in the rocks of western South Dakota, *in* Rich, F. J., ed., Geology of the Black Hills, South Dakota and Wyoming, 2nd ed.: American Geological Institute, p. 154–174.

Sammel, E. A., 1979, Occurrence of low-temperature geothermal waters in the United States, *in* Muffler, L.J.P., ed., Assessment of geothermal resources of the United States—1978: U.S. Geological Survey Circular 790, p. 86–131.

Sass, J. H., and Galanis, S. P., 1983, Temperatures, thermal conductivity, and heat flow from a well in Pierre Shale near Hayes, South Dakota: U.S. Geological Survey Open-File Report 83-25, 10 p.

Sass, J. A., Lachenbruch, A. H., Munroe, R. J., Greene, G. W., and Moses, T. H., Jr., 1971, Heat flow in the western United States: Journal of Geophysical Research, v. 76, p. 3391–3401.

Schoon, R. A., and McGregor, D. J., 1973, Geothermal potentials in South Dakota: South Dakota Geological Survey Report of Investigations 110, 76 p.

Sorey, M. L., Reed, M. J., Foley, D., and Renner, J. L., 1983, Low-temperature geothermal resources in the central and eastern United States, *in* Reed, M. J., ed., Assessment of low-temperature geothermal resources of the United States—1982: U.S. Geological Survey Circular 892, p. 51–65.

Volk, R. W., 1972, The Denver Basin and the Las Animas Arch, *in* Mallory, W. W., ed., Geologic Atlas of the Rocky Mountain Region: Denver, Colorado, Rocky Mountain Association of Geologists, p. 281–282.

Manuscript Received by the Society September 26, 1990

ACKNOWLEDGMENTS

Most of the temperature-gradient measurements for this study used equipment supplied by the Geothermal Laboratory at Southern Methodist University, which also provided thermal conductivity measurements for heat-flow calculations. Six temperature logs in deep wells in western Nebraska were made by Hank Heasler of the University of Wyoming. Others who contributed in the data collection are Duane Eversoll and Marvin Carlson of the Conservation and Survey Division, University of Nebraska; David Becker, Larry Carter, and Jim McCauliffe of the University of Nebraska at Omaha; Merlin Tipton and Robert Schoon of South Dakota Geological Survey; Kenneth Harris and Sidney Anderson with the North Dakota Geological Survey; and Francis Howell with the University of North Dakota. Support for this work was provided in part by U.S. Department of Energy grants DE-AS0779ET27205, DE-FC07-79ID12030, and DE-FG07-85ID12606, and National Science Foundation Grant EAR-8417305.

Chapter 28

Late Quaternary glacial isostatic recovery of North America, Greenland, and Iceland; A neotectonics perspective

John T. Andrews
Department of Geological Sciences, and Institute of Arctic and Alpine Research, Campus Box 450, University of Colorado, Boulder, Colorado 80309

INTRODUCTION

The objective of this chapter is to provide information on the magnitude of elevation changes and rates of deformation that are associated with the recovery of North America, Greenland, and Iceland from the late Quaternary ice loads (Fig. 1). These topics are addressed in detail in various chapters of *The Quaternary Geology of Canada and Greenland* (Fulton, ed., 1989). In particular, the various regional treatments contain much of the basic information on which this present chapter is based. Andrews and Peltier (1989) also provide a survey of both the observational data and glacial isostatic theory. Of the major glaciated areas of North America, only Alaska has a paucity of studies on relative sea-level movements during the last deglacial cycle from the region directly influenced by an ice load (see Mann, 1986). However, Clark (1977) has evaluated the effect of recent changes in glacier thickness on the sea-level history of southeast Alaska.

Neotectonics associated with glacial unloading of the Earth's surface are a different set of processes than movements associated with normal faulting or with the movements of plate margins, although there may be connections between them (e.g., Clark, 1982; Quinlan, 1984; Adams, 1989a, b). Glacial isostatic recovery affects large areas (ca. 14×10^6 km^2 of North America) and is characterized most commonly by spatial and temporal coherence of the response (e.g., Peltier, 1976; Clark, 1980; Quinlan, 1985; Quinland and Beaumont, 1982). Glacial isostatic observations have been used on this *predictive* basis for several decades to constrain appropriate models of the Earth's rheology (e.g., Walcott, 1970; Peltier and Andrews, 1976). Thus, neotectonics associated with the last deglaciation of North America, Greenland, and Iceland primarily consists of a viscoelastic adjustment with observed half-lives of a few hundred to a few thousand years. The processes involved in the recovery from the ice load include the elastic flexing of the lithosphere (Walcott, 1970), the displacement of material from the aesthenosphere, and the gravitational attraction of the water surface to the ice sheets (Clark, 1976). Many of these studies are associated with changes in relative sea level; full discussions of both the methodology and importance of such approaches are presented by van de Plassche (1986) and Devoy (1987).

In the broader arena of plate tectonics (see e.g., Saemundsson, 1986), the application of an ice load to northern North America, Greenland, and Iceland is not simply one of scale of the load. There are substantial differences in the style of Cenozoic tectonics between these areas, and there is a substantial difference in observed neotectonic activity between, for example, the coast of southeastern Alaska and the passive trailing margin of eastern Canada. Some of the regional variations in Quaternary faulting are outlined in various chapters in this volume. It is implicit in this chapter that the *observed* late Quaternary vertical movements in areas within the glaciated margin (Fig. 1) are dominated by isostatic readjustments. Superimposed on these processes are longer-term tectonic adjustments. As the rate of motion associated with glacial isostatic recovery goes to zero, the importance of these other processes will become increasingly apparent as an explanation of changes in level. This may already be happening in areas beyond the former ice margins, and in areas of active plate subduction (see chapters in this volume).

History of observations

The early history of observations on raised marine and glacial lake shorelines in North America is presented in Andrews (1974). During the nineteenth century, explorers in the Canadian and Greenland Arctic commented on the occurrence of Quaternary fossils above high tide and the presence of raised beaches. For example, Kane (1856, p. 308) remarked that evidence in northwestern Greenland ". . . .enables me to assert positively the interesting fact of secular elevation of the crust . . . ," and Hall (1865, p. 572) noted that the local peoples (Innuits) of Frobisher Bay, Canada, understood that sea level was once significantly higher than present because they had observed marine shells ". . . away up on the mountains."

Shaler's (1874) analysis of the recent changes of sea level

Andrews, J. T., 1991, Late Quaternary glacial isostatic recovery of North America, Greenland, and Iceland; A neotectonics perspective, *in* Slemmons, D. B., Engdahl, E. R., Zoback, M. D., and Blackwell, D. D., eds., Neotectonics of North America: Boulder, Colorado, Geological Society of America, Decade Map Volume 1.

along the Maine coast clearly illustrated that he understood the weight of the last ice sheet caused the crust to sag. The exchange of ideas with the northwest European research community also occurred, even in the days before airliners. In 1892, the famous Swedish geologist De Geer (1892) presented a paper in Boston "On Pleistocene changes of level in Eastern North America".

Field work on both glacial lake shorelines and raised marine beaches continued during the early part of the twentieth century (e.g., Fairchild, 1918), but it was not until 1941 that a serious attempt was made to interpret the North American data in a geophysical context (Gutenberg, 1941). The Second World War restricted research, and the next major phase of investigation was led by Canadian researchers working in the Arctic (e.g., Bird, 1954; Craig and Fyles, 1960; Sim, 1960; Loken, 1962). They began investigating the raised marine records in Arctic Canada and were able to numerically date events because of the development of the radiocarbon method. The well-preserved stratigraphic records of North America and Greenland (e.g., Washburn and Stuiver, 1962), resulted in the publication of elevation/age diagrams (see Figs. 2 and 3) and an appreciation of the spatial variability of postglacial rebound.

From the 1970s to the present, field data on glacial isostatic recovery became an essential tool for investigating the rheology of the Earth. This included the development of geophysical models that were constrained by broad geographic coverage of elevation/age information (Walcott, 1970, 1972; Peltier, 1976, 1986; Peltier and Andrews, 1976, 1983; Quinlan and Beaumont, 1982; Cathles, 1975). These data (e.g., Blake, 1975, 1976) enabled initial estimates to be made of the rate of glacial isostatic rebound, or more accurately, the rate of change in relative sea level. Values of greater than 10 m/100 yr (i.e., 100 mm/yr!) were derived from the postglacial delevelling curves at sites within Hudson Bay and along the coast of the Canadian Arctic mainland at 8 ka (Fig. 3). This compares with a maximum of approximately 5 m/100 yr at 6 ka. The tide-gauge record from Churchill, on the west shore of Hudson Bay, suggests a rate of <1 m/100 yr at present, but at other sites in the bay the rate is probably >1 m/100 yr (cf., Barnett, 1970; Andrews, 1970; Peltier, 1986; see Figs. 4 and 5).

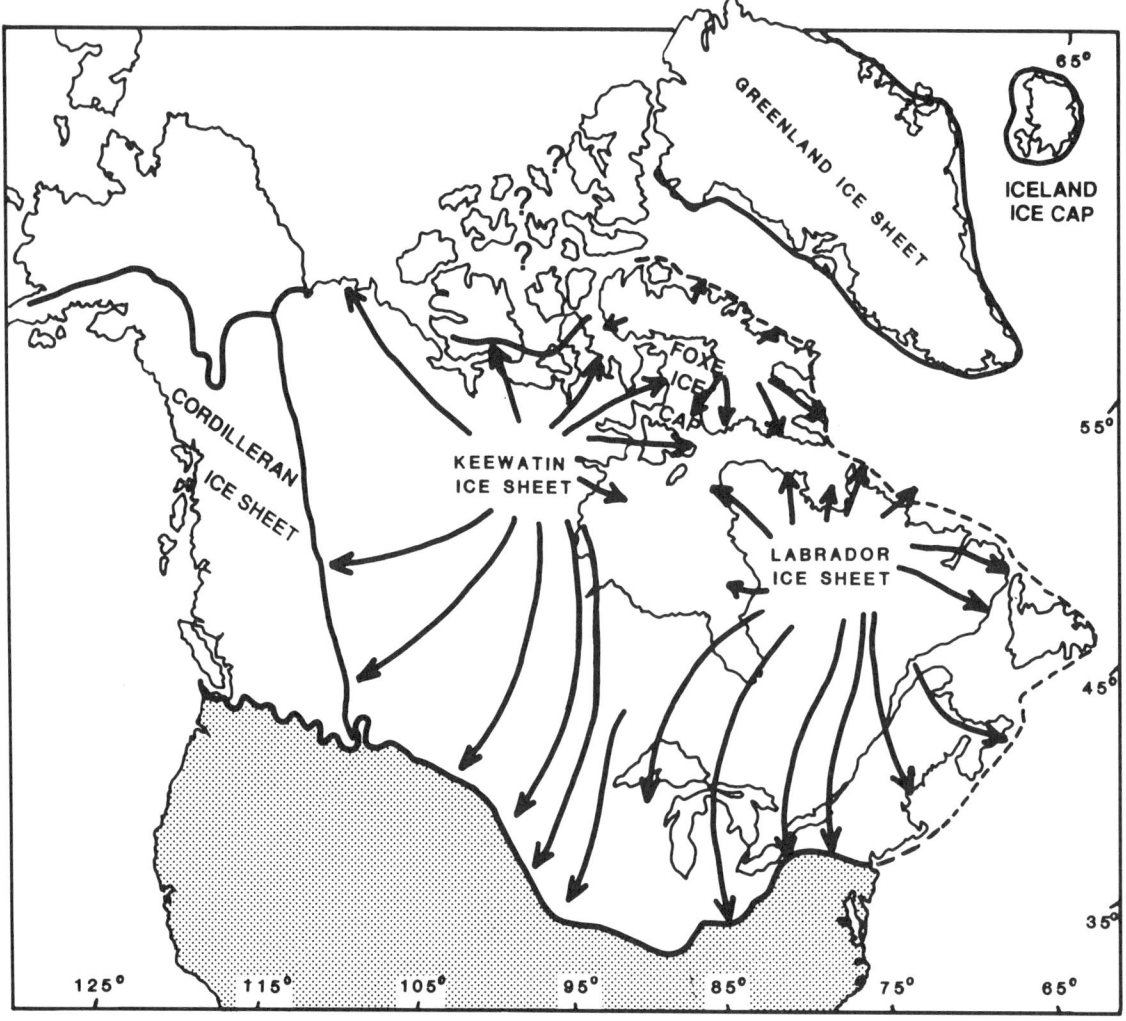

Figure 1. Extent and possible late Wisconsinan configuration of late Quaternary ice masses over North America, Greenland, and Iceland.

Chronology of glacial unloading: North America, Greenland, and Iceland

Glacial isostatic recovery was affected by the manner and rate of unloading (i.e., deglaciation) of the major ice sheets that affected North America, Greenland, and Iceland. Maps showing the retreat of the margins of the North American Wisconsin ice sheet have been available for 15 years (Bryson and others, 1969; Prest, 1969), and a new compilation has just been issued (Dyke and Prest, 1987). Comparable maps have not been published for Greenland and Iceland, although there are now sufficient radiocarbon dates on deglaciation to discuss the chronology of deglaciation (Funder, 1989; Einarsson, 1978; Hjort and others, 1985). However, the relation between the area of the late Pleistocene ice sheets and spatial and temporal variations in thickness, though simple for ideal ice sheets (Bloom, 1971; Paterson, 1972), may be complicated, especially for the Laurentide Ice Sheet (e.g., Fisher and others, 1985; Andrews, 1987). Figure 6 shows the change in average thickness for the Laurentide Ice Sheet based on data in Paterson (1972, Table 5). These estimates were based on a reconstruction of the ice sheet that projected a dome over Hudson Bay (a matter of some controversy (e.g., Fisher and others, 1985; Boulton and others, 1985; Denton and Hughes, 1981; Dyke and others, 1982). Figure 6 represents a *maximum* estimate of average

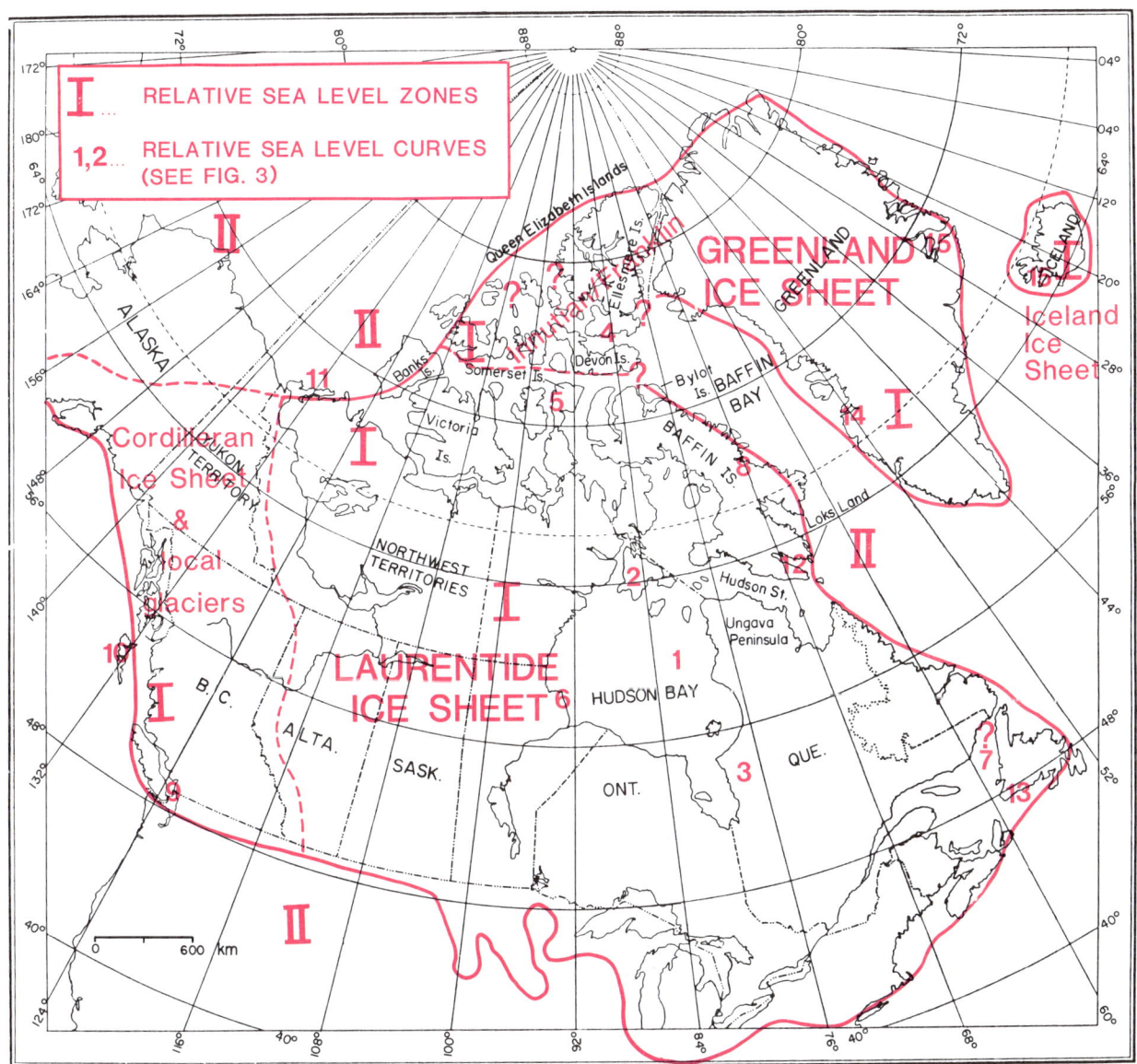

Figure 2. Map of the different relative sea-level zones (after Clark and others, 1978) over northern North America and the location of the sea-level curves illustrated on Figure 3. This figure also shows the appropximate extent of grounded glacial ice for the major North American ice sheets (after Dyke and Prest, 1987).

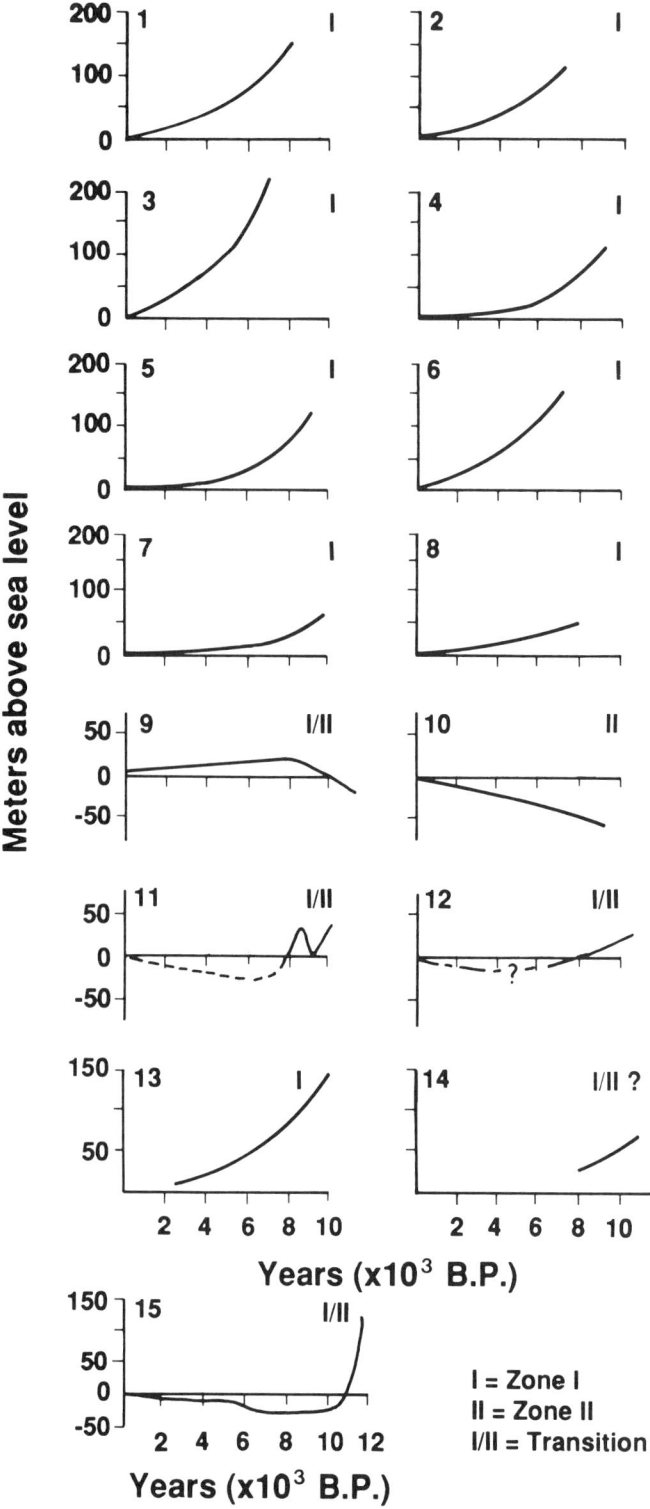

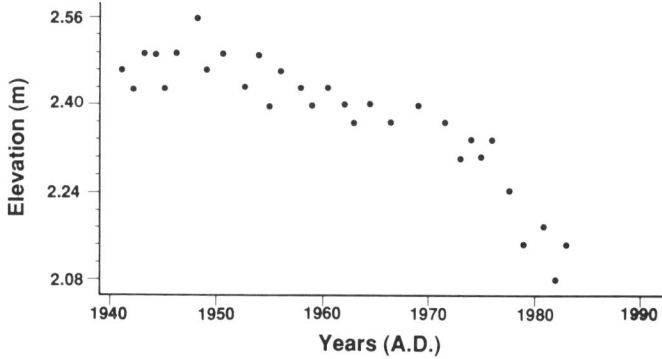

Figure 4. Fluctuations and trend in the average yearly tide-gauge readings from Churchill, Manitoba, eastern Hudson Bay.

Figure 3. Relative sea-level curves from different areas (see Fig. 2 for location): A, relative sea-level curves from Zone I; B, relative sea-level curves for Zone I/II; and C, relative sea level curves for Zone II.

ice sheet thickness during deglaciation; the model does not take into account the evidence for thin ice over the northwest, southwest, and Great Lakes regions of the ice sheet (cf., Matthews, 1974; Beget, 1986, 1987; Fisher and others, 1985; see Fig. 7).

The Cordilleran Ice Sheet started to retreat about 14 to 15 ka, was back to the fiord heads by 11 to 12 ka (Clague, 1983; Clague and others, 1982; Thorson, 1981), and had largely disappeared by 10 ka. The same chronology applies in a general sense to the coastal mountains of southeastern Alaska (Mann, 1986). Very little attempt has been made in coastal southeastern Alaska to reconstruct the actual change of relative sea level during the last 14 kyr or so. Although a number of dates have been obtained on marine molluscs from around Juneau (Miller, 1973), these do not determine the history of sea level. Around Anchorage, the late-glacial marine deposits extend to about 18 m above sea level (a.s.l.; Schmoll and others, 1972; Schmoll and Yehle, 1986) and are referred to the Bootlegger Cove Formation, which has dates associated with it of between 16.3 and 13.9 ka.

The Greenland Ice Sheet probably extended onto its western continental shelf during the late Pleistocene (Kelly, 1985; Funder, 1989) and possibly did so along the eastern margin, although here the evidence is more controversial (Funder, 1989). Some radiocarbon dates (Bjork and Persson, 1981) indicate that retreat started in the northeast by 14 ka, but the late-glacial transgression is marked by fossiliferous muds and near-shore sediments that date primarily between 8 and 10 ka. In places the ice had retreated behind its present margin by 6 ka. A slight readvance may have affected the margin during neoglacial times. The maximum retreat of the margins of the Greenland Ice Sheet during the last 18 kyr has probably amounted to a maximum of 200 km and a change in maximum ice thickness of 100 to 200 m.

The rate and pattern of deglaciation of Iceland has not been addressed in great detail. There is evidence that, in places, ice had retreated to the coast by 12 ka, but about half the island was still ice covered between 10 and 11 ka (Einarsson, 1978). Thereafter, deglaciation proceeded very rapidly. Denton and Hughes (1981) estimated that the maximum ice thickness over Iceland was about 1,500 m.

THE OBSERVATIONAL BASIS

For over 100 years, scientists have noted evidence for the glacial isostatic adjustment of North America in those regions once covered by Laurentide or Cordilleran Ice Sheets (Fig. 1). The data fall into two broad categories: either data on elevation changes at a site with respect to time, or data on the regional deformation of an isochronous surface. Elevation changes are measured with respect to present sea level, or lake level. Measurements and interpretations associated with changes of sea level are documented in chapters in van de Plassche (1986; e.g., Andrews, 1986; Scott and Medioli, 1986; Jardine, 1986; Palmer and Abbott, 1986). Evidence for regional deformation associated with the warping of a former horizontal surface has traditionally been gathered from areas around the margins of the Great Lakes and within the drainage basin of Glacial Lake Agassiz (e.g., Teller and Clayton, 1983; Karrow and Calkin, 1985).

Changes of sea level associated with glacial isostatic adjustments can be dated numerically by the radiocarbon method (e.g., Sutherland, 1986; Mook and van de Plassche, 1986). In areas that have experienced a *fall* in sea level (see later for further elaboration), the age of a particular beach or deposit is commonly assessed by ^{14}C dates on marine molluscs, but driftwood and whalebone have also been used extensively in some areas (cf., Blake, 1975). In areas where sea level has *risen,* the age at which sea level transgressed across terrestrial deposits such as forests or salt marshes is obtained by ^{14}C dating wood or peats (e.g., Heyworth, 1986; Grant, 1980). The great majority of sea-level determinations date from <14 ka; hence, laboratory errors on the

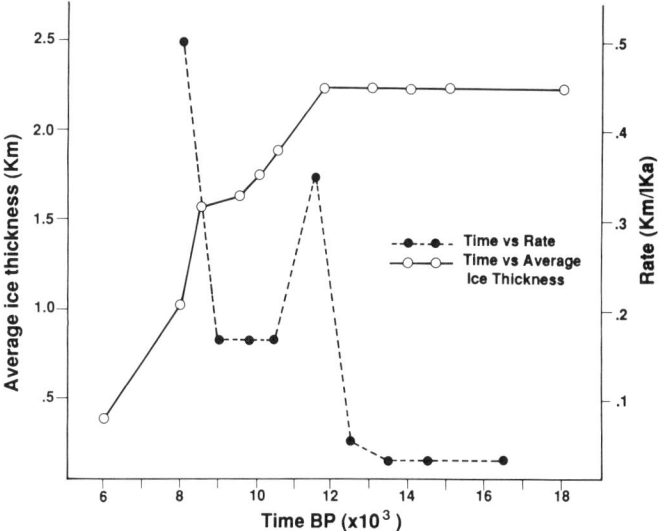

Figure 6. Change in the average thickness of the Laurentide Ice Sheet, and the rate of change, for the period between 18 and 7 ka (data from Paterson, 1972).

radiocarbon dates are usually better than ±0.30 kyr. However, this does not take into account either specific problems associated with the corrections to shell dates (e.g., Sutherland, 1986; Mangerud, 1972) or the divergence between ^{14}C and calendar years. Thus all the ages and rates reported in this chapter are based on ^{14}C years.

Jardine (1986) discussed the problems of measuring the altitude of sea-level indicators. The problems include relating the indicator to a specific level in the annual tidal cycle, the availability of a geodetic benchmark (rarely available throughout much of northern Canada), the accuracy of the survey procedure, and the uniqueness of the indicator's stratigraphic relation to a sea-surface elevation. In the worst-case scenario, altitudes are probably accurate to ±10 m, but at sites where an elevation/age curve is being sought (see Fig. 3 for examples), these errors may be reduced to ±1 m or better by attention to accurate surveying and an appreciation of the appropriate depositional system (Scott and Medioli, 1986; Blake, 1975).

Vertical movements

Clark and others (1978) have distinguished three zones of glacial isostatic response around the margins of the North American ice sheets. In *Zone I* (Fig. 2) there has been continuous crustal rebound, and relative sea level has fallen since site deglaciation. Figure 3A illustrates some typical relative sea-level/age curves from Zone I. Most of the area within the margins of the Laurentide Ice Sheet has a response similar to that illustrated in Figure 3A. At sites within Zone I there exists a "marine limit," which provides evidence on the upper limit to marine inundation (Fig. 8; Bird, 1954; Sim, 1960; Andrews, 1970). The age of this

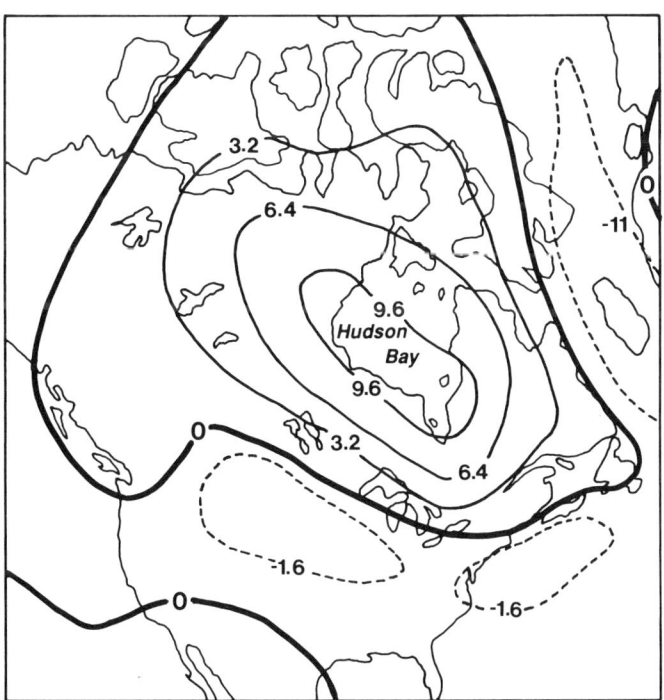

Figure 5. Predicted rates of change of surface elevation associated with glacial unloading of North America (after Peltier, 1986).

feature frequently coincides with the date of ice rewithdrawal or deglaciation. Below the marine limit there are frequently raised beaches, deltas, and other indications of higher sea levels (Fig. 9; Hillaire-Marcel and Fairbridge, 1978). Within Zone I, tide-gauge stations record a fall in sea level during their period of record (e.g., Barnett, 1970; Peltier, 1986; Andrews and Peltier, 1989; see Fig. 4).

The *Zone I/Zone II* transition coincides fairly closely with the outer limit of the former ice sheets. In this area the history of sea level (Fig. 3B) consists of an emergence/submergence cycle. This response results from the interplay between glacial isostatic rebound, the addition of glacial meltwater to the world oceans, and the possible migration of the forebulge (Clark and others, 1978). The observed postglacial sea-level histories from parts of British Columbia, the Canadian Maritimes, the northwestern Canadian Arctic, and parts of northeastern United States reflect the I/II transitional response. In such areas the marine limit may not date deglaciation, and may be formed in response to either regional unloading and/or changes in level of the world ocean.

The sea-level history in *Zone II* is dominated by submergence and reflects, in some complex manner, the rise in sea level caused by the melting of the ice sheets and the addition of about 120 m of water to the oceans (Fairbanks, 1990; see Fig. 3C). The converse of the "marine limit" is the lowest beach *beneath* present sea level. The determination of both the elevation and age of such a feature is difficult, although modern acoustic methods and systematic coring on the continental shelf have added good-quality information (Oldale and O'Hara, 1980).

Regional warping

The neotectonic information present in observations such as Figure 3 represents site-specific histories. These can be used to develop a regional pattern of postglacial changes in elevation by interpolation on each curve and the contouring of the resulting elevations (Fig. 10). Such maps show the isobases on relative sea-level changes over some interval of time. Conversely, if a single former shoreline can be mapped and surveyed over a large geographic area the changes of elevation of such an isochronous surface can be used to further understand glacial isostatic processes (e.g., Broecker, 1966; Andrews and Barnett, 1972) and constrain models of Earth rheology.

In this regard, the deformation of glacial lake shorelines in North America provides a considerable body of information because retreat of both the Laurentide and Cordilleran Ice Sheets resulted in the creation of vast glacial lakes, especially along the southern, southwestern, and western margins of the Laurentide Ice Sheet (Prest, 1970; Fulton, ed., 1984; Teller and Clayton, 1983; Karrow and Calkin, 1985). Surveys of the shorelines left by these lakes result in a "shoreline diagram," such as Figure 11, which plots the elevation of the shorelines in a plane normal to the regional isobases. Diagrams such as Figure 11 have also been constructed on North American marine raised shorelines (e.g., Loken, 1962; Andrews, 1970), but this has been attempted

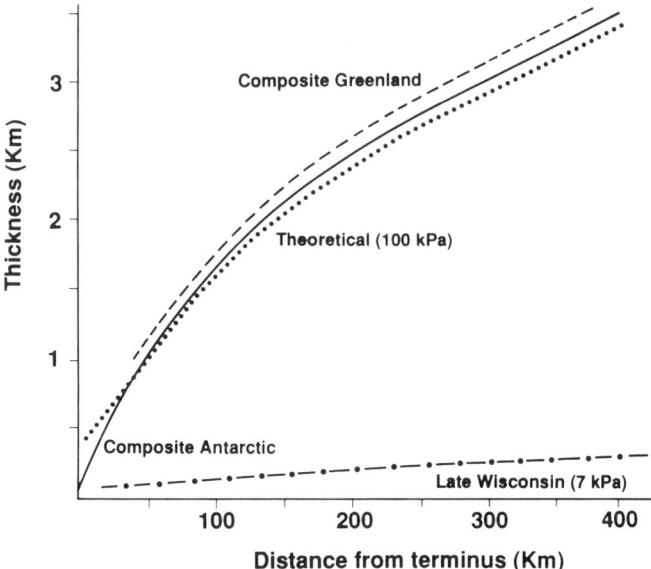

Figure 7. Predicted profiles for the Laurentide Ice Sheet based on a Greenland/Antarctic model versus a "soft bed" model (after Beget, 1987).

for only a few areas. The creation of a mappable lake shoreline is a function of: (1) the span of time that the lake stays at or close to a particular level (this is controlled by the position of the lake outlet and the rate of glacial retreat); (2) length of the fetch; (3) availability of sediment; (4) length of the open-water season; and (5) shoreface gradient. In northern areas, former glacial lake shorelines can be easily recognized and mapped (Fig. 12), but around the Great Lakes and Lake Agassiz, recognition and correlation are more difficult, especially in view of the massive size involved in some of the former lakes.

Dating lake shorelines requires recovery of organic materials collected from beaches or deltas associated with a particular raised shoreline. Wood and peat are frequently dated to provide numeric age estimates.

In lake basins close to the outer limit of the Laurentide Ice Sheet the tilting of the lake basin in an up-glacier direction has resulted in a lake transgression in the distal portions of lake basins. Water-level recorders around the Great Lakes have detected both positive and negative changes of lake level associated with glacial isostatic recovery (cf., Gutenberg, 1941; Peltier, 1986).

NEOTECTONIC RECONSTRUCTIONS

The elements of glacial isostatic changes that can be reconstructed are constrained by the overall objectives of the project. The majority of neotectonic motions included in this volume consist of fault displacements. Such motion is usually episodic, whereas glacial isostatic changes of elevation are probably continuous (e.g., Fig. 3).

In the next section, I have limited the input data to discussions of the following: (1) Elevation and age of the marine limit; and (2) The direction and deformation of lake and marine shorelines per unit time.

Elevation of the marine limit

Within Zone I and in the Zone I/II transition the marine limit is a function of the former ice load, earth rheology, and date of deglaciation (cf., Andrews, 1970; Walcott, 1972). The highest measured marine limits in North America, Greenland, and Iceland occur on the eastern side of Hudson Bay, near the settlement of Great Whale River (Archer, 1968). Elevations of between 230 and 315 m have been measured. Conversely, on the southwest and west sides of Hudson Bay the elevation of the marine limit is <200 m. Marine limits >200 m altitude have also been reported from the region between Montreal and Ottawa, on the southwest coast of the Gulf of Bothnia (Dyke, 1983, 1984), and around Bathurst Inlet on the northern Canadian mainland.

Elevations on the marine limit generally decline toward the margin of the former ice sheet(s). One point along such a transect will display no evidence for marine activity above present; this point represents a "hinge" between sites that have emerged and those that have submerged. Currently, submergence is occurring *inside* the zero isobase. (An isobase is a line that joins sites with equal emergence or submergence over a specific interval of time, say the last 4 or 8 ka; e.g., Andrews, 1970; Dyke, 1974). In the area of present submergence, terrestrial surfaces are being succeeded by salt marsh and/or beach sediments. The zero isobase is not well delimited, but it has been mapped along the outer east coast of Baffin, Bylot, and Devon Islands (Fig. 2), along part of the Arctic mainland coast, and along the Canadian Maritimes and New England coastline.

In British Columbia, the state of Washington, and southeast Alaska, marine limits of >150 and <230 m occur near the fiord heads, and elevations on the limit decline seaward (Clague, 1983; Miller, 1973; Mann, 1986; Thorson, 1981). On the Queen Charlotte Islands (Clague and others, 1982) the marine limit is *not* associated with deglaciation but with an early Holocene marine transgression.

In the High Canadian Arctic, marine limit elevations of >140 m occur on the channels between Axel Heiberg and Ellesmere Islands, and adjacent to northwest Greenland the late Pleistocene (< = 10 ka) marine limit is more than 100 m a.s.l.

In North America, isolines on the marine limit have been presented since the early part of the twentieth century (e.g., Fairchild, 1918; Daly, 1934; Farrand and Gadja, 1962). However, the value of this exercise is limited by the metachronous nature of the limit, and in the last 20 years, research has concentrated on developing isobase maps on glacial isostatic recovery (e.g., Fig. 10).

In North America, the late Pleistocene ice sheets had all virtually disappeared by 6 ka (Bryson and others, 1969; Prest, 1969; Dyke and Prest, 1987). However, the Greenland Ice Sheet has not disappeared, and the amount of retreat between the late-glacial maximum and the present position of the ice sheet's margin is in the range of 100 to 400 km (Funder, 1989). The change in area of the ice sheet was thus only about 500,000 km^2. Relatively, the Greenland Ice Sheet today is equivalent to the Laurentide Ice Sheet at approximately 10 to 11 ka. Funder (1989) presents isolines on the marine limit around the margin of the Greenland Ice Sheet. He delimits three areas of moderate to high Holocene recovery. The first occurs in northwest Greenland and thus matches the high marine limits in northeastern Ellesmere Island (England, 1976). The second zone covers central West Greenland south of Disko Island where the marine limit reaches 120 m and then declines toward the margin of the ice sheet.

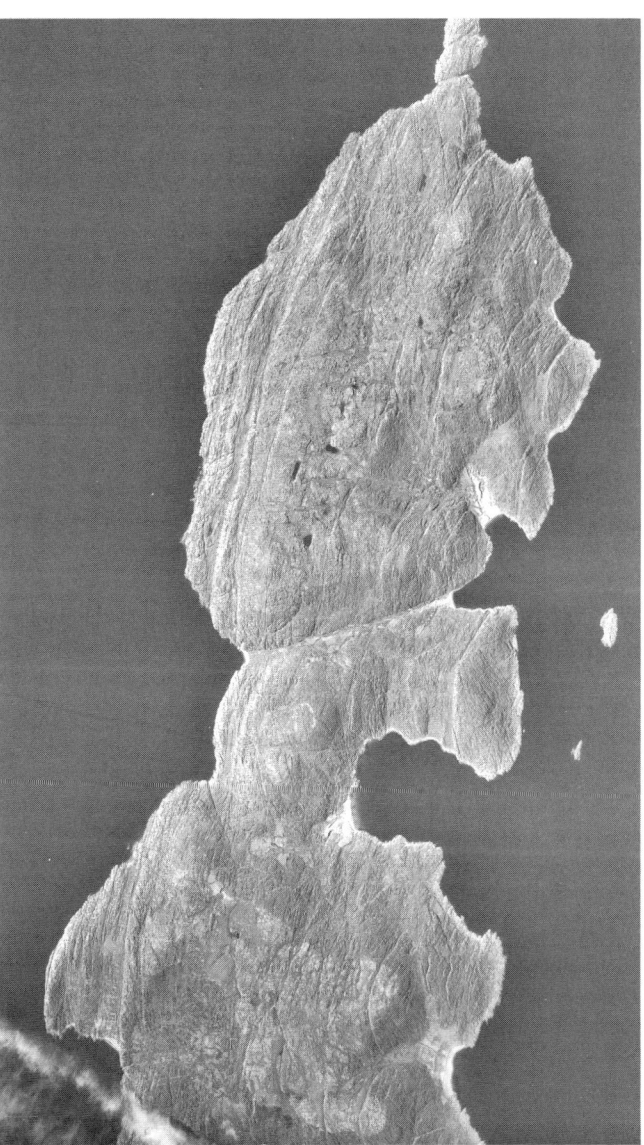

Figure 8. Air photograph (Al6365-12; National Air Photo Library, Government of Canada) of Ottawa Island, Hudson Bay, Northwest Territories. The marine limit is marked by the change in texture between a carbonate-rich till on the uplands to bare bedrock below the marine limit. The limit is about 140 m and dates from about 7.7 ka.

Figure 9. Raised beaches extending toward sea level in a small bay south of Isortoq Fiord, Baffin Island, Northwest Territories. The ridges are formed of shingle and extend from about 100 to 0 m a.s.l.

Finally, marine limits of 90 m are shown in the middle reaches of Scoresby Sund extending northward to Kong Oscors Fjord. Low marine limits of 20 m or so are located in northern West Greenland and along the southern coast of East Greenland and suggest either late deglaciation or a small ice load, probably the former. Compared with Holocene marine limits on eastern Baffin Island, just across Baffin Bay (Fig. 1), the outer West Greenland marine limits are higher. The deglacial chronologies for both West Greenland and Baffin Island are similar (Andrews and Peltier, 1989; Funder, 1989), insofar as dates on deglaciation are usually ≤10 ka and deglaciation of the fiords occurred between 9 and 6 ka. However, most of Baffin Island was deglaciated by 5 ka (Dyke, 1974), whereas the Greenland Ice Sheet stabilized and limited retreat/advanced occurred after 5 ka.

The elevation/age recovery curves from Greenland (e.g., Washburn and Stuiver, 1962; Ten Brink, 1974; Kelly, 1985; Funder, 1989) are characterized by a very rapid initial recovery. Clark (1976) suggested that this is associated with the gravitational attraction exerted by the ice sheet on the adjacent sea surface.

The limits of Iceland's late Quaternary ice sheets are not well known, but it probably extended onto the continental shelf in places. However, some regions in the north and northwest may have been ice free (Denton and Hughes, 1981; Hjort and others, 1985). The ice sheet was approximately 130,000 km^2—one order of magnitude smaller than the Greenland Ice Sheet and two orders of magnitude smaller than the Laurentide Ice Sheet. Nevertheless, the marine limit lies at elevations of between 110 and 32 m (Einarsson, 1966, 1978). Dates on the age of the marine limit suggest that deglaciation of the outer coast of Iceland occurred between 11 and 10 ka.

Late Quaternary deformation: lake and marine shoreline gradients

Figure 13 shows the direction and rate of deformation associated with glacial isostatic deformation in two categories: 10 to 100 and >100 nannoradians/yr. The data on this map are averages for the entire interval since the shorelines were formed.

Although there are over 20 locations plotted on the map, the geographic distribution is strongly biased toward the southern and eastern margins of the Laurentide Ice Sheet. There is a single determination from the southern margin of the Cordilleran Ice Sheet in British Columbia (Fulton and Walcott, 1975), but there are few data on the rates of shoreline deformation across much of western and northern Canada (Fig. 13).

The rate of shoreline deformation depends on the gradient of the rate of glacial isostatic recovery. If the ice sheets had radial symmetry, with thickness (H) being some simple function of distance (L) from the ice margin (i.e., H L$^{0.5}$), and if ice retreat was uniform, then rates of shoreline deformation would be highest at the margins of the ice sheet. However, this idealized relation is complicated by the evidence that ice retreat was highly variable on a continental scale (e.g., Prest, 1969; Dyke and Prest, 1987), and that the ice sheet probably consisted of both "hard" and "soft" beds (Fisher and others, 1985; Boulton and others, 1985; see Fig. 7). Over the Precambrian shield the bed was hard, and the basal shear stress was probably of the order of 50 to 100 kPa (i.e., a "normal" ice sheet profile such as Greenland or Antarctica), whereas in areas underlain by deformable sediments the basal shear stress may have been <10 kPa (Matthews, 1974; Beget, 1987; Fisher and others, 1985).

At the moment the concept of "hard" and "soft" beds is an attractive hypothesis that explains several aspects of the glacial history of the Laurentide Ice Sheet (e.g., Andrews, 1987). It also has implications for regional variations in the rate of shoreline deformation. In areas of "soft" beds with low basal shear stresses the ice would have been thin (see Fig. 7), with a low surface gradient when compared to a Greenland-type model. A simple geographic comparison of the rates of shoreline deformation is impossible because of the variable dates of deglaciation and ages of the shorelines.

Deformation rates of >100 nannoradians/yr occur along

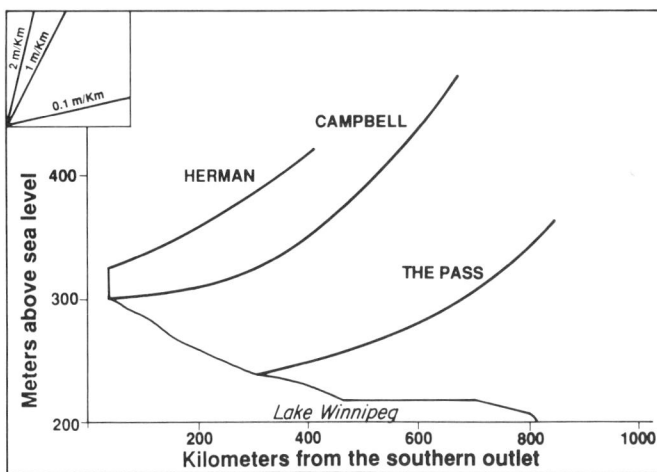

Figure 11. Equidistant shoreline diagram from the southwest margin of the Laurentide Ice Sheet in the vicinity of Glacial Lake Agassiz (from Teller and Thorleifson, 1983).

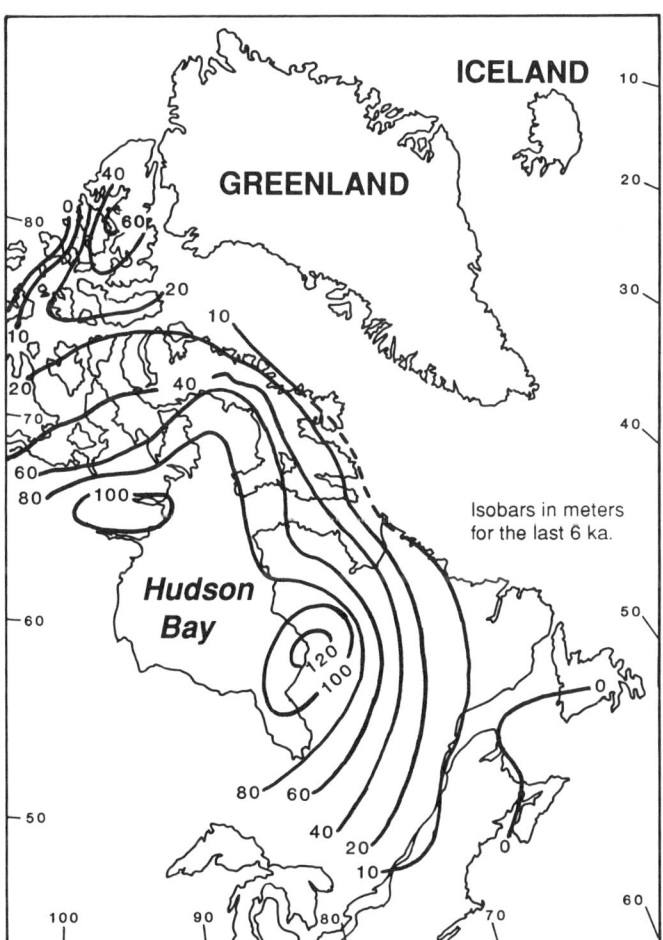

Figure 10. Isobases on relative sea level changes over the last 6,000 years from the region of North America affected by the Laurentide Ice Sheet (after Andrews, 1970).

the eastern seaboard of Canada from Baffin Island to Labrador, but the majority of estimates range between 10 and <100 nannoradians/yr. Shorelines with rates >100 refer to shorelines that were occupied by either the sea or glacial lakes between 8 and 10 ka, and extend across areas with "hard" beds. Hence, these regions were covered by reasonably thick ice. Rates of <100 nannoradians/yr have been measured from coastal and interior Labrador and along the St. Lawrence/Great Lakes/Lake Winnipeg axis, and include mainly "soft" bed regions (Fisher and others, 1985; Boulton and others, 1985; Beget, 1986). I suggest that the deformation rates of glacial shorelines will be useful in testing the validity of the importance of ice-sheet beds on shear stress and ice thickness. However, a rigorous physical analysis of the problem is required.

The direction of tilt of the shorelines can be used to infer the presence of uplift domes within the margins of the former ice sheet (e.g., Andrews and Barnett, 1972). Some interpretations of North American glacial isostatic recovery (Walcott, 1972) indicate that a single center of recovery is located over Hudson Bay, but other interpretations (Andrews, 1970) indicate that at least two domes exist, one northwest and the other east of Hudson Bay. If glacial isostatic recovery is a measure of geographic variations in the ice load, then the directions of shoreline deformation should delimit the geometry of the ice load. Note that there is a critical distinction between an estimate of the location of a zone(s) of maximum ice thickness and the location of an ice divide (cf., Andrews, 1982). Although the two might have coincided, it is not required glaciologically nor isostatically.

The outline of the uplift domes on Figure 13 represents areas where the shoreline tilts intersect. An example from Labrador illustrates the point. Shorelines tilt up to the southwest and westsouthwest in northern and eastern Labrador and to the southeast in southern Labrador; the evidence thus suggests a dome of recovery in central Labrador. Shorelines on Glacial Lake

Figure 12. Glacial lake shorelines in the middle Isortoq Valley close to the present Barnes Ice Cap. Different water levels are marked by shorelines and perched deltas on the valley sides.

Ojibway-Barlowe (Vincent and Hardy, 1979) tilt toward the north-northeast and do not suggest a major zone of uplift over central Hudson Bay; tilts up toward the northwest in southern Quebec do suggest such a center. Although there is considerable uncertainty on the configuration of the ice sheet west of Hudson Bay (Dyke and Prest, 1987; Shilts and others, 1979), there have been relatively few determinations of either the direction or rates of shoreline deformation. However, the opportunity to greatly increase our knowledge in this area does exist. Mapping indicates that glacial lakes were extensive along the west and northwest margins of the Laurentide Ice Sheet and existed in various locations throughout deglaciation (Prest, 1970).

The pattern of tilting of glacial shorelines is more complex than would be expected by the removal of a single-domed ice sheet located over Hudson Bay. The extent to which these data complement, or invalidate, models of current vertical motions induced by deglaciation (e.g., Peltier, 1986) is a matter of considerable interest. Predicted rates of present-day vertical motion (Peltier, 1986; Fig. 5) are based on variations of ICE-1 (Peltier and Andrews, 1976), an ice sheet reconstruction that shows both an ice divide and maximum ice thickness over Hudson Bay. However, the reconciliation of geological and isostatic reconstructions of the Laurentide Ice Sheet (Peltier and Andrews, 1983; Andrews and Peltier, 1989) is far from complete, and interactions between the modellers and the field scientists must remain vigorous.

EVIDENCE FOR EPISODIC RESPONSE

The conventional interpretation of relative sea level curves reflects the interaction between glacial isostatic rebound and the addition of water to the ocean basins. Thus the envisaged processes are relatively constant over time scales of 0.1 to 1 ky. In such a situation the "smooth" curves of Figure 3 are reasonable. However, theory and observations (Quinlan, 1984; Clark, 1982; Oliver and others, 1970; Andrews and Miller, 1985) suggest that rapid changes of sea level in glaciated areas may be associated with unloading.

Daly (1934), in his classic book on the *Changing World of the Ice Age,* proposed two models for the rebound of the crust in glaciated areas. The first model is essentially implicit in much of the earlier sections of this chapter. It postulates that recovery from the load is accommodated by the inward migration of material at great depths. However, his second hypothesis includes a zone of vertical shears or faults that were activated during deglaciation.

There has been a long history of discussion in Scandinavia (cf., Morner, 1980) as to whether faulting occurred during deglaciation. In a detailed study, Sissons and Cornish (1982) showed that the parallel roads of Glen Roy (glacial lake beaches) were faulted with throws of 1 m or less. The "punching" hypothesis favored by Daly (Daly, 1934, Fig. 70) has largely been ignored in North America for the last two decades.

In one of the few papers to explicitly investigate field evidence, Oliver and others (1970, p. 579) noted that ". . . postglacial bedrock faulting, dated by offset striations, is common on outcrops . . ." over a large section of New York and Quebec. Displacements were only a few centimeters at most, but the density of faults caused Oliver and others (1970) to propose that the cumulative displacement could be substantial. Similarly, Adams (1989b) reported small, mainly reverse faults, that displace glacial striations in southeastern Canada and the northeastern United States.

Several workers have examined the relation, or lack thereof, between glacial unloading and seismicity (Oliver and others, 1970; Quinlan, 1984; Hasegawa and Basham, 1989; Adams, 1989b). In some areas, such as coastal British Columbia and Alaska, it is difficult to unravel the glacial isostatic processes from nonrelated tectonic events (e.g., Matthews and others, 1970; Clague and others, 1982; Dragert, 1987). However, there appears to be a growing consensus (cf., Hasegawa and Basham, 1989) that glacial isostatic unloading has reactivated tectonic structures. For example, the association of rapid changes of relative sea level to renewed glacial loading, or with tectonics, has been discussed from the western coast of Canada (e.g., Matthews and others, 1970), and the eastern Canadian Arctic (Andrews and Miller, 1985). Dyke and others (1991) noted an abrupt change in the elevation of the marine limit along the mainland coast of Arctic Canada,, and suggested that it was associated with Holocene block faulting of the Boothia arch. The offset reported by Dyke and others (1991) is of the order of 30 m.

In Iceland, the time of deglaciation is associated with an increase in volcanic activity and the production of volcanic ashes (Einarsson, 1978). Intuitively it might be expected that because of

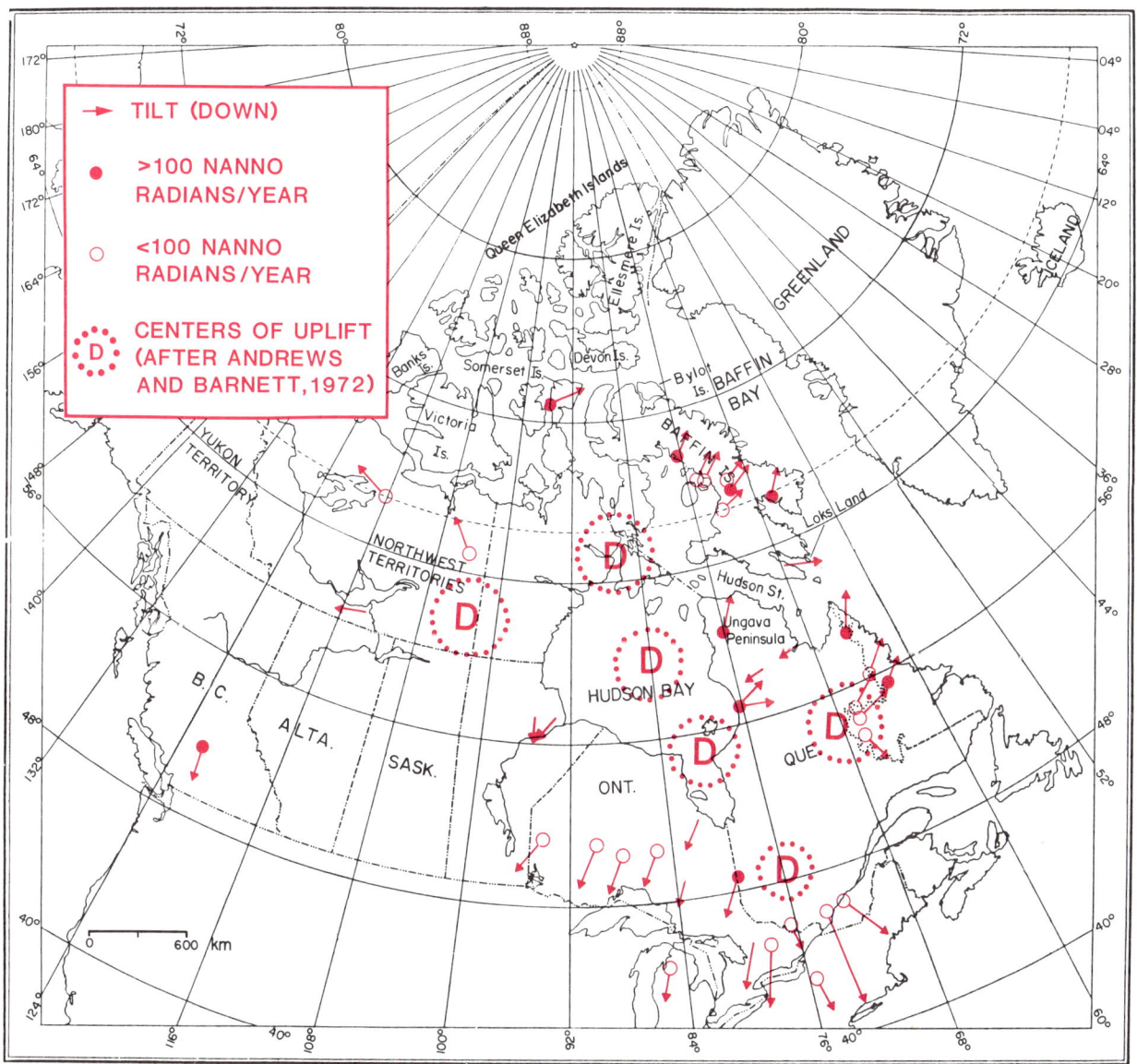

Figure 13. Directions of deformation and average rates of deformation of lake and marine water planes for areas covered by the last North American ice sheet.

its position across the Mid-Atlantic Ridge, Iceland would have responded somewhat differently, than the shield areas of Greenland or Canada. The question of whether a major difference in the production of volcanic rocks between glacial and interglacial periods exists, and whether volcanic activity changed across a glacial/interglacial transition, has not been specifically researched. However, much of the Pleistocene succession in Iceland consists of hyaloclastics that are considered to have been erupted subglacially. If this is a valid interpretation, then it does indeed suggest that the depression of the surface by the ice load caused an increase in surficial magma production (cf., Sjerup and others, 1989).

CONCLUSIONS

Raised marine beaches and warped shorelines (both marine and lacustrine) within the borders of the major ice sheets of North America, Greenland, and Iceland illustrate the glacially induced recovery of these regions. The maximum recorded postglacial change in relative sea level is slightly above 300 m on the east side of Hudson Bay; extensive areas in the southeast and northwest regions of the Laurentide Ice Sheet have recovered at least 200 m since deglaciation. In Greenland, delevelling is less, partly due to the presence of a smaller ice load and the continued existence of the Greenland Ice Sheet throughout the Holocene. A much smaller ice cap substantially depressed Iceland, which may be related to the rheology of the island.

The rate of sea-level change associated with ice unloading can be measured by ^{14}C dating raised marine deposits. The resulting time/elevation graphs indicate that initial rates of sea-level change exceeded 10 m/100 yr in many instances, and most rates of change exceeded 10 m/100 yr in many instances, and most rates of change exceeded 3 m/100 yr. The rates drop exponentially with time, and today the highest estimated rate is between 1 and 2 m/100 yr in southwest Hudson Bay (Weber and others, 1970).

Differential crustal uplift from the ice load caused former shorelines to be tilted toward zones of maximum crustal rebound. The rate of tilting has decreased nonlinearly since deglaciation, but *average* tilts vary between about 10 and 100 nannoradians/yr. Projection of the tilt of these former shorelines toward former ice centers delimited several broad domes of postglacial uplift.

REFERENCES CITED

Adams, J., 1989a, Postglacial faulting in eastern Canada; Nature, origin, and seismic hazard implications: Tectonophysics, v. 163, p. 323–331.
—— , 1989b, Crustal stresses in eastern Canada, *in* Gregersen, S., and Basham, P. W., eds., Earthquakes at North Atlantic passive margins: Dordrecht, Kluwer Academic Publishers, p. 289–297.
Andrews, J. T., 1970, A geomorphological study of post-glacial uplift with particular reference to Arctic Canada: Institute of British Geographers Special Publication 2, 156 p.
—— , ed., 1974, Glacial isostasy: Stroudsberg, Pennsylvania, Dowden, Hutchinson and Ross, 491 p.
—— , 1982, On the reconstruction of Pleistocene ice sheets; A review: Quaternary Science Reviews, v. 1, p. 1–30.
—— , 1986, Elevation and age relationships; Raised marine deposits and landforms in glaciated areas; Examples based on Canadian Arctic data, *in* van de Plassche, O., ed., Sea-level research; A manual for the collection and evaluation of data: Norwich, United Kingdom, Geo Books, p. 67–96.
—— , 1987, The late Wisconsin glaciation and deglaciation of the Laurentide Ice Sheet, *in* Ruddiman, W. F., and Wright, H. E., Jr., eds., North America and adjacent oceans during the last deglaciation: Boulder, Colorado, Geological Society of America, The Geology of North America, v. K-2, p. 13–37.
Andrews, J. T., and Barnett, D. M., 1972, Analysis of strandline tilt direction in relation to ice centres and postglacial crustal deformation, Laurentide Ice Sheet: Geografiska Annaler, v. 54A, p. 1–11.
Andrews, J. T., and Miller, G. H., 1985, Holocene sea level variations within Frobisher Bay, *in* Andrews, J. T., ed., Quaternary environments; Eastern Canadian Arctic, Baffin Bay, and West Greenland, Boston, Massachusetts, Allen and Unwin, p. 585–607.
Andrews, J. T., and Peltier, W. R., 1989, Quaternary geodynamics, *in* Fulton, R. J., ed., Quaternary geology of Canada and Greenland: Geological Survey of Canada Geology of Canada 1, p. 542–572.
Archer, D. R., 1968, The upper marine limit in the Little Whale River, New Quebec: Arctic, v. 21, p. 153–160.
Barnett, D. M., 1970, An amendment and extension of tide gauge data analysis for Churchill, Manitoba: Canadian Journal of Earth Sciences, v. 7, p. 626–627.
Beget, J., 1986, Influence of till rheology on Pleistocene glacier flow in the southern Great Lakes area, U.S.A.: Journal of Glaciology, v. 32, p. 235–241.
—— , 1987, Low profile of the Laurentide Ice Sheet: Arctic and Alpine Research, v. 20, p. 81–88.
Bird, J. B., 1954, Postglacial marine submergence in central Arctic Canada: Geological Society of America Bulletin, v. 65, p. 457–464.
Bjorck, S., and Persson, T., 1981, Late Weischelian and Flandrian biostratigraphy and chronology from Hochstetter foreland, northeast Greenland: Meddelelsser om Gronland Geoscience 5, 19 p.
Blake, W., Jr., 1975, Radiocarbon age determinations and post glacial emergence at Cape Storm, southern Ellesmere Island: Geografiska Annaler, v. 62A, p. 1–71.
—— , 1976, Sea and land relations during the last 15,000 year in the Queen Elizabeth Islands, Arctic Archipelago: Geological Survey of Canada Paper 76-1B, p. 201–207.
Bloom, A. L., 1971, Glacial-eustatic and isostatic controls of sea level since the last glaciation, *in* Turekian, K. K., ed., The late Cenozoic glacial ages: New Haven, Connecticut, Yale University Press, p. 355–380.
Boulton, G. S., and Smith, G. D., Jones, A. S., and Newsome, J., 1985, Glacial geology and glaciology of the last mid-latitude ice sheets: Geological Society of London Journal, v. 142, p. 447–474.
Broecker, W. S., 1966, Glacial rebound and the deformation of proglacial lakes: Journal of Geophysical Research, v. 71, p. 4777–4783.
Bryson, R. A., Wendland, W. M., Ives, J. D., and Andrews, J. T., 1969, Radiocarbon isochrons on the disintegration of the Laurentide Ice Sheet: Arctic and Alpine Research, v. 1, p. 1–14.
Cathles, L., 1975, The viscosity of the Earth's mantle: Princeton, New Jersey, Princeton University Press, 386 p.
Clague, J. J., 1983, Glacio-isostatic effects of the Cordilleran ice sheet, British Columbia, *in* Smith, D. E., and Dawson, A. G., eds., Shorelines and isostasy: London, Academic Press, p. 285–319.

Clague, J. J., Harper, J. R., Hebda, R. J., and Howes, D. E., 1982, Late Quaternary sea levels and crustal movements, coastal British Columbia: Canadian Journal of Earth Sciences, v. 19, p. 597–618.

Clark, J. A., 1976, Greenland's rapid postglacial emergence; A result of ice-water gravitational attraction: Geology, v. 10, p. 310–312.

——, 1977, An inverse problem in glacial geology; The reconstruction of glacier thinning in Glacier Bay, Alaska, between A.D. 1910 and 1960 from relative sea level data: Journal of Glaciology, v. 18, p. 481–503.

——, 1980, The reconstruction of the Laurentide Ice Sheet of North America from sea level data; Methods and preliminary results: Journal of Geophysical Research, v. 85, p. 4307–4323.

——, 1982, Glacial loading, A cause of natural fracturing and a control of the present stress in regions of high Devonian-shale gas production; Albuquerque, New Mexico, Sandia National Laboratories SAND-81-2474C, 18 p.

Clark, J. A., Farrell, W. E., and Peltier, W. R., 1978, Global changes in postglacial sea level; A numerical calculation: Quaternary Research, v. 9, p. 265–287.

Craig, B. G., and Fyles, J. G., 1960, Pleistocene geology of Arctic Canada: Geological Survey of Canada Paper 60-10, 21 p.

Daly, R. A., 1934, The changing world of the ice age: New Haven Connecticut, Yale University Press, 271 p.

De Geer, G., 1892, On Pleistocene changes of level in eastern North America: Proceedings of the Boston Society of Natural History, v. 25, p. 454–477.

Denton, G. H., and Hughes, T. J., 1981, The last great ice sheets; New York, John Wiley and Sons, 484 p.

Devoy, R.J.N., ed., 1987, Sea surface studies: New York, Croom Helm, 649 p.

Dragert, H., 1987, The fall (and rise) of central Vancouver Island; 1930–1985: Canadian Journal of Earth Sciences, v. 24, p. 689–697.

Dyke, A. S., 1974, Deglacial chronology and uplift history; Northeastern section Laurentide Ice Sheet: Boulder, University of Colorado Institute of Arctic and Alpine Research Occasional Paper 12, 73 p.

——, 1983, Quaternary geology of Somerset Island, District of Franklin: Geological Survey of Canada Memoir 404, 32 p.

——, 1984, Quaternary geology of Boothia Peninsula and northern District of Keewatin, central Arctic Canada: Geological Survey of Canada Memoir 407, 26 p.

Dyke, A. S., and Prest, V. K., 1987, Late Wisconsinan and Holocene history of the Laurentide Ice Sheet: Geographie Physique et Quaternaire, v. 41, p. 237–263.

Dyke, A. S., Dredge, L. A., and Vincent, J.-S., 1982, Configuration of the Laurentide Ice Sheet during the late Wisconsin maximum: Geographic Physique et Quaternaire, v. 36, p. 5–14.

Dyke, A. S., Morris, T. F., and Green, D.E.C., 1991, Postglacial tectonic and sea level history of the central Canadian Arctic: Geological Survey of Canada Bulletin 397, 56 p.

Einarsson, T., 1966, Late and post-glacial rise in Iceland and sub-crustal viscosity: Jokull, v. 3, p. 157–166.

——, 1978, Jardfraedt (Geology): Reykjarvik, Iceland, Mal of Manning, 240 p. (translation of parts of the text provided by Aslaug Giersdottir).

England, J. H., 1976, Postglacial isobases and uplift curves from the Canadian and Greenland high arctic: Arctic and Alpine Research, v. 8, p. 61–78.

Fairchild, H. L., 1918, Postglacial uplift of northeastern America: Geological Society of America Bulletin, v. 29, p. 187–234.

Farrand, W. R., and Gadja, R. T., 1962, Isobases on the Wisconsin marine limit: Canada Department of Mines and Technical Surveys Geographical Branch Geographical Bulletin 17, p. 5–22.

Fisher, D. A., Reeh, N., and Langley, K., 1985, Objective reconstructions of the late Wisconsin Laurentide Ice Sheet and the significance of deformable beds: Geographic Physique et Quaternaire, v. 39, p. 229–238.

Fulton, R. J., ed., 1984, Quaternary stratigraphy of Canada; A Canadian contribution to IGCP Project 24: Geological Survey of Canada Paper 84-10, 210 p.

——, ed., 1989, Quaternary geology of Canada and Greenland: Geological Survey of Canada Geology of Canada 1, 839 p.

Fulton, R. J., and Walcott, R. I., 1975, Lithospheric flexure as shown by deformation of glacial lake shorelines in southern British Columbia, in Whitten, E.H.T., ed., Quantitative studies in the geological sciences: Geological Society of America Memoir 142, p. 163–173.

Funder, S., ed., 1989, Quaternary geology of Greenland, in Fulton, R. J., and Funder, S., eds., Quaternary geology of Canada and Greenland: Geological Survey of Canada Geology of Canada 1, p. 741–839.

Grant, D. R., 1970, Recent coastal submergence of the Maritime Provinces, Canada: Canadian Journal of Earth Sciences, v. 7, p. 676–689.

——, 1980, Quaternary sea-level change in Atlantic Canada as an indication of crustal delevelling, in Morner, N.-A., ed., Earth rheology, isostasy, and eustasy: New York, John Wiley and Sons, p. 201–214.

Gutenberg, B., 1941, Changes in sea level, postglacial uplift, and mobility of the Earth's interior: Geological Society of America Bulletin, v. 52, p. 721–772.

Hall, C. F., 1865, Arctic researches and life amongst the Esquimaux: New York, Harper and Brothers, 595 p.

Hasegawa, H. S., and Basham, P. W., 1989, Spatial correlation between seismicity and postglacial rebound in eastern Canada, in Gregersen, S., and Basham, P. W., eds., Earthquakes of North Atlantic passive margins: Dordrecht, Kluwer Academic Publishers, p. 483–500.

Heyworth, A., 1986, Dendrochronological dating in van de Plassche, O., Sea-level research; A manual for the collection and evaluation of data: Norwich, United Kingdom, Geo Books, p. 561–568.

Hillaire-Marcel, C., and Fairbridge, R. W., 1978, Isostasy and eustasy of Hudson Bay: Geology, v. 6, p. 117–122.

Hjort, C., Ingolfsson, O., and Nordahl, N., 1985, Late Quaternary geology and glacial history of Hornstrondin, northwest Iceland; A reconnaissance study: Jokull, v. 35, p. 9–29.

Jardine, W. G., 1986, Determination of altitude, in van de Plassche, O., ed., Sea-level research; A manual for the collection and evaluation of data: Norwich, United Kingdom, Geo Books, p. 569–590.

Kane, E. K., 1856, The second Grinnell Expedition in search of Sir John Franklin, 1853, 1954, 1855, vols. 1 and 2: Philadelphia, Pennsylvania, Childs and Peterson, 464 p. and 467 p.

Karrow, P. F., and Calkin, P. E., eds., 1985, Quaternary evolution of the Great Lakes: Toronto, Geological Association of Canada Special Paper 30, 258 p.

Kelly, M., 1985, A review of the Quaternary geology of western Greenland, in Andrews, J. T., ed., Quaternary environments; Eastern Canadian Arctic, Baffin Bay, and western Greenland: Boston, Massachusetts, Allen and Unwin, p. 461–501.

Loken, O. H., 1962, The late glacial and postglacial emergence and deglaciation of northernmost Labrador: Canada Department of Mines and Technical Surveys Geographical Branch Geographical Bulletin 17, p. 23–56.

Mangerud, J., 1972, Radiocarbon dating of marine shells, including a discussion of apparent age of recent shells from Norway: Boreas, v. 1, p. 143–172.

Mann, D. H., 1986, Wisconsin and Holocene glaciation of southeast Alaska, in Hamilton, T. D., Reed, K. M., and Thorson, R. M., eds., Glaciation in Alaska; The geologic record: Anchorage, Alaska Geological Society, p. 237–265.

Matthews, W. H., 1974, Surface profiles of the Laurentide Ice Sheets in its marginal areas: Journal of Glaciology, v. 13, p. 37–43.

Matthews, W. H., Fyles, J. G., and Nasmith, H. W., 1970, Postglacial crustal movements in southwestern British Columbia and adjacent Washington State: Canadian Journal of Earth Sciences, v. 7, p. 690–702.

Miller, R. D., 1973, Gastineau Channel Formation, A composite glaciomarine deposit near Juneau, Alaska: U.S. Geological Survey Bulletin 1394-C, 20 p.

Mook, W. G., and van de Plassche, O., 1986, Radiocarbon dating, in van de Plassche, O., ed., Sea-level research; A manual for the collection and evaluation of data: Norwich, United Kingdom, Geo Books, p. 525–560.

Morner, N.-A., 1980, The Fennoscandian Uplift; Geological data and their geodynamical implications, in Morner, N.-A., ed., Earth rheology, isostasy, and eustasy: New York, John Wiley and Sons, p. 251–284.

Oliver, J., Johnson, T., and Dorman, J., 1970, Postglacial faulting and seismicity in New York and Quebec: Canadian Journal of Earth Sciences, v. 7, p. 579–590.

Oldale, R. N., and O'Hara, C. J., 1980, New radiocarbon dates from the inner continental shelf off southeastern Massachusetts and a local sea-level rise curve for the past 12,000 yr.: Geology, v. 8, p. 102–106.

Paterson, W.S.B., 1972, Laurentide Ice Sheet; Estimated volumes during the late Wisconsin: Reviews of Geophysics and Space Physics, v. 10, p. 885–917.

Palmer, A.J.M., and Abbott, W. H., 1986, Diatoms as indicators of sea-level change, *in* van de Plassche, O., ed., Sea-level research; A manual for the collection and evaluation of data: Norwich, United Kingdom, Geo Books, p. 457–488.

Peltier, W. R., 1976, Glacial isostatic adjustments; 2. The inverse problem: Geophysical Journal of the Royal Astronomical Society, v. 46, p. 669–706.

—— , 1986, Deglaciation-induced vertical motion of the North American continent and transient lower mantle rheology: Journal of Geophysical Research, v. 91, no. B9, p. 9099–9123.

Peltier, W. R., and Andrews, J. T., 1976, Glacial-isostatic adjustment; 1. The forward problem: Geophysical Journal of the Royal Astronomical Society, v. 46, p. 605–646.

—— , 1983, Glacial geology and glacial isostasy, Hudson Bay, Canada, *in* Smith, D. I., and Dawson, A. G., eds., Shorelines and isostasy: New York, Academic Press, p. 285–319.

Prest, V. K., 1969, Retreat of Wisconsin and Recent ice in North America: Geological Survey of Canada Map 1257A, scale 1:5,000,000.

—— , 1970, Quaternary geology in Canada, *in* Douglas, R. J., ed., Geology and economic minerals in Canada, 5th ed.: Canada Geological Survey Economic Geology Report 1, p. 676–764.

Quinlan, G., 1984, Postglacial rebound and the focal mechanisms of eastern Canadian earthquakes: Canadian Journal of Earth Sciences, v. 21, p. 1018–1023.

—— , 1985, A numerical model of postglacial relative sea level change near Baffin Island, *in* Andrews, J. T., ed., Quaternary environments; Eastern Arctic, Baffin Bay, and western Greenland: Boston, Massachusetts, Allen and Unwin, p. 560–584.

Quinlan, G., and Beaumont, C., 1982, The deglaciation of Atlantic Canada as reconstructed from the postglacial relative sea-level record: Canadian Journal of Earth Sciences, v. 19, p. 2232–2248.

Saemundsson, K., 1986, Subaerial volcanism in the western North Atlantic, *in* Vogt, P. R., and Tucholke, B. E., eds., The western North Atlantic: Boulder, Colorado, Geological Society of America, The Geology of North America, v. M, p. 69–86.

Schmoll, H. R., and Yehle, L. A., 1986, Pleistocene glaciation of the upper Cook Inlet basin, *in* Hamilton, T. D., Reed, K. M., and Thorson, R. M., eds., Glaciation in Alaska; The geologic record: Anchorage, Alaska Geological Society, p. 193–218.

Schmoll, H. R., and Szabo, B. J., Rubin, M., and Dubrovolny, E., 1972, Radiometric dating of marine shells from the Bootlegger Cove Clay, Anchorage area, Alaska: Geological Society of America Bulletin, v. 83, p. 1107–1114.

Scott, D. B., and Medioli, F. S., 1986, Foraminifera as sea-level indicators, *in* van de Plassche, O., ed., Sea-level research; A manual for the collection and evaluation of data: Norwich, United Kingdom, Geo Books, p. 435–456.

Shaler, N. S., 1874, Preliminary report on the recent changes of level on the coast of Maine: Boston Society of Natural History Memoir 2, p. 321–334.

Shilts, W. W., Cunningham, C. C., and Kasycki, C. A., 1979, Keewatin Ice Sheet; Re-evaluation of the traditional concept of the Laurentide Ice Sheet: Geology, v. 7, p. 537–541.

Sim, V. K., 1960, Maximum postglacial marine submergence in northern Melville Peninsula: Arctic, v. 13, p. 178–193.

Sissons, J. B., and Cornish, R., 1982, Differential glacio-isostatic uplift of crustal blocks of Glen Ray, Scotland: Quaternary Research, v. 18, p. 268–288.

Sjerup, H.-P., and 6 others, 1989, Quaternary tephra-chronology of the Icelandic Plateau, north of Iceland: Journal of Quaternary Science, v. 4, p. 109–114.

Sutherland, D. G., 1986, A review of Scottish marine shell radiocarbon dates; Their standardization and interpretation: Scottish Journal of Geology, v. 22, p. 145–164.

Teller, J. T., and Clayton, L., eds., 1983, Glacial Lake Agassiz: Geological Association of Canada Special Paper 26, 451 p.

Teller, J. T., and Thorliefson, H. N., 1983, The Lake Agassiz–Lake Superior connection, *in* Teller, J. T., and Clayton, L., eds., Glacial Lake Agassiz: Geological Association of Canada Special Paper 26, p. 261–290.

Ten Brink, N. W., 1974, Glacio-isostasy; New data from West Greenland and geophysical implications: Geological Society of America Bulletin, v. 85, p. 219–228.

Thorson, R. M., 1981, Isostatic effects of the last glaciation in the Puget Lowland, Washington: U.S. Geological Survey Open-File Report 81-370, 100 p.

van de Plassche, O., ed., 1986, Sea-level research; A manual for the collection and evaluation of data: Norwich, United Kingdom, Geo Books, 618 p.

Vincent, J.-S., and Hardy, L., 1979, The evolution of Glacial Lake Barlow and Ojibay, Quebec and Ontario: Geological Survey of Canada Bulletin 136, 18 p.

Walcott, R. I., 1970, Isostatic response to loading of the crust in Canada: Canadian Journal of Earth Sciences, v. 7, p. 716–726.

—— , 1972, Late Quaternary vertical movements in eastern North America: Reviews of Geophysics and Space Physics, v. 10, p. 849–884.

Washburn, A. L., and Stuiver, M., 1962, Radiocarbon-dated postglacial delevelling in northward Greenland and its implications: Arctic, v. 15, p. 66–73.

Weber, P. J., Richardson, J. W. and Andrews, J. T., 1970, Post-glacial uplift and substrate age at Cape Henrietta Maria, southeastern Hudson Bay, Canada: Canadian Journal of Earth Sciences, v. 7, p. 317–325.

Manuscript Accepted by the Society October 4, 1990

ACKNOWLEDGMENTS

The time required for this contribution has been provided by a grant from the National Science Foundation's program on Surficial Processes. Discussions with colleagues at the University of Colorado and the Geological Survey of Canada have provided material for this chapter. Specific appreciation is given to the various authors noted in the text who have contributed chapters to other related Decade of North American geology volumes. Aslaug Geirsdottir was most helpful for the Icelandic literature. Stephanie Green assisted in drafting figures and proofreading. I thank W. H. Matthews and J. Adams for their constructive reviews.

Index

[Italic page numbers indicate major references]

Abiquiu Reservoir, 235
Adak Island region, 29, 31, 35, 38, *39*, 42
Adak network, *29*, 36
Adams Run Seismic Zone, 301, *302*
Adel, Oregon, 93, 101
Adirondack Mountains, 282, 285, 429
Adirondack–Western Quebec province, 287
aftershocks, *32*, 35, 43, 49, 62, 83, 85, 93, 114, 142, 144, 147, 156, 162, 164, 167, 200, 204, 249, 266, 280, 282, 285, 311, 316, 325, 329, *331*, 336, 415
  activity, 114, 115
  sequence, 82, 89, 115, 263, 330, 335
  swarm, 163
  zones, 32, 34, 36, 61, 101
Agustin fault, 332
Alabama, *303*
Alabama Piedmont, 303
Alabama-Tennessee Seismic zone, 304
Alaska, 4, 32, 54
  central, *60*
  earthquakes, 381, *387*
  heat-flow measurements, *446*
  northeast, *60*, 62
  seismicity, *47*
  southeast, *50*, 390, *446*, 479
  southern, *51*
  stress, 354, *387*
  western, *63*, 381
Alaska-Aleutian arc, 26, *29*
  earthquakes, *32*
  See also Aleutian arc
Alaska Range, 54, 55
Alaska Tsunami Warning Center, 47
Alberta, 74
Albion, Maine, earthquake, 290
Albuquerque area, *237*
  volcanoes, 233, 237
Albuquerque Basin, 237, 462
Albuquerque-Belen Basin, *462*, 464
Alcoa-Maryville area, earthquake, 298
Aleutian arc, 52, 54, *354*
  main thrust zone, *32*
  outer rise region, *31*
  seismicity regions, *29*, *31*
  trench region, *31*
  upper plate region, *41*
  Wadati-Benioff zone, *31*, *38*
Aleutian Islands, 26, 387
  central, 29, 31, 32, 42, 393, 395
  eastern, 29, 392, 393
  seismic networks, 29
  volcanoes, 392
  western, 29, 42, 392, *394*
Aleutian megathrust, 51
Aleutian subduction province, 6, *354*, 361

Aleutian Trench, 42, 48, 51, 54, 390
alignments
  cinder-cone, *402*, 411
  volcanic, 416
alluvial deposits, 249
Alma well, 376, 386
Alpha Rise, 441
Amarillo uplift, 254
Amarillo-Wichita-Arbuckle uplift, 6
Amatitlan, Guatemala, 333
Amchitka region, 29, 39, *40*, 42
Amlia fracture zone, *34*, 393
Anacapa Dume fault, 145
Anadarko basin, 254
Anahim Volcanic Belt–Wells Gray volcanic lineament, 454
Andes, central, 313
Andreanof Islands, 42, 393
  earthquake, 42
anisotropy, rock, 341
Anna, earthquake, 253
anomalies
  gravity, 7, 73, 77, 256, 271, 454
  heat-flow, 433, 470
  magnetic, 7, 271, 424, 425, 427
  stress, 377
  thermal, 470
Antilles arc, 315, *316*
Antilles subduction zone, 314
Antilles Wadati-Benioff zone, 316
Anza, California, 143
Anza Gap, 143
Appalachian detachment, 298
Appalachian décollement, 305
Appalachian Highlands, 292, 294, *304*
  earthquakes, *304*
Appalachian Mountains, 429
  Canada, 439
  northern, 261, *265*
  southern, 304
Appalachian Province, 287, *304*
aquifer, confined, 469
Arbuckle Mountains, 254
arches, *16*
Arctic Canada, 271, 474
Arctic mainland coast, 479
Arctic Ocean
  heat flow, 427
  margin, *271*
Ardsley, New York, earthquake, 284, 290
Arenal volcano, 328
Arizona, 4
Arotic Islands, southern, 437
Asheville-Hendersonville area, 298
ashflow tuffs, 208
Ashley River fault, 302
Aspy fault, *16*
asthenosphere, 71, 454, 473
  counterflow, 280
  drag, 280

Atlantic Coastal Plain, 429, *305*
Atlantic margin, 437, 438
  heat flow, *424*, *439*, 441
Attica
  earthquake, 287
  faulting, 276
Augustine Volcano, 60

Baffin Bay, 437, 480
  earthquake, 261, *269*, 271, 273, 374
Baffin Island, Canada, 269, 349, 375, 380, 383, 479, 480
  earthquake, 352
Baffin Shelf, 352
Baja California, 362, 401, 407, *409*
  northern, 310, 403, 432
Bandelier Tuff, 236
Banning fault, 135, 137, 139, 146
Banning segment, 140
Banning strand, 136, 137
Barbados sediment prism, 427
Barter Island, 60, 63
Bartlett Springs fault, 111, 112, 120
basalts
  flood, 90
  flows, 232
Basin and Range Province, 4, 85, 95, 96, 109, *112*, *126*, 190, 381, 433, 458, 463
  earthquakes, 168
  heat flow, 430, 432, *459*
  southern, 136, *147*, 409, 417, *459*
Basin and Range subprovince, 6, 353
Basin and Range–Middle Rocky Mountain transition, 188
Basin and Range–Rio Grande rift stress province, 413
basins
  extensional development, 190
  intraplate, *352*
  pull-apart, 310, 311
Bath, earthquake, 282
Battle Mountain region, 430, 433
Bay of Fundy, 439, 440
Bear Lake fault, 205
Bear Lake Valley, 197
Beaufort Sea, *73*, 271, 374, 381, 395, 396
  coast, 396
  earthquakes, *73*, 381
Beaufort Shelf, 60, 62
Beaverhead fault, 210
Belen-Socorro area, 231
Belize, 401
Bell arch, *16*
Benioff zone, 77, 85, 101
Berdeleben fault, 64
Bering Sea, 354, 387, 394
  heat flow, *427*
  margin, 397
Beringa Island, 29, 43

488 *Index*

Berlin, New Hampshire, earthquake, 290
Bermuda Rise, 425
Big Bear, earthquake, 146
Big Bend, 136, 137, 145
Black Hills, 428, 469
Blanco fracture zone, 78
Blue Mountains, 85, 94, 285
Blue Mountains province, 77
Boconó fault zone, *314*
Boconó-Morón trend, 315
Bonnet Plume basin, 73
Book Cliffs, 213
Bootheel lineament, 9
Boothia arch, *16*
Boothia Peninsula, 270
  earthquake, 270, 273
Boothia uplift, *16*, 271
Borah Peak, earthquake, 168, 185, 192, 193, 194, *201*, 206, 210, *220*, 241
boreholes, 446
  elongations, 389, 411, 416
  offset, 370
Borrego Mountain, earthquake, 143
Boulder Dam, 212
Bowers Ridge, 34
Bowman, earthquakes, 302
Bowman Seismic Zone, 301, *302*
Brawley Seismic Zone, *142*
breakouts, 413
  oil well, *389*
  wellbore, *342*, *368*, 382, 383
Brevard zone, 300
British Columbia, 71, 383, 479, 480
  coast, 446
  southeastern, 455
  southern, 452
  southwestern, 452, 455
Brooks Range, 60, 62, 395, 397
Brookwood, Alabama, 304
Buchan Gulf, 269
Burgos basin, *413*
Burlington/Niagara Falls area, 268
Byam Martin Channel, 271
  swarm, 374

Cache Valley, earthquake, 203
Cajon Pass, California, 139, 359
  drillhole, 342
Calaveras fault, 111, 112, 120
Caliente/Clover, earthquake, *163*
California, 4
  central, *107*, 109, *110*
  eastern, *112*
  earthquake catalog, 109
  faults, 407, 409
  northern, 99
  seismicity, 99, *107*, *110*
  seismotectonics, *133*
  southern, *133*, *135*, 432
  structural provinces, 99
  tectonic divisions, *135*
Campeche-Chiapas-Petén stress province, 416, 417, 418
Canada, *16*, 390
  Appalachian region, 439
  Beaufort Sea, *73*
  Cascadia subduction zone, 70
  continental margin, 266
  crustal stress, *367*
  eastern, *261*, 430
  heat flow, 430, *437*, *438*, *441*
  Interior Platform, 74
  northeastern, 268
  northern Cordillera, 72
  offshore, 69
  Queen Charlotte transform region, 71
  St. Elias region, *71*
  seismicity, 69, *261*
  seismotectonics, *261*
  southeastern, *262*, *266*
  southern Cordillera, 73
  stress, *367*, *380*, *382*, *383*, 390
  western, 69
Canadian Arctic, 271, 474
Canadian Cordillera, 353, 433
  heat flow, 445, 446, *451*
  heat generation, *451*
Canadian margin, 395
Canadian Maritimes Coast line, 479
Canadian Shield, 262, 275, 370, 374, 377, 383, 437, 438, 442
  heat flow, *430*
Canadian Shield Superior Province, 440
Cape Ann, earthquake, 286, 287
Cape Elizabeth fault, 282
Cape Mendocino, 104, 107, 109, 115
Caracas, earthquake, 315
CARB-NOAM plate boundary, *333*, 335
carbonates, 467
Caribbean plate, *313*, *314*, *323*, 335, 336
Caribbean Sea, heat flow, *427*
Carolina Piedmont, 298
Carolina slate belt, 303
Carrizo Plain, 139
Cascade arc, 85
Cascade convergence province, *357*
Cascade Range, 77, 96
  eastern, 90
Cascade volcanic arc, 357, 433
  heat flow, 431
Cascade volcanic centers, northern, 442
Cascadia Convergence stress province, 381
Cascadia megathrust, 94
Cascadia subduction zone, 70
  earthquakes, 70
Castle Mountain fault, 49, 51, 57, 59
catalogs
  earthquake, 79, 82, 94, 109, 135, *195*, 285, 291, 292
  network, 327, 329
  seismic, 77, 94, 134, 154, 234
cauldrons, 241
Cayman Trough, 427
Cedar Mountains, 114
  earthquake, *159*
Celmira-Ballena fault, 357
Centennial Tectonic Belt, 188
Central America
  active deformation, *401*
  seismograph networks, *324*
  stress, 355, *401*, 416
  strong-motion networks, *324*
  tectonics, *323*
  upper-crustal earthquakes, *333*
  upper-crustal seismicity, *323*
  volcanic chain, *324*, 328, 329, 333, 335
  western, *401*, *415*, 416, *418*
Central American subduction zone, *313*, *324*, *355*
Central American Wadati-Benioff zone, *313*
Central Medasedimentary Belt, 263
Central Nevada Seismic Belt (CNSB), 153, *165*, 175
Central Virginia Seismic Zone, 294, *297*, 304
Centralia fault, 252
Cerrillos, earthquakes, 232
Cerro Prieto fault, 407, 409
Cerro Prieto Seismic Zone, 142, 143
Cerro Prieto transform segment, 409
Chalfant Valley, earthquake, *132*, 163, 164
Challis, Idaho, 201
Chapala graben, 414
Charleston (Missouri), earthquake, 9
Charleston (South Carolina), earthquake, 6, 7, *16*, 286, 300, 301, 303
Charlevoix, Quebec, *15*, 261, *263*, 275
Charlevoix–La Malbaise region, 15
Charlevoix Seismic Zone, 279, *287*, *376*
Charlotte slate belt, 303
Chatham Strait, 72
  fault, 51, 72
Chatham Strait–Denali fault system, 51
Cheraw fault, *17*
Chiapas, northwestern, 417
Chiapas fold belt, 356, *415*, 417
Chiapas-Petén fold belt, 416, 418
Chibougamou, 440
Chicoutimi, Quebec, 275
Chihuahua, Mexico, 403, 409, 413, *462*
  heat flow, *457*
  northeastern, 464
Chile, earthquake, 47
Chixoy-Polochic fault, *324*, 332, *333*, *335*, 336, 337
Chugach Mountains, 65
Chugach–St. Elias fault, 51, 54, 71
Chukchi Sea, 64
  earthquake, 64
Churchhill County, Nevada, 162
Cincinnati-Findlay arch, *253*
cinder cones, 403, 414
Clark Canyon Dam, 212
Clarks Hill Reservoir, earthquakes, 304
Clarkston Valley, 197, 212
  earthquake, 192, *197*, 203, 212
Clear Lake, 110
  volcanic center, 432
Clover Mountains, earthquake, 154

clusters
  epicenter, 109
  seismicity, 51, 57, 113, 114, 115
CNSB. *See* Central Nevada Seismic Belt
Coachella Valley, 140
Coachella Valley segment, 137
Coahuila, 411, *413*
  fold belt, 413
coal extraction, 213
Coalinga, earthquake, 111, 115, 118, *126*, 128, 135
Coast Mountains, 51, 74
  earthquakes, 74
Coast Plutonic Complex, 445, 452, 454
Coast Ranges, 77, 109, *111*, *119*, *126*, 128
  northern, *102*
Coastal Plain province, 292, 300
  earthquakes, 294, *302*, *303*
  Seismic Zones, *301*
Coatzacoalcos-Minatitlán basin, 416
Cochrane Falls, 268
Cocos-Nazca-Caribbean triple junction, 313
Cocos Plate, 324, 336, 349, 424
  subduction, *311*
Cocos Ridge, 313
Colima graben, 414
Colorado, 5, *17*
Colorado lineament, 254
Colorado Plateau, 190, 191, 205, 213, 219, 229, 230, 232, *353*
  heat flow, *432*, *457*
  interior region, *457*, 464
  peripheral regions, *458*
  uplift, 457
Columbia Basin, heat flow, 430
Columbia Bay, earthquakes, 57
Columbia Plateau, 77, *90*, 95, 96
Columbia River Basalt Group, 90, *92*, 95, 96
Comalcalco basin, 417
compression, 359, 361, 390, 395, 397
Concord fault, 111
conductivity, 429
Continental Borderland, 133, *144*
continental margin, southeastern, *266*
continental shelves, 424
convection, 424
convergence, 311
  crustal, 145
  plate, 344
  rate, 77, 94
Cook Inlet, 47, 54, 55, 57, 59, 60
Cook Inlet Volcano, 55
Cooper River basin, 59
Cordillera, 353, 433, 438
  earthquakes, *72*, *74*
  heat flow, 434, *441*, *454*
  northern, *72*
  southern, *73*
  western, 346, *353*, 433
  See also Canadian Cordillera, Mexican Cordillera
Cordilleran extension province, *353*, 357, 381

Cordilleran Ice Sheet, 477, 478, 480
  retreat, 476
Cordilleran stress province, 353
Cordilleran Thermal Anomaly Zone, 430, 433
Cornwall, earthquake, 262
Coso geothermal field, 147
Costa Rica, 343
  focal mechanisms, *328*
  microseismicity, *328*
  seismograph networks, *326*
Cottage Grove fault, 252
Cottage Grove-Shawneeton-Rough Creek fault zone, 253
Couer d'Alene mining district, 213
Cove Fort, Utah, 213
Covington, earthquake, 303
Coyote Lake, earthquake, 120
creep, 111, 119, 120, 128, 142, 143, 147, 165
Criner fault, 7, 14, 15
Crownpoint, earthquake, 234
crust
  basement, 300
  brittle, *119*, 169, 172
  continental, 95, 294
  lower, 119, 190, 206, 361, 416, 452, 454
  mid-, 90, 92
  oceanic, 94, 140, 452, 454, 455
  shortening, 71
  spreading, 70
  stress, *367*
  upper, 5, 74, 90, 92, 107, 135, 145, 190, 205, 256, 297, 323, 328, 359, 417, 429, 446, 452, 459
crystalline basement, 109
Cumana, Venezuela, 315

Dakota aquifer, 428, 467, 470
Dakota Group, 467
Dakota Sandstone, 428, 430, *470*
Dall City, earthquake, 62
Dall Mountain fault, 49, 62
Dalton segment, 71
Dead Sea transform, 360
Death Valley, 148
  earthquake, 154
Death Valley-Furnace Creek fault zone, 164, 172
Decaturville dome, 253
deformation, 7, 119, 153, 169, 171, 185, *220*, 314, 411, *480*
  active, *401*, 418
  compressional, *126*
  contemporary, *191*
  ductile, 144
  extensional, 128
  glacial lake shorelines, 478
  neotectonic, 413
  rates, *193*, 473, 481
  regional, 477
  seismic, 139
  shoreline, 480
  tectonic, 148
deglaciation, 374, 382, 476, 479, 480, 482
  Iceland, 476, 483
dehydration, 454

Denali fault system, 49, 51, 57, 59, 71, 72
Denali-Totschunda-Duke River fault system, 51
Denver, Colorado, 213
Denver Basin region, 353, *461*
Deschutes River, 91
Devon Island, 271, 479
Dillon, Montana, 197
displacement, 107
Dixfield, Maine, earthquake, 282, 290
Dixie Valley, earthquake, *114*, *162*, 220
Dixon Island, 16
Dobbs Ferry fault, 290
Dog Valley fault zone, 163
drillhole elongations, *401*, 409, 414
Duke River fault, 49, 57, *59*
Duke River segment, 71
Dulce, earthquake, 236
Dunnage Zone, 439
Durango, 403, 411

earthquakes, 29, 70, *99*
  background, 204, 207
  by year. *See* earthquakes by year
  catalog, *79*, *82*, *94*, 109, 35, *195*, 285, 291, 292
  causative, 12
  clusters, 64, 73, 110, 163, 187, 209, 262, 268
  crustal, 83, *89*, 94, 95, 309, 342
  cycle, *32*
  databases, *21*
  deep, 135, 279, 379
  depth distributions, *166*, 265
  epicenters, 7
  focal depths, *118*
  hazard, *94*, 145
  historical, 7, 17, 22, 26, 47, 60, 72, 73, 154, 157, *165*, 192, *197*, 212, 247, 263, 265, 267, 301, 305, 314, 316, *333*, 336. *See also* earthquakes by year
  induced, 74, 265
  intermediate-depth, 38, *10*, *313*, 415
  intra-slab, *94*
  intraplate, 311, 342
  large, 2, 32, 36, 42, 47, 48, 51, 53, 71, 115, 135, 153, *157*, *165*, 185, *197*, 214, 310, 333, 335, 368
  low-magnitude, 5, 9, 102
  magnitude, 7, 22, 48, 82
  main thrust-zone, *52*
  mantle, 309, 314
  midplate, *245*
  models, *220*
  moderate, 113, 115, 128, 137, 144, 145, 145, 153, 154, 185, 204, 207, 208, 214, 263
  monitoring, 26, 27
  normal-faulting, 355, 370
  nucleation, 218
  occurrence patterns, 107, *213*
  off fault, *126*
  offshore, 267
  overriding plate, 52, *57*

490 Index

parameters, *21*
pattern, *185*
potential, *300*
pre-1970, 79
prehistoric, 10, 147, 195, 302, 379
regional belt, 185
reverse, 401
shallow, 57, 60, 82, 83, 85, 89, 90, *94*, 101, 185, 266, 279, 313, 314, 379, 415
small, 71, 128, 135, 144, 145, 154, 168, 204, 207, 212, 248, 368
strike-slip, *42*, 255, 310, 401
subcrustal, 85
subduction, 71, *94*, 312
subsea, 52, *54*
swarms, 51, 61, 85, 89, 101, 114, 118, 142, 145, *148*, 163, *203*, 205, 208, 210, 212, *214*, 231, *232*, 237, *238*, 242, 255, 271, 285, 298, 302, 330
tectonic, 212, 327
thrust, 31
thrust-fault, *311*, 352
tsunamigenic, 47
upper-crustal, 327, *333*
volcanic, *327*
volcanic axis, 52, *60*
volcanic-chain, 333
See also specific locations
earthquakes by year
 (1556), 255
 (1610), 335
 (1663), 263
 (1690), 316
 (1727), 279
 (1733), 334, 335
 (1765), 334
 (1766), 315
 (1774), 335
 (1775), 279, 286, 298
 (1785), 334
 (1788), 32
 (1791), 263
 (1809), 335
 (1811), 6, 7, 9, 10, 249, 251
 (1812), 6, 7, 9, 10, 137, 249, 314
 (1816), 317, 334, 335
 (1817), 266
 (1843), 251, 316
 (1847), 32
 (1848), 298
 (1849), 32
 (1850), *157*
 (1852), 153
 (1853), 315
 (1855), 266
 (1857), 110, 111, 136, 137, 148
 (1858), 32
 (1860), 153, *157*, 263
 (1866), 6, 7
 (1868), 231
 (1869), *157*
 (1870), 263
 (1872), 94, 110, 113, 127, 147, 148, 153, *163*, *172*, 194
 (1873), *82*, 95
 (1874), 298
 (1875), 294, 304
 (1876), 298
 (1880), 298
 (1884), *197*, 279, 298
 (1886), *16*, 164, 286, 300, 301, 302, 303
 (1887), 311, 411
 (1892), 115
 (1895), 9, 251
 (1897), *197*, 294, 304
 (1899), 53, 54
 (1903), *157*
 (1904), 48, 62, 266, 279
 (1905), 203
 (1906), 102, 110, 111
 (1908), 154
 (1909), 74, *82*, 203
 (1912), 59, 313
 (1913), 300
 (1914), 303
 (1915), *157*, 194, 335
 (1916), 154, 298, 303, 304
 (1917), 32
 (1918), 71, 74, 232
 (1920), 51, 73, 154, 203, 271, 313
 (1921), 203
 (1925), 192, *197*, 203, 212, 263
 (1926), 74
 (1927), 51
 (1928), 203
 (1929), 62, 71, 203, 261, *267*, 271, 279, 286, 287, 315
 (1930), 145, 333
 (1931), *312*, 314, 329, 333, 336
 (1932), *159*, 311
 (1933), 144, 261, *269*, 271, 352
 (1934), 193, *197*, 335
 (1935), 203, 262, 263, 269
 (1937), 48, 143
 (1938), 32, 392
 (1939), 83, 212, 264
 (1940), 73, 135, *140*, 279, 285
 (1942), 74, 143
 (1943), 59
 (1944), 203, 262, 265, 279
 (1945), 203, 334
 (1946), 32, 35, 71, 83, 135, 147
 (1947), 147, 154, 203
 (1949), 51, 71, 83, 85, *89*, 94
 (1952), 72, 134, 135, 139, 148, 203, 236
 (1954), 143, 157, *162*, 220
 (1955), 73
 (1957), 32, 34, 392, 393
 (1958), 51, 59
 (1959), 185, 194, *200*, 203, 206, 210, 216, 220, 334
 (1960), 47
 (1961), 231
 (1962), 101, 203, 316
 (1963), 374
 (1964), 32, 35, *47*, 48, *52*, 53, 392
 (1965), 31, 32, 42, 43, 85, 253, 271, 330, 335, 393
 (1966), 114, 120, 154, *163*, 203, 233, 236, 287
 (1967), 253, 255, 315
 (1968), 62, 93, 101, 143, 144, 253
 (1969), 144, 255, 269
 (1970), 71, 139, 237, 269
 (1971), 59, 134, 135, *145*, 236, 237, 269, 334
 (1972), 41, 73, 74, 236, 253, 255, 314, 323, 329, 330, 333, 335, 336, 374
 (1973), 51, 145, 266, 298, 299, 311, 313, 326, 329
 (1974), 120, 304
 (1975), 73, 101, 103, 126, 147, 203, 204, 208, 216, 255, 262, *267*, 349
 (1976), 71, 303, 325, *330*, 335, 336, 415
 (1977), 302
 (1978), 145, 262, 311, 352
 (1979), 49, *142*, 145, 146, 147, 264, 282, 302, 311, 344
 (1980), 89, 101, 114, *115*, 118, 127, 253, 296, 313
 (1981), *143*, 144, 312
 (1982), 205, 261, 262, 266, 280, 281, 282, 285, 287, 290, 302, 335
 (1983), 111, 115, 118, *126*, 128, 135, 185, 192, 193, 194, *201*, 206, 210, *220*, 241, 255, 262, 265, 266, 279, 280, 282, 300, 302, 329
 (1984), 94, 120, 265, 266, 284
 (1985), 62, 72, 284, 290, 302, 311, 330
 (1986), 32, *36*, 42, 132, 137, 139, 144, 146, 253, 264, 268, 304, 325, 330, 335, 393
 (1987), 49, 54, 73, 143, 145, 330, 390
 (1988), 16, 49, 54, 147, 206, 275, 290
 (1989), 7, *17*, 132, 276
 (1990), 71
 (1991), 290
East Bear Lake fault, 197
East Greenland, 480
East Mountain area, 213
East Pacific Rise, *311*, 424
El Asnam, earthquake, 344
El Pilar fault, *315*
El Salvador, 323
 focal mechanisms, *330*
 microseismicity, *330*
 seismograph network, *325*
 strong-motion network, *325*
El Vado Reservoir, 235
elevation changes, 473
Elk Lake, earthquakes, 85, 89
Elk Mountains, 461
Ellef Ringnes Island, 271
Ellesmere Island, 479
elongations
 borehole, 389, 411, 416
 drillhole, *401*, 409, 414
Elsinore area, 203
 fault, 133, *144*
Elysian Park fault system, 145
epicenters, 7, 10, 16, 29, 50, 55, 59, 73, 79, 101, *109*, *111*, 142, 144, 187, 197, 201, 204, 209, 231, 236, 240, 248, 270, 277, 293, 298, 329, 336

episodic response, *482*
erosion, 454
Española Basin, 236, 462
Estancia basin, 237
Eureka low, 430
Excelsior Mountains, 164
 earthquake, 168
Explorer Plate, 70, 445
Explorer ridge, 78
extension, *241*, 356, 357, 359, 375, 383, 396, 427, 431, 445, 454, 459
 crustal, 346
 intraplate, 185, 191
 regional, 49

Fairbanks, Alaska, 48, *60*
Fairview Peak, earthquake, 114, *162*
Fairview Peak–Dixie Valley sequence, 126
Fairview Range, 162
Fairweather fault, 49, 51, 72
Fallon, earthquake, *162*
faults
 active, 16, 300, 323, 383
 detachment, 190
 displacements, 478
 extensional, 211
 indicators, *389*
 intraplate, *7*
 listric, *220*
 low-angle, *220*
 major active zones, *7*
 minor, *17*
 normal, 31, 42, 54, 63, 65, 70, 89, 94, 102, 109, 127, 140, 142, 147, 153, 162, 168, 171, 175, 188, *190*, *193*, 208, 210, 215, 217, *220*, 249, 253, 269, 271, 310, 312, 317, 349, 353, 355, 362, 374, 380, 389, 392, 394, 397, 411, 413, 416, 418, 463, 473
 reactivated zones, 7
 reverse, 9, 13, 60, 90, 134, 139, 140, 144, *145*, 235, 254, 294, 302, 313, 314,\ 324, 344, 346, 349, 389, 392
 reverse-slip, 9
 rift, *263*, 271
 slip, *344*, *407*, 414
 strike-slip, 60, 63, 70, 133, 139, 143, 146, 168, 171, 175, 253, 294, 298, 302, 311, 314, 317, 324, 330, 346, 389, 394, 396, 407, 415, 418
 surface, 7, *194*, *233*, 249
 thrust, 9, 254, 266, 324, 374, 389, 454
 transform, 79, 133, *310*, 314, 359, 361, 362, 363, 407
 wrench, 172
 See also specific locations
Ferland, Quebec, 15, *16*
fjords, 446, 479
Flaming Gorge Dam, 212
Flathead Lake, Montana, 203, *212*, 214, 381
Flexure Bay, 16

flows
 basalt, 232
 fluid, *428*
 ground-water, *428*
fluids, hydrothermal, 462
focal depth, *25*, *29*, *118*, 145, 153, *166*, 204, 206, 208, 209, *218*, 240, 241, 263, 264, 292, *293*, 304, *306*, 313, 381
focal mechanisms, 29, 35, *38*, 41, 51, 54, 60, 63, 70, 74, *85*, 89, 91, 95, 102, 104, *119*, 128, 135, 139, 142, 144, 146, 148, 163, *168*, *191*, 200, 202, 205, 222, 255, *263*, 280, 284, *290*, 300, 302, 304, 312, 316, *328*, 335, 341, *342*, 346, 353, 362, 367, *368*, 374, *376*, 379, 381, 387, *401*, 409, 411, 414, 416
folds
 active, 370
 intraplate, *16*
Foothills fault system, 103
Foreland fold and thrust belt, 72, 74, 445
foreshocks, 89, 147, 157
Fort Sage Mountain fault, 163
Fort St. John, British Columbia, 74
Fort Tejon, 137
 earthquake, 110, 136, 137
Foxe Basin, 270
fracture pressures, oil-well, 369
fractures, open, 341
fracturing, hydraulic, 344, *369*
Franciscan block, 112, 128
Franklin Straight, 17
Fraser Valley, 454
Fredericton graben, 439
frequency-magnitude data, 115, 118
Front Range, 428, 460, *461*
Fuego volcano, 328
Furnace Creek fault zone, 353

Gabbs Valley, 162
Gallatin fault system, 209
Gallina-Archuleta arch, *235*
Galway Lake, earthquake, 147
Garibaldi Volcanics, 442, 451, 454, 455
Garlock fault, 146, *147*, 165, 172
Gaza, New Hampshire, earthquake, 282, 290
Gentry Mountain area, 213
Georgia, *303*
Georgia Piedmont, 305
Georgia Strait, southern, 71
Geothermal Map of North America, *423*
geothermal energy, 427
geothermal systems, *428*
Geysers geothermal field, 112, 119, 432
Giles County, earthquake, 294
Giles County Seismic Zone, *296*, 298, 304
Glacial Lake Agassiz, 477
Glacial Lake Ojibway-Barlowe, 482
glacial unloading, 473, 483
 chronology, *475*

glaciation, effects, *375*
Glacier Bay region, 72
Glen Canyon Dam, 212
Glen Roy, 482
Goat Mountain, earthquake, 147
Goat Rocks, 89
Gold King fault, 157
Goochland County, earthquake, 294
Goodnow, earthquake, 262, 282
Gorda Basin, *101*
Gorda Escarpment, 101
Gorda Plate, 107
 subduction, *104*
Gorda ridge, 78, 101
Goshen Valley, 204
grabens, 230
 rift, *233*
gradients
 lake, *480*
 marine shoreline, *480*
Graham County, 460
Graham Island, northern, 71
Gran Quivera National Monument, 237
Grand Banks, 376
 earthquake, 261, *267*, 271, 286
Grand Banks basin, 353
Granduc site, 446
Great Basin, *166*, 171, *175*
 earthquakes, *167*
 seismicity, *153*
 southern, 153, 191
 western, 164, 191
Great Lakes, 477, 478
Great Plains, 229, 230, 232, 437 *440*
 aquifers, 467
 heat flow, *430*, 432, *470*
 northern, 467
 temperature data, *467*
Great Plains stress province, 353, *413*
Great Salt Lake, 197
Great Valley, 109, 110, *115*
Great Valley block, 128
Great Valley–Sierra Nevada block, *126*
Great Whale River 479
Green Valley fault, 111, 112, 120
Greenland, *17*, *473*, *475*, 484
Greenland Ice Sheet, *476*, *479*, 484
Greenville fault, 111
Gregory County, South Dakota, 470
Grenville basement, 265, 300
Grenville Province, 262, 275
Grenville terrain, 430
Grinnell Peninsula, 271
Grizzly Valley fault zone, 163
ground water, 462, 469
 recharge, *461*
Guatemala, 323, 401
 earthquakes, 317, 325, *330*, 333, 336
 focal mechanisms, *330*
 microseismicity, *330*
Guatemala City graben, 415, *417*
Guatemalan volcanic arc, *417*
Guaymas Basin, 424
Guerrero, 415, 417
Gulf Coast, 255, 258, *352*, *428*, 430
Gulf Coast Basin, 430
Gulf Coast province, 346, 413, 416, 417

Gulf coastal plain, 413
Gulf of Alaska, 32, 48, 54, 65, 389, 390, 396
  eastern, 391
  earthquakes, 49, 390
Gulf of California, 133, 140, 310, 401, *407, 409*, 418, 424, 431
  faults, 310
  plate boundary, 407, 409, *418*
  transform system, 363
Gulf of California province, *362, 409*, 413, 417
Gulf of Campeche, 417
Gulf of Honduras, 324
Gulf of Mexico
  coastal plain, 416
  continental margin, 411, 417
  earthquake, 352
  heat flow, *427*
Gulf of St. Lawrence, 439, 440
Gusfat-Lougheed Arch, 271
Gutenberg-Richter relation, 214

Hansel Valley, earthquake, 168, 193, *197*
Hawley Ridge, 36
Hayes Volcano, 54
Hayward fault, 111
hazards, seismic, 157, 165, 194, 313
Healdsburg–Rodgers Creek–Macaama fault, 111, 112, 120
heat flow, 6, 144, *190*, 203, 214, 218, 219, 237, 358, 426, *438, 445, 457*
  anomalies, 433, 470
  aquifer effects, 428
  Arctic Ocean, *427*
  Atlantic margin, *424, 439*, 441
  Atlantic Ocean, *424*, 441
  Baffin Bay, 441
  Basin and Range Province, 430, 432, *459*
  Bering Sea, *427*
  Canada, *437*
  Canadian Cordillera, *445, 446, 451*
  Canadian Shield, *430*
  Caribbean Sea, *427*
  Cascade volcanic arc, 431
  Chihuahua, Mexico (northern), *457*
  Colorado Plateau, *432, 457*
  Columbia Basin, 430
  conductive, 442, 452
  continental, *427*
  Cordillera, 434, *441, 454*
  database, 424
  eastern Canada, 430
  eastern continental shelf, *438*
  eastern margin, 442
  estimates, 439
  Great Plains, *430, 432, 467, 470*
  Gulf of Mexico, *427*
  histograms, 440
  Iceland, *427*
  Labrador Sea, 441
  measurements, 257, 358, 427, 430, 442, 446, 457, 459, 462
  Mexico (Chihuahua), *457*
  Middle Rocky Mountains, 432, 458
  model, 468
  New Brunswick, 442
  North America, *433*
  Northern Rocky Mountains, 430
  Nova Scotia, 442
  ocean, *424*
  Pacific Ocean, *424*, 441
  patterns, *423*
  Peninsular Ranges, 432
  profiles, *441*, 454
  province, defined, 452
  Rio Grand rift, 430, *432*
  Salton Trough, 430
  Sierra Nevada Mountains, 432
  Snake River Plain, Yellowstone area, 431
  Southern Rocky Mountains, 430, *432*
  tectonic settting, 438
  terrestrial, *437*
  United States, *429, 430, 457*
  western margin, 442
  Wyoming Basin, 432, 458
heat generation, *429*, 440
  Canadian Cordillera, *451*
heat sink, 454
heat transfer, 414, 424
Hebgen fault, 200
Hebgen Lake region, 188, 192, 210, 216
  earthquake, 185, 194, *200*, 203, 206, 216, *220*
Hebgen Valley area, 203, 204
Hecate Strait, 71, 455
Helena, Montana, 214
  earthquakes, *203*
Hennessey Shale, 14
Heron Reservoir, 235
High Canadian Arctic, 479
High Cascades, 77
Hispaniola, 316
  western, *316*
Home Bay, 269
Homestead Valley, earthquake, 147
Honduras, northwestern, 333
Honduras Depression, 324, 337
Honey Lake fault, 102
Hosgri fault, 145
hot spot, 188, 263, 313
hot springs, 211, *428*
Hudson Bay, 437, 474, 479, 481, 484
Hudson Strait, 270
Hueco graben, 411
Huslia, earthquake, 63, 65
hydrothermal activity, 208
hydrothermal fluids, 462
hypocenters, 85, 89, 93, 111, 118, 144, 232, 237, 263, 264, 302, 306, 311, 336, 388

Iapetus Rift faults, *276*
ice
  glacial, 17
  loads, 473, 479
  sheets, 473, *475*, 479, 482
  unloading, 484
Iceland
  deglaciation, 476, 483
  glacial isostatic recovery, *17, 473*, 484
  glacial-unloading chronology, *475*
  heat flow, *427*
  ice sheets, 380
Idaho, 5, *201*
Idaho batholith, 203
Idaho Seismic Zone, 188
Iliamna Volcano, 55, 57
Illinois basin, 252, 346, 352
Imperial fault, *140*, 407
Imperial Valley, *140*
  earthquake, 135
Imperial/Cerro Prieto fault system, 361
indicators
  fault, *389*
  thrust earthquake, 391
  volcanic, *389*
Indio segment, 137, 140
injection, fluid, *213*
Insular Belt, 452, 454
Interior Lowlands, 429
Interior Platform, *74*
  earthquakes, *74*
Intermountain Seismic Belt (ISB), 381, 452
  central, 185, 192, 193, *203, 206*
  earthquake record, *194*
  earthquake recurrence, *214*
  northern, 185, 188, *203, 211*, 216
  seismicity, *185, 220*
  seismotectonics, *188*
  southern, 185, 192, 193, *203, 204*
  space-time patterns, *215*
intraplate measurements, 401
Inyan Kara Formation, 467, 470
Ipala graben, 334, 415
Ipala stress province, *417*
Irazu volcano, 329
Irondale, earthquake, 303
Iroquois Falls, 268
ISB. *See* Intermountain Seismic Belt
isostatic recovery, glacial, *17, 473*
Isthmus of Panama, 314
Isthmus of Tehuantepec, earthquakes, 313

Jackson Lake dam, 194
Jalpatagua fault, 330
James Bay, 268
Jasper, Alabama, 304
Jeanne d'Arc Basin, 376
Jemez lineament, *233, 236*
Jemez Mountains caldera, 235, 236
Jocotan-Chamelecon fault, 332, 335
Jornada del Muerto, 463, *464*
Juan de Fuca Benioff zone, 85
Juan de Fuca Plate, 70, 77, 78, 83, 85, *89*, 94, 95, 96, 133, 349, *357*, 381, 424, 445, 451, 454, 455
Juan de Fuca Ridge, 361, 424
Jura Mountains, 343

Kalamath Mountains, 431
Kamchatka Trench, 395
Kankakee arch, *253*
Kayak Island, 53
Kenai Peninsula, 53, 57
Kennedy, Nevada, 157
Kentucky
  earthquakes, 296

eastern, *294*
Kentucky Plateau, 293
Kentucky River fault system, 11, *13*
Kern County, earthquake, 134, 135, 148
Kern River fault, 147
Kerr Dam, 212
Kigluaik fault, 64
kimberlites, 263
kinematics, *119*
Kirkland Lake, 263
Klamath Mountains, *102*
Klamath Mountains province, 77
Kluane Lake, 72
Knoxville, Tennessee, 298
Kobuk fault, 49, 65
Kobuk Trench, 65
Kodiak Island, 47, 53, 54, 55, 57, *392*
Komandorsky Islands region, *32*
Kong Oscors Fjord, 480
Kuril Trenches, 395
Kuril-Kamchatka subduction zone, 42, 43
Kuril-Kamchatka trench, 43

La Babia fault, 413
La Jencia fault, 241
La Malbaie area, 346
Labrador, 263
 southern, 481
Labrador Sea, 261, 263, 426, 437
 earthquakes, *268*
Labrador Shelf, 352
Lake Agassiz, 478
Lake Champlain, 263
Lake Chelan, 90
Lake County uplift, 9
Lake Erie, 276
Lake Jocassee, 301, *302*
Lake Mead, impoundment, 154, 163, *212*
Lake Ontario, 276
Lake Sinclair, 304
Lake Superior, 427
 tectonic zone, *254*
Lake Tadoussac, Quebec, *15*
Lake Timiskaming, earthquake, 262
lakes, pluvial, 17
Lakota Formation, 470
Lancaster, earthquake, 284
Las Animas Arch, *17*
LaSalle anticline, 252
Lassen Peak, 101, 102, 115, 128
Last Chance fault zone, 163
Laurentian Channel, 267
Laurentian Slope Seismic Zone, 266
Laurentide Ice Sheet, 475, 477, 478, 484
lava, 454
LCZ. *See* Lewis and Clark Zone
leak-off pressures, 374
leak-off tests, *372*
Lemhi fault, 210
Lesser Antilles, earthquakes, 316
Lesser Antilles subduction zone, 314, 315
Lewis and Clark Zone (LCZ), *211*
Likely fault, 102
Lillooet River 442

lineament, *241*
liquefaction, 7, 9, 16, 143, 197
Lithoprobe I profile, 454
lithosphere, 70, 339, 380, 424, 445, 446, 453, 454
Little San Bernardino Mountains, 145
Lituya Bay, 51
Livermore, earthquake, 115
Loma Prieta, California, earthquake, *132*
Lone Pine fault, 172
Long Beach, earthquake, 144
Long Valley Caldera, 109, 148, 163
Long Valley–White Mountains region, 115
 seismicity cluster, *113*, 118
Los Angeles basin, 145, 361
Los Muertos Basin, 464
Los Muertos graben, 411
Lost River fault zone, 194, 201, 210
Lost River Range, 194, 201
Louderback Mountains, 157, 162
Lower St. Lawrence zone, *376*, 379
Lytle Creek, earthquake, 139

Maacama fault zone, 102, 112
Mackay, Idaho, 200
Mackenzie Mountains, 72, 73
Macuspana basin, 417
Madison fault, 200
Madison Range, 200
Madison Valley, 203
Magdalen Basin, 439
magmas, 208, 232, 237, *242*, 402, 462
 movement, 335
magmatism, 413, 445
mainshock, 42, 62, 65, 82, 89, 91, 93, 115, 142, 144, 162, 167, 202, 335
Malibu, earthquake, 145
Mammoth, Nevada, *163*
Mammoth Lakes, earthquake, 118, 127
Mammoth zone, 24
Managua, Nicaragua, 333
 earthquakes, *314*, 323, 329, 333, 335, 336
Manic, earthquakes, 265
Maniwaki, earthquake, 262
Manix, earthquake, *147*, 154, 168
Manix fault, 147
Mansfield, earthquake, 303
mantle, 397
 active, 396
 upper, 256, 459
margins
 continental, 75
 convergent, 29, 54
 passive continental, 277
marine limit, elevation, *479*
Maritime Provinces, 429
Marked Tree, 9
McKenzie Delta, 354
Meade thrust, 205
Meager Mountain geothermal area, 451
Mednyy Island, 29
Meers-Duncan-Criner fault zone, *13*, 15

Meers fault, 6, 13, *14*
megashear, 171
melts, partial, 208
Mendocino Escarpment, 102
Mendocino triple junction, 107, *115*, 133
Mercury fault, 376
Mesa Central, 411, 413
Mesilla Basin, 464
meteorite impact, 265
Mexican Cordillera, 353
Mexican Ridges fold belt, 411
Mexican Wadati-Benioff zone, 313
Mexico
 central, *311*, *414*, *418*
 geothermal, *423*
 heat flow, *457*
 northeastern, *409*
 northern, *310*
 northwestern, *407*, 418
 southern, *415*, *416*, *418*
 stress, 355, *401*, 416
Michoacan gap, *312*
microearthquakes, 29, 39, 59, 61, 70, 71, 72, 79, *85*, 112, 120, 154, 163, 167, 168, 208, 264, 310, 330
 data, 77, 249
 swarms, 282, 285
microseismicity, 35, 96, 111, 135, 139, 143, 144, 145, 146, 210, 217, *328*, 332, 336, 411
 Costa Rica, *328*
 rates, 101
Mid-Atlantic Ridge, 374, 427
Mid-continent Stress Province, 255
Middle America, seismotectonics, *309*
Middle American arc, 314
Middle American trench, 311, *316*, 323, 336, 355, 414, 417
Middle Rocky Mountains, heat flow, 432, 458
Middleton Island, 53
Middleton Place–Summerville Seismic Zone, *301*, 305
Midplate province, 5, *6*, 17, 346, *349*, 353, 363, *380*, 382, 383
Milton-Freewater, earthquake, *83*
Mimbres Basin, 464
Minto Flats Seismic Zone, *62*
Minto lineament, 62
Miramichi Highlands, 266
 earthquakes, 261, *266*, 281, 285
 mainshock, *266*
Mission Creek fault, 135
Mission Creek strand, 136
Mississippi Delta, 417, 430
Mississippi Embayment, 7, 249, 251, 256
Mississippi Valley, central, 247
models
 Charleston (South Carolina), *306*
 crustal rebound, 482
 fore-arc sliver, *336*
 frictional, 360
 heat-flow, 457, 468
 hotspot-frame, 313
 kinematic, 409
 normal-faulting earthquakes, *220*

relative plate motion, *335*
seismotectonic, *306*
tectonic, 42
Modoc Plateau, *102*
Mohawk Valley fault zone, 163
Mohorovičic discontinuity, 119
Mohr failure, *359*
Mojave desert, 134, 136, *146*
Moncton, earthquake, 266
Monroe uplift, 7
Montana, 5, *197, 200*
Monte Cristo Valley, 159
Monteregian Hills, 263, 277
Monterrey-Torreón transverse ranges, 413
Monticello Reservoir, 201, *302*
Moodus, Connecticut, 282
Morenci lineament, 241
Moresby Island, 455
Morgan Hill, earthquake, 120
Motagua
  earthquake, 415
  fault, 317, *324*, 330, *331, 333, 335*, 337, 415, 416
Motagua-Polochic plate boundary zones, 401
Motagua-Polochic transform system, 416, 417, 418
Motagua transform system, 355
Mount Ranier, 89
Mount St. Helens, 85
Mt. Shasta, 102
Murray Transform, 145
Murre fault, 376

Nacimiento uplift, *235*
Nahanni, earthquakes, *73*
Nares Strait, 270
Nayarit, 407
Nazca plate, *314*
Nemaha uplift, *254*
neotectonics, *473*
  defined, 1, 3
networks
  catalogs, 327, 329
  microearthquake, 247
  seismic, 29, *195*, 204, 277, 291
  seismograph, 47, 75, 234, 318, 323, *324*
  strong-motion, *324*
  telemetered, 157
Nevada
  seismic patterns, *153*
  seismicity, *153*
  western, *157*
Nevada Seismic Zone, 175
Nevada Test Site, 163, 187, 195
Nevada Transverse Zone (NTZ), 154, 163
  southern, 168
New Brunswick, 266
  earthquake, 280, 287
New England coastline, 479
New England-New Brunswick Appalachians zone, 287
New England Seamount Chain, 263, 277
New Madrid, earthquakes, 6, 7
New Madrid fault zone, 249

evolution, *256*
structure, *256*
New Madrid rift, 11
New Madrid Seismic Zone, 7, 245, 276, 346, 352
  recurrence rates, *251*
New Mexico
  north-central, *234*
  seismicity, *229*
  southern, 413
New York-Alabama lineament, 300
New York City area, earthquakes, *284*
New York-eastern Pennsylvania-New Jersey zone, 287
New York State Zone, western, 287
Newark basin, 282, 284
Newberry, South Carolina, 294
Newcastle Formation, 467
Newfoundland, 439
Newport-Inglewood fault, 133, 144
Nicaragua, 323
  focal mechanisms, *329*
  microseismicity, *329*
  seismograph network, *326*
  strong-motion network, *326*
Nielsen Island, 440
Nootka fault zone, 71
North America
  Geothermal Map of, *423*
  glacial isostatic recovery, *17, 473*, 484
  glacial-unloading chronology, *475*
  heat flow, *423, 433*
  heat-flow map, 433
  Seismicity Map of, *21*
  stress provinces, *349*
  tectonic stresses, *339, 345*
North American-Caribbean plate boundary, 317
North American Plate, 29, 48, 51, 60, *65*, 77, 95, 107, 133, 136, 261, 310, *311*, 313, 316, 324, 335, 341, 354, 374, 375, 390, 397, 401, 414, 417
North Atlantic, 425
North Carolina, *298*
  eastern, *300*
  western, *300*
North Cascades province, 77
  earthquake, 77, *82*, 95
North Gower, earthquake, 262
North Palm Springs, earthquakes, 137
North Slope, 62, 65, 389
Northern Rocky Mountains province, 188
  heat flow, 430
northern California, seismicity rate, 99
Northwest Territories, Canada, *16*
Nova Scotia (northern), Canada, *16*
Novarupta volcano, 60
NTZ. See Nevada Transverse Zone
nuclear explosions, 164, 187
nuclear testing, underground, 154, 168
Nuevo León, 413

Oaxaca, 415, 417
  earthquake, *312*
oceanic measurements, 441

Oceanside, earthquake, 144
Okanogan Highlands, 85
Olinghouse fault, 157
Olympic Mountains, 77
Omenica Crystalline Belt, 381, 452
Oregon
  earthquake catalog, 79, 94
  northwestern, *89*
  seismicity, *77*
  southeastern, *93*
  tectonic provinces, *85*
Oroville, California, 101
  earthquake, 103, 126, 349
Ortigalita fault, 112
Ossippe, earthquakes, 285
Otoskwin River, 440
Ottawa graben, 276
Ottawa River, 262, 263, 265
Ottawa Valley, 263
Ottawa-Timiskaming rift, 268
Ouachita frontal thrust, 258
Ouachita Mountains, 254, 429
Ouachita Seismic Zones, 254
Ouachita system, 254
overcoring methods, *369*
Owens Valley, 147, 148, 353
  earthquake, 110, 113, 127, *163, 172*, 194
  fault, 147
Owens Valley gap, southern, 164
Ozark arch, 253
Ozark Plateau, 429
Ozark uplift, 253, 255, 258

Pacaya Volcano, 336
Pacific coast, *324*, 361
  seismicity, 48
Pacific margin, 49
Pacific Northwest stress provinces, 353, *357*
Pacific Ocean, heat flow, *424*, 441
Pacific Plate, 29, 43, 48, 51, 54, *65*, 107, 133, 136, 310, 349, 355, 388, 390, 391, 392, 396, 397, 401
Pahute Mesa, 168
Pajarito fault, *236*
paleoliquefaction, 9, 16
paleoseismicity, defined, 4
Paleozoic Rift System, 276
Palisades Dam, 212
Palisades Reservoir, 213
Palmdale, California, 137
  earthquake, 139
Palmdale Bulge, 146
Palos Verdes fault, 144
Pamplona fault zone, 71, 391
Panama, *314*
Paris, Idaho, 197
Park Range, 460
Parkfield, earthquake, 120, 128
Pasco basin, 90, 91
Passamaquoddy Bay, earthquake, 265
PDE. See Preliminary Determination of Epicenters
Peace River Arch, 376
Pedernal Hills uplift, 237
Peel Sound, 17
Pender Island, earthquake, 89

Peninsular Ranges, 431, 432
  heat flow, 432
permeability trends, 370
Petén fold belt, 417
Picton, faulting, 276
Piedmont province, 292, 297, 300, 302
  earthquakes, 294, 303, 306
Pierre Shale, 470
Pima County, 460
Pinal County, 460
Pine Valley, 203
  earthquake, 154
Pinto Mountain fault, 146
Pitaycachi fault, 411, 413, 418
plate boundary provinces, 354, 357
plate convergence, 29, 31, 51, 77, 344
plate flexure, 355
plate interfaces, 309
plate interiors, 309
plate margins, movements, 473
plate motions, relative, 339
plate tectonics, 5
Pleasant Valley, earthquake, 114, 168, 194
Pocatello Valley, earthquake, 203, 204, 216
Point Conception, 145
Point Mugu, earthquake, 145
Polochic fault, 317, 415
Port Royal, Jamaica, 317
Post Oak Conglomerate, 14
Potrillo field, 233
potash mining, 213
Powell Lake, 446
Prairies Basin, 428
Prairies envelopes, 442
Precambrian Sheild, 442
Preliminary Determination of Epicenters (PDE), 21, 29, 48
Presidio graben, 411
Prince of Wales Island, 16
Prince Patrick Island, earthquake swarms, 271
Prince William Sound, 51, 52, 53, 55, 57, 65, 391
  earthquakes, 54, 65
prism, accretionary, 356
profiles, heat-flow, 441, 454
provinces, plate interior, 349
Puerto Rico, 316
Puerto Rico Trench, 316
Puget Sound, 71, 94
  northern, 82
Puget Sound basin, 83, 85, 95, 96
Puget–Willamette Lowland, 78
Purcell Mountains, 442
Pyramid Lake fault zone, 157

quarry-floor buckles, 370
Quatemala fault, 415
Quaternary, 193
  late, 473, 480
Quebec, western, 261, 262
Quebec-Maine border, 266
Queen Charlotte–Fairweather fault system, 51, 71, 390
Queen Charlotte fault, 49, 51, 71, 75, 361, 363, 445, 446, 455

Queen Charlotte Islands, 71, 446, 479
  southern, 455
Queen Charlotte province, 361, 380, 381
Queen Charlotte Sound, 455
Queen Charlotte transform region, 71
quiescence, seismic, 110, 111, 112, 115, 119, 135, 208, 210

radioactivity, 433
  high crustal, 429, 460
  surface rocks, 429
  waste, 157
Raft Batholith, 452
Raft River 213
Rainbow Mountain, 162
  earthquake, 162
Ramapo fault system, 282
Rampart–Dall City area, 62
Rangely oil field, 213
Rat Island, earthquake, 31, 42, 43
Raymond Hill fault, 145
rebound
  crustal, 477
  postglacial isostatic, 17, 473, 484
reconstructions, neotectonic, 478
Red Canyon fault, 200
Reelfoot complex, 7
Reelfoot Lake, 7, 9
Reelfoot Rift, 256
Reelfoot Rift–New Madrid Seismic Zone, 7, 11
Reelfoot scarp, 7, 9
Reidel shears, 172
rheology, 218, 473, 474, 479
Richardson Mountains, 72, 73
Rinconada fault, 112
Rio Grande depression, 237
Rio Grande rift, 353, 409, 460, 462
  central, 236
  heat flow, 430, 432, 462
  seismicity, 229
  southern, 464
Rio Grande Rift Province, 4
Rio Grande River, 411
Rio Grande Valley, 232
  ancestral, 241
  earthquakes, 231
Rivera Plate, 133
  subduction, 311
Rivera-Cocos–North American triple junction, 311
rockbursts, 212, 213, 304
rocks
  basement, 218, 377
  crystalline, 218, 370
  igneous, 277
  plutonic, 452
  sedimentary, 212, 218
  silica-rich, 218
Rocky Mountain Trench, 74, 442
Rocky Mountains, 219, 353
  heat-flow data, 461
  Middle, 458
  Southern, 230, 433, 460, 461, 465
Rome trough, 13
Roosevelt Hot Springs, 213
Rough Creek fault zone, 11
Round Valley, earthquake, 114

ruptures
  seismic, 310, 313, 316
  surface, 153, 154, 157, 193, 197, 204, 245

Saddle Mountains, 93
  anticline, 90
Saguenay
  earthquake, 275
  graben, 276
St. Donat, earthquake, 262
St. Elias region, 71, 445
  earthquakes, 49, 54, 71
St. Francis dome, 253
St. Francois Mountains, 349
St. Genevieve–Cottage Grove fault zone, 10
St. Helens Seismic Zone (SHZ), 85, 89
St. Lawrence, lower, 261, 265
  earthquakes, 265
St. Lawrence Platform, 265, 370
St. Lawrence Rift fault system, 376
St. Lawrence River, 263, 264
St. Lawrence Seismic Zone, 15, 265, 276
St. Lawrence Valley–Charlevoix–LaMalbaise-Saguenay zone, 15
Salcha, earthquake, 60
Salcha Seismic Zone, 60
Salinian block, 112, 128, 432
Salt Basin graben, 411
salt domes, 427
Salton Sea, 140, 409
Salton Trough, 140, 143, 427
  heat flow, 430
Salton Trough–Cerro Prieto region, 361
San Andreas discrepancy, 109, 128
San Andreas fault stress province, 357, 409
San Andreas fault system, 26, 102, 107, 110, 111, 119, 120, 128, 132, 133, 136, 145, 146, 148, 341, 357, 359, 361, 363, 409, 432
San Andreas transform boundary, 109
San Ardo oil field, 112
San Benito fault, 362
San Bernardino, California, 137
San Bernardino Mountains, 145, 146, 147
San Bernardino segment, 137, 140
San Bernardino Valley, 311
San Cruz Island, fault, 145
San Diego Trough fault, 144
San Diego Trough–Bahia–Soledad fault zone, 144
San Felipe fault zone, 235, 237
San Fernando
  earthquake, 135
  fault, 145
San Francisco
  earthquake, 102, 110, 145
  volcanic field, 458
San Francisco Bay, 119, 132
San Gabriel Mountains, 143, 145
San Gorgonio Pass area, 147
San Gregorio–Hosgri fault system, 112
San Isidro, 329

San Jacinto fault, 133, 135, *143*
San Jacinto Mountains, 144
San Joaquin fault, 148
San Joaquin Valley, 148
San Juan Basin, *458*
San Juan Islands, 82
San Juan Mountains, 460
San Juan volcanic field, *458*, 461
San Luis Potosi, 403, 411
San Luis Reservoir, 112
San Marcos fault, 413
San Miguel volcano, 330
San Pablo Bay, 111
San Rafael Swell, earthquake, 206
San Salvador, earthquake, 330, 333, 335
San Vicente, El Salvador, 333
Sanak Island, 35
sand, vented, 12
sand blows, 251, 302
sand dikes, 12
Sandwich fault zone, 253
Sangre de Cristo Mountains, 460, 461
Sangre de Cristo uplift, 236
Santa Ana Mesa, 237
Santa Barbara, earthquake, 144, 145
Santa Barbara Channel area, 145
Santa Catarina fault, 334
Santa Cruz Mountains, earthquake, *132*
Santa Maria Volcano, 336
Santa Monica Bay, earthquake, 145
Santa Monica fault, 145
Santa Monica Mountains, 145
Santa Susana fault, 145
Santa Ynez Mountains, 145
Saskatchewan, southern, 74
Sauguenay region, *16*
  earthquake, 16
Savannah River Plant, earthquake, 301, 302
Sawatch Range, 460, 461
Scoresby Sound, 480
Scotian Shelf, 372, 376, 383, 386
Scotian Shelf basin, 353
sea floor, age, 427
sea level
  changes, 473, 477
  submergence, 478
  tectonics, 483
sediments, bottom, 446
segmentation, 35
seismic activity, annual, 115
seismic belt, east-west, 154
seismic catalog, 77, 94, 134, 154, 234
seismic gaps, *32*, 51, 71, 143, *164*, 208, 210, 216, 269, *311*
seismic hazards, 157, 165, 194, 313
seismic network, 29, *195*, 204, 237, 277
seismic quiescence, 110, 111, 112, 115, 119, 135, 208, 210
seismic slip, 190, 208, 313
seismic stations, *79*, 235
seismic strain, 165
seismicity, *204, 217, 233, 237, 249, 281, 285, 291, 297*, 483
  Alabama, *303*
  Alaska (continental), *47*

Aleutian Arc, 29
background, 212, 217, *268*
California, *99, 107*
Canada, *69*, 261
clusters, 51, 57, 113, 114, 115
cross sections, *118*
eastern Kentucky, *294, 298*
Fairbanks area, 60
focal depths. *See* focal depth
Georgia, *303*
Great Basin, *153*
halo, 297, 304
historical, *109, 292,* 298, 299
induced, *163,* 212
instrumental, *293*
Intermountain Seismic Belt (ISB), *185, 220*
mining-related, *231*, 254
network, *294*, 298
Nevada, *153*
New Mexico, *229*
North Carolina, *298*
nuclear-induced, 164
Oregon, *77*
Pacific coast, 48
patterns, *107*, 109, *110*, 115, 128, *153*, 204, *234, 247, 255*, 287
rates, *99*, 288
reservoir-induced, *212*, 235, 294, 301, *302*, 305
Rio Grande rift, 229
South Carolina, *300, 306*
tectonic-induced, 294
Tennessee, *298*
trench, 31
upper-crustal, *323*
variations, *34*
Virginia, 294
volcanic zone, *335*
Washington, *77*
West Virginia, *294*
Seismicity Map of North America, *21*
seismograph network, 47, 75, 107, 234, 261
seismotectonics, *85*, 313, *324*
  California (southern), *133*
  Canada (eastern), *261*
  defined, 4
  Middle America, *309*
  United States (central), *245*
Sevier Desert region, *190*
Sevier Valley, Utah, 169, 175, 191, 205
Seward Peninsula, 47, 49, 64, 65, 354, 396, 397
Shanshi Province, earthquake, 255
shear strain, 128
Shirshov Ridge, 34
shorelines
  glacial lake, 473, 474, 478, 479, 480, 481, 482
  raised marine, 473, 474, 479, 480
shortening, 145
  crustal, *126*, 255, 355, 359
  horizontal, *389*
Shoshone, Idaho, 203
Shumagin gap, 32, 34, 36
Shumagin Islands region, 29, *32*, 35, 36, 38, *39*, 392

Shumagin network, 29, 35, 39
SHZ. *See* St. Helens Seismic Zone
Siberia, eastern, 387
Sierra lineation, southern, 147, 148
Sierra Madre fault, 145
Sierra Nevada, 101, *102*, 109, 126, 136, *147*, 427, 431, 432, 433
  block, 114, 128
  eastern, *126*, 175
  escarpment, *112*
  fault, 148
  foothills, *115*
  heat flow, 432
Sierra Nevada–Great Basin boundary zone (SNGBZ), 153, *163*, 165
Sierra Nevada–Great Valley block, 110
Sigsbee Knolls, 427
silt layers, 16
Sinaloa, 407
Siouxon Peak, earthquake, 89
Sitka, earthquake, 51
slickensides, 94
slip
  aseismic, 32, 316
  fault, *344*
  reverse, 132
slump, submarine, 267
Snake Plain aquifer, 428
Snake River Plain, *220*
  heat flow, 431
  Yellowstone area, 431, *432*
Socorro, New Mexico, 232, *238*
  earthquakes, *238*
  magma body, 237, *242*
Soda Springs, Idaho, 205, 207
solutions
  fault-plane, *240*
  focal-mechanism, 294
Somerset, Colorado, 213
Soná-Azuero fault, 357
Sonoma Range gap, 164
Sonora, 407, 411, 413
  earthquake, 311, 411
  northeastern, 411
  northwestern, 409
South America, northern, *314*
South American craton, 315
South American plate, 315, 316
South Carolina, *300*
South Carolina Coastal Plain, 300
South Carolina Piedmont, 305
South Carolina–Georgia Seismic Zone, 301
South Dakota, south-central, 469
South Fork Mountain fault zone, 102
South Seattle, earthquake, 89
Southern Appalachian Seismic Zone, 294, 298
Southern Great Plains province, *353*
Southern Rocky Mountains, heat flow, 430, *432*
Sovanco fracture, 70
splitting, shear-wave, 370
spreading
  ocean-floor, 140
  rate, 409
  ridge, 361
  sea-floor, 140, 269, 362

spreading centers, 133, 310, 311, 339, 359, 407
stations
  seismic, *79*, 235, 277, 279
  seismograph, 261
statistics, frequency-magnitude, 115, 118
Steamboat Hot Springs area, 163
Stillwater Range, 159, 162
  earthquake, *162*
  gap, 164
Stone Mountain, earthquakes, 298
strain
  geodetic, 370
  rates, *192*, 242
stress, 6, 31, 32, 40, 72, 95, 294, *354*, *401*
  age, *382*
  Alaska, *387*
  anisotropy, 369, *383*
  anomalies, 377
  Canada, *367*, *380*, *382*, *383*, 390
  causes, *383*
  Central America (western), *401*
  compressive, 288
  crustal, *255*, *367*
  directions, 346, *395*
  eastern United States, *346*
  fields, *191*, 255, 262, 263, *280*, *339*
  horizontal orientation, 346, 369, 370, 401, 417
  indicators, *341*, *346*, *362*, *387*, 390
  intraplate regimes, *280*, *417*, *418*
  magnitudes, 171, 346, *370*, *371*
  maximum horizontal compressive (MHC), *387*
  measurements, *341*, *344*, *409*, *411*, *413*, *414*, *418*
  Mexico, *401*
  North America, *349*
  North American Plate, *341*
  Pacific Northwest, 353
  patterns, 370, 396
  plate-boundary, *390*
  plate-interior, *395*
  postglacial, 375
  provinces, *346*, *349*, *353*, *380*, *419*
  quality-ranking system, *341*, *362*
  refraction, *376*, *383*
  renewal mechanisms, *382*
  sources, *417*
  southeast Canada, *346*
  tectonic, 144, 163, 311, *339*, *341*, *345*
  tensor, *375*
  thermoelastic, *341*
  trajectories, *376*, *419*
subduction, 34, 42, 71, *89*, 94, *95*, *104*, 133, 314, 357, 392, 455
  aseismic, *146*, 311
  Cocos Plate, *311*
  Rivera Plate, *311*
  tectonics, 77
  zones, 316
submergence, 479
subsidence, 7, 9, 53, 220
Suisun Bay, 119
Summit, New York, earthquake, 290

Superior Province, 437, *440*, 442
Superstition Hills, earthquake, 142
Suprise Valley fault, 103
Sur-Nacimiento fault zone, 112
Sutton, earthquake, 59
Sverdrup Basin, *271*, 437
  earthquake, 271, 273
Swan fracture zone, 324

Tabasco, 417
  western, 416
Tacana volcano, 330
Tamaulipas, 413
Tamayo fracture zone, 409
Tampico-Misantla basin, *413*
Taos volcanic field, 233
tectonics, 95, *107*, 134, *168*, *323*
  active, 3
  Central America, *323*
  defined, 1
  plate, 5
  subduction, 77
tectonism, 7, 49, 254, 316, 354
  extensional, 311, 413
Tehuantepec Fracture Zone, 313
teleseismic mechanisms, 39
temperatures
  seasonal changes, 341
  subsurface, *467*
Tennessee, *298*
  eastern, *299*
  western, *300*
Tennessee Appalachians, 293
tests, leak-off, *372*
Teton fault, 188, 194, 208, 209
Texas–Louisiana Gulf Coast stress province, 352
Tépic-Zacoalco graben, 409, 414
thermal conductivity measurements, 446
thermal subprovinces, 430
Tijeras fault zone, 237
Tilaran, earthquake, 326, 329, 335
Timiskaming, earthquake, 262, 263
Tintina fault, 49, 62
Tintina Trench, 73
Tiptonville dome, 7, 9
Tiscapa fault, 329, 330
Tobin Range, 159
Toledo caldera, 237
Toombs County, earthquake, 303
Tosco fault, 362
Totschunda fault, 49, 59
Trans-Mexican Volcanic Belt, 313, 353, *356*, 403, 407, 409, 413, *414*, 416, *417*
Transition fault zone, 51, 72
Transverse Ranges, 107, 133, 136, *144*
  central, *145*
  eastern, *146*, 361
  western, *145*, 361
triple junction, 43, 109
  Kuril-Kamchatka subduction zone, 43
  Mendocino, 107
  Rivera-Cocos–North American, 311
Trousers Lake, earthquake, 266
Truckee, California, 102

earthquake, 115
tsunami, 32, 367, 317
  warnings, 47
tuffs, ashflow, 208
Turrialba volcano, 329

Unalaska gap, 34
Unalaska Island, 29, 31, 32, 394
Ungava, Quebec, *17*, 170
  earthquakes, 7, *17*, 270, 273, 276
Union County, earthquake, 300
United States
  central, *245*
  earthquake activity, *277*
  eastern, *429*
  geothermal regime, 464
  heat flow, *429*, *430*, *457*
  northeastern, *277*, *286*
  Seismic Zones, *286*
  seismicity, *291*
  seismotectonics, *245*
  southeastern, *291*
  southwestern, *457*, 464
  western, *430*
  See also specific states
United States Cordillera, 353
uplift, 7, 9, 17, 53, 71, 128, 230, 237, *241*, 253, 255, 270, 454, 457, 484
  domes, 481
upper-crustal seismicity, Central America, *323*, *333*
Utah, 5

Vacaville/Winters, earthquake, 115
Valdez area, 47
Valles caldera, 237
Valley and Ridge physiographic province, 299
Vancouver Island, 70, 71, 445, *454*
  central, 454
  western, 381
Velarde fault, *236*
Velarde graben, 236
velocity
  Caribbean plate, *313*
  P-wave, 111, 190
Venezuela, 317
  earthquake, 314
Ventura basin, 145
Ventura-Winnemucca zone, 153
Veracruz, 417
  southern, 416
Verdi, California, 102
Viking graben, 376
Villa de Reyes graben, 413
Virginia, *294*
Virginia City, earthquake, 203
Virginia Piedmont, 300, *306*
Vizcaino Peninsula, 407
volcanic arcs, 354, 433
  calc-alkaline, 414
volcanic ash, 483
volcanic centers, 109, 115, *428*, 442
volcanic chain, Central America, *324*, 328, 329, 333, 335
volcanic eruptions, 208
volcanic fields, 411
volcanic vents, 209, *344*

volcanics, 459, 460, 462, 483
    alignments, 416
    indicators, *389*
volcanism, 65, 77, 95, 140, 147, 175, 194, 214, *233*, 356, 357, 397, 427, 430, 432, 445
volcanoes, 34, 51, 314, *324*, 389, 402, 429
    Arenal, 328
    Augustine, 60
    Cook Inlet, 55
    flank, 344
    Fuego, 328
    Hayes, 54
    Iliamna, 55, 57
    Irazu, 329
    Novarupta, 60
    San Miguel, 330
    Tacana, 330
    Turrialba, 329
    Wrangell, 57

Wabash Valley, Indiana, *12*, 276
Wabash Valley fault zone, 10, *12*, 15, 252
Wadati-Benioff zone, 95, 104, 313, 314, 316
    Aleutian, *31*, *38*, 51, *54*, *57*, 65
    Antilles, 316
    Central American, *313*
    earthquakes, 52, *54*
    Mexican, 313
    Wrangell, 49, *54*, *57*, 65
Wagner basin, 409
Walker Lane, 175
Walker Lane belt, 6, 153, 163, 171, 172, 353
Walker Pass, earthquake, 135, 147
Wallace, Idaho, 213
Wallace Creek, 137
Warner province, *103*
Warner Valley, 93
warping, 7

crustal, 285
regional, *478*
Wasatch fault, 188, 194, *204*
Wasatch Front, 188, *193*, 194, *204*, 216, 218
Wasatch Plateau, 213
Wasatch Range, 194
Washington, 479
    earthquake catalog, *79*, 94
    northwestern, *89*
    seismicity, *77*
    southwestern, *89*
    tectonic provinces, *85*
Washita Valley fault, 7
washouts, 389
waveforms, 315
waves
    body, 246
    seismic, 164, 245
    surface, 245, 246
    tidal, 47
Waynesville, North Carolina, 298
wedge, accretionary, 314, 354, 417
wellbore breakouts, stress-induced, *243*, *368*, 382, 383
wells, hydrocarbon, 372
West Greenland, 480
West Virginia, *294*
Western Canadian Basin, 372
Western Cascades Range, 431
Western Cascades province, 77
Western Cordilleran province,s *353*
Westmorland, earthquake swarm, 143
Wet Mountains, 460, 461
White Lake basin, 452
White Mountains
    fault, 132
    magma series, 277
    seismic gap, 113, 132, *164*
White Wolf fault, 148
Whittier Narrows, earthquake, *145*
Wichita frontal fault system, 6, *13*, 254

Wichita uplift, 254
Wiggins uplift, 7
Williston Basin area, 74, 461
    stratigraphic formations, *468*
    temperature profile, *468*
Winston-Salem, North Carolina, 298
Wonder, earthquake, *157*
Woodstock fault, 302
World-Wide Seismographic Station Network (WWSN), 79
Wrangell arc, 54
Wrangell volcanoes, 57
Wrangell Wadati-Benioff zone, 49, *54*, *57*, 65
Wrightwood, California, 137
WWSN. *See* World-Wide Seismographic Station Network
Wyoming, 5
Wyoming Basin, heat flow, 432, 458
Wyoming craton, 432

Yakataga area, 396
Yakataga gap, 65
Yakutat Bay, 47, 54
Yakutat block, 391
Yakutat terrane, 49, 51, 66
Yellowstone caldera, 190, 203, 207, 209, 210, 219, 220
Yellowstone hot spot, 190, 207, 210, *220*
Yellowstone Lake, 209
Yellowstone National Park, 24, 288, 192, 203, *208*, 216, 427
    earthquake, 203
Yellowstone–Hebgen Lake region, 209
Yellowstone-SRP volcanic system, 188
Yucatan Peninsula, 417
Yukon Territory, 71, 72, 354, 387
Yuma County, 460

zones, glacial isostatic response, *477*, *478*, 479

Typeset by WESType Publishing Services, Inc., Boulder, Colorado
Printed in U.S.A. by Malloy Lithographing, Inc., Ann Arbor, Michigan

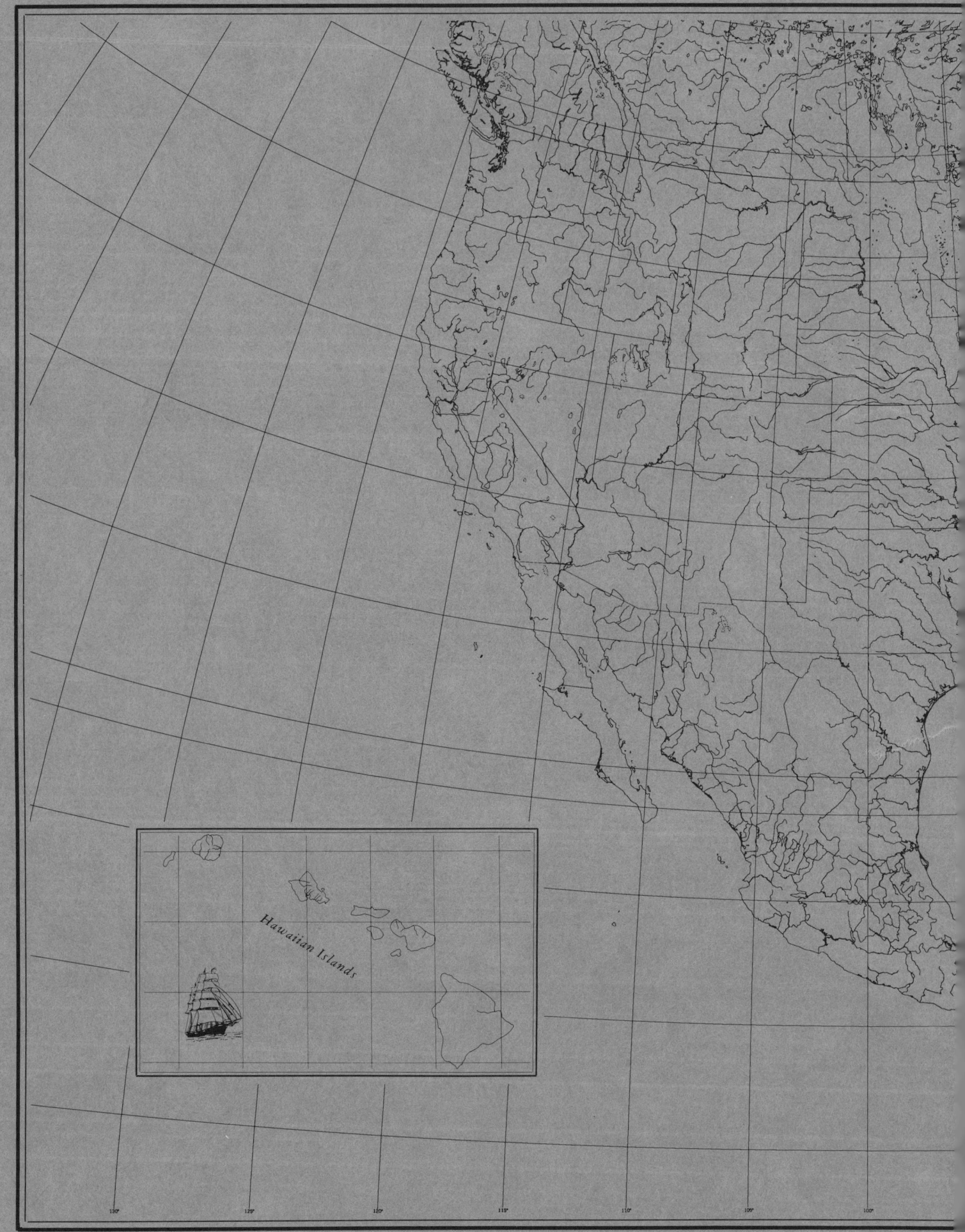